15th International Conference and Exhibition on Fire and Materials 2017

San Francisco, California, USA
6 – 8 February 2017

Volume 1 of 2

ISBN: 978-1-5108-4674-6

Printed from e-media with permission by:

Curran Associates, Inc.
57 Morehouse Lane
Red Hook, NY 12571

Some format issues inherent in the e-media version may also appear in this print version.

Printed by Curran Associates, Inc. (2017)

For permission requests, please contact Grayson Franks Ltd
at the address below.

Grayson Franks Ltd
1 West Yard
Guilford Grove Greenwich
London SE10 8JT
United Kingdom

Phone: 44 208 692 5050
Fax: 44 208 692 5155

office@intersciencecomms.co.uk

Additional copies of this publication are available from:

Curran Associates, Inc.
57 Morehouse Lane
Red Hook, NY 12571 USA
Phone: 845-758-0400
Fax: 845-758-2633
Email: curran@proceedings.com
Web: www.proceedings.com

Fire and Materials 2017

15th International Conference

San Francisco, California,
USA

6th – 8th February 2017

Published by

interscience communications

London, UK

www.intersciencecomms.co.uk

The organisers wish to thank and acknowledge the companies for their sponsorship

American Chemistry Council — North American Flame Retardant Alliance

FTT Fire Testing Technology

Kingspan

FM Global

Proceedings of the Fire and Materials 2017 Conference

held at the Hyatt Centric Hotel Fisherman's Wharf Hotel, San Francisco, California, USA
6 – 8 February 2017

A conference organised by
Interscience Communications Ltd
1 Burt Ash Lane
Bromley, Kent BR1 4DJ
England

Fire and Materials 2017
1004 pp

Fire and Materials 2017 Programme Committee

Fire and Materials 2017 Scientific Committee

To select a paper click on title, or use "Page Thumbnails" on the left hand navigation bar

TABLE OF CONTENTS

ELECTRO-TECHNICAL

FIRE SAFETY REQUIREMENTS FOR WIRES AND CABLES

Marcelo M. Hirschler
GBH International, US
(2 Friars Lane, Mill Valley, CA, 94941)

ABSTRACT

Wires and cables are among the products which are most often regulated for their fire safety. A key reason is that they are often found in concealed spaces, including (particularly in North America) in plenums and other air-handling spaces.

In the US, the federal government has not issued fire safety requirements for wires and cables in buildings. However, most fire safety requirements for wires and cables in buildings are in the National Electrical Code (NEC, NFPA 70), widely adopted throughout the country. NEC fire safety requirements for wires and cables address various categories: (1) power cables, (2) wires and cables associated with power-limited circuits (including signaling cables, fire alarm systems and optical fiber cables) and all types of communications cables and circuits, (3) wires and cables associated with information technology equipment (4) fire resistive cables. The more severe fire safety requirements correspond to the first two categories, which use cable tray fire tests oriented horizontally (for plenum cables, measuring flame spread and smoke, see figure below) or vertically (for riser and tray cables, measuring primarily flame spread).

In the US both major organizations issuing codes, including building codes, call upon the NEC (NFPA 70) for requirements on electrical safety (including fire safety). Interestingly, regarding fire safety regulation of products (including wires and cables) in plenums, the NEC is subservient to a consensus standard known as NFPA 90A (Standard for the Installation of Air-Conditioning and Ventilating Systems).

Specific US fire safety requirements have been issued by the FAA for aircraft and by the USCG for vessels that do not sail in international waters. IMO regulations apply to vessels sailing in international waters, including some fire tests for wires and cables. US federal agencies have issued no wire and cable fire safety rules for land surface or underground transportation. Fire safety of intercity and interstate passenger trains is regulated by the FRA, which do not include wires and cables. Highway passenger vehicles (including automobiles) are regulated by NHTSA and mass transportation (buses and underground trains) is regulated by FTA, but wires and cables are not covered by either. However, NFPA 130, a widely-adopted consensus standard, regulates regulate the fire safety of wires and cables in trains and underground systems.

In the European Union, the EC regulates the harmonization of legislative, regulatory and administrative provisions for all member states. The EC issued the Construction Products Regulation, which includes "Safety in case of fire" as one of the 7 basic requirements. The fire classification system for construction products is EN 13501, which contains standards with the information for requirements for many building products, including wires and cables, covered in EN 13501-6. Tests and classification systems apply to all countries but each determines requirements for any specific application, via interpretative documents. Categories range from A (best) to F (no performance determined) and measurements are made of flame spread and heat release as well as optional additional measurements of smoke emission and acidity.

In the EU, the fire safety of all materials in trains (including wires and cables) is covered in EN 45545-2 (Railway applications - Fire protection on railway vehicles - Part 2: Requirements for fire behavior of materials and components) while the fire safety requirements of the FAA apply to all aircraft worldwide.

This work discusses the major fire tests used to regulate fire safety of wires and cables in the US and EU.

NATIONAL ELECTRICAL CODE

The National Electrical Fire Code (NEC, NFPA 70)[1] applies to the installation and removal of electrical and optical fiber wires and cables in public and private premises, including buildings, structures, mobile homes, recreational vehicles, and floating buildings. It does not apply to installations used to supply electricity in ships, railway rolling stock, aircraft, or automotive vehicles other than mobile homes and recreational vehicles. It also does not apply to installations underground in mine, to installations of communications equipment under the control of communications utilities located or to installations under the control of an electric utility. Virtually all US jurisdictions, and both major building codes, namely the International Building Code (IBC)[2] and NFPA 5000 (NFPA Building Construction and Safety Code)[3] use the NEC as the code with the requirements for fire safety of wires and cables.

Power cables: In the NEC Chapter 3 (Wiring Methods and Materials) contains the fire safety requirements for all power cables. Articles 300 (General Requirements for Wiring Methods and Materials) and 310 (Conductors for General Wiring) contain most of the necessary details for constructions but do not detail the fire test requirements that these wires and cables (or that the corresponding insulations or jackets) must meet. The key table that describes all the relevant power cables by type is Table 310.104(A) Conductor Applications and Insulations Rated 600 Volts, which describes all the acceptable types of cables and its corresponding insulations and outer coverings (including nonmetallic jackets). In the vast majority of cases either the insulation or the outer covering is required to be "flame-retardant" (unless it is fluorinated or non-combustible). The requirements for a material to be "flame-retardant" are not contained in the NEC but there is a requirement that all cables must be "listed" and the NEC contains each of the "Product Safety Standards" (usually Underwriters Laboratories standards) to which each type of cable must be listed. Several types of cables have their own articles: armored cable (article 322), flat cable assemblies (article 324), integrated gas spacer cable (article 326), medium voltage cable (article 328), metal clad cable (article 330), mineral insulated cable (article 332), nonmetallic sheathed cable (article 334), power and control tray cable (article 336), service entrance cable (article 338) and underground feeder cable (article 340). The fire test requirements for materials are mostly based on the UL 94 series of tests[4] and those for the cables themselves are mostly (but not exclusively) based on the UL VW-1[5] vertical wire test (Fig. 1).

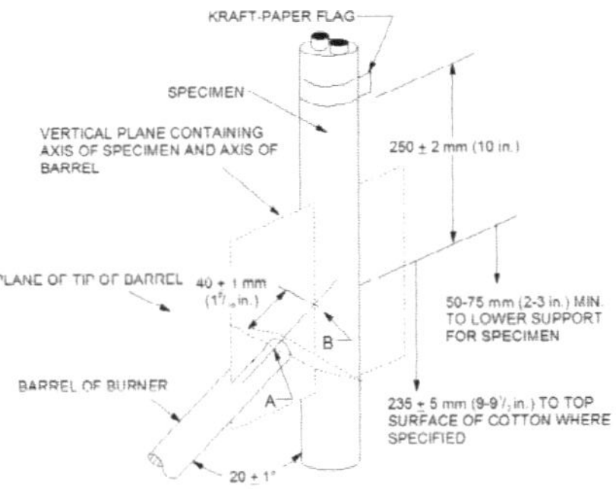

Figure 1: UL VW-1 (500 W Intensity) Test Layout

Cables pass the test if, after five 15 s applications of a small flame, they did not burn more than 0.25 m, did not drip enough to ignite a swab of cotton placed on the floor, or continued to burn, in the absence of the igniter, for over 1 min. The test is quite mild, particularly when large diameter cables are used.

Cables for power limited and communications circuits: These cables are those with the most severe reaction-to-fire fire test requirements in the NEC. They are covered in articles 725 (remote-control, signaling, and power-limited circuits that are not an integral part of a device or of utilization equipment), 760 (fire alarm systems), 770 (optical fiber cables) and articles in Chapter 8 (i.e. article 800 on communications circuits, article 820 on coaxial cable distribution of radio frequency signals typically employed in community antenna television systems, article 830 on network-powered broadband communications systems and article 840 on premises-powered broadband communications systems), where the fire performance of the cable itself, and not that of the component materials, is described.

These cables are covered by a hierarchy of 4 fire tests (Fig. 2; in increasing order of severity): UL VW-1 (see above), UL 1685[6], UL 1666[7] and NFPA 262[8], and cables that are listed as complying with any test are explicitly allowed to substitute for a cable required to comply with a less severe test. Cables complying with the VW-1 test are known as "limited use cables" permitted to be installed in dwellings and raceways.

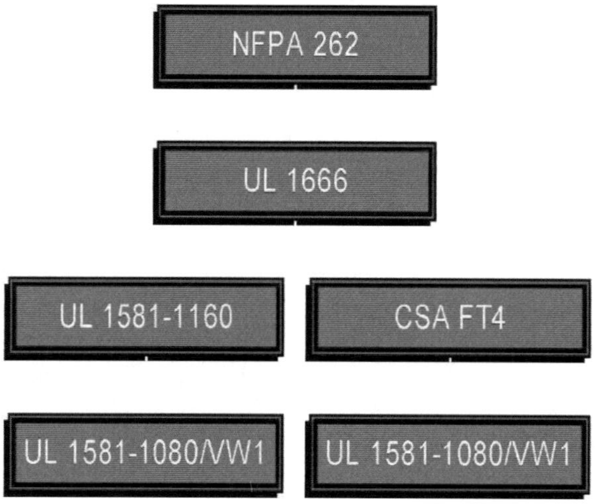

Figure 2: Fire test hierarchy for fire performance of power limited cables in the NEC

UL 1685 is a vertical cable tray test and contains two fire tests, both at a 20 kW intensity: the traditional UL 1581-1160[9] cable tray test with a Venturi burner perpendicular to the cable tray and a version of the CSA FT4[10] test with the same burner at a 20° angle to the horizontal (Fig. 3), making it significantly more severe. The UL 1581 version of the test involves a vertical cable tray, 2.4 m high and 0.3 m wide, loaded with cables and exposed to a gas flame for 20 min. The cables pass the test if the length of char (i.e. wire unprotected by pliant plastic) runs all the way to the top of tray. The burner is sited horizontally (i.e. perpendicular to the tray) at a height of 0.45 m above the bottom of the tray, from the back. The CSA version has a more severe cable loading and cables must exhibit a char length of < 1.5 m to pass the test. There is no difference (in the US) in the applications open to cables meeting either of the vertical cable tray tests (they are permitted to be installed in cable trays) but the Canadian Electrical Code recognizes only the CSA FT4 test and not the alternate one. Cables in this category are allowed to meet a "limited smoke" requirement also, with different requirements for each one of the tests.

The UL 1666 test, for riser cables, is a vertical cable tray test, where cables are exposed in a 5.8 m high concrete shaft divided into two compartments at the3.7 m level, with a 0.3 x 0.6 m opening between the compartments. The ignition source is a gas flame of 145 kW, on for 30 min. The cables will pass the test if there is no "flame" at the top of the bottom compartment during the test. This test is considerably less severe than the plenum cable test (NFPA 262), even though the incident heat is much higher. Risers are "vertical runs penetrating one or more floors or vertical runs in a shaft".

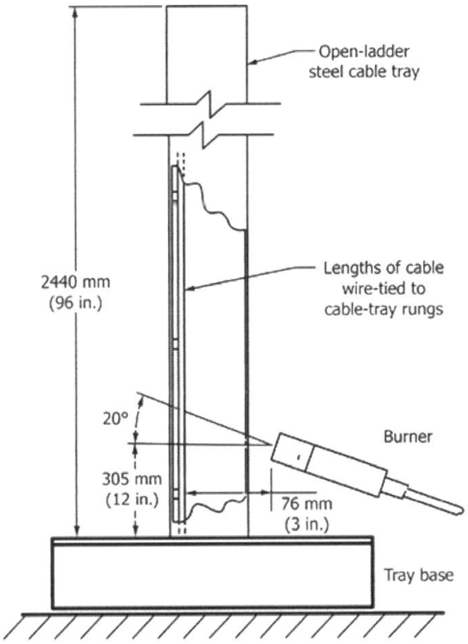

Figure 3: UL 1685/CSA or CSA FT4 (20 kW Intensity) Test Layout

The toughest fire performance requirements, as well as the only smoke requirements, are those for plenum cables. They must pass requirements based on the NFPA 262 test (Fig. 4), a variation on the famous ASTM E84 Steiner tunnel test[11]. This involves loading a horizontal 7.6 m long, 0.3 m wide tunnel with cables and exposing them to a gas flame of 88 kW for 20 min (compared to 10 min for ASTM E84), under an air flow rate of 73 m/min. In order to "pass" the NEC requirements, the cables need to have a flame spread distance of less than 1.5 m, beyond the gas flame itself, a peak smoke optical density not exceeding 0.5 and an average optical density not exceeding 0.15, both measured in the exhaust duct. Note that these requirements are very different from the typical ASTM E84 requirements in codes.

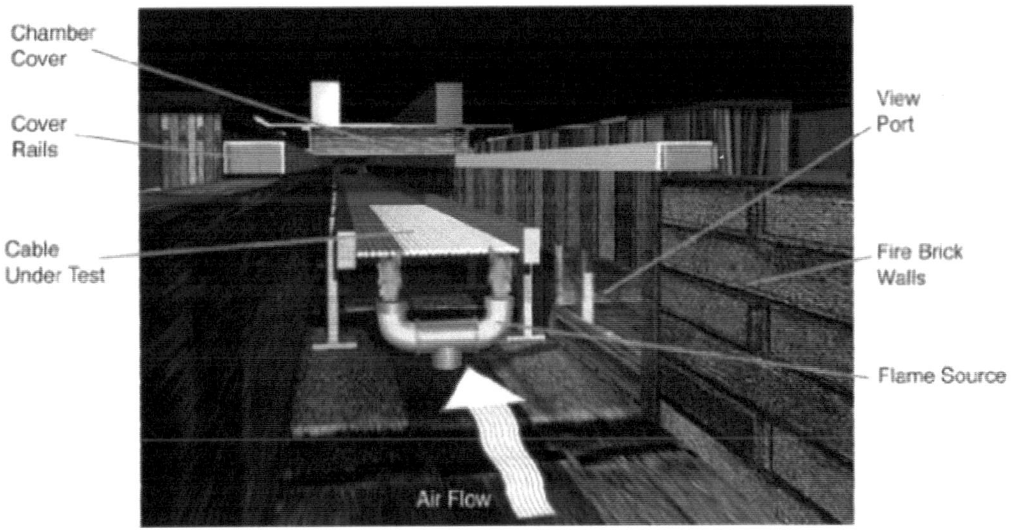

Figure 4: NFPA 262 (88 kW Intensity) Test Layout

It is interesting to note that fire test requirements for all products (including wires and cables) contained in plenums are under the jurisdiction of NFPA 90A[12] (air conditioning standard) and not of the NEC, as determined by NFPA's Standards Council. Plenums are defined in the NEC and in NFPA 90A as "A compartment or chamber to which one or more air ducts are connected and that forms part of the air distribution system".

It is also important to understand what types of ducts and plenums are identified in the NEC and the corresponding terminology because the designations are not obvious. NEC Section 300.22 covers three types of spaces: "Ducts for dust, loose stock, or vapor removal", "Ducts specifically fabricated for environmental air" and "Other spaces used for environmental air (plenums)". Only the latter are the spaces typically considered plenums and are those in which plenum cables are permitted to be installed. Small amounts of plenum cables are permitted in "ducts specifically fabricated for environmental air", as long as they are "necessary to connect to equipment or devices associated with the direct action upon or sensing of the contained air, and the total length of such wiring methods or cabling systems" does not exceed 1.2 m. This is an important recent addition to the NEC which recognizes that small lengths of plenum cables are often needed to provide access to equipment contained within ducts. The other types of ducts (ducts for dust, loose stock, or vapor removal) are not allowed to contain cables.

It must be noted that any wires and cables installed within a totally enclosed, non-ventilated, insulated busway having no provisions for plug-in connections or within electrical metallic tubing, flexible metallic tubing, intermediate metal conduit, rigid metal conduit without an overall nonmetallic covering, flexible metal conduit, or, where accessible, surface metal raceway or metal wireway with metal, are permitted in any application (including plenums). The same permission applies to mineral insulated (Type MI) cables without an overall nonmetallic covering, mineral clad (Type MC) cables without an overall nonmetallic covering, armored (Type AC) cables or listed prefabricated cable assemblies of metallic manufactured wiring systems without a nonmetallic sheath. This means that if cables are hermetically enclosed in metal there is no requirement for them to comply with plenum cable fire tests to be allowed into plenums.

Mechanical codes (both the International Mechanical Code[13] (IMC) and the Uniform Mechanical Code[14] (UMC) contain the same requirements for plenum cables either directly or by reference to the NEC.

All power limited cables are required to be marked to indicate the level of fire performance that the cable is listed to. Plenum cables are marked with a P, riser cables with an R, limited use cables with an X and tray cables need no additional marking, by default. All articles addressing these cables contain the required markings, and also tables with permitted substitutions. Table 1 shows the markings. Article 725 covers Class 1, Class 2, and Class 3 Remote-Control, Signaling, and Power-Limited Circuits, wherein Class 2, Class 3 and power limited circuit cables meet fire tests and the power limited circuit cables are intended for installation in cable trays outdoors and are only listed to the tray cable test. Article 760 covers fire alarm cables, which can be power limited (FPL) or not (NPLF), and they must meet at least a tray cable fire test. Optical fiber cables are covered by article 770 and they can be conductive (OFC) or non-conductive (OFN) and must also at least meet a tray cable fire test. Article 800 covers communications circuit cables and they tend to be permitted to substitute for almost any other power limited cable which complies with the same fire test or a less severe one. Article 820 covers coaxial cable distribution of radio frequency signals typically employed in community antenna television (CATV) systems and article 830 covers network-powered broadband communications systems that provide any combination of voice, audio, video, data, and interactive services through a network interface unit. Article 830 addresses both medium power circuit cables (BM), which are not allowed to be used in plenums and must meet at least a cable tray fire test, and low power circuit cables (BL), which cover the full gamut of fire performance.

A cable which has been listed to a more severe fire test is always permitted to substitute for one of the same type listed to a less severe fire test. However, cables are also permitted to substitute for other cables based not just on the fire test to which they are listed but also other physical and electrical properties.

Table 1 – NEC Markings of Power Limited Cables				
	Plenum	**Riser**	**Tray**	**Limited Use**
Article 725	CL2P, CL3P	CL2R, CL3R	CL2, CL3, PLTC	CL2X, CL3X
Article 760	FPLP, NPLFP	FPLR, NPLFR	FPL, NPLF	
Article 770	OFNP, OFCP	OFNR, OFCR	OFNG, OFN, OFCG, OFC	
Article 800	CMP	CMR	CM	CMX
Article 820	CATVP	CATVR	CATV	CATVX
Article 830	BLP	BMR, BLR	BM, BL	BLX

Wires and cables associated with information technology (IT) equipment: These IT area cables are covered by article 645 of the NEC and most of the fire safety requirements are extracted from NFPA 75[15]. The default type of cables needs to comply with the tray cable fire test requirements (i.e. UL 1685). When wires and cables are installed in ceiling cavity plenums they must meet plenum cable requirements. The same applies if they are installed in "other spaces used for environmental air" but only if "that air space is used for the movement of air". When wires and cables are installed in raised floor plenums they must meet plenum cable requirements if the air space below the raised floor is not sprinklered, but if it is sprinklered the cables need only to comply with tray cable requirements. Power cords must comply with the less severe UL 60950-1[16] or UL 62368-1[17] but are limited to 4.5 m and must be connected directly to IT equipment.

Fire resistive cables: Fire resistive cables are those that will continue operating during a fire, irrespective of their contribution (or lack of it) to the fire itself. They can be listed as individual cables or the cables can be listed as a part of electrical circuit integrity systems.

Electrical circuit integrity systems are those consisting of components and materials intended for installation as protection for specific electrical wiring systems, with respect to the disruption of electrical circuit integrity upon exterior fire exposure. These systems have been investigated for their ability to remain electrically functional during a fire exposure and after the impact, the erosion and cooling effect of a water hose stream test. The fire resistance ratings apply only to the entire system assembly, constructed using the combination of components and materials specified in the individual system. Components and materials are designated for use in a specific individual system for which corresponding ratings have been developed, and are not intended to be interchanged between systems. No fire resistance ratings are assigned to individual system components or materials, and listed electrical circuit integrity systems can contain cables that are not themselves fire resistive cables. Electrical circuit integrity systems are intended to be fastened to a concrete or masonry wall or a concrete floor-ceiling assembly. The fire resistance rating of the assembly used for the fire test (in accordance with UL 2196[18]) must be at least equal to the rating of the electrical circuit integrity system, to ensure that the complete electrical circuit integrity system will survive during the test. Components or materials cannot be interchanged between listed systems.

Electrical circuit integrity systems must be listed to UL 1724[19]. UL 2196, which is the fire resistance test (with a standard time-temperature curve, in accordance with ASTM E119[20]) contained within UL 1724, is a test where either the cable system or the fire resistive cable itself is embedded into a complete building assembly (such as a wall) to be constructed around the electrical circuit. For wall assemblies, the sample is typically 3 m x 3 m and the average temperature ***rise*** on the unexposed surface of the wall must be <139°C with no individual temperature ***rise*** exceeding 181°C and no flaming on the unexposed surface *(Note that this refers to temperature rise and not actual temperature)*. After the fire test the assembly must comply with a hose stream test, which involves water being delivered at high pressure and high flow rate. For acceptable performance, the water stream cannot penetrate through the building assembly. Voltage and current are applied to the cable or assembly during the fire exposure portion of the test, and the electrical and visual performance of the cable or assembly is monitored. Neither the cable nor the assembly is electrically energized during the hose stream part of the test, but it is visually inspected and electrically tested after the hose stream test and fails the test if it cannot continue transmitting a signal.

Fire-resistive cables are insulated electrical cables intended for installation as specified in the individual electrical circuit integrity systems or for installation on their own. The cables must have been tested, and listed, for their ability to remain electrically functional during a fire exposure (to UL 2196) and after the impact, erosion and cooling effect of a water hose stream test.

UL 2196 also includes a fire exposure representing a hydrocarbon pool fire, similar to the exposure associated with the time temperature fire curve in UL 1709[21], but that exposure is not required for cables or assemblies regulated by the NEC. There are also two hose stream exposures defined in UL 2196, the normal exposure and a low impact hose stream, used for circuit integrity (CI) cables only.

Listed fire resistive cables or listed electrical circuit integrity systems are required as wiring for fire pumps (NEC article 695), for emergency systems (NEC article 700) and for critical operation power systems (NEC article 708). In every case the option exists also for encasing the cable in 50 mm of concrete. The fire resistance rating of these cables or systems must be a minimum of two hours. Listed fire resistive optical fiber cables must be part of a listed electrical circuit integrity system with a two-hour minimum fire resistance rating. Circuit integrity (CI) cables (with the low impact hose stream) that are listed in accordance with UL 2196 are allowed in the following applications: Class 1 or Class 2, control, signal or power-limited circuits (NEC article 725), fire alarm cables (NEC article 760), optical fiber cables (NEC article 770) and communications circuits (NEC article 800), all of which have limited power. The 2014 edition of the NEC first added an article (article 728) regulating the installation of fire resistive cables.

Note that any cable that has been listed to one of the reaction-to-fire fire tests described above is entitled to be additionally tested as a fire resistive cable or a circuit integrity cable and be designated as such.

PASSENGER TRAINS: ABOVE GROUND AND UNDERGROUND

In the US, the safety of passenger trains is regulated by the Federal Railroad Administration (FRA) for inter-city and inter-state transport. In the US Code of Federal Regulations, the final rule for fire safety was issued in June 2002 (49 CFR 238)[22] and it has not been amended since. That rule does not contain any information on requirements for fire safety of electrical or optical wires and cables. However, most operators and authorities having jurisdiction of trains and underground transport comply with NFPA 130[23], which does have corresponding fire safety requirements for wires and cables.

Chapter 8 of NFPA 130, on vehicles, contains the fire safety requirements for wires and cables used to regulate the fire safety of most trains and underground in the US, Canada and Singapore. It states that all wires and cables shall comply with the CSA FT4 requirements contained in UL 1685 for both cable char height, total smoke released and peak smoke release rate. As an alternate, wires and cables are allowed to comply with the plenum requirements based on NFPA 262 (as shown above). The reason to allow this is that (as also allowed in the NEC) if you are already listed to a more severe fire test (and one that contains also smoke requirements) there is no need to retest to a less severe fire test. The same requirements apply to wires and cables in trainways (meaning the area where the trains operate) and in enclosed stations.

NFPA 130 adds also that wires and cables used for heat, smoke, or other detection system must comply with the UL 1685 shown above and one of the following: (a) be capable of maintaining circuit integrity for 15 minutes when tested in accordance with IEC 60331-11[24], (b) demonstrate that, if circuit integrity is tested during the UL 1685 vertical flame fire test, a current continues operating for at least 5 minutes during the test or (c) use cables listed as a circuit integrity cable in accordance with the NEC.

In Europe, fire safety requirements for passenger trains are covered by EN 45545-2[25], Railway Applications - Fire protection on railway vehicles, which contains also the requirements for wires and cables. It starts by stating that products that comply with the requirements of class A1 (meaning basically that they are non-combustible) in EN 13501-1[26] need no further testing. The additional requirements for wires and cables are a function of the type of cable and the application. However, basically all cables for both interior and exterior applications must meet requirements based on several tests: IEC 60332-1-2[27] (small scale vertical flame test), IEC 60332-3-24[28] (vertical cable tray test, using a 20.5 kW Venturi burner, with maximum flame spread of no more than 2.5 m), smoke obscuration based on the 3-meter cube test in IEC 61034-2[29] and smoke toxicity requirements based on the test contained in EN 50305[30]. The pass/fail requirements are slightly different for interior and exterior applications. Interestingly there are also requirements for printed circuit boards without any attached technical equipment and for small electrotechnical products including low power circuit breakers, overload relays, contactors, contactor relay, switches, control or signaling switches, terminals and fuses. Printed circuit boards must meet an oxygen index of $\geq$ 28 based on ISO 4589-2[31] (equivalent to ASTM D2863[32]) or a minimum glow wire temperature of 850°C based on IEC 60695-2-11[33]. Small electrotechnical products must comply with a vertical flammability rating of V-0 based on IEC 60695-11-10[34] (a 50 W flame, similar to that in UL 94[35]).

SHIPS

The fire safety of ships that engage in international trade and fly the flag of a country that has signed the International Convention for Safety of Life at Sea (SOLAS)[36], which includes the USA, must comply with the regulations of the International Maritime Organization (IMO). The IMO regulations are detailed in the SOLAS book, periodically amended by "Resolutions" of the IMO committees, and ratified by the signatory states. Details of the fire issues are given in the IMO Fire Test Procedures Code[37], also reissued regularly. Some special vessels are regulated separately: high speed craft that is never too far from shore is regulated by the IMO High Speed Craft Code[38]. All ships that sail in US waters must comply with the requirements specified by the US Coast Guard (USCG), laid out in US Federal Government - Coast Guard: Title 46, Shipping, Code of Federal Regulations, Parts 1–199[39], and in NVIC (US Coast Guard Guide to Structural Fire Protection)[40]. However, neither the IMO FTP code nor the US Coast Guard NVIC contain any requirements for fire safety of wires and cables. Thus, non-US ships have no mandatory fire safety requirements for wires and cables but there are some voluntary requirements in various standards.

Fire safety requirements for wires and cables in US ships are included in 46 CFR Subchapter J[41] (with details in subpart 111.60, Wiring Materials and Methods). However, as will be seem below, there is a significant amount of overlap and some confusion.

The Coast Guard is the authority having jurisdiction over ships engaging in international trade and sailing into US waters or US ports; such ships must comply with IMO regulations and need not also comply with separate Coast Guard requirements. With the instructions from the US federal government in the 1990s that standards should, whenever possible, be delegated to private organizations, the Coast Guard and NFPA agreed to develop NFPA 301[42], which applies to passenger vessels carrying more than 12 passengers, cargo and tank vessels, and towing vessels 12 m or more in length and greater than 500 hp; it does not apply to military ships (although military ships must comply with Coast Guard requirements). NFPA 302[43] applies to boats of less than 300 gross tons used for pleasure or commercial purposes that meet one of the following conditions: (a) motor craft; (b) boats that use cooking, heating, or auxiliary appliances; or (c) boats that have permanently installed ignition source(s); it does not apply to personal watercraft. In fact, passenger vessels carrying up to 12 passengers are covered by NFPA 302. The market for ships in the US that do not need to be built to comply with IMO requirements is actually quite large, because it encompasses all the ships sailing through rivers (e.g., Mississippi River), lakes (e.g., Great Lakes), and coastal waters (without reaching into international waters), as well as ships sailing in amusement parks. NFPA 301 does not become a requirement unless someone chooses to meet it, and that rarely happens, because it contains more severe sprinkler requirements than the US Coast Guard does.

Typically, electrical installations in ships in the USA have always been regulated by the requirements included in 46CFR Subchapter J and by the recommendations of IEEE 45[44]. Recent editions of IEEE 45 have taken the detailed requirements for wire and cable out of IEEE 45 and placed them into IEEE 1580[45]. The critical requirement is that all cables approved to IEEE 1580 must meet the flame spread requirements of the vertical cable tray test in IEEE 1202[46], which is identical to CSA FT4. IEEE 1580 does not require all cables to meet the corresponding smoke obscuration criteria, but it requires that cables that are marked as "limited smoke" must meet the same flame spread and smoke release requirements of CSA FT4. IEEE 1580 also contains material requirements for some “low smoke” materials for uses as cable insulation and cable jacketing, and, by reference, smoke obscuration criteria based on MIL C24643A[47]. The military specification MIL C24643 includes several requirements for low smoke and halogen-free cables. The 46CFR states that each marine shipboard cable must meet all of the construction and identification requirements of IEEE 45 or MIL-C-24643A (both incorporated by reference), and the respective flammability tests contained in them and be of a copper stranded type. This means that they must meet the CSA FT4 test requirements (as stated above), for flame spread. If cables are constructed to IEC 60092-353[48] they must meet the flammability requirements of Class A of IEC 60332-3-22[49]. This test is, again, a cable vertical flame spread test requiring a flame spread of less than 2.5 m and differing from the one that applies to trains in Europe (as shown above) simply in the type of cables and in the mounting. It is interesting to note that the committees associated with IEEE 45 and IEEE 1580 are not very active as these documents have not been updated very recently; the latest version of IEEE 45 is dated 2004 and the one for IEEE 1580 is dated 2010 and neither are standards but just guidance documents as recommended practices.

AIRCRAFT

Fire safety of aircraft is covered by the Federal Aviation Administration (FAA), which regulates all aircraft operating in the US and, virtually, all other aircraft throughout the world. Regulation is covered in Title 14 in the Code of Federal Regulations. All of the aircraft fire test requirements are published in a Fire Test Handbook[50]; found on the FAA website, with regular updates. Wires and cables need to meet a relatively mild exposure to a Bunsen burner, at a 60° angle, for 30 s (Figs. 5 and 6; Handbook, Chapter 3). However, most of the wires and cables used in aircraft (as initially installed by the air frame manufacturers) exhibit a fire performance that significantly exceeds the test requirements. The entertainment systems (usually introduced late into the aircraft by the airlines), typically simply meet the FAA requirements. The FAA also has an ignition test intended for use in determining the resistance of high-temperature electrical wire used in designated fire zones to damage due to flame and vibration; Handbook, Chapter 14).

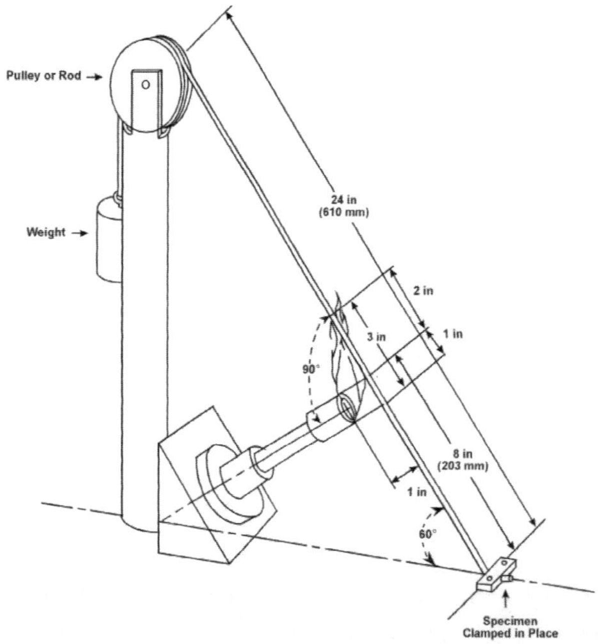

Figure 5: FAA 60-Degree Electrical Wire Bunsen Burner Fire Test Setup

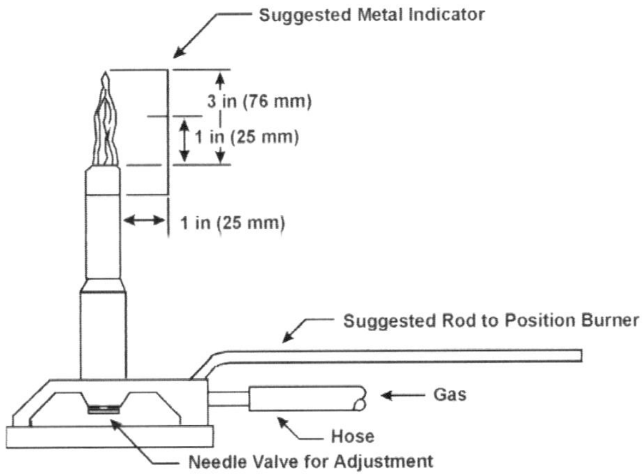

Figure 6: FAA Wire Test Burner Plumbing and Burner Flame Height Indicator

The FAA has been pursuing, for many years, the development of a new test for materials concealed outside of the passenger cabin, and key products in that location are wires and cables. One test that is under development for that purpose is based on the flooring radiant panel (ASTM E648[51]), because that would be consistent with the FAA approach regarding thermal and acoustic insulation, which is also contained in concealed aircraft spaces and is regulated by an ignition test based on ASTM E648 (Handbook Chapter 23). The FAA also has a smoke obscuration test for wire and cable, using the NBS smoke chamber (similar to ASTM E662[52], Handbook, Chapter 19). The FAA prefers to develop test methods that are based on a fire hazard assessment and that are of specific use to the aircraft industry, and to work with the interested parties in the industry to complete the final test modifications and improvements. They generally do not use standards developed by other organizations but modify them for aircraft use.

HIGHWAY VEHICLES

Such vehicles are passenger road vehicles, including private cars, trucks (or lorries), buses and school buses. There are almost no mandatory fire safety requirements at all for such vehicles. The National Highway Safety Transit Administration (NHTSA) has as its only mandatory fire safety test one that applies a small burner to interior materials (FMVSS 302[53]). This test is very mild and applies only to materials found within 13 mm of the passenger compartment (which typically does not include wires and cables). It has been replaced in most public transport applications by the FTA guidelines[54]; however, these guidelines (which are not requirements) also contain no information about electrical cables. Some manufacturers of passenger road vehicles have some of their own cable specifications, which helps obtain a certain degree of fire safety, but this area needs much progress.

CABLES IN THE EU

The EU Construction Products Regulation (CPR)[55], issued in 2011, lays down harmonized rules for the marketing of construction products in the EU. It covers any cable product intended to be incorporated into construction works (fixed installation), including both buildings and civil engineering works and subject to performance requirements on reaction and/or resistance to fire, as of July, 1st 2016. The CPR introduced requirements for declarations of performance and CE marking of cables for construction works in terms of fire performance. CPR replaced the Construction Products Directive (CPD), issued in 1988[56], which provided guidelines without actually requiring member states to comply with them.

The CPR developed Euroclasses, which are uniform assessment methods or classifications that are intended to reflect "real-life environments for building products", based on reaction-to-fire performance. The classes developed for the reaction-to-fire performance of cables require measurements of flame spread and heat release, and recommend additional declarations based on flaming droplets, smoke emission and smoke corrosivity of emissions. Based on the fire performance of the cables, the cables are placed into the Euroclass system as an indicator of expected behavior during a fire. The Euroclass system provides rankings based on various criteria but the required criteria are those based on heat release and flame spread, which indicate the expected contribution to expanding a fire (ranging from "no contribution" to "high contribution," with several levels in between). The seven "Reaction-to-fire performance classes for electric cables" (Euroclasses; A_{CA}, $B1_{CA}$, $B2_{CA}$, C_{CA}, D_{CA}, E_{CA} and F_{CA}) are shown in Table 2. Heat release and flame spread are the classification criteria, but smoke production, flaming droplets and acidity are additional classification criteria (not required).

Table 2 – Classes of Reaction-to-Fire Performance for Electric Cables

Class	Test Method(s)	Classification Criteria	Additional Classification
A_{CA}	ISO 1716[57]	PCS $\leq$ 2.0 MJ/kg	
$B1_{CA}$	FIPEC$_{20}$[58] Scen 2 and	FS $\leq$ 1.75 m and THR$_{1\ 200s}$ $\leq$ 10 MJ and Peak HRR $\leq$ 20 kW and FIGRA $\leq$ 120 W/s	Smoke production and Flaming droplets/particles and Acidity
	IEC 60332-1-2	H $\leq$ 425 mm	
$B2_{CA}$	FIPEC$_{20}$ Scen 1 and	FS $\leq$ 1.5 m and THR$_{1\ 200s}$ $\leq$ 15 MJ and Peak HRR $\leq$ 30 kW and FIGRA $\leq$ 150 W/s	Smoke production and Flaming droplets/particles and Acidity
	IEC 60332-1-2	H $\leq$ 425 mm	
C_{CA}	FIPEC$_{20}$ Scen 1 and	FS $\leq$ 2.0 m and THR$_{1\ 200s}$ $\leq$ 30 MJ and Peak HRR $\leq$ 60 kW and FIGRA $\leq$ 300 W/s	Smoke production and Flaming droplets/particles and Acidity
	IEC 60332-1-2	H $\leq$ 425 mm	
D_{CA}	FIPEC$_{20}$ Scen 1 and	THR$_{1\ 200s}$ $\leq$ 70 MJ and Peak HRR $\leq$ 400 kW and FIGRA $\leq$ 300 W/s	Smoke production and Flaming droplets/particles and Acidity
	IEC 60332-1-2	H $\leq$ 425 mm	
E_{CA}	IEC 60332-1-2	H $\leq$ 425 mm	
F_{CA}	No performance determined		

Notes to Table 2: PCS: gross calorific potential; FS: flame spread (damaged length); THR: total heat released from test start until end of test, excluding the contribution from ignition source; HRR: heat release rate, excluding the contribution from ignition source; FIGRA: FIre Growth RAte index defined as the highest value of the quotient between HRRsm30 excluding the contribution of ignition source and time; H: flame spread. The test from the FIPEC project was subsequently standardized as EN 50399[59].

The FIPEC test is a vertical cable tray test that involves basically the same apparatus as IEC 60332-3, except that it adds measurements for heat and smoke release in the exhaust duct (Fig. 7). There are two versions of this test, namely Scenario 1(Scen 1) and Scenario 2 (Scen 2). Scen 1 involves a test procedure very similar to that in IEC 60332-3, using the 20.5 kW Venturi burner. It differs from the test method in IEC 60332-3 in two major ways: (a) the airflow into the test chamber was increased from 5,000 L/min to 8,000 L/min, so as to improve the response of the heat release measurement and (b) all cables with a diameter greater than 5 mm are mounted individually with a spacing of one cable diameter between each cable. Cables with a diameter less than 5 mm are mounted in bundles (non-twisted) of 1 cm diameter with a 1 cm distance between each bundle. Scen 2 is a test procedure which uses the same IEC 60332-3 burner but with an increased heat input, at a level of 30 kW. Moreover, a non-combustible backing board is mounted on the rear of the cable ladder to provide potential re-radiation. The airflow into the test chamber and the cable mounting procedure is identical to that in Scen 1.

The EU CPR also contains fire resistance requirements for those electric cables that require circuit integrity characteristics. Thus, cables used to ensure the continuity of the power and/or signal supply of safety installations must meet these tests. The various fire resistance classes are based on the survival times in terms of the circuit integrity, and the corresponding times are: 15, 30, 60, 90 and 120 minutes. Classifications of PH 15, PH 30, PH 60, PH 90 and PH 120 apply to small diameter power or signal cables or systems (< 20 mm diameter and with conductor sizes $\leq 2.5\ mm^2$) and P 15, P 30, P 60, P 90 and P 120 apply to the remainder of the cables.

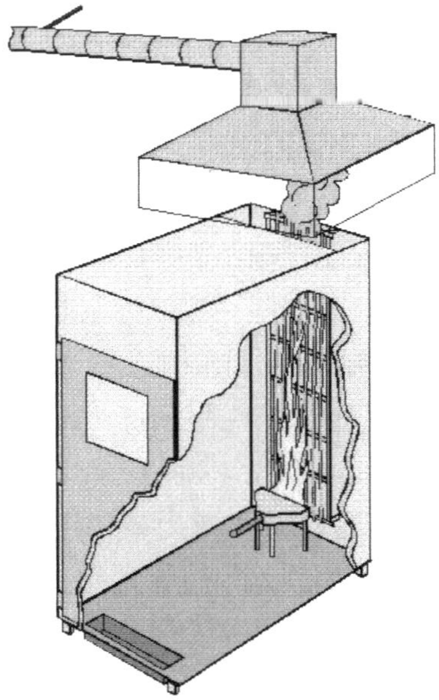

Fig. 7 – Layout of the FIPEC – EN 50399 Vertical Cable Tray Test

CONCLUSIONS

Statistics by NFPA (for the US)[60] indicate that electrical wire and cable insulation represents the item first ignited in homes in 5% of fires and 5% of fire deaths and fires in homes represent approximately ¾ of all fires. It can be assumed that electrical fires in the EU will be relatively similar in this regard. However, when fires involving electrical cables occur, they can result in significant loss of life and of property, due to the presence of these products throughout each building or vehicle. Adequate regulations for the built environment can be found in both the US (via the NEC) and the EU (via the CPR). It appears that the emphasis, which is mainly on flame spread, is, albeit slowly, moving toward increased use of heat release and fire hazard assessment. This means that tests can be chosen in order to obtain more valid fire safety engineering test results, which can then be used as input into fire models.

In the area of transportation, statistics indicate that electrical cables are a critical and growing combustible component. In recent years, this has led to more emphasis being placed on fire safety requirements for electric cables in various public transportation sectors: ships, trains, and aircraft. However, the largest gaps are in terms of those ships regulated exclusively by SOLAS and, of course, all highway vehicles. The fire safety record of ships is fairly reasonable, but that of highway vehicles is definitely very poor (and they typically represent about ¾ of all vehicle fires and an even larger fraction of vehicle fire fatalities).

Smoke obscuration is rarely a requirement for cables in building environments, with the notable exception of plenum cables in the US, which must meet a very severe flame spread and smoke test. In most other building environments, both in the US and the EU, smoke requirements tend to be voluntary.

Smoke obscuration requirements are included in a variety of transportation vehicle environments, to ensure that egress can be effected in an orderly and safe manner.

A fire hazard assessment developed as a result of heat release data should be able to assess a new wire and cable product being considered for use and to conclude whether the new product considered is safer, in terms of predicted fire performance, than the one in established use. The result of such assessments will be the ability to design, with a high degree of confidence, electrical systems which offer excellent fire protection.

REFERENCES

[1] National Electrical Code (NEC, NFPA 70), National Fire Protection Association, Quincy, MA, US.
[2] IBC (International Building Code), International Code Council (ICC), Washington, DC, US.
[3] Building Construction and Safety Code (NFPA 5000), National Fire Protection Association, Quincy, MA, US.
[4] UL 94, "Standard for Tests for Flammability of Plastic Materials for Parts in Devices and Appliances", Underwriters Laboratories, Northbrook, IL, US.
[5] UL VW-1 (vertical-wire) flame test in ANSI/UL 1581, "Reference Standard for Electrical Wires, Cables and Flexible Cords", Underwriters Laboratories, Northbrook, IL, US.
[6] UL 1685, "Standard for Vertical-Tray Fire-Propagation and Smoke-Release Test for Electrical and Optical-Fiber Cables", Underwriters Laboratories, Northbrook, IL, US.
[7] UL 1666, "Standard for Test for Flame Propagation Height of Electrical and Optical-Fiber Cables Installed Vertically in Shafts", Underwriters Laboratories, Northbrook, IL, US.
[8] NFPA 262, "Standard Method of Test for Flame Travel and Smoke of Wires and Cables for Use in Air-Handling Spaces", National Fire Protection Association, Quincy, MA, US.
[9] UL Vertical-Tray Flame Test in ANSI/UL 1581, "Reference Standard for Electrical Wires, Cables and Flexible Cords", Underwriters Laboratories, Northbrook, IL, US.
[10] CSA FT4, Vertical Cable Tray Test in CSA 22.2, "Test Methods for Electrical Wires and Cables (C22.2 No. 0.3)", Canadian Standards Association, Rexdale, Ont., Canada.
[11] ASTM E84, "Standard Test Method for Surface Burning Characteristics of Building Materials", ASTM International, West Conshohocken, PA, US.
[12] NFPA 90A, "Standard for the Installation of Air-Conditioning and Ventilating Systems", Quincy, MA, US.
[13] IMC (International Mechanical Code), International Code Council (ICC), Washington, DC, US.
[14] UMC (Uniform Mechanical Code), International Association of Plumbing and Mechanical Officials (IAPMO), Ontario, CA, US.
[15] NFPA 75, "Standard for the Fire Protection of Information Technology Equipment", Quincy, MA, US.
[16] UL 60950-1, "Information Technology Equipment - Safety - Part 1: General Requirements", Underwriters Laboratories, Northbrook, IL, US.
[17] UL 62368-1, "Audio/video, information and communication technology equipment - Part 1: Safety requirements", Underwriters Laboratories, Northbrook, IL, US.
[18] UL 2196, "Standard for Tests for Fire Resistive Cables", Underwriters Laboratories, Northbrook, IL, US.
[19] UL 1724, "Outline of Investigation for Fire Tests for Electrical Circuit Protective Systems", Underwriters Laboratories, Northbrook, IL, US.
[20] ASTM E119, "Standard Test Methods for Fire Tests of Building Construction and Materials", ASTM International, West Conshohocken, PA, US.
[21] UL 1709, "Standard for Rapid Rise Fire Tests of Protection Materials for Structural Steel", Underwriters Laboratories, Northbrook, IL, US.
[22] Federal Register, Tuesday June 25, 2002, Dept. of Transportation, Federal Railroad Administration, 49 CFR Part 238 et al., Federal Register, Vol 67, No. 122, pp. 42892-42912, "Passenger Equipment Safety Standards; Final Rule."

[23] NFPA 130, "Standard for Fixed Guideway Transit and Passenger Rail Systems", Quincy, MA, US.
[24] IEC 60331-11, "Tests for electric cables under fire conditions - Circuit integrity - Part 11: Apparatus - Fire alone at a flame temperature of at least 750°C", International Electrotechnical Commission, Geneva, Switzerland.
[25] EN 45545-2, "Railway applications - Fire protection on railway vehicles - Part 2: Requirements for fire behavior of materials and components", European Committee for Standardization (CEN), Brussels, Belgium.
[26] EN 13501-1, "Fire classification of construction products and building elements. Classification using test data from reaction to fire tests", European Committee for Standardization (CEN), Brussels, Belgium.
[27] IEC 60332-1-2, "Tests on electric and optical fibre cables under fire conditions - Part 1-2: Test for vertical flame propagation for a single insulated wire or cable - Procedure for 1 kW pre-mixed flame", International Electrotechnical Commission, Geneva, Switzerland.
[28] IEC 60332-3-24, "Tests on electric and optical fibre cables under fire conditions - Part 3-24: Test for vertical flame spread of vertically-mounted bunched wires or cables - Category C", International Electrotechnical Commission, Geneva, Switzerland.
[29] IEC 61034-2, "Measurement of smoke density of cables burning under defined conditions - Part 2: Test procedure and requirements", International Electrotechnical Commission, Geneva, Switzerland.
[30] EN 50305, "Railway applications. Railway rolling stock cables having special fire performance. Test methods", European Committee for Standardization (CEN), Brussels, Belgium.
[31] ISO 4589-2, "Plastics - Determination of burning behaviour by oxygen index - Part 2: Ambient-temperature test", International Organization for Standardization, Geneva, Switzerland.
[32] ASTM D2863, "Standard Test Method for Measuring the Minimum Oxygen Concentration to Support Candle-Like Combustion of Plastics (Oxygen Index)", ASTM International, West Conshohocken, PA, US.
[33] IEC 60695-2-11, "Fire hazard testing - Part 2-11: Glowing/hot-wire based test methods - Glow-wire flammability test method for end-products", International Electrotechnical Commission, Geneva, Switzerland.
[34] IEC 60695-11-10, "Fire hazard testing - Part 11-10: Test flames - 50 W horizontal and vertical flame test methods", International Electrotechnical Commission, Geneva, Switzerland.
[35] UL 94, "Standard Tests for Flammability of Plastic Materials for Parts in Devices and Appliances", Underwriters Laboratories, Northbrook, IL, US.
[36] International Convention for the Safety of Life at Sea (SOLAS), with periodic Amendments (Latest One May 2011), Consolidate Edition, International Maritime Organization, London, UK, 1974, http://www.imo.org/.
[37] International Code for Application of Fire Test Procedures (FTP Code), International Maritime Organization, London, UK, 2010.
[38] International Code of Safety for High-Speed Craft (HSC Code), incorporated into SOLAS as chapter X in 1994, International Maritime Organization, London, UK.
[39] Code of Federal Regulations, US Federal Government - Coast Guard: Title 46, Shipping, CFR, Parts 1–199, Subchapter J: Electrical Engineering (Parts 110–113).
[40] US Coast Guard, NVIC September 1997, "US Coast Guard Guide to Structural Fire Protection", Washington, DC, US.
[41] 46 CFR Subchapter J, "Electrical Engineering", http://tinyurl.com/jyws8r8, Code of Federal Regulations, US.
[42] NFPA 301, "Code for Safety to Life from Fire on Merchant Vessels," National Fire Protection Association, Quincy, MA, US.
[43] NFPA 302, "Standard for Fire Protection Standard for Pleasure and Commercial Motor Craft," National Fire Protection Association, Quincy, MA, US.
[44] IEEE 45, "Recommended Practice for Electric Installations on Shipboard," Institute of Electrical and Electronics Engineers, Piscataway, NJ, US.
[45] IEEE 1580, "Recommended Practice for Marine Cable for Use on Shipboard and Fixed or Floating Platforms," Institute of Electrical and Electronics Engineers, Piscataway, NJ, US.
[46] IEEE1202, "Flame-Propagation Testing of Wire and Cable," Institute of Electrical and Electronics

Engineers, Piscataway, NJ, US.
[47] MIL-C-24643A, "Military Specification, General Specification for Cables and Cords, Electric, Low Smoke, for Shipboard Use," Department of Defense, Philadelphia, PA, US.
[48] IEC 60092-353, "Electrical installations in ships – Part 353: Power cables for rated voltages 1 kV and 3 kV", International Electrotechnical Commission, Geneva, Switzerland.
[49] IEC 60332-3-24, "Tests on electric and optical fibre cables under fire conditions - Part 3-22: Test for vertical flame spread of vertically-mounted bunched wires or cables - Category A", International Electrotechnical Commission, Geneva, Switzerland.
[50] Federal Aviation Administration, "Aircraft Materials Fire Test Handbook," DOT/FAA/AR-00/12, http://www.fire.tc.faa.gov/handbook.stm.
[51] ASTM E648, "Test Method for Critical Radiant Flux of Floor-Covering Systems Using a Radiant Heat Energy Source," ASTM International, West Conshohocken, PA, US.
[52] ASTM E662, "Test Method for Specific Optical Density of Smoke Generated by Solid Materials," ASTM International, West Conshohocken, PA, US.
[53] FMVSS 302, "Motor Vehicle Safety Standard No. 302, Flammability of Materials - Passenger Cars, Multipurpose Passenger Vehicles, Trucks and Buses," National Highway Traffic Safety Administration, Washington, DC, US.
[54] Federal Transit Administration, "Recommended Fire Safety Practices for Transit Bus and Van Materials Selection; Notice," Federal Register, Vol. 57, No. 8, January 13, 1992, pp. 1360–63.
[55] European Union Construction Products Regulation (CPR), Regulation (EU) No 305/2011, April 4, 2011, European Commission, Brussels, Belgium.
[56] European Union Construction Products Directive (CPD), Council Directive 89/106/EEC), December 21, 1988, European Commission, Brussels, Belgium.
[57] ISO 1716, "Reaction to fire tests for products - Determination of the gross heat of combustion (calorific value)", International Organization for Standardization, Geneva, Switzerland.
[58] Grayson, S.J., van Hees, P., Vercellotti, U., Breulet, H. and Green, A., "Fire Performance of Electric Cables - New Test Methods and Measurement Techniques", Final Report on the European Commission SMT Programme Sponsored Research Project SMT4-CT96-2059, Interscience Communications, London, UK, 2000.
[59] EN 50399, "Common test methods for cables under fire conditions - Heat release and smoke production measurement on cables during flame spread test - Test apparatus, procedures, results", European Committee for Standardization (CEN), Brussels, Belgium.
[60] Ahrens, M., "Home Structure Fires – September 2016", National Fire Protection Association, Quincy, MA, US, 2016.

EFFECTS OF CABLE TRAY CONFIGURATION ON FIRE SPREAD

Pascal Zavaleta, Sophie Bascou, Sylvain Suard
Institut de Radioprotection et de Sûreté Nucléaire (IRSN), PSN-RES, SA2I, Cadarache, St Paul-Lez-Durance Cedex, 13115, France

ABSTRACT

Fires involving electrical cables are one of the main fire hazards in Nuclear Power Plants (NPPs). The aim of this work is to study the impact of cable tray configuration on fire spread over multiple cable trays. Two real-scale cable tray fire tests were thus carried out as part of the OECD PRISME-2 Project (CORE campaign) for studying the effect of a protected cable tray (CORE-2) and slanted ladder cable trays (CORE-3) on the main fire characteristics. By considering new configurations, these tests completed the preliminary CFSS tests of the PRISME-2 project which involved five horizontal ladder cable trays. The CORE-2 test implemented the same horizontal trays set-up except that the lower tray was a protected cable tray. Compared with the five ladder cable trays set-up, the protected cable tray delayed the ignition of about 16 min and led to decrease the total mass loss (TML). Furthermore, the fire growth rates and the heat release rate (HRR) were slightly higher for the CORE-2 test. The CORE-3 test used five horizontal ladder cable trays 1 m long followed by five slanted ladder cable trays 2 m long with a 30 degree angle. This configuration shortened the ignition of about 4 min, led to similar TML and increased the fire spread over the cable trays meaningfully and the HRR peak compared with the horizontal ladder trays configuration. Finally, the other fire characteristics such as the average effective heat of combustion or the gas and soot yields were not affected by the presence of the lower protected cable tray or the ladder cable trays orientation (horizontal or slanted).

Keywords: Protected cable tray; Slanted ladder cable tray; Ignition; Fire spread

NOMENCLATURE

CDG	Carbon dioxide generation
CO	Carbon monoxide
CO_2	Carbon dioxide
C_{soot}	Mass concentration of soot (kg/m^3)
$\Delta H_{c,eff}$, EHC	Effective heat of combustion (MJ/kg)
E_{CO_2}	Energy produced by mass unit of generated CO_2 (13.3 MJ/kg)
E_{O_2}	Energy produced by mass unit of consumed O_2 (13.1 MJ/kg)
HFFR	Halogen free flame retardant
HRR	Heat release rate (kW)
M_a	Molecular mass of ambient air (0.029 kg/mole)
M_i	Molecular mass of gas species i = CO_2, CO and H_2O (kg/mole)
ML	Mass loss (kg)
MLR, $\dot{m}$	Mass loss rate (kg/s)
$\dot{m}_i$	Mass flow rate of species i = CO_2, CO and soot (kg/s)
$\dot{m}_{O_2}^C$	Mass consumption rate of O_2 (kg/s)
$\dot{Q}$	Heat release rate (kW)
$\dot{Q}_{burner}$	Fire power of the gas burner (kW)
$\dot{Q}_{CDG}$	Heat release rate assessed from the CDG calorimetry method (kW)
$\dot{Q}_{OC}$	Heat release rate assessed from the OC calorimetry method (kW)

OC	Oxygen consumption
O_2	Oxygen
P	Pressure (Pa)
R	Ideal gas constant (8.314 J/mole/K)
t_0	Starting time of the power law equation describing the growth of the fire (s)
t_1	Fire growth characteristic time (s)
T	Gas temperature (K)
TML	Total mass loss (kg)
X_i	Molar fraction of gas species i = O_2, CO_2, CO and H_2O (mole/mole)

Greek characters

α	Fire growth rate (kW/s^2)
ϕ	Oxygen depletion factor
θ	Volumetric expansion factor (1.105)
$\dot{v}$	Volumetric flow rate in the exhaust network (m^3/s)

Superscripts

$\bar{}$	Average
$\dot{}$	Per unit time (s^{-1})
ex	Exhaust network
G	Generated
in	Incoming air

INTRODUCTION

Several hundred kilometers of electrical cables are present throughout nuclear power plants (NPPs). Power cables are used for instance in the process rooms for supplying electricity to the pumps, turbines, transformers or heaters, or in the switchgear room that contains numerous electrical cabinets connected to multiple cable trays[1]. Furthermore, many cable trays containing both instrumentation and control cables are also found in the cable spreading room and in the main control room. Instrumentation cables are used for digital or analogic transmission for various types of transducers, while control cables serve for example for controlling valves or operating relays and contactors.

A serious cable fire occurred at the Browns Ferry NPP in 1975[2] resulting in the loss of the emergency core cooling system of unit 1. Ever since many efforts have been made on the most recent nuclear installations, all over the world, to enhance the prevention of cable fires, for instance by using flame retardant materials in the cable composition. Nevertheless, nearly seventy fire events from NPPs involving electrical cables as fuel were recorded in the current OECD FIRE Database[3] between the late 1980s and the end of 2014. The study of cable fire is therefore a major concern for fire safety analyses in NPPs. Many cable tray fire programmes were carried out over the last decades[4,5,6,7,8,9] and also more recently[10,11,12,13,14,15]. All these last ones implemented ladder cable trays either horizontally or vertically oriented. However, the NPP tray installations may also feature slanted ladder cable trays or protected cable trays. These last ones are used for the fire protection of electrical cables involved in safety trains circuits. Nevertheless, no available programme has yet studied the impact of such cable tray configurations on fire spread.

Two real-scale cable tray fire tests were thus carried out as part of the OECD PRISME-2 Project[16] (CORE campaign) for studying the effect of a protected cable tray (CORE-2) and slanted ladder cable trays (CORE-3) on the main fire characteristics. By considering new configurations, these tests completed the CFSS-1 to CFSS-4 tests[13] (PRISME-2) project which investigated fire spread over five horizontal ladder cable trays according to various cable-types. The CORE-2 test implemented the same horizontal trays set-up except that the lower tray was a protected cable tray. This last one was a metallic non-perforated and covered cable tray. Furthermore, the CORE-3 test involved five horizontal ladder cable trays 1 m long followed by five slanted ladder cable trays 2 m long with a 30 degree angle. The

CORE-2 and CORE-3 tests used the same cable-type, containing a halogen free flame retardant (HFFR), as the CFSS-2 test and they were conducted in open atmosphere under a large-scale calorimeter.

Firstly, this paper gives a description of the cable tray fire sources, the large-scale calorimeter of the SATURNE facility and the instrumentation used for the tests. Then, the assessments of the main fire characteristics of the CORE-2 and CORE-3 tests, such as the mass loss rate (MLR), the heat release rate (HRR) and the effective heat of combustion (EHC) are showed in detail. Furthermore, the evaluations of additional fire characteristics such as the ignition delay, the fire growth rate and the fire duration, all derived from the HRR, are also proposed in this section. In the last part of this paper, the effects of the cable tray configuration are highlighted from comparative analysis of the previous fire properties with the ones obtained for the horizontal ladder trays set-up of the CFSS-2 test. Finally, the evaluation of the gas and soot yields, proposed in this last part, complete the above comparative analysis.

DESCRIPTION OF THE FIRE TESTS

Cable tray fire sources

The three cable tray fire sources used for the preliminary CFSS-2 test, the CORE-2 and CORE-3 tests had the same following characteristics:

- Five ladder cable trays 3 m long, 0.45 m wide, spaced from each other by 0.3 m and set up against an insulated side wall (Figure 1),
- Every tray was filled with 32 samples of 2.4 m long of the same cable-type (its mass per unit length is about 570 kg/km). This cable-type contains Poly(ethylene/vinyl acetate) (EVA) and Polyethylene (PE) as polymeric materials and aluminium hydroxide (ATH) as halogen free flame retardant (HFFR),
- The cable samples were packed loosely along the five trays.

The differences between the three fire sources are listed hereafter:

- For the CORE-2 test, the lower cable tray was a metallic non-perforated and covered cable tray (Figure 2),
- For the CORE-3 test, the fire source was composed of five horizontal ladder cable trays 1 m long, followed by five slanted ladder cable trays 2 m long, with a 30 degree angle (Figure 3). The cable samples length was 0.8 m along the first part while the one along the slanted trays was 1.6 m.

Figure 1: The five horizontal ladder cable trays used for the CFSS-2 test.

Figure 2: The cable tray set up for the CORE-2 test.

Ignition source

A square sand burner (300 x 300 mm^2), located 0.2 m below the lower tray, was used as an ignition source (Figure 2). This gas burner was supplied with propane gas and its fire power was 80 kW for all the tests. Furthermore, the gas burner was centred under the first cable tray for both the CFSS-2 and CORE-2 tests while it was centred beneath the horizontal part of the lower tray for the CORE-3 test (Figure 3). The ignition source was stopped when the HRR exceeded 400 kW (without taking into account the contribution of the gas burner). Given that, this value is five times higher than the fire power of the gas burner (80 kW), the cable tray fire source is thus considered as ignited and the fire as self-sustained. The ignition delay (t_{ig}) of the cable tray fire source is thus defined as the time to reach a HRR of 400 kW.

Large-scale calorimeter and instrumentation

The tests were carried out under a large-scale calorimeter (Figure 4) located in a facility, named SATURNE. Numerous openings are located at the top of this facility over each wall. They provide enough air to maintain the oxygen concentration at 21 % around the fire source, as in open atmosphere.

Figure 3: The cable tray set up for the CORE-3 test.

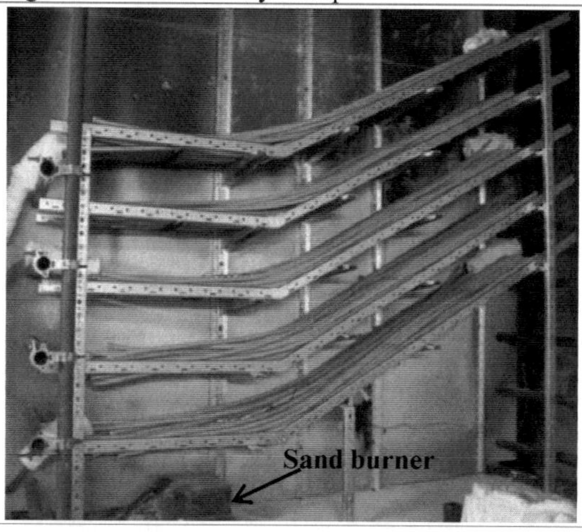

Figure 4: The large-scale calorimeter of the SATURNE facility.

The fire source is centred under the large-scale hood (4.5 x 4.5 m^2 side by side with a 3 m inner diameter, Figure 4). This last one is connected to an exhaust ventilation network which collects all the combustion products released by the fire. Gas species concentrations (O_2, CO_2 and CO) as well as soot mass concentrations, gas temperature, pressure and volume flow rate were measured in the exhaust network. Two weighing devices were located below the cable trays (Figure 4) and allowed measuring the mass loss of materials contained in cables during the fire. For all the tests, five thermocouples were positioned along each of the trays directly above the cables (or just above the metallic cover of the lower tray for the CORE-2 test). All the thermocouples were spaced from each other of 0.5 m. In addition, for the CORE-2 test, three thermocouples were also located inside the protected cable tray just above the cables. They were positioned at the centre and at 0.5 m from the centre in both South and North directions.

FIRE CHARACTERISTICS

Mass loss and mass loss rate

The mass loss (ML) is deduced from the mass measurement carried out with the weighing devices. In addition, the mass loss rate (MLR) is assessed from time derivation of the ML. Both the ML and the MLR are shown in Figure 5 for the CORE-2 and CORE-3 tests. This figure also indicates the total mass losses (TML) of 73 kg and 93 kg as well as the MLR peaks of 78 and 92 g/s, for the CORE-2 and CORE-3 tests respectively.

Figure 5: ML and MLR.

CORE-2 test | CORE-3 test

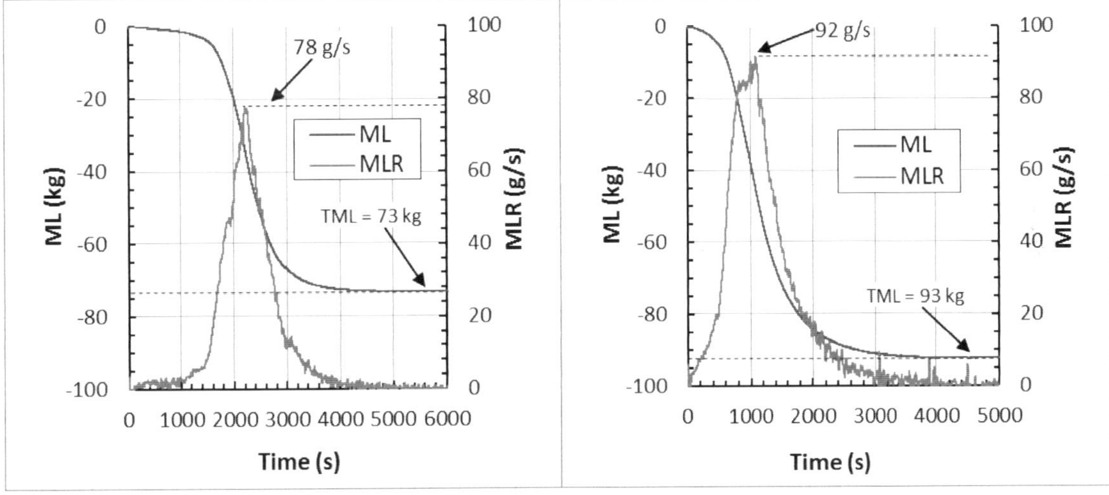

Methods used for the assessment of the heat release rate

The fire heat release rate (HRR) can be assessed with two chemical methods, which are the carbon dioxide generation (CDG) and the oxygen consumption (OC) calorimetry methods.

For the CORE-2 and CORE-3 tests, conducted in open atmosphere, negligible fraction of carbon in the fuel is converted to CO or soot instead of CO_2, as confirmed by their yields assessed in a further section. Consequently, the HRR may be assessed from the CDG calorimetry method ($\dot{Q}_{CDG}$) as follows[17]:

[1] $$\dot{Q}_{CDG} = E_{CO_2}\dot{m}^G_{CO_2}$$

- E_{CO_2}: Energy produced by mass unit of generated CO_2 (13.3 MJ/kg);
- $\dot{m}^G_{CO_2}$: Mass flow rate of generated CO_2 (kg/s).

In addition, $\dot{m}^G_{CO_2}$ is evaluated from:

[2] $$\dot{m}^G_{CO_2} = \dot{m}^{ex}_{CO_2} - \dot{m}^{in}_{CO_2}$$

- $\dot{m}^{ex}_{CO_2}$: Mass flow rate of CO_2 in the exhaust network (kg/s);
- $\dot{m}^{in}_{CO_2}$: Mass flow rate of CO_2 in the incoming air (kg/s).

Finally, it is shown that $\dot{m}^G_{CO_2}$ may be assessed as follows:

[3] $$\dot{m}^G_{CO_2} = (X^{ex}_{CO_2} - X^{in}_{CO_2}) \frac{M_{CO_2} p^{ex}}{RT^{ex}} \dot{V}^{ex}$$

- $X^{ex}_{CO_2}$: Molar fraction of CO_2 in the exhaust network (mole/mole);
- $X^{in}_{CO_2}$: Molar fraction of CO_2 in the incoming air (mole/mole);
- M_{CO_2}: Molecular mass of CO_2 (0.044 kg/mole);
- p^{ex}: Pressure in the exhaust network (Pa);
- R: Ideal gas constant (8.314 J/mole/K);
- T^{ex}: Gas temperature in the exhaust network (K);
- $\dot{V}^{ex}$: Volumetric flow rate in the exhaust network (m^3/s).

Moreover, the HRR is also evaluated from the OC calorimetry method ($\dot{Q}_{OC}$). For negligible fraction of carbon in the fuel converted to CO or soot instead of CO_2, the HRR may be assessed from the OC calorimetry method ($\dot{Q}_{OC}$) as follows[18]:

[4] $$\dot{Q}_{OC} = E_{O_2} \frac{\phi}{1+\phi(\theta-1)} \dot{m}^{ex} \frac{M_{O_2}}{M_a} (1 - X^{in}_{H_2O} - X^{in}_{CO_2}) X^{in}_{O_2}$$

With

[5] $$\phi = \frac{X^{in}_{O_2}(1-X^{ex}_{CO_2}) - X^{ex}_{O_2}(1-X^{in}_{CO_2})}{(1-X^{ex}_{O_2}-X^{ex}_{CO_2})X^{in}_{O_2}}$$

and

- E_{O_2}: Energy produced by mass unit of consumed O_2 (13.1 MJ/kg);
- ϕ: Oxygen depletion factor;
- θ: volumetric expansion factor (1.105);
- $\dot{m}^{ex}$: mass flow rate in the exhaust network (kg/s);
- M_{O_2}: Molecular mass of O_2 (0.032 kg/mole);
- M_a: Molecular mass of the ambient air (0.029 kg/mole);
- X^{in}_i: Molar fraction of gas species (i = CO_2, H_2O and O_2) in the incoming air (mole/mole);
- X^{ex}_i: Molar fraction of gas species (i = CO_2 and O_2) in the exhaust network (mole/mole).

Heat release rate for the CORE-2 test

The HRR assessments from the CDG ($\dot{Q}_{CDG}$) and OC ($\dot{Q}_{OC}$) calorimetry methods are pointed out in Figure 6 (including the gas burner contribution to the HRR) and show very good agreement. So, the final CORE-2 HRR ($\dot{Q}$, Figure 7) is evaluated from the average values of $\dot{Q}_{CDG}$ and $\dot{Q}_{OC}$, minus the fire power of the gas burner ($\dot{Q}_{burner}$), as follows:

[6] $$\dot{Q} = (\dot{Q}_{CDG} + \dot{Q}_{OC})/2 - \dot{Q}_{burner}$$

The main stages of the CORE-2 fire test are the ignition stage, the fire growth stage, the fully developed fire stage and the fire decay stage. They are described hereafter based on the HRR assessment (Figure 7).

Ignition stage (27 min 34 s)

For the CORE-2 test, the gas burner provided a fire power of 80 kW for 1721 s (28 min 41 s) and led to ignite the cable tray fire source at 27 min 34 s (t_{ig}) when the HRR reached 400 kW (Figure 8).

Fire growth stage (from 27 min 34 s to 35 min 54 s)

It is assumed that the following power law equation can describe the growth of the fire[24]:

[7] $$\dot{Q} = \frac{1000}{t_1^2} \cdot (t - t_o)^2$$

Or

[8] $$\dot{Q} = \alpha \cdot (t - t_o)^2$$

With

[9] $$\alpha = \frac{1000}{t_1^2}$$

- α: fire growth rate (kW/s^2);
- t_0: starting time of the power law equation describing the growth of the fire (s);
- t_1: fire growth characteristic time (s). t_1 is defined as the time necessary after t_0 to reach a HRR of 1000 kW (s).

Figure 8 points out that the best power law equation that fits with the HRR growth of the CORE-2 fire is $y = 0.0032 \cdot (t - 1320)^2$. So according to equations [7]-[9], the fire growth rate (α) is 0.0032 kW/s^2, t_1 = 558 s (9 min 18 s) and t_o = 1320 s (22 min).

Figure 6: HRR assessed from the CDG and OC calorimetry methods (CORE-2 test).

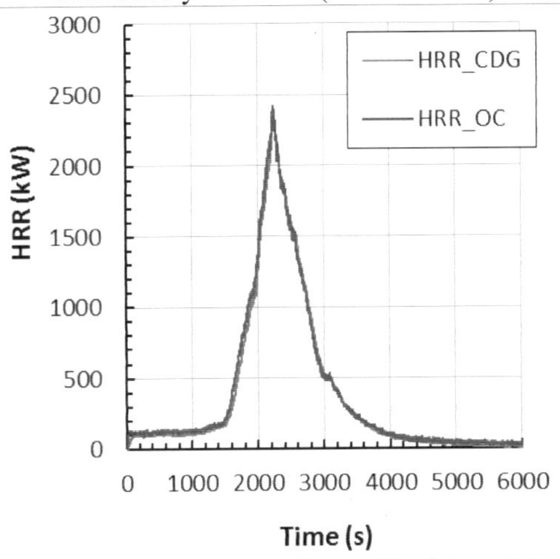

Figure 8: Ignition and fire growth stages of the CORE-2 test.

Figure 7: HRR and the main fire stages of the CORE-2 test.

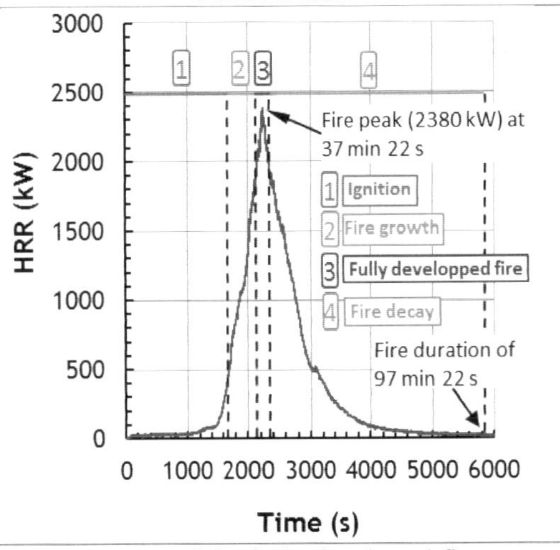

Figure 9: Start of the fully developed fire stage at about 36 min (CORE-2 test).

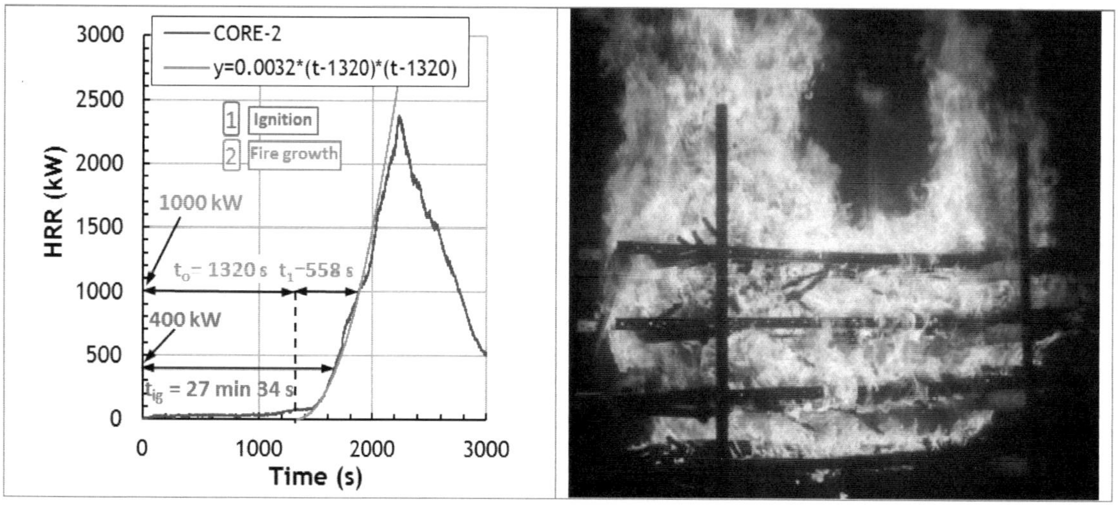

Fully developed fire stage (from 35 min 54 s to 38 min 55 s)

It is considered for the CORE-2 test that the fully developed fire stage started when the HRR exceeded the value of 2 MW. Indeed the fire nearly involved all the cable tray surface area for such value of the HRR (Figure 9). Moreover, it is supposed that this stage ended when the HRR reached for the last time the previous value. According to the HRR of the CORE-2 test (Figure 7), the fully developed fire stage thus started at 35 min 54 s (2154 s) and ended at 38 min 55 s (2335 s). The HRR peak (2380 kW) was reached during this stage at 37 min 22 s.

Fire decay stage (from 38 min 55 s to 97 min 22 s)

The fire decay stage for the CORE-2 fire started after the fully developed fire stage at 38 min 55 s (Figure 7). Furthermore, the fire is supposed to be extinguished once its fire power is lower than 25 kW. This value is indeed similar to the uncertainty of the HRR assessed from the chemical methods considered in this study. The CORE-2 fire is therefore extinguished at 97 min 22 s since this time matches with a HRR of 25 kW (Figure 7). As it is considered that the CORE-2 fire starts with the gas burner operation (t = 0 s), then its fire duration is therefore of 97 min 22 s.

Heat release rate for the CORE-3 test

For the CORE-3 test, the HRR evaluations from the CDG ($\dot{Q}_{CDG}$) and OC ($\dot{Q}_{OC}$) calorimetry methods are also very relevant until t = 2000 s (Figure 10). Beyond this time, the two calorimetry methods show different values. In particular, the OC method gives unrealistic values, greater than zero for t > 4077 s, which is the time when the fire is over according to the CDG method (see further for more details). Analysis of the experimental results indicates a slight drift of the oxygen analyzer occurring in the last stage of the CORE-3 test. Indeed, this analyzer pointed out after the test an oxygen concentration in the exhaust duct (20.78 %vol) slightly lower than the one measured before the test (20.91 %vol). This lower oxygen content is therefore interpreted as a continuing fire by the OC method, thus associating it with a fictive heat release.

Thus, it is proposed to evaluate the CORE-3 HRR ($\dot{Q}$, Figure 11) only from the CDG method, minus the fire power of the gas burner:

[10] $$\dot{Q} = \dot{Q}_{CDG} - \dot{Q}_{burner}$$

The same fire stages as for the CORE-2 are now described hereafter for the CORE-3 test (Figure 11).

Ignition stage (7 min 45 s)

For the CORE-3 test, the gas burner provided a fire power of 80 kW for 468 s (7 min 48 s) and allowed the ignition of the cable tray fire source at 7 min 45 s (t_{ig}) when the HRR reached 400 kW (Figure 12).

Fire growth stage (from 7 min 45 s to 13 min 38 s)

Figure 12 also exhibits that the best power law equation which fits with the HRR growth of the CORE-3 fire test is $y = 0.0073 \cdot (t - 255)^2$. So according to equations [7]-[9], the fire growth rate (α) is 0.0073 kW/s^2, t_l= 371 s (6 min 11 s) and t_o = 255 s (4 min 15 s).

Fully developed fire stage (from 13 min 38 s to 20 min 53 s)

As for the CORE-2 test, it is assumed that the fully developed fire stage started when the HRR exceeded the value of 2 MW. Indeed the fire spread almost over all the cable tray surface area for such value of the HRR (Figure 13). This stage is also supposed ended when the HRR was lower than 2 MW. Given the HRR of the CORE-3 test (Figure 11), this stage therefore started at 13 min 38 s (818 s) and finished at 20 min 53 s (1253 s). The HRR peak (2550 kW) was achieved during this stage at 17 min 23 s.

Fire decay stage (from 20 min 53 s to 66 min 57 s)

The fire decay stage for the CORE-3 fire test started after the fully developed fire stage at 20 min 53 s and ended at 66 min 57 s (Figure 11). The fire is assumed to be extinguished once its fire power is lower than 25 kW (as for the CORE-2 test). The CORE-3 test is thus over at 66 min 57 s since this time matches with a HRR of 25 kW (Figure 11). As it is considered that the CORE-3 fire starts with the gas burner operation (t = 0 s), then its fire duration is therefore of 66 min 57 s.

Figure 10: HRR assessed from the CDG and OC calorimetry methods (CORE-3 test).

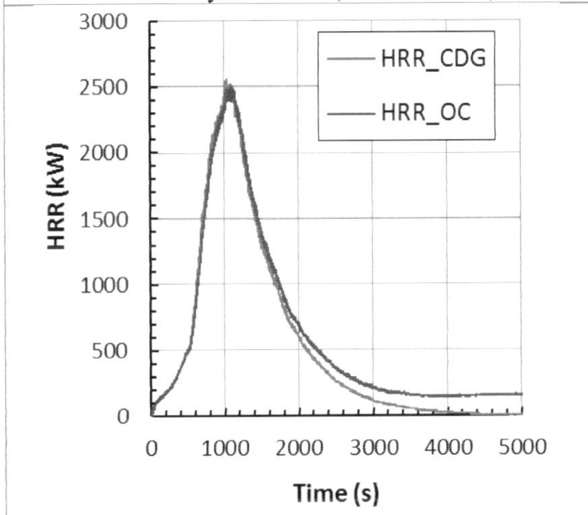

Figure 11: HRR and the main fire stages of the CORE-3 test.

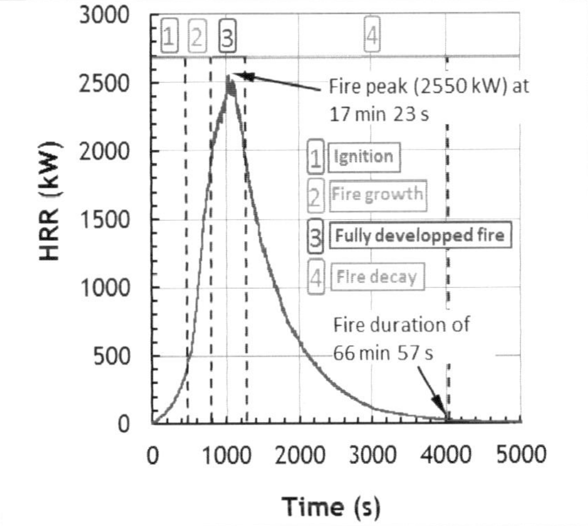

Figure 12: Ignition and fire growth stages of the CORE-3 test.

Figure 13: Starting of the fully developed fire stage at about 13.5 min (CORE-3 test).

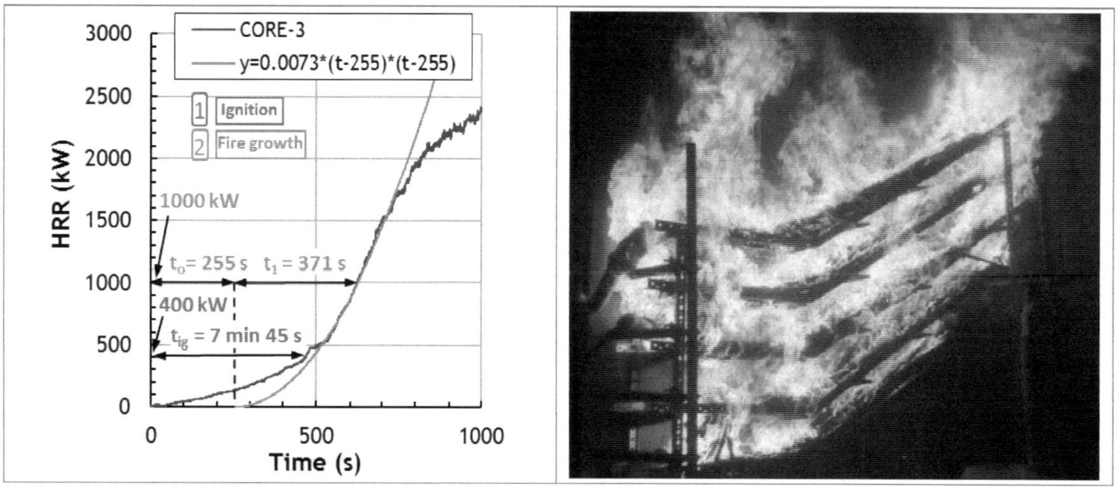

Effective heat of combustion

The combustion behaviour of a fire may be characterized by the effective heat of combustion ($EHC\ or\ \Delta H_{c,eff}$) defined as the ratio of HRR ($\dot{Q}$) to MLR ($\dot{m}$)[18]:

[11] $$\Delta H_{c,eff} = \frac{\dot{Q}}{\dot{m}}$$

Figure 14 therefore points out the EHC evaluated from equation [11] and also the MLR assessed above for the CORE-2 and CORE-3 tests. First, this figure highlights that the EHC is quite low in the early stage of these two tests. Indeed, the highly endothermic dehydration of the flame retardant ATH[19,20], contained in the cable-type studied, delays the burning of the polymeric materials (EVA and PE blend) in particular during the first stage of the tests. Then the EHC continuously increases given the fire spread over all the cable trays area which leads to the progressive consumption of the ATH flame retardant. The EHC after the fire peak (ranging from 30 to about 40 MJ/kg) is indeed consistent with the EHC of the EVA (about 35 MJ/kg[21,22]) and the PE (43 MJ/kg[23]).

Figure 14 indicates that the EHC shows fast and steep fluctuations for the CORE-2 test when $t > 3000$ s and for the CORE-3 test when $t > 1920$ s. These variations are caused by non-negligible fluctuations of the MLR as well as rather low values of this last one (lower than 15 g/s and 18 g/s for the CORE-2 and CORE-3 tests, respectively). Indeed, the EHC is evaluated according to the inverse of the MLR (equation [11]). So it is suggested to assess the average EHC ($\overline{\Delta H_{c,eff}}$) for $1600\ s < t < 3000$ s when MLR > 15 g/s for the CORE-2 test and for $425\ s < t < 1920$ s when MLR > 18 g/s for the CORE-3 test (Figure 14). These MLR values are higher than 20 % of the MLR peak for the two tests since these last ones are of 78 g/s for the CORE-2 test and of 92 g/s for the CORE-3 test (Figure 14). The average EHC is thus assessed at 28.6 MJ/kg for the former test and at 29.2 MJ/kg for the latter test.

Figure 14: EHC and MLR

CORE-2 test | CORE-3 test

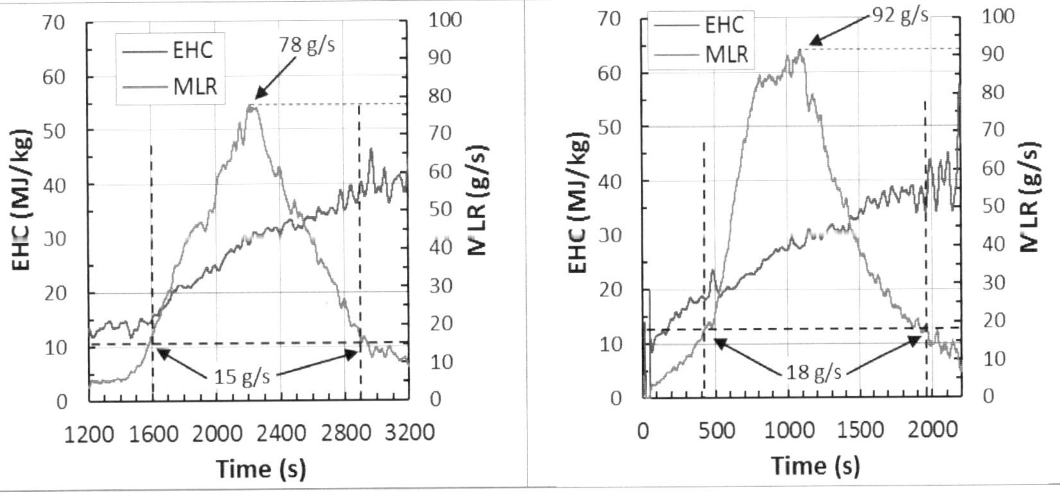

EFFECT OF THE CABLE TRAY CONFIGURATION

The effect of the cable tray configuration on the fire characteristics previously determined are highlighted from comparison with the ones obtained for the CFSS-2 test[13] (five horizontal ladder cable trays). These last ones are summarized in Table 1.

This table firstly indicates similar TML for the CORE-3 (93 kg) and CFSS-2 (97 kg) tests. Indeed, for these two tests the fire involved entirely the five ladder cable trays which were filled with the same cable loading (32 cable samples per tray). By contrast, the TML is lower for CORE-2 (73 kg). For this test, the three gas temperatures measured inside the lower protected cable tray are clearly lower (Figure 15) than the maximal temperatures measured along the burning upper trays (as seen further in Figure 19). Indeed, the cable samples contained inside the lower protected tray showed damages only over a limited portion (Figure 16). These outcomes suggest that the mass loss for this first tray is significantly lower than for the first ladder cable tray of the CORE-3 and CFSS-2 tests, thus leading to the lower TML pointed out above for the CORE-2 test.

The HRR as well as the ignitions delays of CORE-2, CORE-3 and of CFSS-2 tests are summarized in Figure 17. This figure and Table 1 firstly highlight that the protected cable tray delays of about 16 min the ignition of the CORE-2 fire source compared with the five ladder cable trays (CFSS-2). The non-perforated and covered lower tray, located just 0.2 m above the gas burner (Figure 2), acted as a protection screen for the upper ladder cable trays. Compared with the CFSS-2 test (five horizontal ladder cable trays), the average gas temperatures measured along each of the four upper cable trays were indeed significantly lower before the ignition of the CORE-2 cable tray fire source occurring at 27 min 34 s (Figure 19). These temperatures are evaluated as the average of the measurements carried out by the five thermocouples located along each cable tray.

By contrast, the slanted cable trays configuration shortened by about 4 min the ignition of the CORE-3 fire source (Figure 17) compared with the five horizontal ladder cable trays (CFSS-2). The total heat required for the ignition of the cable tray fire source is obtained from the product of the fire duration of the ignition source by its fire power (80 kW). A total heat of 37 MJ was therefore required to ignite the CORE-3 fire source vs 60 MJ for the CFSS-2 test (about one and a half times higher) and 138 MJ for the CORE-2 test (nearly four times larger).

Table 1: Main fire characteristics of the CFSS-2, CORE-2 and CORE-3 tests.

Test	CFSS-2 (horizontal ladder cable trays)	CORE-2 (protected cable tray)	CORE-3 (slanted ladder cable trays)
TML (kg)	97	73	93
Peak of MLR (g/s)	78	78	92
Ignition delay	11 min 57 s	27 min 34 s	7 min 45 s
Fire growth rate (kW/s^2)	$2.2 \cdot 10^{-3}$	$3.2 \cdot 10^{-3}$	$7.3 \cdot 10^{-3}$
Peak of HRR (MW)	2.20	2.38	2.55
Fire duration (min)	72	97	67
$\overline{\Delta H_{c,eff}}$ (MJ/kg)	28	28.6	29.2

Figure 15: Gas temperature inside the lower protected tray (CORE-2 test).

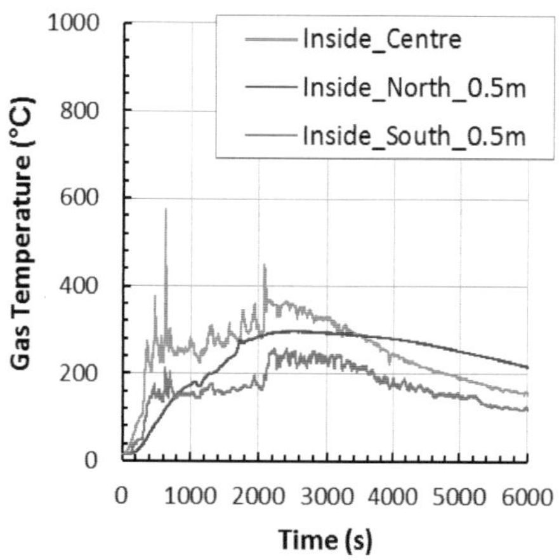

Figure 16: cable samples contained in the protected lower cable tray after the CORE-2 fire.

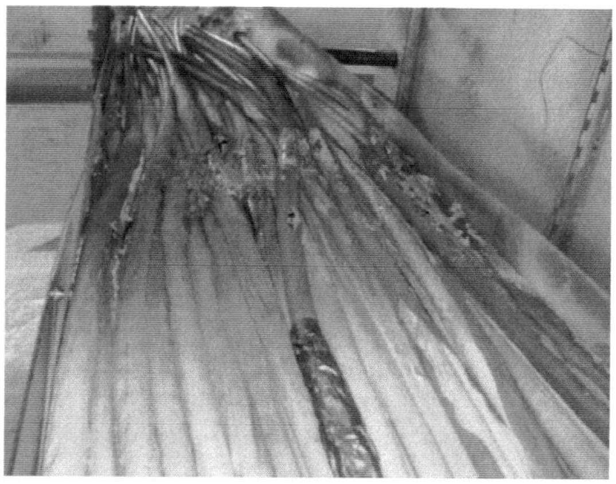

In addition, the best power law equations fitting with the HRR of these three tests are also summarized in Figure 18. This figure and Table 1 indicate that the fire growth rate of the CORE-2 test (0.0032 kW/s^2) is lightly higher than for the CFSS-2 test (0.0022 kW/s^2). It means that the fire spreads slightly faster over the four upper ladder cable trays for the CORE-2 test than over the five ladder cable trays for the CFSS-2 test. Furthermore, Figure 18 points out that the slanted cable tray configuration leads to clearly increase the fire growth rate. This last one is indeed more than three times higher for the CORE-3 test (0.0073 kW/s^2) than for the CFSS-2 test (0.0022 kW/s^2). This outcome is consistent with the faster increase of the average gas temperatures for the CORE-3 test compared with the CFSS-2 test, except for the lower cable tray (Figure 20). Compared with the horizontal cable tray configuration (CFSS-2 test), the slanted tray orientation (30° angle) of the second part of the CORE-3 fire source could indeed favors the pre-heating of the cables by the plume. This effect is indeed supposed to increase the spread rate of the fire as pointed out by previous cable tray fire experiments[12]. In addition, compared with the CFSS-2 test, the slight increase of the fire growth rate for the CORE-2 test could also be due to a higher pre-heating of the cables before they ignite, as a consequence of the much longer operation of the gas burner for the latter test (28 min 41 s vs 12 min 24 s for the former test).

Figure 17: Ignition delays (t_{ig}) for the CFSS-2, CORE-2 and CORE-3 tests.

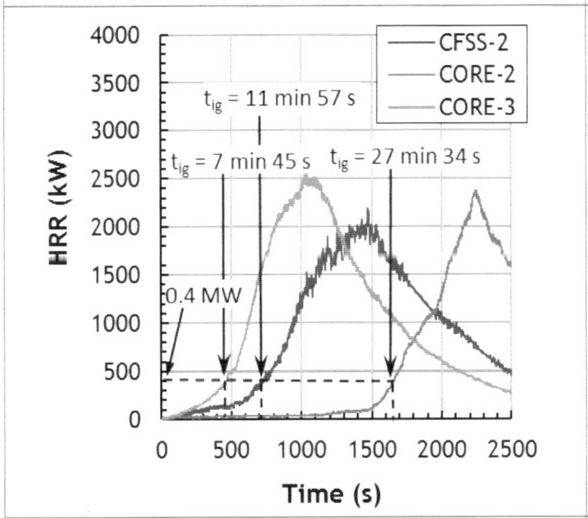

Figure 18: Fire growth rates for the CFSS-2, CORE-2 and CORE-3 tests.

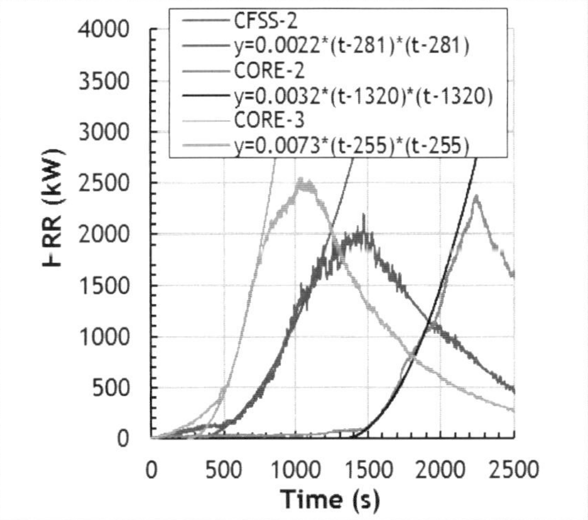

Figure 19: Average gas temperatures along the cable trays (CORE-2 and CFSS-2 tests).

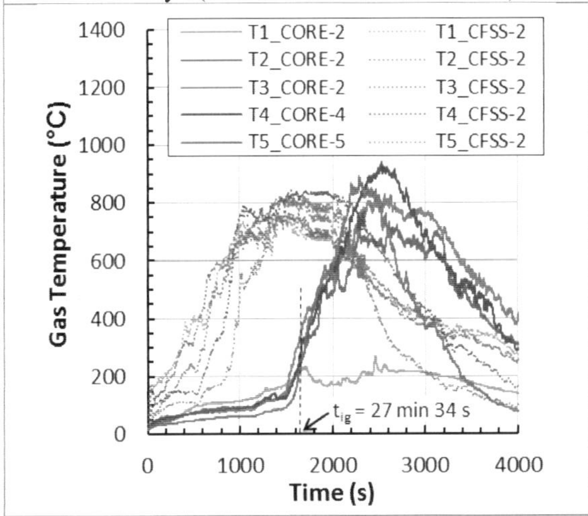

Figure 20: Average gas temperatures along the cable trays (CORE-3 and CFSS-2 tests).

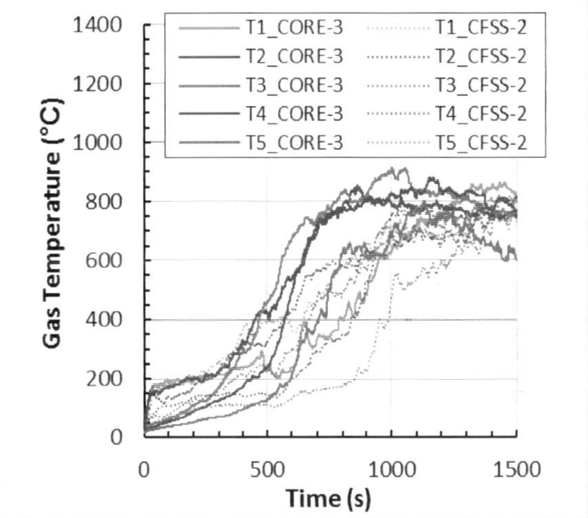

In addition, the HRR peak for the CORE-2 test (2.38 MW, Table 1) is slightly higher than for the CFSS-2 test (2.20 MW, Table 1) and the one obtained for the CORE-3 test (2.55 MW, Table 1) is the highest compared with the former tests. This last outcome suggests for these tests that the HRR peak may increase slightly when the fire growth rate increases (Table 1). Moreover, the fire duration, that includes the gas burner operation time, is equivalent for the CORE-3 fire (about 67 min, Table 1) and the CFSS-2 fire (about 72 min, Table 1). Otherwise the fire duration is significantly higher for the CORE-2 fire (about 97 min) due to the much higher time needed by the gas burner for its ignition.

Furthermore, the gas and soot yields (Y_i) are given in Table 2 for the CFSS-2, CORE-2 and CORE-3 tests. Y_i is defined as the ratio of m_i^G to TML:

[12] $$Y_i = \frac{m_i^G}{TML}$$

- m_i^G: Total mass of gas or soot species (i = CO_2, CO and soot) generated by the cable tray fire (kg).

For CO_2, $m^G_{CO_2}$ is evaluated according to:

[13] $$m^G_{CO_2} = \int_{t=0}^{t_{end\_fire}} (\dot{m}^G_{CO_2} - \frac{\dot{Q}_{burner}}{E_{CO_2}}) dt$$

Where $\dot{m}^G_{CO_2}$ is obtained from equation [3]. The mass flow rate of CO_2 generated by the gas burner ($\frac{\dot{Q}_{burner}}{E_{CO_2}}$) is removed from $\dot{m}^G_{CO_2}$ for proper evaluation of the total mass of CO_2 only generated by the cable tray fire ($m^G_{CO_2}$).

Furthermore, for CO and soot, as their generation by the gas burner is supposed negligible, m^G_i is therefore assessed as follows:

[14] $$m^G_i = \int_{t=0}^{t_{end\_fire}} \dot{m}^G_i \, dt \qquad \text{for i = CO and soot}$$

Moreover, these species are also assumed negligible in the incoming air ($\dot{m}^{in}_i = 0$, for i = CO and soot). So, $\dot{m}^G_{CO}$ is obtained from equation [15]:

[15] $$\dot{m}^G_{CO} = \dot{m}^{ex}_{CO} = X^{ex}_{CO} \frac{M_{CO} P^{ex}}{R T^{ex}} \dot{V}^{ex}$$

and $\dot{m}^G_{soot}$ from equation [16]:

[16] $$\dot{m}^G_{soot} = \dot{m}^{ex}_{soot} = C^{ex}_{soot} \dot{V}^{ex}$$

With:

- X^{ex}_{CO}: Molar fraction of CO in the exhaust network (mole/mole);
- M_{CO}: Molecular mass of CO (0.028 kg/mole);
- C^{ex}_{soot}: Mass concentration of soot in the exhaust network (kg/m^3).

Table 2 first points out that the CO_2 yields are much higher than the CO and soot yields for the three tests. This outcome validates the assumptions considered for the HRR assessment from the CDG and OC calorimetry methods, as previously discussed. In addition, the CO_2, CO and soot yields are similar for the three tests (Table 2). Finally, the evaluations of the average EHC for the CFSS-2, CORE-2 and CORE-3 fires give close values (from 28 to 29 MJ/kg, Table 1). The CO_2, CO and soot yields as well as the average EHC are thus not affected by the presence of the lower protected cable tray or the ladder cable trays orientation (horizontal or slanted).

Table 2: CO_2, CO and soot yields for the CFSS-2, CORE-2 and CORE-3 tests.

Test	Y_{CO_2} (g/g)	Y_{CO} (mg/g)	Y_{soot} (mg/g)
CFSS-2	2.23	27	15
CORE-2	2.17	34	19
CORE-3	2.20	21	12

CONCLUSION

Two real-scale cable tray fire tests were thus carried out as part of the OECD PRISME-2 Project (CORE campaign) for studying the effect of a protected cable tray (CORE-2 test) and slanted ladder cable trays (CORE-3 test) on the main fire characteristics. These tests completed the preliminary CFSS-1 to CFSS-4 tests of the PRISME-2 project which investigated fire spread over five horizontal ladder cable trays according to various cable-types. The CORE-2 and CORE-3 tests used the same cable-type, containing a halogen free flame retardant (HFFR), as the CFSS-2 test and they were conducted in open atmosphere under a large-scale calorimeter.

The CORE-2 test implemented the same horizontal trays set-up as for the CFSS-2 test except that the lower tray was a metallic protected cable tray. This last one was a non-perforated and covered cable tray which delayed the ignition of about 16 min compared with the CFSS-2 test. This lower cable tray indeed acted as a protection screen for the upper ladder cable trays. In addition, the total mass loss (TML) was lower for the CORE-2 test (73 kg) than for the CFSS-2 test (97 kg). Indeed, the cable samples contained inside the lower protected cable tray showed damages only over a limited portion. Furthermore, the fire growth rate (0.0032 kW/s^2) for the CORE-2 test was slightly higher than for the CFSS-2 test (0.0022 kW/s^2). The longer gas burner operation (about 29 min for the CORE-2 test vs about 12 min for the CFSS-2 test) could indeed lead to a greater pre-heating of the electrical cables before they ignite. This effect is indeed supposed to increase the spread rate of the fire.

The CORE-3 test involved the same horizontal cable trays set-up 1 m long as for the CFSS-2 test, followed by five slanted ladder cable trays 2 m long, with a 30 degree angle. Compared with the horizontal trays set-up (CFSS-2 test), the slanted trays configuration shortened by about 4 min the ignition of the multiple cable trays and also clearly increased the fire growth rate. This last one is indeed assessed as more than three times higher for the CORE-3 test (0.0073 kW/s^2) than for the CFSS-2 test (0.0022 kW/s^2). The slanted cable trays orientation (30° angle) of the second part of the CORE-3 fire source could indeed favor the pre-heating of the cables by the plume and therefore the spread rate of the fire. The heat release rate (HRR) peak was also shown higher for this test (2.55 MW) compared with the CFSS-2 test (2.2 MW). This outcome, also observed for the CORE-2 test, suggests for these tests that the HRR peak may increase when the fire growth rate increases. Furthermore similar TML was obtained for the CORE-3 test (93 kg) and the CFSS-2 test (97 kg). Indeed, for these two tests, the fire involved entirely the five ladder cable trays which were filled with the same cable loading (32 cable samples per tray).

Finally, the other fire characteristics such as the average effective heat of combustion or the gas and soot yields were not affected by the presence of the lower protected cable tray or the ladder cable trays orientation (horizontal or slanted).

Further experiments involving similar cable tray fire sources will be characterized in a confined and mechanically-ventilated facility. These investigations will allow for highlighting the effect of confined and ventilated conditions on fire spread over these fire sources.

ACKNOWLEDGMENT

This work was performed by IRSN in the framework of the OECD PRISME-2 project with the support of the partners: ENGIE and Bel V (Belgium), CNSC (Canada), VTT (Finland), DGA and EDF (France), GRS (Germany), NRA and CRIEPI (Japan), CSN (Spain), SSM (Sweden) and ONR, an agency of HSE (United Kingdom).

REFERENCES

[1] U.S. NRC and EPRI (2007), Verification and validation of selected fire models for nuclear power plant applications, NUREG-1824, U.S. Nuclear Regulatory Commission, Washington, DC

[2] U.S. NRC (1975), Cable Fire at Browns Ferry Nuclear Power Station, NRC Bulletin BL-75-04, U.S. Nuclear Regulatory Commission, Washington, DC, USA

[3] Organisation for Economic Co-operation and Development (OECD), Nuclear Energy Agency (NEA), July 2011, "OECD FIRE Database" (OECD FIRE DB 2010:1)

[4] Sumitra, P.S., 1982, Categorization of Cable Flammability. Intermediate-Scale Fire Tests of Cable Tray Installations, Interim Report NP-1881, Research Project 1165-1, Factory Mutual research Corp., Norwood, MA, USA

[5] Nowlen, S.P., 1989, A Summary of Nuclear Power Plant Fire safety Research at Sandia National Laboratories, 1975-1987, NUREG/CR-5384, SAND89-1359, Sandia National Laboratories, Albuquerque, New Mexico, USA

[6] Hirschler, M.M., 1994, Comparison of Large- and Small-scale Heat Release Tests with Electrical Cables, Fire and Materials, 18, p61-76

[7] Mangs, J., Keski-Rahkonen, O., 1997, Full-scale fire experiments on vertical and horizontal cable trays. In: VTT publication 324. VTT Technical Research Centre of Finland, Esppo, Finland

[8] Grayson, S. J., Van Hees, P., Green, A.M., Breulet, H., Vercellotti, U., 2001, Assessing the Fire Performance of Electric Cables (FIPEC), Fire and Materials, 25, p49-60

[9] Tewarson, A., 2003, Characterization of the fire environments in central offices of the telecommunications industry, Fire and Materials, 27, p131-149

[10] Passalacqua, R., Cortes, P., Taylor, N., Beltran, D., Zavaleta, P., Charbaut, S., 2013, Experimental characterisation of ITER electric cables in postulated fire scenarios, Fusion Engineering and Design, 88, p2650-2654

[11] McGrattan, K., Lock, A., Marsh, N., Nyden, M., Bareham, S., Price, M., Morgan, A.B., Galaska, M., Schenck, K., Stroup, D., 2012, Cable Heat Release, Ignition, and Spread in Tray Installations During Fire (CHRISTIFIRE). Volume 1: Horizontal Trays, NUREG/CR-7010, U.S.NRC

[12] McGrattan, K., Bereham, S., 2013, Cable Heat Release, Ignition, and Spread in Tray Installations During Fire (CHRISTIFIRE). Phase 2: Vertical Shafts and Corridors, NUREG/CR-7010, Vol. 2, U.S.NRC

[13] Zavaleta, P., Charbaut, S., Basso, G., Audouin, L., 2013, Multiple Horizontal Cable Tray Fire in Open Atmosphere, Thirteenth international conference of the Fire and Materials, San Francisco, USA, p57-68

[14] Zavaleta, P., Audouin, L., 2016, Fire spreading from a real open-doors electrical cabinet to overhead multiple cable trays into a confined and mechanically-ventilated, INTERFLAM2016, Fourteenth international Fire Science & Engineering conference, University of London, UK, p1563-1574

[15] Li, L., Huang, X., Bi, K., Liu, X., 2016, An enhanced fire hazard assessment model and validation experiments for vertical cable trays, Nuclear Engineering and Design, 301, p32-38

[16] Audouin, L., Prétrel, H., Zavaleta, P., OECD PRISME 2 Fire Research Project (2011-2016) – Current Status and Perspectives, 13th International Post-Conference Seminar on Fire Safety in Nuclear Power Plants and Installations

[17] Tewarson, A., 2002, Generation of Heat and Chemical Compound in Fires, SFPE handbook of Fire Protection Engineering, ISBN:087765-451-4, Third edition, Section three, Chapter 4

[18] Janssens, M., 2002, Calorimetry, SFPE handbook of Fire Protection Engineering, ISBN:087765-451-4, Third edition, Section three, Chapter 2

[19] Delfosse, L., Baillet, C., Brault, A., Brault, D., 1989, Combustion of Ethylene-Vinyl Acetate Copolymer Filled with Aluminium and Magnesium Hydroxydes, Polymer Degradation and Stability, 23, p337-347

[20] Zilberman, J., Hull, T. R., Price, D., Milnes, G. J., Keen, F., 2000, Flame Retardancy of Some Ethylene-Vinyl Acetate Copolymer-Based Formulations, Fire and Materials, 24, p159-164

[21] Wang, L., Jiang, J., Jiang, P.,Yu, J., 2011, Synthesis, characteristic of a novel flame retardant containing phosphorus and its application in poly(ethylene-co-vinyl acetate), Fire and materials, 35, p193-207

[22] Sonnier, R., Viretto, A., Dumazert, L., Longerey, M., Buonomo, S., Gallard, B., Longuet, C., Cavodeau, F., Lamy, R., Freitag, A., 2016, Fire retardant benefits of combining aluminum hydroxide and silica in ethylene-vinyl acetate copolymer (EVA), Polymer Degradation and Stability, 128, p228-236

[23] SFPE handbook of Fire Protection Engineering, ISBN:087765-451-4, Third edition, Appendix C, Table C.3

[24] ISO/DIS 16733-1 - Fire safety engineering – Selection of design fire scenarios and design fires – Part 1: Selection of design fire scenarios, ISO/DIS 16733-1:2014(E)

FIRE-INDUCED RE-RADIATION UNDERNEATH PHOTOVOLTAIC ARRAYS ON FLAT ROOFS

J. Steemann Kristensen[1], B. Merci[2] & G. Jomaas[1,3]
[1]Technical University of Denmark, Dept. of Civil Engineering, 2800 Kgs. Lyngby, Denmark.
[2]Ghent University, Dept. of Flow, Heat and Combustion Mechanics, 9000 Ghent, Belgium.
[3]University of Edinburgh, BRE Centre for Fire Safety Engineering, Edinburgh EH9 3JL, UK.

ABSTRACT

The impact of the reflection of fire-induced heat from a gas burner was studied experimentally to gain knowledge on the interaction between photovoltaic (PV) panels and a fire. The heat flux was measured in a total of eight points at the same level as the top of the gas burner. The gas burner was placed underneath the centre of a PV panel and the eight points were measured in symmetrical pairs of two at four different distances from the burner. The heat release rate from the gas burner was increased stepwise every four minutes. The measurements were made underneath a PV panel installed in a geometry similar to a commercial East-West orientated mounting system and was compared to a baseline test without the re-reflection from the PV panel. A significant increase of the received heat flux was noticed and the trend indicated an ascending percentage-wise difference as a function of an increased heat release rate.
Contrary to the basic view factor theory, the received heat flux was higher underneath the most elevated part of the PV panel, and this occurred due to two important flame related reasons: 1) the deflection of the flame towards the most elevated part of the panel, resulting in an increased amount of radiation from the flame towards the surface; 2) A non-homogeneous distribution of the temperature on the PV panel surface, due to the deflected flame, and thereby a non-homogeneous emission from the heated PV panel.
Finally, it was seen that two similar tests conducted with respectively a brand new PV panel and a PV panel tested for the fourth time, showed very comparable results, except during the period when the thin combustible film underneath the new PV panel was burning. This resulted in a higher heat flux during that period and implies that the results presented herein are conservative in that they are lower than what can be expected in case of a real fire hazard, where the PV panel is by definition involved in the fire for the first time. It can be concluded that PV panels can have a significant contribution in roof fires, as they stimulate fire spread over the roof on which they have been mounted. These findings emphasise that the risk related to the installation of PV panels is not only associated with the increased fire load and possibility of ignition, but largely also with the changed fire dynamic surroundings of the roof construction.

INTRODUCTION

An increased focus on renewable energy, a desire to gain good publicity and especially the possibility of decreasing the energy cost, in a world where the demands of energy increase, are important reasons why photovoltaic (PV) panels become more and more popular. For companies with large flat roof constructions on buildings it can even be cost-beneficial, because it allows for utilization of the unused space on the top of their properties. An initiative along these lines is the recent suggestion to replace all Dutch asbestos roofs with PV panels [1].

However, the installation of PV arrays introduces new and not yet well-studied fire related risks to the roof construction, which is of great interest for all parties involved in PV related power, from the manufacturer of the panels to the owner of the facilities and their insurance companies. A risk assessment related to the

Italian national fire services guidelines by Cancelliere [2] divides current studies into three main domains: (1) ignition, (2) propagation of the fire and (3) the safety of personnel, including fire fighters.

The installation of large PV arrays and thereby large power generating DC systems introduces a risk of electronic malfunctions and thereby a possibility of ignition and fire. Wohlgemuth and Kurtz [3] divided the causes of fire hazards into three categories: (1) hot spots, (2) high series resistance and (3) arching. Pandian et al. [4] studied the consequence of shading faults on PV panels and concluded that the temperature of the PV panel is a potential hazard, especially in the hydrocarbon industry. Solar America Board for Codes and Standards (Solar ABCs) and UL studied the Bakersfield fire [5] and suggested that a current of 1000 ampere in a small, unprotected 10 AWG wire was the source of ignition. In general, it seems to be accepted to assume an electric malfunction to be the main cause of PV related fires, which is why there exists extensive studies in the electronics industry (e.g., Zhao et al. [6], who made a fault analysis of PV arrays to avoid safety hazards).

Each element added to a roof construction can be seen as an addition to the fire load. A PV panel's reaction to flames is tested in the European standard IEC 61730-2:2016 [7] and the American equivalent UL 1703 [8]. Despinasse and Krueger [9] found that the current tests are based on non-easily reproducible test methods and are in the development of a method where they test the PV panel's reaction to a propane flame. Despite the fact that parts of a PV panel are combustible, the current authors have earlier concluded that the PV panel, as a fuel load, does not represent a large fire hazard for a roof construction, because combustible materials represent a minor amount of the panel's total mass [10].

The safety for fire fighters has been addressed by SFPE [11], UL [12] and The Fire Protection Research Foundation [13]. The risk of being electrocuted during fire extinguishment actions has furthermore been studied by Tommasini et al.[14], where they concluded that panels with a voltage up to 1000 V DC could be extinguished safely, by respecting safety distances. Lastly, the Solar ABCs have concluded that the installation of PV panels on sloped roof constructions reduced the fire related properties of the subjacent material [15].

Based on the approach that the PV panels do not contribute with a significant fire load to the roof construction, it is of interest to study why large fires involving PV installations are occurring on regular basis. This increase in PV-related fires, which can, for example, be seen from data based on incident reports from the Italian National Fire Corporation [16], indicates that the installation of PV arrays changes the boundary conditions of the fire by reflecting a significant fraction of the heat back towards the surface of the roof construction instead of into open air. The matter of a possible propagation of fire, without regard to the actual source of the fire, strongly depends on how the surrounding environment responds to the ignition, which once again depends on the critical heat flux for ignition of the adjacent materials. The reflected heat would be added to the heat flux from the fire, resulting in a increased total heat flux towards the top of the roof construction, and when or if the total heat flux exceeds the critical heat flux for the subjacent material, it will ignite. Thus, the propagation of a fire beyond the point of ignition, say from a faulty wire, basically depends on the ratio between the critical heat flux of the unignited material versus the heat flux received from the surroundings. The hypothesis herein is therefore that a propagation of a fire can only occur in a (fire proofed) roofing membrane if the membrane's critical heat flux is exceeded by the sum of the heat from the ignition source or burning roofing membrane and the heat reflected by the PV panels.

The main aim of the experiments presented herein is therefore to examine how the installation of a PV panel influences the energy release from a realistic flaming fire source, which was represented by a gas burner. The gas burner made it possible to define different steady state heat release rates (HRR), and these well-controlled HRRs enabled the possibility of a comparison between the heat flux received at the same spot, with and without the reflection from the PV panel.

THEORY

A fire on a roof construction would influence the surroundings with the three types of heat transfer: conduction, convection and radiation. Without external impacts, such as wind, it is expected that only the conduction and a fraction of radiation will affect the nearby roofing materials, whereas the convection and most of the radiation will be transferred into the open air. By adding a PV installation above the fire on the roof construction, large parts of the radiation will either be absorbed by the panel or reflected back towards the surface of the roof construction that is located underneath the panel. The absorbed heat, together with the heat received from convection, will increase the temperature of the panel resulting in an enhanced amount of emission according to the Stefan-Boltzmann law.

Based on the view factor theory and an assumption of a homogenously distributed temperature over the entire PV panel surface, it is possible to calculate the radiative heat transfer between the solid surface *AC* and the point *D* in Fig. 1. The view factor in point *D*, as a function of the distance between *B* and *D* can be seen in Fig. 2 and Fig. 3 for respectively different angles, θ, and elevations, h_E, of the solid. The graphs indicate that the received heat flux by radiation emission from the heated panel decreases as a function of an increased elevation or a decreased angle. It is furthermore noticed that the view factor is largest underneath the least elevated part of an inclined panel, as expected.

In case of an East-West orientated PV installation, a convectional contribution of heat from the flow of hot air underneath the installation is expected, possibly combined with radiation from the smoke. Herein, the different heat sources are treated separately, because it is the combined additional heat that represents increased hazard to the subjacent surface, and the main objective is to examine whether or not the additional contribution of heat is significant for the propagation of a possible fire.

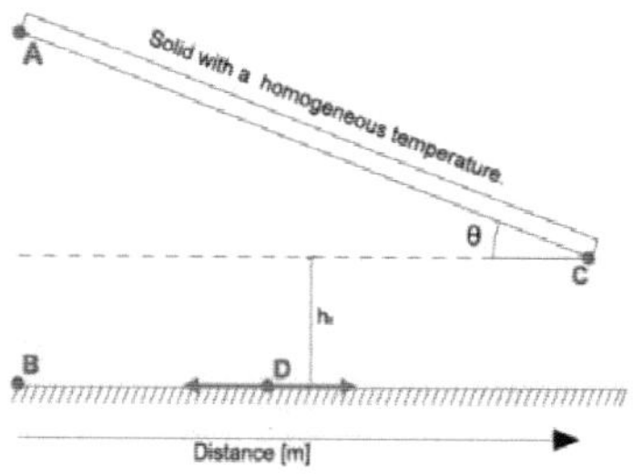

Figure 1 - Simplified set-up of a solid with the angle θ and the elevation h_E above a surface.

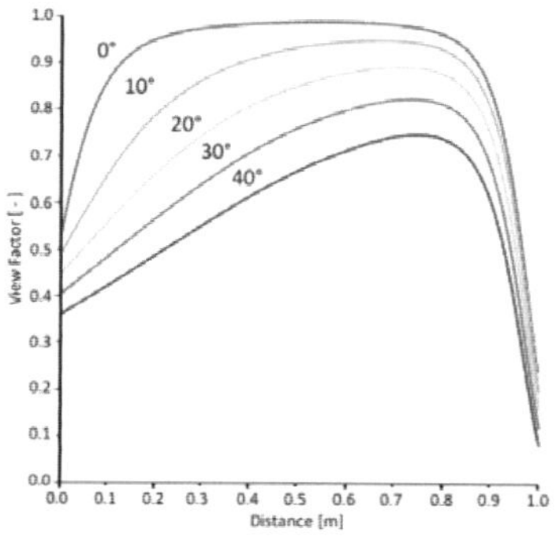

Figure 2 – Theoretical view factor as a function of the distance between point B and D, the angle θ and a fixed elevation of $h_E = 0.1m$ in Fig. 1.

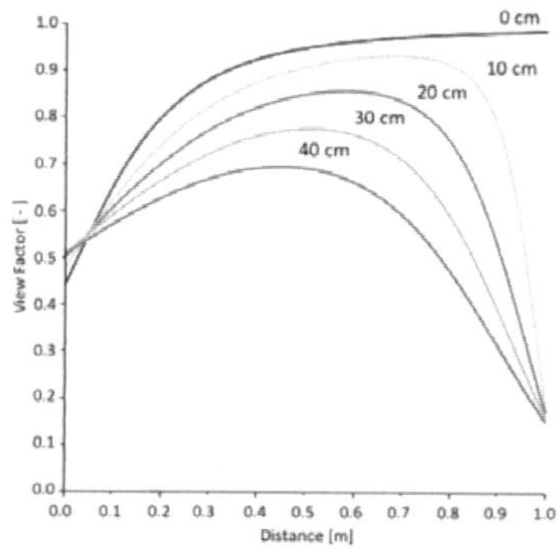

Figure 3 – Theoretical view factor as a function of the distance between point B and D, the elevation h_E and a fixed angle of θ = 10° in Fig. 1.

EXPERIMENTAL SETUP AND APPROACH

The setup seen in Fig. 4 consist of a Bosch Rexroth modular system, which made it possible to change the elevation and angle of the PV panels by adjusting h_L and h_R. Two PV panels with dimensions of $1.0\ m \times 1.7\ m$ were tested. The panels were poly-crystalline panels designed with a front layer of glass and polymer backsheet encasing the PV cells. The main panel was stabilized by a $40\ mm$ aluminum frame and the panel could be connected to a PV system via two connection cables. Because the objective was to test the PV panel's influence on the heat flux received at the subjacent surface it was decided to cut both connection cables and the thereby avoid an additional, but non-reproducible, fire load from the panels.

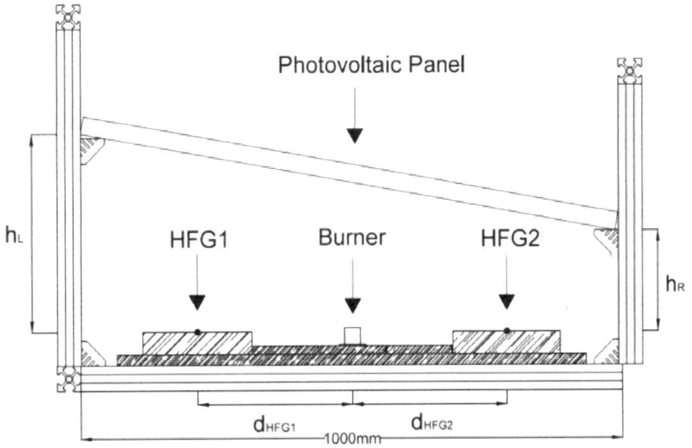

Figure 4 - Schematic of the setup used to study the reflection. The hatched areas underneath the Burner and the Heat Flux Gauges (HFG) are made up of non-combustible insulation.

A scaled sandbox gas burner was placed in the center of the setup, using LPG with minimum 90% propane and maximum 10% butane as fuel. Due to uncertainty with respect to the exact composition of LPG, the HRR has been calculated using the heat of combustion (ΔH_C) for propane. It is noted that the error is small, given the very similar ΔH_C values for propane and butane (respectively 46.3 $MJ\ kg^{-1}$ [17] and 45.7 $MJ\ kg^{-1}$ [17]). The gas flow was controlled by a Bronkhorst flow meter with a maximum flow rate of 10 l/s and thereby thus a maximum HRR of 14 kW from the burner. Due to the sensitivity of the used Heat Flux Gauges (HFGs), the maximum HRRs were not used.

The HFGs were water cooled Hukseflux SBG01 thermal sensors with a working range around 5 kW/m^2. As seen from Fig. 4, the HFGs were installed in a 40 mm thick board of non-combustible insulation to avoid unintended heating from the sides. The insulation boards with the HFGs could be moved horizontally on top of the subjacent second layer of non-combustible insulation – making it possible to measure the heat flux in different positions for each test. The sensitive part of the HFGs were furthermore elevated to the same height as the top of the burner to avoid shadowing effects. Both the water supply to the HFGs and the gas supply to the burner were placed underneath the insulation to prevent influence from the heat.

Test matrix

An overview of the conducted tests is shown in Tab. 1, where the 17 tests have been divided into five subgroups. The difference between the results in tests series A and B was used to examine the difference between the heat flux received at the same distances from the gas burner, with and without the influence of the PV panel. In test series B the panel was placed in a geometry similar to the mounting system used for an East-West orientated array and the elevation and angle of the panel were changed in respectively test series D and E.

Table 1 – Schematic overview of the experiments conducted. All dimensions are in cm.

A: Baseline Test – No PV

Test #	h_L	h_R	d_{HFG1}	d_{HFG2}
1.1	-	-	-28.5	28.5
1.2	-	-	-22	22
1.3	-	-	-16	16
1.4	-	-	-11	11

B: Standard Mounting System

Test #	h_L	h_R	d_{HFG1}	d_{HFG2}
2.1	32	9	-28.5	28.5
2.2	32	9	-22	22
2.3	32	9	-16	16
2.4	32	9	-11	11

C: Similar to test 2.4 – with new PV

Test #	h_L	h_R	d_{HFG1}	d_{HFG2}
3.1	32	9	-11	11

D: Changed height

Test #	h_L	h_R	d_{HFG1}	d_{HFG2}
3.2	27	4	-11	11
3.3	27	4	-28.5	28.5
3.4	37	14	-11	11
3.5	37	14	-28.5	28.5

E: Changed height

Test #	h_L	h_R	d_{HFG1}	d_{HFG2}
4.1	27	9	-28.5	28.5
4.2	27	9	-22	22
4.3	37	9	-16	16
4.4	37	9	-11	11

Approach

Each test was made with an increase of the HHR by 1 kW for approximately 4 minutes each – a slight difference of the time intervals occurred because it was a manual process. All of the tests were running with 0 kW for an initial start-up period to ensure the heat flux received from the surroundings was constant. The PV panels were kept at room temperature for at least 40 minutes between each test to avoid the heat from a previous test to influence the following test. Because of the sensitivity of the used HFGs and because only two new PV panels were available for the tests, it was decided to stop the tests if the received heat flux underneath the PV panels exceeded 7.5 kW for a continuous period. The decision was primarily decided to protect the HFGs but also to avoid structural damage of the panels, including broken glass, which could influence the results. The tests in series B were carried out with the same PV panel whereas the tests in series D and E were conducted with two different panels.

RESULTS AND DISCUSSION

Influence of the PV panel

The actual experimental setup can be seen in Fig. 5. It is important to notice the deflection of the flame towards the most elevated part of the PV panel. The consequence of the deflected flame can be seen in Fig. 6 where the received heat flux is plotted as a function of time, and therefore on the HRR from the gas burner.

In Fig. 6, it is seen that the highest heat flux is measured by HFG1, positioned underneath the most elevated part of the panel (see Fig. 5), which means that the view factor theory alone (Fig. 2 and Fig. 3) cannot explain the results. Furthermore, Figure 6 shows that the increase in HRR from the gas burner results in an

Figure 5 - Photo of the setup. Notice how the flame is deflected upwards underneath the PV panel.

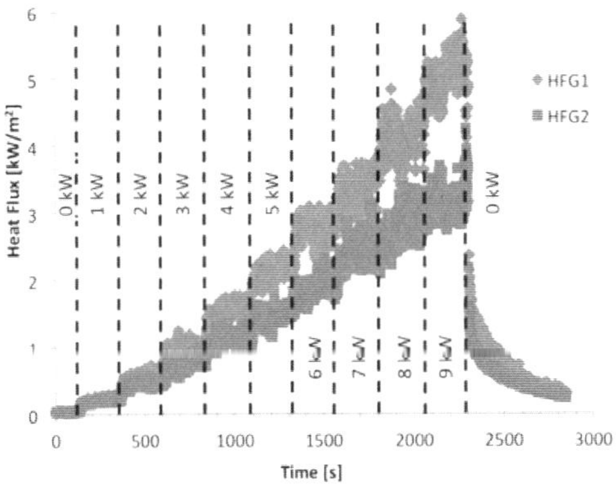

Figure 6 - Raw data from test 2.2: the received heat flux [kW/m²] is plotted as a function of time [s]. The vertical black dotted lines indicate the changes of the HRR over time.

increase of the received heat flux on the subjacent surface. In addition, the fluctuations increase for increased HRR, and these fluctuations in the measured heat fluxes probably relate to fluctuations of the flame length from the gas burner. From the last period in the measurements (Fig. 6) it is seen that the emission from the PV panel, due to its increased temperature, is limited to approximately *2 kW/m²* immediately after interruption of the gas flow (and thereby the almost immediate disappearing of the flame). As the panel cooled down, the measured radiation emission descends towards it original state of approximately *0 kW/m²*.

In order to study the distribution of the heat flux measurements as a function of the location, the measurements were averaged. To that purpose, the averaging was performed over 2 minutes per imposed HRR, once a new steady state had been obtained after increasing the HRR. These averaged values are then assumed to be constant, representative for that imposed HRR. If this heat flux exceeds the critical heat flux for the subjacent material, it is assumed to result in ignition and thereby a propagation of the fire.

The average received heat flux was calculated for all HRRs and all positions of the HFGs, making it possible to plot the measured heat flux as a function of the distance to the gas burner for test series A and B, as seen from Fig. 7. From subfigure A, which is the baseline test without the PV panel, it can be noticed that the measured heat flux is almost symmetrically distributed around the centre of the gas burner which is placed at a distance of 0 cm. The measurements at the left side of the gas burner are slightly higher than the right side of the burner, though. This is probably due to a limited occasional draught towards the left side of the setup in the laboratory. The difference is small, but will be kept in mind for the conclusion. The measurements from test series B, with a PV panel installed in a geometry similar to commercial available mounting system, are shown in subfigure B. A significant increase of the heat flux is observed at all locations, when compared to the measurements without PV panel. As mentioned above, all the measurements underneath the most elevated part of the panel are higher than the measurements at the same distance from the burner underneath the least elevated part of the panel. This indicates the importance of the heat transfer by the flame, as mentioned above. This observation, along with the very fast decrease of the measured heat flux when switching off the burner (Fig. 6, final period in the experiments) emphasises that the emission from the panel, compared to the heat transfer from the flame, is only a limited contributor to the total heat flux towards the subjacent surface.

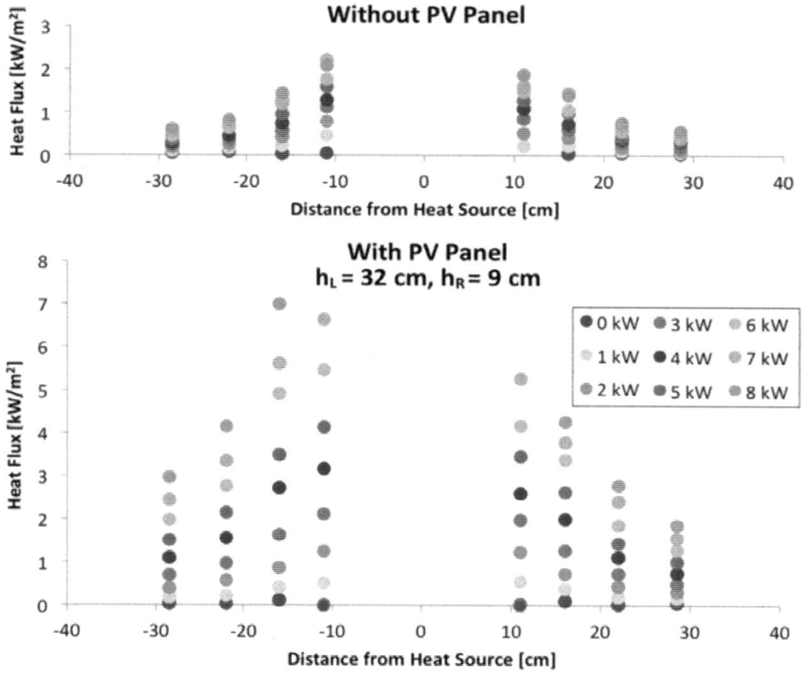

Figure 7 – Averaged measured heat fluxes for test series A (top) and B (bottom). The Heat Flux received as a function of distance to, and HRR from, the heat source. Each colored dot defines a HRR and the gas burner is placed at the distance of 0 cm from the heat source.

With the assumption that the amount of convective heat is limited due to the distance of the panel from the roof surface (Fig. 5) and the fact that there is no accumulation of hot smoke, the differences between the measured heat fluxes at both sides of the gas burner are related to the radiation from the deflected flame from the gas burner. The influence of the flame can also be seen in Fig. 8, where the percentage-wise increase of the received heat flux is plotted as a function of the HRR and the distance from the gas burner, as seen from equation 1.

$$I(D,\dot{Q}) = \frac{\dot{q}''_{PV}(D,\dot{Q}) - \dot{q}''_{Base}(D,\dot{Q})}{\dot{q}''_{Base}(D,\dot{Q})} \times 100\% \qquad (1)$$

Where:

$I(D,\dot{Q})$: is the percentage-wise increase of the heat flux *[%]*

D: is the distance from the gas burner *[cm]*

$\dot{Q}$: is the HRR from the gas burner *[kW/m²]*

$\dot{q}''_{PV}(D,\dot{Q})$: Is the heat flux received for a given distance and HRR in test series B *[kW/m²]*

$\dot{q}''_{Base}(D,\dot{Q})$: Is the heat flux received for a given distance and HRR in test series A *[kW/m²]*

A consistent observation for the plot is that the percentage-wise increase of the heat flux continues to grow for the three most distant positioned HFGs at the left side of the gas burner (i.e., underneath the most elevated part of the PV panel at distances of -28.8 cm, -22 cm and -16 cm), whereas the increase stabilises for the other positions (at values between 180% to 250% of the heat flux that would have been received without the panel). The continued growth, as a function of HRR, for the HFG positioned at the distances of

-16.0 cm, -22.0 cm and -28.5 cm probably occurs due to a HRR-related increase of the deflected flame length. The view factor for radiation from the flame increases significantly as the flame becomes longer, in particular when it reaches a position such that the heat flux meters were directly underneath the flame. Once this took place, the received heat flux hardly increased with further increase of the HRR from the burner, again confirming that the flame radiation was the dominant factor. This also explains why the heat fluxes at the right hand side stabilised quickly and did not change significantly: the view factor from the flame hardly changed with an increased HRR from the burner. Only the emission from the panel increased, due to higher temperatures of the PV panel, caused by the higher HRR from the burner. Visual observations from the experiments supports this.

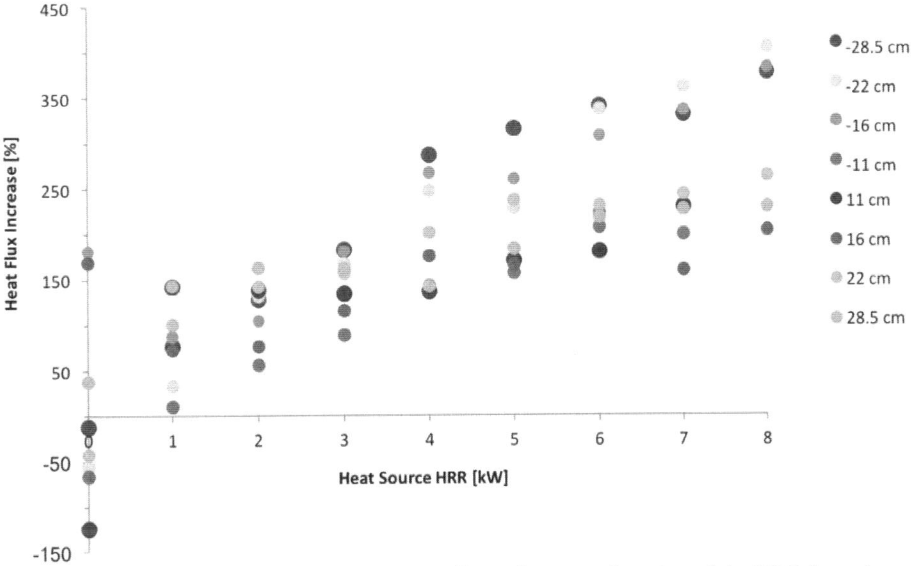

Figure 8 – Overview of the percentage-wise increase of heat flux, as a function of the HRR from the gas burner, measured in respectively the baseline test (test series A in Tab. 1) and underneath the PV panel placed in a standard mounting system (test series B in Tab. 1).

Reproducibility of tests

The performance of the similar tests 2.4, with a reused panel, and 3.1, with a brand new panel (Tab. 1), made it possible to assess the effect of reusing panels in the experiments. Based on the raw data measurements it was not possible to verify the similarity using a 2-sample z-test with a 10 % level of significance. Figure 9 therefore plots a 2-dimensional scatter diagram of paired data sets for the heat flux as received by the two heat flux gauges.

The diagram illustrates a good linear relation between the two tests, with two clearly deviating points, encircled by green circles, for the HRR of 4 kW. This corresponds to a period where the PV panels backing material burns, so there is an additional HRR. Ignoring these 2 points, the slope of the two trend lines indicate almost no difference between the heat flux underneath the new PV panel, compared to the panel tested for the fourth time. This confirms once again that the heat transfer from the flame is dominant.

However, during the burning period, the average heat fluxes measured underneath the brand-new PV panel was around 35 % higher. Based on these observations, it can be concluded that the current measurements are representative for periods while the PV panels are not burning, and therefore are on the lower side of what can be expected during an actual fire (where the PV panels are by definition brand new with respect to fire). The increase occurred due to the combustion of the thin film on the backside of the PV panel resulting in a short term growth of the heat flux.

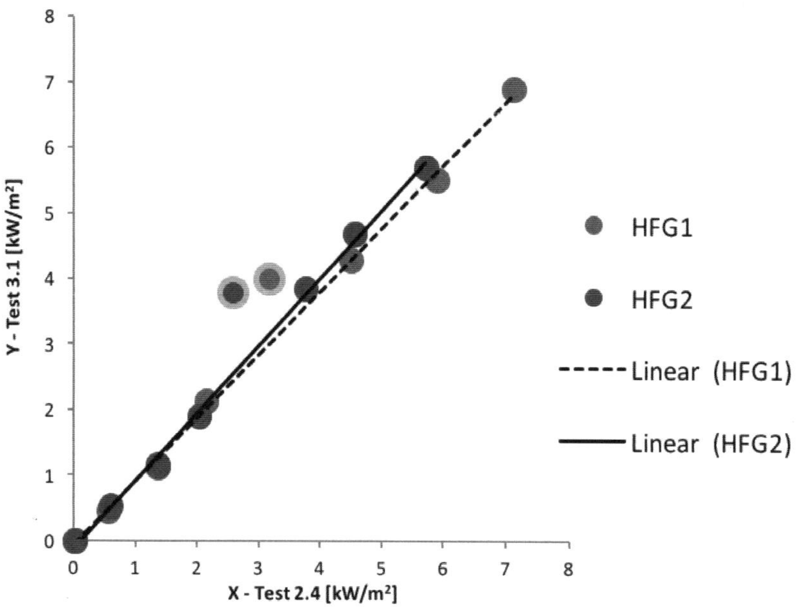

Figure 9 - Two-dimensional scatter diagram of the paired data set from test 2.4 and 3.1, with and without the paired data for a HRR of 4 kW. The two green lines encircle the two measurements made with a HRR of 4 kW and the two linear trend lines ignores these two measurements.

CONCLUSION

Based on experiments, where the received heat flux from a gas burner was measured at the nearby surface with and without the influence of a photovoltaic panel installed above the flame, it was seen that the installation of the PV array significantly increased the heat flux received on the subjacent surface. Contrary to the expected distribution of heat from view factor calculations, it was observed that the heat from, and the deflection of, the flame defined the area of the maximum heat flux. For a fire-induced heat source placed underneath an angled PV panel the maximum heat flux thereby occurred under the most elevated part of the panel, which is where the flame deflected.

PV panels can have a significant contribution in a fire hazard, as they stimulate fire spread over the roof on which they have been mounted. The current findings emphasise that the risk related to the installation of PV panels is not only associated with the increased fire load and possibility of ignition – but largely also with the changed fire dynamics on the roof construction.

Finally, the results could form the basis of a discussion regarding the level of fire related properties for roofing materials used underneath PV installations on flat roof constructions – as an improved reaction to fire performance of the roofing membranes could decrease the possibility of propagation underneath the arrays.

ACKNOWLEDGEMENTS

The authors would like to thank the Otto Mønsted Foundation who funded Bart Merci as a visiting professor at the Technical University of Denmark. The PV panels used in the experiments were sponsored by IKEA Services AB as a part of a larger research project. The authors would furthermore like to thank research associate Bjørn Skjønning Andersen at the Technical University of Denmark, and Richard Clemenceau, laboratory intern from ISTIA - École d'igénieurs de l'Université d'Angers (France), for their assistance in the DTU Fire Laboratory.

REFERENCES

1 Solar Magazine, "'Koppel de vervanging van alle asbestdaken aan zonnepanelen'," 6 June 2016. [Online]. Available: https://solarmagazine.nl/nieuws-zonne-energie/i11486/koppel-de-vervanging-van-alle-asbestdaken-aan-zonnepanelen#. [Accessed 7 June 2016].

2 P. Cancelliere, "PV electrical plants fire risk assessment and mitigation according to the Italian national fire services guidelines," *Fire and Materials,* p. 355–367, 5 December 2014.

3 J. H. Wohlgemuth and S. R. Kurtz, "How Can We Make PV Modules Safer?," in *IEEE Photovoltaic Specialists Conference*, Austin, 2012.

4 A. Pandian, D. John Thiruvadigal and S. Sakthivel, "Fire Hazards and Overheating Caused by Shading Faults on Photo Voltaic Solar Panel," *Fire Technology,* vol. 52, p. 349–364, 6 June 2015.

5 B. Brooks, "The Ground-Fault Protection BLIND SPOT: A Safety Concern for Larfer Photovoltaic Systems in the United States.," [Online]. Available: http://solarabcs.org/about/publications/reports/blindspot/.

6 Y. Zhao, J-F de Palma, J. Mosesian, R. Lyons and B. Lehman, "Line–Line Fault Analysis and Protection Challenges n Solar Photovoltaic Arrays," *IEEE TRANSACTIONS ON INDUSTRIAL ELECTRONICS,* vol. 60, no. 9, 12 June 2012.

7 International Electrotechnical Commission - IEC, *IEC 61730-2:2016,* 2016.

8 Underwriters Laboratories Inc., *UL 1703 Standard for Flat-Plate Photovoltaic Modules and Panels,* Underwriters Laboratories Inc..

9 M.-C. Despinasse and S. Krueger, "First developments of a new test to evaluate the fire behavior of photovoltaic modules on roofs," *Fire Safety Journal,* vol. 71, pp. 49-57, 2015.

10 J. S. Kristensen, "Fire risk assessment of solar cell array installations on large buildings: How to protect the building in case of fire?," Lyngby, Denmark, 2015.

11 C. C. Grant, "Harnessing the Sun: Solar Power and Fire Protection Engineering," [Online]. Available: http://www.sfpe.org/?page=2014_Q3_4&hh. [Accessed 1 11 2016].

12 R. Backstrom and D. A. Dini, "Firefighter Safety and Photovoltaic Installations Research Project," 2011.

13 C. C. Grant, "Fire Fighter Safety and Emergency Response for Solar Power Systems," The Fire Protection Research Foundation, Quincy, MA, 2010.

14 R. Tommasini, E. Pons, F. Palamara, C. Turturici and P. Colella, "Risk of electrocution during fire suppression activities involving photovoltaic systems," *Fire Safety Journal,* vol. 67, pp. 35-41, 2014.

15 L. Sherwood, B. Backstorm, D. Sloan, C. Flueckiger, B. Brooks and A. Rosenthal "Fire Classification Rating Testing of Stand-Off Mounted Photovoltaic Modules and Systems," Solar America Board for Codes and Standards Report, 2013.

16 Luca Fiorentini, L. Marmo, E. Danzi and V. Puccia, " Issue 99: Fires in Photovoltaic Systems: Lessons Learned from Fire Investigations in Italy", "SFPE emerging TRENDS enewsletter," 2015. [Online]. Available: http://www.sfpe.org/?page=FPE_ET_Issue_99&hhSearchTerms=%2299+and+fires%22. [Accessed 19 October 2016].

17 V. Babrauskas, Ignition Handbook, Issaquah: Fire Science Publishers, 2003, 2014, p. 28.

MODELING THE CONDITIONS PRODUCED DURING A LITHIUM BATTERY REACTION

Gerard G. Back
JENSEN HUGHES, INC.

BACKGROUND

Lithium batteries have become the industry standard due to their higher energy densities as compared to legacy batteries (up to 100 times the energy density). Lithium batteries are a family of cells that consist of a lithium-based anode (negative terminal) and a variety of different types of cathodes (positive terminal) and electrolytes [1]. Lithium batteries are grouped into two general categories; primary and secondary batteries. Primary batteries are comprised of single-use cells containing metallic lithium anodes and cannot be recharged. Secondary lithium-ion batteries are comprised of rechargeable cells containing an intercalated lithium compound for the anode and cathode. Non-rechargeable batteries (primary batteries) are referred to throughout the industry as "Lithium" batteries, and rechargeable lithium batteries (secondary batteries) are commonly referred to as "Lithium-ion" batteries. The modeling described later in this paper was performed for rechargeable/Lithium-ion batteries.

Single lithium-ion batteries are commonly referred to as "cells" and have an operating voltage that ranges from 3.6 V– 4.2 V. The cell consists of a negative electrode (anode) and positive electrode (cathode) that serve as a host for the lithium ions (Li+). Lithium ions move from the anode to the cathode during discharge and are intercalated into the cathode (i.e., inserted into voids in the crystallographic structure of). The ions reverse direction during charging as shown in Figure 1.

The layers of anode (copper foil coated with a specialty carbon) and cathode (typically aluminum foil coated with a lithiated metal oxide or phosphate) are separated by a microporous polyolefin film referred to as a separator. An electrolyte composed of an organic solvent and dissolved lithium salt provides the media for lithium ion transport. A cell can be constructed by stacking alternating layers of electrodes (typical for high-rate capability prismatic cells), or by winding long strips of electrodes into a "jelly roll" configuration typical for cylindrical cells. The lithiated metal oxide or phosphate coating on the cathode defines the "chemistry" of the battery.

Figure 1. Generalized Lithium-Ion Cell Components/Configuration

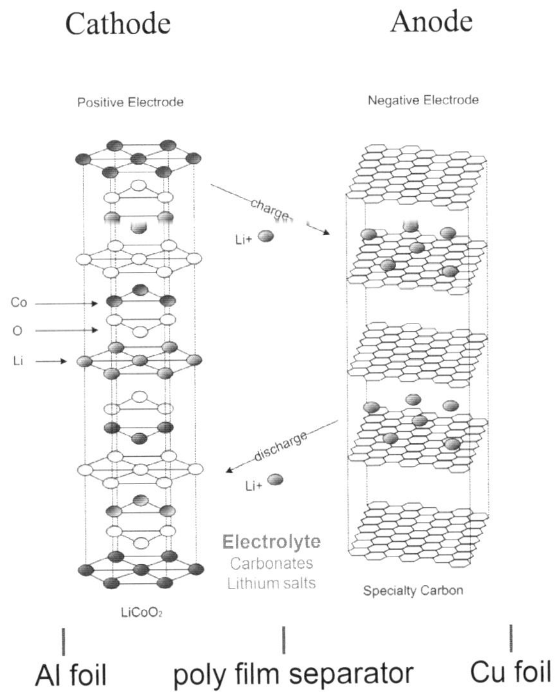

The electrolytes are typically a mixture of organic carbonates such as ethylene carbonate or diethyl carbonate [1,2]. The flammability characteristics of common carbonates used in lithium-ion cell electrolytes in a neat state are provided in Table 1. The mixture ratios of the compounds vary depending upon desired cell properties (e.g., a cell designed for low-temperature applications will likely contain a lower viscosity electrolyte than one optimized for room temperature applications).

Table 1. Flammability Characteristics of Typical Electrolyte Compounds

Electrolyte Component	**Molecular Formula**	**Flash Point** °C	**Boiling Point** °C	**Auto-Ignition Temperature** °C	**Heat of Combustion** kJ/ml
Diethyl Carbonate (DEC)	$C_5H_{10}O_3$	25	126	445	–20.9
Di-Methyl Carbonate (DMC)	$C_3H_6O_3$	18	91	458	–15.9
Ethylene Carbonate (EC)	$C_3H_4O_3$	145	248	465	–17.2
Ethyl Methyl Carbonate (EMC)	$C_4H_8O_3$	25	107	440	–19.2
Propylene Carbonate (PC)	$C_4H_6O_3$	135	242	455	–20.1

The electrolyte solution contains solvated lithium-ions, which are provided by lithium salts, most commonly lithium hexafluorophosphate ($LiPF_6$). $LiPF_6$ decomposes at elevated temperatures to release hydrofluoric acid (HF) during a battery casualty.

There are three basic cell designs; coin-shaped cells, cylindrical-shaped cells, and prismatic-shaped cells. Coin-shaped cells (sometimes referred to as button cells) are low current cells used in watches, calculators, and remote car locks to name a few. The cylindrical-shaped cells incorporate the same design parameters that have been the standard for alkaline cells for years (A, AA, AAA, C, and D cells). Prismatic-shaped cells include standard 9V battery designs (although these use cylindrical alkaline cells) and a variety of "pouch" cells that are typically used in tablets and cell phones. Examples of these cell designs are shown in Figure 2.

The severity of a runaway battery reaction is in part related to the buildup and release of pressure from inside of the cell. Cells with a means of releasing this pressure (i.e., pressure relief vents) typically produce less severe reactions than cells that serve to contain the pressure and rupture due to high pressure (i.e., unvented cylindrical cells). As a result, the cell construction can be a major variable pertaining to the severity of a battery mishap.

Figure 2. Typical Cell Designs

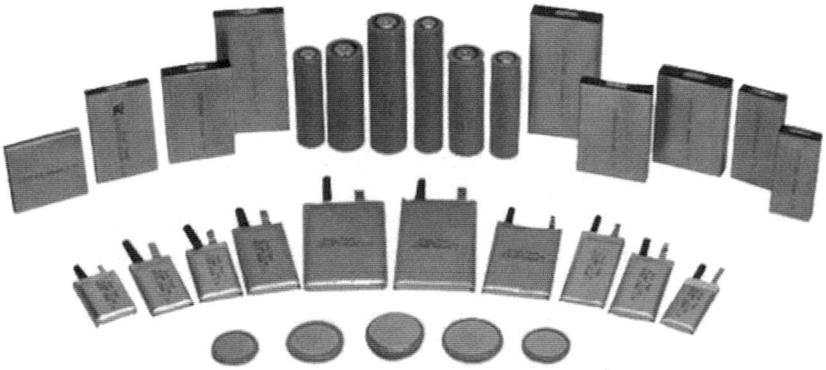

Since very few portable electronic/electrical devices operate at 3.6 V–4.2 V, numerous cells are connected in packs to provide the desired voltage and capacity. Connecting cells in parallel increases pack amperage and discharge capacity while connecting cells in series increases pack voltage. As an example, a 24 V lithium-ion battery pack typically has six cells connected in series. Examples of cylindrical cells and battery packs produced with cylindrical cells are shown in Figure 3. It needs to be noted that many larger battery packs have built-in circuitry used to monitor charging and controlling the discharge cycle of the battery (i.e., shut down the battery when the charge level drops below a specified value).

Figure 3. Example Battery Packs

The form factor cell, "18650" (similar in size and shape to an AA battery) is the "workhorse" of the lithium-ion battery industry and is used in a majority of commercially available battery packs. The "18" refers to the diameter of the cell in and the "65" refers to the length of the cell, both in millimeters. All of the packs shown in Figure 3 are produced with 18650 cells.

LITHIUM-ION BATTERY HAZARDS

A lithium-ion battery is different from legacy battery types (e.g., nickel metal hydride, nickel cadmium, and lead acid) in a number of ways. From a safety and fire protection standpoint, the higher energy densities coupled with the flammable organic electrolyte creates a number of new challenges with regard to the use, storage, and handling of these batteries. Studies have shown that physical damage, electrical abuse (i.e., short circuits and overcharging) and exposures to elevated temperature can cause a runaway reaction within a battery/cell (referred to throughout the industry as thermal runaway). The reaction is the result of an internal short circuit within the battery that vaporizes the organic electrolyte which can lead to an explosion and/or fire [2-4]. Manufacturer's defects such as imperfections and/or contaminants in the manufacturing process can also lead to thermal runaway under certain conditions.

JENSEN HUGHES personnel have conducted over 15 years of research for the U.S. Navy, assessing the fire hazards associated with the spectrum of lithium and lithium-ion batteries for use onboard navy ships and submarines [3,4]. During these programs, a majority of the commercial off the shelve batteries as well as the custom made military batteries were tested for a number of characteristics including their susceptibility to physical damage, heating and over charging. In all, over 250 types of batteries and battery configurations were assessed during these programs. The characteristics measured included likelihood of ignition, gaseous reaction products and heat release rate to name a few. Literally millions of dollars of batteries have been tested and destroyed during these programs.

The data shows that the severity of the reaction is a function of a number of parameters including battery size, chemistry, construction and the battery state of charge (SOC). During a thermal runaway reaction within a cell, a significant amount of gas is produced that causes an increase in pressure within the cell housing. The mechanism for release of this pressure is dependent on the rate of the reaction and the configuration of the cell housing.

For example, at a low SOC, the event may consist of a small amount of liquid/aerosol sprayed through the overpressure vents surrounding the positive terminal. The vented products are flammable and can be ignited if exposed to a flame. At a high SOC (fully charged batteries), the venting is much more forceful and typically involves a much larger volume of liquid/aerosol. The electrolyte liquid/aerosol is again flammable and can be ignited forming small torch-like fires at the overpressure vent ports near the positive terminal. There can be a stronger, sometimes substantial, pressure pulse associated with this electrolyte release. In some instances, depending on the battery construction, flaming debris can be expelled from the vent ports on the battery.

Photographs of venting scenarios are provided in Figure 4. The photograph on the left shows a venting scenario that has not been ignited. The photograph on the right shows the conditions that could occur if the vented products are ignited.

Figure 4. Single Cell Venting Reactions

Smoker | Flare

Occasionally, the pressure release ports fail to operate correctly (or the battery is not equipped with vent ports), causing a build-up of pressure inside the cell casing until the casing fails. When this occurs, the cell literally explodes, expelling the contents throughout the test chamber and producing a substantial pressure pulse. These explosive type reactions can vary from firecracker type bangs to as loud as shotgun blasts or greater. As with the slower venting scenarios, the contents are flammable and can be ignited. Since the entire electrolyte content of the battery is released at once, a substantial fireball can occur if the products are ignited. The high-pressure releases may eject burning or incandescent material that may start nearby fires or ignite the vapor also ejected.

Photographs of explosive type events are provided in Figure 5. The photograph on the left shows an explosive event that was not ignited. The photograph on the right shows a similar reaction where the products were ignited. The fireball in the right side photo is about 1.5 m in diameter and was sustained for about a minute.

In almost every battery casualty test, the same hazardous components were observed; flammable by-products (aerosols, vapors and liquids), toxic gases and flying debris (some burning) and in many instances, sustained burning of the electrolyte and casing material.

Figure 5. Single Cell Explosive and Flaming Reactions

Explosion | Fireball

During a venting reaction (i.e., no ignition of the vented products), the products consist primarily of electrolyte constituents. For most batteries, the products typically consist of 30% Nitrogen (N_2), 20% Carbon dioxide (CO_2), 20% Carbon monoxide (CO), 20% Hydrogen (H_2) and 10% hydrocarbons (C_xH_x). These gases are flammable and have a Lower Explosive Limit (LEL) on the order of 10% by volume (the hydrogen and hydrocarbon percentages may vary with chemistry and state of charge). As a result, the venting of larger battery packs has the potential to produce explosive concentrations and significant fireballs and overpressures if the gases are logged before ignition [3,4].

The volume of vented gas has been determined to be proportional to the electrical energy stored in the cell/battery. For most chemistries, about 0.3 m^3/MJ (10 ft^3/MJ) of vented gas is released during a non-burning reaction [3,4].

For the burning scenario, the electrolyte burns fairly efficiently producing primarily CO_2 and water as the by-products. For most batteries, the products typically consist of 30% Nitrogen (N_2), 50% Carbon dioxide (CO_2) and 20% Water Vapor (H_2O). The burning reaction also tends to liberate the fluorine from the lithium salt (typically $LiPF_6$) dissolved in the electrolyte. The fluorine typically reacts with hydrogen to form hydrogen fluoride (HF). This HF production is also proportional to the electrical energy stored in the cell/battery.

The total combustion energy is typically on the order of about 10 times the electrical energy potential (energy stored in the cell/battery) of the battery. The individual shrink wrap on the cells tends to double this value (i.e., 20 times the electrical energy potential of the battery). Battery packs with thinner plastic casings are about five times this value (i.e., 50 times the electrical energy potential of the battery) and thicker plastic pelican type housings can increase the combustion energy by 10 times this value (i.e., 100 times the electrical energy potential of the battery) [3,4].

The venting and burning rates of multiple-cell packs are a function of the propagation/rate from cell to cell within the battery pack and the contribution of the battery pack casing material. The propagation rate from cell to cell ranges from seconds to minutes depending on the type of cell (chemistry and construction) and battery pack configuration. As a result, some of the smaller packs are consumed in less than a minute while some of the larger packs react/burn for over an hour.

MODELING

To supplement the hazard characterization testing conducted by the USN, a number of modeling exercises were performed to assess the impact of the various parameters associated with a given application. The models were calibrated using baseline empirical data and were then used to conduct sensitivity assessments of the range of application parameters.

The primary model used during these assessments was the Fire Dynamics Simulator (FDS) augmented using a number of solid heat transfer models [5]. FDS is a computational fluid dynamics (CFD) model that has been under development/refinement at the National Institute of Standards and Technology (NIST) Building Fire Research Laboratory (BFRL) for over a decade. Similar to other CFD models, FDS requires the computational domain to be divided into smaller volumes or grid cells in which the mass and energy conservation equations are applied and solved. The size selection of these cell volumes is typically driven by the characteristics of the computational domain (i.e., the volume of the space being modeled) and a compromise between speed and accuracy of the computation.

The modeling exercise/example presented in this paper consisted of developing and tweaking of a model to predict the range of hazards associated with the deployment of an MK 18 Unmanned Underwater Vehicle (UUV) on a USN surface combatant [6]. The UUV was stowed and launched from a Conex Box (referred by the USN as a MILVAN) located in a large open space (i.e., hangar, vehicle storage area or well deck) onboard ship.

The MK 18 Mod 2 UUV is an automated, unmanned underwater sensor platform used for low-visibility underwater missions for both military and civilian purposes. During survey work the vehicle follows a pre-programmed course and contains a number of sensors to gather data which are uploaded once the vehicle return to the base of operations, and trained personnel then conduct post-mission analysis. The mission objectives include hydrographic or environmental mapping, target location and assessment or mine countermeasures operations. An illustration of the MK 18 UUV is provided as Figure 6.

Figure 6. MK 18 Unmanned Underwater Vehicle (UUV)

The MK 18 UUV is powered by large trays (two or three depending on the mission) of lithium-ion batteries located near the center of the vehicle. Each tray consists of five or six pairs of battery modules, depending on configuration. Each module is made up of 63 E-One Moli ICR-18650 cells (4.2V, 2.4 Ah Lithium Cobalt Oxide ($LiCoO_2$)) arranged in parallel groups of nine cells and seven groups in series (9P7S). Loaded at full capacity, the MK 18 contains 2268, 18650 cells. Photographs of the battery configurations are provided in Figure 7.

Figure 7. MK 18 Battery Configurations/Photos

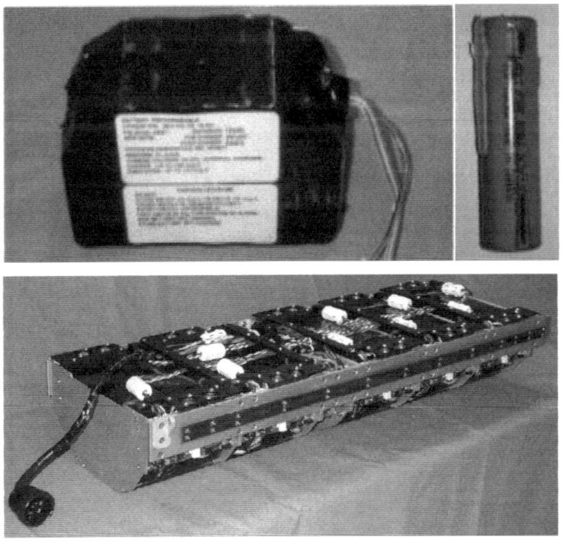

The initial focus of the modeling exercise was to accurately predict the conditions produced inside of the UUV during a battery reaction and the propagation of the reaction through the large number of cells inside of the vehicle. The model was developed and initially calibrated using empirical data from two all-up vehicle tests.

After a number of iterations and additional small scale (cell and module level) tests, the model was able to accurately predict the gas temperature inside of the UUV, the UUV surface temperature, the temperatures of all of the cells located in the UUV and the amount of gas venting from the UUV out the pressure relief port located at the back of the vehicle. A snapshot of the internal gas temperatures predicted by the model one minute into the reaction is provided as Figure 8.

Figure 8. MK 18 Internal Gas Temperature (1:00 into the Reaction)

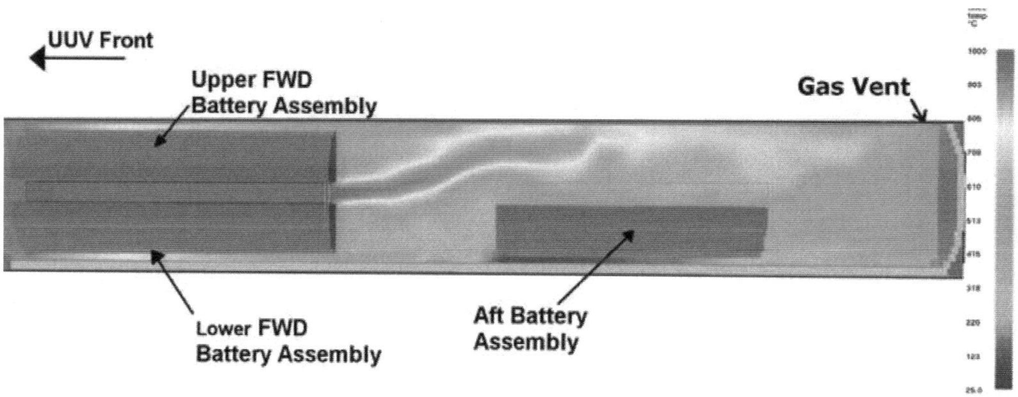

During the exercise, the gases released from the individual cells were modeled as the average of the vented gas products measured during the small-scale tests and mentioned previously (i.e., 30% Nitrogen (N_2), 20% Carbon dioxide (CO_2), 20% Carbon monoxide (CO), 20% Hydrogen (H_2) and 10% hydrocarbons (C_xH_x)). The vented gas temperatures ranged from 400°C to 800°C depending on the battery chemistry and SOC (i.e., the lower temperatures were used for the more benign phosphate based chemistries and the higher temperatures were used for the more reactive Manganese, Cobalt chemistries). The gases were always assumed to vent from the positive terminal of the battery.

Individual cells/batteries were modeled as solid cylinders with the average properties of the cell components (i.e., steel outer shell, copper and aluminum current collectors). Cell to cell propagation was modeled to occur when the adjacent cells reached surfaces temperature from 100°C to 200°C depending on the battery chemistry (i.e., the higher temperatures were used for the more benign phosphate based chemistries and the lower temperatures were used for the more reactive Manganese, Cobalt chemistries).

After a number of iterations and tweaks, the model was able to match the empirical data within a reasonable degree of accuracy. The model was then used to conduct sensitivity assessments on the initial battery reaction location and for a range of battery chemistries. The model was validated using two final all-up vehicle tests in which the predicted condition and measured condition were almost identical.

The final analysis included the modeling of the flammable gases released from the UUV into the MILVAN during the battery reaction and the development of an exhaust gas extraction system to prevent the development of an explosive gas mixture in the space. The MILVAN provides storage for two assembled MK 18 UUVs and storage for one unassembled UUV (i.e., in its shipping container). The MILVAN is a custom built ISO container that is 6.1 m long, 2.7 m wide and 2.6 m tall. An illustration of the MILVAN is provided as Figure 9.

Figure 9. MILVAN Isometric Cut-away View

The flammable gas concentrations predicted by the model for the two configurations are shown in Figure 10. The snap-shot at the top shows the gas concentrations in the MILVAN in the absence of a ventilation system. The green and yellow colored regions correspond the gas concentration in the flammable range. The ignition of these gases would produce significant overpressures within the MILVAN and structural damage of the enclosure. The lower snap-shot shows the same battery reaction within the UUV with a rapid exhaust gas extraction system (2000 CFM supply and exhaust) installed in the MILVAN. During this scenario, the gases are only in the flammable range in the area adjacent to the pressure relief port on the UUV. The ignition of these gases would produce a small fireball at the aft end of the UUV which would be thermally managed by the deluge sprinkling system installed in the enclosure. These predictions were also in good agreement with two full-scale validation tests conducted in the MILVAN.

Figure 10. Single Cell Explosive and Flaming Reactions

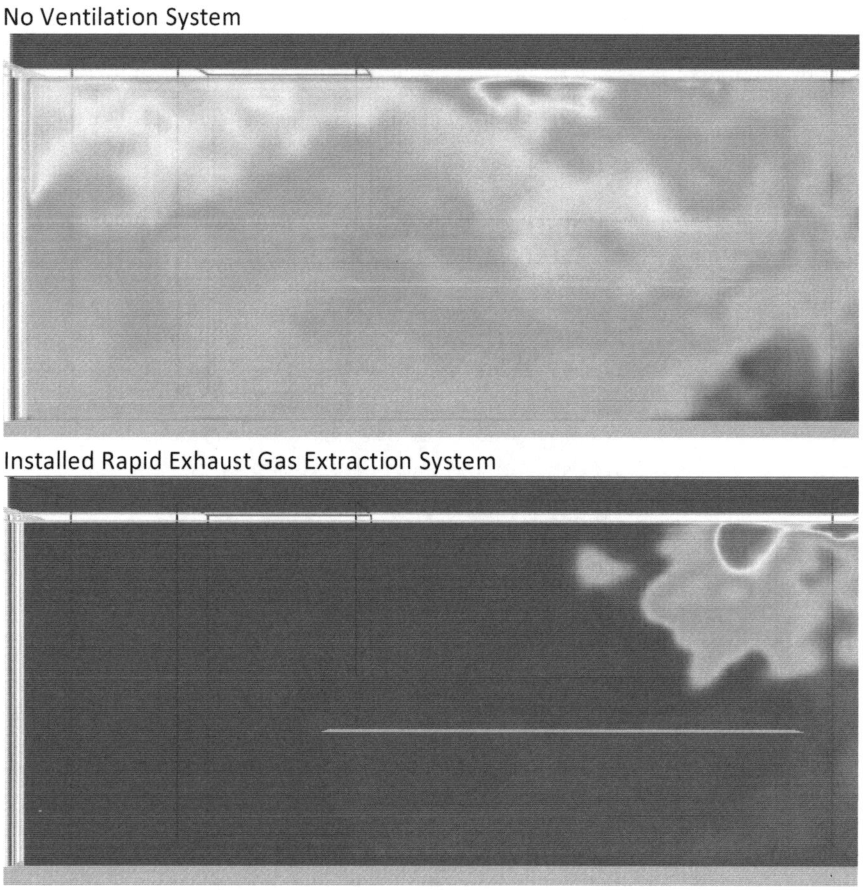

SUMMARY AND CONCLUSIONS

Over the past 15 years, JENSEN HUGHES has supported the US Navy in assessing the fire hazards associated with the spectrum of lithium (non-rechargeable) and lithium-ion batteries. During these programs, a majority of the commercial off the shelf batteries as well as the custom made military batteries were tested for a number of characteristics including their response to physical damage, electrical abuse (i.e., short circuits and overcharging) and exposures to elevated temperature. In all, millions of dollars of batteries (i.e., 250 types of batteries and battery configurations) have been assessed during these programs.

The reaction characteristics measured during these programs included the likelihood of ignition, types and quantities of gaseous reaction products and the heat release rate of the ensuing fire. The data suggests that the severity of the reaction is a function of a number of parameters including battery size, chemistry, construction and the charge level of the battery, to name a few. In addition, the geometry of the battery pack is also a key parameter.

The empirical data was used to develop the input parameters for an FDS model, and to calibrate and validate the accuracy of the model predictions. By the end of the program, the FDS model was found to be able to accurately predict the following parameters:

- The gas temperature inside UUV during a battery reaction,
- The UUV surface temperature,
- Heat transfer to adjacent cells/batteries,
- Cell to cell propagation,
- Module to module propagation,
- Tray to tray propagation,
- The amount and temperature of the gas vented from the UUV, and
- The gas concentrations in the MILVAN during the reaction.

REFERENCES

1. Linden, D., Reddy, T. Handbook of batteries. Third Edition, McGraw-Hill Series, New York, NY, 2002.
2. Mikolajczak, C., et al. Lithium-ion batteries hazard and use assessment. National Fire Protection Research Foundation, Quincy, MA, 2010
3. Back, GG, Winchester, CS., Williams, FW. Lithium battery fire test data summary report. NRL Ltr Rpt Ser 6104/0003, Naval Research Laboratory, Washington, DC, November 14, 2012.
4. Williams, FW, Back GG. Lithium battery fire tests and mitigation report. NRL/FR/6104-14-10-262, Naval Research Laboratory, Washington, DC, August 25, 2014
5. McGrattan, K., McDermott, R., Hostikka, S., Floyd, J. Fire dynamics simulator (Version 6.1.2) user`s guide. National Institute of Standards and Technology, Gaithersburg, MD, 2013.
6. Schwartz, J., Skelding, R., Winchester, C., Banner, J. Safety test report of the MK18 Mod 2 UUV battery module and the MILVAN mitigation and containment unit. NSWCCD-61-TR–2013/1, Bethesda, MD, 2013.

FIRE TESTING

PARTICLE FINITE ELEMENT MODELLING AND ADVANCED EXPERIMENTS ON DRIPPING V-0 POLYPROPYLENE

Melissa Matzen[1], Julio Marcelo Marti[2], Eugenio Oñate[2], Sergio Idelsohn[2], Bernhard Schartel[1]*

[1] Bundesanstalt für Materialforschung und –Prüfung (BAM), Unter den Eichen 87, 12205 Berlin, Germany

[2] International Centre for Numerical Methods in Engineering (CIMNE), Technical University of Catalonia (UPC), Gran Capitan, s/n, 08034 Barcelona, Spain

ABSTRACT

Melt flow and dripping of polymeric materials can be both beneficial and detrimental during fire. It reduces flame spread and result in extinction, as mass and heat are removed from the actual pyrolysis zone. In contrast, melt flow and dripping can provide an additional ignition source, additional process of flame spread and has the potential to start a pool fire. In the vertical UL 94 test, a well adjusted dripping behaviour of flame retarded polypropylene (PP-FR) resulted in a non-flaming dripping V-0 classification. For the polymer samples and their drops collected in UL 94, the decomposition and viscosity was investigated. Particle finite element method (PFEM) was successfully used to simulate the material behaviour in the UL 94 test and increased the understanding of the complex behaviour of polymeric materials during fire.

INTRODUCTION

In small-scale fire tests, such as oxygen index and UL 94 burning chamber, the sustained ignition of the specimen as a reaction to an ignition source is based on complex interactions of different phenomena in the gas phase and the condensed phase. The main phenomena controlling burning are pyrolysis of the polymer, oxidation of the volatiles, mass transport to the flame and heat transport to the pyrolysis front. Additional to the main phenomena there a several sub-phenomena and all interact or compete with each other. With respect to fuel release, e.g. gasification competes with melt flow, dripping, and charring. Further, charring e.g. affects the heating of the polymer by forming protective layer that shields the polymer from the flame and therefore influences the decomposition and the melting of the polymer.

Melt flow and dripping often lead to a supplementary process of flame spread, provide an additional ignition source and even harbour the potential to start a pool fire. On the other hand, melt flow and dripping is used as a flame retardant mode of action in commercial products e.g. foils. It reduces flame spread and can result in extinction, as it removes mass and heat from the actual pyrolysis zone. A well-adjusted melt flow and dripping behaviour is crucial to pass for instance UL 94 burning chamber, FMVSS 302, all kinds of Bunsen burner type tests, glow wire test.

The presence of flame retardants affects the fire properties in numerous folds [1]. They can increase the char yield, work as flame inhibitor, plasticiser, reinforcing filler, build up a protection layer, induced the decomposition or a flow limit. All these modes of action interact with each other. The importance of melt flow and dripping was recently shown in different studies [1-9]. In the work reported in [2], PC/ABS blends were subjected to the test flame in the UL 94 scenario and the fire behaviour investigated experimentally and numerically by particle finite element method (PFEM). The combination of an induced flow limit with flame inhibition resulted in a non dripping V-0 classification in the UL 94 test. PFEM was developed at the International Centre for Numerical Methods in Engineering CIMNE in Barcelona, Spain.

In this work the dripping behaviour of HB classified PP and dripping V-0 classified PP-FR has been investigated and simulated with PFEM. The experimental results were compared with the numerical prediction. The viscosity, thermal transition, thermal and thermo-oxidative decomposition of the flame-retarded PP and corresponding PP has been studied. The results are analysed, compared and simulated to explain the different dripping behaviours.

EXPERIMENTAL

Materials

A commercial polypropylene was compounded with a flame retardant. The sheets 100 mm x 100 mm x 3 mm in size were prepared by a melt pressing process (temperature 190 °C, pressure of 20 bar). The bar specimens were injection-moulded using an industrial-scale injection moulding machine (T_{max} = 210 °C).

Characterisation methods

The thermal and thermo-oxidative stability of the PP was investigated by TGA (Netzsch-TG 209 ASC F1 Iris, Selb, Germany) under 30 ml min^{-1} nitrogen flow and 30 ml min^{-1} synthetic air flow, respectively (heating rate 10 °C min^{-1}; temperature range 30 °C to 900 °C). The kinetics of the material was studied by TGA under 30 ml min^{-1} nitrogen flow at a heating rate of 2, 5, 10, 20 °C min^{-1}. DSC (Netzsch 204 F1, Selb, Germany) was carried out under nitrogen atmosphere (50 mL min^{-1}; temperature range - 40 °C to 250 °C; heating rate 10 °C min^{-1}). The specific heat capacity was measured according to ISO 11357-4. Thermal conductivity was tested with a hot disk on two samples of 3 mm thickness. Rheological analyses of the PP and its molten drops were conducted using a plate-plate rheometer (Rheometer MCR 501, Anton Paar, Ostfildern, Germany) in oscillation mode at 190 °C and 170 °C to 300 °C, respectively. The angular frequency was 100 - 0.1 rad s^{-1} and 0.1 rad s^{-1}, respectively.

Figure 1. Scheme of the experimental setup for: (a) modified UL 94; and (b) temperature measurement.

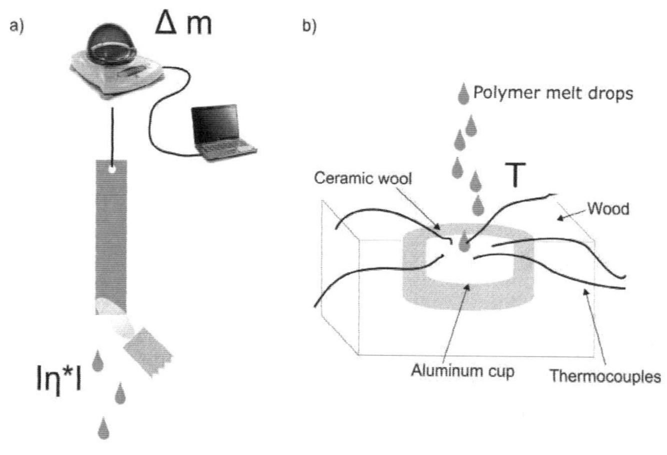

The flammability was determined by the UL 94 test in accordance with IEC 60695-11-10. The melt flow and dripping behaviour was investigated by exposing the bar specimens to the calibrated UL 94 test flame. The sample was hung on a digital mass balance and the test conducted in line with the vertical UL 94 test regulation. The total mass loss was recorded for each sample as a function of time. For the drop temperature measurement (Figure 1), the drops were collected 50 mm beneath the sample in an isolated container equipped with thermocouples and recorded with an IR-camera. The fire behaviour under forced-flaming conditions (Cone Calorimeter, Fire Testing Technology Ltd., Grinstead, UK) was assessed according to ISO 5660. A micro combustion calorimeter by FTT (FAA Micro Calorimeter, Fire Testing Technology Ltd., Grinstead, UK) was used (heating rate 1 °C s^{-1}; 30 °C to 900 °C).

Numerical simulation method

The melt flow and dripping behaviour of polymeric material in a fire scenario is a complex interaction of physical and chemical processes. It involves phenomenon such as fluid flow, heat transfer, material decomposition and flame chemistry. The computational method requires therefore different modules to deal with the polymer as a structure and the polymer as a fluid. The PFEM was developed by the International Centre for Numerical Methods in Engineering (CIMNE) in Barcelona, Spain. The focus of PFEM is the analysis of fluid–structure interaction problems, complex multidisciplinary problems, large continuous deformations, coupled thermal effects, fragmentation, separation of fluid droplets, and complicated conditions between different fluids or between fluids and solids [8-13]. The details of the code is given elsewhere [10].

Model set up

The simple UL 94 test set up is a complex scenario for the quantification of the heat transfer to the specimen. The high temperature applied on the specimen leads to a large temperature gradient on a limited area. Additional edge effects are taking account [15]. A sub-model is implemented to describe the total heat intake consisted of heat intake due to the test flame, and the heat feedback due to combustion of the pyrolysis gases. The estimation of the heat intake by the test flame are detailed reported in the literature [16] and evaluated [2]. A maximum heat intake of 150 kW m^{-2} was set and applied as an exponential decreased function. For the heat feedback as result of combustion of pyrolysis gases a sub- model was used. In the UL 94 test the flame is maintained twice for 10 ± 0.5 s. To simplify the simulation just the first heat application of 10 s is considered. In 3D simulation, the model was discretised by 5718 nodes, forming 21983 four-noded tetrahedrons.

RESULTS AND DISCUSSION

Thermal Transition, Thermal and Thermo-Oxidative Decomposition

Melt flow and dripping of polymeric material during exposure to an external heat source are a complex interaction of physical and chemical processes. The physical processes, such as glass transition and melting, as well as chemical, such as molecular weight reduction and charring, caused by thermal and thermo-oxidative decomposition play a major role with respect to melt flow and dripping. Thus, the thermal transition, thermal and thermo-oxidative decomposition of flame retarded PP-FR and non flame retarded PP were investigated. PP exhibited under nitrogen a single step decomposition with no residue left. In contrast, PP-FR decomposed in three steps with a reduced decomposition temperature ($T_{5\ wt.-\%}$) compared to PP.

UL 94 characteristics

PP achieved in the horizontal UL 94 test the classification of HB and failed in the vertical test. PP burned almost completely with a small residue left. The bar specimen exhibited melt flow and dripping behaviour with flaming drops. In the vertical UL 94 test, PP-FR achieved a UL 94 classification of non flaming dripping V-0.

The mass loss of PP and PP-FR was investigated by exposing bar specimens to a Bunsen burner type ignition source in guidance with the UL 94 test. The dripping behaviour of PP was characterised by a small burning droplets at the beginning, followed by a constant flaming melt flow. The flammability, melting and dripping behaviour of PP-FR differs totally from the behaviour of PP: it extinguished immediately after the first flaming, the second application of the burner for 10 s led to a self-extinguishing due to non-flaming drops. The flame retardant in PP-FR changed the UL 94 classification from HB for PP to a non-flaming dripping V-0.

Temperature of drops

Melt flow and dripping in the UL 94 test can be related to both melting and decomposition. To get an indication about the decree of decomposition of the temperature of the drops was investigated. The temperatures of the drops were measured in the vertical UL 94 set up. Instead of the cotton on the bottom an isolated with thermocouple equipped container was placed 50 mm beneath the sample. The experiment is also recorded by an IR camera. The IR images of the burning bar specimen are shown in Figure 2. In general, the temperature decreased with increasing distance to the burning bar specimen. In comparison to PP the temperature of PP-FR was about 100 °C lower.

Figure 2. IR image: burning melt flow of PP and drop of PP-FR (left to right).

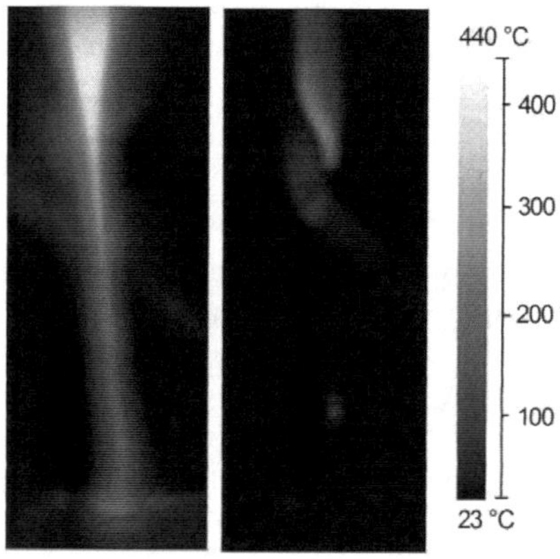

Rheology

The UL 94 test protocol explicitly involves the observation of melt dripping. The viscosity is the controlling factor to adjust the dripping behaviour. For unlinked polymers, the pertinent viscosity for dripping is consistent with the ordinary steady-flow viscosity [16], means dripping and melt flow occurs at low shear rates. The rheological characteristics of PP, PP-FR and its drop material were studied to understand and quantify the effect of the flame retardant in PP-FR and the effect of the flame. For modelling polymeric material in fire scenarios well-estimated input parameters are important for the correct estimation of the material behaviour. Some temperature-dependent material properties in fire situations have been investigated by Stoliarov et al. [17, 18]. The viscosity of PP and PP-FR as a function of temperature was measured. With increasing temperature the viscosity of both materials was decreasing. PP-FR exhibited a higher sensibility to temperature than PP.

Further material properties used for modelling

To describe the materials along with the differences between them it is important to determine the characteristics such as density, thermal conductivity, specific heat capacity, activation energy, Arrhenius coefficient and heat of complete combustion of volatiles of PP and PP-FR.

Numerical Investigation

In Figure 3 the images of PP-FR simulated with PFEM are shown. The images are taken between 0 s and 30 s. For the first 10 s an external heat flux is numerically applied, which is noted in the increase of the temperature of the bar specimen. The calculated temperature of the pyrolysis zone is above the decomposition temperature measured in the TGA. After the dripping the bar specimen is cooling down to about 120 °C. The calculated mass loss of PP-FR correlated to the mass loss in the experiment.

Figure 3. Images of the PFEM simulation for PP-FR.

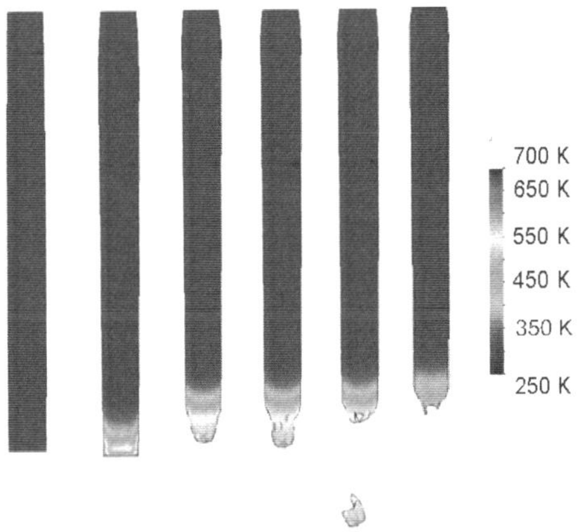

CONCLUSION

The effect of a flame retardant in PP-FR on the dripping behaviour compared to PP has been investigated by experimental and computational approach (PFEM). The UL 94 classification, dipping behaviour, viscosity, thermal transition, thermal and thermo-oxidative decomposition were highly dependent on the flame retardant. The UL 94 classification changed from HB for PP to V-0 for PP-FR. The study of the material enabled PFEM to predict the dripping behaviour in the UL 94 scenario for PP and PP-FR was correct and thus increases the understanding of the complex behaviour of polymeric materials during fire.

REFERENCES

[1] Matzen, M., B. Kandola, C. Huth, and B. Schartel, *Influence of Flame Retardants on the Melt Dripping Behaviour of Thermoplastic Polymers.* Materials, 2015. **8**(9): p. 5621-5646.

[2] Kempel, F., B. Schartel, J.M. Marti, K.M. Butler, R. Rossi, S.R. Idelsohn, E. Onate, and A. Hofmann, *Modelling the vertical UL 94 test: competition and collaboration between melt dripping, gasification and combustion.* Fire and Materials, 2015. **39**(6): p. 570-584.

[3] Ohlemiller, T. and K. Butler, *Influence of polymer melt behavior on flammability.* NIST report NISTIR 6588, 2000.

[4] Wang, Y., J. Jow, K. Su, and J. Zhang, *Dripping behavior of burning polymers under UL94 vertical test conditions.* Journal of Fire Sciences, 2012. **30**(6): p. 477-501.

[5] Kandola, B.K., M. Ndiaye, and D. Price, *Quantification of polymer degradation during melt dripping of thermoplastic polymers.* Polymer Degradation and Stability, 2014. **106**(0): p. 16-25.

[6] Zhang, J., T.J. Shields, and G.W.H. Silcock, *Effect of Melting Behaviour on Upward Flame Spread of Thermoplastics.* Fire and Materials, 1997. **21**(1): p. 1-6.

[7] Casu, A., G. Camino, M. De Giorgi, D. Flath, V. Morone, and R. Zenoni, *Fire-retardant mechanistic aspects of melamine cyanurate in polyamide copolymer.* Polymer Degradation and Stability, 1997. **58**(3): p. 297-302.

[8] Butler, K.M., E. Oñate, S.R. Idelsohn, and R. Rossi, *Modeling polymer melt flow using the Particle Finite Element Method*, in *11th International Interflam Conference*2007, Interscience Communications: London, UK.

[9] Oñate, E., R. Rossi, S.R. Idelsohn, and K.M. Butler, *Melting and spread of polymers in fire with the particle finite element method.* International Journal for Numerical Methods in Engineering, 2010. **81**(8): p. 1046-1072.

[10] Onate, E., S.R. Idelsohn, F. Del Pin, and R. Aubry, *The particle finite element method - an overview* International Journal of Computational Methods, 2004. **1**(2): p. 267-307.

[11] Idelsohn, S.R., E. Onate, and F. Del Pin, *The particle finite element method: a powerful tool to solve incompressible flows with free-surfaces and breaking waves.* International Journal for Numerical Methods in Engineering, 2004. **61**(7): p. 964-989.

[12] Onate, E., S.R. Idelsohn, M.A. Celigueta, and R. Rossi, *Advances in the particle finite element method for the analysis of fluid-multibody interaction and bed erosion in free surface flows.* Computer Methods in Applied Mechanics and Engineering, 2008. **197**(19-20): p. 1777-1800.

[13] Idelsohn, S.R., J. Marti, A. Limache, and E. Onate, *Unified Lagrangian formulation for elastic solids and incompressible fluids: Application to fluid-structure interaction problems via the PFEM.* Computer Methods in Applied Mechanics and Engineering, 2008. **197**(19-20): p. 1762-1776.

[14] Wang, Y., F. Zhang, C.M. Jiao, Y. Jin, and J. Zhang, *Convective Heat Transfer of the Bunsen Flame in the UL94 Vertical Burning Test for Polymers.* Journal of Fire Sciences, 2010. **28**(4): p. 337-356.

[15] Hamins A, B.M., Dillon SE, *Characterization of candle flame.* journal of Fire Protection Engineering, 2005. **15**: p. 265-285.

[16] Franklin, A.G. and R.J. Krizek, *Complex viscosity of a kaolin clay.* Clay and Clay minerals, 1996. **17**: p. 101-110.

[17] Stoliarov, S.I., N. Safronava, and R.E. Lyon, *The effect of variation in polymer properties on the rate of burning.* Fire and Materials, 2009. **33**(6): p. 257-271.

[18] Li, J. and S.I. Stoliarov, *Measurement of kinetics and thermodynamics of the thermal degradation for non-charring polymers.* Combustion and Flame, 2013. **160**(7): p. 1287-1297.

PREDICTING UL 94 RATINGS FOR PLASTICS BASED ON MCC MEASUREMENTS

Mahmood Tabaddor, PhD, Priyansu Panda & Thomas Fabian, PhD
Commercial & Industrial, UL LLC

INTRODUCTION

As manufacturers are always under pressure to be more innovative and faster to market with new product offerings, relying on complicated testing of physical samples can be a competitive disadvantage. However, for any manufacturer, from material-scale to system-level, there are many physical tests that must be conducted to help determine performance. With the growing use of Machine Learning (ML), there are now possibilities that physical tests can be replaced with computer algorithms which can provide high accuracy in predicting performance and so would be an economic boon to manufacturers. Alternatively, if the outcome of a resource-intensive physical test can be predicted based on the results of a simpler physical test, then ML can help learn that relationship and provide tremendous value to manufacturers. Yet the challenge is to understand how to build and validate these ML-based models.

For one set of manufacturers, plastic manufacturers, trying to predict the performance of new materials is challenging. The UL 94[1] test for flammability is one such test. The outcome of the testing is a flammability rating, one of four that are possible. Sample preparation time, amount of materials and length of testing time all motivate the search for a simpler test that could be predictive of the outcome of the UL 94 test. One candidate test is the Micro-scale Combustion Calorimetry (MCC)[2]. MCC is a thermal analysis technique that applies controlled heating to a sample to high temperatures driving combustion. The MCC test does not require molded flame bars typical of the UL 94 test; samples on the order of milligrams are sufficient. Since both tests, UL 94 and the MCC, push the materials into a state of combustion, there might be a technical basis for a link between the outcome of the MCC test and the UL 94 for a material, the first requirement for using ML-based approach.

Brief Literature Survey

Several studies have been carried out in the last few years looking at a possible relationship between MCC and UL 94. In one study[3], the authors visualized a few MCC measurements and UL 94 ratings to uncover a potential relationship. Figure 1, from this study, shows a plot of two MCC variables with the data colored by the resulting UL 94 rating. In this sample set, only two out of the four possible ratings from UL 94 are plotted. Nevertheless, the data does suggest that one or more of the MCC measurements may help the outcome of the UL 94 test. The authors then proceed to develop a flammability index based on three of the MCC parameters but with no stated link to the UL 94 rating.

In other studies[4,5,6], the authors have looked at the relationship between MCC measurements and the associated UL 94 ratings. Figure 2 shows one example from a study[4] where a single parameter, the heat release rate, measured by the MCC test is shown versus UL 94 rating. Again, the data analysis is suggestive of a potential relationship between UL 94 rating category and the MCC test. For this specific example, once again a threshold appears that may help separate one category, V0, from all the other categories. Yet there is overlap and the parameter cannot help distinguish between all four of the UL 94 categories.

Figure 1 (from reference 3)

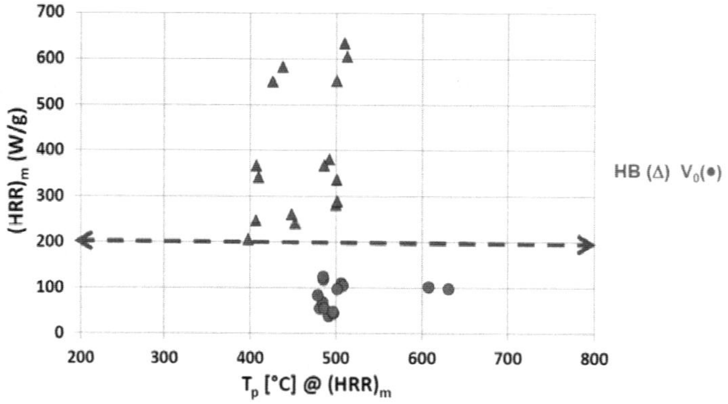

Figure 2 (from reference 4)

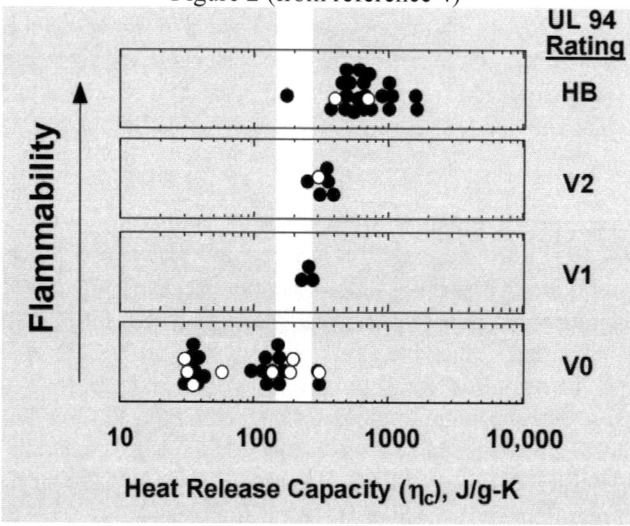

ML-based Modeling Approach

There is a long history in using ML-based algorithms, such as neural networks[7], to predict the properties of polymers. However, these sorts of tools were only available to academics with the time to handle and troubleshoot these complicated algorithms and immune to the pressures of practical product development. Over the last few years, with the boost in cheaper and powerful computational machines and advances in ML algorithms, ML-based modelling is now a more accessible tool for practical product development. Yet, there is a need for a careful and expert oriented process to ensure the modelling process does not generate misleading and incorrect models. Unlike the clichéd garbage in, garbage out, the key concern is that even with good inputs, the modelling process can be carried out incorrectly as the ML algorithms are vastly more complicated than the traditional linear analysis that has been the work horse of the regression in statistical analysis.

Figure 3 shows a schematic of the key steps in an ML-based modelling building process. Without going into details there are several key steps. One is exploratory data analysis. In this step, the data is reviewed, cleaned and visualized to ensure that an understanding of the quality and potential gaps are identified early. Some visualizations look for correlations amongst the list of candidate input variables. Any correlations can help simplify the list of potential input parameters and lead to a more efficient modelling process. Also there is quite a bit of learning that can happen in this stage even before a model is even built. Once the data is assessed and streamlined, then the next key step is to divide the data into two sets (for more advanced analysis, three sets). The purpose of one set, the training set, is to help the ML algorithm to learn all it can about potential patterns within the data. This is where model parameter tuning and selection occur. Then once a model(s) is selected, the accuracy of the model is determined by trying to predict the outcomes given in the validation set. This separation process is one of the keys to the integrity of the model building process. The division of the total data set into a training and validation sets must meet certain requirements not detailed here.

Figure 3

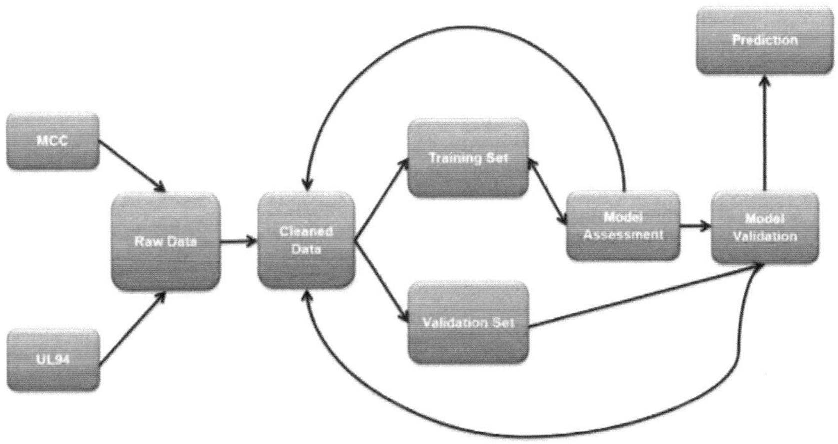

For this project, since the intent is to predict a UL 94 rating, classification ML algorithms will be chosen. The target, or the outcome, that is being predicted is simply the following ratings: V0, V1, V2 and Fail. The inputs are all the measurements from the MCC test and also the thickness of the sample during the UL 94 test. Not all MCC parameters are expected to be predictive of the UL 94 rating but this is part of the learning process. The ML process will help filter those inputs that are not strong predictors of the outcome of interest.

Finally, there is another challenge specific to predicting the UL 94 rating. The rating assignment is based on what we call a "bad apple" approach. A single bad reading can lead to the overall assigned lowered rating from the sample of 5 that are measured. So basically, if only one sample performs sufficiently poor it will override the performance of the other samples and determine the final rating. This is not something that can be expected to be predicted accurately unless there is a reason to believe that the MCC would more sensitive to that individual sample than all the others. We would expect that the model can predict the overall behaviour of all the samples of materials (such as an averaged value), not that of an individual sample the deviates from the others.

To address this concern, two different methods of assigning UL 94 ratings will be examined and the effect on the accuracy of the model will be shown.

THE DATASET

Data was gathered on 39 independent samples. This is generally a small test size for ML based learning but will likely represent the reality of many similar projects. Still, it is a good start and will help identify the potential in ML modeling and determine if there is value in investing in building the dataset size. The 57 samples were manufactured using 25 unique grades of polymers, including both thermoset and thermoplastics, 13 different generic resin types and different thicknesses. MCC data was generated suing Method B with 5 replicates. Measurements included PHRR(W/g), T@PHRR(°C), Onset HRR (10) Temp(°C), Onset HRR(25)Temp(°C), HOC(kJ/g), Residue(%), and Heat Release Capacity(J/g-K).

For the UL 94 test, multiple thicknesses were used with final ratings covering all four categories and 5 replicates for each sample conditioning type. For the UL 94 rating test, 5 samples were tested as received (AR) with no prior conditioning. 5 other samples were conditioned by oven aging (OA) prior to running the UL 94 test. No films were tested. Hence the dataset was divided into two sets: one for AR and one for OA (Figure 4). Each receives its own UL 94 rating. Hence, a different model would need to be built for each one. The samples for the MCC test do not undergo any conditioning and so a model between MCC and AR UL 94 data would be expected to perform better. The UL 94 rating shown follows the procedure described in the standard.

Figure 4

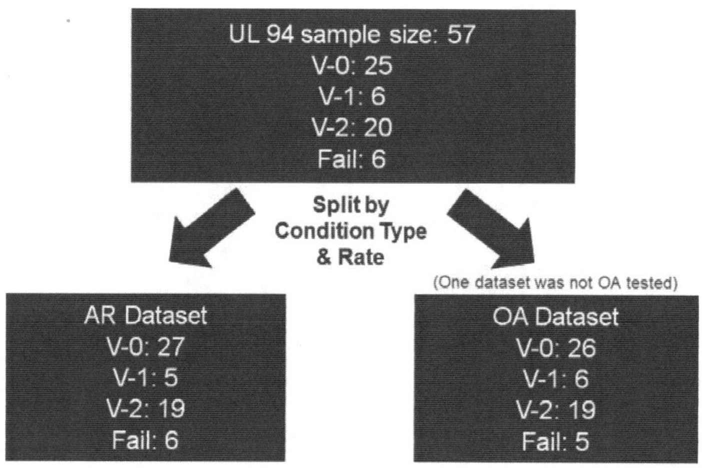

Exploratory Data Analysis

Exploratory data analysis (EDA) is an approach to visualize datasets especially those with many input parameters that define a hyperspace. The objectives of EDA are to suggest hypotheses about the causes of the observed data, support in modelling of the data, and guidelines for future data collection. One such visualization tool is a scatterplot (Figure 5) where all the input variables are plotted against each other in pairings. This helps show the myriad of possible relationships between the parameters.

From this scatterplot, certain patterns are observed and quantified for their strength. For the MCC attributes, there is a strong positive linear correlation between PHRR and HRC. There is a similarly strong positive correlation between OnsetHRR25Temp and OnsetHRR10Temp. Finally, there is a somewhat strong negative linear correlation between HOC and Residue. Looking at the first pairing, PHRR and HRC, the two attributes are perfectly correlated. This is due to the fact that one parameter is derived from the other, under constant heating rates. In building the model, it is not necessary to keep both attributes and the list of inputs to the model is reduced by one.

Figure 5

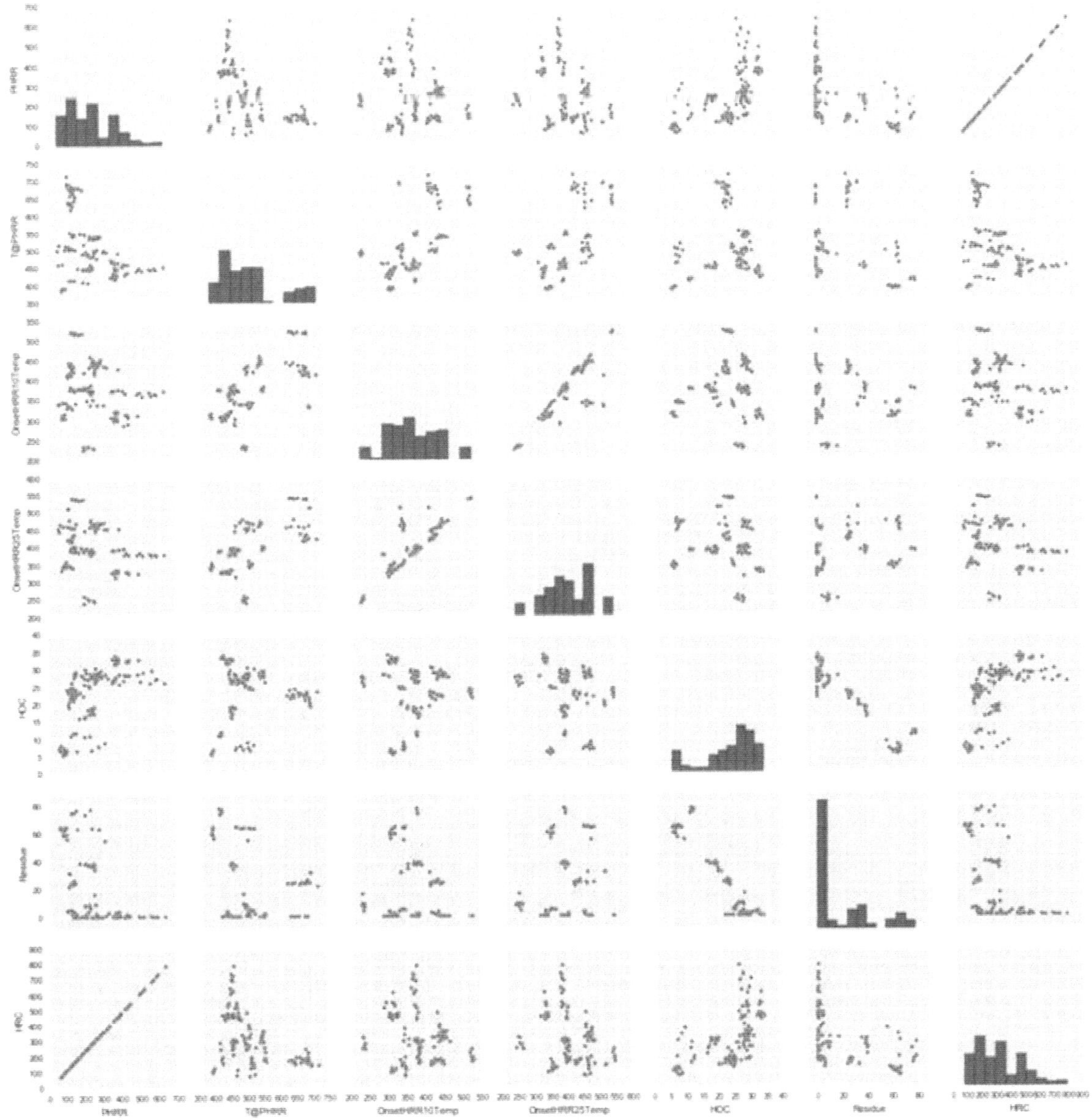

Figure 6 shows the presence of a negative correlation between HOC and Residue. However, by color coding the data to show whether the sample was milled (blue) or bulk form (yellow), then it becomes obvious that the strong correlation holds for the bulk form of the sample. Though there is a relatively strong negative correlation, both attributes are kept for the model building process. A similar analysis was conducted for all the pairings shown to understand where gaps in the data exist and which attributes might be useful to keep in building the model.

Figure 6

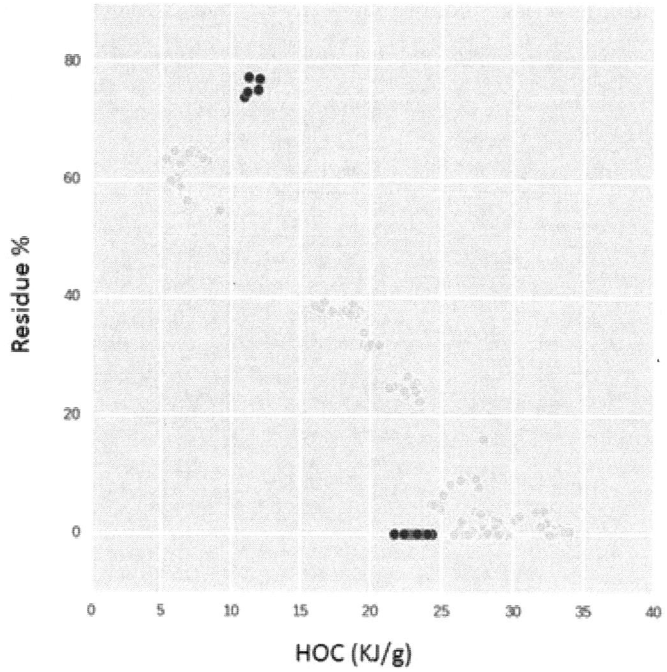

MODELING

The EDA helped pare down the list of potential MCC measurements to only the PHRR, T@PHRR, HOC, Residue, OnsetHRR10Temp and OnsetHRR25Temp. In addition, the thickness from the UL 94 test is a necessary input as it is well known that sample thickness affects assigned rating. For the output, the model needs to predict one of four possible UL 94 ratings: V-0, V-1, V-2, and Fail. Hence the ML algorithms that are described as multi-class classifiers (with a cross validation approach) will be the basis for the modelling building. Specifically, the Random Forest classifier was the algorithm of choice as it performed better than a number of other classifiers. The details of this process are not presented here other than to note that using more than one model is a best practice. There is no single universal algorithm that will work best with all datasets.

Before moving ahead with the modelling building process, it is important to upfront agree on an evaluation metric, one that will help establish the accuracy level of the model and be the basis for deciding on whether a model is ready for release. For this classifier, accuracy will be evaluated on the basis of proportion of true results (both true positives and true negatives) among the total number of cases examined. Tuning of the hyper-parameters of the algorithm required that a grid search technique was used on the training set along with a k-fold cross validation.

Modeling Results

Figure 7 shows what is called a confusion matrix. It is a very simple and compact table that compares the different predictions versus the actual ratings for UL 94 for the AR samples. In this case, the overall accuracy of the model is approximately 52%. This is better than the case for the OA samples which produced a model with an accuracy of 50%. It may not seem like a big difference but for such a small dataset, it does demonstrate that there is likely a stronger relationship between AR samples from UL 94 rating to MCC. This was expected since the MCC samples are not subject to any conditioning.

Figure 7

		Predicted Rating			
		V-0	V-1	V-2	Fail
Actual Rating	V-0	384	110	95	36
	V-1	95	0	20	0
	V-2	82	10	296	57
	Fail	37	3	80	0

From the confusion matrix, it can be seen that one area that challenged the model was accurately predicting V-1 and Fail. The model performed much better for V-0 and V-2. Part of the error may be due to the test method and how a single sample results in a rating change, always lowering the rating. V-0 is considered the best rating followed by V-1, V-2 and finally Fail. The other part of the error may be due to the small number of samples that earned a V-1 or Fail rating in the larger dataset.

Another positive feature of the ML-based algorithms is that the classifier provides a relative ranking of the importance of all the input variables. Figure 8 shows the attribute rankings with thickness having the largest effect. This was expected and shows how a non-material level property influences the predictions. Next is the Residue as the most prominent MCC measurement followed by PHRR, HOC and OnsetHRR10Temp being basically equally impactful. This sort of ranking can help lead to learning as an explanation on why this ranking might be correct needs to take place.

Figure 9 shows the confusion matrix for an alternative method for determining the UL 94 rating on the AR samples. As mentioned earlier, the normal UL 94 rating could be determined based on a single sample, the "bad apple". The theoretical foundations of predictive modeling would conclude that trying to model such a scenario will be unsuitable. So to investigate this issue, a new UL 94 rating was assigned to the sample grouping based on a "majority rules" approach. In this case, the rating that was most dominant amongst the 5 samples was selected. This did result in a change in the UL 94 ratings for some of the samples. The accuracy for the "majority rules" was 67% versus 52% for the "bad apples". This suggests that the current method in which the UL 94 rating is determined will be a limiting factor in developing a highly accurate relationship between MCC data and UL 94 ratings.

Figure 8

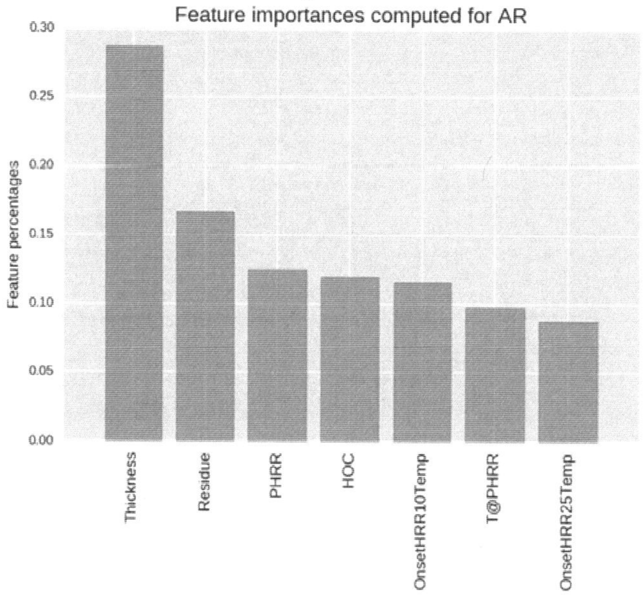

Figure 9

		Predicted Rating			
		V-0	V-1	V-2	Fail
Actual Rating	V-0	485	95	95	15
	V-1	80	0	0	0
	V-2	35	0	391	19
	Fail	22	3	65	0

SUMMARY AND RECOMMENDATIONS

Progress in ML makes it a very practical tool to help manufacturers gain insight into their product development. In this paper, the power of ML is demonstrated with the example of finding a relationship between MCC measurements and UL 94 to help predict the subsequent UL 94 rating. The methodical approach used in the model building uncovered relationships between different MCC attributes and strengthened the choice of the MCC as a surrogate for the UL 94 rating. However, gaps in the data and the quirk of the how the rating is assigned may also suggest some limits to building a very accurate model.

REFERENCES

1. **UL 94.** *Standard for Tests for Flammability of Plastic Materials for Parts in Devices and Appliances.* 2013. Edition 6.
2. **UL.** *Microscale Combustion Calorimetry (MCC) Testing for Polymeric Materials.* s.l. : Performance Materials, 2014.
3. *Micro-scale evalution of flammability for cable materials.* **Yang, HE et al.** s.l. : Proceedings for 62nd International Wire & Cable Symposium, 2013. Vol. Proceedings.
4. *Standard Test for Flammability Screening of AC Materials.* **Lyon, RE.** s.l. : International Aircraft Materials Fire Test Working Group, 2006.
5. *Correlations between Microscale Combustion Calorimetry and Conventional Flammability Tests for Flame Retardant Wire and Cable Compounds.* **Lyon, HE et. al.** s.l. : 56th International Wire & Cable Symposium, 2007.
6. *Assessing Material Consistency Using MCC.* **Fabian, Tom.** s.l. : International Aircraft Materials Fire Test Working Group Meeting, 2015.
7. *On the use of computational neural networks for the prediction of polymer properties.* **Sumpter, BG et al.** 3-4, s.l. : Journal of Thermal Analysis and Calorimetry, 2005, Vol. 46.

MODIFICATIONS TO CONE CALORIMETRY INSTRUMENTATION AND PROCEDURES TO FACILITATE GENERATION OF FIRE SAFETY ENGINEERING DATA IN STANDARDISED TESTS

Sean Gregory, Sandra Pasantes, Fire Testing Technology Ltd, UK
Stephen J Grayson, Interscience Communications Ltd, UK

The cone calorimeter developed by Babrauskas and standardised in ISO 5660 and ASTM E1354 was standardised to measure heat and smoke release from simple combustible materials such as wood, plastics and building products when burnt in air. More recently standardisation has focused on further measurement of other parameters in the cone calorimeter such as testing materials that are thermally mobile whilst testing; measuring parameters in vitiated oxygen conditions; and most recently replacement of the light sources in the optical transmission systems used for smoke measurement.

ISO 5660-1 typically investigated materials with peak HRR of 250-2000 kW/m^2. Researchers either use the results directly or as data for numerical models. More recently researchers, the authors and standardisation groups have developed techniques to measure much lower levels of heat release and these are very recently standardised in ISO 5660-4.

Measuring Critical Ignition flux:

The minimum heat flux supporting sustained ignition is a fundamental parameter that will influence fire initiation and development and hence is very important to determine. Measurement and quotation of this property is both important in defining the fire properties of materials and as data for fire safety engineering studies. ISO TC92 SC1 workers realised that this could be more easily, quickly and conveniently measured in the cone calorimeter (ISO 5660-1) or even in the fire model of that apparatus (as standardised in ISO 1372 and ISO 17754), than in other apparatus and this has led to the development of a standardised iterative measurement technique for determination of this parameter.

Critical heat flux for ignition is the minimum heat flux needed to sustain ignition. An effective Critical Heat Flux value can be determined empirically using the cone and igniter assembly section of the apparatus (as in ISO 13927 or ISO 17554), by exposing specimens to different heat fluxes and measuring the ignition times for sustained flaming. It is determined by trial and error with repeated tests in search of the heat flux for no ignition occurring within 15 min duration. Specimens are tested without retainer frames and the minimum heat flux is also determined by trial and error and is first determined to a coarse resolution of 5 kW/m^2, and then more finely to 1 kW/ m^2.

The lowest value of heat flux at which sustained ignition is achieved is determined, and then the highest value at which ignition is not achieved. The minimum heat flux for ignition is defined as the average between the lowest heat flux at which there is ignition, and the highest heat flux at which there is no ignition for 15 min.

Now this parameter is standardised in the ISO 5660 series and in ASTM 1354, we expect it to be widely quoted in manufacturer literature.

Measuring low heat release rates in cone calorimeter

Some countries now want to regulate low heat release construction products. Both Japan and Canada have already issued standards and regulate using heat release measurements based on cone calorimeter data generated using a protocol and apparatus similar to that outlined in ISO 5660-1. The Japanese building product classification performance limits have shown that the test protocol and apparatus specifications outlined in ISO 5660-1 can lead to "within standard allowable" errors that constitute a significant fraction of the allowable performance levels. This necessitated a better low heat release rate data measurement method. The authors previously reported their work [1, 2] towards development of the low heat release rate measurement system developed for ISO TC92 SC1 since 2005. This was principally focussed to develop a more sensitive technique to increase the signal to noise ratio in the oxygen depletion system, and to tightening both laboratory techniques and analyser specification so that these low levels of heat release can be reliably measured. The most significant advances required are a larger cone heater developed for testing larger specimens and an exhaust with a direct coupling between the plenum and the cone heater to prevent any combustion product leakage. This development was finalised in 2016 and the ISO Technical Specification defining this work (ISO TS 5660-4), published in 2017. This standardised instrument should give countries struggling to measure the low levels of heat release specified by their regulations the tool to do this.

Wider use of the large cone heater from ISO TS 5660-4

The cone heater developed in the ISO TS 5660-4 apparatus, which gives a uniform heat flux over the entire surface area (150 × 150mm) of a specimen to a very tight specification, as shown in Figure 1, is seen as an invaluable resource for other fire studies and standardization groups. Thermal mapping of the areas under the new cone heater has shown that all zones are very thermally uniform over a wide surface area and over a significant depth. This is very useful for measuring specimens that distort, intumesce or collapse during testing. It also gives the opportunity for researchers to test bigger components or finished products in the cone calorimeter environment. We see particular value in assessing the critical ignition flux of these larger specimens/products that can be accommodated in this larger geometry and constant heat flux zone.

The importance of Critical Ignition flux data and ease of its determination in this cone heater apparatus was recently recognize and procedures for its determination added to cone calorimeter standards (ISO 5660-1 and ASTM 1354). This larger unit gives and excellent facility to determine this parameter for say electro technical products or other intermediate sized components.

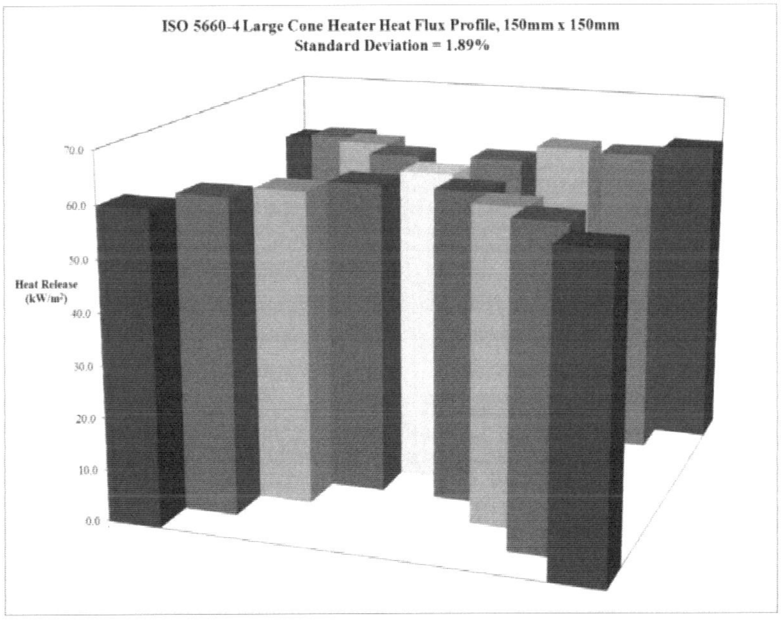

Figure 1: Large Cone Heater Heat Flux Profile (150 × 150mm)

Direct coupling of the larger cone to the ducting

Work on finalising this standard was arrested for some time after Hallman and Hermesky [3] discovered some low buoyancy combustion product systems that were not captured by the conventional ISO 5660-1 ductwork. This led to the authors' successful redesign of the apparatus to directly link the cone to the ductwork as shown in Figure 2.

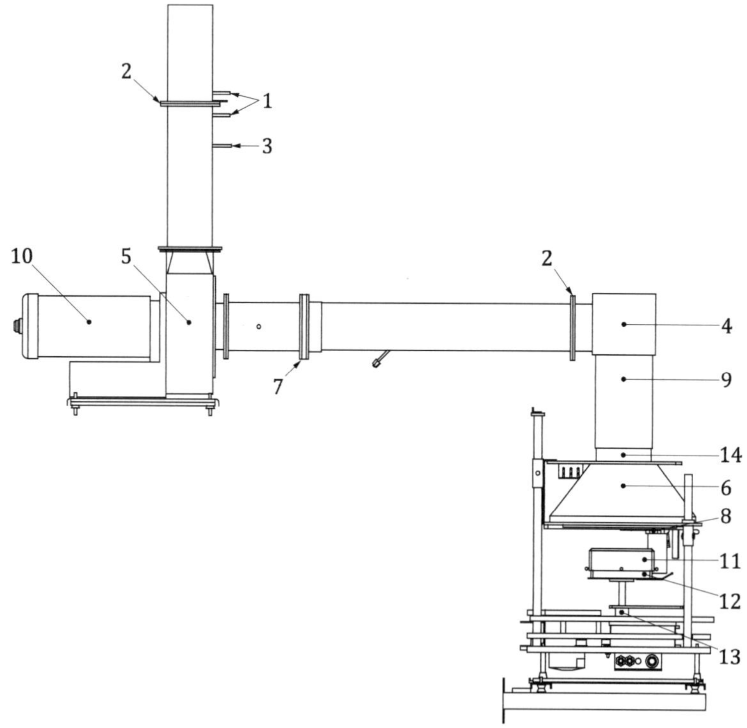

1	pressure ports	8	spark plug
2	orifice plates	9	vertical connection
3	thermocouple	10	blower motor
4	plenum with extension piece	11	retainer frame and specimen
5	blower	12	specimen holder
6	heater	13	weighing device
7	gas sampling probe	14	expansion flange

Figure 2: ISO 5660-4 schematic showing linking of larger format cone to ductwork

The system was proven by a mini-reproducibility study between Fire Testing Technology Ltd (UK), Armstrong World Industries (USA) and Tokyo System Vac (Japan) who each examined low heat release materials both in the ISO 5660-1 apparatus and in the new larger format (ISO 5660-4) apparatus, under a 50 kW/m^2 exposure. The results for the tests on mineral fibre ceiling tiles and on plasterboard from the three laboratories and the averages are shown in Tables 1-4. These show that the Peak HRR and Coefficients of Variance (CoV) were lower for specimens tested according to the ISO 5660-4 than those tested according to ISO 5660-1. This demonstrates the success of increasing the signal has led to better resolution and improved repeatability and should facilitate better measurement of low levels of heat release in bench scale calorimetry.

Table 1: Test results for mineral fibre tested to ISO 5660-1 configuration

	Lab 1	Lab 2	Lab 3	Average
THR (MJ/m^2)	6,83	5,70	5,48	6,00
THR SD (MJ/m^2)	2,23	0,79	0,84	0,73
THR CoV (MJ/m^2)(%)	32,64	13,86	15,31	12,13
Peak HRR (kW/m^2)	20,43	14,65	25,13	20,07
Peak HRR SD (kW/m^2)	4,66	0,54	1,44	5,25
Peak HRR CoV (kW/m^2)(%)	22,80	3,72	5,72	26,14

Table 2: Test results for plasterboard tested to ISO 5660-1 configuration

	Lab 1	Lab 2	Lab 3	Average
THR (MJ/m^2)	4,40	3,30	2,70	3,47
THR SD (MJ/m^2)	1,47	1,48	1,10	0,86
THR CoV (MJ/m^2)(%)	33,48	44,85	40,74	24,87
Peak HRR (kW/m^2)	95,78	95,79	109,11	100,23
Peak HRR SD (kW/m^2)	6,59	6,89	4,51	7,70
Peak HRR CoV (kW/m^2)(%)	6,88	7,19	4,14	7,68

Table 3: Test results for mineral fibre tested to ISO 5660-4 configuration

	Lab 1	Lab 2	Lab 3	Average
THR (MJ/m^2)	6,70	7,80	5,40	6,63
THR SD (MJ/m^2)	0,40	0,49	0,26	1,20
THR CoV (MJ/m^2)(%)	5,97	6,28	4,78	18,11
Peak HRR (kW/m^2)	16,39	15,30	18,70	16,79
Peak HRR SD (kW/m^2)	0,35	0,86	0,92	1,74
Peak HRR CoV (kW/m^2)(%)	2,16	5,62	4,95	10,33

Table 4: Test results for plasterboard tested to ISO 5660-4 configuration

	Lab 1	Lab 2	Lab 3	Average
THR (MJ/m^2)	3,55	5,10	2,20	3,62
THR SD (MJ/m^2)	0,35	0,37	0,14	1,45
THR CoV (MJ/m^2)(%)	9,96	7,25	6,43	40,12
Peak HRR (kW/m^2)	104,76	101,20	94,41	100,12
Peak HRR SD (kW/m^2)	0,68	12,20	6,97	5,26
Peak HRR CoV (kW/m^2)(%)	0,65	12,06	7,38	5,25

Measuring heat and smoke release in vitiated conditions

Considerable work has been done [4,5] in developing an apparatus and measurement techniques to assess the heat release from materials in vitiated oxygen atmospheres and this has now become a significant point of interests for those studying combustion product gas levels for tenability studies. This is now being standardised in ISO TC92 SC1 as ISO 5660-5. Work was initially focussed on heat release rate measurements at oxygen vitiated levels down to 15% using a box enclosure developed by Grayson and Upton at FTT for Mikkola [3]. More recently the same apparatus has been developed for studies at lower oxygen levels by Werrel et al [4] and Marquis et al [5] and the same systems have also been used for pyrolysis studies. This development has led to a modification of the apparatus and its usage; in particularly, the addition of a chimney to the chamber and in some cases linked directly to the exhaust system to prevent after burning of combustion products after leaving the vitiated chamber.

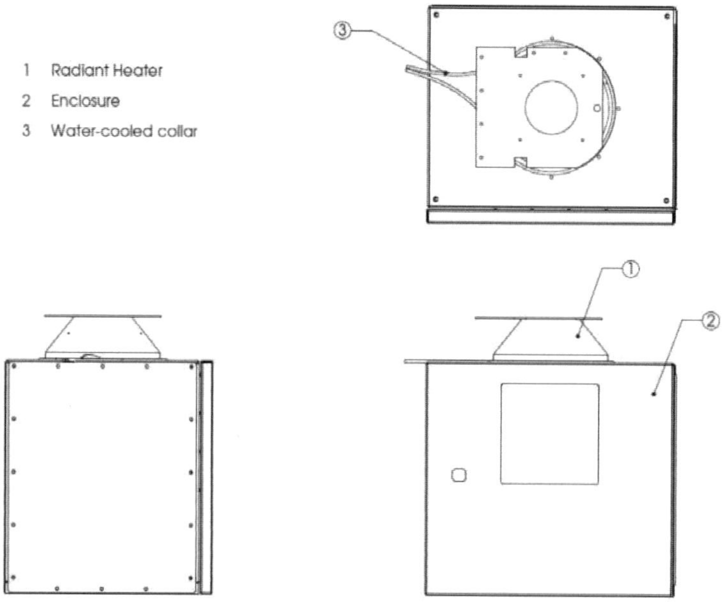

Figure 3: Schematic of the vitiation chamber

Figure 3 show the basic stainless steel chamber that houses the specimen and the load cell. Premixed air-nitrogen mixtures are fed into the base of the chamber via a baffle system and the specimen is exposed to heat from a conical furnace housed in the roof of the chamber. The weighing device is fitted with water cooled shielding as the temperature in the chambers can be high.

A 600mm long chimney can be mounted on top of the top-plate of the conical heater to limit potential secondary burning when combustion products leaving the chamber mix with ambient air, as shown in Figure 4. If secondary burning is observed above the chimney this indicates that a direct link to the cone ducting is needed for correct measurement.

Generally when testing in the range of 15% to 20.95% oxygen and with an enclosure gas supply rate lower than the exhaust flow rate, the enclosure and chimney should not be linked directly to the exhaust hood. When testing with an enclosure gas supply rate equal to or greater than the exhaust flow rate, a setup that directly links the enclosure and chimney to the exhaust hood may be used. When undertaking Anaerobic Pyrolysis experiments these can be carried out in either unlinked or linked apparatus depending on the gas supply rate or in the enclosure alone without the oxygen depletion measurement equipment (as in ISO 13927 or ISO 17554).

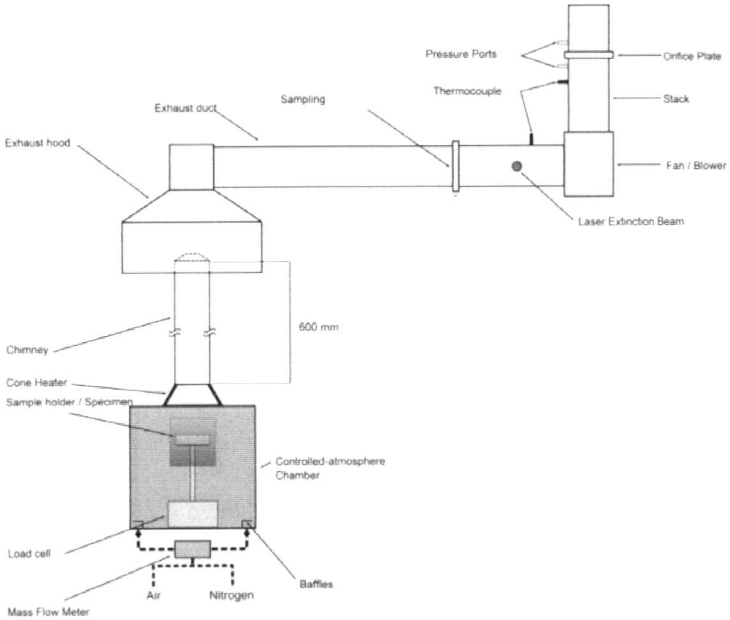

Figure 4: Schematic of the vitiated cabinet in place in the ISO 5660 cone calorimeter

Replacement of white light sources in optical determination of Smoke

Most fire calorimeters and other fire test equipment have used tungsten filament or other white light sources for determining smoke opacity. The latter are becoming increasingly difficult to source so replacements are needed. Lasers have been used extensively, especially in cone calorimetry but these are quite expensive. LED and diode lasers are now being studied as economical alternatives to the lasers and white light systems.

ISO TC92 SC1 is now developing a test method for measuring the smoke density by LED and other new light sources and studies are underway to examine the equivalence of some systems in a customised apparatus in Japan.

Using diode lasers as light sources in fire testing

Wilkins [7] recently carried out a project at Fire Testing Technology Ltd, using diode lasers and compared their performance to standard Helium Neon lasers of the sort used in the cone calorimeter and to date thought to exceptionally stable. The preliminary findings were presented to a recent ISO TC92 committee to show how promising these sources were for white light and laser replacement.

Diode lasers are very robust, stable and available with a large range of wavelengths. They are physically very small so easily handled and housed and very cost effective compared to lasers.

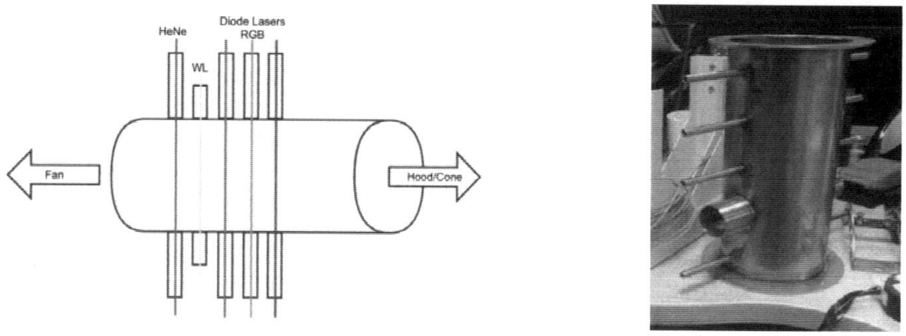

Figure 5: Schematic and Photograph of Wilkins set up

Figure 5 shows a schematic and a photograph of the experimental cone calorimeter duct section Wilkens used for testing the white light, the HeNe laser and the 3 diode lasers (Red, Blue and Green). This consisted of a specially engineered duct insert for the cone calorimeter duct that replaced the usual laser mounting section and allowed all the units to be housed on the cone simultaneously.

Figure 6 shows the noise and drift of a traditional Helium Neon laser used in ISO 5660-1 and many other calorimeters of the type chosen by Babrauskas because of its excellent stability and optical properties. Figures 7,8,9 show the noise and drift of a 650nm, a 532nm and a 405nm diode laser respectively. The diode lasers all have considerably superior noise and drift performances to the Helium Neon and the white light systems.

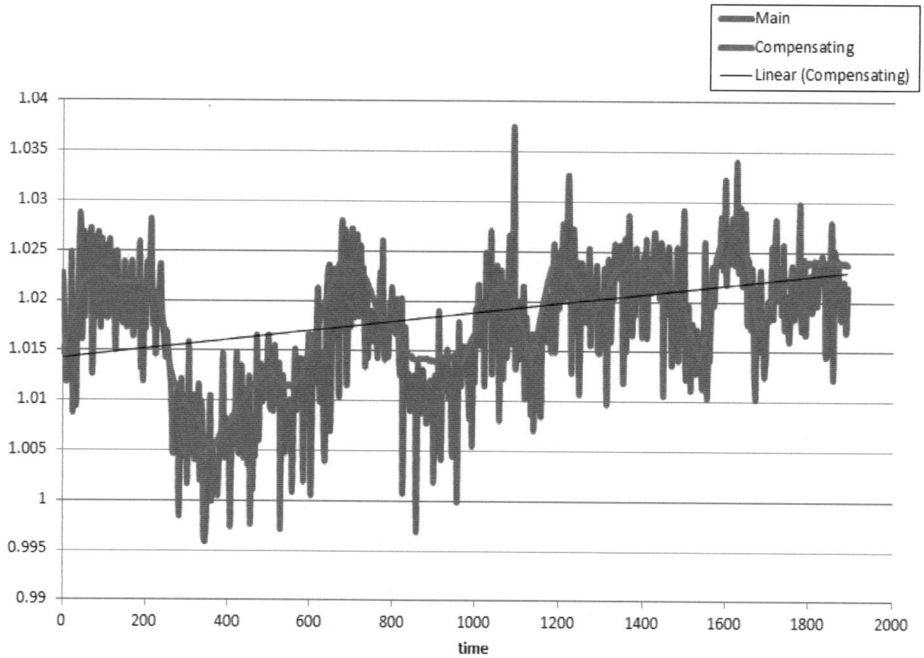

Figure 6: Noise and drift on ISO 5660-1 Helium Neon Laser

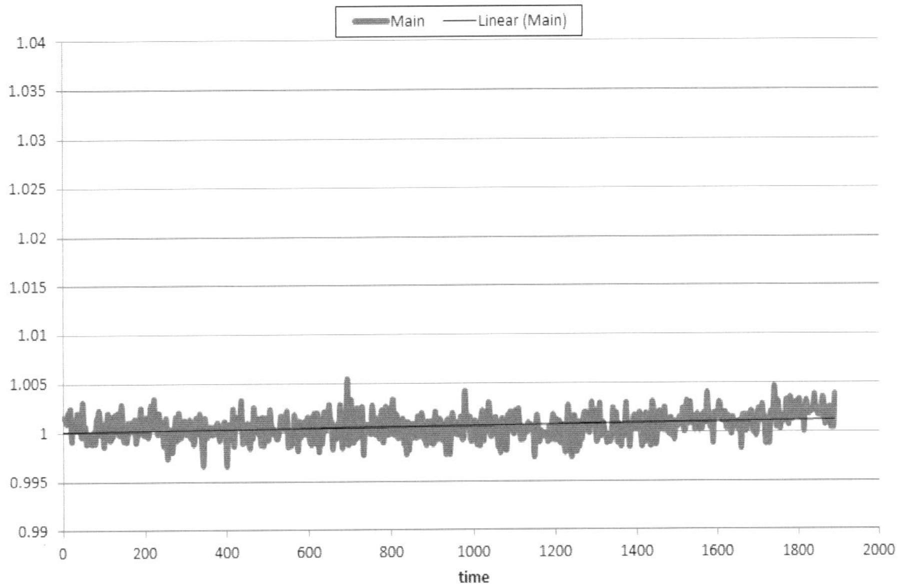

Figure 7: Noise and drift on 650 nm red diode laser

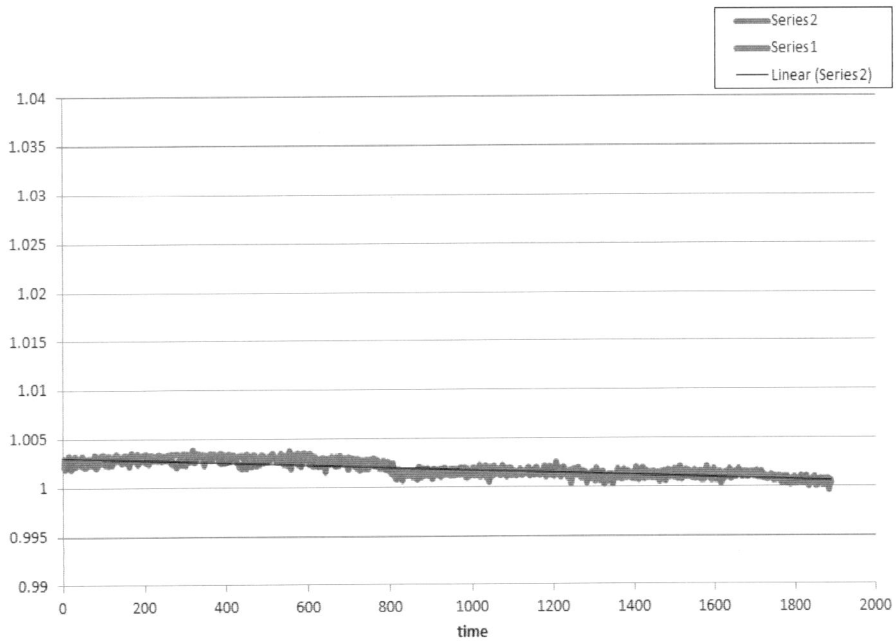

Figure 8: Noise and drift on 532 nm green diode laser

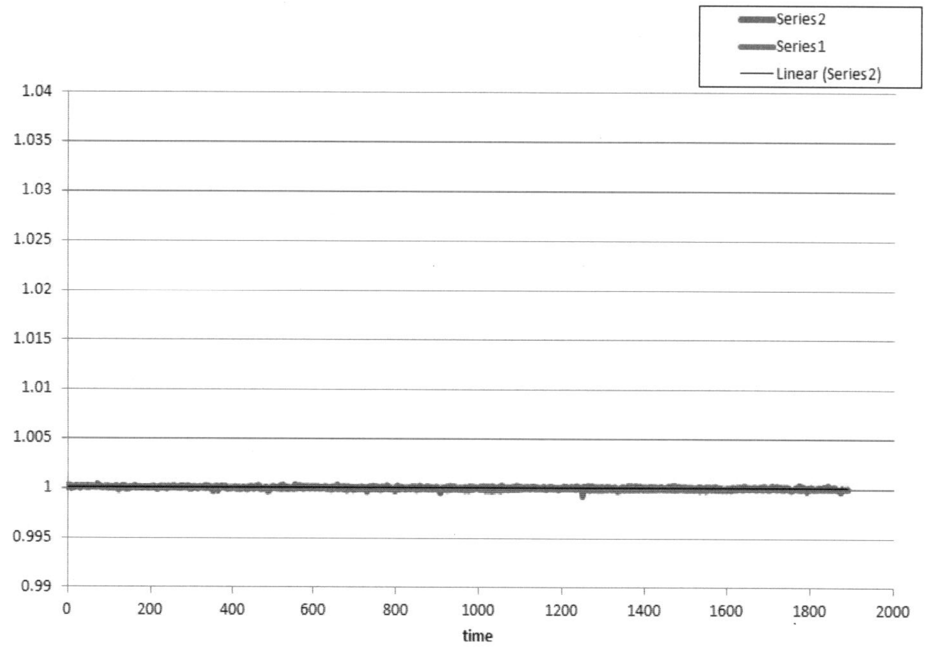

Figure 9: Noise and drift on 405 nm blue diode laser

Work is now required to identify which of these or other laser diodes systems give similar performance and responses to white light system in test conditions.

References:

1 S Gregory, A Green, S J Grayson, S Kumar, and A Cornelissen. Use of cone calorimeter for testing materials with low heat release rates. Fire and Materials 2005, 63-75 Interscience Communications London UK

2 S Gregory, A Green, S Pasantes, S Grayson and S Kumar. Use of cone calorimeter for testing materials with low heat release rates. Fire and Materials 2013, 263-272 Interscience Communications London UK

3 R A Halmann and M Hermesky private communication (within ISO TC92 SC1 WG5)

4 M Werrel, J H Deubel, S Krüger, A Hofmann and U Krause. The calculation of the heat release rate by oxygen consumption in a controlled-atmosphere cone calorimeter. Fire and Materials 2013, 204-226 Interscience Communications London UK

5 D.M. Marquis, E. Guillaume, and D. Lesenechal. Accuracy (trueness and precision) of cone calorimeter tests with and without a vitiated air enclosure. In Proceedings of the 9th Asia-Oceania Symposium on Fire Science and Technology, Hefei, CN, 2012. Procedia Engineering.

6 E Mikkola. Effects of Oxygen concentrations on cone calorimeter results. Proceedings Interflam 1996

7 K Wilkins. Private communication (within ISO TC92 SC1 WG5)

DYNAMIC STUDY OF POLYMER MASS LOSS UNDER HIGH AND FAST RADIATIVE HEAT FLUXES

Guillaume Rambaud[1], Mathieu Gillet[2]
[1]CEA, DAM, GRAMAT, F-46500 Gramat, France
[2]DGA Aeronautical systems, Department MTO, Odeillo, France

ABSTRACT

There are not much articles in literature about behaviour of polymers exposed to heat fluxes above 300 kilowatts per square meters. Moreover, there are fewer ones for heat loads applied in few seconds.
This paper proposes experimental tests to well understand pyrolysis processes of polymers exposed to high (above 500 kW/m^2) and fast (ranging from 1 to 5 s) radiative heat loads. For that, mass losses are measured during heat loads with classical scale. Until now, classical scales with high mass sensibility but with bad time response are not able to measure dynamic mass loss during first step of thermal decomposition. A new scale, based on load cell, was developed to increase time and mass characteristics. This device was tested on polyoxymethylene (POM), a simple polymer.
The tests were achieved using a 45 kW solar concentrator, capable of delivering heat fluxes up to 6 MW.m^{-2} with fully parameterized shapes, lasting from 0.3 second to several hours. Tests were realized on 50 mm in diameter cylinder with thickness ranging from 1 to 2 mm. They were exposed to solar radiative heat flux ranging from 500 to 2250 kW/m^2.
Comparisons between modelled (FDS v6.2) and experiment results were realised on both side temperature samples and mass losses. Mass losses are well compute by FD v6.2. Front side temperatures are also well compute for heating step, but discrepancy appears for the cooling step, after maximum temperature were reached.

KEYWORDS

Solar furnace, high heat flux, POM, mass loss, load cell

INTRODUCTION

The flaming ignition of solids is based on two physical mechanisms. First of all, since flames are a gas phase phenomenon, for a solid to be able of a flaming ignition it must react to heat by breaking down and releasing fuel gas. This pyrolysis process is the principal topic of this paper. Under certain conditions, combustible vapors resulting from the pyrolysed solid ignite in a flaming mode. This paper deals with modelling the thermal degradation of solid materials exposed to high radiative heat fluxes above 300 kW.m^{-2}, which is not accessible to many experimental facilities. As a result, characterization of materials subjected to those levels of thermal radiation is poorly covered by literature.
In this paper, we focus on tests performed on a single thermal degradation process polymer, polyoxymethylene Delrin (POM-Delrin) which was exposed to radiative heat fluxes generated by a solar furnace. Thermo-physical properties were characterized by classical experimental benches. Kinetic parameters were calculated from thermo-gravimeter tests. Because of the specific spectral distribution of the radiative heat source (solar furnace), it was necessary to change emissivity and applied heat flux to take it into account. With these parameters, FDS simulations were made and compared to experiments realized with solar furnace for high radiative heat flux.

EXPERIMENTAL

In order to study thermal degradation and flammability of POM under high and very quick thermal flux, several tests have been realised. For these tests front side samples were exposed to heat, whereas back side samples were not and were exposed to environment. Both side temperatures were measured during tests by two pyrometers. High speed camera (500 fps) was used to detect ignition time and the load cell to measure mass losses. To avoid destruction by flames, load cell is located under samples. Load cell mass resolution is around 10 mg and its time response is around 10 ms.
50 mm POM cylinders, 2 mm thick were exposed to heat flux ranging from 500 kW/m^2 to 2250 kW/m^2 with exposed times of 4900 ms and 600 ms respectively.
The test device used to generate heat aggressions on POM samples is a 45 kW double reflection solar furnace with a parameterized flux (Fig. 1).

Fig. 1: general view of the Main Solar Furnace's elements

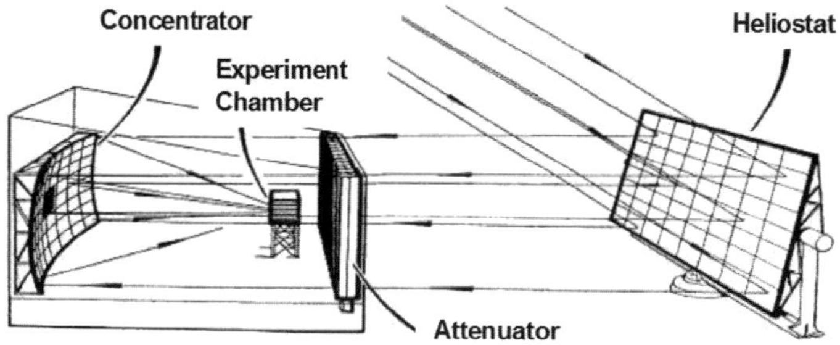

This device is composed of four main elements:

- The heliostat, a mobile plane reflector that follows the sun's course across the sky throughout the day (Fig. 2). It is 13.2 m high and 17.4 m wide (total surface is 230 m^2), and made of 638 square mirrors, 60 cm wide each. It is automatically oriented towards the sun, thus reflecting direct sunlight towards the concentrator.

- The solar radiation attenuator, based on a flux modulator (Fig. 3). It is composed of 20 computer-controlled vertical blades that can open (or close) in 150 milliseconds. Their opening ratio determines the amount of reflected sunlight that reaches the concentrator, thus the final radiative power on target.

- A 100 m^2 concentrator (Fig. 4) made of 384 spherical mirrors installed on a Davis-Cotton type structure, with a 10.75 m focal distance. The energy it concentrates in the focal plane has a circular shape, and is quasi-homogeneous in a 25 mm radius (Fig. 5). The maximal power density can reach 7 MW/m^2 in the focal zone, up to 12 MW/m^2 on a smaller area thanks to a secondary concentrator

- The experiment chamber (Fig. 6) where the tested samples are installed, in the concentrator's focal plane.

Fig. 2: heliostat

Fig. 3 : attenuator

Fig. 4: concentrator

Fig. 5: power distribution in focal plane of the concentrated flux

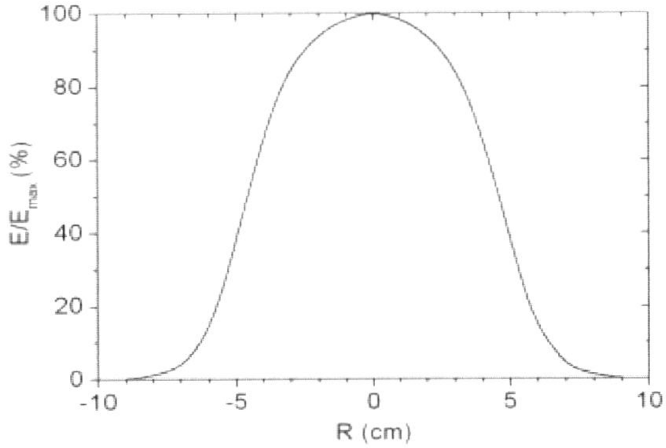

Fig. 6: experiment chamber

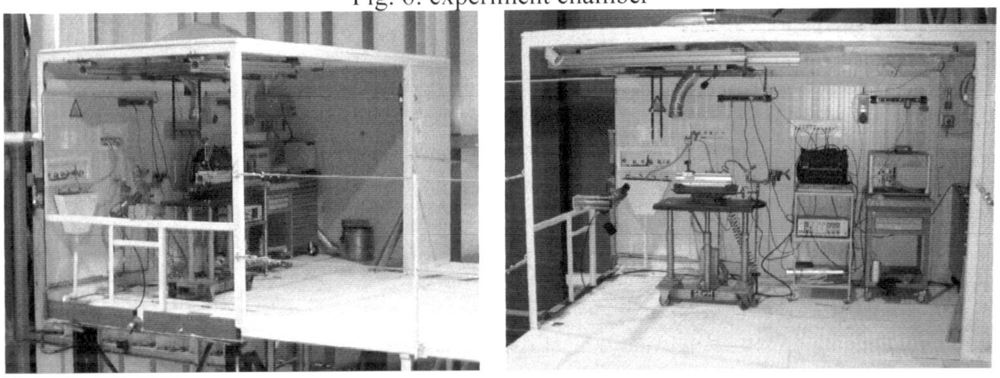

The main specificity of this test facility resides in its ability to generate completely parameterized flux shapes, thanks to its attenuator associated with 2 horizontal blinds that can reduce the illuminated area on the concentrator and make the full dynamics of the vertical shutters available to adjust the radiative power on target with optimal accuracy for low levels (Fig. 7 and Fig. 8).

Fig. 7: examples of shutters positions

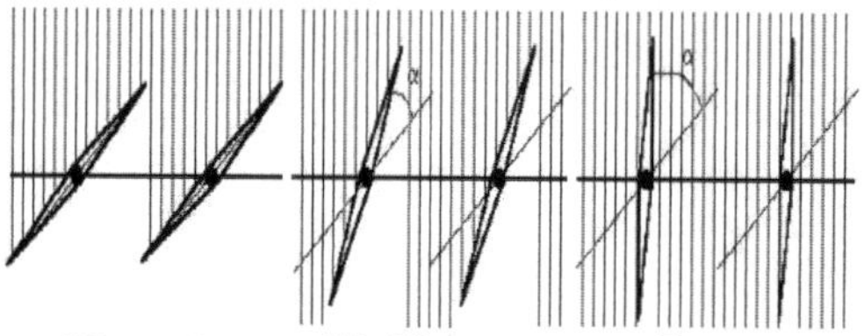

Fig. 8: horizontal blinds positions

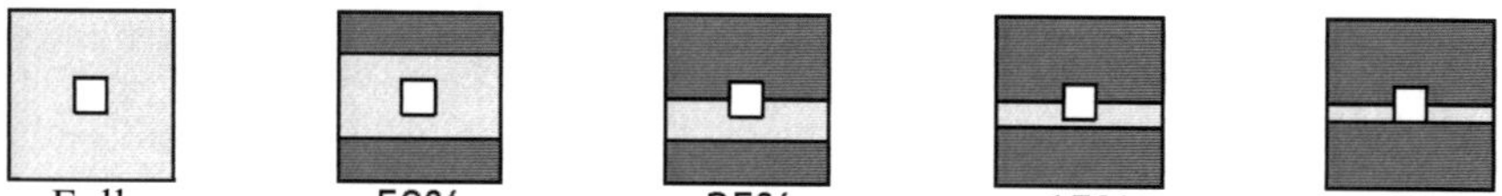

Pyrometers

- An "Ultrakust" infrared pyrometer (8 – 12 µm band) measures the back face temperature. High emissivity paint is used to coat the measured spot on the POM samples, which allows a direct temperature display.
- An "Impac 140" infrared pyrometer (narrow band centered on 3.9 µm) aims at the sample's front face. This device was chosen because of its fast response time (1.5 ms) and specific spectral range, which reduces the perturbations due to smoke release.

High speed video

Color high speed video is recorded at 500 fps by a Weinberger Visario G2 camera. A mirror was used to split the picture and showing on the one picture, the front face reflected by the mirror, and a direct view of the diaphragm and sample's side.

High speed load cell

Experiments were realised with an original set-up designed from a load cell sensor. The sensor was selected for its mass resolution around 10 mg and its time response around 10 ms. Testing of these two design features were realised with an aluminium ball and a drop of water which was released on a plate link to the load sensor (Fig. 9 and Fig. 10). High speed camera with 25000 fps was used to film falls. Comparisons was realised on load cell output signal and high speed camera pictures. Time response of load cell was deduced from the contact time of ball or drop of water on the plate of load cell and variation in output signal. Results are shown in Table 1.

These results show that the load cell response time is relatively quick, even for quite small mass. However, we note a bigger response time for test realised with drop of water due to its squeezable aspect against aluminium ball which is rigid.

Fig. 9: picture of aluminium ball on load cell plate at contact (*on left*); corresponding output load cell sensor signal during the test (*on right*)

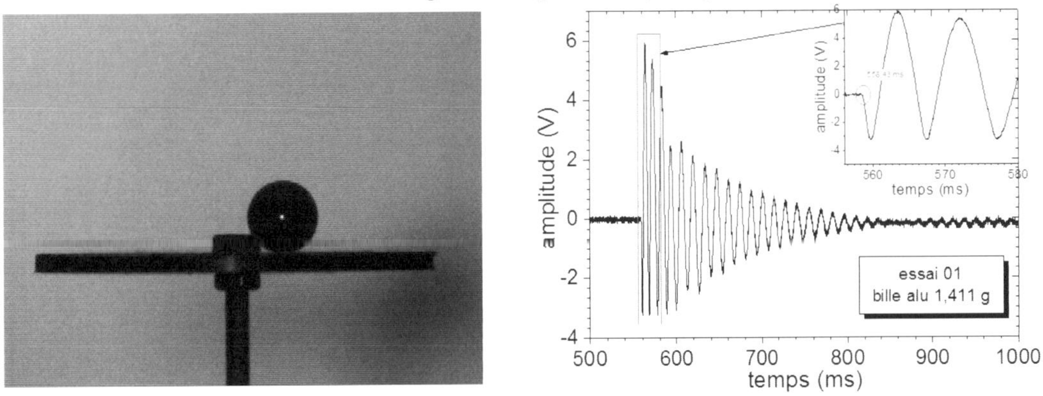

Fig. 10: picture of drop of water at the contact with the plate link to the load cell (on left-***a***), and start of output signal variation (on left-***b***); corresponding output signal (on right)

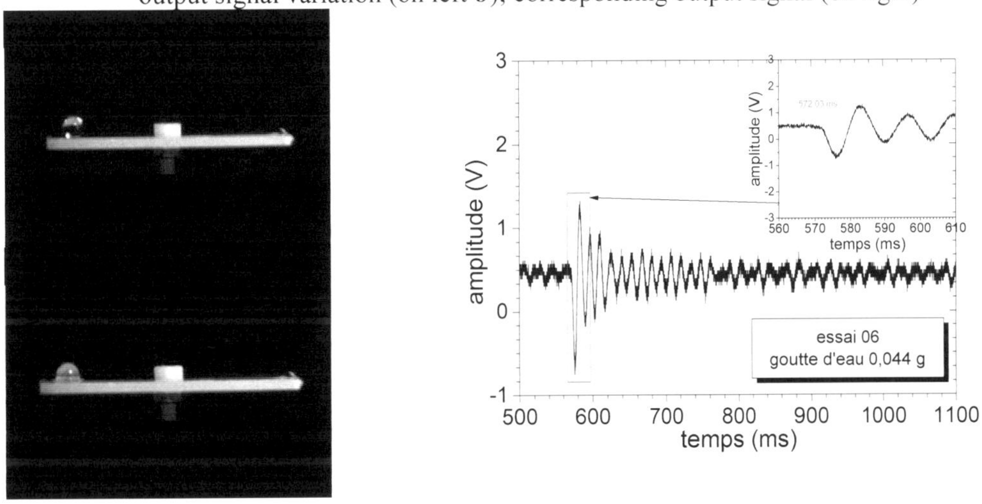

Table 1 : time difference between output signals recorded and high speed camera pictures filmed for the two validation tests.

	mass	diameter	contact time from VNR	contact time from output signal	time difference
aluminium ball	1,411 g	10,00 mm	558,30 ms	558,43 ms	0,13 ms
drop of water	0,044 g	4,6 mm	571,40 ms	572,03 ms	0,63 ms

Experimental load cell was use in the experiment chamber of the solar furnace (Fig. 11). Shutters were used quite quickly to produce desired experimental shapes. This kind of test generates vibrations in the infrastructure which are transported to the load cell apparatus. Because of sensor sensitivity and sample holder which is composed by an aluminium ring linked to the cell sensor with a 20 mm aluminium rod to protect cell sensor from heat flux, coming vibrations are amplified and created a lot of noise. To reduce that, two kinds of shock absorbers were used for working in horizontal mode and in vertical mode (Fig. 11). Impact on output signal can be seen on graphs of Fig. 12. The first one was realised without sample and comparing output signal with and without shock absorbers. The second one, on right, was realised with POM sample (50 mm in diameter cylinder and 2 mm thick). This graph compares output signals for tests realised with vertical working shock absorbers without horizontal working shock absorber and tests realised with vertical and horizontal working shock absorbers. The second one shows also the load cell efficiency and the short response time of it. The last step to reduce noise was to use mathematical treatments. Before that, fast Fourier transform method was used on output signal to find particular frequencies in the overall noise. This analyse was realised for a real test. For this test, a 50 mm in diameter cylinder with 2 mm thick POM sample was exposed to a quite slow triangular shape heat flux obtained in 4.75 s with a linear rising shape from 0 to 550 kW/m^2. The frequency study was realised on several particular steps of output signal (Fig. 13). From these graphs, the main important information is that the particular frequency 14 Hz which appears during shutters opening step and shutters closing step, disappears during mass loss step.

Fig. 11: load cell device in its experimental use (on left). Overall view of experimental set up (on right)

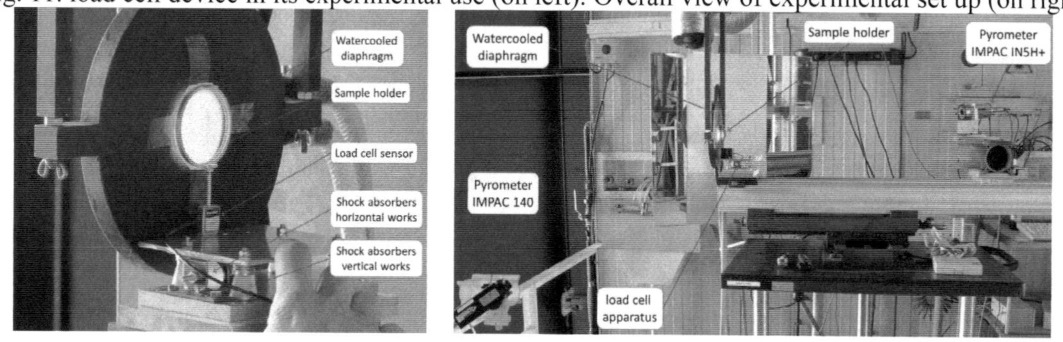

Fig. 12 : comparison of output signal of load cell with and without shock absorbers

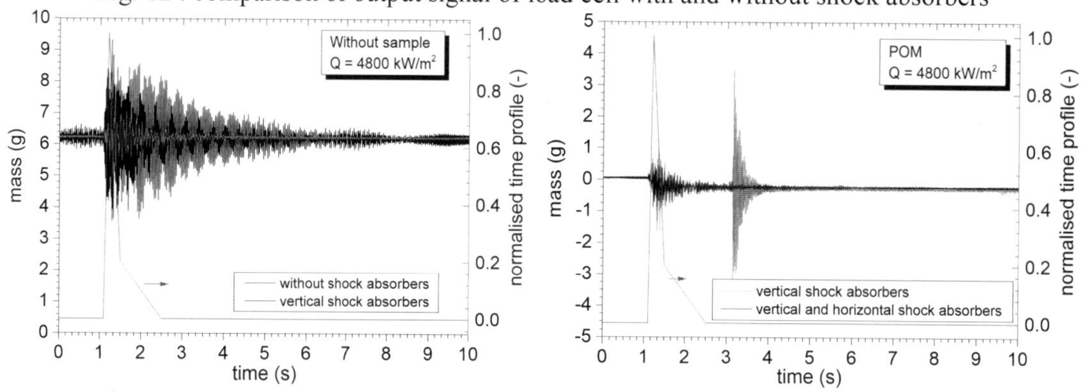

Fig. 13: frequency study of output signal delivered by load cell for particular steps.

1- stabilization before applying heat flux *2- opening shutters*

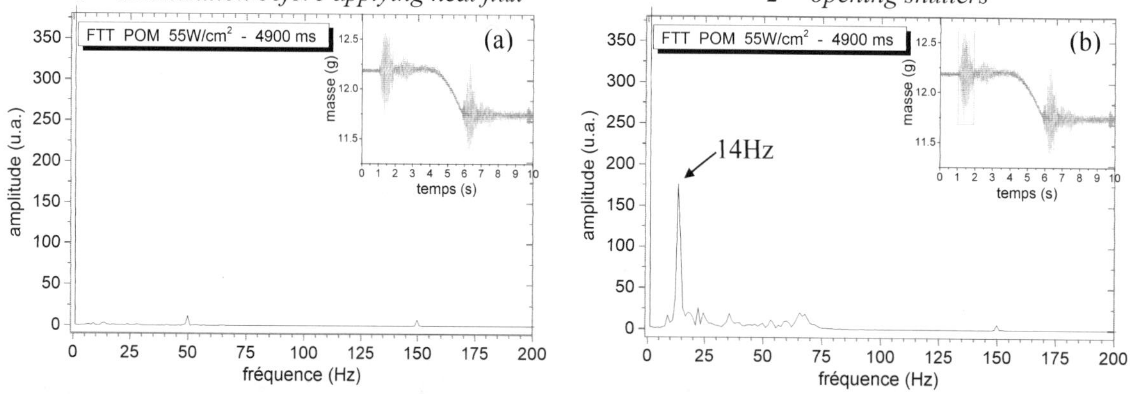

3- starting to apply heat flux *4- mass loss step*

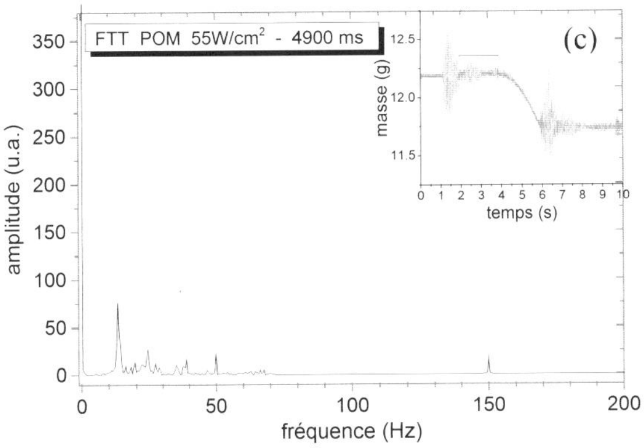

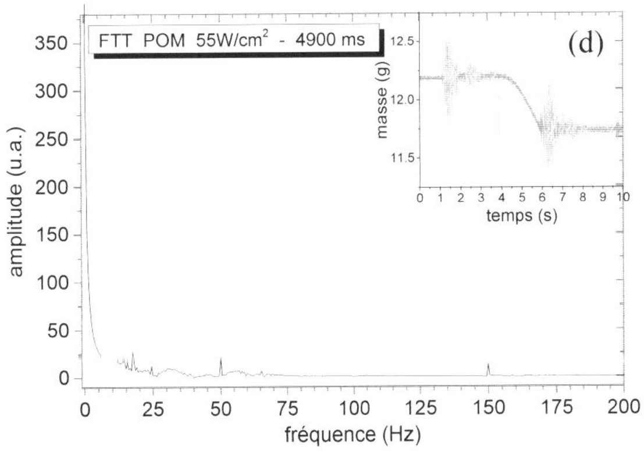

5- closing shutters

6- stabilization to the end of test

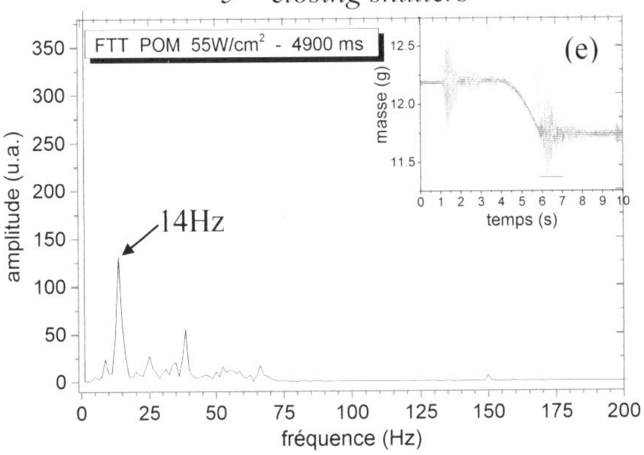

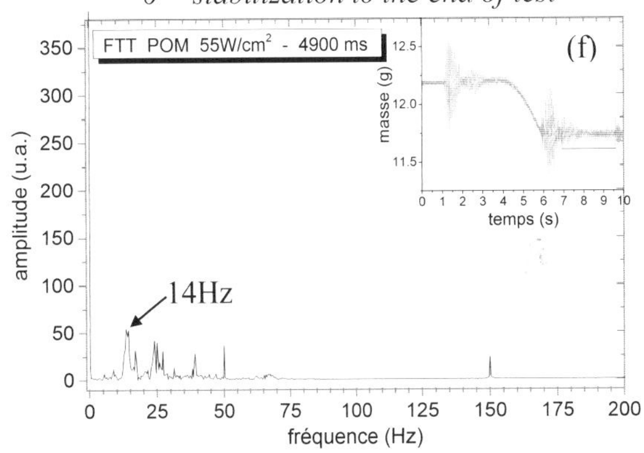

From this analysis, fast Fourier transform treatment with a 12 Hz low pass band filter was used on output signal. This treatment was compared to another mathematical treatment, call Savitzky-Golay method, based on a polynomial regression of the data points in the moving window. This mathematical treatment was realised in a window length of 600 pts with a polynomial order 2. Results are shown in Fig. 14. For the FFT treatment, some oscillation appears, whereas these oscillations don't appear with the other treatment. It could be also seen that oscillations which appear at the shutters opening and closing steps are linked to jolts. Nevertheless, due to the size of the experimental test device, erase or reduce these oscillations is hard to realise. For next analysis, mass evolution in function of time will be represented after Savitzky-Golay treatment. For experimental tests describes in Table 2, static final mass losses measured with the load cell, were compared to laboratory scales measures. Results are presented in Fig. 15. On this graph, ideal case corresponds to a perfect equality between the two devices. These results show that mass losses are quite well located around ideal case line. The difference between results and ideal case is smaller for small mass losses than for biggest ones.

Fig. 14 : mathematical treatments of load cell output signal for thermal degradation test of POM under high radiative heat flux

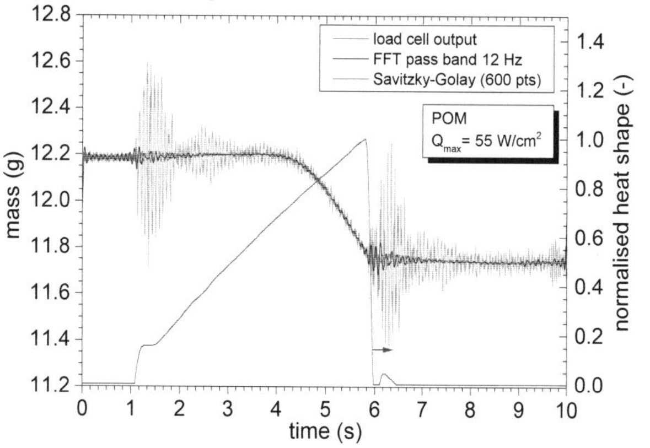

Fig. 15: difference between a laboratory scales and load cell sensor on mass variation at the end of the exposure

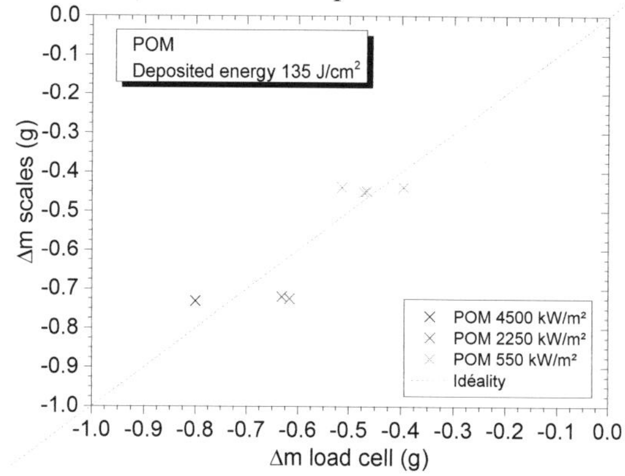

Table 2: time profile and heat flux applied for tests

test n°	Q_{max} *kW/m²*	rise time *ms*	time length *ms*	decay time *ms*	deposited energy *kJ/m²*
1	550	4750	4900		
2	2250	1050	1200	150	1350 kJ/m²
3	4500	450	600		

MODELLING

The numerical model is a one-dimensional simulation of pyrolysis corresponding to experiments realized with the solar furnace. The experiments have been modelled thanks to FDS (Fire Dynamics Simulator) v6.2 software[1]. With this software, the solid phase of the model was computed by solving the heat conduction equation. The decomposition process of the solid phase was computed from the solid phase species conservation equation and the rate of reaction was modelled by an Arrhenius law. All of solid phase is converted in formaldehyde. Beyond the method to characterize absorption coefficient of POM, all other method to identify parameters listed below (Table 3) can be found in reference [2].

Table 3: thermo-physical parameters used to model thermal degradation of POM

Thermal conductivity	ramp
Heat capacity	
Density	1420 g/dm³
Pyrolysis gas	formaldehyde (CH_2O)
Heat of combustion of gas	17000 J/g
Heat of decomposition of solid	1500 J/g
Emissivity	0.98
Activation energy (thermal degradation law)	197000 J/mol
Pre-exponential factor (thermal degradation law)	$5.52.10^{12}$ s^{-1}
Absorption coefficient (solid phase)	10000 m^{-1}

Because of the spectral distribution of solar furnace and characteristics of POM, a part of incident heat flux is absorbed on the solid boundary of POM sample. The remains heat is absorbed in depth, in a thin layer that we identified. To do that, transmittance and reflectance for different POM sample thicknesses were characterized with the spectrophotometer JASCO V670 for wavelengths between 200 nm to 2500 nm. Results have shown that for 0.4 mm thick, POM starts to be transparent at this spectral band. This depth was introduced in this equation $(1-\rho)\, I_0 \exp(-\tau x)$, where ρ is the reflected heat flux, I_0 the incident heat flux, x the depth in sample and τ the absorption coefficient. By dichotomy, this coefficient was identified and introduced in FDS software with the keyword ABSORPTION_COEFFICIENT.

Computational domain shown on Fig. 16 is 2 m in x,y directions and 4 m in z-direction. It is composed by a solid domain which is 0.2 m in x,z directions. The y-direction depending of sample thickness is equal to 0.002 m. The mesh of gas domain is divided in 20 uniform cells in x,y-directions and in 40 cells in z-direction. Mesh cells of solid domain are uniformly ten times smaller than mesh cells of gas domain.

Fig. 16: computational domain

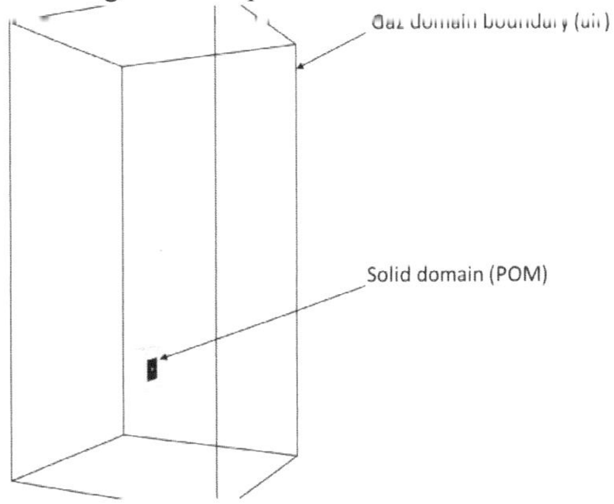

Because the spectral distribution of solar furnace is different from a classical flame spectral distribution, absorptivity and emissivity of POM are not equal. But with FDS, it's not possible to differentiate them. The way we have chosen to take that into account is to modify emissivity and external heat flux to correspond to the experiment. Moreover, for the spectral distribution of solar furnace, POM absorptivity is quite high, around 96 % but because of high level of heat flux, a significate error could be done on the amount of radiative energy.

RESULTS

Modelling was done and compared to experimental tests realised with the solar furnace. POM samples were exposed to several triangular time profiles set up with different maximum heat fluxes, different time lengths and only one deposited energy. To perform that, we have chosen to modify rise times and not decay times (Table 2). 50 mm in diameter and 2 mm thick POM samples were used for those tests.

For the slower experiment (Fig. 17) corresponding to the total time length 4900 ms, the modelling and experiment mass losses begin and stop decreasing at the same instant. And total mass losses are exactly the same.

For the intermediate experiment (Fig. 18) corresponding to the total length time 1200 ms, a small difference appears at the beginning of mass loss step and disappears at the end of mass loss step. As for the above-mentioned test, both modelled and measured final normalised mass are equal.

For the quicker experiment (Fig. 19) corresponding to the total time length 600 ms, load cell sensor begins to show its limits. The number of measured points during the mass loss step is quite small, around 20. Nevertheless, measured normalised mass loss and modelled normalised mass loss are relatively in good agreement.

To summarize, comparisons between modelling and experiments show a very good agreement of mass loss. Note that experiment mass losses were mathematically treated. From these comparisons, two points are important. First, modelling can reproduce mass loss for polymers exposed to quick and high radiative heat fluxes. Second, the capability of the load cell to follow mass loss during these amounts of heat deposited in a very quick time length has been demonstrated.

Fig. 17: POM normalised mass modelled and measured for triangular profile n°1 (Q_{max} = 550 kW/m^2 ; total time length = 4900 ms).

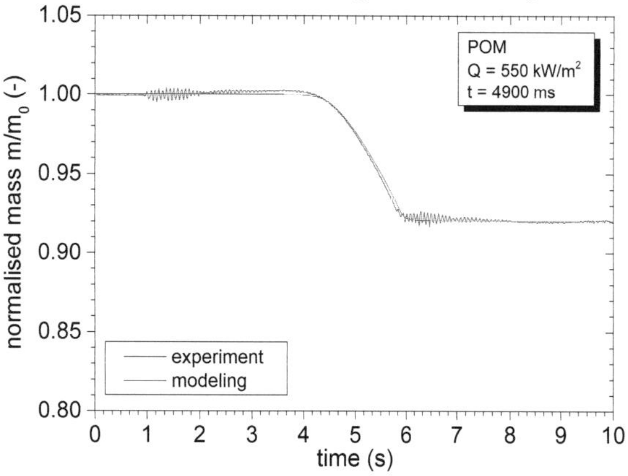

Fig. 18: POM normalised mass modelled and measured for triangular profile n°2 (Q_{max} = 2250 kW/m^2 ; total time length = 1200 ms).

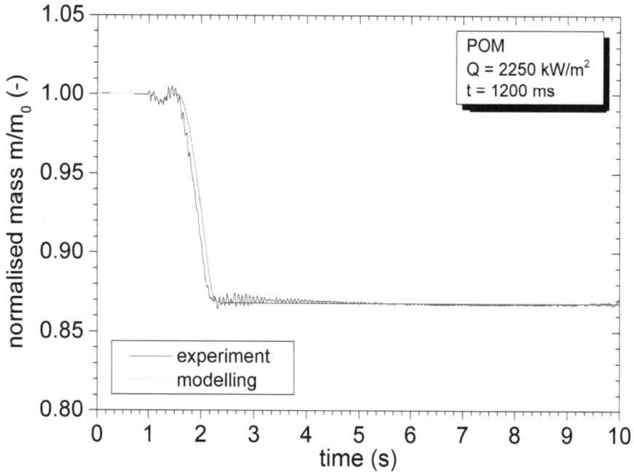

Fig. 19: POM normalised mass modelled and measured for triangular profile n°3 (Q_{max} = 4500 kW/m^2 ; total time length = 600 ms).

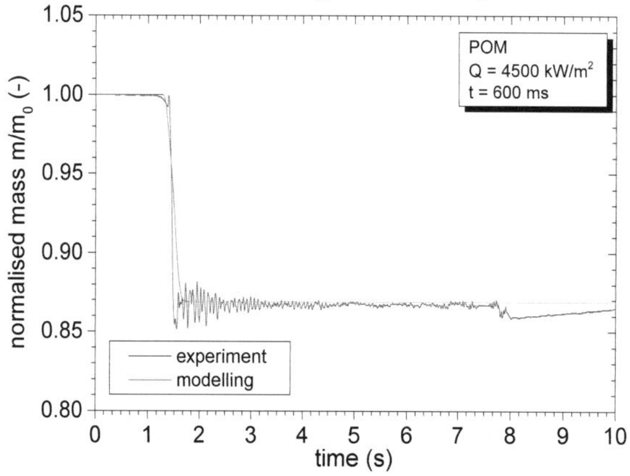

During experiments, front and back sides temperatures were measured respectively with an IMPAC140 and an ULTRAKUST infrared pyrometers. Note that, temperatures below 70 °C can not be measured by IMPAC140.

Temperatures were modelled with FDS v6.2 and comparisons with experimental temperatures are shown on Fig. 20.

For the slower experiment, modelled temperatures and measured temperatures have the same shape until the maximum temperature has been reached. Back face temperature is quite better modelled than the front face temperature, particularly during the cooling step where the gap between modelled temperatures and measured temperatures increases.

For intermediate and faster experiments (Q=2250 kW/m^2 and Q=4500 kW/m^2), the gap between modelled and experimental temperatures, appearing during the cooling step, is bigger than the slower experiment. But, during heating step, front face temperatures are well computed by modelling. For these two experiments, back face temperatures are also well computed during heating step but not during cooling step.

Several reasons could explain these differences. First, for measured temperatures, emissivity was set to 1, whereas POM modelled emissivity is 0.98, for taking into account spectral distribution of solar furnace. Moreover, emissivity could change with thermal degradation because of bubbling. Note that, for this polymer, no char appears during flaming.

Then, pyrolysis gas appearing principally after reaching the maximum temperature, could create a shield between POM front face and pyrometer. This could stop a part of radiative signal emits by exposed sample. So, front side pyrometer sees a part of signal emits by POM front face and a part of signal emits by smoke. The last one, cooler than solid, could explain why pyrometer temperature decreases so quickly after maximum temperature were reached. Otherwise, during early stage of rising step, no gas was created from POM, and so, pyrometer sees only exposed side and that's why modelled temperatures are well fitted. Note that, modelled temperature is set in FDS with the keyword WALL TEMPERATURE which computes temperature from solid domain.
At the end of test, when a large part of smoke has gone out, front face modelled temperature is closed to experimental temperature, although experimental temperatures are always higher.
From a numerical point of view, several reasons could also explain gaps between experimental and modelled front side temperatures. Owing to heat flux was highly and quickly applied, convection heat transfer of exposed side could be less compute than back side which is not exposed to the same dynamic heat flux. How emissivity and absorptivity are compute with FDS could also explain differences between modelled and experimental temperatures. First, with FDS, emissivity is fixed for each material and could not be changed during modelling.
With the amount of heat applied and the surface temperature reached, radiative heat flux is the main part of heat loss by sample. Then, incident heat flux absorbed by sample is a function of absorptivity but net heat flux (incident heat minus heat loss) depends also of emissivity. This difference could introduce an error on heat loss by sample which could be corrected if absorptivity and emissivity were not indissociable.

Fig. 20: front and back faces temperatures modelled with FDS v6.2 and measured with pyrometers

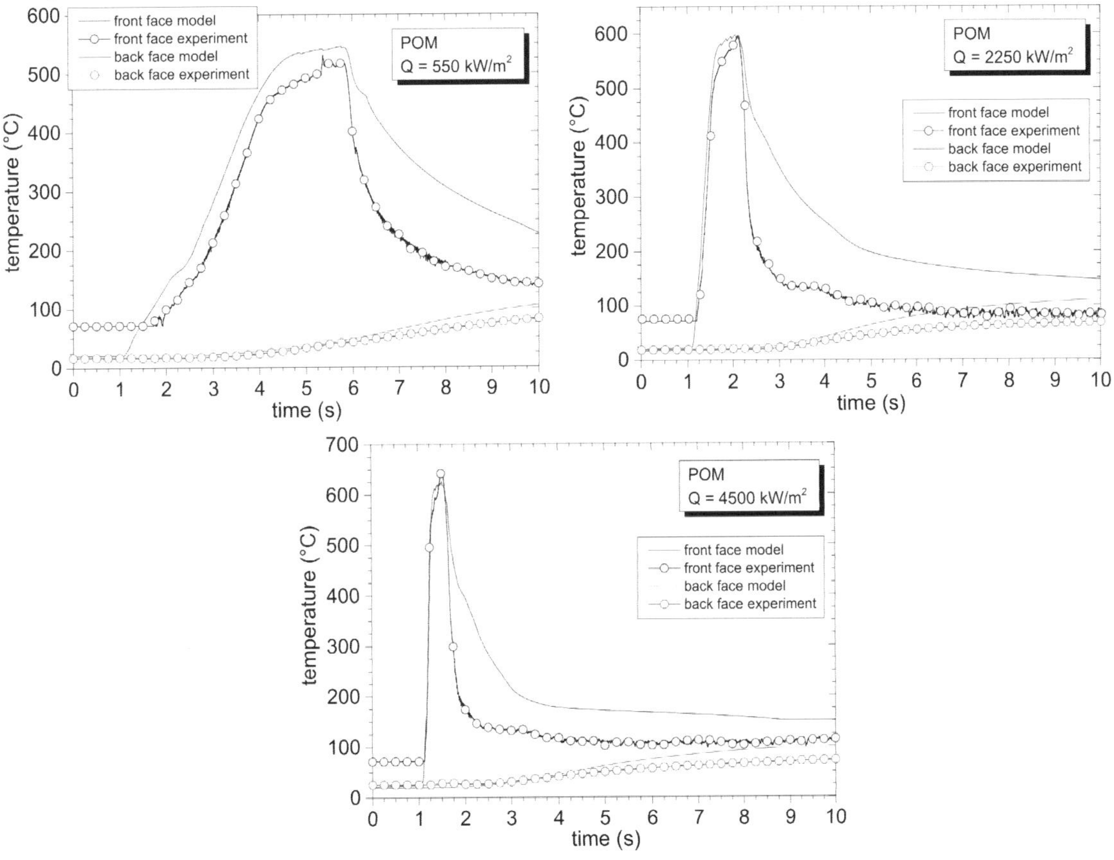

CONCLUSION

These tests have shown the good characteristics of the new load cell to reproduce mass loss for fast and high radiative heat fluxes during dynamic and static steps. Moreover, these tests have demonstrated the capabilities of FDS 6.2 to replicate pyrolysis of simple polymer under triangular heat flux shapes ranging from 500 kW/m^2 to 2250 kW/m^2 for several exposure times. FDS v6.2 modelling was implemented with thermo-physical properties identified with classical experimental benches. The pyrolysis law was modelled by an Arrhenius law. Spectral distribution of solar (heat source used), was taking into account by modifying emissivity and heat flux applied on exposed side of POM samples. Tests realised on POM samples were well computed by modelling. Mass losses were well followed by load cell and well computed by FDS. Results on front and back sides temperatures are also quite well for heating step and less good for cooling step due to smoke or emissivity. So it could be interesting to modify how the model computes absorptivity and emissivity to take into account spectral distribution of solar furnace or other heat source with a special spectral distribution where a large part of heat is not in the infrared.

REFERENCES

[1] K. McGrattan, R. McDermott, S. Hostikka, J. Floyd, C. Weinschenk, K. Overholt, Fire dynamics Simulator (Version 6), User's Guide; NIST Special Publication 1019-6, 2013.

[2] M. Gillet, G. Rambaud, Pyrolysis of solid materials exposed to high thermal radiative heat flux, Fire and Materials conference, 2015

MATERIAL FLAMMABILITY

CALORIMETRIC STUDY OF STANDARD COMMODITY FUELS

Gaurav Agarwal*, Marcos Chaos[1], and Yi Wang
FM Global, Research Division, 1151 Boston-Providence Turnpike, Norwood, MA 02062, USA

ABSTRACT

Standard commodities are representative fuels that are used in the fire protection community to evaluate the risks posed by hazards present at a given occupancy. This work focuses on obtaining detailed flammability data and combustion behavior of standard commodity fuels, including Class 3, CUP (cartoned unexpanded plastic) and CEP (cartoned expanded plastic) commodities using the Fire Propagation Apparatus (FPA). The constituent materials of the aforementioned standard commodities, *viz.*, single wall corrugated cardboard, unexpanded polystyrene cups, expanded polystyrene trays, and paper cups, were investigated. Tests were also conducted for mixtures of component materials in the ratio representative of the respective standard commodities. FPA combustion tests were designed to accurately measure the combustion efficiency, and convective and radiant fraction of the tested samples. The global flammability and combustion properties of the fuel mixtures were found to be reasonably well approximated by the weighted combination of the properties of the individual fuel constituents.

INTRODUCTION

In the fire protection community, the risk exposure of industrial/commercial facilities is evaluated by conducting large-scale fire tests using representative fuels called standard commodities. The standard commodity fuels used in such tests include cartoned paper commodity (Class 3), cartoned unexpanded plastic (CUP), cartoned expanded plastic (CEP), among others[1]. Such standard commodities, as shown in Figure 1, are solid combustible fuels that typically have multiple material compositions and/or complicated fuel geometries with multiple surfaces that can be potentially exposed to fire and water[2, 3].

Figure 1: Pallet load of standard commodities[Δ]

(a) Class 3 (b) CEP (c) CUP

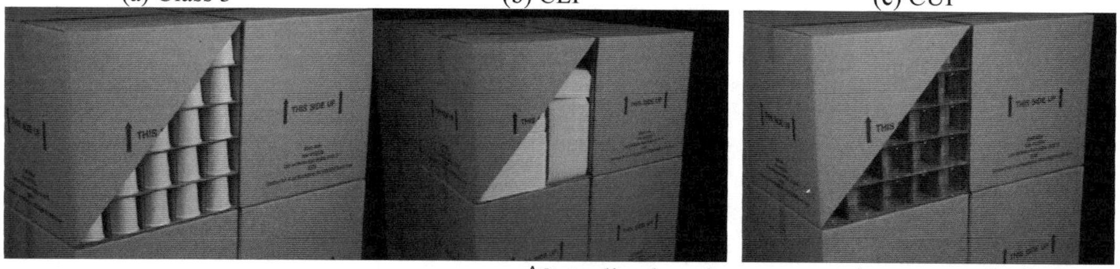

[Δ]Outer linerboard cut open to view commodity inside

The fire tests used to evaluate standard commodities include free burn and sprinkler protection tests. These fire tests exemplify the burning characteristics and determine the protection guidelines for commodities stored in industrial/commercial facilities[1-3]. The overall hazard of the commodity significantly depends on the heat release rate (HRR) measured during such fire tests. The HRR, and therefore the burning behavior, of standard commodities depends on the flammability characteristics of its component fuels, including their

[1] Presently at Lawrence Livermore National Laboratory, Livermore, CA 94550, USA

thermochemistry parameters, chemical heat of combustion, and the convective and radiant fractions of the chemical energy released. Such flammability data are challenging to obtain for standard commodities because the fuels are a mixture of multiple solid fuels. Detailed measurements of the flammability data of standard commodities will aid in better understanding their burning characteristics, and will also further the development of comprehensive computational fluid dynamics (CFD) models for simulating the burning behavior of such standard commodities[4-6].

Previous literature studies have considered the flammability behavior of standard commodities by varying approaches. Hamins and McGrattan[7] conducted bench-scale cone calorimeter experiments[8] on a unit cell of CUP commodity to provide heat release rate (HRR), heat of combustion, and chemical combustion efficiency data. The derived global flammability properties of CUP were used in CFD modelling of rack-storage fires, and the HRR calculated from bench-scale tests was used to extrapolate the burning behavior of a pallet load of CUP commodity[9]. Agarwal et al.[10] developed a complementary experimental and modeling approach to investigate the burning behavior of CUP standard commodity. Bench-scale flammability tests were designed for a unit cell of CUP commodity, and a two-fuel combustion model was developed to simulate the global burning behavior of a CUP unit cell. Both aforementioned studies[7, 10] calculated HRR by the weighted sum of the thermochemical parameters for each of the constituent fuels of CUP, *viz.* corrugated cardboard and unexpanded polystyrene (PS). Model-effective flammability parameters determined from Agarwal et al.[10] were implemented by Gupta et al.[6] in an open source CFD modeling toolbox (FireFOAM[11]) to simulate and validate the burning behavior of rack-storage-scale CUP fires[2]. The modeling approach used by Gupta et al.[6] assumed some global flammability parameters, including heat of combustion and radiant fraction, of CUP also to be the weighted sum of the properties of the constituent fuels of CUP. However, the validity of the weighted sum nature of flammability parameters of CUP constituents, and the constituents of other standard commodities, has not been established.

This work focuses on obtaining detailed flammability data of standard commodity fuels and their component materials by investigating their combustion behavior using a Fire Propagation Apparatus (FPA). The FPA setup, described in the standards[12, 13], is modified to measure the convective fraction of samples. The radiant fraction and the chemical combustion efficiency have been derived from the chemical heat of combustion and the convective fraction measurements. Smoke, carbon-di-oxide, and carbon-monoxide yields of fuel samples are also measured in the study.

EXPERIMENTS

Materials and External Laboratory Analyses

The materials tested in this study include single-wall corrugated cardboard, unexpanded PS cups, expanded PS trays (10" x 10"), and paper cups. These materials are the constituents of CUP, CEP and Class 3 standard commodities. Figure 2 shows the four materials evaluated in this study.

Figure 2: Materials tested in the study (a) Single-wall corrugated cardboard, (b) Unexpanded PS cup, (c) 10" x 10" Expanded PS tray, and (d) Paper cup.

(a) (b) (c) (d)

Heat of combustion, proximate analysis, and ultimate analysis (elemental composition) tests were conducted on each of the aforementioned four materials. Table 1 details the tests performed on each sample. Test methods are present in the open literature[14-19]. Table 2 provides the summary of test results.

Table 1: Methods for laboratory analysis of materials.

Property/Evaluation		Method
Gross heat of combustion	$\Delta H_{c,gross}$ (kJ/g)	ASTM D-5865-10a [17]
Proximate analysis	Loss on drying (mass %)	ASTM D3173-11 [14]
	Volatile (mass %)	ASTM D3175-11 [15]
	Ash (mass %)	ASTM D3174-12 [16]
	Fixed Carbon (mass %)	Calculation by difference
Ultimate analysis	C (mass %)	ASTM D5291-16 [19]
	H (mass %)	ASTM D5291-16 [19]
	N (mass %)	ASTM D5291-16 [19]
	O (mass %)	Calculation by difference
	S (mass %)	ASTM D4239-83 [18]
	Cl (mass ppm)	ASTM D4327-11 [20]

Table 2: Summary of results from laboratory analyses of standard commodity materials

Property/Evaluation	Corrugated Cardboard	Paper Cups	Unexpanded PS Cups	Expanded PS Trays
$\Delta H_{c,gross}$ (kJ/g)	17.98	20.05	41.51	37.12
Loss on drying (mass %)	4.45	4.63	0.22	0.73
Volatile (mass %)	77.11	89.57	98.68	98.81
Ash (mass %)	4.81	0.52	0.34	0.62
Fixed Carbon (mass %)	13.63	9.91	0.98	0.57
C (mass %)	45.27	48.78	91.92	91.98
H (mass %)	6.56	6.79	7.84	7.83
N (mass %)	< 0.5	< 0.5	< 0.5	< 0.5
O (mass %)	46.63	43.86	0.00	0.23
S (mass %)	--	0.011	< 0.0066	< 0.0070
Cl (mass ppm)	--	307	< 27	< 50

-- Not measured

FPA tests

FPA combustion tests were designed to measure the chemical heat of combustion and the convective fraction of the tested samples[12, 13]. FPA is a versatile device that is used to investigate the flammability characteristics of materials through ignition, combustion, pyrolysis, and vertical fire propagation tests that can be done in controlled gas environments (e.g., oxygen reduced air, oxygen enhanced air, or air in presence of extinguishing agents).

The FPA setup, described in the ISO and ASTM standards[12, 13], was modified to measure the convective fraction of samples. In the standard FPA[12, 13], gas temperature in the exhaust duct is measured using a type-K thermocouple, located approximately six duct diameters downstream of the mixing orifice plate in the central axis of the exhaust duct. The accuracy of type-K thermocouples is ±2.2°C and its test range is -200°C to 1200°C. In this study, a type-T thermocouple was used, which has a narrower operating range (-50°C to 400°C) and a better accuracy of ±0.1°C from 0°C to 50°C. However, the final accuracy of the type-T thermocouple was ±0.5°C after accounting for the resolution limits of the data acquisition system. Furthermore, a new measurement of ambient temperature was designed. Three aspirated type-T thermocouple probes were evenly distributed and installed along the circumference of the lower edge of the FPA collector hood, with the probes extending approximately 5 cm from the edge. This location was chosen as it corresponded with the largest inflow air velocity, as measured by a handheld anemometer. The probes were aspirated in order to eliminate the influence of indirect radiation heat from the FPA heaters and test flames. Lastly, three layers of 3 mm thick ceramic blanket insulation (by Cotronics Corp) were wrapped around the outer perimeter of the exhaust duct to minimize heat losses from the duct.

Samples were placed in aluminum sample holders of either 102 mm (4 inch) diameter or 114 mm (4.5 inch) diameter, and 25 mm (1 inch) deep. Combustion tests were conducted under natural convection at 30 kW/m^2 external heat flux. Samples were ignited using a 10 mm long ethylene/air pilot flame, located approximately 10 mm above the surface of the samples. The pilot was turned off and the igniter was removed once the sample ignited. The sample holder was placed on a load cell (0-4500 g range, ±0.01% FS accuracy, ±10 mg noise). The MLR data were obtained by taking the derivative of the transient mass loss data using Savitsky-Golay filters[21-23]. Gaseous combustion products were collected in an exhaust duct system and the data were used to determine HRR through calorimetric methods[24]. A smoke measurement system based on laser attenuation[25] was used to determine the smoke extinction coefficient[12, 13].

Component materials of the standard commodities, *viz.*, single wall corrugated cardboard, unexpanded PS cups, expanded PS trays, and paper cups, were investigated. Furthermore, combustion tests were conducted for mixtures of component materials in the ratio representative of the respective standard commodities. Tests were also conducted for acetone and black PMMA samples, which are well characterized in the literature[24], to provide validation for the approach used. Benchmark experiments were performed with the radiant heaters operating alone, without fuel inside sample holders, at 30 kW/m^2 external heat flux to calculate the duct temperature increase contributed by the radiant heaters. Table 3 shows the test matrix for FPA combustion tests; the sample preparation methods and experimental descriptions.

Table 3: Test matrix for FPA combustion tests.

Material	Description	Heat flux (kW/m^2)
Benchmark	No sample; 102 mm diameter and 25 mm deep Aluminum dish; 3 repeats.	30
Acetone	100 ml acetone in 102 mm diameter and 25 mm deep Aluminum dish; 1 test.	0
PMMA	102 mm diameter and 13 mm deep cylindrical sample in 102 mm diameter and 25 mm deep Aluminum dish; 1 test.	30
Cardboard	102 mm diameter single-wall cardboard sheets compressed in 102 mm diameter and 25 mm deep Aluminum dish. 3 repeats.	30
Cardboard	114 mm diameter single-wall cardboard sheets compressed in 114 mm diameter and 25 mm deep Aluminum dish$^{\Delta}$. 1 test.	30
Paper cups	Shredded paper cups in 102 mm diameter and 25 mm deep Aluminum dish. 3 repeats.	30
PS cups	Shredded unexpanded PS cups in 102 mm diameter and 25 mm deep Aluminum dish. 3 repeats.	30
PS trays	114 mm diameter unexpanded PS tray cutouts compressed in 114 mm diameter and 25 mm deep Aluminum dish$^{\Delta}$. 3 repeats.	30
Cardboard + paper cups	50% cardboard and 50% paper cup mixture by mass. Shredded samples mixed in 102 mm diameter and 25 mm deep Aluminum dish. 1 experiment.	30
Cardboard + PS cups	30% cardboard and 70% PS cup mixture by mass. Shredded samples mixed in 102 mm diameter and 25 mm deep Aluminum dish. 1 experiment.	30
Cardboard + PS trays	25% cardboard and 75% PS tray mixture by mass. Shredded samples mixed in 114 mm diameter and 25 mm deep Aluminum dish$^{\Delta}$. 1 experiment.	30

$^{\Delta}$A larger size aluminum dish allowed larger fuel mass to be tested in FPA combustion tests.

CALCULATION PROCEDURE

Thermochemistry data

The empirical chemical formulae of fuels are calculated while assuming a known number of carbon atoms in a monomer of polymers tested in this study. Chemical formula for cardboard and paper cups is assumed to consist of six carbon atoms due to cellulose ($C_6H_{10}O_5$, CAS 9004-34-6) being its major component. Styrene (C_8H_8, CAS 100-42-5) is the monomer of PS based products, and hence eight carbon atoms are assumed in its monomer chain. Complete combustion of an oxygenated hydrocarbon fuel to carbon-di-oxide can be expressed as Eq. 1.

$$C_xH_yO_z + \left(x + \frac{y}{4} - \frac{z}{2}\right)O_2 \rightarrow xCO_2 + \frac{y}{2}H_2O \qquad [1]$$

The gross heat of combustion of fuels were obtained from bomb calorimeter tests. However, the "lower" or "net" heat of combustion is more relevant for fire applications, and is expressed in Eq. 2.

$$\Delta h_{c,net} = \Delta h_{c,gross} - \frac{y}{2}\Delta h_{vap} \qquad [2]$$

The $\Delta h_{c,gross}$ and $\Delta h_{c,net}$ are respectively the gross and net heat of combustion on mole basis. The $\Delta h_{c,vap}$ is the heat of vaporization of water (= 44 kJ/mol[26]). Mass based heat of combustions are expressed as:

$$\Delta H_{c,net} = \Delta H_{c,gross} - \left(\frac{y}{2}\right)\frac{\Delta h_{vap}}{M_f} \qquad [3]$$

Here, M_f is the molecular weight of fuel.

For charring materials, *viz.* cardboard and paper cups, the heat of char combustion is assumed to be 32.8 kJ/g[27]. The minimum possible char mass fraction, μ, yielded during pyrolysis is calculated by the volatile fraction reported from the proximate analysis (Table 3). Using these values, the net heat of combustion per unit mass of volatiles ($\Delta H_{c,vol}$) is calculated as per Eq. 4. It is noted that for solid fuels, the char mass fraction and therefore the volatile gas composition, are strongly dependent on the external heat flux.

$$\Delta H_{c,vol} = \frac{\Delta H_{c,net} - \mu\Delta H_{c,char}}{(1-\mu)} \qquad [4]$$

Carbon monoxide and soot formation are considered in cases of incomplete combustion , so that complete oxidation to CO_2 is hindered[28]. Only CO formation thermochemistry is assumed in this work, and the fuel is assumed to undergo following reaction during incomplete combustion:

$$C_xH_yO_z + \left(\frac{x-z}{2} + \frac{y}{4}\right)O_2 \rightarrow xCO + \frac{y}{2}H_2O \qquad [5]$$

The chemical heat release rate is determined by either CO-CO_2 generation (Eq. 6) or O_2 consumption calorimetry (Eq. 7) [24].

$$\dot{Q}_{ch} = \Delta H^*_{CO_2}\dot{G}_{CO_2} + \Delta H^*_{CO}\dot{G}_{CO} \qquad [6]$$

$$\dot{Q}_{ch} = \Delta H^*_{O_2(CO_2)} \dot{C}_{O_2} - \left(\Delta H^*_{O_2(CO_2)} - \Delta H^*_{O_2(CO)}\right) \frac{\psi_{O_2(CO)}}{\psi_{CO}} \dot{G}_{CO} \quad [7]$$

Here, $\dot{G}_{CO_2}$ and $\dot{G}_{CO}$ are the generation rates of CO_2 and CO, respectively, and $\dot{C}_{O_2}$ is the consumption rate of O_2. The following parameters are defined to calculate the HRR by either calorimetry methodology:

(a) ψ_{CO_2} (g of CO_2/g of fuel): stoichiometric yield for the complete conversion of fuel to CO_2.
(b) ψ_{CO} (g of CO/g of fuel): stoichiometric yield for the complete conversion of fuel to CO.
(c) $\psi_{O_2(CO_2)}$ (g of O_2/g of fuel): stoichiometric oxygen to fuel mass ratio for complete conversion to CO_2.
(d) $\psi_{O_2(CO)}$ (g of O_2/g of fuel): stoichiometric oxygen to fuel mass ratio for complete conversion to CO.
(e) $\Delta H^*_{CO_2}$ (kJ/g of CO_2): net heat of combustion per unit mass of generated CO_2, assuming complete conversion to CO_2.
(f) ΔH^*_{CO} (kJ/g of CO): net heat of combustion per unit mass of generated CO, assuming complete conversion to CO.
(g) $\Delta H^*_{O_2(CO_2)}$ (kJ/g of O_2): net heat of combustion per unit mass of O_2 consumed, assuming complete conversion to CO_2.
(h) $\Delta H^*_{O_2(CO)}$ (kJ/g of O_2): net heat of combustion per unit mass of O_2 consumed, assuming complete conversion to CO.

Molar balance of Eqs. 1 and 5 yield the following parameters:

$$\psi_{CO_2} = xM_{CO_2} / M_f \quad [8]$$

$$\psi_{CO} = xM_{CO} / M_f \quad [9]$$

$$\psi_{O_2(CO_2)} = (x + y/4 - z/2)M_{O_2} / M_f \quad [10]$$

$$\psi_{O_2(CO)} = (x/2 + y/4 - z/2)M_{O_2} / M_f \quad [11]$$

Thereby, the species-based heat of combustion are calculated as follows:

$$\Delta H^*_{CO_2} = \frac{\Delta H_{c,net}}{\psi_{CO_2}} \quad [12]$$

$$\Delta H^*_{CO} = \frac{\Delta H_{c,net}}{\psi_{CO}} - \Delta H_{CO \to CO_2} \quad [13]$$

$$\Delta H^*_{O_2} = \frac{\Delta H_{c,net}}{\psi_{O_2(CO_2)}} \quad [14]$$

Here, $\Delta H_{CO \to CO_2}$ is the heat generated, per unit mass of CO, in the reaction $CO + (1/2)\, O_2 \rightarrow CO_2$ (10.1 kJ/g of CO[26])

Chemical combustion efficiency, and convective and radiative fraction

The chemical HRR for tests listed in Table 2, where a single fuel constituent is present, are calculated using CO-CO_2 generation calorimetry, i.e. Eq. 6. For a single fuel constituent, chemical HRR calculated through CO-CO_2 generation calorimetry is more accurate in comparison to O_2 consumption calorimetry due to better resolution of the CO_2 and CO FPA gas analyzers. The chemical HRR for combustion tests with fuel mixtures are calculated using O_2 consumption calorimetry, i.e. Eq. 7, because the $\Delta H^{*}_{O_2(CO_2)}$ parameter for different fuel constituents are of approximately same value, whereas the $\Delta H^{*}_{CO_2}$ and ΔH^{*}_{CO} parameters vary greatly depending on the fuel constituent in the mixture undergoing combustion[24]. Although in Eq. 7, the $\Delta H^{*}_{O_2(CO)}$ parameters vary for fuels with relatively low combustion efficiencies, the second term in the R.H.S of Eq. 7 is small and therefore neglected in this work.

To calculated chemical HRR per Eqs. 6 and 7, the CO-CO_2 generation rates and O_2 consumption rate are calculated from the procedures and formulations mentioned in the established FPA standards[12, 13]. The chemical heat of combustion is calculated as per Eq. 15, by taking the ratio of the net integrated chemical HRR and cumulative fuel mass loss over the combustion period.

$$\Delta H_{c,ch} = \int \dot{Q}_{ch} dt \Big/ m_{loss} \qquad [15]$$

The combustion efficiency is calculated as a ratio of chemical and theoretical net heat of combustion.

$$\chi = \Delta H_{c,ch} \big/ \Delta H_{c,net} \qquad [16]$$

Under conditions of enthalpy balance, the combustion efficiency measured for a fuel is the sum of its convective fraction and the radiant fraction[24, 29].

$$\chi = \chi_{conv} + \chi_{rad} \qquad [17]$$

The convective HRR is determined from the excess temperature of the gaseous products in the sampling duct, mass flow rate of the gaseous products, and the specific heat capacity of products as per Eq. 18.

$$\dot{Q}_{conv} = \dot{m}_{d} C_p \left(T_d - T_{amb}\right) \qquad [18]$$

As previously described, the convective HRR release rate is corrected by subtracting the benchmark convective HRR contributed by the radiant heaters at the heat flux used in the experiments. Benchmark experiments were performed with the radiant heaters operating alone, without fuel inside sample holders, at 30 kW/m^2 external heat flux to calculate the duct temperature increase contributed by the radiant heaters.

The convective fraction, χ_{conv}, is calculated by taking the ratio of the net integrated convective HRR over the combustion period of the fuel and the theoretical net heat of combustion, as shown in Eq. 19 . The radiant fraction, χ_{rad}, is calculated from Eq. 16 by taking the difference of χ and χ_{conv}.

$$\chi_{conv} = \int \dot{Q}_{conv} dt \Big/ \Delta H_{c,net} \qquad [19]$$

The smoke generation rates (SGR) and smoke yield, y_s, is calculated as per procedures and formulations mentioned in established FPA standards[12, 13].

RESULTS

Thermochemistry constants

Thermochemistry data of the four materials selected for this study are reported in Table 4. The reported data can be used as an archived reference for large-scale tests involving these materials.

Table 4: Derived thermochemistry data for standard commodity materials

Fuel → / Derived data ↓		Corrugated Cardboard	Paper Cups	Unexpanded PS Cups	Expanded PS Trays
Chemical Formula		$C_6H_{10.4}O_{4.7}$	$C_6H_{9.95}O_{4.05}$	$C_8H_{8.13}$	$C_8H_{8.12}O_{0.02}$
MW (g/mol)		157.10	146.88	104.28	104.51
Heat of combustion (kJ/g)	Gross	17.98	20.05	41.51	37.12
	Net	16.53	18.56	39.79	35.41
	Volatile	13.65*	16.90*	39.89	35.46
ψ_{CO_2} *(g/g)*		1.68	1.80	3.38	3.37
ψ_{CO} *(g/g)*		1.07	1.14	2.15	2.14
$\psi_{O_2(CO_2)}$ *(g/g)*		1.28	1.41	3.08	3.07
$\psi_{O_2(CO)}$ *(g/g)*		0.67	0.75	1.85	1.84
$\Delta H^*_{CO_2}$ *(kJ/g)*		9.83	10.32	11.79	10.51
ΔH^*_{CO} *(kJ/g)*		5.34	6.12	8.42	6.41
$\Delta H^*_{O_2(CO_2)}$ *(kJ/g)*		12.93	13.18	12.93	11.54

*Assuming the minimum possible char fraction to be 0.15

Due to their inherent lignocellulosic nature, the thermochemistry data of corrugated cardboard and paper cups are expected to be comparable, and also similar to that of wood materials listed in the SFPE handbook[24]. However, from Table 4, the net heat of combustion of paper cups is noticeably higher (by 10%, approximately) than the net heat of combustion values of cardboard and other lignocellulose materials mentioned in the SFPE handbook[24]. The use of plastic (polyethylene) in the manufacturing of paper cups accounts for the observed higher heat of combustion.

The derived thermochemistry data for PS cups are consistent with the PS thermochemistry data reported in the SFPE handbook[30], *viz* the net heat of complete combustion, chemical formula, and the combustion calorimetry data. Gross and net heats of combustion for PS trays are ~11% lower than the similar parameters for PS cups. This difference can be attributed to the higher ash and oxygen content of PS trays (Table 2). Nonetheless, the data for PS trays are consistent with data for expanded PS found in Ref.[24].

Specific HRR plots

The specific HRR plots, i.e. HRR normalized on the basis of initial mass of fuels, for the tests conducted in this study are shown in Figure 3.

Figure 3(a) show plots for combustion of cardboard sheets, shredded paper cups, and shredded mixture of cardboard and paper cups in 50:50 ratio by mass. Specific HRR profile predicted from a weighted sum of individual cardboard and paper cup specific HRR profiles in 50:50 mass ratio, is also included. It is observed that the specific HRR profiles for cardboard sheets and paper cups are similar; the higher specific HRR peak for paper cup is consistent with its higher net heat of combustion in comparison to that of cardboard (Table 4). The experiment for 50:50 mixture of shredded cardboard and paper cup exhibits an earlier and higher ignition peak, and termination of the combustion process than that of constituent fuels. This is due to the presence of cardboard in shredded form providing easier ignition, and higher exposed surface area of shredded cardboard leading to higher specific HRR peak. The profile for predicted specific HRR of cardboard and paper cup 50:50 mixture deviate slightly from the experimental profile of mixture due to aforementioned reasons.

Figure 3: HRR normalized on the basis of initial mass for fuels and mixtures (a) Cardboard and paper cup (b) Cardboard and unexpanded PS (c) Cardboard and expanded PS

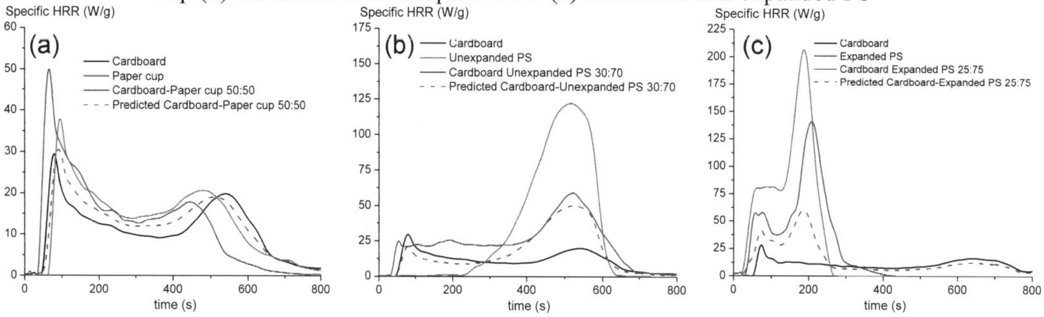

Combustion plots of cardboard sheets, shredded unexpanded PS cups, and shredded mixture of cardboard and PS cups in 30:70 ratio by mass, are shown in Figure 3(b). The specific HRR profile of a predicted weighted sum for cardboard and unexpanded PS in 30:70 mass ratio was generated from their individual specific HRR data. Unexpanded PS combustion profile reveals significant difference from the cardboard combustion profile. Firstly, the ignition of unexpanded PS occurs at a notably later time than cardboard ignition; this is due to higher specific heat capacity and heat of pyrolysis of PS in comparison to cardboard, as well as the energy required to melt PS before ignition[24]. Secondly, the peak specific HRR of unexpanded PS is approximately four times that of cardboard due to the higher net heat of combustion of PS cup in comparison to cardboard (Table 4). Regardless, the predicted specific HRR profile of fuel mixture match sufficiently well with the specific HRR profiles of the cardboard and unexpanded PS fuel mixtures.

Similar plots for cardboard and expanded PS tray tests are shown in Figure 3(c). These tests were conducted using 114 mm diameter dishes to accommodate sufficiently high mass of expanded PS in the sample holder, unlike the plots shown in Figures 3(a) and (b). The specific HRR profile of expanded PS is greatly different than that of cardboard and even unexpanded PS. For expanded PS, the ignition occurs earlier than that for cardboard and the rate of increase of HRR is also quite higher than that for unexpanded PS in Figure 3(b). The peak specific HRR of expanded PS is approximately seven times higher than that of cardboard. The peak specific HRR of expanded PS is almost twice higher than that of unexpanded PS in Figure 3(b) and the duration of combustion is almost half; a part of difference in this observation is attributed to the difference in dish size for combustion tests of two types of PS leading to higher heat feedback in expanded PS combustion case. The relatively faster ignition and combustion of expanded PS, in comparison to unexpanded PS, can be attributed to lower heat of gasification of expanded polymer products in comparison to granular polymer products[24]. Due to early ignition and high HRR from expanded PS, the combustion behavior of cardboard in cardboard and expanded PS 25:75 mixture run is also affected. From Figure 3(c), it can be observed that the presence of high heat from expanded PS combustion lead to early combustion of cardboard and its char; therefore the mixture combustion experiment completed earlier and was lacking in char combustion peak, in comparison to the predicted run of the weighted sum of cardboard and expanded PS mixture. These observations reveal that the energy released from certain fuel constituents in a fuel mixture can affect the combustion behavior of other constituents in the mixture; this thermal interaction between constituents of a mixture shall be resolved while predicting the burning behavior of a fuel mixture.

Global calorimetric properties

Table 5 shows the list of global calorimetric properties calculated for the fuels tested in this study. Average values of CO_2 yield, y_{CO_2}, CO yield, y_{CO}, and smoke yield, y_s, calculated from the experiments conducted in the study are listed in Table 5; all yields are calculated based on the fuel mass lost during combustion period. Table 5 reports the literature values for the materials in square brackets, which were obtained by Khan et al.[24] by conducting FPA tests at 50 kW/m² external heat flux. Not all materials tested in this study have been reported in the mentioned literature[24]; e.g. global calorimetric properties of corrugated cardboard and paper cup are not reported, and therefore instead, calorimetric values of wood from the aforementioned literature source[24] are listed in Table 5. The values listed in round brackets for mixture fuels are calculated by weighted sum of the literature values of constituent materials of the mixture.

Table 5: Global calorimetric results: combustion yields

Material	y_{CO_2} (g/g)	y_{CO} (g/g)	y_s (g/g)
Acetone	2.08 [2.14]	0.002 [0.003]	0.000$^{\Delta}$ [0.014]
PMMA	1.82 [2.12]	0.007 [0.010]	0.008 [0.022]
Cardboard	1.03 [1.27]~	0.046 [0.004]~	0.000$^{\Delta}$ [0.014]~
Paper cups	0.99 [1.27]~	0.046 [0.004]~	0.000$^{\Delta}$ [0.014]~
PS cups	1.83 [2.33]	0.074 [0.060]	0.138 [0.164]
PS trays	2.13 [2.33]	0.085 [0.060]	0.117 [0.164]
Cardboard + paper cups (50:50)	1.03 (1.27)	0.036 (0.004)	0.000$^{\Delta}$ (0.014)
Cardboard + PS cups (30:70)	1.61 (2.01)	0.066 (0.043)	0.085 (0.119)
Cardboard + PS trays (25:75)	1.71 (2.07)	0.049 (0.046)	0.074 (0.127)

~Property taken from wood results.

$^{\Delta}$Small amount of smoke generated was not accurately resolved by the laser system.

The experimental yield values calculated for materials are repeatable, with standard deviation, for example, for cardboard CO_2 and CO yields being approximately 2% for three repeat tests. Acetone tests were conducted in absence of external heat flux, and the various yield values obtained are consistent with the literature values[24]. For the rest of the experiments conducted under external radiation, the obtained experimental yields are within a reasonable range of the literature[24]. The experimental results, except PS, show slightly higher CO yields, and also show lower CO_2 yields in comparison to literature[24]; these are attributed to lower chemical combustion efficiency of the fuels at tested 30 kW/m² external heat flux, in comparison to literature values[24] obtained at 50 kW/m² heat flux. The experimental smoke yields, obtained from the laser attenuation system, are slightly lower than the literature values[24], which were obtained from a white-light based turbidimeter system[31]. The experimental yields for mixture fuels obtained in the study exhibit weighted additive nature while comparing to the values noted in round brackets.

Table 6 reports the average values of chemical heat of combustion, and combustion, convective, and radiant efficiencies of the tested fuels and mixtures. Similar to Table 5, the literature values[24] for fuels are noted in square brackets, and for mixture fuels, the values listed in round brackets are calculated by weighted sum of the literature values of constituent materials of the mixture.

Table 6: Global calorimetric results: chemical, convective and radiative efficiencies

Material	$\Delta H_{c,ch}$ (kJ/g)	χ	χ_{conv}	χ_{rad}
Acetone	27.6 [27.9]	0.93 [0.94]	0.72 [0.68]	0.21 [0.26]
PMMA	21.7 [24.2]	0.86 [0.96]	0.60 [0.66]	0.26 [0.30]
Cardboard	10.5 [12.4]~	0.64 [0.73]~	0.42 [0.46]~	0.22 [0.27]~
Paper cups	11.5 [12.4]~	0.62 [0.73]~	0.41 [0.46]~	0.21 [0.27]~
PS cups	23.7 [27.0]	0.60 [0.69]	0.29 [0.28]	0.30 [0.41]
PS trays	26.2 [27.0]	0.74 [0.69]	0.28 [0.28]	0.46 [0.41]
Cardboard + paper cups (50:50)	10.6 (12.3)	0.60 (0.73)	0.38 (0.46)	0.22 (0.27)
Cardboard + PS cups (30:70)	20.6 (22.9)	0.63 (0.70)	0.35 (0.33)	0.28 (0.37)
Cardboard + PS trays (25:75)	23.8 (23.6)	0.78 (0.70)	0.37 (0.32)	0.40 (0.37)

~Property taken from wood results.

The experimental chemical heat of combustion values for solid fuels are slightly lower than literature values due to relative higher level of incompleteness of combustion at 30 kW/m^2 external heat flux. For the acetone test, the chemical heat of combustion and combustion efficiency from experiments are close to those reported in the literature[24]; this is due to fact that heaters were not used in acetone combustion tests in the literature measurements[24] as well. For all solid fuels, with the exception of PS trays, the chemical efficiency are calculated to be lower than those reported in the literature[24]; there is no mention of chemical efficiency specifically for expanded PS in the literature[24]. The measured chemical efficiency is repeatable, with standard deviation, for example, for cardboard combustion efficiency being approximately 2.5% for three repeat tests. For the three fuel mixtures tested in the study, the chemical heat of combustion, and hence the combustion efficiency values, measured in the study exhibit weighted additive nature. This observation is in an agreement with the approach used in a recent fire modeling work[6], where chemical heat of combustion of the CUP commodity was assumed to be the weighted sum of that of its constituents.

The calculated convective and radiant fraction of acetone are close to the literature values[24]. The experimental uncertainty in the measurement of convective fraction, and therefore radiant fraction, is generally higher than the uncertainty for other global calorimetric properties. For example, for cardboard combustion, the standard deviation in convective fraction measurement is close to 17%, thereby taking the uncertainty in the measurements to ±34% of the average values. The higher uncertainty in convective fraction measurements is due to the correction of heating effect of the exhaust duct from indirect radiation from heaters, and also due to variation in the HVAC cycles of the room in which the tests were conducted. These can be resolved in the future experiments by blocking off indirect radiation from heaters to duct, and stabilizing the room ambient air temperature by better controlling the HVAC system.

The convective and radiant fraction of all solid fuels, except for PS trays, are calculated to be slightly lower than the literature values[24], as the combustion efficiency from the tests conducted in the study are lower due to the lower external heat flux used in the experiments. Although it is generally accepted that the radiant fractions for well-ventilated buoyant turbulent diffusion fires for liquid and gaseous fuels, of diameter ranging 0.1-2 m[29, 32, 33], are constant and can be treated as material flammability properties, it has been shown in recent studies[34] that the radiant fraction and the combustion efficiency of solid fuels can show an increasing trend with an increase in applied heat flux to the sample surface. Regardless, the convective and radiant fraction for the fuels reasonably match with results mentioned in the literature[24] that were taken at 50 kW/m^2 external heat flux. The measured radiant fractions are shown to positively correlate with the smoke yield (generally unburnt soot) of the fuels, as also noticed in other literature studies[29, 33, 36]. The radiant fractions of the fuel mixtures were calculated to be with reasonable estimates of the weighted additive radiant fractions of the individual fuel constituents. This observation validates the approach taken in the recent CFD modeling work[6] where the radiant fraction of the CUP commodity was assumed to be weighted sum of the radiant fraction of cardboard and unexpanded PS constituents.

CONCLUSIONS

In this study, detailed flammability data of standard commodity fuels and their component materials were obtained using the FPA. The commodities investigated in this study included cartoned paper commodity (Class 3), cartoned unexpanded plastic (CUP), cartoned expanded plastic (CEP). FPA combustion tests were conducted for component materials of these standard commodities, *viz.*, single wall corrugated cardboard, unexpanded polystyrene cups, expanded polystyrene trays, and paper cups. Tests were also conducted for mixtures of component materials in the ratio representative of the respective standard commodities. Additionally, the gross heat of combustion, proximate analysis, and ultimate analysis tests were conducted for the aforementioned four component materials and their thermochemistry data were obtained.

FPA combustion tests were designed to accurately measure the combustion efficiency, and convective and radiant fraction of the tested samples. The CO_2, CO, and smoke yield of samples were also measured during these tests and reported in this study. The aforementioned global flammability parameters for mixture fuels were calculated to be within reasonable estimates of the weighted additive nature of the individual fuel constituents. This observation is consistent with the approach taken in a recent CFD modeling work[6] where global calorimetric parameters, including the radiant fraction and the chemical heat of combustion, of the CUP commodity was assumed to be a weighted sum of the global calorimetric parameters of cardboard and unexpanded PS constituents.

REFERENCES

1 Xin, Y.; Tamanini, F., Assessment of Commodity Classification for Sprinkler Protection Using Representative Fuels. Fire Safety Science 2008, 9, 527-538.

2 Thumuluru, S.; Xin, Y.; Gupta, A., An Experimental Study on the Freebirn Characteristics of Complex Fuels, 14th International Conference and Exhibition on Fire Science and Engineering (Interflam), Interscience Communications Ltd.: London: Nr Windsor, UK, 2016; pp 891-902.

3 Xin, Y.; Wang, Y.; Chaos, M.; Dorofeev, S. B., Complex Fuel Experiments Using the Characteristic Fuel Unit (CFU) Method for Cartoned Commodities, 14th International Conference and Exhibition on Fire Science and Engineering (Interflam), Interscience Communications Ltd.: London: Nr Windsor, UK, 2016; pp 903-914.

4 Wang, Y.; Meredith, K.; Zhou, X.; Chatterjee, P.; Chaos, M.; Ren, N.; Dorofeev, S. B., Numerical Simulation of Sprinkler Suppression of Rack Storage Fires. Fire Safety Science 2014, 11, 1170-1183.

5 Ren, N.; de Vries, J.; Meredith, K.; Chaos, M.; Wang, Y., Firefoam Modeling of Standard Class 2 Commodity Rack Storage Fires, Proceedings of the Fire and Materials Conference 2015, 14th International Conference, San Francisco, USA, 2015.

6 Gupta, A.; Agarwal, G.; Meredith, K.; Thumuluru, S.; Xin, Y.; Chaos, M.; Wang, Y., Modeling of

Large-Scale Fires in Cartoned Plastic Commodity Rack-Storage, 14th International Conference and Exhibition on Fire Science and Engineering (Interflam), Interscience Communications Ltd.: London: Nr Windsor, UK, 2016; pp 915-926.

7 Hamins, A.; McGrattan, K. B., *Reduced-Scale Experiments to Characterize the Suppression of Rack-Storage Commodity Fires, Report No. NISTIR 6439*. National Institute of Standards and Technology: 1999.

8 Standard Test Method for Heat and Visible Smoke Release Rates for Materials and Products Using an Oxygen Consumption Calorimeter. In *ASTM E1354-14e1*, ASTM International: West Conshohocken, PA, 2014.

9 McGrattan, K. B.; Hamins, A.; Stroup, D., *Smoke and Heat Vent, Draft Curtain Interaction – Large Scale Experiments and Model Development, Report No. NISTIR 6196-1*. National Institute of Standards and Technology: 1998.

10 Agarwal, G.; Gupta, A.; Chaos, M.; Meredith, K. V.; Wang, Y., Experimental investigation and inverse modeling of the flammability behavior of cartoned plastic commodity. Proceedings of the Combustion Institute 2016.

11 *FireFOAM, available at <https://github.com/fireFoam-dev/fireFoam-2.2.x>.*

12 Reaction to Fire tests – Measurement of Material Properties Using a Fire Propagation Apparatus. In *ISO 12136*, International Organization for Standardization: Geneva, Switzerland, 2011.

13 Standard Test Methods for Measurement of Material Flammability Using a Fire Propagation Apparatus (FPA). In *ASTM E2058-13a*, ASTM International: West Conshohocken, PA, 2013.

14 Standard Test Method for Moisture in the Analysis Sample of Coal and Coke. In *ASTM D3173-11*, ASTM International: West Conshohocken, PA, 2011.

15 Standard Test Method for Volatile Matter in the Analysis Sample of Coal and Coke. In *ASTM D3175-11*, ASTM International: West Conshohocken, PA, 2011.

16 Standard Test Method for Ash in the Analysis Sample of Coal and Coke from Coal. In *ASTM D3174-12*, ASTM International: West Conshohocken, PA, 2012.

17 Standard Test Method for Gross Calorific Value of Coal and Coke. In *ASTM D5865-10a*, ASTM International: West Conshohocken, PA, 2010.

18 Sulfur in the Analysis Sample of Coal and Coke Using High Temperature Tube Furnace Combustion Methods, Method C, Annual Book of ASTM Methods. In *ASTM D4239-83*, ASTM International: West Conshohocken, PA, 1992.

19 Standard Test Methods for Instrumental Determination of Carbon, Hydrogen, and Nitrogen in Petroleum Products and Lubricants. In *ASTM D5291-16*, ASTM International: West Conshohocken, PA, 2016.

20 Test Methods for Anions in Water by Chemically-Suppressed Ion Chromatography. In *ASTM D4327*, ASTM International: West Conshohocken, PA, 2011.

21 Savitzky, A.; Golay, M. J. E., Smoothing and Differentiation of Data by Simplified Least Squares Procedures. Analytical Chemistry 1964, 36, (8), 1627-1639.

22 Staggs, J. E. J., Savitzky–Golay Smoothing and Numerical Differentiation of Cone Calorimeter Mass Data. Fire Safety Journal 2005, 40, (6), 493-505.

23 Khan, M. M.; de Ris, J. L., Operator Independent Ignition Measurements. Fire Safety Science 2005, 8, 163-174.

24 Khan, M. M.; Tewarson, A.; Chaos, M., Combustion Characteristics of Materials and Generation of Fire Products. In The SFPE Handbook of Fire Protection Engineering (5th Ed.), Chapter 36, Gottuk, D.; Hall Jr., J. R.; Harada, K.; Kuligowski, E.; Puchovsky, M.; Torero, J. L.; Watts Jr., J. M.; Wieczorek, C. J., Eds. Springer: New York, 2016; pp 1143-1232.

25 Mulholland, G. W.; Johnsson, E. L.; Fernandez, M. G.; Shear, D. A., Design and testing of a new smoke concentration meter. Fire and Materials 2000, 24, (5), 231-243.

26 Linstrom, P. J.; Mallard, W., NIST Chemistry webbook; NIST standard reference database No. 69. 2001.

27 Chaos, M. Derived Thermochemistry Values for Corrugated Cardboard; FM Global: Norwood, MA, 4 February 2013, 2013.

28 Brohez, S.; Delvosalle, C.; Marlair, G.; Tewarson, A., The Measurement of Heat Release from

Oxygen Consumption in Sooty Fires. Journal of Fire Sciences 2000, 18, (5), 327-353.
29 Tewarson, A., Combustion efficiency and its radiative component. Fire Safety Journal 2004, 39, (2), 131-141.
30 Tewarson, A., Generation of Heat and Chemical Compounds in Fires. In The SFPE Handbook of Fire Protection Engineering (2nd Ed.), Section 3, Chapter 4, The National Fire Protection Association: Quincy, MA, 1996; pp 3-53–3-124.
31 Newman, J. S.; Steciak, J., Characterization of Particulates from Diffusion Flames. Combustion and Flame 1987, 67, (1), 55-64.
32 Markstein, G. H., Relationship between smoke point and radiant emission from buoyant turbulent and laminar diffusion flames. Symposium (International) on Combustion 1985, 20, (1), 1055-1061.
33 De Ris, J.; Cheng, X.-F., The Of Role Of Smoke-point In Material Flammability Testing. Fire Safety Science 1994, 4, 201-312.
34 Zeng, D.; Chaos, M.; Khan, M. M.; Dorofeev, S. B., Radiation Characteristics of Corrugated Cardboard Flames. Fire Safety Science 2014, 11, 97-110.
35 Delichatsios, M. A., Smoke yields from turbulent buoyant jet flames. Fire Safety Journal 1993, 20, (4), 299-311.
36 Zeng, D.; Chaos, M.; Wang, Y., Radiative Characteristics of Fires with Different Sooting Propensities, 14th International Conference and Exhibition on Fire Science and Engineering (Interflam), Interscience Communications Ltd.: London: Nr Windsor, UK, 2016; pp 1405-1414.

INFLUENCE OF PLASTIC FORMULA ON FLAMMABILITY PROPERTIES

Thomas Fabian, Julie Shih, Lillian Chen, Scott MacLeod, Renate Messing, Paul Moore
UL LLC, USA

ABSTRACT

Two series of polyamide 6,6 were investigated to explore the influence of colorant and polymer molecular weight on thermal degradation and flammability properties. A commercially available natural color PA66 and six different colored samples made from the natural color PA66 pellets and commercially available, polyamide-based standard color concentrate pellets constituted the colorant influence study while two commercially available PA66 of differing molecular weight along with two differing blends of the respective base PA66 polymers constituted the molecular weight influence study. Molded test specimens were characterized for thermal degradation behavior (TGA) and material combustibility by limited oxygen index (LOI), UL 94 vertical flame test (UL 94), and microscale combustion calorimetry (MCC). Despite limiting the study samples were to a single condensation polymer chemistry, six colorants using the same carrier chemistry, and four molecular weights, several important findings expected to translate to other polymer systems were made:

1. Few correlations between the different test methods and measured flammability properties were observed
2. UL 94 afterflame times were generally longer for “Oven-Aged” samples than the “As-Received” samples, however, the differences were not always statistically significant
3. MCC measurements by the two ASTM standard defined methods were different such that Method A (anaerobic conditions) tended to yield simpler heat release rate profiles, higher temperatures at combustion onset and temperatures corresponding to the peak heat release rate, lower peak heat release rates and heats of combustion, and greater residue
4. Color was found to have a greater influence than polymer molecular weight for properties measured at molecular level such as thermal degradation (TGA) and microscale combustion calorimeter (MCC)
5. Color influenced the measured flammability properties at statistically significant levels, however, the degree of influence varied for the different properties and varied by color
6. Molecular weight was more influential for test methods in which sample flow was possible such as UL 94 and LOI than molecular level tests

Keywords

Flammability, LOI, material properties, microscale combustion calorimeter, polyamide, TGA, UL 94.

INTRODUCTION

Flammability properties of polymeric materials are most frequently assessed by UL 94 flame testing. The resulting UL 94 flame test categorical ratings (HB, V-2, V-1, V-0, 5VB, and 5VA for bulk plastics) demonstrate the acceptability of the material usage at the tested thicknesses in electrical and electronics applications. In addition to test specimen thickness, UL 94 ratings are known to vary with material properties and composition including color and additive / filler concentrations. Consequently, different polymer grades must be evaluated separately along with representative colors.

This approach has served the electrical and electronics application market well since UL 94 was first published in September 1972. From a product development viewpoint, however, the categorical nature of UL 94 ratings is not as insightful for discerning the influence of formulation compositional changes on material flammability as continuous response measurements such as heat release rate. Heat release rate measurements such as ASTM E 1354 / ISO 5660 cone calorimeter and the more recent ASTM D 7309 microscale combustion calorimeter (MCC) are limited in that they do not facilitate observation of polymer dripping much less formation of flaming droplets that could accelerate flame spread. Yet droplets that ignite the cotton indicator below a UL 94 test specimen are the distinction between a UL 94 V-1 and V-2 rating. Thus it is questioned whether MCC or cone calorimeter measurements can distinguish material formulas differing in aspects that have been observed to yield differing UL 94 ratings while maintaining similar fuel loads such as polymer molecular weight and anti-drip agent concentration.

UL's R&D and Performance Materials divisions collaborated on a series of investigations to explore the influence of colorant, polymer molecular weight, and concentrations of various additives on thermal degradation and flammability properties. This Paper covers the colorant and molecular weight investigations in which two series of samples based polyamide 6,6 (PA66) were tested by thermogravimetric analysis (TGA), UL 94, limited oxygen index (LOI), and MCC.

SAMPLES

All of the samples were based on polyamide 6,6. Natural color and six different colored samples were made by UL's Performance Materials division from natural color PA66 pellets and commercially available, polyamide-based standard color concentrate pellets. Both the natural color and colored samples were melt blended (nominally 1 wt% color concentrate loading) the same way to minimize chemistry variation introduced by the blending process. Four different molecular weight samples were made by UL's Performance Materials division by melt blending different ratios of two PA66 materials. All four molecular weight study samples were processed the same way to minimize chemistry variation introduced by the blending process. Samples and descriptions are summarized in Table 1.

Table 1: Sample ID and descriptions.

Study	Sample ID	Description
Color	160-1	Natural, NC
	160-2	White, WT
	160-3	Black, BK
	160-4	Red, RD
	160-5	Gray, GY
	160-6	Yellow, YL
	160-7	Blue, BL
Molecular Weight	152-1	$\overline{Mw}$ = 37.1 kDa
	152-2	$\overline{Mw}$ = 42.6 kDa (25:75 blend of 152-1 and 152-5)
	152-3	$\overline{Mw}$ = 45.3 kDa (50:50 blend of 152-1 and 152-5)
	152-4	$\overline{Mw}$ = 66.8 kDa

Three basic sample shapes were inmolded using a Shinwa Seiki S-80CN II injection molding machine:
1. UL 94 flame bars – 125 × 13 × 3.0 mm thick
2. ISO 527 Type 1a tensile bars – 4.0 mm thick (2 cavity, end-gate)
3. Plaques – 150 × 100 × 3.0 mm thick

No mold release agent was used during the formation of any samples.

EXPERIMENTAL

Molded samples were characterized for thermal degradation behavior (TGA) and material combustibility by limited oxygen index (LOI), UL 94 vertical flame test (UL 94), and microscale combustion calorimetry (MCC).

Thermogravimetric Analysis (TGA)

Thermal decomposition behavior of the samples was characterized using a TA Instruments Q500EGA model TGA following the test procedures outlined in ASTM D 3850 *Standard Test Method for Rapid Thermal Degradation of Solid Electrical Insulating Materials by Thermogravimetric Method (TGA)*. TGA test specimens weighing 20 ±0.5 mg[1] cut from the center of flame bars (no outer flame bar edge included in specimens) were heated from standard lab temperature to 850 °C at 20 K/min under nitrogen or air purge gas. One trial was conducted per sample by each purge gas.

Limited Oxygen Index (LOI) Tests

Limited oxygen indices of the samples were measured using a CSI Limited Oxygen Index measurement apparatus. All tests were conducted in accordance with ASTM D 2863 *Standard Test Method for Measuring the Minimum Oxygen Concentration to Support Candle-Like Combustion of Plastics (Oxygen Index)*. LOI test specimens (80×10×4 mm) were cut from the middle section of the molded ISO 527 Type 1a tensile bars and conditioned at standard laboratory conditions (23 °C and 50% r.h.) for at least 40 hours prior to testing. Three replicates were conducted per sample.

UL 94 Flammability Tests

UL 94 flammability ratings of the samples were evaluated in accordance with UL 94 50W Vertical Burning Test. Ten test specimens per sample were preconditioned at each of the two prescribed conditions prior to evaluation:

a) "As-Received": a minimum of 48 hours at 23 ±2 °C and 50 ±5% r.h.
b) "Oven-Aged": in an air-circulating oven for 168 ±2 hours at 70 ±2 °C and then cooled for at least 4 hours in a desiccator at 23 ±2 °C and <20 % r.h.

One end of a vertically oriented test specimen was exposed to a 20 mm high blue (methane) flame for 10 seconds. The flame was removed and the time for the afterflame to cease, t_1, was measured. The flame was then reapplied for an additional 10 seconds and removed. The time for the second afterflame to cease, t_2, and the time for afterglow to cease, t_3, was measured. The test was conducted 300 mm above absorbent cotton.

Microscale Combustion Calorimetry (MCC)

Combustion characteristics of the samples were measured using Govmark MCC-3 Microscale Combustion Calorimeter units. All tests were conducted in accordance with either ASTM D 7309 Method A or Method B. Molded sample test specimens weighing 2.10 ±0.10 mg were cut from flame bars (no outer flame bar edge included in specimens) with one specimen harvested per flame bar. Specimen laden sample pans were heated in the MCC's precision controlled furnace at a rate of 0.8 °C per second to 900 °C under nitrogen (Method A) or dry 20:80 (v/v) oxygen-nitrogen atmosphere (Method B). The combustor temperature was set at 900 °C and fed with dry 20:80

[1] The sample mass of gray sample tested under air was decreased to 11 mg due to several failed attempts using the greater mass.

(v/v) oxygen-nitrogen gas mixture. Five replicates were conducted for each sample in each method.

RESULTS

Thermogravimetric Analysis (TGA)

Overlays of the TGA weight loss profiles for the seven color and four molecular weight samples tested in both nitrogen and air environments are shown in Figure 1.

Figure 1: TGA measured thermal degradation behavior in nitrogen (top row) and dry air (bottom row).

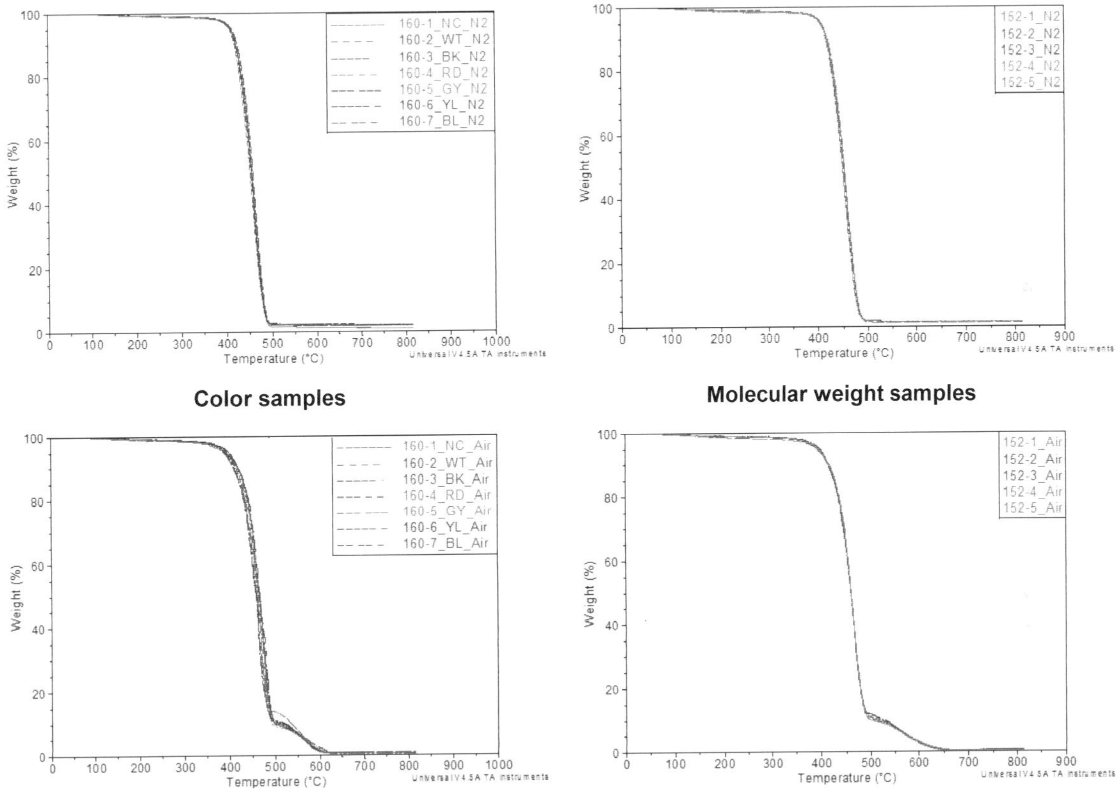

Limited Oxygen Index (LOI) Tests

LOI tests results for the seven color and four molecular weight samples are plotted in Figure 2.

Figure 2: LOI values for color and molecular weight samples. Error bars are s.d.

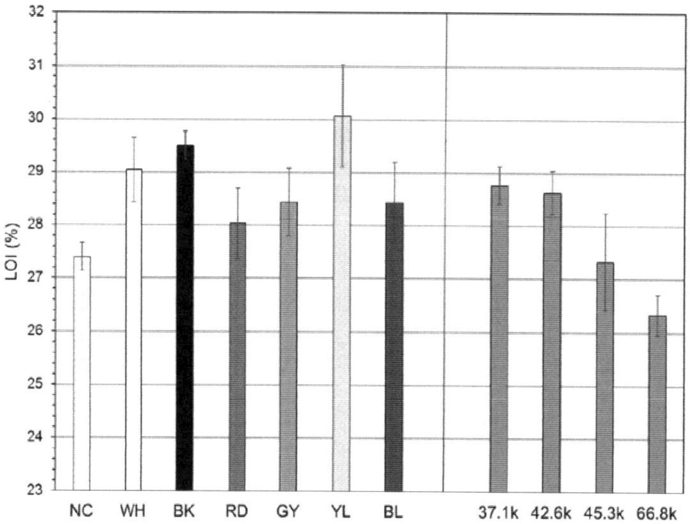

UL 94 Flammability Tests

UL 94 flame ratings of the seven color and four molecular weight samples were determined from UL 94 Table 8.1. Resulting flame ratings and average afterflame burn times for the ten replicates per precondition are listed in Table 2.

Table 2: UL 94 flame ratings and afterflame times for color and molecular weight samples.

Sample	UL 94 Flame Rating	"As-Received" Afterflame Time (s)			"Oven-Aged" Afterflame Time (s)		
		t_1	t_2	$t_1 + t_2$	t_1	t_2	$t_1 + t_2$
160-1 **NC**	V-2	2.8 ±2.0	2.1 ±1.4	4.9 ±2.2	5.6 ±4.0	8.5 ±10.2	14.1 ±12.9
160-2 **WT**	Fail	3.6 ±3.1	1.7 ±0.9	5.3 ±3.0	4.1 ±3.7	11.4 ±13.6	15.5 ±17.0
160-3 **BK**	V-2	1.0 ±0.0	2.0 ±1.2	3.0 ±1.2	2.5 ±2.5	2.4 ±1.3	4.9 ±2.6
160-4 **RD**	V-2	1.7 ±1.9	2.2 ±1.3	3.9 ±2.8	4.0 ±4.0	2.7 ±3.4	6.7 ±7.2
160-5 **GY**	Fail	4.0 ±3.5	3.9 ±2.9	7.9 ±5.8	6.4 ±6.4	**14.5 ±12.3**	**20.9 ±13.4**
160-6 **YL**	V-2	1.1 ±0.3	2.3 ±0.5	3.4 ±0.5	3.6 ±3.6	1.8 ±1.9	5.4 ±5.2
160-7 **BL**	V-2	2.7 ±2.6	3.4 ±1.6	6.1 ±4.1	1.0 ±1.0	1.1 ±0.3	2.1 ±0.3
152-1 **37.1k**	V-2	1.2 ±0.4	2.4 ±1.3	3.6 ±1.2	8.3 ±1.5	10.5 ±13.5	18.8 ±13.6
152-2 **42.6k**	Fail	1.5 ±1.0	2.7 ±1.3	4.2 ±1.7	13.1 ±5.9	**23.3 ±12.4**	**36.4 ±12.2**
152-3 **45.3k**	Fail	1.1 ±0.3	3.3 ±2.5	4.4 ±2.4	**10.1 ±9.1**	**10.7 ±14.0**	**19.7 ±15.4**
152-5 **66.8k**	V-2	3.0 ±2.7	3.6 ±1.2	6.6 ±2.4	2.5 ±2.3	6.7 ±4.6	9.2 ±4.7

Note: Values in **bold red** are minimum values

Microscale Combustion Calorimetry (MCC)

Overlays of the MCC heat release rate profiles for the seven color and four molecular weight samples measured by Methods A and B are shown in Figure 3.

Figure 3: MCC measured heat release rate profiles by Method A (top row) and Method B (bottom row).

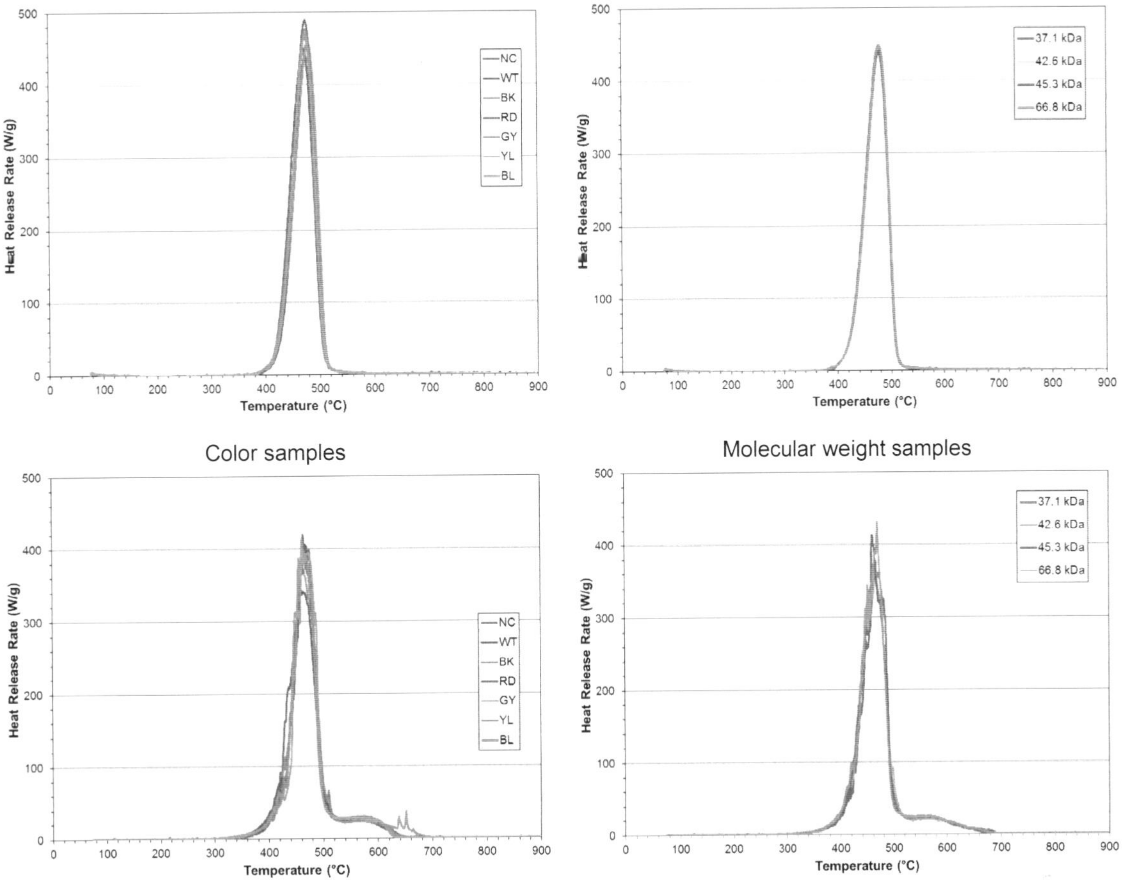

ANALYSIS OF COLOR INFLUENCE

TGA degradation behavior of the seven color samples were very similar under the nitrogen atmosphere and similar under the dry air atmosphere, Figure 1. A small deviation was observed for the GY sample tested under dry air atmosphere in that it exhibited approximately 4 % less weight loss during the main degradation stage than the other colors.

Analysis of the temperatures at the onset of degradation and temperatures at 3%, 5%, 10%, and maximum weight loss, Figure 4, revealed that under the nitrogen atmosphere, the WT, GY, and YL samples consistently degraded at lower temperatures than the NC sample with the WT sample being the most significantly different. The BK and RD samples degraded at comparable temperatures to the NC sample, sometimes lower, sometimes higher. Only the BL sample consistently degraded at higher temperatures than the NC sample. In contrast to the nitrogen atmosphere test results, only the BK sample degraded at higher temperatures than the NC sample under dry air atmosphere. All of the other colors degraded at lower temperatures with GY and YL being the most significant.

All of the colors yielded greater residue than the NC sample under the nitrogen atmosphere (1.8 to 2.4 % versus 0.9 % for NC). The more complete combustion expected for testing under a dry air atmosphere resulted in less residue than under the nitrogen atmosphere, less than 50% of residue observed under the nitrogen atmosphere. In contrast to the nitrogen atmosphere test results, only the BK sample yielded less residue than the NC sample under the dry air atmosphere.

Figure 4: Temperatures at degradation onset and at 3%, 5%, 10%, and maximum weight loss for color samples under nitrogen (left) and dry air (right) atmospheres.

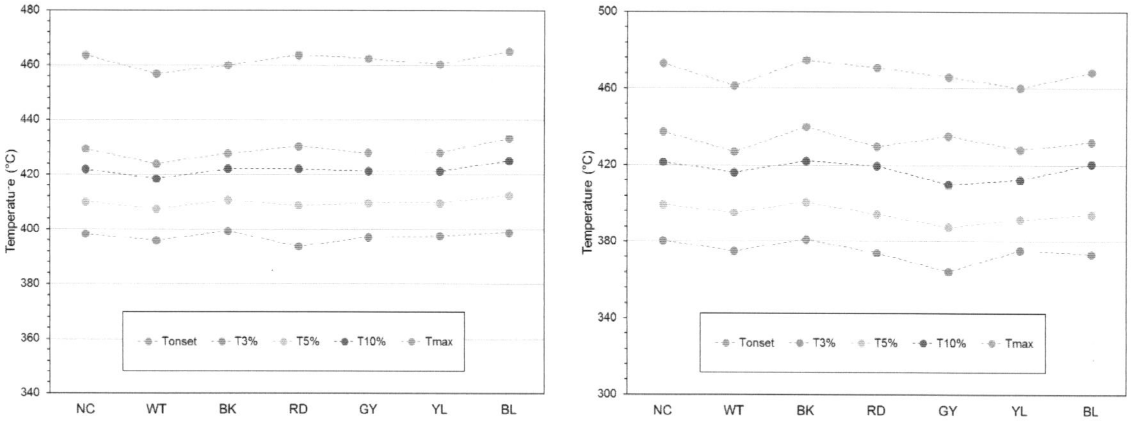

Addition of color to the NC PA66 was found to result in greater LOI values than for the NC sample, Figure 2. Rank ordering the seven PA66 samples based on LOI indicated YL, BK and WT samples were the most significantly impacted. T-test analysis (two-tailed, homoscedastic Student's t-test at α of 0.05) of the results, Table 3, indicated the samples with the three greatest LOI values (WT, BK, and YL) were statistically significantly greater than for the NC samples. Furthermore, the two highest LOI samples, BK and YL, were also statistically significantly greater than for the second lowest LOI sample, RD. Based on the vastly different LOI results for the two samples with the lowest noncombustible content, BK and NC, the amount of noncombustible content did not appear to be the source of observed LOI differences.

Table 3: T-Test results comparing LOI values measured for different color samples. Values in bold red are significant at an alpha value of 0.05.

Sample	WT	BK	RD	GY	YL	BL
160-1 **NC**	**0.013**	**0.001**	0.201	0.060	**0.010**	0.088
160-2 **WT**		0.287	0.126	0.301	0.190	0.341
160-3 **BK**			**0.024**	0.055	0.380	0.081
160-4 **RD**				0.493	**0.039**	0.528
160-5 **GY**					0.070	1.000
160-6 **YL**						0.081

As expected, color was found to influence UL 94 flammability test results, Table 2. The two materials with the most residue (by TGA in dry air and MCC Test Method B), GY and WT, had the worst flammability performance and did not meet the requirements for the least conservative UL 94 vertical burning flammability rating of V-2. The GY color samples consistently had the longest afterflame times. The other five color samples resulted in the same V-2 UL 94 flammability rating. The BK samples consistently had shorter afterflame times than the NC samples which was somewhat surprising considering that the color black has greater emissivity than the light color of the NC sample.

The effect of color on UL 94 flammability performance, however, differed for the "As-Received" and "Oven-Aged" samples. Analysis of T-test results (two-tailed, homoscedastic Student's t-test at α of 0.05) comparing the respective afterflame times measured for the different colors under the two conditions indicated that color was slightly more influential on "As-Received" t_1 afterflame times than "Oven-Aged" afterflame times but more influential on the other afterflame times in the "Oven-Aged" state. Seven color sample combinations did not have statistically different afterflame burning times regardless of the condition:

NC and WT	BK and RD
WT and GY	RD and YL
GY and NC	YL and BK
NC and RD	

Interestingly, the NC, WT, and GY samples were statistically indistinguishable yet only the NC sample met the afterflame requirements for a V-2 rating suggesting it was a borderline passing material for a V-2 UL 94 rating. Individual test results for these three samples (one NC test specimen failed versus two test specimen failures for each WT and GY) supported this conclusion.

Combustion behavior of the seven color samples were very similar to each other as measured by the two MCC test methods, Figure 3. Method A resulted in a single peak whereas Method B resulted in a main peak and a trailing higher temperature peak of lower intensity. The higher temperature peak observed in the Method B results is consistent with burning char. Small heat release rate spikes were observed in the "char" peak of some BK samples tested by Method B. These spikes were suspected to result from brief flaming combustion of the char.

The six colored samples had lower temperatures at peak heat release rate, lower heats of combustion, and more residue than the NC sample by Method A. These differences were statistically significant for every color with the exception of the temperature at peak heat release rate for the BK sample (478.3 °C versus 478.6 °C for NC). Temperatures at combustion onset were within 5 °C for all seven PA66 samples. WT, GY, and YL samples had lower temperatures at combustion onset than the NC samples while the other three colors, BK, RD, and BL had greater temperatures. The rank order of these results closely paralleled the rank order of temperatures at 10% weight degradation ($T_{10\%}$) measured by TGA under nitrogen. Although the observed temperatures at combustion onset for the colored samples were nominally different than for the NC sample, only the temperature at combustion onset for the BL sample was statistically significantly different from the NC sample (415.8 °C versus 413.2 °C for NC). Peak heat release rates of the WT and BK samples were lower than the NC sample whereas the other colors were greater. Of these, only the YL, BL, and RD samples, were statistically significantly different from the NC sample. In summary, no two color samples were the same at a statistically significant level for all five assessed characteristics.

Beyond the similar trends in the temperature at combustion onset and the TGA under nitrogen $T_{10\%}$ measurements, comparison of the resulting key MCC characteristics to LOI and UL 94 results did not suggest any strong correlations. LOI results indicated the NC and YL samples were the extreme performers. This behavior was paralleled by MCC heat of combustion results, however, the LOI and heat of combustion results for the other samples were considerably different. For example, the BK sample had an LOI value similar to the YL yet the BK heat of combustion was statistically different from the YL sample and statistically indistinguishable from the NC sample. MCC temperatures at combustion onset grossly follow t_1 afterflame times measured for "Oven-Aged" UL 94 samples, however, results for the NC sample are completely incongruous suggesting the correlation is tenuous at best.

In contrast to the absence of any general trends observed for the Method A measured MCC characteristics, Method B results for these formulations suggested an inverse relationship between temperature at combustion onset and residue, Figure 5. Additionally, the WT sample tended to have more extreme results as compared to the other samples. It had the lowest temperature at combustion onset, the lowest peak heat release rate and corresponding temperature, and heats of combustion and residue that were on the higher end.

The six colored samples had lower temperatures at combustion onset than the NC sample. These differences were statistically significant for every color with the exception of the BK sample (394.3 °C versus 397.3 °C for NC). As observed for the Method A results, the rank order of Method B temperatures at combustion onset closely paralleled the rank order of temperatures at 10% weight degradation ($T_{10\%}$) measured by TGA under dry air.

The NC sample exhibited greater peak heat release rate than all of the colored samples except the BL sample (419.7 W/g versus 417.4 W/g for NC). Of these, only the WT, BK, and YL samples were different at a statistically significant level. Temperatures at peak heat release rate were within 14 °C for all seven PA66 samples. RD and GY samples had greater temperatures at peak heat release rate than the NC samples, BL samples was within 0.1 °C of the NC samples, and the other colors, BK, WT, and YL had lower temperatures. Although the observed temperatures at peak heat release rate for the colored samples were nominally different than for the NC sample, only the temperature at peak heat release rate for the WT sample was statistically significantly different from the NC sample (457.8 °C versus 466.6 °C for NC). Heats of combustion for the YL and BL samples were lower than the NC sample whereas the other colors were equal or greater. Of the six colored samples, only the lowest heat of combustion material, the YL sample, was statistically significantly different from the NC sample. In summary, no two color samples were the same at a statistically significant level for all five assessed characteristics.

In contrast to the Method A MCC results which did not parallel LOI results, there appears to be a weak inverse correlation between peak heat release rates measured by Method B and LOI. No MCC characteristic trended similarly to UL 94 ratings or afterflame times.

Figure 5: Inverse relationship between residue and temperature of combustion onset as measured by MCC Method B. Error bars are s.d.

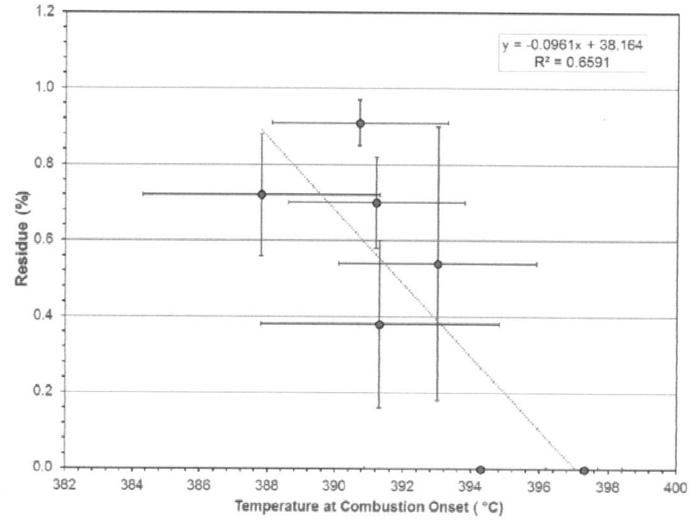

ANALYSIS OF MOLECULAR WEIGHT INFLUENCE

TGA degradation behavior of the molecular weight samples in both nitrogen and dry air atmospheres were respectively similar, Figure 1. Analysis of the temperatures at the onset of degradation and temperatures at 3%, 5%, 10%, and maximum weight loss, Figure 6, did not indicate any clear molecular weight influences on degradation rate under either nitrogen or dry air atmospheres. At best, the temperature at degradation onset in dry air was weakly related to molecular weight:

$$\text{Tonset (DA)} = 0.1325\ \text{Mw} + 424.86,\ r^2 = 0.7672$$

Residue after TGA testing was comparable for the different molecular weight samples under the respective atmosphere and, as expected, testing under dry air resulted in less residue (~1.45 % for nitrogen versus ~0.5 % for dry air).

Figure 6: Temperatures at degradation onset and at 3%, 5%, 10%, and maximum weight loss for molecular weight samples under nitrogen (left) and dry air (right) atmospheres.

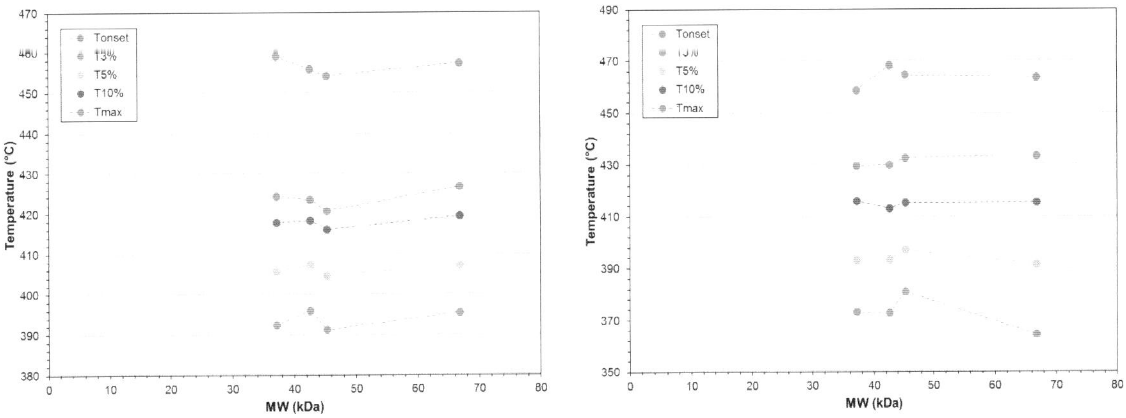

In contrast to thermal degradation, molecular weight was found to have an inverse linear relationship with LOI as seen in Figure 7. The relationships between LOI and weight-average molecular weight were:

$$\text{LOI} = -0.0813\ \text{Mw} + 31.665,\ r^2 = 0.842$$

T-test analysis (two-tailed, homoscedastic Student's t-test at α of 0.05) of the results indicated the LOI values measured for the worst LOI performer, 66.8k, was statistically significantly lower than the two lowest molecular weight samples, 37.1k and 42.6k (25:75 blend).

Figure 7: LOI as a function of weight-average molecular weight. Error bars are s.d.

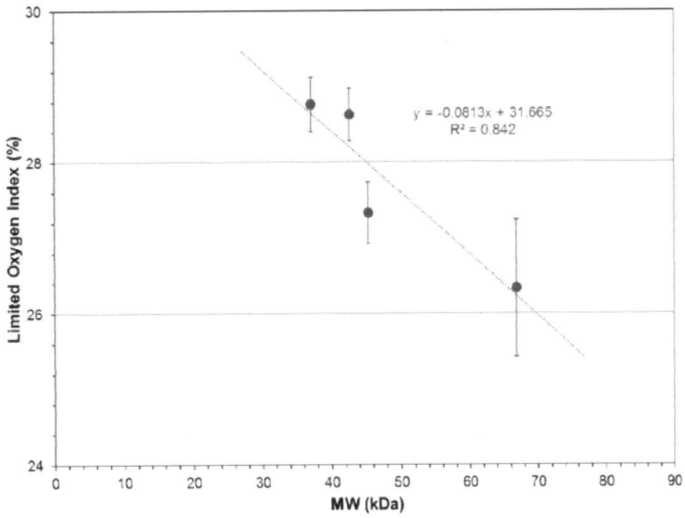

The influence on UL 94 vertical flame test ratings was unclear at best for these molecular weight samples. Both of the base samples, 37.1k and 66.8k, yielded the literature published V-2 ratings but the two blends did not meet the requirements for the least conservative UL 94 vertical burning flammability rating of V-2. Because the molecular weights of the two base samples bracketed the molecular weights of the blends, it was suspected that the diminished flammability performance of the blends was related to the mixing of a broader range of polymer chain lengths (molecular weights).

UL 94 afterflame times give some hint at possible molecular weight influences,
Figure 8. The "As-Received" preconditioned samples in particular exhibited a linear relationship between molecular weight and afterflame time:

$$\text{"As-Received" } t_1 \text{ (s)} = 0.0641 \text{ Mw} - 1.372, \ r^2 = 0.8923$$
$$\text{"As-Received" } t_2 \text{ (s)} = 0.0366 \text{ Mw} + 1.2439, \ r^2 = 0.7581$$
$$\text{"As-Received" } t_1{+}t_2 \text{ (s)} = 0.1007 \text{ Mw} - 0.1281, \ r^2 = 0.9995$$

Although the relationship with molecular weight was positive for afterflame time whereas the negative relationship observed for LOI behavior, the net impact on flammability was in the same direction: greater flammability with higher molecular weight.

Figure 8: UL 94 afterflame times for "As-Received" (left) and "Oven-Aged" (right) samples as a function of weight-average molecular weight. Error bars are s.d.

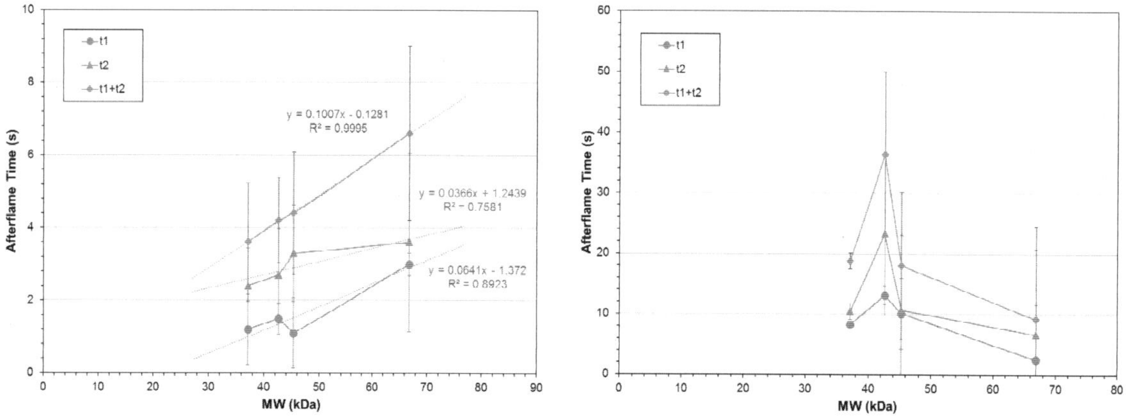

For the most part there was good correlation between the t_1 and t_2 afterflame times such that samples exhibiting shorter t_1 afterflame times also exhibited shorter t_2 afterflame times. The one exception to this trend was the 50:50 blend (45.3k) in the "As-Received" condition where the blend exhibited the shortest average t_1 afterflame time of all the samples but the second longest average t_2 afterflame time.

Comparison of the T-Test results for the three afterflame times revealed that the "Oven-Aged" condition was more differentiating than the "As-Received" condition. In the "As-Received" condition the only material with afterflame times that differed at a statistically significant level (two-tailed, homoscedastic Student's t-test at α of 0.05) from the other samples was the highest molecular weight sample, 66.8k, whereas in the "Oven-Aged" condition all of the sample combinations except the lowest molecular weight (37.1k) and the 50:50 blend (45.3k) differed for at least one of the afterflame times.

Although the afterflame times of the lowest molecular weight (37.1k) and the 50:50 blend (45.3k) samples were statistically indistinguishable, the lowest molecular weight sample met the afterflame requirements for a V-2 rating whereas the 50:50 blend did not. This suggested the lowest molecular weight sample was a borderline passing material for a V-2 UL 94 rating. Individual test results for these samples supports this conclusion (one lowest molecular weight test specimen failed versus three 50:50 blend test specimens).

Combustion behavior of the four molecular weight samples were very similar to each other as measured by the two MCC test methods, Figure 3. Method A resulted in a single peak whereas Method B resulted in a main peak and a trailing higher temperature peak of lower intensity. The higher temperature peak observed in the Method B results was consistent with burning char.

MCC characteristics for the five samples measured by Method A were remarkably consistent suggesting any potential molecular weight effect to be minimal, at least for this polymer type and molecular weight range. Temperatures at combustion onset were within 3.5 °C for all the samples, peak heat release rates within 7.5 W/g, temperatures corresponding to the peak heat release rates within 2 °C, heats of combustion within 0.2 kJ/g, and residue within 0.1 %.

The lowest molecular weight sample (37.1k) had the highest peak heat release rate (by at least 3 W/g) and temperature at combustion onset (by at least 2 °C) and second highest temperature corresponding to the peak heat release rate, Figure 9 through Figure 11. In contrast, the highest molecular weight sample (66.8k) exhibited the lowest temperatures at combustion onset and at the peak heat release rate and the second lowest heat release rate. Of these MCC characteristics, only the differences in temperature at combustion onset were statistically significant. The temperatures at combustion onset exhibited by the lowest molecular weight sample (37.1k) were different from all of the other samples at a statistically significant level. Heats of combustion for the four molecular weight samples were not different at a statistically significant level nor were the residue with the exception of the highest molecular weight sample (66.8k) from the 25:75 blend (42.6k). In summary, MCC testing by Method A did not differentiate these molecular weight samples.

As with the MCC results measured by Method A, MCC Method B measured results were minimally influenced by molecular weight, if at all. Temperatures at combustion onset were within 3.0 °C for all the samples, temperatures corresponding to the peak heat release rates within approximately 6 °C, heats of combustion within 1.0 kJ/g, and residue was negligible at no greater than 0.03 %. The range of measured peak heat release rates was greater by Method B than Method A but all of the test specimens were within 46 W/g,

The lowest molecular weight sample (37.1k) tended to yield the same type of extremes for various MCC characteristics by Method B as by Method A. It had the highest temperature at combustion onset (by at least 1.5 °C) and the highest peak heat release rate (by at least 11 W/g), Figure 9 through Figure 11. The 25:75 blend (42.6k) lowest peak heat release rate and temperature at combustion onset by Method B in contrast to the highest molecular weight sample (66.8k) that did by Method A. The differences in these three MCC characteristics for the four samples were not statistically significant with the exception of the peak heat release rates of the 25:75 blend (42.6k) versus the lowest and highest molecular weight samples (37.1k and 66.8k respectively). The heats of combustion for the three lower molecular weight samples were not statistically distinguishable, however, the highest molecular weight sample (66.8k), which exhibited the highest heat of combustion (by 0.6 kJ/g), was statistically significantly different from the other samples. Of the four samples, only the two which yielded the high and low residue extremes, the lowest molecular weight (37.1k) and 25:75 blend (42.6k) samples respectively, were different at a statistically significant level. In summary, no one sample differed from the other samples for all five assessed characteristics at a statistically significant level. More importantly, many combinations of samples were statistically indistinguishable by all five MCC characteristics suggesting MMC Method B testing to also be unsuitable for distinguishing different molecular weight materials.

Comparison of the MCC Methods A & B results to TGA results did not reveal the same correlation between MCC measured temperatures at combustion onset and TGA measured temperatures corresponding to a prescribed weight loss as observed for colored PA66 samples. Surprisingly the rank order of residues measured by MCC Method A was the exact opposite of that observed for TGA in nitrogen, though this may have been coincidental given the non-significant distinction between MCC measured residues.

Beyond the common trends with TGA, comparison of the resulting key MCC characteristics to LOI and UL 94 results did not suggest any strong correlations. LOI results indicated a clear molecular weight effect but neither MCC method did. None of the UL 94 afterflame time rank orders paralleled any of the MCC measured properties. Interestingly, the two samples that did not meet the requirements for a UL 94 V rating exhibited the lowest heats of combustion by MCC, granted the MCC measured heats of combustion for the different molecular weight samples did not differ at statistically significant levels.

Figure 9: Temperatures at combustion onset (left) and peak heat release rate (right) measured by MCC Methods A and B as a function of weight-average molecular weight. Error bars are s.d.

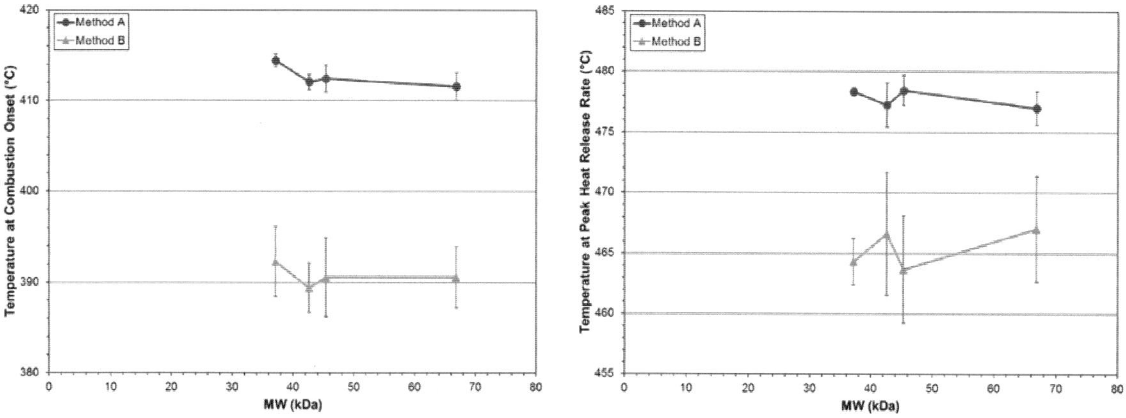

Figure 10: Peak heat release rates (left) and heats of combustion (right) measured by MCC Method A and B as a function of weight-average molecular weight. Error bars are s.d.

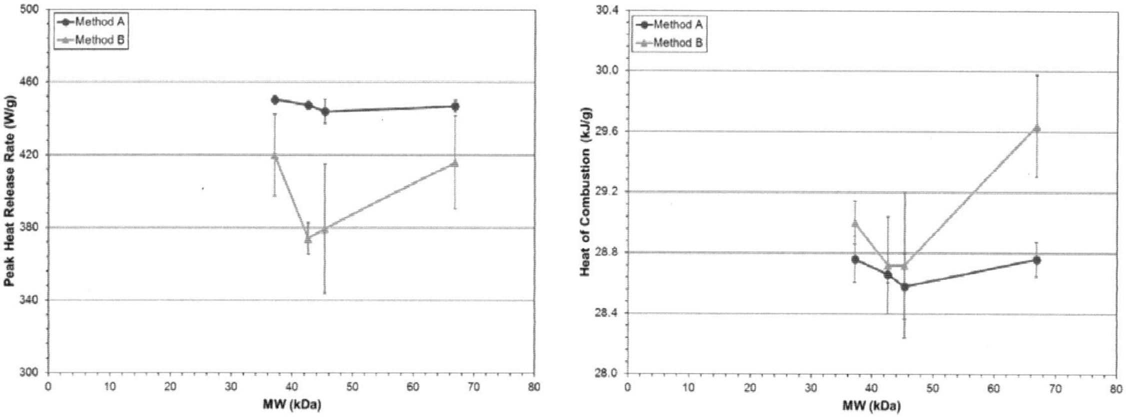

Figure 11: Residue measured by MCC Method A and B as a function of weight-average molecular weight. Error bars are s.d.

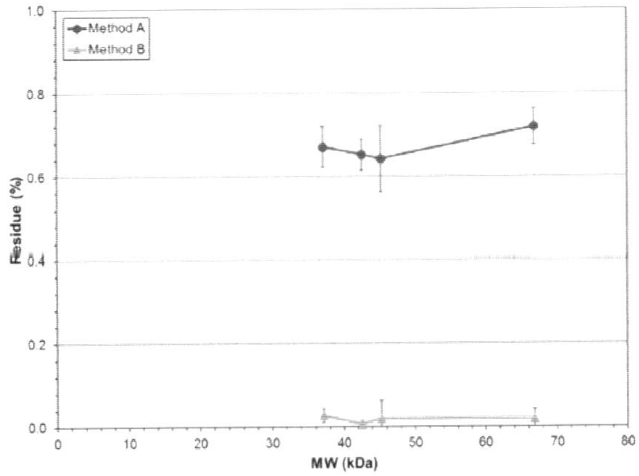

SUMMARY

Two series of polyamide 6,6 were investigated to explore the influence of colorant and polymer molecular weight on thermal degradation and flammability properties. A commercially available natural color PA66 and six different colored samples made from the natural color PA66 pellets and commercially available, polyamide-based standard color concentrate pellets constituted the colorant influence study while two commercially available PA66 of differing molecular weight along with two differing blends of the respective base PA66 polymers constituted the molecular weight influence study. Although the study samples were limited to a single condensation polymer chemistry, PA66, six colorants using the same carrier chemistry, and four molecular weights spanning approximately 37 to 67 kDa, several important findings expected to translate to other polymer systems were made:

1. Few correlations between the different test methods and measured flammability properties were observed.
2. Consistent with materials that absorb moisture, UL 94 afterflame times were generally longer for "Oven-Aged" samples than the "As-Received" samples. However, the differences in afterflame times were not always statistically significant.
3. MCC measurements by the two ASTM standard defined methods were different. Heat release rate profiles measured by Method A (anaerobic conditions) resulted in a single peak whereas Method B (aerobic conditions) resulted in a main peak and a trailing higher temperature peak of lower intensity corresponding to char combustion. Temperatures at combustion onset were higher by Method A, peak heat release rates were lower by Method A and the corresponding peak temperatures were higher. Heats of combustion were similar for the two methods but typically lower for Method A and residue was greater for Method A.
4. Color was found to have a greater influence than polymer molecular weight for properties measured at molecular level such as thermal degradation (TGA) and microscale combustion calorimeter (MCC).
5. Color influenced the measured flammability properties at statistically significant levels, however, the degree of influence varied for the different properties and varied by color.
6. Molecular weight was more influential for test methods in which sample flow was possible such as UL 94 and LOI than molecular level tests.

A METHODOLOGY FOR PREDICTING AND COMPARING THE FULL-SCALE FIRE PERFORMANCE OF SIMILAR MATERIALS

Chad M. Lannon
Holmes Fire LP., San Francisco, USA

Stanislav I. Stoliarov
University of Maryland, Department of Fire Protection Engineering, College Park, USA

James M. Lord
Bureau of Alcohol, Tobacco, Firearms and Explosives, Fire Research Laboratory, Ammendale, USA

Isaac T. Leventon
National Institute of Standards and Technology, Fire Research Division, Gaithersburg, USA

ABSTRACT

In some cases, fire investigators may perform reconstructive fire testing to identify the cause, origin, and progression of a particular fire. Accurate reconstruction of the fire requires the laboratory structure to be outfitted with building materials that perform similarly to the original materials found in the field. Therefore, a method was developed to assist fire investigators with selecting building materials for reconstructive fire testing based on the modeled performance of materials recovered from the fire scene. This method requires milligram- and gram-scale testing to determine material properties and a numerical model that is used to interpret the test results and simulate the room corner scenario, which represents full-scale material performance. The model consists of a two-dimensional pyrolysis solver, ThermaKin2D, coupled with an empirical flame heat feedback model, which was developed in this study. A criterion was also developed that characterizes relative fire performance of each material during this full-scale fire scenario. This method was validated using full-scale room corner tests. Overall, this method is expected to improve a fire investigator's ability to perform accurate fire reconstruction, and to assist them in quantifying the uncertainty in their test results.

INTRODUCTION

An effective method in determining the cause, origin and progression of a fire is to perform reconstructive fire tests. A reconstructive fire test strives to recreate a fire scenario by building a full scale replica of the structure in a laboratory, and using it to test hypotheses. When building these structures, it is often necessary to substitute materials that can be purchased at building supply stores for the original materials that were found during the fire investigation because the original materials were either destroyed or there is an insufficient quantity remaining to conduct large scale tests. Accurate reconstructive fire testing requires these materials to behave similarly to the original materials during the likely fire scenario. Typically, materials are chosen for use in reconstruction tests based on the information available about the construction of the building, or through post-fire visual examination of the remaining building materials. The fire forensic field needs a quantitative methodology for comparing various materials and predicting how they perform relative to one another in a full-scale fire scenario so that substitute materials can be better selected for reconstructive fire testing.

The goal of this project was to develop a systematic procedure that allows fire investigators to obtain the material properties of several similar materials so that they can be implemented into a model which predicts and compares their performance in a full-scale fire scenario. The full-scale fire scenario chosen for this project was the NFPA 286 room corner test [1]. The room corner test is a standard full-scale test used to evaluate the flammability of combustible walls and or ceiling materials and measure their overall contribution to the fire growth during a compartment fire.

A two dimensional pyrolysis model, ThermaKin2D was chosen to simulate upward flame spread along the corner-walls during the room corner test. ThermaKin2D has been designed to simulate milligram-scale decomposition and bench-scale gasification, as well as flame spread [2]. This model requires an extensive set of material properties. Therefore, a procedure was developed in this work to obtain these properties through inverse modeling of milligram- and bench-scale tests. Tests were also performed in an open-corner wall assembly to characterize the heat feedback from a spreading corner-wall fire and a fire produced by the room corner ignition source. A flame model was developed to represent gas-phase heat feedback from these two aspects of the room corner test and coupled with ThermKin2D as a surface boundary condition to produce a unified model of material degradation and upward flame spread. The ThermaKin2D room corner model was then validated against the room corner tests and criteria were established to compare performance of similar materials.

Material Description

Medium density fiberboard (MDF), an engineered wood product, was selected for detailed study in this work because it is widely used in both modern and legacy construction and thus, the Bureau of Alcohol, Tobacco, Firearms and Explosives Fire Research Laboratory (ATF FRL) has identified it as common material utilized during reconstructive fire testing. Four types of MDF wall paneling were selected from building supply stores. Each MDF sample has the same trade name and similar composition, but they are produced by three different manufactures in varying thickness (τ), densities (ρ), and finish. Gypsum wallboard was also used in this study because NFPA 286 requires the base surface of the walls, ceiling and floor to be composed of gypsum wallboard [1]. The τ of Sample A, B, C, D and the gypsum wallboard was 6.2×10^{-3} m, 4.1×10^{-3} m, 3.6×10^{-3} m, 3.2×10^{-3} m and 12.6×10^{-3} m, respectively. The ρ of Sample A, B, C and D was 1089 kg m^{-3}, 889 kg m^{-3}, 889 kg m^{-3}, 1080 kg m^{-3} and 481 kg m^{-3}, respectively.

Modeling

A two-dimensional pyrolysis model, ThermaKin2D, was utilized to obtain the material properties of the four samples and predict their full-scale performance during the room corner test. ThermaKin2D is a flexible numerical solver that was designed to simulate the gaseous fuel production of a pyrolyzing material subjected to heat. ThermaKin2D is able to simulate one and two dimensional pyrolysis by solving mass and energy conservation equations, while conservation of momentum is introduced implicitly. In ThermaKin2D, reactants and their products are represented by components; each component is characterized by physical, thermal, optical and chemical properties. Each component must be specified as either a solid, liquid or a gas. These components undergo chemical reactions, which may occur in series or in parallel. Physical structures simulated in ThermaKin2D are rectangular and defined by initial temperature, material composition, length, and thickness. A highly flexible thermal boundary condition formulation implemented in ThermaKin2D provides a capability to construct empirical flame heat feedback submodels within the framework of the pyrolysis model. A detailed description of the ThermaKin2D governing equations and flame submodel formulation can be found in [2, 3].

MATERIAL PROPERTY DETERMINATION PROCEDURE

During decomposition, the virgin sample may undergo many reactions that produce various species, including condensed-phase residue in the form of char. All of these species have material properties that characterize heat and mass transport through the sample. The relevant chemical reactions and their respective properties are the pre-exponential factor (A), activation energy (E), heat of effective combustion of gaseous volatiles (ΔH_{eff}), stoichiometric yield (θ), and heat of reaction (h), the thermal properties are ρ, heat capacity (c), and thermal conductivity (k) and the optical properties are the absorption coefficient of radiation (α) and emissivity (ε). In order to simulate decomposition through pyrolysis modeling, these properties must be specified for all species involved in each reaction.

A procedure was developed that used widely available standard milligram and bench-scale experiments to determine all of these properties for each MDF sample. The micro-scale combustion calorimeter (MCC) was the milligram-scale experiment selected because it enables the user to obtain the maximum amount of properties while requiring the least amount of material [4]. The MCC test decouples condensed-phase pyrolysis and gas-phase combustion and measures heat release rate (HRR) based on oxygen consumption. MCC testing was used to determine each sample's char yield with respect to the initial virgin sample (θ_C). A model was constructed in ThermaKin2D to simulate the MCC test and an optimization procedure was performed where the kinetic parameters; A and E were manually changed until the models predictions match the experimental HRR results. The cone calorimeter was the bench-scale experiment used to determine each sample's ΔH_{eff} and char density (ρ_C). Testing occurred at a range of incident heat flux conditions (q''_{inc}) and heat release rate per unit area (HRRPUA) was measured. A cone calorimeter model was also developed in ThermaKin2D and an optimization procedure was performed where the remaining thermal, optical, and chemical properties were manually changed until the models predictions matched all the experimental HRR results.

Milligram-scale - MCC Tests

Five MCC experiments were performed on each sample in accordance with the relevant standard [4]. During MCC testing, a 3.5×10^{-6} kg $\pm$ 0.15×10^{-6} kg sample was inserted into the pyrolyzer and linearly heated in a nitrogen environment at 1 K s^{-1}. The MCC measured the HRR normalized by the initial mass (HRR/m_0) as a function of the pyrolysis temperature. The sample's initial (m_0) and final mass (m_f) were measured by a microbalance to determine θ_c. θ_c for Sample A, B, C and D was 17.5 ± 3.1 %, 14.4 ± 6.3 %, 17.6 ± 0.4 % and 11.6 ± 2.4 %, respectively. The uncertainty was calculated as two standard deviations of the experimental mean ($2\sigma_{mean}$).

Milligram-scale - MCC Model

Current research methods define how MDF pyrolyzes by conducting thermogravimetric experiments and prescribing apparent reaction schemes with corresponding kinetic parameters that best fit the mass loss rate data [5, 6]. These semi-global reaction schemes represent many chemical reactions that occur during decomposition. One of the goals of this project was to determine the minimum complexity needed to accurately simulate MDF pyrolysis and thus reduce the number of variables needed to parameterize the pyrolysis model. Two reaction schemes were considered. The simplest reaction scheme is the 1 Reaction Model, shown in Equation 1. It defines MDF decomposition by a single, first order chemical reaction:

$$Sample \rightarrow \theta_C * Char + (1-\theta_C) * Gas \qquad [1]$$

Equation 2 presents the second reaction scheme (2 Reaction Model), where MDF decomposition is modeled by a series of two, consecutive, first order reactions. In this model, the first reaction produces a solid *Intermediate* component and gaseous volatiles. Then the *Intermediate* component undergoes a second reaction to produce solid char and more gaseous volatiles. The stoichiometric coefficients (θ_I) and (θ_{CI}) were defined such that the model's final char yield matched the MCC test results.

$$\begin{aligned} &Sample \rightarrow \theta_I * Intermediate + (1-\theta_I) * Gas_1 \\ &\qquad \Rightarrow Intermediate \rightarrow \theta_{CI} * Char + (1-\theta_{CI}) * Gas_2 \end{aligned} \qquad [2]$$

The MCC experiment is a thermally thin heating scenario, where the effects of heat and mass transport are minor and the rate of pyrolysis is controlled by the sample's chemical reactions and kinetic parameters: *A* and *E*. A thermally thin model was implemented in ThermaKin2D to represent this scenario and simulate sample pyrolysis according to a 1 K s^{-1} heating rate. Using the MCC model, each sample's kinetic parameters were determined independently for both reaction schemes by conducting an optimization procedure according to Li's technique [7]. Figure 1 compares experimentally measured and model predicted HRR/m$_0$ curve based on the final set of kinetic parameters for the 1 Reaction (solid line) and 2 Reaction Models (dashed line). The accuracy of the HRR/m$_0$ fit was calculated based on the coefficient of determination (R^2). For all MDF samples, the 1 Reaction Model was able to obtain a fit between 0.91 and 0.93, while the 2 Reaction Model obtained a fit between 0.96 and 0.99. Experimental data points represent an average of the five tests and error bars indicate $2\sigma_{mean}$. Overall, the 1 Reaction Model produce a good fit to the experimental data and the 2 Reaction Model produces a nearly perfect fit.

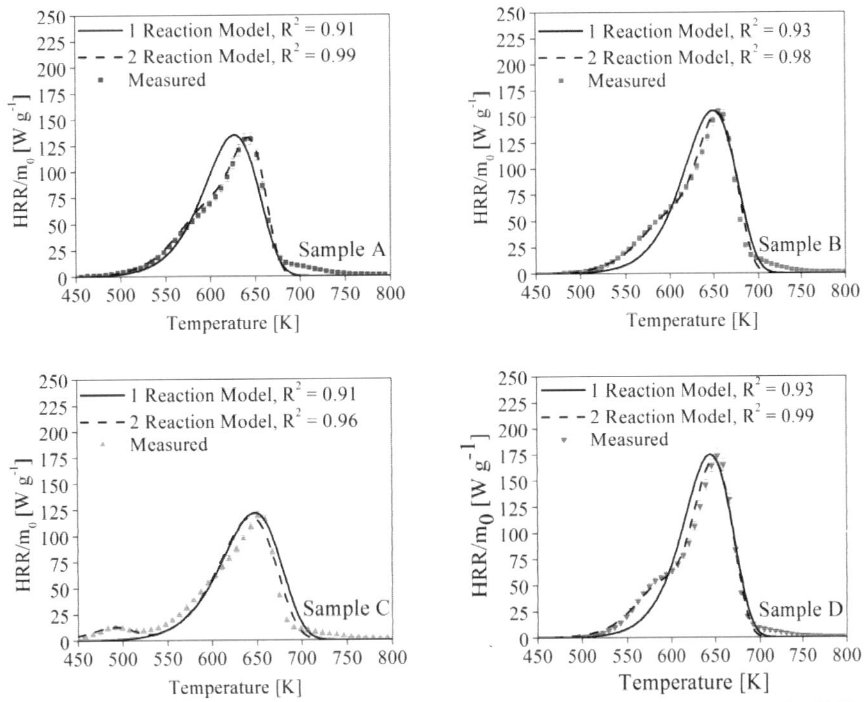

Figure 1. Experimentally measured & model predicted HRR/mo during MCC tests of MDF samples with a heating rate of 1 K s^{-1}

Bench-scale - Cone Calorimeter Tests

Cone calorimeter experiments were performed on each sample according to ASTM E1354 [8]. All samples were tested three times at a range of incident heat fluxes - q''_{inc} equal to 20 kW m^{-2}, 50 kW m^{-2}

and 80 kW m^{-2} which represents the thermal conditions materials may be subjected to during the room corner test. If the sample did not ignite at 20 kW m^{-2}, then it was tested at 25 kW m^{-2}. In bench-scale and full-scale testing conducted in this work, MDF samples were stapled to gypsum wallboard to replicate common installation practices. Stapling the MDF samples to the gypsum wallboard prevented swelling during cone calorimeter experiments and delamination from the walls during full-scale tests. A set of five equidistant staples joined the MDF samples and gypsum wallboard. Prior to cone calorimeter testing, the MDF-gypsum specimen, was placed in the sample holder, on top of a 3.0×10^{-2} m layer of insulation and encapsulated by an edge frame to prevent sample swelling. The exposed square surface area was 88.4×10^{-4} m^2. Preparing the sample in this manner was found to be nonintrusive compared to using a mesh wire grid, an alternative option provided in ASTM E1354.

Throughout testing, HRRPUA was measured based on the principle of oxygen consumption calorimetry. ΔH_{eff} was calculated as the total heat released divided by the total mass lost from ignition to the moment when flaming combustion became unstable and transitioned to smoldering combustion. ΔH_{eff} for Sample A, B, C and D was measured to be 11.3×10^6 J kg^{-1}, 11.7×10^6 J kg^{-1}, 10.3×10^{6}J kg^{-1} and 12.9×10^6 J kg^{-1}, respectively. After these experiments, the char mass was measured and its structure was geometrically analyzed to determine the ρ_c. The ρ_c for Sample A, B, C and D was 191 kg m^{-3}, 128 kg m^{-3}, 382 kg m^{-3} and 247 kg m^{-3}, respectively.

Bench-scale - Cone Calorimeter Model

Cone calorimeter experiments were simulated in ThermaKin2D using a one dimensional transport assumption. In the model, the top surface boundary condition prior to ignition was represented by a radiative q''_{inc} of 20, 25, 50 or 80 kW m^{-2} and convective heat loss due to air entrainment across the sample based on a convective heat transfer coefficient and ambient temperature of 10 W m^{-2} K^{-1} and 300 K. In order to mimic specimen setup, gypsum wallboard was located under the MDF and it's back surface was treated as an adiabatic boundary due to the insulation. Mass transport of the gaseous volatiles produced within the sample was specified to occur at a rate that would not inhibit their flow through the material. Once the mass loss rate reached the critical mass flux ($\dot{m}''_{crit}$) which was material dependent, ignition occurred and flaming combustion was simulated by specifying an additional radiative heat flux of 20×10^3 W m^{-2} to the top surface boundary [9]. After ignition occurs, it was assumed that the flames covered the entire surface and therefore the convective heat loss at the surface was turned off. The simulations were performed using the grid size and time step of 2.5×10^{-5} m and 0.005 s, respectively. When the grid size and the time step were increased and decreased by a factor of two, the model results did not change, thus indicating convergence.

This cone calorimeter model was used to perform an optimization procedure where the remaining material properties could be determined: k, c, ε and α, h and $\dot{m}''_{crit}$ for the virgin sample, intermediate component and char. Beginning the optimization procedure required an initial guess for the unknown material properties, as well as a realistic lower and upper bound for which each property could be altered. This optimization procedure was developed with the intention that it could be applied to any type of material, not just MDF. Therefore, the initial guess and adjustable range reported in Table 1 for all unknown properties was based on a comprehensive literature review of polymer material properties by Stoliarov et al. [10]. In this literature review, Stoliarov et al. reviewed 97 polymers and reported the average thermal, optical and chemical material property values, as well as the minimum and maximum range. The minimum range for k, ε, and h and maximum range for α were altered to better reflect the property values reported in literature for wood, lignocellulosic materials, charring polymers and MDF products [7, 11, 12, 13]. In this study, a positive value for h indicates that the reaction is exothermic.

The two other properties that were adjustable during this optimization procedure, but not based on Stoliarov's literature review were $\dot{m}''_{crit}$ and k of char. Equation 3 is an expression for $\dot{m}''_{crit}$ derived by Lyon and Quintiere [14].

$$\dot{m}''_{crit} = HRR^{*}/\Delta H_{eff} \quad [3]$$

$\dot{m}''_{crit}$ is a function of each sample's ΔH_{eff} and HRR at incipient ignition (HRR^{*}). Based on their literature review, HRR^{*} was 21± 6 kW m^{-2} for a range of polymers. During the optimization procedure, ΔH_{eff} was kept constant, while HRR^{*} was adjusted within the range to determine $\dot{m}''_{crit}$.

k of char was based on the radiation diffusion approximation which assumes that conductive heat transfer through the char is driven by radiation. This approximation was shown to accurately describe thermal conductivity of another lignocellulosic material, cardboard [12]. In this work, the initial value and adjustable range of char's k reflected the average, minimum, and maximum coefficient reported for cardboard by McKinnon [12]. Overall, all material properties were specified to be temperature independent except for k of char.

Table 1. Cone Calorimeter Parameters: Initial Property Value and Adjustable Range

Property	Average	Minimum	Maximum
	Virgin Sample		
k [W m^{-1} K^{-1}]	0.24 [10]	0.07 [11]	0.42 [10]
c [J kg^{-1} K^{-1}]	2300 [10]	1100 [13]	2900 [10]
ε [-]	0.88 [10]	0.70 [12]	0.94 [10]
α [m^2 kg^{-1}]	3 [10]	0.6 [10]	100 [7]
$\dot{m}''_{crit}$ [kg s^{-1} m^{-2}]	$21 \times 10^3/\Delta H_{eff}$ [14]	$15 \times 10^3/\Delta H_{eff}$ [14]	$27 \times 10^3/\Delta H_{eff}$ [14]
h [J kg^{-1}]	-1.3×10^6 [10]	0 [9]	-2.5×10^6 [10]
	Intermediate Component		
k [W m^{-1} K^{-1}]	0.24 [10]	0.07 [11]	0.42 [10]
c [J kg^{-1} K^{-1}]	2300 [10]	1100 [13]	2900 [10]
ε [-]	0.88 [10]	0.70[12]	0.94 [10]
α [m^2 kg^{-1}]	3 [10]	0.6 [10]	100 [7]
h [J kg^{-1}]	-7×10^5 [10]	0 [9]	-2.5×10^6 [10]
	Char		
k [W m^{-1} K^{-1}]	$5\times10^{-10}\times T^3$ [12]	$1\times10^{-10}\times T^3$[12]	$1\times10^{-9}\times T^3$[12]
α [m^2 kg^{-1}]	10 [10]	0.6 [10]	100 [7]

In order to limit the amount of unknown properties which need to be determined, c and ε of char, c of gaseous volatiles and all the properties of gypsum wallboard were not adjustable, but based on comprehensive studies. Li reported that the average c and ε of char is 1500 J kg^{-1} K^{-1} and 0.86, respectively, while the c of gaseous volatiles was 1.8 J kg^{-1} K^{-1} [7]. Rahmanian studied gypsum wallboard extensively and found that k and c were 0.1 W m^{-1} K^{-1} and 950 J kg^{-1} K^{-1} [15]. While, the optical properties, α and ε were specified to be 1000 m^2 kg^{-1} and 0 so that the back surface of gypsum wallboard was represented as an adiabatic and non-transparent.

The optimization procedure began for each sample by implementing the average values reported in Table 1 as the initial guess for any unknown properties, as well as the measured and estimated properties from MCC testing and simultaneously modeling the various incident heat flux exposures. The overall goal of the optimization procedure was to accurately predict the time to ignition and shape of the HRRPUA curve. A sensitivity analysis performed by Stoliarov et al. demonstrated which material properties were the most influential in dictating the time to ignition and shape of the HRRPUA curve during cone calorimeter experiments [10]. Based on this sensitivity analysis, the adjustable parameters were manually altered, while remaining within the minimum and maximum bounds, until the models predictions best fit the experimental results at all incident heat flux conditions. Since ThermaKin2D calculates mass loss rate per unit area; this value was multiplied by the sample's average ΔH_{eff} to produce HRRPUA, which was compared to experimental data from tests. A total of 8 separate optimization procedures were performed in order to characterize all four MDF samples when both the 1 and 2 Reaction Model was utilized. Figure 2 compares the experimentally measured and model predicted HRRPUA for Sample A, B, C, and D in the cone calorimeter at an incident heat flux of 20 kW m^{-2} or 25 kW m^{-2}, 50 kW m^{-2} and 80 kW m^{-2}, respectively using each sample's final set of optimized properties. The colored shaded area represents the range of experimental data points.

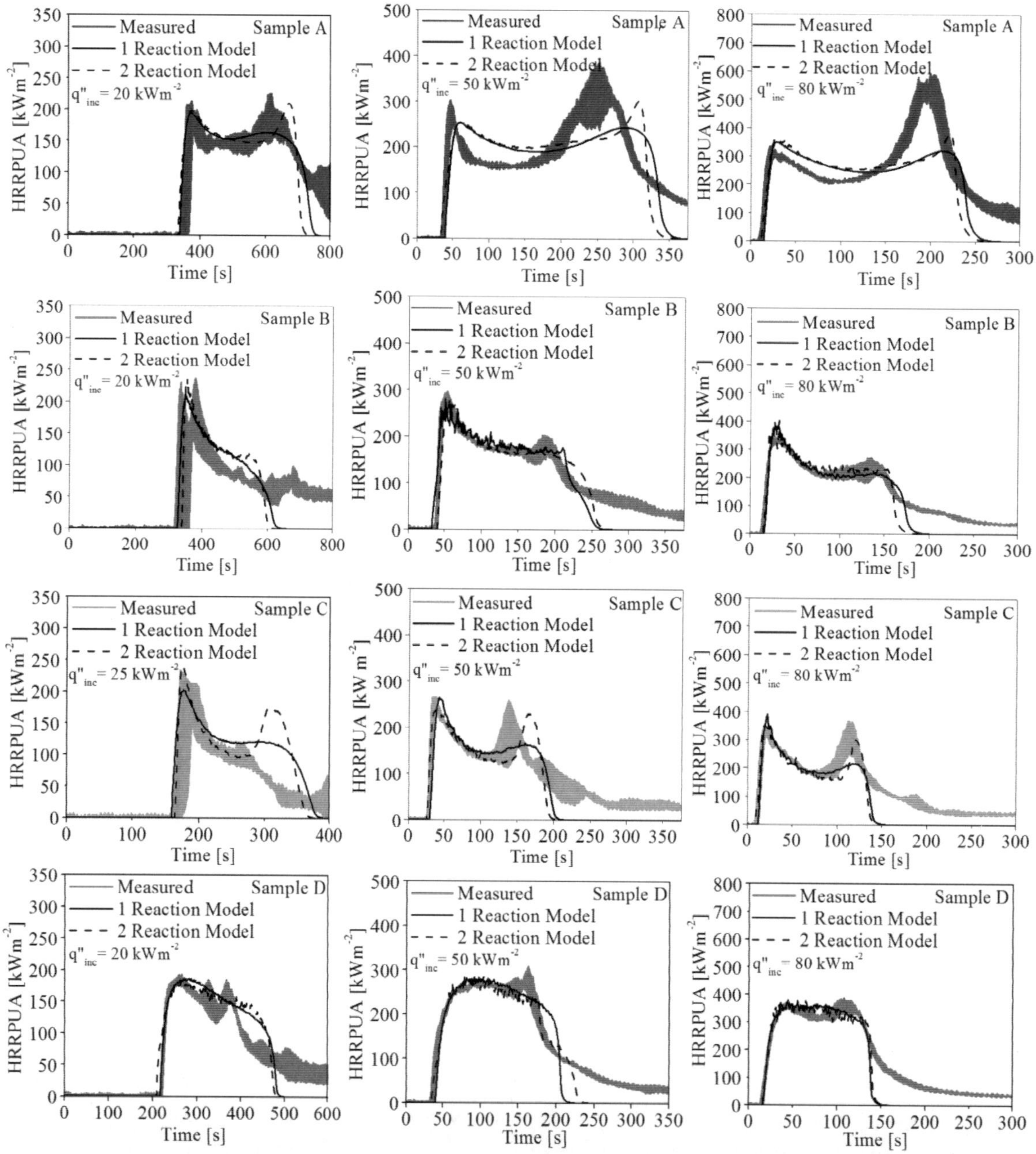

Figure 2. Experimentally measured and model predicted HRRPUA during cone calorimeter tests of MDF samples with an q''_{inc} exposure of 20×10^3 or 25×10^3 W m^{-2}, 50×10^3 W m^{-2} and 80×10^3 W m^{-2}

The quality of each model fit was quantified based on the relative difference between the HRR model predictions and corresponding experimental measurements [9]. The 1 and 2 Reaction Models were able to achieve an average relative difference of 14 and 15 % for all the samples under all radiant exposures, respectively. Therefore, a decision was made to utilize the 1 Reaction Model for all simulations of full-scale experiments because it provided comparable accuracy to the 2 Reaction Model when simulating cone calorimeter tests, while requiring fewer model inputs and lower computational cost. Table 2 provides each sample's complete set of material properties needed to fully parameterize their 1 Reaction Model.

Table 2: Complete Set of Sample Properties for the 1 Reaction Model

		Virgin Material Properties			
Parameter	**Sample A**	**Sample B**	**Sample C**	**Sample D**	**Source**
τ [m]	6.2×10^{-3}	4.1×10^{-3}	3.6×10^{-3}	3.2×10^{-3}	Measured
ρ [kg m^{-3}]	1090	889	889	1080	
k [W m^{-1} K^{-1}]	0.2	0.13	0.095	0.22	Optimized
c [J kg^{-1} K^{-1}]	2900	2900	2900	2650	
ε [-]	0.82	0.71	0.725	0.91	
α [m^2 kg^{-1}]	20	5	7	3	
$\dot{m}''_{crit}$ [kg s^{-1} m^{-2}]	2.22×10^{-3}	2.33×10^{-3}	2.64×10^{-3}	2.13×10^{-3}	
h [J kg^{-1}]	-3.0×10^5	-4.0×10^5	-1.5×10^5	-1.2×10^6	
θ_C [%]	17.5	14.4	17.6	11.6	Measured
A [s^{-1}]	1.44×10^7	5.81×10^7	4.80×10^6	2.80×10^8	Optimized
E [J mol^{-1}]	1.04×10^5	1.15×10^5	1.019×10^5	1.2225×10^5	
ΔH_{eff} [J kg^{-1}]	11.3×10^6	11.7×10^6	10.3×10^6	12.9×10^6	Measured
		Char Material Properties			
ρ [kg m^{-3}]	191	128	382	255	Measured
k [W m^{-1} K^{-1}]	9.5×10^{-10}×T^3	2.5×10^{-10}×T^3	1.25×10^{-10}×T^3	4.0×10^{-10}×T^3	Optimized
α [m^2 kg^{-1}]	35	10	15	13	
c [J kg^{-1} K^{-1}]	1500	1500	1500	1500	Literature [7]
ε [-]	0.86	0.86	0.86	0.86	

OPEN CORNER-WALL TESTS

An open corner-wall assembly, consisting of two 2.4 m × 1.2 m perpendicular walls with a partial ceiling was constructed under the 1 MW rated hood at the ATF FRL. This assembly was used to conduct two sets of experiments: burner exposure and flame spread tests. These tests characterized the heat feedback from the square burner used as the ignition source in the room corner tests and a spreading corner-wall fire so that a corner-wall model could be constructed to simulate the room corner tests. During both sets of tests, the 30.5×10^{-2} m square, 35.4×10^{-2} m tall propane gas burner designed in accordance with NFPA 286 was positioned on the floor in the corner of the assembly [1]. A mass flow controller fed CP grade propane to the burner at a rate that would achieve the desired fire size based on propane having a ΔH_{eff} equal to 46.4×10^6 J kg^{-1} [16]. Heat flux was measured using six water-cooled Schmidt-Boelter heat flux gauges that were 2.54×10^{-2} m in diameter with a 100 kW m^{-2} capacity. These gauges were arranged vertically along the right corner-wall, flush with the wall surface, separated from one another by 0.25 m and placed 0.10 m from the corner of the wall to ensure that they were always located within the flame region. Flame propagation and heat transfer to the walls was assumed to be symmetric with respect to the corner. During testing, heat flux gauges were cooled with water at an average temperature of 313 K, which has been shown to mitigate condensation on the gauge surfaces [17]. The 1 MW hood collected the products of combustion. The hood flow rate and O_2, CO_2 and CO concentrations were measured to determine HRR with a reported uncertainty of 10 % at 50 kW and 5% at 500 kW [18]. Two video cameras simultaneously recorded each test: the center-view video camera was positioned at a 45° angle to the corner-wall assembly and recorded flame propagation along both walls, while the side-view video camera faced only the right wall and it was used to estimate flame width along a single wall. Still-frame images were captured at 5 s intervals by the side-view video camera to determine the flame width at various heights along the flame height based on the distance from the corner to the edge of the luminous flame region.

Burner Characterization

Three tests were performed for the burner exposure tests, where the noncombustible gypsum walls were exposed to a propane gas burner producing a 40 kW fire for 300 s and the total heat flux from the burner fire to the surface of the corner-walls ($q"_b$) was measured. Figure 3 displays the $q"_b$ measurements from the 40 kW burner fire. The left image graphs shows that the $q"_b$ increases over the first 30 s of the test before reaching a relatively steady value, $q"_{steady}$. The right graph shows the $q"_{b,steady}$ measurements versus the height above the burner, y. $q"_b$ was the highest in the locations near the burner and decreased with increasing height above the burner.

Heat transfer from the propane gas burner was represented in ThermaKin2D using Equation 4, which defines an expression that that best fit $q"_{b,steady}$ measured during these experiments. This expression defines $q"_{b,steady}$ in terms of kW m^{-2}. In ThermaKin2D simulations, this boundary condition (Equation 4) was scaled by a factor γ, which varies between 0 and 1 throughout the duration of burner exposure. γ is prescribed to increases linearly from 0 to 1 over the first 30 s and then remain equal to 1 until the burner flame is turned off. The $q"_{b,steady}$ model is shown to accurately compare to the experimental measurements in the left image of Figure 3.

$$q"_{b,steady} = \begin{cases} -64.5 * y + 53.9, & 0 < y \leq 0.7\ m \\ -9.9 * y + 17.2, & 0.7\ m < y \leq 1.7\ m \end{cases} \qquad [4]$$

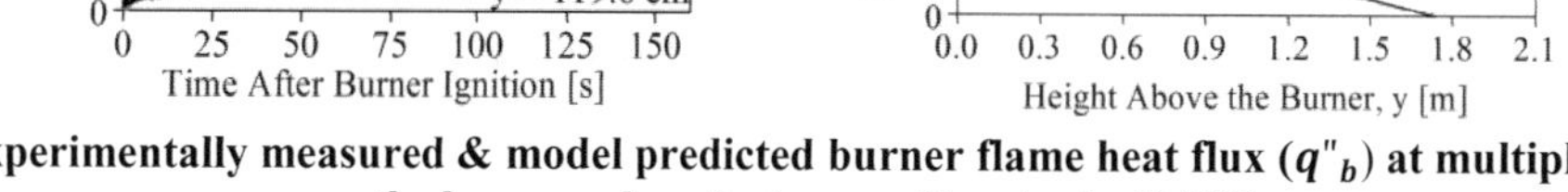

Figure 3. Experimentally measured & model predicted burner flame heat flux ($q"_b$) at multiple heights above the burner when the burner fire size is 40 kW

Flame Spread and Heat Feedback

During flame spread testing, 0.61 m wide and 2.4 m tall MDF samples were stapled to the corner-walls and exposed to a 25 kW fire produced by the propane gas burner. Samples A, B, C and D were ignited by exposing them to the burner for 345 s, 227 s, 293 s and 250 s, respectively. The burner HRR and time of exposure were selected to minimize contribution of the igniter to the fire growth while ensuring uniform ignition at the base of the sample. One flame spread experiment was performed on each sample in the open corner-wall assembly where HRR, flame width and total heat flux from a spreading flame to the corner-walls (q''_{flame}) were measured at various heights.

Corner-wall flame spread was assessed once the fire from the propane burner was turned off and unassisted flame spread occurred over the sample. During these tests, the flame width remained relatively constant, which indicated that lateral flame spread was negligible. The mean single side flame width of Samples A, B, C and, D was determined to be 0.25 m. It is assumed that the flame width is symmetric with respect to the corner. Therefore, the average total flame width along both corner-walls was 0.50 m. Fire size, (Q') was determined by normalizing the measured HRR by the average total flame width, 0.50 m. The total heat flux measurements were analyzed only after the propane burner was turned off and until flames reached the ceiling.

During testing, the corner-wall fire was initially unsteady and thin after the propane gas burner was turned off. As the fire spread vertically, the flames thickened and became increasingly turbulent which corresponded to an increase in measured q''_{flame} at each location. Once the flames cover the heat flux gauge, the q''_{flame} measurements reached a relatively steady value between 35 kW m^{-2} and 50 kW m^{-2}. The average peak steady state total flame heat flux ($q''_{f,steady}$) during corner-wall flame spread, but prior to ceiling flame spread was 42.0 kW m^{-2}. $q''_{f,steady}$ was found to be independent of the material or position of the heat flux gauge.

Typically, flame height (y_f) is defined based on visual observations of the flame tip height or continuous flame region [19, 20]. In this study, y_f was defined quantitatively by a measure of flame heat flux, as the distance from the base of the flame to the highest location where the measured heat flux is within 85 % of $q''_{f,steady}$. Flame height during corner-wall fire growth, y_f was developed based on Leventon's flame height model for a vertically-oriented sample, Equation 5 [21].

$$y_f = a(Q')^b + c \quad [5]$$

Equation 5 defines the y_f as a function of Q' and was developed to fit the experimental data using three empirical constants a, b, and c. In this equation y_f and Q' are in terms of m and kW m^{-1}. The model fit was $R^2 = 0.93$ when the empirical factors, a, b, and c equal 0.165, 0.50 and -0.686, respectively. Figure 4 plots the y_f experimental data and model as a function of the Q' during corner-wall upward flame spread. Sample C was not included because the fire self-extinguished once the propane burner was turned off.

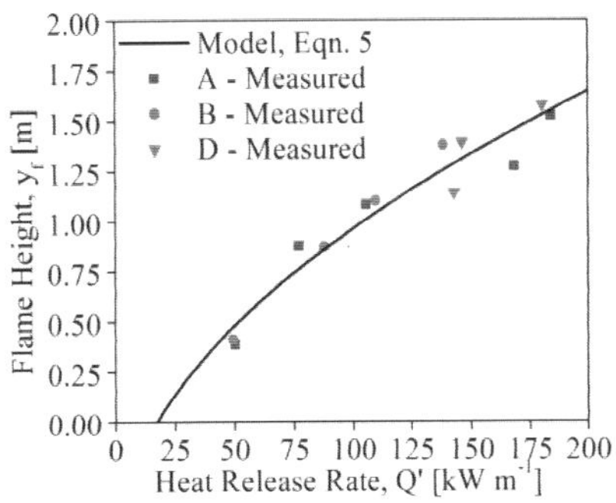

Figure 4. Flame height as a function of fire size during corner-wall flame spread

A model describing heat feedback from corner-wall flames produced during MDF combustion was also developed using the framework of Leventon's wall flame heat feedback model, Equation 6 [3]. Leventon's model was first developed to describe the heat feedback during flame spread along small-scale PMMA samples. This model was then generalized by Korver to describe the behavior of wall flames supported by a wide range of materials [22]. In Equation 6, the total heat flux from the flame to the surface is steady, $q''_{f,steady}$ over the flame height region and decays beyond the flame height region as a function of $q''_{f,steady}$, an empirical constant α_f and dimensionless height (y^*), Equation 7.

$$q''_{flame} = \begin{cases} q''_{f,steady} & y_b \leq y_f \\ (\alpha_f \times q''_{f,steady})\left(e^{-\ln(\alpha_f)\times(y^*)^2}\right), & y_b > y_f \end{cases} \quad [6]$$

$$y^* = \frac{y_b + y_0}{y_f + y_0} \quad [7]$$

Equation 7 defines the y^* as a function of y_f (Equation 5), the height above the base of the flame, y_b, and the last empirical constant y_0. In these equations, y_b and y_0 are in terms of m, q''_{flame}, and $q''_{f,steady}$ are in terms of kW m^{-2} and y^* was dimensionless.

This model was utilized to fit the heat flux gauge data at the various vertical heights as a function of Q'. The corner-wall flame heat feedback model achieves the best fit of the experimental heat flux data when the empirical constants, y_0 and α_f equal 4.5 m and 49.82, respectively. Figure 5 plots the experimentally measured and model predicted flame heat flux at six measurement locations: 0.196 m, 0.446 m, 0.946 m, 1.196 m, 1.446 m and 1.696 m, respectively. The mean absolute error of the model and the experimental measurements was calculated to be 2.4 kW m^{-2}. Measurements of q''_f could not be obtained when $Q' < 50$ kW m^{-1}. Therefore, the corner-wall flame heat feedback model contains greater uncertainty for small to intermediate sized flames, observed immediately after ignition. The burner and corner-wall flame heat feedback models were implemented in ThermaKin2D. An assumption was made in both cases that the heat feedback was radiative in nature.

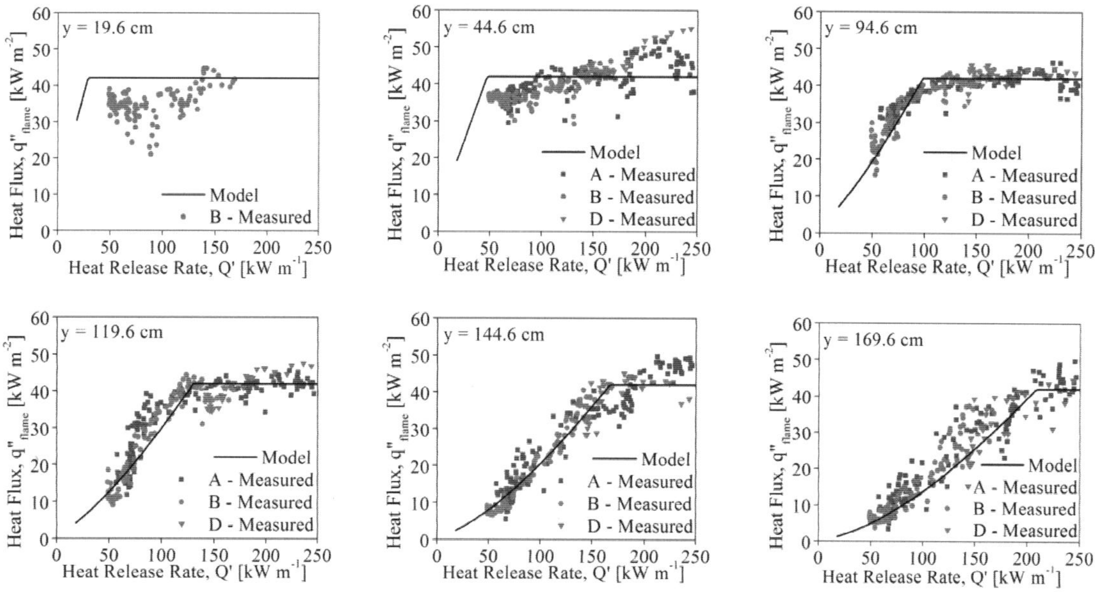

Figure 5. Experimental measured and model predicted flame heat flux during the open corner-wall flame spread tests at various heights above the burner

FULL-SCALE

Room Corner Tests

Room corner tests were performed at the ATF FRL according to the standard room corner test, NFPA 286 [1]. The room corner compartment was 2.4 m × 3.6 m wide and 2.4 m tall, with a 0.8 m × 2.0 m opening. Gypsum wallboard was installed along the walls and ceiling. The MDF samples were stapled to all gypsum walls, except for the front wall where the door was located. The same square propane gas burner used during the open corner-wall tests was also used during the room corner tests. NFPA 286 requires the gas burner to produce a 40 kW fire for the first 5 minutes and a 160 kW fire for the next 10 minutes [1]. The ignition source specified during NFPA 286 subjects the walls and ceiling material to extreme thermal conditions throughout the entire test, which drives flame spread and controls fire development within the compartment. The goal of room corner tests conducted in this project was to isolate the ignition source and focus solely on flame spread so the fire performance of these similar materials could be compared. Therefore, a 40 kW fire was produced from the burner until ignition occurred and then the burner fire was turned off and the fire was allowed to propagate on its own. Preliminary testing revealed that all the samples ignited once they were exposed to the burner fire for 165 s. The room corner assembly was positioned under the 4 MW at the ATF FRL so that compartment HRR could be measured. A video camera located inside the compartment was utilized to identify the moment when flames had spread upwards along the entire corner-wall and began to spread horizontally along the ceiling. Two room corner tests were performed for Samples A, B, and D, while only one test was performed using Sample C due to time constraints.

Once the propane gas burner was turned off during room corner testing, the resulting corner-wall flame would either continue to propagate upward towards the ceiling and the compartment would eventually transitioned to flashover or the sample would not sustain flame spread and the fire would self-extinguish. The 4 MW hood was able to measure compartment HRR with a combined standard uncertainty of 42 %, 4.6 % and 2.6 % at 50 kW, 500 kW and 1100 kW, respectively [23]. Time to ceiling flame spread was based on video recordings within the compartment and time to flashover was defined as the moment when the flames extended beyond the door [1]. Both tests of Sample B and D resulted in compartment flashover. For Sample B and D, it took on average 292 s and 247 s for flames to spread to the ceiling and only 470 s and 378 s for the compartment to transition to flashover. During the first test on Sample A, flames spread to ceiling after 532 s and compartment flashover was achieved at 717 s. During the second test for Sample A and the only test for Sample C, the fire self-extinguished prior to spreading to the ceiling and therefore, compartment flashover never occurred. Sample D and B were the most hazardous materials because flashover occurred the quickest for these samples, while Sample A and C were the least hazardous.

In characterization of material contribution to the fire growth, the most critical aspect of the room corner test was the time it took for flames to spread to the ceiling. The HRR values of 220 kW m^{-1} approximately corresponded to the point when the flame reached the ceiling. All experiments where the fire size exceeded 300 kW m^{-1} resulted in flashover.

Room Corner Model

Room corner tests were modeled in ThermaKin2D by simulating flame spread along the corner-walls of the compartment. This model was approximated as two-dimensional and the corner-walls of the assembly were configured by having the 2.05 m vertical MDF sample positioned on top of gypsum wallboard. The complete set of material properties reported in Table 2 were implemented to represent Sample A, B, C and D. The burner was modeled using the heat flux formulation (Equation 4) described above. After 165 s of the burner exposure, the corner-wall flame model, represented by Equation 5, 6 and 7 was turned on. The presence of the flame was controlled in this model by $y_f > 0$ condition. Local $\dot{m}'' >$

$\dot{m}''_{crit}$ condition was used to determine the position of the base of the flame. The cell size of the x and y grids in the model were 5×10^{-5} m and 0.01 m, respectively, while the time step was 0.01 s. Increasing or decreasing the integration parameters by a factor of 2 did not change computation results, thus indicating convergence. Figure 6 shows the experimentally measured and model predicted room corner results. The results shown here do not include the HRR contribution from the propane burner.

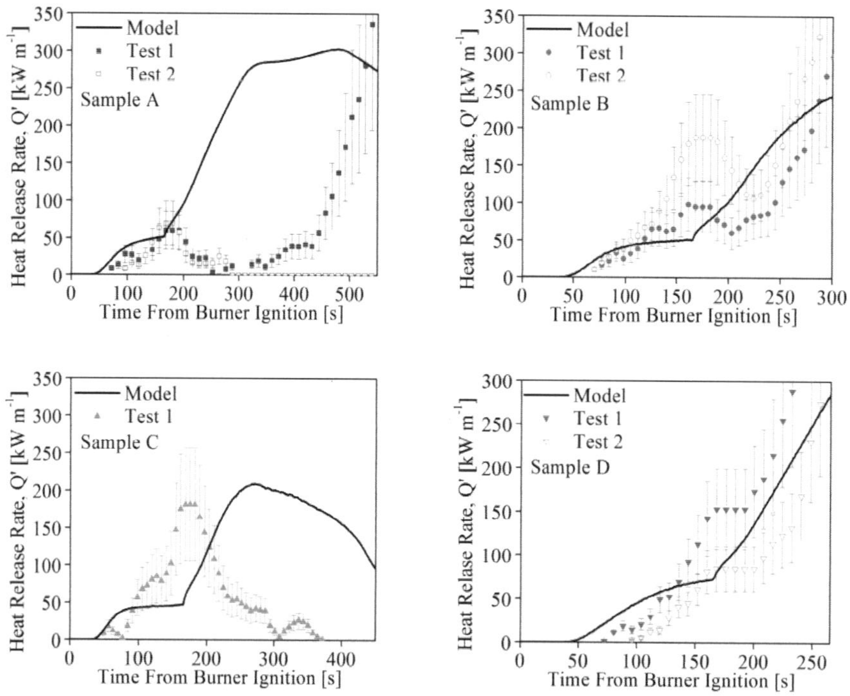

Figure 6. Experimentally measured & model predicted Q' during room corner tests and simulations. The Q' produced by the gas burner fire was removed from the experimental measurements

The model produced excellent predictions for Sample B and D, which were among the more flammable materials analyzed in this study. The model captured the general shape of the HRR curve for Sample C, although the time to peak HRR was significantly over predicted. For Sample A, the model correctly predicted the fire size late in the experiment but failed to predict the shape of the HRR history. Both Sample A and C were the materials that exhibited borderline spread/no spread behavior. This behavior made it difficult to obtain repeatable experimental data on these materials and model this data with a high degree of accuracy.

Comparison Criterion

Relative performance of these similar materials in the room corner scenario was quantified based on the average fire growth rate (FGR). FGR was calculated as the slope of the line beginning when the fire from the gas burner was turned off to the point when Q' reached 300 kW m^{-1}. As was discussed above, this critical value of Q', Q'_{crit} was a threshold the achievement, which indicated that the fire will proceed to grow until compartment flashover occurred. Equation 8 is the expression for FGR where,

$$FGR = \frac{Q'_{crit} - Q'_{b,off}}{t_{crit} - t_{b,off}} \quad [8]$$

Here, $Q'_{b,off}$ and $t_{b,off}$ are the fire size and time when the burner was turned off, and t_{crit} is the time when Q'_{crit} was reached. For those experiments where the value of 300 kW m^{-1} was not reached, Q'_{crit} and t_{crit} were assigned the values corresponding to the maximum Q' achieved during the experiment. If the maximum Q' did not reach 220 kW m^{-1} (i.e. the flame never reached the ceiling), the material was declared to self-extinguish. Table 3 compares the FGR values calculated for the experimental results and model predictions.

Table 3: FGR Computed from Experiments and Simulations

Sample	Fire Growth Rate (FGR)	
	Experimental Mean	**Model**
Sample A	0.7	0.9
Sample B	1.2	1.3
Sample C	Self-extinguish	Self-extinguish
Sample D	2.2	2.1

Experimental mean is the average FGR values for each test. Since Sample A had one room corner test that self-extinguished and another that resulted in flashover, the FGR is only based on the test where flashover occurred. The experimental and model predicted FGR values compare well. Much like the experiments, the model qualitatively predicts that FGR is the highest for Sample D, followed by Sample B, then Sample A and Sample C self-extinguishes without flashover.

CONCLUSION

A new procedure was developed for fire investigators that can be used to compute and compare how similar materials perform during the room corner test, a full-scale compartment fire test. This procedure requires fire investigators to determine the material properties of exemplar and candidate samples through MCC and cone calorimeter testing and modeling. Once the material properties are determined, they can be implemented into a ThermaKin2D model that simulates corner-wall flame spread during the room corner tests. This model relies on two submodels that represent heat feedback from the room corner ignition source, the gas burner fire and a spreading corner-wall flame. It is expected that the corner-wall flame model developed in this work can be generalized for a wide range of materials. Lastly, the FGR criterion can be used to rank how hazardous these materials are relative to one another. This procedure is expected to improve a fire investigator's ability in performing accurate reconstructive fire tests because a fire investigator will be able to select candidate materials for their reconstructive fire test that performs most similarly to the exemplar material found during the fire investigation.

REFERENCES

1. NFPA 286, Standard Methods of Fire Tests for Evaluating Contribution of Wall and Ceiling Interior Finish to Room Fire Growth, National Fire Protection Association, Quincy, MA, (2006).
2. Stoliarov, S. I., Leventon, I. T., Lyon, R. E., Two-Dimensional Model of Burning for Pyrolyzable Solids, Atlantic City, NJ, (2013).
3. Leventon, I.T., Li, J., Stoliarov, S.I., A Flame Spread Simulation Based on a Comphrensive Solid Pyrolysis Model Coupled with a Detailed Empirical Flame Structure Representation, Combustion and Flame, vol. 162, pp. 3884 - 3895 (2015).
4. ASTM D7309: Standard Test Method for Determining Flammability Characteristics of Plastics and Other Solid Materials Using Microscale Combustion", ASTM International, West Conshohocken, Pa.
5. Li, K, Huang, X, Fleischmann, C., Rein, G., Ji, J., Pyrolysis of Medium Density Fibreboard: Optimized Search for Kinetic Scheme and Parameters via Genetic Algorithm Driven by Kissinger's Method, Energy & Fuels, vol. 28, pp. 6130 – 6139, (2014)

6. Becidan, M., Várhegyl, G., Hustad, J.E., Skreiberg, O, Thermal decomposition of biomass wastes. A kinetic study, Industrial and Engineering Chemistry Research, vol. 46, no. 8, pp. 2428–2437, (2007).
7. Li, J, A Multiscale Approach To Parameterization of Burning Models for Polymeric Materials, University of Maryland, PhD Dissertation (2014).
8. ASTM E1354-14: Standard Test Method for Heat and Visible Smoke Release Rates for Materials and Products Using an Oxygen Consumption Calorimeter", ASTM International, West Conshohocken, Pa.
9. Lannon, C., A Methodology for Determining the Fire Performance Equivalency Amgonst Similar Materials During a Full-scale Fire Scenario Based on Bench-scale Testing, Department of Fire Protection Engineering, University of Maryland, Masters Thesis (2015).
10. Stoliarov, S. I., Safronava, N., Lyon, R.E., The effect of variation in polymer properties on the rate of burning, Fire and Materials, vol. 33, no. 6. pp. 257–271, (2009).
11. Gronli, M.G., A Theoretical and Experimental Study of the Thermal Degradation of Biomass., The Norwegian University of Science and Technology, Trondheim, Norway, (1996).
12. McKinnon, M. B., Development of a Model for Flaming Combustion of Double-Wall Corrugated Cardboard, Department of Fire Protection Engineering, University of Maryland, Masters Thesis (2012).
13. Li, K. Y. , Fleischmann, C.M., Spearpoint, M.J., Determining thermal physical properties of pyrolyzing New Zealand medium density fibreboard (MDF), Chemical Engineering Science, vol. 95, pp. 211–220, (2013).
14. Lyon, R.E, Quintiere, J.Q, Criteria for piloted ignition of combustible solids, Combustion and Flame, vol. 151, no. 4, pp. 551–559, (2007).
15. Rahmanian, I., Thermal and Mechanical Properties of Gypsum Boards and Their Influences on Fire Resistance of Gypsum Board Based Systems, University of Manchester, (2011).
16. Babrauskas, V., SFPE Handbook of Fire Protection Engineering, 4th ed. Quincy, MA: National Fire Protection Association, (2008).
17. Bureau of Alcohol, Tobacco, Firearms and Explosives Fire Research Laboratory, Laboratory Instruction: Heat Flux Transducers.
18. Bureau of Alcohol, Tobacco, Firearms and Explosives Fire Research Laboratory, Technical Reference: Fire Products Collector - 1 MW Square TR011a.
19. Lattimer, B. Y., Sorathia, U., Thermal Characteristics of Fires in a Noncombustible Corner, Fire Safety Journal, vol. 38, no. 8, pp. 709–745, (2003).
20. Hasemi, Y., Yoshida, M., Takashima, S., Kikuchi, R.,Yokobayashi, Y., Flame length and Flame Heat Transfer Correlations in Corner-Wall and Corner-wall-ceiling Configurations, Interflame'96, (1995).
21. Leventon, I.T., Prediction of Upward Flame Spread over Polymers, Department of Fire Protection Engineering, University of Maryland, PhD Dissertation (2016).
22. Korver, K.T., A Generalized Model for Wall Flame Heat Flux During Upward Flame Spread On Polymers, Department of Fire Protection Engineering, University of Maryland,Masters Thesis (2015)
23. Bureau of Alcohol, Tobacco, Firearms and Explosives Fire Research Laboratory, Technical Reference: Fire Products Collector - 4 MW TR011c.

SCALE REDUCTION: HOW TO PLAY WITH FIRE?

Serge Bourbigot, Johan Sarazin, Pierre Bachelet, Fabienne Samyn, Maude Jimenez & Sophie Duquesne
R_2Fire group/UMET – UMR/CNRS 8207, ENSCL, University of Lille, Avenue Dimitri Mendeleïev – Bât. C7a, CS 90108, 59652 Villeneuve d'Ascq Cedex, France

INTRODUCTION

Fire testing for evaluating the reaction and resistance to fire of materials is generally time-consuming and expensive. Nevertheless, those tests are required to evaluate materials in many sectors such as building or transportation. Due to the complexity of fire phenomenon, full-scale tests are still the main and the most credible tool for investigating fire-related issues. However, the cost of those tests significantly increases with scale. The purpose of this paper is to examine the scale reduction of different tests and to mimic fire scenarios at the small scale.

In this paper, the scaling reduction of some fire tests of interest is presented. It deals with the Steiner tunnel (ASTM E-84), furnace test (UL 1709), single burning item (SBI) and the burner test (ISO 2685). It covers different types of materials (composites and steel) and several applications (composite and steel for building and carbon fiber reinforced composite (CFRP) for aviation). In all cases, the fire protection is provided with an intumescent coating.

The fire protection of materials like CRP or steel can be achieved with reactive insulation coatings which react when exposed to fire increasing their thermal insulation properties, thereby protecting the underlying substrate. It happens with intumescent coatings which expand and form an insulative layer at the surface of the substrate upon heating. The intumescence process results from a combination of charring and foaming at the surface of the substrate[1,2]. The result of this process is the formation of a multicellular (alveolar) barrier, thick and non-flammable, which protects the substrate or residual material from heat or flame action. The charred layer acts as a physical barrier which slows down heat and mass transfer between gas and condensed phase. The formation of an intumescent char is a complicated process involving several critical aspects: rheology (expansion phase, viscoelasticity of char), chemistry (charring) and thermophysics (limitation of heat and mass transfer)[3]. Hence, the formation of the intumescent barrier must be adapted to the fire scenario. It is considered in the different fire tests presented in this following.

STEINER TUNNEL (ASTM E-84)

The use of composites in civil engineering structural components is increasing thanks to their light weight, high specific strength, and stiffness properties[4]. In this section, we consider polyethylene terephthalate (PET) foam sandwich structured composites used for making the ceiling of railway stations. We investigated the potential benefit of using intumescent coatings for improving the resistance to fire of PET foam-based composite. The goal is to show if better performance in terms of fire retardancy and smoke emission can be obtained using an appropriate commercial intumescent coating for protecting the composite against fire. The composite to be tested is a PET foam block (ARMAform, Armacell Benelux SA) covered with glass woven fabrics (Gurit) impregnated by an epoxy resin (Ampreg 21FR, Gurit). Different commercial intumescent coatings have been selected, including epoxy-based [Chartek8 of International Paint, Firetex M93 of Sherwin Williams (formely Leigh's paint) and XE 2458/1 of Axson], polyester-based [Polycor 2330 BR, Cray Valley] and latex-based [Avicoat FS/DS of Arabian vermiculite, Firefree 88 of Firefree coatings, Intuflam of Lurie and Pyroplast ST200 of Rutgers Organics] resins. A mass of 1kg of wet paint per square meter has been applied.

The associated test for evaluating our materials is the Steiner Tunnel (ASTM E-84). Steiner tunnel is a large scale test which is referenced as a method to assess flame spread and smoke density. It is the primary test method for evaluating resistance to fire of interior finish materials. The apparatus consists of a tunnel-like enclosure measuring 8.7 × 0.45 × 0.31 m (25 × 1½ × 1 ft). The test specimen is 7.6 m (24 ft) long and 0.46 m (18 in.) wide and is mounted in the ceiling position. It is exposed at one end, designated as the burner end, to an 88-kW (5000-Btu/min) gas burner. There is a forced draft through the tunnel from the burner end with an average initial air velocity of 1.2 m/s (240 ft/min). The measurements consist in flame spread over the surface and smoke obscuration in the exhaust duct of the tunnel. The test duration is 10 minutes, flame spread is determined visually through windows built into the tunnel and an optical cell mounted at the tunnel exhaust measures the smoke density. A flame spread index (FSI) is calculated on the basis of the area under the curve of flame tip location versus time. The FSI is 0 for an inert board, and is normalized to approximately 100 for red oak flooring. A smoke developed index (SDI) is calculated on the basis of the area under the light obscuration vs. time curve, and is equal to 100 for red oak flooring. To make fast evaluation of our materials, we developed a 'mini' Steiner tunnel (reduction of the Steiner tunnel at 1/8 scale) which is fully described in Figure 1. The burner is fed with methane and the flow rate is 0.74 l/min (gas velocity = 1.9 m/s). It gives a temperature at the surface of the evaluated sample of about 980°C. Temperatures as a function of time (it is not required in the ASTM E84 but it gives information about the efficiency of the intumescent coating as heat barrier) are measured during the test using 6 thermocouples (hereafter called Th.x with $1 \leq x \leq 6$), small windows permit to follow flame spread and smoke is collected in the chimney to measure its opacity with a smoke density analyzer including a halogen light.

Figure 1. (a) Schematic view of the lab-scale Steiner tunnel and (b) its internal construction (side view)

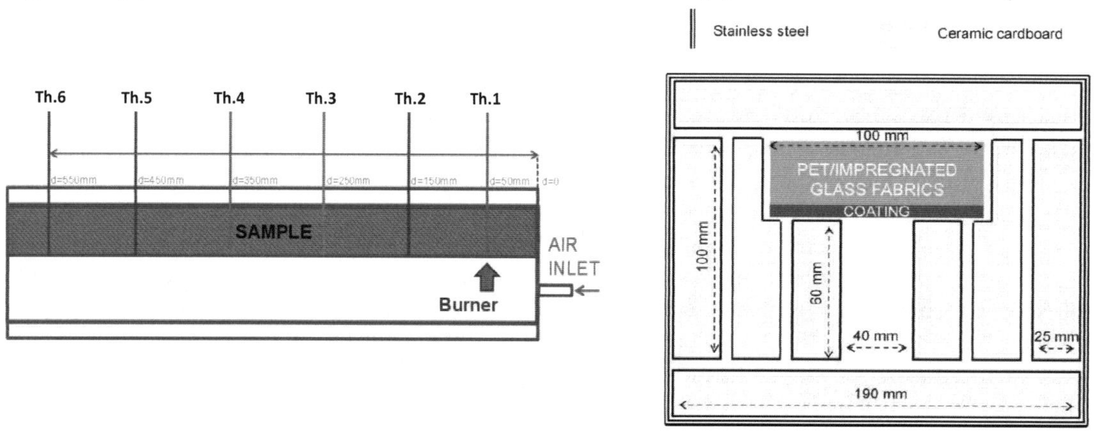

(a) (b)

The temperature/time curves (curves not shown)) are very promising and allow us to discriminate the different coatings. Avicoat, Pyroplast ST200 and Intuflam give the best barrier properties. Visual observation of the residues indicates that not all the coatings have developed an intumescent structure. Avicoat and Intuflam do not swell but they provide the best performance forming a non-expanded char at the surface. In our conditions, thermal constraint is not high enough to make the expansion of the material. At the end of the test, we can observe that the PET foam is intact when protected with Avicoat which indicates that the efficiency of this coating is really high.

Regarding the good results obtained with Avicoat in terms of fire barrier properties, a specimen coated with this reference has been tested under large scale ASTM E84 conditions in the evaluation center of Intertek in Elmendorf-USA. The Avicoat coating was about 700 μm thick and it exhibited good adhesion to the composite when carrying the ASTM D3359B test. The description of the specimen indicates that a charred aspect is observed on 85% of the total length (Figure 2-a). Then, a discoloration occurs. The picture of a section of the specimen however clearly shows that no intumescent phenomenon appeared during the test. The observation of the sample tested on the small scale tunnel also shows a charred aspect without intumescence; the coating is only distorted and cracked (Figure 2-b). These results suggest that, on the one hand, the standard conditions are reproduced: the results of the small scale test can be extrapolated for the evaluation of the thermal barrier effect (time-temperature curves). On the other hand, the thermal constraints seem to influence the behavior of the coating as follows: the flame constraint and the conditions of the tunnel test lead to no intumescence, only to the formation of a crust.

Figure 2: (a) Avicoat coating protecting PET foam after the large scale Steiner tunnel and (b) after the mini Steiner tunnel test

(a)

(b)

FURNACE TEST (UL 1709)

The protection of metallic materials against fire has become a very important issue in the construction and petrochemical industries, as well as in the marine and military fields. Structural steel loses a significant part of its load-carrying ability when its temperature exceeds 500°C. Prevention of the structural collapse of a building is crucial to ensure the safe evacuation of people and it is a prime requirement of building regulations in many countries.

We have developed a small scale furnace test to evaluate the fire performance of intumescent coatings in cellulosic and hydrocarbon fires. This test was designed to mimic the ISO834 (or ASTME119) and the UL1709 normalized temperature/time curves, respectively related to cellulosic fire and to hydrocarbon fire. The lab-scale furnace exhibits an internal volume of 40 dm^3 (Figure 3). Refractory fibers (stable up to 1300°C) cover the different faces of the furnace. The furnace was equipped with two gas burners (20 kW propane burners). The gas pressure was fixed at 1.8 bars and the flow was regulated in order to mimic the UL1709 curve (in this work). A temperature probe inside the furnace regulates the temperature and a K-type thermocouple allows the furnace temperature profile to be registered. The temperature at the back side of the plate is measured as a function of time using a pyrometer (temperature measured in the center of the plate). The backside of the plate is coated with black paint (Jeltz) having a constant emissivity (0.92) and thermally resistant up to 800°C in order to get reliable temperature measurements with the pyrometer.

Figure 3. Furnace set up to mimic cellulosic and hydrocarbon fire scenario

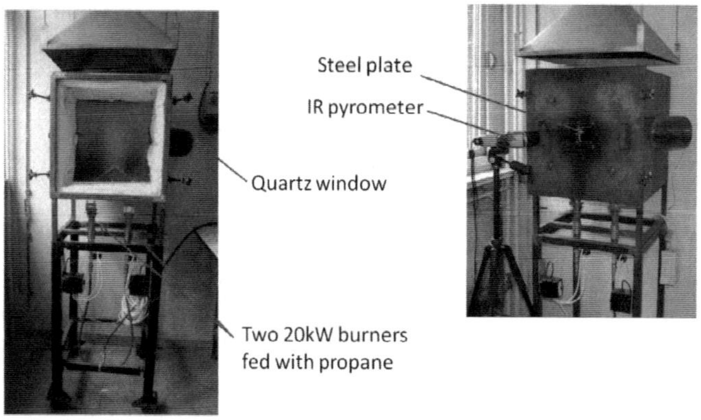

Novel intumescent paints based on silicone as polymeric matrix (hereafter called Intu-1 and Intu-2)[5] were developed in the lab and were evaluated with the small furnace on steel plates according to UL-1709 temperature curve (Figure 4). The failure temperature has been taken at 500°C. Uncoated steel plate reaches 500°C at 240s while the coated steel with intumescent coatings (thickness on steel plate = 1 mm) reaches the failure temperature at times higher than 980s. It shows that intumescence is an efficient way to protect steel since the time to failure temperature is 5 times longer than that with virgin steel. Our intumescent paints have been compared to a commercial paint and it appears that they are much more efficient at shorter times (t<500s) suggesting a faster development of the intumescence at the beginning of the test.

Figure 4. Temperature measured on the backside of steel plates coated with intumescent paints as a function of time in the conditions of UL 1709 in a small furnace

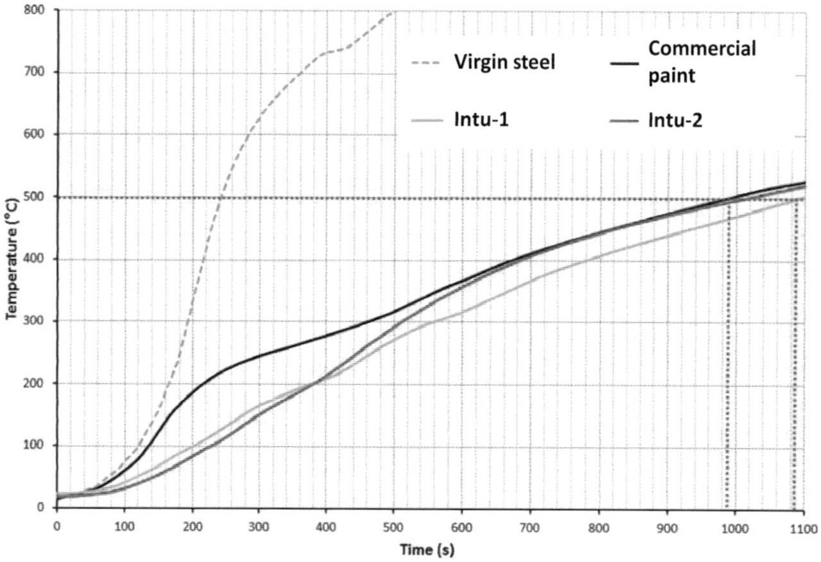

SINGLE BURNING ITEM (SBI, EUROCLASS)

The SBI test method is planned to assess the performance of building products in a (real scale) room corner scenario. In the pursuit of our efforts to make scale reduction, we have developed a SBI at 1/3 scale (Figure 5). It measures all parameters as in the large scale SBI including heat release rate (HRR), total heat release (THR), FIGRA, smoke opacity and SMOGRA. Wooden panel protected by an intumescent paint was evaluated with our bench-scale SBI. The results show that HRR is dramatically reduced when the panel is protected by the intumescent coating (Figure 6 – (a)). The intumescent charred layer develops upon burning and provide the protection (Figure 6 – (b)).

Figure 5. (a) Mini-SBI test; (b) Cross-section of parameters measurement (smoke opacity, O_2 analyser and pressure)

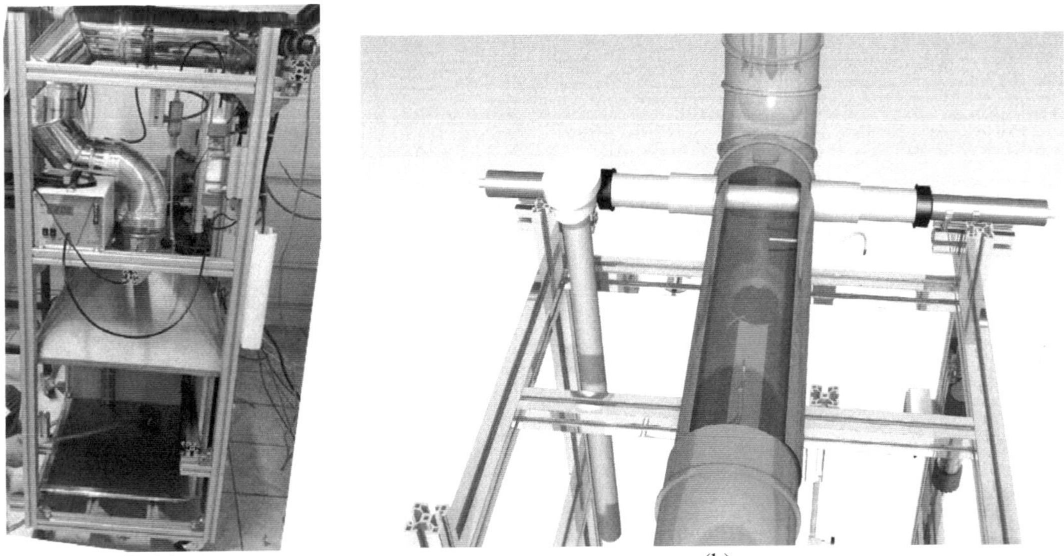

(a) (b)

Figure 6. (a) HRR as a function of time of wooden panels at the bench-scale SBI; (b) wooden panel protected by an intumescent paint burning at the bench scale SBI

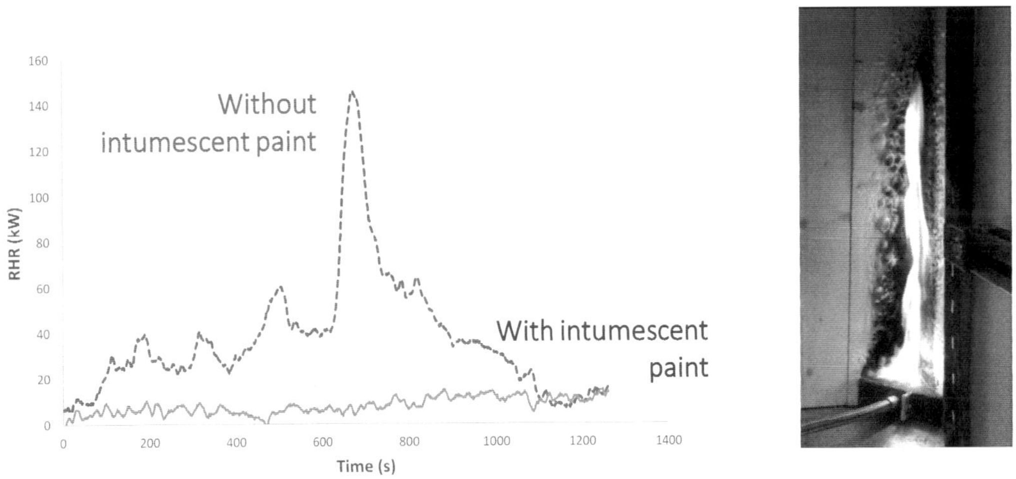

(a) (b)

Our mini SBI was designed to be correlated with the values of the large scale SBI. It gives good prediction for the wooden panels as shown in Table 1.

Table 1. Comparison of the classification of wooden panels between the large scale SBI and the bench scale SBI

	Wooden panel (SBI)	**Wooden panel (mini SBI)**	**Wooden panel (intumescent)**
THR @600s [MJ]	15.7 (+/-2)	18.2 (+/-3)	3.2 (+/-1)
FIGRA [W/s]	440 (+/- 79)	524.3 (+/- 85)	164 (+/- 25)
TSP [m²]	47 (+/- 10)	66.8 (+/- 10)	0
SMOGRA [m²/s²]	3 (+/-1)	10.9 (+/-3)	0
Rating (Euroclass)	Ds2d0	Ds2d0	Cs1d0

BURNER TEST (ISO 2685)

Composite materials are increasingly being used in the design of aircraft. However, the recent application of carbon fiber reinforced polymer (CFRP) in aircraft structure has introduced potential fire threats and it raises concerns regarding fire safety[6,7].

The fire resistance of components, equipment and structure located in 'fire zones' (e.g. compartments containing main engines and auxiliary power units) is conventionally evaluated according to the international standard ISO 2685. The test is established on a pass/fail basis thereby an element is considered to pass its fire test if it is capable to sustain its design function after the standard flame exposure for a specific lapse of time. In this research we are interested in the fire behavior of CRFP undergoing fire as defined in ISO 2685. Even if ISO 2685 is a medium-scale test, it is time-consuming and expensive, especially when CFRP samples must be produced for this purpose. In this paper, a small scale bench was developed to mimic ISO 2685 and to evaluate CFRPs with and without fire protection (intumescent coating).

The physical scaling method has been applied in this work to reduce the ISO 2685 test. Despite its simplifications, the scaling technique permits to improve the understanding of the fire dynamic involved in large scale fire and in fire testing[8]. Virtual benches were modeled (not presented here) considering a reduction by 3 (except for the sample thickness) of ISO 2685 on steel and aluminum plates. Dimensionless numbers were determined by the well-known Froude scaling technique and were used in the heat diffusion equation associated with the appropriate boundaries for making simulations. Comparisons of temperature distribution on the backside of the plates were done between the ISO bench and the virtual reduced one (Figure 7-(a)). The temperature distribution looks similar (similar shapes) but the measured and simulated temperatures are significantly different. Nevertheless, temperature distribution becomes similar (thermally thin sample) as that of the ISO bench when the duration of the test is decreased and heat transfer (convection and radiation) is increased (Figure 7-(b)).

Figure 7. Comparison of the temperature distribution on the backside of steel plates measured during ISO 2685 test and by simulating ISO 2685 on reduced virtual bench with 1/3 reduction (a) and with 1/3 reduction changing the duration of the test and the boundaries (b) (Note the temperatures are in °C).

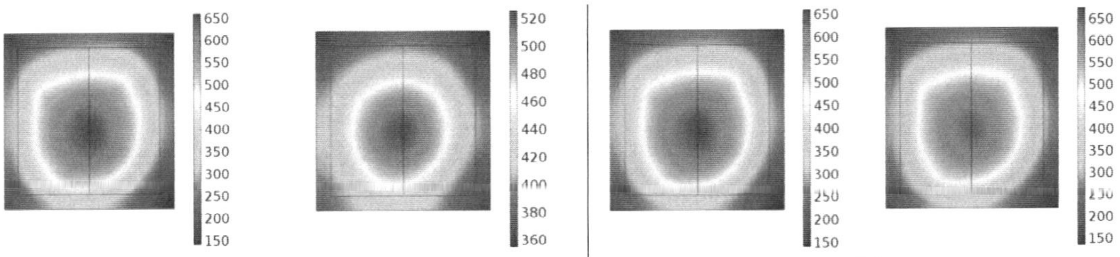

(a) (b)

The main conclusions are temperature differences measured during the full scale ISO test compared to the virtual bench are mainly due to scaling changes. On the other hand, it was shown temperature distribution in the flame, the design of the burner and boundaries conditions had a minor influence on temperature changes.

A reduced scale bench was built up and was equipped with propane burner, insulated holder for the samples and with infrared camera and bichromatic pyrometers for measuring temperatures on the backside of the sample (Figure 8-(a)). CFRP sample prepared by Safran Company was coated with a commercial intumescent paint and the temperatures measured on the backside of the sample was significantly reduced compared to virgin CFRP. As an example, temperature measured in the center of the sample is 345°C in the stationary state with intumescent coating while it reaches 485°C without intumescent coating (Figure 8-(b)).

Figure 8. Reduced scale bench equipped with temperature rake for measuring flame temperature and with gauge for measuring heat flux impacting the sample (right); sample holder insulated from the frame (left); the two frames (right and left) are on the rail moving back and forth for calibration's purpose (a) – Intumescent coating formed after fire testing on the reduced scale ISO 2685 test (b).

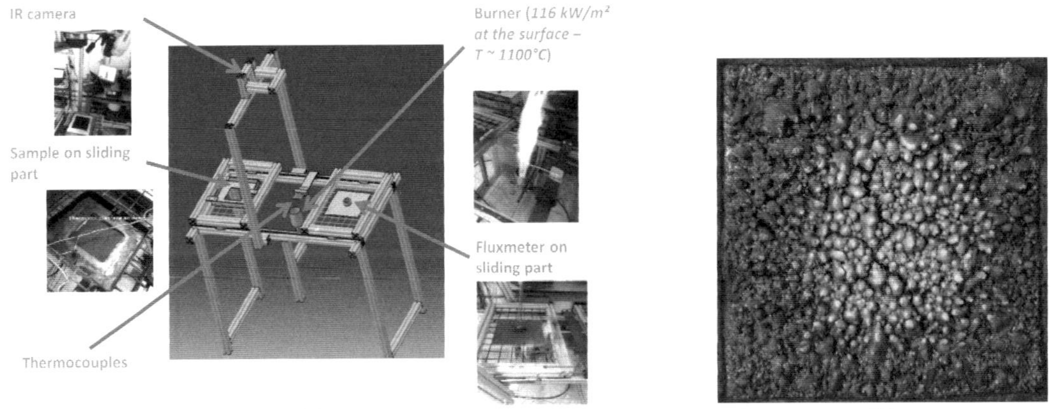

(a) (b)

CONCLUSION

Scaling reduction permitted to successfully build several bench-scale tests mimicking different fire scenarios and providing appropriate and reliable results. It was shown that prediction of fire performance at the large scale can be made. Those benches were used to examine the fire protection of several materials (steel, wooden panel and CFRP) by intumescent coatings. Intumescent coating offers highly efficient way to fire protect the substrates and the intumescence concept is therefore a method of choice in the field of fire protection.

REFERENCES

1. Bourbigot S, Le Bras M, Duquesne S, Rochery M. Recent Advances for Intumescent Polymers. Macromol Mater Eng 2004;289:499-511.
2. Alongi J, Han Z, Bourbigot S. Intumescence: Tradition versus novelty. A comprehensive review. Prog Polym Sci 2015;51:28-73.
3. Jimenez M, Duquesne S, Bourbigot S. Multiscale Experimental Approach for Developing High-Performance Intumescent Coatings. Ind Eng Chem Res 2006;45:4500-8.
4. McCrary-Dennis MC, Okoli OI. A review of multiscale composite manufacturing and challenges. J Reinf Plast Compos 2012;31:1687-711.
5. Bourbigot S, Gardelle B, Vandereecken P, Duquesne S, inventors; Dow Corning, assignee. Protecting substrates against damage by fire patent WO2013150121 (A1). 2013 13 October 2013.
6. Sikoutris DE, Vlachos DE, Kostopoulos V, Jagger S, Ledin S. Fire burnthrough response of CFRP aerostructures. Numerical investigation and experimental verification. Applied Composite Materials 2012;19:141-59.
7. La Delfa G, Luinge JW, Gibson AG. Integrity of composite aircraft fuselage materials under crash fire conditions. Plast, Rubber Compos 2009;38:111-7.
8. Quintiere JG. Scaling applications in fire research. Fire Safety Journal 1989;15:3-29.

HAZARDOUS MATERIALS

ANALYSIS OF FIRE BARRIERS WITH RESPECT TO FIRES WITH COMBUSTIBLE GASES AND LIQUIDS

Patrick van Hees[1], Jonathan Wahlqvist[1], Blanca Andres[1,2], Konrad Wilkens[1,2], Abhishek Bhargava[1,2], Karlis Livkiss[1,2]

[1]*Lund University, Division of Fire Safety Engineering, Lund, Sweden.*
[2]*Danish Institute of Fire and Security Technology, Hvidovre, Denmark*
**Corresponding author email: patrick.van_hees@brand.lth.se*

ABSTRACT

Fires including flammable gases and/or liquids differ from typical residential fires where the fuel is mainly composed of solid combustibles. The traditional ISO 834 curve was established to represent the time-temperature curve for a fully developed fire. However for combustible gases the type of thermal exposure might be substantially different and therefore other curves have been developed such as the hydrocarbon curve (HCC) and the jetfire exposure. Knowledge on how test results from traditional ISO 834 fire test can be used to estimate the behaviour under HCC or jetfire conditions is limited. The Swedish Contingency Agency MSB initiated research on this topic to find correlation or guidance rules. For this reason, both literature studies, statistical studies, calculation and fire tests were performed. Fire tests were performed at different scales and levels, and this paper focuses mainly on a real scale fire test conducted in a 2-container set-up where one of the containers doors was replaced by a typical fire barrier. It was observed that the exposure in a real fire is different with respect to pressure, temperature levels and temperature distributions. While the real fire test had a long period of lower room temperatures before it fully developed, the unequal temperature distribution as well as the radiative heat transfer from the pool fire resulted in a fire resistance, which was close to the rating of the wall according to ISO 834. Taking into account the results of the other fire test as well as calculations, it could be concluded that heat exposure from hydrocarbon fires will lead to a substantial decrease of fire rating. On top of this it was observed that certain materials do not resist sufficiently to, for example a jet fire exposure. Hence it is important to perform a full risk analysis when choosing an appropriate fire barrier for fires, which can include combustible gases and liquids.

KEYWORDS: Combustible gases and liquids, Fire Barriers, Fire Resistance.

BACKGROUND

Storage of combustible liquids and gases often lead to different types of fires than traditional cellulosic fires. The real scale fire exposure for fire barriers can therefore be different than traditional ISO 834 exposure[1]. International test standards are available to determine fire resistance according to other exposures. These correspond better with combustible liquids and gases. Examples of such exposures are the HCC curve (EN13633-2)[2] or the jetfire exposure (ISO 22899)[3,4]. However few results are available on similar type of fire barriers subjected to different exposures and to real scale tests.

For this reason the Swedish Contingency Agency MSB requested the University of Lund to conduct a research project in order to obtain an answer to a number of research questions such as:

- How should a fire barrier be constructed to protect an object when there is a fire outside? What literature is available?
- What is the real time-temperature curve for combustible gases and liquids and how does the room influence the results?
- Which full-scale test results exist under real scale conditions and what are the temperature and pressures inside the room or on the outside?
- How does a fire barrier tested according to ISO834 react when exposed to other conditions such as HCC and jetfires?
- How can pressure relief valves be dimensioned and what guidance rules are available?
- How should fire barriers be constructed to protect areas outside the room of origin?

This paper will focus mainly on the experimental part of the study i.e. pool fires inside rooms. For more information on the overall research and the detailed answers to the research questions, the project reports[5,6] are recommended. This paper should be considered as an extended summary.

METHODOLOGY

The overall methodology, used in order to obtain an answer to the research questions, consisted of a literature review, statistical studies, calculation and a number of orientation tests as given in figure 1.

Figure 1. Overview of the overall methodology.

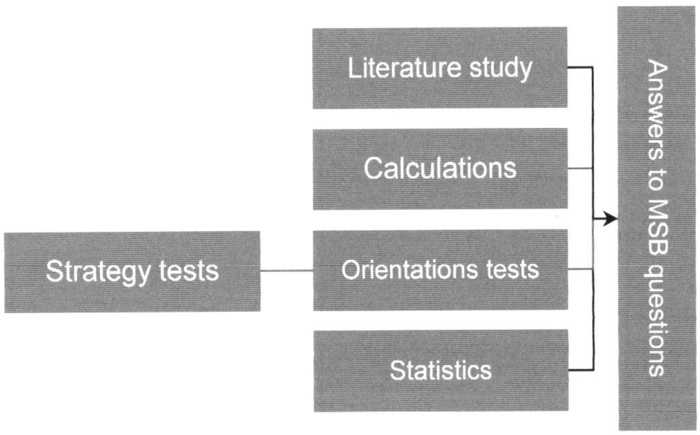

The testing strategy, which is the main focus in this paper, is given in figure 2. The choice of the tests and the number of tests where chosen subject to available resources.

Figure 2. Overview of testing strategy

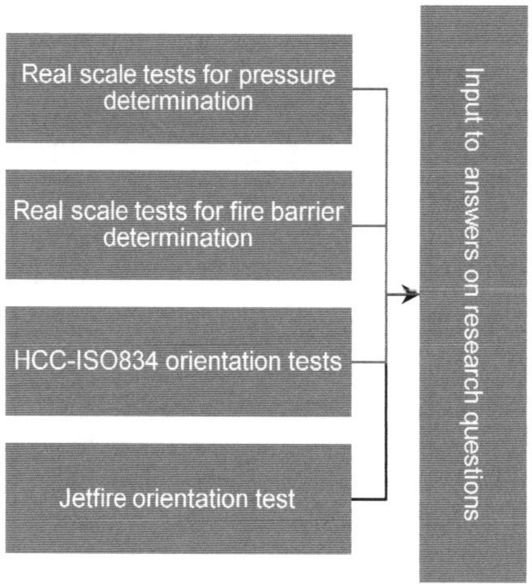

This paper focuses on the results of the fire tests. For the results of the other methods in the methodology the final report may be consulted[6].

CONDUCTED TESTS AND RESULTS

Comparative fire tests of elements according to ISO 834 and HCC Curve EN1363-2

In these tests four elements with size of 0.5x0.5 m where mounted in a cellular concrete wall. The elements were exposed according to an ISO 834 and HCC fire curve in a medium scale fire resistance furnace of 1.5m^3, see Figure 3. The elements were built up according to the Swedish guidelines for fire resistance of walls[7]. A description of the 4 elements is given in Table 1. For smaller elements only thermal isolation is the criteria, which can be investigated correctly due the size of the specimens. Integrity and stability should be checked at larger scale. Through temperature comparison however it could be observed that for the HCC curve the results for thermal insulation were reduced by approximately 50%[6,8]. The tests were later also used to validate thermal heat transfer models[8]. Figure 4 gives and overview of the temperature as a function of time for the 4 constructions.

Figure 3. Picture before and after the test in the medium scale fire furnace

Figure 4 Comparison of temperature measurements between the different constructions from Table 1 (HCC, ISO 834 curves and real scale)[6,8]

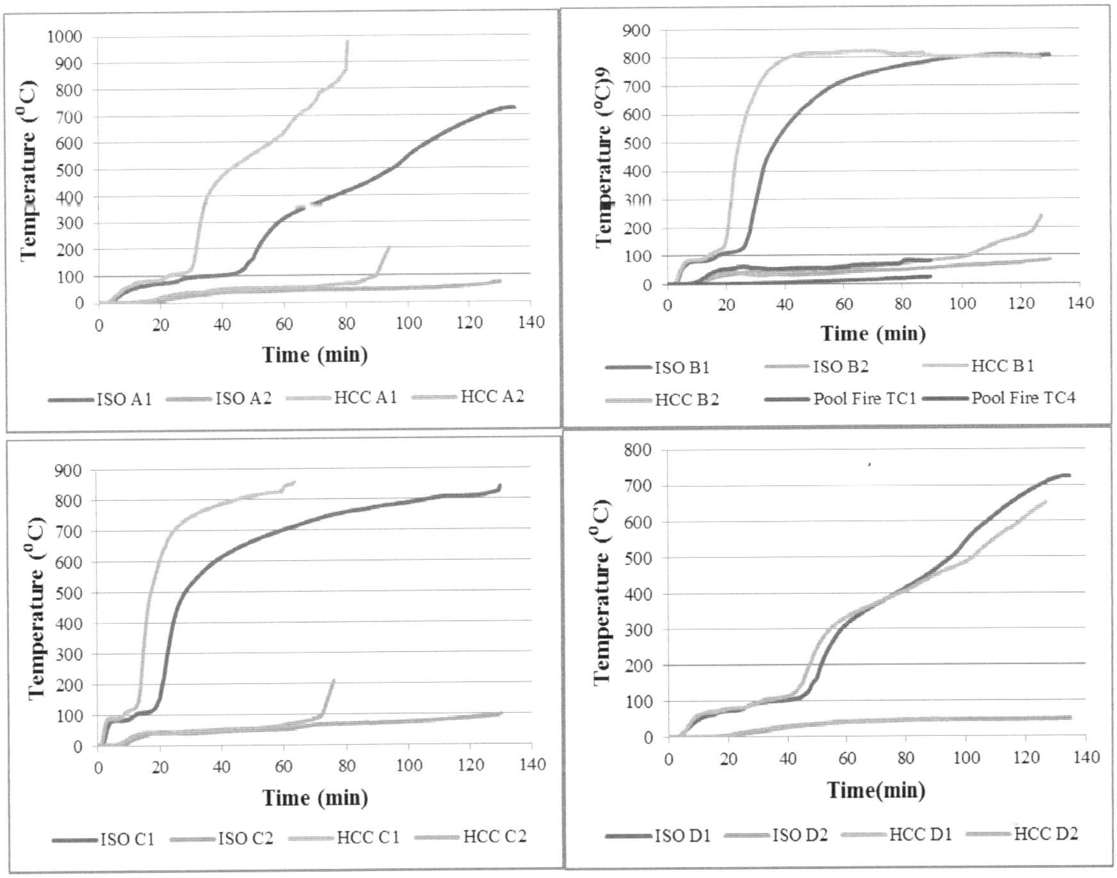

Legend to figure 4:

ISOx1: thermocouple on the inside of protections board (exposed side) for test assembly x during the ISO test
ISOx2: thermocouple on the unexposed side of test assembly x during the ISO test
HCCx1: thermocouple on the inside of protections board (exposed side) for test assembly x during the HCC test
HCCx2: thermocouple on the unexposed side of test assembly x during HCC test
Poolfire TC1: thermocouple on the unexposed side of the fire barrier during the real scale test position 1 (upper part)
Poolfire TC4: thermocouple on the unexposed side of the fire barrier during the real scale test position 4 (lower part)

Table 1. Tested constructions[6,8]

Assembly Tabulated value (ISO834)[a)]	Construction	Non-Exposed side	Wooden Studs	Exposed side	Insulation
A-EI60	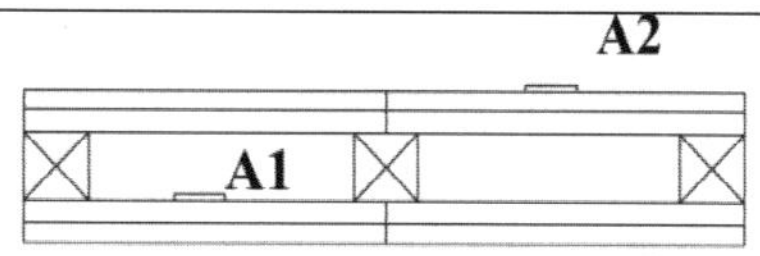	13 and 15 mm gypsum	45 mm	13 and 15 mm gypsum	None
B-EI60		13 mm gypsum	70 mm	10 mm masterboard and 13 mm gypsum	Mineral wool 95 mm
C-EI60	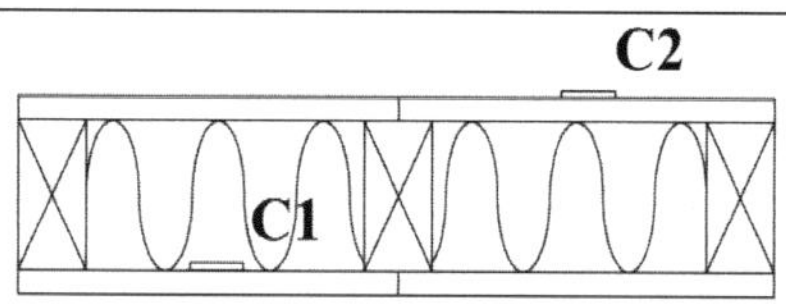	15 mm gypsum	95 mm	15 mm gypsum	Glass wool 95 mm
D-EI90	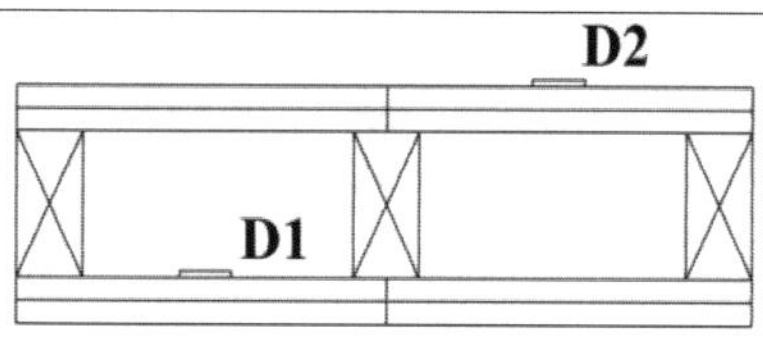	2 x 15 mm gypsum	95 mm	2 x 15 mm gypsum	None

a) According to Swedish guidelines[7]

Fire tests with a limited jet fire

A reduced scale jet fire test was developed for product development and a number of typical fire barrier constructions were tested as given in Figure 5. It was observed that certain lightweight constructions do not resist a jetfire exposure although they could meet ISO 834 requirements. This work was conducted as a master thesis project[9].

Figure 5 Picture of a reduced scale jet fire test.

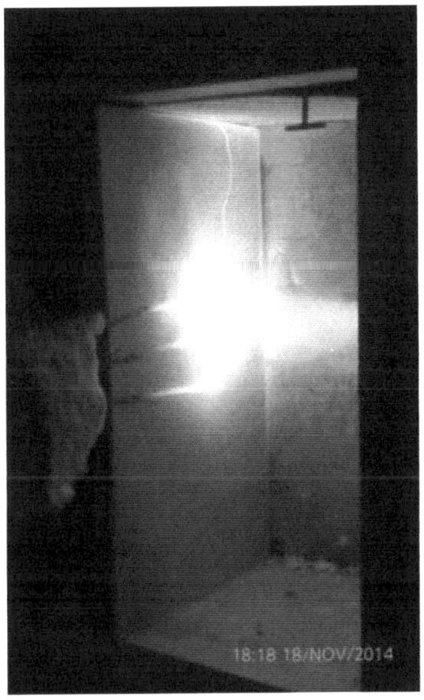

Real scale fire tests for pressure determination[10]

All tests were conducted on site at the Ringhals (Swedish nuclear power plant) fire brigade training field. The setup consisted of two standard 20 foot containers that had been connected to each other with a fire door, a fire damper and a floor drain (in each compartment connected through a pipe underneath the containers), see Figure 6 and 7. All of the connections could be either opened or closed which was used to create different smoke spread scenarios. The inside of the fire room had been insulated using mineral board (mounted using steel studs) and 70 mm of glass fiber insulation, the non-fire container was un-insulated. The inner dimension of the insulated container was 5.60x2.14x2.30 m (length, width, height). The inner dimension of the un-insulated container was 5.90x2.35x2.39 m (length, width, height).

The fire source consisting of a pool fire in two sizes (300x300 and 435x435 mm) was placed in the middle of the room, an array of thermocouples was placed in the centerline of the room 2 m from the center of the fire source with a spacing of 20 cm between, starting 20 cm from the floor (Figure 8). Measurement points for oxygen levels were placed at the edge of the pool vessel at a height of 30 cm and in the thermocouple array at a height of 1.8 meters over the floor. A differential pressure measurement was made by a small pipe inserted through the container wall and welded to get an airtight connection. A connection for a fan was placed in one of the lower corners of the fire room (see Figure 7c and 8), this was used to connect a fan standing outside of the containers in some scenarios where mechanical ventilation was of interest. Bi-directional probes were placed by the fire damper opening (see Figure 7 middle) and the floor drain opening in the non-fire compartment to be able to measure the flow induced by the fire in the cases where these were in open position. Thermocouples were also placed at the same positions with the addition of one thermocouple in the fan inlet pipe.

The experimental setup was very tightly sealed since the effect of the fire pressure was of prime interest. All corners, joints and screw holes were sealed with fire resistant joint filling to prevent leaking in other places than those that were being monitored.

Figure 6 a, b. Overview of the experimental setup on site (left) c. the connecting pipe between the floor drains in the two rooms (fire and non-fire room) (right picture).[10]

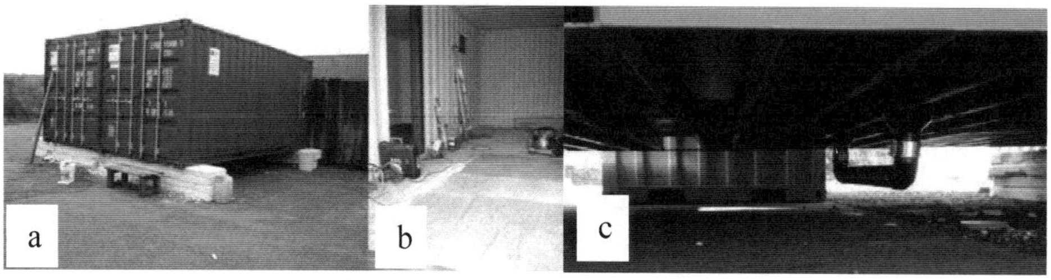

Figure 7 The placement of the fire damper, the left picture is from inside the fire room (a), the middle picture is in the non-fire room where the bi-directional probes and coupled thermocouples were placed (b). The right picture is a close up picture of the connection pipe (200 mm) where the fan was connected (c). The picture is taken from inside the fire room; the fan was placed on the other side of the wall (outside).[10]

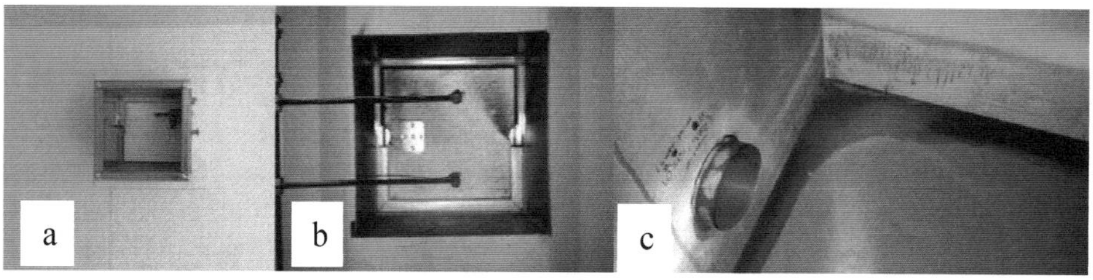

In total 15 different test set-ups were selected and they can be in detailed found in references 5 and 10. Testing parameters were different type openings, fire size, fan speed and damper mode. In this paper test nr 6 is presented as an example for the overpressure. For this test, the fan inlet was plugged, the floor was drain opened up and was filled with water (approximately 10-15 cm) at both ends of the pipe (both fire room and non-fire room), the fire damper was closed and the fire door was closed. A 435x435 mm^2 pool fire containing heptane was used. Figure 9 gives the pressure in the room as a function of time. For many tests overpressure between 200-400 Pa were observed, sufficiently to empty e.g. the water drains.

Figure 8 3D overview of the experimental setup. The fan connection is in the top left corner, the fire source is placed in the middle of the left room, the floor drain is placed close to the fire and its position is mirrored in the other compartment. The fire door and fire damper placement can also be seen in the middle wall.[10]

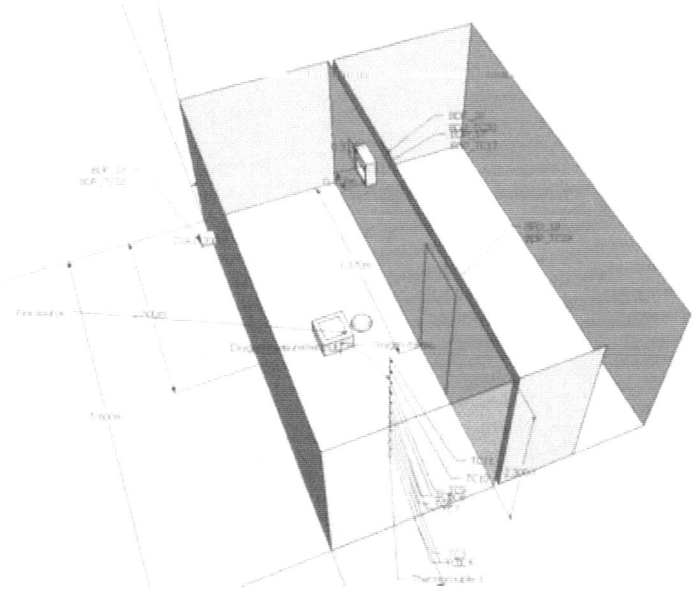

Figure 9 Pressure (Pa) inside the room as a function of time(s) for test nr 6.[10]

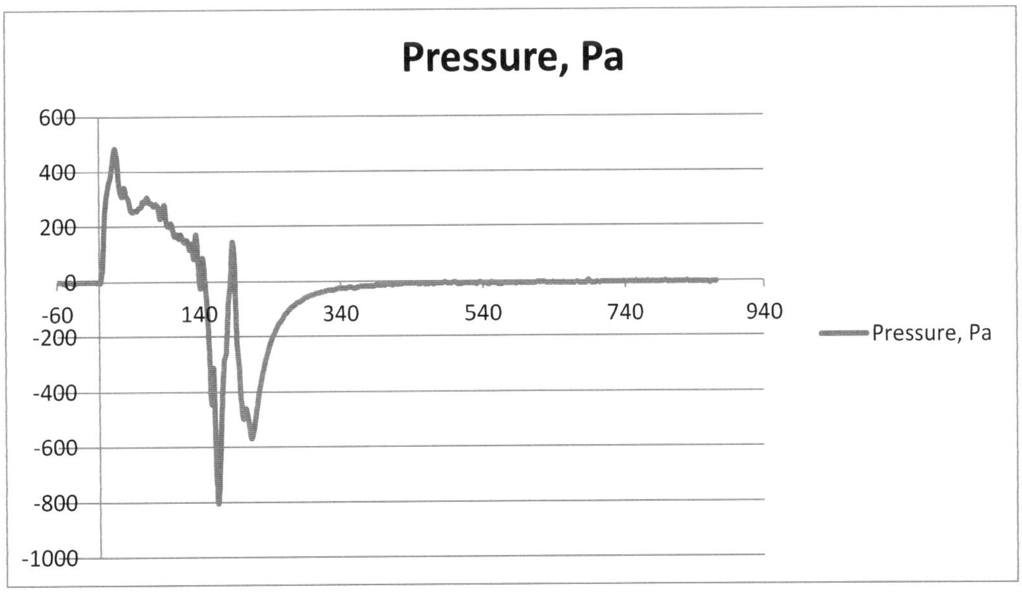

Real scale fire test for fire barrier evaluation[6,8]

In this test 100 litres of heptane was burned in a typical barrel used for storage (Figure 15). The barrel was placed inside a shipping container and first filled with water. The container was connected to another container by means of a standard door opening. The front opening of the first container, containing the barrel, was closed by means of a fire barrier with a one-hour fire rated wall. This wall was even protected on the inside by an extra thermal protection layer of fibre silicate board and corresponds with wall nr B in the medium fire resistance furnace tests. The second container had an opening in the front to the outside to allow fire gases to escape from the first container through the door and through the second container to the outside (see figure 10). The barrel was placed close to the fire barrier wall to be evaluated. The test results showed much more local exposure and uneven temperature distribution (see Figures 11, 12a and 14). Despite the fact that the fire in the room only obtained flashover levels in the last 30 minutes and that the levels before that were much lower than in ISO 834 (see Figure 13), the tests resulted in a fire resistance time very close to exposure with ISO 834 i.e. 92 minutes. The time temperature curve of the upper thermocouples on the unexposed side followed mainly the HCC curve in the medium scale furnace while the lower thermocouples followed the ISO 834 curve (see Figure 4). All tests were conduced at the Ringhals (Swedish nuclear power plant) fire brigade training field.

Figure 10 a. Overview of the experimental set-up. b. Picture during the test.[6,8]

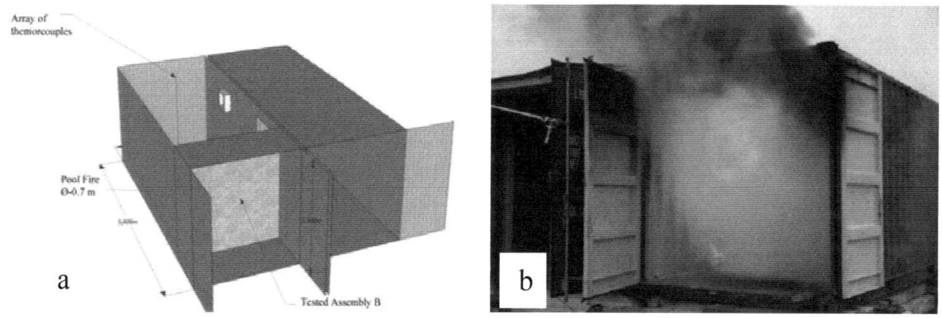

Figure 11 a. Uneven thermal exposure as seen by a thermal camera on the right. b. Thermal hot spot at the end of the test at 90 minutes.[6,8]

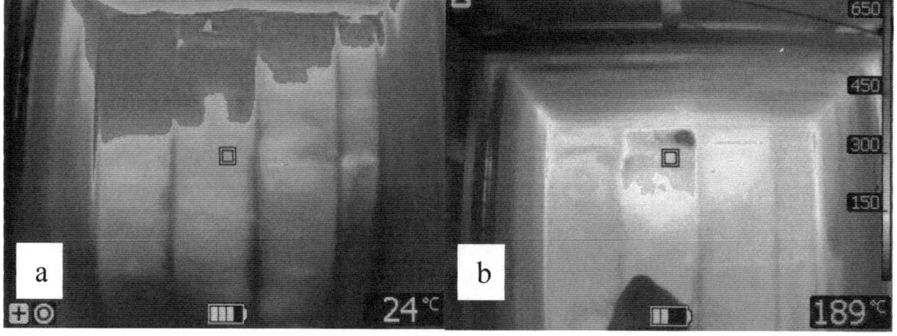

Figure 12 a. Uneven thermal exposure in the corners observed by a thermal camera on the unexposed side. b. Unexposed side of the fire barrier at the end of the test at 90 minutes.[6,8]

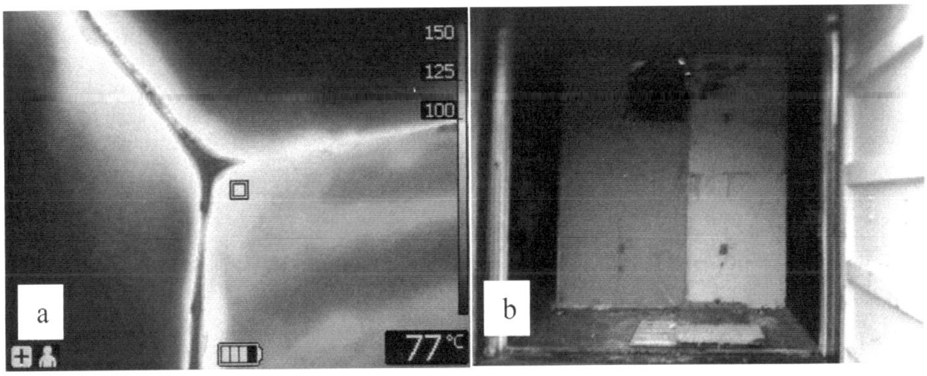

Figure 13. Temperature as a function of time inside the fire room for the real scale fire tests in comparison with ISO 834 and HCC curve, at 60 minutes the fire became more turbulent as the inner structure started burning[6,8]

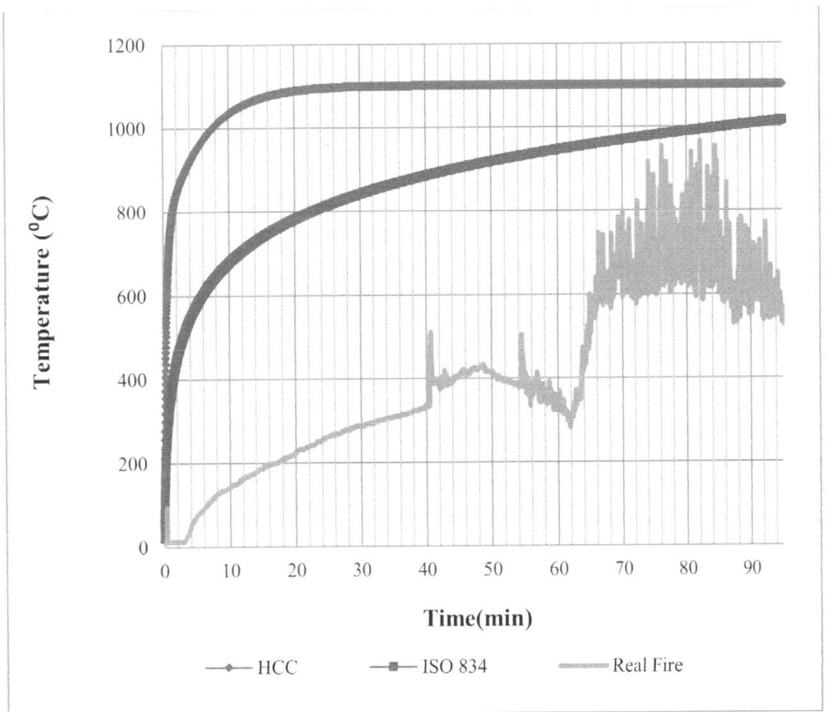

Figure 14. Temperature as a function of time on the unexposed side of the fire barrier, showing the large differences between the thermocouples on the upper part (1,2) and the lower part (4,5).[6,8]

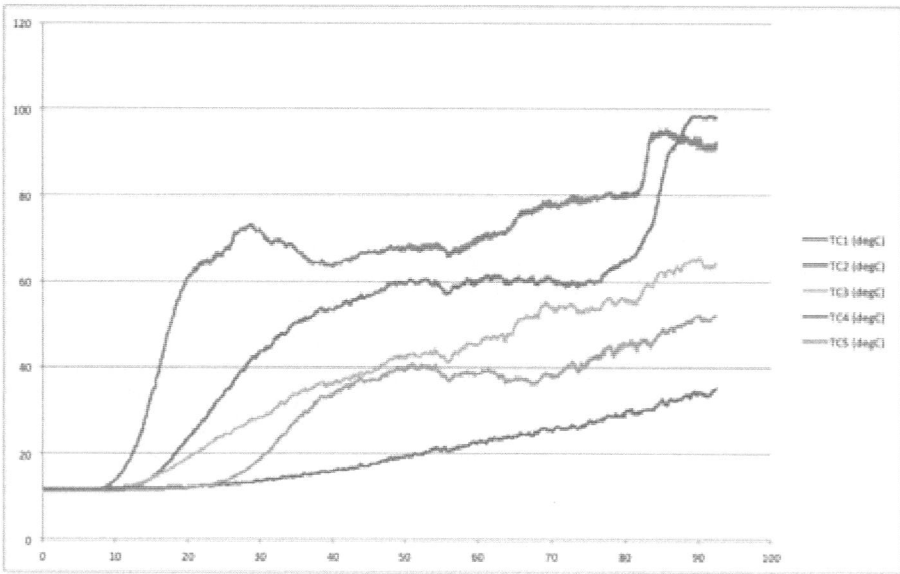

Figure 15 a. Picture of the fire source at the beginning of the test. b. During the test. c. Thermographic picture of it during the test.[6,8]

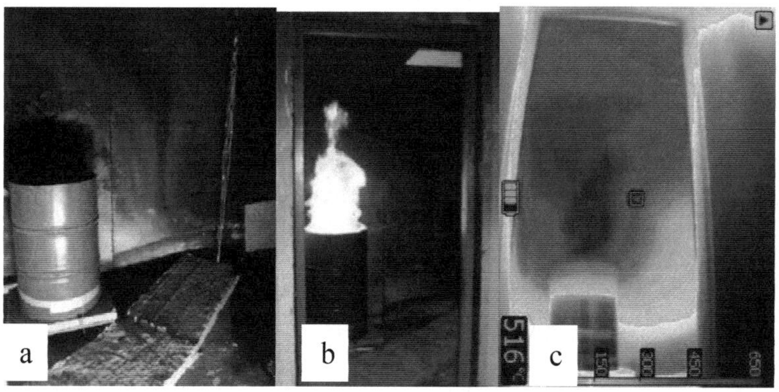

CONCLUSIONS

The results of this study show very clearly that the behaviour of fire barriers is different with respect to fire resistance time depending on the type of exposure. While this might be obvious, it was clear in this project that few comparative data sets were in fact available in the literature. Therefore a number of orientation tests in a fire resistance furnace were performed as well as real scale pool fire tests inside a container set-up with the same type of constructions. The comparison between HCC and ISO 834 tests showed a decrease in time to critical temperature on the non-exposed side by approximately 50%. The tests on a reduced scale jet-fire tests revealed that certain light-weighed constructions do not resist against a jet-fire exposure. The real scale tests revealed that the thermal exposure is different than in the standardised test methods. It is mostly the local exposure and unevenly distributed temperature profiles, which seem to subject the fire barriers wall to different scenarios. Furthermore, it was observed in the tests that the pressure inside a closed room was substantially higher than in standardised tests. Finally a number of safety factors were given to the regulators. An overall observation was that much more knowledge and structured fire research is needed for determining the properties of fire barriers subject to non-standard, different fire exposures.

ACKNOWLEDGEMENTS

The authors would like to thank Ringhals AB, DBI Danmark and Promat International Belgium for providing their testing facilities. For the tests at Ringhals the authors would like to thank Fredrick Carlstein (Ringhals AB) and Tommy Magnusson (Ringhals AB). For the tests at DBI the authors thank Christian Andersen (DBI), and Dan Lauridsen (DBI). For the tests at Promat RTC the authors thank Christoffer Schön (LTH), Joris Jacobs (Promat RTC) and Kees Both (Promat RTC). Finally the authors would like to thank the Swedish Contingency Agency (MSB) for the funding of this project and especially David Gardsjö for the fruitful discussions during the project.

REFERENCES

1. CEN, EN 1363-2:1999, Fire resistance tests Alternative and Additional procedures, CEN Geneva 1999
2. ISO 834, ISO 834-1:1999 Fire-resistance tests -- Elements of building construction -- Part 1: General requirements, Geneva 1999, (1999)
3. ISO, ISO 22899-1: 2007-12 (E), (2007), Determination of the resistance to jet fires of passive fire protection materials - Part 1: General requirements, Geneva, (2007)
4. ISO, ISO/TR 22899-2:2013, Determination of the resistance to jet fires of passive fire protection -- Part 2: Guidance on classification and implementation methods, Geneva, (2013)
5. van Hees, P., Bjarne Husted B., Johansson, N., Brandteknisk avskiljning vid hantering av brandfarliga gaser och vätskor – fas 1, LTH report nr 3176, LUTVDG/TVBB--3176—SE, Lund, (in Swedish), (2013).
6. van Hees, P., (2014), Brandteknisk avskiljning vid hantering av brandfarliga gaser och vätskor – fas 2, LTH report nr 3187, LUTVDG/TVBB--3187—SE, Lund, (in Swedish), (2014)
7. Bengston, S., Frantzich, H., Jönsson, R. & Marberg, P.-A., 'Brandskyddshandboken.', Lund University, (2012)
8. Andres B., A., van Hees, P., Experimental and thermal analysis of wall assemblies exposed to standard and parametric fires, Confab conference First international conference on structural safety under fire and blast, Glasgow September 2015, (2015)
9. Schön, C., Jet and hydrocarbon fires exposure to passive fire protection. An investigation into the suitability of the interpretation of the laws regulating risks with jet and hydrocarbon fires and its consequential compliance concerning fire separation, LTH report, Lund, (2016)
10. Wahlqvist J., Andres B., Bhargava A., Livkiss K., Wilkens K., van Hees P., Experimental study on enclosed fires with various interconnections between compartments, LTH report, (2013)

Indoor Fire Testing of Transportation Container Systems to Meet DOT Hazardous Material Classification Requirements

Debbie Finfrock, Walt Gill, Anay Luketa, and Brandon Servantes
Sandia National Laboratories, Albuquerque, NM*

INTRODUCTION

Assigning a shipping container and its explosive contents to the proper Division in Hazard Class 1 (HD) for transportation purposes requires conducting an actual fire test with one or more packages. The procedures for making the assignment are given in Department of Defense Ammunition and Explosives Hazard Classification Procedures, TB 700-2 [1]. Details for conducting the test are given in External Fires, Munition Test Procedures, STANAG 4240 [2]. The requirements, while intentionally straightforward, can be difficult to meet in an outdoor setting.

In particular, the requirement stating "A quantity of fuel sufficient to maintain a fully developed fire for 150% of the estimated time required to cause all explosive material to react" is rendered difficult-to-meet due to the presence of wind. While both pretesting with mockups and the employment of computer simulation modeling are normally used to estimate the required time, these approaches have limited value in outdoor setups. Heat absorbed by the package from the fire can vary as much as a factor of 4 depending on wind speed and direction at any instance during the fire. This large uncertainty in heat input translates to large uncertainty in pre-determining the expected time-temperature histories in the package interior.

A tacit assumption in other requirements concerning thermocouple readings during the fire, the location of the packages in the setup, and the size of the fuel source, is that the packages are engulfed in flame during the entire test. This rarely occurs in outdoor testing as the packages, flames and wind all interact in a coupled manner that produces random, large-scale, time-varying flow features. This coupling is somewhat allayed by following the requirements, but ultimately conformance is a post-test qualitative assessment of video records and thermocouple traces. The programmatic risk from non-conformance to these requirements is that of finding unconsumed explosives in the packaging during post-test analysis and having to repeat the test – with a longer burn, a larger fire, or better wind conditions. In any event, the result is increased cost and time delays both for the shipping container development and for the program depending on component shipping. At Sandia, we have chosen to mitigate this risk by moving the testing indoors.

Here, we present a two-test series to qualify a container and contents for HD 1.4 classification that has been done in the FLAME facility as an example of how moving indoors completely eliminated the programmatic risk discussed above. Pre-test modeling with a high fidelity fire model fitted with virtual thermocouple instrumentation shows how the fire and packages were configured to conform to requirements. The full dress

* Sandia National Laboratories is a multi-program laboratory managed and operated by Sandia Corporation, a wholly owned subsidiary of Lockheed Martin Corporation, for the U.S. Department of Energy's National Nuclear Security Administration under contract DE-AC04-94AL85000.

rehearsal fire conducted to validate the modeling and to determine the time-temperature history of internal package locations is treated. Finally, results from the actual qualification fire will be presented to show how instrumentation, such as infrared (IR) video, was used to demonstrate full compliance with requirements.

REQUIREMENTS

The overall strategy in the classification fire test is to expose the shipping container system to a fire that will ultimately react all of the contained explosive material and to observe the effects of that reaction on the package and its surroundings; be it an energetic event with blast and shrapnel or a benign event with no effect. The fire is to be typical of a transportation accident where all of the explosives are consumed.

To insure the fire is typical, the TB 700-2 document addresses the test setup by specifying prototype packages under test be placed on a rack above a pool of liquid hydrocarbon fuel and meet the following conditions:

a) The packages are to be in their transportation and storage configuration (e.g., pallet or unit load) with a total volume of at least 5.3 cubic feet [ft^3, or 0.15 cubic meter (m^3)] or a minimum of three packages, whichever is the greater.

b) The fuel basin filled with liquid fuel (i.e., JP-4, JP-5, Jet A-1 (JP-8), and AVCAT (NATO F-34 or F-44)), or commercial kerosene (Class C2 /NATO F-58) at least 1 meter (m) larger than the stack in every direction.

c) A quantity of fuel sufficient to maintain a fully developed fire for 150% of the estimated time required to cause all the explosive material to react.

d) A metal grid to support the packages approximately 0.3 – 1 m above the fuel surface at the onset of the test.

There are other requirements for instrumentation associated with the confirmation of package performance but are not germane to the present discussion. With the exception of requirement "c," these configuration requirements are intended to insure the packages are fully engulfed and are exposed to high temperature flames.

The STANAG 4240 document further imposes requirements on the performance of the pool fire and describes a valid test:

e) An average flame temperature of at least 800°C, as measured by all valid thermocouples (sample rate > 1.0 Hertz) at the test-item without contribution from the burning munition, will be considered a valid test. This temperature is determined by averaging the temperature from the time the flame reaches 550°C until all munition reactions are completed.

Issues with the requirements in outdoor testing

Requirement "e" will typically force the metal grid system to the 1 m height, as the lower region of a pool fire in no-wind conditions is generally composed entirely of fuel vapor and is unable to burn. However, pushing the test item up in the fire renders it vulnerable to the effects of wind. Fuel vapor is heavier than air, and a slight breeze will push it out from underneath the test stand, resulting in flames occurring on and attaching to the downwind side of the test item. The temperatures on the downwind side can be quite high due to enhanced mixing. Although the heat flux from the flames to the package surface can be very high as a result, the overall average heat input to the package can be low, leading to an extended period of time required to heat the internal package contents to a desired level (typically 400°C to 600°C) that insures the complete utilization of the explosive material.

Following the pool size requirements in TB700-2 will alleviate some of the wind effect but will not eliminate it. Wind barriers near the pool would be highly effective, but are not allowed due to a requirement for an 11-m diameter cleared area around the fire for blast, shrapnel and heat flux observations. Figure 1 shows the observed variation in heat flux from the fire to an engulfed object in a fire set-up as described in the standard with wind barriers beyond the 11 m limit. The solid line indicates the nominal evolution of flux absorbed by the package surface as the package heats up. The range in observed absorbed flux was gathered from similar outdoor pool fires over a period of 20 years. The wide range is due to varying wind conditions from fire to fire and brings out the point that the outdoor fires considered here are not repeatable. This in turn makes meeting requirement "c" difficult, if not impossible.

Figure 1

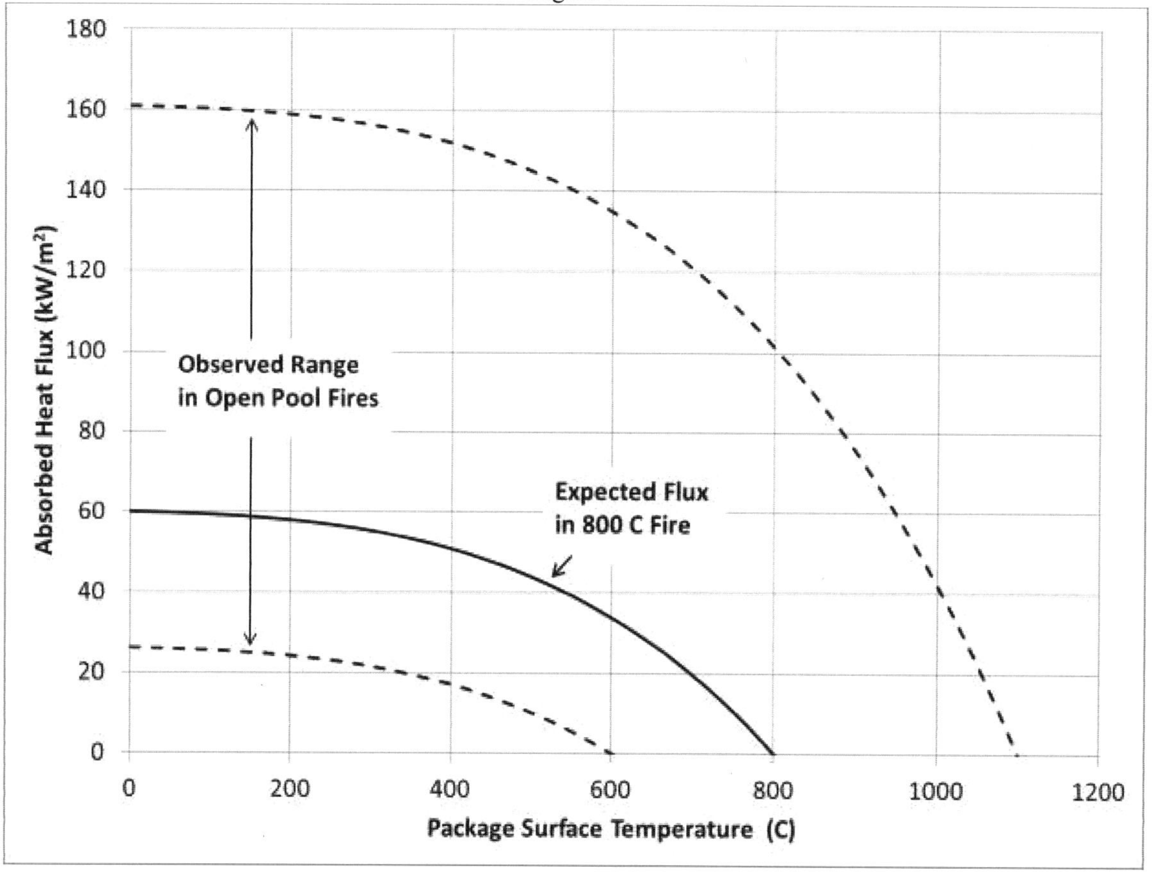

The poor definition of the heat flux at the boundary of the package limits the use of modeling for determining the time required to reach desired internal temperatures. Installing internal instrumentation violates the package integrity and can compromise the interpretation of the test outcome. Running the test with a dummy instrumented package to determine the internal temperature response followed by a test with prototype package for the 150% time extension is not viable because of the inability to reproduce the same heating conditions from test to test.

Mitigation of the issues

It is understood the reason behind the 150% time requirement is to insure that all the explosive material in the package has reacted and no further possibility of some characterizing event can occur (blast, fragment, or sudden heat flare). In practice, one estimates the time and accepts the risk that post-test, unexpended explosive may be found and the test will have to be repeated. At Sandia, we have recognized the source of the problem is the presence of wind, and have elected to mitigate the risk by moving the fire indoors whenever possible.

Our indoor facility (Figure 2), FLAME, is a seven-story ~20 m-diameter test cell that has water-cooled walls and well controlled/characterized airflow equipment. Systems to allow jet fuel, methanol, and other liquid fuels as well as hydrogen, methane, and other gas fuels are part of the design. The facility has served as an international resource for development and validation of applicable high-fidelity, full physics fire models. Because of this, fires in the facility are highly repeatable and well understood.

The dimensions of the cell allow for performing the qualification fires for packages that would fit in a 3 m diameter pool fire (~ 55 gallon drum size). The amenability to modeling allows package testers to design the fire fixtures and make accurate predictions of the package and thermocouple response. An example of how this capability was used to advantage is given in the following sections of this paper.

Figure 2

EXAMPLE OF A QUALIFICATION TEST

Description of package and test set up

The package under test was a nominal nine-gallon (34-liter) steel drum containing explosive parts and organic foam fill. The drum was fitted with a standard locking lid and gasket assembly. Two burn-out vents were installed to prevent pressure build-up during the fire from the decomposing foam fill material. The configuration of the explosive components and total quantity were such that explosive limits of the burn facility were not exceeded.

Figure 3 shows the drums installed in the test set-up. In accordance with the requirements, four drums are placed on a steel grill above a 3-m diameter fuel pool. Witness screens for recording potential projectiles, Bikini blast gauges for detecting potential overpressure release, and heat flux gauges for capturing potential heat flares are shown deployed in the clear area around the fire. Cameras, both visual and IR, were used to record the packages' response.

Type K, Inconel-sheathed, mineral-filled thermocouples were installed on the drum sides, tops, and bottoms to demonstrate compliance with the 800° C flame temperature requirement. In Figure 4, the thermocouples can be seen banded to the drum sides and bent over to extend off the surfaces.

Figure 3.

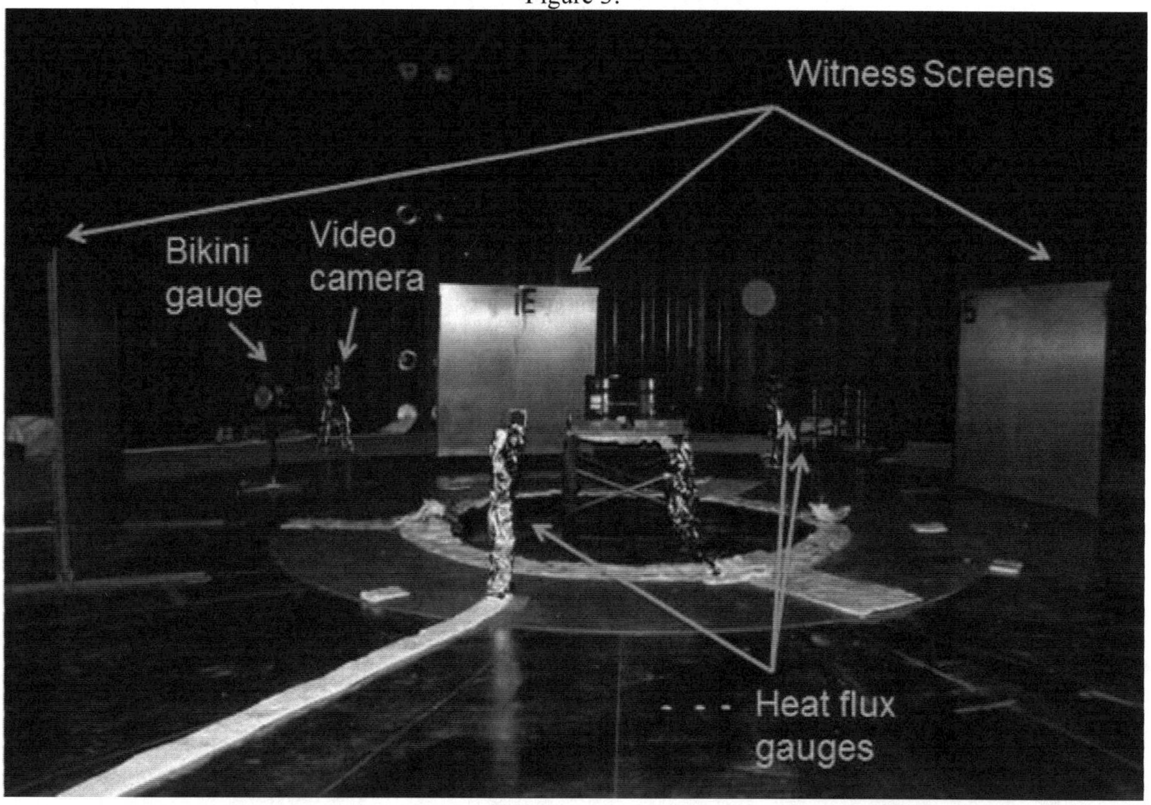

Figure 4.

In a hydrocarbon fuel pool fire, flame temperature is the temperature registered by a thermocouple, and is a strong function of location. To ensure compliance with the 800°C flame temperature requirement, computer simulations were completed prior to the test set-up in order to select the proper height of the steel grill above the pool.

Computational simulations

The objective of the simulations was to determine if average thermocouple temperatures within 40 to 60 millimeters (mm) of the surface of the containers reach at least 800°C. The simulation was performed using Fire Dynamics Simulator (FDS), a computational fluid dynamics open source code developed by NIST. Simulation is routinely used within Sandia's Fire Science and Technology Department to both facilitate the design of fire experiments and to explore a greater range of parameter space, beyond that covered by experiments. Once experimental data is obtained, validation is often performed in order to gain confidence in the predictive capability of the simulations.

The FLAME facility allows for the study of fire environments in an enclosed facility where boundary conditions can be controlled and characterized. The facility is cylindrical with a diameter of 18.8 meters and has a sloped ceiling of approximately 18 degrees where the height transitions from 12.2 m to 14.6 m. The outlet is at the center of the ceiling and is 4.9 m in diameter and exits to a rectangular 3.0 by 3.7 m duct. The walls of the facility are water cooled to prevent radiation interactions. The burner at the center of the facility can burn gaseous or liquid fuel up to 3-m in diameter and is surrounded by a 12.7 m diameter steel plate. Air is provided from a ring at the floor of the facility and air and combustion products are extracted at the outlet at a rate to replicate a quiescent outdoor pool fire.

Figure 5 shows a cut-away of the FLAME facility and the test set-up created for the simulations. Shown is the 3-m diameter center burner with a 1.68 m square platform supporting four transport containers (shown in orange) above the burner. The containers are 0.46 m in height, 0.41 m in diameter, and are arranged to form a square as seen from above. Five simulations were performed, three of which explored different heights of the platform of 0.46 m, 1 m, 1.3 m in order to determine if elevation within the fire would affect thermocouple temperatures surrounding the containers, and one to assess the effect of allowing the containers to become fully heated such that their surfaces radiate providing additional heat transfer to the thermocouples. Early simulations used only three drums. Due to a subsequent update on the number of containers to be tested, a final calculation was performed with four drums.

Figure 5

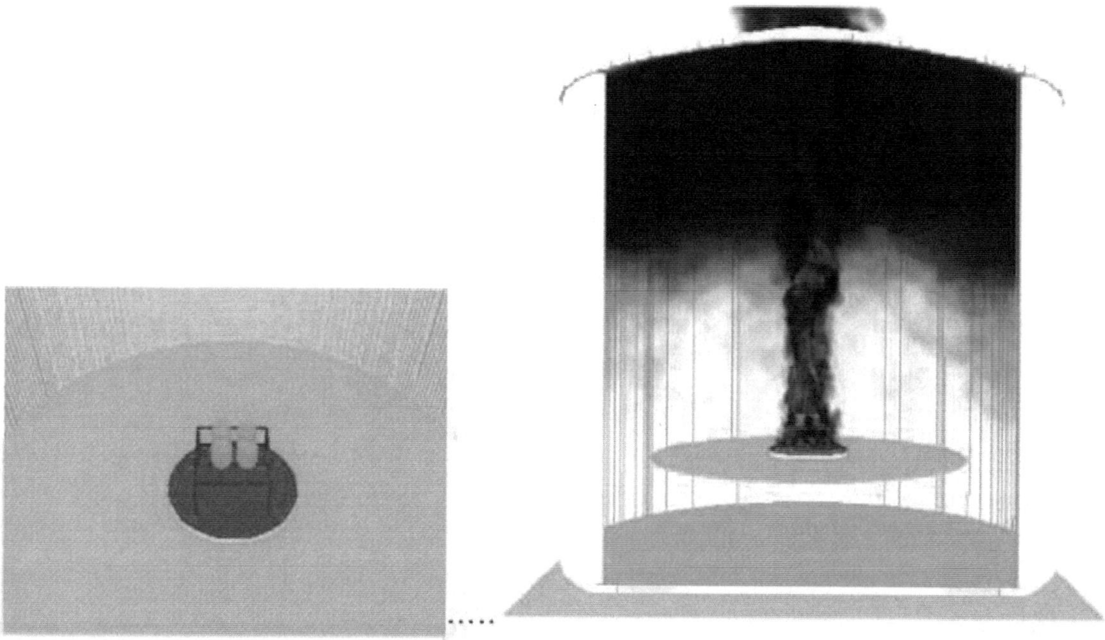

A mesh size of 12 million elements was used and a cell size of 0.05 m was used to resolve regions encompassing the fire. Each simulation was performed for 60 seconds of real time and took about 3 days of computational time on 36 processors. The heat of combustion and fuel mass flux was specified as 46,000 kilojoule per kilogram (kJ/kg) and 0.038 kilogram per meter squared-second (kg/m^{2}s), respectively, which provides a heat release rate of about 12.4 megawatts (MW). For three of the simulations, the containers were not allowed to heat up since the simulation time was not of sufficient duration. Thus, the containers were essentially at room temperature. For one of the simulations, the containers were assigned a very low thermal conductivity. This has the effect of minimizing heat conduction into the container and allowing the surface temperature to rise to a value that is representative of steady-steady heat transfer conditions. This strategy was used to represent the heating up of the containers, since it would be computational prohibitive to perform a calculation for 3600 seconds, representative of the duration of a test (~1 hr).

The thermocouple model within FDS was used to evaluate temperatures. A thermocouple model was required since actual thermocouples do not measure gas temperature due to their inherent heat transfer pathways. Thus, in order to have an equitable comparison with experimental measurement, simulation thermocouple temperatures were evaluated. Thermocouples will typically read about 200 to 300°C below gas temperature [3, 4]. Thus, for specification requirements, the temperature criteria should refer to thermocouple temperatures rather than gas temperatures.

For the simulations with three containers, 36 thermocouples were placed around the containers within a distance from 40 to 60 mm from their surfaces. For the simulation with four containers, 120 thermocouples were evaluated. Figure 6 shows example results where the platform height of 0.46 m above the pool is predicted not to meet the required flame temperature criteria.

Figure 6

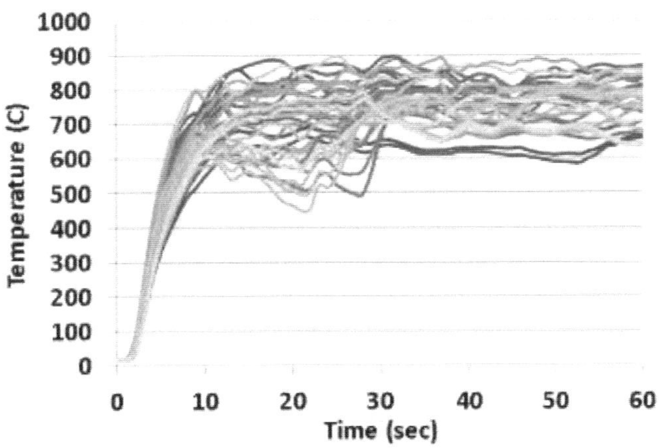

Table 1 provides the average among all thermocouples during steady-state for each case, as well as their range and standard deviation. The results indicate that the height of the platform affects thermocouple temperature. The height of 0.46 m results in the lowest average thermocouple temperature, since the simulated packages were more within the cooler core. Raising the platform to 1.3 m results in the highest average temperature, however, there are still several thermocouples below 800°C. Including drum surface re-radiation brings the thermocouple average above the requirements at 1 meter.

Table 1: Average thermocouple temperatures, ranges, and standard deviations for all cases.

Number of containers	*Platform height above pool (m)*	*Average temperature of all thermocouples (°C)*	*Average range (°C)*	*Standard deviation (°C)*
3	0.46	754	45	16
3	1	784	45	15
3	1.3	798	43	14
3	1*	904	51	15
4	1*	906	80	23

*Drums contributing to radiation to thermocouples.

Based on the variation in temperature shown from these results, it should be noted that in an experiment, the placement of thermocouples can result in very different outcomes. There is a potential to bias results by clustering thermocouples in certain locations. Thus, testing requirements should include a sufficient number of thermocouples to adequately map the surface of the containers. As mentioned previously, these simulation cases were performed with containers that stay at ambient temperature and were not allowed to heat up. When the containers were theoretically allowed to heat up and re-radiate to the thermocouples, which is representative of actual conditions, all thermocouples temperatures were above 800°C. The case where the platform was at a height of 0.46 m would similarly result in higher temperatures if the containers were allowed to heat up, but potentially not all thermocouples would rise to the required temperature, since some were as low as nearly 600°C. Based on the results from these simulations, a platform height of 1 m was judged to provide sufficient conditions to result in an average thermocouple temperature of 800°C.

Fire Test Results

Two actual fire tests were performed. In the first mock test, no explosives were installed in the drums; however, the organic foam fill was included. Thermocouples were routed into the locations where components would be installed and attached to small metal masses sized to mimic the heat capacity of the components. Prior to the test, explosive subject matter experts established 600°C as being the critical temperature for ensuring all the explosive material would be consumed. As can be seen in Figure 7, the internal temperatures reached 600°C approximately 10 minutes after the start of the fire. Also, comparing the thermocouple readings for the model and experiment clearly demonstrates that one cannot expect one-to-one correspondence, however, on average (on space and over time) model predictions can be used to select acceptable test configurations.

Figure 7

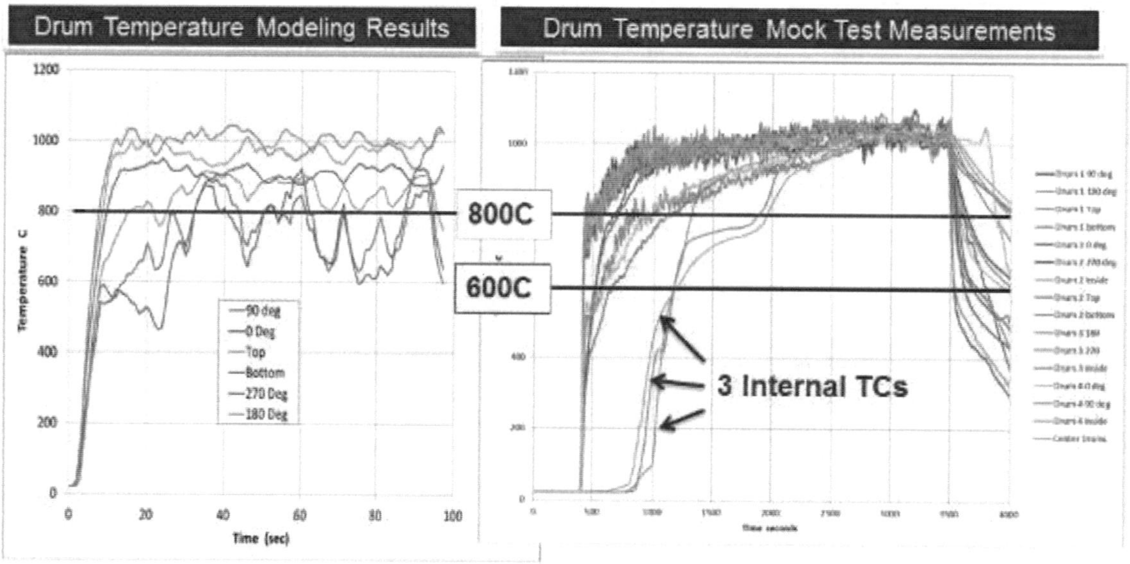

800°C External STANAG Requirement
600°C Internal Explosive Safety Requirement

In the second test, explosives were installed in the drums. No internal thermocouples were used so as not to compromise the integrity of the package walls. The burn was conducted for 30 minutes, which was greater than the required 150% time for explosive consumption. Figure 8 shows a wide view of the fire, a close up view of the steel grill engulfed in flames, and a wide IR view. The IR view demonstrates full engulfment of the packages by the flames. No energetic events were observed during the course of the burn. Post-test examination of the drums confirmed all explosives were consumed.

CLOSURE

We have shown there is advantage to performing HD classification fires indoors. All requirements of TB 700-2 and STANAG 4240 were met with a low programmatic risk by being out of the wind. This was possible because the candidate package contained a small amount of explosive and fit in a 3 m diameter pool. Larger sized packages and larger explosive quantities will still have to be conducted outdoors as the risk of facility damage weighs against the programmatic risk of having to repeat the tests.

The required test time was determined by performing two tests – an instrumented mock test followed by a test with the prototype package. Repeatability of the indoor fire allowed this approach. As modelling capabilities advance to the point of confidently coupling a package response model to the fire model, the need for a preliminary mock fire will be eliminated. Amenability to modelling achieved by the well-controlled boundary conditions in the indoor setting is the enabling technology in this approach.

Figure 8

REFERENCES

[1] Department of Defense Ammunition and Explosives Hazard Classification Procedures, TB 700-2

[2] External Fires, Munition Test Procedures, STANAG 4240

[3] Brundage, A.L., Donaldson, A.B., Gill, W., Kearney, S.P., and Nicolette, *Thermocouple Response in Fires, Part 1: Considerations in Flame Temperature Measurements by a Thermocouple*, Journal of Fire Sciences, vol 29 – May 2011

[4] Brundage, A.L., Donaldson, A.B., Gill, W., Kearney, S.P., and Nicolette, *Thermocouple Response in Fires, Part 2: Validation of Virtual Thermocouple Model for Fire Codes,* Journal of Fire Sciences, vol 29 – May 2011

CHARACTERISATION OF A VEHICLE FIRE ON A 7 TONNE LPG ROAD TANKER

Lorraine Jenney

Spadeadam Testing and Research Centre, DNV GL, UK

INTRODUCTION

Since 2013, there have been two incidents involving 7 Te LPG road tanker vehicle fires in the UK. The first event occurred in 2013 on the A1 in Northumberland near Belford; several Fire and Rescue Services attended the scene and an area of 1000 m radius was cordoned off with an exclusion zone resulting in long delays on the main route between the North East of England and Scotland. The Fire and Rescue Service attempted to keep the tank cool to reduce the likelihood of the 7 Te LPG tank failing and resulting in a Boiling Liquid Expanding Vapour Explosion (BLEVE). A second tanker was involved in an incident in August 2015, this incident occurred on the M56 near Chester resulting in a 1600 m exclusion zone being put in place around the vehicle. The exclusion was put into place as the Fire and Rescue Service could not approach the tanker due to safety concerns. As the Fire and Rescue Service could not tackle the vehicle fire it was allowed to burn out before approaching, this caused the M56 and the nearby railway line to be closed for 6 hours in both directions[2]. On both occasions there were no casualties and neither incident lead to a BLEVE event. However, in both cases the Fire and Rescue Service had to put into place large exclusion zones as little is known about the possible consequences of the vehicle fire impinging on the 7 Te LPG vessel.

More information is required by the Fire and Rescue Services to help best tackle similar incidents in the future. Statistics show that incidents like these, while uncommon, do occur. In March 2016, Bolton News reported that a road tanker carrying a flammable liquid had a near miss when the brakes stuck and caught fire whilst driving along the A666 in the Midlands. This could easily have led to a similar incident as the two LPG tankers and set the entire vehicle alight[3]. While the news article does not state that the tanker was carrying LPG, it highlights that these incidents have an increasing probability as more vehicles use the roads to transport hazardous products such as LPG[9].

The worst-case scenario resulting from a road tanker crash would be for the vessel on the back of the vehicle to fail and for BLEVE to occur. The exact definition of a BLEVE is often debated in research papers. The general description is that the event is a sudden, almost instantaneous, opening of a vessel and the explosive release of the contents[4]. BLEVE events can have catastrophic consequences which, in over 80 separate events, have claimed more than 1000 lives between 1940 and 2005. Not only have these events caused a large death toll they have also left thousands more people seriously injured[5]. Studies into previous incidents show that a fire heating the outside of the vessel, known as 'hot BLEVEs', caused 36% of BLEVE events. This is considerably higher than the second most common cause of failure that is mechanical damage to the skin, causing 22% of vessel failures, or 'cold BLEVEs'. The fatalities and injuries recorded were caused by; rocketing fragments from the failure of the vessel, the blast wave from the initial explosion and thermal radiation from the ignited fireball.

RESEARCH METHODOLOGY

The vehicle was located on a remote test pad of the Spadeadam Testing and Research Centre in Cumbria, UK. The vehicle was positioned on a concrete pad with a steel bund under the wheels of the road tanker. Within this outer bund, an inner bund contained fuel to represent spilled fuel in a real event. The fuel bund was water filled, with a layer of diesel on the top to simulate a full fuel tank spill under the chassis, an additional volume of diesel was included to be representative of the propane contained in the pipework under the vessel. The diesel was poured into the bund under the road tanker to stop an unplanned BLEVE of the fuel tank or rupture of the propane pipework, the fuel tank was left without the fill cap and all pipework was drained prior to testing. The vehicle was ignited at three locations; in the engine bay, in the cab and in the diesel filled bund. The engine bay was ignited first and when fully engulfed the cab and the bund were simultaneously ignited to try to create the worst-case fire by gaining the highest heat fluxes concurrently. The fires were ignited using an electronic ignitor, petrol soaked rags and magnesium filings.

The instrumentation was set up to record the heat flux around the outer surface of the vessel using water-cooled calorimeters. Thermocouples were used to measure the vessel wall temperature and the flame temperature.

The calorimeter locations were chosen to provide information on the heat load distribution on the vessel from the vehicle fire. The fire from the chassis and cab provide a heat load at one end of the vessel and under the entire length of the vessel. The aluminium crash bars protect the hose boxes and fill points under the centre of the vessel; the LPG contents of these components are all potential fuel to the vehicle fire. As the majority of the fuel for the fire is below the centreline of the vessel the 20 calorimeters were all located on the bottom half of the vessel. A line of calorimeters ran along the bottom dead centre (BDC) line of the vessel, a second row were located along a line 25 % of the way up the height of the vessel and a third row along a line 50 % of the height up the vessel, Figure 1.

Type K stainless steel sheathed thermocouples, 3 mm diameter, were fitted into the vessel wall to a depth of 3 mm off the outer surface of the vessel to measure the temperature of the wall material. The wall temperature measurements help to understand the heat gain from the fire and with the characterisation of the fire when commissioning the burner array. These wall temperature thermocouples were located adjacent to all calorimeter locations and designated the same reference number.

At eight locations along the BDC of the vessel, a 1.5 mm diameter hole was drilled through the wall. 1.5 mm diameter stainless steel sheathed Type K thermocouples were inserted through the wall with the tip protruding 1 mm from the outer surface. These thermocouples measured the temperature immediately outside the vessel during the test. These thermocouples were located adjacent to all calorimeter locations and designated the same reference number.

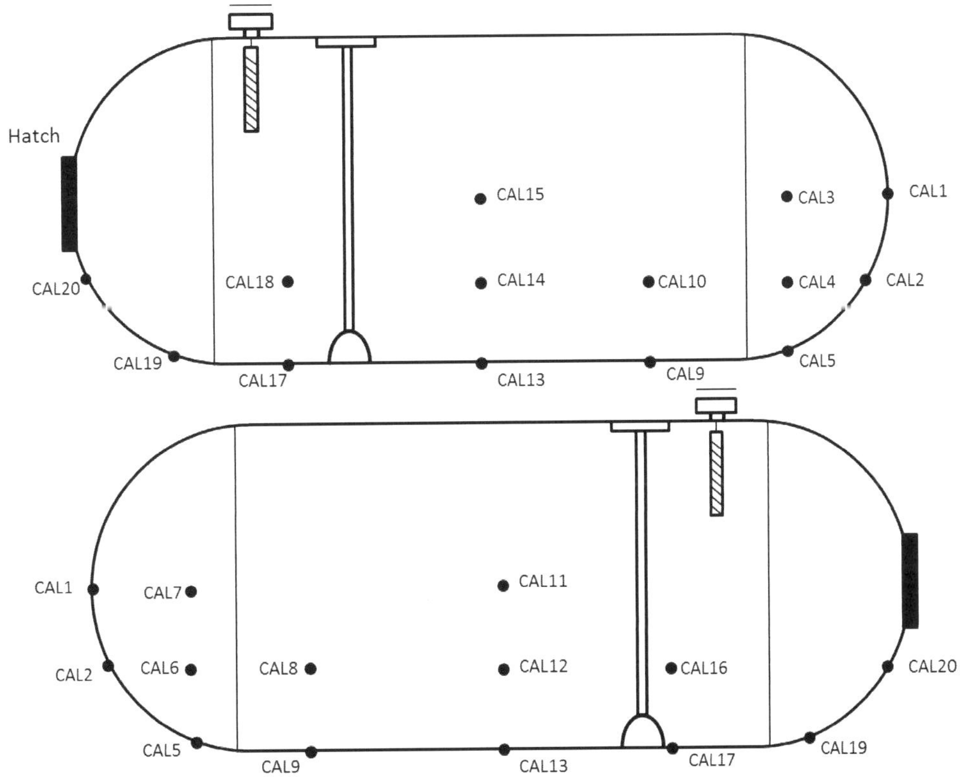

Figure 1- Locations of calorimeters, wall temperatures and flame temperatures

The vehicle was setup in a large bund, 6 m x 6 m, to capture any diesel or oil spills. A secondary 1.5 x 1.5 m bund was located under the centre of the vehicle by the diesel fuel tank. The bottom 150 mm of the bund was filled with water to prevent buckling from the heat during the fire and 300 litres of diesel was poured on top. The volume of diesel was calculated based on the volume of the fuel tank, engine pipework and the volume of the hoses under the vessel, which are usually filled with propane during normal transportation of the vehicle. The volume of propane was replaced with the same volume of diesel for ease of testing. It was expected that over the 2.25 m^2 area of the bund the diesel would burn at a regression rate of 1.5 mm.min^{-1}, this gave a predicted burn time of 40 minutes. The mass burning rate was calculated using Eq. 1.

$$\dot{m}'' = \frac{\rho_l \times R \times 10^3}{60} \qquad \text{Eq. 1}$$

Eq. 1 gives a mass burning rate of 0.022 kg.m^{-3}.s^{-1} using the density of diesel (ρ_l) as 885 kg.m^{-3} and the regression rate (R) of 1.5 mm.min^{-1}.

In the interest of safety prior to the test, all of the tyres on the vehicle were depressurised and holes drilled into the sidewalls to prevent an accumulation of pressure and subsequent explosion during the fire. The tyres provide high heat flux to the sides of the vessel so it was important to ensure this was directed onto the vessel walls by keeping the tyres attached to the vehicle. The batteries were removed from the vehicle to prevent harmful acids being released once engulfed in the fire. All of the fuel, coolant and oil tanks had caps removed to prevent the build-up of pressure and limit the likelihood of an explosion.

Three ignition devices were located around the vessel to ensure that all parts of the vehicle and the diesel pool fire ignited at the same time to generate the highest heat flux levels. The first device ignited was located inside the engine bay. This used a container of petrol and rags ignited using a remotely activated spark, the container was located beneath hoses and cabling to ensure the engine bay fully ignited. The second device was located inside the cab of the vehicle under the driver's steering wheel, this too was a petrol-filled container ignited in the same way as the engine bay. Once the engine bay and the cab were fully engulfed, the diesel pool fire was remotely ignited from outside the exclusion zone and upwind of the fire.

RESULTS AND DISCUSSION

Vehicle Fire

The vehicle fire took place on 26th August at Spadeadam Testing and Research Centre. The vehicle was set up on the remote Test Site West location to allow for an exclusion zone of 50 m to the immediate area to protect personnel from the radiative heat from the fire, and a larger 200 m exclusion zone downwind of the fire was imposed due to the smoke hazard from the diesel pool fire and the burning vehicle.

A large plume of black smoke formed from the diesel pool fire and from the large volumes of rubber and plastic on the vehicle, Figure 2. The downwind exclusion zone allowed the smoke to disperse safely from the area.

Figure 2- Fully engulfed cab and diesel pool fire under vehicle

The wind was predominantly from the west throughout the test pushing the flames to the east side of the vehicle.

Figure 3 shows the difference between the downwind and upwind sides of the vehicle and vessel post-experiment. The downwind side was covered in a layer of soot up the entire vessel with the paint burnt, while the upwind side of the vessel shows far less damage with hazard signs still visible on the side. The heat from the diesel pool fire and the chassis fire has completely melted the aluminium crash barrier on the downwind side of the vehicle, which shows the temperatures experienced here were more than 660 °C, the melting point for aluminium. The diesel tank located on the downwind side of the vessel also burst during the fire and all of the plastic housings for the batteries entirely melted away. The tyres on both sides of the vehicle were left with only the wire inserts as the rubber had burnt away during the experiment.

The roller-shutter boxes containing 80 m of hose were left closed on both the upwind side of the vessel and the back of the vessel prior to the experiment. Both of these boxes and hoses remained intact during the fire and the hoses remained visibly undamaged. The hoses on the downwind side under the vessel were also still in place at the end of the experiment however they had suffered a large amount of damage from the flames.

Figure 3- Downwind (LHS) and Upwind (RHS)

The cab of the vehicle and the engine bay burnt out during the experiment. The windscreen melted during the experiment and flowed down the front of the engine. All of the seats and the fabric around the cab burnt. All of the flammable parts of the engine ignited and the heat of the fire caused aluminium parts to melt, including the steps up to the cab on the downwind side of the vehicle.

At the time of experiment initiation, the ambient air temperature was 15°C with a relative humidity of 77%. The wind was in a West North West direction at 20.4 km/h (5.7 $m.s^{-1}$).

The temperature measured in the flame in Figure 4 shows that the peak measured temperature was 540 °C at location FLT_17. FLT_17 was located to the south edge of the diesel bund and in close proximity to the two rear inner tyres. The videos show that the rear inner tyres were alight within 10-15 minutes of the ignition of the diesel pool fire which is prior to the time of peak temperature measurement. FLT_09 also saw a peak temperature of over 500 °C, this thermocouple was located close by the front set of tyres indicating that the peak temperature is from a combination of the diesel pool fire and the burning tyres. During the experiment the diesel in the 1.5 m x 1.5 m bund spilled out into the larger bund, this meant that the majority of the diesel had burnt after 30 minutes. The chassis and the tyres were still smouldering under the vessel until around 90 minutes when the tyres on the upwind side of the vehicle caught fire causing a series of flame temperature peaks in the data at FLT_09 and FLT_05, both close to the front tyres.

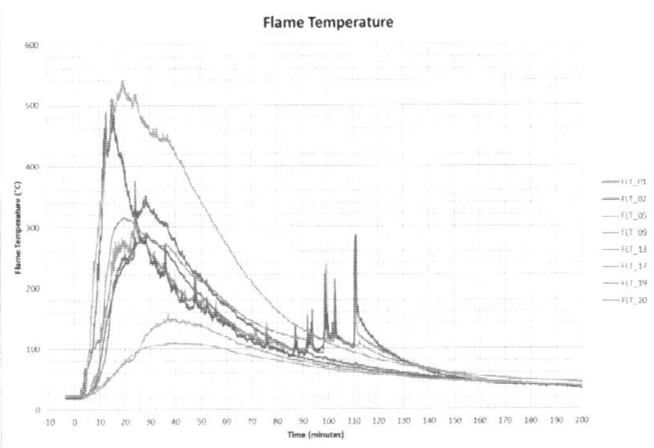

Figure 4- Flame Temperature

The wall temperature of the vessel peaked at 803 °C at thermocouple WT15 located on the mid-height line of the downwind side of the vessel, see Figure 5. The next three highest wall temperature measurements were WT10, WT14 and WT18, which are below WT15 on the downwind side of the vessel at one quarter of the vessel height. All three of these thermocouples measured temperatures in the steel of greater than the 600 °C. The ultimate tensile strength of the vessel would begin to diminish significantly past this temperature[8]. This implies that when the tank is at a lower fill level such as the 50% or 25% it is possible that the metal temperatures will be sufficiently high to result in a BLEVE of the vessel.

After around 16 minutes, the initial peaks from the diesel pool fire begin to the decrease. This is due to the diesel spilling out of the initial bund and covering a larger area increasing the rate at which the fuel was burning. Due to the slope of the bund all the fuel funnelled to the downwind side of the vehicle; this reduced the amount of flame around the vessel. After 40 minutes, much of the downwind vessel was at nominally 200-300 °C.

After 90 minutes, the temperature in the vessel increases again due to the front tyres, hoses and cables burning under the vessel.

The average temperature measured in the flame at each location shown in Table 2. The average temperature taken between 7.5 minutes and 8.8 minutes from the start of the test; where the heat flux was nominally steady on the calorimeters. The temperature measured in the flame continued to increase and peaked at around 27 minutes after the rear tyres started to burn.

Table 1 Temperature within the Flame at Steady State

Position	Temperature (°C) at steady state
FLT_01	42.1
FLT_02	35.6
FLT_05	52.7
FLT_09	148.5
FLT_13	103.0
FLT_17	238.8
FLT_19	33.9
FLT_20	28.9
Average	**85.4**

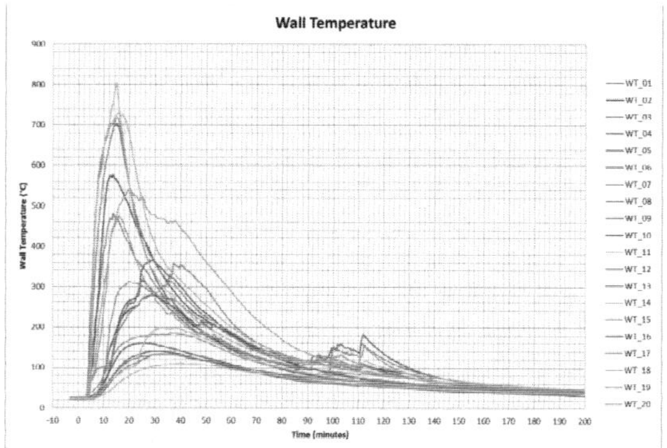

Figure 5- Wall Temperature

The average wall temperatures at each location are shown in Table 2. The average temperature was taken between the same nominally steady state heat flux region of 7.5 minutes and 8.8 minutes from the start of the test. These wall temperatures continued to increase and peaked at around 13.5 minutes, after this point the majority of the diesel pool, engine bay and drivers cab had burnt away, only the underside of the chassis and tyres continued to burn. WT_10 which was located above the front tyre on the downwind side of the vessel began to show temperatures that are close to the 600°C threshold.

Table 2 Wall Temperature Average Values at Steady State

Position	Temperature (°C) at steady state
WT_01	43.48
WT_02	35.60
WT_03	269.16
WT_04	294.54
WT_05	40.43
WT_06	29.73
WT_07	36.06
WT_08	44.58
WT_09	142.79
WT_10	569.78
WT_11	36.77
WT_12	56.48
WT_13	98.67
WT_14	458.91
WT_15	562.34
WT_16	36.30
WT_17	202.44
WT_18	586.29
WT_19	47.39
WT_20	27.28
Average	**181.0**

The calorimeter data shows a spike in the heat flux readings of over 650 $kW.m^{-2}$, see Figure 6. This is likely an anomaly as it is higher than could be reasonably expected in a pool fire or vehicle fire. This could be due to the instrument cabling being exposed to high radiative heat and being damaged. After 80 minutes all of the calorimeters return to zero, the calorimeters do not show a spike in the heat flux around 90 minutes as seen on the thermocouples when the tyres ignite downwind. The data for each location has been analysed focusing on the first 80 minutes.

Generally, the peak heat flux on the vessel is around 200 kW.m^{-2} and occurs in the first 10 minutes from the start of the experiment. This is consistent with previous published values for surface emissive power from pool fires on land[7].

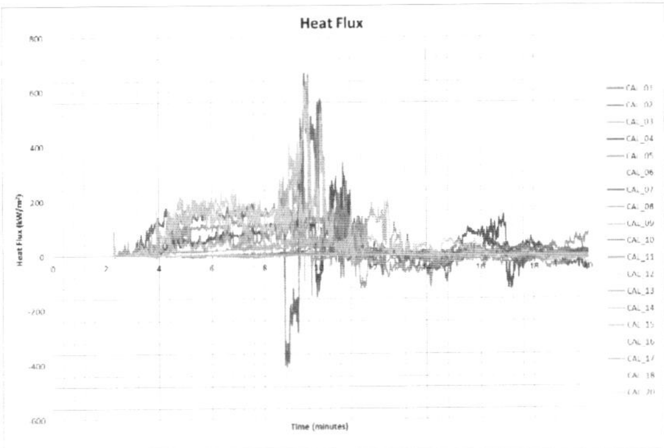

Figure 6- Measured heat flux for first 40 minutes

The calorimeters along the BDC of the vessel recorded negligible readings in comparison to those further up the vessel wall. The most likely cause of the comparatively low temperatures measured at some calorimeter locations is much of the material underneath the vehicle is metal and non-flammable. This material is acting as a protective barrier to radiant heat from the pool fire and tyre fire. The air gap between the vessel and the chassis is also allowing the wind to flow between the fire and the vessel wall, which could also be reducing convective heat flux on the centreline of the low parts of the vessel.

The upwind calorimeters show limited heat flux on the vessel. At the time of testing the wind speed was 5.7 m.s^{-1} in a WNW direction. The wind was pushing most of the flame to the east side of the vessel. The peak in heat flux registered on some sensors at around 20 minutes onwards coincides with the front tyres taking fire. In a lower wind speed the flame from a diesel spill, or in this experiment from the diesel bund, would be more evenly distributed up both sides of the vessel.

The downwind side of the vessel saw much higher heat flux in the first 20 minutes of the fire compared to the upwind side. The heat flux on the downwind side of the vessel was around 200 kW.m^{-2} this is around the expected value for a pool fire. If there was no wind during a vehicle fire then the total heat gain of the vessel could be higher as the flames will be more uniformly distributed across the vessel. More of the flame would be in contact with the vessel wall and this would transfer more heat into the vessel increasing the wall temperature and increasing the likelihood of failure. This must be considered when developing a propane burner to replicate the vehicle fire.

The average heat flux over the entire vessel during the nominally steady state period between 7.7 minutes and 8.5 minutes is shown in Table 3. The average heat flux across the vessel was 44.3 kW.m^{-2} over all measurement locations, this heat flux could have been greater if the test had been carried out with a lower wind speed. During test three, calorimeters, CAL 13, CAL 17 and CAL 19 were damaged due to the heat damaging the instrument cables, for the table below these results have been removed. CAL 06 and CAL 16 were removed due to electrical noise.

Table 3 Calorimeter Averages at Steady State

Calorimeter	Heat Flux (kW.m^{-2})
CAL_01	6.2
CAL_02	2.8
CAL_03	72.6
CAL_04	97.0
CAL_05	6.3
CAL_07	0
CAL_08	0.8
CAL_09	33.1
CAL_10	159.9
CAL_11	0
CAL_12	4.6
CAL_14	108.1
CAL_15	122.8
CAL_18	160.1
CAL_20	0.0
Average	44.3
Total	**768.2**

The results from the vehicle fire have been used to build a representative propane burner system to produce similar heat flux onto the vessel. Propane is a good source of fuel to replicate the vehicle fire as the surface emissive power of LPG on land according to Table 6.2 (from Ref. 2) is 48 $kW.m^{-2}$ which is similar to the average value seen in the vehicle fire. Commercial propane was used as the fuel in the propane burner, this has a small butane content.

Propane Burner Commissioning

The data collected from the vehicle experiment was used to create a burner array to replicate the vehicle fire to conduct further tests into the possibility of a BLEVE without the need to set fire to a vehicle and to make the tests more repeatable, with the advantage that experiments can also be stopped and started at will. Only the calorimeter data is used for comparison between the two arrays as the wall temperature is time dependent and the flame temperature is related to the fuel. The initial burner array was fabricated with 40 nozzles in four rows of 10 spaced 600 mm apart. The centre two rows of nozzles were pointed directly upwards and the outer row of nozzles were angled at 45° to the vertical. All of the nozzles were positioned to be inside the footprint of the original chassis. See Figure 7. To fuel the burner array liquid propane was used at saturation pressure (nominally 7.5 barg) resulting in an average propane flow rate of around 125 l/min.

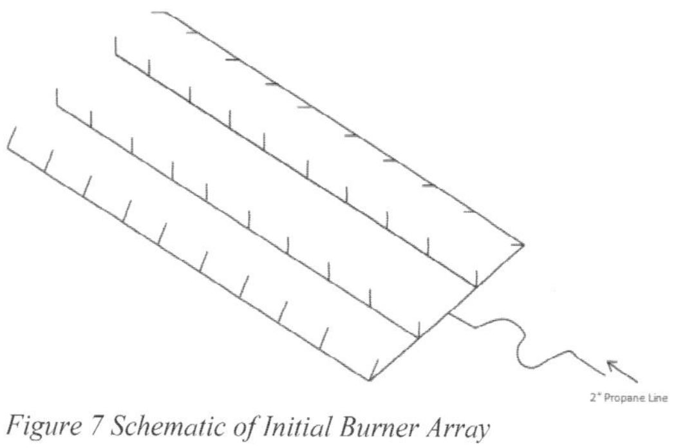

Figure 7 Schematic of Initial Burner Array

The nozzles were fabricated from a 6.35 mm (¼") OD stainless steel, with an internal diameter of 3 mm and a length of 200 mm. Each nozzle was welded onto 25 mm (1") OD stainless pipe which in turn was connected to a 50 mm (2") NB propane supply line.

The initial burner array trials produced more heat flux on the vessel than was seen in the vehicle fire and therefore not representative; see Table 4. Photographs showed that the ends of the vessel were subjected to almost constant flame coverage not seen in the vehicle fire. The calorimeters around the centreline of the vessel also saw large heat fluxes. In the vehicle fire, the majority of the underside of the vessel and the front end of the vessel were shielded from the pool fire and the cab fire by the steel wall of the cab and the pipework and hoses on the chassis.

For the second commissioning run the two end rows of nozzles were welded closed and the centre two rows of nozzles had alternate nozzles welded closed, see Figure 8. A steel mesh was also fitted above the nozzles under the vessel to try to limit the flame thickness on the underside of the vessel.

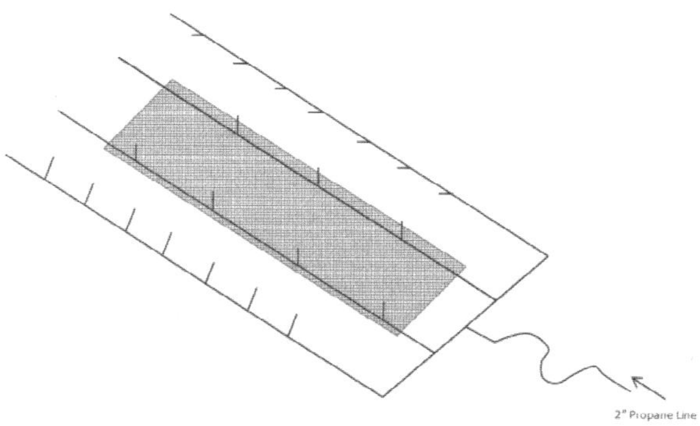

Figure 8 Schematic of Burner Commissioning Run 2

The steel mesh was welded in place using steel tubes secured into the ground. The mesh was positioned 100 mm from the top of the centre nozzles.

Figure 9 Modifications to the Burner Array Commissioning Run 2

The photograph from Commissioning Run 2 shows that the ends of the vessel are no longer totally engulfed by the flames, this reduced the heat flux on the ends of the vessel. However, the total heat flux on the vessel increased due to the mesh helping to spread the flame to cover both sides of the vessel. The centreline of the vessel still showed heat flux readings far greater than the initial vehicle fire, see Table 4.

Figure 10 Commissioning Run 2

For Commissioning Run 3 a solid steel plate was welded on top of the mesh to create a solid barrier between the nozzles and the vessel as shown in Figure 11, this was used to reduce the level of heat flux on the bottom of the vessel. The sheet used was 3 mm thick plate, the plate was thin enough to deflect the flame away from the vessel whilst still allowing some heat to be transferred through the steel.

Figure 11 Steel Plate for Commissioning Run 3

Commissioning Run 3 was much more comparable to the vehicle fire, again summarised in Table 4.

Table 4 Comparison of calorimeter readings for burner commisioning trials

Calorimeter	Vehicle	Commissioning Run 1	Commissioning Run 2	Commissioning Run 3
CAL_01	6.2	100.4	DNR	DNR
CAL_02	2.8	107.5	DNR	DNR
CAL_03	72.6	142.2	218.9	33.3
CAL_04	97.0	91.4	239.5	93.3
CAL_05	6.3	146.3	256.7	1.9
CAL_06	DNR	87.4	234.8	81.3
CAL_07	0	43.4	108.7	84.8
CAL_08	0.8	72.8	199.7	1.3
CAL_09	33.1	105.5	DNR	DNR
CAL_10	159.9	125.5	251.5	123.7
CAL_11	0	3.8	37.4	65.4
CAL_12	4.6	31.5	117.4	81.3
CAL_13	DNR	DNR	DNR	DNR
CAL_14	108.1	140.0	301.2	11.3
CAL_15	122.8	DNR	DNR	DNR
CAL_16	DNR	43.6	97.3	13.8
CAL_17	DNR	DNR	DNR	DNR
CAL_18	160.1	108.1	261.9	3.5
CAL_19	DNR	DNR	DNR	DNR
CAL_20	0.0	121.1	DNR	DNR
Average	**44.3**	**91.9**	**193.7**	**49.6**

The average of the steady state calorimeter measurements across all calorimeters is used to compare between the vehicle fire and commissioning runs. The vehicle fire and Commissioning Run 3 are similar. The propane burner gives a more uniform heat flux across the vessel when compared to the vehicle fire with localised high heat flux.

The temperatures measured in the flame around the bottom of the vessel on Commissioning Run 3 are, in the majority, similar to those seen in the vehicle fire at around 500 °C, see Figure 12. The wall temperatures in Figure 13 are also similar to those seen on the vehicle fire with two measurement locations reaching the failure threshold of around 600 °C, the test was stopped after a short period of time, however it is likely that the other thermocouples would reach similar temperatures. If this propane burner arrangement was allowed to run for 25 minutes as was the duration of the significant heat flux seen in the vehicle fire then it is likely, depending on the fill level of the vessel that the material of the vessel could fail leading to a rupture or BLEVE of the vessel.

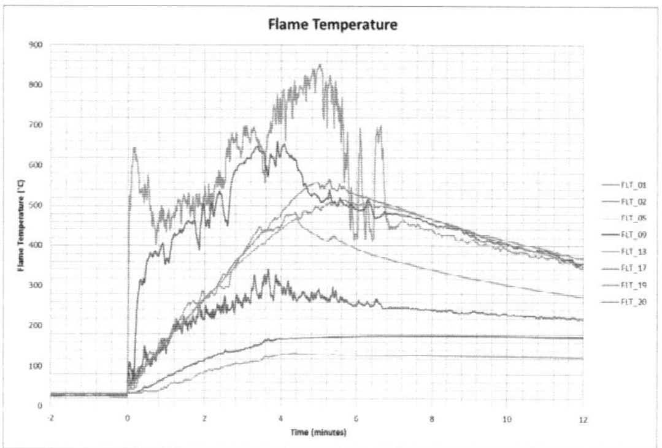

Figure 12 Flame Temperatures on Commisioning Run 3

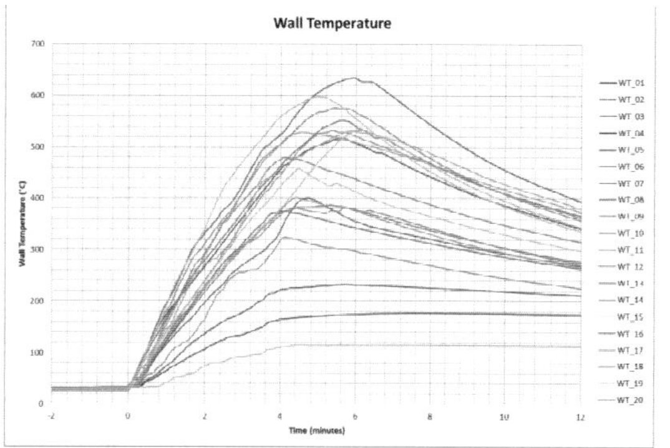

Figure 13 Wall Temperatures on Commissioning Run 3

The heat flux graph for all the measurement positions shows that while the highest heat fluxes are lower than the peaks seen in the vehicle fire it is much more uniform across the vessel than in the vehicle fire, see Figure 14. As the table of results and the graphs of heat flux and temperature show similar values to the vehicle fire, it can be concluded that the propane burner arrangement in Commissioning Run 3 is suitable, with the findings of this paper to replicate the vehicle fire and for use in the continuation of the test program into the possibility of a BLEVE event with 7 Te road tankers when subjected to fire.

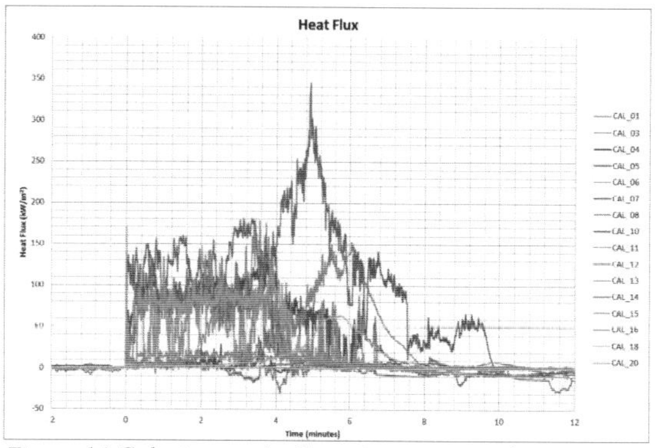

Figure 14 Calorimeters in Commissioning run 3

Concluding Remarks

This paper concludes that there is a strong possibility that a road tanker fire on a LPG road tanker could cause a BLEVE event. The literature surrounding the possibility of a BLEVE incident has been reviewed and shows the consequences of a BLEVE event could be catastrophic. A road tanker fire has been characterised and shows temperature rises in the vessel steel up to the region of 600 °C where it is known that vessel failures will occur in pressurised vessels. A propane burner system has been built and optimised to replicate a vehicle fire and the associated heat flux observed on the walls of the vessel and will be used to confirm if LPG vessel failure can occur in simulated vehicle fires in future work in 2017.

References

[1]Jeffrey, S. (2013, December 11). Retrieved from North East Chronical Live: http://www.chroniclelive.co.uk/news/north-east-news/lpg-tanker-fire-closed-a1-6391715

[2]BBC News. (2015, August 11). BBC News. Retrieved from BBC News: http://www.bbc.co.uk/news/uk-england-manchester-33854932

[3]Proctor, I. (2016, March 24). The Bolton News. Retrieved from The Bolton News: http://www.theboltonnews.co.uk/news/14378038.Brakes_on_tanker_carrying_flammable_liquid_burst_into_flames_on_A666/

[4]Cunningham, A. M. (1994). The boiling liquid expanding vapour explosion . Journal of

Loss Prevention in the Process Industries, Volume 7, Issue 6, 474-480.

[5]Abbasi, T. A. (2007). The Boiling Liquid Expanding Vapour Explosion (BLEVE): mechanism, consequence, assessment management. Journal of Hazards, 489-519.

[6]HSE. (1998). Pressure Test Safety. Buxton: HSE Health and Safety Executive .

[7]HSE. (1992). Oil and Gas Fires: Characteristics and Impacts. Buxton: HSE.

[8]K. Moodie, L. C. (1988). Fire Engulfment tests on a 5T LPG Tank. Journal of Hazardous Materials, 55-71.

[9] Ahmed, Fahmeda. (2015, 5 May) Calor. Retrieved from Calor Website: https://www.calor.co.uk/news/new-report-supports-strong-forecasting-for-increased-lpg-take-up-as-an-automotive-fuel-to-2050/

FIRE DYNAMICS

LIMITING OXYGEN CONCENTRATION OF THIN MATERIALS IN OPPOSED FLOW UNDER MICROGRAVITY CONDITIONS

Shuhei Takahashi* and Keisuke Maruta
Gifu University, Japan

INTRODUCTION

It is expected that the manned space missions such as long-term stay in space stations, exploring Moon and Mars, will be more active in feature. In such manned space activities, the suppression of fire hazard is one of the most important issues. The kind of materials used in space environments is and will be widely spread and new developed materials may be adopted for space missions. In the present circumstances, the flammability of the materials launched to space has been checked by the pass/fail flammability test, for example, NASA STD 6001B test[1]. In NASA STD 6001B test, the sample is set vertically in the corresponding condition in which the material will actually be used, and it is ignited at the bottom end to observe the upward spreading flame and its behaviour.

The NASA STD 6001B test is a pass/fail test; therefore, if the material is used in a different environment, the uses have to carry out the flammability test again although the material is the same one. Additionally, the repeatability and the reproducibility of the NASA STD 6001B are poor due to the nature of upward spreading flame and the use of a chemical ignitor for ignition, and the uses need to repeat the same flammability test for several times to obtained the result. Moreover, the facilities which can conduct NASA STD 6001B are limited in a few institutions in the world, so that it is not convenient for uses who intend to adopt some novel materials for a space mission. Furthermore, there remains a question whether the NASA STD 6001B really predict the fire hazard of the material in microgravity because the test is conducted in normal gravity condition. For example, NASA has proposed an index test (ISO/TS 16697), based on NASA STD-6001B, but Olson et al. reported that some materials had lower upward limiting oxygen index (ULOI) in reduced gravity conditions than in normal gravity conditions[2]. On the other hand, in many plastic manufacturing companies, index flammability tests are often used to confirm the quality of the products. The ISO 4589-2 is a typical index test for plastic materials, which uses downward flame spread over the sample, and its good repeatability and reproducibility are desirable for giving index to the material tested. The oxygen index (OI) obtained by ISO 4589-5 does not guarantee the fire safety, but it is also often regarded as a measure of fire resistance of the material and is helpful for choosing material for a certain purpose.

The Japan Aerospace Exploration Agency (JAXA) is now carrying out the International Space Station orbital experiment project on fire safety in reduced gravity environments, named FLARE; Flammability Limits At Reduced-g Experiment. The FLARE project is an international cooperative project including, JAXA, NASA, ESA and many universities and institutions. In the FLARE project, we will investigate the behaviour of flame spread over flat materials or electric wire insulators, and their ignition characteristic in microgravity environments in order to clarify the mechanism and features of flame spread and ignition of solid materials in microgravity, which is one of the objectives. Another objective of the FLARE project is to establish an evaluation method to give an index to the material in reduced gravity environment, which is to propose to ISO as a new standard as ISO 4589-4. The new developed standard will be helpful for screening materials used in space antecedent to the pass/fail test.

The ISO 4589-2 give OI as the minimum oxygen concentration in which downward flame spread occurs over the sample. In the FLARE, we start from ISO 4589-2, which is of course conducted in normal gravity conditions, and will improve it so that it can predict the limiting oxygen concentration (LOC) in reduced gravity environments. The concept for predicting the LOC in microgravity is as follows. Figure 1 shows the typical flammability map of a flat material in an opposed flow[3]. The horizontal axis is the opposed flow velocity and the vertical axis is the oxygen concentration. It is found that the flammability limit is a U-shaped curve and flame spread is suppressed both in high and low opposed flow velocity region, and eventually blow-off extinction and radiative extinction occur, respectively. The important issue is that the LOC in the reduced gravity environments can be different from the LOC obtained by ground-based flammability test because the buoyant flow exists in normal gravity. For example, if we conduct downward flame spread test (ISO 4589-2) in normal gravity, where the opposed (upward) buoyant flow is about 30-40 cm/s, the blow-off extinction was observed and the corresponding oxygen concentration is recorded as LOC of the material (LOC_{1g}). However, the bottom peak of the limiting curve may located at lower opposed flow region than buoyant flow, and in such a situation, the LOC in reduced gravity can be much lower than the LOC_{1g}. (see Fig. 1) Thus, there is a gap (we define it as ΔO_2) between LOC_{1g} obtained in normal gravity and the minimum limiting oxygen concentration (MLOC) in reduced gravity or microgravity condition where the buoyant flow is much slower or is not present. The extensive studies in decades have reported that the flammability of a solid material more readily increases[4-11] in certain microgravity environments than under normal gravity conditions.

If we can predict the ΔO_2 without conducting microgravity experiments, it would reduce the cost for screening the materials for space use. In the present paper, we focus on the flame spread over a thermally thin material in an opposed flow[12-21] to introduce a simplified model which can predict theΔO_2 by scale analysis, and then we show the predicted MLOC for PMMA (Acryplen: Mitsubishi Rayon) and NOMEX (NOMEX HT90-40: DuPont). The predicted results are compared with experimental results obtained by normal gravity experiments and parabolic flight experiments.

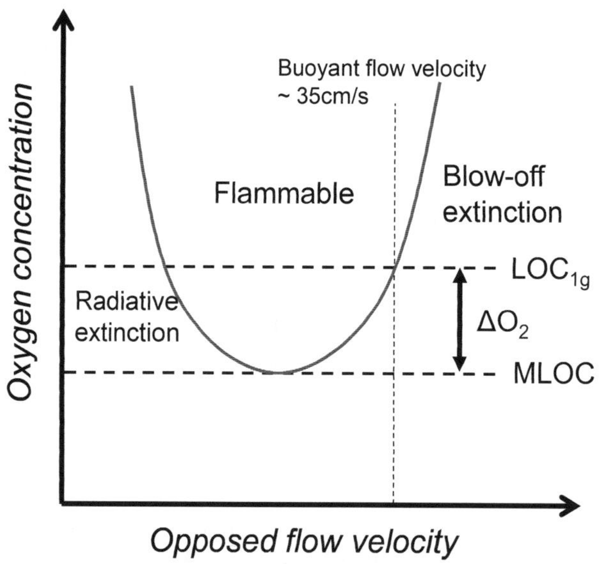

Fig. 1 Flammability map of a thin material in an opposed flow.

SCALE ANALYSIS

First, we set up the heat balance equation by scale analysis[22-26] to interpret the behaviour of flame spread over a thermally thin material in an opposed flow shown in Fig. 2.

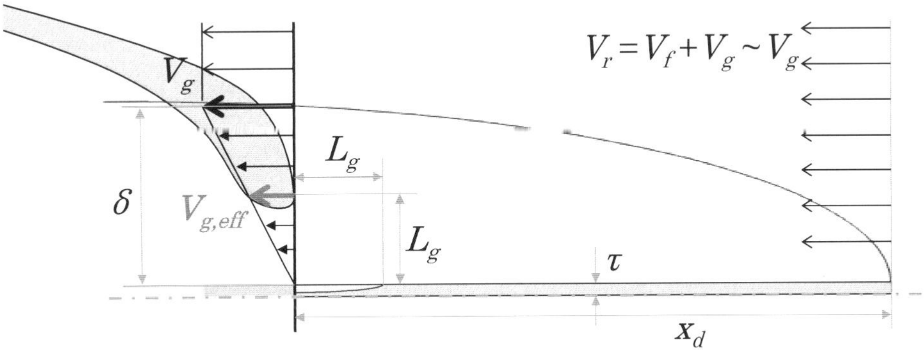

Fig. 2 Schematic of two-dimensional flame over a thin material in an opposed flow.

When the radiative heat loss is negligible small and the chemical reaction rate is sufficiently fast (thermal regime), the flame spread rate is determined by the heat flux due to the conduction driven by the temperature gradient in the gas-phase preheat zone.

$$V_{f,th}\rho_s c_s L_{sy}(T_v - T_\infty) \sim \lambda_g \frac{(T_f - T_v)}{L_{gy}} L_{gx} \qquad [1]$$

Assuming that the sample in thermally thin and the flame spread rate is much smaller than the flow velocity, the preheat zone lengths in gas-phase and solid-phase are expressed as $L_{sx} \sim L_{gx} \sim L_{gy} \sim \alpha_g / V_g$, and $L_{sy} = \tau$, respectively. Inserting them into Eq. 1, we obtain the flame spread rate in the thermal regime as

$$V_{f,th} \sim \frac{\lambda_g}{\rho_s c_s \tau} F \text{ ,where } F = \frac{T_f - T_v}{T_v - T_\infty}. \qquad [2]$$

The result in Eq. 2 is identical to the de Ris's expression.[20] When the opposed-flow velocity is very low, L_g becomes large and the radiative loss from the sample surface cannot be neglected (microgravity regime). In such conditions, we have to add the term, Q_{rad}, accounting the radiative heat loss into Eq. 1.

$$Q_{rad} \sim \varepsilon(1 - \alpha_{abs})\sigma\left(T_v^4 - T_\infty^4\right)L_g W \qquad [3]$$

On the other hand, when the opposed flow velocity is very high, the characteristic residence time become comparable to the characteristic chemical reaction time, and eventually blow-off extinction occurs. In such conditions, it is expected that the Damkohler number, Da, should be close to unity.

$$Da \equiv \frac{t_{res}}{t_{chem}} = \frac{L_g / V_g}{\rho_g / \dot{\omega}_g} \sim \frac{\alpha_g}{V_g^2} \rho_g Y_O A \exp(-E / RT_f) \qquad [4]$$

The Da is determined by the ratio of the characteristic residence time, t_{res}, to the characteristic chemical time, t_{chem}, but it can also be considered as the ratio of the total combustion heat of a pyrolysis gas in the preheat zone to the heat required to sustain a temperature gradient in the gas-phase preheat zone.

$$Da \equiv \frac{L_g / V_g}{\rho_g / \dot{\omega}_g} \sim \frac{L_g{}^2 \dot{\omega}_g \ \Delta h_c}{\rho_g c_g L_g V_r (T_f - T_v)} = \frac{L_g{}^2 \dot{\omega}_g \Delta h_c}{L_g \lambda_g (T_f - T_v) / L_g} \qquad [5]$$

Hence, we made a model in which the heat conduction from the flame to the solid material was reduced by the factor of $(1-1/Da)$. Using this concept and Eqs. 1 to 4, we obtained the heat balance equation including both the radiative heat loss effect and finite kinetic effect as follows,

$$V_f \rho_s c_s L_{sy} W (T_v - T_\infty) + \varepsilon (1 - a_{abs}) \sigma \left(T_v{}^4 - T_\infty{}^4 \right) L_g W \sim \left(1 - \frac{1}{Da} \right) \lambda_g \frac{(T_f - T_v)}{L_{gy}} L_{gx} W \qquad [6]$$

Introducing a non-dimensional flame spread rate, η (= $V_f / V_{f,th}$), we obtain a simple non-dimensional form,

$$\eta + R_{rad} + \frac{1}{Da} = 1 \ \text{ or } \ \eta + \frac{1}{Da_R} + \frac{1}{Da} = 1 \qquad [7]$$

In Eq. 7, R_{rad} is the non-dimensional factor expressing the radiative heat loss. Note that R_{rad} can be regarded as the reciprocal of so-called radiative Damkohler number, $1/Da_R$.

$$R_{rad} = \frac{1}{Da_R} = C_1 \frac{\varepsilon (1 - a_{abs}) \sigma (T_v^4 - T_\infty^4)}{\rho_g c_g V_r (T_f - T_v)} \qquad [8]$$

In the kinetic regime, the opposed flow velocity is so high that the effective flow velocity, $V_{g,eff}$, seen by the flame font should be different from the bulk flow velocity, V_g, due to the boundary layer on the sample. Assuming the linear velocity profile in the boundary layer[15], the $V_{g,eff}$ is estimated as $V_{g,eff} = V_g L_g / \delta$ using the relations $\delta \sim x_d / Re_x^{1/2}$. This concept means that the effective flow velocity varies according to the location from the sample edge. Hence, if we introduce certain blow-off point, $x_d = x_{d,b}$, as a fixed value, we obtain the modified form of Da as

$$V_{g,eff} \sim \frac{V_g}{\mathrm{PrRe}_x{}^{1/2}} \propto V_g^{\frac{1}{2}} \ \text{ and } \ Da = C_2 \frac{\alpha_g}{V_g} \rho_g Y_O A \exp(-E / RT_f) \qquad [9].$$

Equation 7 implies that if the value of $R_{rad} + 1/Da$ is close to unity, the heat balance cannot be held and extinction occurs. The two non-dimensional parameters contain C_1 and C_2 as empirical constant, respectively. In order to evaluate one of the empirical constants, C_1, we conducted microgravity experiment with 4.5s drop tower in advance with using PMMA film as the base-material. The details of the drop experiment are described in Ref. 27, and we evaluated that $C_1 = 2.76$ for Eq. 8 from the fact the radiation extinction occurred at the condition where opposed flow velocity was 3cm/s and the oxygen concentration was 21%. The other empirical constant, C_2, for Eq. 9 was also determined as 0.28 from the fact that the LOC for the downward spread in 21% oxygen concentration was 17.0%. In the following discussion, we will use the same C_1 and C_2 for both PMMA and NOMEX.

EXPERIMENTAL SETUP

In order to validate the developed simplified model, we conducted following ground-based experiments and microgravity experiments for PMMA, the base-material, and NOMEX sheet, typical fire resistant material.

Downward flame spread experiment

In order to investigate the flame spread behaviour in the kinetic regime, we conducted downward flame spread experiment (blow-off experiment) with thin PMMA film and NOMEX sheet. Figure 3 shows the schematic of the wind tunnel and the glovebox used. The dimensions of the sample are 8 cm in length, 2 cm in width and 0.125mm in thickness. The sample was held by sample holder which consisted of two reverse U-shaped thin plates made of stainless steel. The thickness of the holder plate was 0.5 mm, and the holder was located at the centre of the wind tunnel. The wind tunnel was made by acrylic resin and the flame spread could be observed from outside. The cross-section of the duct was 14 cm x 14 cm, and the length of the duct was 35 cm. The wind tunnel had a motored fan at the top and three rectifying honeycomb plates inside. The rotation of the fan was controlled by a PWM driver and we could create upward flow with desired velocity ranging from 40 to 130 cm/s. The hot wire anemometer was set inside the duct 3cm apart from the sample surface to measure the bulk velocity of the upward flow. The wind tunnel was installed in the glovebox whose dimensions were 1 m x 1 m x 1 m (1000L). We can vary the oxygen concentration inside the glovebox.

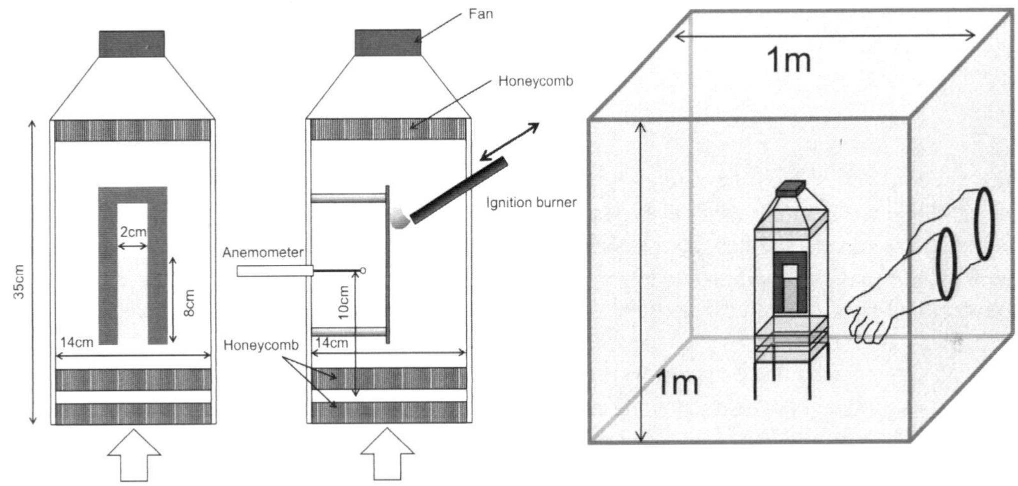

Fig. 3 The wind tunnel and the glovebox used for blow-off extinction experiment.

The ignition was done at the top end of the sample by a butane diffusion burner through a hole in the wall of the wind tunnel. The hole had a lid and after we ignited the sample, we remove the burner and the lid closed. When the upward flow velocity condition was more than 60 cm/s, we ignited the sample at 60 cm/s and then increased the flow velocity to the setting value. In such situations, the sample length was adjusted so that 8cm sample length was left after changing the velocity. The change of the oxygen concentration inside the glove box was less than 0.1% during the blow-off test. In the blow-off experiments, we recorded the flame spread by CCD camera, measured the location of the flame and specified the oxygen concentration and flow velocity in which blow-off extinction occurred.

Parabolic flight experiment

The flame spread under microgravity environments were observed in 20 s parabolic flight experiments carried out by Diamond Air Service. Figure 4 shows the schematic of the test chamber used for these flight experiments[25]. This test chamber was also 35 L in volume, in which a rolled PMMA film or

NOMEX sheet was placed, which is similar to the setup in the work of Ferkul et al.[28] The chamber had a suction fan for creating a flow ranging from 0 to 20 cm/s in the test section. The dimensions of the test section were 14 cm × 14 cm × 13 cm, and honeycomb plates were installed on both the upstream side and the downstream side of the test section. The sample sheet in the test section was 8 cm in length and 2 cm in width. The downstream end of the sample was ignited by the Ni-Cr wire 10 s before the onset of microgravity, and the flame that spread upstream was recorded with a CCD camera and IR camera.

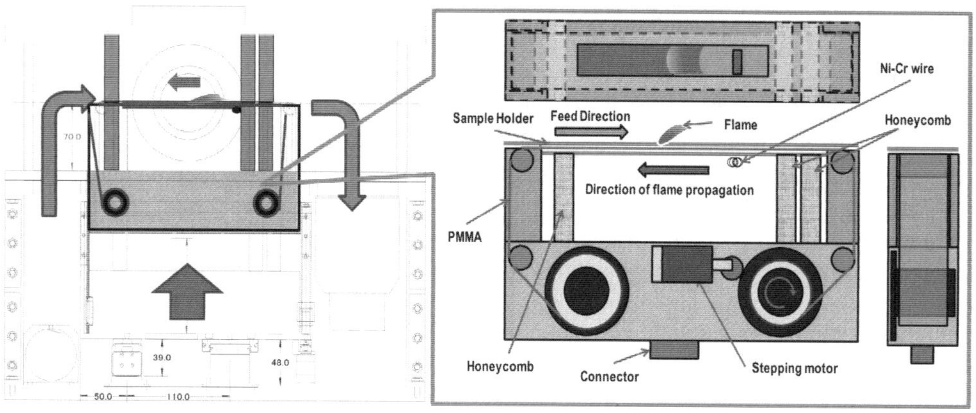

Fig. 4 Schematic of experimental apparatus for parabolic flight experiment.

Under the conditions in which the initial oxygen concentration was lower than the LOC_{1g}, we could not ignite the sample before the onset of microgravity. Hence, under such conditions, we ignited the sample at the onset of microgravity. When the sample was ignited right at the onset of microgravity, the time remained for flame spread was less than 20 s. The typical duration of heating for ignition was about 5 s. If the flame survived until the end of the microgravity environment, we judged it as being flammable; else, we judged it as being extinct. Then, we defined the MLOC as the minimum LOC at which the flame survived. The spread rate V_f was defined as the average spread rate in 5 s.

Molecular sieves and zeolites were packed inside the test chamber to absorb the water and carbon dioxide generated during the combustion process. A pressure transducer, a thermocouple, and an oxygen sensor were also installed to monitor the pressure, temperature, and oxygen concentration, respectively. We wound the rolled sheet to feed a new sample area after the completion of combustion. After the complete combustion of the 8-cm-long sample sheet, the oxygen concentration in the chamber decreased by 0.8%, but the actual decreased concentration of oxygen depended on the consumed length of the sample in the experiment. After filling an oxygen/nitrogen mixture in the chamber, we repeated the combustion test without adding consumed oxygen, and so, the oxygen concentration in the chamber decreased gradually during the experiment. The initial pressure was 101 kPa; it decreased to 97 kPa after a series of parabolic flight experiments.

RESULTS AND DISCUSSION

In the scale analysis, we clarified that the flammability limit of a thin material in an opposed flow was interpreted by the combination of the radiative heat loss factor, R_{rad}, and the Domkohler number, Da. The Da has significant effect in high opposed flow region (where V_g > 60cm/s) whereas R_{rad} becomes important in low opposed flow region (V_g < 10cm/s). Therefore, the accuracy of predicting blow-off limit has relatively large effect on determining MLOC than that of radiative extinction limit. Then, we first observed the downward flame spread with a forced upward flow in normal gravity.

Downward flame spread experiment

We show the result of downward blow-off tests with a forced upward flow. Figure 5 shows the typical temporal behaviour of the downward flame spread over PMMA. The ambient oxygen concentration was 20.9 %, and the forced flow velocity was 80 cm/s. In the earlier stage, the flame has a certain flame length and the luminous flame propagated at $V_f \sim 2.5$ mm/s. In the middle stage, the flame propagated with the steady rate, but the flame length decreased in time. After 20 s, the spread rate suddenly decreased to 0.46 mm/s and the flame length became very short. In this situation, the flame was located behind the thin sample edge with small amount of molten PMMA.

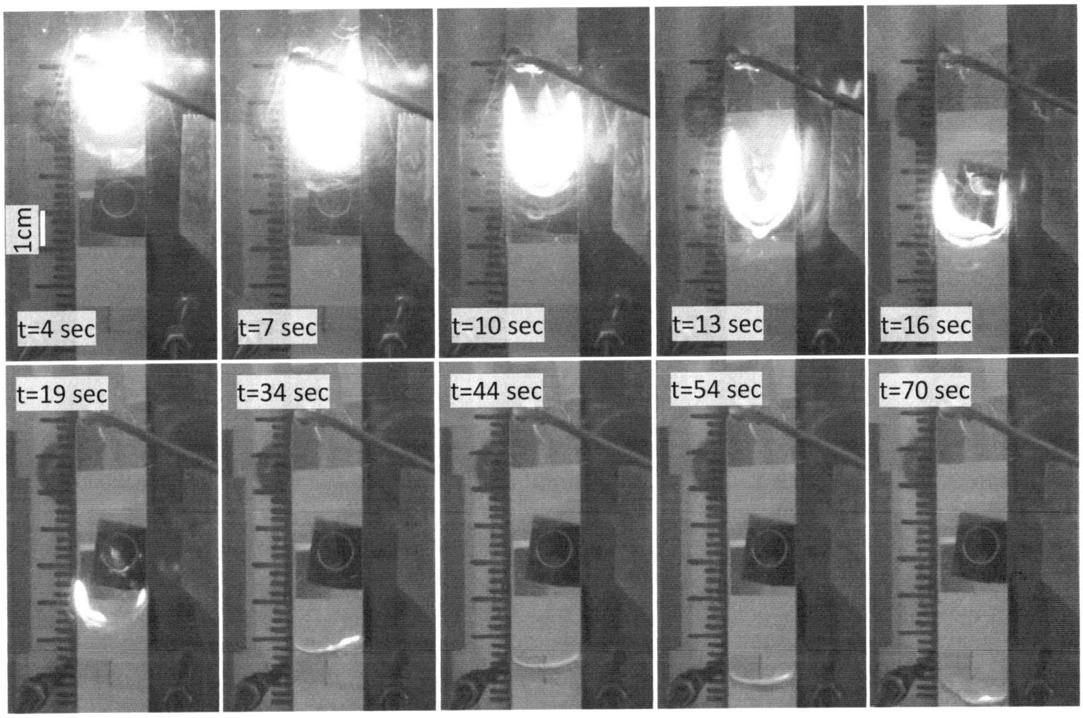

Fig. 5 Temporal behaviour of downward flame spread along a PMMA sheet. The thickness of the sample is 0.125mm. The bulk velocity of the upward wind is 80 cm/s, and the ambient oxygen concentration is 20.9%.

The flame location vs. time obtained from in Fig. 5 is plotted in Fig. 6a. The downward flame spread over PMMA film into two modes; one is surface spread mode and the other is wake spread mode. In the wake spread mode, the supply of oxygen is caused by the strong shear between the bulk flow and wake zone, therefore, this mode can be achieved only in kinetic regime where V_g is relatively high. Thus, in the wake spread mode, the characteristic residence time estimated by L_g/V_g is expected to be shorter than the characteristic chemical reaction time. Therefore, in the present paper, we defined the blow-off occurred if the flame could not support the surface spread mode. For example, in the experiment shown in Figs. 5 and 6, we defined that the blow-off occurred at $x_{d,b} = 2.3$ cm and $V_g = 80$cm/s. On the other hand, as MOMEX did not create molten product, the wake mode spread was not observed for NOMEX, and the flame perfectly quenched when it could not support the surface spread mode.

Then, we investigated the blow-off location, $x_{d,b}$ with varying the force flow velocity, V_g. It is found that the location of blow-off and the forced flow velocity has a linear relation, and the blow-off velocity is quite different depending on the location where blow-off occurs. If $x_{d,b}$ changed from 3cm to 6cm, the blow-off velocity increased about 25-30%. Bhattacharjee et al.[29] reported that the effective velocity increased 27% when the sample length changed from 6cm to 3cm in a similar condition, and the value is consistent with our experimental results. In Ref. 29, they emphasized the importance of evaluation of effective flow velocity seen by a flame in the kinetic regime. Therefore, in the kinetic regime, Da should be estimated with taking account of the velocity distribution of the boundary layer.

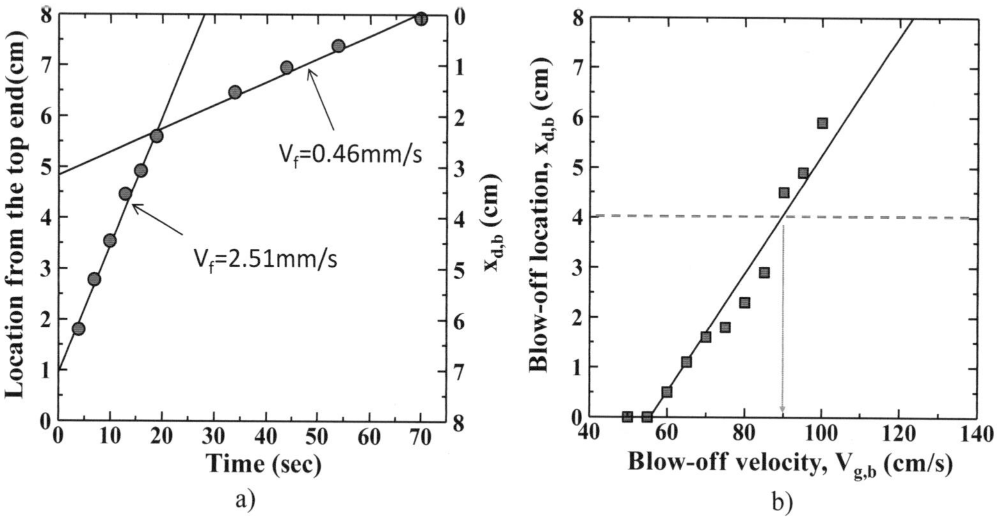

Fig. 6 The behaviour of blow-off extinction in a forced flow in 20.9% O_2 level; a) the location of the flame vs. time from the ignition at V_g = 80 cm/s, b) the blow-off location with varying forced upward flow velocity.

The results of downward spread experiment show that the forced velocity when the blow-off occurs differs according to the location where the blow-off occurs. Therefore, it is important to determine the definition of blow-off criteria. Hence, in this paper, we define the blow-off forced velocity, $V_{g,b}$, as the forced flow velocity where the 4cm of the sample (half of the sample length) is burnt and blow-off occurs. Therefore, for example, for PMMA film, the $V_{g,b}$ is 90 cm/s for the oxygen concentration of 20.9 %.

Using the results of downward spread experiment, we can draw the limiting line for PMMA. In order to calculate R_{rad} and Da, we have to know proper activation energy, E, and pre-exponential factor, A, for the blow-off of NOMEX as well as the adiabatic flame temperature, T_f, pyrolysis temperature, T_v, properties of ambient gas. However, it is very difficult to find such E and A in the literatures or measure/calculate them in contrast with other quantities. Additionally, materials like NOMEX may require illatively large decomposition energy and the characteristic pyrolysis reaction time may be much greater than characteristic gas-phase reaction. In such conditions, the relation $t_{py} >> t_{chem}$ holds and the expression of Damkohler number should be different from Eq. 9. Assuming that the pyrolysis reaction is oxidative reaction and the pyrolysis reaction undergo at T_{py} which is proportional to T_f, thus, $T_{py} = \beta T_f$ ($\beta < 1$), on the surface, we can express the reaction rate, ω_s, and Da using reduced activation energy E'.

$$\omega_s \sim \rho_s \rho_g Y_O A \exp(-E/RT_{py}) = \rho_s \rho_g Y_O A \exp(-E/R\beta T_f)$$

and

$$Da \equiv \frac{t_{res}}{t_{py}} = \frac{L_g / V_{g,eff}}{\rho_s / \dot{\omega}_s} \sim \frac{\alpha_g}{V_g} \rho_g Y_O A \exp(-E'/RT_f) \text{ where } E' = E/\beta \qquad [10]$$

It is found that the form of Eq. 10 is identical to Eq. 9; therefore, we can handle such materials in the same manner if we know proper *E'* and *A*.

In order to obtain proper *E* and *A* for the material considered, we propose an evaluation method. Figure 7 shows the concept for determining the two parameters for blow-off limit. We choose two blow-off conditions; one is the blow-off condition in natural convention only, which corresponds to the LOC_{1g}, and the other is the blow-off condition in which a certain forced flow velocity, $V_{g,b}$, is given to measure the limiting oxygen concentration, LOC_b. Assuming that the buoyant flow velocity is 35 cm/s, the parameters can be calculated with LOC_{1g}, $V_{g,b}$ and LOC_b.

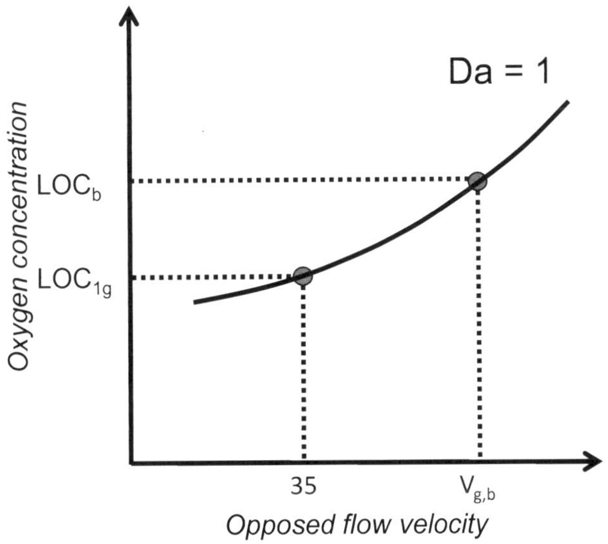

Fig. 7 Concept of the evaluation method for determining proper *E* and *A* for blow-off.

Figure 8 shows the measured LOC_b with varying the forced flow velocity and the calculated *E* and *A* for PMMA by the proposed method above. In the experiment, the blow-off condition was investigated with increasing the forced flow velocity at a fixed oxygen concentration in the glovebox. It is found that the LOC_b increases with the forced velocity, which is consistent with the theory in Eq. 9. The calculated *E* and *A* with LOC_{1g} and each LOC_b are shown in Fig. 8b. The *E* and *A* in Ref. 21, which were 1.5×10^5 and 1.36×10^9, are also shown as the blue line and the red line, respectively. When the forced flow velocity is close to the buoyant flow velocity, 35 cm/s, the calculated parameters are apart from the values in Ref. 21. This is because the forced flow merged with the buoyant flow and the substantial flow velocity became larger than the force flow itself. However, when the flow velocity is more than 70 cm/s, the calculated *E* and *A* converges at the referred values, respectively. For example, if we choose the condition of $V_g = 80$ cm/s as the reference, we obtain *E* and *A* as 1.46×10^5 and 1.10×10^9, which are almost the same as the values in Ref. 21. Thus, it is found that the proposed method can be a helpful tool to determine the blow-off limit of the material considered.

The *E* and *A* for blow-off phenomena of NOEMX have not been found in the literatures, however, we can estimate proper values for them in the same manner adopted in Fig. 8. The results for NOMEX sheet are shown in Fig. 9. The calculated *E* (or maybe *E'*) converges on a line near $E = 9.5\times10^5$ when the forced flow velocity is more than 70 cm/s. Hence, we adopted the condition where the forced flow velocity is 80 cm/s for LOC_b and obtained $E = 9.36\times10^5$ and $A = 1.37\times10^{24}$ for MOMEX, respectively.

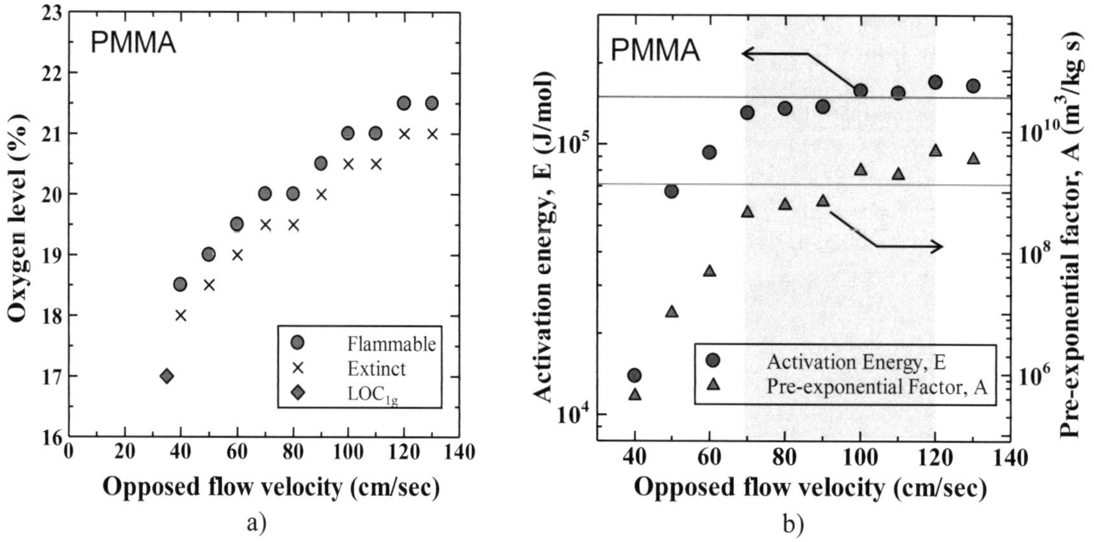

Fig. 8 The result of blow-off experiments for PMMA; a) the LOC_{1g} and LOC_b with varying the forced flow velocity, b) the calculated activation energy, *E*, and the pre-exponential factor, *A*. The two lines in Fig. 8b represent the *E* and *A* in Ref. 21, respectively.

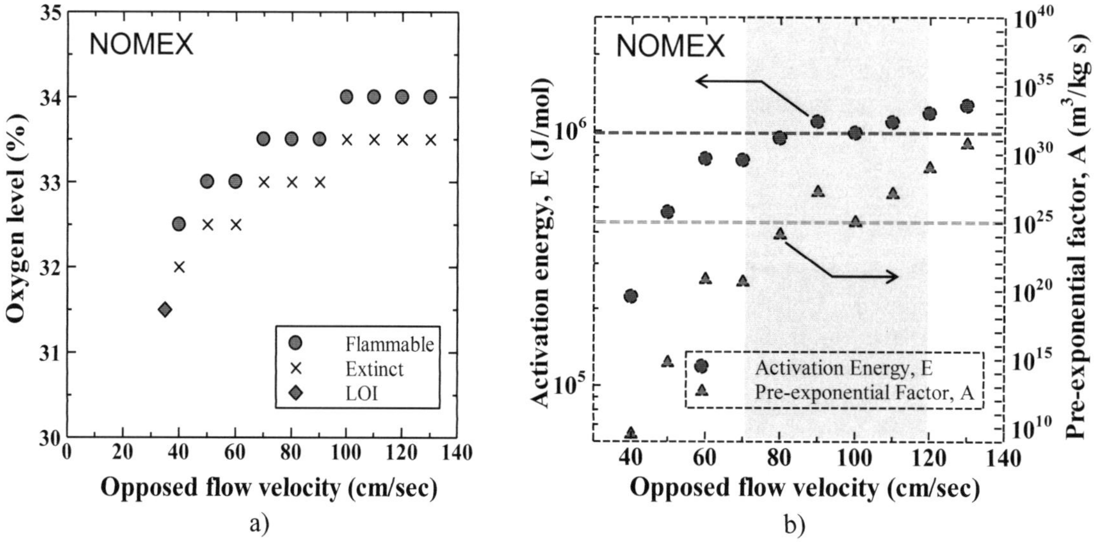

Fig. 9 The result of blow-off experiments for NOMEX; a) the LOC_{1g} and LOC_b with varying the forced flow velocity, b) the calculated activation energy, *E*, and the pre-exponential factor, *A*.

Parabolic flight experiment

Then, we show the results of parabolic flight experiment for PMMA and NOMEX in Fig. 10 with the flammability limiting curves predicted by the properties in Table 1. The thermodynamic properties in gas-phase were calculated at a fixed temperature, which was equal to the pyrolysis temperature, T_v. The characteristic pyrolysis temperature, T_v, was selected as the temperature where the mass loss rate was the maximum in thermogravimetric analysis under air. The solid circle mean the conditions in which the flame spread continued during the microgravity period. The rectangles and open circles are the results in downward flame spread experiments with natural convection only and with a forced flow, respectively. It is found that for both materials, the MLOC is less than LOC_{1g}, which is the characteristic trend in microgravity environment. In the experiment, the ΔO_2 for PMMA and MOMEX is 2.1% and 2.0%, respectively. When PMMA was used as the sample, the MLOC was observed at lower flow velocity region of V_g < 10 cm/s. On the other hand, for NOMEX, the MLOC was located at relatively higher velocity region of V_g = 10~20 cm/s. The difference is due to the difference in the characteristic pyrolysis temperature, T_v.

Table 1 The properties used in the simplified model.

	PMMA[21]	NOMEX[30, 31]
Pyrolysis temperature, T_v (K)	670	730
Enthalpy of combustion, Δh_c (kJ/kg)	25900	25530
Activation energy, E (J/mol)	1.50×10^5	9.36×10^5 *
Pre-exponential factor, A (m³/kgs)	1.36×10^9	1.37×10^{24} *

* Calculated value in Fig. 9

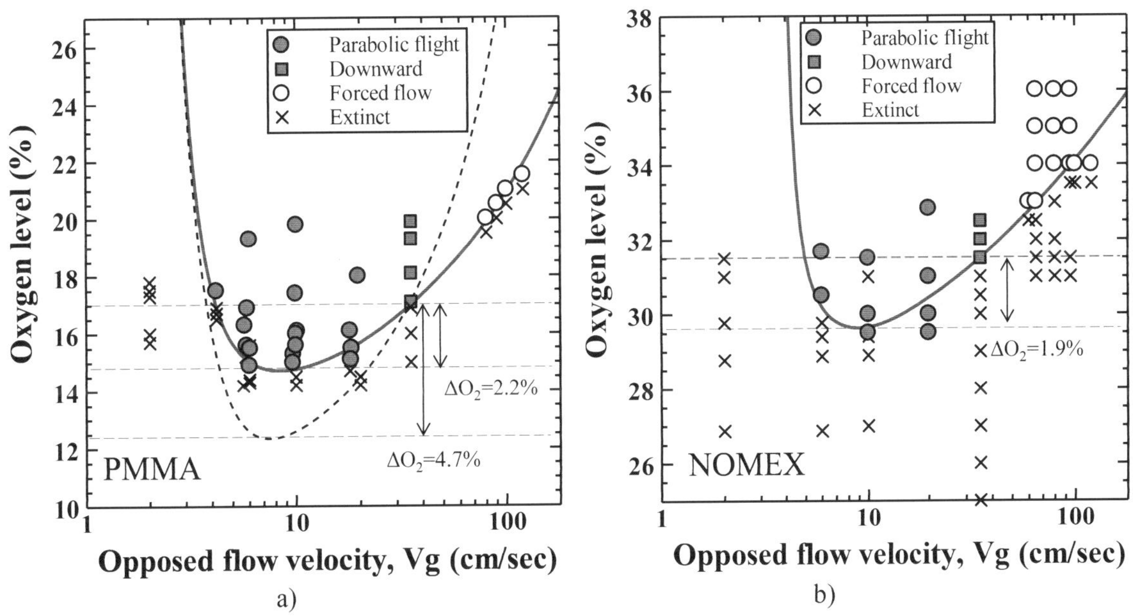

Fig. 10 The flammability map with the predicted limiting curve by the simplified model; a) for PMMA film and b) for NOMEX sheet. The prediction of original model with Eq. 7 is shown as the dashed line in Fig. 10a.

The predicted curves well agree with the experimental results and succeed in capturing the feature of their limiting curves. The predicted ΔO_2 for PMMA and MOMEX is 2.2% and 1.9%, which are closed to the corresponding experimental result. For comparison, we also show the prediction with original blow-off limiting curve shown in Eq. 4 as the dashed line, which did not include the effect of boundary layer. The dashed line cannot present the blow-off extinction in higher flow velocity region and obviously overestimates the ΔO_2 in mild flow region. Thus, it is found that the accurate expression of the blow-off limit is necessary for predicting the MLOC.

The comparison between the obtained experimental results and the predictions with is shown in Table 2. Especially for NOMEX, although the limiting curve and the MLOC are calculated with the empirical parameters obtained by downward spread experiments, the result quantitatively agrees with the actual MLOC. This means that the behaviour of flame spread over NOMEX sheet in microgravity environment has been predicted without conducting microgravity experiments. Even though the result of downward spread test will not necessarily guarantee the fire safety, the proposed simplified model and blow-off test method have a potential for a screening test to choose the candidate material used in space activity.

Table 2 Measured and predicted MLOC and ΔO_2 for PMMA and NOMEX.

		PMMA	NOMEX
LOC_{1g}		17.0 %	31.5 %
MLOC	Experiment	14.9 %	29.5 %
	Prediction	14.8 %	29.6 %
ΔO_2	Experiment	2.1%	2.0 %
	Prediction	2.2%	1.9 %

CONCLUSION

In order to predict the flammability limit of thin materials in an opposed flow under microgravity, a simplified model was developed by scale analysis. The scale analysis revealed the two important parameters, R_{rad} and Da, which explained the suppression of flame spread in low flow velocity region and high flow velocity region, respectively. The developed model states that the flammability limiting curve is expressed as the conditions in which $R_{rad} + 1/Da = 1$, and the minimum limiting oxygen concentration (MLOC) of the material considered corresponds to the minimal point of the curve. Therefore, the MLOC or ΔO_2, which is the difference between MLOC and LOC_{1g}, can be calculated without conducting microgravity experiments provide that the necessary properties foe the parameters are known. Among the properties, the activation energy, E, and the pre-exponential factor, A, for expressing blow-off phenomena are rarely available in literature. Hence, we also proposed an evaluation method to determine these E and A by downward blow-off experiment. The experimental results for PMMA and NOMEX showed that the proper E and A could be obtained by conducting the blow-off test providing a forced flow of 70-120 cm/s. The predicted flammability limiting curves for both materials well agreed with the results obtained by the parabolic flight experiments, and it was expected that the proposed method could be a useful tool for screening the materials used in microgravity environments.

ACKNOWLEDGEMENTS

This study was conducted as part of the FLARE project supported by JAXA. We are also thankful to the staff of Diamond Air Service for their excellent technical support to parabolic flight experiments.

NOMENCLATURE

A	Pre-exponential factor	$V_{g,eff}$	Effective flow velocity seen by the flame front
a_{abs}	Absorption coefficient of gas, $a_{abs} = 0$	$V_{g,b}$	Blow-off flow velocity
C_1	Empirical constant for R_{rad}, $C_1 = 2.67$	V_f	Flame spread rate
C_2	Empirical constant for Da	$V_{f,th}$	Flame spread rate in thermal regime
c_g	Specific heat of gas	V_r	Velocity relative to flame, $V_r = V_g + V_f$
c_s	Specific heat of solid	W	Sample width
Da	Damkohler number	x_d	Length from the sample edge
Da_R	Radiative Damkohler number	$x_{d,b}$	Blow-off length

E Activation energy
E' Reduced activation energy for pyrolysis, $E' = E/\beta$
Δh_c Heat of combustion
L_g Diffusion length
Re_x Length Reynolds number
R_{rad} Radiation loss factor, $R_{rad} = 1/Da_R$
T_f Adiabatic flame temperature
T_v Pyrolysis temperature
T_∞ Ambient temperature
t_{chem} Characteristic chemical time
t_{py} Characteristic pyrolysis time
t_{res} Characteristic residence time
V_g Opposed flow velocity
α_g Thermal diffusivity of gas, evaluated at T_v
β Correction factor for pyrolysis temperature
δ Boundary layer thickness
ε Surface emissivity, $\varepsilon = 1$
λ_g Gas-phase conductivity evaluated at T_v
λ_s Solid-phase conductivity
η Non-dimensional spread rate, $\eta = V_f/V_{f,th}$
ρ_g Gas density evaluated at T_v
ρ_s Solid density
τ Fuel half-thickness
ω_g Gas-phase reaction rate
ω_s Pyrolysis reaction rate

REFERENCES

[1] NASA-STD-6001 B: Flammability, Offgassing, and Compatibility Requirements and Test Procedures (2011).

[2] S. L. Olson and P.V. Ferkul, "Evaluating Materials Flammability in Microgravity and Martian Gravity Compared to NASA's Normal Gravity Materials Flammability Testing", 42st International Conference on Environmental Systems (ICES), 15-19 July 2012, San Diego, CA.

[3] K. R. Sacksteder, J. S. Tien, Proc. Combust. Inst. 25 (1994) 1685-1692.

[4] H. Shih and J. S. Tien, Proc. Combust. Inst. 28 (2000) 2777-2784.

[5] K. Prasad, Y. Nakamura, S. L. Olson, O. Fujita, K. Nishizawa, K. Ito, T. Kashiwagi, Proc. Combust. Inst. 29 (2002) 2553-2560.

[6] T. Kashiwagi, K. B. McGrattan, S. L. Olson, O. Fujita, M. Kikuchi, K. Ito, Proc. Combust. Inst. 26 (1996) 1345-1352.

[7] S. L. Olson, P. V. Ferkul, J. S. T'ien, Proc. Combust. Inst. 22 (1989) 1213-1222.

[8] S. L. Olson, F. J. Miller, Proc. Combust. Inst. 32 (2009) 2445-2452.

[9] S. L. Olson, T. Kashiwagi, O. Fujita, M. Kikuchi, K. Ito, Combustion and Flame 125 (2001) 852-864.

[10] A. Kumar, H. Y. Shih, J. S. T'ien, Combustion and Flame 132 (2003) 667-677.

[11] C. Jiang, J. S. T'ien, H. Shih, Proc. Combust. Inst., 26 (1996) 1353-1360.

[12] A. Kumar, H. Y. Shih, J. S. T'ien, Combustion and Flame 132 (2003) 667-677.

[13] K. R. Sacksteder, J. S. Tien, Proc. Combust. Inst., 25 (1994) 1685-1692.

[14] C. Jiang, J. S. T'ien, H. Shih, Proc. Combust. Inst., 26 (1996) 1353-1360.

[15] I. S. Wichman, Combustion and Flame 50 (1983) 287-304.

[16] S. Rybanin, Proc. Combust. Inst., 27 (1998) 2791-2796.

[17] L. K. Honda, P. D. Ronney, Proc. Combust. Inst., 28 (2000) 2793-2801.

[18] I. T. Leventon, S. I. Stoliarov, Proc. Combust. Inst., 34 (2013) 2523-2530.

[19] Y. Kudo, A. Ito, Proc. Combust. Inst., 29 (2002) 237-243.

[20] J. N. de Ris, Proc. Combust. Inst. 12 (1969) 241-255.

[21] A.C. Fernandez-Pello, S.R. Ray, I. Glassman, Proc. Combust. Inst., 18 (1981) 579-589.

[22] S. Bhattachrjee, J. West, and R. A. Altenkirch, Proc. Combust. Inst. 26 (1996) 1477-1485.

[23] S. Bhattacharjee, R. Ayala, K. Wakai, and S. Takahashi, Proc. Combustion Inst., 30 (2005), 2279-2286.

[24] S. Bhattacharjee, S. Takahashi, K. Wakai, C. Paolini, Proc. Combust. Inst. 33 (2011) 2465-2472.

[25] S. Takahashi, T. Ebisawa, S. Bhattacharjee, T. Ihara, Proc. Combust. Inst. 35 (2015) 2535-2543.

[26] S. Takahashi, K. Tsuboi, K. Maruta, Interflam 2016 1 (2016) 733-738.

[27] S. Takahashi, S. Bhattacharjee, T. Ihara, and K. Wakai, J. Japan Society of Microgravity

Application, Vol. 24, No. 3 (2007) 225-230.
[28] P. V. Ferkul, J. Kleinhenz, H. Shih, R. Pettegrew, K. Sacksteder, J. T'ien, Microgravity Science and Technology: XV/2 (2004) 3-12.
[29] S. Bhattacharjee, W. Tran, M. Laue, C. Paolini, Y. Nakamura, Proc. Combust. Inst. 35 (2015) 2631-2638.
[30] R. N. Walters, S. M. Hackett, R. E. Lyon, Fire and Materials 24, 5 (2000) 245–252.
[31] S. Villar-Rodil, A. Martinez-Alonso, J.M.D. Tascon, J. Analytical and Applied Pyrolysis 58-59 (2001) 105-115.

EFFECT OF OXYGEN DEPLETION ON THE RADIATIVE CHARACTERISTICS of BUOYANT TURBULENT DIFFUSION FLAMES

Dong Zeng*, Yi Wang
FM Global, Research Division
1151 Boston-Providence Turnpike, Norwood, MA 02062, USA

ABSTRACT

In many fire scenarios (such as a fully developed compartment fire), gaseous fuel burns in in vitiated environments. Changes of radiative and convective heat transfer from the flame to the environment or solid fuel surfaces due to vitiation need be better understood. This work presents an experimental study on the radiative characteristics of buoyant turbulent ethylene diffusion flames burning in a nitrogen-diluted atmosphere. The experiment was performed on a 15-cm diameter pool-like ethylene fire enclosed in a water-cooled compartment. The ambient oxygen concentration was controlled by adding nitrogen to the air flow while maintaining the oxygen flow rate at ten times of the stoichiometric oxidant requirement of the flames. The fuel supply rate was constant while the ambient oxygen concentrations were reduced from atmospheric condition to 8 vol %. At different ambient oxygen concentrations, the radiant fraction, combustion efficiency, flame shape, and temperature along the central axis of the flame were measured. In addition, the radiative power per unit height was determined by measuring the radiative heat flux of flame sections using a slit radiometer, which facilitates the interpretation of radiant fraction. The global soot volume fraction and temperature were also determined by coupling measured spectral flame emission and RADCAL modeling. It was found that the radiant fraction reduces with decreasing ambient oxygen concentrations, which was mainly attributed to reduced soot concentration.

INTRODUCTION

In an underventilated fire condition, the composition of the air in an enclosure where fuel burns can change as oxygen is diluted and displaced by combustion product. Similarly, in the case of commonly used fire protection systems such as sprinkler, water mist, and gaseous agent systems, air can be replaced by water vapor or other suppression agent that results in a low oxygen environment. Characteristics of fire in a vitiated air environment could be different from those in an air condition and have been investigated both experimentally [1-6] and numerically [7, 8]. Among those, the aspect of radiative characteristics is the focus of the current study as the radiative heat flux is dominating heat transfer in intermediate- to large-scale fires.

Santo and Tamanini [1] studied the flame properties of polymethyl methacrylate (PMMA) pool fire in nitrogen diluted atmosphere. They found that the total radiation reduces with the decreased oxygen volume fraction from 21% to 18%, while the Schmidt temperature, or assumed constant flame temperature of a gray, homogeneous flame, does not change significantly. Santo and Delichatsios [2] measured the effect of combustion products, e.g., CO_2 and H_2O on the burning of pool fires. It was shown that vitiation conditions had an insignificant effect on the heat release rate, but decreased radiative power per unit flame height, which was attributed to the reduction of soot formation due to a lower adiabatic flame temperature with increased vitiation. Orloff *et al.* [3] measured the radiant fraction of buoyant turbulent jet flames of hydrocarbon fuels and showed that the radiant fractions decreased with lower oxygen concentrations. Tewarson and coworkers [4] studied the influence of oxygen concentration on the fuel parameters of small pool fires of several liquid and solid fuels. They found that the combustion efficiency reduced, flame radiation decreased, and convective fraction increased when the oxygen concentration decreased from that of air. They also noted that inversely, flame radiative fraction increased and convective fraction decreased as oxygen concentration increased. Nasr *et al.* [5] investigated the burning process of heptane pool fires at different ventilation flows. Reduction of oxygen concentration was accompanied by a lower radiant fraction, which had a faster reduction rate from the air condition to the oxygen concentration of 18% and decreased slowly with the further reduction of oxygen concentration. Such a variation contributed to a faster reduction of the radiative heat flux toward the fuel surface than the convective heat flux. Overall, the heat flux toward the fuel surface reduced with decreased oxygen concentration. White *et al.* [6] studied turbulent flames generated from a line burner burning methane and propane with a co-flow of air mixed with varying dilution levels of nitrogen. It was shown that the radiant fraction decreased linearly for both fuels with declining oxygen concentration until the rapid onset of extinction. Such a decrease was attributed to diminishing soot radiation based on the flame color and luminosity analysis.

Recently, several modeling studies attempted to simulate the heat flux effect of flames in the oxygen depleted environments. In general, the empirically-fitted correlation or experimental data were applied in the simulation. Wahlqvist and van Hees [7] utilized a correlation of mass loss rate with oxygen level developed from the experimental data [9]. Vilfayeau [8] modeled the line burner experiment of White *et al.* [6]. The variation of radiative characteristic with different oxygen concentrations was treated using the different radiant fraction measurements in the simulation. These simplified modeling approaches show the importance of developing a gas phase radiation model in the vitiated gas environment.

The current study aims to investigate the radiative characteristics of low strain, buoyant, turbulent flames of ethylene fuel. For this purpose, a compartment facility was developed to provide a well-controlled burning environment. The characteristics of flame burning in the reduced oxygen environment leading to an extinction condition were measured with the focus of the radiative aspects of flame.

EXPERIMENTAL

Apparatus and Instrumentation

Global flame characteristics including radiant fraction, radiation power profile, spectral emission, puffing frequency, and combustion efficiency were measured on axisymmetric, buoyant, turbulent pool-like diffusion flames. A schematic of the experimental apparatus is shown in Figure 1. A water-cooled enclosure provided control of the ambient O_2/N_2 concentration ratio and minimized the ambient draft disturbances that may contribute to a less quiescent environment. Air mixed with additional nitrogen was supplied by gas reservoirs and controlled by rotameters at about 10 times the stoichiometric oxidant requirement of the flames. Air and nitrogen were mixed in a static mixer (Koflo 1.5-40-6-2) before being supplied to the enclosure. The enclosure had a cross section of 1.22 × 1.22 m^2, and a height of 1.83 m. The air-nitrogen mixture was supplied to the lower chamber of the enclosure through a distribution box. Uniform ambient flow was achieved by passing the gas mixture through a plenum followed by several screens and a layer of 2.54 cm layer of fine sand. During a test, the air flow rate was fixed and the enclosure oxygen concentration was monitored and controlled by adjusting the nitrogen flow rate.

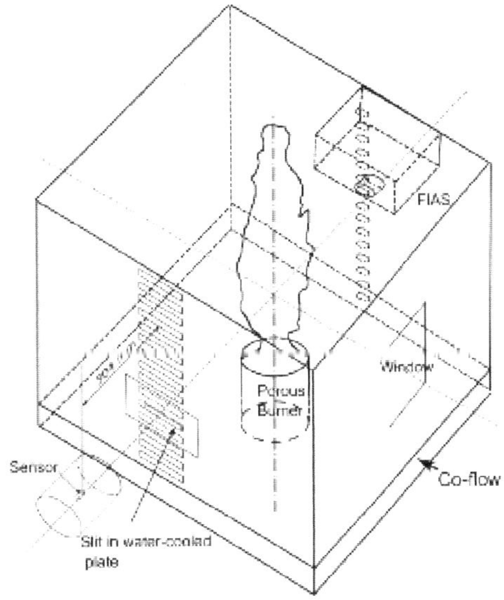

Figure 1. Schematic of the apparatus and instrumentation.

The gaseous fuel was supplied from a steel round burner of 15.2 cm in diameter; uniform exit velocities over the burner surface were ensured by using two layers of coarse and fine sand (2.54 and 5.08 cm thickness, respectively). The fuel flow rate was maintained constant by a mass flow controller (Sierra C100M-DD-3-OV1-SV1-PV2-S0). With the reduction of ambient oxygen concentration, the flame with a relatively small heat release rate tends to blow off before the global extinction, as a result of heat loss to the burner. Such extinction is not desirable as it does not represent the typical extinction behavior encountered in large-scale fire suppression. Therefore, the flame base near burner needs to be compensated for such a heat loss, e. g. White et al. [6] used the oxygen injection near the burner rim. In the current work, the flame was anchored by 36 pre-mixed ethylene/air pilot flames surrounding the burner. Each premixed flame was adjusted to be approximately 2.5 cm long. The total heat release rate of pilot flames is approximately 1 kW.

A characterization test was conducted to verify that the oxygen concentrations of co-flow surrounding the flame were maintained at the prescribed level. The oxygen concentration at the heights above the burner of 5.1, 30.5, 55.9, and 81.3 cm, and the radial distance of 22.9 cm from burner axis was measured for a steady state flame similar to those examined in the test. The sampling locations virtually encapsulated the largest size of flames encountered in the study and represented the co-flow environment of flame. The steady state oxygen concentrations of these locations were consistent, within a 1.3% difference from the corresponding oxygen concentration in the inlet condition. This indicated that the co-flow and enclosure system effectively controlled the oxidizer environment and that ambient air infiltration had a negligible effect on flame.

Experimental procedures

The water-cooled enclosure was positioned under a fire products collector (FPC). The calorimetry measurements were conducted in the same fashions as those in the = Fire Propagation Apparatus test [10, 11]. The chemical heat release rate is determined using CO/CO_2 generation calorimetry [12]

$$\dot{Q}_C = \frac{\dot{m}_{FPC}}{M_{FPC}}[\Delta H^*_{CO_2} M_{CO_2}(X_{CO_2} - X^0_{CO_2}) + \Delta H^*_{CO} M_{CO}(X_{CO} - X^0_{CO})] - \dot{Q}_{C,p} \quad (1)$$

where $\dot{Q}_C$ is the chemical heat release rate (kW), $\dot{m}_{FPC}$ is the mass flowrate through the FPC duct (kg/s), M_{FPC} is the molecular weight of FPC duct flow, $\Delta H^*_{CO_2}$ and ΔH^*_{CO} (kJ/g) are the net heat of complete combustion to CO_2 and CO per unit mass of generated CO_2 and CO; X_{CO_2} and $X^0_{CO_2}$ are measured mole fractions of CO_2 in the FPC duct and in the ambient air, respectively, and X_{CO} and X^0_{CO} are measured mole fractions of CO in the FPC duct and in the ambient air, respectively. $\dot{Q}_{C,p}$ is the chemical heat release rate of the pilot flames (kW).

The combustion efficiency is evaluated by:

$$\chi = \frac{\dot{Q}_C}{\dot{m} \cdot \Delta H_c} \tag{2}$$

where $\dot{m}$ is the fuel flow rate (g/s), and ΔH_c is the theoretical heat of combustion (kJ/g).

The water-cooled enclosure prohibited the use of a single radiometer to measure the radiation fraction. Instead, a slit-radiometer traversed vertically through the flame height and measured the radiation power profile. The slit-radiometer viewed the flames in the water-cooled enclosure by horizontal slots of 1.27 mm high, 12.7 cm wide, and 5.08 cm apart in the vertical direction. Care was taken to ensure the radiation measurement was not affected by the presence of the enclosure. The radiometer, slit plates, and all viewing surfaces of the enclosure were water-cooled and painted with flat black paint and kept free of any soot deposit. The radiometer is equipped with a thermopile sensor (Dexter Research Center, Inc., Model 1M, BaF2 window, and sensitive area 1 mm diameter). The radiometer was calibrated by means of a blackbody source of a 2.54 cm diameter aperture. For the determination of the effective slit height Δz, the slit-radiometer was calibrated at a proper distance of 91.4 cm from a 6.35 mm diameter source aperture. The radiometer traversed vertically with the field of view scanning through the source aperture. Shown in Figure 2, the distance between the points corresponding to the half values of the maximum signal is deemed as the effective slit height. For radiative power measurement, iso-tropic emission was assumed. The irradiance q_{rs} measured by the slit-radiometer sensor at a distance R_s from the flame with an effective slit height of Δz yields a measurement of radiative power per unit flame height:

$$S(z) = \frac{4\pi R_s^2 q_{rs}}{\Delta z} \tag{3}$$

The integration of radiative power per unit height along the flame yields the total radiative power. Therefore, the radiant fraction is:

$$\chi_r = \frac{\int_0^{L_f} S(z)dz}{\dot{m} \cdot \Delta H_c} \tag{4}$$

where L_f is the flame height (m)

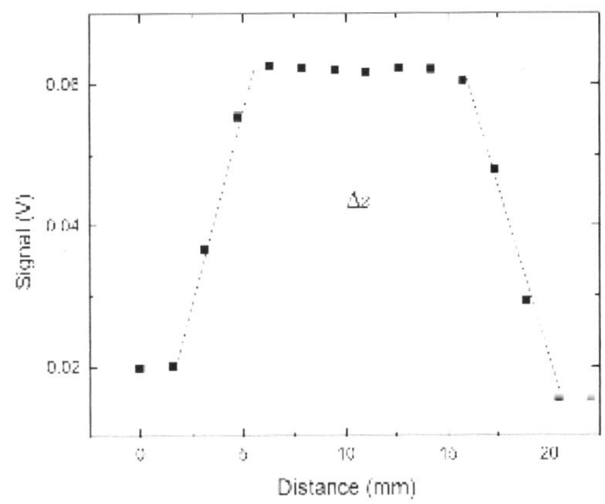

Figure 2. Slit-radiometer slit height calibration.

Line-of-sight (LOS) spectral intensity measurements were conducted with a fast infrared array spectrometer (FIAS) for the diametric path at different heights above the burner. Shown in Figure 1, the FIAS viewed the flames through openings of 1 cm diameter. The FIAS's sampling rate is 400 Hz; its spectral range is 1–5 μm (20 nm resolution), approximately, covering the near-infrared soot continuum radiation bands and gas radiation from water vapor and carbon dioxide. Experimental uncertainties (95 % confidence) are on the order of 10% [13]. At each measurement location, 12,000 instantaneous radiation intensity samples (corresponding to 30 seconds) were obtained. Collected data were time averaged for analysis. Both the slit-radiometer and FIAS were traversed vertically from the burner surface in 50.8 mm increment until the signal level reached nearly zero.

Perpendicular to the FIAS's LOS, flame images were recorded by a video camera (Redlake MotionXtra HG-XR) at a frame rate of 25 frames per second. Variable exposure settings were applied to facilitate high contrast between flame and background. For each test, a total of 3000 flame images were recorded. The flame images were post-processed based on the RGB color model [14] to identify the instantaneous flame height. The mean flame height is determined based on the 50% of intermittency of flame, as defined by Zukoski et al. [15]. Figure 3 shows the probability density function (PDF) and cumulative density function (CDF) of instantaneous flame heights of ethylene flame burning in air co-flow. The flame height corresponding to the 0.5 of CDF value is reported as the mean flame height of this test condition.

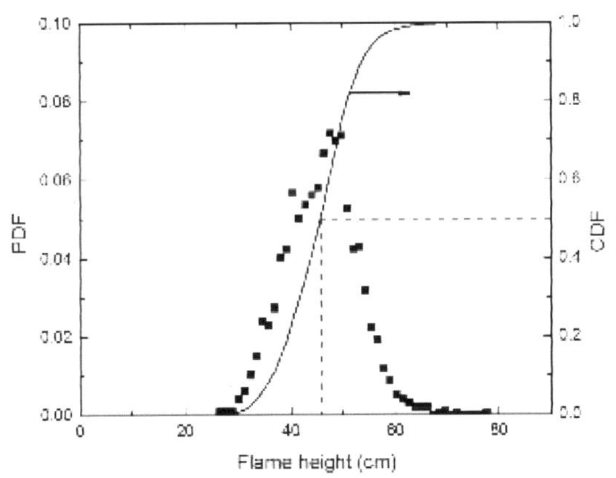

Figure 3. Example of mean flame height determination.

The flame heights can be evaluated using Heskestad's correlation [16]:

$$\frac{L_f}{D} = fn(\frac{c_{p,\infty} T_\infty}{g \rho_\infty^2 (\chi_c \Delta H_c / r)^3} \frac{\dot{Q}^2}{D^5}) \quad (5)$$

where L_f is the flame height; D is the burner diameter; $c_{p,\infty}$ is the specific heat of air; T_∞ and ρ_∞ are the ambient temperature and density, respectively; χ_c is the convective fraction; g is the acceleration of gravity; ΔH_c is the heat of combustion; r is the mass stoichiometric ratio of air to fuel.

The presence of strong periodic pulsations in buoyant diffusion flame affects many flame characteristics. In this study, the puffing frequency was determined by analyzing the dominant frequency of radiant power of flame slice from the slit-radiometer measurement, following Grant and Jones [17], Portscht [18]. The radiant power of a horizontal flame slice at 2 burner diameter (2D) height above burner was applied. In Figure 4, the time series of the radiant power signal is shown in the upper panel. They are transformed into the frequency domain as shown in the lower panel. The frequency corresponding to the maximum power is the puffing frequency.

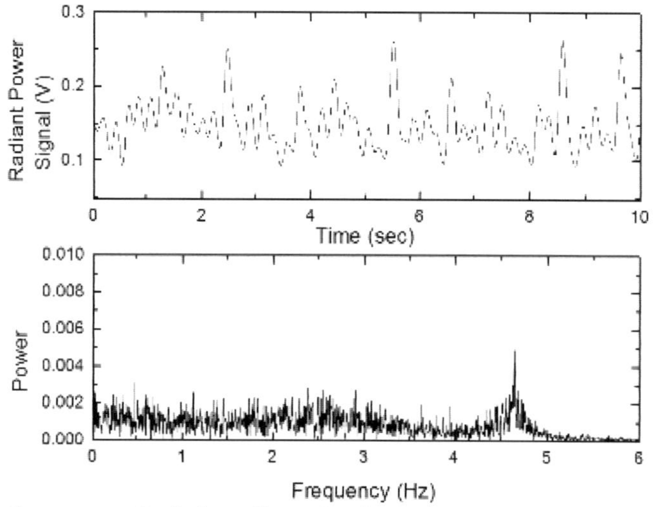

Figure 4. Puffing frequency of ethylene flame in air environment; upper panel: time series of radiant power of a horizontal flame slice; lower panel: data in frequency domain, showing the maximum frequency of 4.65 Hz.

EXPERIMENTAL RESULTS AND DISCUSSION

The experimental results are reported and discussed in the following sections. Flame images of different oxygen concentration condition are presented first. Variation of combustion efficiency and CO/CO_2 yields with oxygen concentration in the oxidizer stream are also reported. The effect of oxygen concentration on flame height and flame puffing frequency is discussed next. Last, the radiative characteristics of ethylene flames in co-flow with different oxygen concentrations are reported and discussed.

Flame Images

Figure 5 shows the flame images captured under oxygen concentration conditions ranging from 10 vol% to that in air. These flame images of different oxygen concentrations were captured using a Nikon D800 camera with the same settings including the shutter speed of 1/25 s, the aperture size of F5, and the ISO value of 200. For the flame image acquired in the 10 vol% oxygen environment, the flame emission in the visible range was small and the image was post processed to increase the visibility of flame.

The flame is luminous with the oxygen concentrations from that in air to 12 vol% as a result of soot incandescent emission. Qualitatively, the flame luminosity did not change significantly from air condition to 14 vol%, then became less luminous at 12 vol% oxygen concentration. Visual observation showed that there were blue flame strips in the flame of 12 vol% condition. In the case of 10 vol% oxygen concentration, soot emission is nearly absent, and flame emission is dominated by luminescence. When the oxygen concentration is 12 vol% and less, flames are weaker and show periodical extinction. Compared to the methane and propane flames examined by White et al. [6], which transitioned from yellow to blue flames at 17 vol% and 15%, respectively, ethylene flame exhibits a lower transition oxygen concentration (~11 vol%), which is attributed to the higher soot formation capability of ethylene.

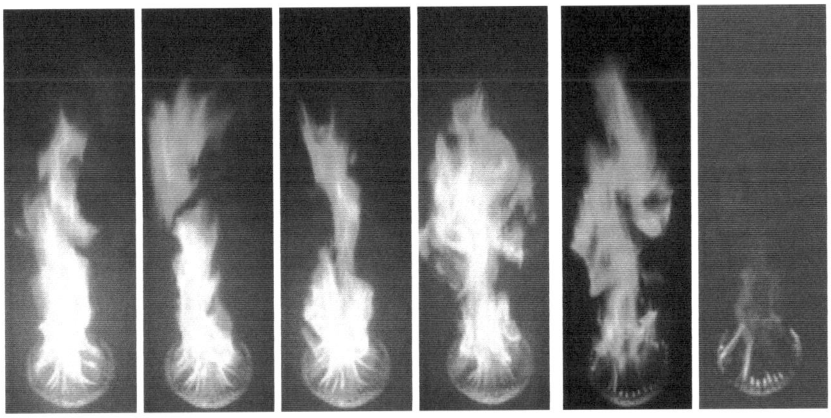

Figure 5. Flame images under different oxygen conditions, from left to right: air, 18%, 16%, 14% 12%, and 10%.

Combustion Efficiency of Buoyant Turbulent Ethylene Flame

The left panel of Figure 6 shows variation of combustion efficiency (χ) of buoyant turbulent ethylene diffusion flames with decreased oxygen concentration in the oxygen/nitrogen mixture environment. The error bar in both panels of Figure 6 shows the fluctuation of data due to the characteristics of buoyant turbulent flame. Combustion efficiency is 0.86 in the air environment. χ decreases slightly from 0.86 to 0.8 on average when oxygen concentration decreases from the air condition to 11 vol%. The further decrease of oxygen concentration is accompanied by a drastic reduction of χ. During this stage, the flame becomes increasingly unstable and the upper part of the flame shows extinction behavior and χ reduces to 0.12 at the oxygen concentration of 8.2 vol%, which is lower than the LOI values of CH_4 and C_3H_8 [19]. The right panel of Figure 6 shows the CO/CO_2 yields as a function of oxygen concentration. The CO_2 yield trend with oxygen concentration variation coincides with the combustion efficiency, shows a small decrease from the air condition to 11 vol% oxygen concentration, and then quickly decreases with further reduction of oxygen concentration. On the other hand, CO yield maintains near zero level initially, and then increases to a maximum value of 0.09 at an oxygen concentration of 9 vol%, before it reduces slightly with the further reduction of oxygen concentration.

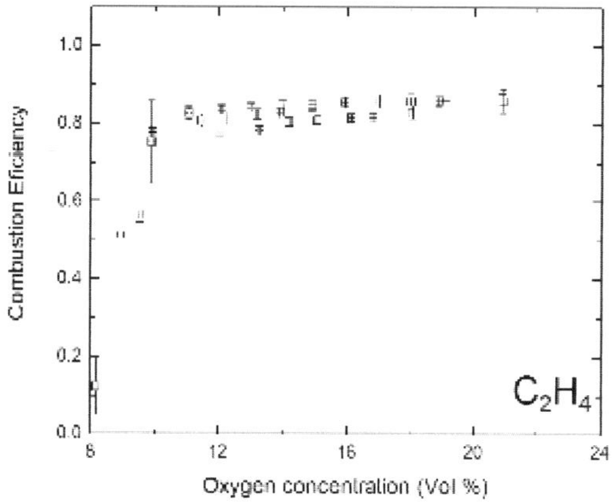

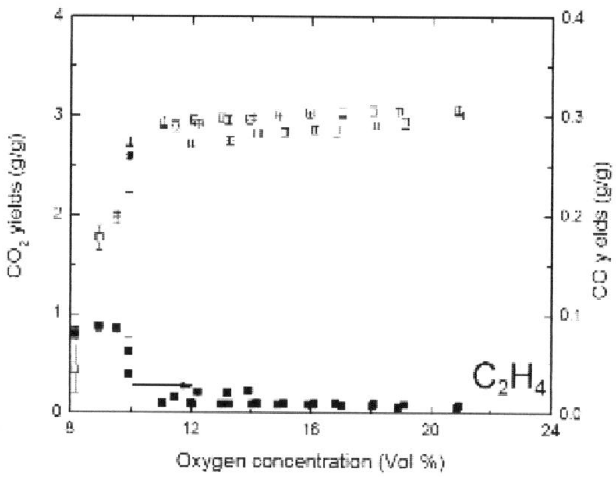

Figure 6. Combustion efficiency (left panel) and CO/CO_2 yields (right panel) of buoyant turbulent ethylene diffusion flame under different oxygen/nitrogen mixture environment, error bar represents standard deviation of time series.

Flame height and puffing frequency

The correlation of flame height in Eq. (5) can be rearranged as the following: r can be represented as r_{O_2} / Y_{O_2} for the reduced oxygen environment, r_{O_2} is the mass stoichiometric ratio of oxygen to fuel; $\dot{Q}$ is expressed as $\dot{Q}\chi$; χ_c can be represented by $\chi - \chi_r$, where χ_r is the radiant fraction. Therefore Eq. (5) can be expressed as:

$$\frac{L_f}{D} = fn(N) = fn(\frac{c_{p,\infty}T_\infty}{g\rho_\infty^2 Y_{O_2}^3}\frac{1}{[(\chi - \chi_r)\Delta H_c / r_{O_2}]^3}\frac{(\dot{Q}\chi)^2}{D^5}) \quad (6)$$

The first term in the RHS of Eq. (6) represents variables of environment, the second term represents fuel parameter, and the third term represents chemical heat release rate. Consequently, in the current test condition with a fixed theoretical heat release rate, $L_f / D \propto \chi^2 / [Y_{O_2}(\chi - \chi_r)]^3$. Figure 7 shows the relation of the non-dimensional parameter N, or that in the function of RHS in Eq. (6) with the normalized flame height. In the same figure, the dependence of L_f / D on oxygen concentration is also shown. The flame height increases continually with decreased oxygen concentration until oxygen concentration is approximately 15 vol%. The further decreased oxygen is accompanied by a lower flame height. This may be attributed to 1. Decreased soot formation tendency and flame becomes increasingly less luminous, as discussed in the next section; 2. In these relatively low oxygen molar fraction environments, flame becomes increasingly unstable. In the relation of L_f / D with N, the data of oxygen volume fraction larger than 0.15 is fitted linearly, as shown in Figure 7, with an R-square value of 0.95.

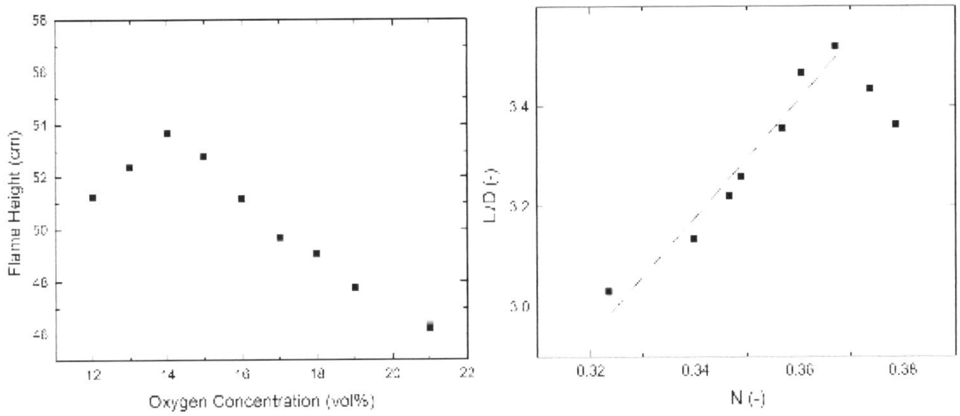

Figure 7. Left panel: ethylene flame height versus N and oxygen concentration; right panel: normalized ethylene flame height versus N.

The combustion efficiency only slightly reduces from ~0.86 to ~0.8 on average when the oxygen volume fraction declines from 0.21 to 0.15. The increased flame heights in this range shows that flame requires more residence time to entrain ambient oxidizer to achieve these combustion efficiencies. The effects of the increased convective fraction of fire (see Figure 9), and simultaneously increased diluent fraction in the oxidizer conflicts with each other. A higher convective fraction can increase the flame temperature while the increased diluent can reduce the flame temperature. To further understand the relative importance of these factors, the puffing frequency is analyzed next.

Figure 8 shows the puffing frequency as a function of oxygen concentration in the co-flow environment. Puffing is an oscillatory behavior that can affect the air entrainment rate into the flame, flame height, combustion efficiency, and radiative characteristic of flames. In the pool-fire like scenario encountered in the current study, buoyant plume gases accelerate upward which results in a toroidal vortical structure within one burner diameter above the fuel source. Such a vortex decays on the flame surface near the burner lip. In the meantime, buoyant fuel gas accumulates inside the flame envelope and subsequently accelerates to form the next structure [20, 21]. Puffing frequency can be described by [21]:

$$f = C\left(\frac{\rho_\infty}{\rho_f} - 1\right)\sqrt{\frac{g}{D}}[(1+\frac{1}{Ri})^{1/2} - \frac{1}{\sqrt{Ri}}]^{-1} \quad (7)$$

where C is a constant, ρ_f is the flame density, Ri is the Richardson number, defined as $\frac{(\rho_\infty - \rho_f)gD}{\rho_\infty U^2}$, and U is the fuel velocity. For the air condition, Eq. (7) is simplified as $f = 1.68D^{-0.5}$ [21], which yields a puffing frequency of 4.3 Hz, similar to the measured value (4.65 Hz) in the air condition. With the decrease of oxygen concentration in the co-flow environment, the puffing frequency reduces. In the Eqn. (7), the Ri number has a relatively large value and the change of the $\frac{\rho_\infty}{\rho_f} - 1$ term mainly contributes to the trend shown in Figure 8. Therefore, the flame density has to increase with decreased oxygen concentration. This shows that the dilution effect of increased nitrogen diluent is more substantial than the other factors and results in the reduced flame temperature with increased oxygen concentration.

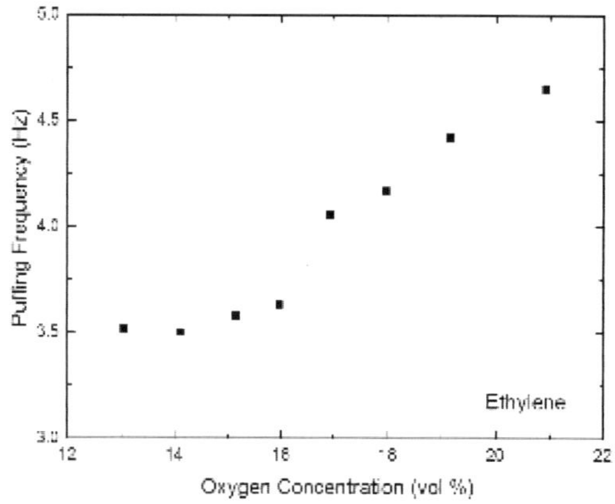

Figure 8. Flame puffing frequency of ethylene flames at different oxygen concentrations.

Radiative Characteristics

The left panel of Figure 9 shows the radiant power distribution of ethylene flames burning in co-flow with different oxygen concentrations. These radiant power distributions are similar in that all have a large power near the burner surface, then decrease due to neck-in effect before they reach the maximum power at approximately 1.5~2 of burner diameter height. The radiant power distribution extends to a larger height when oxygen concentration decreases, consistent with the flame height variation shown in Figure 7. Overall, the radiant power decreases with lower oxygen concentration. The right panel of Figure 9 shows the radiant fraction computed using Eq. (3) and (4). The radiant fraction decreases from 0.33 to ~0.1 when oxygen concentration reduces from ambient to 10 vol%, consistent with the observation with methane and propane [6].

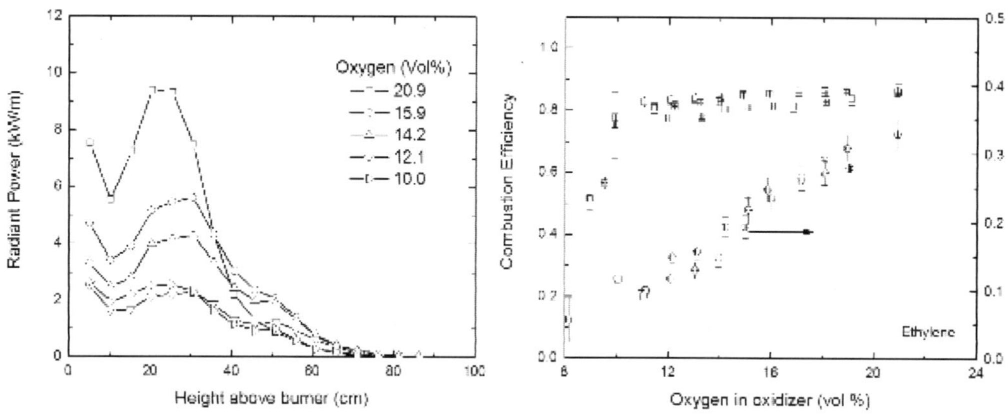

Figure 9. Left panel: Radiation power distribution of Ethylene flames at different oxygen concentrations; right panel: radiant Fraction of Ethylene flames at different oxygen concentrations, error bar represents standard deviation of time series.

Figure 10 shows the LOS emission spectra of the diametric line at 20.3 cm height above burner, or 1.33 burner diameter. These spectra are averaged values of 30 s time series. For spectra in the atmosphere with 14 vol% or higher oxygen concentration, soot broadband emission is evident. Within this range, soot broadband spectral intensity reduces with decreased oxygen concentration. In the meantime, gas emission, such as CO_2 at the wavelength of 4.3 μm and H_2O at the wavelength of 2.5 μm, is relatively invariant and is mainly responsible for radiation emission when oxygen concentration is less than or equal to 12 vol%.

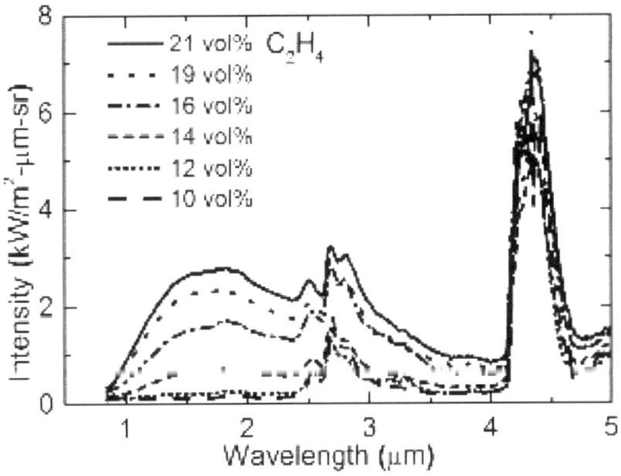

Figure 10. Spectral emission of ethylene flames at different oxygen concentrations.

Figure 11 shows the PDF of LOS spectral intensity of the diametric line of 30.5 cm or 2D height above burner, at the wavelength of 1.99 μm, which is dominantly generated from soot emission. The data of oxygen concentration from 14 vol% to air condition is plotted. PDF of intensity acquired from air condition shows a bimodal characteristic, including a broadly distributed intensity from 1 to 9 $kW/(m^2 sr\mu m)$. In the mean time, there exists a large probability near zero intensity, an indication of flame intermittency at the measured location. When oxygen concentration decreases to 17 vol%, the intermittency reduces and PDF shifts toward smaller intensity values. With the further decrease of oxygen concentration, e. g. 14 vol%, the PDF further shifts toward smaller intensity with an increase of intermittency. Such a variation of PDF with oxygen concentration is the result of interaction between flame geometry and soot emission capability. In the air condition, flame height is lower (see Figure 7), and there is a higher probability of intermittency due to the increased oscillatory behavior in the upper flame region. For the same oxygen concentration condition, soot emission intensity has a larger value when flame presents. When oxygen concentration decreases, intermittency decreases due to flame height increases. In the case of the oxygen concentration of 14 vol%, it seems that the soot emission capability of flame further reduces which results in PDF shifts to smaller intensity values and increased intermittency compared to those with oxygen concentration in the range of 16~19%.

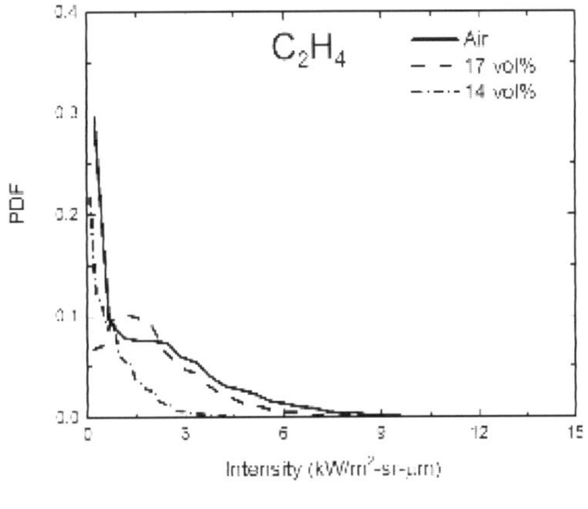

Figure 11. PDF of diametric line LOS emission intensity at the wavelength of 1.99 μm.

Figure 11 indicates that flame soot emission capability decreases with a lower oxygen concentration in a co-flow environment. The soot broad band emission can be described using Planck's law:

$$I_\lambda = \varepsilon_\lambda \frac{C_1}{\lambda^5} \frac{1}{\exp(\frac{C_2}{\lambda T}) - 1} \tag{8}$$

where I_λ is the spectral intensity (kW/m^2-µm-sr), ε_λ is the spectral emissivity, λ is the wavelength (µm), T is the temperature (K), and C_1 and C_2 are the first and second radiation constants (1.191×10^5 kWµm^4/m^2/sr and 1.439×10^4 µmK, respectively). The flame emissivity values can be expressed by the spectral absorption coefficient (κ_λ) and effective flame pathlength (l) [22-24]:

$$\varepsilon_\lambda = 1 - \exp(-\kappa_\lambda l) \tag{9}$$

For soot broadband radiation, the spectral absorption coefficient can be derived by invoking the small particle limit (e.g., Rayleigh limit) of Mie theory [22-25]. The flame emission typically peaks between 1~3 µm, which is over an order of magnitude larger than reported soot particle diameters [26]. At the Rayleigh limit, the absorption coefficient is independent of particle size and inversely proportional to wavelength:

$$\kappa_\lambda = 6\pi \operatorname{Im}(\frac{m_{\bar{\lambda}}^2 - 1}{m_{\bar{\lambda}}^2 + 1}) \frac{f_v}{\lambda} = \frac{36\pi n_{\bar{\lambda}} k_{\bar{\lambda}}}{(n_{\bar{\lambda}}^2 - k_{\bar{\lambda}}^2 + 2)^2 + (2 n_{\bar{\lambda}} k_{\bar{\lambda}})^2} \frac{f_v}{\lambda} \tag{10}$$

Where f_v is the soot volume fraction, and $n_{\bar{\lambda}}$ and $k_{\bar{\lambda}}$ are the optical constants (i.e., $m_{\bar{\lambda}} = n_{\bar{\lambda}} + i k_{\bar{\lambda}}$, where $m_{\bar{\lambda}}$ is the complex index of refraction) of soot evaluated at different wavelengths. The narrow-band radiation model RADCAL [27] was utilized and κ_λ was evaluated using the Dalzell and Sarofim optical constants incorporated in RADCAL. For each diametric line at different flame heights, the values of mean radiation temperature and soot volume fraction were determined through optimization, using the RADCAL model [27], in order to match the measured spectral radiation intensities. Only those wavelength ranges attributable to soot emission were considered in the model. Therefore, the wavelength ranges of 1~2.31 µm were considered. The optimal values of scalars were achieved by minimizing the mean squared error between the calculated and measured mean intensity within these wavelength ranges. To reflect the effect of soot emission, the mean spectra were recalculated by rejecting the spectra with no significant soot band emission, e. g. $I_{1.99\mu m} > 0.2$ (kW/m^2-µm-sr). In the derivation of soot volume fraction, a fixed pathlength of 10 cm is assumed for all cases.

The left panel of Figure 12 compares the effective soot temperature of flames at different heights above burner generated in the oxygen concentration ranging from 15 ~ 21 vol%. The temperatures of all cases are within the range of 1340~1400 K, which shows the soot temperatures maintain at similar range even though the oxygen concentration reduces, and correspondingly, the overall flame temperatures shift to smaller values. This finding is consistent with that of Santo and Tamanini [1] which seems to indicate that the soot formation/oxidization reactions are favored in a fixed temperature range. This condition does not vary appreciably when oxygen concentration is changed. The right panel of Figure 12 shows the soot volume fraction variation with different flame heights and oxygen concentrations. For those with 19 vol% and 21 vol% cases, soot volume fractions are similar, and those with 21 vol% oxygen shows a lower soot volume fraction at higher height e. g. larger than 30 cm, which is attributed to a shorter flame height (see Figure 7). However, overall, soot volume fraction decreases with reduced oxygen concentration. Therefore, when oxygen concentration decreases in the oxygen/nitrogen co-flow environment, the flame temperature distribution shifts toward lower values as a result of increased nitrogen dilution and inversely, to a lesser extent, decreased radiant heat loss. In the meantime, the soot formation/oxidization reactions are relatively favored in a certain temperature range. The overall lower temperature distribution in reduced oxygen concentration contributes to the lower soot volume fraction.

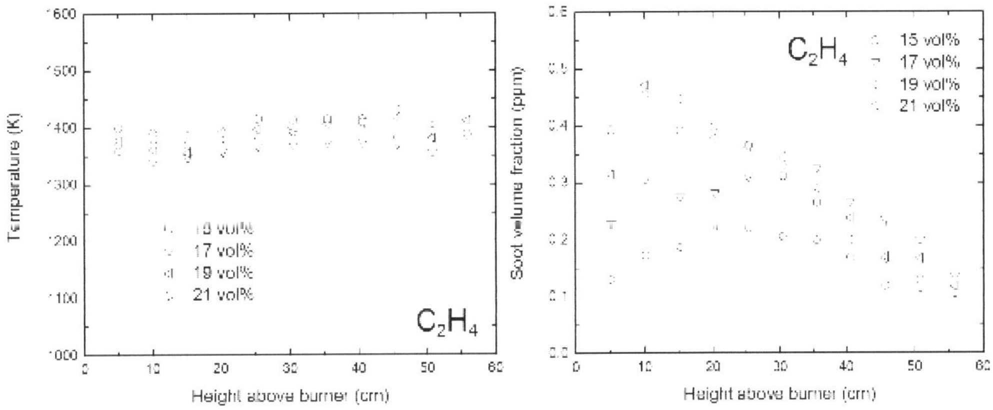

Figure 12. Left panel: soot temperature of Ethylene flames at different oxygen concentrations, right panel: soot volume fraction of Ethylene flames at different oxygen concentrations

CONCLUSIONS

The characteristics of buoyancy-driven, turbulent ethylene flame in nitrogen/oxygen mixture oxidizer were investigated experimentally. For the purpose of this study, an experimental apparatus was developed to provide a controlled gas environment. Combustion efficiency decreases slowly with lower oxygen concentration until it promptly reduces when global extinction begins. In the global extinction stage, the CO yield significantly increases. The luminous flame heights show an increasing trend when oxygen concentration decreases from air condition to oxygen concentration of 14 vol%, then decreases with the further reduced oxygen concentration. The increase trend is mainly attributed to the effect of increased nitrogen dilution which contributes to a lower global flame temperature. Such a lower global flame temperature at reduced oxygen condition also contributes to a smaller puffing frequency. The radiant power distribution of flames in different oxygen concentrations is measured and interpreted for their radiant fraction. Radiant fraction generally decreases with lower oxygen concentration mainly as a result of reduced soot emission. Radiation power distribution shows the similar trend with different oxygen concentrations, and also demonstrates the effect of flame elongation. The spectral emission of flames in different oxygen environments shows reduced soot broadband emission with lower oxygen concentration. Detailed analysis of probability function density shows such a reduction of soot broadband emission is the result of the complex interaction of flame fluctuation and soot emission capability of flame. Furthermore, the intermittency filtered spectra are analyzed using the RDACAL model for its emission temperature and soot volume fraction. It was found that the emission temperature is relatively independent of ambient oxygen concentration. Instead, reduced oxygen concentration mainly lowers the soot volume fraction and correspondingly reduces radiant fraction.

The current study focuses on ethylene fuel. Our ongoing study is investigating several other fuels with different radiative characteristics than ethylene. In the meantime, local radiation measurement such as temperature and soot volume fraction can further facilitate the understanding of the mechanisms by which oxygen depletion affects flame radiation. Current work is targeted to these directions.

ACKNOWLEDGEMENTS

The authors would like to thank Mr. Robert Tabinowski of FM Global for his efforts in conducting the experiments.

REFERENCES

1 Santo, G. and Tamanini, F., *Influence of oxygen depletion on the radiative properties of PMMA flames.* Symposium (International) on Combustion, 1981. **18**(1): p. 619-631.

2 Santo, G. and Delichatsios, M.A., *Effects of vitiated air on radiation and completeness of combustion in propane pool fires.* Fire Safety Journal, 1984. **7**(2): p. 159-164.

3 Orloff, L., De Ris, J., and Delichatsios, M.A., *Radiation from buoyant turbulent diffusion flames.* Combustion Science and Technology, 1992. **84**(1-6): p. 177-186.

4 Tewarson, A., Lee, J.L., and Pion, R.F., *The influence of oxygen concentration on fuel parameters for fire modeling.* Symposium (International) on Combustion, 1981. **18**(1): p. 563-570.

5 Nasr, A., Suard, S., El-Rabii, H., Garo, J.P., Gay, L., and Rigollet, L., *Heat feedback to the fuel surface of a pool fire in an enclosure.* Fire Safety Journal, 2013. **60**: p. 56-63.

6 White, J.P., Link, E.D., Trouvé, A.C., Sunderland, P.B., Marshall, A.W., Sheffel, J.A., Corn, M.L., Colket, M.B., Chaos, M., and Yu, H.Z., *Radiative emissions measurements from a buoyant, turbulent line flame under oxidizer-dilution quenching conditions.* Fire Safety Journal, 2015. **76**: p. 74-84.

7 Wahlqvist, J. and van Hees, P., *Implementation and validation of an environmental feedback pool fire model based on oxygen depletion and radiative feedback in FDS.* Fire Safety Journal, 2016. **85**: p. 35-49.

8 Vilfayeau, S., Large eddy simulation of fire extinction phenomena, 2015, UNIVERSITY OF MARYLAND, COLLEGE PARK.

9 Peatross, M. and Beyler, C. *Ventilation Effects On Compartment Fire Characterization.* in *Proceedings of the Fifth International Symposium on Fire Safety Science.* 1997.

10 Reaction to fire tests -- Measurement of material properties using a fire propagation apparatus, 2011, International Organization for Standardization, Geneva, Switzerland,

11 Standard Test Methods for Measurement of Synthetic Polymer Material Flammability Using a Fire Propagation Apparatus (FPA), 2009, West Conshohocken, PA,

12 Tewarson, A., *Generation of Heat and Chemical Compounds in Fires*, in *SFPE Handbook of Fire Protection Engineering*, P.J. DiNenno, Editor. 2008. p. 3-124.

13 Zheng, Y., Barlow, R.S., and Gore, J.P., *Measurements and calculations of spectral radiation intensities for turbulent non-premixed and partially premixed flames.* Journal of Heat Transfer, 2003. **125**(4): p. 678-686.

14 Patel, P. and Tiwari, S., *Flame detection using image processing techniques.* International Journal of Computer Applications, 2012. **58**(18): p. 13-16.

15 Zukoski, E.E., Cetegen, B.M., and Kubota, T., *Visible structure of buoyant diffusion flames.* Symposium (International) on Combustion, 1985. **20**(1): p. 361-366.

16 Heskestad, G., *Engineering relations for fire plumes.* Fire Safety Journal, 1984. **7**(1): p. 25-32.

17 Grant, A.J. and Jones, J.M., *Low-frequency diffusion flame oscillations.* Combustion and Flame, 1975. **25**: p. 153-160.

18 Portscht, R., *Studies on characteristic fluctuations of the flame radiation emitted by fires.* Combustion Science and Technology, 1975. **10**(1-2): p. 73-84.

19 White, J.P., Measurement and Simulation of Suppression Effects in a Buoyant Turbulent Line Fire, in Department of Fire Protection Engineering2016, University of Maryland: College Park, Md.

20 Cetegen, B.M. and Ahmed, T.A., *Experiments on the periodic instability of buoyant plumes and pool fires.* Combustion and Flame, 1993. **93**(1): p. 157-184.

21 Malalasekera, W.M.G., Versteeg, H.K., and Gilchrist, K., *A review of research and an experimental study on the pulsation of buoyant diffusion flames and pool fires.* Fire and

Materials, 1996. **20**(6): p. 261-271.
22 Yuen, W. and Tien, C., *A simple calculation scheme for the luminous-flame emissivity.* Symposium (International) on Combustion, 1977. **16**(1): p. 1481-1487.
23 Tien, C. and Lee, S., *Flame radiation.* Progress in Energy and Combustion Science, 1982. **8**(1): p. 41-59.
24 Bard, S. and Pagni, P., *Carbon particulate in small pool fire flames.* Journal of Heat Transfer, 1981. **103**(2): p. 357-362.
25 Dalzell, W.H. and Sarofim, A.F., *Optical constants of soot and their application to heat-flux calculations.* Journal of Heat Transfer, 1969. **91**(1): p. 100-104.
26 Newman, J.S. and Steciak, J., *Characterization of particulates from diffusion flames.* Combustion and Flame, 1987. **67**(1): p. 55-64.
27 Grosshandler, W.L., RADCAL: A Narrow Band Model for Radiation, NIST technical note 1402, 1994.

A METHOD FOR MEASUREMENT OF SPATIALLY RESOLVED RADIATION INTENSITY AND RADIATIVE FRACTION OF LAMINAR FLAMES OF GASEOUS AND SOLID FUELS

Catherine Hamel[ab], Fernando Raffan-Montoya[b], Stanislav Stoliarov[b]
[a] AcuTech Consulting Group
[b] Department of Fire Protection Engineering, University of Maryland, College Park

ABSTRACT

This work introduces a new method for determination of the radiation intensity and radiative fraction for axisymmetric laminar diffusion flames for both solid and gaseous fuels by using a modified DSL-R camera employed to collect monochromatic (900 nm) images and a Schmidt-Boelter heat flux gauge. The high spatial resolution provided by the images of the camera allows for a multi-emitter treatment of the 2-6 cm flames. The flame's radius and intensity are extracted from the images and presented as two curves that are functions of the flame-axis position. Each point on the flame sheet is discretized at pixel-level resolution and treated as a differential emitting surface. Radiation transport equations are formulated and solved numerically to compute a function that relates the camera's readings to the total radiation heat flux detected by the gauge. The calculation yields spatially resolved radiation intensity information. Integration of this intensity over the flame surface divided by the total heat release rate yields the global radiative fraction. In this work, polyethylene (solid fuel) is studied and three gaseous fuels (methane, propane and acetylene) are studied to validate the methodology.

INTRODUCTION

Radiation from flames is an important topic of study because this mode of heat transfer dominates many fire growth scenarios[1]. The fraction of combustion energy lost to the environment through radiation, which is termed the radiative fraction, χ_r, is an important parameter used extensively in engineering calculations as well as in state-of the-art computational fluid dynamics fire models[1,2]. Therefore, having a complete understanding of the radiative fraction for a wide range of fuel types is imperative for accurate fire growth predictions.

The radiant properties of diffusion flames have been studied extensively by various researchers in the past four decades[3,4,5,6]. These studies have utilized various methods of measuring the radiation of flames of both gaseous and solid fuels. For gaseous fuels, methods for determining the radiative fraction have been developed for both laminar and turbulent flames. Markstein[3] conducted a set of experiments on turbulent flames of multiple gaseous fuels using a circular 12.7mm diameter nozzle and a radiometer placed 3m from the burner axis. Assuming spherical isotropy and using a theoretical total heat release rate definition, Markstein found the radiative fraction to be a constant value across a wide range of heat release rates (0-40 kW). In the same study, Markstein studied laminar flames burning from a 4.1mm diameter burner with a radiometer placed 300 mm from the burner axis. Unlike the turbulent flames, Markstein found the radiative fraction of the laminar flames to be dependent on the heat release rate, showing an increase to the half-power with an increase in the theoretical total heat release rate[3]. Markstein also related the radiative fraction of a laminar flame to the smoke point by finding a linear relation between the radiative power at the smoke point and the theoretical total heat release rate at the smoke point. Markstein's study therefore introduced the smoke point as an important scaling parameter for gaseous fuels that can be used for laminar-to-turbulent radiation predictions[3]. Additionally, the radiative fractions from turbulent flames of various solid fuels have been studied by Tewarson[4] using the fire propagation apparatus (FPA). In these studies, the radiation losses were not measured directly but rather calculated based on measurements of the convective losses. Tewarson concluded that the use of a fixed, global radiative fraction is erroneous due to a dependence on fire size, ventilation conditions, and chemical distributions in the flame[4].

The current literature reveals a lack of studies on the radiation from laminar flames of solid fuels and spatially-resolved measurements of flame radiation intensity. This study aims to fill this void with a unique method that measures the spatially-resolved radiation intensity and the radiative fraction of laminar flames of solid fuels without relying on convective measurements or empirical relationships. Additionally, a scaling parameter between laminar and turbulent flames is desirable (unlike Markstein's study[3], the smoke point cannot be used due to smoke point measurements being challenging for solid fuels). Therefore, this study provides a method for obtaining the maximum radiation intensity value, which is introduced as a potential scaling parameter between laminar and turbulent flames of solid fuels.

METHODOLOGY

The method for measuring the spatially-resolved radiation intensity and the radiative fraction of axisymmetric laminar flames is comprised of two sets of experiments, with one measuring the radiation heat release $\dot{Q}_{rad}$ and the other measuring the total heat release $\dot{Q}_{tot}$. Taking the ratio of these two heat release rates gives the radiative fraction as:

$$\chi_r = \frac{\dot{Q}_{rad}}{\dot{Q}_{tot}} \quad [1]$$

In the calculation of $\dot{Q}_{rad}$ [kW], a spatially-resolved radiation intensity $I_f(z)$ [kW/m^2-srad] is also calculated. The method assumes a relationship between the image intensity along the centerline of the flame and the radiation intensity, and it is related with a spatially-resolved function. The method also assumes that the flame is represented by an axisymmetric laminar flame sheet that is optically thin (meaning it absorbs no radiation) and radiates from both the inside and outside of the flame sheet.

Experimental Setup and Procedure

The total heat release $\dot{Q}_{tot}$ is measured in the Milligram-scale Flaming Calorimeter (MFC) apparatus, which is shown in Figure 1. The MFC utilizes oxygen-consumption calorimetry to measure the effective heat release rate and the corresponding effective heat of combustion of both gaseous and solid fuels. For solid fuels, a platinum coil

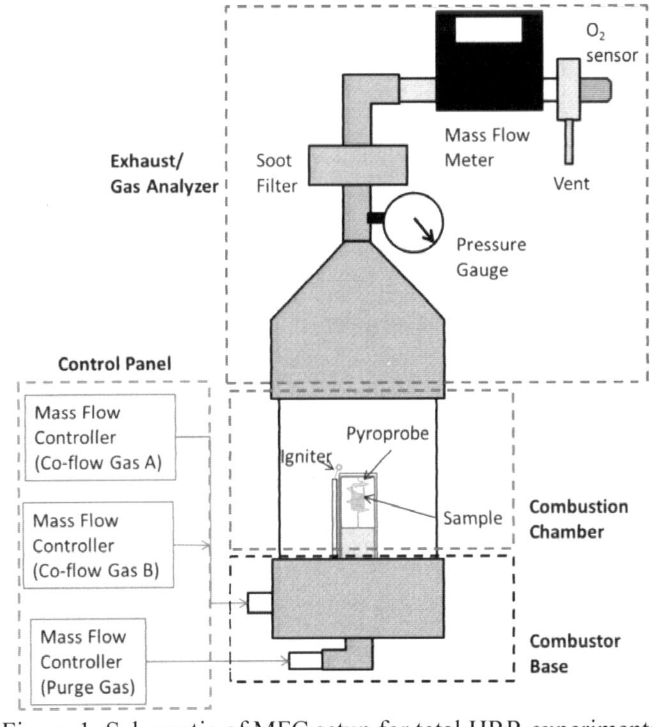

Figure 1. Schematic of MFC setup for total HRR experiments.

pyroprobe is used to heat a 30 mg solid sample at a prescribed heating rate until gaseous volatiles are produced and a laminar flame is established. For gaseous fuels, the mass flow rate of the fuel is controlled with a digital flow controller and data is acquired for 30 seconds of the steady flame before suppressing the flame. Previous work by Raffan-Montoya et al.[7] provides details on the experimental and analysis procedures of the MFC measurements.

The radiative heat release rate $\dot{Q}_{rad}$ is measured in an altered setup of the MFC apparatus that allows for unobstructed optical access to the flame. This altered setup is shown in Figure 2. A Nikon D800 DSLR camera equipped with a 900 nm band pass filter is placed 19 cm from the burner and acquires images of the flame. The sensor of the camera was altered by MaxMax LLC to extend the spectral sensitivity to the infrared region. Additionally, the incident heat flux at various locations is measured by a Schmidt-Boelter water-cooled heat flux gauge (with a sensor diameter of 7 mm) that can be traversed in both the vertical and radial directions. The method assumes that only the radiative contributions of the heat flux are measured by the gauge, and a thermocouple is consequently placed at the same location of the heat flux gauge in order to monitor for and avoid potential convective contribution. In the actual setup, the camera is rotated 90° from the location shown in Figure 2 such that the vector normal to the face of the camera lens is perpendicular to the vector normal to the heat flux gauge. Additionally, the entire setup shown in Figure 2 is placed inside a draft hood that protects the setup from ambient light and environmental drafts.

The radiation heat release experiments begin by taking a series of calibration images of the burner to determine a number of geometric definitions, which include a pixel-to-millimeter conversion factor, the location of the origin in Cartesian coordinates (which is placed at the center of the outlet of the burner), and the location of the heat flux gauge. A co-flow of 1.8 SLPM of air is passed through a pool of glass beads and aids in establishing a steady flame. A data acquisition system (DAQ) records the heat flux gauge reading in addition to the thermocouple data and the DAQ is started prior to igniting the flame in order to record an ambient noise reading.

After the DAQ is started, a flame is established by piloted ignition, followed by the acquisition of 16 images

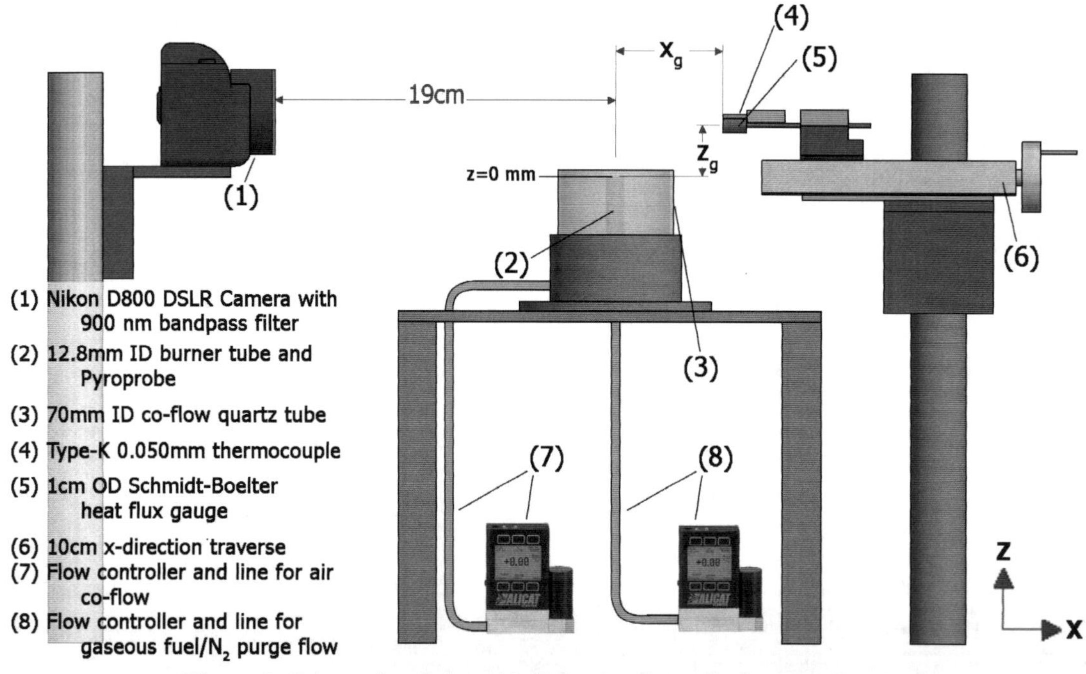

Figure 2. Schematic of altered MFC setup for radiation experiments.

by the camera. The flame is suppressed and DAQ data continues to be acquired for 30 seconds after flame extinction. The data set of each trial consists of 16 images of the flame, heat flux data and thermocouple data. For each fuel, two to three flame heights are studied, with a minimum of three heat flux gauge placements per flame height. At each specification of fuel, flame height, and gauge placement, three trials are done.

Parameter Determination

The images of the flame and the heat flux gauge data provide a means of obtaining important information about the flame geometry and radiation. The two parameters obtained from the flame images are the radius profile of the flame, $R(z)$, and the normalized intensity along the centerline of the image, $I_c(z)$. Additionally, the mean heat flux at a particular heat flux gauge height is found as q''_{gauge}.

The radius profile and flame height is found by first defining the flame region as any pixel location that has a greyscale intensity value greater than or equal to 7% of the maximum greyscale value of the image. The sensitivity of the method to this 7% threshold value is minimal across flame heights and fuel types. The flame height is found for each image and is defined as the maximum z-location in the image where the pixel value exceeds the 7% threshold value. The flame height location of all images are averaged to obtain the mean flame height, h_f, and the uncertainty of this value is taken as two standard deviations (2σ) of the mean value.

Each image is split along the z-axis into two half-images and the radius profile of each half-image is defined as the series of pixel locations above the threshold that are furthest from the z-axis. The set of profiles from the three trials are converted to physical units by dividing by the pixel-to-millimeter conversion factor and are then averaged together to determine a mean radius profile. This mean radius profile is fit with a sixth-order polynomial function, $R(z)$, which can be subsequently operated on. The uncertainty is resolved along the z-axis and is taken as 2σ of the mean.

In a similar fashion, the normalized centerline intensity profile is determined from the flame images. The centerline is chosen as a representative location for each height that describes the intensity of any element on the flame sheet at that z-location, due to the camera capturing true line-of-sight intensity along the centerline of the flame. The images are first normalized by the maximum greyscale value of a 16-bit image (65535) to create an array of pixels with values between 0 and 1, and are then normalized by the exposure time of the image acquisition, allowing for comparison between fuels and resulting in units of [1/s]. The column of pixels along the z-axis are taken as the "centerline intensity" for all images and are averaged together to obtain a mean centerline intensity profile. This mean experimental profile is fit with a Gaussian function, $I_c(z)$, which can be subsequently operated on as in the case of the radius. Again, the uncertainty is taken as 2σ of the mean.

The final parameter obtained from the radiation experiments is the mean heat flux at each heat flux gauge location. The heat flux values recorded during the image acquisition period are first averaged together. In order to account for the heat flux contributions of a heated apparatus, the background heat flux is taken as the heat flux obtained 5-seconds after the extinction of the flame, and this background flux is removed from the mean heat flux during image acquisition. The resulting value is taken as the heat flux at the gauge and the values of the three trials at each gauge location are averaged to obtain a mean heat flux at that gauge location, q''_{gauge}.

Calculation of Radiation Properties

With the parameters known from the radiation experiments, the numerical representation of the geometric configuration must be established. First, the 2-dimensional radius profile of the flame is swept around the z-axis in 2π radians to obtain an axisymmetric flame sheet. The result of this operation is shown in Figure 3. This flame sheet is resolved with a differential step size of $\delta z = 1$ pixel (0.02 mm). Each point on the flame sheet is defined in cylindrical coordinates with a differential area of:

$$\delta A_p(z,\theta) = R(z)\delta\theta\sqrt{\left(\frac{dR(z)}{dz}\delta z\right)^2 + (\delta z)^2} \qquad [2]$$

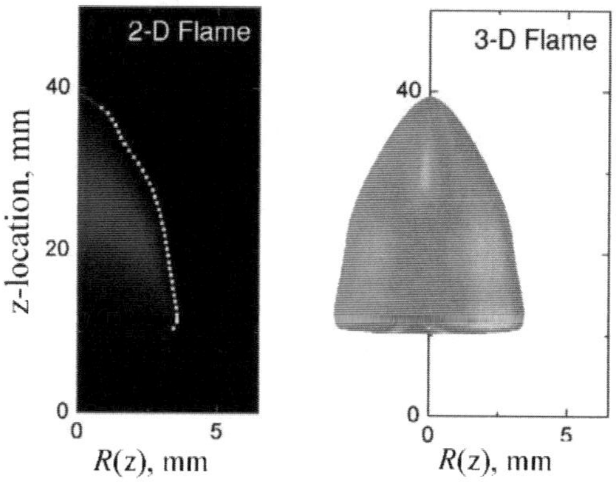

Figure 3. 2-dimensional and 3-dimensional presentations of the flame sheet.

where $\delta\theta$ is a differential angle that is defined as $\delta\theta = 2\pi R(z)/\delta z$. This definition of the flame is easiest described in cylindrical coordinates, but the radiation equations are defined in Cartesian coordinates. Therefore, the location of every point of the flame can be converted from cylindrical coordinates to Cartesian coordinates using the set of equations:

$$x_p = R(z)\cos\theta; \quad y_p = R(z)\sin\theta; \quad z_p = z \qquad [3]$$

where (x_p, y_p, z_p) defines a point on the flame sheet in Cartesian coordinates, with an origin set at the center of the top of the burner outlet, as shown in Figure 2.

Additionally, due to the axisymmetric assumption and Equation 3, the flame surface can be defined by one function $F(x_p, y_p, z_p)$ as:

$$F(x_p, y_p, z_p) = x_p{}^2 + y_p{}^2 - R^2(z_p) = 0 \qquad [4]$$

where $F(x_p, y_p, z_p)$ is a symbolic function that may be mathematically manipulated.

With the flame sheet defined, the important geometric vectors that describe radiation from the flame sheet to the heat flux gauge are the unit vector normal to the differential surface at each point of the flame sheet, $\vec{n}_p$, the unit vector normal to the gauge face, $\vec{n}_g$, and the unit vector pointing from the flame surface to the gauge, $\vec{n}_{pg}$. These vectors are shown in Figure 4. First, the unit vector normal to the differential flame surface at the location (x_p, y_p, z_p) can be found using calculus, which defines the gradient of a scalar function defining a surface as the normal vector to said surface[8]. Mathematically, the normal vector is found by:

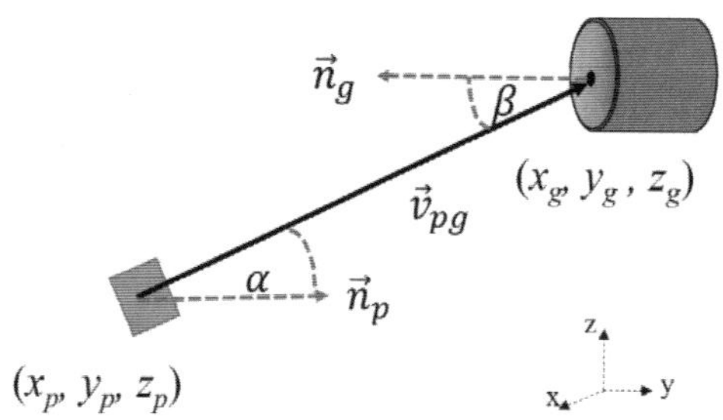

Figure 4. Geometric consideration between differential surface and gauge face.

$$\vec{v}_p(x_p, y_p, z_p) = \nabla F(x_p, y_p, z_p) = \nabla\left[x_p{}^2 + y_p{}^2 - R^2(z_p)\right] \qquad [5]$$

and the unit vector pointing normal to the point is found by normalizing by the magnitude of $\vec{v}_p$, or:

$$\vec{n}_p(x_p, y_p, z_p) = \frac{\vec{v}_p}{\|\vec{v}_p\|}. \qquad [6]$$

The vector $\vec{n}_p(x_p, y_p, z_p)$ can be found at any point on the flame sheet. Next, the unit vector normal to the face of the heat flux gauge, $\vec{n}_g$, has a trivial definition due to the heat flux gauge being placed in the yz-plane and running parallel to the y-axis, therefore only having a vector component in the y-direction. Thus, the unit vector $\vec{n}_g$ is given as $\vec{n}_g(x_g, y_g, z_g) = 0\hat{\imath} - 1\hat{\jmath} + 0\hat{k}$.

The gauge face is assumed to be a point with infinitesimally small surface area, pointing in the direction of $\vec{n}_g$. The last vector that needs to be calculated is the vector connecting a point (x_p, y_p, z_p) on the flame sheet to the heat flux gauge. Using vector algebra, this vector is found by taking the difference of the two locations:

$$\vec{v}_{pg}(x_p, y_p, z_p) = (x_g - x_p)\hat{\imath} + (y_g - y_p)\hat{\jmath} + (z_g - z_p)\hat{k} \qquad [7]$$

The unit vector corresponding to $\vec{v}_{pg}$ is then found by normalizing by the magnitude of the vector:

$$\vec{n}_{pg}(x_p, y_p, z_p) = \frac{\vec{v}_{pg}}{\|\vec{v}_{pg}\|} \qquad [8]$$

With the geometry of the flame-gauge interaction defined, a mathematical description of the heat flux to the gauge can be formulated. The heat flux detected by the gauge, $\dot{q}''_{gauge}$ can be defined in terms of the spatially-dependent radiation intensity $I_f(z)$ as:

$$\dot{q}''_{gauge} = \sum_{A_f} \frac{I_f(z)|\vec{n}_p \cdot \vec{n}_{pg}|(\vec{n}_g \cdot \vec{n}_{pg})\delta A_p}{\|\vec{v}_{pg}\|^2} \qquad [9]$$

where A_f is the surface area of the entire flame sheet defined by F. The dot product term $\vec{n}_p \cdot \vec{n}_{pg}$ is placed in absolute values in order to mathematically account for the heat flux gauge accepting radiation from both sides of the flame sheet. It is assumed that the radiation intensity $I_f(z)$ can be related to the centerline intensity $I_c(z)$ by a relation function $c(z)$ that will be determined through inverse modelling. The assumed definition of the radiation intensity is given as:

$$I_f(z) = I_c(z)c(z) \qquad [10]$$

Using Equations 6-11, Equation 10 can be expanded as:

$$\dot{q}''_{gauge} = \sum_z \sum_\theta \frac{I_c(z)c(z)|\vec{n}_p \cdot \vec{n}_{pg}|(\vec{n}_g \cdot \vec{n}_{pg})R(z)\delta\theta\sqrt{\left(\frac{dR(z)}{dz}\delta z\right)^2 + \delta z^2}}{(R(z)\cos\theta - x_g)^2 + (R(z)\sin\theta - y_g)^2 + (z - z_g)^2} \qquad [11]$$

and the only unknown of the equation is the spatially-dependent relation function $c(z)$.

This relation function is inverse modelled, or 'calibrated' such that the percent difference between the predicted value of $\dot{q}''_{gauge,pred}$ and the measured $\dot{q}''_{gauge}$ is minimized. Therefore, the coefficients of the relation function are iterated until the quantity $|\dot{q}''_{gauge} - \dot{q}''_{gauge,pred}|/\dot{q}''_{gauge}$ is minimized for each of the three gauge locations (at one particular flame height). Each $c(z)$ function is thus unique to each fuel type and flame height. Additionally, the choice of $c(z)$ is validated by predicting the heat flux at various "spot-check" gauge locations that differ from the three calibration locations. Once a valid choice of $c(z)$ is found, the radiation intensity profile can be found using Equation 11 and the corresponding radiation heat release rate $\dot{Q}_{rad}$ is found by integrating over the flame sheet area:

$$\dot{Q}_{rad} = 2\pi \sum_z \sum_\theta I_f(z)\delta A_p = 2\pi \sum_z \sum_\theta c(z) I_c(z) R(z) \delta\theta \sqrt{\left(\frac{dR(z)}{dz}\delta z\right)^2 + (\delta z)^2}$$
$$= 4\pi^2 \sum_z c(z) I_c(z) R(z) \sqrt{\left(\frac{dR(z)}{dz}\delta z\right)^2 + (\delta z)^2} \quad [12]$$

Because of the axisymmetric assumption of this method, Equation 13 becomes a simple 1-dimensional integration along the z-axis. With $\dot{Q}_{rad}$ known, the calculation of the radiation parameters is complete and the final unknown needed to find the radiative fraction is the total heat release, $\dot{Q}_{tot}$, per Equation 1.

Determination of Total HRR $\dot{Q}_{tot}$ and Radiative Fraction χ_r

The total heat release rate can be defined in various ways depending on the definition of the heat of combustion, Δh_c. In some cases, Δh_c can be measured, while in other instances it can be calculated from the heats of formation. In the current study, the effective heat of combustion is experimentally measured using the calorimetry capabilities of the MFC and the complete heat of combustion is also calculated by assuming stoichiometric, complete combustion reaction and the heats of formation of the reactants and products are used. The calculation of the complete $\Delta h_{c,com}$ is given as:

$$\Delta h_{c,com} = \sum\left(n\Delta H_f^\circ\right)_{prod} - \sum\left(n\Delta H_f^\circ\right)_{react} \quad [13]$$

where n is the number of moles of the species in the assumed stoichiometric equation and the products are assumed as only carbon dioxide and gaseous water vapor. This is a theoretical heat of combustion corresponding to complete combustion reactions. In reality, the combustion efficiency is lower than unity and the effective heat of combustion $\Delta h_{c,eff}$ represents the actual heat release per unit mass consumed. The effective $\Delta h_{c,eff}$ is found experimentally in the MFC apparatus by measuring the mass flow rate at the outlet of the combustion chamber and the oxygen concentration of the outlet flow. With these two measurements, in addition to the mass flow rate and oxygen concentration prior to combustion, the total mass of oxygen consumed can be found. An empirically-calculated constant describing the amount of energy released per gram of oxygen consumed is taken as 13.1 ± 0.6 kJ/g_{O2}, and this is coupled with the mass of oxygen consumed to find the instantaneous heat release rate (HRR). Integrating the HRR curve gives the total heat released in [J] and dividing by the total amount of mass burned gives $\Delta h_{c,eff}$ in [kJ/g_{fuel}].

The total heat release rate is found by multiplying the complete or effective heat of combustion by the mass flow rate of fuel:

$$\dot{Q}_{tot} = \dot{m}_f \Delta h_c \quad [14]$$

The mass flow rate of gaseous fuels is explicitly obtained by multiplying the volumetric flow rate set by the digital flow controller by the density of the fuel at standard conditions (25°C, 1atm). For solid fuels, the mass flow rate is found by averaging the instantaneous HRR during the image acquisition period (taken from the MFC experiment) and dividing by the measured heat of combustion to get an estimated mass loss rate. Either the complete $\dot{Q}_{tot}$ (using $\Delta h_{c,com}$) or the effective $\dot{Q}_{tot}$ (using $\Delta h_{c,eff}$) can be calculated.

With $\dot{Q}_{tot}$ known, Equations 1, 13, and 14 are used to find the radiative fraction, χ_r. Taking a ratio of the radiative and total heat releases, the radiative fraction is known and can be calculated using either the total or effective heat release rate.

RESULTS

The methodology explained above is applied to four fuels, including one solid state fuel and three gaseous fuels. Polyethylene (PE) is chosen as the solid fuel. In order to validate the method against previous studies' literature values, three well-studied gaseous fuels were also studied: methane, propane, and acetylene. These three fuels provide a wide range of radiation properties, with methane being a very low radiator and acetylene being a very high radiator. In accordance with the methodology, two to three flame heights are studied for each fuel. The test matrix for the current study is shown in Table 1. The flame heights are controlled either by the volumetric flow controllers in the case of the gaseous fuels, or the heating rate of the pyroprobe heater for the solid fuels (a higher heating rate corresponds to a higher flame height). The results of these fuels are presented here.

Table 1. Test matrix of current study.

Gaseous Fuels	Propane			Methane			Acetylene		
Flow rate (sccm)	56.0	69.6	106.0	191.3	205.0	216.0	94.8	117.8	179.8
Gauge location, z_g (cm)	1.5 2.5 3.5	1.5 3.5 4.9	2.5 3.5 4.9	2.4 3.4 4.9	2.4 3.4 5.0	2.4 3.4 5.0	1.4 2.4 4.9	1.4 2.4 4.9	2.4 3.4 4.9

Solid Fuel	Polyethylene	
Heating rate (°C/s)	8.0	15.5
Gauge location, z_g (cm)	0.9 1.4 1.9	1.5 3.4 5.0

Radius Profiles, $R(z)$ and Flame Height, h_f

The mean radius profiles, along with the accompanying uncertainty and the polynomial fit, are shown in Figure 5 for all fuels and flame heights. Open symbols show the mean radius value along with the uncertainty (uncertainty shown with error bars and is on the order of the symbol size). The maximum uncertainty of all flames is 12.0% for the tip of the methane flame, and the mean uncertainty amongst all fuels and flame heights is less than 5.0%. The measured flame height is also shown with horizontal dashed

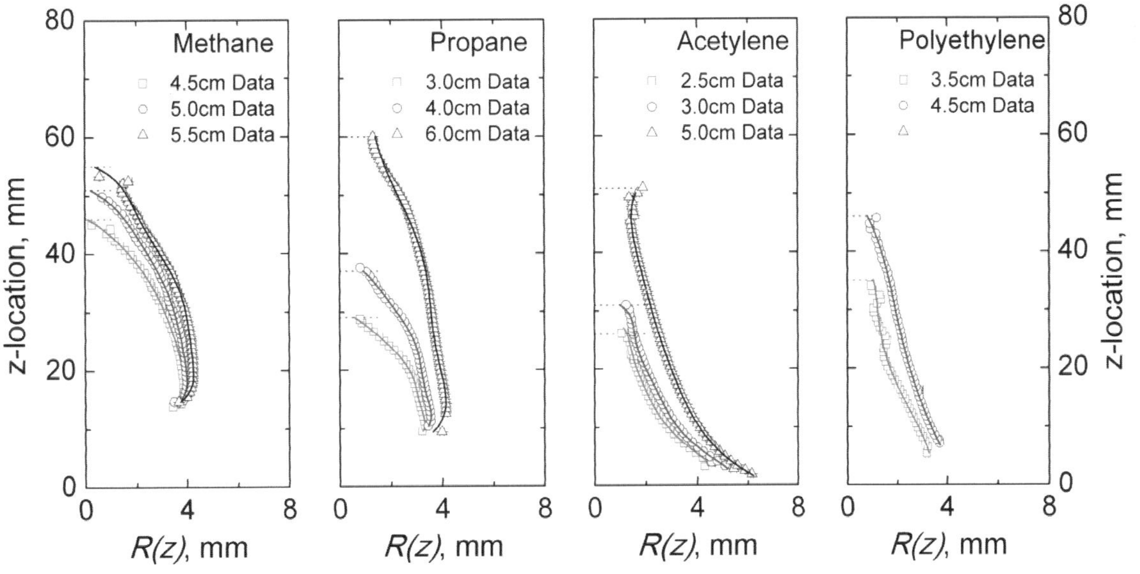

Figure 5. Mean radius values (open symbols) with accompanying uncertainty (error bars shown and on the order of the symbol size). The radius polynomial function $R(z)$ is shown with a solid line and the horizontal dashed lines indicate the flame height.

lines, and it is seen that all flames are between 2 and 6cm, which is the desired height for the MFC apparatus. Additionally, the 6th-order curve fits described the flame shape well for all fuel types and flame heights.

In should be noted that the flame heights of the PE tests were chosen such that the taller flame exceeds the smoke point and the shorter flame falls below the smoke point. Additionally, all of the acetylene flames exceeded the smoke point. Acetylene's radius profiles differ from the rest of the fuels in shape by extending down to $z = 0$ mm. This is due to acetylene producing soot almost immediately after exiting the burner tube. Conversely, the other fuels have a "lift-off" region of roughly 10 mm that corresponds to a low-emission region of the flame that radiates at lower wavelengths than 900 nm due to low soot-production towards the base of the flame. This finding is consistent with the observations of Rankin et al.[5] who observed this lift-off region for laminar ethylene diffusion flames from a 7.6 mm circular burner using an infrared camera.

Centerline Intensity Profiles, $I_c(z)$

The mean centerline intensity profiles and the uncertainty are shown in Figure 6 for all fuels and flame heights. The Gaussian curve fits are also shown with solid lines. The uncertainty of the profiles is highest at 5.8% for the 5.5cm methane flame and the mean uncertainty of all flames fell below 3%. It should be noted that the methane profiles are plotted on a different scale than the three other fuels.

Figure 6 shows that methane has a significantly lower centerline intensity than the other fuels, with a peak intensity value falling an order of magnitude lower than the peaks of the other fuels, which is consistent with the understanding of methane as a lowly-radiative fuel[1]. Additionally, acetylene has the highest peak intensity, following it being a highly sooting and radiative fuel. The acetylene profiles also have a peak intensity at 10-30% of the flame height, which is much lower than the other fuels that peak at 50-70% of the flame height. This agrees with the flame radius profiles shown in Figure 5 that show acetylene flames with a wider, sootier base than the other fuels. The polyethylene and propane profiles have similar shape and peak intensity values.

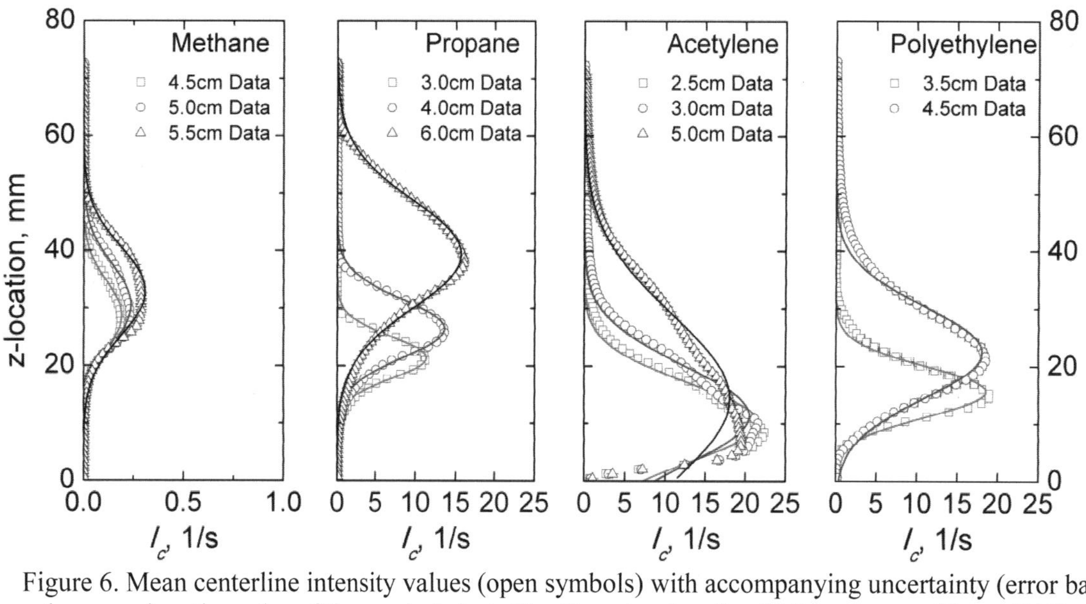

Figure 6. Mean centerline intensity values (open symbols) with accompanying uncertainty (error bars shown and on the order of the symbol size). The Gaussian function fit I_c is shown with a solid line.

Heat Flux Gauge Values, $\dot{q}''_{gauge}$, and Relation Function, $c(z)$

With the profiles of $R(z)$ and $I_c(z)$ known, the relation functions can be optimized using the mean heat flux gauge readings. It was found that a shallow convex 2nd order polynomial best predicted flames that were below the smoke point and a linear fit best predicted flames that exceeded the smoke point. The relation functions were optimized using the three heat flux gauge positions of each flame height, and the validity of the choice was confirmed with "spot-checks" predictions of heat flux gauge positions outside of the three optimization positions.

Figure 7 shows the measured heat flux values along with the uncertainty for one flame height of each fuel with open symbols and error bars. The predicted heat flux values that are found using the optimized $c(z)$ profiles and Equation 12 are shown with solid symbols. These values are plotted against the normalized gauge height position (normalized by the flame height) in order to show the strength of the predictions within the flame region. It is shown that within the flame region ($z_g/h_f < 1$), the chosen relation functions predict the heat flux gauge values within an average deviation of 7.1%, 4.6%, 5.4%, 7.7% for methane, propane, acetylene and polyethylene, respectively. These deviations fall within the 8.0% uncertainty of the $\dot{q}''_{gauge}$ measurements. Additionally, the relation functions are able to predict the heat flux at "spot-check" values within 8.0%, despite these locations not being part of the optimization data set. The relation function is able to predict the heat flux at any location, including the regions beyond the flame height, as can be seen in Figure 7. The highest prediction for acetylene occurs at 1.6 times the flame height and the prediction is well outside of the experimental uncertainty. However, the radiation intensity is minimal at a location far from the flame region, and therefore the deviation in this region is ignored.

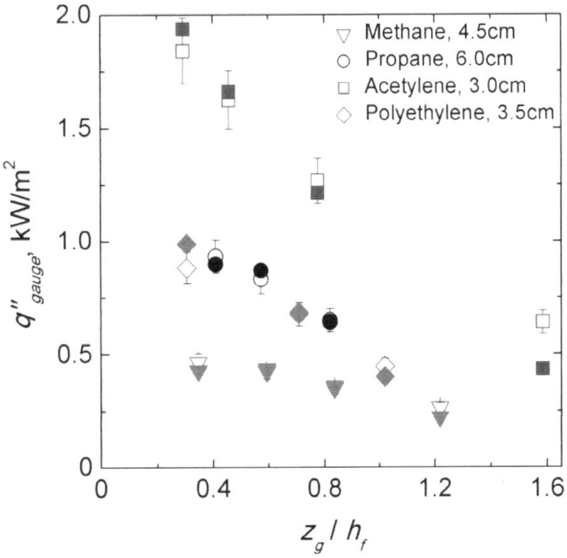

Figure 7. Measured (open symbols) values of heat flux and predicted values (filled symbols) found using the optimized relation functions. Plotted against gauge height normalized by the flame height.

Spatially-resolved Radiation Intensity Profiles, $I_f(z)$, and Total Radiation Heat Release Rate, $\dot{Q}_{rad}$

Using Equation 11 and the determined $I_c(z)$ and $c(z)$ profiles, the radiation intensity profiles $I_f(z)$ are found. Figure 8 shows the resulting profiles, and the maximum uncertainty is found to be 8.0% at the flame tip after an uncertainty propagation analysis. The radiation intensity profiles show that one again methane has the lowest radiation emission and acetylene has the highest. The 3.5cm flame of the polyethylene flame (which was below the smoke point) has similar radiation intensity behaviour to the 3.0cm flame of propane.

It is observed that the peak radiation intensity is nearly constant across different flame heights for the gaseous fuels. The polyethylene flames do not peak at the same intensity value, likely the result of one flame being below the smoke point and the other being beyond the smoke point. If the peak radiation intensity value is constant across heat release rates (for flame heights that are all below/all above smoke point), this value can potentially be used as a predictor of radiative fraction for turbulent flames in a similar fashion to Markstein's use of the smoke point[3].

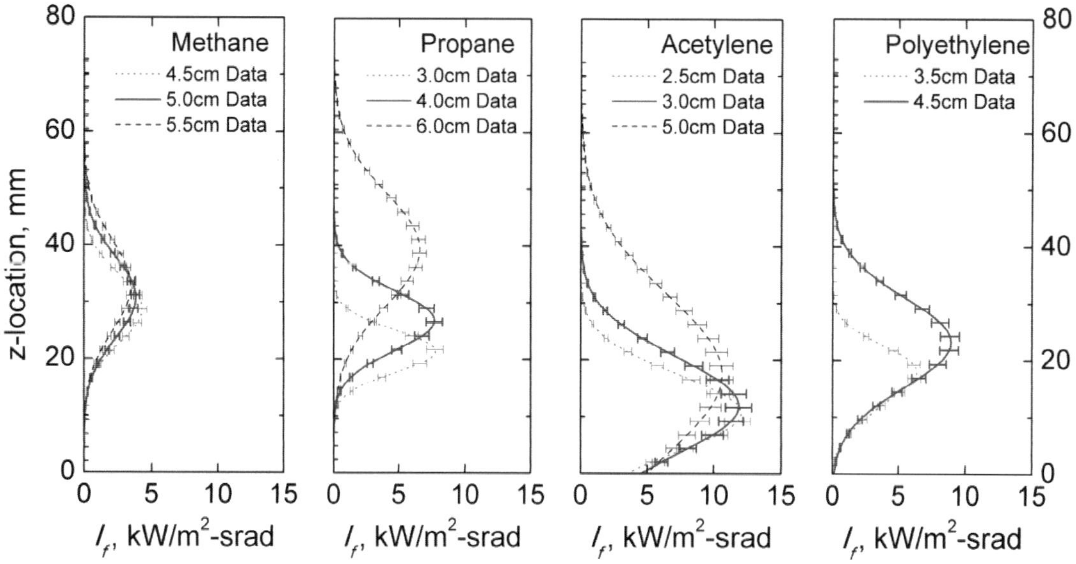

Figure 8. Radiation intensity profiles of all fuels and flame heights.

Using Equation 13, the $I_f(z)$ profiles are integrated over the flame sheet area to obtain the total radiation heat release rate, $\dot{Q}_{rad}$, which are shown in Figure 9 with an uncertainty of 8.0%. Once again, acetylene has the greatest radiation heat release rate, methane has the lowest radiation, and polyethylene and propane show similar radiation losses for comparable flame heights.

Total Heat Release Rate, $\dot{Q}_{tot}$

The total heat release rate value $\dot{Q}_{tot}$ are found using Equation 15. The values for $\Delta h_{c,eff}$ is measured in the MFC using Method A[7] and the complete heat of combustion, $\Delta h_{c,com}$ is found using Equation 14. The $\Delta h_{c,eff}$ values are shown in Table 2. For all fuels except acetylene, it is expected that the effective heat of combustion would be measured close to the complete heat of

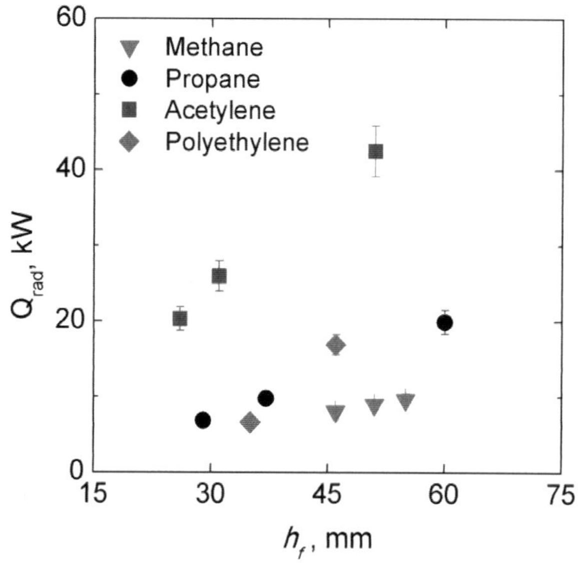

Figure 9. Measured (open symbols) values of heat flux versus predicted values (filled symbols) using the optimized relation functions.

Table 2. Net and complete heat of combustion values for all fuels.

Fuel	Flame Height [cm]	Effective Δh_c [J/kg]	Std. Error (+/-) [J/kg]	Complete $\Delta h_{c,com}$ [J/kg]
Methane	5.0	**51**	0.3	50
Propane	3.0	**46**	0.4	46
	4.0	**47**	0.3	
Acetylene	2.5	**35**	0.1	48
Polyethylene	3.5	**44**	0.4	44 [4]

combustion due to fairly high combustion efficiencies in the MFC setup. This is the case for methane, propane and PE, as seen in Table 2. However, for acetylene, which is a highly sooting flame and therefore has a low combustion efficiency, the effective heat of combustion is noticeably lower than the complete heat of combustion. Due to a lack of heat of formation data for polyethylene, the complete heat of combustion for PE could not be calculated. Therefore, for PE $\Delta h_{c,eff}$ is compared with data from Tewarson, who measured the heats of combustion in the oxygen bomb calorimeter[4], which measures close-to-complete combustion. Tewarson's heat of combustion value falls within the uncertainty of the $\Delta h_{c,eff}$ value for PE. The uncertainty of $\Delta h_{c,eff}$ values are taken as 5.0% in accordance with the standard MFC procedure[7].

With the heat of combustion known, the total heat release rate is found using Equation 15. Two Q_{tot} values are found based on differing definitions of the heat of combustion (effective versus complete). These values are shown in Table 3. An important observation is that the PE flames produce a noticeably lower $\dot{Q}_{tot}$ than the other fuels for flames of similar height. The reason for this is that during the experiments of the solid fuels, a purge flow of 50sccm of nitrogen is flowed through the burner tube in order to carry the volatiles of the pyrolized fuel into the combustion region. This purge flow therefore stretches the flame and results in lower heat release for a similar flame height when compared to gaseous fuel counterparts.

Table 3. Net and complete heat of combustion, total heat release rate, and radiative fraction of all flames.

	Propane			Methane			Acetylene			Polyethylene	
Flame Height [cm]	3.0	4.0	6.0	4.5	5.0	5.5	2.5	3.0	5.0	3.5	4.5
$\Delta h_{c,eff}$ [J/kg]	46	47	-	-	51	-	35	-	-	44	-
$\Delta h_{c,com}$ [J/kg]	46	46	46	50	50	50	48	48	48	44	44
Q_{tot} [W] (effective)	78.9	98.1	149.3	106.4	114.2	120.4	59.7	74.2	113.2	36.3	49.0
Q_{tot} [W] (complete)	78.6	97.7	148.8	103.8	111.4	117.4	67.7	84.1	128.3	39.1	52.8
χ_r (effective)	8.9	10.2	13.7	7.8	8.1	8.2	35.0	36.0	38.7	18.8	35.5
χ_r (complete)	8.9	10.3	13.8	8.0	8.3	8.4	30.9	31.8	34.1	17.4	32.9

Radiative Fraction, χ_r

Using Equation 1, the radiative fraction is found for all flames, and the results are listed in Table 3. Additionally, the radiative fraction corresponding to the complete heat release rate are shown in Figures 10 and 11. An uncertainty propagation analysis showed that the uncertainty of the χ_r values is on the order of 10% of the value.

The χ_r values corresponding to a complete heat of combustion are compared to the literature value of Markstein[3], who has the most comprehensive data set for radiation from laminar diffusion flames. Markstein defines the radiative fraction in terms of a complete heat release rate, and thus the complete $\Delta h_{c,com}$ values are used in the comparison. The results for the fuels that did not exceed the smoke point at any flame height (propane and methane) are shown in Figure 10. The data of Markstein[3] is plotted with open symbols, the data of the current study is shown with closed symbols and the accompanying uncertainty.

The data of the current study fall roughly 20% lower than Markstein's data for both methane and propane. The reasoning for this discrepancy was investigated and a possible reason is proposed here. In Markstein's study, the radiometer is placed at a distance of 30cm from the burner axis and a spherically isotropic assumption is made during the calculation of the total radiative intensity, $\dot{Q}_{rad}$:

$$\dot{Q}_{rad} = 4\pi S_y{}^2 \dot{q}''_{gauge} \qquad [15]$$

where S_y is the radial separation distance between the burner axis and the face of the gauge. This assumption disregards any spatial deviations in radiation properties. In order to test whether this assumption leads to the discrepancy between Markstein's data and the current study, Markstein's assumptions and method are replicated and applied to the current data. The current method has the capability of predicting the radiant heat flux at any location using the relation function and the centerline intensity, so the heat flux at a distance of 30 cm (to replicate Markstein's method) is estimated and Equation 16 is used to find a new radiative fraction. The data corresponding to this replication is labeled "4π- Method" in Figure 10 and fell within the uncertainty of Markstein's data (not reported, but estimated to be 10%) for both propane and methane. Therefore, assuming spherical isotropy for laminar flames makes a noticeable difference in the radiative fraction value.

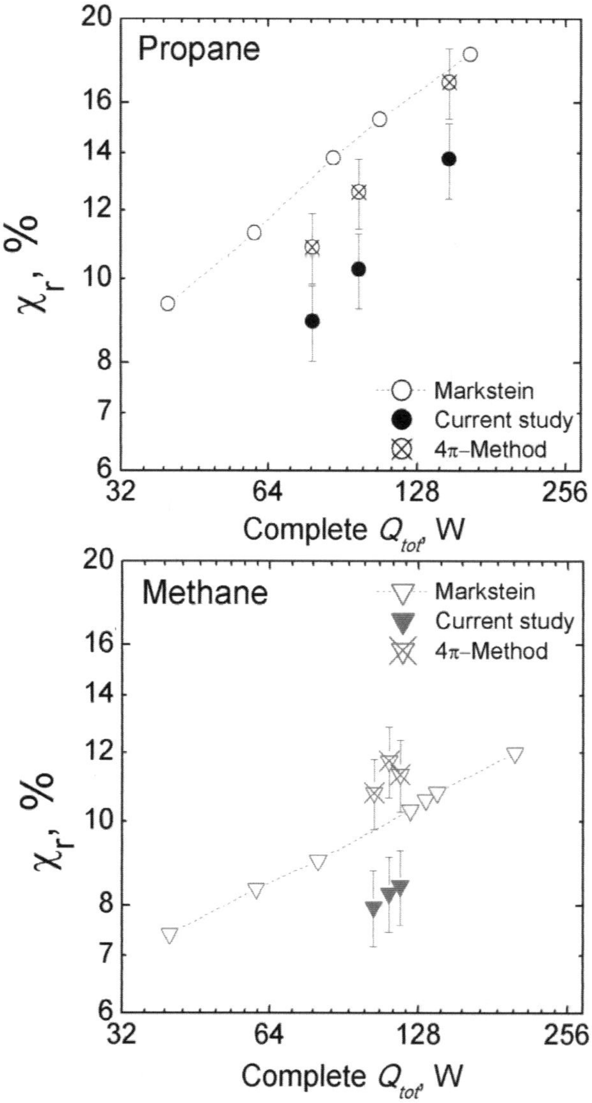

Figure 10. Radiative fraction values from the current study and previous studies for methane and propane. Data shown on log2-log2 plot.

Similarly, the radiative fractions of fuels that did exceed the smoke point for at least one flame height (acetylene and polyethylene) are shown in Figure 11. Although Markstein did not measure the radiative fraction of either acetylene or PE, a family of highly-sooting flames was studied[3] and the corresponding values of radiative fraction are shown in Figure 11 with a shaded region for flames below and above the smoke point (SP). The results from the current study for polyethylene are shown with the corresponding uncertainty and behave as expected. For the 3.5cm flame, which did not exceed the smoke point, the radiative fraction falls below the region of Markstein's data for highly sooting flames below the smoke point. For the 4.5cm flame, which did exceed the smoke point, the radiative fraction increased significantly and overlaps with Markstein's region of highly-sooting flames above the smoke point. For acetylene, the current study overlaps with the region of highly-sooting flames above the smoke point from Markstein's data. Additionally, averaged data from Sivathanu & Gore, who determined radiation losses from laminar acetylene/air flames using soot production models and species concentration measurements[6], aligns with both Markstein's data and the current study's data.

When the current study's data is replicated with the 4π- Method, the acetylene values increase but not as significantly as in the case of the methane and propane flames. This indicates that the spherical isotropic assumption is better suited for flames that are highly sooting and radiative. For flames that are not highly radiative (like methane and propane), the spherical isotropy assumption potentially neglects important spatial deviations along the height of the flame and therefore results in over-prediction of the radiative fraction.

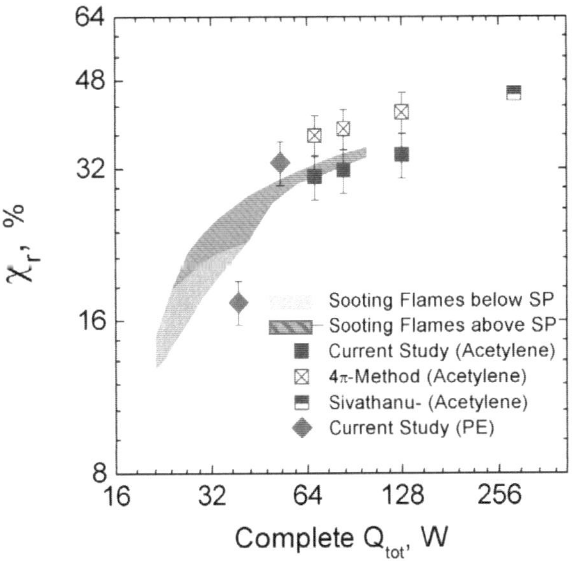

Figure 11. Radiative fractions of current study and previous studies for acetylene and polyethylene. Data shown on log2-log2 plot.

CONCLUSIONS AND FUTURE WORK

In this study, two configurations of the Milligram-scale Flaming Calorimeter were used to measure the radiation and total heat release rates from the laminar flames of both solid and gaseous fuels. The spatially-resolved radiation intensity were determined and used to compute the radiative fraction for four fuels. The results of the study show that this method produces values consistent with previous studies, particularly when the assumptions of the previous studies are replicated. The results also indicate that an assumption of spherical isotropy, as in the study of Markstein [3], may lead to over-predicted radiative fraction measurements, particularly for fuels that do not produce large amounts of soot, such as propane and methane. Additionally, the data indicates that for flame heights of a particular fuel that are all below (or all above) the smoke point, the value of the maximum radiation intensity is nearly constant within a range of $\dot{Q}_{tot}$.

The authors provide a number of recommendations for future work that would further bolster this methodology. The first recommendation is to study a more diverse set of polymeric fuels in order to validate the method across a range of fuel types. These fuels should include charring fuels and heating rates should be chosen to study the flames above and below the smoke point. Additionally, the number of tests can be reduced and the method accelerated by the additional of multiple gauges measuring the heat flux simultaneously.

By altering the definition of the flame to a radiating volume (instead of a flame sheet) and changing the accompanying mathematical description of the radiation, this method can be used with turbulent flame experiments as well. Therefore, an additional recommendation is to find a similar scaling factor between laminar and turbulent flames for solid fuels as Markstein used the smoke point for gaseous fuels. Because the smoke point is extremely difficult to measure for solid fuels, it would be useful to find another parameter that could serve the same purpose. The current results indicate that the maximum radiation intensity value is a possible scaling parameter, due to it being nearly independent of $\dot{Q}_{tot}$.

ACKNOWLEDGEMENTS

This work was sponsored by the Federal Aviation Administration, Grant #12-G-011. The collaboration between the Federal Aviation Administration and the Department of Fire Protection Engineering is invaluable and this work is made possible by their support.

REFERENCES

[1] Klote, J. H., 2012. Fire Science and Design Fires. In: *Handbook of Smoke Control Engineering.* Atlanta: ASHRAE, pp. 149-170.

[2] McGrattan, K., McDermott, R., Hostikka, S. & Floyd, J., 2010. *Fire dynamics simulator (version 5) user's guide NIST special publication 1019-5,* Gaithersburg, Maryland, USA: NIST Building and Fire Research Laboratory.

[3] Markstein, G., 1984. Relationship between smoke point and radiant emission from buoyant turbulent and laminar diffusion flames. *Proceedings of the Twentieth Symposium (International) on Combustion,* pp. 1055-1061.

[4] Tewarson, A., 2004. Combustion efficiency and its radiative component. *Fire Safety Journal,* 39(2), pp. 131-141.

[5] Rankin, B., Blunck, D. & Katta, V., 2012. Experimental and computational infrared imaging of bluff body stabilized laminar diffusion flames. *Combustion and Flame,* 159(9), pp. 2841-2843.

[6] Sivathanu, Y. & Gore, J., 1994. Coupled radiation and soot kinetics calculations in laminar acetylene/air diffusion flames. *Combustion and Flame,* 97(2), pp. 161-172.

[7] Raffan-Montoya, F., Ding, X. & Stoliarov, S., 2015. Measurement of heat release in laminar diffusion flames fueled by controlled pyrolysis of milligram-sized solid samples: Impact of bromine- and phosphorus-based flame retardants. *Combustion and Flame,* 16(12), pp. 4660-4670.

[8] Kreyszig, E., 1999. Gradient of a Scalar Field Directional Derivative. In: *Advanced Engineering Mathematics.* 8 ed. New York: John Wiley & Sons, Inc., p. 449.

TEMPERATURE AND VELOCITY FIELD MEASUREMENTS FOR A PURE BUOYANCY DRIVEN PLUME IN A ENCLOSURE

Alexander Belt, Lukas Arnold, Leonie Rommeswinkel, and Anna Tscherniewski

Institute for Advanced Simulation
Forschungszentrum Jülich, Germany

a.belt@fz-juelich.de, l.arnold@fz-juelich.de

INTRODUCTION

This paper discusses the thermal and fluid dynamics of buoyancy driven flow conditions. In the field of fire safety engineering plumes are important for transport and mixing processes, which must be precisely represented by computer models. Enclosure fires can be computed by zone models or field models, i.e. computational fluid dynamics (CFD) models. Whereas CFD codes, which solve the full set of Navier-Stokes equations, are the most sophisticated. Regarding the computing effort, the conservation equations need to be simplified, e.g. by modeling turbulence, to make CFD codes applicable for fire safety analysis. For this purpose, assumptions are made, which need to be validated by means of experimental data. Therefore, to achieve representative flow conditions, the non-intrusive 2D particle image velocimetry (2D-PIV) measurement technique is applied to the experimental investigation discussed in this paper. It is applied to measure the flow field along a plane perpendicular to the surface of a heat source, see Figure 1. In experiments performed by Meunders (2016)[9] environmental as well as geometrical impacts have been identified, e.g. the cooling of the operating computer, the laser system or the mounting of the heating coil, that had a strong impact on the flow conditions. To ensure the applicability of the experimental data for model validation purposes experimental conditions should be as undisturbed as possible. Thus the heating source used in Meunders has been redesigned.

The field of view (FOV) of the camera, that is used to capture the tracer particles, is smaller than the total area of interest, i.e. laser illuminated region inside the enclosure. Therefore, the camera needs to be mounted on a traverse unit along the z-axis. As the increased gas temperature is driving the plume, a number of thermocouples (TC) is placed into the enclosure, see Figure 1, to monitor the plume temperatures. Two lines (L1, L2) of thermocouples are positioned vertically to measure the temperature stratification and three horizontal lines (L3, L4, L5) monitor the temperature inside the plume in a plane parallel to the laser sheet of the PIV system. To check the influence of the TCs on the flow conditions in the plume, experiments were conducted with and without the corresponding TCs. The results presented in this study show the instantaneous and averaged velocity fields of the plume for various power outputs of the heating bock. Furthermore, the flow fields measurements are presented for the disturbed (by TCs) and the undisturbed flow field. By means of the temperature measurements, an analysis regarding the symmetry of the plume and the Rayleigh number *Ra* with respect to the turbulent kinetic energy are presented.

To the authors' knowledge there are only a few PIV applications in the context of fire experiments. Zhou et al.[21] used PIV for a small scale application to measure the surrounding air and near the base of pool fires. Tieszen et al.[17] demonstrated the application of PIV in large scale pool fire experiments. Sun et al[15] conducted a comparison of a thermal PIV (TPIV) technique with traditional PIV for a small scale pool fire of alcohol. Suadr at al.[14] investigated small scale compartment fires with PIV.

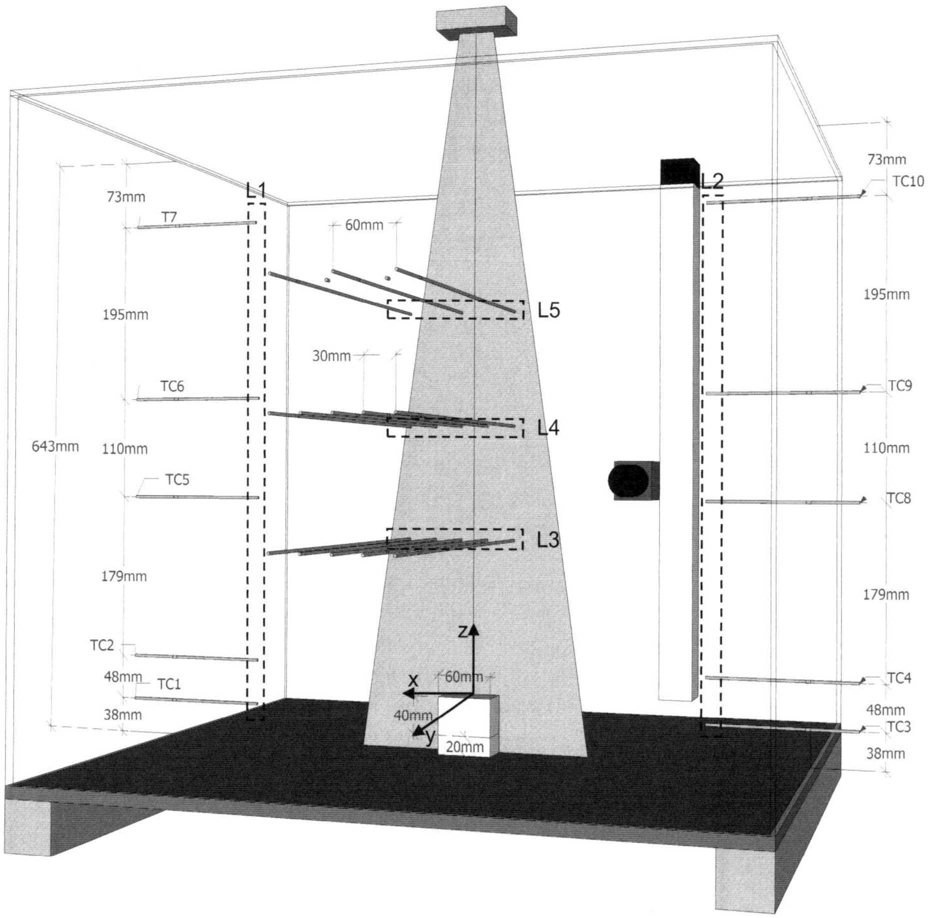

Figure 1: Setup of the open plume experiment. The green sheet indicates the laser sheet, at which the flow measurement is carried out. The thermocouple (TC) positions and their group lines (L) are indicated. From right to left line L3 contains TCs 11-15, L4 contains TCs 16-20, and L5 contains TCs 21-23. The exact positions of the thermocouples in these lines is shown in Figure 6.

Bryant et al.[3] conducted PIV measurements of the velocity field in the doorway of a standard size room containing a fire. His work was motivated by the need for experimental data over the spatial scales of a full-scale enclosure fire combined with the spatial resolution required to provide validation data for CFD models.

One of the most prominent CFD models in fire safety science is the Fire Dynamics Simulator (FDS)[7,8]. Its turbulence treatment is based on the large eddy simulation (LES) approximation. FDS is separated into sub models, e.g. hydrodynamic model, which numerically solves a form of the Navier-Stokes equations appropriate for low-speed, thermally-driven flow, or the combustion model. Modeling the overall fire dynamics in CFD codes is very complex as it combines many physical and chemical processes. Therefore, this study reduces the complexity by substituting the fire source through a static heat source inside an enclosure. A rectangular copper block instrumented with an electrical heating resistor and a thermocouple to monitor the inner temperature is used as a heating device. The instrumentation is at the block's bottom and is lead out through the floor, so there is no disturbance of the flow field. As the copper block heats up the surrounding air, natural convection is induced in the enclosure. The surface temperature of the heating block is the driving force for the flow. It is controlled by the electrical voltage and current provided to the heating resistor. Neglecting environmental changes, e.g. the temperature around the enclosure, the heating power of the copper block is the only parameter changed in between experiments. Because the flow properties, here laminar versus turbulent, are important, the experimental matrix conducted in this study, is selected to cover both regimes and their transition region.

FLOW PROPERTIES

Turbulence plays an important role in fire dynamics, e.g. in mixing processes and combustion. An indicator for the turbulence intensity is the turbulent kinetic energy k. Given an instantaneous velocity field $v = (v_x, v_y, v_z)$, the Reynolds decomposition $v = \langle v \rangle + v'$ distinguishes the mean $\langle v \rangle$ and fluctuating $\vec{v}'$ components. The mean velocity is computed by time averaging and in case of PIV measurements over N frames with the corresponding instantaneous velocities $\vec{v}_i$ at the ith frame as

$$\langle \vec{v} \rangle = \frac{1}{T}\int_0^T \vec{v}\,\mathrm{d}t \approx \frac{1}{N}\sum_{i=1}^{N} \vec{v}_i \quad . \qquad [1]$$

The turbulent kinetic energy is the kinetic energy of the fluctuations, therefore is directly given by

$$k = \frac{1}{2}\langle \vec{v}'^2 \rangle = \frac{1}{2}\left(\langle v_x'^2 \rangle + \langle v_y'^2 \rangle + \langle v_z'^2 \rangle\right) \quad . \qquad [2]$$

The above statement, Equation 2, is true for 3D velocity fields, however within this study a 2D-PIV system was used. Consequently, the determination of the 3D characteristics of the turbulence can not be considered. With the assumption of local isotropy the 2D turbulent kinetic energy is formulated by

$$k = \frac{3}{4}\left(\langle v_x'^2 \rangle + \langle v_z'^2 \rangle\right) \quad . \qquad [3]$$

The Rayleigh number Ra characterizes buoyancy driven flows. It indicates weather the flow is laminar, transient or turbulent.[16] It depends on the dimensionless Grashof number Gr and the Prandtl number Pr, i.e. $Ra = Gr \cdot Pr$. Where Gr is the fraction of buoyancy and viscous forces, while Pr is the ratio of kinematic viscosity and thermal conductivity. The Rayleigh number is given by

$$Ra = \frac{\rho g \beta (T_{HS} - T_{top}) L^3}{\kappa \nu} \qquad [4]$$

ρ	density
β	thermal expansion coefficient
κ	thermal diffusivity
ν	kinematic viscosity
T_{top}	temperature at the surface of the heat sink, here the ceiling
T_{HS}	temperature at the surface of the heat source, here the heating block surface
L	characteristic length scale (distance between heating block and ceiling).

The fluid properties (ρ, κ, ν) are calculated as described by Stelzer[13] under the assumption of dry air $(21\%\ O_2, 79\%\ N_2)$ at the average temperature T_{avg}, see Equation 5, and at ambient pressure ($p_a = 100\,\mathrm{kPa}$).

$$T_{avg} = \frac{T_{top} + T_{HS}}{2} \qquad [5]$$

The thermal expansion coefficient is given by

$$\beta = \frac{1}{T_\infty} \quad , \qquad [6]$$

where T_∞ is the environmental gas temperature. In Figure 2 the gas temperature w.r.t. height for the lowest and the highest power setting are shown, compare with the experimental matrix given in Table 1.

The temperature at the lowest position is the temperature at the surface of the heating block, which was experimentally determined in previous experiments to avoid disturbances of the flow conditions during the experiments. The temperature at higher positions is given by the measurement of the central thermocouples of lines L3-L5, i.e. TC13, TC18 and TC22 (see Figure 1), but the highest position, which is the average of TC07 and TC10.

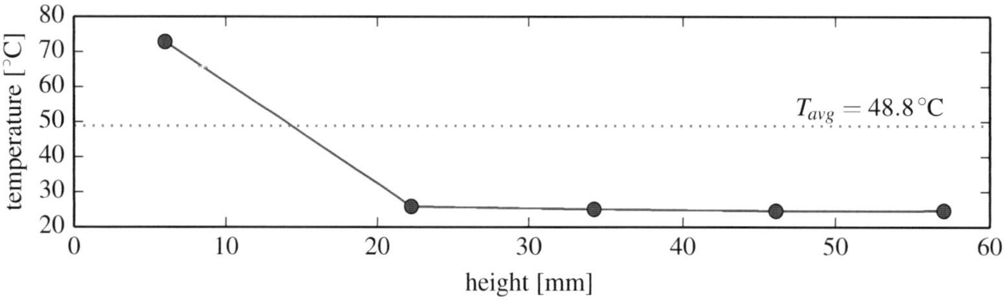

(a) Gas temperature at 50 V power setting.

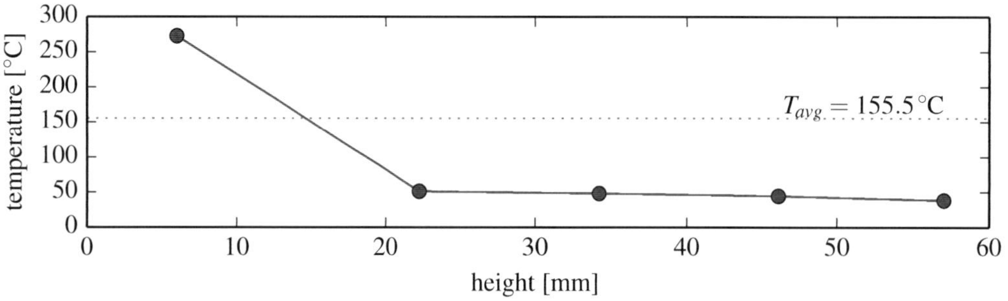

(b) Gas temperature at 150 V power setting.

Figure 2: Temperature plot for the lowest and highest electrical power setting. The lowest height position (6 mm) represents the measurement directly at the heating block, whereas the other ones are form the thermocouples of line groups L3-L5 and the average of TC07 and TC10. The average temperature (Equation 5) is indicated.

EXPERIMENTAL SETUP

In order to confine the particles and to ensure defined boundary conditions, an enclosure made of polymethyl methacrylate (PMMA) slabs, also known as acrylic glass, is used. It is 735 mm wide, 575 mm deep, and 650 mm high (inner dimensions). A schematic representation of the whole experimental setup including the enclosure is shown in Figure 1. The PMMA slabs for the walls and ceiling are 5 mm thick, and for the floor a 12 mm thick board of plywood is used. Two holes in the floor are used as particle inlet and exhaust port respectively. For measuring the gas temperature 3 mm Ni-CrNi sheath thermocouples have been used. A two point calibration, as described by Bernhard[2] has been conducted for all thermocouples. Boiling water and the average of all thermocouples at ambient conditions have been used as reference points. The heat source has been constructed with a Cartesian grid with 10 mm resolution in mind. This is advantageous for recreating the setup in CFD models with Cartesian grids, such as FDS. In this way the geometry can be represented without having to neglect any geometrical details. The dimensions of the heat source are 60 mm × 60 mm × 40 mm (l×w×h), see the technical drawing in Figure 3.

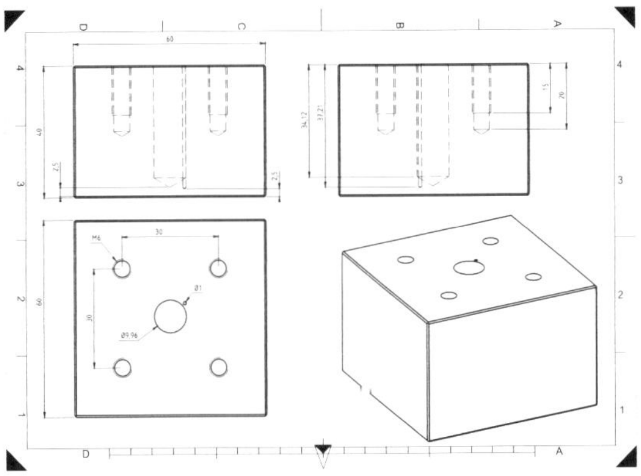

Figure 3: Technical drawing of the heating block. All lengths are given in millimeter.

A heating coil and a thermocouple are embedded inside the block of copper. The borehole for the thermocouple has a diameter of 1 mm and the one for the heating coil 9.96 mm. Both boreholes end in a distance of 2.5 mm from the top surface of the heating block, so the temperature measurement is in the vicinity of the surface. For mounting purposes four threaded holes (M6) are arranged in a square. To establish constant boundary conditions a constant voltage was applied to the heating coil, which lead to very constant surface temperatures. As surface temperatures that lead to oxidation processes of the copper are expected, the surface of the copper block was coated with chalk spray, which has a constant emissivity ($\varepsilon = 0.9$). To measure the total energy output P_{HS} of the heat source, an ampere meter and a voltmeter are connected in series and parallel to the heat source respectively. It is assumed that all electrical power is converted to thermal energy. Copper has a very high conductivity and thus quickly equalizes temperature differences. Therefore a homogeneous surface temperature can be expected. For illumination purposes a Dual-Nd:YAG Laser is used. As can be seen in Figure 1, the laser inlet is at the ceiling of the enclosure. The light sheet is aligned perpendicular to the surface of the heating block. The light sheet is manually aligned to the x-axis, so that the light sheet is in the x-z-plane. As tracer particles, di(2-ethylhexyl)sebacate (DEHS)[18] was used and generated with the TSI Aerosol Generator Model 3079[19]. It is a non-soluble, colorless and odorless liquid that boils at approximately 232 °C to 249 °C at 100 kPa atmospheric pressure. The lifetime of a particle with 0.3 µm diameter is given as four hours, but this will depend upon ambient conditions such as temperature and pressure. The camera, a PCO2000, is mounted on a traversing unit.

EXPERIMENTAL PROCEDURE

To provide different flow properties the electrical power provided to the heating block is varied. The heating coil, heats up the copper block. This results in different surface temperatures and therefore different convective power outputs to the plume. The electrical power was set to five values and is the only parameter that is actively varied within the experiments, see Table 1.

Table 1: Experimental Matrix

electrical voltage [V]	electrical current [A]	electrical power [W]	resulting block surface temperature [°C]
50	0.21	10.5	72.7
75	0.34	25.5	116.7
100	0.44	44.0	166.9
125	0.56	70.0	218.3
150	0.69	103.5	272.7

The PCO2000 is a CCD (charge coupled device) camera with a pixel resolution of 2048×2048[10].

The camera is calibrated once at the beginning of the experimental procedures. The extension of the used calibration target is 200 mm×200 mm (w × h). It has a well defined pattern of white dots, thus the effect of distortion can be measured and compensated. The distance between the camera and the calibration target and the orthogonal perspective lead to a field of view (FOV) of 197.3 mm×197.3 mm. This field of view is not large enough to capture the full plume and therefore the camera is mounted on a traversing unit. After the calibration, the target is removed from the enclosure and electrical power is supplied to the heating block. To ensure steady state conditions the temperatures at different positions in the enclosure were monitored. Each experiment was conducted at least twice. The first PIV measurement was started after about 30 minutes and the second pass after about 90 minutes. For each measurement the acquisition started at position 1 of the traversing stage, compare Figure 4. The PIV setup worked at 5 Hz and 155 double frames were captured for each position. Thus the measurement at one position takes 31 seconds. After that, the camera was traversed 60 mm upwards to the next position with a traversing speed of 5 mm s^{-1}. Consequently it takes 12 seconds from one position to the next.

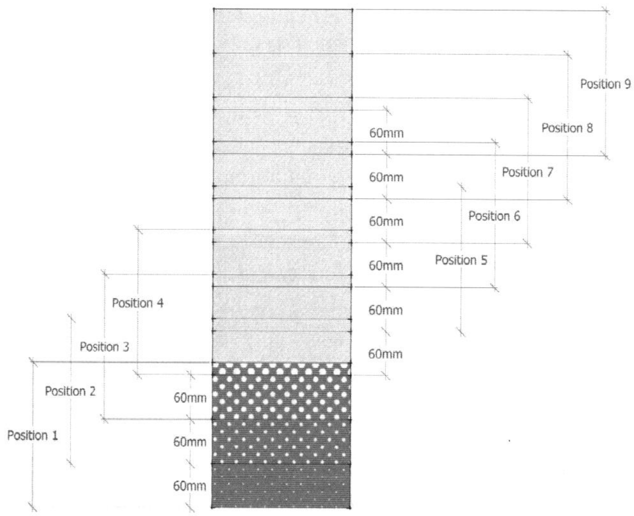

Figure 4: Scematic representation of the traversing positions of the camera. The different pattern in the FOV in position1 highlight the overlap of the different camera positions

As the FOV of the camera is larger than 180 mm×180 mm (w × h) the overlap from one position to the next is more that 66%. In total the camera is traversed eight times. The increment from one position to the next was kept constant. Thus the resulting FOV is about 660 mm×180 mm and the total measurement takes more than 6 minutes.

RESULTS

The temperature measurements within the plume as well as the velocity data are used for the evaluation of the flow and temperature conditions. First, the symmetry of thermal boundary conditions is shown, followed by the presentation of the thermocouple data and the velocity field. This section closes with the analysis of the turbulent kinetic energy.

Symmetry of Thermal Stratification

The flow conditions are driven by density differences, due to changes in gas temperature. To observe weather there are symmetrical conditions within the enclosure or not a set of thermocouples is used to observe the temperature at the left (L1) and the right (L2) wall, see Figure 1. In Figure 5 the temperature curves of the opposing walls are displayed for the five different power settings. The readout frequency of the temperature measurements is 1 Hz. The values from opposite sides are in a good agreement, i.e. temperature difference of less than 1 K, for all power settings, thus the thermal conditions can be assumed as symmetrical.

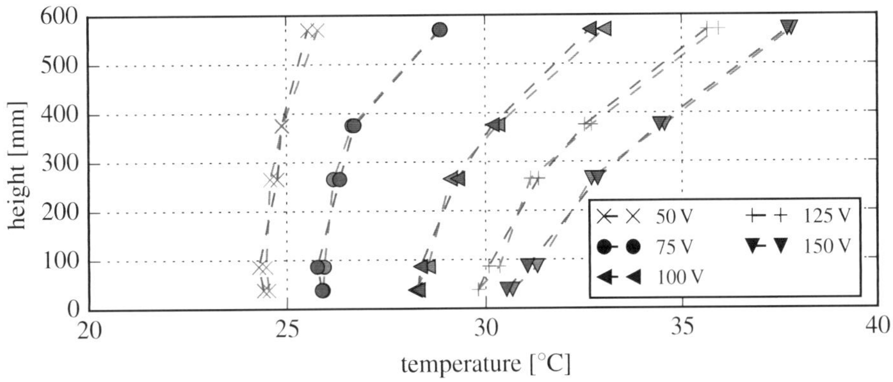

Figure 5: Temperature behaviour besides the enclosure walls, TC1-TC3 ($z = 38$ mm), TC2+TC4 ($z = 86$ mm), TC5+TC8 ($z = 265$ mm) TC6+TC9 ($z = 375$ mm), TC7+TC10 ($z = 570$ mm). Red markers represent TCs in positive and the blue one in negative y-direction.

Temperature Measurement in the Plume

For the lowest power setting laminar flow conditions are expected, which transit to a turbulent flow at higher power outputs. Laminar flow will be streamlined, whereas due to increasing sheer stresses and fluctuations the flow becomes distorted in the turbulent case, see Schlichting[12]. The ambient air entrainment into the plume is therefore increased for a turbulent plume, which has a significant impact on the plume temperature.

For the temperature instrumentation TCs are mounted from the backside of the enclosure. Because of their own weight the TCs are not exactly mounted in a symmetrical spatial arrangement. By means of the calibrated CCD camera their exact positions have been determined, see Figure 6.

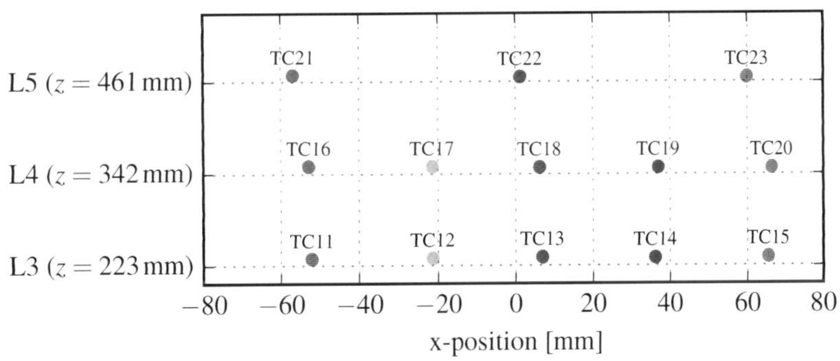

Figure 6: Positions of the thermocouples in lines L3 to L5.

It must be noted, that the thermocouple lines L3 and L4 are disarranged about 6 mm to the right and L5 is almost centered. Therefore, a symmetry of the temperature measurements within the plume can not be expected. In Figure 7 the measured plume temperatures are presented. The marker represent the mean temperature over 5 minutes. The mean values and the standard deviations for the line central thermocouples (TC13, TC18, TC22) are listed in Table 2.

For the lowest power setting (50 V) a small increase of the temperature (TC13) with respect to its neighbors was observed. Furthermore, negligible temperature fluctuations were measured at higher positions. In the second case (100 V) there is a significant increase of temperature at the center of each line (TC13, TC18, TC22) and a smaller increase at only one of the thermocouples next to it at the two lower lines (TC12, TC17). The likely reason for that asymmetry is the disarrangement of the thermocouples. For a time interval of 50 minutes, data not shown here, a small increase of the overall temperature behavior was observed but only small local fluctuations of the temperature, thus one can assume, that there is no global movement of the plume. At the mid and high positions, a small increase of thermocouple at the edges was observed. What substantiates the assumption, that the plume widens. The highest power setting (150 V) shows a similar behavior as the previous case, but the temperature amplitudes are higher and a broader plume can be deduced. In the 100 V power case, temperature fluctuations in the range of 2 °C at the central position (TC13) were observed. This fluctuation range decreases to 1 °C in L4 and about 0.5 °C in L5. For the highest power (150 V) setting, the fluctuation range decreases from 3 °C (L3) to 1 °C (L4) and to 0.5 °C in L5.

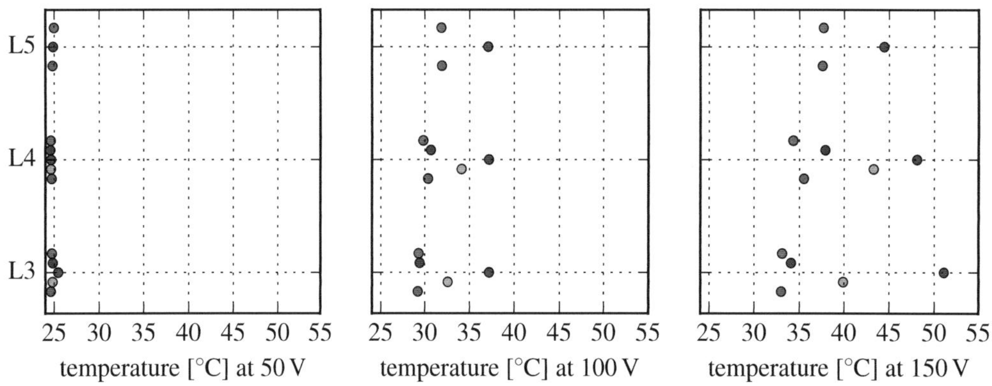

Figure 7: Mean temperature (5 minutes) for three different power setting in the thermocouple lines inside the plume (L3, L4, L5). The marker positions in this plot are shifted to prevent overlap. Compare TC positions with Figure 6.

Table 2: Line Central TC (TC13, TC18, TC22) Temperatures – Means μ and Standard Deviations σ

electrical voltage [V]	$\mu_{L3} \pm \sigma_{L3}$ [°C]	$\mu_{L4} \pm \sigma_{L4}$ [°C]	$\mu_{L5} \pm \sigma_{L5}$ [°C]
50	25.44 ± 0.078	24.64 ± 0.052	24.87 ± 0.055
75	27.25 ± 0.226	26.88 ± 0.174	27.25 ± 0.198
100	37.16 ± 0.559	37.15 ± 0.349	37.06 ± 0.245
125	45.92 ± 1.054	44.10 ± 0.444	42.01 ± 0.154
150	51.10 ± 0.863	48.09 ± 0.244	44.45 ± 0.172

Velocity Field

Each FOV (1-9) was evaluated with a standard multi-pass Fourier transformation cross-correlation method. By means of the multi-pass method, the interrogation window size is reduced from initially

128×128 pixels (1 pass) to the final size of 48×48 pixels (3 passes) with an overlap of 50%. The computed data is filtered by a median filter as described in Westerweel (2005)[20]. For the analysis of the total FOV the flow velocity is averaged in the overlapping areas. Figure 4 demonstrates that even the fourth position contributes to the first one. In Figure 8 the averaged z-velocity fields for three different power outputs, same as in previous section, with and without the instrumentation with thermocouples are presented. Velocity profiles along the x-direction across the plume at four selected heights (111.5 mm, 282.5 mm, 401.5 mm, 552 mm) are presented besides. A qualitative comparison of the two power settings 50 V and 100 V reveals a very good agreement of the velocity measurement with and without the thermocouples. The high power case (150 V) the velocity plot for the experiment with thermocouples looks inhomogeneous. This is caused by a low particle distribution. The fluid was seeded at the beginning of each test series, but the particles depose over time. As the particle density was not checked for each experiment, the low value was determined too late. By means of all velocity profiles, except for the lowest power (50 V) without TCs, the plume is centered at the z-axis.

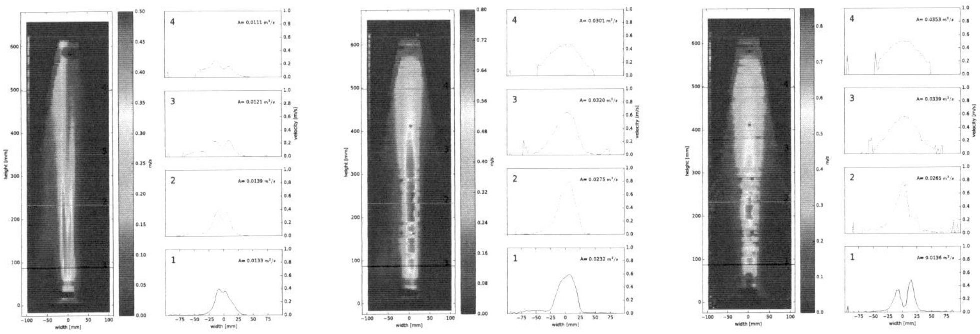

(a) 50 V with TC (b) 100 V with TC (c) 150 V with TC

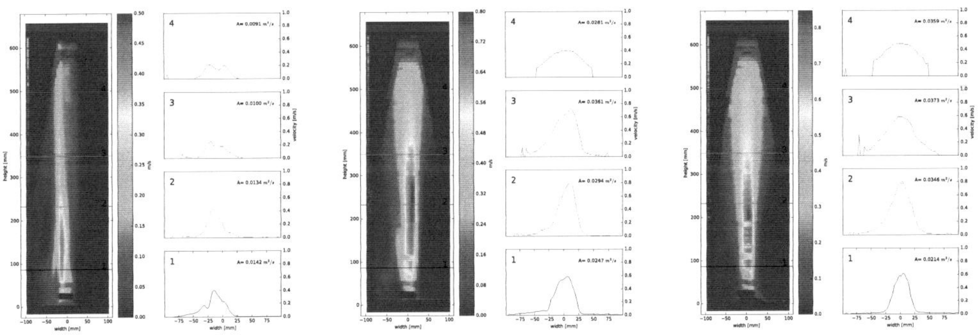

(d) 50 V without TC (e) 100 V without TC (f) 150 V without TC

Figure 8: Left part of sub figures: velocity field (z-component). Right part: slices along the x-axis at selected positions.

Turbulent Kinetic Energy

The turbulent kinetic energy and the instantaneous velocity profiles are used for the evaluation of the plume regarding its turbulence intensity. The TKE is determined by the velocity fluctuations. Figure 9 illustrates the instantaneous velocity profiles for the lowest and the highest power setting at camera position 5 at a fixed z-value (398 mm). It presents the mean velocity as well as 155 instantaneous profiles. To highlight the manifoldness of the profiles five profiles are displayed colored. Within the acquisition time of 31 seconds, the lowest power setting shows very diverging velocity profiles. This is due to the dynamics, i.e. slow panning, of the plume at the lowest power setting and its averaging blurs the velocity profile. Thus the mean is not mandatory representative for the instantaneous flow conditions.

At the highest power setting (150 V), the instantaneous velocity profiles fluctuate around the mean.

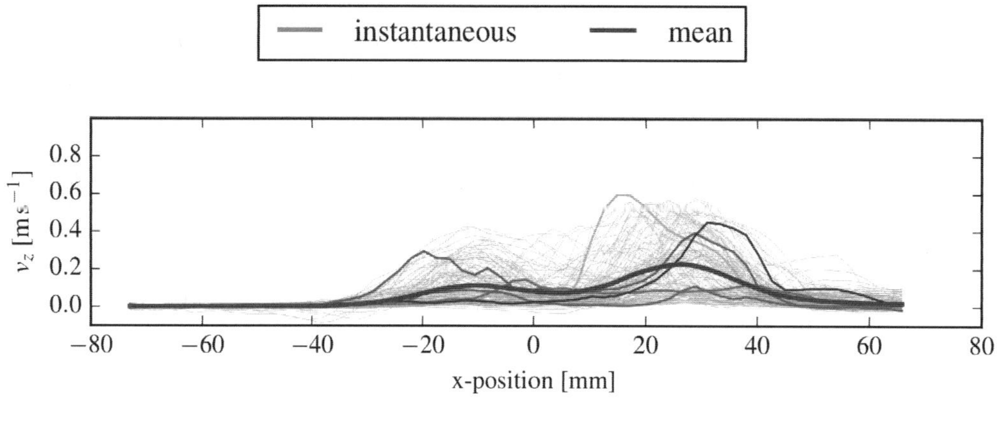

(a) Velocities at 50 V power setting.

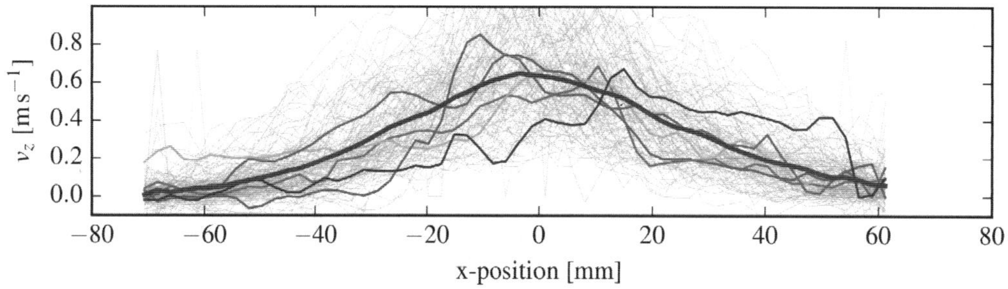

(b) Velocities at 150 V power setting.

Figure 9: Instantaneous and mean velocities at given hight, here $z = 398\,\mathrm{mm}$, for all 155 frames at 50 V (a) and 150 V (b) power setting. Selected frames are displayed in color, the thick blue line indicates the total average and the thin gray lines illustrate all individual frames.

The convergence of the velocity profiles towards the mean can be defined as the distance δ, here a form of the L_2 norm, of intermediate means $\langle v_z \rangle_n$ that cover n frames to the total mean $\langle v_z \rangle_N$ of $N = 155$ frames, see Equation 7,

$$\delta(n) = \int_{\text{line}} \left(\langle v_z \rangle_N (x) - \langle v_z \rangle_n (x) \right)^2 \mathrm{d}x \quad \text{with} \quad \langle v_z \rangle_n (x) = \frac{1}{n} \sum_{i=1}^{n} v_{z,i}(x) \qquad [7]$$

and $v_{z,i}$ being an instantaneous velocity profile. Figure 10 shows the observed convergence. While the low power setting shows a small distance at the beginning, the high power case quickly reduces its distance from a higher initial value. It saturates at about 40 frames and drops to zero at the final stage. In contrast to that, the distance in the low power case continuously drops which indicates, that there is a lot of dynamics and convergence is achieved on longer scales as in the high power case.

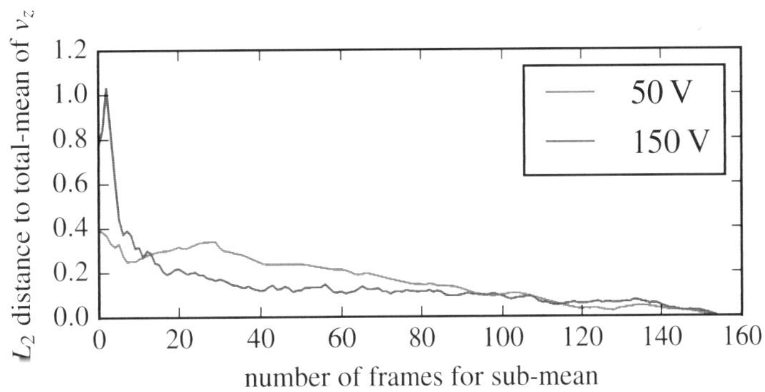

Figure 10: Distance of intermediate means to the total mean as a function of frames used for the intermediate mean, see Equation 7.

In Figure 11 the TKE fields are presented. The fields are evaluated the same way as the velocity fields and the slices are in the same positions. For the low power case the TKE is negligible at the lower positions and slowly develops over the height. The high power case shows significant TKE at the bottom, directly above the heating block. Next to the block the profile is very tight and it widens with the height. The local maximum is decreasing from the bottom to the height, but the line integral, see plots on the left side of the sub-figures of Figure 11, stays nearly constant.

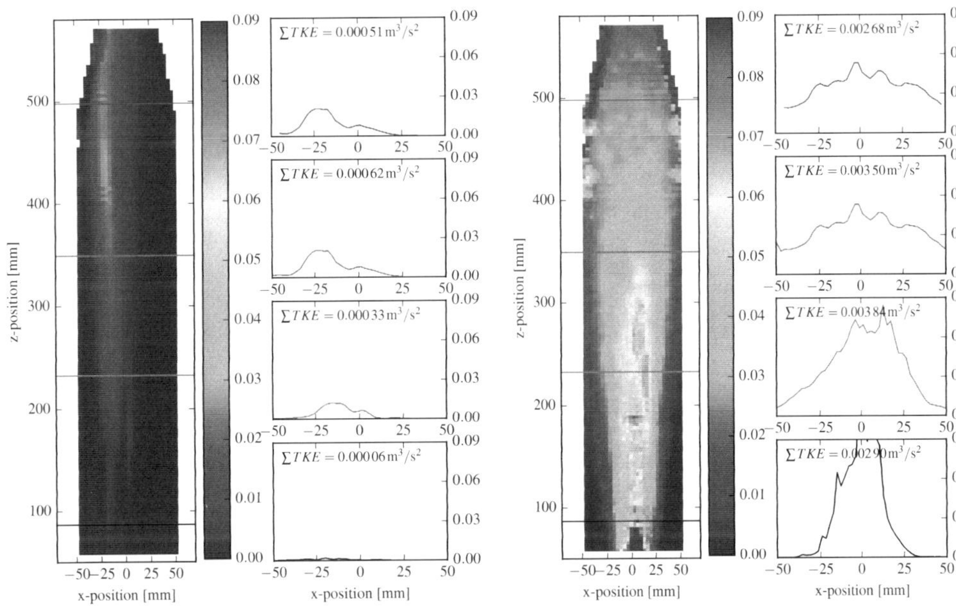

(a) 50 V (b) 150 V

Figure 11: Turbulent kinetic energy contour plots.

Finally in Figure 12 the local TKE along the z-axis is presented for all power settings. High power settings (100 V to 150 V) have an increased TKE just above the heating block. Considering the heating block and its geometry, the flow velocity along the vertical walls of the heating source is increasing with the surface temperature. At the upper edge, flow separation occurs. This phenomenon is similar to a backward facing step, as it has be investigated by e.g. Armaly et al.[1] or in Hudy[5]. This has a significant impact on the flow conditions above the heating block because vortex shedding can occur.

The transition from laminar to turbulent flow at a heated vertical wall occurs at $10^8 < Ra < 10^9$. In the considered cases the Rayleigh number is below 4×10^5, considering the block's walls. Thus the TKE next to the block is based on the flow separation at the edges. The flow conditions at the vertical walls of the heating block have not been investigated within this study and reflections at the upper surface of the block avoided measurements right next to the surface. Therefore a quantitative analysis regarding the flow field in these regions could not be conducted. After a peak at about 75 mm the TKE strongly decreases for the three high power cases. In range height range 100 mm to 320 mm the TKE slowly increases again. Afterwards it decreases, which already was displayed in the TKE fields (Figure 11). For the both test cases with lower power settings the TKE is very low up to a height of 400 mm. For the low power setting (75 V) the TKE increases slightly with increasing height, whereas it stays at a low level for the lowest power setting (50 V), thus turbulence can be excluded for both lowest power settings.

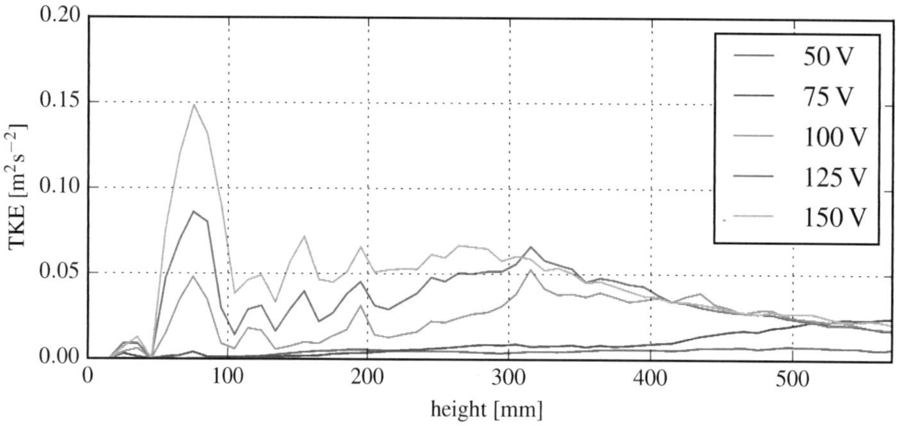

Figure 12: Turbulent kinetic energy (TKE) along the z-axis.

As described above, the average temperature between the surface of the heating block and the ceiling is used for the computation of gas properties, see Equation 5. Assuming a Gaussian temperature distribution within the plume, the temperature is decreasing from the centre to the boundary of the plume. Thus the mean temperature level of the plume volume is much lower than it is by averaging between the bottom and the top. For a better estimation of the mean temperature of the gas within the plume a linear interpolation was done for the temperature of the heating block, the temperature measurements at the z-axis and the average of TC07 and TC10. The estimated averaged temperature was used for the calculation of the gas properties within the plume, which were needed for the calculation of the Rayleigh number. In Figure 13 the global Rayleigh number is plotted over the temperature difference between the heating source and the ceiling. By means of the conducted experiments and the TKE analysis it can be concluded, that the case at 50 V is below a critical Rayleigh number, the 75 V case is transient and all other test cases are turbulent. However, it is not clear weather the turbulence is induced by the vortex shedding or the temperature difference between the block and the ceiling. Thus for a similarity approach the geometry of the heating block, the height of the plume, and the temperature difference of the gas has to be considered.

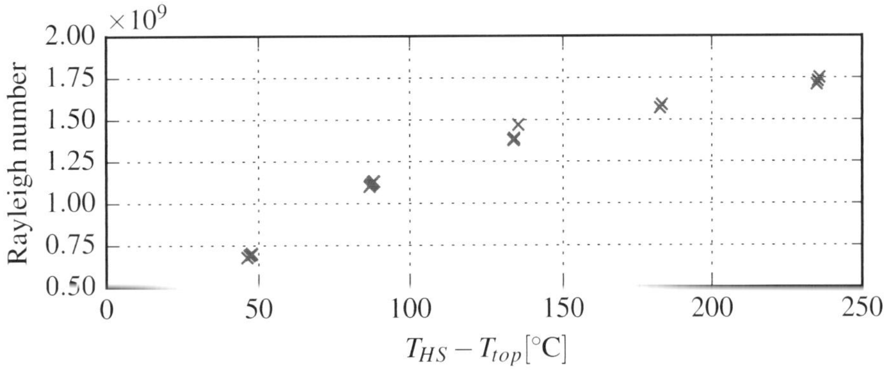

Figure 13: Rayleigh number over the temperature difference between the surface of the heating source (T_{HS}) and the average of the temperature of TC07 and TC10 (T_{top}) for all experiments conducted in this work.

CONCLUSIONS AND FUTURE WORK

A comparison of velocity fields of the experiments performed with and without thermocouples showed, that both are in a good agreement. So the influence on the flow conditions was not further considered. By means of temperature measurements next to the walls, thermal symmetrical conditions have been determined for the environment of the plume. The determination of the exact TC positions in the plume showed, that they are not mounted in the symmetry axis. Thus a statement regarding the symmetry of the plume could not be made by means of the temperature measurements. But these measurements showed first indicators for turbulence next to the block for the higher power settings and a decrease of the turbulence along the z-axis, what was not expected, if one considers the Rayleigh number, which increases with the height of the plume. The temperature information at different positions in the plume provides the opportunity for a better estimation of the averaged fluid properties. The TKE analysis showed that for the lowest two power settings pure laminar conditions are present, whereas for the three upper power settings unintended TKE was observed next to the heating source. Furthermore, the TKE field analysis showed, that the TKE is not decreasing as it was recognized in Figure 12 and the temperature measurements in the plume. The TKE spreads along the height of the plume.

The analysis of the instantaneous velocity showed, that the plume significantly changes its inclination in the laminar case, whereas the turbulent plume keeps its straight alignment. Only velocity fluctuations around the mean velocity profile have been observed. Thus the averaging process for the laminar conditions blurs the velocity information, which spreads the plume and therefore limits the information contained in the mean.

This experimental setup provides detailed data, which can be used for model validation purposes. But the simplification of the heating source, although the combustion model can be excluded, creates other challenges, e.g. the vortex shedding. Following experiments will be redesigned in the way that the block will be embedded in the floor.

As the flow conditions for buoyancy driven flows in a homogeneous distributed gas mixture are driven by temperature gradients, the temperature is a very important quantity. The spatial resolution of the TC instrumentation does not provide a holistic temperature distribution. For more detailed information, a higher spatial resolution is necessary, however, more thermocouples will significantly disturb the plume. Therefore, a non intrusive measurement technique for the temperature distribution, e.g. filtered Rayleigh scattering, might be an option for future experiments.

REFERENCES

[1] B. F. Armaly, F. Durst, J. C. F. Pereira, B. Schönung: *Experimental and theoretical investigation of backward-facing step flow* J. Fluid Mech.,vol. 127, pp. 473-496, (1983)

[2] F. Bernhard: *Technische Temperaturmessung physikalische und messtechnische Grundlagen, Sensoren und Messverfahren, Messfehler und Kalibrierung Handbuch fuer Forschung und Entwicklung, Anwendungspraxis und Studium*, Springer, Berlin, (2004)

[3] R. A. Bryant: *Application of Stereoscopic PIV to measure the flow of air into an enclosure containing a fire*, Exp Fluids 47 (2009)

[4] P. E. Ciddor: *Refractive index of air: new equations for the visible and near infrared* Appl. Optics 35, 1566-1573, (1996)

[5] L. M. Hudy: *Simultaneous Wall-Pressure and Velocity Measurements in the Flow Field Downstream of an Axisymmetric Backward-Facing Step*, PhD thesis, Michigan State University, Department of Mechanical Engineering, (2005)

[6] LaVision: *DaVis 8.3 Software*, Product Manual, Göttingen, (2016)

[7] R. J. McDermott: *A velocity divergence constraint for large-eddy simulation of low-Mach flows*, Journal of Computational Physics, 274 (2014)

[8] K. McGrattan, R. McDermott, C. Weinschenk, K. Overholt, S. Hostikka, J. Floyd: *Fire Dynamics Simulator User's Guide* NIST Special Publication 1019 Sixth Edition (2016)

[9] A. Meunders: *A study on buoyancy-driven flows: Using particle image velocimetry for validating the Fire Dynamics Simulator* PhD Thesis, ISBN 78-3-95806-173-6, Wuppertal, (2016)

[10] PCO: *PCO2000*, Datasheet, Kehlheim, (2008)

[11] H. Pitsch: *Turbulence*, CEFRC Combustion Summer School, (2014)

[12] H. Schlichting, K. Gersten: *Grenzschicht-Theorie* 10. überarbeitete Auflage, Springer, ISBN-10 3-540-23004-1, Berlin, (2006)

[13] J.F. Stelzer: *Stoffwertalgorithmen*, Karl Thiemig AG, München (1984)

[14] S. Suard, A. Koched, H. Pretrel, L. Audouin: *Numerical simulations of fire-induced doorway flows in a small scale enclosure*, International Journal of Heat and Mass Transfer, 81 (2015)

[15] L. Sun, X. Zhou, S. Mahalingam, D. R. Weise: *Experimental investigation of the velocity field in buoyant diffusion flames using PIV and TPIV algorithm* Proc Int Assoc Fire Saf Sci (2005)

[16] A. Thess: *Wärmeübertragung durch freie Konvektion: Grundlagen*, VDI Wärmeatlas, Chapter F1, 753-757 (2013)

[17] S. R. Tieszen, T. J. O'Hern, R. W. Schefer, E. J. Weckman, T. K. Blanchat: *Experimental study of the flow field in and around a one meter diameter methane fire* Combust Flame 129 (2002)

[18] TOPAS: *Di-Ethyl-Hexyl-Sebacate*, Data sheet, Dresden

[19] TSI Incorporated: *Atomizer Aerosol Generator Model 3079*, (2012)

[20] J. Westerweel, F. Scarano: *Universal outlier detection for PIV data*, Experiments in Fluids, 1096-1100, Springer-Verlag (2005)

[21] X. C. Zhou, J. P. Gore, H. R. Baum: *Measurements and prediction of air entrainment rates of pool fires*, Proc Combust Inst 26, (1996)

UPHOLSTERED FURNITURE

EXPERIMENTAL AND NUMERICAL INVESTIGATION ON FIRE BEHAVIOUR OF FOAM/FABRIC COMPOSITES

Konrad Wilkens Flecknoe-Brown[1,2], Karlis Livkiss[1,2] & Patrick van Hees[2]
[1] Danish institute of fire and security technology, Denmark
[2] Division of fire safety engineering, Lund University, Sweden
Corresponding author email: Konrad.wilkens@brand.lth.se

ABSTRACT

Upholstered furniture is a major contributor to the fuel load in a fire compartment. Modelling the fire behaviour of upholstered furniture would support performance based fire safety engineering, by allowing the specification of realistic fire scenarios. Modelling upholstered furniture composites in cone calorimeter test conditions is undertaken to validate fundamental assumptions about the fire behaviour of these combinations, without including complexities of a real sized system (e.g. constructional details of mattresses or sofas). Kinetic parameters were taken from micro combustion calorimeter tests and thermal properties were taken from literature and optimised using individual material cone calorimeter tests. A novel methodology of sample preparation was proposed for the cone calorimeter tests in an attempt to increase the one dimensionality of heat transfer. Modelling showed mixed results when compared with experiments. The results also indicated the models incapacity to capture material-material interactions, such as melting of polyester on the top of flexible polyurethane slab or shielding created by cotton residue to protect flexible polyurethane slab from heat exposure.

KEYWORDS: Upholstery, Polyurethane, Fabrics, Cone Calorimeter, Heat release rate, Simulation

INTRODUCTION

Upholstered furniture is often the first item to ignite and a major contributor to the fuel load within a fire compartment[1,2]. For this reason predicting the fire behaviour of these products would be beneficial for performance based fire safety design.

Flexible polyurethane foam and fabric combinations form the basis of most upholstered furniture in both residential and commercial settings. The contribution of these materials to the overall fire load within an enclosure can be substantial. Flexible polyurethane foams present a large combustible mass and the fabric may enhance or lessen the reaction to fire properties of the composite. Having the ability to model these material combination is therefore of significant interest to the fire safety engineering community and for product development.

A multi-scale approach for studying the fire behaviour of building systems has been proposed and implemented by researchers[3–5]. The multi-scale approach follows a stepwise procedure; obtaining the material properties from a milligram scale and performing modelling by slowly increasing the physical dimensions of the samples. Before predicting reaction to fire behaviour of the full system (i.e. couch or mattress), the prediction of cone calorimeter test results is performed as an intermediate step validating the fundamental model assumptions. However, both testing and modelling of layered composites is challenging even in cone calorimeter test scenarios[4,6]. Testing challenges include: ensuring valid boundary conditions for modelling purposes and avoiding edge effects (e.g. heat loss around edges). Modelling challenges include obtaining appropriate material properties, and treatment of the interactions between different material layers in fire.

This study had a number of objectives. First was to investigate the feasibility of modelling upholstered furniture composites with Fire Dynamics Simulator (FDS) using input parameters obtained from individual materials and parameter optimisation. Second was to propose a novel specimen preparation method for these composites, with the specific purpose of model validation using cone calorimeter tests.

MATERIALS

The materials used in this study were chosen for their widespread use within building content, specifically upholstered furniture, which is a large contributor to the fire load within many building fire compartments[2,7]. Non-fire retarded flexible polyurethane foam (FPUF) was chosen, as it is a common form of cushioning agent found in these types of products. The fabrics; cotton, polyester and a cotton/polyester blend (PC blend), were also selected, as they constitute a large fraction of the materials used as coverings for upholstered furniture. All materials used for this study are commercially available and have been used in previous research[8–10]. Table 1 provides the basic properties of the materials used in this study.

Table 1 – Materials[8,9]

Material	Composition (mass %)	Density
FPUF	Approx. 32% Toluene Diisocyanate TDI, 68% polyol	21 kg/m^3
Cotton	100% Cotton	0.160 kg/m^2
Polyester	100% Polyester	0.400 kg/m^2
PC blend	25% Polyester 75% Cotton	0.183 kg/m^2

EXPERIMENTAL

The experimental program consisted of number of Micro Combustion Calorimeter (MCC) experiments and Cone Calorimeter experiments. MCC tests were used to study reaction kinetics of the investigated individual materials and to obtain reaction kinetics parameters activation energy E_a, pre-exponential factor A and reaction order n. Cone calorimeter experiments of individual materials were used for validation and optimisation of material input parameters using Fire Dynamics Simulator (FDS) version 6.5.2 [11]. Later, cone calorimeter tests with the layered composites were performed and modelled with FDS using the optimised model input parameters. Time to peak heat release rate (PHRR), PHRR and total heat released (THR) were analysed to assess the repeatability of the tests methodology. Time to PHRR for the experiments is defined as time starting from the visual ignition to the PHRR. An overview of all the performed tests in this study is presented in **Table 2**.

MCC Experiments

Micro scale experiments performed in this study on the individual materials that were used in the layered composite experiments were undertaken using the MCC developed by the US Federal Aviation Administration (FAA)[12]. The MCC employs pyrolysis-combustion flow calorimetry (PCFC), which principally works by separating out the two main processes in flaming combustion; solid-phase pyrolysis of the specimen and the gas-phase combustion of the pyrolysis gases. This is accomplished by using a two-stage rector, where a specimen is pyrolysed in one chamber under controlled atmospheric and heating conditions; the evolved pyrolysis gases then travel into a second furnace chamber, mix with oxygen and are then combusted at high temperature.

The MCC measures the rate of oxygen consumed by the conversion of pyrolysis gases to heat through the combustion process. This is then transformed into HRR through the principles of oxygen consumption calorimetry[13]. Samples tested in the MCC are typically in the order of milligrams.

The experimental procedure for the MCC tests followed ASTM D7309-11[14]. Testing of all samples was performed in triplet to get a measurement of the repeatability of heat release rate outputs. The tests included in this study followed ASTM D7309 – Method A: Controlled Thermal Decomposition, in which specimens are pyrolysed under anaerobic conditions, i.e. in a nitrogen atmosphere. A heating rate of 1K/s or 60K/min was used as per recommendations in ASTM D7309-11 and FAA documentation [14,15]. The test method and results for fabrics used in this study has previously been reported by Wilkens Flecknoe-Brown et. al.[10].

Cone Calorimeter Experiments

FPUF, three types of fabrics and FPUF/fabric composites were tested as a part of this study. When individual fabrics and foam/fabric composites were tested, a novel sample preparation method was used compared to the standard method[16] with a sample holder and edge frame. This was done in order to reduce the heat exchange around the perimeter of the sample – typically in the standard configuration this may occur through the metallic edge frame. FDS is limited to one dimensional heat transfer within solids, thus the standard specimen preparation may not be appropriate for FDS model validation purposes. Limiting heat losses around the perimeter of the specimen and keeping each layer of the samples cross-section to a single material, creating a one-dimensional material system. The new method used the same sample holder, however the edge frame was substituted with ceramic and stone wool frame for the insulation purposes. This was done in order to limit heat losses through the edge frame.

The tested FPUF dimensions were 100x100x25 mm. The tested fabric dimensions were 110x110 mm to completely cover the FPUF. Five unexposed sides of FPUF sample was wrapped in one layer of aluminium foil. The sample was placed in the sample holder with a 25 mm thick layer of ceramic insulation at the bottom. An edge frame made from a single 50mm thick stone wool piece with a cut out of 100x100mm in the centre and placed around the specimen. Ceramic wool insulation was stuffed in between the FPUF and edge frame, to further ensure the insulation around the perimeter of the specimen. Whenever the fabric was used in combination with the FPUF, it was placed on the top of the upper surface of FPUF and fixed to the frame with a number of metallic staple pins pushed through the stone wool. When a fabric specimen was tested alone, ceramic wool insulation was used as a backing layer. An example of specimen preparation is presented in Figure 1.

Figure 1 – Sample preparation

Each sample of tested FPUF and fabrics was weighted, resulting with average mass of 5.41g, 3.29g, 2.74g and 6.78g for FPUF, cotton, PC blend and polyester respectively. Apart from the sample preparation, the experimental procedure followed closely to ISO 5660[17]. Each day before the tests calibration and maintenance procedure was performed, including exchange of gas analyser dying agent, gas calibration and HRR rate calibration with methane. The heat flux from the cone heater was measured before each change of the heat flux.

When FPUF was tested alone, three repeated tests under two different heat fluxes (25 and 35 kW/m^2) were performed using the large cone calorimeter [18]. The test samples with exposed surface area of 0.1x0.1m were used in large edge frame and the void between the edge frame and the tested foam was filled with ceramic insulation. The insulation around the perimeter was provided in a very similar manner compared with the individual fabric and composite tests done for this study. One of benefits of using the large cone calorimeter is it ability to provide more uniform heat flux distribution over the exposed specimen surface.

Table 2: Overview of performed tests

Specimen	Apparatus	Exposure	Repeated tests
Cotton	MCC,	1 K/s	3
	standard cone calorimeter	35kW/m^2	3
Polyester	MCC,	1 K/s	3
	standard cone calorimeter	35kW/m^2	3
CP blend	MCC,	1 K/s	3
	standard cone calorimeter	35kW/m^2	3
FPUF	MCC	1 K/s	3
	Large cone calorimeter	25 and 35kW/m^2	3
FPUF/fabric	Standard cone calorimeter	25, 35 and 50 kW/m^2	3 (for each fabric)

Cone Calorimeter Individual Material Test Results

Each type of fabric and FPUF was tested separately. For FPUF three repeated tests under two different heat fluxes (25 and 35 kW/m^2) were performed resulting in 6 tests in total. For individually tested fabrics three repeated tests were performed under one external heat flux 35 kW/m^2, resulting in 9 tests in total.

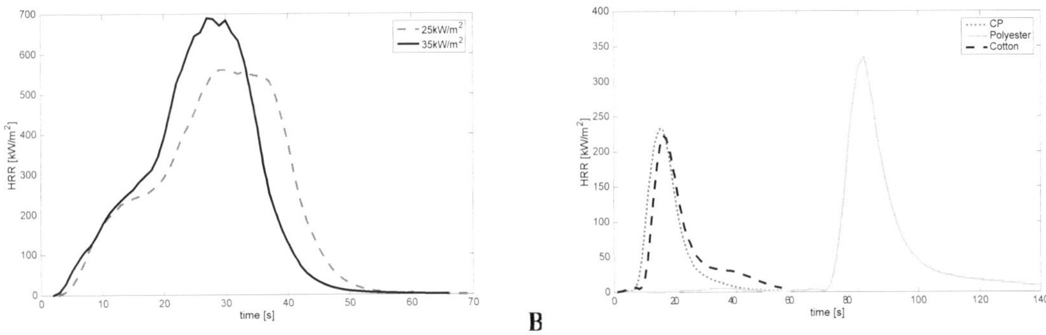

A B

Figure 2 – averaged experimental result (from the 3 repeated tests) for all materials at different heat fluxes (a) FPUF (25-35kW/m^2), (b) Fabrics (only 35kW/m^2)

Table 3 – Quantitative results at different heat fluxes for FPUF

Cone heat flux [kW/m^2]	25		35	
	time to PHRR (from ignition) [s]	PHRR [kW/m^2]	time to PHRR (from ignition) [s]	PHRR [kW/m^2]
FPUF (test 1)	24	687	22	586
FPUF (test 2)	30	811	19	574
FPUF (test 3)	29	641	21	556
FPUF average THR [kJ]	14044±308		14193±174	

Table 4 - Quantitative results at 35kW/m^2 for fabrics

Cone heat flux [kW/m^2]	**35**	
	time to PHRR (from ignition) [s]	PHRR [kW/m^2]
C (test 1)	7	233
C (test 2)	6	239
C (test 3)	6	208
C THR [kJ]	3056±305	
P (test 1)	9	356
P (test 2)	11	325
P (test 3)	13	316
P THR [kJ]	5901±184	
CP blend (test 1)	7	250
CP blend (test 2)	9	250
CP blend (test 3)	7	224
CP blend THR [kJ]	2822±65	

Cone Calorimeter: Composite Testing

A test program of FPUF/fabric composites was performed for three different combinations (i.e. cotton, CP blend and polyester each in combination with same type of FPUF). Three repeated tests under three different heat fluxes (25, 35 and 50 kW/m^2) were performed resulting in 27 tests in total.

Based on visual observations, differences were observed in the fire behaviour of the composite for the different fabrics used during the tests. The cotton fabric became black soon after ignition, nevertheless kept its position and shape during the initial stages of the burning. Based on observations it may be suggested that FPUF collapsed under the cotton fabric and pyrolysis gases were diffusing through the fabric as the composite burned. The fabric which was remaining, fell in the opening of the burned out FPUF only at the later stages of the burning and kept smouldering/glowing for about 30 seconds after the flames extinguished. Polyester fabric shrunk and melted before the ignition, forming distributed local pools on the exposed surface of the FPUF. The places where these polyester pools formed collapsed slower during burning. The surface of the sample thus did not recede uniformly, and bumps could be observed. Visually it seemed that the FPUF ignited before the polyester fabric in this test configuration. The CP blend pealed and shrunk around the perimeter of the sample before the ignition. Some of the fabrics' structure was still visible during burning, but in general the CP blend did not maintain its shape for as long as the cotton tests and collapsed together with the FPUF during burning. Observations made for fabrics were consistent with individual fabric tests.

The average HRR of three repeated tests for each incident heat flux level are presented in Figure 3. Composites consisting of CP blend and cotton fabrics resulted in two HRR peaks, while polyester composite resulted in a single peak.

Time to PHRR and PHRR for each of the HRR peaks and THR are presented in Table 5. On average the highest difference in PHRR values were observed for the CP/FPUF combination, tested under at heat flux 50 kW/m^2 (the difference between minimum and maximum values were 22.6 and 132 kW/m^2 for first and second HRR peak respectively). On average the highest differences in time to PHRR and THR were observed for the Cotton/FPUF combination under incident heat flux 25 kW/m^2 (the difference between minimum and maximum values were 2 and 18 seconds for first and second HRR peak respectively and 1685 kJ).

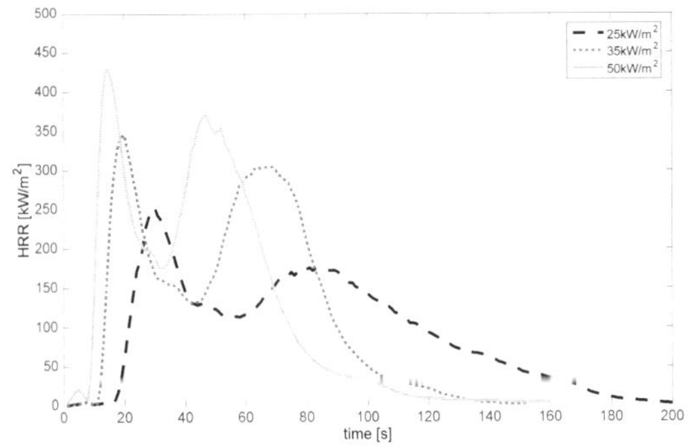

A

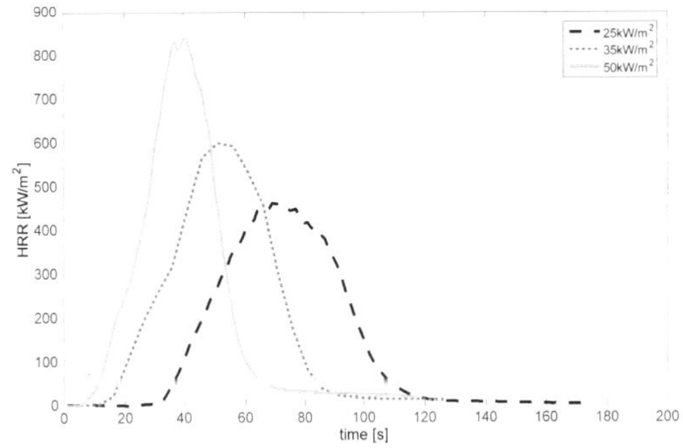

B

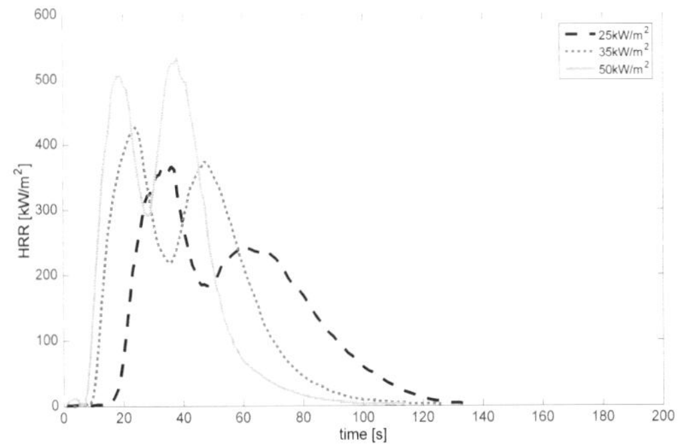

C

Figure 3 – averaged HRR of repeated composite tests (a) cotton, (b) polyester, (c) CP blend

Table 5 - Quantitative results at different heat fluxes for Fabric/FPUF composites

Cone heat flux [kW/m²]		**25**		**35**		**50**	
		time to PHRR (from ignition) [s]	PHRR [kW/m²]	time to PHRR (from ignition) [s]	PHRR [kW/m²]	time to PHRR (from ignition) [s]	PHRR [kW/m²]
Cotton/FPUF (test 1)	1st peak	10	314	9	326	7	418
Cotton/FPUF (test 2)		8	290	8	363	7	402
Cotton/FPUF (test 3)		8	293	8	366	8	481
Cotton/FPUF (test 1)	2nd peak	72	166	63	290	40	375
Cotton/FPUF (test 2)		72	191	50	346	44	382
Cotton/FPUF (test 3)		54	192	57	313	39	408
Cotton/FPUF THR [kJ]		18050±996		18528±458		19113±421	
Polyester/FPUF (test 1)	1st peak	36	449	38	622	31	965
Polyester/FPUF (test 2)		33	493	34	614	29	843
Polyester/FPUF (test 3)		33	486	40	625	31	800
Polyester/FPUF THR [kJ]		22507±678		24184±71		24806±206	
CP blend/FPUF (test 1)	1st peak	19	339	13	404	11	516
CP blend/FPUF (test 2)		18	401	14	452	12	538
CP blend /FPUF (test 3)		17	403	15	457	14	533
CP blend/FPUF (test 1)	2nd peak	41	310	32	393	33	480
CP blend/FPUF (test 2)		39	225	39	386	30	612
CP blend /FPUF (test 3)		48	246	38	353	31	560
CP blend/FPUF THR [kJ]		17654±457		17866±483		18487±860	

MODELLING

Modelling work performed for this study was completed in 3 stages. Micro-scale modelling of the individual materials, 1D cone calorimeter simulations of the individual materials and 1D cone calorimeter simulations of FPUF/fabric layered composites. Optimisation of model input parameters took place in the first 2 stages of the simulation procedure as per **Figure 4**, with the aim being to investigate if these optimised parameters could then be used to model the FPUF/fabric composite tested in the final stage. In the first stage; simulation of MCC test results for all 4 materials was performed to determine the materials kinetic parameters (A, E_a, n). Models were developed based on guidelines provided in the Fire Dynamics Simulator (FDS) user guide [11] and verification guide[19]. The reader is also referred to a previous publication by the authors[10] on this stage 1 of modelling process.

In the second stage, once the individual materials kinetic parameters had been validated against the experimental MCC data, simulation of results from the cone calorimeter experiments on the same individual materials was performed. This stage of simulation differs from the previous 'micro-scale' stage by the inclusion of heat transfer properties, since heat transfer was assumed not to be significant at the micro-scale stage. In particular, properties such as the conductivity, specific heat and material density were expected to play an important role in accurately predicting the cone calorimeter experiments. Experimentally determining the 'true' values for these properties, for all the materials examined in this study is not a trivial task. Even for relatively 'simple' measurements, such as the density of thin fabric materials, a large uncertainty is expected. Attempting to quantify these input parameters for the specific materials investigated was not within the scope of this study, thus initially such properties, when they could not be easily experimentally determined, were approximated with literature values[20–23]. Literature values acted as a baseline from which the heat transfer properties were then optimised systematically (as the kinetic properties were in the first stage, refer[10]) in an attempt to achieve predicted HRR results that reflected the experimental results obtained from the individual material cone calorimeter experiments. Only the input parameters that were not obtained in the micro-scale simulations/experiments were optimised at this stage, other model inputs, e.g. kinetic parameters, were left unaltered. This methodology presents as a double optimisation routine, whereby desired material input parameters are tuned using experimental data at the appropriate scale, e.g. MCC at micro-scale, Cone calorimeter at intermediate scale.

The final stage of the modelling process for this study involved collating the previous simulations stages into a composite model. Using this model, the cone calorimeter experiments performed on the FPUF/fabric combinations discussed in the previous chapters were simulated without further material property input refinements. Two different approaches for modelling at this stage were taken; the first approached modelled the material combinations as different layers, i.e. layer1-fabric, layer2-FPUF as per the experimental setup. The second approach, based on observations during the experimental stage about how the FPUF/fabrics interacted, simulated the composite as one material, with multiple components (i.e. fabric and FPUF components), with each mass fraction representing the various materials involved in the composite.

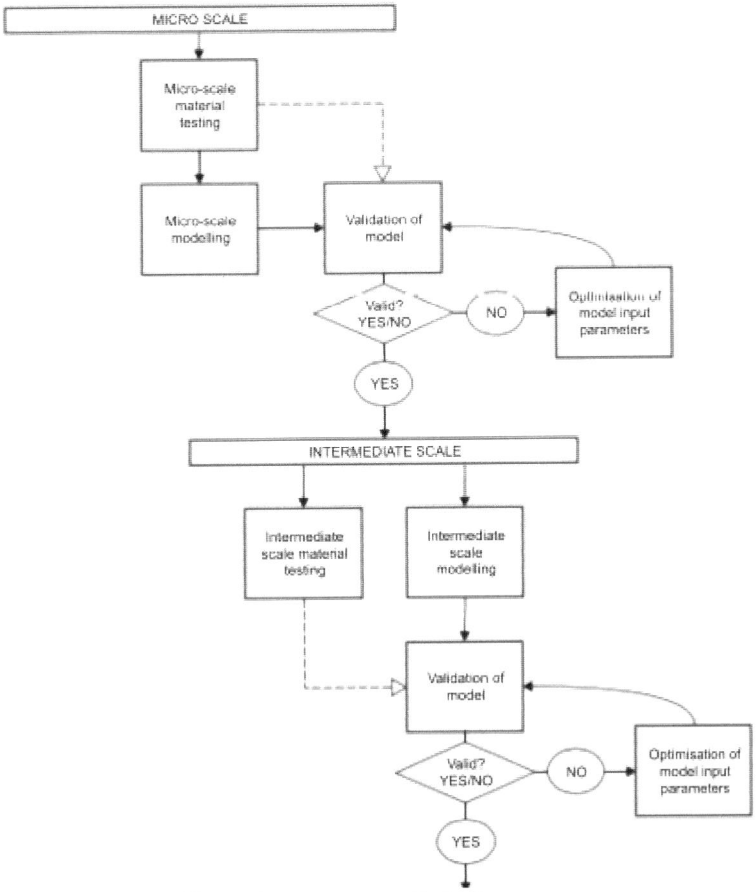

Figure 4 - two-stage optimisation process

Results: Stage 2 – individual material cone calorimeter simulations

Results presented in this section (Figure 5) compare the stage 2 simulations of the individual materials used in this study against the average HRR from the repeated experiments performed in the cone calorimeter.

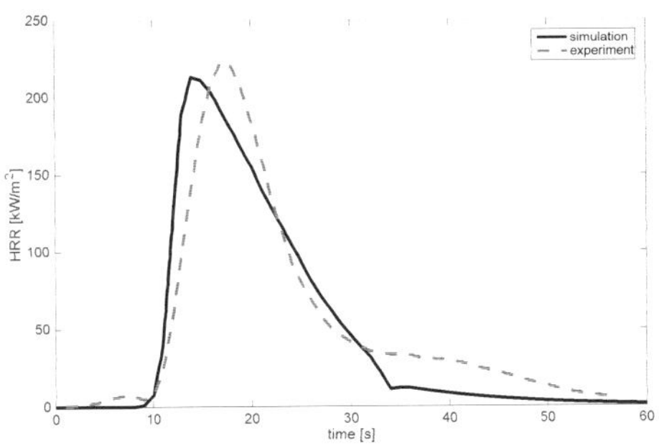

A

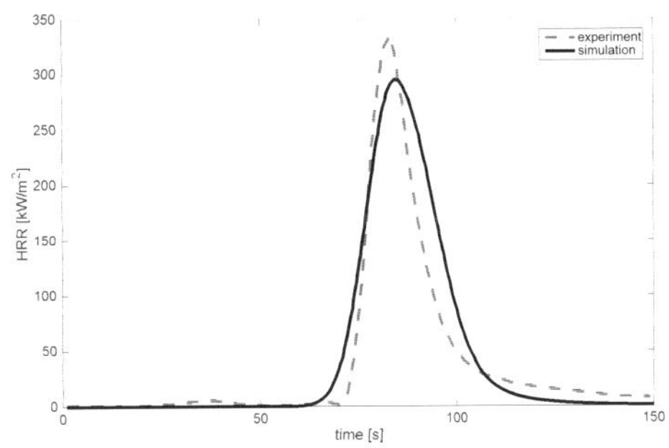

B

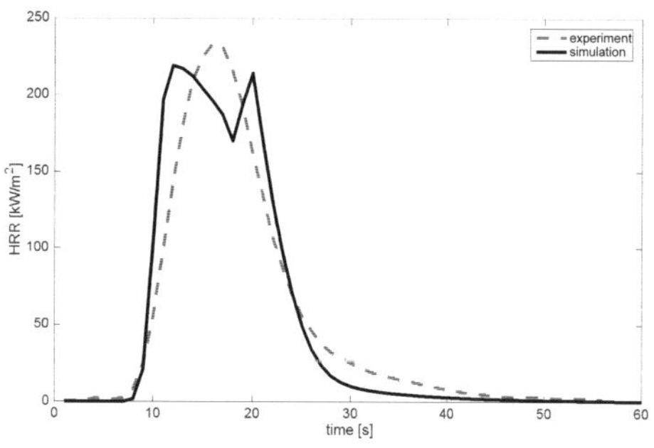

C

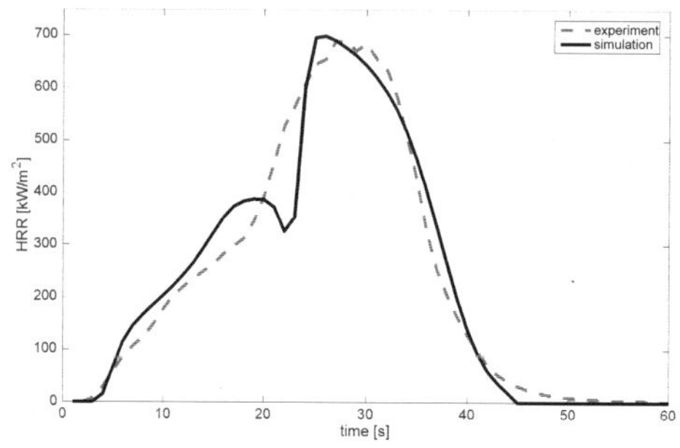

D

Figure 5 – (a) cotton, (b) polyester, (c) blend (PC), (d) FPUF

Results from the stage 2 simulations overall show a reasonable fit with averaged experimental data from the cone calorimeter experiments with the overall shape and peak HRR values showing similarities. Table 6 summaries the results using peak HRR and time-to-peak HRR (from start of test) as the comparison method.

Table 6 – Quantitative results for experimental vs simulation for 35kW/m^2

Material	**Time to peak HRR (s)**		**Peak HRR (kW)**	
	Exp value	Sim value	Exp value	Sim value
Cotton	17	14	223	213
Polyester	83	84	335	295
CP Blend	16	12	233	219
FPUF	27	26	690	699

Both the cotton and the polyester simulations incorporated oxidation of char. The char fraction was determined from the mass leftover after the MCC tests were performed in nitrogen[10]. The PC blend fabric was simulated with two materials (cotton 75% polyester 25%) and oxidation of the residual char, while the FPUF was simulated using two component mass fractions 32% and 68% respectively, assumed to be the main chemicals used to produce the FPUF, i.e. TDI and a polyol (refer Table 1), whose properties were also obtained via MCC measurements.

For the fabric materials, it was decided, due to the uncertainty in the measurement of material thickness that the thickness for all fabric material would be set at 1mm. Thus, in order to achieve the approximately correct total energy content, the density values (kg/m^3) were also part of the optimisation process.

It should also be reiterated at this stage that input values (particularly those effected by heat transfer) for these simulations were highly optimised via the two-stage process described previously to achieve these simulated results to compare with the experimental results. The aim of this process was to investigate if, with these optimised input parameters, the behaviour of the FPUF/fabric composites could be predicted. All input values used in these simulations were constant values. This optimisation may be exemplified by the use of the specific heat (c_p) variable for the polyester simulations being increased excessively (up to 15 kJ/(kg · K)) to delay the 'ignition' time to that observed in the experiments. This is clearly not a physical value, but in this case it is used to account for the melting behaviour of the polyester observed in the cone calorimeter experiments.

Results: Stage 3 – simulation of the composite FPUF/fabric composites

After stage 2 was completed, stage 3 models were produced by combining the material input properties from the stage 2 simulations for the relevant composites. For the layered models, as per the experiments, the material properties were added as separate layers; fabric then FPUF, with separate thicknesses (1mm and 25mm respectively). For the mass-fraction approach, material properties for each layer of the composite were combined into one material, and their mass-fractions were calculated based on the contribution of each materials mass to the overall mass of the composite. Results are presented in Figure 6, Figure 7 and Figure 8.

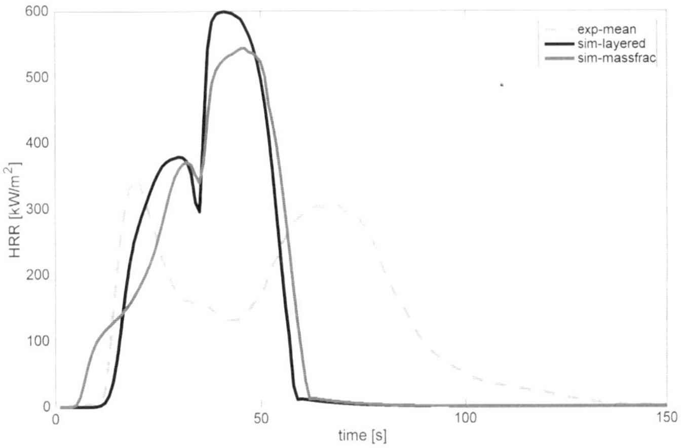

Figure 6 – Cotton/FPUF composite modelling results vs experiment

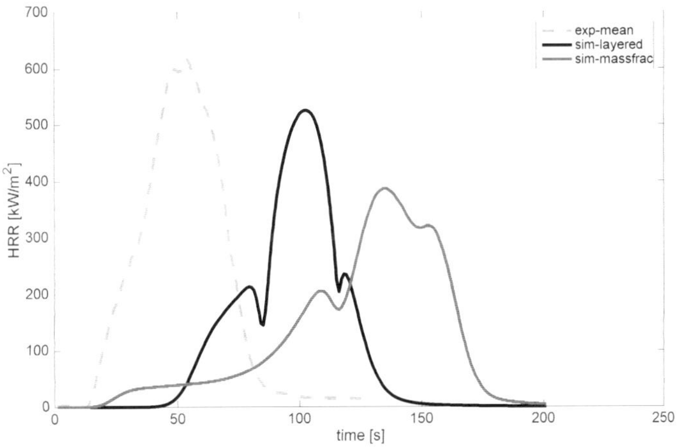

Figure 7 – Polyester/FPUF composite modelling results vs experiment

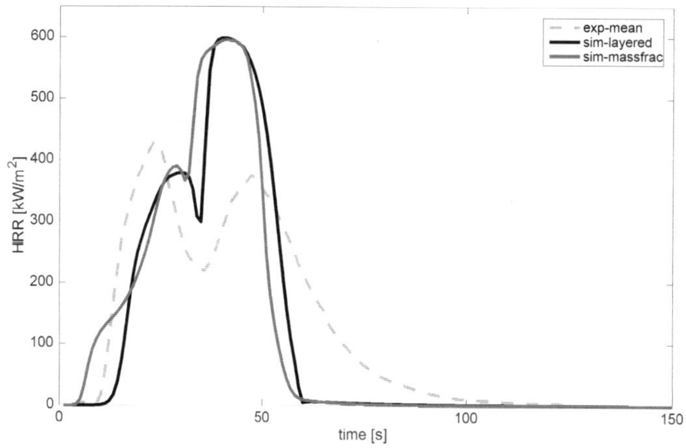

Figure 8 – CP Blend/FPUF composite modelling results vs experiment

The initial results from the composite models show large differences between experimental results and simulations. For the cotton and PC blend simulations two peaks are predicted, and values for the first PHRR have less than a 10% error in height, however, second PHRR values are significantly higher for both sets of simulations, compared to experimental values, more so for cotton/FPUF combination. Burn times are also significantly shorter in the simulations when compared to experiments.

For the polyester-FPUF composite mock-ups, the layered model predicts to some degree the overall shape of the experiments and the main peak HRR value has an error of less than 20%. However, number of peaks and the time to ignition are severely over-predicted in both model cases. A discussion as to the possible causes for the discrepancies observed in these simulation results is discussed in the next section.

DISCUSSION AND CONCLUSION

The main conclusion illustrated by this study is that, even when using highly optimised individual material properties, modelling composite products such as the mock-up upholstered furniture still seems to be out of reach of current material fire behaviour models. The principal reasons for this, in the authors' opinions, are the models' limitations in accounting for the material-material interactions that happen in composites such as the ones used in this study. Current models (e.g. FDS) at this stage can simply not account for the observations that were made in the experiments. A prime example of this is in the case of the composite cotton/FPUF samples. Experimental observations described that after the initial burn off of volatiles from the cotton, the 'char' that was leftover remained in place, and acted as a partial barrier to the volatiles flowing from the FPUF underneath. This effect may also be observed from the extended burn time in the HRR data from the experiments (also to a lesser extent in the blended cases), when comparing them to the simulations. In the layered case simulations, effectively how FDS calculates the results is by first burning the top layer, once the fuel from this layer has been expended, the fuel from the second layer is exposed and begins to be consumed. Hence the double PHRR is observed in the simulations. However, once the first layer is expended, FDS also assumes that the next layer is basically the new top layer and is free to burn without hindrance unlike what was observed in the experiments. This was the reasoning for implementing the second; mass-fraction, simulation method in this study – to attempt to counter this layer-by-layer process. However, this method also has its limitations, as it is now a single mixture of all the materials, FDS's main guide to when pyrolysis of all the material components begins is now the kinetics obtained from stage 1 of this process. Thereby this method also fails to account for such material-material interaction effects.

Due to these observed material-material interactions, which are a significant limitation for current models, in order to attempt any kind of composite simulations, researchers will still require experimental knowledge on how the materials may potentially affect each other in order to attempt to counter these perceived model limitations. An example of this may also be given from this study; in the individual polyester tests and subsequent simulation, it was observed that the time to ignition was delayed extensively due to the material first going through a melting process, before pyrolysis and ignition occurred. To account for this process (melting is also a current model limitation) the specific heat (c_p) of the polyester was increased significantly to help delay the simulated ignition time. This substitution worked well in this case, with the model predicting the cone calorimeter results quite well. However in the composite polyester/FPUF experiments in was noted that as the polyester curled up and began to melt, this exposed the foam to the cone heater, allowing it to begin pyrolysing before the polyester had completed the melting process, thus ignition in this composite stage was likely due to the FPUF. The FPUF ignition may have also then speed up the polyester decomposition process, thus both materials were likely burning at the same stage. Knowledge of this interaction between materials means that the increased c_p value used to account for the delayed ignition in the single material simulations may not have a purpose in the composite model. To test this, the polyester/FPUF model was modified, and the c_p value for the polyester was reduced to a value closer to that found in the literature[23]. Results are shown below in Figure 9.

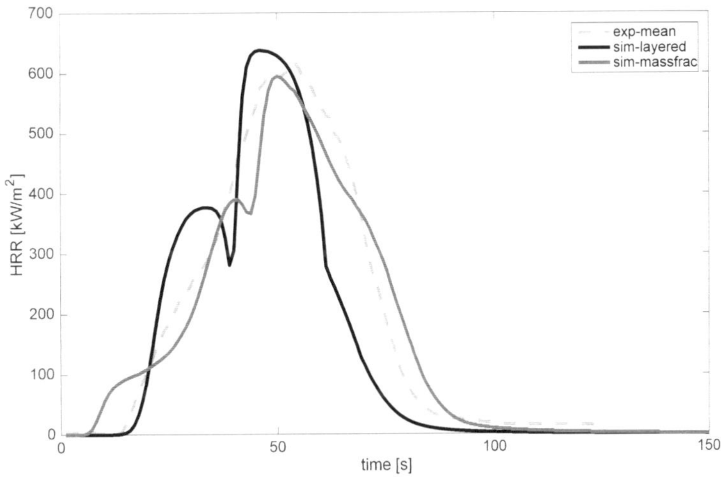

Figure 9 – modified polyester/FPUF simulation results vs experiment

The modified model results shown in Figure 9 are greatly improved in both model cases, greater so for the mass-fraction model case. Main Peak HRR value is within 5% of the experiment, and the overall shape was significantly improved against the initial model simulation (refer Figure 7).

Examples such as this illustrate that with current model limitations; experimentation and modelling must go hand-hand. It also highlights some of the areas where deeper knowledge is required and where current models require further development to help further improve the fire research communities modelling capabilities.

AKNOWLEDGEMENTS

The authors acknowledge the European Union's Seventh Framework Programme (FP7) support under grant no. 316991, and thanks the Interscience Fire Laboratory for use of their MCC, Fire Testing Technology (FTT) for use of their test equipment, Laboratoire national de metrologie et d'essais (LNE) for supplying FPUF samples and the FIRETOOLS group. This study is part of the FIRETOOLS project – a collaboration between Lund University and DBI (Danish Institute of Fire and Security Technology).

REFERENCES

1. Krasny, J., Parker, W., & Babrauskas, V. *Fire behavior of upholstered furniture and mattresses*. (2008).
2. National Fire Protection Association. *White paper on Upholstered Furniture*. (2013).
3. van Hees, P., Guay, F., Bhargava, A., Wilkens, K., Livkiss, K., Andres, B., Lauridsen, D., & Lundstrom, F. V. Simulation of Fire Technical Properties of Products and Construction Barriers to Support Efficient Product Developement in Industry. *Proceedings of 13th international Fire and Engineering Conference Interflam* (2013).
4. Guillaume, E., Camillo, A., & Rogaume, T. *Application and Limitations of a Method Based on Pyrolysis Models to Simulate Railway Rolling Stock Fire Scenarios*. *Fire Technology* **50,** (2013).
5. Marquis, D. M., Pavageau, M., & Guillaume, E. Multi-scale Simulations of Fire Growth on a Sandwich Composite Structure. *Journal of Fire Sciences* **31,** 3–34 (2012).
6. Sundström, B. *Fire safety of upholstered furniture the final report on the CBUF research programme*. (London Interscience Communication, 1996).
7. Ahrens, M. & National Fire Protection Association. *Home Fires that began with Upholstered Furniture. National Fire Protection Association* (2011).
8. Bustamante Valencia, L. Experimental and numerical investigation of the thermal decomposition of materials at three scales: application to polyether polyurethane foam used in upholstered. (2009).
9. Leisted, R. Design Fire for Building Content in Arson Scenarios. (Lund University, Report 5460, 2014).
10. Wilkens Flecknoe-brown, K., Hostikka, S., & Hees, P. van. Pyrolysis modelling of composite fabrics based on individual fabric properties using micro-combustion calorimetry. in *Interflam Conference Proceedings* (Interscience Communications Limited, 2016).
11. McGrattan, K., McDermott, R., Weinschenk, C., Overholt, K., Hostikka, S., & Floyd, J. *Fire Dynamics Simulator User's Guide*. (2015).
12. Lyon, R. E. & Walters, R. N. *A Micro Combustion Calorimetry*. (2002).
13. Janssens, M. Measuring rate of heat release by oxygen consumption. *Fire Technology* 234–249 (1991).
14. International, A. *Standard Test Method for Determining Flammability Characteristics of Plastics and Other Solid Materials Using Microscale Combustion Calorimetry*. (2011). doi:10.1520/D7309-11.Copyright
15. Lyon, R. E., Walters, R. N., Stoliarov, S. I., & Safronava, N. *Principles and Practices of Microscale Combustion Calorimetry*. (2013).
16. ASTM E1474 Standard Test Method for Determining the Heat Release Rate of Upholstered Furniture and Mattress Components or Composites Using a Bench Scale Oxygen Consumption Calorimeter.
17. International Organisation for Standardisation. ISO 5660-1&2 Reaction-to-Fire Tests-Heat Release, Smoke Production and Mass Loss Rate -Part 1&2 (Cone Calorimeter Method). (2002).
18. Gregory, S., Green, A., Pasantes, S., Grayson, S. J., & Kumar, S. Use of the cone calorimeter for testing materials with low heat release rates. *Proceedings of 13th international Fire and*

Engineering Conference Interflam **1,** 871–882 (2013).
19. McGrattan, K., Hostikka, S., McDermott, R., Floyd, J., Weinschenk, C., & Overholt, K. Fire Dynamics Simulator, Technical Reference Guide, Volume 2: Verification. **4,** (2013).
20. Valencia, L. B. Experimental and numerical investigation of the thermal decomposition of materials at three scales: application to polyether polyurethane foam used in upholstered furniture. (2006).
21. Majumdar, A., Mukhopadhyay, S., & Yadav, R. Thermal properties of knitted fabrics made from cotton and regenerated bamboo cellulosic fibres. *International Journal of Thermal Sciences* **49,** 2042–2048 (2010).
22. Stanković, S. B., Popović, D., & Poparić, G. B. Thermal properties of textile fabrics made of natural and regenerated cellulose fibers. *Polymer Testing* **27,** 41–48 (2008).
23. Harper, C. A. *Handbook of Building Materials for Fire Protection*. (McGraw-Hill, 2004).

EXPERIMENTAL EVALUATION OF FIRE BARRIERS EXPOSED TO OPEN-FLAME IGNITION ON FULL-SCALE UPHOLSTERED FURNITURE

Andrew Lock*†, Linda Fansler, & David Miller
U.S. Consumer Product Safety Commission
Rockville, MD 20850

Abstract

U.S. Consumer Product Safety Commission (CPSC) staff has extensive experience evaluating upholstered furniture flammability, experience acquired through ongoing efforts to reduce the fire hazards associated with consumer products. CPSC staff has explored several options for improving the flammability performance of furniture. As part of a recent research effort, CPSC staff evaluated the flammability performance of upholstered chairs constructed with enhanced flammability performing textile materials or "fire-blocking barrier" materials. A preliminary study suggested that fire-blocking barrier materials used in upholstered chairs would reduce the fire growth and energy output of furniture involved in a fire. The fire-blocking barrier materials used in this study were textile barriers made from fibers with enhanced flammability characteristics. The barriers are used in other textile product applications and are commercially available. CPSC staff conducted a limited series of full-scale, open-flame ignition experiments using upholstery fabrics, fire-blocking barrier materials, and simulated aging of the furniture as independent variables. This paper reports the test results, provides statistical tests of differences, summarizes the conclusions, and identifies areas for future research.

A total of 72 chairs were tested, representing 24 different combinations of these test variables. The frame and padding material were standardized across the upholstered chairs. Two cover fabrics and five fire-blocking barriers were used in this test series. The testing was conducted by CPSC staff at CPSC's National Product Testing and Evaluation Center in Rockville, MD. Oxygen consumption calorimetry was used to measure the Heat Release Rate (HRR) and Time-to-Peak-Heat Release Rate (TTPHRR) for the chairs in an open-calorimetry configuration. The ignition scenario for these tests was a BS5852 source 3, which is a 240-mm butane flame, applied to the center back crevice of the chair for 70 seconds.

A summary of the HRR and TTPHRR of the chairs is presented. A first order comparison of the different barriers, cover fabrics, and aging effects is presented first. An Analysis of Variance (ANOVA) is conducted to analyze the contribution of each independent variable and the interactions of the independent variables. The effect of each variable, as well as interactions between variables, is estimated and a determination is made about whether these effects are statistically significant.

Areas for additional study are needed and are being investigated to determine the viability and feasibility of barriers as a potential improvement in flammability performance of upholstered furniture. Areas for further study include, but are not limited to, the added cost of the barriers, the effect of the barriers on the hand (feel) of the furniture, as well as the viability of barriers in different ignition scenarios and construction variants, *i.e.*, polyester fill, instead of polyurethane foam, as

* Corresponding Author: Andrew Lock, alock@cpsc.gov, +1-301-987-2099.
† This paper was prepared by the CPSC staff, and has not been reviewed or approved by, and may not necessarily reflect the views of the Commission.

padding materials, permanent cushions, instead of removable ones, different crevice shapes and surface contact. Preliminary work on the effect of changing the filling material of the back cushion indicates significantly different performance in comparison to foam back chairs.

Introduction

As part of the Upholstered Furniture Flammability project at the CPSC, staff has been working to evaluate the feasibility of solutions to mitigate the Upholstered Furniture Flammability hazard. Upholstered Furniture was the first item ignited in an annual average of 4,900 reported fires between 2011 and 2013, in the United States.[1] Those fires accounted for 390 deaths, 660 injuries, and $238 million in property loss. This serves to illustrate the magnitude of upholstered furniture fire hazards. John Hall has also recently suggested that a different analysis, which accounts for the item that contributes the most to flame spread, whether or not it is the first item ignited, would attribute an additional 130 deaths to upholstered furniture.[2,3]

Following a 2012 validation study by CPSC,[4] staff determined that a barrier provided some improvement in the flammability performance of full-scale upholstered furniture. In 2013, CPSC staff held an Upholstered Furniture Fire Safety Technology Meeting to promote a discussion of fire barrier technologies and the potential benefits barriers could provide toward improving or reducing furniture flammability.[5] As a result of that meeting, staff developed a limited test program to assess fire barrier effectiveness in reducing the fire hazards posed by upholstered furniture flammability. Five fire barriers were selected from commercially available barriers to represent different types of barrier products from several different manufacturers. Staff contracted to have 72 chairs built with these fire barriers using specific cover fabrics, foam, and dimensions. These chairs were subjected to small open-flame and smoldering ignition sources.[‡] In addition, some chairs were also evaluated in a mechanically stressed condition to assess the effect of wear on flammability performance.

Experimental Setup

The chairs that were used in this test series were based on a chair design used by CPSC staff in previous studies.[4] The dimensions of the chair and a representative photo of a chair are presented in Figure 1.

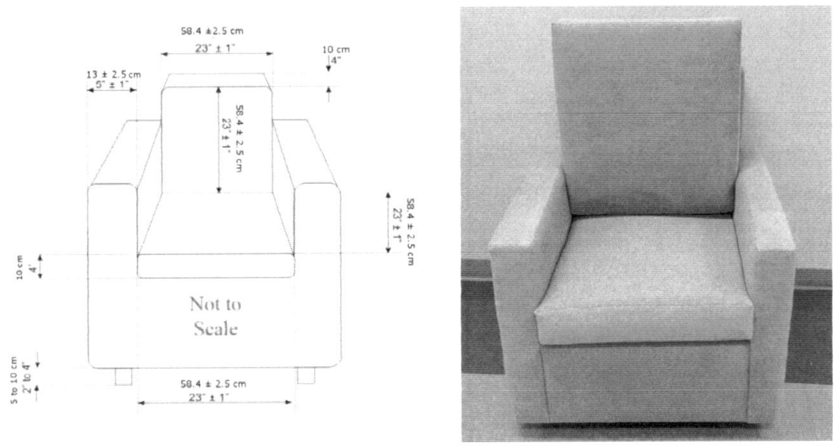

Figure 1. Schematic representation of the dimensions (left) and photo (right) of a test chair used in the test series.

‡ Only the open-flame results are presented here.

The materials, including materials with improved flammability performance, were all commercially available and purchased by the furniture manufacturer. The chairs evaluated with the open-flame ignition source used 18 combinations of materials (1 type of foam, 5 types of fire barriers, plus no fire barrier, and 2 types of cover fabrics). These materials are described subsequently. The materials combinations are listed in Table 1 for the open-flame tests. The cover fabric and barrier descriptions are in Table 2. The chairs were constructed with commercially available non-flame-retardant polyurethane foam. The foam was specified with a density of 1.8 + 0.1 lb/ft,[3] Indentation Load Deflection (ILD): 25% to 30%, Air Permeability: greater than 4.0 ft^3/min and non-flame-retardant (FR). Unfortunately 7 $\pm$ 2% melamine (a common FR chemical) was found in the foam that was used. This was out of spec and may make it difficult to compare with other chairs containing foam without melamine.

Table 1. Chair Material Combinations for Full-Scale, Open-Flame Testing

Combination	Foam	Polyester Batting	Fire Barrier	Cover Fabric	Number of Chairs[6]	
					Mechanically Stressed	Non-Stressed
2	PU	YES	NO	F2	3	3
3	PU	YES	NO	F3	3	3
5	PU	NO	FB1	F2	3	2
6	PU	NO	FB1	F3	3	3
8	PU	NO	FB2	F2	3	3
9	PU	NO	FB2	F3	3	2
11	PU	NO	FB3	F2	3	3
12	PU	NO	FB3	F3	3	2
14	PU	NO	FB4	F2	3	5
15	PU	NO	FB4	F3	4	2
17	PU	NO	FB5	F2	3	3
18	PU	NO	FB5	F3	2	5

Table 2. Upholstery Fabrics for Full-Scale Tests

Fabric Code	Fiber Content	Weight (oz/yd^2)	Fabric Construction
F2	100% cotton	13	Twill (denim)
F3	56% rayon/34% polyester/10% cotton	10	Jacquard

Table 3. Fire Barriers for Full-Scale Tests

Fire Barrier Code	Fiber Content	Construction*	Chemical Analysis[6]
FB1	Cellulose based	Nonwoven	0.6% Phosphorous 0.2% Chlorine
FB2	Fibrous glass/modacrylic/polyester	Needle punched	2.6% Antimony
FB3	Modacrylic/silica	Knit/sock	0.8% Antimony 0.3% Phosphorous 11.0% Chlorine
FB4	Rayon/polyester	High loft	1.0% Phosphorous 0.3% Chlorine
FB5	Rayon/polyester	Densified	1.2% Phosphorous 0.2% Chlorine
FB6**	Polyester	Nonwoven	

*Construction description from manufacturer.

**Conventional material used in chair construction.

The upholstered chair samples were removed from any packaging before conditioning. The upholstered chair specimens were conditioned at a temperature of 20 ± 3 °C (70° ± 5° F) and a relative humidity of 50% ± 5% for 48 hours. The test was started within 10 minutes of removing the specimens from the conditioning area. In the test room, the chairs were placed in a tray on a load cell in the center of the hood. The HRR was measured by oxygen consumption calorimetry. Once ignited, the chairs were allowed to burn until staff determined that no sustained combustion remained, or until staff determined that no more useful information could be collected. Typically, this was after the fire had reached an obvious peak, the remaining HRR had dropped below 200 kW, and staff observed that there were minimal combustibles left burning.

For the open-flame ignition, a British Standard 5852[7] source 3 flame was used. This is a 240 mm butane flame that is applied to the back seat crevice for 70 seconds. The point at which the butane flame was removed from the chair was designated as the start time for the test.

Staff included a mechanical stress variable to study the potential impact of general wear on flammability performance. Staff developed a constant-force pounding procedure based on a prior study[8] that looked at the effect of wear on upholstery fabric chair flammability. A force of 750 + 20 N (169 + 4.5 lbf) was applied vertically downward at the center of the seat cushion. The force was applied through a 250 ± 1 mm diameter plate with a 25 ± 1 mm radius at the edges, at 70 ± 5 cycles per minute. Two hundred thousand cycles were applied mechanically to the cushions to simulate the mechanical stress of long-term use.

Results and Discussion

The experimental results are presented below, along with a statistical analysis of those results. Only the results of the open-flame testing are presented here. A complete data set can be found in the memoranda prepared by CPSC staff.[6,9,10]

For chairs wherein which the fire barrier offered the most effective protection to the filling, the flames were typically smaller, less intense and self-extinguished in a matter of minutes. For chairs with no barrier, or with less-protective barriers, the flame spread on all the chairs ignited by open flame was similar, but this progression generally took significantly longer time for chairs with barriers than without. Aside from those chairs that self-extinguished, the general pattern for flame spread was as follows. When the ignition flame was removed, there were flames on the seat and back cushions. Initially, the flame spread was up the back cushion and towards the corners of the seating area. Next, flames either reached the top of the back cushion or went around the side of the back cushion between the back cushion and the back of the chair. The flame front on the seat cushion moved outward as well. The flames increased in intensity and size and continued to burn as the back cushion and back of the chair became more involved in the combustion process. The back cushion burned through first, followed by the back of the chair. That led to rapid progression with the entire chair soon engulfed in flames. As the seating area became more involved, melt dripping occurred below the chair.

The highest HRR observed in this testing was 2,117 kW from a non-barriered chair. The lowest non-zero peak HRR observed was 469 kW. Essentially, all chairs that burned produced a peak heat release rate (PHRR) of approximately 500 kW or higher. In general, the barriers reduced the intensity of the fire and delayed the onset of the fire.

Figure 2 and Figure 3 show characteristic example cases of HRR curves with and without the five barriers for fabrics F3 and F2, respectively. In each case, the chairs with fire barriers tended to reduce the PHRR (by about 40%-60%) and delay the onset of the fire (by about 8 to 20 minutes). In both, figures the chair without the fire barrier had a much higher PHRR and burned much more rapidly.

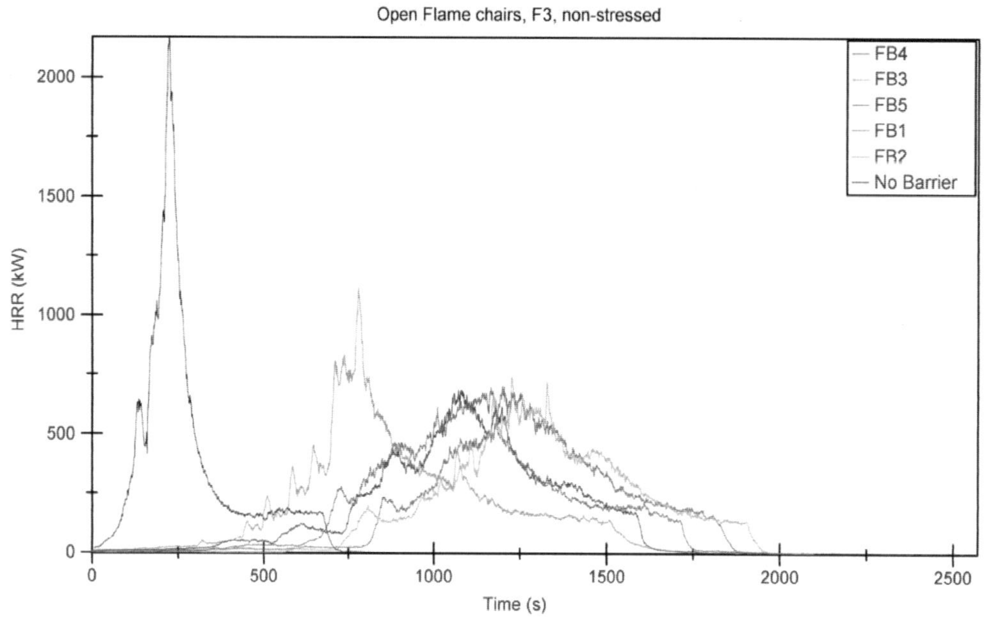

Figure 2. Example comparison of different fire barriers on open flame chairs with cover fabric F3, non-stressed.

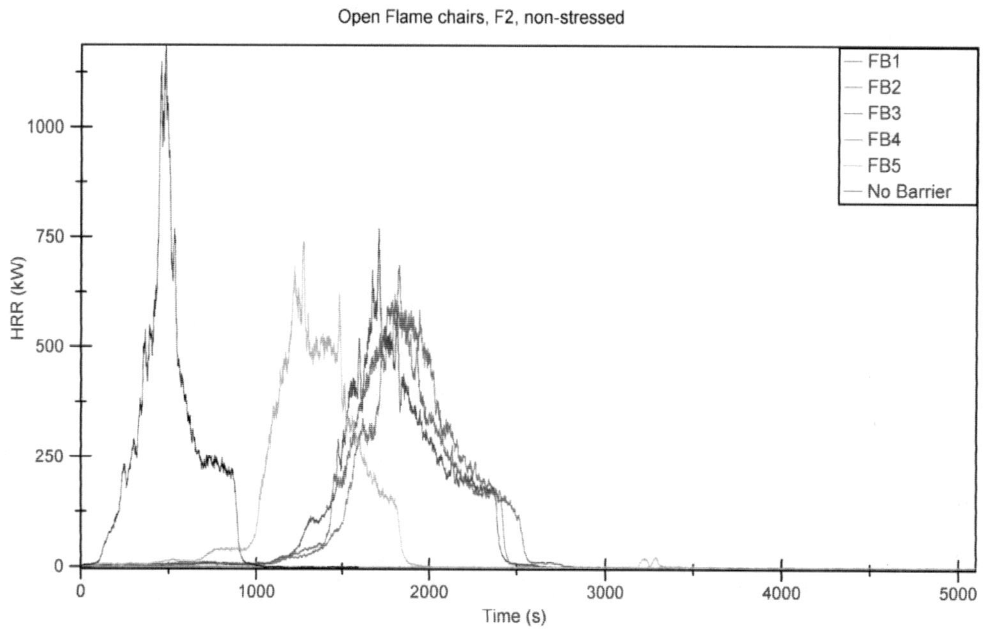

Figure 3. Example comparison of different fire barriers on open-flame chairs with cover fabric F2, non-stressed. Notice that FB2 is zero for the entire test.

The main purpose of the study is to assess the effectiveness of the barriers in reducing the intensity and slowing the progress of full-scale chair fires. Staff evaluated the effect of fire barriers on PHRR to evaluate how fire barriers address intensity of chair fires. Staff decided before the testing began that barriers would be compared using statistical inference to the "no barrier" chairs (FB6) but not to each other. Figure 4 shows the average PHRR by barrier and includes 95 percent confidence intervals. It illustrates the much higher average PHRR for the chairs with FB6.

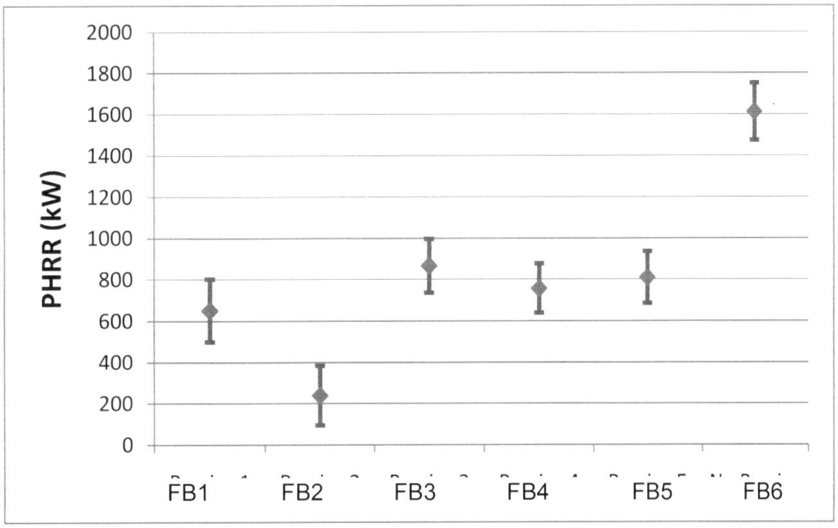

Figure 4. Average Peak Heat Release Rate by Barrier

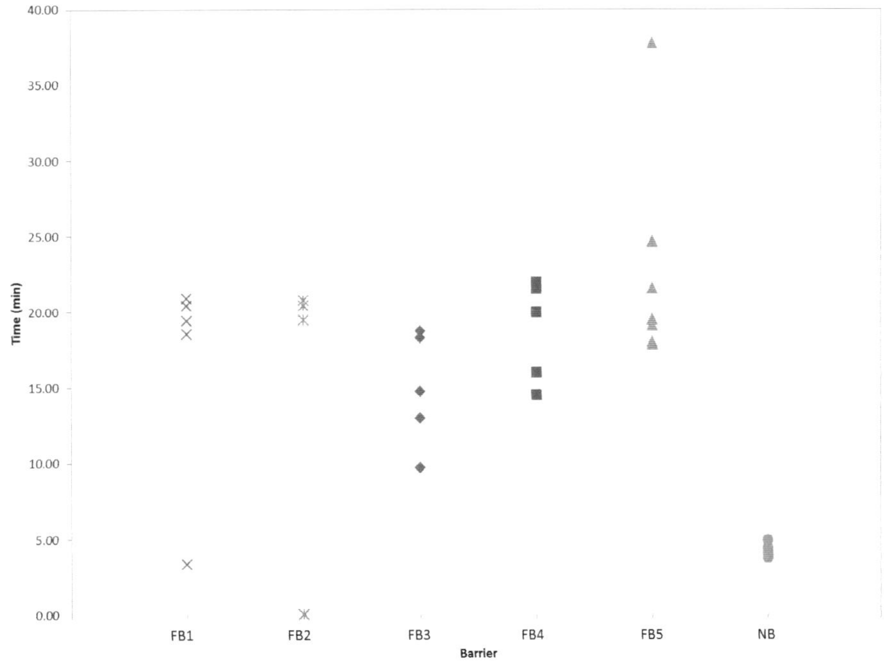

Figure 5. Time to Peak HRR for all open-flame tests on chairs with cover fabric F3. Note that a zero value indicates that no PHRR was observed.

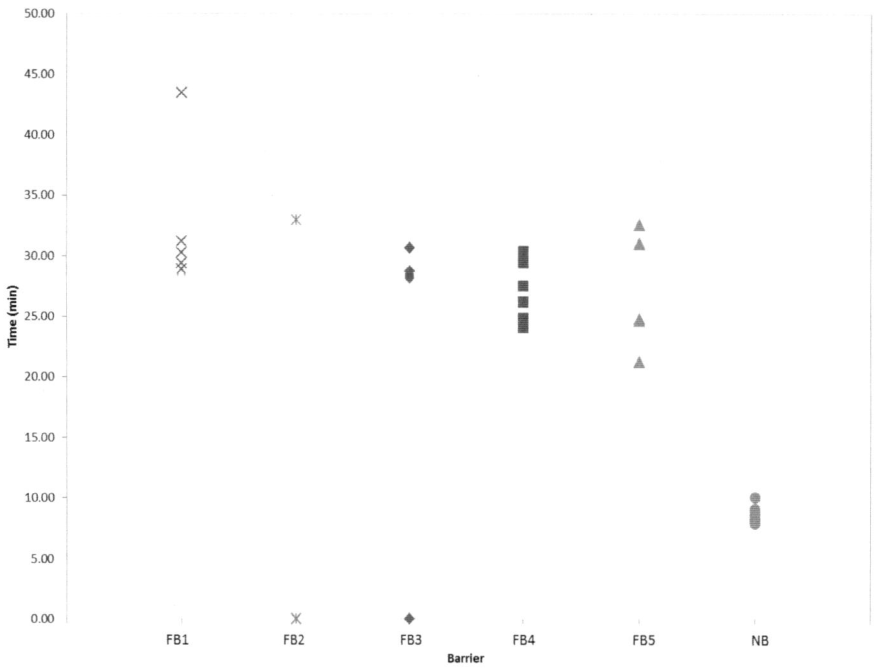

Figure 6. Time to Peak HRR for all open-flame tests on chairs with cover fabric F2. Note that a zero value indicates that no PHRR was observed.

Staff performed an analysis of variance (ANOVA) on the open-flame chairs with PHRR as the dependent variable and barrier, fabric, and stress as the independent variables. Model selection began with the full model. In the full model, there are three main effects (barrier, fabric, and mechanical stress) and four interactions: one three-way interaction (barrier*fabric*stress), and three two-way interactions (barrier*fabric, barrier*stress, and fabric*stress). Interaction terms were eliminated if their p-values were greater than 0.25.

In ANOVA, before evaluating the main effects, the interactions are considered. The three-way interaction was removed from the model because its p-value was 0.387; and barrier*stress was removed because its p-value was 0.497. Both barrier*fabric and fabric*stress were left in the model with p-values below 0.25 but were not significant at the 0.05 level.

The main effects are presented in Table 4; they are the effect that each individual variable has on PHRR (aside from the interactions). The main effects in this model are fire barrier, fabric, and stress. The fire barrier and fabric effects are statistically significant, but the stress effect is not. The barriers had the most significant effect on the PHRR of the chairs, followed by the fabric.

Table 4. PHRR main effects

Effect	Estimated Change in PHRR	95% Confidence Interval	Standard Error	P-value
Fire Barrier	FB1 = -1,120.3	FB1 = (-1,399.8, -840.9)	139.1	< 0.0001
	FB2 = -1,483.2	FB2 = (-1,762.6, -1,203.7)	139.1	< 0.0001
	FB3 = -767.0	FB3 = (-1,026.8, -507.1)	129.3	< 0.0001
	FB4 = -1,008.1	FB4 = (-1,256.9, -759.4)	123.8	< 0.0001
	FB5 = -996.7	FB5 = (-1,245.5, -748.0)	123.8	< 0.0001
Fabric	F2 = -532.4	F2 = (-830.5,-234.2)	148.4	0.0008
Stress	Not stressed = -11.7	Not stressed = (-175.1, 151.7)	81.3	0.89

As with the PHRR, ANOVA was conducted with TTPHRR as the dependent variable and barrier, fabric, and age as independent variables. However, unlike with PHRR, the chairs that did not ignite (7 of 63) were removed from the data. The model selection process led to a model with three main effects: barrier, fabric, and age, and the barrier*fabric and barrier*stressed interactions. Unlike with the PHRR, the TTPHRR barrier*fabric interaction is statistically significant (p-value < 0.05). The barrier*stress interaction was not significant.

The p-value from the ANOVA for the overall barrier*fabric interaction is 0.0172. This assumes normally distributed residuals. The p-values using the bootstrap standard errors and with "No Barrier" (FB6) as the baseline reference level, were statistically significant for FB1 (p-value = 0.02185) and FB3 (p-value = 0.00003)[§]. This implies that the fabric effect is different for chairs with FB1 and FB3 than it is for chairs without barriers.

The main effects, shown in Table 5, are fire barrier, fabric, and age. The effects for fire barrier and fabric are statistically significant. The effect for stress is not statistically significant. The estimates of the effects come from the model. The standard errors that are used to compute the confidence intervals and p-values, are derived from the bootstrap.

Table 5. Time to Peak HRR Main Effects

Effect	Estimate	95% Confidence Interval	Standard Error (Bootstrap)	P-value
Fire Barrier	FB1 = 14.1	FB1 = (9.4, 18.9)	2.4	< 0.0001
	FB2 = 16.0	FB2 = (10.2, 21.8)	3.0	< 0.0001
	FB3 = 10.5	FB3 = (6.8, 14.3)	1.9	< 0.0001
	FB4 = 14.3	FB4 = (10.8, 17.8)	1.8	< 0.0001
	FB5 = 13.7	FB5 = (9.8, 17.6)	2.0	< 0.0001
Fabric	F2 = 4.7	F2 = (1.4, 8.0)	1.7	0.0050
Stress	Not stressed = -0.6	Not stressed (-3.9, 2.7)	1.7	0.7189

The effect of fire barriers on TTPHRR demonstrates the ability of the fire barriers to slow the progress of fires. The barriers are compared to the plain poly chairs (FB6), but not to each other. Figure 7 shows the average TTPHRR by barrier and includes 95 percent confidence intervals. The plain poly chairs clearly reach their peak much sooner than the chairs with barriers.

Mechanical Stressing

As mentioned, mechanical stressing was not found to be statistically significant in the performance of the chairs in this test series. The mechanically stressed chairs' HRR curve was compared to the non-stressed chairs. The mechanically stressed chairs had similar PHRR to the non-stressed chairs. However, in general, the fires seemed to peak earlier for the stressed chairs than for the non-stressed chairs. Specifically, of the 12 combinations tested, on average, seven of the combinations stressed reached the PHRR earlier than the non-stressed combinations. In the remaining five combinations, the non-stressed chairs reached the PHRR earlier than the stressed chairs. HRR plots for Combo 15 are presented below in Figure 8. These two combinations have both different cover fabric and different

[§] The ability to find a statistically significant result for the interaction with FB2, and fabric may have been hampered by having so many FB2 chairs removed from the data because they did not ignite. Although nine of the 63 chairs with known barriers tested had FB2, only three of the 56 that ignited had FB2. Six of the seven chairs that did not ignite were FB2 chairs.

barrier material. In both cases, the mechanically stressed chairs generally peak earlier than the non-stressed; however, there is some overlap. The mechanical stressing generally caused the open-flame chairs to have a lower TTPHRR.

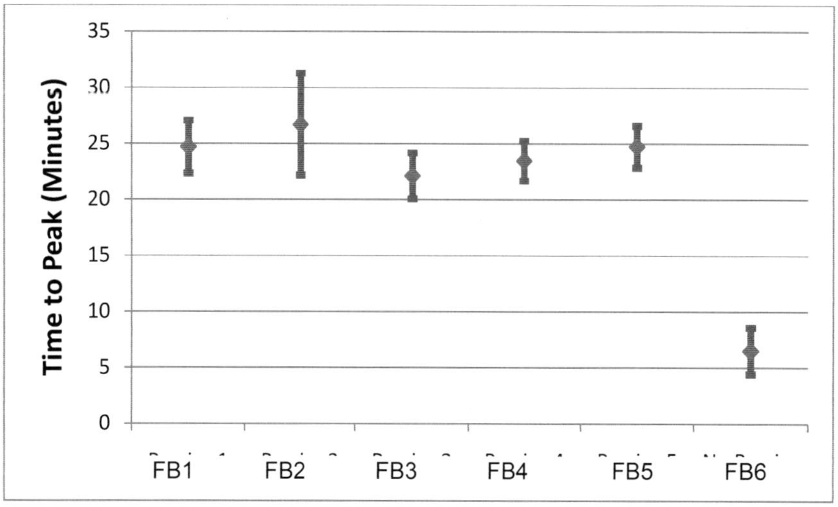

Figure 7. Average time to peak heat release rate by barrier.

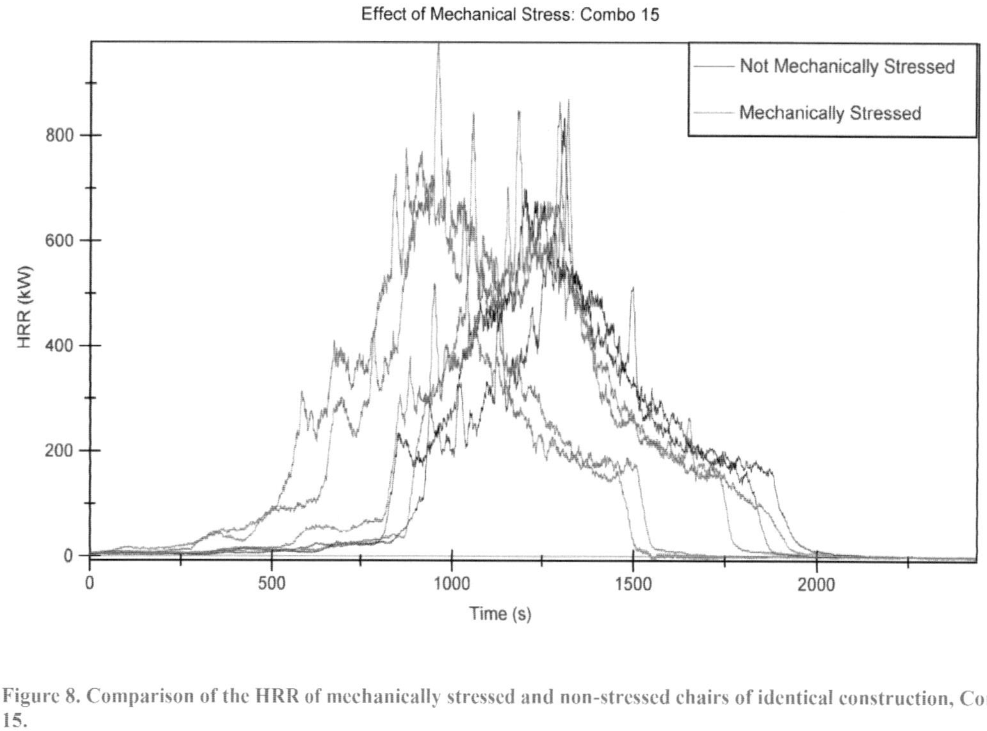

Figure 8. Comparison of the HRR of mechanically stressed and non-stressed chairs of identical construction, Combo 15.

Summary/Conclusions

This paper presented results from a recent test series conducted by CPSC staff to evaluate the performance enhanced flammability performing textile materials or "fire-blocking barrier" materials on the flammability of residential upholstered furniture. In open-flame tests, a set of five fire barriers and plain polyester batting were tested with two different cover fabrics and with and without mechanical stressing as independent variables. The barriers and cover fabrics had the largest effect on the performance of the chairs and mechanical stressing was determined to be statistically insignificant. The fire barriers generally reduced the intensity of the PHRR when compared to polyester batting for all of the open-flame testing conducted. The use of barriers increased the TTPHRR for open-flame tests. This paper did not discuss the smoldering performance of these barriers, and it should not be assumed that the open-flame performance of the barriers is an indication of smoldering performance. The smoldering performance of these barriers is discussed in our full report on this test series.[9]

The next step for CPSC staff is to examine the performance of a wider variety of barrier materials and to investigate the differences in performance with different chair construction. CPSC staff is working on a test program, including chairs with different types of blown in fill in the back cushion, which more closely represents the current residential upholstered furniture market. Those data should be made available in 2017.

References

[1] David Miller, CPSC, "2010-2013 Residential Fire Loss Estimates."
[2] Hall, Jr., J.R., The Smoking-Material Fire Problem, National Fire Protection Association, Quincy, MA, 2012.
[3] Hall, Jr., J.R., Estimating Fires When a Product Is the Primary Fuel but not the First Fuel, with an Application to Upholstered Furniture, Fire Technology, 51, 381–391, 2015.
[4] Memorandum to Dale R. Ray, Project Manager, Upholstered Furniture Project, from S. Mehta, Engineering Sciences, "Upholstered Furniture Full Scale Chair Tests-Open Flame Ignition Results and Analysis," May 2012.
[5] CPSC Meeting logs: http://www.cpsc.gov/en/Newsroom/Public-Calendar/Agendas/Upholstered-Furniture-Fire-Safety-Technology-Meeting-Agenda/.
[6] Memorandum to Andrew Lock, Project Manager, from Matthew Dreyfus, Directorate for Laboratory Sciences, Division of Chemistry, "Chemical Analysis of Fire Barriers," February 2, 2016.
[7] BS 5852:2006, Methods of test for assessment of the ignitability of upholstered seating by smoldering and flaming ignition sources.
[8] Memorandum to Dale Ray, Project Manager, Upholstered Furniture Project, from W.Tao, G. Sushinsky, B. Bhooshan, and David Cobb, Laboratory Sciences, "Cleaning and Wear Effects on Upholstery Fabric Flammability", May 2000.
[9] Memorandum to Andrew Lock, Project Manager, from Linda Fansler and Andrew Lock , CPSC "Summary Report of Open Flame and Smoldering Tests on Chairs" Directorate for Laboratory Sciences, Division of Engineering, January 28, 2016.
[10] Memorandum to Andrew Lock, Project Manager, from David Miller, CPSC Directorate for Epidemiology, Division of Hazard Analysis, "Analysis of Chair Open-Flame and Smoldering Data," January 22, 2016.

Fire-blocking Performance of Laminated Barrier Fabrics: Cone Calorimetric Characterization

Shonali Nazaré*, William M. Pitts, and John R. Shields

Fire Research Division, Engineering Laboratory, National Institute of Standards and Technology, 100 Bureau Drive, MS-8665, Gaithersburg, MD 20899-8665, USA

ABSTRACT

The fire blocking performance of barrier fabrics laminated to upholstery materials was investigated. The results are two-fold. The first involves information gathered directly from cone calorimetry tests conducted on foam/fabric composites to better understand the role played by fire-protective barrier fabrics in protecting underlying flexible polyurethane foam. Systems containing foam covered with upholstery fabric, but no fire-blocking barrier fabric, demonstrate a delay in ignition time when compare to foam alone. Once ignited, the cover fabric is the dominant contributor to the peak heat release rate of the foam/fabric composite while the foam is the largest component of the total heat released. This is in contrast to the foam/fabric composite including a laminated fire-protective barrier, where the peak heat release rate is reduced significantly. Secondly, while the cone calorimetric experiments focused on heat release rates, it was noted that the mass loss of the foam was negatively correlated with the effectiveness of the barrier fabrics. The estimated mass loss of foam during the period following the initial intense burning phase and 600 s is proposed as a means for characterizing the effectiveness of the barrier fabric in reducing the contribution of flexible polyurathane foam (FPUF) to the heat release rate (HRR). Subject to testing of additional laminated fabric combinations, this parameter could prove useful for screening upholstery combinations prior to full-scale testing and may be useful in predicting the performance of full-scale upholstery products.

1. INTRODUCTION

A promising approach for reducing the flammability of residential upholstered furniture (RUF) that avoids adding large amounts of flame retardants either to the flammable flexible polyurethane foam (FPUF) or to the cover fabric (CF) is to incorporate a fire-blocking barrier fabric (BF) into the system. Previous studies [1] have shown that the use of fire-blocking barrier fabrics is a key element in reducing the heat release parameter in calorimetric tests. However, barrier layer inclusion adds extra steps and labor to the upholstering process and may increase cost. Laminating the barrier material directly to the upholstery covering may eliminate much of the additional time and labor costs associated with implementing fire-blocking barrier fabrics.

Fabric lamination is a common process in the textile industry and involves bonding two or more flexible layers of materials with diverse functionalities to produce a single multi-functional product with a desired set of performance properties. Fire-blocking laminates can be made by bonding a highly fire resistant BF to a cover fabric. Such laminates are commonly used in upholstered seating in public places and vehicles of mass transport [2]. Alternatively, thin layers of FPUF containing fire retardant can be laminated with various textile substrates, with the inner layer containing the FPUF. These are commonly used to introduce fire-barriers into institutional mattresses [2]. When the laminates containing a BF are exposed to flame, the outer side of the laminate will burn, while the underlying BF acts to limit the involvement of underlying cushioning in the fire. Laminated barrier fabrics maintain the aesthetics of the exterior fabric while providing better fire performance.

Research efforts investigating barrier effects on flammability [1,3]4,5,6] of upholstered furniture surged in the 1990s during the development and implementation of California Technical Bulletin Cal TB 133 (1991). This bulletin sets a combustion performance standard for seating used in public occupancies. Other work, focusing on incorporating barrier fabrics in residential upholstered furniture and their effects on ignitability of foam/fabric combinations also exists [7]. The Cal TB 133 test criteria include a number of flammability properties following ignition of the upholstered furniture: peak heat release rate (PHRR), total heat release rate (THR), smoke obscuration, carbon monoxide production, and mass loss. The test calls for full-scale testing of prototype chairs which can be expensive and time consuming; making it difficult to test every combination of foam/fabric available, particularly in the RUF market.

To reduce the burden of full-scale testing, bench scale tests have been developed particularly for screening of materials to be tested in full-scale tests [8,9,10,11,12,13]. One of the earliest was proposed by Babrauskas et al [11]. Data gathered from cone-calorimeter experiments were used to predict PHRR and time to flashover on full-size mock-ups of upholstered furniture and wall lining materials. Other work has specifically attempted to correlate cone calorimetry data of FPUF/fabric mock ups to the fire-response of full sized upholstered furniture [1,6, 9,14,15,16, 17]. As efforts were made to associate mock-up, cone calorimetry measurements with full scale responses to fire, several groups have tried to determine the appropriate parameters. In one of the early works, Petrella [10] proposed an arbitrary scale based on cone calorimetry measurements: total heat released (THR) and the ratio of peak heat release rate (PHRR) to time to ignition (TTI) to assess the contribution of residential upholstery furniture to room flashover. Work continues [11,15,18] to predict scale-up response based on flashover potential, the ratio of ignition time to PHRR.

In a recent study, Chow [19] attempted to estimate the heat release rate (HRR) of furniture made with several different upholstery components based on the THR and HRR of the individual elements measured in the cone calorimeter. They were able to combine THR values of component materials to correlate with the measured value of combined materials, however this was not true of the HRR curve. They attributed

this to differences in the ability of the materials to absorb thermal radiation and hence the burning behavior when the upholstery components were combined to represent material assembly in the actual furniture.
It has been shown that the heat release rate as measured in the cone calorimeter is not an adequate predictor of full-scale performance of upholstered products [1, 17]. Numerous studies, including the most recent [20, 21], have shown that the cover fabric has the largest effect on RUF burning behavior. This is primarily due to the fact that these studies considered PHRR as the flammability parameter for comparisons. While the above studies focused primarily on highly flammable furniture materials, and hence the choice of PHRR parameter, utilizing a HRR parameter may not adequately capture the improved flammability performance that fire-blocking barriers can impart to RUF. Forsten [6] attempted to apply CBUF models [13] to predict the performance of furniture showing low or no fire propagation. It was concluded that the average HRR at 180 s after ignition correlated well with full-scale performance. However, the CBUF models failed to predict the THR and time to untenability for upholstery systems incorporating fire-blocking barriers, probably due to the choice of flammability parameters as input data in the models.

The heat release parameter does not adequately assess BF effectiveness in presence of cover fabrics. To capture the effectiveness of a given barrier fabric, a new parameter, mass loss of FPUF at a particular time, is proposed. In the present work, the fire blocking performance of barrier fabrics laminated to the upholstery fabrics has been investigated. Cone calorimetry tests were conducted on composite test specimens to investigate individual and combined effects of upholstery layers in protecting the underlying FPUF. Three different types (knitted, woven and nonwoven) of commercially available BFs were laminated to three types of cover fabrics with varying physical properties. The laminated fabrics as well as the individual component materials (barrier fabrics and cover fabrics) were tested in a FPUF/fabric composite configuration to assess fire-blocking performance. By comparing tests of composites where the cover fabric is changed or even omitted, a better understanding of component contribution to burning behavior can be obtained. The effectiveness of the different laminated BFs and the mechanisms by which they protect the flammable cushioning component will be discussed.

2. EXPERIMENTAL

2.1. Materials

Three upholstery cover fabrics were selected to include different fiber compositions, area densities, and weave patterns. The barrier fabrics were chosen to represent three different fabric structures and air permeabilities. These BFs have been characterized for their thermo-physical properties during earlier studies [22, 23]. The cover fabrics and barrier fabrics were laminated by a commercial manufacturer. With these three different cover fabrics and three BFs, nine combinations of laminated BFs were obtained. The physical properties of the laminated barrier fabrics are also given in Table 1. Pre-cut (100 mm × 100 mm × 50 mm), non-flame retarded FPUF pieces meeting Cal TB 117-2013 [10] requirements were procured from Innocor Foam Technologies, Coldwater, MS, USA.

Note that certain commercial equipment, instruments or materials are identified in this paper in order to specify the experimental procedure adequately. Such identification is not intended to imply recommendation or endorsement by the National Institute of Standards and Technology, nor is it intended to imply that the materials or equipment identified are necessarily the best available for this purpose.

2.2 Cone Calorimetry

Assembly of fabric/foam composite samples
The FPUF block was covered with fabric on the top and partially covered on four sides, leaving the bottom surface uncovered. The fabric was carefully stapled at the corners. The fabric/foam composite sample was placed in an aluminum foil pan of 10 mm depth in order to contain any liquid generated by FPUF pyrolysis

during the test. The aluminum foil pan along with the fabric/foam composite sample was then placed on a sample holder base formed from a calcium silicate board. A retainer steel frame, as described in ASTM D 1354 and ISO 5660, was placed over the fabric/foam sample, thus exposing 88 cm^2 of the top surface. The retainer frame was bolted onto the sides of the sample holder. This provided compression, qualitatively assessed by pressing, typical of cushioning effects found in real RUF. Also, the retainer frame tended to hold the top surface of the composite sample in place, thereby preventing the sample from displacing during initial heat exposure. This sample preparation was slightly different from the CBUF protocol [13] that is often employed in testing of fabric/foam composite specimens. The CBUF sample preparation requires pre-forming of the fabric shell, using glue to hold the sides of the fabric. Further, the CBUF protocol does not include the use of a retainer steel frame to provide compression.

The samples were tested at an incident heat flux of 35 kW/m^2. A spark igniter was used to ignite the test materials. If a sample self-extinguished, and no flaming was apparent, the spark igniter was re-inserted and activated until re-ignition occurred or until 10 mins had elapsed after flame extinction occurred.

Table 1. Sample specification and physical properties of various fabrics. The standard uncertainty in measuring thickness is ± 0.1 mm and that for area density is about ± 5 g/m^2.

Type of Fabric	Sample ID	Fiber Blend	Structure	Thickness, mm	Area density, g/m^2
Barrier fabrics	BF-K	FR rayon/glass fiber/PLA fiber (core spun yarn)	Knitted	1.4	165
	BF-W	Glass fiber core/ FR acrylic fiber	Woven	0.5	170
	BF-NW	FR rayon/glass fiber/PLA	Nonwoven	2.9	290
Cover fabrics	CP	53% cotton/47% polyester	Woven	1.0	382
	P	100 % polyester	Woven	1.4	320
	RP	40% rayon/60% polyester	Woven	1.0	312
Laminated barrier fabrics	BF-K/CP	FR rayon/glass fiber/PLA fiber (core spun yarn)	Knitted: Laminated	2.4	711
	BF-W/CP	Glass fiber core/ FR acrylic fiber	Woven: Laminated	2.0	581
	BF-NW/CP	FR rayon/glass fiber/PLA	Nonwoven: Laminated	3.2	746
	BF-K/P	FR rayon/glass fiber/PLA fiber (core spun yarn)	Knitted: Laminated	2.0	664
	BF-W/P	Glass fiber core/ FR acrylic fiber	Woven: Laminated	1.4	529
	BF-NW/P	FR rayon/glass fiber/PLA	Nonwoven: Laminated	3.0	667
	BF-K/RP	FR rayon/glass fiber/PLA fiber (core spun yarn)	Knitted: Laminated	2.0	647
	BF-W/RP	Glass fiber core/ FR acrylic fiber	Woven: Laminated	1.2	493
	BF-NW/RP	FR rayon/glass fiber/PLA	Nonwoven: Laminated	3.2	694

ID: identification; PLA: polylactic acid; FR: flame retardant.

Note: The thickness of fabrics was determined using a Check-line Digital Thickness Gauge (model 500-JD-A). This device measures changes in capacitance with changes in the thickness of the fabric. The change in voltage due to change in capacitance is converted to a distance. The thickness of fabrics was determined at a low measuring force of 0.6 N ± 0.3 N per unit area

3. RESULTS AND DISCUSSION

3.1 Limitations of cone calorimeter testing

Although widely used for characterization of highly flammable materials, the cone calorimeter has several limitations with regards to time response, sample size, and sample holder configuration. Materials such as those considered in the current study can provide significant experimental challenges associated with the materials' low mass, rapidly changing thickness of the sample, and burning behavior. These same sample properties often lead to rapid changes in HRR over periods on the order of seconds or tens of seconds. Such periods are comparable to the response times of cone calorimeters, with a value close to 5 s being typical. When the two periods are comparable, measured HRRs can be strongly distorted from the actual time behavior. For the sharp HRR peaks characteristic of these experiments, the measured maximum HRR would be expected to underestimate the actual value, while the time required to reach the maximum value would be overestimated [24]. The following results are based on the experimental data as recorded without correcting for time response effects. As a result, the findings represent relative characterizations for foam- fabric composites and should not be considered as absolute values.

3.2 Effects of sample holder

A number of studies [25,26,27,28,29] have reported on influences of retainer frames, insulated frames, and protective wire sample-surface grids on cone calorimetric measurements. In this study, we have focused on the effects of including a retainer frame and placing an insulating ceramic blanket beneath the composite wrapped in aluminum foil on the burning behavior of FPUF.

The effects of insulating layers and the use of a retainer frame on the shape of HRR curves are shown in Figure 1, where averaged HRR curves for three repeated tests are shown for four mounting techniques. The times to ignition (4 s), the effective heats of combustion, and the total mass loss at the end of the test are very similar for all mounting techniques. These results are consistent with observation that there was essentially complete combustion of FPUF during each test.

Figure 1. Averaged heat release data for FPUF with different sample holder configurations. The estimated standard deviation in measuring HRR is ± 10 %.

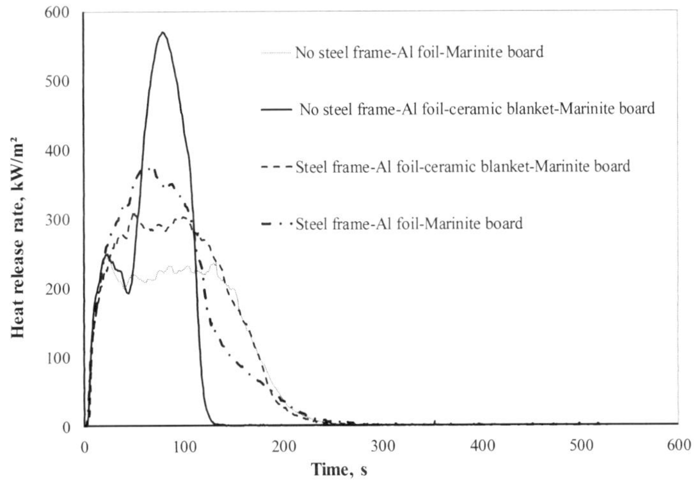

The most striking feature of the HRR curves in Figure 1 is the profile of the FPUF specimens backed with a ceramic blanket and Marinite board in the absence of a retainer frame. These specimens showed distinct two-step burning of FPUF [24, 30, 31]. After the initial surface burning of the FPUF specimen, the HRR drops slightly due to the receding FPUF surface and resultant changes in the heat flux at the surface of the specimen. Similar second sharp peaks in the HRR curves to those observed here have been attributed to burning of the liquid derived primarily from the polyol component of the FPUF [31, 32] once the FPUF has fully collapsed and the polyol component has accumulated at the bottom of the aluminum pan. However, the inclusion of the retainer frame resulted in a single broad peak of reduced intensity. This reduction in PHRR of FPUF in the presence of retainer frame has been generally attributed to the heat sink effects [28] of the frame as well as restricted air entrainment at the sides of the burning specimen [26]. In the absence of a retainer frame, large differences between cases with and without ceramic blanket suggests that heat losses from the bottom of the aliminium foil pan are important as well. The specimens tested with the retainer frame in the presence or absence of the ceramic blanket showed very little difference in burning behavior of FPUF, and this result is in agreement with Valencia et al. [27] who performed similar experiments to compare the effects of boundary conditions on the burning rate of FPUF. The retainer steel frame without ceramic blanket was chosen as the preferred sample mounting technique as it provided a uniform and reproducible cushioning effect to the FPUF and composite samples, believed to be similar to those found in actual upholstered furniture products.

3.3 Flammability of upholstery fabrics

Previous studies [6] have shown that a cover fabric can easily generate a PHRR of over 100 kW in the first 90 s of a full chair burn based on the fact that the chairs self-extinguished and the FPUF was not involved in active burning. In order to study the component contribution, flammability properties of individual cover fabrics were investigated. A summary of cone calorimetric measurements for the three upholstery cover fabrics considered in this study is shown in Figure 2. Cover fabrics (100 mm × 100 mm) were tested as a single layer of fabric placed on aluminum foil. Pieces of ceramic blanket of various thicknesses were utilized to ensure that fabrics were located at the proper height below the cone calorimeter. The retainer steel frame was used to limit configuration changes in flexible textile materials and improve measurement reproducibility. The thermoplastic polyester fabric (P) melted, whereas cotton/polyester (CP) and rayon/polyester (RP) fabrics charred in place before ignition. The time to ignition was found to be inversely proportional to the fabric weight. The total heat released and the effective heats of combustion (EHC) were in proportion to the polyester content in the fabrics. The THR per unit weight, however, can provide an indication of the relative flammability contribution the fabric will make to the upholstered product. In this context, 100 % polyester fabric (P) seems to provide a greater contribution to the total heat release of the upholstered product.

Figure 2. Derived cone calorimetry data for upholstery fabric; solid black bars represent polyester (P), hatched bar represents cotton/polyester (CP) and grey bars represent rayon/polyester (RP) CFs. Error bars represent uncertainties with experimental standard deviations.

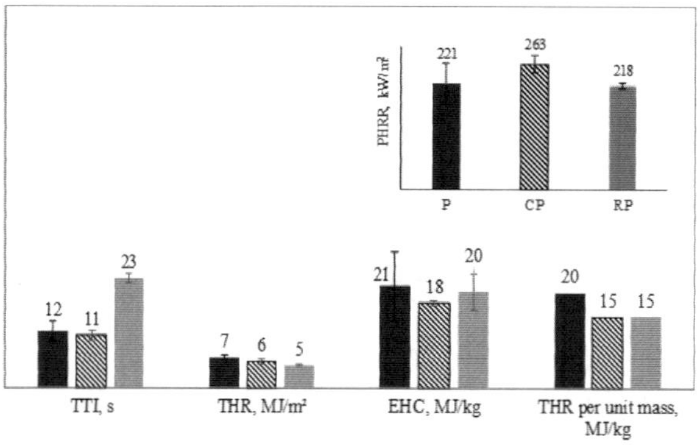

3.4 Fire-blocking performance of upholstery fabrics

The averaged HRR and mass loss (ML) data from three different tests for FPUF-CF combinations are shown in Figure 3 and derived cone calorimetric data are given in Table 2. In addition to slightly reducing the burning intensity of FPUF-CF composites, the only impact of covering the FPUF with upholstery fabric is the increase in ignition times (see Table 2). Once ignited, FPUF-CF composites with all three upholstery fabrics generated higher THR compared to FPUF alone. The THR increases is consistent to the flammable content of the composite specimen. The slightly reduced effective heat of combustion is likely due to incomplete combustion of available fuel.

Figure 3. Averaged heat release and averaged mass loss data for (a) FPUF-P, (b) FPUF-CP, and (c) FPUF-RP composites. The estimated standard deviation in measuring HRR and ML is ± 10 %.

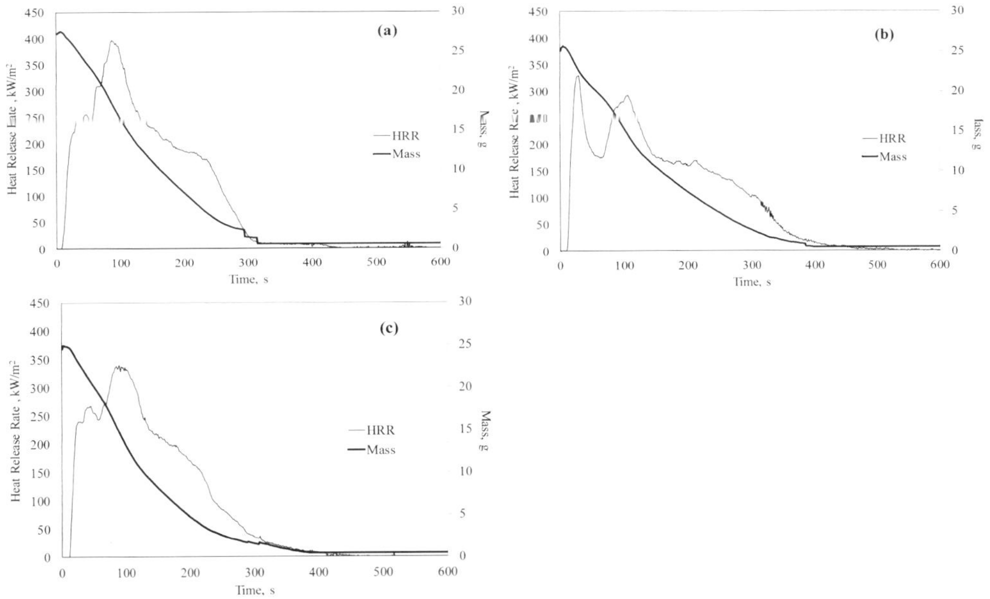

Table 2. Flammability properties of foam/fabric composites under the cone calorimeter at 35kW/m^2. Uncertainties are reported as experimental standard deviations.

Sample ID	TTI, s	PHRR, kW/m²	THR, MJ/m²	EHC, MJ/kg	Mass loss*, %	Avg. HRR _180s, kW/m²
FPUF	4 ± 1	394 ± 57	42.9 ± 0.4	25.4 ± 0.1	99 ± 2	227 ± 10
FPUF/P	10 ± 2	411 ± 46	63.1 ± 1.0	22.4 ± 0.3	92 ± 2	248 ± 4
FPUF/CP	12 ± 1	331 ± 12	59.9 ± 0.4	23.5 ± 0.4	97 ± 2	204 ± 4
FPUF/RP	13 ± 3	352 ± 31	57.9 ± 0.7	23.1 ± 0.2	96 ± 3	238 ± 15
FPUF/K	14 ± 0	168 ± 7	24.9 ± 6.3	11.9 ± 3.2	61 ± 1	39 ± 3
FPUF/W	7 ± 2	125 ± 4	21.5 ± 4.4	16.1 ± 7.0	80 ± 5	31 ± 6
FPUF/NW	10 ± 1	126 ± 4	16.8 ± 3.1	8.8 ± 5.6	63 ± 10	30 ± 8
FPUF/K-P	19 ± 2	279 ± 17	56.6 ± 4.4	19.4 ± 0.7	66 ± 5	70 ± 4
FPUF/W-P	16 ± 1	235 ± 10	53.6 ± 3.6	18.0 ± 1.7	75 ± 10	58 ± 6
FPUF/NW-P	16 ± 3	256 ± 4	49.4 ± 19.5	15.6 ± 2.4	78 ± 18	6 ± 2
FPUF/K-CP	19 ± 2	364 ± 10	52.6 ± 0.9	21.6 ± 4.6	72 ± 3	79 ± 4
FPUF/W-CP	16 ± 1	346 ± 2	47.9 ± 3.4	16.7 ± 1.1	77 ± 4	60 ± 6
FPUF/NW-CP	18 ± 2	351 ± 14	50.7 ± 3.9	15.9 ± 1.4	73 ± 2	63 ± 2
FPUF/K-RP	20 ± 1	326 ± 10	63.9 ± 9.1	17.6 ± 0.8	73 ± 1	72 ± 2
FPUF/W-RP	15 ± 1	325 ± 6	53.9 ± 8.2	17.9 ± 2.3	83 ± 6	56 ± 9
FPUF/NW-RP	19 ± 3	329 ± 1	49.1 ± 0.4	15.6 ± 0.6	78 ± 2	57 ± 0

ID: identification, TTI: time to ignition, FO: flame out, PHRR: peak heat release rate, THR: total heat release rate, EHC: effective heat of combustion.
* Mass loss for complete composite.

All three FPUF-CF combinations have two peaks. The first peak is primarily assigned to the burning of the cover fabric (peak$_{fabric}$), and the second peak is due to combustion of FPUF (peak$_{foam}$) [1, 15]. The separation of these two peaks, however, is determined by the burning behavior of the cover fabric and how well the cover fabric protects the FPUF from the radiant heat. A more prominent separation of the fabric and FPUF peaks can be noted in Figure 3(b). The burning behavior of the cotton/polyester cover fabric (CP) appears to be dominated by the cotton content and, more importantly, the area density of the fabric. The CP charred in place, and the char protected the FPUF from heat and flames. The protective property of this char could be attributed to the area density and weave pattern of the cover fabric. In the case of the thermoplastic fiber fabric in the FPUF-P composites, the FPUF was exposed to the radiant heat as soon as the cover fabric melted. The cover fabric and the FPUF burned almost concurrently and hence the two peaks are not as distinct. Thus, it can be concluded that the cover fabric has some fire-blocking effect, depending on the fiber content, area density, and weave pattern, for FPUF-CF composites exposed to radiant heat flux. Cellulosic fiber containing cover fabrics provide some protection due to their char forming characteristic.

3.5 Fire-blocking performance of barrier fabrics

Woven Barrier Fabric

When initially exposed to radiant heat from the cone heater, weak intermittent flames were observed on the surfaces of FPUF-W composites before sustained flaming ignition was recorded at 7 s ± 2 s. The woven barrier fabric (BF-W) is a thermally thin, highly permeable barrier. This BF has a high heat transfer factor and lower thermal protective index of 1.6 [22] compared to the other two BFs considered in this study. The rapid heat transfer through such a barrier is sufficient to initiate significant thermal decomposition of FPUF. The combustible gases generated due to thermal decomposition of FPUF are easily transported to the region of the ignition source. However, the BF-W contains flame retardant treated fibers that release gases known to operate in the gas phase. These flame retardant chemicals initially interfere with the combustion reactions to delay sustained flaming ignition of the composite specimen. Once the flame retardants were depleted, sustained ignition occurred. The composite specimens burned only briefly for 120 s ± 10 s before the flames extinguished. The spark igniter was re-inserted and energized to reignite any combustible gases being generated. The flame extinguishment, the re-insertion of spark igniter, and removal of spark igniter after re-ignition of the specimen are clearly marked in Figure 4 (a) to 4 (c) showing averaged HRR curves. The spark igniter was inserted twice following the first flame extinguishment for BF-W composites. During this period, weak flaming was observed. This weak burning is corroborated by the very low HRR (HRR < 50 kW/m^2) recorded during this period. As can be noted from the mass loss curve in Figure 4 (a), the composite specimen continued to lose mass after the second flame extinction, indicating that the FPUF continued to pyrolyze. However, pyrolysis gases were not generated in sufficient concentration and/or mixed with the necessary air concentration to support flaming combustion. At the end of the test, the average mass loss of the specimens was 80 % ± 5 %.

Knitted Barrier Fabric

The knitted barrier fabric (BF-K) has a lower permeability as compared to the woven BF-W. From Table 2, it can be noted that these composites ignited after a 14 s exposure. The increased delay in ignition compared to BF-W composite could be due to the combined effects of low gas permeability, retarded combustion reactions due to the presence of FR treated fibers, and the fact that the knitted BF provided a better thermal barrier (thermal protective index, TPI = 1.8) compared to the woven BF-W (TPI =1.6) [22]. The FPUF-K composite initially burned weakly (HRR < 168 kW/m^2 ± 7 kW/m^2), and the flames extinguished within 180 s ± 4 s. After the spark igniter was reinserted, sustained flaming ignition was not recorded for almost 450 s. Occasional, short-lived, flashing flames, similar to those observed during the initial ignition period, were observed on the surface of the specimen. The FPUF under the BF continued to pyrolyze and lose mass (see Figure 4 (b)) during this period. As the BF-K was being heated, the organic fibers on the sheath of the core-spun yarn knitted barrier were completely oxidized and consumed during the initial burning period and subsequent heating. Visual observations at the end of the test indicated that the knitted structure of the glass filament core remained. The gas permeability of such thermally decomposed BF-K is believed to be higher than that of pristine BF-K. With increased gas permeability, the combustible gases could more easily be transported to the spark igniter and flaming re- ignition occurred after 721 s ± 12 s. However, the burning intensity was remarkably low (< 50 kW/m²). The second burning phase yielded a higher THR value for the FPUF-K composite. Mass loss at the end of the tests for this composite was 61 % ± 1 %. The lower mass loss of FPUF-K compared to that of BF-W suggests that more protection was provided by the knitted BF.

Non-woven Barrier Fabric

Similar to the woven and knitted BFs, FPUF-NW initially ignited and burned for a brief period (140 s ± 40 s), yielding a sharp peak in the HRR curve. The spark igniter was re-inserted as soon as the flames extinguished. In this case, it took more than 10 min for the composite to reignite. During this period the underlying FPUF was heated sufficiently to continue pyrolyzing, as indicated by the continuing mass loss. The thermal protective index of the non-woven barrier (TPI = 5.6) is the highest compared to the other two barriers [22] considered in this study. However, as the BF-NW was heated and organic fibers were consumed in the initial burning period, the gas permeability of the BF-NW would be expected to increase. The combustible gases trapped under the non-woven barrier ignited sporadically as and when brief periods of combustible mixtures were present at the spark igniter. While the flaming re-ignition occurred at the spark igniter, unstable flaming was observed inside the sample holder under the barrier. The visual observation of flaming combustion under the BF-NW is reflected in the very small HRR (< 10 kW/m²) recorded during this period. The specimens reignited after 15 min of exposure, and the residual gases burned on the specimen surface with very low intensity before extinguishing. The average mass losses for these composites were the lowest observed.

Figure 4. Averaged heat release and averaged mass loss curves for (a) FPUF-W, (b) FPUF-K, and (c) FPUF- NW composites. The estimated standard deviation in measuring HRR and ML is ± 10 %.

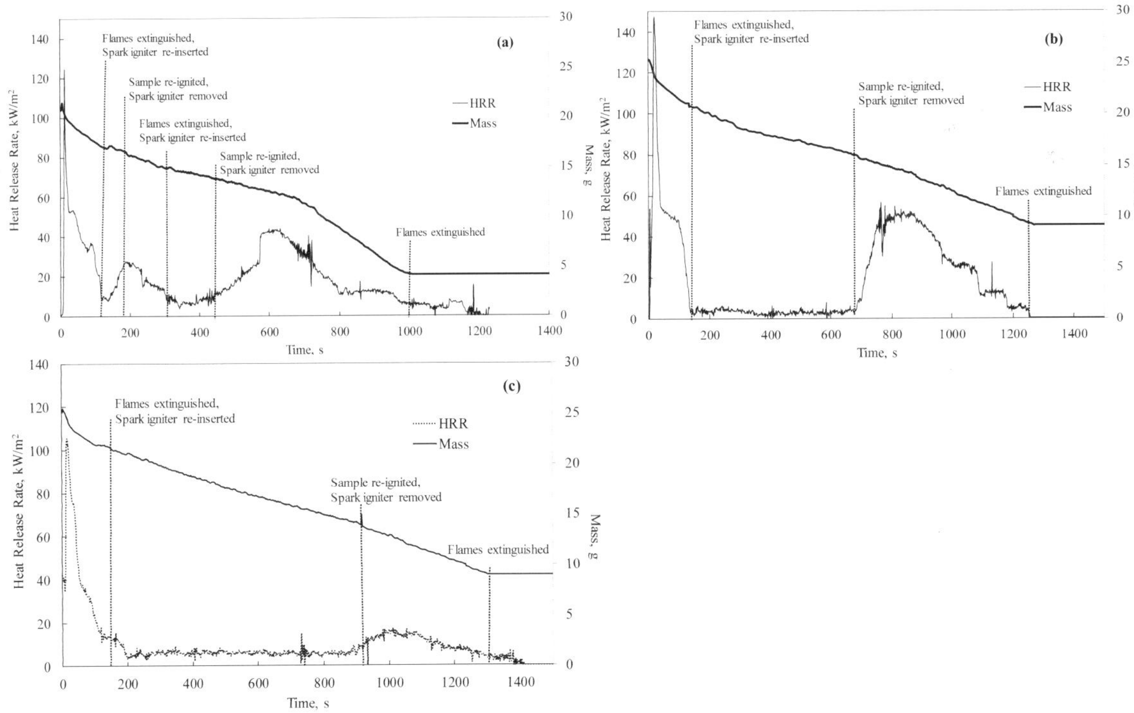

3.6 Fire blocking performance of laminated barrier fabrics

Cone calorimeter results for the nine combinations of BFs and CFs tested with FPUF are given in Table 2. It can be noted that the PHRR values recorded for the various combinations differ only slightly. Comparison of the HRR and ML curves in Figure 5 for the composites covered with BFs laminated with CP fabric suggests that following the initial sharp HRR peaks, the shapes of the curves are dominated by the type of barrier fabric. The timing of the initial characteristic peaks suggests that they are associated primarily with burning of the cover fabrics. Once the cover fabric was consumed, the protective performance of barrier fabrics was largely determined by their mass and heat transfer properties.

Figure 5. Averaged HRR and averaged mass loss curves for (a) FPUF-W-CP, (b) FPUF-K-CP, and (c) FPUF-NW-CP. The estimated standard deviation in measuring HRR and ML is ± 10 %.

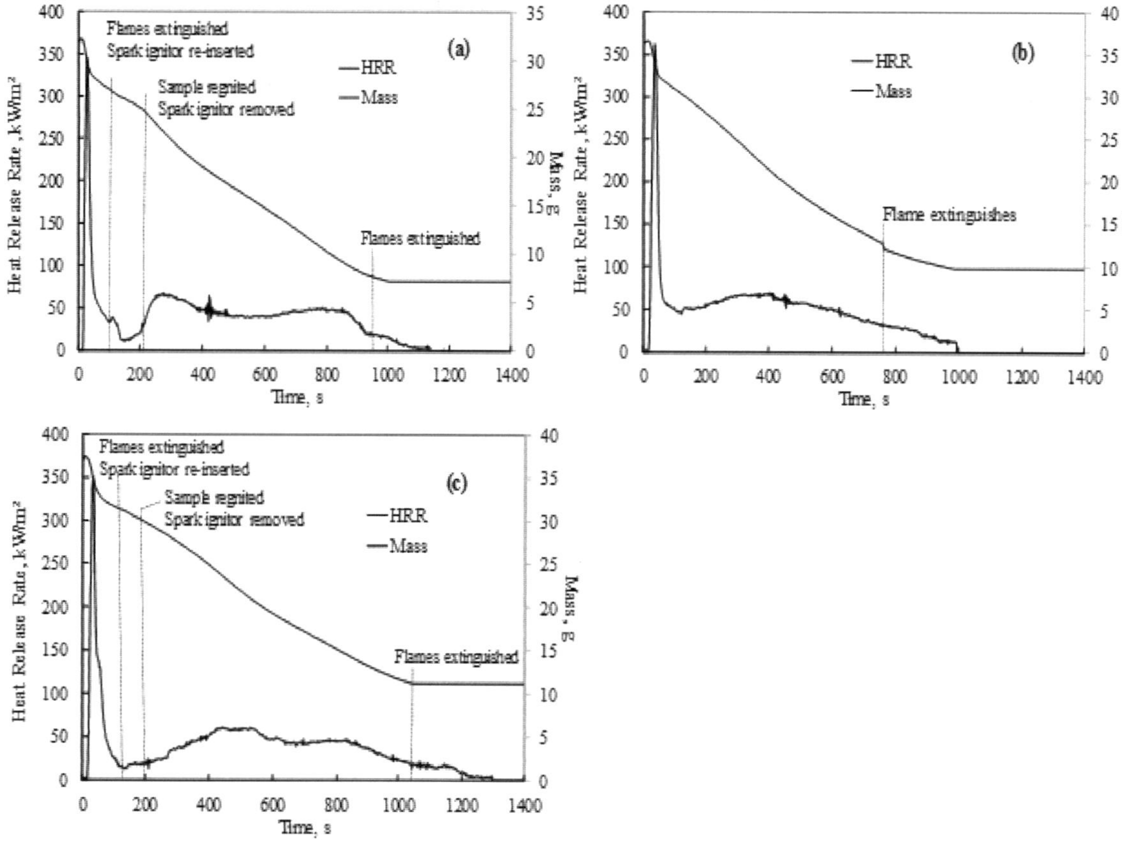

It is clear from Figure 5 that the peaks in the HRR curves do not fully characterize the overall burning behavior of the specimens. However, differences in the fire-blocking performance of laminated barrier fabrics do appear to be captured by the HRR curves following the peaks. The PHRR, although used frequently to compare the fire hazard of such composites [33], does not contain complete information and in some cases could be misleading.

When compared with FPUF composites covered with BFs alone, the laminated barriers are not as effective. Particularly, composite FPUF-K-CP continues to burn after the initial peak generated by the cover fabric. The composite does not self-extinguish as seen in cases of FPUF-W-CP and FPUF-NW-CP. As noted from Figure 5 (a) and Figure 5 (c), the specimen is re-ignited and sustained burning continues for a long time. The covering component on the sides of the FPUF apparently provides sufficient additional fuel for the burning to continue. The HRR is however very low indicating the, the laminated BFs provide significant reductions in flammability as compared to composites with cover fabric alone.

The role of various cover fabrics, barrier fabrics, and laminated barrier fabrics in protecting FPUF can be characterized in terms of changes in flammability properties with and without the covering component. Typical indicators for reaction-to-fire properties such as the PHRR, THR, and mass loss (ML) are considered. The changes in PHRR, THR, and ML of FPUF in the presence of the various covering components are shown in Figure 6 (a) to 6 (c). Negative values indicate that the covering component resulted in improved flammability properties while positive values indicate poorer flammability properties. It can be clearly seen from Figure 6 (a) that the BFs (K, W, and NW) significantly reduced the PHRR as compared to composites with cover fabrics alone. In those cases with the thermoplastic cover fabric (P) alone, the PHRR was slightly increased, suggesting an increased fire hazard of FPUF in combinations with thermoplastic cover fabrics. The laminated barrier fabrics also reduced the PHRR of the FPUF, but not as efficiently as the barrier fabrics alone, clearly due to HRR contributions from the cover fabrics.

The total heat release rate, on the other hand, is related to the amount of material that burned. Comparison of data in Figure 6(b) suggests that apart from BFs, all other covering components in FPUF composites, increased the THR of the specimen. The THR data in Figure 6(b) shows the fire-blocking effects of barrier fabrics in the absence of cover fabric. The BF limits exposure of the FPUF to heat and flames and can lead to extinction of FPUF combustion.

Figure 6 (c) shows changes in ML percent for all composites as compared to FPUF. Negative ML percent values for all composite specimens suggest some char yield, proportionate to the original specimen mass, whereas the FPUF was completely consumed during burning. This total mass loss corresponds to the difference between the specimen and the residual mass. It does not differentiate the mass loss of foam and mass loss of the covering component. And therefore, this parameter is not useful in characterizing the protective performance of barrier fabrics. A more detailed mass loss analysis for assessing protective performance of barrier fabrics is described in the section below.

Figure 6. Changes in (a) PHRR, (b) THR, and (c) Mass Loss % for FPUF composites as compared to FPUF alone tested under 35 kW/m² heat flux. Error bars represent uncertainties with experimental standard deviations.

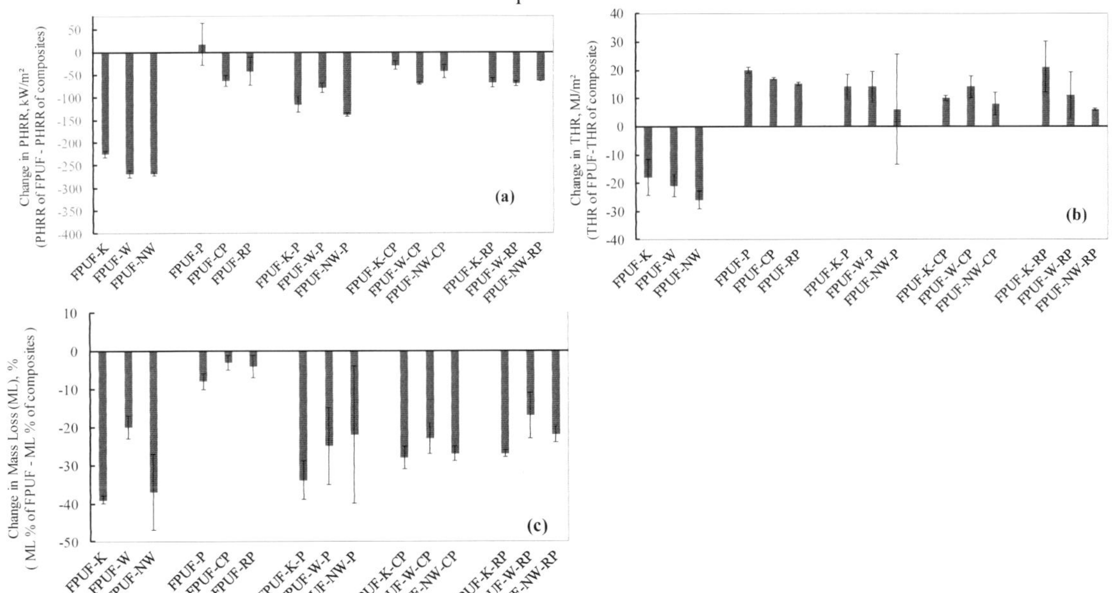

3.7 Mass loss analysis

As discussed above, the peak heat release rate of the FPUF/BF/CF composites in our test configuration is dominated by the burning behavior of the cover fabric. Therefore, the PHRR may not adequately capture or assess the protective performance of barrier fabrics. An alternative parameter that approximates the mass loss of foam, is therefore proposed to capture the effectiveness of barrier fabrics in protecting the underlying FPUF. Since the early flaming is dominated by burning of the laminated fabrics, the mass loss of FPUF was estimated by considering the mass loss following the initial phase of composite burning dominated by the cover fabric. The covering component burning period was assumed to be essentially complete by 60 s. This was assessed from the HRR curves of the covering components (not shown here) and the end of the rapid mass loss rate in Figure 5 (a) to 5 (c). The approximate percent mass loss of foam, excluding any mass loss of cover fabrics was calculated using the following equation:

$$\%\ \text{mass loss of foam} = \frac{(\text{mass at 60 s} - \text{mass at given time}) * 100}{\text{mass at 60 s}}$$

The estimated mass loss of foam at different times during cone calorimetry testing for the various composite combinations are shown in Figure 7 (a) to 7 (c). Uncovered FPUF was completely consumed at 180 s. The cover fabrics protected the FPUF marginally up to 180 s. In contrast, barrier fabrics protected the FPUF fairly well. At 300 s, the average percent of estimated mass loss of foam for composites covered with BFs was 16 % ± 4 %. While BF-K and BF-NW continued to protect the FPUF, BF- W failed by splitting and exposing the underlying FPUF, and the foam was completely pyrolyzed by 600 s. The estimated FPUF mass loss analysis thus confirms that BF-K and BF-NW protected the foam better than BF-W. Comparing the mass loss data for composites tested with laminated barriers (Figure 7 (c)) to those with no barriers (Figure 7 (a)), is consistent with the BFs playing a significant role in protecting the foam.

Figure 7. Percentage of mass loss of foam as a function of time for (a) FPUF and FPUF-CF, (b) FPUF-BF, and (c) FPUF-BF-CF composites. Values are reported with ± 10 % uncertainty.

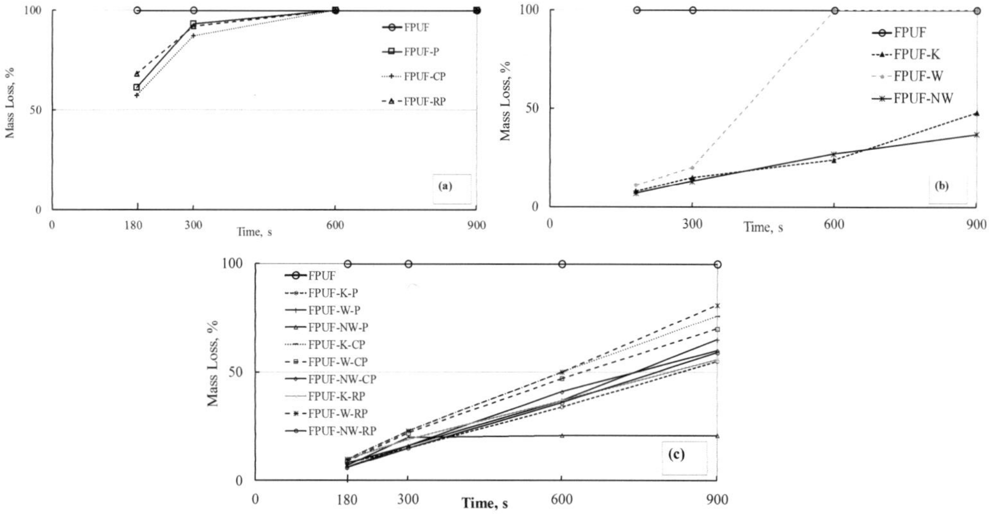

4. CONCLUSIONS

The fire-blocking performance of three different types of barrier fabrics (knitted, woven and nonwoven) laminated to three types of upholstery fabrics with varying physical properties was investigated. Cone calorimetry tests were conducted on composite test specimens to investigate individual and combined effects of upholstery layers in protecting the FPUF. The retainer steel frame without ceramic blanket was chosen as the preferred sample mounting technique as it provided a uniform and repeatable cushioning effect to the sample, similar to those present in actual upholstered furniture products.

Composites with upholstery fabric covered FPUF had longer ignition times than those with FPUF alone. However, once ignited, the cover fabrics provided limited or no fire-blocking effect on FPUF-CF combinations. FPUF covered with thermoplastic cover fabric generated the highest PHRR. It is important to note here that although the thermoplastic cover fabrics resist smoldering ignition and therefore would not require any barrier according to the current Cal TB 117-2013 standard [34], these FPUF/fabric combinations could be extremely hazardous when exposed to flaming ignition sources.

For composites covered with BFs alone, a brief flaming period was observed following ignition, which was followed shortly by flame extinguishment. However, in our experiments, the spark igniter was reinserted and activated to ignite any combustible gases that were being generated. Although the foam continued to pyrolyze, the pyrolysis gases were not generated in sufficient concentration and/or mixed with the necessary concentration to support sustained flaming combustion. Once sustained flaming occurred, only a very small amount of heat was released before flame extinction. Generally, all three barriers tested in this study are highly effective in reducing the flammability of the FPUF.

Generally, in terms of protecting the FPUF, laminated barriers are not as effective as BFs alone, however, BFs are significantly better than the cover fabrics alone. Composites covered with laminated barrier fabrics do not show significant reduction in flammability as assessed in terms of the heat release rate. Moreover, protective properties of laminated barriers deteriorated when compared to FPUF composites tested with BFs only primarily due to the fact that the burning behavior of the cover fabrics plays a dominant role in PHRR of FPUF/fabric composites. However, the estimated mass loss of FPUF during the period following the initial intense burning phase associated with fabric combustion developed above provides a potential quantitative characterization for the fire-blocking performance of BFs. The average estimated mass loss percentage of FPUF at 600 s during these tests provided a quantitative measure that could allow classification of laminated fabric effectiveness. Subject to testing of additional laminated fabric combinations, this parameter could prove useful for screening upholstery combinations prior to full-scale testing and may prove useful in predicting the performance of full-scale upholstery products.

5. REFERENCES

[1] Ohlemiller, T.J. and Shields, J.R., Behavior of mock-ups in the California Technical Bulletin 133 test protocol: Fabric and barrier effects. NISTIR 5653, 1995, National Institute of Standards and Technology, Gaithersburg, MD 20878.

[2] Nazaré, S. and Davis, R., A review of fire blocking technologies for soft furnishings. Fire Science Reviews, **1**:1 (2012).

[3] Eggestad, J. and Johnsen, A.C., Effects of interliners on the ignitability of upholstered furniture. Journal of Fire Sciences, **5**, 152-162 (1987).

[4] Gallagher, J.A., Interliner effect on the fire performance of upholstery materials. Journal of Fire Sciences, **11**, 87-105 (1993).

[5] Fesman, G. and Jacobs, B.A., Evaluating the flammability of upholstered furniture intended for public occupancies. Journal of Fire Sciences, **7**, 414-425 (1989).

[6] Forsten, H.H., Use of small scale testing to predict Cal 133 performance. Fire and Materials, **21**, 153-160 (1997).

[7] Fansler L., Exploratory Tests of Barrier Materials. Memorandum to Dale Ray, Project Manager, Upholstered Furniture Directorate for Economic Analysis. Consumer Product Safety Commission Washington DC, 20207. (October 2004)

[8] Gallagher, J.A., Minimum flux for fire propagation: A new parameter for classification of foam/fabric composites. Journal of Fire Sciences, **10**, 40-57 (1992).

[9] Vanspeybroeck, R., Hees, P.V. and Vandevelde, P., Combustion behavior of Fabric and polyurethane flexible foam mock-up combinations under cone calorimetry test conditions. Fire and Materials, **17**, 167-172 (1993).

[10] Petrella, R.V., The assessment of full-scale fire hazard from cone calorimeter data. Journal of Fire Sciences, **12**, 14- 43 (1994).

[11] Babrauskas, V., Bench-scale methods for prediction of full-scale behavior of furnishings and wall linings, SPFE Technical Report 84-10 (1984) Boston, MA.

[12] Babrauskas V., Baroudi D., Myllymaki J. and Kokkala M., The cone calorimeter used for predictions of the full-scale burning behavior of upholstered furniture. Fire and Materials, **21**, 95-105 (1997).

[13] Fire Safety of Upholstered Furniture- the final report on the CBUF research Programme. Edited by Bjorn Sundstorm, European Commission Measurement Testing Report EUR 16477 EN, 1995.

[14] Price, D., Liu, Y., Hull, T.R., Milnes, G.J., Kandola, B.K. and Horrocks, A.R., Burning behavior of fabric/polyurethane foam combinations in the cone calorimeter. Polymer International, **49**, 1153-1157 (2000).

[15] Kotresh, T.M., Indushekar, R., Subbulakshmi, M.S., Vijayalaxmi, S.N., Krishna Prasad, A.S., Padaki, V.C. and Agrawal, A.K., Effect of heat flux on the burning behavior of foam and foam/Nomex III fabric combination in the cone calorimeter. Polymer Testing, **25**, 744-757 (2006).

[16] Coles, A.R., Flammability of Upholstered Furniture Using the Cone Calorimeter, Fire Engineering Report 2001/1, Canterbury University, Christchurch, New Zealand, (2001).

[17] Janssens M. L., Reducing Uncertainty of Quantifying the Burning Rate of Upholstered Furniture, Document Number 239050, U. S. Department of Justice, Washington, DC, 208 pp, (July 2012).

[18] Ostman, B.A.-L., Nussbaum, R.M., Correlation between small scale rate of heat release and full scale room flashover for surface linings, in : Fire Safety Science- Proceedings of Second International Symposium, Hemisphere Publishing, New York, 823-832 (1989).

[19] Chow, W.K., Assessment on heat release rate of furniture foam arrangement by a cone calorimeter. Journal of Fire Sciences, **20**, 319-328, 2002.

[20] Storesund, K., Steen-Hansen, A. and Bergstrand, A., Fire Sate Upholstered Furniture: Alternative strategies to the use of chemical flame retardants. SP Fire Research Report A15 20124:2. Available from: http://en.spfr.no/media/publikasjoner/upload/2015/rapport-spfr-a15-20124-2.pdf

[21] Fabian, T., Upholstered Furniture Flammability, Research Report, Underwriters Laboratory Inc., (July 31st 2013).

[22] Nazaré, S.; Pitts, W.; Flynn, S.; Shields, J.; Davis, R. D., Evaluating fire blocking performance of barrier fabrics. Fire and Materials, **38**(7), 695-716 (2014).

[23] Nazaré, S.; Pitts, W.; Matko, S.; Davis, R. D., Evaluating smoldering behavior of barrier fabrics. Journal of Fire Sciences, **32**(6), 539-**562 (2014).**

[24] Pitts, W.M. Applied heat flux distribution and time response effects on cone calorimeter characterization of a commercial flexible foam. Fire Technology. **50**, 635-672 (2014). DOI: 10.1007/s10694-011-0235-8.

[25] Tsantaridis, L. and Ostman, B., Communications: Retainer frame effects on cone calorimeter results for building products. Fire and Materials, **17**, 43-46 (1993).

[26] Ezinwa, J.U., Modelling full-scale fire test behavior of polyurethane foams using cone calorimeter data, A thesis submitted to Department of Mechanical Engineering, University of Saskatchwen, Saskatoon, May 2009.

[27] Valancia, L.B. Rogaume, T. Guillaume, E. Rein, G. and Torero, J.L., Analysis of principal gas products during combustion of polyether polyurethane foam at different irradiance levels, Fire Safety Journal, **44**, 933-940, (2009).

[28] Babrauskas, V., Twilley, W.H. and Parker, W.J., The effects of specimen edge conditions on heat release rate. Fire and Materials, **17**, 51-63 (1993).

[29] Nazaré, S., Kandola, B.K., and Horrocks, A.R., The use of cone calorimetry to quantify the burning hazard of apparel fabrics. Fire and Materials, **26** (4-5), 191-199 (2002).

[30] Ravey, M.; Pearce, E.M., Flexible polyurethane foam. I. Thermal decomposition of a polyether-based, water-blown commercial type of flexible polyurethane foam. Journal of Applied Polymer Science, **63** (1), 47-74 (1997).

[31] Kramer, R.H., Zammarano, M., Linteris, G.T., Gedde, U.W., Gilman, J.W., Heat release and structural collapse of flexible polyurethane foam. Polymer Degradation Stability, **95**, 1115-1122 (2010).

[32] Pitts, W.M., Role of two stage pyrolysis in fire growth on flexible polyurethane foam slabs. Fire and Materials, **38**: 323-338 (2014).

[33] Schartel, B. and Hull T.R., Development of fire-retarded materials-Interpretation of cone calorimeter data. Fire and Materials, 31, 327-354 (2007).

[34] State of California Department of Consumer Affairs. *Technical Bulletin Cal-117–2013: Requirements, Test Procedure and Apparatus for Testing the smolder Resistance of Materials Used in Upholstered Furniture.* Bureau of Electronic and Appliance Repair Home Furnishings and Thermal Insulation: Sacramento, CA, USA (2013).

INFLUENCE OF MODERN FURNITURE ON THE FIRE DEVELOPMENT IN FIRES IN HOMES LARGE SCALE FIRE TESTS IN LIVING ROOMS

Anja Hofmann[*], Sven Kaudelka[*], Frederik Rabe[*], Tanja Gnutzmann[*], Andrea Klippel[+]

*Bundesanstalt für Materialforschung und -prüfung (BAM), Unter den Eichen 87, 12205 Berlin, [+]Otto von Guericke University, Postbox 4120, 39106 Magdeburg, Germany

ABSTRACT

About 80 % of all fire fatalities in Germany occur in fires in homes[1]. UK statistics show that living and bed room fires are more often responsible for fatalities than kitchen fires although kitchen fires occur much more often[2]. It has been known for some time that modern materials tend to burn differently from older materials and it has been acknowledged that the amount of combustible plastics in homes has increased significantly over the last decades[3-7]. To investigate the influence of modern furniture and ventilation conditions of fires in homes a series of four large scale tests in two living rooms with adjacent rooms was performed by BAM and the Frankfurt fire service[8]. Two living rooms, one with older furniture and one with modern furniture were tested twice each. Each test started with the ignition of a paper cushion on an upholstered chair. The influence of modern materials on the fire development was investigated as well as the influence of the ventilation on the fire development. Two tests with closed windows and two tests with open windows were performed. Temperatures were measured in the living rooms and the adjacent rooms, gas compositions were measured in the adjacent rooms and videos were taken in all rooms. The ventilation as well as the different materials influenced the fire development regarding temperatures in the rooms and smoke production and composition significantly. The fire in the living room with modern furniture developed faster than the same setting with older furniture. More ventilation (open window) led to higher temperatures in the rooms and faster fire development as well. Smoke gas composition was measured in the adjacent rooms to assess the positive effect for a person being in the adjacent room and not in the room of fire origin. In all settings an upholstered chair was the first burning item. The modern upholstered chair was investigated in three pre-experiments under different conditions to enhance the understanding of the living room fires. The measured mass losses and derived mass loss rates have been used for numerical modelling of the pre-experiments.

INTRODUCTION

Fires in homes are hazardous: about 80 % of all fire fatalities in Germany occur in these fires[1]. UK statistics which are much more detailed than German statistics show that living and bed room fires are more often responsible for fatalities than kitchen fires although kitchen fires occur much more often[2]. It has been known for some time that modern materials tend to burn differently from older materials and it has been acknowledged that the amount of combustible plastics in homes has increased significantly over the last decades[3-7]. A test series of four experiments in living rooms with adjacent rooms has been performed with the Frankfurt fire service and BAM, the German Federal Institute for Material Research and Testing, to investigate on hand the influence of older and modern furniture and ventilation conditions on the fire and smoke development of fires in homes and on the other hand the to assess the differences in conditions for inhabitants who are in the room of fire origin or in the adjacent room. In all four large scale experiments a similar furnishing of the living room has been used, in two experiments older furniture was used and in two experiments modern furniture. The experiments had different ventilation conditions which were realised by open and closed windows, see next section for details. With temperature measurements in the room of fire origin and the adjacent room and gas measurements in the adjacent room and video footage inside the rooms the fire and smoke development could be assessed. In all four experiments an upholstered chair was chosen as first burning item. Additionally, to the series of large scale experiments three pre-experiments with a modern upholstered chair (first burning item in the tests with modern furniture) has been performed. Three chairs identical to the ones which were used in the large scale test series has been tested under different conditions: in a big hall, in a small room with the front wall open and in the same room with a front wall and a door opening in it, details are in section *Experimental investigation of first burning item: upholstered chair*. The mass loss rate was measured in all three tests. These pre-tests were performed to have a better understanding of the fire in the living room, especially with the aim to model the room fires numerically in the future. Often the mass loss rate of the first burning item or the heat release rate is used in numerical calculations of room fires. First step was to model the small room fires (pre-tests) numerically. The measured mass loss rate was used to predict the fire development and the temperatures in the small room tests.

SERIES OF LARGE SCALE FIRE TESTS IN LIVING ROOMS

Test set-up

A series of large scale experiments was performed with the Frankfurt fire service and BAM as part of the TIBRO project which was funded by the German Federal Ministry of Education and Research.

The living room (test room) was furnished with older furniture (about 30 years old) and modern furniture, figure 1. The fire load was 16.6 kg/m² for the modern furniture and 12.6 kg/m² for the older furniture. Each living room had an adjacent room to assess the temperature and smoke development in the adjacent room additionally to the room of fire origin, figure 2.

Figure 1: Living room with older furniture (left), with modern furniture (right)

Figure 2: Test site, building with two living rooms and two adjacent rooms and one measurement room in the middle (left), living room with adjacent room - schematic (right)

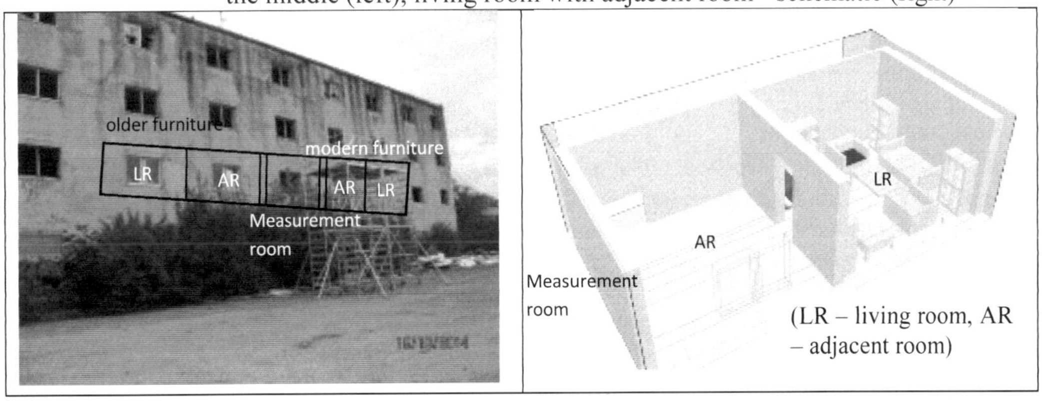

Overview over the measurements and locations gives figure 3. Temperatures were measured in three different locations of the living rooms and the adjacent rooms in three heights (1.25 m, 1.72 and 0.1 m below the ceiling). Gas measurements were done in the adjacent rooms in two different heights (1.25 m and 1.72 m).

Figure 3: Test set-up, location of gas measurements (left hand side) and locations of temperature measurements (right hand side) in vertical projection (top row) and horizontal projection (bottom row)

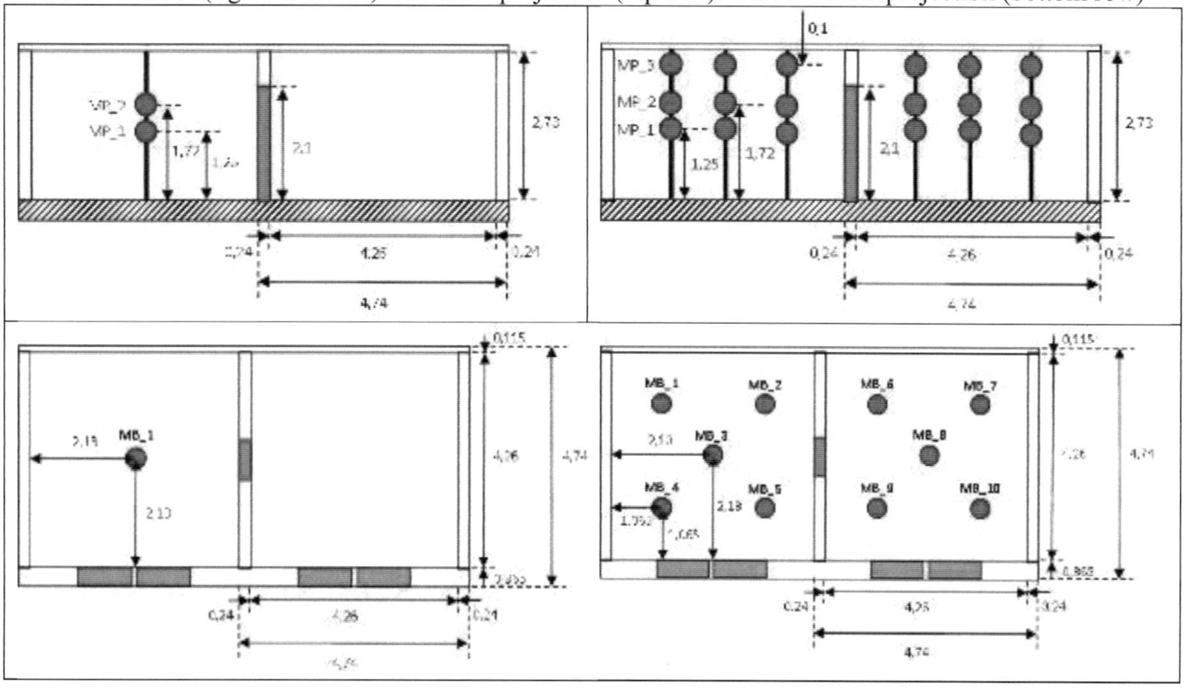

In all four experiments, see table 1, an upholstered chair was ignited by a burning paper cushion. The paper cushion was a 100 gr cushion made of newspaper (similar to the ignition source for train seats according to DIN 5510). Three pre-tests with the modern upholstered chair under different conditions were performed. These tests are described in the section experimental investigations of burning upholstered chair. For the experiment 1_2 with the older furniture one window was open, for the experiment 2_2 with the modern furniture two windows were open. Each window has a size of 1.1 m to 1.4 m and the door between the rooms has a size of 0.8 m to 2.1 m.

Table 1: Overview over test series in living rooms

Experimental set-up	Closed windows	Open windows
Living room, older furniture	**experiment 1_1**	**experiment 1_2**
Living room, modern furniture	**experiment 2_1**	**experiment 2_2**

Results of the experiments

In experiments 1_1, 2_1 and 2_1 only the upholstered chair burnt. Only in experiment 2_2 a fully developed fire was established. In figure 4 the measured temperatures and the measured gas concentrations for the experiments with the older and the modern furniture with closed windows are shown. The temperatures are averaged over the measurement points lying in the hot gas layer and the measurement points lying below the hot gas layer, respectively. Temperatures had been measured in the room of fire origin and the adjacent room. Smoke composition had been measured in the adjacent room in two heights in the centre of the room.

As it can be seen in figure 4 (top row) the room temperatures in the experiment with modern furniture and closed windows are significantly higher than in the experiment with older furniture and closed windows. This is true for the room of fire origin and the adjacent room. Maximum temperature in the hot gas layer in the living room for the older furniture is about 125 °C and for the modern furniture about 300 °C. After 300 s the maximum temperature is already reached for the experiments with the modern furniture and for the experiment with the older furniture the maximum temperature is reached only after 600 s. The gas concentrations of CO, CO2 and HCN are shown in figure 4 (centre and bottom row) for the two measured heights (1.25 m and 1.72 m) in the adjacent room for the experiment with the older furniture and closed windows and the experiment with the modern furniture and closed windows. The gas concentrations for the 1.25 m height are in the same order for both experiments although the slope is steeper for the experiment with the modern furniture. In 1.72 m height the gas concentrations are significantly higher in the experiment with modern furniture and closed windows than in the experiment with the older furniture and closed windows as can also be seen in figure 4 (bottom row).

Figure 4: Experimental results for the experiments with older and modern furniture and windows closed

Comparison between older furniture and modern furniture (experiments 1_1 and 2_1), windows closed

Older furniture	Modern furniture

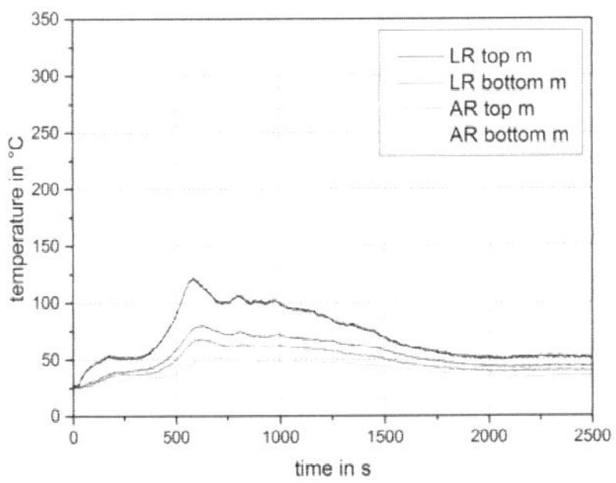

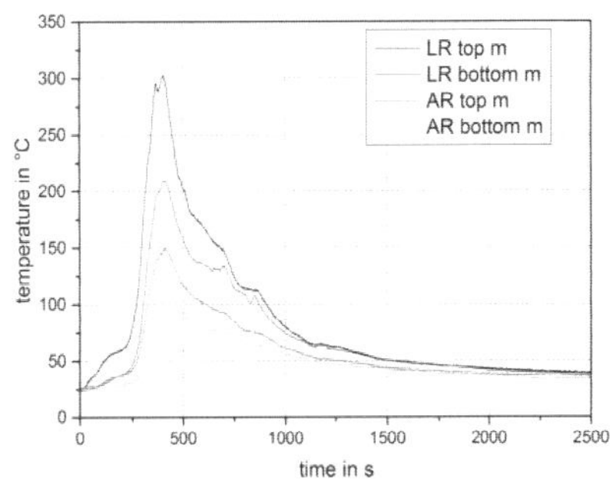

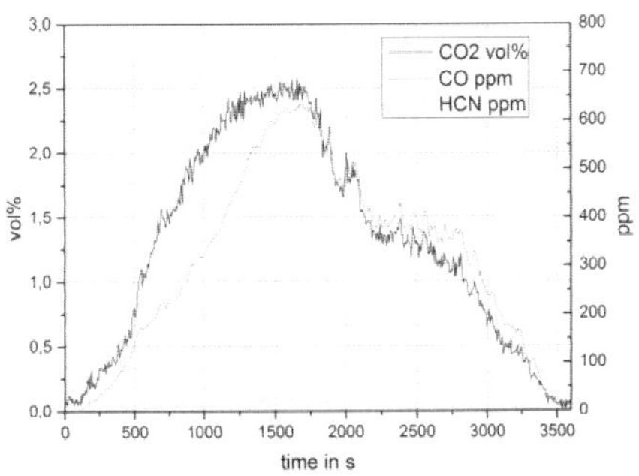

Gas measurements in 1.25 m height

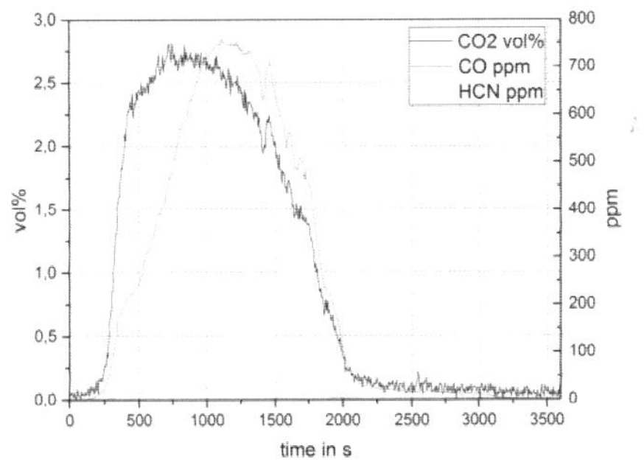

Gas measurements in 1.25 m height

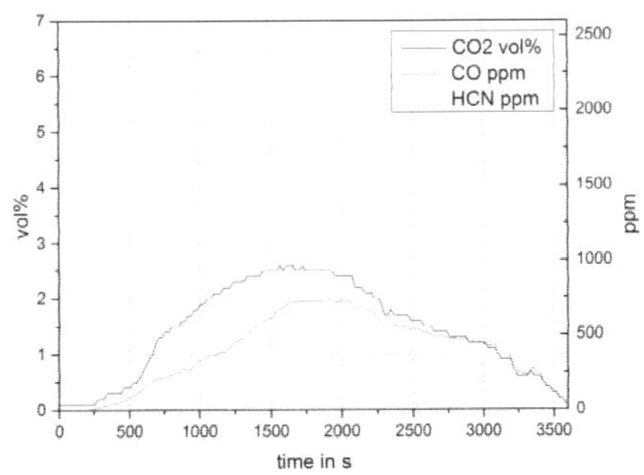

Gas measurements in 1.72 m height

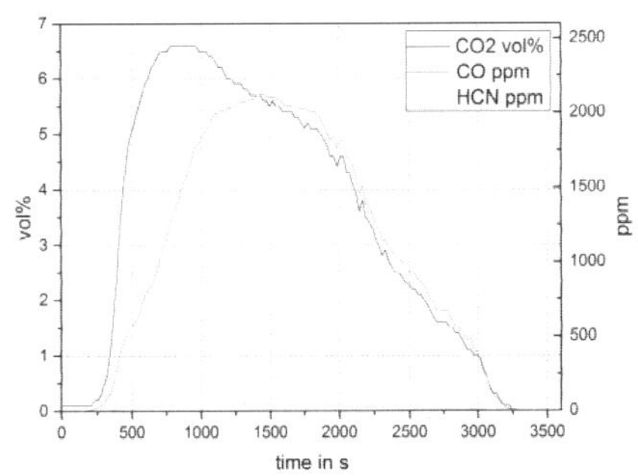

Gas measurements in 1.72 m height

(LR – living room, AR – adjacent room, top – average of 5 thermocouples in 2.63 m height, bottom – average of 10 thermocouples in 1.25 m and 1.72 m height)

Experiments with modern furniture and closed and open windows show the influence of ventilation on the fire development. In figure 5 consequently the experimental results for the living room experiments with modern furniture with closed windows (left hand side) and the two open windows (right hand side) are compared. The maximum temperatures for the experiment with two open windows are above 600 °C for the hot gas layer which indicates fully developed fire in the room. The maximum temperatures for the experiment with closed windows are about 300 °C. The gas measurements in 1.25 m and 1.72 m height in the adjacent room (centre row and bottom row) give significantly higher maximum concentrations of smoke gases (CO_2, CO and HCN are shown) although the first 500 s of both experiments show similar development.

Figure 5: Comparison of experimental results for the experiments with modern furniture and windows closed and modern furniture with two windows open

Comparison between experiments with modern furniture and windows closed with modern furniture and two windows open (experiments 2_1 and 2_2)

Windows closed

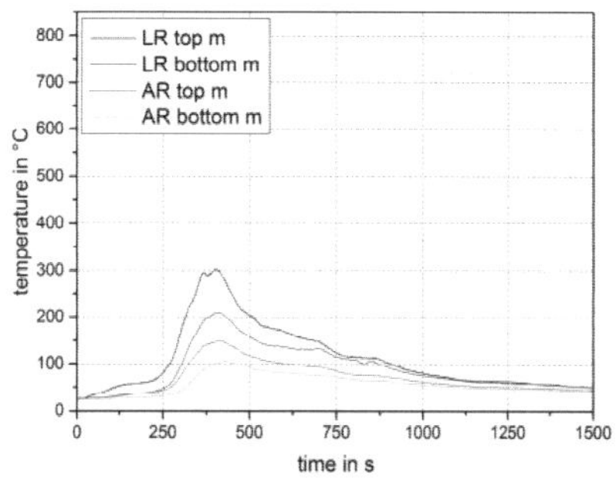

Two windows open

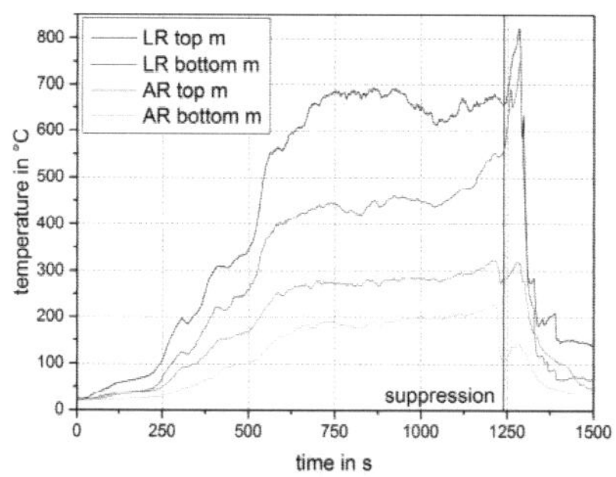

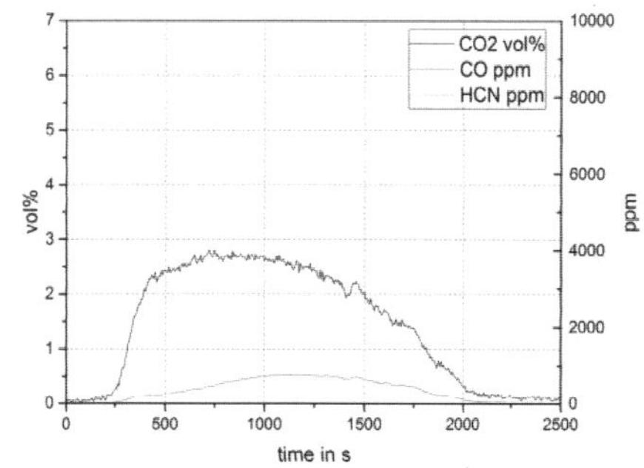

Gas measurements in 1.25 m height

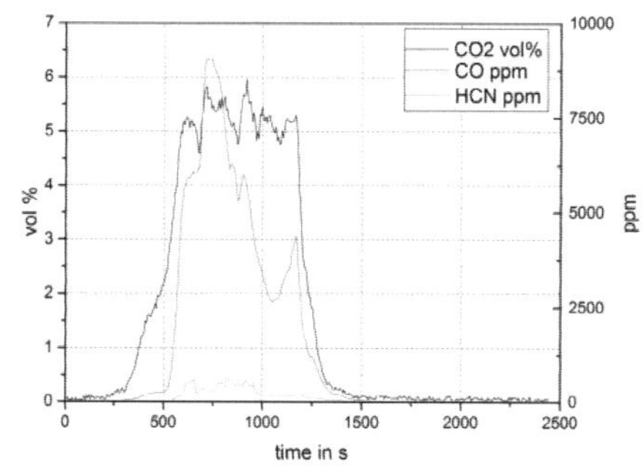

Gas measurements in 1.25 m height

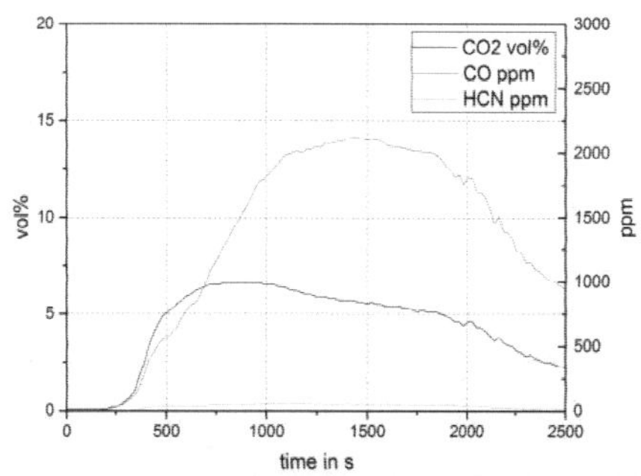

Gas measurements in 1.72 m height

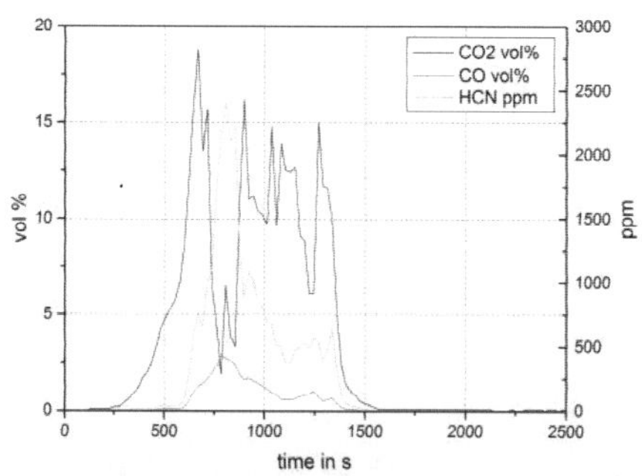

Gas measurements in 1.72 m height

(LR – living room, AR – adjacent room, top – average of 5 thermocouples in 2.63 m height, bottom – average of 10 thermocouples in 1.25 m and 1.72 m height)

That the fire development of both experiments (modern furniture and closed and open windows) is similar can be seen in the temperature development in the rooms as well, figure 6.
Within the first 400 s of both experiments the temperature development in the living rooms (LR, top - hot gas layer and LR, bottom - cold gas layer) is similar. Only after 400 s the temperature development is quite different with significantly higher temperatures in the better ventilated experiment (two windows open). In the experiment with the modern furniture and closed windows only the upholstered chair burnt. The fire couldn't develop further because of the limited ventilation. In the experiment with the modern furniture and two open windows a fully developed fire occurred that was suppressed by the fire service after 1200 s.

Figure 6: Comparison of temperatures for the living rooms and adjacent rooms for the experiments with modern furniture with closed and open windows (experiments 2_1 and 2_2) for the first 600 s of the tests

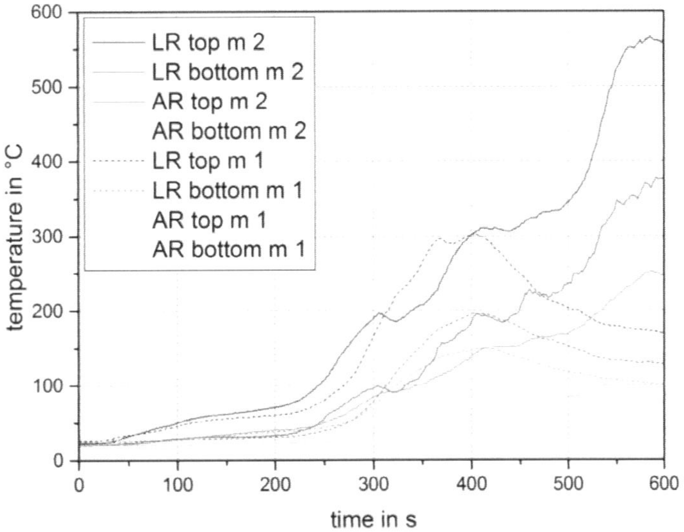

(LR – living room, AR – adjacent room, top – average of 5 thermocouples in 2.63 m height, bottom – average of 10 thermocouples in 1.25 m and 1.72 m height, 1 – experiment 2_1, 2 – experiment 2_2)

In the series of experiments, the kind of furniture and the ventilation conditions changed the fire development significantly. To assess the influence of modern furniture on the fire development in fires nowadays the experiment with the older furniture and open window (experiment 1_2) is compared to the experiment with modern furniture and closed windows (experiment 2_1). We already saw that the modern furniture under the same conditions (closed windows) led to higher temperatures in the rooms and higher amount of toxic smoke gases. The comparison in figure 7 takes the results of the experiment with older furniture and open window (1_2) which develops faster than with closed windows (1_1) and the results of the experiment with the modern furniture and closed windows (2_1) which developed much slower than the experiment with the modern furniture and open windows (2_2).
Although the experiment with the older furniture is better ventilated than the experiment with the modern furniture the temperatures in the living rooms and adjacent rooms are lower for the older furniture than in the experiment with the modern furniture, figure 7 (top row). The maximum measured values for the gas measurements (CO_2, CO and HCN) are in the same range for the measurements in 1.72 height (bottom row). For the measurements in 1.25 m height the maximum values for the older furniture are higher but the slope of the measured gas concentrations is higher for the modern furniture experiment in the beginning of the experiments. To show that in figure 8 the first 600 s of the experiments are shown as an excerpt.

Figure 7: Comparison of experimental results for the experiments with older furniture and one window open and modern furniture and windows closed

Comparison between older furniture and modern furniture (experiments 1_2 and 2_1)

Older furniture, one window open

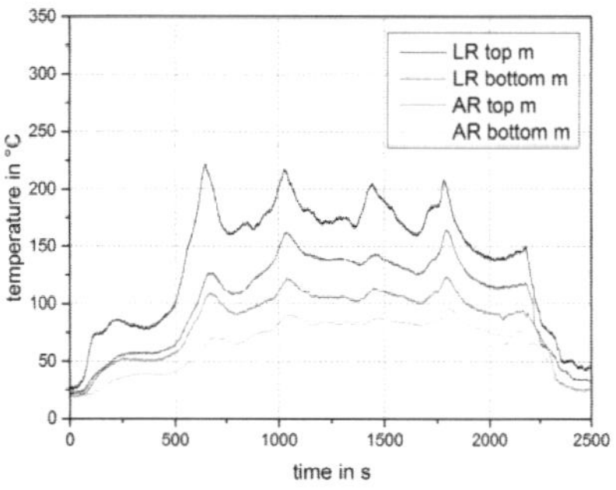

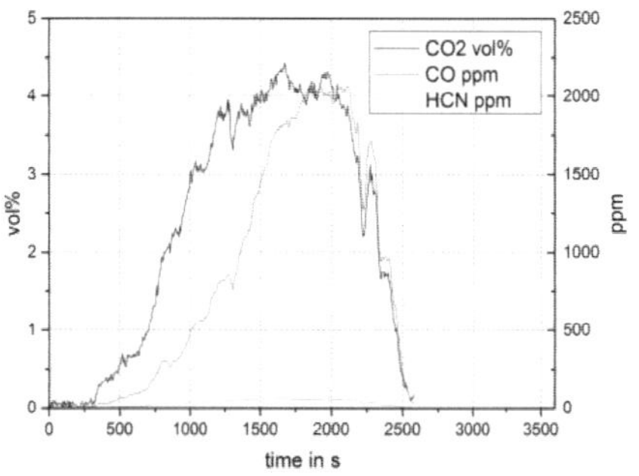

Gas measurements in 1.25 m height

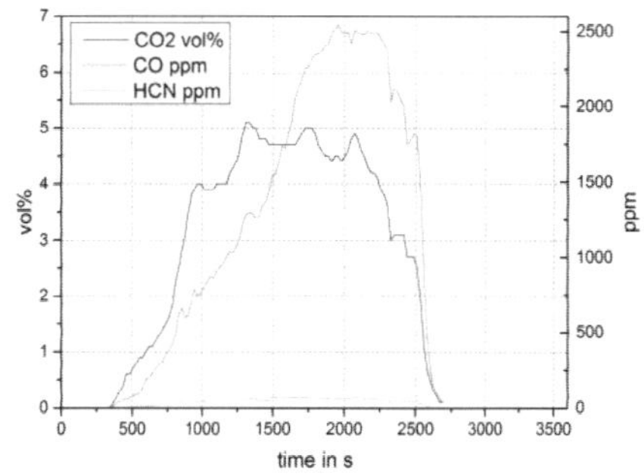

Gas measurements in 1.72 m height

Modern furniture, windows closed

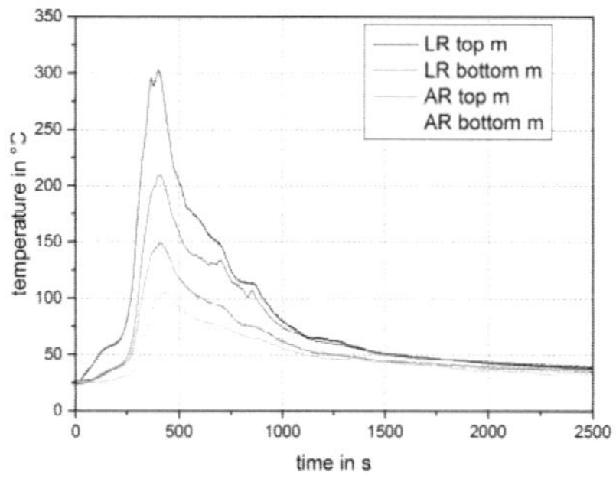

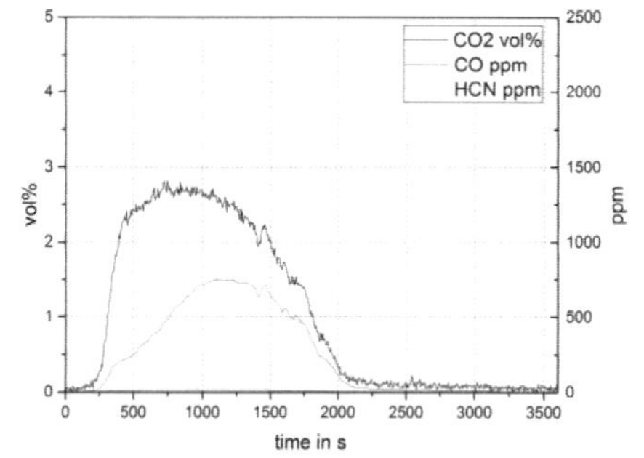

Gas measurements in 1.25 m height

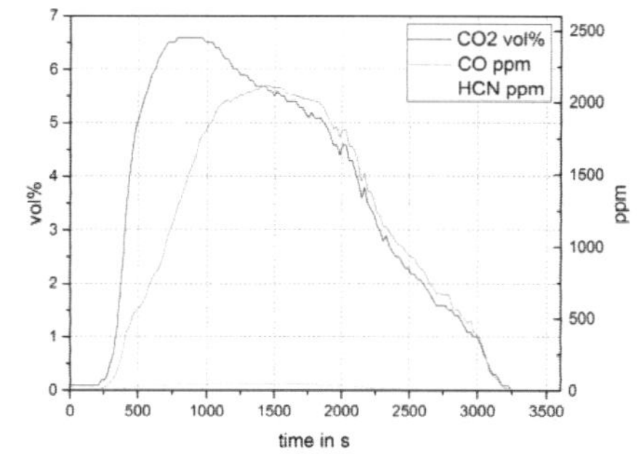

Gas measurements in 1.72 m height

(LR – living room, AR – adjacent room, top – average of 5 thermocouples in 2.63 m height, bottom – average of 10 thermocouples in 1.25 m and 1.72 m height)

Figure 8 shows the comparison between the gas measurements in the experiment with the older furniture and the open window (1_2) and the experiment with the modern furniture and closed windows (2_1) for 1.25 height and 1.72 m height.

Figure 8: Comparison of gas measurements in 1.25 m height (left) and 1.72 m height (right) for the experiments with older furniture and one window open and modern furniture and windows closed for the first 600 s of the experiment

Comparison of gas measurements in different heights for the experiment with older furniture and open window (1_2) and modern furniture and closed windows (2_1)	
Gas measurements in 1.25 m height	Gas measurements in 1.72 m height

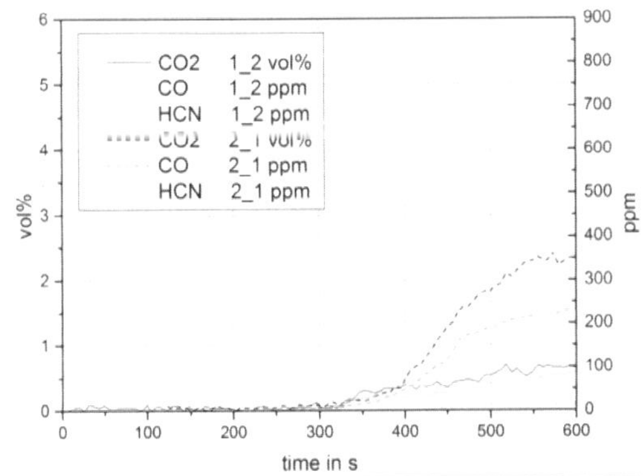

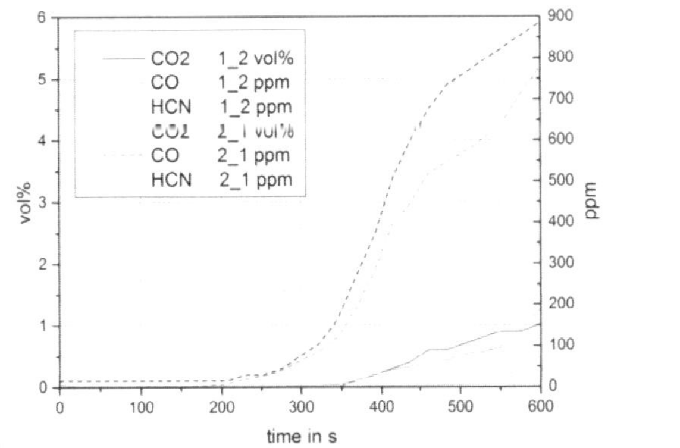

FIRST BUNRING ITEM: MODERN UPHOLSTERED CHAIR

Experimental investigations

Three pre-tests with the modern upholstered chair which was used in experiment 2_1 and 2_2 of the living room tests as first burning item were performed, see table 2. In all three experiments the modern upholstered chair was ignited with a 100 gr paper cushion which was made similar to the standard train seat ignition source (according to DIN 5510) as in the living room experiments as well.

Figure 9 shows the three test set-ups and the measured mass loss rate for all three experiments. In pre-experiment 1 only the mass loss was measured, in pre-experiments 2 and 3 the mass loss, temperatures in the room and smoke gas composition were measured.

Table 2: Overview over pre-experiments with modern upholstered chair

Pre-experiment 1	Pre-experiment 2	Pre-experiment 3
Upholstered chair in big hall	Upholstered chair in little room – no front wall	Upholstered chair in little room – with door opening

Figure 9: Test set-up of pre-experiments and resulting mass losses and mass loss rates

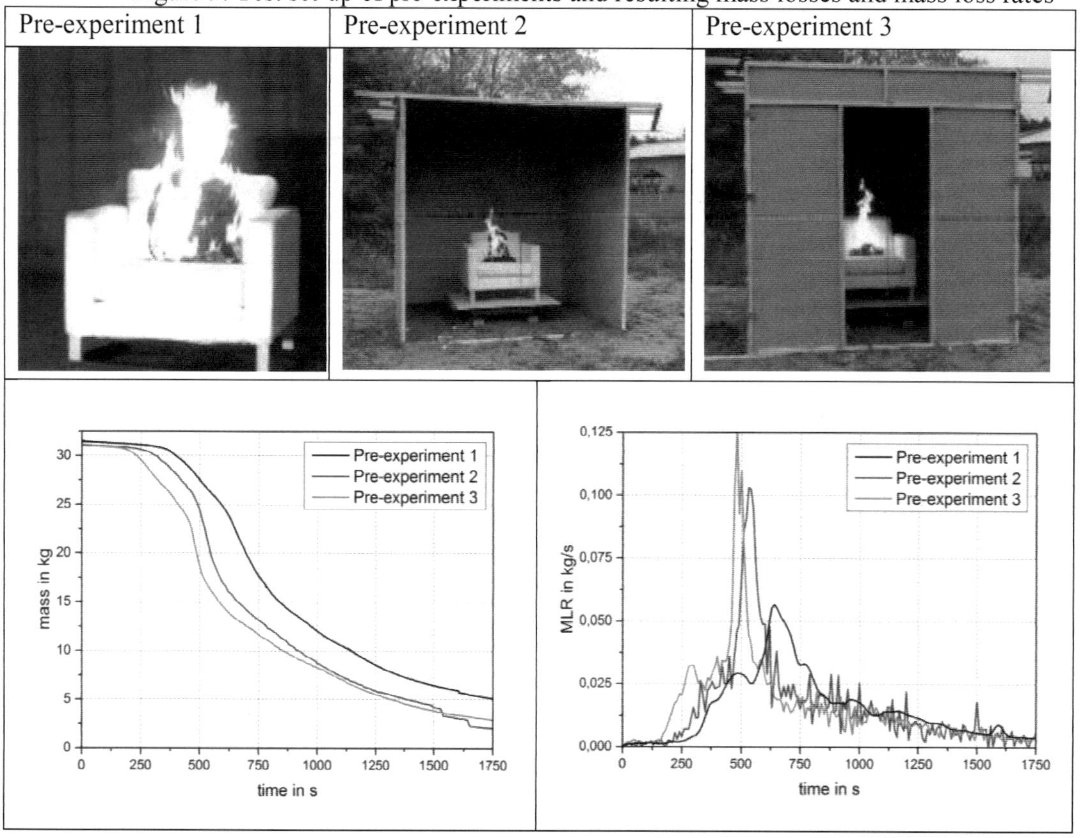

Figure 10 shows the fire development for pre-experiment 2 and 3 after 5 minutes and after 9 minutes in pre-experiment 2 and after 8 minutes in pre-experiment 3. After 5 minutes the fire of pre-experiment 3 is more developed than in pre-experiment 2. After 9 minutes in pre-experiment 2 occurs a short period where the smoke layer burns. In pre-experiment 3 already after 8 minutes flash-over occurred in the small room. This corresponds to the measured temperatures which are shown in figure 11. Temperatures were measured under the ceiling at four locations as shown in figure 11 (left hand side). The average temperatures of the four thermocouples over time are shown in figure 11 (right hand side) for pre-experiment 2 and 3. Temperatures under the ceiling exceed in both cases 600 °C. In pre-experiment 3 the maximum temperatures are significantly higher than in pre-experiment 2: about 900 °C and about 680 °C, respectively. The slope of the temperature rise is also steeper for pre-experiment 3 and the maximum is reached earlier than in pre-experiment 2. Additional temperatures have been measured in front of the chair which are not discussed here.

Figure 10: Observations of fire development in pre-experiments 2 and 3

5 minutes after ignition

5 minutes after ignition

9 minutes after ignition

8 minutes after ignition

Figure 11: Measured temperatures in pre-experiments 2 and 3 (left) with locations of thermocouples (right)

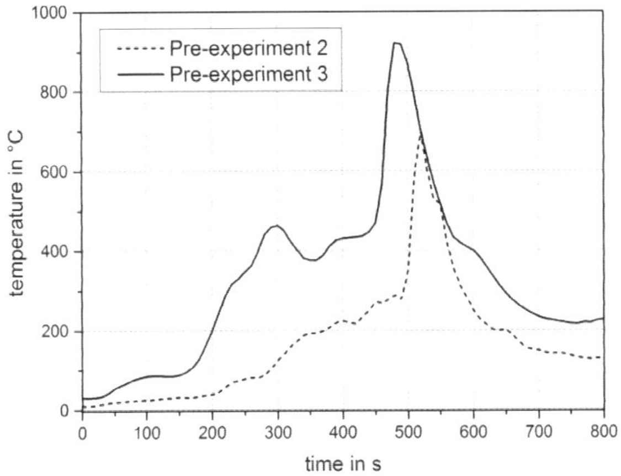

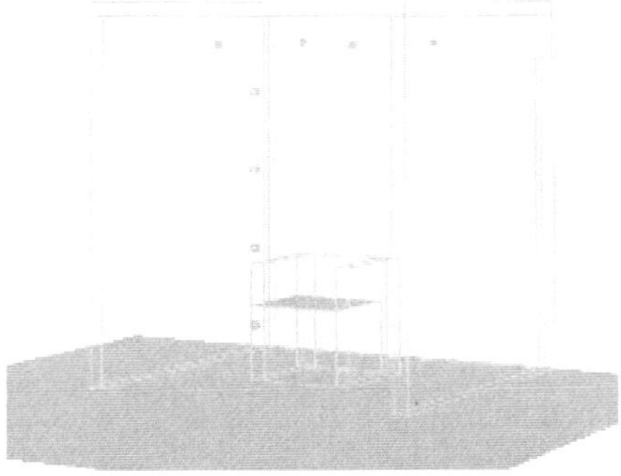

Numerical investigations

The small room configuration has been simulated with the CFD (Computational Fluid Dynamics) software FDS[8,9] (Fire Dynamics Simulator, version 6.01). The small room and the surroundings were modelled with a cubic grid with 0.05 m cell size (side length). The small room has the dimensions 2.5 m x 2.5 m x 2.5 m. The upholstered chair is modelled under the assumption the material is polyurethane foam (NFPA Handbook, Babrauskas[10]) with a density of 38 kg/m³, a specific heat of 1 and a conductivity of 0.05 and a thickness of the material of 0.2 m. The pyrolysis gas is assumed to be C6.3H7.1NO2.1 which burns with a soot yield of 0.1 The seat of the upholstered chair is assumed to be the burning area (0.49 m²). The heat release rate over time curves are calculated from the measured mass loss with an assumed heat of combustion of 25.000 MJ/kg for polyurethane foam.[11] The walls of the small room are assumed to be gypsum plaster with a thickness of 0.1 m. The gypsum plaster is modelled with a density of 1440 kg/m³, a conductivity of 0.48 and a specific heat of 0.84.[12]

As can be seen in figure 12 if the measured mass loss is used to model pre-experiment 2 and 3 it is possible to predict the fire development in the small rooms satisfactorily. The fire development after 5 minutes with a more advanced fire in the room with door opening than in the room without front wall can be predicted well. Also the smoke layer ignition after 8 minutes in the room without front wall and the flash-over after 9 minutes in the room with door opening are captured well by the calculations. The results of the calculation are shown without smoke to make the effects better visible.

Figure 12: Predictions of fire development using mass loss rate from same experiment

	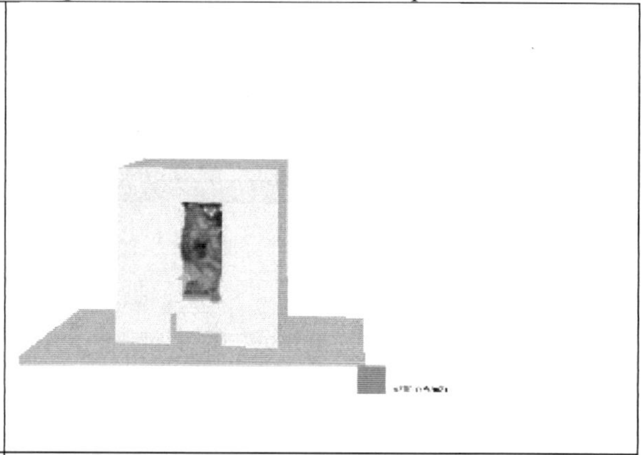
5 minutes	5 minutes
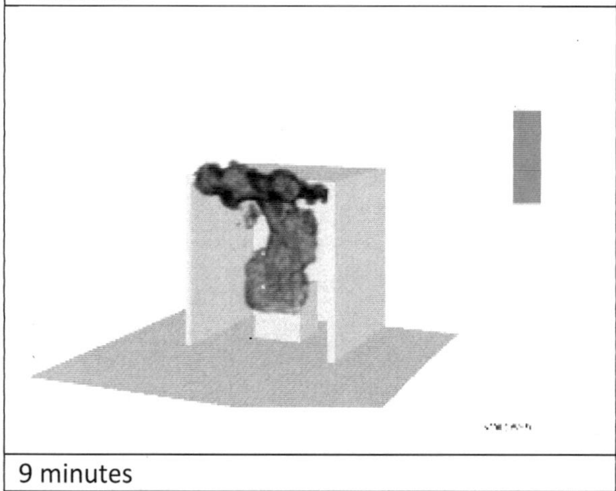	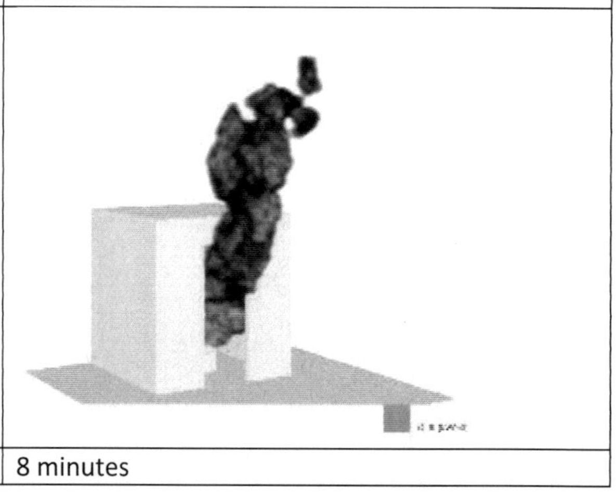
9 minutes	8 minutes

In comparison to the calculations above the small room without front wall was calculated again – the only

difference to the calculations above is the use of the mass loss of the upholstered chair from the experiment in the hall (pre-experiment 1). Grid size, materials and combustion model are the same as above.
In figure 13 the predicted fire development for this case between 9 and 10 minutes is shown. The ignition of the smoke layer does not occur in this model set-up.

Figure 13: Predictions of fire development using mass loss rate from pre-experiment 1 (open hall)

9 minutes	9 minutes 40 s	10 minutes

Figure 14: Predicted temperatures under the ceiling in small room, averaged

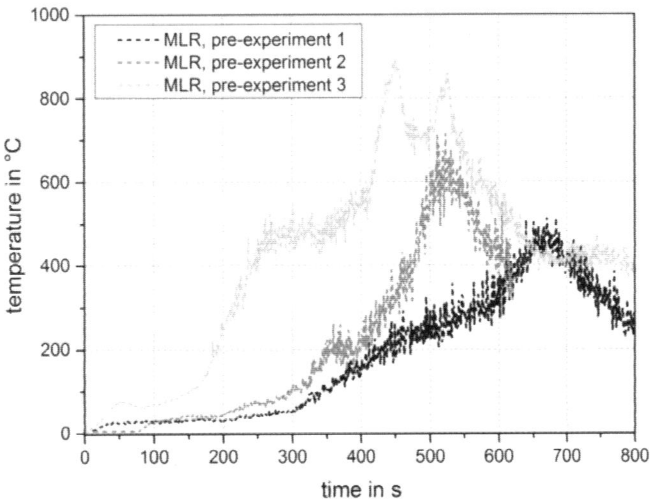

Predicted temperatures in the small rooms under the ceiling correspond to this result. In figure 14 the averaged temperatures over the four thermocouples under the ceiling are plotted over time for the different different calculated cases: small room with front wall and door and use of the mass loss rate (MLR) of corresponding pre-experiment 3 (black line); small room without front wall and use of MLR from corresponding pre-experiment 2 (dotted grey line) and small room without front wall and use of MLR from pre-experiment 1 (grey solid line). Pre-experiment 1 was performed in a big hall with no walls or ceiling nearby. Predicted temperatures for pre-experiment 2 and 3 using the corresponding measured MLR show an exceed over 600 °C under the ceiling that corresponds with the smoke layer ignition in pre-experiment 2 and the flash-over in pre-experiment 3. In the third case where temperatures under the ceiling in the small room without front wall were predicted using the MLR from pre-experiment 1 which was performed in a big hall the temperatures does not exceed 600 °C. This corresponds with the prediction shown in figure 13 that the smoke layer does not ignite.

Discussion of the results

Experimental and numerical investigations of the first burning item of the large scale experiments of the modern living rooms showed that identical chairs burn with significant different mass losses under different room conditions. Three experiments with the identical modern upholstered chair and the same method of ignition with a paper cushion had been performed. In the first test (pre-experiment 1) the upholstered chair was ignited in a big hall with no walls or ceiling close to the item. In the next two tests (pre-experiments 2 and 3) the upholstered chair burnt in the same small room, only difference was the front opening: in pre-experiment 2 the front wall was missing and pre-experiment 3 the front wall was installed with a door opening. The measured mass losses and the derived mass loss rates (figure 9, bottom row) show significant differences although the identical chair and the same ignition method was used. The measured temperatures which were only recorded for pre-experiment 2 and 3 also show significant differences (figure 11, left hand side) as well as the observations of the fire development (figure 10). In the small room with front wall and door the fire development was quicker and the temperatures higher – only in this case flash-over in the room occurred. In the small room without front wall the fire development was slower and only the ceiling smoke layer ignited for several seconds.
The measured mass loss, more precise the derived mass loss rate (MLR) was used in the numerical models to predict the fire development in the small rooms and the temperatures in the room under the ceiling. On the one hand when the MLR of the identical experiment was used for the numerical predictions the fire development could be predicted very accurately. The predicted temperatures under the ceiling are satisfactorily similar to the measured values in shape and magnitude although in the case of the small room with front wall and door the predicted temperatures showed a second peak that was not measured in the experiments as can be seen in figure 15. In figure 15 the predicted and measured averaged temperatures under the ceiling for pre-experiment 2 and 3 are shown.

Figure 15: Predicted and measured temperatures (averaged) under the ceiling in small room

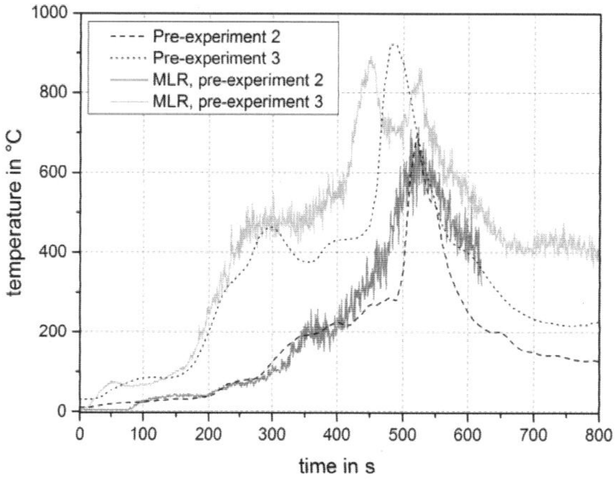

On the other hand when the MLR of pre-experiment 1 (hall) was used to predict the fire development in the small room without front wall the fire development as well as the predicted temperatures are quite different to the observed fire development and measured values: in figure 13 was shown that in this case no smoke layer ignition is predicted although it occurs in the experiment and could be predicted when the MLR rate from the identical experiment is used. Also the predicted temperatures under the ceiling are significantly lower than the measured ones. They stay below 600 °C which is in accordance to the predictions of the fire development.

CONCLUSIONS

A series of four large scale experiments in living rooms with adjacent rooms have been performed. In two experiments older furniture has been used and in two experiments modern furniture has been used. The ventilation conditions have been varied by windows. Two experiments have been performed with closed windows to compare the fire development between the older and the modern furniture. The fires with modern furniture developed significantly faster than the fires with older furniture and produced higher concentrations of smoke gases and higher temperatures in the room of fire origin and the adjacent room. In the room of fire origin and the adjacent room the maximum temperatures were reached after about 300 s for the modern furniture and only after 600 s for the older furniture. Also the concentration of smoke gas components as CO, CO_2 and HCN rises much faster for the modern furniture fires.

The fire develops into a fully developed fire in the experiment with modern furniture and two open windows. After the temperature development in the rooms is similar to the experiment with modern furniture and closed windows for the first 400 s the temperatures rise significantly after that. The maximum temperatures are about 700 °C for the open windows and about 300 °C for closed windows. Also the smoke gas concentrations in the adjacent room are similar for open and closed windows for the first 400 s of the experiments. After that the smoke concentrations of the open window experiment exceed the concentrations of the closed window experiment but for the first 400 s even with closed windows the CO concentrations for instance in 1.72 m height are very high (about 2000 ppm) but they are reached later than in the open windows experiment.

The faster fire development caused by modern furniture becomes even more apparent when the experiment with older furniture and one open window is compared with the experiment with modern furniture and closed windows. Although the fire is more ventilated in the experiment with the older furniture the maximum temperatures in the experiment with the modern furniture are distinctly higher than in the experiment with the older furniture. The smoke gas concentrations are significantly higher for the first 600 s of the experiments for the modern furniture fire in both measured heights.

As it is planned to model the large scale experiments numerically the first burning of the modern furniture was investigated in three pre-experiments to understand the influence of room geometry and ventilation on fire development better. The modern upholstered chair was tested in three different settings: in a hall, a small room without front wall and the same small room but with front wall and door opening. The measured mass losses of the three pre-experiments showed significant differences. These mass losses were used as input in the numerical models of the upholstered chair in the small rooms to see how good the fire development and temperatures could be predicted. Predictions were satisfying when the mass loss of the corresponding experiment was used. In one case the measured mass loss of the first pre-experiment in the hall was used to predict the fire development and temperatures in the small room without front wall. In this case the smoke layer ignition could not be predicted and the temperature in the room were distinctly under the measured temperatures. As measured heat release rates – often measured in open calorimetry – are often used as input in numerical models, for instance for the first burning item, it is important to realise that even with the same chair tested with the same ignition method the results can differ and influence the model predictions.

ACKNOWLEDGEMENTS

We thank the Frankfurt fire service for their technical support and for providing the test facility, without the input of the Frankfurt fire service the tests would not have been possible to perform.

The tests were run as part of the TIBRO project which was funded by the German Federal Ministry of Education and Research (BMBF) with funding reference 13N12176.

REFERENCES

1. https://www.destatis.de/DE/Publikationen/Thematisch/Gesundheit/Todesursachen/Todesursachen.html; with analysis of data also in: A. Hofmann, S. Kaudelka. *Experimentelle und Numerische Untersuchung von kritischen Brandszenarien* (BAM part of final report, TIBRO project, funded by German Federal Ministry of Education and Research (BMBF) with funding reference 13N12176). Unpublished final report, available only in German from project partners and BMBF. 2015
2. https://www.gov.uk/government/uploads/system/uploads/attachment_data/file/456652/Fire_Statistics_Great_Britain_2013-14___PDF_Version_.pdf
3. "Comparison of Modern and Legacy Home Furnishings," UL Experiment, Nov. 2009. Web: 12 Oct. 2012. https://www.ul.com/room_fire/room_fire.html.
4. S. Kerber. *Analysis of Changing Residential Fire Dynamics and Its Implications on Firefighter Operational Timeframes*, Fire Technology, 48, 4: 865-891, Oct. 2012
5. Christian Knaust, Anja Hofmann. *Full Scale Investigations of Fast Spreading Room Fires.* Materials Testing 56 (2014) 1 Carl Hanser Verlag, Germany
6. A. Hofmann, C. Knaust, A. Beard. *High risk items in home fires, fire safety regulations and benefit for home fire safety.* Flame Retardants Conference 2008, 12.-13.2.2008, London UK, ISBN 978-0-95565 48-1 -7
7. A. Hofmann, C. Knaust. A. Beard. *Hazardous fires in children's rooms. Experimental and numerical investigation with comparison to real cases.* Interflam 2007, London, UK, 3rd – 5th September 2007, ISBN 978-0-9556548-0-0
8. A. Hofmann, S. Kaudelka. *Experimentelle und Numerische Untersuchung von kritischen Brandszenarien* (BAM part of final report, TIBRO project, funded by German Federal Ministry of Education and Research (BMBF) with funding reference 13N12176). Unpublished final report, available only in German from project partners and BMBF. 2015
9. K. McGrattan et. al. *Fire Dynamics Simulator User's Guide.* NIST Special Publication 1019. Sixth Edition. USA 2016
10. K. McGrattan et. al. *Fire Dynamics Simulator, Technical Reference Guide* Volume 1: Mathematical Model. NIST Special Publication 1018-1, Sixth Edition, USA 2016
11. NFPA Fire Protection Handbook, 20th Edition, 2008
12. M.J. Hurley. *SFPE Handbook of Fire Protection Engineering.* Fifth Edition. Springer 2016
13. J. Quintiere. *Principles of Fire Behavior.* Second Edition. CRC Press. USA 2016
14. A. Hofmann, Ch. Knaust, S. Krüger, T. Raspe and J. H. Deubel, *Detailed analysis of a smouldering fire scenario at the murder scene, experimental and numerical investigations.* Volume 38, Issue 8, December 2014, Pages: 806–816
15. A. Hofmann and R. Muehlnikel. *Experimental and numerical investigation of fire development in a real fire in a five-storey apartment building*, Fire and Materials, Volume 35, Issue 7, November 2011, Pages: 453–462,
16. S. Krüger, A. Hofmann, A. Berger, N. Gude. *Investigation of smoke gases and temperatures during car fire – large-scale and small-scale tests and numerical investigations.* Fire Mater. 2016, 40, 6: 785-799
17. A. Löhnert (Klippel), N. Monreal, Ch. Knaust, A. Hofmann and U. Krause. *CFD modeling approach of smoke toxicity and opacity for flaming and non-flaming combustion processes.* Fire Mater. 2016, 40, 6: 759-772

COMPARTMENT FIRES

AN ANALYTICAL MODEL FOR FLASHOVER TO REVEAL MECHANISMS

J. G. Quintiere
U of Maryland, USA

ABSTRACT

An approximate model is formulated to reveal the many factors that control flashover in a room fire. Suffice to define flashover as a distinct "jump" or "runaway" in the course of fire growth. The model investigates the instability that underlies the jump for fixed area fires and growing fires. The transient effects are illustrated.

INTRODUCTION

Despite much interest, flashover is a term that is ill defined, and only empirically described. Most can agree that flashover marks a dramatic increase in fire conditions due to the confinement of a room. The name *flashover* applies to this period of change. It can be caused by several situations:

1. The rapid ignition, burning rate and flame spread of materials due to increasing heat flux arising from the heated room. This is depicted in Figure 1. This event is called commonly flashover. Note at the onset of flashover the equivalence ratio is well below 1; the fire is over-ventilated.

2. The accumulation of fuel rich gases and their sudden exposure to air with ignition. This is commonly called backdraft by the fire service and others, but can be regarded as a type of flashover. As a premixed flame can move rapidly through the gas, this event can release a lot of energy in a short time. As a result, the pressure in the compartment can increase, causing a large pulse of flame and possible structural and glass failures.

Figure 1. Stages of compartment fire

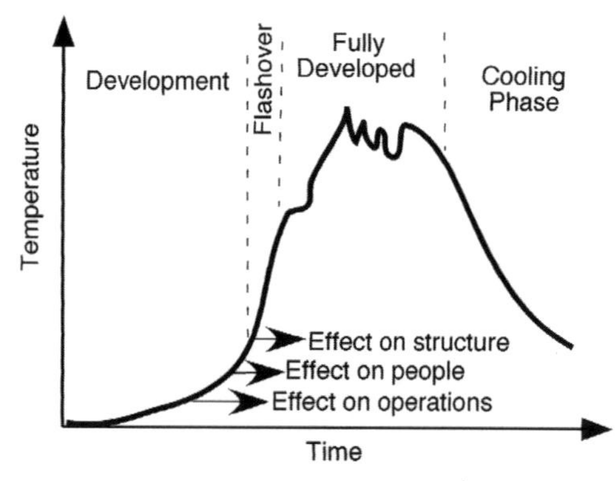

We will regard flashover as that described in 1 and not address backdraft here. Recently it was discussed based on estimations and experimental results [1]. The conclusions from that analysis displayed several aspects of flashover. There is a beginning and end to flashover; it is a transition that can occur over seconds or minutes. It is due to instability. It is much like an ignition process. In ignition, first there is fuel and air then there is a flame. In flashover, first there is a small fire, and then there is a big fire.

For flashover, one type of instability is caused by the firepower suddenly exceeding the heat loss and energy flow rates from the compartment. This imbalance results in a sharp temperature increase in the smoke of the compartment. This form of flashover can result for a fixed area fire or a growing fire. Preheating the surface ahead of the flame to its piloted ignition temperature causes another instability. The sudden ignition of remote fuel surfaces around the main fire is a third way a "jump" can occur in the fire. All of these mechanisms can lead to a relatively sudden increase in the fire, and the fire will seek to reach its maximum, possible state within the room. This fully developed fire state can be fuel-limited (over-ventilated) or air-limited (under-ventilated). All of the fuel gases may not be able to burn within the room. In the under-ventilated state flames will move and seek out available air supply locations. The flames will not fill the room, but the smoke layer will descend to near floor level.

Thomas et al [2] have described the mathematics of the instability for flashover. The Thomas model treats the fire as steady, and ignores the transient process governed by a differential equation in time. The current model considers the unsteady process. The model is similar in spirit to the approximate analytic unsteady model of Graham [4]; however no graphical results were given there. The current analysis will consider the steady-instability formulation, unsteady non-linear results, and analytical solutions to a linearized formation. The effect of spread and remote ignition will be explicitly modeled, and their influence on flashover will be examined.

MODEL FORMULATION

Figure 2 illustrates the room and fire being addressed as a discrete fire and single doorway. The gases in the room form a uniform property gas layer, approximately one-half of the room's volume. The airflow through the doorway is computed by a formula that accounts for the variation in gas temperature. The conductance of the room boundary surfaces is considered constant defined by U_w, without any loss of generality. The generation of fuel is considered small compared to the flow rate of air. The upper room surfaces and the gas enclosed are assumed to be at the same temperature, T.

Figure 2. Model configuration

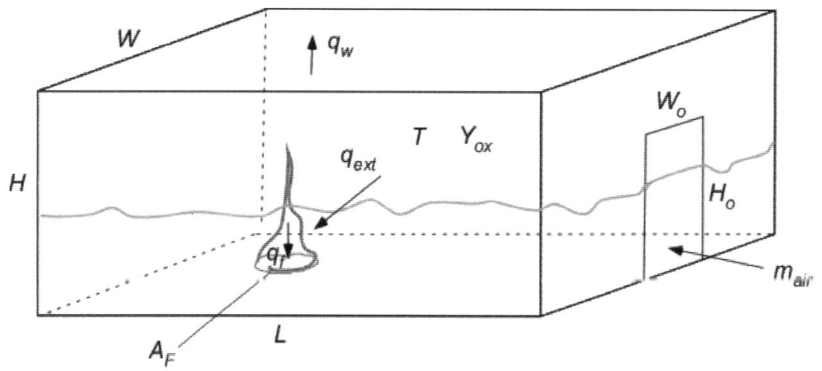

The terms in the figure are defined as follows:

> $\dot{q}''_f$ net flame heat flux, $\dot{q}''_{ext}$ incident external heat flux from room,
> $\dot{q}''_w$ heat flux to boundary, $\dot{m}_{air}$ inlet air flow rate, T smoke layer temperature,
> Y_{ox} smoke layer oxygen mass fraction, A_F area of fuel burning, and F is the view factor from the hot gas to the fuel surface.

Dimensionless variables are used to reduce the number of parameters to a minimum. These are defined:

$$\phi = \frac{T}{T_\infty}, \tau = \frac{t}{t_o}$$

$$t_o = \frac{mc_pT_\infty}{\dot{q}''_f A_{F,o}(\Delta h_c / L)}, \text{ time for fuel mass to equal room gas, } m = \rho_\infty V / 3$$

$$\frac{d\phi}{d\tau} = R - L, \tau = 0, \phi = 1$$

$$R = \left[1 + q_e(\phi^4 - 1)\right]\left(A_F / A_{F,o}\right)$$

$$L = \left(q_o m_o + q_w\right)\left(\phi - 1\right) \qquad [1]$$

$$q_e = \frac{F\sigma T_\infty^4}{\dot{q}''_f}, \text{ external to flame heat flux}$$

$$q_o = \frac{\rho_\infty c_p T_\infty \sqrt{g} A_o \sqrt{H_o}}{\dot{q}''_f A_{F,o}(\Delta h_c / L)}, \text{ doorway enthalpy to fire energy flow rate}$$

$$q_w = \frac{U_w A_w T_\infty}{\dot{q}''_f A_{F,o}(\Delta h_c / L)}, \text{ wall heat loss rate to fire energy flow rate}$$

$$m_o = 0.141 e^{-cw^{2/3}} \sqrt{(\phi - 1)}, c = 0.7, w = W_o / H_o, \text{ dimensionless doorway mass flow rate}$$

The solution requires the integration of a non-linear equation primarily due to the external heat transfer from the room to the fuel in R. A linearized approximation would lead to an exponential solution for ϕ in τ, with a critical parameter governing "runaway" or steady state. Flame spread or discrete remote ignitions will govern the functional form of the fuel burning area, A_F. In previous modeling this area was not explicitly formulated as a function of time; here this will be done.

As one can see there are many variables associated with this problem. We shall try to minimize the number by fixing some without any significant loss in generalization. These will be fixed throughout the numerical results.

Room size: 4 x 4 x 2.44 m high, A_w (floor, ceiling, half wall areas) = 51.5 m^2
Door size: 1 x 2.1 m high, $A_o H_o^{1/2}$ = 3.04 m$^{5/2}$
Construction Conductance, U_w = 0.06 kW/m^{2}K (like gypsum board heated for about 200 s)

CONSTANT AREA FIRE

The first case considered will follow the Thomas et al [2] model and consider the fuel burning area to be fixed. The instability will be revealed by equating $R = L$ and their derivatives in ϕ. This implies that a steady solution exists for Eq. [1], and the R and L curves in ϕ are tangent, such that any slight change in ϕ will cause a "runaway" to a new stable state. The value of ϕ at the tangent is the onset of flashover and the new stable point is such that the fire becomes ventilation-limited or the fuel area must increase.

The results for the tangent flashover onset value for ϕ are found from the following:

$$q_e = \frac{F\sigma T_\infty^4}{\dot{q}_f''} = \frac{1+\left(\dfrac{0.141cw^{2/3}q_o e^{-0.427/\sqrt{\phi-1}}}{2\sqrt{\phi-1}}\right)}{4\phi^3(\phi-1)-(\phi^4-1)}, cw^{2/3}/2 = 0.213 \qquad [2]$$

gives the flashover value for ϕ as a function of q_o and $\dot{q}_f''/F$, and the ratio of wall heat transfer to fire energy, q_w giving the locus for the boundary of stable and unstable solutions is found from

$$q_w = \frac{1+q_e(\phi^4-1)}{(\phi-1)} - 0.141 q_o e^{-0.427/\sqrt{\phi-1}} \quad . \qquad [3]$$

Let us plot these results for three types of fuel packages: a heptane pool fire, a PMMA wall fire, and wood crib fire **all with a fuel area of 2 m^2**. Consider the parameters for each fire type. These are shown in Table 1 as estimates, and are derived from literature values realizing that the wall fire most easily receives radiation from the smoke layer, while the internal surfaces of a burning crib receive the least. But for the crib, the net flame heat flux is augmented by re-radiation from a neighboring surface.

Table 1. Estimated parameters for three fire types of 2 m^2.

Fire type	Net flame heat flux, $\dot{q}_f''$ kW/m^2	View factor, F --	$\Delta h_c / L$ --	q_o --
Heptane Pool	35	0.5	89	0.54
PMMA Wall	20	0.8	15.6	5.4
Wood Crib	25	0.2	2	10.7

Figures 2 and 3 show these results. Figure 2 is very similar to the Thomas et al 2 result, but here the vent flow rate is given by an explicit formula. The figure needs to be interpreted as it is plotted against the net flame heat flux per view factor. As seen from Table 1 only one value pertains to each of the fire types; the large circles indicate these points. However, if one considers just pool fires then the abscissa can vary from about 40 to 100 indicating the onset of flashover in the range of about 500 to 700 °C. Yet to cause the instability for a wall fire of PMMA or a wood crib fire indicates temperatures of about 450 °C to nearly 900 °C. Intuitively it is likely accepted that to cause "flashover" in a discrete wood fire, the feedback temperature would have to be very high, as indicated.

Figure 3 shows the conditions for the heat loss to the boundary construction that leads to possibility for flashover. If the heat loss to the boundary is much smaller than the fire energy produced, flashover is likely. These results show that the conditions for flashover are not associated with some fixed compartment temperature. This is contrary with much of the fire literature that oversimplifies flashover.

Figure 2. Temperature at onset of instability for a fixed area fire, 2 m^2

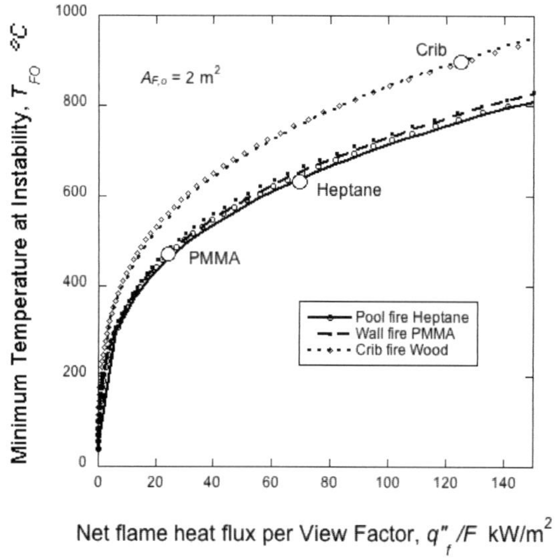

Figure 3. Dimensionless heat loss and fire energy at the instability a. PMMA wall and b. Heptane pool, 2 m^2

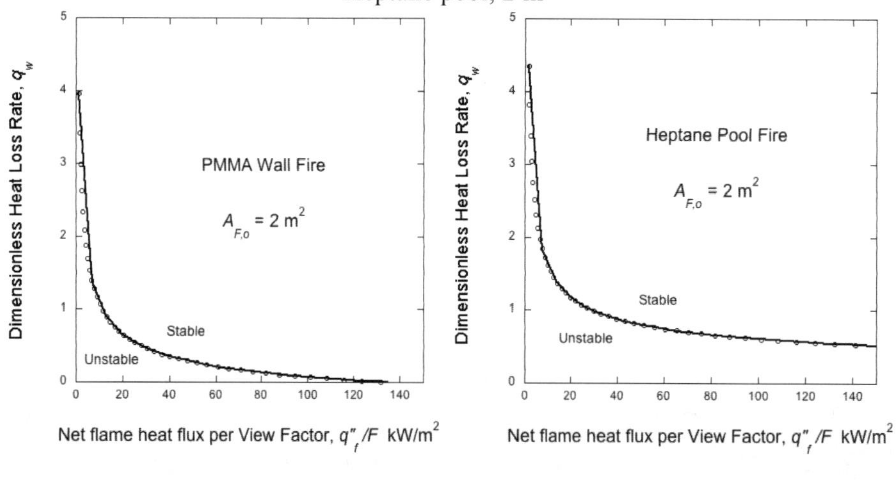

OTHER MODELING RESULTS

The enhancement of the burning rate over its free burning rate at the instability is given as

$$R_{FO} = \left[1 + q_e(\phi_{FO}^4 - 1)\right]\left(A_{F,FO} / A_{F,o}\right). \quad [4]$$

The dimensionless energy release rate when the fire is ventilation-limited is given as

$$R_{VL} = m_o q_o q_{ox}$$
$$\text{where } q_{ox} = \frac{Y_{ox,\infty}\Delta h_{ox}}{c_p T_\infty} = 10.24 \text{ for normal air}. \quad [5]$$

The ratio of the mass loss rate at the ventilation-limit to the mass rate of fuel burned within the room can also be expressed to indicate the rate of fuel available to burn outside the room.

$$\frac{\dot{m}_{VL}}{\dot{m}_{burned}} = \frac{1 + qe(\phi_{VL}^4 - 1)}{R_{VL}}. \quad [6]$$

The dimensionless oxygen mass fraction is found from the differential equation:

$$\frac{dy}{d\tau} + m_o q_o (y - 1) = -R / q_{ox} \text{ with } y = Y_{ox} / Y_{ox,\infty}. \quad [7]$$

Finally the volume of the flame in the room if the fire becomes ventilation-limited is given as

$$V_{flame} / V_{room} = R_{VL} / q_{Turb}$$
$$\text{where } q_{Turb} = \frac{\dot{Q}'''_{Turb} HWL}{\dot{q}''_f A_{F,o}(\Delta h_c / L)},\ \dot{Q}'''_{Turb} = 1200 \text{ kW/m}^3. \quad [8]$$

Let us vary the area of the fire from about 0.25 to 4 m^2 in the specified room. The energy rate R and L curves are plotted against ϕ for the heptane and PMMA pool fires. The parameters pertaining to these fires are given in Table 2.

Table 2. PMMA and heptane pool fires.

Parameter	PMMA pool fire	Heptane pool fire
$\dot{q}''_f$ kW/m^2	60	30
$\Delta h_c / L$ --	15.6	89
$q_e = \frac{F\sigma T_\infty^4}{\dot{q}''_f}$	0.00373	0.00745
$q_o = \frac{\rho_\infty c_p T_\infty \sqrt{g} A_o \sqrt{H_o}}{\dot{q}''_f A_{F,o}(\Delta h_c / L)}$	$3.59/A_{F,o}$	$1.26/A_{F,o}$
$q_w = \frac{U_w A_w T_\infty}{\dot{q}''_f A_{F,o}(\Delta h_c / L)}$	$0.984/A_{F,o}$	$0.345/A_{F,o}$

Figure 4. Energetics of room fires for (a) PMMA and (b) heptane pool fires

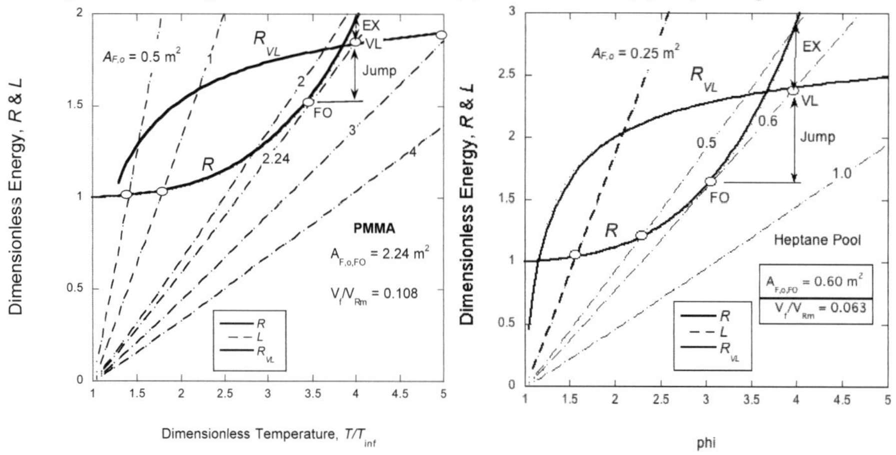

FLASHOVER DYNAMICS

Let us examine the unsteady aspect of flashover and its transition to a fully developed fire. Numerical solutions of Eq. [1] can be developed, but they will likely hide the underlying mechanism to cause flashover. An appropriate linearization of the *R*-term can allow an analytic solution and give some insight to the nature of flashover. The approximation $\phi^4 - 1 \approx 167(\phi - 1)$. Further approximations are given: $m_o = 0.092$ for $\phi = 2$, and a PMMA pool fire is considered with parameters from Table 2. The solution to the linearized governing equation is

$$\phi = 1 + \frac{e^{\lambda\tau} - 1}{\lambda} \qquad [9]$$
$$\lambda = 167 q_e - \left(m_o q_o + q_w\right)$$

It is clearly seen that $\lambda > 0$ leads to a "runaway", and $\lambda = 0$ denote the onset of flashover. With the parameters of Table 2 for a PMMA pool fire, the minimum area to initiate flashover is found as 2.12 m^2 (compared to 2.24 m^2 in Figure 4a for the non-linear solution). The dynamic solution for various fuel areas is plotted in Figure 5. For areas below 2.12 m^2 the solution reaches a steady over-ventilated state. For 1.5 m^2, flashover is not achieved, but the fire does become ventilation-limited. For 2.5 m^2, the fire becomes ventilation-limited at about 6 s. The particular ventilation limited condition is found where $R_{VL} = R$, giving

$$\phi_{VL} = \left(1 + \frac{(3.43 / AF,o) - 1}{0.00373}\right)^{1/4}$$

$$\text{and for } \tau = \left(\frac{A_{F,o}}{7.36}\right) t > \tau_{VL} \qquad [10]$$

$$\phi = 3.61 + (\phi_{VL} - 3.61) e^{-0.179(t - t_{VL})}$$

This example shows that a ventilation-limited condition can occur without flashover, and flashover for a fixed area pool fire can occur in less than 10 seconds with a nearly constant rate of temperature rise. How flashover might be perceived from an observer of these fires might be different than the exact mathematics indicates here.

Figure 5. Dynamics of flashover for a PMMA pool fire

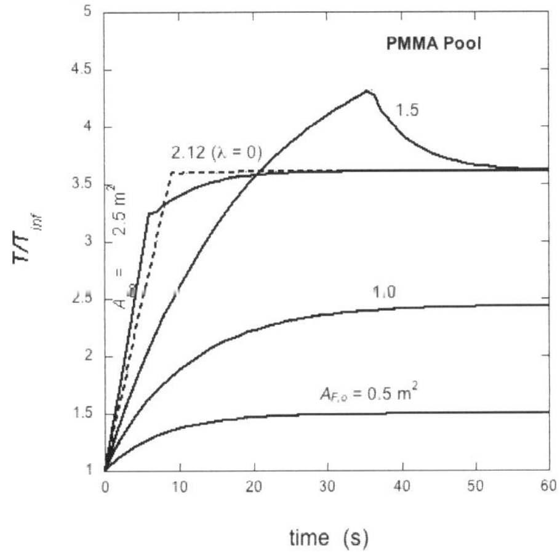

SPREADING FIRES

The dynamics of a spreading fire can be very different from a fixed area fire that responds only to radiation feedback. A spreading fire could spread laterally by opposed flow spread, upward or under a ceiling by wind-aided spread, or by sudden increase inn area due to remote ignition. The feedback instability is still present in all of these as indicated in Figure 4. But that instability analysis cannot reveal the dynamics of the fire. All of these can be explicitly modeled within the construct of Eq. [1]. Now the growth rate of the area must be expressed. This will be done for three cases:

1. a horizontal axisymmetric spreading pool fire,
2. an upward spreading wall fire fixed in width, and
3. an increase in area due to auto (non-piloted) ignition.

The dynamics of these scenarios have many features, and an easy solution cannot be rendered. However, we can examine the minimum conditions that can lead to a very rapid increase in the fire. For the spreading fires, this will occur when the surface temperature ahead of the moving flame reaches its ignition temperature. The surface velocity mathematically would be infinite, but in reality it would achieve the premixed flammability velocity of the order of about 50 to 100 cm/s. For remote ignition, a jump in the area of the fire would occur when the surrounding fuel surfaces achieve their auto-ignition temperatures. For each of these limiting conditions, the minimum "flashover" or sudden increase in fire temperature can be estimated.

MINIMUM TEMPERATURES FOR SPREADING FIRES

Here we are considering "long" heating times before rapid spread occurs. This translates to the equilibrium temperature of the pre-heated surface before spread or ignition. The surface is heated by its flame in spread and in all cases by the room heat transfer feedback. Consider each.

Opposed flow flame spread: Consider an axisymmetric horizontal burning region of radius, r, with an external room heat flux heating both the flame region and the non-burning region at temperature T_s. The flame spreads into a region at T_s based on a local heat flux, and the surface there has been heated by the room external heat flux and the radiation heat flux from the fire. The latter is taken at the radius, r, and is based on a radiation fraction, X_r. The spread velocity and the heat fluxes are described as follows as shown in Figure 7.

spread velocity, $v = d_s / t_s$

time to spread, $t_v = (\pi / 4) k\rho c (T_{ig} - T_s)^4 / \dot{q}_s''$

characteristic spread time, $t_s = (\pi / 4) k\rho c T_{ig}^4 / \dot{q}_s''$

flame spread heat flux, $\dot{q}_s''$

external heat flux, $\dot{q}_{ext}'' = F\sigma(T^4 - T_\infty^4)$

flame radiation heat flux, $\dot{q}_{f,r}'' = (X_r / 4)(\dot{q}_f'' + \dot{q}_{ext}'')(\Delta h_c / L)$

Figure 7. Schematic of heat flux contributions in horizontal spread

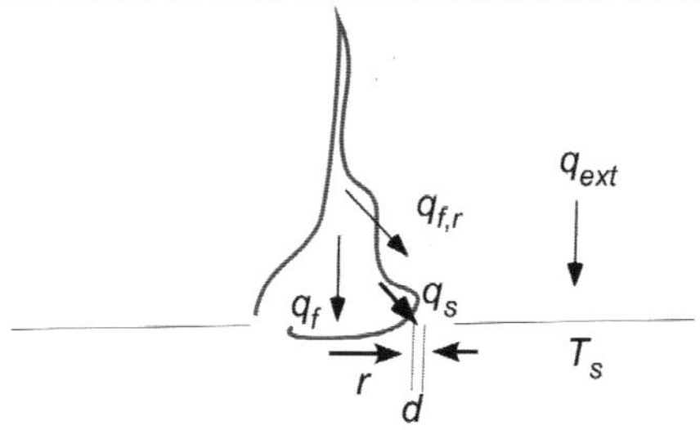

The room temperature that can bring the preheated surface temperature to its ignition temperature can be found as

$$\phi_{FO} = \left(1 + \frac{(\varepsilon_s / F)(\phi_{ig}^4 - 1) + (h_s / F\sigma T_\infty^4)(\phi_{ig} - 1) - (X_r \Delta h_c / 4q_e L)}{(1 + X_r \Delta h_c / 4L)}\right)^{1/4}. \qquad [11]$$

The surface emissivity, ε_s and the convective heat transfer coefficient, h_s are introduced in the equation.

Upward flame spread: Here the flame length is a function of the energy release rate of the fire, and flame radiation away from the flame is negligible. However, there is convective preheating of the surface ahead of the flame. We approximate this as the flame heat flux (an over-estimate). The long term heating result is given as

$$\phi_{FO} = \left(1 + (\varepsilon_s / F)(\phi_{ig}^4 - 1) - 1 / q_e\right)^{1/4}. \qquad [12]$$

Remote ignition: Let us consider this a vertical surface target just adjacent to a symmetric floor flame. The long term heating equation is identical to Eq. [11] but with the ignition temperature taken as is auto-ignition value.

Let us compare these minimum values of room temperature to allow rapid fire growth after a long time in these scenarios with that of the instability temperature of a static pool fire of 2 m^2 in the room of this paper. Figure 8 shows the results in terms of flame net heat flux. The surface emissivity is taken as 1, its convective heat transfer coefficient is taken as 15 W.m2K, the piloted ignition temperature is 350 °C and the auto-ignition temperature is 450 °C. The view factor is 0.5 for the horizontal and remote ignition fires and 1 for upward spread. The results show that flashover is a static fire is likely to occur at a much higher temperature than a spreading fire, and that upward flame spread is the easiest scenario to promote flashover.

Figure 8. Minimum temperatures for fire growth is several scenarios

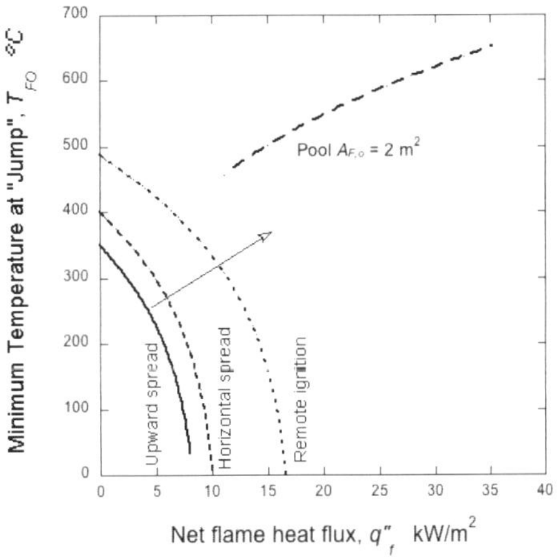

EXAMPLE OF DYNAMIC FIRE GROWTH

Let us consider an example with flame spread in which the area is growing horizontally and asymmetrically with time. The governing equation is [1] but now the area term is given by

$$A_F / A_{F,o} = \left(1 + q_v \int_0^{\tau} \frac{d\tau}{\left(1 - \phi_s / \phi_{ig}\right)^2}\right)^2 \approx \left(1 + \frac{q_v \tau}{\left(1 - 1/\phi_{ig}\right)^2}\right)^2$$

$$q_v = \left(\frac{d}{r_o}\frac{t_o}{t_s}\right), \text{ spread velocity to flow velocity} \qquad [13]$$

$$\phi_s = 1 + \frac{\dot{q}''_{inc}}{\dot{q}''_s}\sqrt{\frac{t_o}{t_s}\tau}$$

$$\dot{q}''_{inc} = F\sigma T_\infty^4\left(\phi^4 - 1\right)\left(1 + X_r \Delta h_c / 4L\right) + \dot{q}''_f \Delta h_c / 4L$$

For illustration, the following representative values are selected: $q_e = 0.004$, $q_o = 4.0$, $m_o = 0.092$, $q_w = 0.9$, $q_v = 0.001$, and $\phi_{ig} = 0.4$. The results are shown in Figure 9. There is an increasing rate of rise in the room temperature. Perhaps the last numerical point might be accepted as the onset of flashover (541 °C). It shows again that flashover is not always so precisely determined, even by the mathematics.

Figure 9. Dimensionless room temperature rise in axisymmetric horizontal flame spread

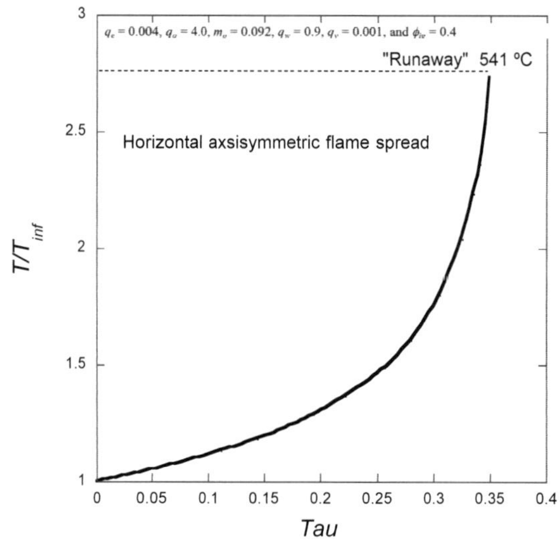

CONCLUSIONS

It has been shown that flashover can be modeled, but its indication is not always obvious. For a fixed area fire, or slowly changing fire, there is a possible instability that triggers a "jump" in the fire conditions. The end state can be all the fuel surfaces in the room contain a flame provided the ventilation allows sufficient air supply. In most room fire cases, the fire jumps to a ventilation-limited state in which the flames will not fill the room. There is not a fixed temperature that defines flashover. The temperature can vary with many factors, especially the nature of fire growth.

REFERENCES

[1] Quintiere, J. G., "Flashover", Proc. of the Eighth International Seminar on Fire and Explosion Hazards (ISFEH8), Hefei, China, 2016.

[2] Thomas, P.H., Bullen, M.L., Quintiere, J. G., and McCaffrey, B.J., "Flashover and Instabilities in Fire Behavior", *Combustion and Flame* 38: 1980, pp. 159-171.

[3] Quintiere, J. G. and Wang, Lei, "A General Formula for the Prediction of Vent Flows", *Fire Safety Journal*, v 44, n 5, p 789-792, July 2009.

[4] Graham, T. L., Simplified Equations for Flashover in Compartment Fire, Proc. of the Seventh International Seminar on Fire & Explosion Hazards (ISFEH7), Providence, RI, 2013.

VERTICAL FLAME SPREAD IN WOODEN CORNERS AS A FUNCTION OF FUEL MOISTURE CONTENT

Arjen Kraaijeveld & Torgrim Log*
Western Norway University of Applied Sciences, Haugesund, Norway

ABSTRACT

January 18th 2014 a severe fire destroyed 40 structures, including a listed building, and threatened Old Lærdalsøyri, one of the largest and best preserved old wooden towns in Norway. One of the key factors in that fire was the very dry indoor climate prior to the fire. The present work therefore investigates fire development in wooden corners as a function of the Fuel Moisture Content (FMC) to investigate how that single parameter influences vertical fire spread from a start fire at the floor level. Test corners made of 18 mm thick pine wood were conditioned to different FMC values and placed on a balance for mass recordings during the fire tests. This thickness ensured semi-infinite boundary conditions, i.e. no influence of the thickness during the fire development. Temperatures at 10 cm intervals were recorded by thermocouples and the fire development was video recorded for fire spread analysis. Cone Calorimeter test specimens were conditioned to varying FMC-values and tested at 20, 30 and 50 kWm^{-2} for establishing time to ignition and theoretically estimating time to flashover.

The he vertical fire spread was very much dependent on the wooden corners FMC values. Drier wood resulted in much earlier high temperatures and Heat Release Rates (HRR). It was, however, observed that the early developing fires declined for some minutes before a delayed new fire development took place. More humid test corners resulted in delayed fire development, but often resulted in similar temperatures as observed for the drier corners. This may be a result of longer period of drying and more pre heating before these fires developed. From a fire risk perspective, the results are quite conclusive. Wood conditioned in air of e.g. 20% Relative Humidity (RH) represents a much more severe fire risk than wood conditioned in accordance to standard test conditions, i.e. 50% RH.

INTRODUCTION

During the previous centuries fire was one of the most severe threats to villages, towns and cities. The wooden constructions were vulnerable to devastating conflagrations. Norway has a rich tradition with respect to wooden constructions, with the 28 remaining Viking Stave Churches, as well as Viking ships preserved in sea mud or at burial sites, being the most known wooden constructions. Due to the humid climate in major parts of the country most wooden constructions have, unfortunately, perished during the centuries due to natural wood deterioration processes. The Hanseatic League, and associated economic trade, resulted in growth of coastal towns and general prosperity. Abundance of fish, wood and other valuables for the continental market represented lucrative merchandises both for the natives and foreign traders. The result was a blooming development of coastal wooden towns from Oslo to Trondheim. The city of Bergen (meaning mountains in German) is the most known Norwegian Hanseatic town – and one of the most haunted with respect to major conflagrations.

In general, the coastal towns of Norway were located very close to the local protected harbour. Given hilly coastal areas this often resulted in very densely built towns, very susceptible for fire spread if one building was ignited. Most towns therefore suffered from severe conflagrations several times each century. In Norway, most conflagrations happened during the period January to April/May. There are examples of two such conflagrations within a few years resulting in complete recession changing vibrant towns to abandoned areas. The most well-known conflagration is the one in Ålesund, January 23rd 1904. The fire brigades were not able to control or extinguish the fire in the first house and storm strength (22-30 m/s) winds spread the fire through the town area. The wind direction changed during the blaze compounding the already challenging situation. That fire was one of the most severe fires in Norway's modern history, killing one person, destroying 850 structures and resulting in 10.000 - 12.000 homeless people. The fact that major conflagrations were most frequent in the period January to April/May, while the windiest season is generally the period August to December, raises some questions as to whether other factors than strong winds may play a role in conflagration risk.

There are still 180 sites with quite well preserved wooden houses from the years 1700-1900 in Norway. Among these, the Bryggen harbor area in Bergen and the quite atypical Røros mountain mining village are on the UNESCO World Heritage List. In order to be preserved and cared for, the best way to protect the majority of these historic building sites is to allow the structures to remain a part of the daily village or city life. This will, of course, expose them to the fire risks of the modern society.

January 18th -19th 2014, a severe fire destroyed 40 structures, including 4 historical buildings, one of these a listed building, and threatened Old Lærdalsøyri, one of the largest and best preserved 17th – 18th century wooden towns in Norway. One of the key factors in this devastating fire was the very low Fuel Moisture Content (FMC) of inhabited (heated) buildings prior to the fire[1]. Storm strength wind was the other key factor[2]. A few days later, the Résidence du Havre nursing home fire[3], Quebec, Canada, killed 32 people and injured 15. The statement describing this fire disaster was: "A dangerous situation quickly turned into a disaster". We all know that dry wood burns better that more humid wood. Dramatic fire development resulting in the fire brigades arriving to late, i.e. at a point where the fire is beyond control, may be related to low FMC of inhabited (heated) buildings in general.

Together with the excessive heating in cold winter weather and candle lights, etc. increasing ignitions probabilities, this dry wood fast fire development may to a large extend explain the higher frequency of conflagrations in the period of January to April/March associated with very low ambient dew point. This is, however, barely mentioned in the literature[4,5] until Log[1] analysed the fire in Lærdalsøyri in 2014. Recently, Kraaijeveld, et al.[6] demonstrated that the time to flashover in ¼ scale wooden compartments is very dependent on the wood FMC. This again depends on the indoor relative air humidity (RH), which becomes very low in cold climate zones due to the low water vapour pressures at low ambient temperatures. Earlier flashover in dryer wooden compartments may partly be explained by faster initial flame spread.

The present paper aims at studying vertical flame spread and potential fire development towards flashover versus wood FMC. Pine (18 mm thick) was used to make representative wooden corners, placed on a balance for mass recordings during the tests. A 3.0 kW methanol fire, resting on a separate balance at the floor level, was used as the ignition source, i.e. similar to the start fire used by Kraaijeveld, et al.[6] The quite transparent methanol flame allowed for studying wood surface carbonization, as well as observing wood pyrolysis luminous flames for flame height recordings. Thermocouples were used to measure temperatures at 10 cm intervals vertically from the methanol fire for temperature development and flame height analysis. The tests were recorded by video for visual post fire analysis. Wood samples were also tested in the ISO 5660 Cone calorimeter for measurements of time to ignition and Heat Release Rate (HRR) at 20, 30 and 50 kWm^{-2} heat fluxes. The main parameter controlled was the wood FMC and the results will be discussed with respect to the recorded observations.

FIRE DYNAMICS

Fire development towards flashover

According to Babrauskas and Peacock[7], the single most important parameter that characterizes fire behavior is the Heat Release Rate (HRR), which may be expressed by:

$$\dot{Q}_c = A_f \cdot \dot{m}_f'' \cdot \chi \cdot \Delta H_c \quad \text{(W)} \qquad [1]$$

where A_f (m^2) is the fuel surface area, $\dot{m}_f''$ (kgm^{-2}s^{-1}) is the released mass flux from the fuel surface, χ is the burning efficiency (≤ 1) and ΔH_c (Jkg^{-1}) is the heat of combustion of the actual fuel. With only a few exceptions, such as solid phase oxidation of char coal and smouldering, the combustion process takes place in the gas phase (flame) where the volatiles (pyrolysis products) from the involved fuel react with entrained oxygen. The heat balance governing the production of volatiles may be expressed by the net heat fluxes to the surface and heat consumed for wood pyrolysis and water evaporation:

$$\dot{Q}_F'' + \dot{Q}_E'' - \dot{Q}_L'' = \dot{m}_f'' \cdot L_V \quad \text{(Wm}^{-2}\text{)} \qquad [2]$$

where $\dot{Q}_F''$ (Wm^{-2}) is the heat flux supplied by the flame, $\dot{Q}_E''$ (Wm^{-2}) is the heat flux from any external heat source (such as distant flames, hot smoke layers or hot objects), $\dot{Q}_L''$ (Wm^{-2}) is the heat losses from the surface and L_V (Jkg^{-1}) is the latent heat of vaporization and pyrolysis. Rewriting Eq. 2 gives:

$$\dot{m}_f'' = \frac{\dot{Q}_F'' + \dot{Q}_E'' - \dot{Q}_L''}{L_V} \quad \text{(kgm}^{-2}\text{s}^{-1}\text{)} \qquad [3]$$

In compartment fires, flames and smoke are partly entrained at the ceiling level. As the smoke layer temperature increases, the heat flux from the smoke layer increases dramatically:

$$\dot{Q}_{E,S}'' = \phi \varepsilon \sigma T_S^4 \quad \text{(Wm}^{-2}\text{)} \qquad [4]$$

where ϕ is the configuration (view) factor, ε is the smoke layer emissivity, σ ($5.67 \cdot 10^{-8}$ Wm^{-2}K^{-4}) is the Stefan-Boltzmann constant and T_S (K) is the smoke layer temperature. The flame emissivity is given by:

$$\varepsilon = 1 - \exp(-K \cdot L) \qquad [5]$$

where K (m^{-1}) is the extinction coefficient and L (m) is the flame thickness in the direction of radiation. At smoke layer temperatures around 550-600°C, the fire generally passes through a stage of sudden increase in fire severity, i.e. flashover. In this phase, cooling the smoke layer to prevent flashover may represent an efficient way to limit further fire development[8]. Post flashover, all combustible surfaces are generally involved in the fire and the mass flow of fuel exceeds the amount being combusted by the air flow into the compartment. The excess fuel then burns as flames out of the ventilation openings giving increased potential for external fire spread.

Fire development towards flashover has been studied by numerous researchers after the connection between back radiation and fire development was acknowledged. The development of the ISO 5660 Cone Calorimeter by Babrauskas at NIST was a break through when it comes to standardized testing. The concept of exposing 10 cm by 10 cm test specimens to variable heat fluxes to record time to ignition, smoke production, heat release rate, etc. very much improved fire testing. As an example, Wang, et al.[9] suggested the following equation for predicting time to flashover based on cone calorimeter testing at 50 kW/m^2 heat flux:

$$t_{\text{FO}} = 0.0716 \frac{t_{\text{ig}}^{0.25} \cdot \rho^{1.72}}{\left(\dot{Q}_{300}''\right)^{1.30}} + 57.4 \qquad [6]$$

where t_{ig} (s) is the recorded time to ignition, ρ (kgm^{-3}) is the test specimen density and $\dot{Q}_{300}''$ (kWm^{-2}) is the heat release rate during the 300 s period after ignition.

Spread of flame in solid fuel fires

Assuming there is no heat production or heat consumption (like evaporating humidity, etc) in a solid fuel, the heat equation is expressed by:

$$\frac{\partial T}{\partial t} = a \cdot \nabla^2 T \quad (\mathrm{Ks^{-1}}) \qquad [7]$$

where t (s) is the time and a (m^2s^{-1}) is the thermal diffusivity given by:

$$a = k/(\rho \cdot C_P) \quad (\mathrm{m^2s^{-1}}) \qquad [8]$$

where ρ (kgm^{-3}) is the density and C_P ($Jkg^{-1}K^{-1}$) is the specific heat capacity. When a fire is heating solid objects there will initially be large heat losses into the solid due to the steep temperature gradients just inside the surface over the "depth of heating" expressed by:

$$\delta = \sqrt{at} \quad (\mathrm{m}) \qquad [9]$$

Assuming that wood ignites at a certain surface temperature, T_{ign}, usually in the range 325-350°C, solving Eq. 7 for a constant heat flux to a semi-infinite solid and eliminating external heat losses, the time to ignition may be estimated by[10]:

$$t_{ign} = \frac{\pi}{4} k\rho C_p \left[(T_{ign} - T_o)/\dot{Q}''_E\right]^2 \quad (\mathrm{s}) \qquad [10]$$

After ignition, the mass flux (Eq. 3) increases due to heat flux from flames, hot smoke and other hot fuel surfaces as well as less heat losses with time due to the increasing "depth of heating".

VERTICAL FLAME SPREAD IN WOOD

For a wooden corner set-up with a given external start fire, such as a pan of burning liquid, different mechanisms may control the fire spread dependent on the wood Fuel Moisture Content (FMC). For high FMC values, drying the wood requires a significant amount of heat. For temperatures below the boiling point of water, the drying mass flux within the solid wood is given by Fick's law:

$$\frac{\partial C}{\partial t} = D_W \cdot \nabla^2 C \quad (\mathrm{kgm^{-3}s^{-1}}) \qquad [11]$$

where C (kgm^{-3}) is the water concentration and D_W (m^2s^{-1}) is the diffusion coefficient of water in wood. The drying time is proportional to the inverse of the thickness squared, i.e. thin objects dry comparably faster (similar to the depth of heating, Eq. 9). Gas phase diffusion is given similar to Eq. 11 where D_W is substituted by the gas phase diffusion coefficient. For temperatures above the boiling point of water, the excessive pressure is the driving force of the drying mass transport. According to Baronas, et al.[11] typical diffusion coefficients of water in wood are in the range $1\text{-}5 \cdot 10^{-10}$ m^2s^{-1}, i.e. two orders of magnitude lower than wood thermal diffusivity ($8 \cdot 10^{-8}$ m^2s^{-1}). In fire scenarios, the heat flow then dictates the wood drying and water generally boils off in an excess pressure driven flow mode.

High FMC will in principle limit the fire development, i.e. fire spread, in several ways. The heat of combustion (ΔH_c) will decrease with added water. The combustion efficiency (χ) will also be reduced.

This would give less heat flux from the flame to the fuel surface ($\dot{Q}_F''$) and less external heat flux ($\dot{Q}_E''$) to any potential new fuel (Eq. 3 and Eq. 4). High FMC also gives increased thermal conductivity, i.e. increased internal heat losses ($\dot{Q}_L''$). Water adds latent heat of vaporization, i.e. increased (L_V). The 1700 fold expansion upon evaporation at 100°C dilutes oxygen levels and increases gas phase heat capacity. These parameters add up to give low Heat Release Rate (HRR) and slow spread of fire for high humidity content wood and may even completely stall the fire spread. On the contrary, the drier the wood, the worse all these parameters are with regard to fire development and fire severity.

Flame spread over liquid surfaces is generally observed to be governed by flashing flame spread. This is due to the surface force driven flow of hot liquid over colder liquid and flashing between the actual (time dependent) location of the fire point and the flash point, followed by the continuous flame front. Upwards fire spread on vertical wood surfaces may not necessarily be that straight forward. For vertical wooden surfaces, the transient heating and associated heat losses, as well as surface volatiles release and char formation, depletes the wood surface layer from the lightest pyrolysis products. This introduces a less combustible layer of high emissivity insulating deeper parts of the wood. External, as well as internal, heat losses may then be quite significant after each flash or short period of vigorous burning. Very dry wood, which ignites fast, experiences particularly large heat losses immediately after the first early major flash. This may significantly delay further combustion. A several minutes pre heating period may follow before the internal heat losses are compensated. A wood construction of (slightly) higher FMC may, however, igniting later may be more preheated before the onset of the first flame flash. It may then experience self-sustained continuous burning even earlier than the driest wood.

Whether the dry wood results in a very fast and severe fire development depends on several factors, such as the size of the start fire and the fire environment. In a compartment configuration, external heat radiation from a developing smoke layer may make the fire development pass the threshold for heat release rate versus heat losses and result in a very fast flashover. To consider this was outside the scope of the present work, but must be kept in mind in compartment fire configurations.

EXPERIMENTS

Pine (18 mm thick) was used as a representative wooden fuel. Corners were made from glued pine hobby boards (18 mm thickness and 300 mm width) cut to 1200 mm height. Two of these plates were put together to make a 90° test corner. Prior to testing the wooden corners were first conditioned in air of ~10% Relative Humidity (RH) for 48 hours. All surfaces were exposed, i.e. 9 mm "drying depth". 10 cm by 10 cm test specimens were preconditioned in the same way for Cone Calorimeter studies. The test specimens were then conditioned to constant relative humidity in the range 20 - 80% RH until equilibrium was achieved. This range corresponded to about 3.5 to 14% Fuel Moisture Content (FMC). Dummy test specimens were treated similarly and used for establishing the dry weight base for FMC calculations prior to, and after oven drying 48 h at 105°C.

Cone calorimeter testing was done in accordance to the ISO 5660 except that the ISO standard requires the test specimens to be equilibrated at 50% RH. The delay from the conditioning to the cone calorimeter testing was kept as low as practicable to avoid test specimens gaining or loosing humidity before testing. Test specimens were exposed to 20 kWm^{-2}, 30 kWm^{-2} and 50 kWm^{-2}, respectively.

The FMC conditioned wooden corners were placed on light weight concrete plates (Jøtul JGFW-5 calcium silicate panel 50 mm thick, density ~225 kgm^{-3}) resting on a balance (DS 30k0.1, 30 kg 0.1g, Kern instruments) for mass recordings during the fire tests. A 3.0 kW (9 x 9 cm square cup burner) methanol start fire, resting on a separate balance (DS 3k0.01, 3kg 0.01 g, Kern instruments), ensured visibility during the early stage of testing due to the nonluminous methanol flame. Type K thermocouples (1.5 mm diameter, MRC HypTeck AS) were located at 10 cm intervals vertically in the center of the corner and about 2 cm from the wood surfaces. Thermocouples exposed to such optical thin flames will, due to radiative losses and later on radiative gain from wood char, not necessarily record correct gas temperatures. For comparison, they will however, be telltales when comparing temperatures recorded during the vertical flame spread for corners of different FMC values.

The vertical flame spread was video recorded for later flame spread analysis. The tests were terminated after about 25 minutes. A corner of noncombustible building materials (Jøtul JGFW-5 calcium silicate panel 50 mm thick, density ~225 kg/m^3) was also tested in the same corner configuration to establish a representative temperature field from the methanol fire without any wood drying or combustion (but still with heat losses to the corner material).

The 10 cm by 10 cm test specimens were tested in a Cone Calorimeter at 50 kWm^{-2} heat flux. Time to ignition was measured as well as parameters needed for estimating time to flashover in Eq. 6 (Wang, et al. 2012). This allowed for comparing the results obtained with recent ¼ scale compartment tests (Kraaijeveld 2016). Cone Calorimeter tests were also performed at 20 and 30 kWm^{-2} representing conditions more relevant for early stage optically thin flames spreading prior to flashover.

RESULTS

Mass loss of the noncombustible corner and mass loss of the methanol start fire are presented in Fig. 1. It is evident that the noncombustible corner did lose some mass, most likely water, during the fire test. The temperatures recorded for this noncombustible corner, as shown in Fig. 1b, are therefore a little lower than it would have been if this corner did not lose any mass at all.

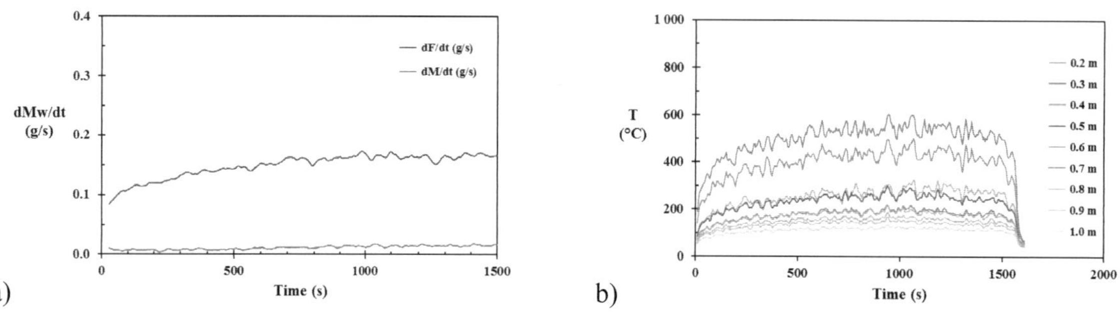

a) b)

Fig. 1 a) Mass loss rate of the noncombustible corner, dM/dt (g/s), and mass loss rate of the methanol fuel (dF/dt) as a function of time.
b) Temperatures recorded for the noncombustible corner as a function of time and height.

Temperatures recorded for representative dry and humid wood corners, respectively 3.5% FMC and 13.5% FMC, are shown in Fig. 2. It is evident that the initial temperature development was faster for the dry wood corners. The average temperature of three dry wooden corners (4.1% FMC), medium humid corners (9.4% FMC) and humid corners (13.4% FMC) recorded at 0.5 m and 1.0 m above the start fire is shown in Fig 3. Trend lines and temperatures recorded for the noncombustible corner are added in Fig. 3. It is evident that there is a general difference in the appearance of the temperature versus time development. The driest corners tend to give high temperatures in the early stage of the fire development and lower temperatures in the later phase of the corner fire. This is opposite for the more humid corners.

The higher initial temperatures for the dry corners are explained by less water content. The lower temperatures in the later phase of fire development may be explained by an early combustion depleting the combustible surfaces from volatiles. After a prolonged time of pre heating (drying) the more humid wooden corners they give comparably higher temperatures. The longer preheating period results in less internal heat flow. When the fire involves this fuel in a later phase it is both dried and preheated.

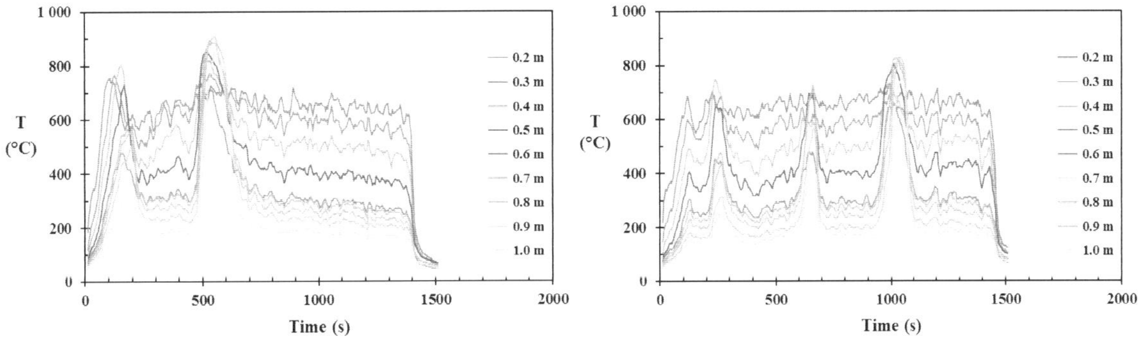

Fig. 2 Recorded temperatures as a function on time: left 3.5% FMC and right 13.5% FMC.

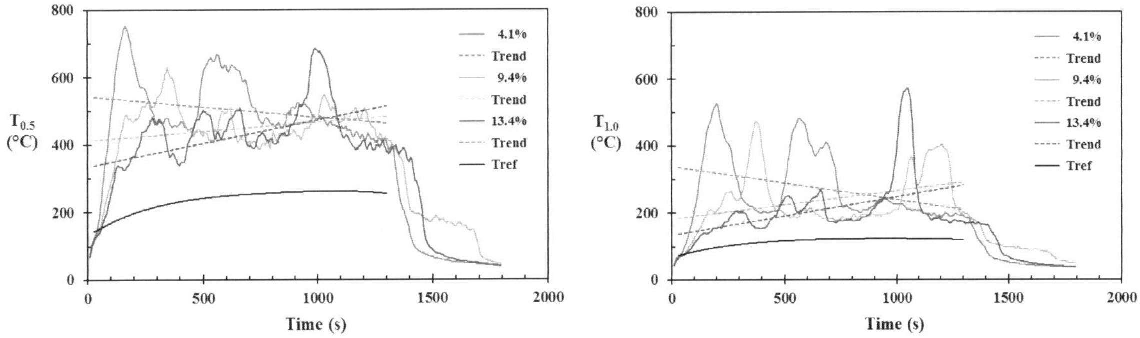

Fig. 3 Temperatures at 0.5 m (left) and 1.0 m (right), linear trend lines and temperatures of the noncombustible reference fire as a function of time.

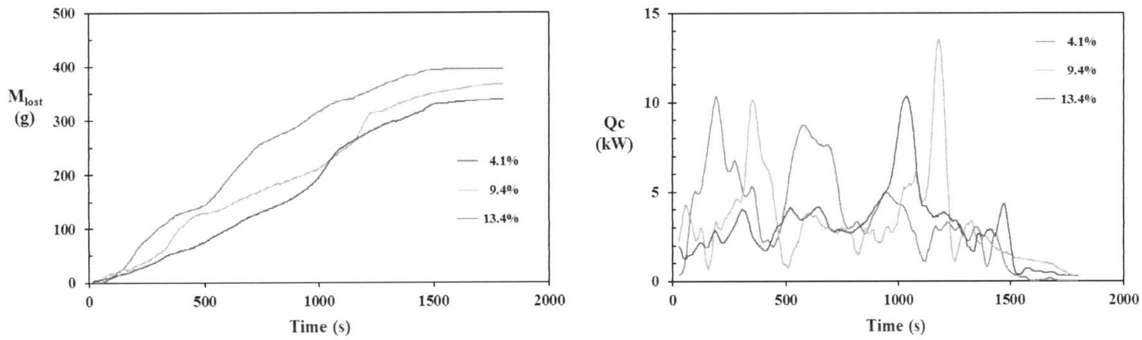

Fig. 4 Dry wood equivalent mass loss (left) and corresponding Rate of Heat Release (left) for 4.1%, 9.4% and 13.4% FMC wooden corners (assuming ΔH_c = 20 MJ/kg and χ = 0.8).

The equivalent dry wood mass loss as a function of time for the respective groups of wood FMC values, i.e. 4.1%, 9.4% and 13.4%, is shown in Fig. 4 together with the calculated Heat Release Rates (assuming $\Delta H_{c,wood}$ = 20 MJ/kg and χ = 0.8). It is clearly seen from Fig. 4 (left) that the dry wood shows significantly higher HRR in the early stage of the fires. This may be an explanation for the earlier flashover observed for dry ¼ scale compartments in the study by Kraaijeveld, et al. (2016). It is clearly seen that the heat production is higher for the dry wooden corners the first 2-3 minutes after start fire ignition. In a compartment configuration this may result in early flashover. If not, the flames may for a while decline and eventually increase again after sufficient wood pre heating.

Time to ignition in the Cone Calorimeter as a function of FMC values for heat flux 20 and 30 kWm^{-2} is shown in Fig. 5. The recorded time to ignition at 50 kWm^{-2} is as shown in Fig. 6 (left). Given representative values of wood thermal properties, as given in Table 1, the corresponding time to flashover calculated by Eq. 6 are given in Fig. 6 (right). It turns out that Eq. 6 results in a very low time to flashover dependency on the FMC values. This is counter to intuitive understanding of dry wood behavior in fires, and quite far from the recent measured time to flashover by Kraaijeveld, et al. (2016) for ¼ scale compartment fires. In that work, a more exponential dependency, as seen in Fig. 7, was obtained.

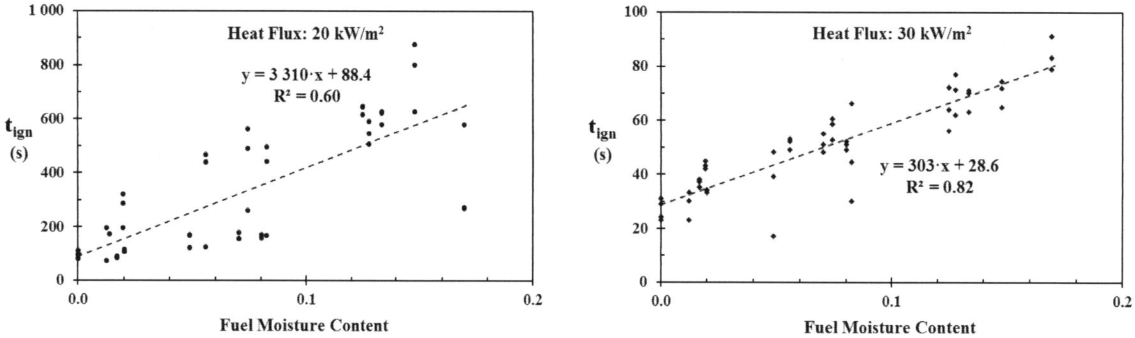

Fig. 5 Time to ignition in the Cone Calorimeter for test specimens of varying FMC values for heat flux 20 kWm^{-2} (left) and 30 kWm^{-2} (right).

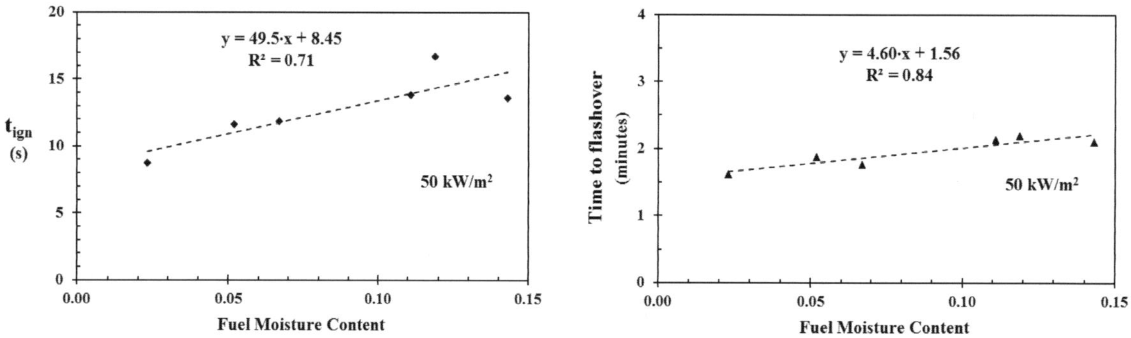

Fig. 6 Time to ignition in the Cone Calorimeter for test specimens of varying Fuel Moisture Content (left) at 50 kWm^{-2} and time to flashover based on Eq. 6 (right).

The Equilibrium Moisture Content (EMC) corresponding to 50% RH (standardized testing) and 20% RH (typical indoor RH values in cold climates) is according to Log[1] 9.3% and 4.5% FMC, respectively. Using these values and the linear equations in Fig. 5 and Fig 6 (left), the ignition at the dry (20% RH) condition happens at 0.82, 0.78 and 0.62 time the respective ignition times at standardized testing using the data recorded at 50, 30 and 20 kW/m^2, respectively. When wood is just ignited, from there on and during the subsequent fire development, there is usually not a constant external heat source other than the start fire. The fire is therefore considered to be self-supplied. The more humid the wood is the less heat supply and the more heat losses due to evaporation of involved water. It may therefore be argued that using the times to ignition obtained at 50 kWm^{-2}, and even 30 kWm^{-2}, may not give a representative picture of subsequent fire development. Probably the Cone Calorimeter results obtained at

20 kWm^{-2} are more representative, evidenced by giving results closer to the flame spread data recorded.

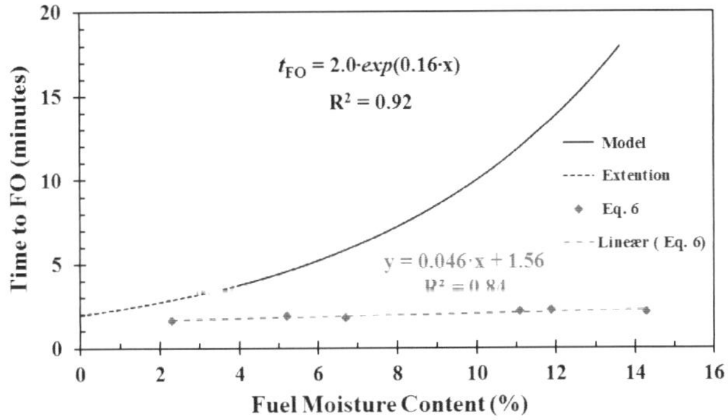

Fig. 7 Recorded time to flashover (FO) in ¼ scale wooden compartments (Kraaieveld, et al. 2016) and time to flashover based on Cone Calorimeter experiments at kWm^{-2} and Eq. 6.

In the upper part of the fire plume flame oscillations are normal. The height of the flame may often be defined as the height at which an average temperature in the range 500 - 600°C is observed. Fig. 8 shows representative flame spread 3 minutes after ignition for 3.5% and 13.5% FMC wood corners. It is clearly seen that the drier wooden corner is more involved in the combustion process.

Fig. 9 shows the temperature development for dry, medium and humid wooden corners until an average temperature of 500°C is observed at 0.5 m and 1.0 m. At 0.5 m height, the driest wooden corners reach 500°C much earlier than the medium and humid wood corners. At 1.0 m height, this trend is not that clear. Two of the medium FMC wooden corners (9.2% and 9.4% FMC) show flames at 1.0 m height about 200 s earlier than the driest corner (3.5% FMC). Drier wood does therefore not necessarily give earlier major flame spread illustrating the variations often experienced in fire testing. The trend is, however, quite clear, i.e. dry wood generally results in significantly faster vertical flame spread.

Average vertical flame spread obtained from the video recordings is shown in Fig. 10. As expected, it is clearly seen that the vertical flame spread is dependent on the FMC values and that the flame is spread faster in the drier corners compared to the more humid corners.

Table 1 Thermo-physical properties and Cone Calorimeter results (50 kW/m²)

FMC	Mass (kg)	Density (kg/m³)	Ign. time (s)	Q300 (MJ/m²)	FO time (s)
0.023	0.075	416.7	8.7	34.2	97.4
0.052	0.079	438.9	11.6	30.3	112.4
0.067	0.074	409.4	11.8	30.4	106.2
0.111	0.084	466.7	13.8	28.0	128.1
0.119	0.085	472.2	16.7	28.3	132.0
0.143	0.088	488.9	13.6	30.4	126.0

Fig. 8 Representative flame spread 3 minutes after ignition (left 3.5% FMC, right 13.5% FMC).

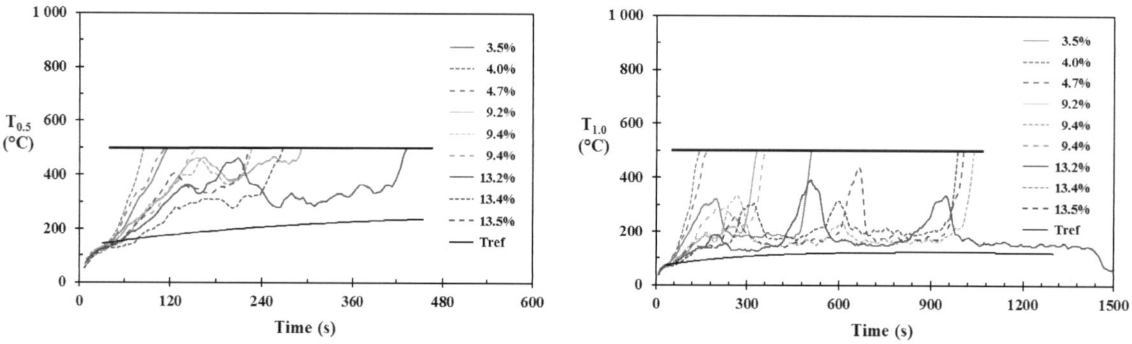

Fig. 9 The development towards continuous flame (defined as $T_{ref} = 500°C$) at 0.5 m (left) and 1.0 m (right) height above the start fire.

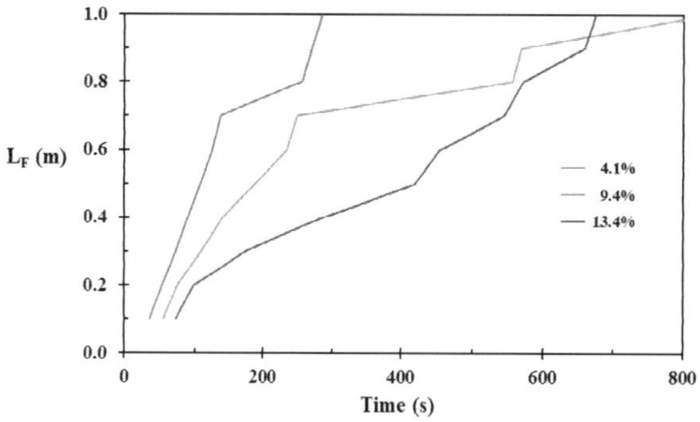

Fig. 9 Vertical flame spread based on video recordings for dry, medium and humid test corners.

DISCUSSION AND CONCLUSION

In the present work vertical flame spread in pine wood corners at varying Fuel Moisture Content was studied. The fire spread was assisted by a 3.0 kW methanol start fire at the base of 1.20 m tall wooden (pine) 18 mm thick corners. The wooden corners and the methanol fires rested on individual balances for separate mass loss recordings and subsequent Heat Release Rate (HRR) calculations. For the wooden corners, the equivalent dry mass, obtained by drying parts cut from the corners at 105°C prior to the fire spread tests, was used for HRR calculations. This set-up worked flawlessly throughout the measurement series.

Though there were large individual variations in fire development, nonlinear fire development as a function of Fuel Moisture Content (FMC), as observed by Kraaijeveld, et al. (2016) for ¼ scale wooden compartments, were also found in the present work. The vertical fire spread was very dependent on the wood corners FMC values.

The cone calorimeter measurements at 50 kWm^{-2} and the use of Eq. 6, did not support this finding. However, Cone Calorimeter testing at 20 kWm^{-2} heat flux showed results more in line with the corner fire tests.

Thin vertical flames have a small flame thickness, i.e. small optical path length. Thermal radiation is therefore very limited and the heat flux is dominated by convection. Even visible flames typically show a heat flux rate of 20 - 30 kWm^{-2}. Cone Calorimeter testing at these heat fluxes therefore gives a better picture of the dependency of vertical flame spread than testing at heat fluxes at 50 kWm^{-2}. In the Cone Calorimeter, the dominant heat flux is from the cone itself. In a compartment configuration, where a minor start fire ignites the wall, the burning wall materials act as the heat supply for convective heat transfer from flames and hot gases to virgin fuel. From the view point of the fire, this is a much more vulnerable fire spread mechanism. Humidity boiled off from the heated wood will then cool the area burning while passing through, dilute the pyrolysis product and act like a heat sink in the flames and the rising hot buoyant plume. These dehumidification processes will then influence the flame spread in a nonlinear way. For a given start fire, a sufficiently high FMC-level resulting in no self-sustained flame spread may exist. Testing this was, however, outside the scope of the present work.

The driest wooden corners quite early peaked in temperatures and (dry wood equivalent) mass loss rates, i.e. Heat Release Rates (HRRs). From that point on, temperatures and mass loss rates relaxed some before new peaks were observed several minutes later. This may be explained by the development of a protective and insulating char layer at lower elevations limiting the heat supply in a situation where the internal heat losses in the wood was still large due to a small depth of heating. A significant period of preheating was then needed for the next peak to develop, i.e. delays of several minutes were observed.

It may be questioned whether the later peaks in fire development could be related to the limited wood panel 18 mm thickness. Given a thermal diffusivity of about $8 \cdot 10^{-8}$ m^2s^{-1} and e.g. 1000 s heating period, the depth of heating $\delta \approx 9$ mm. In addition to this comes the material depth lost due to pyrolysis and combustion, at most 4-5 mm in the present tests. The wooden materials used may therefore be considered semi-infinite during the fire testing. Delayed peaks in fire development can therefore not be attributed to any reflection heat wave caused by limited wood thickness.

The development towards peaks in temperatures and mass loss rates was also observed for the medium (9.4%) FMC and the high (13.4%) FMC values. These peaks did, however, occur later. This may be explained by the fact that less heat was released and an extended drying period required before these wooden corners could achieve self-sustained burning. It was interesting, however, to notice that when delayed temperature peaks were achieved, temperatures and the dry wood equivalent mass loss rates were close to those of the dry wooden corners.

The dry wood fire risk implications are quite clear. A dry wood fire, i.e. very fast developing fire, represents a much higher threat to people, monetary values and the environment. Significantly less time is available for interfering with the developing fire and less time is available for escaping the dangers of the quickly developing fire. Fire fighters have significantly less time to respond before the fire is also threatening neighbour structures. Fires in dry wooden constructions therefore still represent a threat to wooden towns and single buildings, i.e. as it was centuries ago. This was evidenced January 18th 2014 in Lærdalsøyri and 5 days later in L'Isle-Verte, Canada.

In the modern society we do, however, have far better possibilities to monitor developing risks, such as monitoring indoor relative humidity as a proxy for potential fast fire development, as suggested in a recent study[12]. Knowing the risk, it may be far easier to manage the risk by introducing proper risk reducing measures.

REFERENCES

1. Log T (2015) "Cold Climate Fire Risk; a Case Study of the Lærdalsøyri Fire, January 2014", *Fire Technology*, **52** (6) 1825-43, DOI: 10.1007/s10694-015-0532-8

2. DSB Report (2014) "Brannene i Lærdal, Flatanger op på Frøya vinteren 2014, Lærepunkter og anbefalinger / The Fires in Lærdal, Flatanger and Frøya the Winter 2014, Learning Points and Recommendations", Norwegian Directorate for Civil Protection, Dec. 2014, 55 pp, ISBN: 978-82-7768-342-3 (In Norwegian)

3. Delâge C (2015) "RAPPORT DU COMMISSAIRE AUX INCENDIES DU QUÉBEC", Janvier 23 2014 L'Isle-Verte incendie, Québec, Canada (Coroners fire investigation report)

4. Pirsko AR and Fons WL (1956), "Frequency of Urban Building Fires as Related to Daily Weather Conditions", Interim Technical Report AFSWP-866, US Department of Agriculture, Forest Service, Division of Fire Research, Washington, DC.

5. Babrauskas V (2003) "Ignition Handbook", Fire Science Publishers, Issaquah, WA, USA, ISBN 0-9728111-3-3

6. Kraaijeveld A, Gunnarshaug A, Schei B and Log T (2016) "Burning Rate and Time to Flashover in Wooden ¼ scale Compartments as a Function of Fuel Moisture Content", *Proc. 10th Int. Fire Sci.&Eng. Conf, Interflam 2016*, Windsor, UK, 4th – 6th July 2016, 553-558

7. Babrauskas V and Peacock R (1992) "Heat Release Rate: The Single Most Important Variable in Fire Hazard", *Fire Safety J.*, **18**, 255-272, DOI: 10.1016/0379-7112(92)90019-9

8. Log T and Cannon-Brookes P (1995) "'Water Mist' for Fire Protection of Historic Buildings and Museums", *Int. J Museum Mgmt.. and Curatorship*, **13** (3) 283-98, DOI: 10.1016/0260-4779(95)00064-X

9. Wang ZZ, Hu XQ, Jia FC and Galea ER (2013) "A two-step method for predicting time to flashover in room corner test fires using cone calorimeter data", *Fire and Materials*, **37** (6), 457-473, DOI: 10.1002/fam.2139

10. Carslaw HS and Jaeger JC (1959) "Conduction of Heat in Solids" (2nd ed.). Oxford Science Publications, Oxford, ISBN: 0-19-853368-3.

11. Baronas R, Ivanauskasa F, Juodeikienėc I and Kajalavičiusc A (2001) "Modelling of Moisture Movement in Wood during Outdoor Storage", *Nonlinear Analysis: Modelling and Control*, **6** (2) 3-14

12. Log T (2017) "Indoor Relative Humidity as a Fire Risk Indicator", *Building and Environment*, **111**, 238-48, DOI: 10.1016/j.buildenv.2016.11.002

NUMERICAL MODELLING OF THERMAL CONDITIONS DURING FIRES IN CAVE-LIKE GEOMETRY

Virginie Drean(1)*, Gildas Auguin(1), Sharon Sarfati(1), Catherine Ferrier(2), Jean-Claude Leblanc(3), Delphine Lacanette(4), Jean-Christophe Mindeguia(4), Axel Bellivier(5)

(1) Efectis France, France (2) PACEA University of Bordeaux, UMR CNRS 5199, France (3) TRACES University of Toulouse, UMR CNRS 5608, France (4) I2M, University of Bordeaux, UMR CNRS 5295, France (5) Laboratoire Central de la Préfecture de Police, Paris, France

INTRODUCTION

An experimental characterization of fires trying to be representative from those used in Palaeolithic caves is performed. After an initial experimental phase to design these fires as part of the French collaborative research project CarMoThaP (Characterization and Modelling of Thermal Alterations and combustion residues on walls, University of Bordeaux, UMR 5199 CNRS PACEA), the characteristic quantities of the fires (heat release rate, mass loss, gas concentrations, ambient temperatures and temperatures in the flame) are obtained from a calorimeter hood.

The results acquired are then used as input data for numerical simulations performed with the CFD code FDS[1]. The fire model is first validated with a simulation of the calorimetric hood, and then applied to a cavity-like geometry, made in Urgonien limestone, taking into account the restricted ventilation conditions, and allowing predicting the ambient conditions during the fire. Determination of smoke propagation in the cavity, tenability conditions in terms of toxicity and temperature are then investigated.

Furthermore, these calculations[2,3] are likely to allow assessing the conditions in which thermal alterations of limestone (rubification, spalling) and carbonaceous deposits (soot, char) occurred, as observed in some Palaeolithic caves such as Chauvet-Pont d'Arc[4] (France) (Figure 1).

In this paper, the experimental characterization of fires, followed by numerical modelling of fire in a cave-like facility is discussed. Experimental and numerical results are compared and allow further study of thermal conditions in the facility and solicitations on walls leading to the observed thermal alterations.

Figure 1 : Thermo-alterations (rubification, spalling) at vault of Chauvet-Pont d'Arc cave (France)

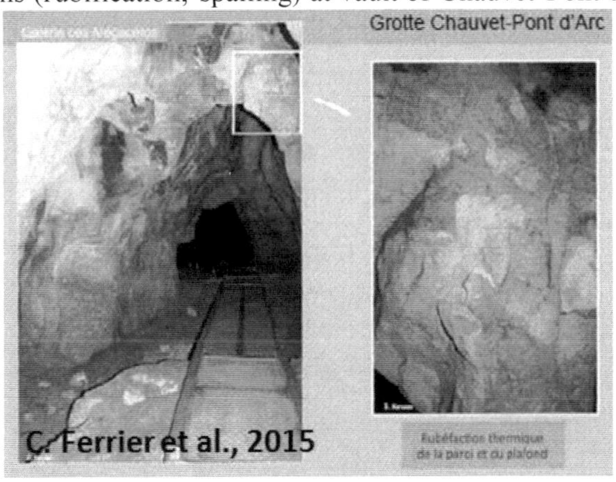

EXPERIMENTAL CHARACTERIZATION OF A FIRE APPLIED IN A CAVE-LIKE CONFIGURATION

Fire presumably representative from those occurred in Palaeolithic caves is experimentally characterized. This fire has been designed to reach the flames heights and temperatures leading to the observed thermal alterations of limestone[5].

At Efectis France lab of Maizières-lès-Metz, the large calorimeter hood provides basic features of fires such as heat release rate and chemical species production rates. A specific instrumentation (thermocouples, flow meters) is also used to determine the thermal stresses representative of a fire (Figure 4).

Numerous results are acquired for gas phase conditions, as ambient temperatures and temperatures in the flames. Wood mass loss during fire is registered as well as oxygen consumption, leading to the fire heat release rate and heat of combustion evaluation.

Preliminary design of fire source

In February 2016, a preliminary test campaign has been conducted in an outdoor platform. The tests were dedicated to design the fire sources used for the overall tests in cave-like conditions performed in the context of the research project CarMoThaP.

The selected fire source must provide thermal stresses on an Urgonien limestone target that led to the thermo-physical alterations observed in the Chauvet-Pont d'Arc cave. This target with dimensions 150 x 150 mm (large x width) is hold 2,5 m above the fire as representative of the vault height in the cave, and insulated by mineral wool on its back and lateral faces.

The fire source is made of Sylvester pine bundles. A total of 30 bundles are calibrated using individual branch with diameters from 1 to 4 cm. The average weight of a typical bundle is close to 5,5 kg, with a length of 80 cm (Figure 2).

Figure 2 : Typical calibrated wood bundle

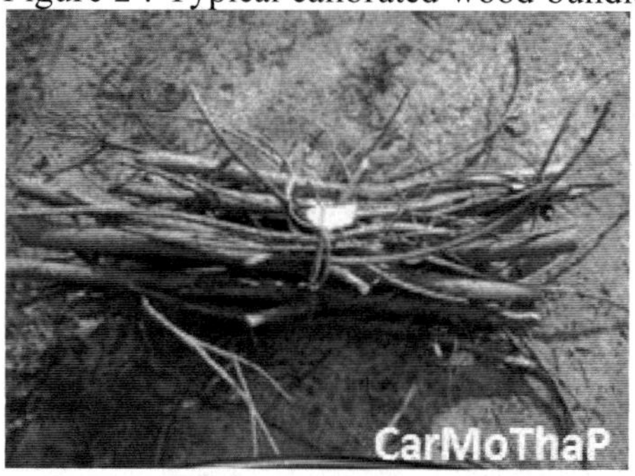

The fire source is then initially constituted with 4 bundles (Figure 3 a). The global dimensions of the fire source are then close to 800 x 800 x 900 mm (L x W x H). When partially consumed, additional bundles are added, almost every 4 minutes. It permits to obtain a flame temperature close to 400°C in the vicinity of the limestone target (250°C target surface temperature), corresponding to one possible thermal criteria leading to alterations of Urgonien limestone[5].

For this tests campaign, thermocouples were located at several heights above the fire, and inside the limestone target (Figure 3 b). The fire mass loss was recorded during the test.

The global methodology used in the cave fire tests is then defined in terms of bundle calibration, fire source disposal and temperature criteria for supplying the fire with additional bundles.

The designed fire source must then be characterized using a calorimeter hood providing basic features of fire such as heat release rate or chemical species production rates, and also to use a specific instrumentation to determine the thermal stresses representative of this fire.

Figure 3 : a) initial fire source with 4 bundles b) Instrumentation: thermocouples, balance and limestone target

Experimental characterization of the fire source

At Efectis France lab of Maizières-lès-Metz, the large calorimeter hood, with internal dimensions of 8,4 x 8,4 m was in March 2016 used to characterize the designed fire source. The main information provided will be used as input data for numerical simulation, and analysed to understand the fire behaviour in controlled ventilation conditions (Figure 4).

A total of 30 pine bundles are calibrated, with a total mass of 125 kg. Samples are selected to evaluate the wood hygrometry and placed in a stove with constant ambient temperature of 60°C. The wood humidity is close to 20%.

Wood mass loss during fire is recorded as well as oxygen (O_2) consumption and production rates of carbon dioxide (CO_2) and monoxide (CO), leading to the evaluation of the fire heat release rate and heat of combustion evaluation.

Urgonien limestone targets with dimension 150 x 150 x 100 mm (length x width x depth) similar to those used in the design phase of the fire source is hold 2,5 m above the fire to estimate the residual alterations after fire. Lateral and back faces were insulated with mineral fiber.

Figure 4 : Fire characterization testing under calorimeter hood

A specific instrumentation with thermocouples and heat flux meters is implemented to determine the thermal stresses inherent to this selected fire. Numerous results are acquired for gas phase conditions, as ambient temperatures and temperatures in the flames (Figure 5, Figure 6, Figure 7).

Gas phase instrumentation is constituted with nineteen 1,5 mm diameter Inconel thermocouples implemented above the fire, and at different locations radially to the fire. The Urgonien limestone targets are instrumented with four 3 mm diameter Inconel at 3 and 5 mm depth from the target exposed surface, and four K5 type thermocouples located on exposed and unexposed target faces. Two total flux meters with measure range of 0-50 kW/m² are disposed 1 m and 1,5 m radially from the fire, at 0,5 m above the mean fire height (i.e. 1,5 m from ground).

The experimental results achieved allow the understanding of the fire behavior and its growth when additional bundles are added during the test such as gas temperature above the fire increased. Limestone target surface temperatures are observed to be greater than 250°C, and lead to thermo-alterations.

Figure 5 : Instrumentation location above the fire and experimental results

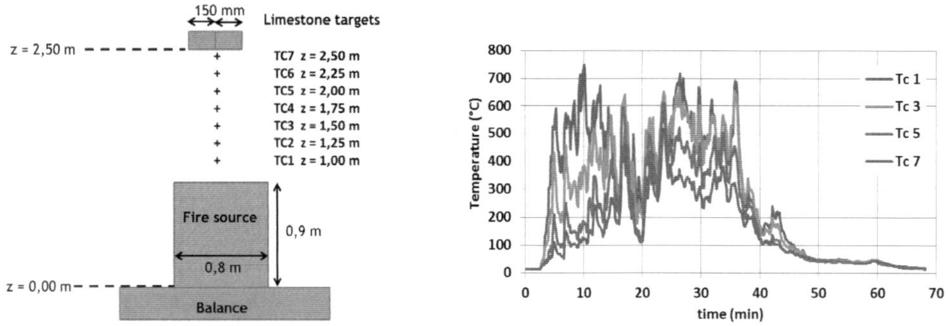

Figure 6 : Instrumentation location radially from the fire and experimental results

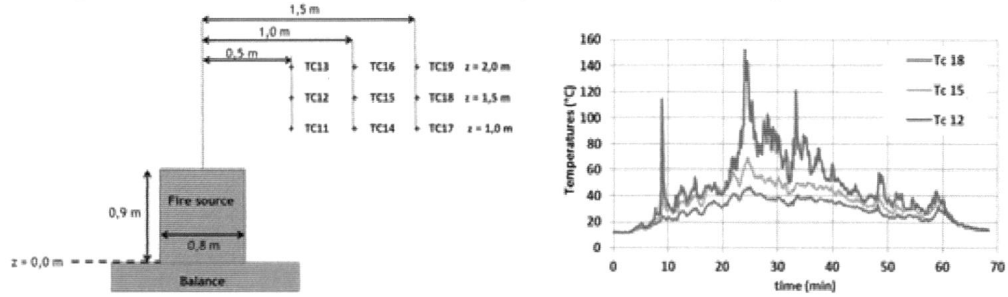

Figure 7 : Total heat flux radially from the fire and experimental results

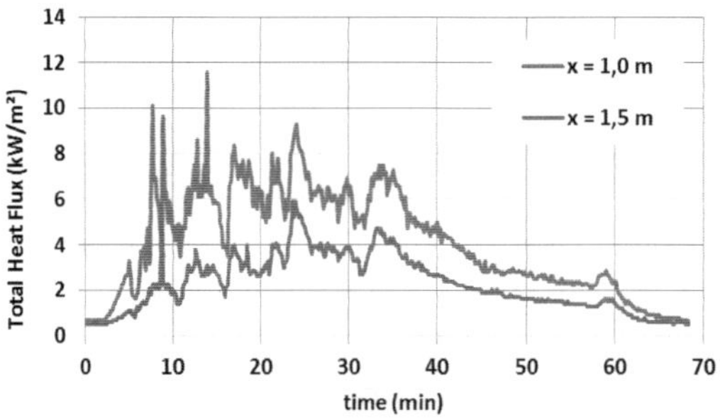

The heat release rate of the fire is evaluated thanks to oxygen O2 and CO/CO2 concentration, temperature conditions and volume flow conditions in the hood chimney venturi (Figure 8). The heat of combustion of the characterized fire is deduced from its mass loss rate. The measured value is close to 13

MJ/kg. This value is in agreement with bibliography commonly used for wood combustion (12-20 MJ/kg).

Figure 8 : Experimental heat release rate of the fire source

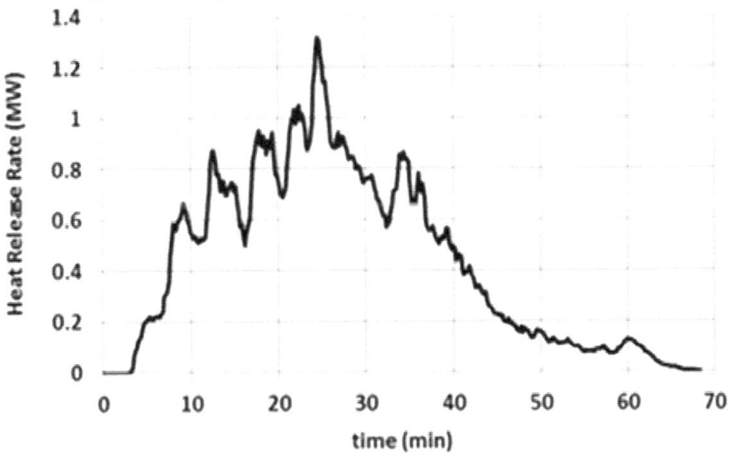

Each experimental measurement is then used to validate the numerical model of the fire source in the hood facility, before being implemented in a cave-like geometry.

Experimental fire applied to the cave-like conditions

The validated fire source and test protocol were applied in a Rupelian limestone quarry, located in Lugasson, South West of France, in May 2016. The aim of this last test campaign is to collect data for the hot gas temperatures in the volume of the cavity, the thermomechanical behavior of the wall material of the cavity and the nature of the thermal alterations of the wall.

The career facility is shaped in "L", with dimensions close to 9 x 9 m (L x W). Its height varies depending on the location in the cave, from 1,3 m to 2,5 m. Its walls are made of Rupelian type of limestone.

The test protocol in terms of fire source design (30 calibrated pine bundles), and additional supply to ensure gas temperature high enough to led to limestone alterations is slightly modified to the one validated during fire characterization under hood. A supply of 2 bundles is applied, although 4 bundles were used before.

The instrumentation location is schematized on Figure 9. It consists in thermocouple trees above and in vicinity of the fire, and at different locations inside the quarry and at the entrance, thermocouples at surface and in depth of an Urgonien limestone target above the fire, velocity and oxygen probes at the quarry entrance. Thermocouples were positioned mainly in the higher part of each thermocouple trees to capture the hot gas layer, and to quantify the stratified flows in the restricted ventilation configuration of the quarry.

Thereby, numerous results are acquired for gas phase conditions, as ambient temperature, temperature in the flames vicinity and at entrance, and flow velocity going out the gallery.

Acquired results for thermocouple trees located near the fire (A0 and A1), in the gallery (A4) and at the entrance (A9) are indicated on Figure 10 and Figure 11. Stratification of the flow is observed in the quarry and the temperature of the hot layer is up to 200°C. The hot gas layer height appears close to 1,3 m from the ground. At the cave entrance, the cold layer temperature is close to ambient, and the hot gases escaping from the gallery have temperature between 50 and 250°C.

Surface and in depth temperature on limestone targets above the fire are recorded and allow further analysis of the observed thermal alteration.

The results of this test will allow the validation of a fire model. It is also helpful to better understand the behavior of a fire in a cave-like geometry, and allows studying thermo-alterations of limestone..

Figure 9 : Measurement location in the limestone quarry

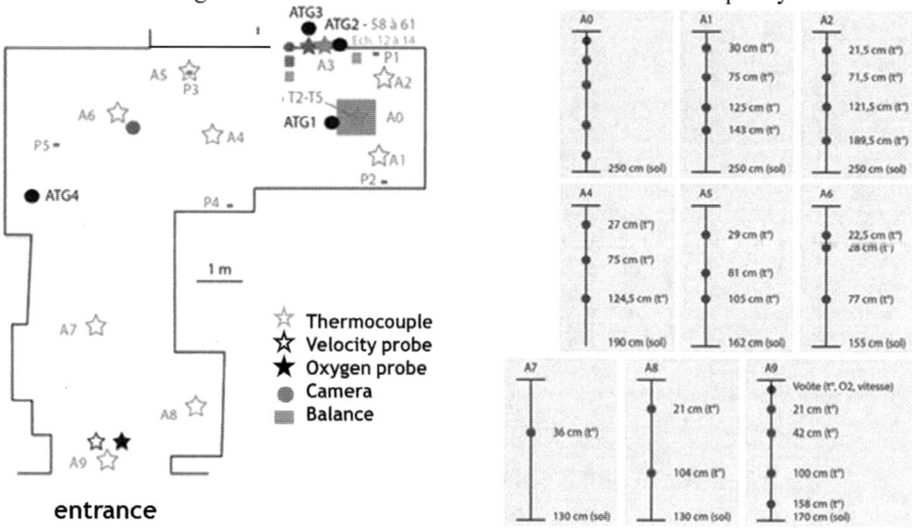

Figure 10 : Gas temperatures above and near fire in the limestone quarry

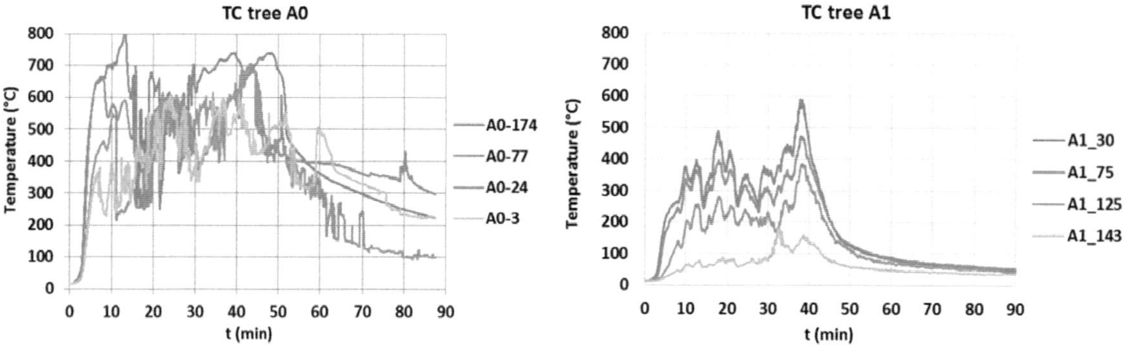

Figure 11 : Ambient gas temperatures in the quarry

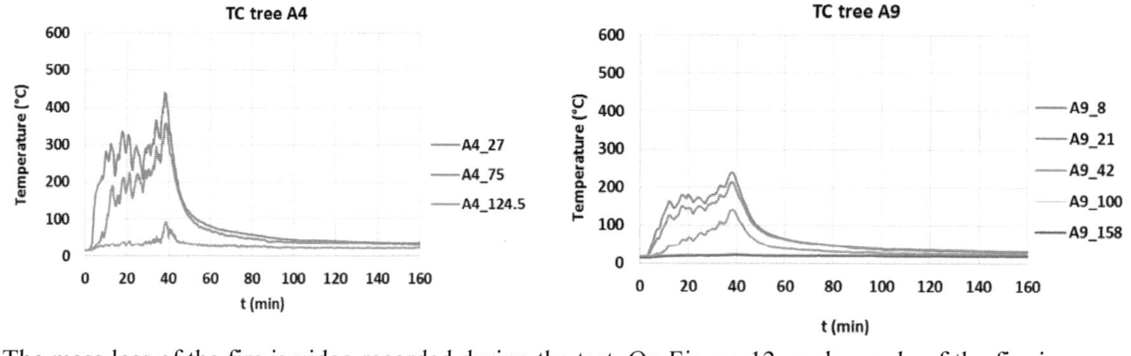

The mass loss of the fire is video recorded during the test. On Figure 12, each supply of the fire is indicated by a vertical green line. The evaluated heat release rate is compared with one characterized under the calorimeter hood, and similarity is observed. When evaluated using the mass loss of the fire, artefact are observed due to the local supply of the fire. However, the fire source is in the range of 1MW, as evaluated previously under the calorimeter hood. The repeatability of the test protocol is then observed and the main differences are due to restricted ventilation of the quarry.

Figure 12 : Mass loss of fire source – Evaluated heat release rate

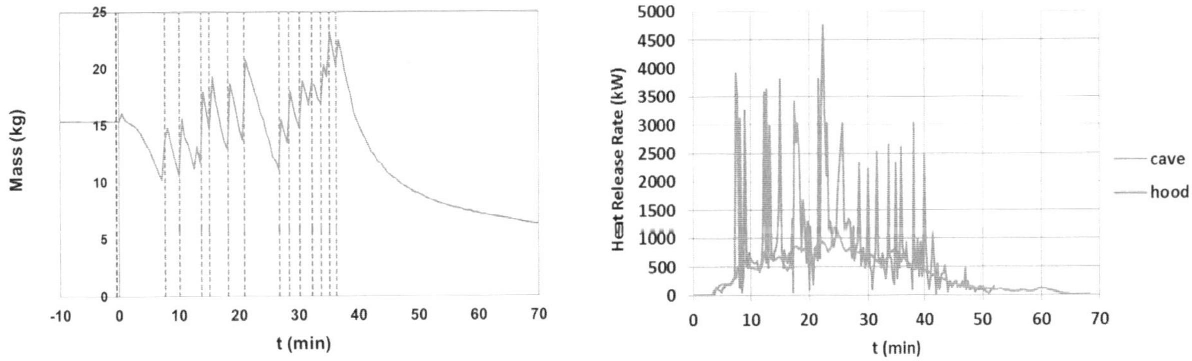

NUMERICAL MODELLING OF FIRE THERMAL SOLICITATIONS

The fire characteristics defined during experiment are used as input of the numerical simulations.

First, the fire load model is validated using a numerical simulation of the calorimeter hood. Then, the cave-like geometry is reproduced and the numerical model is validated in restricted ventilation conditions based on the comparison between numerical and experimental results obtained during fire test (Figure 13). The thermal solicitations received at walls are evaluated and allow qualitative prediction of the thermo-alterations of limestone during fires.

Computational code

The FDS[1] code version 6 is used to numerically predict the fire combustion and hot gases development in restricted ventilation conditions.

FDS is a computational fluid dynamics (CFD) code developed by the NIST and based on a finite volumes formulation for a regular and Cartesian grid. This tool allows 3D modeling of the computational domain. The code includes a combustion model with a mixture fraction approach and a large scale model (LES) for the description of turbulent flows. The hydrodynamic part of the system is similar to Navier-Stokes equations governing incompressible flow, by assuming a low Mach number formulation for decoupling the pressure of the associated velocity field calculation. The radiative transfer is described for a medium gray with a source term corresponding to the blackbody radiation. Absorption coefficients are based mainly on the soot radiation spectrum. Additional model for the soot deposition on walls is implemented in this version of FDS.

Numerical validation of the fire source model

The fire source model is firstly validated using a numerical simulation of the calorimeter hood used during the fire characterization test campaign and detailed in previous section of the document.

The calorimeter hood, with an enclosure of dimensions 8,4 x 8, 4 x 5 m (length x width x height under chimney Venturi), is modelled as shown on Figure 13. Cells with size 0,1 x 0,1 x 0,1 m are used, refined enough to capture the main combustion and ventilation phenomena. Thermal properties of walls (cellular concrete and gypsum board) and steel structure are included. Lateral ventilation openings and door are modelled with same conditions as during the real test campaign to ensure similar ventilation conditions in the hood. External domain with ambient temperature and pressure conditions is simulated far enough from hood to avoid the exterior open pressure boundary conditions effects.

The fire characteristics defined during experiment are used as input of the simulations, such as fire dimensions and heat release rate (Figure 8). The numerical instrumentation is implemented to follow the test configuration for thermocouples and flux meters (Figure 5, Figure 6 and Figure 7). Oxygen concentration, local pressure and temperature in the hood chimney are calculated as in the real facility.

Figure 13 : Numerical modelling of the experimental large calorimeter hood facility

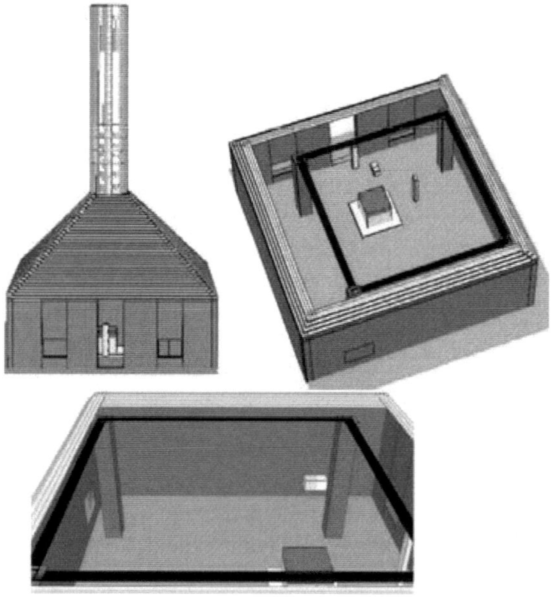

The numerical results achieved for gas temperatures above the fire at 1,5 m and 2,5 m from ground (TC3 and TC7, corresponding to 0,5 m and 1,5 m above the top of the fire) are compared with experimental data on Figure 14.

The numerical model allows the prediction of the thermal conditions in the fire area. The magnitude and the evolution versus time of temperature are fairly well reproduced.

An under prediction of the temperature for TC3 at the beginning of the fire during 8 minutes is observed. This is due to the fire ignition and propagation from a bundle to the other ones not well reproduced by numerical simulation. However, the maximum values of temperatures are not reached at this time, and temperatures are well evaluated from the beginning at higher altitudes above fire (TC7).

From 20 to 38 minutes after ignition, temperatures of 500 to 600°C are reached at TC3 location, and 300 to 350°C at TC7, 2,5 m above ground, corresponding to the altitude of the experimental quarry vault.

Figure 14 : Thermocouples above the fire - experimental (EXP) and numerical (FDS) results

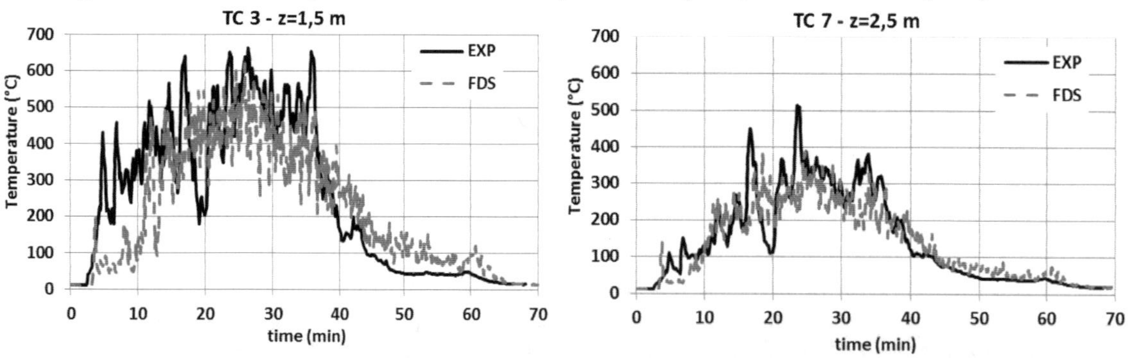

The numerical results achieved for gas temperatures radially at 0,5 m and 1,5 m (TC12 and TC18) from the fire at 1,5 m high from ground, corresponding to man height, are compared with experimental data on Figure 15.

The numerical model allows the prediction of the thermal conditions radially from the fire area. The order of value and the evolution versus time of temperature is fairly well reproduced for both locations. The maximum values of temperatures are close to 150°C locally (TC12). These peak values can be related to additional supply of the fire source.

From 20 to 38 minutes after ignition, temperatures of 80 to 100°C are reached at TC12 location. These values at man height are high to ensure safety conditions near the fire. At 1,5 m radially from fire, temperatures close to 40°C (TC18) are registered, low enough to permit human standing in this area.

Figure 15 : Thermocouples radially from the fire - experimental (EXP) and numerical (FDS) results

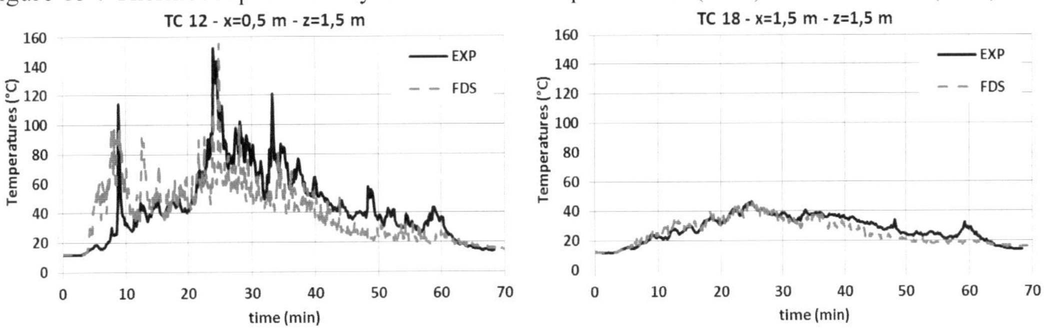

The numerical results achieved for total heat flux at 1 m and 1,5 m radially from the fire at 1,5 m high from ground are compared with experimental data on Figure 16.

The numerical model allows the prediction of the thermal stresses radially from the fire area with some over evaluation in the range of 2 kW/m². However, the order of value and the evolution with time of total heat flux is fairly good reproduced for both locations. The maximum values of heat flux are close to 10 kW/m² and 6 kW/m² respectively at 1 m and 1,5 m radially from fire. As observed for temperature, these values at man height are high to ensure safety conditions near the fire in this specific configuration.

Figure 16 : Flux meters at 1 m and 1,5 m radially from the fire - experimental (straight) and numerical (dotted) results

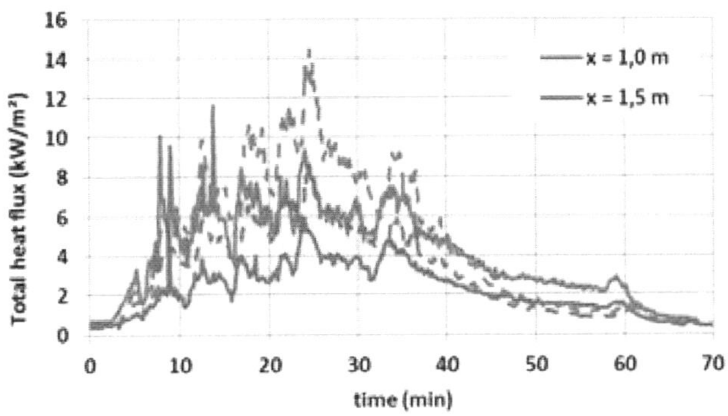

The numerical results achieved for gas phase quantities in the hood chimney (Venturi) are compared with experimental data on Figure 16. The numerical fire model is then validated regarding the values predicted in the hood Venturi. This model will allow the understanding of the fire behavior and its growth when additional bundles are added during the test in the cavity-like geometry.

Figure 17 : Oxygen concentration (%) and gas temperature in the hood chimney - experimental (straight) and numerical (dotted) results

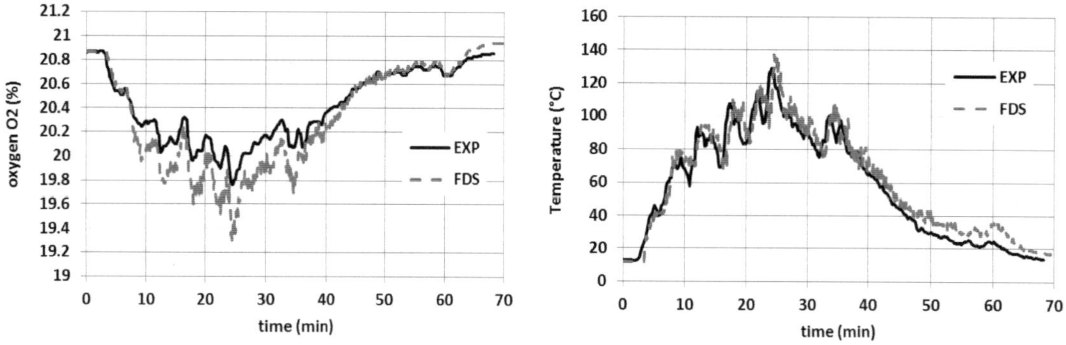

Evaluation of thermal conditions in the restricted ventilation quarry geometry

Once validated, the fire source model is implemented in the limestone quarry geometry detailed in previous section of the document.

The quarry facility, shaped in "L", with internal dimensions close to 9 x 9 m (L x W) is modelled as shown on Figure 18. The variation of its height from 1,3 to 2,5 m depending on the location in the quarry, is also simulated. Cells of 0,05 x 0,05 x 0,05 m are used, refined enough to capture the main combustion and ventilation phenomena, and to allow the complex geometry modelling. Thermal properties of walls (Rupelian type of limestone) are included, and roughness close to 5 cm is considered. The main ventilation opening is the quarry entrance, with dimensions of 2,5 m x 1,7 m (L x H) and is modelled to ensure similar ventilation conditions in the quarry as during the real test campaign. External domain with ambient temperature and pressure conditions is simulated far enough from the entrance to avoid the exterior open pressure boundary conditions effects.

The fire characteristics defined during experiment are used as input of the simulations, such as fire dimensions and heat release rate (Figure 12). However, the shape of the fire is kept constant, while it varies during test due to addition of the bundles.

The instrumentation location schematized on Figure 9 is implemented numerically: thermocouple trees above and in vicinity of the fire, and at different locations inside the quarry and at the entrance, thermocouples at surface and in depth of limestone target above the fire, velocity and oxygen probes at the quarry entrance.

Figure 18 : Numerical modelling of the experimental quarry facility

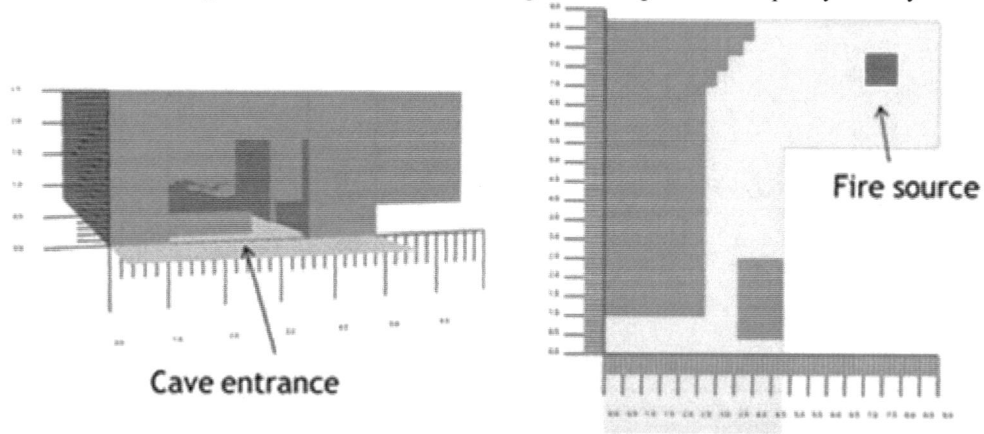

The numerical results achieved for gas temperatures above the fire (thermocouples tree A0) from 3 cm to 174 cm under the vault (corresponding to 2,5 m to 0,8 m above the ground) are compared with experimental data on Figure 19.

The predicted temperatures are close to the acquired experimental values. The maximum temperature reached above fire with simulation is close to 700°C, while a local increase of temperature to 800°C is observed experimentally at 10 minutes, mainly due to position of additional bundle to supply the fire.

The code fails to reproduce the temperature after 38 minutes of test, when the fire is no more supplied (decrease of heat release rate as observed on Figure 12). The temperature above fire stays relatively constant and high after 38 minutes experimentally because of the incandescent embers, not simulated numerically.

Figure 19 : Thermocouple tree above the fire (A0) - experimental (EXP) and numerical (NUM) results

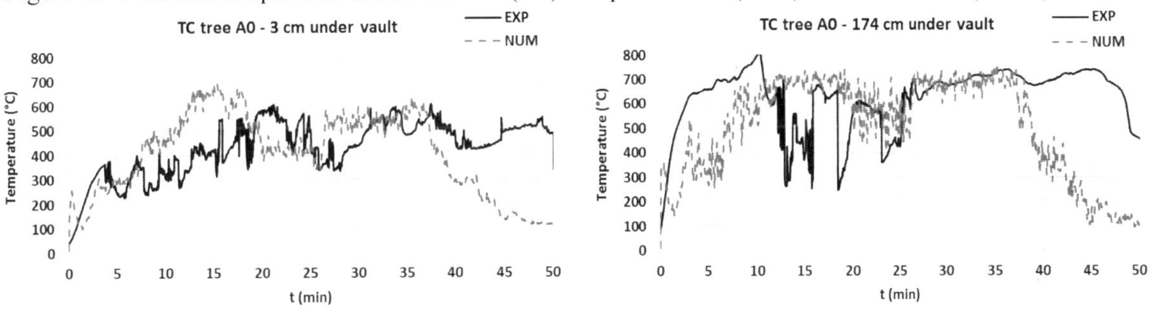

The numerical results achieved for gas temperatures in the vicinity of the fire (thermocouples tree A1) from 30 cm to 143 cm under the vault (corresponding to 2,2 m to 1 m above the ground) are compared with experimental data on Figure 20.

For this location in the quarry, the predicted temperatures are very close to experimental ones. The numerical simulation captures the main combustion and dynamics conditions in the fire vicinity. The flow stratification is correctly reproduced, with gas temperatures lower than 200°C at 143 cm from vault. The layer height seems close to 1,2 m from ground. High thermal solicitations exist in this part of the quarry, detrimental for human tenability.

Figure 20 : Thermocouple tree in the vicinity of the fire (A1) - experimental (EXP) and numerical (NUM) results

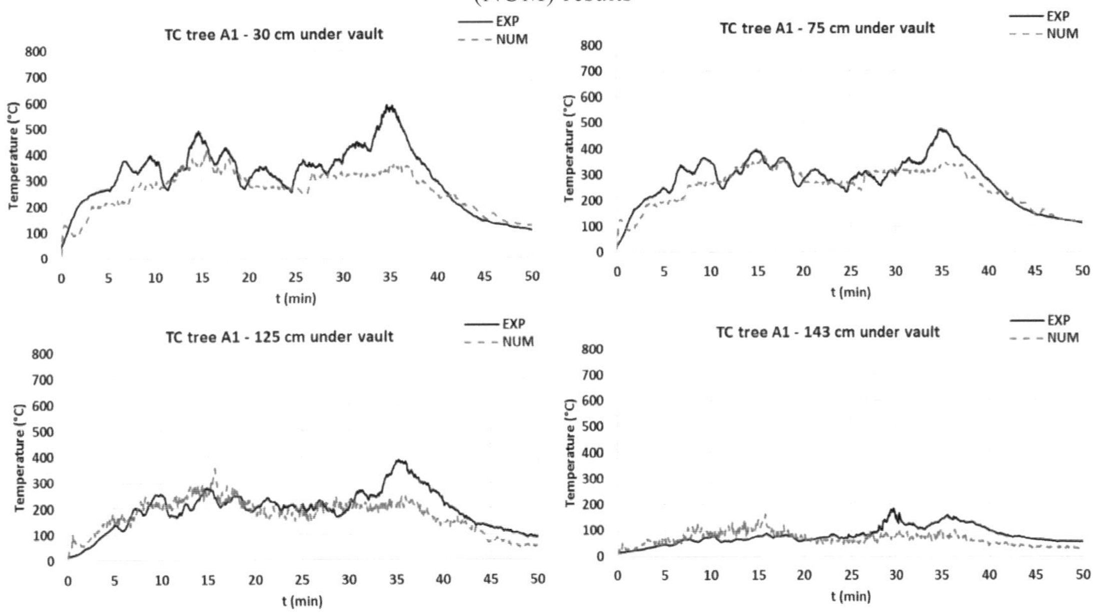

The numerical results achieved for gas temperatures in gallery (thermocouples tree A4) from 27 cm to 124,5 cm under the vault (corresponding to 1,6 m to 0,7 m above the ground) are compared with experimental data on Figure 21.

The predicted temperatures are very close to experimental ones for the hot gas layer (27 cm under vault); under estimation appears at the end of the fire, from 30 to 38 minutes of test. The cold gas layer is also correctly reproduced (124,5 cm under vault) numerically. However, despite the flow stratification is successfully captured by the code, the hot gas layer height appears overestimated by FDS: the temperature calculated 75 cm under vault (corresponding to 1,2 m above ground) is lower than experimental one. Again, high thermal solicitations exist in this part of the quarry, detrimental for human tenability conditions 1 m above ground.

Figure 21 : Thermocouple tree in the gallery (A4) - experimental (EXP) and numerical (NUM) results

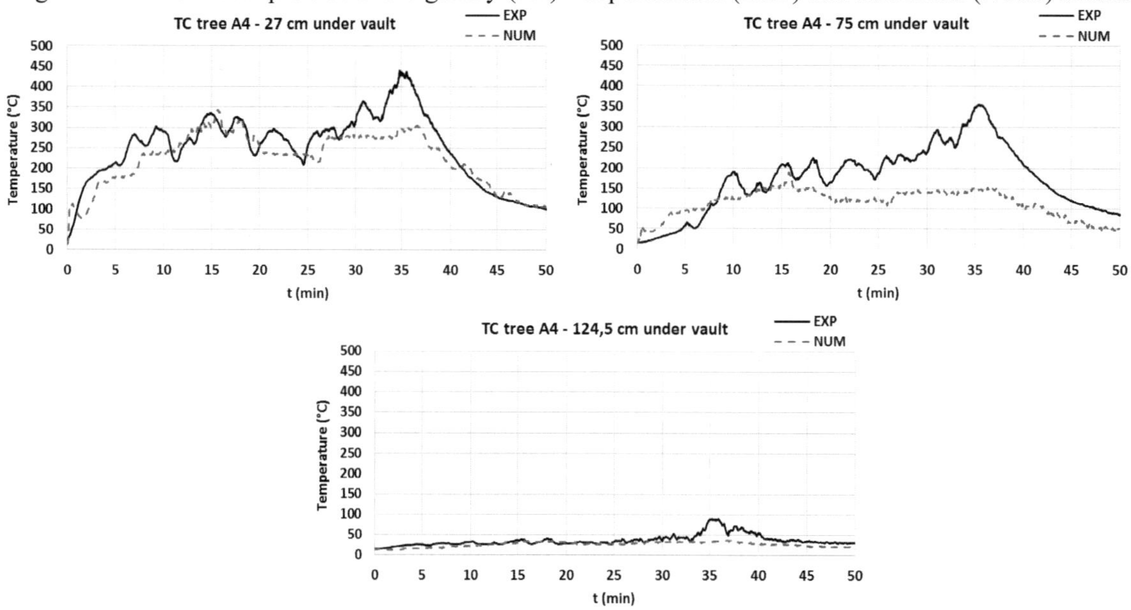

The numerical results achieved for gas temperatures at gallery entrance (thermocouples tree A9) from 8 cm to 100 cm under the vault (corresponding to 1,6 m to 0,7 m above the ground) are compared with experimental data on Figure 22.

As observed for other locations in the gallery, the predicted temperatures are very close to experimental ones for the hot gas layer (8 and 21 cm under vault). The cold gas layer is also correctly numerically reproduced (100 cm under vault). At the gallery entrance, the numerical model captures the flow stratification successfully. High thermal solicitations exist at the quarry entrance, detrimental for human tenability 1,2 m above ground, with escaping gas temperature close to 200°C.

Figure 22 : Thermocouple tree at the gallery entrance (A9) - experimental (EXP) and numerical (NUM) results

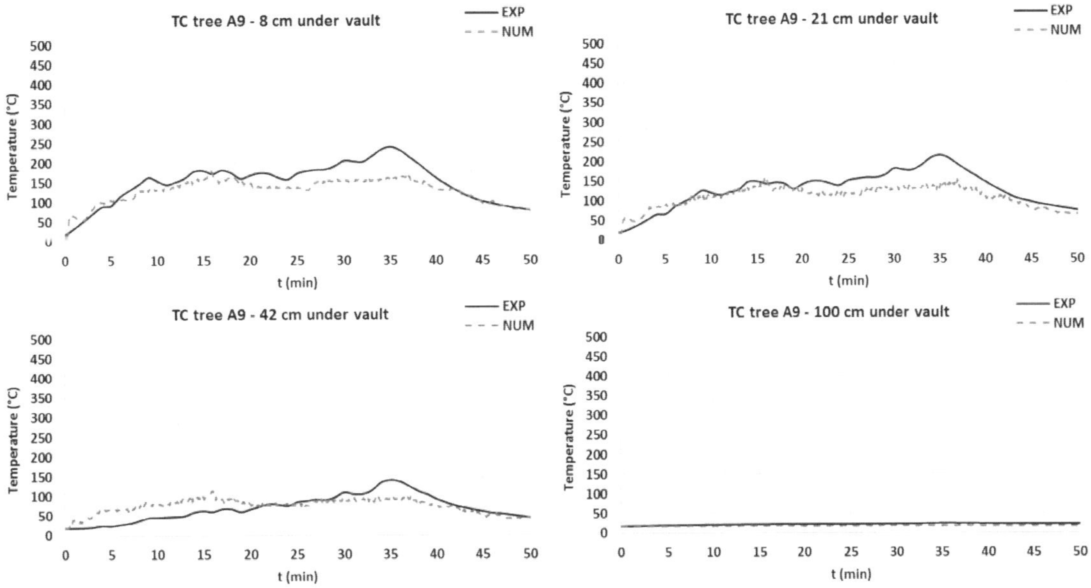

The thermal conditions in the quarry are correctly reproduced numerically at each instrumented locations. The code allows capturing the flow stratification and dynamics in this restricted ventilation configuration. Temperature fields evaluated numerically at time t=15 minutes are indicated on Figure 23. Plan and transverse view of the gallery are provided. The flow stratification is clearly observed.

Figure 23 : Numerical temperature fields for plan view (z = 1,3 m) and in the transverse axis of the gallery

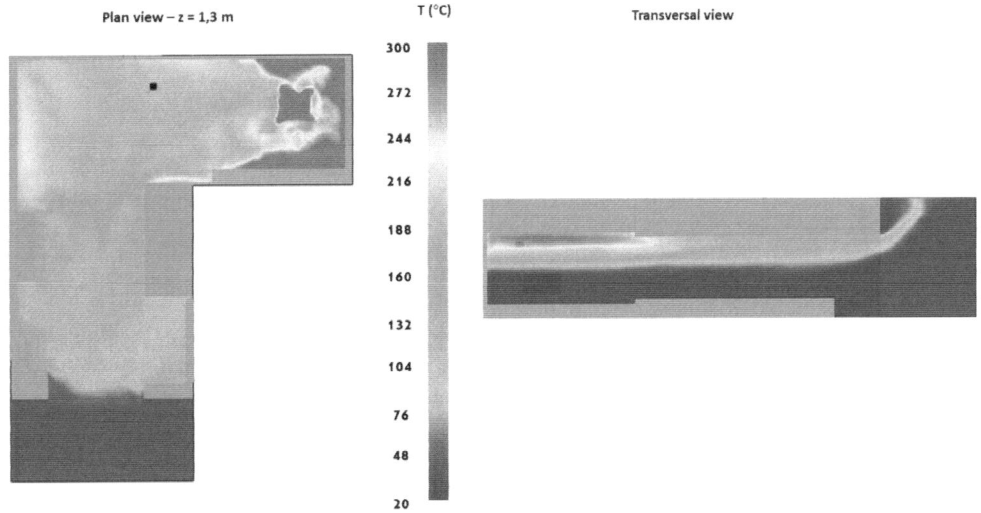

The last validation of the numerical model concerns the evaluation of the flow velocity escaping at the quarry entrance. Numerical and experimental results are compared on Figure 24 until 28 minutes of test. After this time, the velocity probe performance was lost because of temperatures higher than those allowed by the device. A correct evaluation of the flow velocity by simulation is observed, with magnitude close to 1,5 m/s.

Figure 24 : Gas velocity at the gallery entrance (A9) - experimental (EXP) and numerical (NUM)

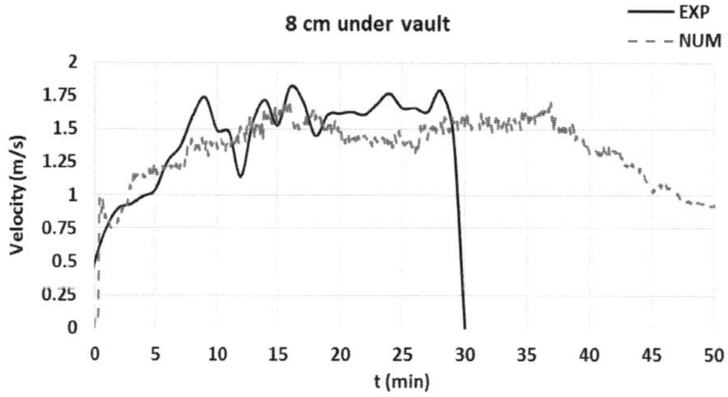

Regarding the high thermal conditions in the quarry during the fire, human tenability seems very limited. An additional numerical evaluation of the toxicity in the quarry is performed. Carbon monoxide concentration calculated at time t=15 minutes at altitude z = 1,3 m from ground is indicated on Figure 25. Lethal values higher than 150 ppm are observed in the quarry.

Figure 25 : Carbon monoxide concentration evaluated numerically at z = 1,3 m in the gallery

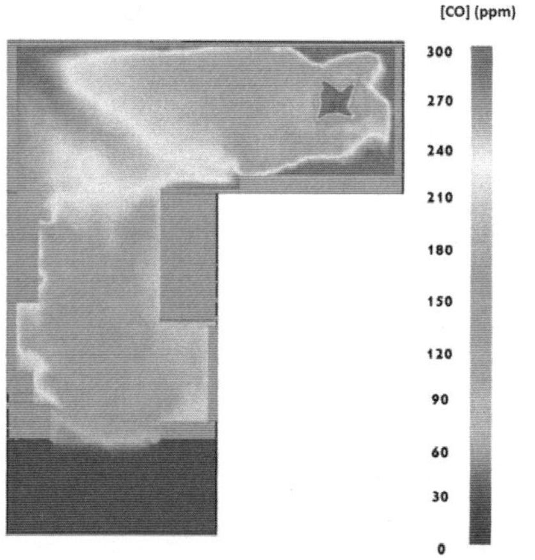

Finally, because the numerical model can predict with confidence the thermal conditions in the quarry, gas temperature measured at the surface of the limestone sample suspended 2,5 m above the fire is used to evaluate the material temperature.

Experimental and numerical gas temperatures at the sample surface are compared on Figure 26, and a correct reproduction is observed with the simulation. The evaluated sample surface temperature is indicated on Figure 26. Values higher than 250°C are reached after 10 minutes of fire, corresponding to one possible thermal criteria leading to alterations of limestone[5]. Then, this indicates that the designed fire should lead to limestone thermal alterations in the quarry.

Figure 26 : Limestone temperature above the fire (A0) - experimental (EXP) and numerical (NUM)

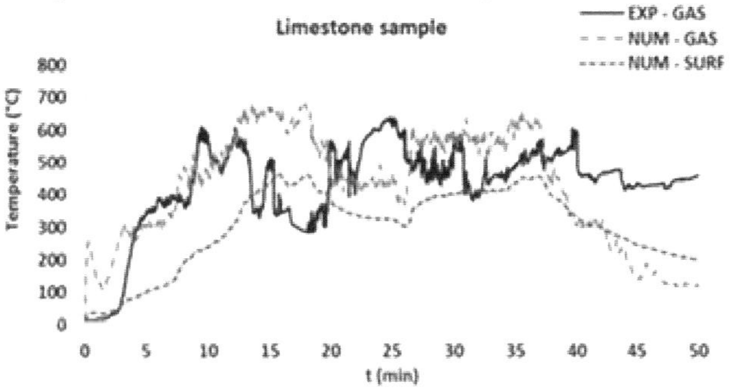

CONCLUSIONS AND PERSPECTIVES

An experimental and numerical characterization of fire in a quarry-like facility has been addressed. The numerical model was first validated by comparison with results acquired from calorimeter hood experiment. Then, it was applied in the quarry facility modelled with the code FDS.

The predicted temperatures are close to the experimental ones acquired for each location in the quarry. The numerical simulation captures the main combustion and dynamics conditions and the flow stratification is correctly replicated in this restricted ventilation configuration. The cold gas layer is also correctly reproduced numerically. High thermal solicitations exist in each part of the quarry, including at the entrance, detrimental for human tenability. The numerical evaluation of the flow velocity leaving at the quarry entrance is correct with magnitude close to 1,5 m/s. The evaluated Urgonien limestone sample surface temperature shows values higher than 250°C, corresponding to one possible thermal criteria leading to alterations of limestone and indicates that the designed fire should lead to limestone thermal alterations in the experimental quarry.

Finally, the designed fire seems adapted for studying thermal loads imparted to Palaeolithic caves, despite detrimental human tenability. The proposed numerical model can predict with confidence the thermal conditions in a cave-like facility, and allows further study of thermal conditions and solicitations on walls leading to the thermal alterations observed in Palaeolithic caves. Additional tests are planned at the end of 2016 with same methodology and fire source that detailed in this paper. The cave facility will be used to evaluate the reproducibility of such thermal conditions, and enhance the understanding of the fire behaviour in restricted ventilation conditions.

The research was supported by the Nouvelle Aquitaine region (2015-1R40205). The authors acknowledge SDIS 33, the owner of the quarry Christian Boucher, and Jean-Luc Vidal for the wood supply.

REFERENCES

1 Fire Dynamics Simulator (Version 6), www.nist.gov

2 Lacanette D., Ferrier C., Mindeguia J.C., Debard E., Kervazo B. 2014 - 3D numerical simulation of a fire in a simplified gallery of the Chauvet-Pont d'Arc cave, 17th World UISPP Congress, Burgos, Spain, September 2014

3 Drean V., Ferrier C., Leblanc J.C., Mindeguia J.C., Lacanette D., Bellivier A., Auguin G., Sarfati S., Modelling of thermal conditions during fire in paleolitic caves, CarMoThaP project, 22th journées du GDR CNRS 2864 Incendie, June 2016, Paris, France

4 Ferrier C., Debard E., Kervazo B., Brodard A., Guibert P., Baffier D., Feruglio V., Gély B., Geneste J.-M., Maksud F. - Les parois chauffées de la grotte Chauvet-Pont d'Arc (Ardèche) : caractérisation et chronologie. Paléo, n°25, 2014

5 Mindeguia J.C., Sirieix C., Brodard A., Guibert P., Ferrier C., Malaurent P., Puyo-Lacanette D. - Simulation of the thermomechanical impact of an experimental fire in limestone gallery-Implications for the study of Palaeolithic cave Chauvet-Pont d'Arc (Ardèche). 21ème Congrès Français de Mécanique, Bordeaux, France, août 2013

RABOT2012 MULTI-COMPARTMENT FIRE TESTS A COMPARATIVE STUDY WITH CFD

Andrea Klippel* & Frederik Rabe* & Alexander Klippel & Anja Hofmann[a] & Ulrich Krause*

*OvGU Otto von Guericke University, Faculty of Process and Systems Engineering, Universitätsplatz 2, 39106 Magdeburg, Germany
[a] Bundesanstalt für Materialforschung und –prüfung (BAM), Unter den Eichen 87, 12205 Berlin, Germany

Keywords: Computational Fluid Dynamics, full scale fire tests, smoke layer height, validation, combustion process

ABSTRACT

The application of Computational Fluid Dynamics (CFD) in fire safety concepts is becoming increasingly important. CFD can be used for performance based design for buildings of particular type and use. Safety parameters can be calculated with CFD to predict and achieve certain acceptance criteria. Quality and validation of CFD simulations have been discussed since CFD programs became available. The wide range of applications of CFD makes validation even more complicated, but for specific tasks there is no standardized solution. Modeling of fire tests depends on the discretization method and the possibility of application of the used models/reactions on different scenarios. These scenarios can be varied widely and after an accident it is not always possible to retrieve all boundary conditions such as fire load, ventilation conditions (windows open/closed) and location of first ignition. Experimental data from the multi-compartment fire tests RABOT2012 were used for the validation of a method to model the evolution and propagation of fire smoke and its properties with Computational Fluid Dynamics.
The RABOT2012 fire tests, which were done by the Ghent University Belgium, give a detailed description of a specified scenario along with measured data of temperature, smoke layer height, flow velocities and heat fluxes. Therefore it is suitable to validate an adjustable model for flaming and non-flaming combustion, which was implemented in ANSYS CFX R15.0. The aim of the numerical study was to test if the used modeling methods and post analysis can cover the combustion process and the experimental results of RABOT2012. ANSYS CFX was used to compute the generation and spread of smoke from an arrangement of furniture through the interconnected rooms. With the chosen fluid models and sub models and the implemented expressions in CFX a good agreement between the calculated temperatures and spread of smoke with the measured data were achieved for the first 10 minutes of the fire.

INTRODUCTION

In fire safety engineering CFD is often used for performance based design for buildings of particular type and use. Different CFD programs and various models for predicting a combustion process are available for the user. In order to accurately model a combustion process, it is important to know which stage of fire development you want to cover with your modeling. A complete or incomplete combustion causes a significantly different composition of products in fire smoke.
Sometimes no statistically representative data sets are available for fire safety engineers to validate their numerical models which are used for modeling fire and smoke spread in complex geometries. When CFD calculations are carried out to predict and achieve certain acceptance criteria by using safety parameters, the user is responsible for choosing suitable models and evaluation methods, as there is no standardized solution for fire scenario modeling. When simulating fire and smoke propagation the processes are subjects of fundamental physics. At the same time three-dimensional turbulence flow problems are only solvable using methods of computational fluid dynamics and its underlying

approximations. Verification of physical models as well as the validation of the models are an important component when working with CFD applications and has been discussed since the programs became available. Modeling of the combustion process is still subject of current research, e.g. Hiremath et al. [1] and Kempel et al. [2] as well as in this paper. The prerequisites for a successful evaluation of smoke modeling with ANSYS CFX was to find a as precisely as possible documented fire experiment in a real compartment with interconnected rooms. Evaluation criteria were the temporal and spatial development of temperatures and smoke layer heights in the fire room and in the interconnected rooms as well. For a reliable validation process the data base of one fire test is insufficient. But the fire tests were done by researchers of Ghent University and the data base of the one test configuration available was used to see if the fire and modeling methods achieve acceptable agreements between the experimental und numerical results or if there is a significant deviation.

The RABOT2012 fire tests give a detailed description of a specified scenario along with measured data of temperatures, smoke layer height, flow velocities and heat fluxes. Therefore it is suitable to validate an adjustable model for flaming and non-flaming combustion, which was implemented in ANSYS CFX R15.0. The use of a public available dataset offers the opportunity of a comparable validation. Additionally the temperature of the upper layer and the smoke layer height are compared to empirical correlations as done for fast spreading room fires by Knaust et al. [4]. Empirical correlations can lead to fast estimations of temperatures and smoke layer heights for simple scenarios and should be used at least additionally, if applicable to the scenario. CFD gives the opportunity of modeling multi-compartment fire scenarios with evaluation of temperatures, toxicity and visibility but is difficult to validate and time-consuming.

EXPERIMENTAL DATA

All experimental data has been taken from the RABOT2012 fire tests, which were done by the Ghent University, Belgium. Four large-scale multi-compartment fire tests were conducted in an apartment in one of the 'RABOT' towers in the city of Ghent (Belgium) in September 2012. The tests include determination of smoke layer heights by video recordings and thermocouple data as well as heat release rates of the used materials. Characterization of the tests and experimental results are published [5, 6]. The website of the RABOT2012 fire tests [7] hosts a publicly available video dataset of the RABOT2012 fire tests and as well as thermocouple data and information on geometry and fire load.

One objective of the RABOT2012 fire tests was to apply and evaluate a multi-view video fire analysis algorithm for flame spread and smoke layer height detection within a multi-compartment setup like the RABOT fire tests. For this work data from Test 1 of total 4 fire tests was used.

The geometry is shown in Figure 1. The setup consists of a bedroom with two sofas (polyurethane) and a bookshelf with books. The fire was ignited in the corner of the left sofa (see Fig. 1) by means of a lighter and six small wood cribs soaked in heptane. Next to the bedroom a living room was located equipped with additional cameras and thermocouples. The living room was connected to a corridor, which leads to the staircase and was open during the tests. All ceilings have a height of 2.5 m and the doors have a height of 2 m. The floor covering in the rooms was vinyl layer on top of a 30 cm thick concrete slab.

The ceiling linings were composed of:

- 2 cm thick gypsum plasterboard and
- a 30 cm thick concrete slab.

The wall linings were composed of:

- thin wall paper, 2 cm thick gypsum plasterboard,
- 3 cm thick fibre insulating board and
- a 30 cm brick layer.

Figure 1. Geometry of Test 1, RABOT2012 [5]

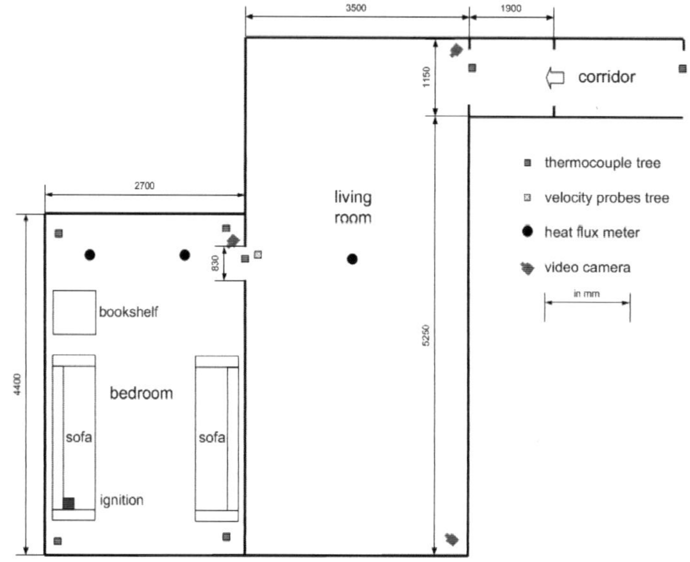

A total of three thermocouple trees were located in the doorways between bedroom and living room, living room and in the exit doorway. Nine thermocouples were attached to every tree, starting 30 cm above the floor and then 20 cm apart each. Determination of smoke layer height by temperature was done with the least-square method (see Figure 2) according to He et al. [8] and equations 1-3.

Figure 2. Determination of smoke layer height by thermocouples [5, 6, 7]

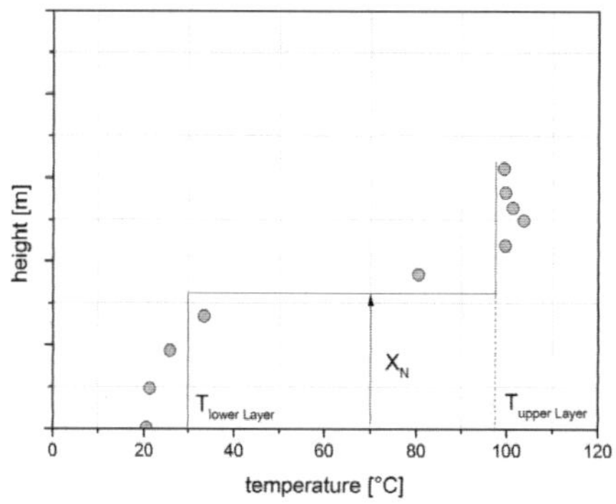

$$T_{upper\ Layer} = \frac{1}{H_D - X_N} \int_{X_N}^{H_D} T dz \qquad [1]$$

$$T_{lower\ Layer} = \frac{1}{X_N} \int_{0}^{X_N} T dz \qquad [2]$$

$$\sigma^2 = \frac{1}{X_N} \int_0^{X_N} \left(T - T_{lower\ Layer}\right)^2 dz + \frac{1}{H_D - X_N} \int_{X_N}^{H_D} (T - T_{upper\ Layer})^2 dz \qquad [3]$$

where H_D is the height of the ceiling [m], X_N is height of smoke free layer [m] and T is temperature [°C].

Figure 3. Average temperature of lower layer (left) and upper layer (right) of bedroom, living room and corridor [5, 6, 7]

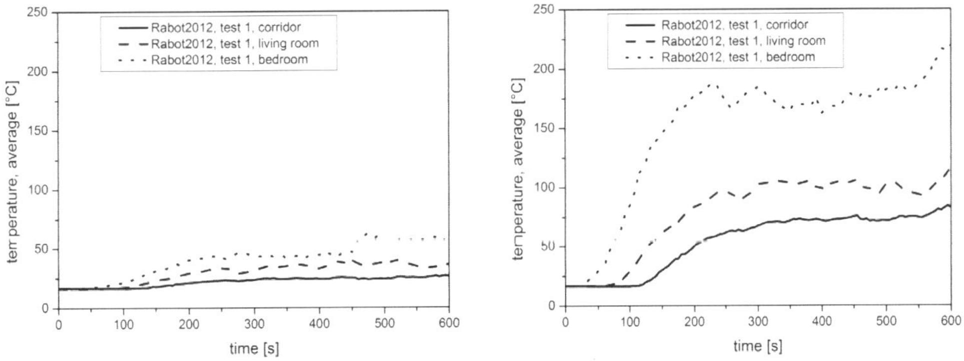

Determination of smoke layer height by video recording was done by an algorithm which uses the Discrete Wavelet Transform (DWT) based on evaluation of Video Energy Lines see Fig. 4 and Fig. 5 [6].

Figure 4. Schematic of video-based smoke layer height determination [6]

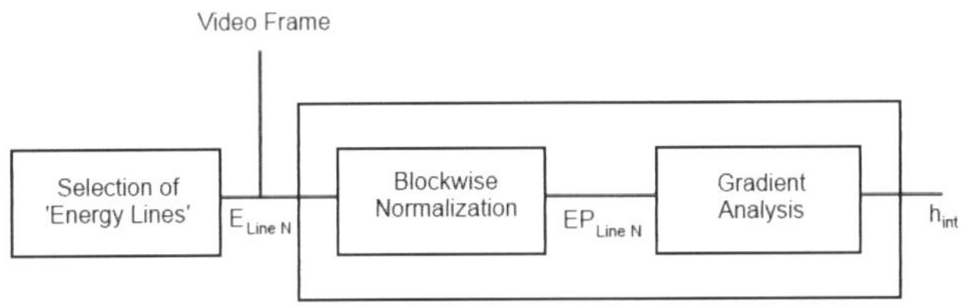

Figure 5. DWT-based energy line(s) (left), Energy profile (middle), Detected smoke layer height (right) [6]

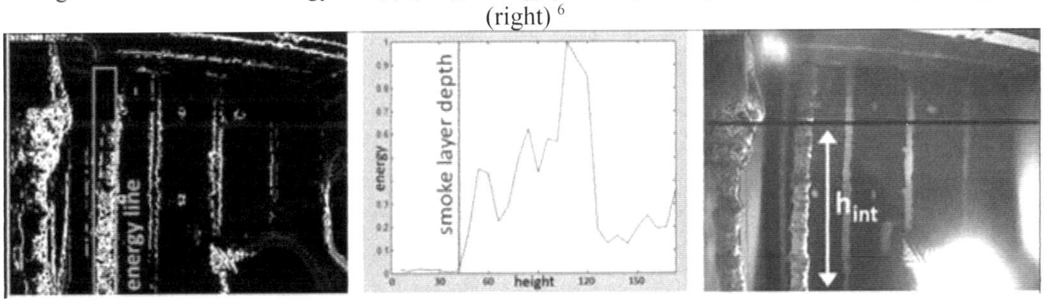

Problems can arise in one camera but can (most probably) be compensated by others. The paper [6] proposes to analyse the within- and between-variance (see eq. 4 and eq. 5) of multi-view smoke layer height estimations. Within-variance means indication on the accuracy of a camera's measurements and between-variance means the certainty of the overall measurements [6].

$$\sigma_w(t) = \frac{1}{N}\sum_{j=1}^{N}\left(h_{int}(t, LINE_i) - \overline{h_{int}(t, LINE_{1...N})}\right)^2 \qquad [4]$$

$$\sigma_b(t) = \sum_{i=1}^{M}\left(h_{int}(t, CAM_i) - \overline{h_{int}(t, CAM_{1...N})}\right)^2 \qquad [5]$$

where σ is the standard deviation and h_{int} is the height.

Average smoke layer heights determined video-based and by thermocouples are shown in Fig. 6 for bedroom, living room and corridor. Thermocouple trees were located in the doorways with a height of 2.0 m.

Figure 6. Average smoke layer height of bedroom, living room and corridor, video-based (left), by thermocouples (right) [6, 7]

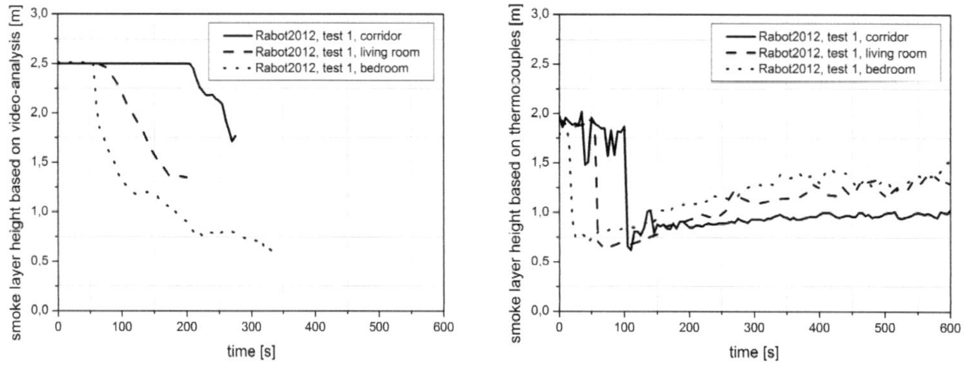

Measured temperatures, smoke layer heights by temperature and video-based are shown in results compared to simulated data. Smoke layer heights determined by video cannot be compared to those determined with thermocouples due to the fact that video based smoke layer heights were measured in the room with a height of 2.5 m and smoke layer heights by temperature were measured in doorways with 2 m in height.

MODELING WITH EMPIRICAL CORRELATIONS OF RABOT2012 FIRE TESTS

In order to compare the measured data and numerical data empirical correlations for upper layer temperatures and smoke layer height are used as well. The empirical equations are used to estimate the temperatures of the upper layer and smoke layer height in the bedroom (fire room) of the test.

Upper layer temperature

With the heat release rate it is possible to estimate the upper layer temperature. The expression is used to compare to experimental and numerical data. Upper layer temperature can be calculated up to 600°C according to an empirical correlation of McCaffrey, Quintiere and Harkleroad [9, 11], which depends on the heat release rate and some geometric and material constants. If the fire occurs near a wall the temperature can be multiplied by a factor of 1.3 [9, 11].

$$T_{upper\ layer} - T_{\infty} = 6.85 \left(\frac{\dot{Q}^2}{A_0 \sqrt{H_0} h_k A_T} \right)^{1/3} \qquad [6]$$

Where $\dot{Q}$ is the heat release rate [kW], A_0 is the area of the opening [m^2], H_0 is the height of the opening [m], h_k is the effective heat transfer coefficient and A_T the total compartment surface area [m^2].

The transient, effective heat transfer coefficient h_k can be calculated according to eq.7, if the considered time is below the time factor t_F can (Eq. 8) [9, 11].

$$h_k = \sqrt{\frac{k\rho c_p}{t}} = \sqrt{\frac{b^2}{t}} \qquad [7]$$

where k is the thermal conductivity of the compartment, ρ is the density of the material [kg/m^3], c_p is

the specific heat capacity [J/(kg K)] and t is fire duration [9, 11].
For consideration of the time depending reaction of material the time factor t_F can be used [9, 11].

$$t_F = \frac{d^2}{a\pi} \approx 1.9\ hours \qquad [8]$$

Where d is the thickness of the wall [m] and a is the temperature conductivity [m^2/s]. The used constants for the bedroom of the fire test can be seen in Table 1. Almost two hours for t_F means that the transient constant h_k should be used [9, 11].

Table 1. Material and geometric properties of the bedroom (fire room)

parameter	*values*
area of opening A_0	1.6 m^2
height of opening H_0	2.0 m
height of ceiling H	2.5 m
surface area A_T	53.5 m^2
temperature T_∞	290 K
wall thickness d	0.3 m
density ρ	1.2 kg/m^3
specific heat capacity c_p	1000 J/(kg K)
b^2	58000 $W^2s/(m^4K^2)$
temperature conductivity	4.10e-07 m^2/s
floor area A_F	12.8 m^2
gravity g	9.81 m/s^2
heat release rate	300 kW

The comparison of measured upper layer temperatures and calculated ones is shown in Fig. 7. There is general agreement of the temperatures but a very simplified curve.

Figure 7. Average temperature of upper layer in bedroom by thermocouples and empirical calculation[7]

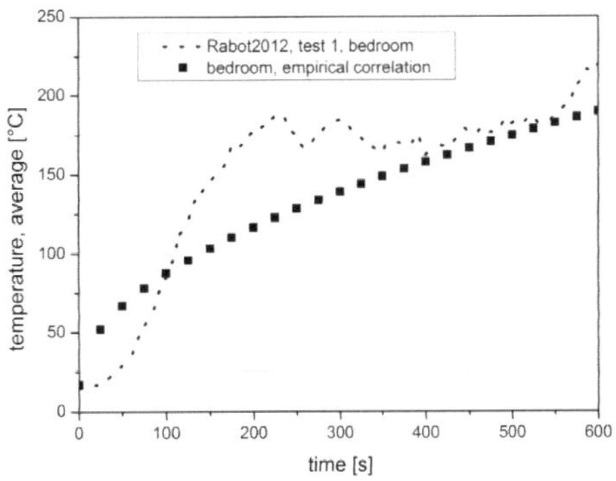

Calculation of the smoke layer height can be done according to Quintiere [11]. Smoke layer height h_{int} presents height of ceiling *H* minus height of smoke free layer X_N. Dimensionless height y is expressed by height of lower layer X_N normalized by the room height *H* and given with eq. 9.

$$y = \left[1 + \frac{2 * 0.21}{3} \dot{Q}^{*1/3} \tau\right]^{-3/2} \qquad [9]$$

With time constant $\tau = t\sqrt{\frac{g}{H}\frac{H^2}{A_F}}$, where t is time, H is the room height and A_F is area of the floor and the modified heat release rate:

$$\dot{Q}^* = \frac{\dot{Q}}{\rho c_p T_\infty \sqrt{g} H^{5/2}} \quad [10]$$

The comparison between video-based smoke layer determination and empirical calculation shows a mostly conservative result for the empirical approach in the fire room (see Fig. 8).

Figure 8. Video-based smoke layer height and empirical calculation of smoke layer height[7]

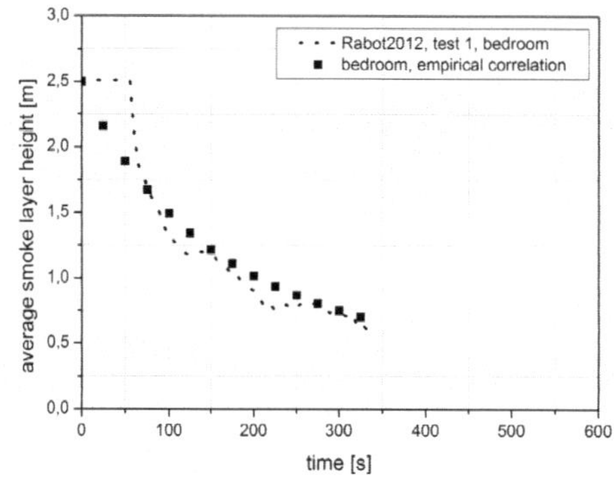

Results of empirical calculations

Empirical calculations of upper layer temperature and smoke layer height can be used as conservative estimation for the bedroom (fire room). These simple approaches cannot easily be used for multi-compartment fires but should be used wherever possible in addition to CFD calculations, e.g. to evaluate the temperature range in the fire room.

NUMERICAL MODELING

The underlying physical phenomena in a combustion process are not only very complex but also three-dimensional and non-linear. For the prediction of fluid flow field variables in a combustion process e.g. temperature, pressure, and velocities the use of a CFD Code is required. The numerical simulation of turbulent fluid flows as well as the heat and mass transfer and chemical reactions are based on physical models of fluid mechanics. The fundamental equations of numerical fluid computation are the continuity equation (conservation of mass), the Navier-Stokes equations (conservation of momentum), the energy equation (conservation of energy) and the turbulence equations [12].

For discretisation ANSYS CFX uses the finite volume method to treat the transport phenomena like convection, diffusion and source. It allows the conversion of the integrated governing equations for each control volume in a system of algebraic equations, which can be solved with iterative methods.

All local conservation equations (mass, momentum and energy) of a compressible fluid can be written as a system of equations as follows [15]:

mass
$$\frac{\partial \rho}{\partial t} + div(\rho \boldsymbol{u}) = 0 \quad [11]$$

x-momentum
$$\frac{\partial(\rho u)}{\partial t} + div(\rho u \boldsymbol{u}) = -\frac{\partial p}{\partial x} + div(\mu\ grad\ u) + S_{Mx} \quad [12]$$

y-momentum
$$\frac{\partial(\rho v)}{\partial t} + div(\rho v \boldsymbol{u}) = -\frac{\partial p}{\partial y} + div(\mu\ grad\ v) + S_{My} \quad [13]$$

z-momentum
$$\frac{\partial(\rho u)}{\partial t} + div(\rho w \boldsymbol{u}) = -\frac{\partial p}{\partial z} + div(\mu\ grad\ w) + S_{Mz} \quad [14]$$

energy
$$\frac{\partial(\rho i)}{\partial t} + div(\rho i \boldsymbol{u}) = -p\ div\ u + div(k\ grad\ T) + \Phi + S_i \quad [15]$$

equations of state	$p = p(\rho, T) and\ i = i(\rho, T)$ $p = \rho RT\ and\ i = C_v T$ (for a perfect gas)	[16]

where $\frac{\partial(\rho u)}{\partial t}, \frac{\partial(\rho v)}{\partial t}, \frac{\partial(\rho u)}{\partial t}$ express the rates of increase of x, y and z-momentum, t is time, ρ is the density, p is pressure, T is temperature, S_M is momentum source, μ is dynamic viscosity and Φ stands for the dissipation function.

In the present study Reynolds averaging was used to consider the turbulence. The selected turbulence model was the shear stress turbulence model (SST). The SST model is an eddy-viscosity model, which combines the k-ω (near-wall region) and the k-ε (in the outer part of the flow) models[13].
Radiation is considered by the Monte Carlo (MC) approach. The MC model is essentially based on photons histories (that means photons are tracked through the domain and for every photon a history will be written) to simulate the underlying physical interactions between photons and their environment [13]. Due to the multi-compartment setup a flaming combustion reaction for polyurethane with an eddy-dissipation-model was chosen out of a set of reaction equations for non-flaming and flaming combustion. The Eddy Dissipation model is an approach to consider the influence of the turbulence [13]. When reactants mix at the molecular level, they instantaneously form products [13]. For the simulation the eddy dissipation combustion model was used because no information were available about detailed reaction kinetics for the combustion process in RABOT2012.
Further numerical basics are skipped for brevity but can be found in literature by Ferziger et al. [12] or in the documentation of ANSYS CFX R15 [13] or in Versteeg et al. [15].

Figure 9 shows the geometry of the fire test as implemented in ANSYS CFX R15.0. The bedroom, where the fire starts, can be seen on the left. In the living room next to the bedroom a section of the grid can be seen. A Cartesian grid with 1046280 cells was created with a cell size of 0.05 m. The arrows on the right symbolize an opening, which represents a door, which was open during the fire test.

Figure 9. Geometry of modeled fire test with boundary conditions and mesh

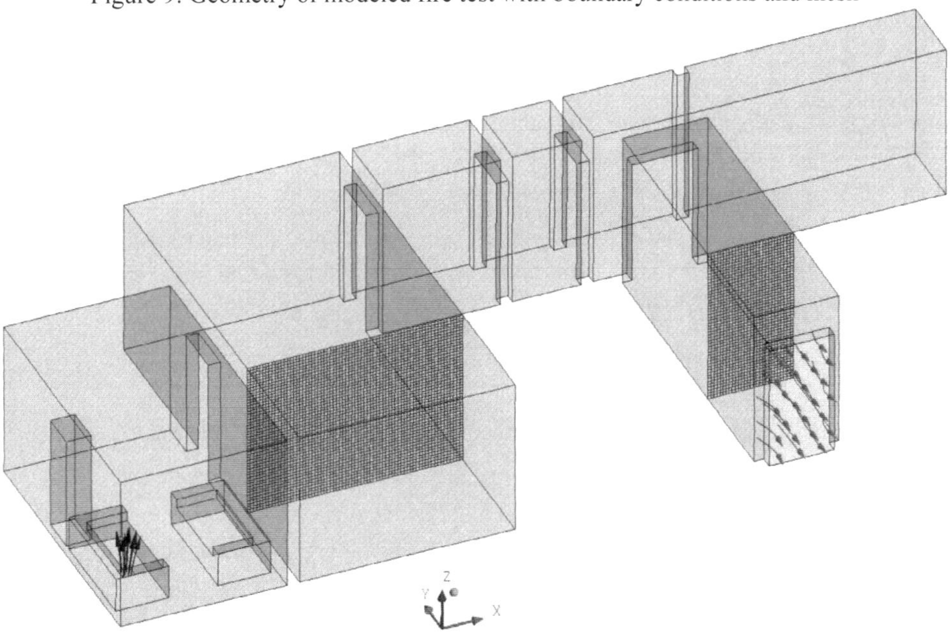

Table 2 gives information about grid, settings and boundary conditions. Total simulation time was 600 s, which is sufficient time for self-evacuation and is above usual fire brigade response times.

Table 2. Boundary conditions and settings of simulation

Fluid Models	***Set-up***
Heat Transfer	Thermal Energy
Turbulence	Shear Stress Transport (SST)
Combustion	Eddy Dissipation (EDM)
Thermal Radiation	Monte Carlo
Solver Control	***Set-up***
Advection Scheme	High Resolution
Transient Scheme	Second order Backward Euler
Convergence Criteria	max. 10E3
Conservation Target	0.01
Basic Settings	***Set-up***
Analysis Type	transient
Time Duration	600 s
Timestep	0.01 s
Mesh	structured with hexahedra
Number of Elements	1 046280
Number of Nodes	1 103634
Material	polyurethane
Reference Pressure	101325 Pa
Buoyancy Model	*buoyant*
Boundary Type	***Set-up***
Inlet	
Flow Regime	subsonic
Relative Pressure	0 Pa
Mass and Momentum	mass flow rate (user function)
Wall	
Mass and Momentum	no-slip wall
Wall Roughness	smooth wall
Outside Temperature	17 °C
Opening	
Flow Regime	subsonic
Relative Pressure	0 Pa
Mass and Momentum	Opening Pressure and Direction

COMPARISON OF EXPERIMENTAL AND NUMERICAL DATA

For evaluating the simulation results measured temperatures and smoke layer heights were used of all three apartment rooms. As acceptance criterion it was assume that the smoke layer should not drop under 2 m. The fire was started in the bedroom and spread to the living room and finally to the corridor.

Smoke layer heights

Since smoke is not a physical property such as pressure, different criteria were used to distinguish lower and upper smoke layer. For describing smoke layer height a temperature criterion (smoke ≥ 50°C) and a mass fraction criterion (smoke mass fraction ≥ 1%) were assumed. As can be seen in Figure 10 smoke layer height is described by 1% mass fraction of smoke as criterion and compared to the video-based results from the experiment and to the empirical correlations. The video-based results and simulated smoke layer heights in the bedroom show a good agreement for the first 200 s. Video-based smoke layer detection stopped when cameras were reached by smoke layer. For the bedroom and the adjacent living room the development of the smoke layer height both simulation criteria are suitable. The smoke layer development in the corridor is over predicted when simulated. Simulated smoke layers quickly drop below 1 m, this could be due to the differences between video-based measurement and simulation of smoke layer with criteria. The video-based smoke layer determination does not give any information about how much smoke is in the air or about species concentration but only about visibility. The following figures show the compared numerical and experimental results for developing smoke layer heights. In addition for the fire room (bedroom) an empirical correlation were done for comparison.

Figure 10. Comparison of simulated smoke layer height over 1% mass fraction of smoke, video-based evaluated smoke layer height from experiment and the empirical correlation in the bedroom of RABOT fire test

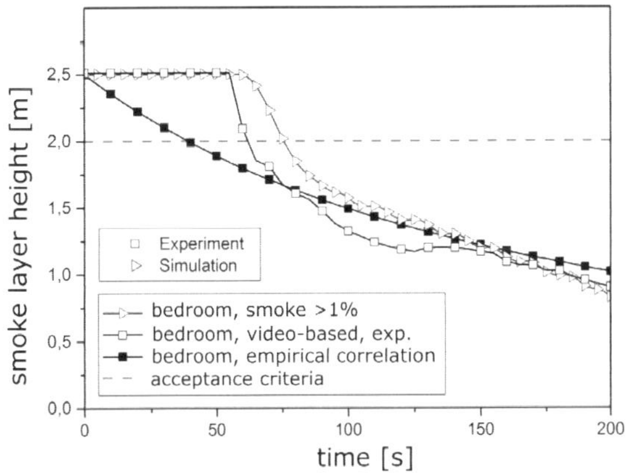

In order to clarify that not only choice of models and scenario can have a huge impact on results, Figure 11 shows the comparison of smoke layer height by mass with video-based results from the fire test. Both curves are in a good agreement for the first 75 s. After that the CFD predicted smoke layer height underestimates the experimentally determined course of the smoke layer height in the living room.

Figure 11. Comparison of simulated smoke layer height over 1% mass fraction of smoke and the video-based evaluated smoke layer height from experiment in the living room of RABOT fire test

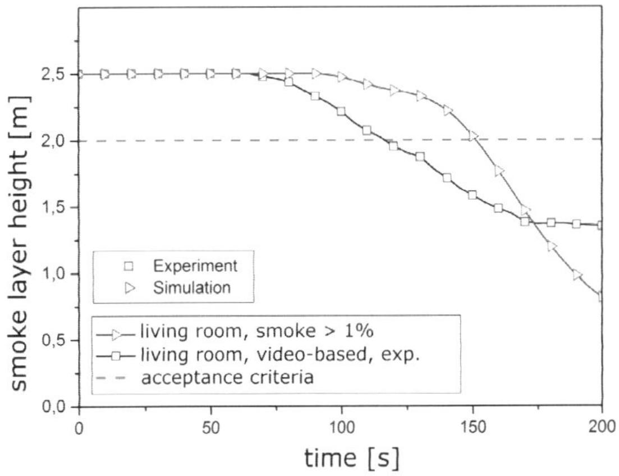

Figure 12 shows the main results of predicted smoke layer height for smoke mass fraction criterion and temperature criterion in comparison with the empirical correlation and the experiment. The developing smoke spread can be described by both criteria in simulations. An upper smoke layer can be detected after approximately 70 s. Measured and simulated data are in an acceptable agreement of the first 200 s, then simulated smoke is quickly predicted in the whole room.

Figure 12. Comparison of simulated smoke layer height for smoke mass fraction and temperature criterion, video-based evaluated smoke layer height from experiment and the empirical correlation in the bedroom of RABOT fire test

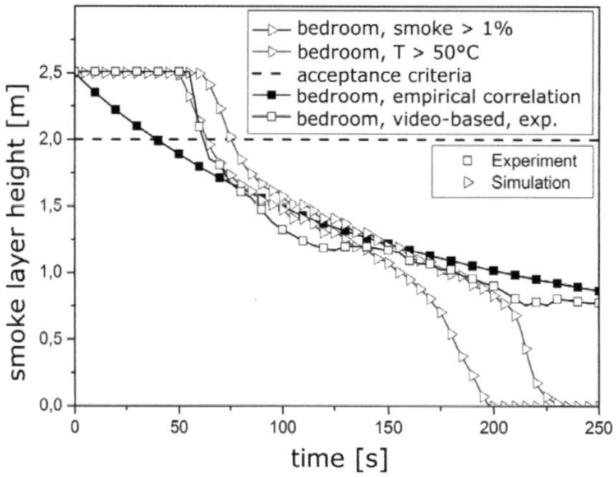

The quantitative results from the diagrams are given in the following tables for specific times. Table 3 compares the time at which the smoke reaches the height of the chosen acceptance criteria for the numerical calculated experimental results. A minimum level for lower layer was set to a height of 2.0 m.

Table 3. Comparison of simulated, measured and empiric calculated times to reach 2.0 m lower layer height.

time when smoke layer drops below 2.0 m				
	Criteria			
Room	*CFD, smoke 1%*	*CFD, T>50°C*	*exp.,video-based*	*empiric calc.*
bedroom	75 s	65 s	65 s	40 s
living room	154 s	145 s	120 s	-
corridor	164 s	157 s	260 s	-

For interests of safety engineering a critical acceptance criteria for lower layer height was chosen at 1.5 m. Table 4 compares the numerical and measured data for times when smoke layer drops below 1.5 m

Table 4. Comparison of simulated, measured and empiric calculated times to reach 1.5 m lower layer height.

time when smoke layer drops below 1.5 m				
	Criteria			
Room	*CFD, smoke 1%*	*CFD, T>50°C*	*exp.,video-based*	*empiric calc.*
bedroom	113 s	100 s	90 s	100 s
living room	170 s	161 s	160 s	-
corridor	173 s	165 s	-	-

Getting a qualitative impression of the CFD calculated smoke spread in RABOT fire test geometry Figure 13 shows the development of a smoke volume which is indicated as an isovolume moving through the domain. The isovolume is described using the variable mode in ANSYS Post with the variable smoke mass fraction in the range above 1 % by mass.

Figure 13. Side view of the qualitative time course smoke development (isovolume of smoke mass fraction ≥ 1 % by mass) through the RABOT geometry

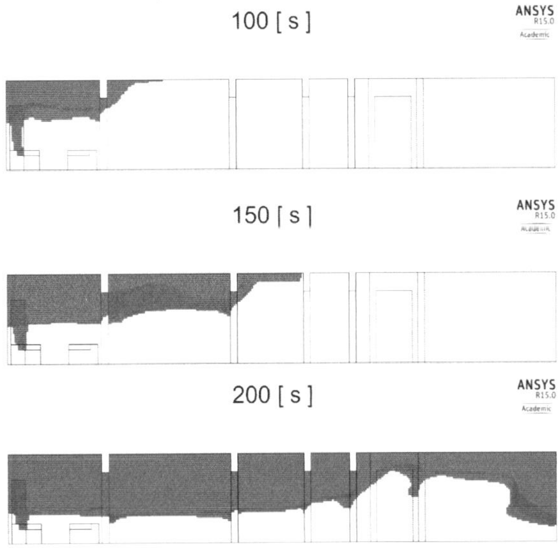

Temperatures

In two rooms temperatures were measured for eight heights between 0.3 m and 1.7 m and compared with simulated temperatures. In addition, three thermocouples positions per room, one near the ground, one in medium height and one under the ceiling, show the absolute temperature differences between measurement and simulation in the first four minutes. In almost all cases the temperature difference were smaller than 20 K. For a better understanding, see Figure 1 again for the geometrical position of the thermocouple trees between the bedroom and living room (thermocouple tree bedroom) and between living room and corridor (thermocouple tree living room). At the same place in the domain the CFD calculated temperatures were evaluated.
In Figure 14 the experimentally determined temperatures are compared again with the simulated temperatures. The three heights are exemplary for the temperatures near the ground, in the middle of the thermocouple tree and under the ceiling. The simulated and measured temperature evaluation show a good agreement at thermocouple tree position between living room and corridor. The maximum temperature deviations are never more than 15 °C.

Figure 14. Comparison of the simulated and measured temperatures between bedroom and living room (left) and between living room and corridor (right) with plotted temperature deviation

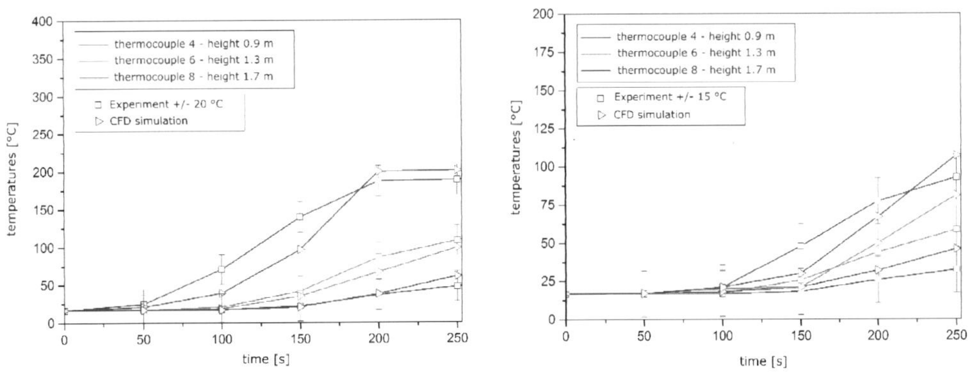

Qualitative prediction of the Fractional Effective Dose

In addition to measured quantities like temperatures and smoke layer heights the Fractional Effective Dose (FED) was calculated numerically and evaluated to account for toxicity. The Fractional Effective Dose model is not available in ANSYS CFX because it's a general code for any fluid or physics applications. The FED model was implemented in ANSYS CFX with user defined expressions using CFX expression language. The basic FED equation is shown in eq. 12 according to Purser et al. [16]. Using the FED model time to compromised tenability or time to incapacitation can be predicted. This depends on the chosen threshold criteria. A detailed description of the model is given in reference 16.

$$\mathrm{FED} = \sum_{i=1}^{n} \sum_{t_1}^{t_2} \frac{C_i}{(Ct)_i} \Delta t \qquad [12]$$

where C_i is the average concentration of a dose related toxicant such as an asphyxiant gas i over the chosen time interval, Δt is the chosen time interval [min] and $(Ct)_i$ expresses the specific exposure dose.

Experimentally determined species concentrations from the RABOT fire test were not available for validation of the FED. So the simulation data are used only for a qualitative assessment of toxicity. The evaluation shows a clear trend of the Fractional Effective Dose. The following Figure 15 shows the smoke distribution through the domain for different time steps. The smoke is described with the mass fraction criterion (smoke $\geq$ 1% by mass) and is indicated with Fractional Effective Dose values in the range between 0 and 0.3. After about 200 s the FED reaches critical values in some areas of the bedroom (fire room). The dark grey indicated areas represent FED values of 0.3. The assumption is if FED reaches values of 0.3 about 11.4 % of the population (statistically estimation) experience compromised tenability. After about 300 s the simulation results clearly show a propagation of critical toxic areas throughout the bedroom into the living room. Even if this are only qualitative statements this investigation shows that spread of toxic gases should be adequately analysed and evaluated for risk assessment and that statistically representative database for species concentrations from real room fire tests are indispensable for any validation processes with CFD.

Figure 15. Numerical calculated time course smoke distribution indicated with FED values $\leq$ 0.3 in the RABOT fire test geometry

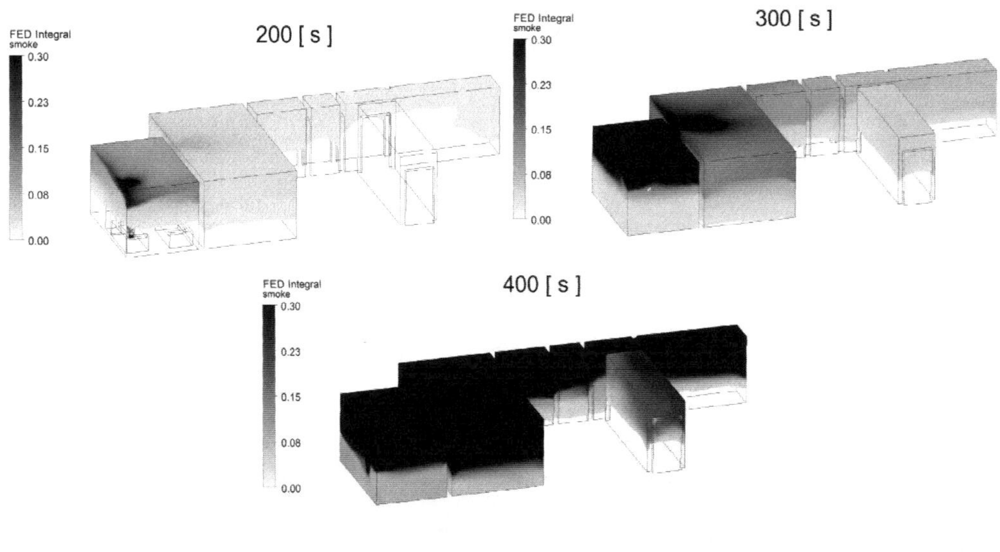

CONCLUSION

CFD methods are widely spread in fire safety engineering but are mainly used for a forecast of a given scenario with no experimental data available. A simplification of the combustion process is always necessary when investigating room fires. Unfortunately even when more detailed approaches are added there are hardly any well documented reference fire tests with a public available dataset. The University of Ghent conducted 4 fire tests in an apartment building in 2012 in Belgium. The fire tests experimental data is not only available to the public but many parameters were recorded, such as temperatures, heat fluxes, smoke layer heights and flow velocities. So one of total four fire tests was used for this comparative study. It was aimed to evaluate if a single fire scenario can be described sufficiently with a general CFD code, ANSYS CFX R15. Smoke layer heights and temperatures were evaluated for all three rooms of the apartment. It could be shown that temperatures and smoke spread e.g. development of smoke layer heights in the interconnected rooms of RABOT can be calculated with ANSYS CFX with acceptable deviation to the experimental measured ones.

For a well done validation process the experimental data base are not enough to be statistically representative for evaluating a real room fire with CFD e.g. physical behaviour of quantities.

Nonetheless for future fire safety engineering with CFD methods it is absolutely necessary to carry out more reference tests which are available to the public. These tests should repeated at least 10 times under identical conditions. Future research should aim on a valid evaluation as well as on a more detailed description of the combustion process that does not only include heat release rates and flame spread velocities but reaction kinetics, air flow conditions, pressure increase, species concentration, transmission measurements, oxygen concentration and ventilation conditions (all quantities measured on several positions in room fire tests).

Unfortunately there is no standardized approach for using CFD, which requests responsible and well educated users who need to choose suitable models and have to decide whether programs and models can be used for the given scenario. Furthermore the numerical results should be considered carefully, wherever possible, should be checked with empirical or analytical approaches if experimental data is not available. This comparative study has clearly shown that for modeling room fires first a good database of real room fires and second much more CFD calculations are needed under variation of turbulence models and turbulence intensities or combustion models e.g. with detailed reaction kinetics. More research studies are required on verification methods of CFD approaches and sub models as well as on validation processes with experimental data when using CFD for fire and smoke modeling for reliable application in fire safety engineering.

Acknowledgements

We would like to thank Bart Merci, Steven Verstockt and Tarek Beji of the Ghent University, Belgium, for providing the 'RABOT2012' data set.

REFERENCES

[1] V. Hiremath, S. Lantz, H. Wang and S. Pope, „Computationally-efficient and scalable parallel implementation of chemistry in simulations of turbulent combustion", Combustion and Flame, No. 159, pp. 3096-3109, October 2012.

[2] F. Kempel, B. Schartel, G. Linteris, S. Stoliarov, R. Lyon, R. Walters and A. Hofmann, „Prediction of the mass loss rate of polymer materials: Impact of residue formation," Combustion and Flame, Bd. 159, No. 9, pp. 2974-2984, September 2012.

[3] P. Roache, „Verification and Validation in Computational Science and Engineering", Socorro, New Mexico: Hermosa Pub, 1998.

[4] C. Knaust und A. Hofmann, „Full Scale Investigations of Fast Spreading Room Fires in press", Materials Testing, Bd. 56, 2014.

[5] T. Beji, S. Verstockt, B. Merci, R. Van de Walle, Abecassis-Empis, C., Krajcovic, M., and Majdalani, A., "RABOT2012 - Presentation of the multi-compartment full-scale ('Rabot') fire tests," 7th International Seminar on Fire and Explosion Hazards, (ISFEH presentation), 2013.

[6] S. Verstockt, T. Beji, B. Merci and R. Van de Walle, „RABOT2012 - Presentation of a multi-view video dataset of the full-scale ('Rabot') fire tests“ in 7th International Seminar on Fire and Explosion Hazards Providence, R.I.; United States, 2013. Verstockt, S., Beji, T., Merci, B., and Van de Walle, R., "RABOT2012 - Presentation of a multi-view video dataset of the full-scale ('Rabot') fire tests," 7th International Seminar on Fire and Explosion Hazards, (ISFEH presentation), 2013.

[7] Multimedia Lab 2010 -Ghent University, Belgium, „RABOT2012“[Online]. Available: http://multimedialab.elis.ugent.be/rabot2012/. [Zugriff am 05 11 2013].

[8] Y. He, A. Fernando, Luo and M., „Determination of interface height from measured parameter profile in enclosure fire experiment“, Fire Safety Journal, Bd. 31, pp. 19-38, 1998.

[9] W. a. P. T. Walton, „Estimating Temperatures in Compartment Fires“ The SFPE Handbook of Fire Protection Engineering, 3 Hrsg., Bd. 1 von 2, Chapter 6, Section 3,, D. D. C. B. a. W. W. P.J. DiNenno, Hrsg., Quincy, MA: National Fire Protection Association and The Society of Fire Protection Engineers, 2003.

[10] U.S. Nuclear Regulatory Commission, NUREG-1824; „Verification and Validation of Selected Fire Models for Nuclear Power Plant Applications“, Bd. Volume 2: Experimental Uncertainty, Rockville, MD, 2007.

[11] J. Quintiere, Fundamentals of fire phenomena, Wiley, 2006.

[12] J. Ferziger and M. Peric, „Computational Methods for Fluid Dynamics“, 3rd Hrsg., Berlin, Heidelberg, New York: Springer, 2002.

[13] ANSYS, Inc., ANSYS CFX-Solver Theory Guide R15.0, Canonsburg, Pa.: ANSYS, Inc., 2013.

[14] Department for Communities and Local Government, Fire incidents response times: England, 2012 to 2013, England: Department for Communities and Local Government, 2013.

[15] H.K. Versteeg and W. Malalasekera, „An introduction to Computational Fluid Dynamics The Finite Volume Method“, Longman Scientific & Technical, Longman Goup Ltd., Copublished in the Unites States with John Wiley & Sons Inc., 1995.

[16] D.A. Purser and J.L. McAllister, „Assessment of Hazards to Occupants from Smoke, Toxic Gases and Heat“, SFPE Handbook of Fire Protection Engineering, DOI 10.1007/978-1-4939-2565-0-63, Society of Fire Protection Engineers 2016.

REPEATABILITY OF UNDERVENTILATED COMPARTMENT FIRE TESTING WITH COMPLEX FUEL PACKAGES

Peter Senez[1,2], Patrick Mulherin[1] & Elizabeth Weckman[2]

There are a number of compartment fire experiments documented in literature, with an emphasis on investigating the fire behavior up to and following flashover. Given the complexity and costs inherent in large scale fire testing, many researchers have chosen to examine fire behaviour by running multiple, different fire scenarios with a single test of each, in lieu of repeated experimentation under the same conditions. However, since fire dynamics is inherently variable due to the complex physical-chemical interactions between the burning fuel and the environment, it is often questioned whether such a single test approach does lead to credible and universal descriptions of the overall behaviour of real fires. This uncertainty is exacerbated when studies are extended to fire conditions for which a dearth of experimentation has taken place such as for the large-scale under-ventilated furniture fires which form the subject of the present study. This paper discusses the overall control and repeatability of a series of controlled, full-scale compartment fire tests to aid interpretation of existing fire data obtained using single tests, as well as in consideration of future test work.

The repeatability of fire development was studied in a series of large-scale fire experiments conducted in a burn facility representative of a two-storey residential house. The furniture load included an upholstered three seat sofa and matching armchair, as well as hardwood coffee and side tables. Three different types of upholstered furniture were constructed using identical frames but different upholstery materials. In each test, the sofa was ignited and the ensuing fire development and spread was evaluated. Three repeat tests were conducted for each of three types of furniture, for a total of 9 instrumented tests. The fires were established under reduced ventilation conditions characteristic of energy-efficient homes.

Various measurements were recorded throughout the structure for each test, including sample mass, gas temperatures, heat flux, carbon monoxide concentrations, oxygen concentrations, and smoke detector activation times. Additionally, quantitative observations were compiled, including detailed documentation of the furniture before and following the fires, video recordings of fire propagation, damage to proximal furniture, and window failure. The figure below shows the layout of the burn house and instrumentation.

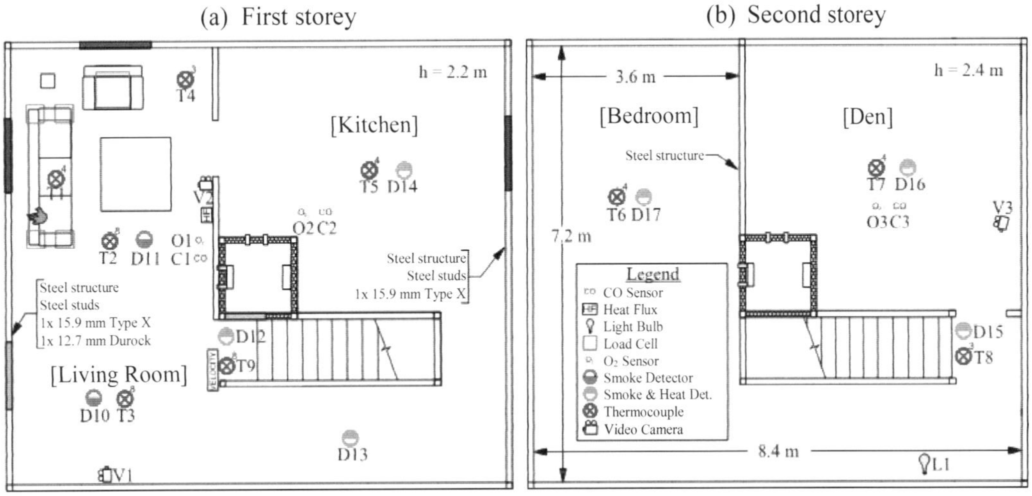

[1] JENSEN HUGHES, BC, Canada
[2] University of Waterloo, Department of Mechanical and Mechatronics Engineering, ON, Canada

In all of the tests, the sofa fires developed with an initial incubation period, a period of rapid fire growth, followed by an abrupt transition to decay. The fires largely self-extinguished within 10 – 15 minutes of ignition. A summary of the main experimental measurements is shown in the table below, along with the deviation in the measurements within each furniture type. Across the three replicate tests on a given type of furniture, the rate of fire growth, peak heat release rate, gas concentrations and compartment temperatures were very repeatable.

Furniture Type	**Consumed sofa mass**	**Average flame spread rate**	**Incubation time**	**Peak HRR**	**Time at peak HRR**	**Peak heat flux**	**Peak 1 Floor temperature**	**Peak 2 Floor temperature**
	(kg)	*(cm/min)*	*(min)*	*(kW)*	*(min)*	*(kW/m²)*	*(°C)*	*(°C)*
Furniture	20.07	66	4.0	1954	7.3	5.7	503	169
Type A	± 4.93	± 6	± 0.1	± 336	± 0.1	± 0.8	± 14	± 9
Furniture	20.29	113	2.6	2242	5.0	11.9	653	207
Type B	± 1.74	± 15	± 0.6	± 233	± 0.6	± 1.0	± 34	± 13
Furniture	19.88	30	5.5	1301	10.9	10.5	483	179
Type C	± 1.61	± 3	± 1.1	± 250	± 1.3	± 1.6	± 42	± 12

There was also consistency in the overall trends of fire development for all three types of furniture, with expected deviations in the details from one furniture type to the other. The figure below illustrates the consistency of the burn damage and fire patterns for the sofa and armchairs in each test series.

(a) Sofas (b) Armchairs

These results demonstrate that overall fire behavior in the compartment was repeatable even with complex the fuel loadings and geometries involved. Damage patterns also support a notion of consistency in fire trend and pattern development in upholstered furniture fires under ventilation limited conditions. Given the consistency of this data, it appears that, in certain cases, only single tests may be required to capture representative fire dynamics for a given fuel source in a compartment and garner results that are reflective of the overall trends in fire behaviour. In particular, running multiple repeats of each full-scale test condition may not be necessary where overall trends in behaviour are of interest and control of the external conditions and variables can be maintained. It should also be noted though that microscopic analysis of results from a specific point within the room will have a much higher degree of variability; thus, investigation of spatially resolved details of the fire environment usually will require repeat tests.

TOXICITY

AN INVESTIGATION INTO THE RELEVANCE OF THE CONTRIBUTION TO TOXICITY OF DIFFERENT CONSTRUCTION PRODUCTS IN A FURNISHED ROOM FIRE

Roy Weghorst*, Kingspan Insulation ltd., Edith Antonatus, BASF SE, Steffen Kahrmann, BASF SE, Christine Lukas, Dow Chemical Company Ltd, Julian Bulk, CURRENTA GmbH & Co. OHG

* Corresponding author: roy.weghorst@kingspan.com

1. Abstract

There is a trend to low energy buildings triggered by tighter energy regulations across the globe. In most cases, this translates into the use of increased insulation layer thicknesses in buildings. The use of high performance insulation products increases with these demands, in order to allow for more efficient building design and increased floor-area. Current research suggests that fires could develop differently in these modern buildings caused by factors including triple glazing and air tightness. It is also emphasised by some that the increased use of organic insulation products generate larger volumes of toxic combustion gases.

This paper presents and evaluates two fire tests conducted in a furnished domestic room only differing in the type of the used wall insulation product to assess its relative contribution to the fire compared to the room content. The test results with regard to heat release rates, smoke and toxic gas emissions show that the organic polyisocyanurate insulation and the mineral fibre insulation behave similarly during the fire and the main threat for occupants emanates from the room contents.

2. Introduction

Building fires, and the evaluation of the toxic hazards from them, are very complex. The risk of building occupants being affected by toxic fumes depends on several factors such as the amount and availability of combustible material, stage of the fire, conditions of the combustion process, etc.

It is known that modern domestic room fires can become fully developed, and flashover can occur after only a few minutes. If an occupant has not escaped by this time, then they would be unlikely to survive. Occupants in other rooms may also be exposed to related smoke while in situ or while evacuating a building, although without themselves suffering the effects of the heat from flashover (because they are separated from the original fire). The question has been raised in recent European discussions about what the role of construction products is in the overall generation of toxic smoke.

In addition to all of the above, generation of toxic combustion products is not simply a material property: the smoke produced by burning construction products, and the resulting hazard, are strongly dependent on the way a product is integrated into the building, the fire scenario (e.g. room size, temperature, ventilation), and exposure time.

This overall complexity makes it difficult to evaluate the smoke toxicity of a single product, in particular a construction product, which generally burns after the building contents, in the later stages of a fire. It has also led international standardisation committees addressing fire safety of building and construction to conclude that, so far, there is no suitable test method for assessing the contribution of individual construction products to the emission of toxic gases linked to real fires. Therefore, the described approach has been chosen to conduct two comparative tests incorporating data generated during the three different phases of a room fire, phase 1 being fire development, phase 2 the flashover (fully developed fire) and phase 3 the decay phase of the fire.

2.1 Objective of the study

The objective of this study was to assess the contribution of the building fabric versus that of the building contents to heat release rate, smoke production and toxicity of fire effluents in the case of a fire in a furnished domestic room. In order to establish the influence of the construction products two test setups were chosen in two rooms constructed with different insulation materials, under the same fire conditions and the same content within the room.

2.2 Scenario

The chosen fire scenario for this comparative testing simulates a waste bin fire (propane burner, 30 kW as defined in ISO 9705) that ignites the curtain and spreads to the armchair (first 5 minutes of the test). The burner was therefore placed in a corner and the curtain was installed just above the burner. After 5 minutes the burner was turned off and the further development of the fire was observed and analysed.

The ignition of the armchair was chosen to be the start of the analysis to minimise variations in fire spread in the early stages of the fire (burning curtains).

3. Description of the test configuration

3.1 Principle

The experimental procedure was conducted in a furnished room insulated and covered with a paper faced gypsum board. The only variation in the two test set-ups was the insulation material used behind the plasterboard. In order to achieve comparable thermal transmittance of the build-ups insulation thickness differed (80 mm vs. 140 mm) to match thermal performance. Both rooms were identically furnished.

3.2 Room set-up

Inner dimensions of the test room without insulation were in accordance with the dimensions given in ISO 9705 (3.6 m x 2.4 m x 2.4 m). The floor, ceiling and walls were made of cellular concrete. The walls, excluding the wall containing the door, were covered with the two different insulation systems. Details for the insulation products are given in table 1.

In order to achieve the same inner volume of the test room in the second test with PIR insulation, an additional inner wall was constructed inside the ISO-room wall (thickness 50 mm) which also consisted of cellular concrete before mounting the insulation system. In both tests the insulation products were placed between wooden battens (distance in both cases 570 mm) which were fixed on the concrete wall. The insulation layer was covered with gypsum cardboard (thickness 12.5 mm) according to EN 520. The gypsum board lining was applied with vertical joints, located on top of the wooden battens, without horizontal joints. A power socket was placed near to the armchair, to create a realistic weak spot in the gypsum layer.

With this build-up two comparable test configurations were achieved.

	Test 1	**Test 2**
Insulation Product	Mineral wool without facing *(according to EN 13162)*	PIR insulation boards with Kraft aluminium facing on both sides *(according to EN 13165)*
Reaction to fire classification according to EN 13501-1 for the product as placed on the market	A1	E
Thermal conductivity according to ($W/*m^{-1}*K^{-1}$)	0.035 *(according to EN 13162)*	0.022 *(according to EN 13165)*
Thickness of insulation product used in test (mm)	140	80

Table 1: Insulation products used in Test 1 and Test 2

3.3 Room contents

Both rooms were fully furnished.

The contents of the rooms were:

- Curtain (fabric): installed width of 80 cm with a curtain rod (installed 15 cm from the wall and 5 cm from the ceiling);
- Armchair (timber and foam filling, 80 cm x 70 cm x 55 cm) with 2 pillows;
- Small table (timber, 55 cm x 55 cm x 50 cm) with a few magazines and a remote control;
- TV bench (timber, 120 cm x 40 cm x 74 cm);
- TV (19 inch);
- Bookcase (timber, 40 cm x 28 cm x 202 cm) with 7 identical books.

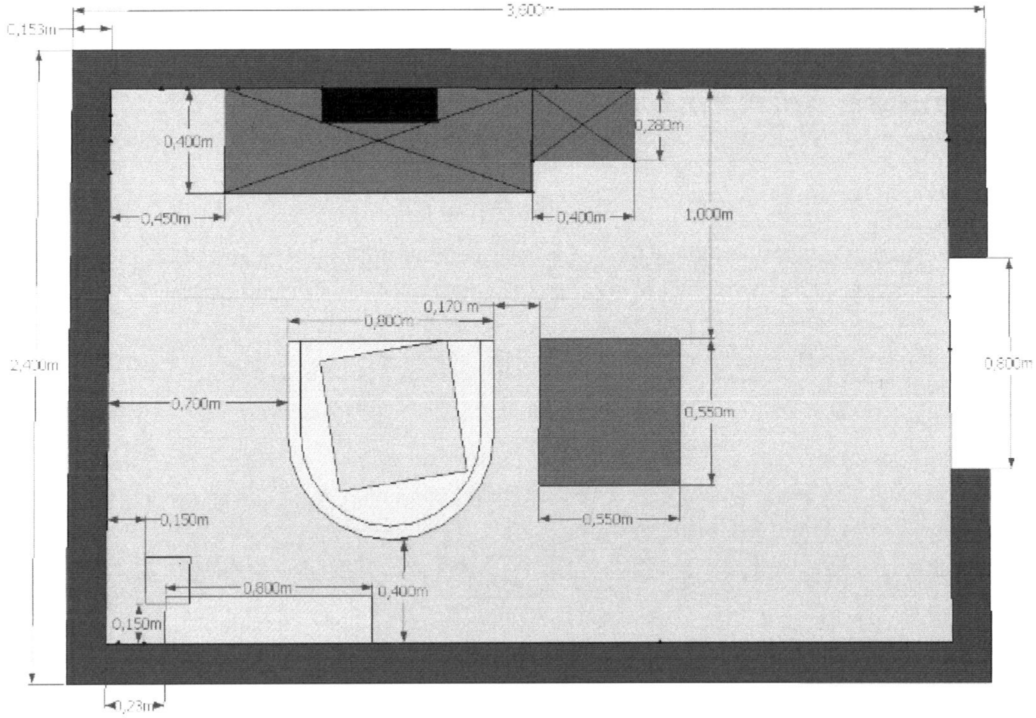

Figure 1: Contents of the tested rooms in relation to the burner

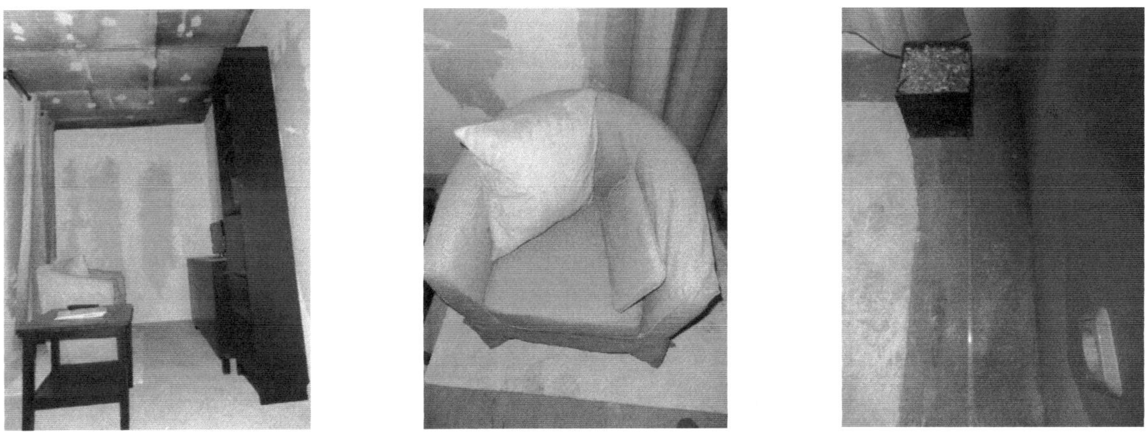

Figure 2: Photos of the room content and the power socket

4. Measurements taken

4.1 Temperature measurements

Thermocouples were installed directly behind the gypsum plasterboard and behind the insulation material in three heights. In addition, a thermocouple tree with seven thermocouples was placed on the left side of the doorway. The temperatures were recorded throughout the entire duration of the test.

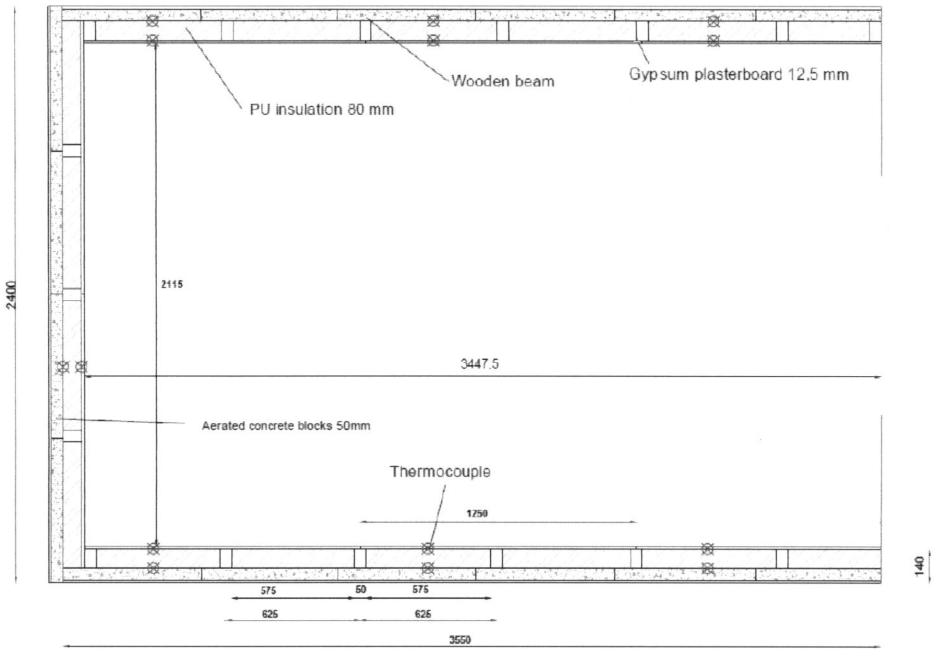

Figure 3: Thermocouple locations at heights of 0.3, 1.2 and 1.8 m from the floor shown in blue

4.2 Heat release rate

The heat release was measured and recorded using the oxygen depletion method according to ISO 9705-1:2016 and the European standard EN 14390:2007. The sampling port was installed in the exhaust duct and oxygen and carbon dioxide were continuously measured which allowed for determination of the heat release rate.

4.3 Smoke obscuration

The smoke obscuration was measured with an optical path system in the exhaust duct using white light as described in ISO 9705. The smoke production rates were measured in m2/s. The calculation was in accordance with ISO 9705 and EN 14390.

4.4 Concentration of effluent gases

The fire effluent gases were measured with an FTIR spectrometer. The sampling and the analysis were done based on ISO 16405 (Room corner and open calorimeter - Guidance on sampling and measurement of effluent gas production using FTIR technique) and ISO 19702 (Guidance for sampling and analysis of toxic gases and vapours in fire effluents using Fourier Transform Infrared (FTIR) spectroscopy).

The following effluent gas components were analysed:

- Carbon monoxide (CO)
- Carbon dioxide (CO_2)
- Hydrogen cyanide (HCN)
- Formaldehyde (CHOH)
- Acrolein (C_3H_4O)
- Sulfur dioxide (SO_2)
- Hydrogen Chloride (HCl)
- NO_x (measured by the sum of N_2O, NO and NO_2)

Gas sampling was installed in the duct work of the smoke exhaust system of the testing facility. This sampling position represents cooled and diluted fire effluents. ISO 16405 states that this sampling position is often preferred over sampling outflowing gases directly at the top of the doorway as matrix effects from the fire effluents are minimized by the dilution. At this position due to cooling and dilution effects all chemical reaction in the fire effluent would have ceased which would not be the case when sampling directly from the immediate vicinity of the fire where further chemical reaction, combustion and decomposition could take place. Using the known volume flow rate in the exhaust duct, the dilution effect has been reversed prior to calculating the Fractional Effective Dose (FED) and Fractional Effective Concentration (FEC).

5. Results and discussion

5.1 Course of tests, time scales for data presentation and heat release

In both tests the burner was shut off after 5 minutes. In the first test, the curtain was ignited by the burner after 5 seconds, fell down and ignited the arm chair after 4 minutes. In the second test the curtain did not cause ignition of the arm chair. Therefore, the arm chair was moved closer to the burner and the burner was started again. In this test the arm chair was only ignited 9:40 minutes after the start of the test. Following the ignition of the armchair in both tests the fire spread to the other contents of the room, subsequently leading to further development of the fire (phase 1).

As the purpose of the test was to analyse the relevance of the contribution to toxicity from the applied insulation products in relation to the contribution from the contents burning in the room, the start of the test, in both cases, was defined as the moment when the arm chair was ignited and hence the fire started to develop in the room. This approach is confirmed by the fact that the time to flashover (phase 2) in the room, caused by burning of the contents of the room only, was 7:24 minutes after ignition of the chair in Test 1, and 6: 55 minutes in Test 2. Therefore, the point of time when the chair ignited is referred to as "start of the test" in the following discussion and in the diagrams.

In both tests flashover occurred about 7 minutes after the start of the test (see Figure 4). A second peak of the heat release rate was observed some minutes later. Subsequently, the fire decayed (phase 3). First cracks appeared in the gypsum plasterboard in both tests after about 20 minutes.

During the decay phase (from the 20th minute onwards) a difference in the heat release rate was observed by a slightly less steep decay in test 2, however, the HRR curves of both tests were below 50 kW after approximately 50 minutes.

The following diagrams illustrate heat release rate versus smoke production and later some of the toxic gas components.

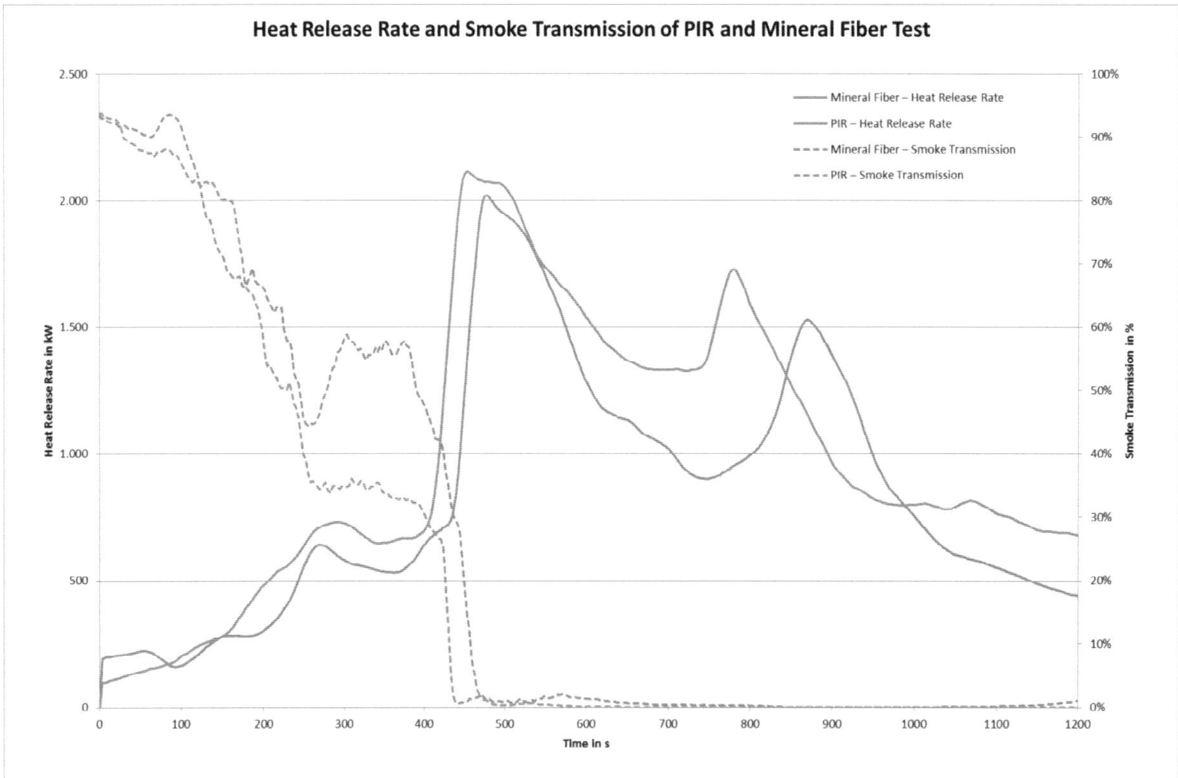

Figure 4: Heat Release rate versus smoke transmission

5.2 Gaseous Effluents

The concentrations of the measured gases also correspond to the course of the heat release rate and the smoke obscuration. The example curves below in Figure 5 show that for CO no significant difference can be seen. For HCN only in a very late phase of the test (more than 20 minutes after the start of the test) there is a slight increase in concentration in Test 2 compared to Test 1. This phenomenon can be attributed to the PIR insulation product which was partly exposed to radiative heat and direct flame impingement from the room's interiors by this time. However, this occurred far after much higher concentrations were reached in both tests due to the burning room contents (very similar in both tests).

The formaldehyde and acrolein data, on the other side, indicate that in the same late phase of the tests the insulation was contributing. Here, after more than 20 minutes, the concentrations were slightly higher in Test 1 – where the mineral fiber insulation had been installed. Also these increases show that the contribution from the insulation product took place very late and the concentrations were significantly lower than those reached during the second (flashover) phase, when the contents of the room were burning intensely.

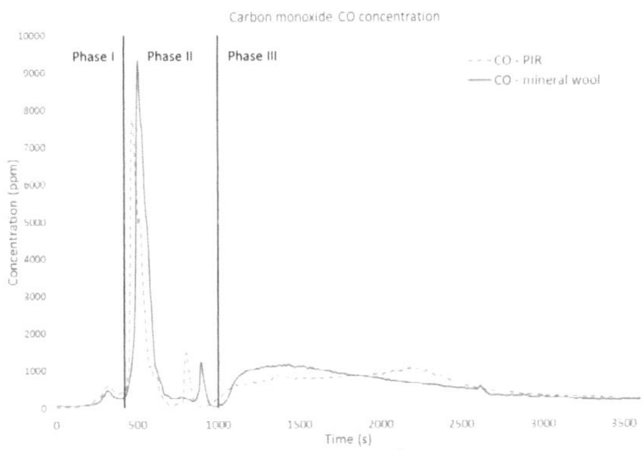

Concentration of Carbon monoxide

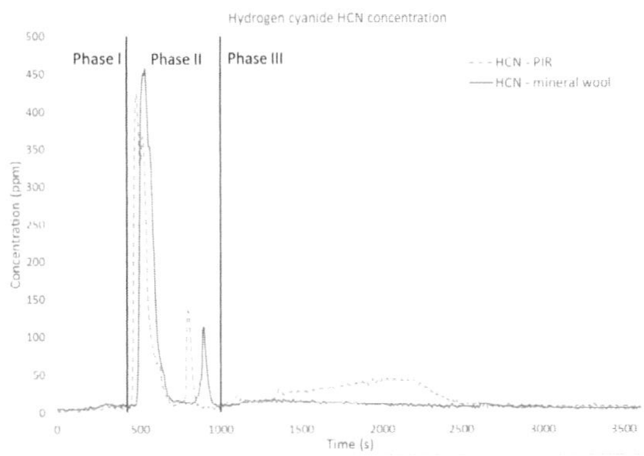

Concentration of Hydrogen Cyanide

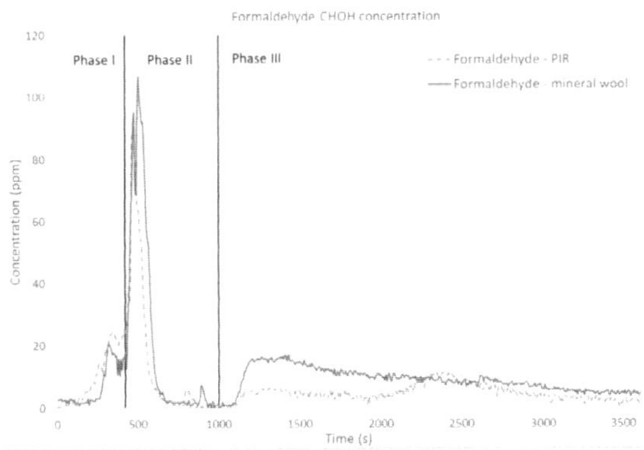

Concentration of Formaldehyde

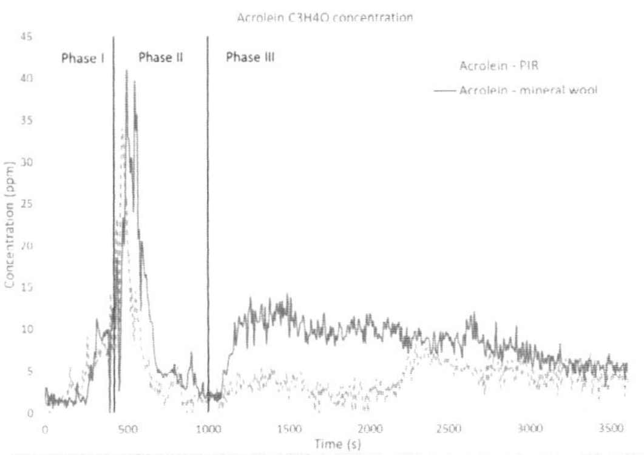

Concentration of Acrolein

Figure 5: comparison of some gas concentrations measured during the test

5.3 Comparative toxicity assessment

The Fractional Effective Dose (FED) and Fractional Effective Concentration (FEC) values were computed according to ISO 13571: 2012 and based on all the fire effluents measured during the fire tests and mentioned earlier. This means that the effluents from the armchair and other furniture contents were the major contributors to both the FED and FEC values, particularly in the fully developed phase of the fire (between 6 and 17 minutes). In detail, FED was calculated based on the concentrations of carbon monoxide (CO), hydrogen cyanide (HCN) and carbon dioxide (CO2) measured during the tests. The data is uncorrected for the gas burner output, but this is consistent in both tests.

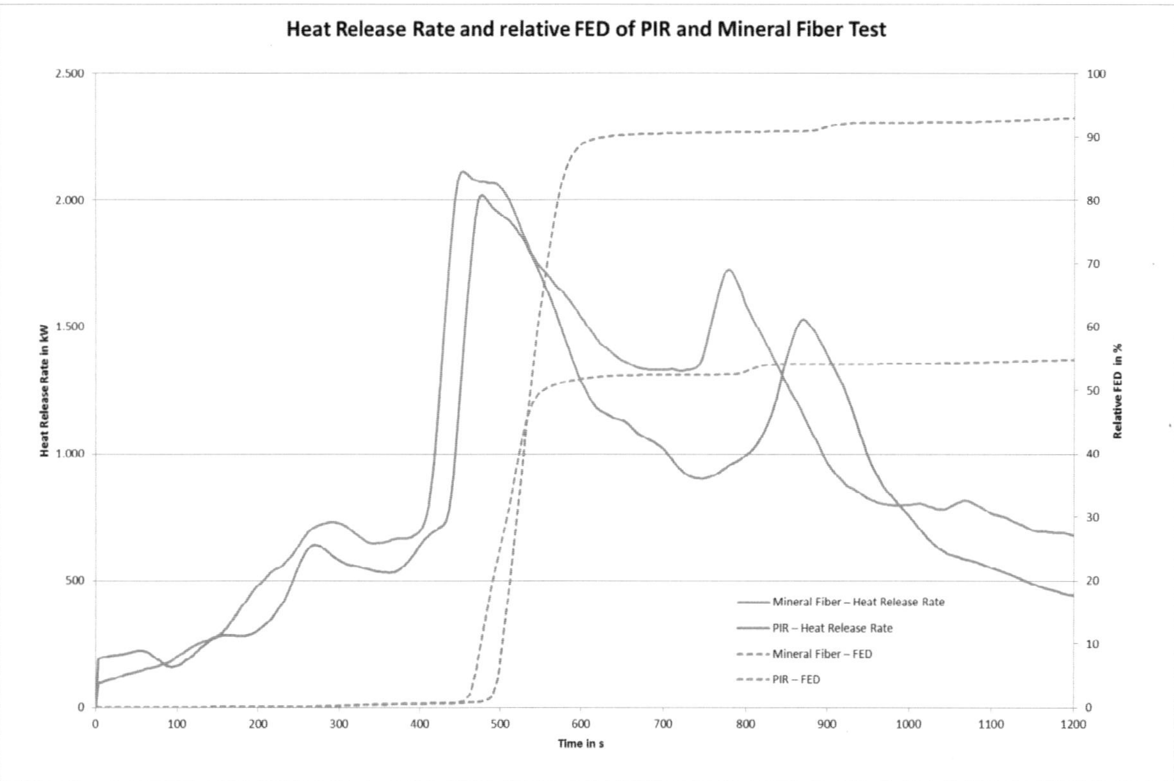

Figure 6: FED comparison

FEC was calculated based on the concentrations of formaldehyde (CHOH), acrolein (C_3H_4O), sulphur dioxide (SO_2), hydrogen chloride (HCl) and nitrous oxides (NO_X). Of course numerous other irritant species can be formed in fires. The range of effluent species selected for analysis covered those species of toxicological significance that could reasonably be expected to be released, based on knowledge of the composition of the material and in consultation with the published documentation for exposure criteria.

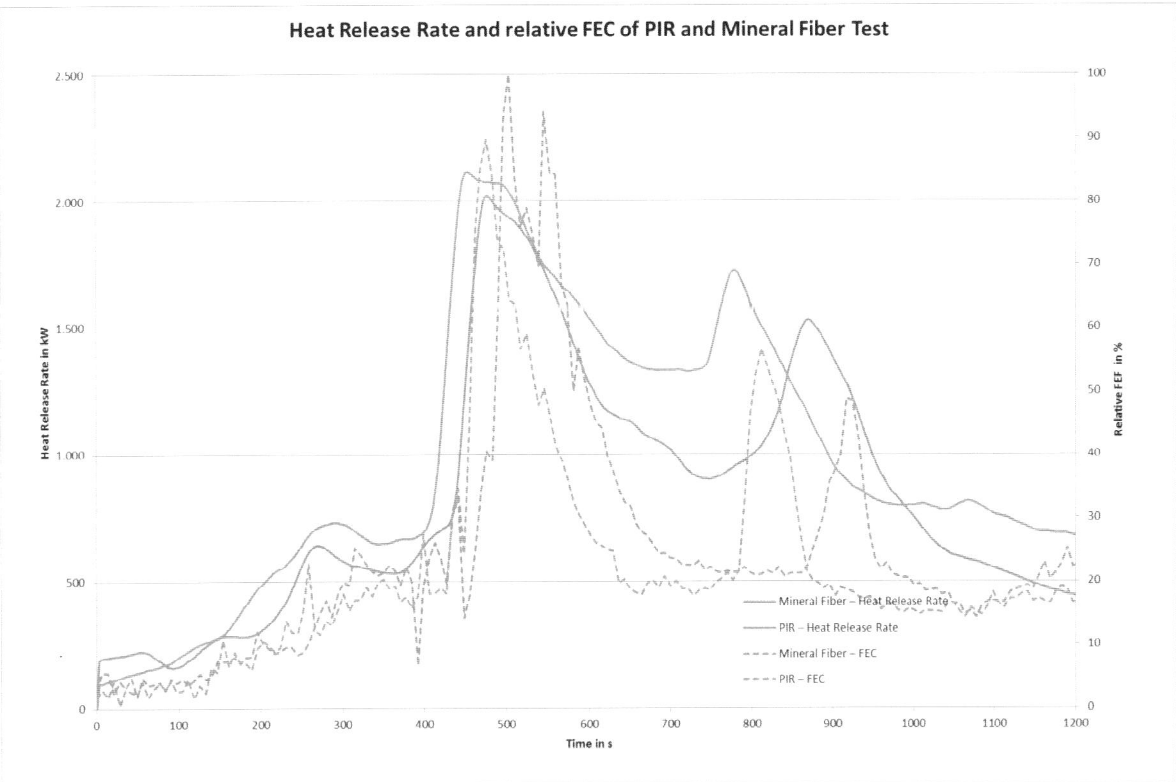

Figure 7: FEC comparison

Note: Toxicity is only to a certain degree a material property. It is strongly influenced by the environment, availability of oxygen, thermal attack, air flow and surfaces available for combustion. The chemistry of the combustion of a given material can therefore proceed along various routes and produce species in very different quantities dependant on conditions to which it is subjected. Such changes would impact the FED and FEC values.

The toxicity evaluation via FED and FEC – as shown in figures 6 and 7 – clearly demonstrates that the major part of the toxic hazard was already reached in the early phases of both tests, where the wall insulating construction products were not yet contributing. Already after less than 10 minutes of testing time, the FED did not change much more. For the FEC in both tests similar developments could be observed. Only about 18 minutes after the start of the test, a slight increase in the FEC occurred in Test 2 (PIR insulation). But also here the increase and the absolute values were significantly lower than the peak values reached in the flashover phase.

6. Conclusions

It is acknowledged that – with regard to domestic fire deaths – smoke inhalation causes or at least contributes to the majority of the fatalities. The results of tests presented here, demonstrate that in the considered domestic fire scenario the toxic potential caused by gaseous emissions basically arises from the burning and consequent smoke emission of the combustible building contents. This occurs significantly before the building envelope has even been affected by the fire. However in the tests presented here, minor contributions to the concentrations of the critical gases due to the involvement of the insulation products in the fire could be observed in a very late phase of the fire. However this was far after substantial FED and FEC values had been reached due to the burning of the contents of the room.

Accordingly, to reduce the risk to occupants induced by smoke gases, it is crucial to detect fires early within a building. This then allows occupants to evacuate before they are exposed to smoke inhalation. Therefore, it is important to investigate suitable fire safety options such as stipulating the installation and maintenance of smoke detectors for all buildings and to improve their design in order to facilitate means for fast and safe evacuation. In this way then early detection of and fast response to fires constitute very effective measures to ensure the safety of occupants.

Acknowledments

We would like to thank PU-Europe for sponsoring this research and the team from Warringtonfire Gent for the good cooperation throughout the entire process. We appreciate the contribution of Beth Dean from Exova Warringtonfire for the FED/FEC calculations and comparison of the gaseous effluents and its relevance in the different stages of fire development in this room fire scenario.

7. References

[1] ISO 9705: 2016 - Reaction to fire tests - Room corner test for wall and ceiling lining products - Part 1: Test method for a small room configuration

[2] ISO 16405: 2015 - Room corner and open calorimeter -- Guidance on sampling and measurement of effluent gas production using FTIR technique

[3] EN 13501-1: 2007+A1:2009: Fire classification of construction products and building elements – Part 1: Classification using data from reaction to fire tests

[4] EN 520: 2004 + A1:2009: Gypsum plasterboards - Definitions, requirements and test methods.

[5] EN 13165: 2012: Thermal insulation products for buildings. Factory made rigid polyurethane foam (PU) products. Specification

[6] EN 13162: 2012+A1:2015: Thermal insulation products for buildings. Factory made mineral wool (MW) products. Specification

[7] ISO 13571: 2012 - Life-threatening components of fire - Guidelines for the estimation of time to compromised tenability in fires

[8] ISO 19702: 2015 - Guidance for sampling and analysis of toxic gases and vapours in fire effluents using Fourier Transform Infrared (FTIR) spectroscopy

POLYMER COMBUSTION PRODUCTS AT CONSTANT FUEL/OXYGEN RATIOS IN THE MICROSCALE COMBUSTION CALORIMETER WITH FTIR

Louise C. Speitel, Richard N. Walters and Richard E. Lyon
Federal Aviation Administration
William J. Hughes Technical Center
Fire Safety Branch ANG-E21
Atlantic City International Airport, NJ 08405

ABSTRACT

The microscale combustion calorimeter (MCC) has proven to be useful for assessing the maximum potential of a material to contribute to a fire. Accurate and reproducible flammability properties of the solid material are obtained by forcing complete pyrolysis of the sample and complete combustion of the pyrolysis products in excess oxygen in a standard test (ASTM D 7309).

The FAA's microscale combustion calorimeter was modified to provide a constant fuel/oxygen mass ratio relative to the stoichiometric fuel/oxygen ratio. This equivalence ratio, Φ, is held constant during the entire course of polymer thermal decomposition in a constant heating rate experiment. A Fourier transform infrared spectrometer (FTIR) was connected for quantitative analysis of the combustion products. In this way, combustion products can be measured under conditions of early stage (over ventilated) to late stage (vitiated) fires, and the effect of flame-retardants on the toxic potency of the smoke can be determined using milligram samples in a 15-minute test.

In this paper we report on this MCC-Φ-FTIR method. A wider mix of combustion gases are generated in vitiated combustion, with more complex FTIR spectra. FTIR methods were developed for this complex mix of spectra to determine gas yields. Over 20 plastics were tested with the MCC-Φ-FTIR technique at equivalence ratios ranging from fuel-lean to fuel-rich. Results are presented for the combustion products of a variety of materials.

INTRODUCTION

There is an ongoing effort to eliminate certain halogenated flame-retardants from consumer products because of the potential for exposure from the natural environment or through contact with flame-retardant articles in the built environment [1-6]. More recent arguments against the use of flame-retardants in consumer products include the inhalation of toxic combustion products generated by burning of flame-retardant plastics commonly used in passenger airplanes, public transportation, home furnishings, construction materials and electronics [7]. Fire fighters are thought to be at particularly high risk from this type of exposure [8,9]. It is well known that the type and quantity of toxic combustion products generated by burning plastics depends on the material composition and the fire environment, and that these toxic combustion products increase as the oxygen concentration in the air decreases at the later stages of a fire [7]. Several bench scale methods are currently in use to measure combustion products of materials over a range of fire conditions. These methods attempt to independently control the burning (mass loss) rate of the sample and the mass flow rate of oxygen to the flame to adjust the fuel/oxygen mass ratio relative to the stoichiometric value, also known as the equivalence ratio, Φ = (mass of fuel/mass of O_2)/(mass of fuel/mass of $O_2)_{stoich}$. In these bench-scale tests, the reported equivalence ratio is a global value for the entire apparatus, because the local fuel/oxygen ratio in the combustion zone (flame) is unknown. In this study the MCC is used to obtain a well-defined local Φ by precise control of the fuel/oxygen ratio entering a pre-mixed combustor. The gaseous combustion products exiting the combustor are analyzed by coupling the MCC to a Fourier transform infrared spectrometer (FTIR) using a heated transfer line. In this way, toxic combustion products can be measured as a function of material composition at well-defined, local equivalence ratios ranging from oxygen rich ($\Phi < 1$), to stoichiometric ($\Phi = 1$) to fuel rich ($\Phi > 1$), thereby extending the utility of the MCC to the study of the toxic potency of smoke from burning polymers and flame-retardant plastics.

MATERIALS

All gases were ultra high purity (UHP) grade obtained from a local supplier along with liquid nitrogen. Samples were additive-free solid polymers obtained from commercial sources, compounded plastics from distributors, or materials obtained from aircraft cabin components/assemblies.

METHODS

Controlling the Equivalence Ratio: In the standard MCC method [10] the sample temperature increases linearly from about 80°C to 1000°C at a constant rate of 1 K/s and the pyrolysis/fuel gases are mixed with excess oxygen and combusted for a few seconds at 1000°C. In the standard test the equivalence ratio varies from $\Phi = 0$ at the beginning and end of the test (fuel pulse) to about $\Phi = 1/2$ at the peak rate of heat release (oxygen consumption), so there is at least twice as much oxygen available at any time in the combustor than is required to completely convert the fuel gases to carbon dioxide, water and acid gases. Figure 1 shows the Φ history during a standard test of polystyrene.

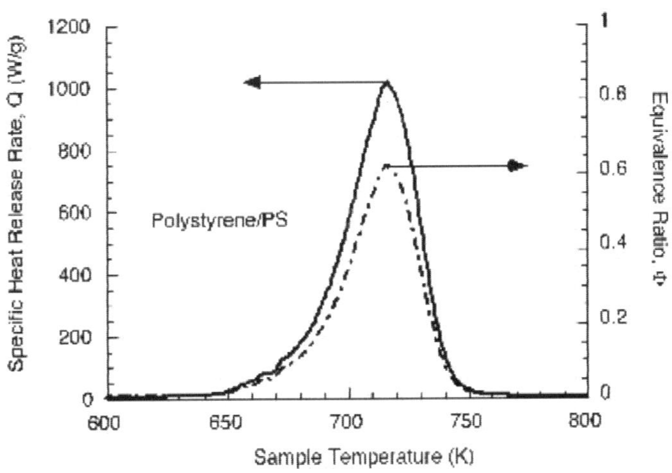

Figure 1. Specific Heat Release Rate Q and Equivalence Ratio Φ for Polystyrene/PS in the Standard Test (ASTM D 7309)

The Microscale combustion calorimeter was constructed in-house according to ASTM D 7309 specifications and tests were conducted according to the standard method with the exception that the concentration of oxygen during the test was set by the equivalence ratios, which were $\Phi = 0.5$, 1.0 and 2.0 for this study. Initial sets of experiments were performed to generate the oxygen consumption data needed to control the oxygen for the subsequent controlled fuel/oxygen tests. These tests were performed under the standard method where the combustor was maintained at 1000°C to ensure complete oxidation of any volatile thermal decomposition products in excess oxygen (see Figure 1). Several tests were run for each material and the data was either averaged or the most representative test was used for the control file of the stoichiometric oxygen consumption per unit mass (oxygen demand) of sample at each temperature throughout the sample decomposition range. The control file was stored as a look-up table and used to control the flow rate of oxygen that would result in the desired equivalence ratio. The oxygen consumption module of the MCC (water removal, flow meter and oxygen sensor) was connected to the outflow of the FTIR to check for leaks, but only the total heat could be obtained from the oxygen consumption data due to distortion of the transient signal in the relatively large volume FTIR gas cell.

Measuring the Combustion Products: The MCC was coupled to an FTIR for gas analysis by a 6-foot (1.83 m) long heated transfer line maintained at a temperature of 165°C. An in-line, heated, stainless steel filter housing was inserted at the end of the heated line prior to the inlet port of the FTIR gas cell. The filter housing contained a 0.2 μm Teflon filter to remove particulates and any possible condensates from the gas stream and was maintained at a temperature slightly below 165°C to trap organics. A plastic tube filled with desiccant was attached to the outlet of the gas cell to remove water from the combustion stream prior to entering the flow meter and oxygen sensor of the MCC for oxygen consumption measurements. Basically, the FTIR was inserted between the combustor and the oxygen consumption module of the MCC. The schematic of The MCC-Φ-FTIR is shown in figure 2.

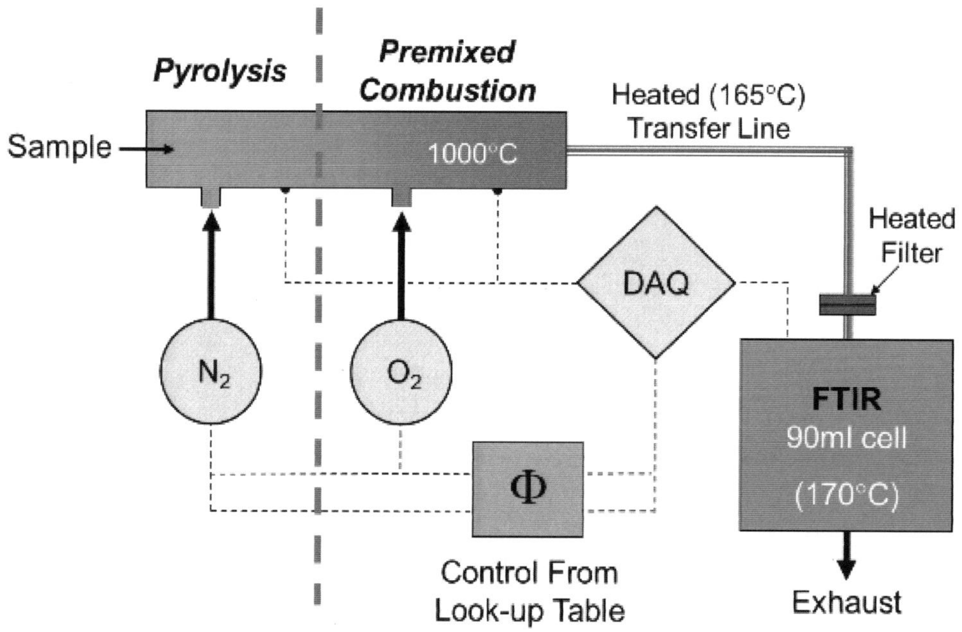

Figure 2. Schematic of the MCC-Φ-FTIR Used for Generating and Measuring Combustion Products of Polymers at Different Fuel/Oxygen (Equivalence) Ratios.

Figure 3 is a photograph of the MCC-Φ-FTIR used in this study.

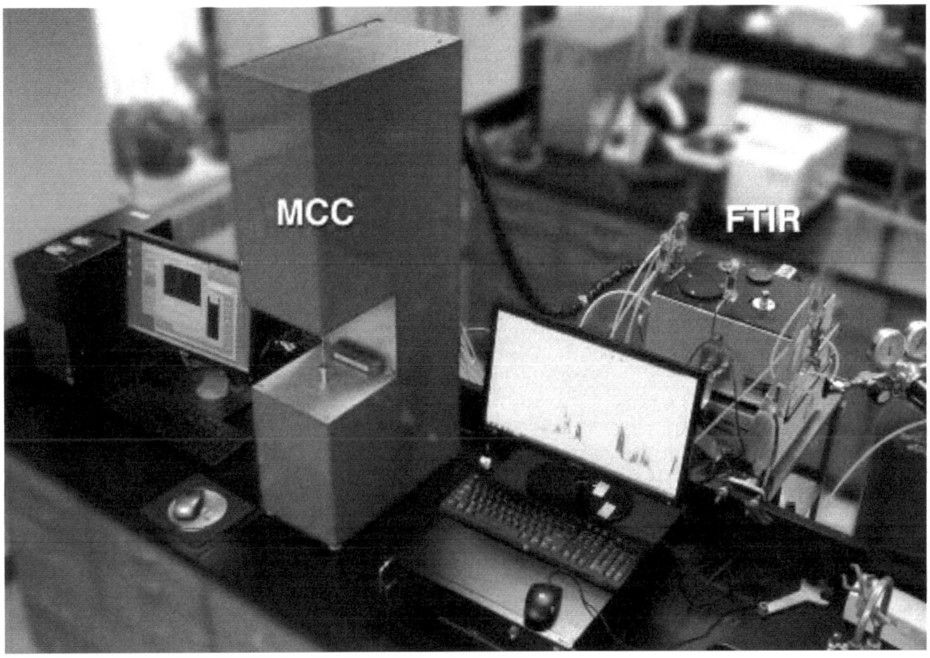

Figure 3. Photograph of the Coupled MCC-Φ-FTIR Used for Generating and Measuring Combustion Products of Polymers at Different Fuel/Oxygen (Equivalence) Ratios.

Two separate methods were employed for the FTIR analysis. The methods are illustrated in figures 4 and 5. The selected regions for quantification are indicated in grey. The differences between the 2 methods are indicated with dashed circles.

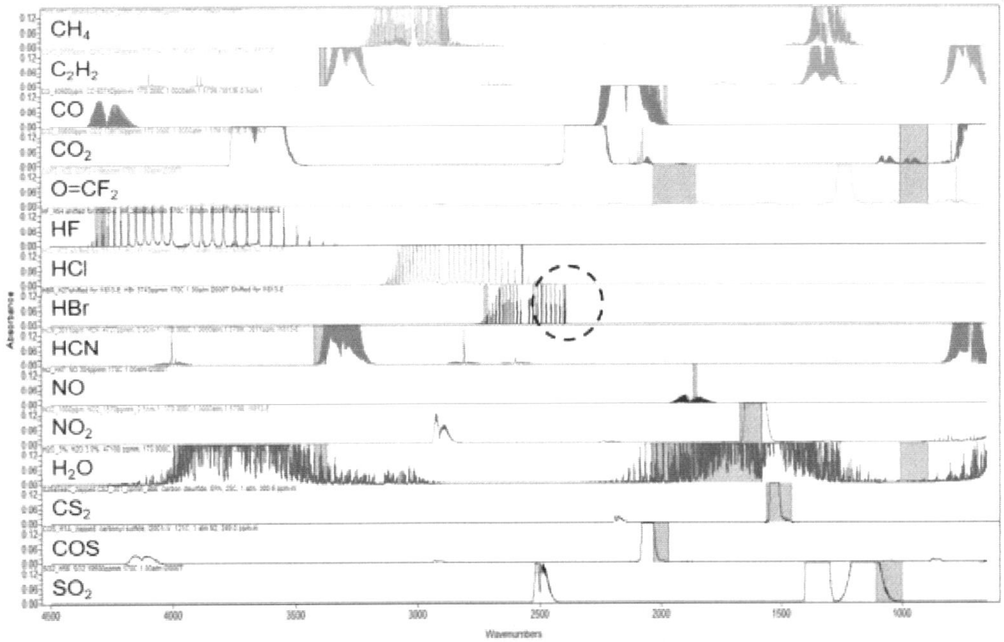

Figure 4: Spectral Regions for FTIR analysis of combustion gases: Method 1.

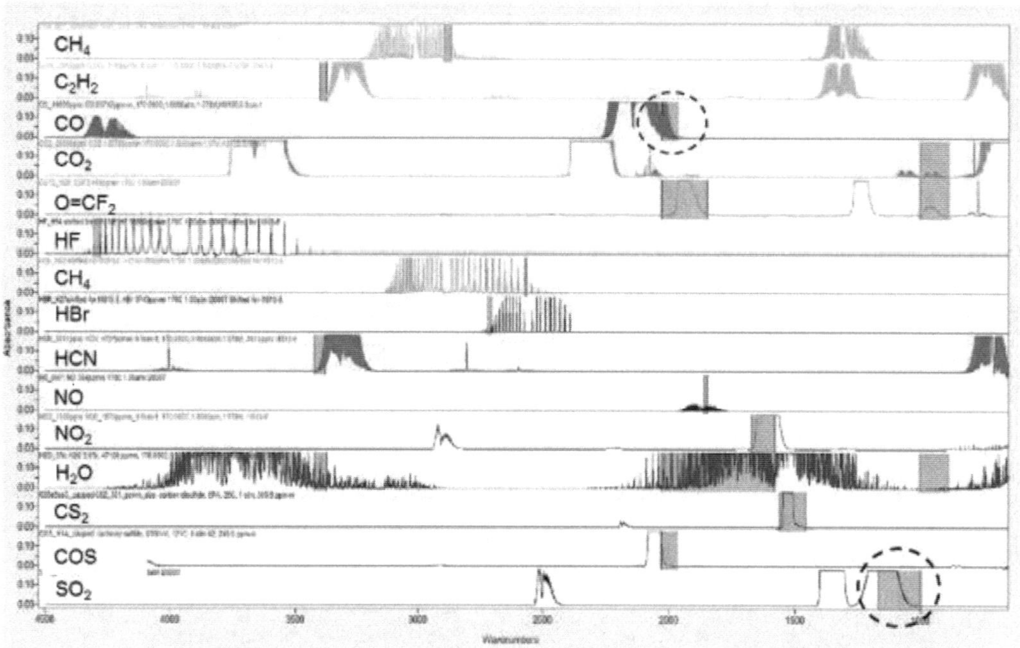

Figure 5: Spectral Regions for FTIR analysis of combustion gases: Method 2.

A Midac Model I1813 FTIR Spectrometer with a 2-meter cell was used for all tests. The optical path of this cell is about 1000 times longer than the 3-mm CO and CO_2 NDIR analyze cells, contributing to a much greater sensitivity and lower limit of detection. The sample cell is nickel, the mirrors are coated with gold, and the windows are zinc selenide (ZnSe). The cell volume is approximately 90 ml. The interferometer, beam splitter, and windows are constructed of ZnSe with a germanium coating for moisture protection. The detector is a sterling engine-cooled Mercury Cadmium Telluride (MCT) detector, providing more than an order of magnitude additional sensitivity. All interface optics are gold-coated for high light throughput and corrosion resistance. The sample cell has a horizontal orientation to minimize buildup of soot on the cell mirrors.

All calibration spectra and test spectra were obtained at unit gain. A triangular apodization, a Mertz phase correction, and a resolution of 0.5 cm^{-1} were used for all calibration and test spectra with the exception of CS_2. The method creates a spreadsheet with the time profiles of the concentration and error (residuals) data for the analyte gases. Gas concentrations are reported as zero for each gas and spectra if the residuals for the spectral region are 30% or greater than the calculated concentration. Cell pressure and temperature are monitored and recorded with each spectrum. The Autoquant Pro software performs piecewise-linear Beer's Law calculations for each analyte in each test spectra to correct the analyte concentration for any pressure and temperature variations from the calibration spectra. However, the cell pressure was always close to atmospheric pressure.

All tests were conducted in an extractive mode at 170°C and 0.5 cm^{-1} resolution with 8 bi-directional scans every 6 seconds, on average. Calibration spectra for most gases were obtained under these conditions except that 256 scans were run. Most calibration spectra were obtained at 170°C using a Midac FTIR: H_2O, CO, CO_2, HCN, C_2H_2, and NO_2 spectra were prepared in-house using the same FTIR as was used for the MCC tests. HF, HCl, HBr, NO, SO_2, $COCl_2$, COF_2 and CH_4 were prepared by Midac Corporation at 170°C using a 10-m path cell. COS and C_3H_8 calibration spectra were prepared by Midac at 121°C. NH_3 and N_2O calibration spectra were prepared by Midac at 25°C. CS_2 was prepared by the US Environmental Protection Agency at 25°C. All calibration spectra prepared on a different FTIR were shifted to match the absorbance frequencies of the MCC-FTIR for the regions used for quantification .

Nineteen gases were analyzed including CO, CO_2, HF, HCl, HBr, SO_2, COS, NO, NO_2, COF_2, NH_3, N_2O, and water. A gas of concern that is considered non-quantitative at this point in time is CS_2. Gases that are in the method but not reported are H_2O, C_2H_2, CH_4, and CS_2. Yields of CO and CO_2 are presented in units of g/kg. All other combustion product yields are presented in units of g/kg. These yields can be used to calculate the concentration of a toxic gas evolved in a volume for a unit mass of a sample and per unit weight loss of sample. Data were collected, analyzed, and plotted using the following commercial software: Midac Autoquant Pro™ software, Operant LLC: Essential FTIR®, and Microsoft® Excel®. Spectral bands for all quantified gases were selected to have minimal common absorbance. Another requirement for spectral band selection is low absorbance so that the calibration is linear over a wide concentration range. Broad non-overlapping bands were selected when possible.

Yields of the gaseous decomposition/combustion products were compared as a function of Φ for a selection of materials. It is hoped that this technique, combined with an appropriate toxicity model, can be used as a part of a toxicity assessment for polymers.

RESULTS & DISCUSSION

Preliminary tests were run during the development of the controlled fuel/oxygen ratios with the MCC configured with additional CO & CO_2 analyzers. These analyzers (Edinburgh Instruments) were placed after the flow meter and oxygen sensor so the MCC oxygen consumption history was well resolved. A series of FAA database thermoplastics were tested to evaluate the algorithm for well-characterized samples in the MCC. In order to expand the parameter space for the study of the effect of material composition (including flame retardants) on incomplete combustion, tests were conducted at different fuel/air (Φ) ratios and combustion temperatures (T_c). The accuracy of the yields from the CO and CO_2 analyzers were checked using a carbon mass balance for samples of known composition that were completely combusted at $T_c \geq 900°C$. A char composition of C_5H_2 was found to best correlate the CO_2 yield of the volatile component of charring polymers. See table 1.

Table 1: Experimental Values Measured in the MCC and Theoretical Values Calculated From the Chemical Composition and Sample Weight.

				O_2 Demand (g/g)		CO_2 Yield (g/g)		Char Yield (%, w/w)	
Polymer	Formula	MW (g/mol)	HR (kJ/g)	Theory	Experiment	Theory	Experiment	Theory	Experiment
POM	CH_20	30.03	14.5	1.07	1.11	1.47	1.59	0.00	0.02
PMMA	$C_5H_8O_2$	100.12	24.7	1.92	1.89	2.20	2.40	0.00	0.03
PP	C_3H_6	42.08	42.9	3.42	3.27	3.14	3.29	0.00	0.13
HDPE	C_2H_4	28.06	43.2	3.42	3.30	3.14	3.25	0.00	0.23
ABS	$C_{15}H_{17}N$	211.31	36.9	2.92	2.82	3.12	3.44	0.00	0.75
PVDF	$C_2H_2F_2$	64.02	28.5	1.25	2.18	1.37	2.86	0.00	1.10
PA66	$C_{12}H_{22}O_2N_2$	226.32	28.9	2.33	2.21	2.33	2.89	1.68	1.16
HIPS	$C_{12}H_{14}$	158.24	38.3	3.13	2.92	3.34	3.50	2.34	2.52
PET	$C_{10}H_8O_4$	192.17	16.8	1.67	1.28	2.29	1.97	13.55	13.02
PVC	C_2H_3Cl	62.48	10.7	1.41	0.82	1.41	1.00	20.84	19.71
PC	$C_{16}H_{14}O_3$	254.28	21.1	2.27	1.61	2.77	2.03	24.41	23.94
PPS	C_6H_4S	108.15	15.5	2.07	1.18	2.44	1.24	46.28	41.97
PEI	$C_{37}H_{24}O_6N_2$	592.61	10.5	2.16	0.80	2.75	1.16	53.71	56.80

The Φ control algorithm for the oxygen and nitrogen flow meters was checked for accuracy by running a test without a sample and comparing the program response to the measured response for a fictitious sample of identical weight. A composite material was selected for this check because it had a complex heat release rate history. The composite sample exhibited three peaks where two of them were overlapping, as shown in Figure 6. The stoichiometric amount of oxygen was calculated from the measured HRR curve. A blank was then run for the composite using the Φ-control file and the output for the mass flow controller along with the measured oxygen concentration were recorded. Figure 6 shows that there is excellent agreement between the experimental and programmed oxygen profiles for this complicated oxygen consumption history.

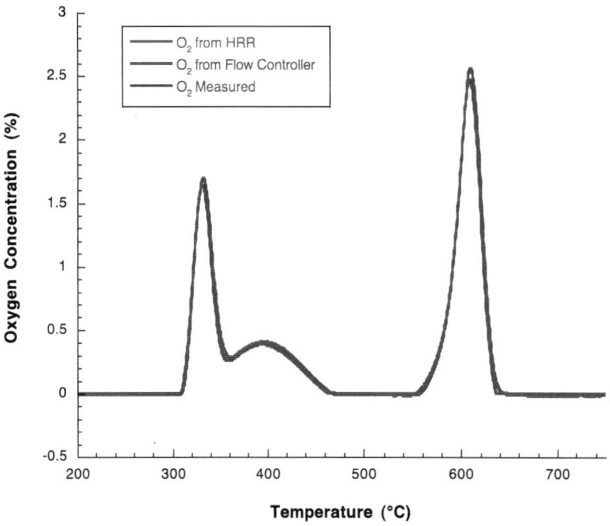

Figure 6: Plot of the Programmed and Measured Oxygen Concentration for Complete Combustion of a Composite Sample for Φ = 1.

Samples of FAA database plastics and aircraft materials were selected for chemical composition and tested at several different Φ values. Initially only O_2, CO and CO_2 were measured using the in-line analyzers and validated with a CO_2 mass balance as described earlier. Figure 7 shows a plot of the CO/CO_2 ratio over a rage of Φ values for these materials. As expected, the CO levels increase drastically with increasing fuel/O_2 ratios.

These data were used to benchmark the CO and CO_2 yields obtained for the same materials in the FTIR method as follows. The CO and CO_2 analyzers were removed and an FTIR connected to the outlet of the combustor as describe previously. Initial results from the comparison of the CO and CO_2 yields (grams of gas species per gram of sample) indicated the FTIR yields of CO were comparable to the values obtained from the in-line CO gas analyzer while CO_2 yields averaged about 30 percent higher. Consequently, further development of the FTIR method is needed.

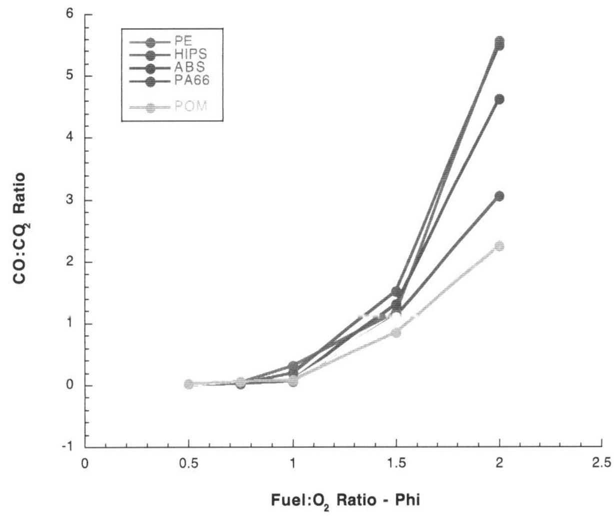

Figure 7: Plot Showing How the CO/CO_2 Ratio Changes Over the $Fuel/O_2$ (Φ) Range.

The MCC-Φ CO and HCN yields were compared to yields obtained with other test methods. It can be seen in figure 8 that the yields obtained with the MCC over a range of Φs fall within the range of the other test methods. The MCC-Φ has the tightest control of the equivalence ratio of the other methods listed. For example, the low CO yields obtained with the controlled atmosphere cone calorimeter and the Fire Propagation Apparatus could be due in part to further oxidation of the hot gases after leaving the stack and before entering the hood. Gases are measured further downstream of the hood in both these test methods.

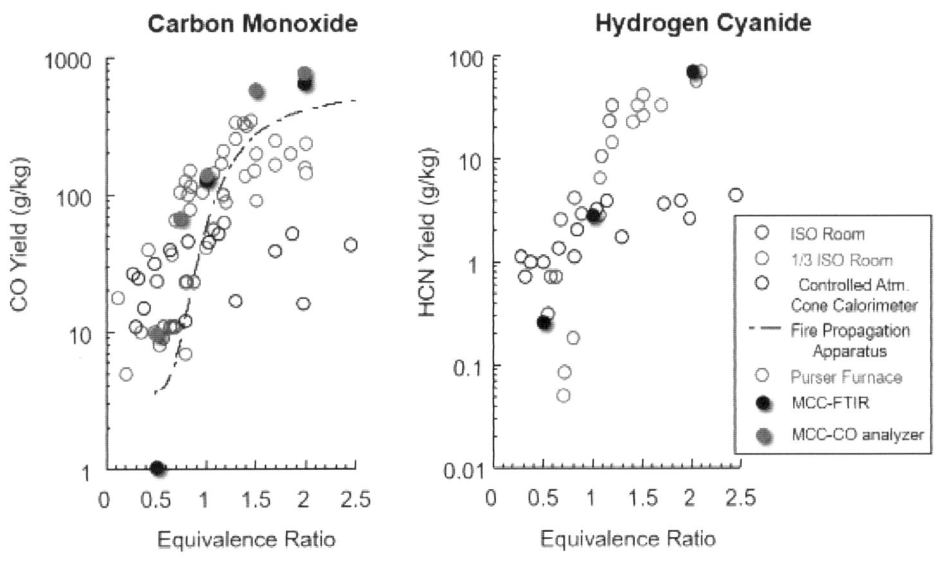

(A. Stec & R. Hull, Fire Toxicity, 2010)

Figure 8. Comparison of Vitiated PA66 Yields Obtained with MCC-Φ and Other Test Methods

Several nitrogen-containing samples were tested at multiple Φs, and the results are shown in Figure 9. Most of the samples produced little or none of the asphyxiant gas hydrogen cyanide (HCN) in well ventilated conditions ($\Phi < 1$), with the exception of a flame-retarded polyphthalamide. In contrast, polyurethane and the non-FR polyphthalamide produced HCN at stoichiometric fuel/oxygen ratio (Φ = 1), while all nitrogen-containing polymers produced HCN under oxygen starved/vitiated (Φ > 1) conditions.

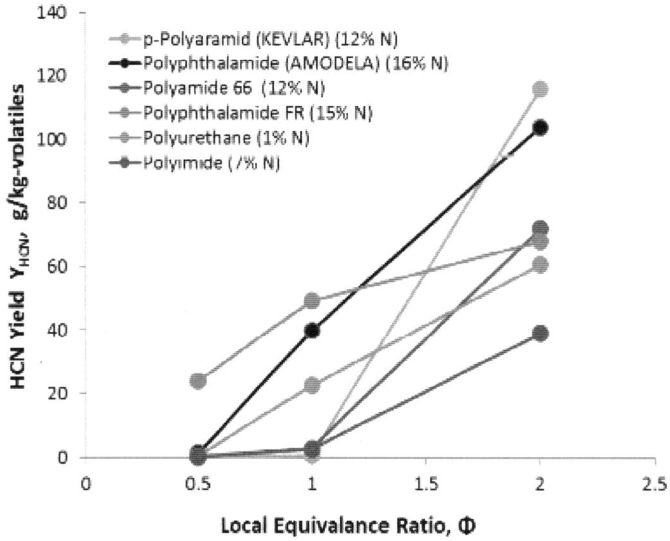

Figure 9. Yields of HCN for Several Nitrogen Containing Polymers.

Yields of HF remained fairly constant throughout the different Φs for PVF, as expected, since oxygen plays no role in HF formation. For example, PVF yielded 262 g-HF/kg-sample. Three different grades of proprietary polyvinylfluoride/PVF films currently used in aircraft had fairly uniform HF yields of 139 g/kg, 213g/kg and 106 g/kg over the range Φ = 0.5-2.0. However, HBr yields for brominated polystyrene increased with Φ with yields of 0.87, 40.8, and 117.3 g/kg at Φ = 0.5, 1, 2 respectively.

Figure 10 shows the yields of toxic combustion products sulfur dioxide/SO_2 and carbonyl sulfide/COS for two sulfur-containing polymers reported on a weight loss basis. Yields of carbon disulfide/CS_2 were high, but there is not enough confidence in the CS_2 values to report them here. Most of 19 gases analyzed in the FTIR method are not reported here because they were not present in the spectra or there was not enough confidence in the results. In the case of CS_2, a very strong IR absorber, an insufficient number of calibration spectra were available to define the shape of the calibration curve. It can be seen from figure 10 that the total yields of sulfurous gases are proportional to the weight percentage of Sulfur in the sample for both PPS and PPSU.

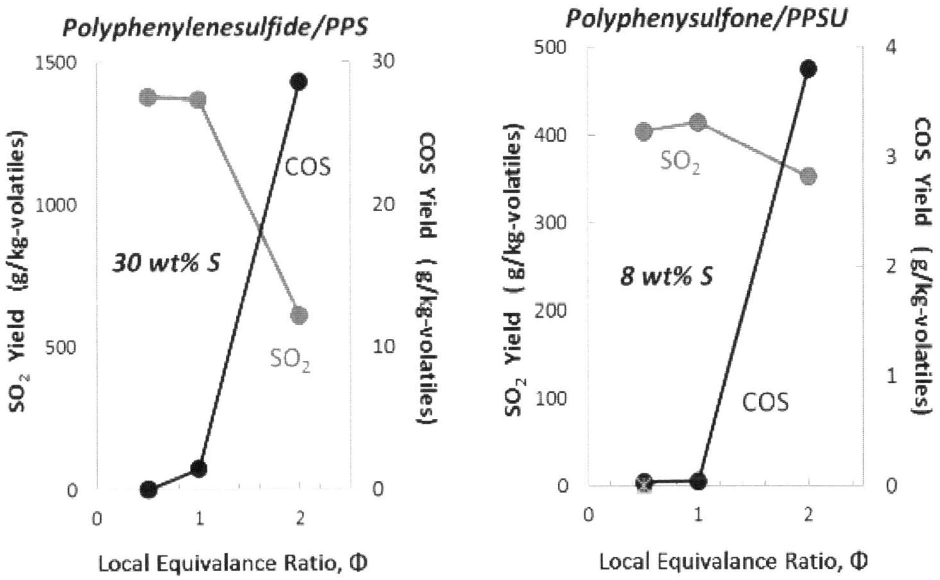

Figure 10. Yields of SO_2 and COS for Sulfur-Containing Polymers at Different Equivalence Ratios.

Figure 11 shows a spectrum for polyphenylenesulfide MCC decomposition products at $\Phi = 2$ with calibration spectra for CS_2, COS and SO_2 stacked below.

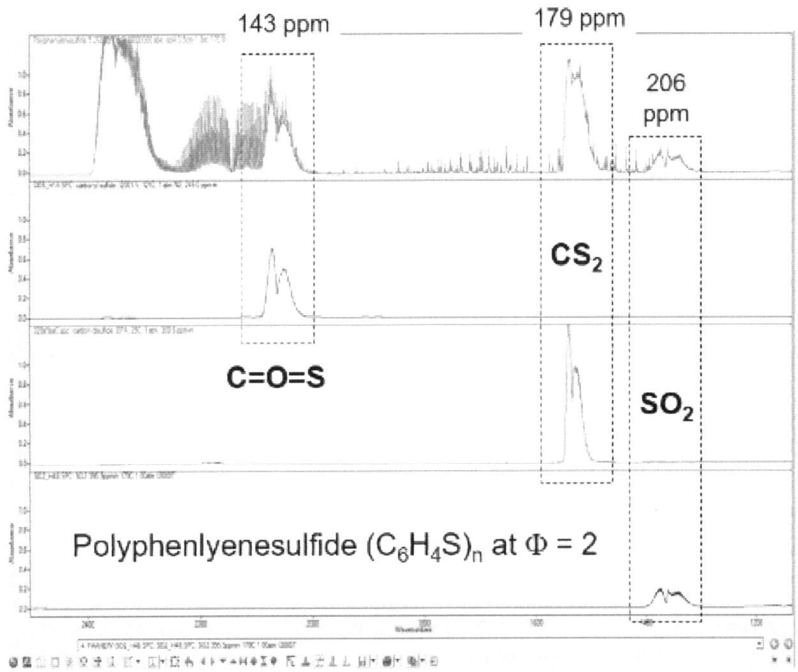

Figure 11. Polyphenylenesulfide Spectrum at $\Phi = 2$ with Stacked Calibration Spectra for CS_2, COS and SO_2

The gas yields obtained by MCC-Φ-FTIR are shown in table 2.

Table 2. Gas Yields for MCC-Φ Tests

Φ	Material	CO (g/g)	CO_2 (g/g)	COF_2 (g/kg)	COS (g/kg)	HBr (g/kg)	HCN (g/kg)	HCl (g/kg)	HF (g/kg)	NO (g/kg)	NO_2 (g/kg)	SO_2 (g/kg)
0.5	ABS	0.000	4.397	0.00	0.00	0.05	0.30	0.00	0.00	36.28	0.00	0.12
1	ABS	0.070	4.197	0.00	0.00	0.17	1.47	0.00	0.00	15.63	0.00	0.08
2	ABS	1.080	1.903	0.00	0.11	0.00	40.14	0.03	0.00	2.03	0.00	0.08
0.5	Polyphthalamide PPA-1002	0.007	3.134	0.00	0.00	0.00	1.36	0.12	0.62	37.57	0.00	0.00
1	PPA-1002	0.348	2.377	0.00	0.20	0.00	38.83	0.09	0.09	11.29	0.00	0.00
2	PPA-1002	0.670	1.213	0.09	0.21	0.00	101.50	0.16	0.00	5.20	0.00	0.00
0.5	Polyphthalamide-FR PPA-FR-1133	0.136	1.425	0.00	0.03	0.00	13.84	0.15	0.00	15.10	0.00	0.06
1	PPA-FR--1133	0.246	1.070	0.03	0.07	0.00	28.51	2.90	0.09	11.42	0.00	0.00
2	PPA-FR -1133	0.372	0.649	0.05	0.03	0.00	38.91	6.41	0.21	6.66	0.01	0.00
0.5	PVC/Acrylic FR-9815E	0.003	0.533	0.60	0.07	13.19	0.10	401.47	12.58	1.54	0.00	0.66
1	PVC/Acrylic FR-9815E	0.046	0.458	0.14	0.02	3.65	0.15	387.16	7.20	1.44	0.04	1.08
2	PVC/Acrylic FR-9815E	0.168	0.230	0.00	0.06	9.89	0.34	378.41	10.28	1.03	0.00	0.20
0.5	Polystrene-butadiene high Impact HIPS	0.000	4.448	0.00	0.01	0.00	0.40	0.14	0.00	0.03	0.00	0.00
1	HIPS	0.042	4.230	0.00	0.01	0.04	0.09	0.07	0.00	0.00	0.00	0.08
2	HIPS	1.095	1.764	0.01	0.04	0.00	0.18	0.05	0.00	0.00	0.00	0.00
0.5	Isoplast Polyurethane	0.000	2.990	0.00	0.00	0.00	0.12	6.09	0.80	43.13	0.00	0.41
1	Isoplast Polyurethane	0.303	2.339	0.00	0.03	0.00	20.96	2.29	0.22	18.54	0.00	0.43
2	Isoplast Polyurethane	0.791	1.116	0.07	0.10	0.00	56.56	0.29	0.00	6.33	0.00	0.69
0.5	Polyetheretherketone PEEK E-1000	0.000	1.649	0.00	0.01	0.00	0.03	0.14	0.11	0.06	0.00	0.22
1	PEEK E-1000	0.103	1.319	0.00	0.02	0.00	0.11	0.54	0.09	0.24	0.00	0.38
2	PEEK E-1000	0.474	0.623	0.00	0.02	0.00	0.05	0.48	0.00	0.05	0.00	0.69
0.5	Polyketone PEK E-1230	0.084	0.911	0.00	0.01	0.00	0.06	0.58	0.00	0.29	0.00	0.06
1	PEK E-1230	0.208	0.606	0.01	0.02	0.00	0.12	0.36	0.09	0.02	0.00	0.06
2	PEK E-1230	0.255	0.487	0.02	0.06	0.00	0.06	0.53	0.00	0.02	0.00	0.00
0.5	Polyaramide KEVLAR	0.000	1.485	0.01	0.00	0.00	0.00	0.03	0.00	38.60	0.00	0.00
1	KEVLAR	0.004	1.452	0.00	0.00	0.00	0.35	0.12	0.00	29.55	0.00	0.36
2	KEVLAR	0.409	0.694	0.01	0.03	0.00	70.61	0.04	0.19	15.78	0.00	0.50

Φ	Material	CO (g/g)	CO_2 (g/g)	COF_2 (g/kg)	COS (g/kg)	HBr (g/kg)	HCN (g/kg)	HCl (g/kg)	HF (g/kg)	NO (g/kg)	NO_2 (g/kg)	SO_2 (g/kg)
0.5	Polyhexamethylene adipamide PA66	0.000	3.169	0.00	0.00	0.00	0.25	0.16	0.00	31.99	0.00	0.47
1	PA66	0.120	2.735	0.01	0.02	0.06	2.80	0.00	0.00	8.07	0.00	1.57
2	PA66	0.634	1.333	0.05	0.02	0.25	71.33	0.15	0.00	1.69	0.00	0.24
0.5	Polyethylene (high density) PE	0.000	4.351	0.00	0.01	0.36	0.58	0.04	0.00	0.58	0.00	0.00
1	PE	0.491	3.319	0.11	0.01	0.00	0.00	0.06	0.00	0.02	0.00	0.23
2	PE	0.933	1.894	0.36	0.14	0.00	0.11	0.07	0.00	0.08	0.00	0.07
0.5	Polyimide	0.000	0.685	0.00	0.00	0.00	0.04	0.02	0.61	23.18	0.00	0.00
1	Polyimide	0.020	0.640	0.00	0.00	0.00	0.99	0.08	0.00	17.41	0.00	0.00
2	Polyimide	0.163	0.397	0.00	0.00	0.00	15.17	0.05	0.00	9.09	0.00	0.00
0.5	Polybromo3styrene	0.012	2.714	0.00	0.00	0.87	0.00	2.90	8.57	1.60	0.00	0.03
1	Polybromo3styrene	0.440	0.786	0.05	0.00	40.78	0.12	10.94	44.41	0.40	0.00	0.16
2	Polybromo3styrene	0.817	0.464	0.98	0.02	117.25	0.10	7.76	36.52	0.45	0.00	0.00
0.5	Polyoxymethylene POM	0.000	2.484	0.00	0.02	0.00	0.23	0.06	0.00	0.06	0.00	0.00
1	POM	0.033	2.218	0.00	0.00	0.00	0.00	0.09	0.00	0.07	0.00	0.04
2	POM	0.470	0.949	0.00	0.02	0.00	0.14	0.00	0.00	0.16	0.00	0.04
0.5	Polyphenylene sulfide PPS	0.001	1.343	0.00	0.01	0.00	0.00	0.07	12.25	1.95	0.00	777.48
1	PPS	0.086	1.215	0.00	0.86	0.07	0.07	0.85	6.20	1.21	0.00	757.49
2	PPS	0.481	0.517	0.01	16.12	0.00	0.31	4.40	8.41	0.56	0.00	306.87
0.5	Polyvinylfluoride PVF	0.000	2.401	0.14	0.02	0.00	0.04	10.88	270.02	0.00	0.01	0.00
1	PVF	0.136	1.931	0.00	0.12	0.12	0.49	2.22	287.25	0.12	0.00	0.00
2	PVF	0.780	0.497	0.00	0.25	0.12	0.12	0.99	230.64	0.00	0.00	0.12
0.5	Polyphenylsulfone PPSU R5200	0.000	1.759	0.00	0.02	0.02	0.11	1.09	0.78	0.02	0.00	241.19
1	PPSU 5200	0.010	1.822	0.00	0.03	0.00	0.10	0.69	0.24	0.02	0.00	236.54
2	PPSU 5200	0.402	0.622	0.00	2.14	0.14	0.15	1.84	0.23	0.01	0.00	198.52
0.5	PPSU R7200	0.019	1.029	0.12	0.00	0.00	0.06	0.03	0.13	0.02	0.00	171.70
1	PPSU R7200	0.120	0.906	0.07	0.23	0.00	0.04	0.33	0.30	0.00	0.00	190.16
2	PPSU R7200	0.271	0.519	0.13	1.29	0.00	0.12	1.04	0.19	0.02	0.00	175.44
0.5	PPSU R7700GY	0.000	1.811	0.00	0.00	0.05	0.11	7.87	0.46	0.04	0.00	256.45
1	PPSU R7700GY	0.047	1.798	0.01	0.07	0.00	0.26	8.03	0.88	0.13	0.00	267.21
2	PPSU R7700GY	0.437	0.634	0.00	1.94	0.00	0.10	7.66	0.23	0.02	0.00	230.79

Φ	Material	CO (g/g)	CO2 (g/g)	COF_2 (g/kg)	COS (g/kg)	HBr (g/kg)	HCN (g/kg)	HCl (g/kg)	HF (g/kg)	NO (g/kg)	NO_2 (g/kg)	SO_2 (g/kg)
0.5	Polyvinylfluoride PVF TCW20BE5	0.000	1.302	0.59	0.02	0.00	0.10	3.97	111.21	0.19	0.00	0.00
1	PVF TCW20BE5	0.028	1.234	0.50	0.01	0.10	0.15	2.62	138.32	0.06	0.00	0.00
2	PVF TCW20BE5	0.369	0.425	0.49	0.06	0.00	0.16	1.74	143.90	0.10	0.00	0.00
0.5	Polyamideimide PAI 4203L	0.000	0.741	0.00	0.03	0.13	0.04	0.73	0.00	22.44	0.00	0.00
1	PAI 4203L	0.001	1.017	0.02	0.02	0.21	0.10	0.67	0.15	17.39	0.00	0.22
2	PAI 4203L	0.131	0.524	0.00	0.07	0.09	9.74	0.14	0.00	12.19	0.00	0.72
0.5	Polyetherimide PEI 9075	0.068	1.257	0.00	0.01	0.00	0.13	0.00	0.00	22.65	0.00	0.00
1	PEI 9075	0.137	1.212	0.01	0.00	0.00	0.48	0.07	0.00	19.58	0.00	0.00
1.5	PEI 9075	0.242	0.968	0.01	0.19	0.00	7.37	0.02	3.72	12.31	0.00	0.05
2	PEI 9075	0.360	0.746	0.06	0.01	0.00	8.46	0.04	0.00	7.83	0.00	0.00
0.5	Liquid crystal polyester G930 LCP G930	0.000	1.286	0.00	0.00	0.00	0.00	0.35	0.00	0.05	0.00	0.21
1	LCP G930	0.020	1.117	0.00	0.00	0.00	0.28	1.36	0.00	0.12	0.00	0.16
2	LCP G930	0.251	0.569	0.01	0.02	0.00	0.07	0.99	1.25	0.03	0.00	0.05

CONCLUSIONS

Preliminary data is presented for a method to measure toxic combustion products generated in a fire over the full range of ventilation conditions (fire stages) using 5-milligram samples in a 15-minute test. This work is part of a larger project at the FAA to understand the relationship between flame retardant additives used in plastics and the toxic potency of the smoke. Future work includes refinement of the FTIR method to further minimize interferences, benchmarking the MCC data with flaming combustion measurements in a fire calorimeter, testing flame-retardant samples. It is hoped that this technique, combined with an appropriate toxicity model, can be used as a part of a toxicity assessment for polymers.

REFERENCES

1. Darnerud, P.O., Eriksen, G.S., Johannesson, T., PLarsen, B., Viluksela, M., Polybrominated Diphenyl Ethers: Occurrence, Dietary Exposure, and Toxicity, Environmental Health Perspectives, 109(1), 49-68 (2001).
2. Hites, R.A., Polybrominated Diphenyl Ethers in the Environment and in People: A Meta-Analysis of Concentrations, Environmental Science & Technology, 38(4), 945-956 (2004).
3. United States Code of Federal Regulations, Title 14: Aeronautics and Space, Volume 1, Chapter I: Federal Aviation Administration, Part 25, Airworthiness Standards: Transport Category Airplanes, Subpart D- Design and Construction- Fire Protection, Section 25.853 Compartment Interiors.

4. Gerecke, A.C., Brominated Flame Retardants in Settled Dust of a Commercial Aircraft, BFR 2007, Amsterdam, April 24-27, 2007.
5. Christiansson, A., Hovander, L., Athanassiadis, I., Jakobsson, K., Bergman, A., Polybrominated Diphenyl Ethers in Aircraft Cabins- A Source of Human Exposure, Chemosphere, 73(10) 1654-1660 (2008).
6. Certain Polybrominated Diphenylethers; Significant New Use Rule and Test Rule, Docket Identification Number EPA- HQ-OPPT-2010-1039, Environmental Protection Agency, 77 Federal Register, 19862, April 2, 2012.
7. Fire Toxicity, A. Stec and R. Hull, Eds., Woodhead Publishing Limited, Oxford, U.K., 2010.
8. K.W. Fent, G.P. Horn and S. DeCrane, Firefighters' Perspective on Flame Retardants, Fire Protection Engineering, 68, 46-48 (2015).
9. T.Z. Fabian, J.L. Borgenson, P.D. Gandhi, C.S. Baxter, C.S. Ross, J.E. Lockey, J.M. Dalton, Fire Technology 50(4), 993-1019 (2014).
10. Standard Test Method for Determining Flammability Characteristics of Plastics and Other Solid Materials Using Microscale Combustion Calorimetry ASTM D 7309-13. American Society for Testing and Materials (International): West Conshohocken, PA, 2013.

USE OF FTIR COMBINED WITH SMALL-SCALE FIRE TESTS AS SCREENING TEST TO TOXICITY TEST

Hideki Yoshioka[1], Tetsuya Hayakawa[2], Satoshi Fujimoto[3], Tomohiro Naruse[4], Xuansu Zhao[5]
Yutaka Tanaike[5], Takafumi Noguchi[5], Koichi Yoshida[6], and Yoshihiro Hase[7]

[1] National Institute for Land and Infrastructure Management (NILIM), Japan
[2] Tokyo System Vac. Inc.(TSV), Japan
[3] Utsunomiya University, Japan
[4] Building Research Institute (BRI), Japan
[5] The University of Tokyo, Japan
[6] Yokohama National University (YNU), Japan
[7] Mitsubishi Plastics, Inc., Japan

INTRODUCTION

According to the Building Standard Law of Japan[1], a non-combustible building material shall not generate smoke or gas that is detrimental to evacuation for twenty minutes in the event of a fire. In cases of quasi-noncombustible materials and fire-retardant materials, this time will be shortened to ten minutes and five minutes respectively. In order to prove that performance, toxicity tests using mice are performed in Japan to evaluate the gas generated from the interior finish materials used at building compartments/rooms when they are set on fire.

At the same time, some negative comments have arisen with regards to animal tests from ethical viewpoints, even though they are one of the promising ways to properly evaluate the toxicity. On the other hand, the fractional effective dose (FED) method is internationally used[2], in standards such as ISO 13344[3] and ISO 13571[4], etc. In this research, FED is calculated based on Fourier Transform Infrared (FTIR) analysis when it is combined with several different types of small-scale fire tests. The results are then compared with the results of mean values of the time for eight mice to stop moving in toxicity tests when identical test specimens are used.

The authors have performed toxicity tests in Japan (both with and without mice) [5] combined with FTIR, an ISO 5659-2 smoke chamber[6,7] combined with FTIR, and an ISO/TS 19700 tube furnace[8] combined with FTIR, so far. And in this paper specifically, the authors try to focus on the correlation among the time for mice to stop moving, the FED calculated from a toxicity test (which is a facility for mice testing), and the FED calculated from a smoke chamber test. ISO/TS 19700 has been already discussed[9,10] in detail and good results have been produced, and there is no doubt that the tube furnace method is a promising way to evaluate toxicity, but also it is said that a tube furnace is more suitable for testing cables and smaller parts[11]. On the other hand, in this paper the authors are interested in examining building interior finish materials which constitute a larger area and consequently a smoke chamber is more suitable for this purpose. Also, now in ISO/TC92/SC1/WG5 meetings, a combination of a smoke chamber and FITR is being discussed and a new draft called ISO/DTS 19021 is being made[12], with much interest in it being shown[13, 14] primarily by the meeting participants etc.

TEST SPECIMENS

As shown in Table 1, the specimens are categorized into three groups, which are wallpaper, core material for a sandwich panel, and wood. There are four kinds of wallpaper — polyester, rayon, vinyl chloride, and silk. There are four kinds of sandwich panel core materials, namely, polyethylene with aluminum hydroxide, polyurethane foam, polyisocyanurate foam, and phenolic foam. There are three kinds of wood specimens: FRT (fire-retardant treated) wood of a noncombustible level, FRT wood of a quasi-noncombustible level, and wood without any fire-retardant treatment. The type of wood is Japanese cedar. In total, eleven kinds of specimens are tested.

Table 1: Test specimens.

Group	Category	Details
A	Wallpaper	(1) polyester, (2) rayon, (3) vinyl chloride, and (4) silk.
B	Core material for a sandwich panel	(5) polyethylene with aluminum hydroxide, (6) polyurethane foam, (7) polyisocyanurate foam, and (8) phenolic foam.
C	Wood (Japanese cedar)	(9) FRT* wood of a noncombustible level, (10) FRT* wood of a quasi-noncombustible level, and (11) wood without any treatment.

* FRT stands for fire-retardant treated.

TOXICITY TEST IN JAPAN

As shown in Figures 1 and 2, the setup for the toxicity test in Japan using mice consists of mainly three parts which are 1) a furnace, 2) a pre-mixing chamber, and 3) a mice exposure chamber[15]. As for 1) the furnace, it abides by JIS A 1321[16], referring to BS 476-6[17] but with some modification. The heating surface area of the specimen is 180 by 180 mm, the heater strength is 1.5 kW, the amount of propane gas is 350 ml/min for the first three minutes, the primary air supply is 3 l/min for the first six minutes, and the secondary air supply is 25 l/min for the first six minutes. Regarding 2) the pre-mixing chamber and 3) the mice exposure chamber, they have the same volume with internal dimensions of 500 × 500 × 500 mm, and both chambers contain a small fan in the ceiling for mixing the gas inside each chamber. There is a small duct combining 2) the pre-mixing chamber and 3) the mice exposure chamber, and there is an exhaust flow from the duct at 6 l/min for the first six minutes. In the regulatory tests, this exhaust rate is 10 l/min, but in this research it was set to 6 l/min.

In the usual classification practice, a test is performed for fifteen minutes, but in this research it is extended to be thirty minutes.

In 3) the mice exposure chamber, eight mice are situated inside each "activity wheel" on each "seat," and a magnet is located in both the activity wheel and seat. Every time a mouse rotates in an activity wheel, both the magnets come into contact with each other and this produces a signal that is recorded in a data logger. During the test, the temperature in 3) the mice exposure chamber is set to 28–30 degrees Celsius, controlled by an electric heater. The mice are female, 4–5 weeks old, and 18–22 g in weight.

The FTIR is located right next to the toxicity test equipment, and the sampling position in 3) the mice exposure chamber is shown by the “x” mark in Figure 1, which is 250 mm from the walls, and 100 mm from the bottom of the chamber. The flow rate for FTIR is 4 l/m.

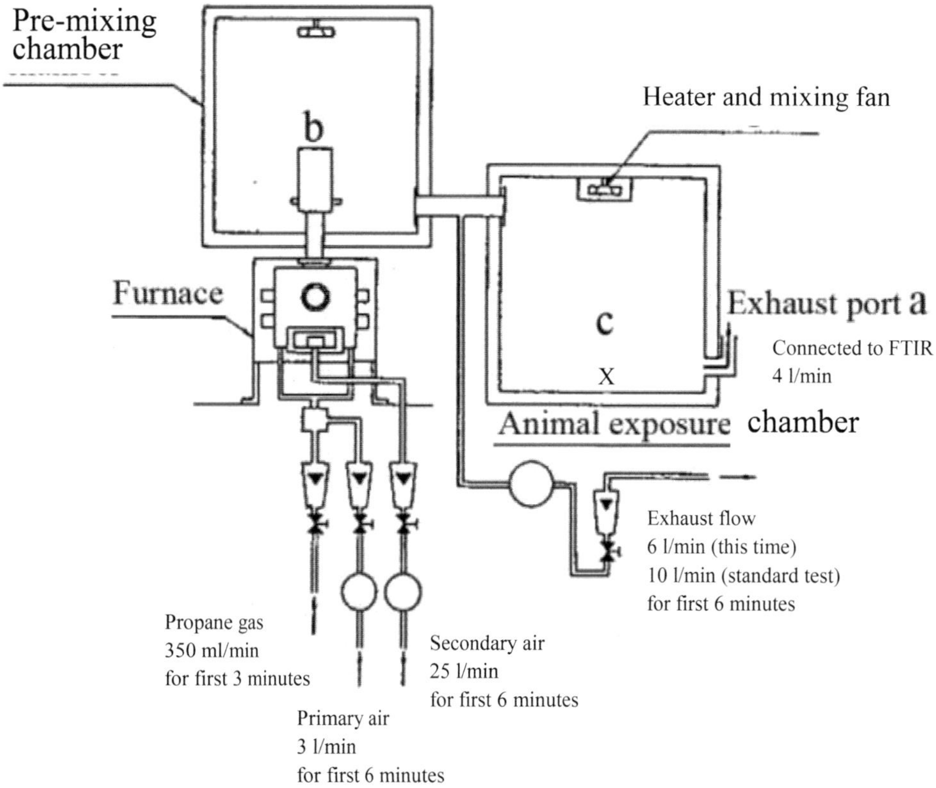

Figure 1: Diagram of toxicity test using mice in Japan.

Figure 2: Test setup (toxicity test combined with FTIR).

Method of calculating average time for mice to stop moving

X_s (average time for mice to stop moving) is defined as in equation 1.

$$X_S = \bar{X} - \sigma \qquad [1]$$

Where $\bar{X}$ (min) is just the simple average value of the time to stop moving among all the eight mice, and σ (min) is its standard deviation.

SMOKE CHAMBER TEST COMBINED WITH FTIR

A test method using FTIR combined with a smoke chamber (ISO 5659-2) is now being developed as ISO/DTS 19021. In this research, the authors tried to generally follow the test procedures described in ISO/DTS 19021 as much as possible, but with different sampling flow rate. This rate is set at 1.5 L/min in the standard, but the value of 4.0 L/min was used for this research as it had traditionally been a more common value. Almost all the other items followed both ISO/DTS 19021 and ISO 5659-2. Basic information on the experiments in this research is described in Table 2, and a still picture is shown in Figure 3.

Table 2: Basic information on smoke chamber test combined with FITR in this research.

Heating conditions for specimen in the smoke chamber	25 kW/m^2, with pilot flame
Sampling probe position in the smoke chamber connected to FTIR	300 mm below the ceiling
Sampling flow rate	4.0 L/min *
All sampling lines	180°C
Test time (data collection time)	30 minutes **

* It is set to be 1.5 L/min in the latest version of ISO/DTS 19021.
** It is set to be 20 minutes in the latest version of ISO/DTS 19021.

Figure 3: Test setup for ISO/DTS 19021 (smoke chamber (ISO 5659-2) combined with FTIR).

EXPERIMENTAL RESULTS

A basic summary of the experimental results is shown in Table 3. FED is calculated according to ISO 13571, using equation 2 based on asphyxiant gases CO and HCN, multiplied by CO_2 induced hyperventilation factor V_{CO2} from equation 3, 30 minutes from the test start. Regarding toxicity tests on the wood specimens (numbers (9), (10), and (11)), FED values and the time for mice to stop moving were recorded at different tests and laboratories, but both specimens and test procedures can be considered as technically almost identical. As for specimen (6), polyurethane foam, in the toxicity test without mice, the FED value is not described in the table as collecting data on HCN did not progress smoothly. As for the FED calculated from the smoke chamber tests on specimens (9) and (10), two pieces were tested and the averaged values are described in Table 3.

$$\mathrm{FED} = \sum_{t_1}^{t_2} \frac{V_{CO_2} \cdot [CO]}{35000} \Delta t + \sum_{t_1}^{t_2} \frac{V_{CO_2} \cdot [HCN]^{2.36}}{1.2 \times 10^6} \Delta t \quad [2]$$

$$V_{CO_2} = e^{\frac{[CO_2]}{5}} \quad [3]$$

Table 3: Summary of experimental results.

Group	Specimen	Toxicity test						Smoke chamber test					Ratio of FED
		Xs [min]	FED	Max CO_2 [%]	Max CO [%]	CO_2/CO ratio	Max HCN [ppm]	FED	Max CO_2 [%]	Max CO [%]	CO_2/CO ratio	Max HCN [ppm]	(Toxicity test)/ (SDC test)
Wallpaper	(1)	-	0.77	3.06	0.09	34.04	23	0.17	2.33	0.03	78.60	6	4.5
	(2)		1.11	2.44	0.14	17.72	34	0.14	1.73	0.03	55.52	8	7.9
	(3)		0.79	2.06	0.12	17.35	29	0.23	3.95	0.04	107.82	4	3.4
	(4)		0.96	1.93	0.09	21.21	92	0.28	1.89	0.04	50.12	40	3.4
Core material for sandwich panel	(5)		0.65	1.13	0.10	11.53	35	0.77	2.73	0.13	21.22	8	0.8
	(6)		-	5.80	0.81	7.12	-	0.74	2.64	0.05	53.01	76	-
	(7)		2.30	7.01	0.10	69.52	93	0.87	3.54	0.08	46.64	75	2.6
	(8)		0.49	1.68	0.06	27.96	35	0.93	1.94	0.13	14.74	49	0.5
Wood (Japanese cedar)	(9)	18	0.86	1.33	0.12	10.92	21	0.14	0.78	0.03	26.00	14	6.4
	(10)	12.2	1.09	1.54	0.16	9.74	8	0.19	0.57	0.04	14.25	19	5.7
	(11)	9.6	6.38	2.21	0.64	3.45	141	2.14	4.09	0.27	14.91	3	3.0

* As for the smoke chamber tests on (9) and (10), two specimens were tested, and the averaged values are shown.

** FED values are calculated at thirty minutes from the start of tests.

Comparison between toxicity test and smoke chamber test regarding FED at 30 minutes

In order to make a scientific comparison, a scaling factor should be multiplied by the FED values obtained from the two different tests, as both specimen size and chamber volume differ from each other. The smoke chamber test could be considered as a closed system in a single chamber except for a small sampling flow from the chamber to the FITR. On the other hand, in the toxicity test, the system used is a moving one with two chambers, and with far more enforced flow from the furnace to the pre-mixing chamber and finally to the mice chamber. For this reason, it is for the time being very difficult to establish how to properly set the scaling factor, and therefore, the authors decided to simply compare the FED values at 30 minutes from both tests without multiplying by the scaling factor.

Both FEDs (at 30 min) calculated from the toxicity test and smoke chamber test (those in Table 3) are plotted in Figure 4. Specimens (1)–(4) which are types of wallpaper, and specimens (9) and (10) which are FRT (fire-retardant treated) wood of various levels, show relatively small FED values, which are 0–2 for the toxicity test, and 0–0.5 for the smoke chamber test. These plots are emphasized with the circle in Figure 3. As for (5) polyethylene with aluminum hydroxide and (8) phenolic foam, they show a small FED (less than 1) from the toxicity test, but a relatively large FED (0.7–1) from the smoke chamber test. The results of these two specimens clearly deviate from the trends shown by other specimens. As for (7) polyisocyanurate foam and (11) wood without any treatment, they show very large FEDs, both in the toxicity test and smoke chamber test, but their plotted positions are not hugely deviating from the linear approximation shown in Figure 4.

Overall, except for specimens (5) and (8), the FED from the toxicity test is far larger than that from the smoke chamber test, and the ratio is somewhere around four times larger with the toxicity test than the smoke chamber test. On the other hand, specimens (5) and (8) show larger FED values with the smoke chamber test than the toxicity test; therefore, it is impossible to make a generalized statement about FED between the toxicity test and smoke chamber test. At least it can be said that for wallpaper and wood specimens, the toxicity tests produce FED values that are more than three times larger than those from the smoke chamber test.

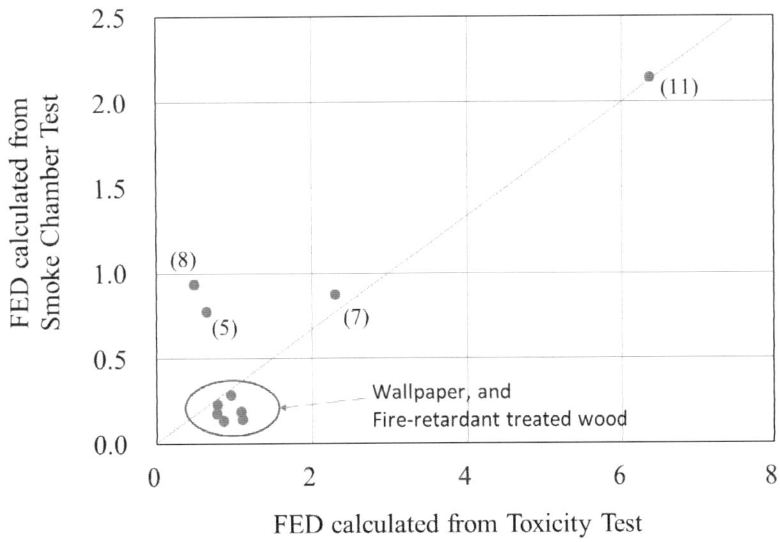

Figure 4: Plot of FED (at 30 min) from toxicity test and smoke chamber test.

Comparison between toxicity test and smoke chamber test regarding CO_2/CO ratio

When comparing toxic gas data from different fire tests, the CO_2/CO ratio is one of the important items as described in ISO 29903[18], and these ratios are plotted in Figures 5 and 6 together with the maximum level of CO. Raw numbers are shown in Table 3.

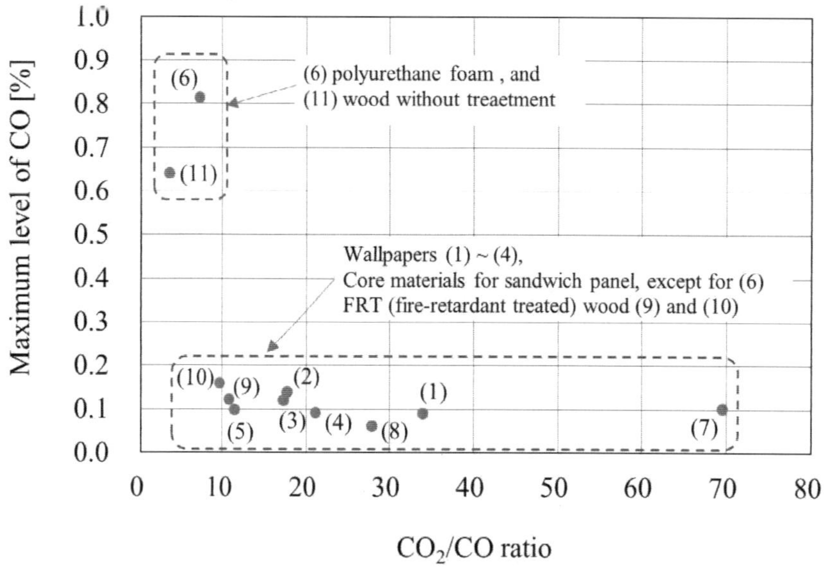

Figure 5: Relationship between maximum level of CO and CO_2/CO ratio for toxicity tests.

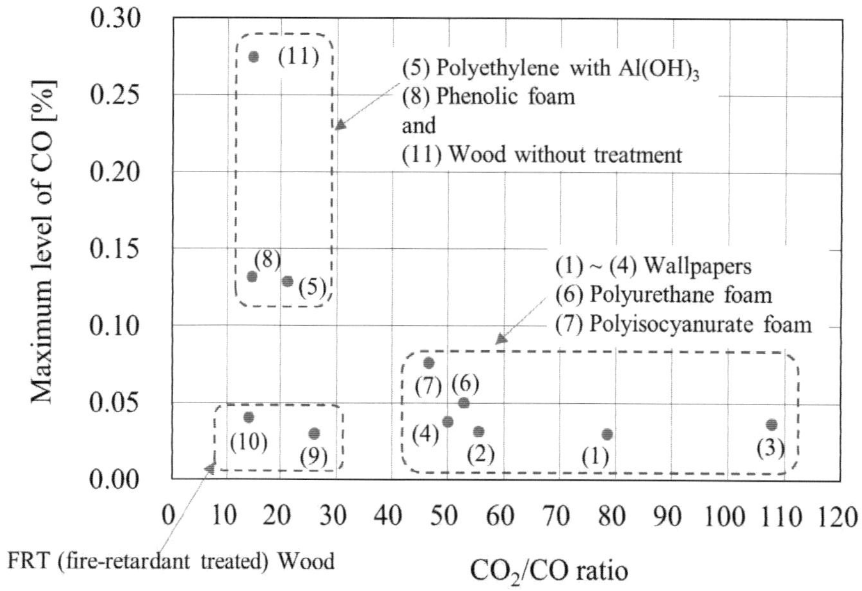

Figure 6: Relationship between maximum level of CO and CO_2/CO ratio for smoke chamber tests

Regarding Figure 5, which shows the toxicity test results, there are two major groups in the plots. One is constituted by (6) polyurethane foam and (11) wood without fire-retardant treatment, where the maximum level of CO is larger than 0.6% and the CO_2/CO ratio is less than 10. The other group is constituted by all the other specimens, where the maximum level of CO is less than 0.2% and the CO_2/CO ratio is larger than 10. There is no unusual data in the grouping results of toxicity tests.

Regarding Figure 6, which shows the results of the smoke chamber test, there are three major groups in the plots. The first group is constituted by (5) polyethylene with aluminum hydroxide, (8) phenolic foam and (11) wood without treatment, where the maximum level of CO is larger than 0.1% and the CO_2/CO ratio is less than 30. The second group is constituted by wallpaper (1)–(4), together with core materials for a sandwich panel (6) and (7), where the maximum level of CO is less than 0.1% and the CO_2/CO ratio is larger than 40. And finally, the third group is constituted by FRT wood types (9) and (10), where the maximum level of CO is less than 0.05% and the CO_2/CO ratio is less than 30. When the maximum value of CO is small, the CO_2/CO ratio tends to be large, and in this regard, the third group could be considered to be an unusual result compared to the other two groups.

Comparing Figures 5 and 6 generally, the toxicity test has a larger maximum level of CO and smaller CO_2/CO ratio than the smoke chamber test, which could indicate that the conditions present in a toxicity test lead to a lower combustion efficiency than that of the smoke chamber test. Also shown in Table 3, toxicity tests generally produce larger FED values than smoke chamber tests. Furthermore, as for the (1)–(4) wallpaper and (7) polyisocyanurate foam, Figures 5 and 6 generally have the same phenomenon, where they have a smaller level of CO and larger CO_2/CO ratio compared with other specimens' results. Also, specimen (11) wood without treatment, has the same phenomenon both in Figures 5 and 6, where it has larger level of CO and smaller CO_2/CO ratio compared with other specimens. On the other hand, as for the plotted positions for the other specimens such as most of the core materials for sandwich panels, and FRT (fire-retardant treated) wood types, there seems to be no specific correlation between Figures 5 and 6.

Comparison between FED at 30 min and X_s (average time for mice to stop moving) in toxicity test

Regarding the results of toxicity tests on wood specimens (9)–(11), Figures 7 and 8 respectively show the relationship between FED at 30 minutes and X_s (average time for mice to stop moving) in minutes, and the relationship between maximum level of CO and X_s (average time for mice to stop moving) in minutes. Specimen (11) wood without treatment has the shortest X_s, while specimen (9) FRT wood of a "noncombustible" level has the longest X_s, and these seem to be valid results. As shown in Figures 7 and 8, the longer X_s is, the larger FED and the maximum level of CO are. Even though there are only three specimens, at least with respect to wooden specimens, the same trend is seen between FED and the maximum level of CO, when they are compared with X_s. When a certain laboratory does not have proper FTIR apparatus, it might be possible to first use CO analyzer data in order to screen the wooden specimens from the viewpoint of toxicity. But it is impossible to generalize this result and it is definitely necessary to increase the number of specimens and the results to find a more concrete correlation between the CO level and X_s from a toxicity test using mice in Japan.

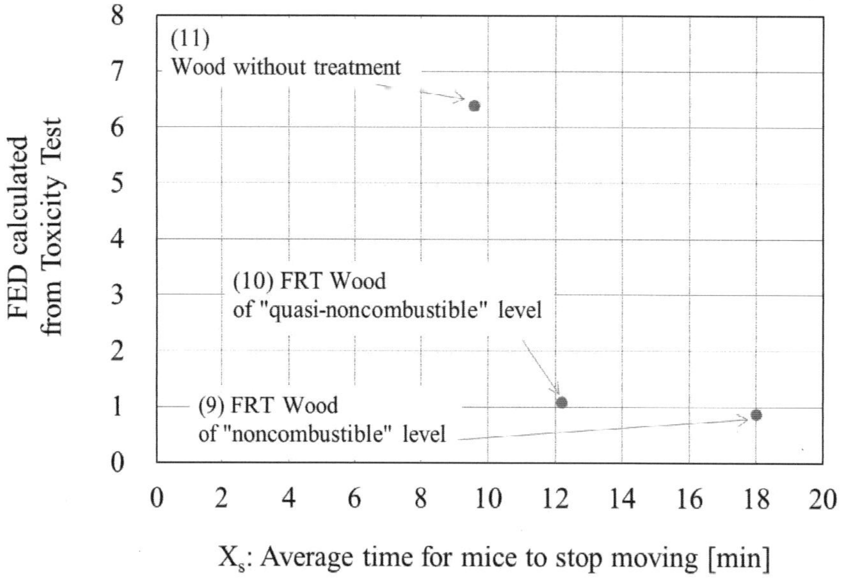

Figure 7: Relationship between FED at 30 min and X_s (average time for mice to stop moving) for wooden specimens in a toxicity test.

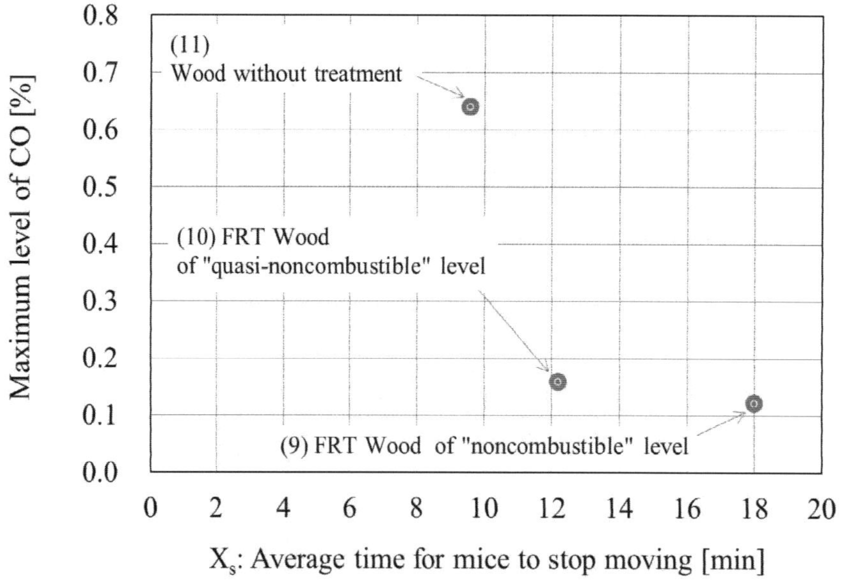

Figure 8: Relationship between maximum level of CO and X_s (average time for mice to stop moving) for wooden specimens in a toxicity test.

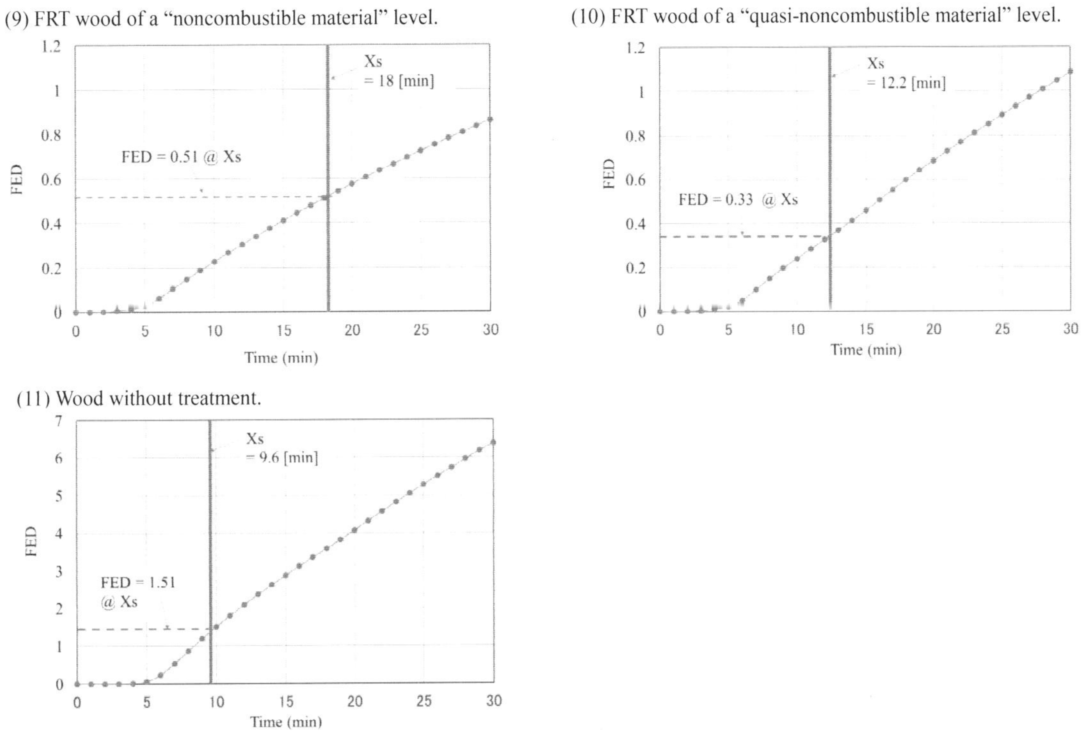

Figure 9: Time series plot of FED on wood specimens for toxicity tests.

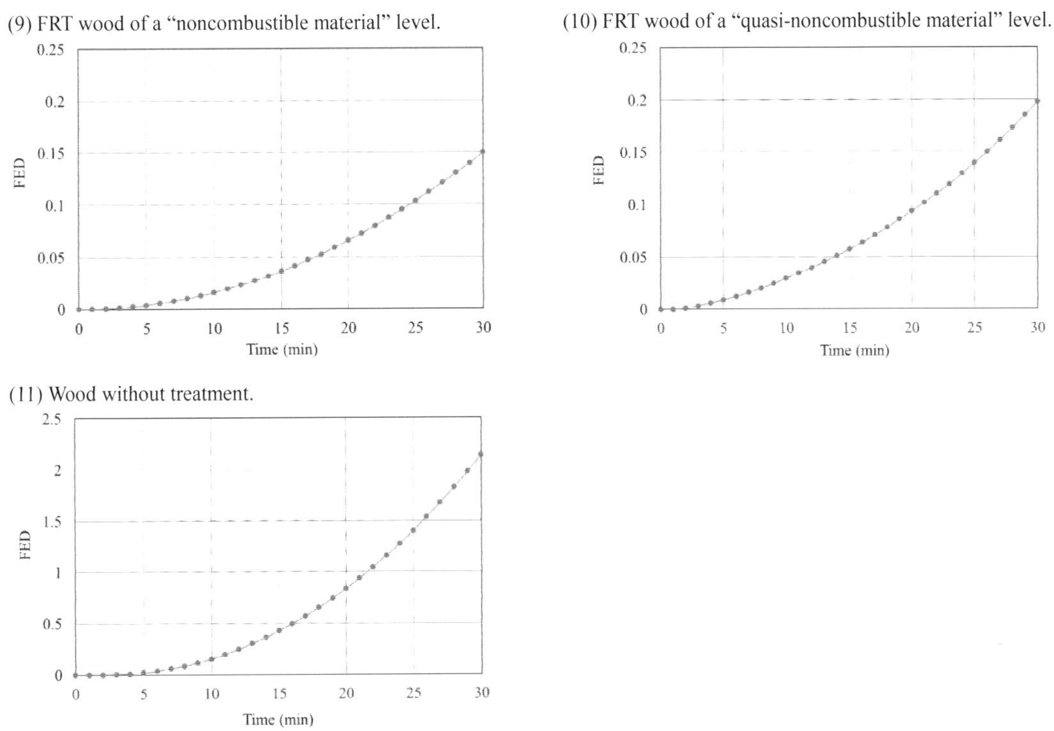

Figure 10: Time series plot of FED on wood specimens for smoke chamber tests.

Time series plots of FED on wood specimens are shown in Figures 9 and 10, for toxicity tests and smoke chamber tests respectively. All the graphs are steadily rising. Figure 9 shows that FED starts to increase from zero around 5 minutes, while Figure 10 shows that FED starts to increase from zero around 3–4 minutes.
In Figure 9, the average time for mice to stop moving is indicated by a vertical straight line, and also, FED calculated at X_s (average time for mice to stop moving) is 0.51, 0.33, and 1.51 for specimens (9), (10), and (11) respectively. No specific correlation has been found between FED and X_s in toxicity tests so far, and it is necessary to increase the number of test specimens to find a more reliable correlation.

CONCLUSION

In order to reduce the number of toxicity tests in Japan using mice while maintaining the criteria as much as possible, the authors are now trying to find any possible alternative methods using gas analysis data based on the FTIR technique and calculations referring to existing equations such as FED from ISO 13571. Among the several existing internationally recognized methods, ISO/DTS 19021, which is a combination of the ISO 5659-2 smoke chamber test and FTIR, was chosen and performed to compare the results with toxicity tests in Japan for identical specimens (wallpaper, core materials for a sandwich panel, and wooden specimens). These specimen targets are primarily interior finish materials for buildings and they account for relatively large areas in real life. The ISO/TS 19700 tube furnace test was also performed by the authors for certain specimens, but the results are not discussed in this paper, as generally it is a method more suitable for cables or relatively small-area materials in real life. But it is one of the promising ways to evaluate the toxicity of specimens and to find a good correlation with large-scale tests.

Regarding the comparison between the toxicity test and smoke chamber test, the former has a forced flow from the specimen furnace to the mixing chamber and then to the mice chamber during the first six minutes. Meanwhile, the latter has basically closed conditions in a single chamber except for the small sampling flow to FTIR. Therefore, it is difficult and complicated to find a suitable scaling factor for comparing these two tests, as the scaling factor simply calculated from the specimen area and the chamber volume could cause a misunderstanding. Therefore, the authors did not consider the scaling factor in this paper, and instead, the CO_2/CO ratio was discussed as it is one of the main principles for comparing the different fire tests as indicated in ISO 29903. The toxicity test has generally a larger maximum level of CO and smaller CO_2/CO ratio than the smoke chamber test for most of the specimens. Furthermore, the toxicity test has generally produced FED values that are more than three times larger than those for the smoke test for most of the specimens, but with exceptions for just two specimens where the toxicity test has lower FED values.

Regarding the comparison between FED and X_s [min] (average time for eight mice to stop moving) in the case of toxicity tests on the wooden specimens, valid results and phenomena were found. FRT (fire-retardant treated) wood of a "noncombustible" level has the longest X_s and the smallest FED at 30 minutes, while wood without treatment has the shortest X_s and the largest FED at 30 minutes. As for the FED value at a time of X_s, there seems to be no coherency among the three different wooden specimens. With respect to the maximum level of CO, the magnitude relation is the same as FED, where FRT wood of a "noncombustible" level has the smallest value, while wood without treatment has the largest value. It might be said that laboratories in Japan having no FTIR facility could use the data on CO for screening purposes. But it is premature to generalize anything on the relationship between FED and X_s [min] in this paper yet, as only a limited number of wooden specimens were tested.

Overall, this paper indicates how to compare the toxicity test in Japan and smoke chamber test, considering X_s [minute] and FED for interior finish specimens of buildings. Discussion was limited due to the limitation on the number of specimens and tests performed, but the authors will continue this work to find any correlation between these different tests with certain criteria and calculations based on the test results.

FUTURE SCOPE

In addition to wooden specimens, other specimens will be tested with mice to find any correlation between X_s (average time for mice to stop moving) and FED. And, FEC values based on irritant gases will also be discussed. Furthermore, a scaling factor will be considered in order to compare the toxicity test and smoke chamber test, as they have different dimensions for both test specimens and chambers.

ACKNOWLEDGMENT

The authors wish to express their sincere gratitude to Mr. Tatsuo Ando, Mr. Masamichi Tamura (The University of Tokyo), Mr. Yuji Hayakawa (TSV), and Dr. Tsuneto Tsuchihashi (General Building Research Corporation of Japan) and Dr. Shuitsu Yusa (Waseda University) for their dedicated contribution to this research. The authors also appreciate the kind help provided by Dr. Eric Guillaume (Efectis), Prof. T. Richard Hull, Prof. Anna A. Stec (The University of Central Lancashire), and Prof. Thomas Rogaume (The University of Poitiers) in gathering the latest international information on toxicity research. Dr. Sean Gregory and Dr. Sandra Pasantes (Fire Testing Technology) are greatly appreciated for their kind help on experimental technique using FTIR apparatus.

REFERENCES

1. Article 108-2, Enforcement Order, Building Standard Law of Japan.
2. J.H. Troitzsh : Smoke Toxicity in Fire Safety Regulations and Tests, 1st Asia-Oceania Symposium

on Fire Safety Materials Science and Engineering, Suzhou, China, 2015.

3. ISO 13344:2015. Estimation of the lethal toxic potency of fire effluents.
4. ISO 13571: 2012. Life-threatening components of fire -- Guidelines for the estimation of time to compromised tenability in fires.
5. Xuansu Zhao, Takafumi Noguchi, Tomohiro Naruse, Hideki Yoshioka, Yutaka Tanaike, Satoshi Fujimoto, Yoshihiro Hase, Tetsuya Hayakawa, and Tatuso Ando: "The influence of gas produced on refuge safety at the time of fire -One consideration of the fire smoke gas in the Smoke Toxicity Test based on FTIR analysis-", The 10th Asia-Oceania Symposium on Fire Science and Technology, Tsukuba, Japan, October 2015.
6. ISO 5659-2:2012. Plastics -- Smoke generation -- Part 2: Determination of optical density by a single-chamber test.
7. Koichi Yoshida, Ken Matsuyama, Kaoru Wakatsuki, Tsukasa Fukaya, Mitsunori Watanabe, and Silvio Messa: "Comparative Assessment on Burning Characteristics of Electric Cables", Fire and Materials conference, 2013.
8. ISO/TS 19700:2016. Controlled equivalence ratio method for the determination of hazardous components of fire effluents -- Steady-state tube furnace
9. J.A. Purser, D.A. Purser, A.A. Stec, C. Moffat, T.R. Hull, J.Z. Su, M. Bijloos, and P. Blomqvist: "Repeatability and Reproducibility Round Robin for the ISO TS19700 steady state tube furnace", Fire Safety Journal 55 (2013) 22–34.
10. Eric Guillaume and Laurent Saragoza: Application of FTIR Analyzers to Fire Gases – Progress in Apparatus and Method Validation for Quantitative Analysis, Fire and Materials conference, 2015.
11. EN 45545-2. Railway applications - Fire protection on railway vehicles - Part 2: Requirements for fire behavior of materials and components.
12. ISO/DTS 19021. Test method for determination of gas concentrations in ISO 5659-2 using Fourier transform infrared spectroscopy.
13. Tuula Hakkarainen: "Accuracy of Fire Toxicity Measurements of Smoke Gases Produced by Train Products", Interflam, University of London, UK, 2013.
14. Kazuma Araya, Koichi Yoshida, Hideki Yoshioka, Tetsuya Hayakawa, and Tsukasa Fukaya: "Influence of Different Ignitors Used in Single Smoke Chamber on Smoke and Gas Measurement", Interflam, University of London, UK, 2016.
15. Tsuneto Tsuchihashi, etc.: "Evaluation Method of Gas Toxicity Based on Combustion Gas Toxicity Test", GBRC Vol.31 No.4, October 2006 (in Japanese).
16. JIS A 1321:1994. Testing method for incombustibility of internal finish material and procedure of buildings.
17. BS 476: Part6: 1989. Fire tests on building materials and structures, Part6. Method of test for fire propagation for products.
18. ISO 29903:2012. Guidance for comparison of toxic gas data between different physical fire models and scales.

FIRE TOXICITY OF AIRCRAFT SEAT COVERS AND SEAT FOAM

Lisa Witty, Abdulaziz A Alarifi, Itohoimo Iniabasi Obot, Gordon E Andrews and Herodotos N Phylaktou
Fire and Explosion Research
School of Chemical and Process Engineering, University of Leeds, Leeds, LS2 9JT, UK.

ABSTRACT

A cone calorimeter with air supply box around the fire test zone was shown to reproduce all the classic features of compartment fires and hence could provide a basis for the determination of toxic gases emitted from materials in fires, under realistic test conditions. Direct hot exhaust gas sampling was used from a small chimney attached to the cone heater outlet and heated sample lines, pumps and filters were used to transport the sample to a heated Temet FTIR, calibrated for all the toxic gases of interest in fire toxicity. Three aircraft seat material were tested, two seat covers of fabric and leather and the seat foam. All the materials were from commercial suppliers to the aircraft industry. Ventilation conditions in the fires were typical of those on aircraft at 9-15 ACH. All three material samples had S and N in their composition, the S from the fire retardants and N from the use of an acrylic fibre in the seat covers and of PUF in the seat foam. As a consequence SO_2 and HCN were major toxic species in all the material fires. In all the fires HF was detected and for the foam and the leather seat cover HCl was detected and these indicate the presence of halogen based fire retardants in the material composition. Thus, most of the toxic gases measured were due to the presence of S, N, F and Cl in the material composition. Other toxic gases detected were the more conventional fire toxic gases CO, formaldehyde, acrolein and benzene. Only in the fabric fire was NO_2 significant. The peak toxicity occurred in the initial flaming combustion and the toxicity order was leather > fabric > foam on both an LC_{50} and $COSHH_{15min}$ toxic assessment basis. Both toxic assessment methods showed the same group of toxic gases were important and only their relative importance changed between the toxic assessment methods. It was concluded that toxic gas assessment should be added to current material test programmes on fire behavior and that the cone calorimeter, modified as in the present work, was a good basis for the toxic gas measurements.

INTRODUCTION

The toxicity of gases produced from the burning of interior aircraft materials, predominantly aircraft seats was investigated. The FAA regulatory requirements for interior furnishings in aircraft are based largely on flammability tests[1] with cabin and cargo liners, and seats subject to further tests that measure the heat release rate and the density of smoke produced [1, 2]. The regulations governing civil aviation in the USA are in the Code of Federal Regulations (CFR) Chapter 1 – Federal Aviation Administration, Department of Transport, Parts 1-1399. The flammability test requirements for aircraft are in Part 25. The smoke test procedures are Part 25 Amendment 25-66, which requires the NBS smoke chamber[1, 2] to be used. The smoke measurement is obscuration and there is no requirement to measure the toxicity of the smoke. However, aircraft manufacturers, recognizing the importance of smoke toxicity, have added toxicity testing requirements of their own[1] and use the NBS smoke chamber for this.

The NBS smoke chamber (ASTM E662) uses a 76mm square 25mm thick test specimen, which is housed vertically with a cone radiant heat source at 25 kW/m^2. It is sealed in a 0.51 m^3 chamber. The smoke is collected in the chamber and its obscuration determined. The weakness of the method is that the fire is in a closed volume and there is no ventilation, which prevents high heat release occurring. This original method has been replaced by an improved version in ISO5659 which has variable radiant heating. The specimen is also mounted on a load cell so fire mass consumption can be determined. However, the issue of the lack of fire ventilation has not been addressed. The IMO and railway materials testing use this method to analyze the products of the fire in the smoke chamber for toxic gases[2]. In the present work a modified cone calorimeter is used with undiluted raw gas sampling, with the ventilation of the fire controlled. This method enables toxic gas yields to be determined as a function of time and, as the fire equivalence ratio varies as the fire develops, the yield can be determined as a function of the fire equivalence ratio[3, 4]. It is considered that this is an improved method to determine the yield of toxic gases from different materials.

Although these tests cover flammability in great detail, the toxicity of smoke produced from the burning of materials is not covered substantially in the Aircraft Fire Tests, nor in test for construction or other transport areas[5]. The FAR25.852 tests only consider the optical density of the smoke and not the products contained within the smoke. These tests ensure that the product being placed in an aircraft do not initiate a fire. However, the tests do not consider the smouldering stage of a fire before it self-extinguishes which can release toxic species, at a higher yield than flaming combustion (NIST, 2002). Also the tests do not consider the material burning in a fire started elsewhere and this is particularly relevant to the toxic gas release from flame retardants. Case studies have shown that toxic gases have a great impact on escape and survivability of passengers and crew on an aircraft and so the current tests cannot be considered satisfactory.

Material developments in recent years in furnishing materials have resulted in more synthetic polymers being used than the natural wood, cotton, leather and wool, than have traditionally been used. Synthetic polymers generally have considerably higher calorific values than the natural materials, similar to that of petroleum [6]. However, these synthetics can be fire retarded during the manufacturing process to incorporate fire retardants in the material unlike natural materials that need surface treatments or coatings to induce fire retardant properties[7]. A synthetic leather product is investigated in this work.

Chaturvedi[8] reviewed work that has been carried out on toxic fumes from aircraft fires. Aviation fire investigations have discovered that the victims can have a very high presence of cyanide in their blood. This prompted research to be carried out on to the toxic levels presented from fires in aircrafts from their furnishings. Chaturvedi[8] investigated 75 different aircraft interior materials including panels, foams, fabrics, floorings and insulations. These were all ranked for their toxicity based upon the relative time to incapacitation and time to death caused by the inhalation of thermal degradation of the interior materials in rats. The materials were thermally decomposed by an isothermal heating regime at 600°C using thermogravimetric analysis. He concluded that the ranking of the most important toxic gases in aircraft internal fires was $HCN>NO_2>HF>HCl$ and that CO was not the dominant toxic gas as it is in building compartment fires.

Alarifi et al.[3] have shown, using the present procedures with the cone calorimeter, that for fires of aircraft fabric seat covers the main toxic gases were SO_2, HCN, CO and NO_2 in order of importance. The SO_2 comes from the fire retardant which was chosen to avoid halogen toxic gas hazards in fires, but simply generated a different toxic gas. The HCN came from the acrylic base of the fabric. They also showed that a floor and wall carpet fire had HCN, CO, NH_3 and SO_2 as the main toxic gases in that order of importance. Andrews et al.[4] investigated the toxicity from aircraft blankets supplied to all passengers and showed that many of these were acrylic with HCN as a key fire toxicity load, with CO, formaldehyde and benzene as the other significant toxic gases. For a non-acrylic blanket, benzene was the most toxic gas, then acrolein, CO and HCN. Also the total toxicity of the non-acrylic blankets was significantly less than the acrylic blankets. This illustrates how undesirable high fire toxicity materials could be screened from use in aircraft if material fire tests for aircraft included the measurement of the toxic gases. Also this work shows that the list of Chaturvedi[8] for toxic gases of significant in aircraft fires should be extended to include SO_2, benzene, acrolein, formaldehyde, ammonia and CO. In only one of the above fires was NO_2 significant as a toxic gas (seat fabric cover), but it was measured in all the tests. Whether HF is present in fire toxic gases depends on the fire retardants used in the materials and it was not present in any of the above fire tests.

PVC covers for electrical cables are another source of HCl in fires and Al-Seyad et al.[9] have shown that the cone calorimeter can be used to determine toxic gas yields from PVC cable fires. They found that the dominant toxic gas was acrolein for both $COSHH_{15\ min}$ impairment of escape and LC_{50} 50% of deaths in 30 minutes criteria. HCl was a significant toxic species but not as important as acrolein and for LC_{50} formaldehyde was more important than HCl, but for $COSHH_{15\ min}$ toxicity HCl was the second most important toxic gas. The important role of the irritant gases acrolein and formaldehyde in PVC cable fires had not previously been reported. The release of HCl in PVC fires act to retard the rate of combustion of the cables and produce high levels of partially burned aldehydes.

Al-Seyegh et al.[10] have also shown in a small compartment fire that mixed fire loads of PVC cables and pine wood cribs resulted in the HCl from the PVC retarding the combustion of the wood and increasing the CO emissions, as well as having the same yield of HCl as in the cone calorimeter tests[10]. There was evidence that the presence of PVC sheathed cables retarded the rate of fire development and decreased the combustion efficiency of the wood, with large emissions of CO and THC and higher acrolein and formaldehyde yields than was found in similar pine crib fires with no PVC present[11]. This was due to the 2% of the total toxicity in these wood/PVC cable small compartment fires.

An alternative way to investigate fire toxicity from practical materials in fires is to use small scale compartment fires,which enable the key features of compartment fires to be reproduced[11,-13]. Gottuck et al.[11], Phylaktou et al.[12] Bundy et al [13] used small compartments of around $1m^3$ with controlled air inlet at the b[3]ottom of the compartment and fire gases out near the top of the compartment, this enabled that variation of fire equivalence ratio with time to be similar to that in full scale compartment fires. The $1.6m^3$ fire compartment of Phylaktou et al.[12] has been used extensively by the authors[11, 13, 15-18] for investigating fire toxicity of various materials as a function of fire equivalence ratio. This investigation of fire toxicity in such small compartment fires often uses dilution of the products of combustion from the compartment and toxic gas analysis of the diluted products. This enables the water vapor of the products of combustion to be kept in the gas phase and for cold line gas sampling to be used[11]. However, direct heated gas analysis is preferable for toxic gas analysis[12, 13, 21, 22] as post fire oxidation of the toxics is avoided and the dilution reduces the concentration which makes measurement more difficult.

Andrews et al.[15] used a 1.6 m^3 compartment fire with a mixed aircraft passenger seat fire load, which was mainly acrylic blankets and in one fire contained the life vest (PVC), in a small compartment fire at aircraft ventilation rates found that the main toxic gases were HCl, HCN, SO_2, CO, Benzene in that order of contribution. Although the PVC from the life vest gave a high HCl toxicity, there was significant HCl toxicity in other fires with no life vest and this was concluded to be due to the use of chlorine based fire retardants. The range of toxics were similar to those found using the cone calorimeter with similar fire loads.

Toxic gases can be measured in full scale fire compartment tests, but this is easier if direct hot gas fire exhaust gas sampling is used[19-21]. Alarifi [19-21] have investigated wooden pallet fires in a standard domestic room fire with one door opening. In this real scale fire the peak CO yield in the pre-flashover region increased from 0.01 at Ø=0.1 to 0.24 at Ø=1.7. Aljumaiah et al.[11] used a 1.6 m^3 compartment experimental rig for pine wood cribs and the peak CO yields varied between 0.14 and 0.25, depending on the air ventilation. This was in good agreement with the full scale experiments for the same Ø. The formaldehyde peak yield was 0.012 in the full scale tests and 0.002 – 0.018 in the 1.6m^3 compartment depending on the ventilation. The acrolein peak yield was 0.005 in the large compartment and 0.003 – 0.008 in the small compartment depending on the ventilation. Results from pine wood using the present ventilation controlled cone calorimeter has CO yields of 0.1 – 0.24, depending on the ventilation conditions. These results for the full scale, medium scale and cone calorimeter compartment fire tests are in good agreement and justify the use of the present cone calorimeter procedures for evaluating toxic emissions from materials used in aircraft interiors.

EXPERIMENTAL METHODS

The materials were tested for toxic gas release using a small 100mm square specimens on a cone calorimeter, modified to enable ventilation controlled combustion and direct undiluted sampling of the products of combustion. The standard cone calorimeter dilution was still used for the determination of the heat release rate (HRR) by oxygen consumption. The cone calorimeter also determines the smoke generation by the obscuration method. The fire products were collected by a hood and duct directly above the cone where O_2 concentrations and smoke density are measured. The cone calorimeter had two modifications for this work: firstly, a 36L cubic compartment with a window was put around the test specimen so that the air flow to the fire could be regulated; and secondly, a chimney was added, the same diameter as the outlet from the cone heater, to enable the sampling of raw fire gas product analysis without external air dilution. Similar enclosures have been used by other researchers for testing materials in an inert or reduced oxygen environment, which is referred to as the Controlled Atmosphere Cone Calorimeter (CACC). In this work metered air was supplied to the chamber to achieve the desired air changes per hour, ACH, of 9 and 15. This simulates the ventilation conditions in the aircraft cabin in accordance with a survey conducted in 1994 on commercial airlines[23]. The air flow at these conditions were 0.11g/s and 0.18 g/s, which convert to 11and 18 g/(sm^2) respectively in terms of air mass flow per surface area of the fire load. At these two air feed conditions the dilution ratio on an air mass flow ratio bases, using the 28.8 g/s of air flow in the diluted exhaust of the cone, was 262/1 and 160/1 respectively. It is this reduction of the toxic gas concentration by over 100/1 that makes direct raw exhaust gas toxicity measurement on the cone calorimeter much easier than for measurements in the diluted gases.

The ratio of the air mass flow to the fuel mass loss rate gave the metered A/F ratio for the fire as a function of time. However, a better resolution of the A/F as a function of time is by carbon balance[24] and this was used to determine the variation of equivalence ratio, Ø, with time in the fire. This was used in all the conversions to toxic gas yields, which were computed from the FTIR gas concentration using Eq. 1.

$$\text{Yield} = K\,C\,(1 + A/F) \qquad (1)$$

Where K is the conversion coefficient, which is the ratio of molecular weight of a specific pollutant to the molecular weight of the exhaust gas. The molecular weight of the exhaust gas is close to that of air and the MW has been used in this work. K is a constant for a particular toxic gas e.g.0.969 for CO. C is the concentration of the component. If this is measured in ppm or % then the equation has to be multiplied by10^{-6} or 10^{-2} respectively.

Raw combustion products from the ventilation controlled fire chamber were sampled by adding a 210mm long exhaust pipe on the conical heater central hole with the same internal diameter (80mm) as the cone upper opening. This acted as a chimney and enabled a gas sample probe to be inserted into this outlet 'chimney', without influencing the entrainment of air into the conical outlet. The direct hot raw gas sampling avoided any oxidation of the toxic gases in the dilution of the fire products in the cone calorimeter. However, tests on pine wood have shown that the main advantage of the raw sampling is that the concentration of the species being measured is much higher than in the diluted exhaust in the cone calorimeter and this enabled more toxic gases to be detected, that would be below FTIR detection limits in the diluted gases. There was little evidence of post dilution oxidation being significant as the yields of various toxic gases was the same in the raw and diluted samples. As discussed above the dilution of the raw gases from the cone exit was 267/1 and 160/1 for the two air flows to the cone calorimeter fire chamber. The FTIR has a limit of detection of about 1- 2ppm for most species and the dilution of the raw gas would prevent species from being measured in the diluted sample below about 300ppm in the raw gas. Many of the toxic gases reported in this work would be undetectable in the diluted sample and hence this is the dominant reason to determine the toxic gases in the raw sample.

Oxygen consumption calorimeter was also used on the raw sample gas as well as the conventional diluted oxygen consumption HRR with the standard cone calorimeter system, which was unaffected by the addition of the cone discharge chimney. This enabled the proportion of the total heat release that was released in the ventilation controlled compartment to be determined. Essentially these modifications of the cone calorimeter convert it into an equivalent apparatus to the flame propagation apparatus, FPA, [25] where a fire sample on load cell is placed in a vertical quartz tube with an external conical heater and metered airflow up the quartz tube. The FPA has been used by Tewarson[26] in producing his gases yield database. These modifications enable the apparatus to meet ISO 19706 fire stages 3a & 3b[27] for under ventilated flaming tests.

The heat flux was set at 40kW/m^2 which represented a surrounding flaming fire, as suggested by the classification of fires in the ISO19706[27]. The raw gas analysis from the modified cone calorimeter chimney was sampled using a single point stainless steel tube into a 190°C heated sample line. This heated sample line was connected to a heated pump and sample filter system and then connected to the FTIR through another heated sample line. Finally the FTIR detection chamber was also fully heated at 190°C. This ensured no loss of toxic species due to condensation or solution in water. SO_2, HCl, NO_2 and many other species such as aldehydes are lost or reduced in concentration if they contact condensed water in the sampling system. The calibration details for the FTIR for the toxic gases in this work are summarized in Table I.

A CE Flash EA2000 elemental analyser was used to determine the HCONS analysis. A Mettler Toledo TGA/DSC 1 instrument was used to perform thermogravimetric analysis for water, volatiles, carbon and ash. The sample was heated in nitrogen on the TGA and the temperature increased to 110°C and held there to determine the weight loss due to water content of the sample. Then the temperature was increased slowly to 700°C and the additional weight loss was the volatile fraction. Then the nitrogen was switched to air and the temperature increased to 950°C and any carbon was burnt and this weight loss was the fixed carbon. Any mass remaining was the ash or inert content of the sample.

A Gasmet FTIR CR-2000 instrument with 190°C heated detection cell was used for toxic gas analysis. The FTIR had a 2 meter path length multi-pass sample cell with a volume of 0.22 l. The FTIR sample cell was coated with rhodium-gold non-porous protective layer resisting corrosion. A liquid nitrogen cooled MCT detector was used with10 spectra scans per second and scans were averaged over 10s. The longer the time average, the more improved was the signal to noise ratio and more species could be resolved in each analysis. The FTIR was calibrated for the 64 species likely to be encountered in combustion processes, including all of the toxic gases of interest in fire toxicity, over a range of concentrations for each gas, as shown in Table I. An example of a recorded spectra with SO_2 identified for the fabric seat cover fire is shown in Fig. 1.

Table I: FTIR Calibration range and Wavelength Analysis Region for the most Important Toxic Gases

Gas name	Formula	No of calibration points	Calibration range [ppm]	Analysis area (cm^{-1})		
				1st range	2nd range	3rd range
Carbon Dioxide	CO_2	10	3000 - 30100			
Carbon monoxide	CO	26	20-9960	2000-2270		
Nitrogen dioxide	NO_2	9	20-4885	1520-1680		
Sulphur Dioxide	SO_2	12	20 - 1000	1250-1450		
Acrolein	CH_2O	5	40-219	1600-1850	2600-2900	900-1100
Formaldehyde	C_3H_4O	5	50 - 500	1650-1800	2600-3100	
Hydrogen cyanide	HCN	6	20-500	600-800	1300-1550	3180-3435
Hydrogen Chloride	HCl	7	20 - 489	2550-3100		
Hydrogen Fluoride	HF	4	33.9 - 100	3600-4200		
Hydrogen Bromide	HBr	6	20 - 592	2300-2700		
Benzene	C_6H_6	5	50-500	600-750	2900-3200	
Hexane	C_6H_{14}	3	50-500	2800-3000	1300-1600	

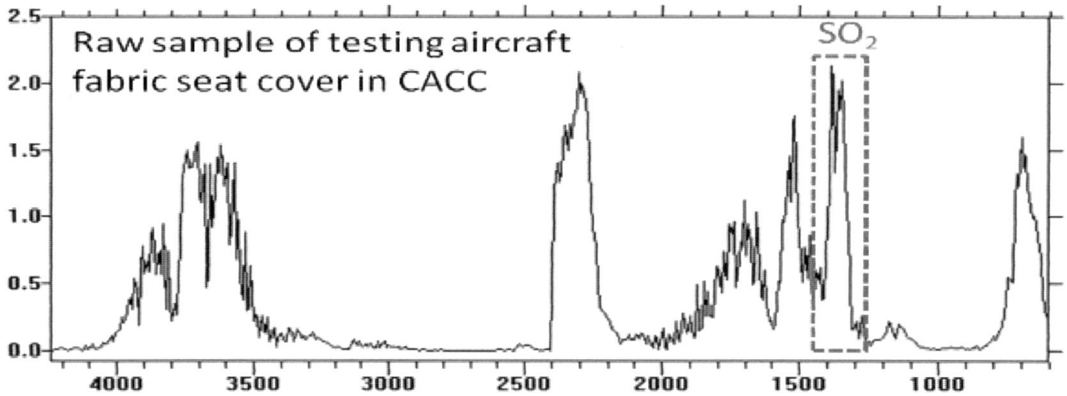

Figure 1 Absorption spectra for a raw CACC hot gas sample analysed by the FTIR showing the SO_2 absorption peak used for SO_2 quantification.

MATERIALS INVESTIGATED

Seat foam and seat leather and fabric[2] coverings were investigated. The seat foam was supplied by Metzeler Schaum UK and was a graphite combustion modified polyether polyurethane foam that was specifically designed for aircraft seating foam. The leather seat coverings were supplied by British Airways furnishing department, but there was only enough sample for one test at one ventilation rate. The seat fabric cover that was used for comparison[2] was provided by Replin Fabrics. All the materials passed the current fire tests for propagation rates and smoke yields. However, none of these tests required the toxic gases to be analysed, so that toxic gas risk assessment could be undertaken. These tests only rate the risk of these materials being the initial source of a fire if there is a pilot ignition of the material, they do not test the behavior of the materials when subject to radiant heating from a fire elsewhere on the aircraft. The present work aimed to determine the toxic gases released if the materials were subject to radiant heating and a 40 kW/m^2 radiant ignition source[27] was used.

The seat fabric was folded to give 9 layers with the normal outer surface exposed to the radiation. The thickness of the 9 layers was 6.84mm and the initial starting mass was 44.8grams. The ignition time was relatively quick at 14 seconds. After ignition, the flame burnt quite fiercely across the whole sample but then stopped burning in the middle relatively quickly and just flamed at the edge where there was an air supply. The foam was placed over foil, in case of melting and dripping which may affect the weight loss. The initial mass of the foam was 17.4 grams and a thickness of 23.7mm. The ignition time was exceptionally fast at 6 seconds. During the test the foam burnt quite brightly initially and then slowed to a single laminar flame in the centre of the sample as shown in Figs. 16 and 17. The leather provided was not sufficient to fold into several layers. It was bigger than the 100mm square sample area and the surplus was folded at the edges to ensure that the cut edge was not exposed to radiation. Like the foam, the sample was lined with foil. The initial mass of the leather was 21.1grams with a thickness of 2.36mm. Ignition occurred quickly at 15 seconds. As soon as the leather ignited, the flame was very large and could be seen through the top of the cone about 40cm above the sample.

The elemental analysis results are shown in Table II and the TGA analysis in Table III. There was significant S in all three materials and this indicates that ammonium sulphate and/or ammonium sulphamate ($NH_4SO_3NH_2$) could have been the flame retardants[28] as both S and N are in the elemental composition. The foam is based on polyurethane which gives it the highest N of the three materials. The high N in the leather seat cover material indicates that it was a composite material with and acrylic base. It will be shown that all three materials had HF in the toxic gases and the seat foam and leather seat

Table II. Elemental analysis of the materials tested and their global chemical formulae

Material	C (wt %)	H (wt %)	N (wt %)	S (wt %)	O (wt%)	Formula
Leather	40.43	6.31	10.56	3.97	38.74	$CH_{1.87}O_{0.44}N_{0.22}S_{0.04}$
Seat Fabric	48.20	6.23	14.98	4.53	26.06	$CH_{1.55}O_{0.59}N_{0.27}S_{0.03}$
Seat Foam	59.52	7.68	5.15	3.74	23.91	$CH_{1.55}O_{0.33}N_{0.07}S_{0.02}$

Table III. TGA analysis of volatile, fixed carbon, water and ash fractions with Stoichiometric A/F

Material	Water %	Volatiles %	Fixed Carbon %	Ash %	Stoichiometric A/F by mass (from data in Table 1)
Leather	NM	NM	NM	NM	8.22
Seat Fabric	4.7	68.3	25.0	2	6.47
Seat Foam	0	80	3	17	8.95

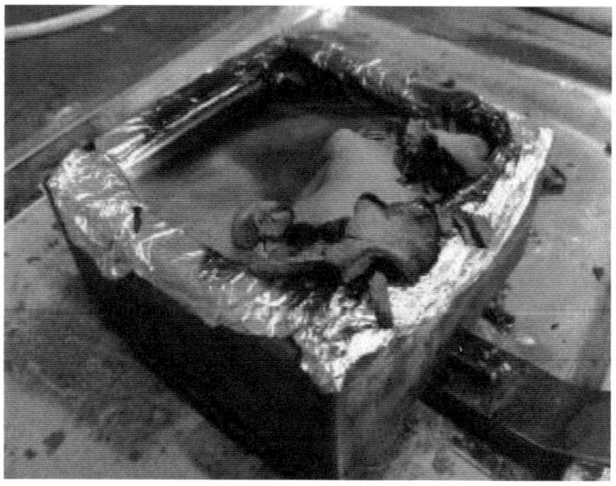

Figure 2 Foam post test at 15 ACH

Figure 3 Leather seat cover post test at 15 ACH

covers had HCl in the toxic gases. This indicates that these materials in addition to Sulphur containing fire retardants also had halogenated fire retardants. The elemental analysis only measured the compounds in Table II. The TGA results in Table III showed that the seat foam released volatiles slowly between 500 and 650K (30% mass loss) and then rapidly between 650 and 700K (50% mass reduction). At the end of the test there was foam remaining, as shown in Fig. 2, which was a mixture of carbon and ash or inerts. Fig. 3 shows the debris at the end of the leather seat fire.

THE CONE CALORIMETER COMPARTMENT FIRE DEVELOPMENT

The enclosure around the 100mm square test fire gives all the features of a compartment fire with air fed at the bottom of the chamber and burnt gases extracted from the ceiling layer of the chamber. This is similar to the 1.6 m^3 fire compartment test rig of Phylaktou[13] and the rigs of Gottuck[12] and Bundy[14]. The leather sample at 9ACH had 75% of the weight loss in a 300s flaming combustion period, after which it self extinguished. The foam burned brightly initially, as in Fig. 4, and then slowed to a single laminar flame in the centre of the foam as shown in Fig. 5.

Figure 4 Foam 15 ACH early in the fire

Figure 5 Foam 15 ACH later in the fire

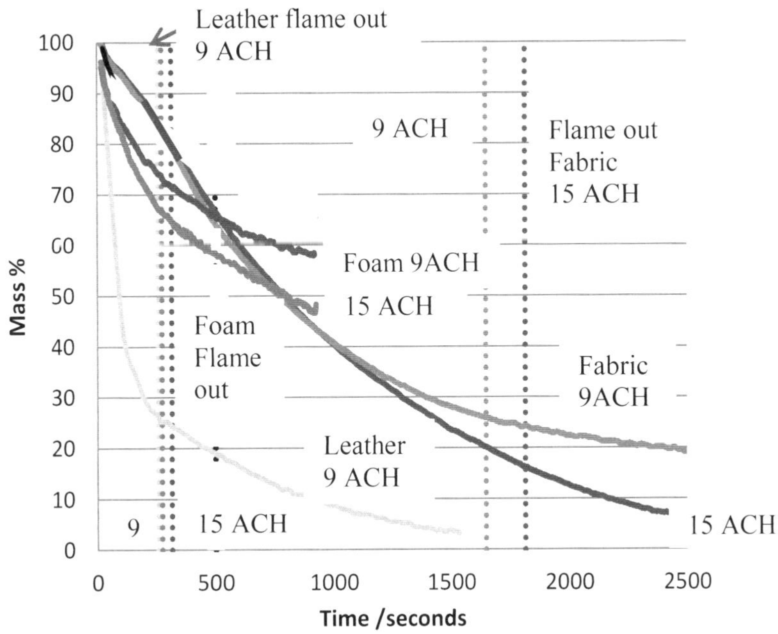

Fig. 6 Mass loss as a function of time in the cone calorimeter fires

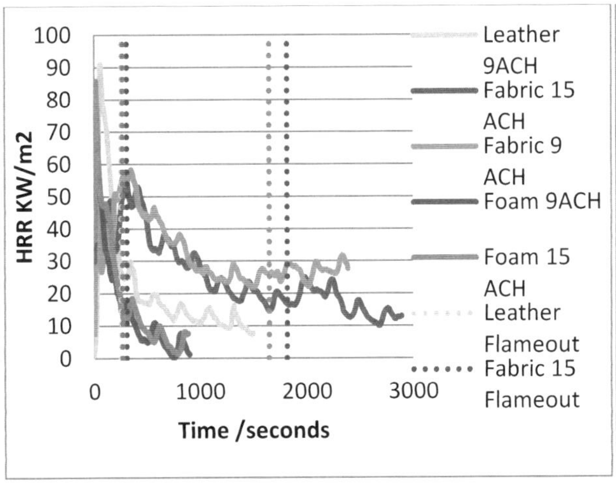

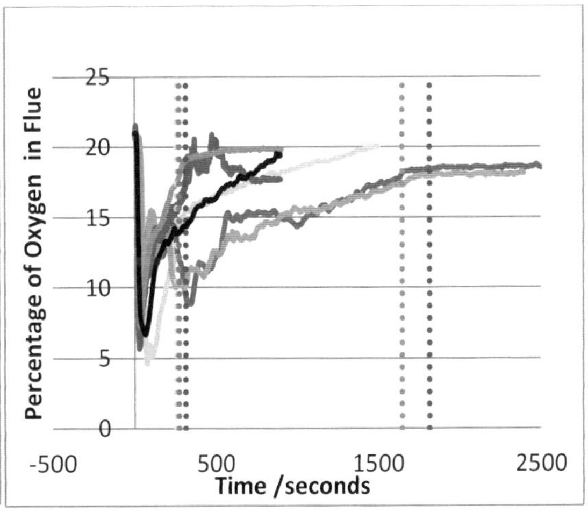

Fig. 7 HRR by oxygen consumption

Fig. 8 Oxygen in the undiluted fire products

The fire weight loss as a function of time for all three materials and for the different ventilation rates are shown in Fig. 6. This shows the fast combustion of the leather foam cover at 9ACH. The foam had the next quickest burn rate with a flame out after 300s at 9ACH, as for the leather cover. The flame period increased to 350s at 15 ACH and the mass loss rate was a little higher at 15 ACH, but it was not proportional to the increase in air flow. The fabric seat cover had the slowest combustion, but the longest period of flaming combustion. There was only a small increase in the flaming combustion time and no difference in the mass loss rate with increase in air flow, until close to the flame out. In the smoldering combustion phase the mass loss was greater for the higher air flow.

The heat release rates, HRR, by oxygen consumption are shown in Fig. 7 and the trends follow those discussed above for the mass loss rate. The oxygen content of the undiluted combustion product gases in the chimney above the cone outlet is shown in Fig. 8. These results show the common features of compartment fires with a fire growth phase to a peak HRR and minimum oxygen, followed by a fire decay phase and for all the materials tested a smoldering combustion phase. The mass loss rates in Fig. 6 are very high and all the materials were burning at both air flows with rich combustion. Both the foam and leather remained rich combustion in the smoldering phase, but the slower burning fabric fires had lean combustion in the smoldering phase. For all the materials in the flaming combustion phase the equivalence ratio, Ø, was about 4 and gave high concentrations of toxic gases.

SPECIFIC OPTICAL SMOKE DENSITY, D_s

The Specific Optical Density, D_s, is a dimensionless measure of the amount of light transmission passed through the smoke of a test sample. The smoke measurement in a cone calorimeter are based on an instantaneous measurement of smoke concentration in a system. The cone calorimeter measures the smoke using a monochromatic light beam, providing data for the extinction coefficient, k, in 1/metres. This can be transcribed to the specific optical density using Eq. 2[29].

$$D_s = \frac{\int k_i v_i dt}{2.303\, A} \qquad [2]$$

where: k_i = Extinction coefficient 1/m (measured using laser absorption)
v_i = Volumetric flow rate through the system $m^3/s = 0.0004 m^3/s$
t = time
A = exposed surface area of the burning sample $m^2 = 0.01 m^2$

This equation can be integrated and the known values put in to produce Eq. 3.

$$D_s = \frac{k_i \times 0.0004 \times t}{0.02303} \qquad [3]$$

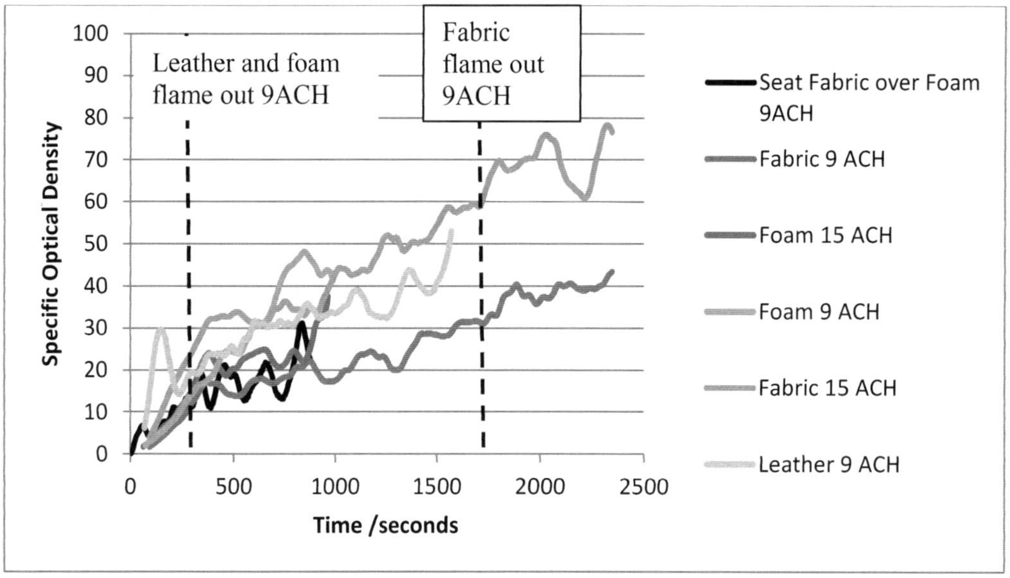

Fig. 9 Specific Optical Density (dimensionless) as a function of time for the three seat materials

Figure 9 shows the measured D_s as a function of time for the three seat materials at the two ventilation rates. In addition test with a dual fire load of seat fabric and foam are shown, which are very close to the separate seat fabric results for 9 ACH and lower than the foam results. Fig. 9 shows that for foam the peak smoke production was after the flame was out. Increasing the ventilation reduced the smoldering foam smoke production. For the leather seat cover the peak smoke production was also after the flame out time. For the fabric seat cover material the flaming smoke production dominated as this lasted much longer than for the other materials. For flaming combustion the increase in ventilation increased the smoke production. There also continued to be smoke production in the smoldering phase of the fabric fires. Fig. 9 shows that at 9 ACH at 800s the ranking of the smoke production is foam > leather > fabric. At 15 ACH only the foam and fabric were tested and the fabric smoke was significantly higher than the foam smoke, mainly because of the longer duration of flaming combustion.

TOXICITY

Table IV Toxic Gas Limits

Toxic Gas	15 Min Exposure Limit COSHH ppm	LC_{50} 30 min ppm	Toxic Gas	15 Min Exposure Limit COSHH ppm	LC_{50} 30 Min ppm
CO	200	3,000	CH_2O	0.3	300
NO_2	5	500	SO_2	5	500
HCN	10	135	HCl	5	3,700
C_6H_6	3		HF	3	2000
C_3H_5O	2	250			

The nine most significant toxic gases in the present work are shown in Table IV with their LC_{50} and $COSHH_{15min}$ exposure limits. The LC_{50} limit is one related to 50% deaths after 30 minutes exposure and the $COSHH_{15min}$ limit is related to no impairment of escape and is similar to the US AEGL 10 minutes impairment of escape limit. The measured toxic gas concentration has been divided by these limits to get the relative toxicity N_{LC50} or $N_{COSHH15min}$ for each gas and then all the N for each gas have been summated to give the total N. This has been done for all 64 gases measured by the FTIR, but the gases in Table IV were the dominant toxic gases in the present work. The importance of each toxic gas has been shown by giving the % of the total N for each gas as a function of time in the cone calorimeter fire. The toxicity results are shown for the three aircraft seat materials in Figs. 9-20. They all show a peak LC_{50} and $COSHH_{15min}$ toxicity in the initial flaming combustion with significant toxicity in the smoldering combustion phase of the fire. The toxic gases of importance for the three materials for 9ACH are summarized in Table V. This shows that for most of the fires that the LC_{50} and $COSHH_{15min}$ identified the same toxic gases of importance, but the range of gases of importance for impairment of escape ($COSHH_{15min}$) was always greater and usually included acrolein, which was only significant for LC_{50} toxicity for fabric seat covers. HF and HCl were also significant for $COSHH_{15min}$ for all three materials but not for LC_{50}. It is clear that all three materials tested had a complex mixture of toxic gases and that CO was never the most important toxic gas on an LC_{50} or $COSHH_{15min}$ basis. For three materials SO_2 was an important toxic gas and the origin of this was the high S in the materials shown in Table II. This was due the use of sulphates in the fire retardants.

For the seat fabric cover there was an initial very high toxic gas peak immediately after ignition. For LC_{50} this was dominated by HCN (75%) with CO and SO_2 as the only other significant toxic gases. For $COSHH_{15min}$ this initial toxic peak was also dominated by HCN (50%) but with SO_2, benzene and CO as the other significant toxic gases. For the seat foam at 9ACH there were two peaks in total toxicity at 30s and 300s. The first peak had, in order of importance, benzene, HCN, CO, SO_2 and acrolein as the most significant toxic gases. Figs. 9-12 show that the transition from flaming to smoldering combustion resulted in an increase in toxicity on an LC_{50} and $COSHH_{15min}$ basis. For $COSHH_{15min}$ the smoldering combustion yielded 7 important toxic gases with acrolein the most important. For LC_{50} the smoldering combustion only had CO and HCN as important toxic gases.

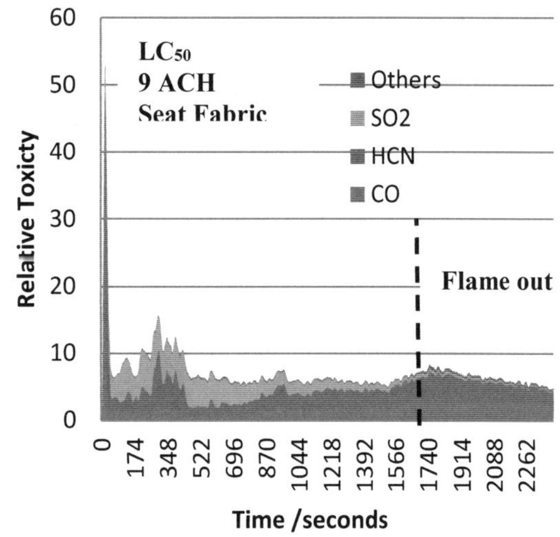

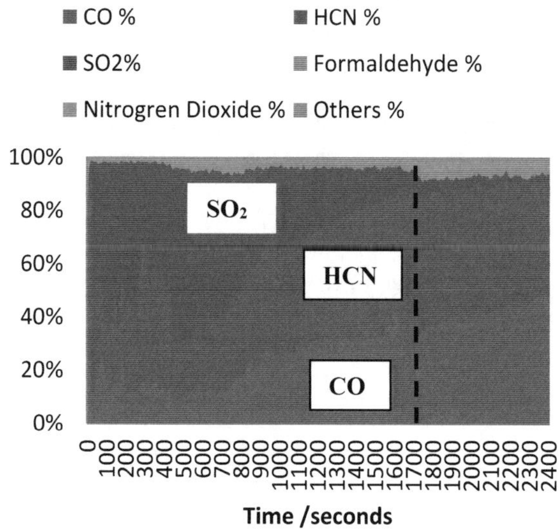

Figure 9 Total relative toxicity, N, and the major species for the seat fabric at 9ACH

Figure 10 Contribution to the LC_{50} toxicity

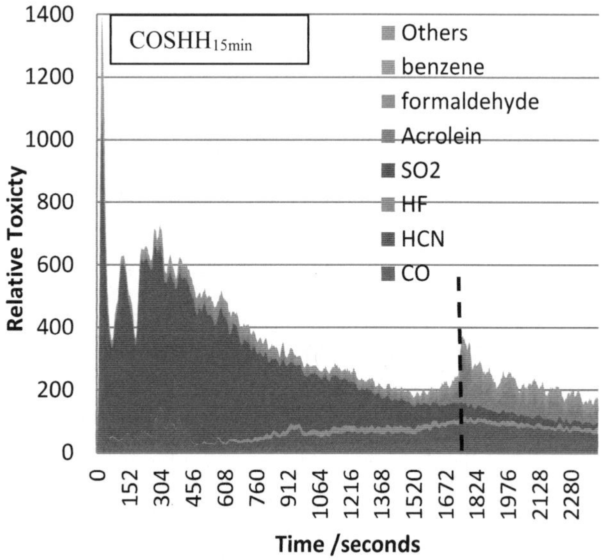

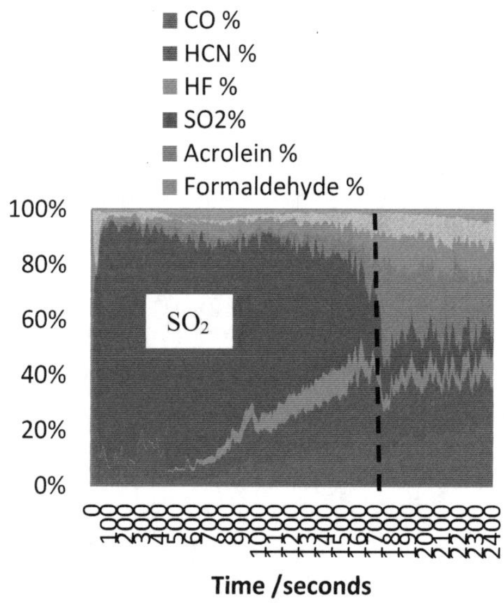

Figure 11 Total relative toxicity, N, at 9ACH and the major species for seat fabric

Figure 12 Contribution to the $COSHH_{15min}$ toxicity

For the foam fires Figs. 13-16 show that there were two main toxic emissions stages in the fire, the initial flaming combustion and then the start of smoldering combustion after the flame out. In both of these stages the most important toxic gases were different. In flaming combustion at the peak toxicity for LC_{50} was 50% from HCN, 45% from CO and 3% from SO_2. In the lower total toxicity later phase of flaming combustion SO_2 was more important at 45% with CO at 25% and HCN at 20%. Following the end of flaming combustion the smoldering combustion had higher toxicity than the later stages of flaming combustion. CO was the dominant LC_{50} smoldering combustion toxic gas at 45% but HCN

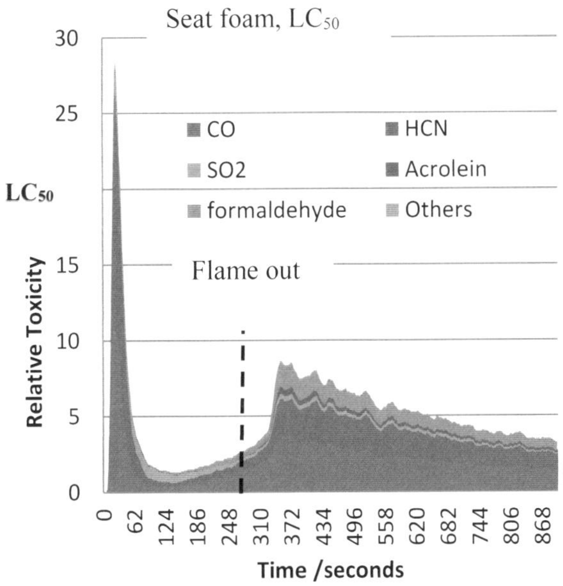

Figure 13 Total relative toxicity, N, and the major species for seat foam at 9ACH

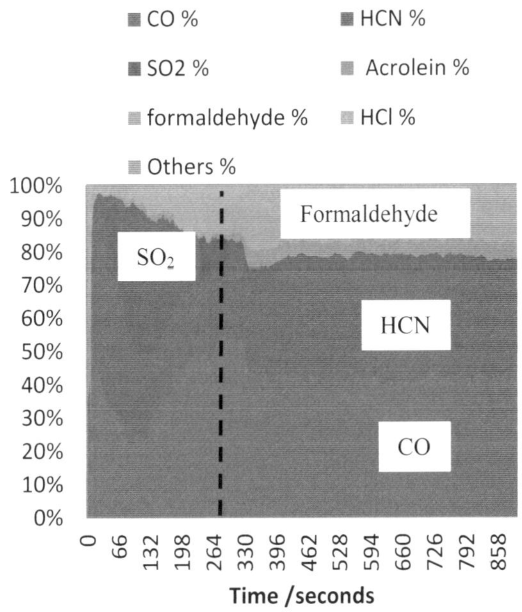

Figure 14 Contribution to the LC_{50} toxicity

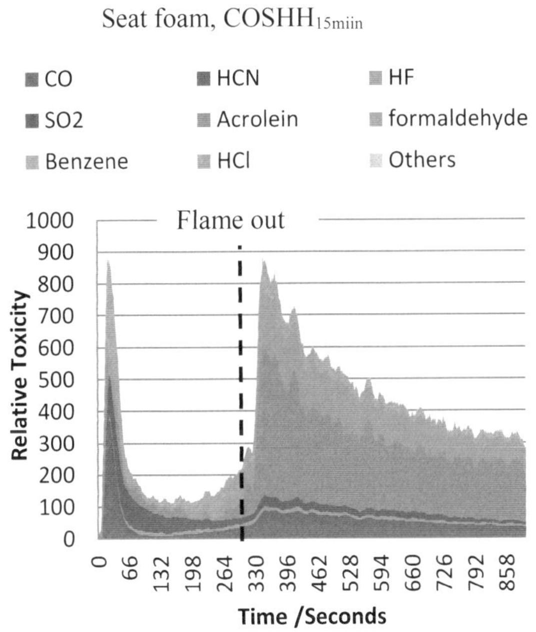

Figure 15 Total relative toxicity, N, at 9ACH and the major species for seat foam

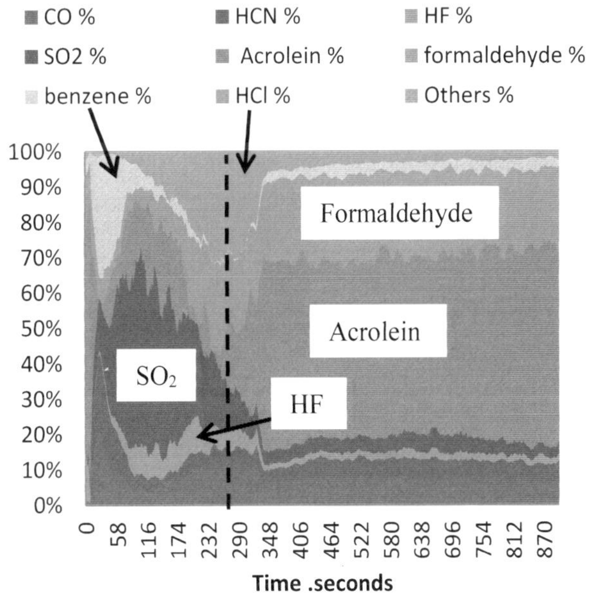

Figure 16 Contribution to the $COSHH_{15min}$ toxicity

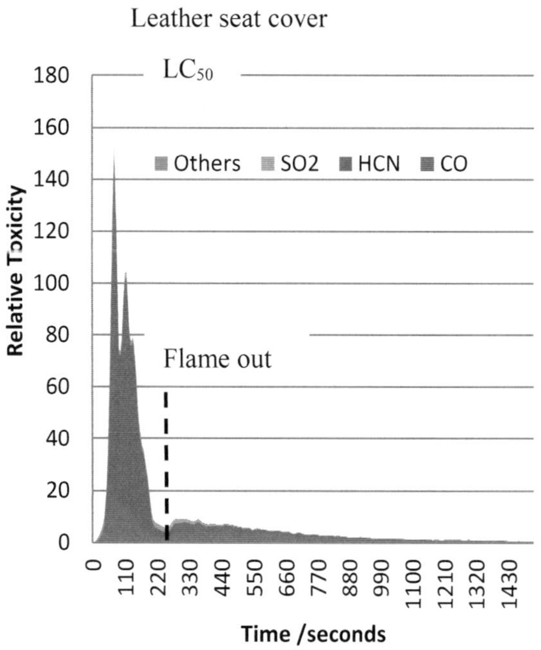

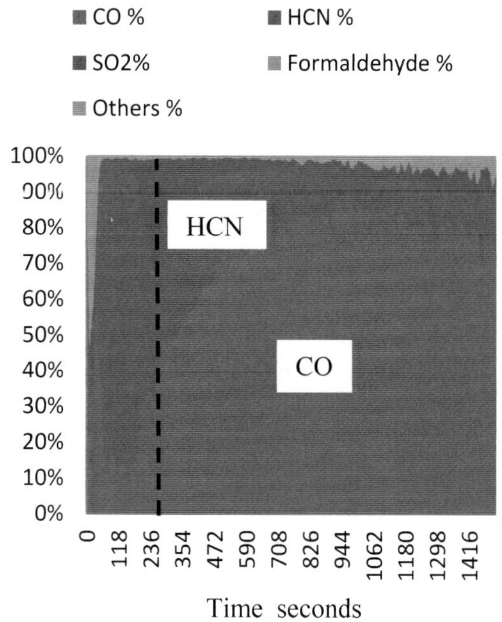

Figure 17 Total relative toxicity, N, and the major species for seat leather cover at 9ACH

Figure 18 Contribution to the LC_{50} toxicity

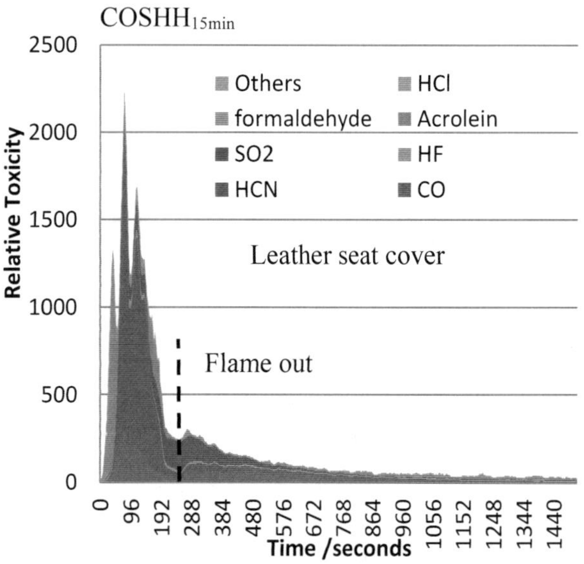

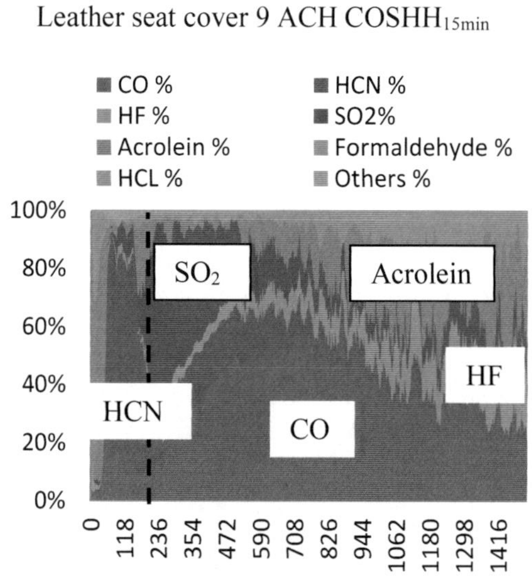

Figure 19 Total relative toxicity, N, at 9ACH and the major species for seat leather

Figure 20 Contribution to the $COSHH_{15min}$ toxicity

increased in importance to 30% with formaldehyde at 20%. For $COSHH_{15min}$ at the peak toxicity, after the start of flaming combustion, benzene was the most important toxic gas (35% of total toxicity), HCN and CO were the next most important (20% each), then SO_2 (15%) and finally formaldehyde (5%).

Table V Significant toxic gases

Toxic Gas	Fabric LC_{50}	Fabric $COSHH_{15min}$	Foam LC_{50}	Foam $COSHH_{15min}$	Leather LC_{50}	Leather $COSHH_{15min}$
SO_2	Yes	Yes	Yes	Yes	Yes	Yes
HCN	Yes	Yes	Yes	Yes	Yes	Yes
CO	Yes	Yes	Yes	Yes	Yes	Yes
Formaldehyde	Yes	Yes	Yes	Yes	Yes	Yes
Acrolein		Yes	Yes	Yes		Yes
HCl			Yes	Yes		Yes
HF		Yes		Yes		Yes
Benzene				Yes		
NO_2	Yes					

Table VI Peak Toxicity N values

Material	LC_{50} 9ACH Flaming	$COSHH_{15min}$ 9ACH Flaming	LC_{50} ACH Smoldering	$COSHH_{15min}$ Smoldering	LC_{50} 15 ACH Flaming	$COSHH_{15min}$ 15 ACH Flaming
Fabric	52	1350	7	350	40	1250
Foam	28	850	8	850	24	900
Leather	150	2200	8	250	NM	NM

The later stages of flaming combustion had lower toxicity that was dominated by SO_2 and formaldehyde. In the smoldering combustion phase for foam fires the $COSHH_{15min}$ toxicity was a high as the initial flaming combustion toxicity but had acrolein at 50% of the total toxicity and formaldehyde as the next most important at 25%. The foam fires had significant HF toxicity across both flaming and smoldering combustion and this indicated that a fluorine based fire retardant had been used.

The leather seat cover material had very high LC_{50} and $COSHH_{15min}$ total toxicity in the initial flaming combustion with much lower toxicity in the smoldering combustion phase. In this initial flaming combustion toxicity, HCN dominated for both toxicity assessments (80%) and this was likely to originate from an acrylic base to the simulated leather material. After the flame out CO was the most important toxic gas for both toxic gas assessment methods.

Table VI summarizes that peak toxic N for the three materials. The value of N is a measure of the dilution of the fire product gases that is required in a building or aircraft before the gases are safe. N is a scalar and can be predicted in fire CFD codes to define a death zone and a safe means of escape zone. Table IV shows that the material with the highest overall toxicity in flaming combustion was leather on both an LC_{50} and $COSHH_{15min}$ basis. The fabric seat cover was the next most toxic and the foam the least toxic. This is not what would be expected as the use of PU foam in the mixed composition foam, would be expected to dominate the toxicity through HCN generation. However, the use of acrylic material base for the fabric and leather seat covers resulted in high levels of HCN from the covers. In the smoldering phase of the fires the foam had the highest toxicity under $COSHH_{15min}$ toxicity assessment, but there was little difference in the three materials under LC_{50} toxicity assessment. The high $COSHH_{15min}$ toxicity was due to high levels of acrolein and the high weighting to acrolein in the $COSHH_{15min}$ assessment relative to the LC_{50}. Table VI also shows that increasing the ventilation of the fire compartment reduced the total toxicity of flaming combustion. This is the opposite trend to that in Fig. 9 for the smoke emissions.

CONCLUSIONS

1. The cone calorimeter with a ventilation controlled air box simulates most of the key features of compartment fires and offers the possibility of small scale material testing for fire toxicity under realistic conditions.
2. Adding a small chimney to the cone calorimeter cone fire product outlet enables direct exhaust gas sampling to be used with heated sample handling and a heated FTIR to determine the toxic gas composition without the problem of sample dilution and the associated oxidation reactions that might occur.
3. Aircraft passenger seat materials, fabric covers, leather covers and seat foam, were investigated using the cone calorimeter. The materials all had a sulphur based fire retardant which gave SO_2 as one of the significant toxic gases. Also all the materials had a significant N content through the use of acrylic fibres as the base of both of the seat cover materials and the foam was a modified PUF with fire retardants added. This fuel N content gave all materials a HCN toxic gas hazard.
4. HF was detected in the toxic gases for all three materials and HCl was detected in the foam and leather fires. The presence of these gases indicate that in addition to sulphur based fire retardants there were halogenated fire retardants.
5. The toxic gases emitted by all the material were very high and were emitted immediately that flaming combustion started and hence were easily generated in the fires. The composition of the materials and the flame retardants used determine the main toxic gases emitted. If toxic gases were measured in material testing for fire propagation then different decisions on the most appropriate materials could be made. At present there is no requirement to make the measurements and so no toxic gas data exists to inform the choice of materials in sensitive applications such as aircraft.

ACKNOWLEDGEMENTS

We would like to thank Replin Fabrics for supplying seat cover fabrics for testing, British Airways furnishing department for supplying the leather seat cover material and Metzeler Schaum UK for supplying 6 pieces of aircraft seat foam for testing. The heated FTIR was provided by the EPSRC.

References

1. Peterson, J.M., 2006. Flammability Tests for Aircraft. Ch. 12 of "Flammability Testing of Materials used in Construction, Transport and Mining" p. 275-301. Ed. V.B., Apte, Woodhead Publishing, Cambridge, England.
2. Grayson, S., 2008. Smoke Opacity Test Methods. p.171-183 of "Hazards of Combustion Products" Eds. Babrauskas, V., Gann, R. and Grayson, S. Interscience Communications, London.
3. Abdulaziz A Alarifi, Lisa Witty, Gordon E Andrews and Herodotos N Phylaktou, 2013. Ignition and toxicity evaluation of selected aircraft interior materials using the cone calorimeter and FTIR analysis. *Interflam2013* Vol. 1 (pp. 37-49). UK: Interscience Communications Ltd.
4. Gordon E. Andrews, Michael Bell, Lisa Tang, Abdulaziz A. Alarifi & Herodotus N. Phylaktou, 2015. Aircraft Blanket Ignition and Toxic Emission in Simulated Aircraft Cabin Fires using the Cone Calorimeter. Proc. Fourteenth International Conference on Fire and Materials 2015,San Francisco, USA. p.734 – 748, 2015. Interscience Communications, London.
5. Babrauskas, V., Gann, R. and Grayson, S. (Eds.), 2008. Hazards of Combustion Products. Interscience Communications, London.
6. Staggs, J., 2011. Polymer and Textile Fires, University of Leeds, UK.

7. Hull, T., Stec, A. et al., 2009. Fire retardancy of polymers . Cambridge: Royal Society of Chemistry
8. Chaturvedi, A. 2012. Aviation combustion toxicology: an overview. *Reviews of Environmental Contamination and Toxicology. 214, 15-40.*
9. Al-Sayegh, W.A., Aljumaiah, O., Andrews, G.E. and Phylaktou, H.N 2015. PVC Cable Fire Toxicity using the Cone Calorimeter. Proceedings of the IAFSS, 10th Asia-Oceania Symposium on Fire Science and Technology, 10th AOSFST, October 5-7, 2015, Tsukuba, Japan.
10. Al-Sayegh, W.A., Aljumaiah, O., Andrews, G.E. and Phylaktou, H.N., 2016.. Wood Crib with PVC Cables: Compartment Fire Toxicity. Proc. 8th International Seminar on Fire and Explosion Hazards, Hefei, China, April. 2016.
11. Aljumaiah, O. Andrews, G.E., Phylaktou, H.N., Mustafa, B. G., Al-Qattan, H. and Shah, V., 2011. Air Starved Wood Crib Compartment Fire Heat Release and Toxic Gas Yields. Fire Safety Science – Proceedings of the Tenth International Symposium, pp. 1263 - 1276 (IAFSS10), Maryland. Ed.: Michael Spearpoint.Publisher:International Association for Fire Safety Science. DOI: 10.3801/IAFFS.FSS.10-1263.
12. Gottuck, D.T., Roby, R.J. and Beyler, C.L., 1992. A Study of Carbon Monoxide and Smoke Yields from Compartment Fires with External Burning. *Twenty Fourth*
13. Phylaktou, H.N., Ledger, J. and Andrews, G.E. Enclosed pool fires in low ventilation enclosures: flame temperatures and global heat loss using gas analysis, edited by Michel Curtat, Fire Safety Science - Proceedings of the Sixth International Symposium, pp.591-603, ISBN 0-925-223-25-5 (2000).
14. Bundy, M., Hamins, A., Johnsson, E.L., Kim, S.C., Ko, G.H. and Lenhert, D.B., Measurement of heat and combustion products in reduced scale ventilation limited compartment fires. NIST TN 1483, July 2007.
15. G.E. Andrews, Hu Li, Alex Hunt, Daniel Hughes, Steven Bond, Paul Tucker, Shahjehan Akram and H.N. Phylaktou, 2008. Toxic Gases in simulated aircraft interior fires using FTIR analysis. Proc. Fifth Int. Conf. on Fire and Explosion Hazards, Univ. Edinburgh, p. 846-855.
16. Aljumaiah, O., Andrews, G.E., Alqahtani, A.M., Husain, B.F., Singh, P. and Phylaktou, H.N., 2010. Air Starved Acrylic Curtain Fire Toxic Gases using an FTIR. Sixth International Seminar on Fire and Explosion Hazards, 11-16 April, 2010, University of Leeds, UK. Proceedings p. 804-816, 2011. ISBN-13:978-981-08-7724-8.
17. Aljumaiah, O., Alshammari, S., Burell, G., Cox, M., Andrews, G.E. and Phylaktou, H.N., 2010. Toxic Emissions from Folded Cotton Towels in a Low Ventilation Compartment. Sixth International Seminar on Fire and Explosion Hazards, University of Leeds, UK. Proceedings p. 792-803, 2011. ISBN-13:978-981-08-7724-8.
18. Gordon E. Andrews, Paul Yeomans, Herodotos N. Phylaktou and Omar A. Aljumaiah, 2015.. Toxic Gases from Compartment Fires with Hanging Cotton Towels and Low Ventilation. Proc. of the First International Fire Safety Symposium, 20-23 April, 2015. Coimbra, Portugal.
19. Alarifi, A.A., Dave, J., Phylaktou, H.N., Aljumaiah, O.A. and Andrews,G.E. "Effects of Fire Fighting on a Fully Developed Compartment Fire: Temperatures and Emissions". Fire Safety Journal 68 (2014): 71-80. DOI 10.1016/j.firesaf.2014.05.014 *Fire Safety.Journal°68 (2014): 71-80.*
20. Alarifi, A.A., Phylaktou, H.N., Andrews, G.E., Dave, J. and Aljumaiah, O.. Developing and Post-Flashover Fires in a Full Scale Room: Thermal Environment, Toxic Emissions and Effects of Fire-fighting Tactics., 2015. Proceedings of the 2nd IAFSS, Second European Symposium on Fire Safety Science, ESFSS, Nicosia, Cyprus.
21. Alarifi, A.A., Phylaktou, H.N. and Andrews. G.E., 2015. Heated Raw Gas Sampling with Heated FTIR Analysis of Toxic Effluents from Small and Large Scale Fire Tests. Proceedings of the IAFSS, 10th Asia-Oceania Symposium on Fire Science and Technology, 10th AOSFST, Tsukuba, Japan.

22. Alarifi, A.A., Phylaktou, H.N. and Andrews, G.E., 2015. Toxic Gas Analysis from Compartment Fires using Heated Raw Gas Sampling with Heated FTIR 50+ Species Gas Analysis. Proc. of the First International Fire Safety Symposium. Coimbra, Portugal.
23. Foundation, F.S., *Special Double Issue: Guidelines Enable Health Authorities to Assess Risk of Tuberculosis Transmission Aboard Aircraft.* Cabin Crew Safety, 1998.
24. Alarifi, A.A., Phylaktou, H.N. and Andrews, G.E., 2015. Heated Raw Gas Sampling with Heated FTIR Analysis of Toxic Effluents from Small and Large Scale Fire Tests. Proc. of the IAFSS, 10th Asia-Oceania Symposium on Fire Science and Technology, 10th AOSFST, Tsukuba, Japan.
25. ASTM, *ASTM E2058 -13 Standard Test Methods for Measurement of Synthetic Polymer Material Flammability Using a Fire Propagation Apparatus (FPA),* 2006, ASTM International: West Conshohocken, PA.
26. Tewarson, A., *Generation of Heat and Chemical Compounds in Fires*, in *SFPE Handbook of Fire Protection Engineering*. 2002, NFPA: Quincy, Massachusetts.
27. ISO19706, B., *BS ISO 19706-2007 Guidelines for assessing the fire threat to people*. 2007.
28. Touval, I., 1993. Antimony and other Inorganic Flame Retardants. In: Kirk-Othmer Encyclopedia of Chemical Technology, 4th Ed., John Wiley,Vol.10, pp.936-954.
29. Peacock, R and Bukowski, R. (1998). *Evaluation of Passenger Train Car Materials in the Cone Calorimeter.* Available: http://fire.nist.gov/bfrlpubs/fire98/PDF/f98012.pdf.

REAL-TIME MEASUREMENTS OF PARTICULATE AND TOXIC-GAS CONCENTRATIONS

Fumiaki Takahashi [1], Paul S. Greenberg [2], Susana Carranza [3], Gary W. Hunter [2], Michael, J. Kulis [2], & Gordon Berger [4]
[1] Case Western Reserve University, USA; [2] NASA Glenn Research Center, USA;
[3] Makel Engineering, Inc., USA; [4] Universities Space Research Association, USA

ABSTRACT

Time-resolved concentration measurements of particulates and gaseous species have been made in a 326 litter-smoke chamber and in a 66 m^3-burn room. Smoke from a pyrolizing wood specimen in a cylindrical furnace (250 – 400 °C), diluted with cold air in the smoke chamber, was characterized using an optical aerosol sensor and solid-state microsensors for O_2, CO_2, CO, and hydrocarbons with reference instruments: a personal dust monitor, a scanning mobility particle spectrometer, and an optical CO gas analyzer. The particulate size distribution showed a peak of ultrafine aerosol particles (mode < 100 nm) at an early stage and then shifted to larger size (100 – 500 nm). The mass concentration of particles peaked at a few minutes after the heat exposure and decreased monotonically. The concentration of hydrocarbons generally peaks at the beginning and CO and CO_2 continued to increase afterward. In the burn-room experiment, different fuels (wood pallet, mattress, and sofa) were burned individually in a controlled mode. The measurements of the heat flux and temperature and continuous gas sampling were made. The temporal variation of the heat flux correlates well, in general, with that of the gas temperature, showing an initial peak and a gradual decay, typical of small-scale burn tests. The variation in the mass concentration of particles correlates, to some extent, with the CO concentration. The results provide insight into the smoke characteristics and assist in developing a hand-held real-time particulate and toxic-gas detector.

INTRODUCTION

There are high-priority needs [1,2] for improvements in personal protective equipment (PPE) and environmental monitoring system to reduce incidence of firefighter exposure of chemicals and particulates during structure fire overhaul activities and in all aspects of wildland fire. Wildland firefighters do not wear respiratory protection despite intermittent high-level and long-term low-level exposure to smoke. Effective assessment of the post-fire environment requires not only the detection of multiple toxic gases and vapors, but also (and perhaps more importantly) assessing the presence and makeup of suspended particulates that may contain significant additional health risks including polycyclic aromatic hydrocarbons (PAHs) [3]. The challenge is to integrate and adapt the technologies effectively and reliably to make the firefighter aware of dangerous situation in their work environment.

Removal of respiratory protection during post-fire (overhaul) activities could expose firefighters and investigators to a variety of unknown toxic gases, vapors, and airborne particulates, but current practice relies largely on the CO concentration [4]. The overhaul phase, which lasts for an average of 30 minutes, is the firefighting stage in which fire suppression is complete and firefighters are searching the structure for hidden fire or hot embers [5]. During the overhaul phase of a structure fire, when there is little or no smoke in the environment, firefighters is most likely to remove their respiratory facepiece and work in this environment with lesser level of (or without) respiratory protection. However, ultrafine (< 100 nm) smoke particles during overhaul can be too small to be visible by naked eye, suggesting that "clean" air is not really that clean [6].

Exposure of firefighters to particulates and toxic gases have long been studied in various types of fires: structure [4-20], vehicle [21-22], aircraft [23-26], wildland [27-33], wildland-urban-interface (WUI) [34], prescribed [35], training [36-37], and experiment [38]. The studies include different phases of firefighting: "startup, knockdown, and overhaul" in structure/vehicle fires or "direct attack, line holding, and mop-up" in wildland/prescribed fires.

Fire smoke is a complex mixture of gases, vapors, liquid and solid aerosols, produced by thermal decomposition (or pyrolysis), smoldering, and flaming combustion of materials. It includes carbon monoxide, irritants (aldehydes), volatile organic compounds (VOCs), polycyclic nuclear aromatic hydrocarbons (PAHs), ultrafine (< 100 nm) particles, and respirable (< 3.5 μm) particulates. Toxic chemicals, including those known to be carcinogens, may be absorbed in the aerosols. Gaseous effluents may condense on protective equipment and exposed skin [6]. The type and amount of chemicals and particulates in smoke vary, depending on the fuel, oxygen availability, the reaction temperature, and the mode of combustion. Contemporary structure firefighting environments contain a variety of toxicants due to synthetic materials and chemicals (e.g., plastics, fabrics, flame retardants, and wood preservatives).

The toxicants concentration measurements [4,5] during the overhaul in a number of actual structure fires, a handful of toxicant species were found, in particular fires, to exceed the standard occupational exposure limits [39,40]: the NIOSH Recommended Exposure Level – Short Term (REL – ST), the Occupational Safety and Health Administration (OSHA) Permissible Exposure Limit (PEL), the short-term exposure limit (STEL), its ceiling values, and even Immediately Dangerous to Life and Health (IDLH). Among 25 fires investigated [5], in particular fires, the species that exceeded the published ceiling values were acrolein (C_3H_4O), CO, formaldehyde (CH_2O), glutaraldehyde ($C_5H_8O_2$); and those exceeded STEL include benzene (C_6H_6), nitrogen dioxide (NO_2), and sulfur dioxide (SO_2) (in addition to the above species). Among 38 fires [4], in particular fires, the species that exceeded IDLH include NO_2, acrolein, CO, arsenic (As) and mercury (Hg); and those exceeded REL – ST or PEL were glutaraldehyde, hydrogen chloride (HCl), hydrogen cyanide (HCN), ozone (O_3), particulates (<10 μm), and benzene (in addition to the above species).

On the other hand, many studies on smoke exposure and the resulting adverse health effects among wildland firefighters have long been performed by the US Forest Service [27-32]. Exposure to wildland smoke may be one of the least understood risks of wildland firefighting. Wildland firefighters can be exposed to significant levels of CO and respiratory irritants (including formaldehyde and acrolein), (benzene), and respirable particulate matter [32-34]. The greater concern with their exposure is the combination of irritants in smoke from wildland fires. Correlations among the pollutants are recommended to be used to predict exposure to several pollutants when resources allow only one (such as CO) to be monitored. In a WUI fire, both forest fuels and structural fuels burn, creating a noxious mixture of chemicals in a firefighter's breathing space [34]. Unlike structural firefighters, wildland firefighters are not equipped with a self-contained breathing apparatus (SCBA) to protect them from inhaling structural smoke, working in these hazardous smoke conditions.

Numerous studies have demonstrated adverse health effects associated with the respiratory uptake of airborne particles. These studies address both animal and human subjects, and span controlled laboratory exposure and actual conditions encountered in the field. Elevated rates of coronary disease and morbidity are observed in firefighters engaged in suppression and overhaul activities [13,41,42] and have been associated with exposure to particulates [20]. Measurements of wood smoke particles in wildland fires have demonstrated as a mechanism for cellular injuries [43].

A compact real-time measurement device for simultaneously monitoring particulates and toxic gases is needed for firefighters during structural fire overhaul and wildland firefighting. As an initial effort, this study aims to combine and test a NASA-developed optical particulate sensor and micro-fabricated gas sensors for O_2, CO, and hydrocarbons to validate appropriateness of the technology transfer to the fire service use. This paper reports the experimental results of the smoke characteristics obtained in the laboratory and a burn room.

EXPERIMENTAL METHODS

The particulate and micro-fabricated gas sensors are tested in the NASA Glenn Research Center's Gases & Aerosol from Smoldering Polymers (GASP) Laboratory[44]. The entire GASP facility is contained in polycarbonate boxes attached to lab ventilation for safe operation. The GASP facility causes thermal decomposition and oxidation of the specimen by heating samples. Filtered air from a cylinder flows (4.4 l/min) through a tube furnace (3.45 cm i.d.) to push the decomposition byproducts into a 326 litter-smoke chamber (73.5 cm L x 58.3 W x 76.1 cm H). The smoke chamber includes pass-through for smoke/byproduct inlet, gas outlets to some of the analytical instruments. Smoke is generated by inserting a wood specimen (0.5 - 2 g) placed on a ceramic boat into the tube furnace at a set temperature (250, 300, 350 and 400 °C). The wood specimen for the laboratory experiment is sampled from a wood pallet for the burn-room experiment and used without any pre-conditioning (i.e., drying or steaming).

The particulate and micro-fabricated gas sensors are also tested in a 66 m^3-burn room (6.4 m L x 4 m W x 2.6 m H) at the Fire Training Academy of the Cleveland Division of Fire. Three different fuels (wood pallet, mattress, and sofa) are used to burn individually in a controlled mode. Three wood pallets are set in a roof shape with straws on the bottom pallet and placed at the center of the room. A single-size mattress is placed flat on two layers of wood pallets with straws. A sofa without cushions is placed on the floor. The fuel is ignited by a propane torch and extinguished with a water spray after a desired period (< 10 min). A HD video camera records the entire burning processes. Two LED lighting fixtures are used to illuminate the burning fuel. A vertical sensor support stand is placed ≈2.7 m away from the center of the room. A Gardon-type heat flux transducer (Medtherm 100 kW/m^2) and two copper gas sampling tube (3.86 mm i.d.) outlets (one for particulates and another for gas sensors) are facing toward the fire at 1.5 m above the floor. Six K-type thermocouples are place along the support stand at different heights (15 cm, 61 cm, 1.1 m, 1.5 m, 2.0 m, and 2.4 m). The real-time data are acquired throughout the ignition, burning, extinguishment, and post-fire periods using laptop computers.

Researchers at the NASA Glenn Research Center (GRC) have been collaborating with Case Western Reserve University to develop suitable sensor technologies for monitoring firefighter particle exposure. Initially being tested is a small light scattering device. This sensor was originally developed for advanced fire detection and environmental monitoring on the International Space Station and eventually other spacecraft. The Multi-Parameter Aerosol Scattering Spectrometer (MPASS)[45] is an extremely small robust package, and utilizes solid-state and integrated optical technologies similar to those developed for optical storage media such as Compact Discs. Several detection channels are employed, and can be optimized to characterize various moment quantities of interest, such as total aerosolized mass or surface area. In contrast to single-particle counters, the MPASS measures the scattering from an ensemble of particles. For this reason, the measurement accuracy depends on the composition of the particles (their refractive index), as well as the particle size distribution itself. Therefore, a device of this type can be better optimized for specific applications by better understanding the nature of the aerosols that are encountered. Part of the present effort involves more thoroughly characterizing the various particulate environments that are associated with firefighter exposure.

Efforts on chemical species gas sensors at the NASA GRC produced Smart Fire Detection Systems[46] able to measure a range of chemical species, including CO, NO_x, CO_2, O_2, and hydrocarbons. Multiple hardware platforms are available and can be designed for different functionalities. The presence of the microprocessor significantly enables the "smart" attributes of the hardware system and allows reprogramming of a range of functions. The Smart Fire Detection Systems include an array of miniaturized chemical microsensors. This microsensor technology is fabricated using Microelectromechanical Systems (MEMS) fabrication techniques to minimize size, weight, and power consumption. Three micro-fabricated sensor platforms are used: resistance-based sensors, Schottky diodes, and electrochemical cells. These basic platforms can be tailored for a range of measurements by proper selection of materials within the microstructure. A multiparameter measurement approach is used with sensors intended to have limited cross-sensitivity (orthogonality) allowing quantitative measurements. Such multiparameter measurements combined with quantitative readings significantly increases the overall awareness of the environment. A previous demonstration of this approach suggested a notably decreased false alarm rate in aircraft cargo bay fire detection. Future development is intended to widen the range of combustion species detected through designing selective sensors based on these microsensor platforms. Combined with particulate detection, this work aims to provide the firefighter with a new set of smart tools to understand the environment in a hazardous fire event.

Several commercial off-the-shelf (COTS) instruments provide reference values for the MPASS and micro-fabricated gas sensors. A tapered element oscillating microbalance (TEOM)[47] personal dust monitor (Thermo Scientific, Model PDM-3600) measures the aerosol mass concentration directly. The MPASS is placed directly in line with the TEOM, such that both instruments sample the aerosol identically. The TEOM dust monitor was used during both the laboratory and burn-room testing. A scanning mobility particle spectrometer (SMPS)[47], which consists of an Electrostatic Classifier (TSI, Model 3080) and an Ultrafine Condensation Particle Counter (UCPC, TSI ,Model 3776), directly measures the number concentration as a function of particle size (15 – 660 nm), wherein the corresponding surface area can be calculated. A CO analyzer (Vista Photonics, Model COMA 1) based on multipass tunable diode laser spectroscopy measures the CO concentration for a reference. The SMPS and COMA were used in the laboratory only. In addition to the reference instruments, a thermogravimetric analyzer (TGA, TA Instruments Model 2950) determines the thermal decomposition characteristics of the wood in air.

RESULTS AND DISCUSSION

Figure 1 shows the TGA results for the wood used for both the laboratory and burn-room testing. The wood specimen is pyrolyzed under air at a heating rate of 20 °C. There are two distinct peaks of the weight loss rates at 400 °C and 490 °C. These peaks fall into the following description[48],

> *"The degradation reaction of lignin is an exothermic reaction, with peaks occurring between 225 ℃ and 450 ℃; temperatures and amplitudes of these peaks depend on whether the samples were pyrolyzed under nitrogen or air. All wood components end their volatile emissions at around 450 ℃. "Above 450 ℃, the remaining wood residue is an activated char that undergoes further degradation by being oxidized to CO_2, CO, and H_2O until only ashes remain. This is referred to as afterglow."*

Thus, the former temperature peak may be due to the degradation reaction of lignin and the latter peak may be that of activated char being oxidized.

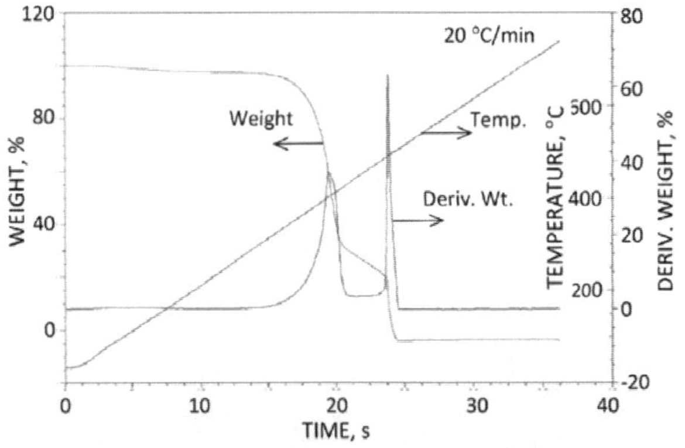

Figure 1 TGA analysis of the wood in air

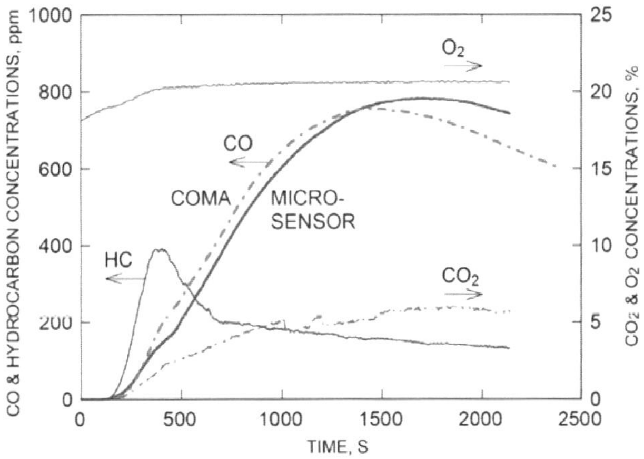

Figure 2 Concentrations of O_2, CO_2, CO, and hydrocarbons in the smoke chamber. Furnace temperature: 400 °C.

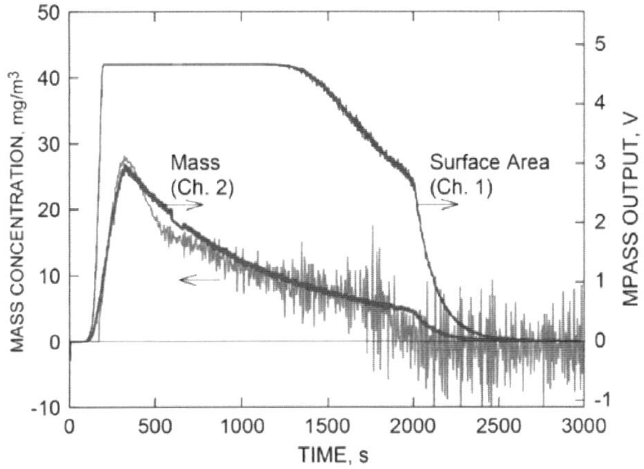

Figure 3 TEOM mass concentration of particulate and MPASS outputs in the smoke chamber. Furnace temperature: 400 °C.

Figure 2 shows the temporal variations in the measured gas concentrations of O_2, CO_2, CO, and hydrocarbons in the smoke chamber at the furnace temperature of 400 °C. The initial weight of the wood specimen before the test is 2.012 g. The air flow through the tube furnace was turned on and then the wood specimen was pushed into the midpoint of the heated cylindrical furnace (t = 0). White smoke was observed to come out from the exit of the furnace during an initial period. The oxygen concentration in the smoke chamber was initially lower than that in air as a result of the previous run and increased as the fresh air flowed in. The concentrations of CO_2, CO, and hydrocarbons increased rapidly at the beginning. White smoke was observed to exhaust from the tube exit of the furnace. The concentration of hydrocarbons peaked during the early stage (t = 400 s), when the O_2, CO_2 and CO concentrations showed changed their slopes slightly. The CO_2 and CO concentrations continued to increase steadily until their maxima (t > 400 s). The microsensor CO concentration reading generally followed the COMA analyzer value, but it was somewhat lower probably because of the calibration (for > 200 ppm), since the bottled dry air was used in this experiment, instead of the moist ambient air.

Figure 3 shows the TEOM mass concentrations of particulate and MPASS outputs in the smoke chamber at the furnace temperature of 400 °C. The channels 1 and 2 of the MPASS output provide surface area and mass, respectively. The surface area signal (Ch. 1, blue) saturated due to excessive particulate loading in the

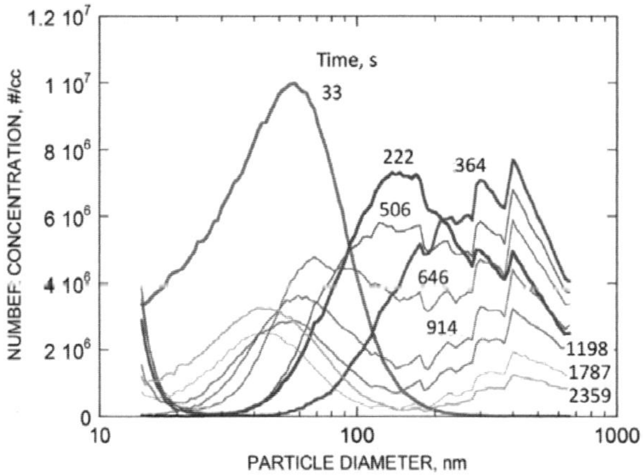

Figure 4 Particle size distributions during the laboratory experiment. Furnace temperature: 400 °C.

environment. The scale of the mass signal (Ch.2, green) was adjusted to match with the peak of the TEOM signal (red), which occurred during the early stage (t = 400 s) coincident with the peak of the hydrocarbon concentration (see Fig. 2). The TEOM signal peak was within the device's published measurement range (< 200 mg/m3). The MPASS signal variations for total aerosolized mass follows the TEOM mass concentration qualitatively. As the mass concentration decreased down to very low (near zero) values after the fire extinguishment, the TEOM signal fluctuated significantly with a sampling period of 5 s, probably because of the discrete nature of highly dispersed relatively large dust particles at later time, as described below. On the other hand, the MPASS exhibited clean signals even at the near-zero values, presumably because it measures light that is scattered from an ensemble of particles within a 1mm^3-order measuring volume (as opposed to a particle mass difference within the sampling time).

Figure 4 shows the particle size distributions at different elapse times in the smoke chamber at the furnace temperature of 400 °C. At the very beginning (t = 33 s) after the specimen was inserted into the tube furnace, there is a single peak (mode: 57.3 nm) with slightly skewed toward larger diameters from a logdnormal distribution. The particle size distribution at an early time for the furnace temperature of 350 °C (not shown) exhibited more symmetric peak closer to the lognormal. At later times (> 3 min. in Fig. 4), peaks shifted to a larger range (100 – 500 nm), which contributed to the peak of the mass concentration shown in Fig. 3.

The burn-room testing of the particulate and micro-fabricated gas sensors was conducted for three types of fuels: (1) wooden pallets (Fig. 5a), (2) a mattress (Fig. 5b), or (3) a couch without cushions. The results for

(a) (b)

Figure 5 Photographs of the fuel: (a) wood pallets and (b) mattress.

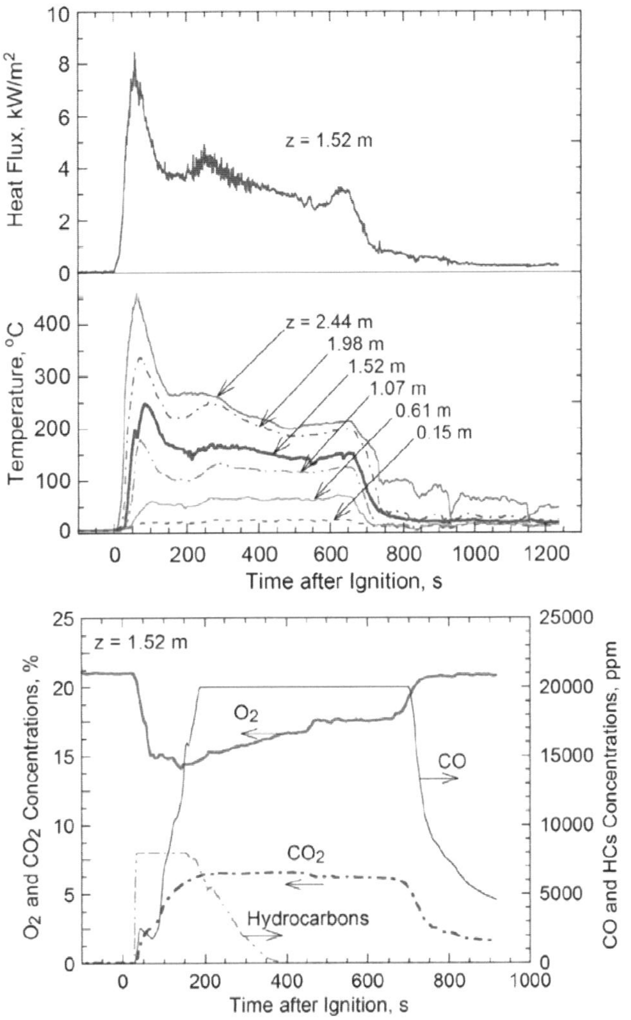

Figure 6 (a) Heat flux, temperatures, and (b) concentrations of O_2, CO_2, CO, and hydrocarbons in the burn room fire testing. Fuel: wood pallets.

the last item have been reported elsewhere[49]. Figure 6 shows the heat flux, temperatures, O_2, CO_2, CO, and hydrocarbons concentrations with burning wood pallets. The heat flux and temperature at ~2.7 m away from the edge of the pallet and z = 1.52 m above the floor reached peaks of~8 kW/m^2 and 250 °C, respectively, at ~1 min. after ignition and decreased rapidly as water was sprayed over the fire at ~11 min. after ignition. The O_2 concentration decreased to the minimum of ~14% in the initial flaring period (< 3 min.), during which the heat flux and temperature peaked, the hydrocarbon concentration signal increased rapidly, and saturated at 8000 ppm. The CO concentration level also increased rapidly and saturated at 2 % for the most of the testing time. The CO_2 concentration increased until the initial flaring period and maintained its high level (~6 %) until extinguishment. The TEOM results indicated (not shown) that the flow rate dropped from 2 l/min to ~0.45 l/min due to an excessive concentration (> 500 mg/m^3) of particulates clogged the filter and thus the measurement failed during the burning test. To avoid clogging, sample gases were diluted with the ambient air to reduce the concentration 1/10 for the successive runs.

Figure 7 shows the heat flux, temperatures, O_2, CO_2, CO, and hydrocarbons concentrations with a burning mattress (94 cm × 188 cm). The heat flux and temperature at ~0.8 m away from the edge of the pallet and z =

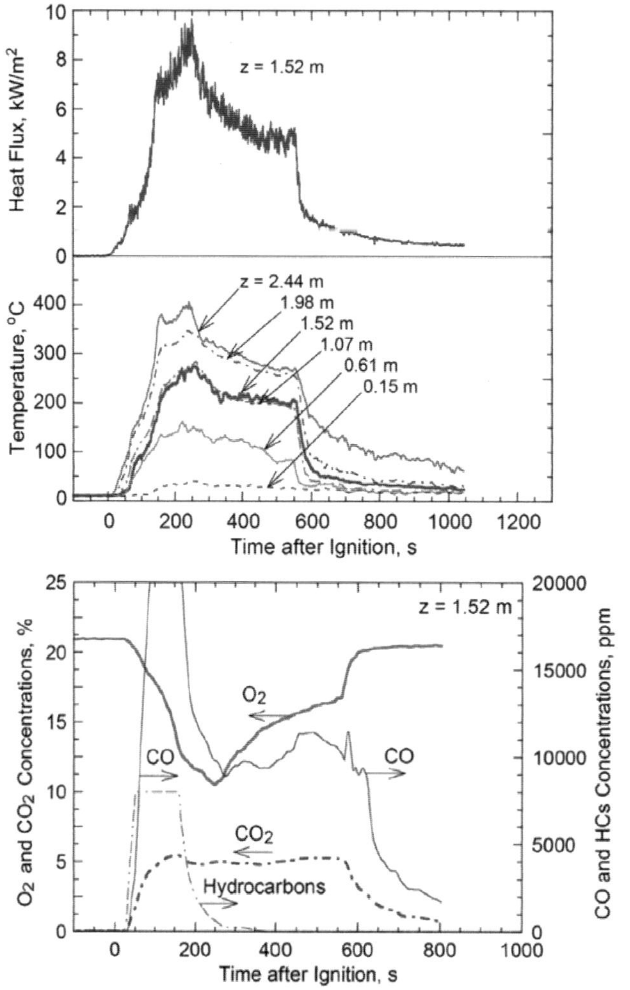

Figure 7 (a) Heat flux, temperatures, and (b) concentrations of O_2, CO_2, CO, and hydrocarbons in the burn room fire testing. Fuel: mattress.

1.52 m above the floor reached peaks of~9 kW/m^2 and 270 °C, respectively, at ~4 min. after ignition and decreased rapidly as water was sprayed over the fire at ~9 min. after ignition. The O_2 concentration decreased to the minimum of ~10% in the initial flaring period (< 4 min.), during which the heat flux and temperature peaked. The CO and hydrocarbon concentrations increased rapidly and saturated at 2 % and 8000 ppm, respectively for ~3 min. The CO_2 concentration increased during the initial flaring period and maintained its level at ~5 % until extinguishment.

Figure 8 shows the TEOM mass concentration of particulate and MPASS outputs in the burn room fire testing of a burning mattress. Both the surface area (Ch. 1, blue) and mass (Ch. 2, red) signals saturated due to excessive particulate loading, which exceeded 2.5× the device's published measurement range (< 200 mg/m3). The scale of the MPSS signal (green) was adjusted to match with the variations of the TEOM signal (red). The MPASS signal variations for total aerosolized mass follows the TEOM mass concentration qualitatively. The result indicates that, after a proper calibration, MPASS can be used as a mass concentration sensor to be incorporated into a hand-held real-time particulate and toxic gas detector for firefighters.

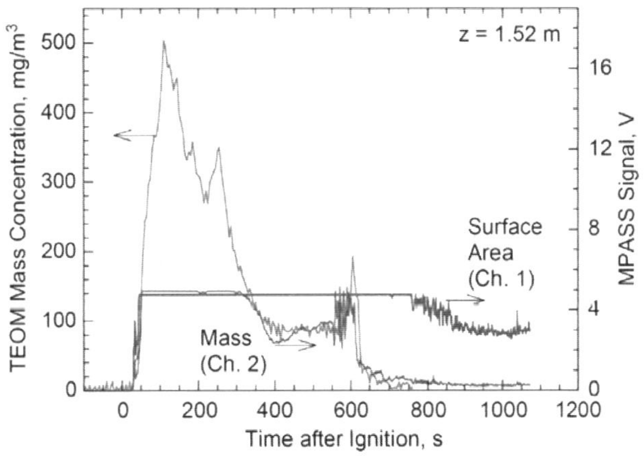

Figure 8 TEOM mass concentration of particulate and MPASS outputs in the burn room fire testing. Fuel: mattress.

CONCLUSIONS

Protecting firefighters from exposures to particulates and various chemicals is highly complex. To understand the conditions encountered in the firefighting environment and to develop a tool to accomplish it, simultaneous measurements of concentration levels of particulates and multiple toxic-gas components have been made in the laboratory and the burn room. The NASA-developed MPASS particulate sensor and microsensors for O_2, CO_2, CO, and hydrocarbons exhibited promising results in a comparative measurement with reference instruments. The MPASS has demonstrated superior performance in generating a clean output under low signal levels. In addition to particulates and these species, sensors for detecting low levels of irritants (acrolein and formaldehyde) are currently under development in this study.

ACKNOWLEDGMENTS

This work was supported by the U.S. Department of Homeland Security, Federal Emergency Management Agency, Assistance to Firefighters Grant Program, Fire Prevention and Safety grants (Nos. EMW-2012-FP-01284 and EMW-2014-FP-00688). Assistance in conducting the experiment, coordinated by Batt. Chief Sean DeCrane, at the Fire Training Academy of Cleveland Division of Fire is acknowledged.

REFERENCES

[1] Anon. (2011). Report of the 2nd National Fire Service Research Agenda Symposium, National Fire Academy, May 20 - 22, National Fallen Firefighters Foundation (NFFF), p. 50.

[2] Anon. (2013). National Occupational Research Agenda (NORA), National Public Safety Agenda for Occupational Safety and Health Research and Practice in the U.S. Public Safety Sector, 29 Oct. 2013 Revision.

[3] IAB (2012). Evaluation of Hazards in the Post-Fire Environment, https://iab.gov/Uploads/evaluation_of_hazards_in_the_post-fire_environment.pdf.

[4] Weiss, D.C., Miller, J.T. (2011). A study on chemicals found in the overhaul phase of structure fires using advanced portable air monitoring available for chemical speciation, State of Oregon Governer's Fire Service Policy Council and Tualatin Valley Fire & Rescue, February 25.

[5] Bolstad-Johnson, D., Burgess, J., Crutchfield, C., Storment, S., Gerkin, R. and Wilson, J. (2000). Characterization of Firefighter Exposures during Fire Overhaul. *AIHAJ*, 61(5), pp.636-641.

[6] Fabian, T., Borgerson, J.L., Kerber, S.I., Gandhi, P.D., Baxter, C.S., Ross, C.S., Lockey J.E. and Dalton, J.M. (2010). Firefighter Exposure to Smoke Particulates. Final Report, Underwriters Laboratories, Inc.

[7] Pryor, A.J., Johnson, D.E., and Jackson, N.N. (1975). Hazards of Smoke and Toxic Gases Produced in Urban Fires. *JFF/Combustion Toxicology* 2(2), pp.64-112.

[8] Gold, A., Burgess, W. and Clougherty, E. (1978). Exposure of firefighters to toxic air contaminants. *American Industrial Hygiene Association Journal*, 39(7), pp.534-539.

[9] Treitman, R.D., Burgess, W.A. and Gold, A. (1980). Air contaminants encountered by firefighters. *American Industrial Hygiene Association Journal*, 41(11), pp.796-802.

[10] Lowry, W.T., Juarez, L., Petty, C.S. and Roberts, B. (1985). Studies of Toxic Gas Production during Actual Structural Fires in the Dallas Area. *Journal of Forensic Sciences*, 30(1), p.10965J.

[11] Brandt-Rauf, P.W., Fallon, L., Tarantini, T., Idema, C. and Andrews, L. (1988). Health hazards of fire fighters: exposure assessment. *Occupational and Environmental Medicine*, 45(9), pp.606-612.

[12] Jankovic, J., Jones, W., Burkhart, J. and Noonan, G. (1991). Environmental Study of Firefighters. *Annals of Occupational Hygiene*, 35(6), pp.581-602.

[13] Burgess, J.L., Nanson, C.J., Bolstad-Johnson, D.M., Gerkin, R., Hysong, T.A., Lantz, R.C., Sherrill, D.L., Crutchfield, C.D., Quan, S.F., Bernard, A.M. and Witten, M.L. (2001). Adverse Respiratory Effects Following Overhaul in Firefighters. *Journal of Occupational and Environmental Medicine*, 43(5), pp.467-473.

[14] Austin, C.C., Wang, D., Ecobichon, D.J. and Dussault, G. (2001). Characterization of Volatile Organic Compounds in Smoke at Experimental Fires. *Journal of Toxicology and Environmental Health, Part A*, 63(3), pp.191-206.

[15] Fabian, T.Z., Borgerson, J.L., Gandhi, P.D., Baxter, C.S., Ross, C.S., Lockey, J.E. and Dalton, J.M. (2010). Characterization of Firefighter Smoke Exposure. *Fire Technology* 50(4) pp.993-1019.

[16] Baxter, C.S., Ross, C.S., Fabian, T., Borgerson, J.L., Shawon, J., Gandhi, P.D., Dalton, J.M., Lockey, J.E. (2010). Ultrafine Particle Exposure During Fire Suppression—Is It an Important Contributory Factor for Coronary Heart Disease in Firefighters?. *Journal of Occupational and Environmental Medicine*, 52(8), pp.791-796.

[17] Crewe, R.J., Stec, A.A., Walker, R.G., Shaw, J.E.A., Hull, T.R., Rhodes, J., and Garcia-Sorribes, T. (2013). Experimental Results of a Residential House Fire Test on Tenability: Temperature, Smoke, and Gas Analyses. *Journal of Forensic Sciences*, 59(1), pp.139-154.

[18] Fent, K.W., Eisenberg, J., Snawder, J., Sammons, D., Pleil, J.D., Stiegel, M.A., Mueller, C., Horn, G.P., and Dalton, J. (2014). Systemic Exposure to PAHs and Benzene in Firefighters Suppressing Controlled Structure Fires. *Annals of Occupational Hygiene*, 58(7), pp.830-845.

[19] Fent, K.W., Evans, D.E., Booher, D., Pleil, J.D., Stiegel, M.A., Horn, G.P., and Dalton, J. (2015). Volatile Organic Compounds Off-gassing from Firefighters' Personal Protective Equipment Ensembles after Use. *Journal of Occupational and Environmental Hygiene*, 12(6), pp.404-414.

[20] Baxter, C.S., Hoffman, J.D., Knipp, M.J., Reponen, T., Haynes, E.N. (2014). Exposure of Firefighters to Particulates and Polycyclic Aromatic Hydrocarbons. *Journal of Occupational and Environmental Hygiene*, 11(7), pp.D85-D91.

[21] Fent, K.W., Evans, D.E. (2011). Assessing the risk to firefighters from chemical vapors and gases during vehicle fire suppression. *Journal of Environmental Monitoring*, 13(3), p.536.

[22] Evans, D.E., and Fent, K.W. (2015). Ultrafine and respirable particle exposure during vehicle fire suppression. *Environ. Sci.: Processes Impacts*, 17(10), pp.1749-1759.

[23] Pane, G.A. (1988). Toxic smoke inhalation and aircraft fires. *The American Journal of Emergency Medicine*, 6(1), p.82.

[24] Chaturvedi, A.K., and Sanders, D.C. (1996). Aircraft fires, smoke toxicity, and survival. *Aviation Space Environmental Medicine* 67(3) pp.275-278.

[25] Gandhi, S., Lyon, R. and Speitel, L. (1999). Potential Health Hazards from Burning Aircraft Composites. *Journal of Fire Sciences*, 17(1), pp.20-41.

[26] Anon. (2013). Smoke, Fire and Fumes in Transport Aircraft. 2nd Ed., Royal Aeronautical Society, London, U.K.
[27] Reinhardt, T.E. (1991). Monitoring Firefighter Exposure to Air Toxins at Prescribed Burns of Forest and Range Biomass. Research Paper PNW-RP-441, USDA Forest Service, October.
[28] Reinhardt, T. E., and Ottmar, R. D. (1997). Smoke Exposure among Wildland Firefighters: A Review and Discussion of Current Literature. General Technical Report PNW- GTR-373, USDA Forest Service, February.
[29] Sharkey, B. (1997). Health Hazards of Smoke. USDA Forest Service, November.
[30] Reinhardt, T. E., Ottmar, R. D., and Hallet, M. J. (1999). Guide to Monitoring Smoke Exposure of Wildland Firefighters. General Technical Report PNW- GTR-448, USDA Forest Service, March.
[31] Reinhardt, T. E., and Ottmar, R. D. (2000). Smoke Exposure at Western Wildfires. PNW-RP-525, USDA Forest Service, July.
[32] Reinhardt, T.E., Ottmar, R.D. (2004). Baseline Measurements of Smoke Exposure at Among Wildland Firefighters. *Journal of Occupational Environmental Hygiene*, 1, pp.593-606.
[33] Austin, C. (2008). Wildland Firefighter Health Risks and Respiratory Protection. IRSST Report R-572.
[34] Hvenegaard, S. (2012). Health impact of smoke exposure in wildland-urban interface fires: a literature review. FPInnovations, Wildland Operations Research, March.
[35] Reinhardt, T. E., and Ottmar, R. D. (2000). Smoke Exposure among Firefighters at Prescribed Burns in the Pacific Northwest. PNW-RP-526, USDA Forest Service, October.
[36] Fent, K.W, Musolin, K., and Methner, M. (2013). Evaluation of Chemical Exposures during Fire Fighter Training Exercises Involving Smoke Simulant. National Institute for Occupational Safety and Health, HETA 2012-0028-3190, July.
[37] Kirk, K.M. and Logan, M.B. (2015). Firefighting Instructors' Exposures to Polycyclic Aromatic Hydrocarbons during Live Fire Training Scenarios. *Journal of Occupational and Environmental Hygiene*, 12(4), pp.227-234.
[38] Austin, C.C., Wang, D., Ecobichon, D.J., and Dussault, G. (2001). Characterization of Volatile Organic Compounds in Smoke at Experimental Fires. *Journal of Toxicology and Environmental Health, Part A* 63(3), pp.191-206.
[39] Anon. (2014). Documentation for Immediately Dangerous To Life or Health Concentrations (IDLHs). National Institute for Occupational Safety and Health (NIOSH). http://www.cdc.gov/niosh/idlh/default.html.
[40] Anon. (2006). TLVs and BEIs Based on the Documentation of the Threshold Limit Values for Chemical Substances and Physical Agents & Biological exposure Indices. American Conference of Governmental Industrial Hygienists (ACGIH).
[41] Kales, S.N., Soteriades, E.S., Christophi, C.A., Christiani, D.C. (2005). Emergency Duties and Deaths from Heart Disease among Firefighters in the United States. *New England Journal of Medicine* 356, pp.1207-1215.
[42] Geibe, J.R., Holder, J., Peeples, L., Kinney, A.M., Burress, J.W., Kales, S.N. (2008). Predictors of on-duty coronary events in male firefighters in the United States. *American Journal of Cardiology* 101, p.585.
[43] .Leonard, S.S., Wang, S., Shi, X., Jordan, B.S. Castranova, V., Dubick, M.A. (2000). Wood smoke particles generate free radical and cause lipid peroxidation, DNA damage, NFκB activation and TNF-α release in macrophages. *Toxicology* 150, pp.147-157.
[44] Kulis, M.J., Meyer, M.E., Mudget, P.D., Berger, G.M., and Pilgrim, J.S. (2014). Heating of Printed Circuit Board, Wire Insulation and Electronic Components for Fire Signature Sensor Evaluation. 44th International Conference on Environmental Systems, Tucson, Arizona, July 13-17.
[45] Greenberg, P.S., Fischer, D.G. (2010). Advanced Particulate Sensors for Early Warning Fire Detection. 40th International Conference on Environmental Systems, Barcelona, Spain, July 11-15, AIAA 757995.

[46] Hunter, G.W., Xu, J.C., Biaggi-Labiosa, A.M., Ward, B., Dutta, P., and Liu, C.C. (2011). Smart Sensor Systems for Spacecraft Fire Detection and Air Quality Monitoring. 41st Int. Conference on Environmental Systems, July, AIAA 766637.

[47] Kulkarni, P, Baron, P.A., Willeke, K. (Eds.) (2011). Aerosol Measurement: Principles, Techniques, and Applications 3rd Edition, John Wiley & Sons, Hoboken, New Jersey.

[48] White, R.H. and Dietenberger, M.A. (April 2010). Fire Safety of Wood Construction, in Wood Handbook—Wood as an engineering material. General Technical Report FPL-GTR-190. U.S. Department of Agriculture, Forest Service, p.18-1.

[49] Takahashi, F., Greenberg, P.S., Hunter, G.W. (2014). Firefighter Particulate and Gas Sensor Technology, *Fire Safety (FS)-World*, Fall edition, 2014.

FIRE AND THE ENVIRONMENT

OFFICE EQUIPMENT FIRE SAFETY: THE IMPLICATIONS OF ENVIRONMENTALLY PREFERABLE PURCHASING PROGRAMS ON FLAME RETARDANT USAGE AND FIRE SAFETY

Tim Earl, GBH International, USA

ABSTRACT

In the United States, computers and other office equipment cause or contribute to over 1,300 fires each year, resulting in approximately $70 million in damages annually, in addition to lost business revenue. Flame retardants have been an integral part of these products for many years, most prominently in plastic housings and printed circuit boards.

As the amount of electronic office equipment sold in the US continues to increase, for both commercial and residential use, the manufacture and disposal of this equipment has come under scrutiny with regard to health and environmental concerns. New standards have been developed to encourage the use of more environmentally friendly materials in these products. The most prominent examples of these standards are the IEEE 1680 Family of Standards for Environmental Assessment of Electronic Products.

The Green Electronics Council manages EPEAT - Electronic Product Environmental Assessment Tool. Companies may register individual products with EPEAT, which has three performance levels: Bronze, Silver, and Gold.

"Environmentally preferable purchasing programs" affect the marketplace as large purchasers, such as the US federal government, demand that suppliers provide EPEAT-registered products.

Many of the EPEAT standards explicitly prohibit or restrict materials which are key ingredients in some of the most effective flame retardants for these applications. This situation has created a challenge for researchers to develop new flame retardants to maintain fire safety while also complying with material restrictions. This is exacerbated by some attempts to prohibit new flame retardants before they even reach the market, without studies demonstrating harm to human health. Flame retardant manufacturers are thus discouraged from spending time and money researching alternatives for fear of never recouping those costs with a marketable product. The end result is a potential weakening of fire safety requirements to comply with restrictions on the use of certain substances.

This paper will detail the process by which these voluntary standards have been developed and grown into de facto requirements through widespread adoption. It will then discuss the effect this has on the use of existing flame retardants, the search for alternatives and the implications for fire safety.

OFFICE EQUIPMENT/ELECTRONICS AND FIRE SAFETY

Consumer electronics and office equipment present an often overlooked fire hazard in the typical home or office. According to a 2013 NFPA report[1], there were an estimated 730 reported home fires and 630 reported non-home fires involving office equipment each year from 2007-2011, resulting in $70 million in damages and 11 deaths annually.

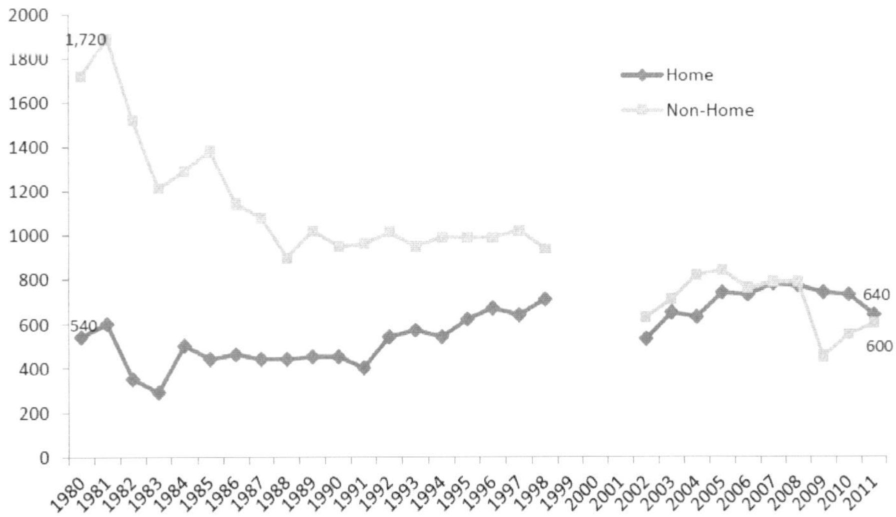

Figure 1: Home and Non-Home Fires Involving Office Equipment, by Year
Structure Fires Reported to U.S. Fire Departments
(Gap from 1999-2001 due to uncertainty caused by low participation)
Source: *Home and Non-Home Fires Involving Office Equipment*, NFPA, 2013[1]

Computers are the most common pieces of office equipment involved in fires, representing 48% of home office equipment fires and 36% of office fires involving office equipment. When computer-related equipment is added, the numbers become 79% and 67%, respectively of office fires involving office equipment. Although details of ignition causes are often not available, wire or cable insulation and housings or casings are most frequently identified as the first items ignited.

Televisions and Computer Monitors

Televisions and computer monitors are often lumped together in the same category, although there can be notable differences in the materials used. In a study conducted by Dr. Juergen Troitzsch[2], the author noted that TV backplates were usually made of polystyrene, whereas PC monitors were often made of ABS, PC/ABS, or PVC. However, over time the demarcation line between TV set and monitor has been blurred to such an extent that they can reasonably be placed in the same category. Furthermore, flat panel TV sets have become a common fixture in modern offices, particularly as presentation displays in conference rooms. As such, TVs should be considered office equipment.

UL 94[3] is a simple Bunsen burner test used to assess the flammability of plastic materials used in appliances and electronics. A small specimen is exposed to the burner flame for 10 seconds. Classification is based on the afterglow time and whether any particles or drops ignite a piece of cotton underneath the specimen. Based on those results, the material is assigned a rating of V-0, V-1, or V-2, with V-0 having the highest performance. Materials unable to pass the vertical test may be tested in a horizontal test and receive an HB rating, which indicates much worse fire performance. There is an additional classification for materials which fail even the HB test, called B (for burn).

TV enclosures in the US are required to have a UL94 V-0 rating, the highest level of performance. In Europe, the requirements are much less restrictive and the vast majority of TV enclosures are only HB rated.

Fig 2: UL94 Test

The Troitzsch study included small scale and full scale tests for several different TV sets sold in the US, Europe, and Japan (which has similar requirements to the US). The results showed that a better UL 94 classification resulted in much better full scale results, with UL 94 HB-rated sets exposed to a small ignition source leading to fully developed room fires in a very short period of time.

A 2000 study by Safety Engineering Laboratories[4] in the US came to the same conclusions as the Troitzsch paper and also included oxygen consumption calorimetry data showing a 33% reduction in heat release rate for UL 94 V-0 sets when compared to UL 94 HB-rated sets.

A 2001 study by the UK Consumer Affairs Directorate[5] concluded that TV fires were 30% less likely in the US than in the UK, when normalized to the number of TV sets in each nation.

Lithium Ion Battery Fires

Lithium ion batteries are ubiquitous in modern electronics. A typical business professional is likely to carry a laptop computer, tablet, and smartphone, all of which are powered by lithium ion batteries. The fire hazard of these batteries is well known, first gaining widespread coverage in 2006 when Apple[6], Dell[7] and Sony[8] recalled over 4,000,000 laptop batteries after several well-publicized fires, including one at Los Angeles International Airport.[9]

More recently, Samsung recalled approximately 1 million Galaxy Note 7 smartphones after several reported fires.[10] Following this recall, The Federal Aviation Administration (FAA) issued an unprecedented total ban on carrying these phones on aircraft in the United States.

The problem is even more serious with lithium polymer (LiPo) batteries, which are essentially lithium ion batteries in a soft package instead of a rigid body. This makes them much lighter, which is why they are used in hobby drones and self-balancing scooters (so-called hoverboards). However, the soft packaging provides no impact protection and makes them more susceptible to damage which can lead to fires, such as a January 2016 fire in Nashville, Tennessee, which destroyed a large single family home.[11] As a result of such fires, Underwriters Laboratories (UL) published a hoverboard standard, UL 2272.[12] All self-balancing scooters in the US must now comply with that standard.

While LiPo batteries are not generally used in office equipment, some tablet and smartphone manufacturers are starting to incorporate them as they design thinner devices.

ELECTRONICS AND THE ENVIRONMENT

The prevalence of consumer electronics in our lives and their seemingly increasing rate of obsolescence has led some groups to closely scrutinize these devices, focusing on their end of life management. Specific issues identified by one such group, the Electronics Takeback Coalition, include:

- Limited recyclability, resulting in most e-waste going to landfills
- The presence of toxic (presumed or proven) materials in consumer electronics
- Exporting materials to countries with less restrictive laws regarding disposal[13]

The second point will be discussed in detail, as it is a source of conflict with fire safety regulations.

Products of Combustion and Toxicity

Opponents of flame retardants frequently claim that the smoke produced when flame-retarded materials burn is more toxic than smoke from non-FR materials. A 2000 paper by Troitzsch[14] of the Fire and Environment Protection Service in Germany looked into this, specifically investigating the impact of pollutants and the role of brominated flame retardants in two major fires in Germany. The data showed that decomposition products from flame retardants did not play a role in the acute toxicity of the products of combustion. It found that the toxicity of PAHs (polynuclear aromatic hydrocarbons) was orders of magnitude higher than that of flame retardants or their products.

Disposal of electronics is addressed by most environmental standards, so incineration is an item of concern. This was addressed back in 1987 in a report by the New York State Energy Research and Development Authority.[15] The report concluded that the release of dioxins from burning waste was most heavily influenced by temperature, and modern incinerators operate at temperatures high enough to virtually eliminate dioxin release. A German study[16] also found that increased chlorine and bromine in the waste stream did not lead to increased emissions of dioxins and furans.

The Impact of Fire on the Environment

One counterpoint to environmental concerns with flame retardants is the negative impact of fires on the environment. This topic was discussed in a Fire-LCA (Life Cycle Assessment) model for televisions published by SP, Swedish National Testing and Research Institute.[17] The study concluded that "in the case of a number of key emission species there is a markedly higher total emission over the whole life cycle from the NFR (not fire retarded) TV than from the FR (fire retarded) TV in the present day scenario" and this is "ample evidence that the avoidance of large numbers of fires is important from an environmental point of view quite apart from the safety considerations discussed above."

The study also found that PAHs, and dibenzodioxins, and furans were produced at high levels during fires, making them a significant source of total LCA emissions. PAHs in particular were singled out as the greatest potential threat to health and the environment. Thus, the contention by some organizations that flame retardants are more hazardous than the fires they prevent is not supported by the facts, at least in this case study.

EPEAT (Electronic Product Environmental Assessment Tool) - Background

In 2005, the EPA solicited applications for a $300,000 grant to implement the EPEAT system. The grant's stated goal was "to provide funding for the completion and pilot testing of a system to assist institutional purchasers in identifying and selecting environmentally preferable electronic products."[18] The grant was awarded to the Green Electronics Council (GEC).

GEC partnered with The Institute of Electrical and Electronics Engineers (IEEE) to produce the first three EPEAT standards:

- IEEE 1680.1 for PCs and Displays[19]
- IEEE 1680.2 for Imaging Equipment[20]
- IEEE 1680.3 for Televisions[21]

IEEE 1680.4 for Servers is currently under development and will be discussed later in this paper.

These are collectively known as the IEEE 1680 family of standards. They contain required and optional criteria in several areas including:

- Energy conservation
- Reduction/elimination of environmentally sensitive materials
- Materials selection
- Responsible end-of-life management
- Corporate performance
- Packaging

Meeting all of the required criteria makes a product eligible to be listed as EPEAT Bronze. Silver and Gold listings require earning a certain number of optional points.

Manufacturers must undertake an onerous process to comply with these standards. First, they must join EPEAT and pay an annual EPEAT Participation Fee to the GEC. The amount of the fee depends on the company's sales of the applicable products. For example, as of November 2016, a company which sold $5M - $10M of computer products (slates, notebooks, desktops, integrated systems, work stations, thin clients, and displays) in 2015 would pay a fee of $3,800, which a company with $1B - $2.5B of such sales would pay $38,200. (The maximum fee is $152,000 for a company selling over $50B).[22]

But that fee only allows a company to participate. In order to list a product on the EPEAT registry, manufacturers must work with an EPEAT Conformity Assurance Body (CAB) to declare that the product is in conformance with the appropriate EPEAT standard.

While listing a product on the EPEAT registry is optional, large purchasers of office equipment are heavily encouraged by many environmental interest groups to require their suppliers to be EPEAT registered. The US federal government is the largest organization to make such a demand.

In January 2007, US President Barack Obama issued Executive Order 13423 – Strengthening Federal Environmental, Energy, and Transportation Management.[23] This Executive Order contained several directives for federal agencies intended to increase sustainability, including a requirement that 95% of electronic products be EPEAT-registered where applicable.

Several US states and foreign governments, hospitals, universities, and large corporations also require EPEAT registry for electronic office equipment. Thus, these new optional standards have become de facto requirements for manufacturers, as they would give up a large share of the market if they did not comply.

EPEAT – Requirements

These standards come into potential conflict with fire safety through limits or prohibitions on certain substances, most notably bromine and chlorine.

IEEE 1680.1 (PCs): This standard has two different restrictions on flame retardants and one on PVC:

- (Required) Paints, coatings, plastics, rubbers, and seals shall be free from flame retardants containing SCCPs (not more than 0.1% by weight), 10 carbon atoms to 13 carbon atoms, minimum 48% chlorine by weight, unless it can be shown that the SCCPs are present above this threshold due to the use of recycled content.
- (Optional) Plastic parts >25 g shall be free from flame retardants (not more than 0.1% of total weight) that are classified as dangerous substances under European Council Directive 67/548/EEC. Many flame retardants have not been tested to this directive, so this creates an additional testing burden. It is also a very open ended directive which is not based on evidence of harm. (For example, most of the criteria say the substance "may cause" some sort of harm with no reason given for this classification).
- (Optional) Plastic parts >25 g shall not have PVC. While this is not as direct as a restriction on flame retardants, PVC is often used because of its superior fire performance compared to other thermoplastic materials.

IEEE 1680.2 (Imaging Equipment): External plastic casings greater than 25g shall not contain more than 1000 ppm bromine or chlorine "attributable to brominated flame retardants, chlorinated flame retardants, and PVC." Bromine or chlorine in excess of those amounts that is not attributable to FRs or PVC is allowed but must be documented. The allowable thresholds increase to 3,000 ppm for parts containing 25% or more postconsumer recycled content.

IEEE 1680.3 (TVs): Bromine and chlorine restrictions in this standard match those of IEEE 1680.2

IEEE 1680.4 (Servers) (Proposed): This standard is still under development. There has been some compromise between manufacturers who rely on these substances and NGOs seeking to reduce or eliminate their use. As such, a direct prohibition on bromine and chlorine is not likely to appear in this standard.

UL 110 (Standard for Sustainability for Mobile Phones)

In 2012, Underwriters Laboratories published UL 110, Standard for Sustainability for Mobile Phones.[24] The UL standards development process differs significantly from IEEE and some other Standards Development Organizations (SDOs) but still meets the requirements of American National Standards Institute (ANSI) for a consensus process. UL 110 offers two levels of achievement, Certified and Platinum.

UL 110 has no mandatory restriction on bromine or chlorine. There are, however, two optional sections regarding these substances:

- Restriction of Bromine and Chlorine in Mobile Phone Printed Circuit Board – points are awarded if the combined total of bromine and chlorine in the printed circuit board resin plus reinforcement matrix is less than 1,500 ppm and neither bromine nor chlorine exceeds 900 ppm.
- Restriction of Bromine and Chlorine in All Materials of Mobile Phone – Optional points are awarded if the product as a whole meets the requirements above.

Note that in all of these standards, the restrictions are on any substance which contains bromine and chlorine, with no consideration for the specific chemistry of the compound. While this approach has been heavily criticized for not having a sound scientific basis, with each new standard the previous standards are cited as justification for this approach.

Typical Bromine Loading in Electronics

Flame retardants are used in electronics to meet the UL 94 requirements where applicable. Most of these flame retardants are bromine-based. Typical bromine loading for these materials follow:
ABS Plastic (May be used in housings and fans)

- For a UL 94 V-0 rating (most common), bromine content is typically 8-13%.
- For a UL 94 V-2 rating (less common, burns more), bromine content is typically 4-7%.

HIPS Plastic (May be used in housings)

- For a UL 94 V-0 rating (most common), bromine content is typically 8-13%.
- For a UL 94 V-2 rating (less common, burns more), bromine content is typically 3-7%.

Printed Wiring Boards / Printed Circuit Boards

- PWBs are required to meet UL 94 V-0.
- Bromine content is typically 15-18%.

PBT Plastic (May be used in connectors)

- For a UL 94 V-0 rating (most common), bromine content is typically 7-13%.

Polyamides (May be used for connectors and internal parts)

- For a UL 94 V-0 rating (most common), bromine content is typically 10-15%.

These bromine loading levels are much higher than the levels permitted by any of the current EPEAT standards, and even higher than the elevated levels that may be allowed (in 1680.4) for materials with higher post-consumer recycled content. As a result, registering products on EPEAT means discontinuing the use of brominated flame retardants in these applications.

THE FUTURE FOR ELECTRONIC SUSTAINABILITY STANDARDS AND FLAME RETARDANTS

There are indications that some of the more stringent requirements established by EPEAT may be eliminated or relaxed in the near future.

Since the federal government purchasing mandate was established by Executive Order and not an act of Congress, it is subject to repeal at any time. While this often occurs after a change of administration, the sitting president may choose to do so as well, which is what happened in this case.

In March of 2015, President Obama signed Executive Order 13693, Planning for Federal Sustainability in the Next Decade,[25] which revoked Executive Order 13423 and established new environmental policies.

The reference to EPEAT was removed and replaced with a more open ended purchasing mandate. Beginning October 1, 2016 federal agencies are directed to purchase environmentally preferable products or services that:

a) meet or exceed specifications, standards, or labels recommended by EPA that have been determined to assist agencies in meeting their needs and further advance sustainable procurement goals of this order; or
b) meet environmental performance criteria developed or adopted by voluntary consensus standards bodies consistent with section 12(d) of the National Technology Transfer and Advancement Act of 1995 (Public Law 104–113) and OMB Circular A–119;

This second option opened up the possibility of competing standards from other SDOs. This has caused some concern for NGOs who heavily invested in the EPEAT process and do not want to see standards developed which they view as less stringent.

An article by Barbara Kyle of the Electronics Takeback Coalition[26], written after Executive Order 13693 was signed, illustrates the challenge in dealing with organizations which do not fully understand the consensus process for development and amendment of codes and standards. The article criticizes the American National Standards Institute (ANSI) essential requirements as not actually reflecting consensus. In fact, the ANSI process is well-respected, and US representation on international committees (such as IEC and ISO) is through ANSI.

The article further explains how EPEAT stakeholders left IEEE for the server standard and went through NSF International to develop the server standards. According to Ms. Kyle, the EPEAT stakeholders did not consider the IEEE (ANSI-accredited) process to be fair, and they thought NSF offered a more level playing field. In contrast, when the NSF committee for a server standard was established, some industry groups objected to the NSF system of classifying participants, stating that it gave more votes to NGOs and thus was not balanced. Fearing that they would be overwhelmed on what they considered an unequal playing field, many manufacturers chose not to participate in the NSF committee.

Meanwhile, IEEE continued with development of IEEE 1680.4 for servers (with higher participation than the NSF committee) until the US Department of Energy initiated a mediation process to bring the two development efforts together. At the time of writing, a joint IEEE/NSF committee continues to work on the EPEAT server standard. The joint agreement stipulates that, if the process takes too long, one organization can pull out and publish their own standard. As such, this issue is not yet resolved.

The disagreement over what constitutes a consensus process and a balanced committee is likely to continue with every new standards activity.

ASTM E60 Practice for The Development of Multi-Attribute Sustainability Standards For Building Products

Recognizing the number of new sustainability standards under development and the demand for more, ASTM Committee E60 on Sustainability began work on a "standard for standards" to provide guidance on the process. This draft practice lists requirements for the SDO and the process itself. The intent is to provide a framework for organizations to follow so that the standard they develop will be accepted.

If successful, the practice could help to alleviate concerns of "green washing" (disinformation disseminated by an organization so as to present an environmentally responsible public image) by some and committee imbalance by others. This practice is still early in the draft stage and it addresses building products only. However, it could serve as the foundation for similar standards covering other areas, including electronics. Such standards could help resolve the problem of organizations arguing over the validity of a standard due to balance and process concerns.

CONCLUSION

Halogenated materials continue to face a high level of scrutiny from environmental and health advocates. Whether out of scientific ignorance or a simple desire for expediency, recent regulatory efforts target all compounds with any bromine or chlorine in them, regardless of the specific chemistry. Thus, even flame retardants which have been proven to be safe (such as polymeric materials that cannot leach out) are under attack.

So-called environmental leadership standards have been developed with the stated purpose of allowing manufacturers to demonstrate going above and beyond regulatory requirements for sustainability (a term which itself seems to have been expanded to include human health). Some of the standards or proposed standards even state that they expect only a small percentage of manufacturers to be able to comply. However, once any of the standards is published, its proponents then lobby large purchasers, including governments at all levels, to require compliance with these standards from their suppliers. Thus the voluntary standards become requirements for any company wishing to maintain their market share.

In the office and consumer electronics market, EPEAT standards have restricted bromine and chlorine at the elemental level, eliminating some of the most effective flame retardants from that market. Recently, there has been some movement away from this prohibition as the US federal government no longer requires agencies to purchase EPEAT-registered products, and newer standards are less restrictive.

It is important for flame retardant manufacturers to be aware of this regulatory activity and be involved in the process.

REFERENCES

1. Home and Non-Home Fires Involving Office Equipment, National Fire Protection Association, Quincy, MA, September 2013.
2. Fire Safety of TV-Sets and PC-Monitors, Dr. Juergen Troitzsch, European Flame Retardants Industry Panel, European Flame Retardants Association
3. UL 94, Tests for Flammability of Plastic Materials for Parts in Devices and Appliances, Underwriters Laboratories, Northbrook, IL.
4. Television and Electronic Device Burn Demonstrations, Safety Engineering Laboratories, Inc., June 2000.
5. Causes of Fires Involving Television Sets in Dwellings, Consumer Affairs Directorate, Department of Trade and Industry, London, April 2001.
6. https://www.cpsc.gov/Recalls/2006/Apple-Announces-Recall-of-Batteries-Used-in-Previous-iBook-and-PowerBook-Computers-Due-To-Fire-Hazard/
7. https://www.cpsc.gov/Recalls/2006/Dell-Announces-Recall-of-Notebook-Computer-Batteries-Due-To-Fire-Hazard/
8. https://www.cpsc.gov/Recalls/2007/Sony-Recalls-Notebook-Computer-Batteries-Due-to-Previous-Fires/
9. https://www.cnet.com/news/flaming-lax-laptop-was-a-thinkpad/
10. https://www.cpsc.gov/Recalls/2016/samsung-recalls-galaxy-note7-smartphones
11. https://www.washingtonpost.com/news/innovations/wp/2016/02/19/hoverboard-fires-are-no-joke-two-kids-almost-died-as-their-house-burned-down/
12. UL 2272, "Outline of Investigation for Electrical Systems for Self-Balancing Scooters", Underwriters Laboratories, Northbrook, IL.
13. http://www.electronicstakeback.com/resources/problem-overview/
14. Fire Gas Toxicity and Pollutants in Fires, The Role of Flame Retardants, J. Troitzsch, 2000.
15. Results of the Combustion and Emissions Research Project at the Vicon Incinerator Facility in Pittsfield, Massachusetts: Final Report, New York State Energy Research and Development Authority, 1987
16. Effects of Combusting Plastics in WTE Plants on Dioxin Formatin and Ash Quality, J. Vehlow, H. Hunsinger, F.E. Mark, 2003.
17. Fire-LCA study of TV sets with V0 and HB enclosure material, M. Simonson, C. Tulin, H. Stripple, SP, Sweden, 2002.
18. http://www.federalgrants.com/Electronic-Product-Environmental-Assessment-Tool-EPEAT-3422.html
19. IEEE 1680.1 - IEEE Standard for Environmental Assessment of Personal Computer Products, Including Notebook Personal Computers, Desktop Personal Computers, and Personal Computer Displays, IEEE, New York, NY, 2009.
20. IEEE 1680.2 - IEEE Standard for Environmental Assessment of Televisions, IEEE, New York, NY, 2012.
21. IEEE 1680.2 - IEEE Standard for Environmental Assessment of Imaging Equipment, IEEE, New York, NY, 2012.

22. http://www.epeat.net/documents/manufacturer-resources/fees/PC-EPEAT-Participation-Fees_7-1-16_to_6-30-17.pdf
23. Executive Order 13423 – Strengthening Federal Environmental, Energy, and Transportation Management, Federal Register Vol. 72, No. 17, January 26, 2007.
24. UL 110, Standard for Sustainability for Mobile Phones, Underwriters Laboratories, Northbrook, IL.
25. Executive Order 13693 – Planning for Federal Sustainability in the Next Decade, Federal Register Vol. 80, No. 57, March 25, 2015
26. http://web.archive.org/web/20160430120835/http://www.electronicstakeback.com/2015/04/06/new-executive-order-is-the-grim-reaper-for-epeat/

ASSESSMENT OF THE ENVIRONMENTAL IMPACT OF WAREHOUSE FIRES AND FIRE SERVICE RESPONSE

Francine Amon and Jonatan Gehandler
SP Fire Research, SP Technical Research Institute of Sweden

ABSTRACT

This work is an extension of a previous feasibility study in which a tool was developed to examine the environmental and economic consequences of warehouse fires. The present work provides guidance for future environmental evaluations of structure fires and fire response based on experience derived from the previous work. Only the environmental aspects of the tool are addressed in this work- the economic assessment part is not included. A quantitative risk assessment (QRA) approach was used to predict the amount of damage caused by a warehouse fire with and without fire service response; the results were then applied to life cycle assessment (LCA) models to evaluate the environmental consequences. The present work extends the scope of the previous feasibility study to include feedback from users and peers, along with observations and experiences collected during the development of the tool. This information is presented in terms of potential increased accuracy, functionality, and usability of the tool, but it is also intended to provide guidance for similar research in evaluating the environmental aspects of structure fires and fire service response.

INTRODUCTION

Concern for the health of the natural environment is growing as human population grows and as new levels of contamination of scarce resources are discovered. Most fires in the built environment contribute to contamination of air and possibly also to surface water, groundwater, sediment, and soil [1-3]. Firefighting operations can also impact these same receptors [4]. Fire debris and ash often contain many harmful constituents, depending on the fuel and burning conditions of the fire [5]. Also, there is a heavy environmental impact associated with replacing damaged structures and contents.

A tool has been developed that estimates the environmental and economic impact of fire on communities and compares a (theoretical) case of no fire service response to an actual case that includes fire service response [6]. The tool targets warehouse fires as a first step toward understanding whether such a tool could have practical value for the fire service to illustrate environmental benefits resulting from fire interventions. Further development of the Environmental and Economic Assessment (Enveco) tool could also aid the fire service in their pre-planning activities to understand the environmental trade-offs of different fire attack strategies for specific situations, such as adopting a defensive response (letting the fire burn while protecting the surroundings) versus using foam or water to extinguish the fire.

The tool is capable of assessing the impact of replacing the damaged portions of the warehouse structure and contents, including adjacent structures and contents if the fire spreads. It also includes the impact of the fire effluent coming from the burned structure(s) and contents and from firefighting operations. Finally, the impact of the fire service response (equipment and suppression media) is included in the real fire case, but not in the baseline case.

Depending on the intended use of such a tool, there are many methods for estimating environmental impact as well as predicting the fate of a structure fire if the fire service does not respond. In the following text the basic approach of the Enveco tool is outlined and analyzed with respect to feedback received from potential users of the tool and from peers via workshops and other events. Also, observations made during the feasibility study phase of the tool that could have a bearing on the functionality of a more advanced version of the tool are examined in order to provide guidance for further work in this area. In particular, the assumptions made, along with their effect on the sensitivity of the model input and uncertainty of the results are discussed.

APPROACH

The intent of the Enveco tool was to estimate the savings in environmental (and economic) impact resulting from the actions of the fire service when responding to a warehouse fire. Since environmental impact is not a familiar concept for many people, it was useful to frame the results as a comparison between a fictitious baseline case in which the fire service exists but does not respond and the real fire case under study to show the savings due to fire service intervention. As shown in Figure 1, a QRA approach was used to predict the amount of damage caused by a warehouse fire if there is no fire service response; the results were then applied to LCA models to evaluate the environmental consequences.

Figure 1. Diagram of Enveco tool components (the economic part of the tool is not included).

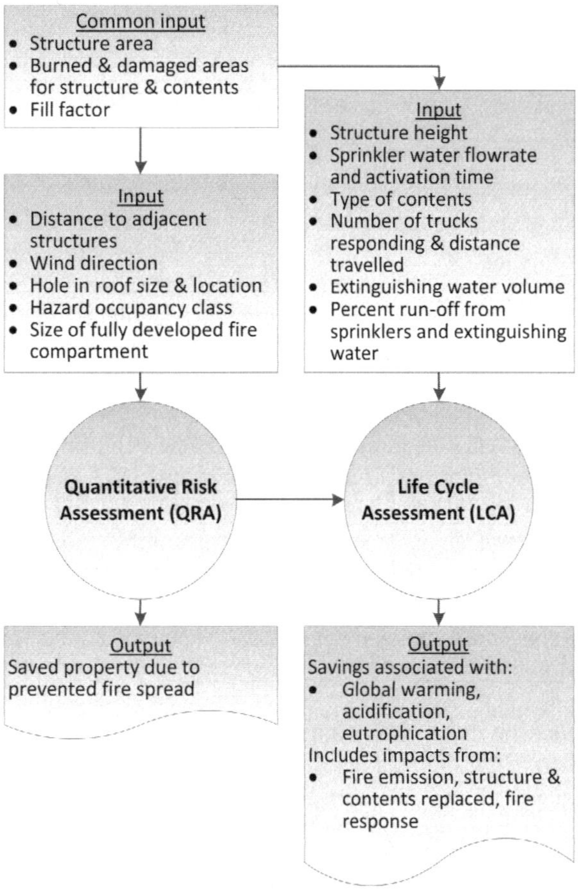

Many simplifying assumptions were made for both the QRA and LCA models. These assumptions were:

- The fire is well developed by the time the fire service arrives, requiring a defensive strategy to be adopted.
- The fire service uses only water for suppression (from outside the structure) and to protect adjacent structures.
- The adjacent structures are of similar structure type and have similar contents.
- It is assumed that the level of fire safety for buildings is the same whether or not the fire service responds to the fire.
- The fire could potentially spread to other structures via a hole in the roof of the burning warehouse.
- The heat flux of the contents of the warehouse can be represented by the hazard occupancy classes defined in NFPA 13 [7].
- Fire spread occurs via piloted ignition in the downwind direction when an adjacent surface is exposed to heat flux of at least 18 kW/m^2.
- Fire spread occurs via spontaneous ignition in the non-downwind directions when an adjacent surface is exposed to heat flux of at least 33 kW/m^2.
- The contents of the warehouse can be represented by general categories of materials for the LCA.
- The warehouse has a single floor and 10 m high walls.

The QRA modeled the main mode of fire spread from large warehouse fires, i.e. by radiation from flames ejected through a partially collapsed roof. The input to the QRA model is based on information taken from the real fire case in the comparison. The details of both the QRA model and the LCA methodology are given in [6].

In light of experience gained from the Enveco tool feasibility study and from feedback collected since the feasibility study, the QRA and LCA methodologies used are briefly outlined in the following sections and then analyzed with respect to the potential for improving the functionality, accuracy, and usability of the tool. It is hoped that these analyses will also provide useful information to other research in this general area of study.

QRA methodology

The QRA methodology was based on the recommended approach of Ingason in [8], which is in turn based on the work of Heskestad [9, 10] and Mudan and Croce [11] and is applied to warehouse fires.

Response to warehouse fires is frequently defensive because the fire can grow rapidly in large open spaces having high fuel loads and the conditions can quickly become unsafe for firefighter entry [12]. The QRA requires a fully developed compartment/warehouse fire and basic data about the fire conditions at the time fire spread is most likely to occur. This method uses deterministic information from the known fire, such as the orientation and location of adjacent structures, the size and location of breaches in the burning warehouse roof, and wind/weather conditions. It also uses distributions and probabilistic parameters, e.g. the diameter and position of the hole in the roof to inform a predictive model of fire spread potential for the theoretical baseline case in which the fire service does not respond.

Flame height, flame tilting and radiation to adjacent targets are calculated in a spreadsheet using Monte Carlo simulations. In the Monte Carlo simulation 10 000 simulations are run. For each run, the deterministic parameters are assigned their constant value and each uncertain parameter is assigned a random value base on its probabilistic distribution. The uncertain parameters have an equally plausible uniform distribution between two limits.

The heat flux of the warehouse contents is estimated from the hazard occupancy class defined in NFPA 13 [7] and ranges from 0.1 – 0.3 MW/m^2 for the least severe Light Hazard class to 1.3 – 2.5 MW/m^2 for the most severe Extra Hazard (Group 2) class.

To summarize, the following input is required to estimate the risk for fire spread from a warehouse to an adjacent building:

- Fully developed fire for warehouse-
  - Length and width of the compartment/warehouse on fire (constants).
  - HRR per unit area (uniform distribution).
  - Upper and lower limit for the roof hole diameter (worst case for each fire simulation, i.e. a diameter from approximately 10 m up to half the side length, and position (north, west, south, east, anywhere) of the hole in the roof (uniform distributions).
- Wind direction (north, west, south and east) (constant probability).
- Distances from adjacent structure (up to four: north, west, south and east) to burning warehouse (constant).
- Target material ignition limit:
  - Piloted ignition (in wind direction): 18 kW/m^2 (1.7 kW/ft^2) (constant).
  - Spontaneous ignition (in all other directions): 33 kW/m^2 (3.1 kW/ft^2) (constant).

Default values are provided for the user in cases where the input data is highly uncertain or unknown; the only inputs without default values are the warehouse/compartment dimensions.

The output of the QRA is the probability of the fire spreading to adjacent structure(s). It is assumed that the adjacent structure(s) are the same size as the burning warehouse. The probability represents the area of the adjacent structure that could burn, based on the size of the burning warehouse. As an example, if the burning warehouse is 10 000 m^2 and the probability of fire spread is 10 %, the amount of property saved is 1 000 m^2 of warehouse.

Analysis of QRA methodology

Feedback from the US fire service and other parties with an interest in the Enveco tool generally addressed the issue of data input. There is not usually an opportunity to measure structure dimensions and distances to adjacent structures during a fire response, especially when it is a fully involved warehouse fire. It is also not always a simple matter to come back to the site and collect data after the fire has been extinguished. The heat flux from the burning warehouse is potentially difficult to estimate and could be prone to large variations between users of the tool. Essentially, the tool must be as automated as possible for the fire service to get good value from it.

Future work to address these issues could include an exploration of online mapping tools that automatically generate distances between structures, structure sizes and types (zoning information for residential, industrial, commercial, agricultural structures, etc.) given the address of the burning structure. Local weather data, including wind speed to calculate the effects of tilting flames, could probably also be automatically input directly into the tool.

A diagram of the relationship of the burning warehouse to adjacent structures, which is the most important input specific to the incident site, is shown in Figure 2. This basic diagram is included in the tool for user reference, but could serve as a starting point for designing an automated tool to generate the necessary information if existing methods are not available.

The assumption that the adjacent structures are the same size as the burning warehouse can be eliminated if the size of the threatened structures is known. This is also information that can possibly be obtained from online mapping sources.

Likewise, information about the hazard occupancy of the burning structure may be available in electronic municipal records or through fire service pre-planning activities and could potentially be exported to the tool. The difference here is that some effort would be required to set up the connections necessary to access this information and it may be difficult to do.

The model can be modified to calculate the probability of fire spread based on the worst case with respect to the size and location of the hole in the roof, given a range of possible values, or it can use the average probability. These modifications could result in a reduction in accuracy.

The above automation ideas would leave virtually no input necessary from the user except the address of the burning structure, although it would be necessary for someone to check the values to ensure they are reasonable.

Figure 2. Definition of point of origin and coordinate system (x, y) to determine distances to adjacent buildings.

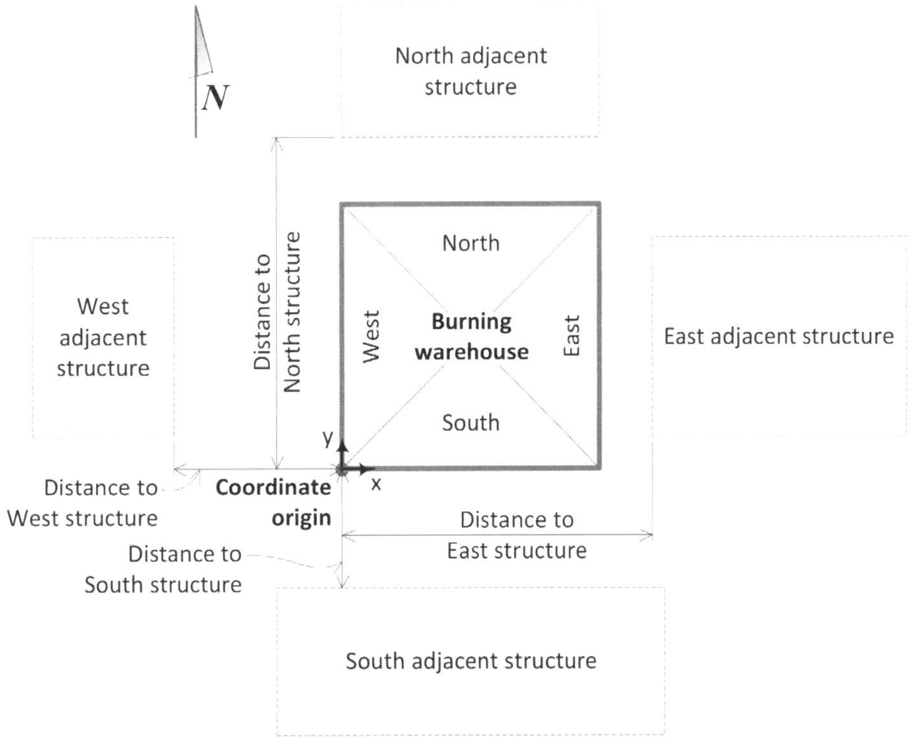

Another request from future users of the tool was to enable it to handle incipient fires in which an aggressive attack might be appropriate. In this situation the fire service could use the tool to estimate fire damage as a function of response time and resources. There are political issues related to this request, e.g. the tool could be used to support decisions about the density and size of fire stations, which is outside the original scope of this work. However, fire spread *within* the structure could be quite important for structure fires in which people are threatened.

To achieve this goal an estimate of fire growth would need to be built into the QRA, along with a method of accounting for the time between the ignition of the fire and the arrival of the fire service. Both these activities are non-trivial, have high uncertainty, and the results would likely be very sensitive to the input. At this time no known viable solutions have been proposed.

The current QRA model does not include materials with extreme flame temperatures such as magnesium. It covers fires for which the ratio of radiated heat to total heat release lies in the range between 0.2 and 0.3. In a future model this ratio could be made into an input parameter or tied to the hazard occupancy class of the structure.

The last significant request related to the QRA was to allow the Enveco tool to analyze more than one case at a time. Generalized values cannot currently be used without severe degradation of the results due to the importance of the proximity of the burning structure to adjacent structures, coupled with the local weather. It would be necessary to completely change the QRA approach to enable the analysis to include multiple cases.

LCA methodology

In general, LCA-based environmental impact methods can be used to assess a wide range of environmental impact categories, for example: global warning, eutrophication, resource depletion, ecotoxicity of soil and water bodies, etc, depending on which impact assessment method is considered important for the goals of the LCA. In this work, LCA is applied to the production of the warehouse and its contents, and also includes the system of fire service response and fire emissions to the environment.

Characterizing the fire effluent is a challenge because it is dependent on the fuels involved, the ventilation conditions, and suppression activities. For this reason it is difficult to predict the composition, concentration, and fate of pollutants coming from the fire. Measurements made by Blomqvist et al., in which fire effluent to air from fire tests of furnished rooms [13] and fire effluent to water from various non-industrial structural fire test results [14] were are used to represent the fire effluent for both the warehouse structure and its contents.

The amount of pollutants emitted to air is based on the percent of the warehouse footprint area in which the fire consumed the warehouse structure and contents. The amount of PAH and VOC in sprinkler water run-off to the environment (in mg) is calculated from user input of the sprinkler water discharge rate, length of operation time, and the estimated percent of sprinkler water run-off that escaped into the environment.

Replacement of the burned warehouse is modelled from cradle to gate. The gate is positioned at the end of the construction stage, meaning that the model includes extraction of raw materials from the earth, transport of the materials to refining facilities, additional processing of the materials in one or more facilities, transport of the finished materials to distribution centers (warehouses), transport to the construction site, and finally construction of the warehouse. It is assumed that a fire will occur in both cases, therefore the LCA model stops at the end of the construction phase and does not include the use or the non-accidental end of life phases because they would be the same in both cases and therefore do not contribute to the comparison.

The environmental impact associated with the existence of the fire department is shared in both cases, but the response to the warehouse fire is only included in the real fire case. Therefore, the impacts associated with parking lots, training facilities, maintenance, energy use, waste streams, structures and contents of the fire station(s) are the same for both cases and can drop out of the analysis. The actual equipment that is used to respond to the fire (quantity and type of vehicle and distance travelled) is assigned to the real fire case but isn't present in the baseline case, so it will appear in the comparison. The production and maintenance of the trucks is not included in the LCA model because these aspects of the truck life cycle would be the same for both cases.

The following input is required by the Enveco tool to estimate the environmental impact of a warehouse fire:

- Description of warehouse and contents-
    - Length and width of the fire compartment and overall warehouse on fire (shared with QRA).
    - Storage rack height.
    - Fill factor (percent of warehouse in which contents are actually stored).
    - Burned area of structure and contents (contributes to fire effluent).
    - Damaged area of structure and contents (that must be replaced).
    - Presence of sprinkler system, overall discharge rate, and length of time it functioned.
    - Mix of contents.

- Description of fire service response-
  - Number of trucks responding.
  - Distance from fire station to incident.
  - Amount of water used.
  - Amount of water that escaped into the environment.

Default values are provided for the user in cases where the input data is uncertain or unknown; the only inputs without default values are the warehouse/compartment dimensions (which could possibly be obtained from online mapping algorithms) and the sprinkler activation time. The default values either come from the literature or are calculated based on other input values using relationships obtained from the literature.

The output is given for global warming (in kg CO_2 equivalents), acidification (in kg SO_2 equivalents), eutrophication (in kg N equivalents), ozone depletion (in kg CFC-11 equivalents), smog (in kg O_3 equivalents), ecotoxicity (in CTUe equivalents), and energy used (in MJ). This output is broken down into categories of fire emissions, structure replaced, contents replaced, and fire department.

Analysis of LCA methodology

Similar to the feedback about the QRA, users do not want to input any more data than is absolutely necessary. This is particularly true for the LCA part of the Enveco tool. Default values were provided to aid users in this respect; these values can be discarded if better information is available. Most of the default values have been derived from national statistics and literature; a future improvement to the tool would be to regionalize the input as much input as possible. This could make a significant difference in the LCA results that depend on energy usage and structure construction styles.

Metal building system (MBS) construction was assumed for the warehouse structure, however, it is possible to provide a selection of construction styles for the user. Doing this would require additional user input. A sensitivity analysis was conducted using three types of construction (MBS, structural insulated panels, and tilt-up concrete panels) and the results showed that the largest percent change was approximately 12 %. Compared with the stated 15 % uncertainty of the LCA modeling program results, this level of sensitivity was considered acceptable. If the tool is expanded to include other types of structures, for example residential or commercial structures, selection of the construction styles/materials would be much more important due to the wider range of materials and the closer proximity of the structures.

A function that could be implemented in the tool is the ability to choose fire protection systems other than sprinklers. Fire protection system requirements generally depend on the hazard occupancy of the structure and its intended use. These systems could affect both the amount of damage to the structure/contents and the environmental impact of the system itself, for example if certain types of foam are released into natural water and soil systems or wastewater treatment facilities. The amount of water discharged by sprinkler systems could have a high impact, largely with respect to replacement of structural materials and contents because of water damage. Compartmentation could have an effect on the types of contaminants found in the fire effluent that vary with the completeness of combustion, although this must be balanced against the overall amount of smoke produced by the fire.

The most uncertain input data is the warehouse contents. A warehouse fire was chosen as the initial application of the tool because it simplified some of the choices for predicting fire spread in the baseline case. The downside of using a warehouse fire is the wide range of possibilities for contents. Nearly anything can be stored in a warehouse. The Enveco tool currently has eight categories of materials to represent the warehouse contents (plastic, paper, wood, textile, soft furniture, chemicals, dry food, and electronics), but these categories do not adequately capture the environmental impact of replacing products if they have been damaged by fire. For example, if the warehouse was used to store garden supplies there could be many products that are made from plastic, wood, textile, etc and these finished products would have a very different environmental impact than their raw materials. In other words, the individual material impacts cannot be summed to estimate the overall impact of the product. Replacement of the damaged warehouse structure and its contents has the largest impact by far, compared with fire emissions and fire service activities.

When analyzing the sensitivity of the warehouse contents it was found that the results varied as much as 87 % when the mix of contents was randomly changed for individual model runs. This very high sensitivity is dampened considerably when using the results in a comparison, where the impact of the mix of contents only applies to the difference in content volume between the baseline case and the real fire case. However, there is much room for improvement in characterization of the warehouse contents. It may be possible to improve the accuracy of the assessment of the contents by tying them to the hazard occupancy class, permitting processes, or other information in public records that relate to the use of the warehouse. Characterization of the structure contents will be less difficult for certain other structure fires (e.g. residences) because there is much more content consistency than in warehouses.

Ideally, the fire effluents used in the Enveco tool would be directly related to the construction style of the warehouse structure and its actual contents. The fire effluents used in the LCA model were selected based on previous work in which furnished rooms or non-industrial structures were burned [13, 14]. The fidelity of the fire effluent will improve as more information becomes available from fire tests and models. The fire response consists of driving the fire trucks to and from the incident and possibly includes contamination of soil and water from fire water run-off. Considering the assumptions made concerning water usage, both the fire response and the fire effluent contribute several orders of magnitude less environmental impact than replacement of the warehouse structure and its contents.

There are many uncertainties associated with the components of the Enveco tool due to non-specific input data, but many of them become less severe when used in the comparison between the theoretical baseline case of no fire service intervention and the real case where the fire service intervenes. This is because some portion of these uncertain parameters used in the models are applied to both scenarios in the comparison and are therefore partially cancelled out of the results.

Feedback from users included adding a mechanism by which the effects of reduced availability of the fire service while responding to a large warehouse fire could be incorporated into the results. If most of the fire service resources are occupied, other fire incidents may go unheeded or the response time could be much longer than normal. This situation could make a large difference in the results. At this time no known viable solutions have been proposed.

Another request was to include additional structure types (single family residences, commercial buildings, apartment buildings, mixed types, etc.) and locations (wildland area, suburban neighborhood, central business district, etc.). In some ways the Enveco tool would be easier to apply to other structure types, for example when identifying the structure contents. This expansion of the tool would require a much larger database that can handle a wider range of construction styles and materials and would also need to be regionalized to account for climate and transport differences. It may be possible to use online mapping tools to facilitate some of the modifications that would enable this functionality without adding to the user input burden.

The LCA does not include the effects of salvaging damaged materials or preparing the site for new construction. This could be important if the site requires decontamination due to particularly toxic fire effluents (negative impact) or if materials can be recycled or provide energy through incineration (positive impact). It is possible to develop methods of including these effects, possibly by using literature data to estimate the amount of salvaged material and by using the hazard occupancy class to estimate the amount of decontamination effort.

Lastly, the mix of career versus volunteer firefighters is not addressed in the LCA calculations. There could be differences in distances traveled, in the types and amount of vehicles and fire suppressant used. These differences are not likely to affect the results significantly because the impact of the fire service response is very much lower than the impact of replacing structure materials and contents.

CONCLUSIONS

The Enveco tool was intended to help the fire service understand and share information about environmental (and economic) savings to the communities they protect. The goal of this work was to analyze feedback from peers and potential users regarding the further development of the Enveco tool in terms of improving its accuracy, functionality, and usability. Observations made by the authors during the feasibility study were included for this purpose as well.

The primary message from potential fire service users was that too much input is required, and some of the input is difficult to collect. In most cases default values are available if the user doesn't have specific knowledge about the fire incident being studied. Improvements could be made to incorporate online mapping algorithms into the tool that automatically insert much of the user input, given the address of the incident.

If information in public and/or private records can be accessed by the tool it may be possible improve its accuracy with respect to heat flux (which is a highly sensitive parameter in the fire spread calculations) and the characterization of warehouse contents. Automatic population of the input fields would also improve the usability of the tool. Ideally, updates to database(s) connected to the tool would also serve to keep the tool as current as possible. The functionality of the tool can also be extended more easily as new information becomes available.

Many of the fire service personnel that provided feedback on the tool requested functionality that would enable it to handle incipient fires in which an aggressive attack might be appropriate. This would be a major change to the tool and would involve estimating the fire growth and accounting for the time between the ignition of the fire and the arrival of the fire service. Developing this functionality while maintaining a reasonable level of uncertainty would be difficult; however, it would add a substantial amount of value to the tool for the fire service and would be an important feature if the tool was adapted to apply to residential and commercial structures.

Feedback from several peers indicated that expanding the tool to include different structure types, particularly residential structures, is an important improvement. Since residential fires are the most common structure fires in the US it is a natural next step to include them in the tool. Doing this would require a much larger LCA database that can handle a wider range of construction styles and materials and would also need to be regionalized to account for climate and transport differences.

Even though there are many uncertainties associated with the components of the Enveco tool due to non-specific input data, many of them become less severe when used in the comparison between the theoretical baseline case of no fire service intervention and the real case where the fire service intervenes. This is because some portion of these uncertain parameters used in the models are applied to both scenarios in the comparison and are therefore partially cancelled out of the results.

ACKNOWLEDGEMENTS

Without financial support from the Fire Protection Research Foundation and the National Fire Protection Association this work would not have been possible. Thank you very much both for funding this work and for the excellent dissemination opportunities after it was finished. We would also like to acknowledge the valuable assistance provided by Mai Tomida and Brian Meacham of Worchester Polytechnic Institute in collecting and analyzing much of the supporting information upon which this work is based.

REFERENCES

1. Palm, A., et al., *Assessing the environmental fate of chemicals of emerging concern: a case study of the polybrominated diphenyl ethers.* Environmental Pollution, 2002. **117**(195-213).

2. Alaee, M., *Recent progress in understanding of the levels, trends, fate and effects of BFRs in the environmen.* Chemosphere, 2006. **64**: p. 179-180.

3. Lönnermark, A., et al. *Emissions from Fire - Methods, Models and Measurements.* in *Interflam 2007.* 2007. Royal Holloway College, University of London, UK: Interscience Communications.

4. Noiton, D., J. Fowles, and H. Davies, *The Ecotoxicity of Fire-water Runoff. Part II. Analytical Results.* 2001, New Zealand Fire Service Commission. p. 23.

5. Blomqvist, P. and M. Simonson-McNamee, *Large-scale generation and characterisation of fire effluents*, in *Fire Toxicity*, A. Stec and R. Hull, Editors. 2010, Woodhead Publishing Limited. p. 461-514.

6. Amon, F., et al., *Development of an Environmental and Economic Assessment Tool (Enveco Tool) for Fire Events.* 2016, Springer.

7. NFPA, *NFPA 13: Standard for the Installation of Sprinkler Systems.* 2015, NFPA: Quincy, MA.

8. Ingason, H., H. Tuovinen, and A. Lönnermark, *Industrial fires - A literature survey*, in *SP Report 2010:17.* 2010, SP Technical Research Institute of Sweden: Borås, Sweden.

9. Heskestad, G., *Luminous Heights of Turbulent Diffusion Flames.* Fire Safety Journal, 1983. **5**: p. 103-108.

10. Heskestad, G., *A reduced-scale mass fire experiment.* Combustion and Flame, 1991. **83**: p. pp 293-301.

11. Mudan, K.S. and P.A. Croce, *Fire Hazard Calculations for Large Open Hydrocarbon Fires*, in *The SFPE Handbook of Fire Protection Engineering*, P.J. DiNenno, et al., Editors. 1995, The National Fire Protection Association.

12. Dinaburg, J. and D. Gottuk, *Fire Detection in Warehouse Facilities*, in *Fire Detection in Warehouse Facilities.* 2012, Springer New York. p. 1-59.

13. Blomqvist, P., L. Rosell, and M. Simonson, *Emissions from Fires Part II: Simulated Room Fires.* Fire Technology, 2004. **40**(1): p. 59-73.

14. Blomqvist, P., A. Lönnermark, and M. Simonson, *Miljöbelastning vid bränder och andra olyckor-Utvärdering av provtagning och analyser*, in *The Environmental Impact of Fires and Other Accidents. An evaluation of samples and analyses.* 2004, Räddningsverket: Borås. p. 52.

WOOD STOVE CHIMNEY SMOKE MONITORING AND OPACITY AND ODOR REDUCTIONS

Noel Putaansuu IAAI-CFI, CFII, CFEI

INTRODUCTION

Wood heating has been used since the Stone Age. Humans have found that a few pieces of dry wood can make a camp fire or be placed into a box made from metal, brick and stone with a channeled combustion air supply and vented outdoors. Modern urban property lots place neighboring structures in close proximity and wood smoke is at times problematic for the neighboring residents. Code enforcement fines are complaint driven and often only levied when a person complains to officials about a neighbor's emitted smoke. Code enforcement uses trained eye or instrumentation to determine if the opacity is more than 20%.

There are accounts of heating related smoke and the decline of air quality in history and recently:

- Significant chimney design in history is accredited to Louis Savot and Benjamin Thomas (Count Ruthford) 1796.
- Great Smog of London 1952.
- Washington State regulates wood stoves in 1995 limiting emissions.

There are an estimated 9.3 million wood stoves in operation in the United States alone. These wood fuel stoves provide heat for residences using an abundant inexpensive renewable resource. The newer (1995) wood stoves have smoke reducing technology that ends at the top of the wood stove.[1] Smoke from wood stoves has prompted restrictions in urban and rural regions. Burn bans during inversion weather patterns restrict or prevent operation.[2,3] As a Fire Science community we need to ensure wood stoves are an optional available heat source in the future. Using off the shelf consumer level components, levels of opacity and odor reduction are possible.

Problem Defined

Wood stoves are used by people who are able to collect wood. This low cost fuel when dried can meet a major portion of the heating needs of a structure. Wood is a plentiful renewable resource. However, the wood stove industry has products that only partially complete the combustion of the wood fuel. Wood stoves are used by a wide cross section of society, although, people who are trying to save money may be the more predominant users. Keeping these wood stoves operating will ensure that this heating option is available in the future. The primary impediment is the vented smoke odor and opacity. The code enforcement is driven primarily by neighboring residence complaints when the vented wood smoke drifts onto a neighbors property.[2] The problem is that no technology is available to measure, monitor and reduce emissions at the top of the stack where the pipe is exposed and smoke is venting to the atmosphere.

In the northern latitudes where wood is in good supply, cities tend to locate in coastal regions and transportation hubs often in valleys with hills and mountains around the perimeter. This topography combined with a temperature inversion weather system has the tendency to hold the smoke emitted by wood stoves down low in the valley, the result being lower than desirable air quality.

Impact

With the problems associated with smoke, wood stoves are threatened with restricted use. If wood stoves are restricted to the point of not being useful, this heating option will diminish. During a power outage or other catastrophic event that disrupts our infrastructure, a wood stove may be the only viable option for heating a structure. In addition, if wood stoves don't consume the waste wood, the wood may decompose releasing Methane and Carbon dioxide without the containment benefit of providing heat to those in need. Methane is also considered a higher initial impact greenhouse gas, as much as 100 times, more than Carbon dioxide.[4]

CURRENT TECHNOLOGY

Technology is constantly changing. As industry develops new technology devises, cost often diminishes as capability improves allowing the practical application in areas not previously open for development and widespread use. Below are technologies and terms that can be applied to wood stove smoke reductions.

Microprocessors: Also called Micro controllers. Arduino is a family of consumer level microprocessors with integrated networking and analog to digital conversion. Arduinos can monitor sensors and activate relays and servomechanisms as well as be programed to record and process data.

Thermoelectric Power Generation: A thermoelectric generator, or TEG (also called a Seebeck generator), is a solid state device that converts heat directly into electrical energy through a phenomenon called the Seebeck effect.[5]

Thermal Oxidation: A thermal oxidizer (also known as thermal oxidiser, or thermal incinerator) is a process unit for air pollution control in many industrial facilities that decomposes hazardous gases at a high temperature and releases them into the atmosphere.[6]

Opacity Measurement: ASTM E 84, ASTM E662, ASTM E 906 describe smoke measurement methods. It is essentially directing a light beam across a duct or chamber and using calibrated neutral density filters to measure the incident light on a photo sensor. To measure **Optical obscuration,** a light beam is passed through the smoke and a detector opposite measures the light received. The more smoke particles that are present between the two, the less light impacting the sensor will be measured.

Temperature Measurement: Thermocouples (TC) and resistant temperature detectors (RTD) are common and proven instruments for measuring temperature and relaying to microprocessors.

Gas Sensors: Combustion gas sensors including for Carbon Monoxide and Oxygen, are available and can provide signals to microprocessors.

Servomechanism: A device which provides position control.[7]

Secondary Air Injection: A vehicle emissions control strategy introduced in 1966, wherein fresh air is injected into the exhaust stream to allow for a more complete combustion of exhaust gases. [8]

Lidar (also called LIDAR, LiDAR, and LADAR): is a surveying method that measures distance to a target by illuminating that target with a laser light. [9]

Smoke Minute: This is a term created for the smoke readings that exceed the 20% opacity for more than one minute. In another jurisdiction that has another criteria, this term may have a different meaning.

HYPOTHETICAL SOLUTION

If the vented smoke odor and/or opacity could be real time monitored this instantaneous information would allow for the fuel and combustion air control changes to reduce the emission opacity levels.

If the measured vented opacity were recorded, the operator would be able to use the feedback to control emissions. This record of emissions could be used to defend against a fine or infraction. Summarized smoke production data could be stored and used to calculate pollution levels contributed by wood stoves aiding in future design innovation.

If the temperature of the fire box and the stack were monitored, the control of the wood stove could be improved with the feedback of information. The operator could then regulate the heat level to achieve a more comfortable room temperature with less room temperature extremes that would require restricting combustion air, which increase pollutants.

If passive thermal oxidizers were used, the impact may produce a measured emission reduction of smoke from a wood stove vent. A passive oxidizer would be defined as gravity powered by the buoyant hot exhaust gases with no powered fans or mechanisms.

If active controls were integrated into a thermal oxidizer, combustion air supplied by powered fans could be controlled to achieve optimal conditions for oxygen content and temperature. An active thermal oxidizer would include any system that used powered fans, actuators and feedback control loops using information from sensors.

Discussion

There are no off-the-shelf products for measuring the smoke opacity of residential wood stove smoke vents. The enforcement officials use a trained human eye or other methods such as LIDAR to levy fines up to $1000 for smoke opacity that exceeds 20% for six minutes of operation in any hour.[2] It is likely that the human eye method is subject to light levels, physical background color and human factors that diminish accuracy. A wood stove operator may have no idea what opacity level his vent is producing. At night even an outside trained viewer may not be able to estimate the opacity of the smoke emitted within reasonable accuracy.

Using off the shelf electronics such as light emitters and light sensors; a calibrated optical measurement system can be produced to monitor smoke opacity. This has been done in fire laboratories on regular basis for decades. Using this technology on the consumer level is now economically viable. (Figure 4.)

Using simple passive or active designs, thermal oxidizers can be used to reduce smoke opacity. Using the Three (3) Ts of conventional thermal oxidizers: Temperature, Time, Turbulence; simple and effective smoke reduction can be accomplished.

Connecting the sensors to readily available collection and analysis hardware allows the information to be summarized and stored in a database thereby providing information for weather and air quality predictions.

Opacity Measurement in Series / Sequence

Once an opacity measurement system (OMS) is developed that is reliable and repeatable, two can be installed in sequence with a smoke reduction technology installed between the first and second OMS units to provide feedback on the performance of the smoke reduction technology being evaluated. This will be the foundation for innovation as end users can attempt many forms of smoke reduction technologies and feedback results can be uploaded to the internet Cloud. Competition would encourage design contests promoting innovation.

POTENTIAL SMOKE AND ODOR REDUCTION TECHNOLOGIES

Thermal Oxidizer

Thermal oxidizers are used to convert unburned fuel, aromatic compounds, soot or undesirable odors out of an exhaust stream prior to releasing to the atmosphere. Thermal oxidizers work by providing supplemental heating of the smoke gases, combining combustion air, providing turbulence and holding the gases for a period of time.

Secondary Air Injection

Secondary air injection is similar to a thermal oxidizer but without a supplemental heating component. This technology is used in the automotive industry. Applying this technology to the upper stack would be achievable with little cost and effort. This could be passive with draft induced air or active with fans pumping air.

Warm Temperature Thermal Oxidization

Typically commercial thermal oxidizers operate between 590 °C (1,094 °F) and 650 °C (1,202 °F) and hold the gases for 0.5 seconds to achieve the desired opacity and odor reductions. These commercial sized thermal oxidizers use large gas burners and have large combustion air fans. During operations of large commercial oxidizers the ground nearby will often shake and hearing protection is often necessary. An example of "warm temperature thermal oxidization". Is a one cylinder motorcycle, a 2005 BMW F650 GS that has two exhaust chambers; Aspirated Air Injection is used on the left side while the exhaust

exits on the right side.

Figures: 1 upper left, 2 upper right, 3 left

Exhaust pipe is on left side (Figure 1). Gases enter left chamber and cross to right chamber (Figure 2). Small port on left chamber pulls air during negative pulse (Figure 3).

The motorcycle industry is faced with increasingly restricted emission standards.[9] In 2016, ten years after the manufacture of the pictured F650 the industry appears to be phasing out the large one cylinder motorcycles possibly due to them not meeting the more restrictive emissions standards as well as other market forces. If BMW determined that thermal oxidation (simply injecting air into the exhaust stream) could improve emissions, then the same practice may also be effective in wood stove emissions.

Controlled rain/snow cover

Typical chimney caps prevent rain and snow from entering the chimney. The cap often disrupts the thermal column of rising hot gases. The venting of a smoke stream that has no cap or cover to alter the flow will allow the hot gases to rise higher up into the air above the stack before dissipating. While not in use the cover would need to reposition to prevent rain or snow from entering the stack. The cover should not seal the stack opening and always allow a portion of the gases to vent so that when starting a wood stove the hot gases would flow out and trigger the reposition of the rain cover by the servomechanism.

System electrical power generation

Wood stoves produce sufficient heat in the form of hot gases to power a warm temperature active thermal oxidizer by way of thermoelectric generators. When fired to a high level, stack temperatures above 300 °C (600 °F) are common. Design temperature for components must be able to accommodate stack temperatures of 538 °C (1000 °F).[10] Using hobby level optical sensors, cross flow heat exchangers and electricity generated from the wood stove hot gases, a self powered opacity reduction and monitoring device could be implemented.

Prototype Opacity Measurement

Using hobby level fabrication, Arduino sensors, light emitters, photo-sensors and thermocouples a functioning system was developed.

By giving the operator a meter to measure the opacity, the operator could make adjustments to the air flow and fuel to adjust the opacity released. It is analogous to putting speedometers in vehicles, operators could then better control the speed of the vehicle to accommodate speed limits.

The technology exists to measure the opacity in situ. Most fire laboratories have visible smoke opacity measurement instrumentation. The cost has finally lowered, becoming feasible for residential use.

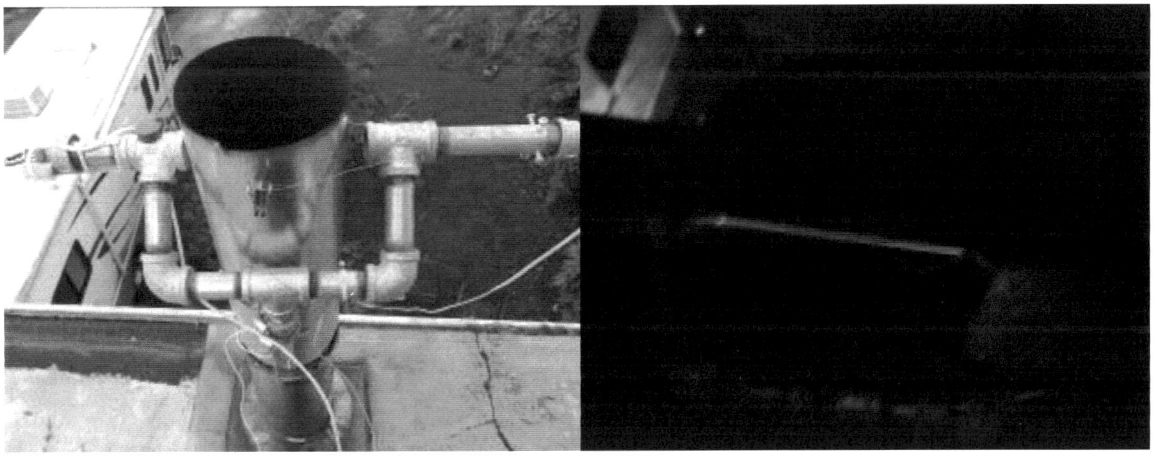

Figures 4 & 5 prototype optical bench and laser cross section.

Charts

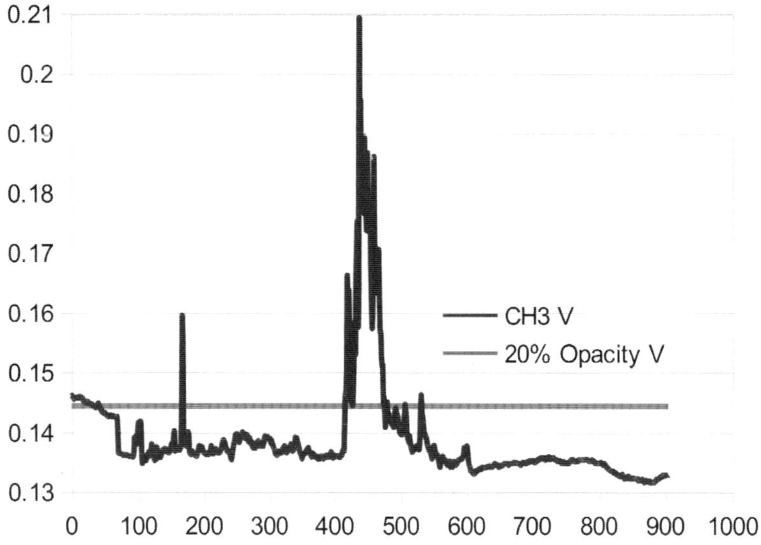

Figure 6: 900 minutes of operation. 20% Opacity neutral Density Filter as reference during the first 45+ minutes. Fire started near 100 minutes.

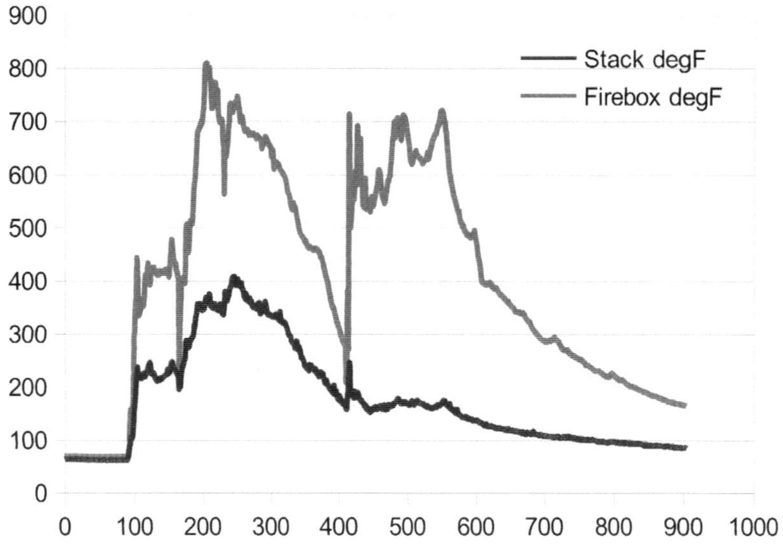

Figure 7: Temperature of fire box and stack

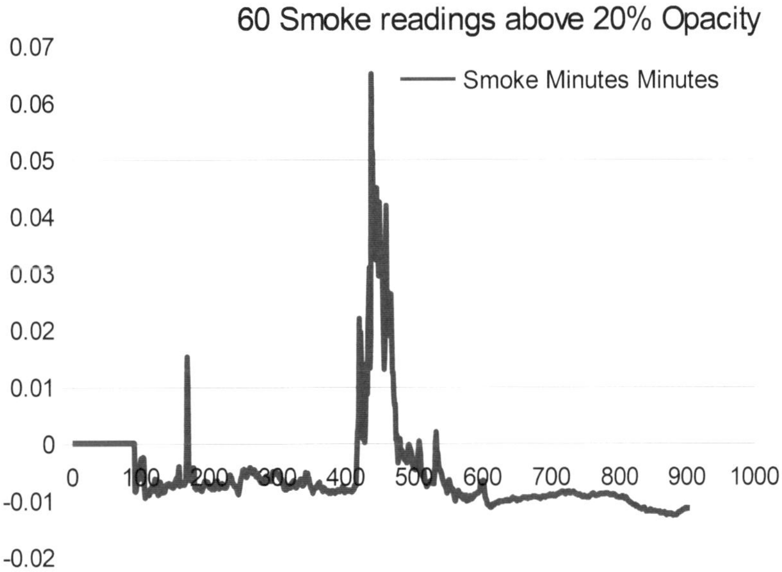

Figure 8: Calculation for emission levels that exceed 20 % Opacity. The term Smoke Minute is each minute the opacity output exceeds 20%. Chart and data indicate 60 Smoke Minute readings between 400 and 500 minutes. By Washington state law where only 6 Smoke Minutes are permitted each hour. This smoke emission is out of compliance

Enforcement

The inception of code enforcement is a complaint from a neighbor. They may notice that the odor of a neighbor's wood stove is blowing onto their property causing them discomfort. They then call the pollution control authority and an enforcement agent may visit the property. If the agent views smoke coming from the stack during a burn ban, fines from $750-$1400 can be levied in Washington State depending on region. Other states have or are considering fine systems for smoke production of wood stoves.

Considering that there is a segment of the population that is limited in income and using wood stoves and an accumulated wood supply to heat their house for a reduced cost, a fine of this magnitude is discouraging to wood stove operators. With the threat of a fine, wood stove users may decide that a wood stove is not needed or is no longer viable and go without heat or use more costly heating methods.

Discussion

The charts in Figures 6, 7 and 8 present a wood stove monitored by the prototype optical bench. (Figure 4) The optical bench was calibrated using a 0.1 Optical Density filter or 20% Opacity. The filter was then removed and a baseline was established for 0% Opacity. A fire was started and when the room reached a comfortable temperature the air flow control was pulled out to reduce the combustion air and heat output. The opacity then increased and exceeded the opacity limit of 20% for 60 minutes. These over 20% opacity readings that have more than one minute duration can be termed "smoke minutes". Only six (6) smoke minutes are allowed in Washington State once the stove is at operating temperature. Stack temperature measured by a thermocouple as the chimney pipe exits the masonry chimney was over 100 °C (212 °F) during open air flow control position and under 100 °C (212 °F) when the air control was pulled all the way out (restricted). Fire box temperature measured by a thermocouple in the top of the wood stove chamber was in the 370 °C (700 °F) range, during open flow and restricted combustion air flow until the fuel was consumbed.

The above charted data in Figures 6, 7 and 8 was one set of data recorded, a day with an afternoon temperature average of 10 °C (50 °F). When the day's temperature drops near and below 0 °C (32 °F) higher wood stove temperatures for both fire box and stack are likely to have operating temperatures of 650 °C (1200 °F) for the fire box and 204 °C (400F) for the post combustion stack exhaust gases are not uncommon. Stack temperatures of 538 °C (1000 °F) are target design values.[10]

Conclusion

Wood stoves have been in use for decades and remain popular. They use a local renewable fuel that may otherwise be a waste material. Wood stoves are a viable heating option during power outages, as well as in rural cabins and suburban neighborhoods. The low cost of the fuel can save money on heating a structure.

The wood stove production of smoke is the primary problem and is affecting air quality as population density increases. Washington State laws set a precedent for wood stove emission opacity limits and a fine structure for enforcement. In the future, Washington State may determine the fines are a revenue source and begin monitoring more wood stove users and levying an additional number of fines.

By using an opacity measurement method, a wood stove operator can benefit from the feedback and reduce emissions thereby reducing the potential for fines. By incorporating active controls and smoke reduction technology, wood stoves would be usable with less potential for irritating bystanders and thus the enforcement of fines.

Fresh air with oxygen can be preheated and proportioned into the high temperature oxygen depleted smoke stream and mixed to oxidize the unburned fuel gases and products that create odor. In my opinion, even if the results are small (in the 10%-25%) range the benefits of cleaner air are significant and worthy. I also believe that greater reductions are possible with refinements in design.

By logging the information collected such as opacity, temperature, oxygen level and other parameters over time, the data can be uploaded to the secure internet Cloud into a database for predictive modeling of air quality. With an oxygen sensor and mass flow sensor a rudimentary oxygen consumption calorimeter for wood stoves can be developed.

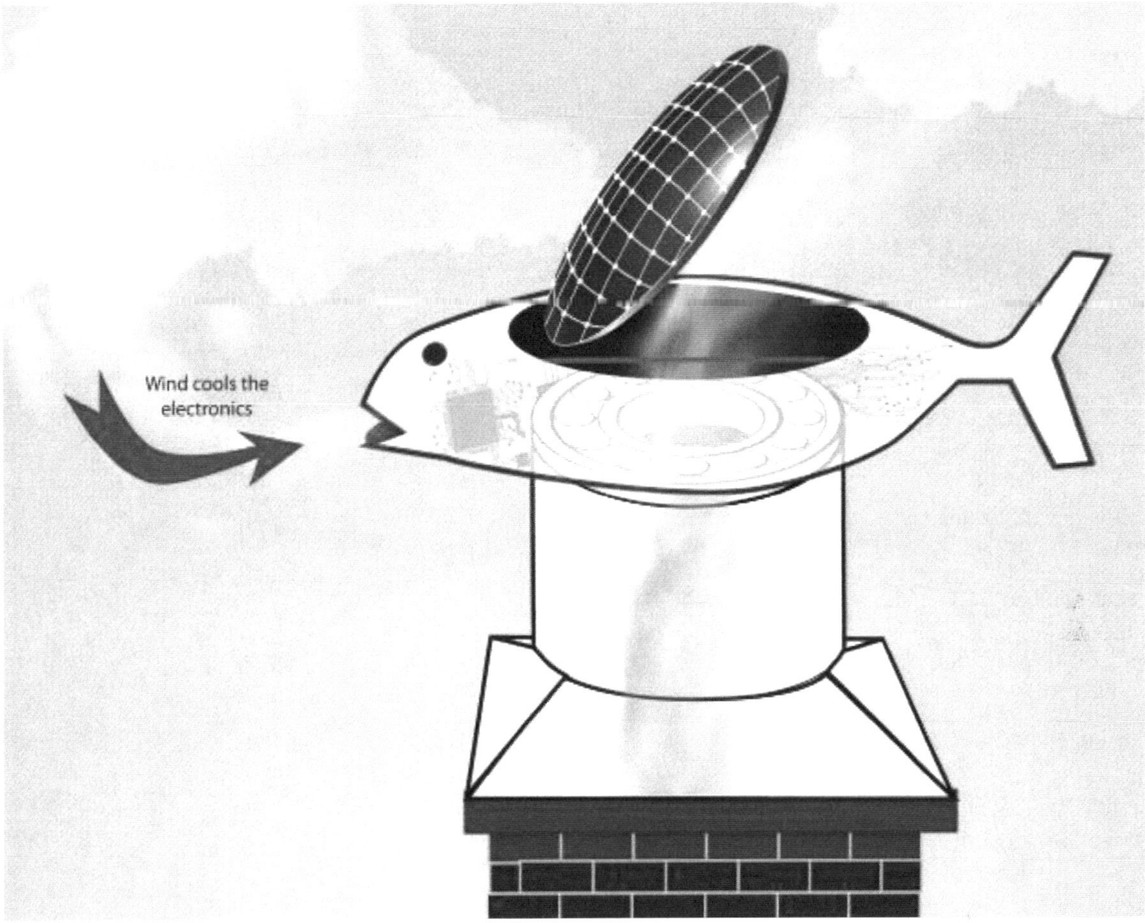

Figure 9: Future chimney top concept diagram. Diagram prepared by Sandra Putaansuu

Credits:
Research and Author: Noel Putaansuu CFI, CFEI, CFII – email:nputaansuu@gmail.com
Technical Assistance: Thayer Putaansuu
Graphics: Sandra Putaansuu email:poustergraphics@gmail.com

Peer Reviewers:
Reid Larson
Gerard Schaefer P.E.

References

1.
http://www.forgreenheat.org/resources/history.html
1995:Washington State requires all stoves sold in the state emit no more than 4.5 grams per hour.
2.
http://www.pscleanair.org/priorities/woodheating/Pages/woodburninglaw.aspx
Wood Burning and the law, Washington State
3.
http://www.pscleanair.org/Pages/default.aspx
Washington State Burn Ban
4.
https://en.wikipedia.org/wiki/Greenhouse_gas
comparison of Carbon Dioxide and Methane as greenhouse gases
5.
https://en.wikipedia.org/wiki/Thermoelectric_generator
Heat to Electricity Conversion
6.
https://en.wikipedia.org/wiki/Thermal_oxidizer
Thermal Oxidization
7.
https://en.wikipedia.org/wiki/Servomechanism
Servomechanism
8.
https://en.wikipedia.org/wiki/Secondary_air_injection
Secondary Air Injection
9.
https://en.wikipedia.org/wiki/Lidar
LIDAR
10.
http://canadamotoguide.com/2016/09/19/the-end-of-big-single-cylinder-thumpers/
Reference to potential phasing out of big single cylinder motorcycles.
11.
UL Standard for Safety for Factory-Built Chimneys for Residential Type and Building Heating Appliances, UL 103, Eleventh Edition, Dated October 15, 2010
The maximum continuous flue-gas outlet temperatures do not exceed 1000°F (538°C).

Other resource links

http://www.dailymail.co.uk/news/article-3972966/Inconvenient-truth-wood-burning-stove-bad-environment-health.html
Inconvenient truth about your wood-burning stove: They can be bad for the environment AND your health
By DAVID DERBYSHIRE FOR THE DAILY MAIL
PUBLISHED: 19:18 EST, 25 November 2016 | **UPDATED:** 19:24 EST, 25 November 2016

www.smokelesschimney.com
A future resource for wood stove owners

FLAME RETARDANTS

EPOXY FLAME RETARDANT APPROACHES – A REVIEW FOR TRANSPORTATION AND NEW FLAME RETARDANT ADDITIVES

Alexander B. Morgan*, Vladimir Benin, Mary L. Galaska
Center for Flame Retardant Material Science, University of Dayton, Dayton OH 45469 USA

ABSTRACT
Epoxies are increasingly incorporated into fiber-reinforced composites in use for transportation (automobile, bus, train, aircraft) due to the variety of benefits they provide over traditional metal materials including light weight, resistance to corrosion, and sometimes ease of manufacture. However, because of their fundamentally different properties when compared to metal, they present a very different fire threat to transportation than that encountered with metals. Specifically, epoxy composites cannot be extinguished with foams, they have very different structural failure characteristics under fire exposure, and they tend to be much more flammable than metals. Therefore, epoxy composites require different fire safety approaches than those typically used with vehicles, some of which is known, but other aspects of composites flammability are not as well known outside smaller academic and fire safety engineering circles. In this paper, a quick review of epoxy composite flammability will be discussed, including fire-fighting measures and known ways of providing fire protection. Fire protection measures including intumescent paints, integrated burn resistance barriers, and flame retardant chemistry will also be discussed. Finally, some newer flame retardant additives developed in our labs which react directly into the epoxy for lowered heat release will be discussed, along with an outlook for what other technologies, or combinations of technologies, are needed for epoxy flame retardant success.

INTRODUCTION

Epoxy Composite Flammability

Fiber reinforced polymer composites are a class of materials that bring mechanical strength and lighter weight, as well as resistance to rust/corrosion over existing metallic materials. These property benefits over metals allows them to bring fuel efficiency to vehicles, as well as maintenance cost savings over existing metal structures which require paint/depaint/corrosion protection efforts on a regular basis. For these reasons, these composites are being increasingly used in aircraft, automotive, mass transport (bus, rail, shipping), and in some applications, buildings. The range of chemical compositions for these composites can vary greatly due to the wide range of polymers and fiber types available. Common polymers used in composites include polyolefins (polyethylene, polypropylene), polyesters, polyamides (nylons), engineering plastics (such as polyetherimide, polyetheretherketone), and thermoset resins (vinyl esters, epoxies, bismaleimides, polyimides). Fiber types are not as varied, with fiberglass and carbon fiber making up most of the volume for composite use. The composites growing the fastest in use are epoxy and carbon fiber composites, with the largest growth for these composites occurring in high performance automotive and aircraft. Some well-known examples of these epoxy-carbon fibers are their use in "hypercars" such as the Lamborghini Sesto Elemento,[1] increasing use for parts and components in mass production automotive,[2] and the carbon fiber used extensively on the Airbus A380 and Boeing 787 Dreamliner. The increase in fiber-reinforced composite use is not expected to abate anytime soon due to the benefits these composites bring, and likely will increase as new carbon fiber capacity and newer composite manufacturing methods become available as well. However, the composites are not an equal replacement when it comes to fire protection. Composite structures can fail well before they ignite, as polymers lose structural rigidity once they are heated past their glass transition temperature, leading to in some cases catastrophic fire / structural losses.[3,4,5,6,7,8] Further, fire-fighting approaches for carbon fiber are somewhat different than they are for metal structures, which require different tools to penetrate the structure and put the fires out.[9] Further, there has been some recent reports finding that the carbon fibers in a fire can break loose and present additional inhalation hazards that would not be seen in a fire involving metal structures.[10,11] Clearly, fiber-reinforced composite is not the same as metal, and presents a higher heat release / fuel load wherever it used, and it does require the use of fire protection.

Epoxy Fire Protection Approaches

Fire protection for fiber-reinforced composites generally falls into two categories: fire protection barriers and flame retardant additives. Sometimes the two are combined, but in general the two are used exclusively depending upon end-use applications. Fire protection barriers are used most often for composites where mechanical strength of the composite must be kept high, and adding anything to the composite (such as flame retardant additives), can sometimes weaken composite mechanical properties. The barrier approach is used often with structural composites. Flame retardant additives are used when barriers cannot be used, or, the flame retardant needs to be integrated into the polymer. The additive approach is used most often with composites that will be exposed, or, have other things bonded/connected to them, such as epoxy-based circuit boards or structural composites which may only have aesthetic fascia attached to them. Some specifics on some of the specific chemistries are discussed below.

Fire Protection Barriers:

The use of a fire protection barrier is a proven approach for allowing fiber-reinforced composites to resist fires for set periods of time, or, to have some flame retardant performance in specific fire risk scenarios. The fire protection barrier can be an insulation batting or metal barrier that protects the composite from heat/flame. Examples of this include mineral wool insulation and metal shrouds / plates / panels that cover the composite. The barrier can also be something more directly connected to the composite, such as an intumescent paint or intumescent flame retardant/polymer film with is "co-cured" (co-polymerized) directly onto the outer layers of the composite during composite manufacture. Intumescent paints are very commonly used to provide fire protection to composites, but, can debond or be scratched open which can limit their effectiveness.[12,13] However, they are easily applied to composites and do not affect the composite mechanical properties in their end-use applications. The use of co-cured fire protection barriers where the flame retardant and/or intumescent is covalently bonded to the top layer of the composite has been covered in the open literature recently, and it shows some great promise in providing fire protection which is integral to the composite.[14,15,16,17] These coatings, since they are integrated into the top layer of the composite, cannot be scratched off or easily removed, and so provide durable fire protection when compared to paints. Further, because they are only located on the outer/top layers of the composite, they tend to have no negative effects on composite mechanical properties. However, they do require some additional manufacturing steps and control of resin viscosity during composite manufacture so the integrated top protection barrier is not ruined or damaged during composite manufacture.

Flame Retardant Additives:

Where coatings and other fire protection barriers are not appropriate for use, or, there is a desire for additional fire protection should the barrier fail, flame retardant additives are often used for fiber-reinforced composites. The chemistries for these additives are highly varied, and are covered in several good review papers.[18,19,20] The term "additive" refers to any flame retardant chemical which is "added" to the composite, and is used to cover both chemicals which do not react with the polymer during composite manufacture, as well as those that have functional groups which can react / copolymerize with the other monomers used to make the fiber-reinforced composite. If we consider epoxy resins used in composites, an example of an non-reactive additive would be aluminium diethylphosphinate, while an example of a reactive additive would be tetrabromobisphenol A diepoxide. There are pros and cons to the use of additives, which depend upon how the additive does/does not react with the polymer during composite manufacture. Non-reactive additives are easily added to a material and tend to not interfere with polymer processing / polymerization kinetics, but, they can present other problems including "filtering out" as resin + flame retardant passes through fiber reinforcement during manufacturing if the additive is insoluble in the monomers. Further, these additives can migrate out of a material over time if they are not reacted in, resulting in loss in fire protection performance over time, as well as chemicals migrating into the environment which presents other problems if the additives are persistent, bioaccumulative, and/or toxic (PBT). Reactive flame retardant additives on the other hand cannot ever migrate out of the composite because they are covalently reacted into the polymer during composite manufacture, but, they can cause changes in polymer viscosity and processing which results in manufacturing changes that can add cost to the final product. When considering the two approaches with fiber-reinforced composites, the reactive approach is the better one to utilize because it eliminates any and all PBT issues, and, it ensures fire protection over the lifetime of the product. With this in mind, we have focused on reactive flame retardants which eliminate the PBT issues once they are incorporated into the polymer during manufacture. In this paper, we present some newer work conducted in our labs utilizing phosphorus based flame retardants which have epoxy functional groups that can react with the epoxy during epoxy composite manufacture. The chemical structures of the materials we evaluated are shown below in Figure 1.

Figure 1. Epoxy Functionalized Phosphorus Flame Retardants

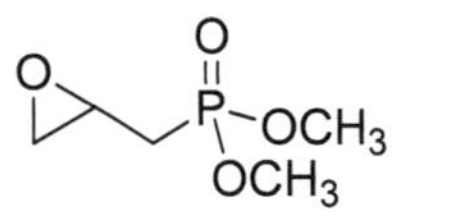

Dimethyl Epoxyphosphonate (DMEP)

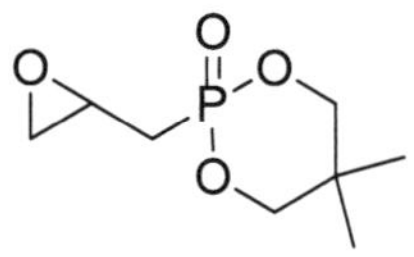

Epoxycyclophosphonate (ECPh)

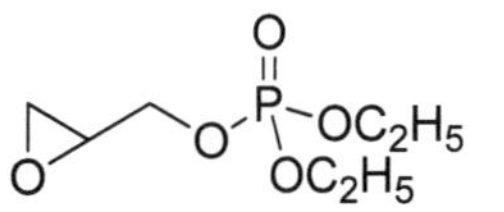

Diethyl Epoxyphosphate (DEEP)

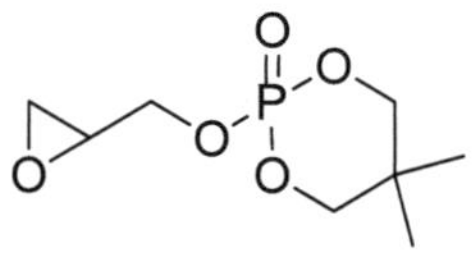

Epoxycyclophosphate (ECP)

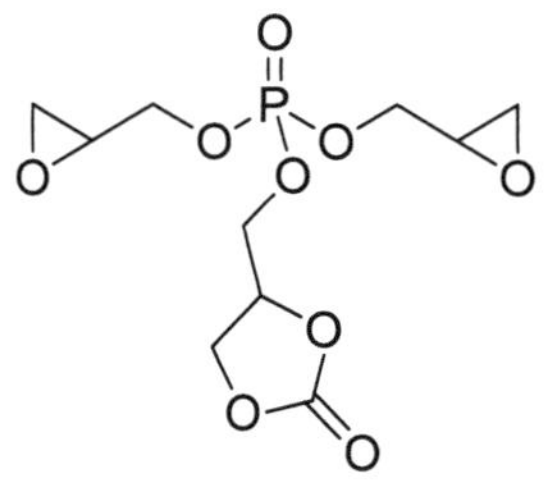

Diepoxy Carbonate Phosphate (DECP)

EXPERIMENTAL SECTION:

Epoxy Sample Preparation

The epoxy used for this effort was an epoxy Novolac (D.E.N. 438) originally sold by Dow Chemical and now sold by Olin. Curing agent (dicyandiamide) was purchased from Acros Organic and was used as received. The catalyst was 2-methylimidazole, and was purchased and used as received from Aldrich Chemical. The five epoxy functionalized flame retardants listed in this paper (DMEP, DEEP, DECP, ECPh, ECP) were synthesized and characterized as per previously published work.[21] 2.37g of the epoxy was weighed into a disposable Aluminum pan toward one side. The curing agent and the catalyst were weighed into the same pan to the opposite side. A few drops of ethanol was added to the catalyst and stirred till dissolved. Then some ethanol (~5 ml) was added to the curing agent and stirred. The curing agent does not dissolve fully in the ethanol, but will appear to "wet out" and disperse better. Then the pan was placed on a hot plate and gently heated till the epoxy thinned. The epoxy was stirred into the liquid catalyst and mixed a long time (~15-20 minutes) till it appeared clear and homogeneous. Then the curing agent was mixed in and stirred until the mixture appeared cloudy. If the epoxy sample was noted to be getting too hot (starting to gel too fast without full mixing of the additives) the sample was rapidly removed from heating and allowed to cool prior to further processing. Upon complete mixing, the sample was placed in a vacuum oven for 60 minutes at a temperature of 70°C to pull off the ethanol. Then it was cured at 170°C for one hour. Batch weights for the samples made in this paper are shown below in Table 1, which was based on the information in Table 2.

Table 1. Batch weights for a 2.5g sample batch:

Batch #	Polymer DEN 438 (g)	DMEP (g)	ECPh (g)	DEEP (g)	ECP (g)	DECP (g)	Curing agent (g)	Catalyst (g)	Total batch weight (g)
1	2.308	0.329					0.125	0.005	2.767
2	2.308		0.417				0.125	0.005	2.855
3	2.308			0.417			0.125	0.005	2.855
4	2.308				0.446		0.125	0.005	2.884
5	2.308					0.625	0.125	0.005	3.063

Table 2. Flame Retardant Molecular Weight and Chemical Information / Formula Calculations.

FR	Molecular weight (g/mol)	% P	Grams to yield 0.25g for a 10 g batch (2.5% P)	Grams to yield 0.0625g for a 2.5g batch (2.5% P)
DMEP	166	19	1.316	0.329
ECPh	206	15	1.667	0.417
DEEP	210	15	1.667	0.417
ECP	222	14	1.786	0.446
DECP	310	10	2.50	0.625

Micro Combustion Calorimeter (MCC) Testing:

The samples provided by UDRI were tested with the MCC at 1°C/sec heating rate under nitrogen from 150 to 700°C using method A of ASTM D7309-12 (pyrolysis under nitrogen). Each sample was run in triplicate to evaluate reproducibility of the flammability measurements.

RESULTS AND DISCUSSION

For initial small scale flammability performance screening, we utilized the micro combustion calorimeter (MCC) as per the ASTM D7309-12 standard. MCC is also known as pyrolysis combustion flow calorimetry (PCFC),[22] and has been shown to be a very useful tool in measuring the inherent heat release of polymers with and without flame retardant structures.[23,24,25,26,27,28,29,30] The technique does have its limitations,[31] but, relative to the goals of composite fire safety, it does measure heat release reduction and char formation. Increased char formation will suggest that the composite will be able to hold onto more mass that is not converted into heat from burning, and that increased char yield should relate to reductions in heat release. Specifically, more of the carbon-based polymer fuel will be converted into high thermal stability carbon/semi-graphitic char which cannot be easily pyrolyzed and combusted further. This char will not be available for further combustion. Lowered heat release from the remaining available polymer fuel in the composite will ensure that when the composite is burning, it is burning with lower intensity that allows for ease of extinguishment, lowered flame spread, and the potential for self-extinguishing behavior depending upon the intensity of the impinging flame / heat source.

Measured results from the MCC are shown in Table 3and cover the following measurements:

- Char yield: This is obtained by measuring the sample mass before and after pyrolysis. The higher the char yield, the more carbon/inorganic material left behind. As more carbon is left behind, the total heat release should decrease.

- HRR Peak(s): This was the recorded peak maximum of heat release rate (HRR) found during each experiment. The higher the HRR value, the more heat given off at that event. This value roughly correlates to peak heat release rate that would be obtained by the cone calorimeter.
- HRR Peak(s) Temp(s): The temperatures at which peak HRRs were measured.
- Total HR: This is the total heat release for the sample, which is the area under the curve(s) for each sample analysis.

Along with the epoxy control and the five epoxy + flame retardant samples, a polystyrene standard is given as an internal control as per the ASTM standard. The data in Table 3 shows that all of the phosphorus based flame retardants combined with the epoxy yielded notable increases in char yield, and significant decreases in peak heat release rates. However, all of the flame retardant samples also show an earlier onset of heat release and a multi-peak heat release rate (HRR) curve, indicating that the flame retardants are changing the decomposition chemistry of the base epoxy. The exact mechanisms of chemical change and heat release reduction cannot be elucidated exactly from the MCC data, but, the results do make it clear that the reduced heat release is due to enhanced char formation, as indicated by the improved char yield simultaneously measured with the lowered peak HRR values. Some of the peak HRR values are quite a bit lower than that of the control, indicating that the flame retardant not only enhances char yield and lowers the total fuel (Total HR) available for burning, but, it also lowers the intensity of burning vs. the non-flame retarded material. The results show good potential for these flame retardants and merit further consideration for scale-up and combination with fibers for composite testing.

Table 3. Epoxy + Phosphorus Based Flame Retardants

Sample	Char Yield (%)	HRR Peak(s) Value (W/g)	HRR Peak(s) Temp(s) (°C)	Total HR (kJ/g)
Epoxy control	26.36	352	441	17.4
	26.18	337	445	17.4
	26.31	330	440	17.2
ECPh	39.82	145, 100, 67	368, 426, 537	14.1
	40.03	142, 97, 66	373, 424, 537	13.9
	39.80	131, 108, 69	371, 426, 528	14.2
ECP	41.39	112, 92, 60	359, 430, 504	13.1
	42.01	115, 97, 58	358, 424, 524	12.8
	41.89	107, 90, 59	345, 425, 523	13.1
DMEP	43.49	119, 108, 55	365, 415, 529	13.2
	43.28	107, 113, 57	368, 419, 529	13.2
	43.78	121, 109, 56	366, 425, 527	12.7
	43.36	117, 110, 53	368, 414, 533	13.1
	43.89	124, 108, 53	366, 417, 527	12.8
DEEP	40.97	68, 163, 44	361, 437, 529	13.0
	41.28	83, 156 42	360, 415, 537	13.3
	43.02	72, 125, 52	360, 421, 526	12.9
	42.77	76, 137, 48	359, 414, 531	12.9
	44.02	66, 120, 51	354, 417, 526	12.5
DECP	41.27	112, 114, 30	358, 406, 550	12.6
	40.75	137, 110, 29	366, 415, 552	12.6
	40.69	129, 111, 30	361, 419, 553	12.7
PS std 5-13-16	0.03	1388	447	38.8
	-0.08	1387	447	38.9
	-0.08	1401	446	38.9

Heat release rate curves and char pictures for the samples tested in this paper are shown below in Figures 2-8. Some samples show good reproducibility of measurement, while others show some scatter in the reproducibility of the material, suggesting that some improvement in mixing of flame retardant and epoxy may still be needed for those materials. Along with the HRR curves, char pictures are shown, and all of the samples show minimal flow of the char during thermal decomposition, suggesting very low levels of deformation during burning.

Figure 2: Heat Release Rate Curve (left) and char picture (right) for DEN 438 + ECPh.

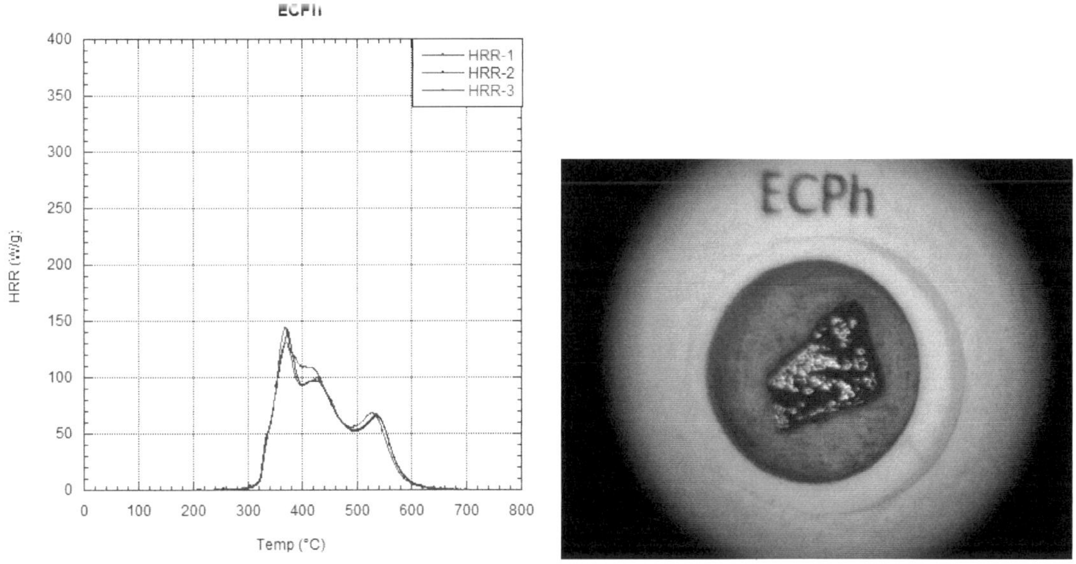

Figure 3: Heat Release Rate Curve (left) and char picture (right) for DEN 438 + ECP.

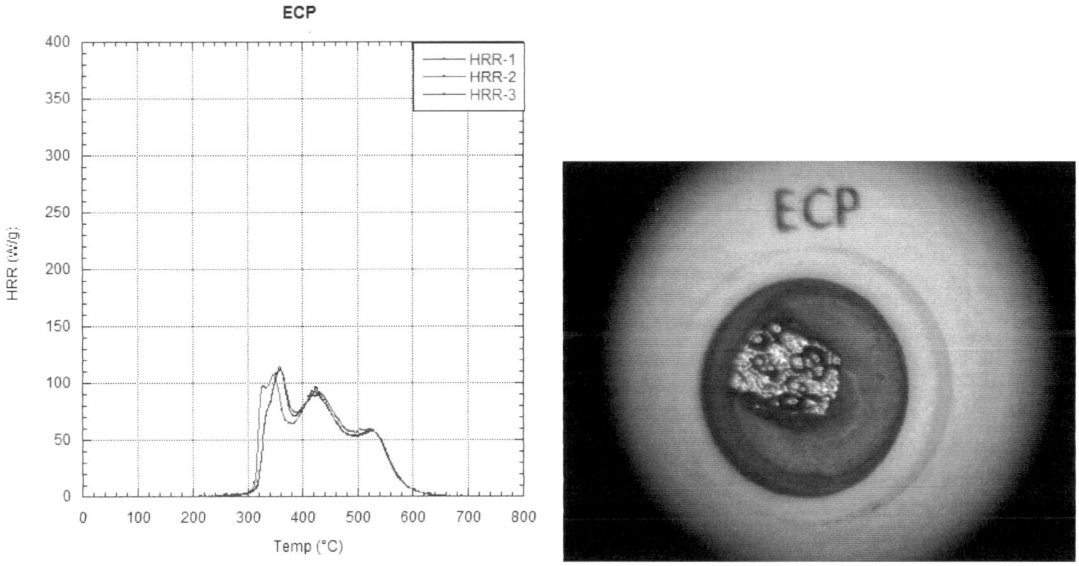

Figure 4: Heat Release Rate Curve (left) and char picture (right) for DEN 438 + ECP2.

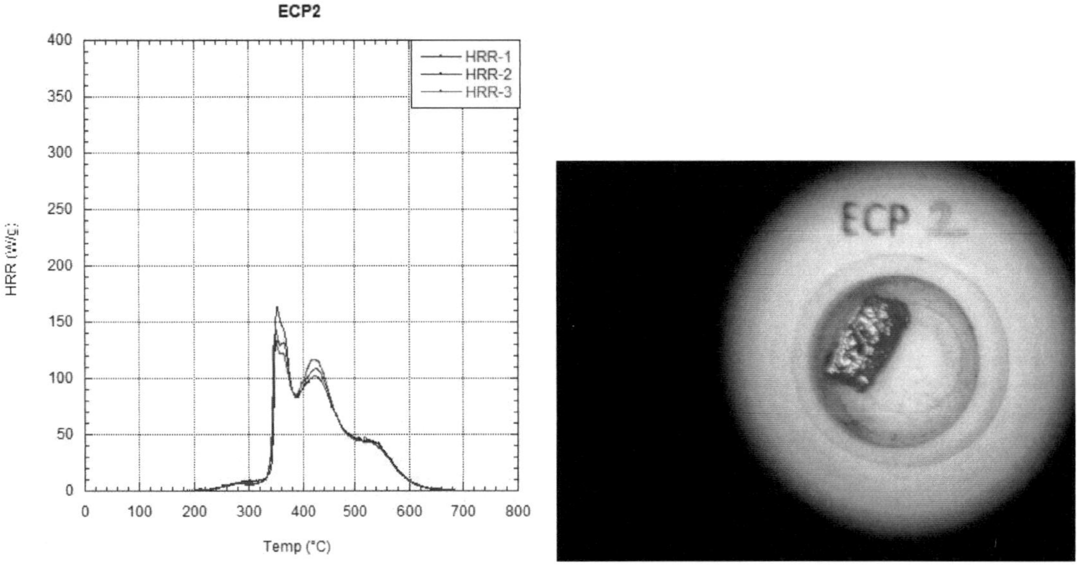

Figure 5: Heat Release Rate Curve (left) and char picture (right) for DEN 438 + DMEP.

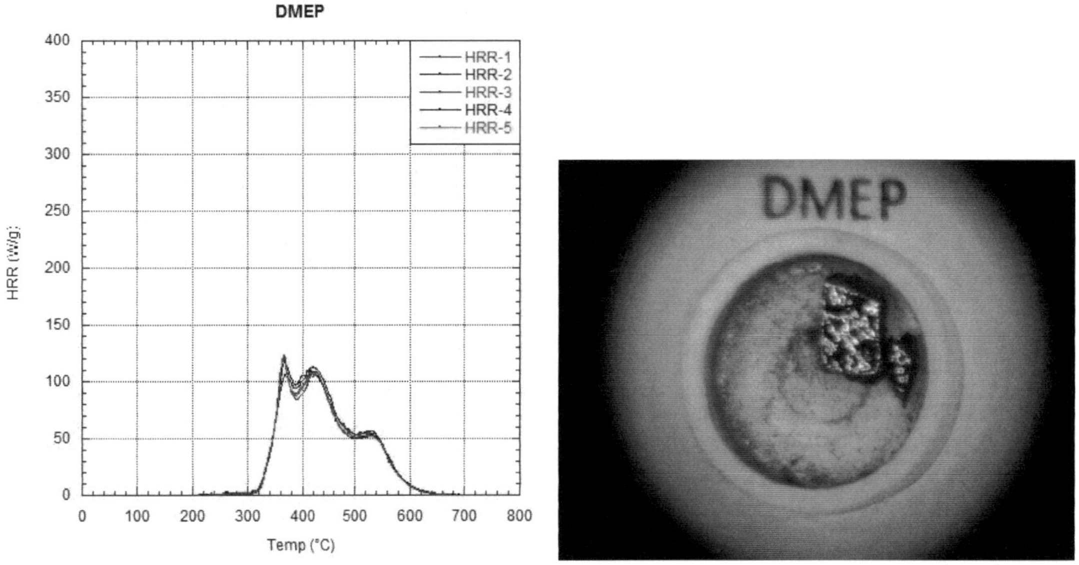

Figure 6: Heat Release Rate Curve (left) and char picture (right) for DEN 438 + DEEP.

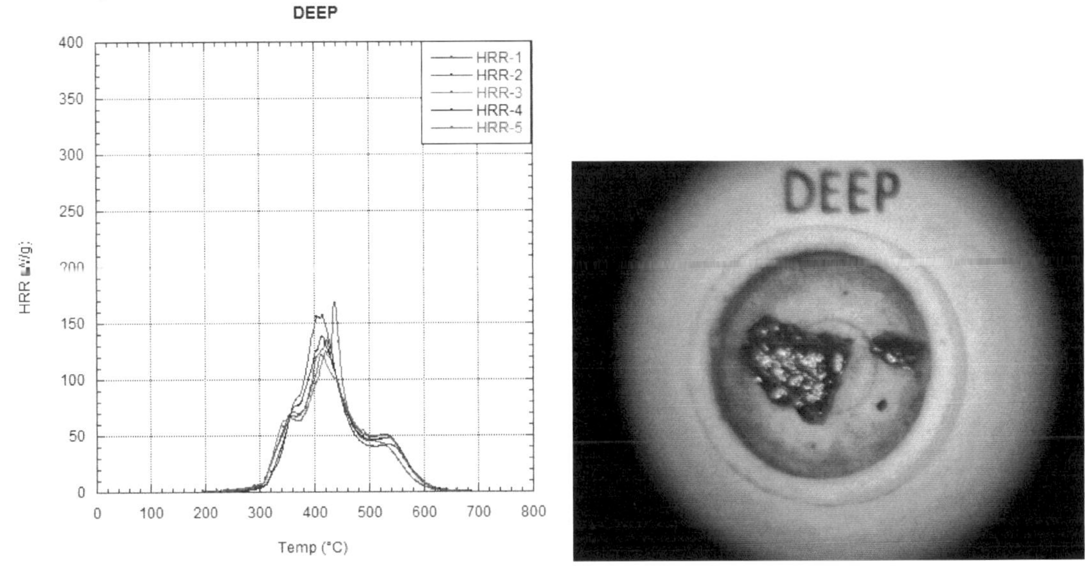

Figure 7: Heat Release Rate Curve (left) and char picture (right) for DEN 438 + DECP.

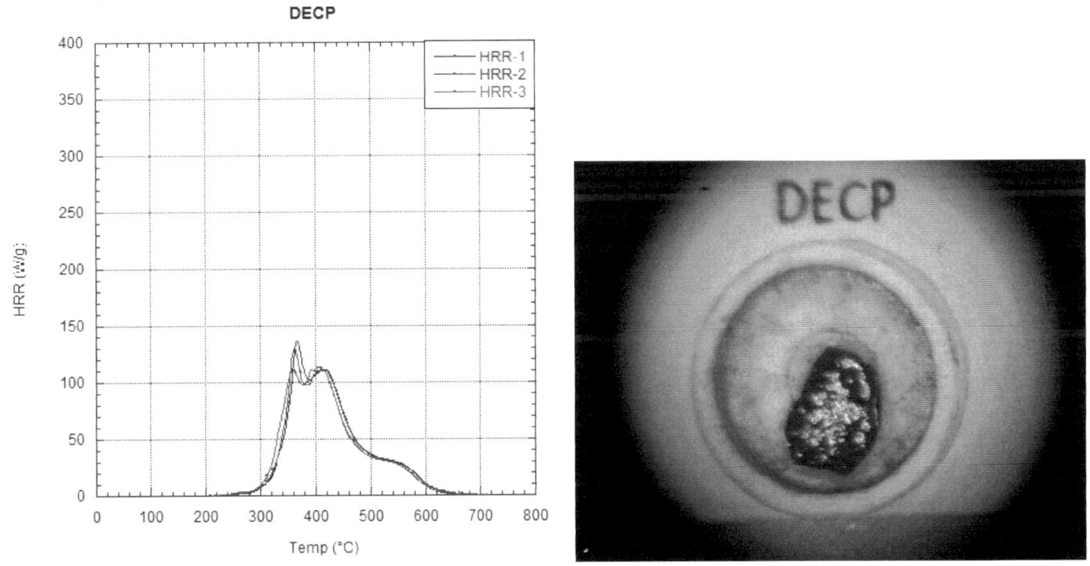

Figure 8: Heat Release Rate Curve (left) and char picture (right) for DEN 438 control.

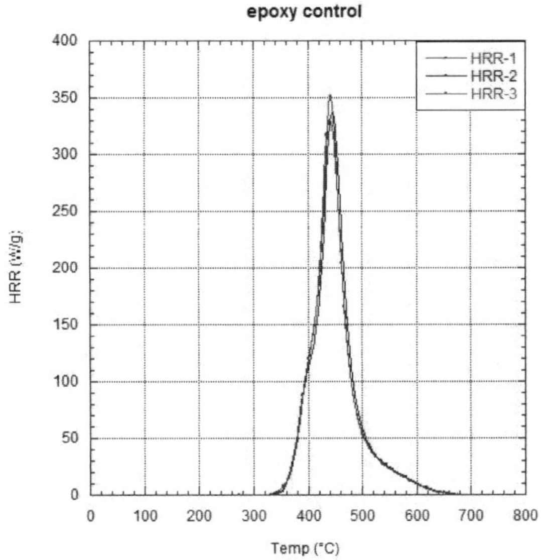

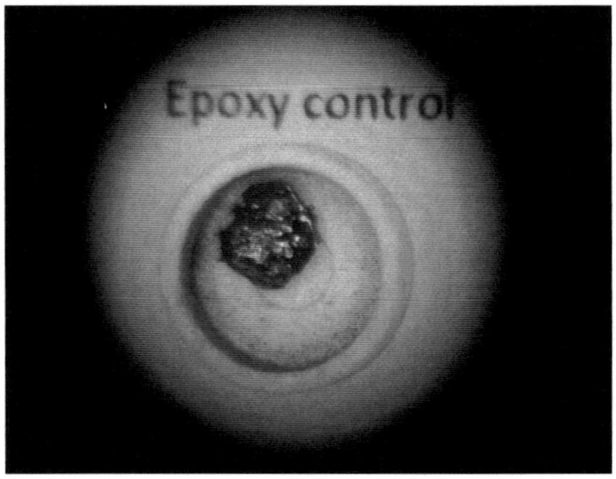

CONCLUSIONS

In this paper, we have shown the effects of new reactive flame retardants for epoxy, with all of the additives showing significant reductions in peak HRR and total HR compared to the control epoxy. The increase in char yield combined with lowered heat release suggests that the epoxy-based flame retardants presented in this paper are condensed phase / char formation flame retardants. Additional studies on these flame retardants, including their effect on glass transition temperature and mechanical properties, compatibility with fiber reinforcement, and composite manufacturing methods, are recommended to see if the technologies in this paper would yield self-extinguishing behaviour or other enhanced flame retardant effects in other tests. Further, the effect of these flame retardants on mechanical properties (positive, negative, or no effect) needs to be validated as well since just about all composite applications require fire safety and mechanical strength. Relative to transportation applications, reduced heat release should make it easier to extinguish composite fires, but, full scale fire testing to verify this. Still, the results are highly promising when compared to the epoxy control and show the value of these potential reactive phosphorus-based flame retardants for epoxies.

REFERENCES

[1] https://en.wikipedia.org/wiki/Lamborghini_Sesto_Elemento (last accessed 11/05/16)

[2] https://www.speautomotive.com/SPEA_CD/SPEA2010/pdf/TS/TS8.pdf (last accessed 11/05/16)

[3] "Fire Properties of Polymer Composite Materials" Eds. Mouritz, A. P.; Gibson, A. G. Springer-Verlag, The Netherlands, 2006. ISBN 978-1-4020-5355-9.

[4] "Effect of stress on the fire reaction properties of polymer composite laminates" Elmughrabi, A. E.; Robinson, M.; Gibson, A. G. *Polym. Degrad. Stab.* **2008**, *93*, 1877-1883.

[5] "Modeling composite high temperature behavior and fire response under load" Gibson, A. G.; Browne, T. N. A.; Feih, S.; Mourtiz, A. P. *Journal of Composite Materials* **2012**, *46*, 2005-2022.

[6] "Fire behaviour of composite laminates" Dodds, N.; Gibson, A. G.; Dewhurst, D.; Davies, J. M. *Composites Part A* **2000**, *31*, 689-702.

[7] "Flammability Properties of Aircraft Carbon-Fiber Structural Composite" Quintiere, J. G.; Walters, R. N.; Crowley, S. DOT/FAA/AR-07/57 Report, October 2007.

[8] "Wrinkling of sandwich polymer matrix composite panels under transverse thermal gradients" Gu, Pei; Asaro, R.J. *Fire Safety Journal* **2008**, *43*, 151 – 160

[9] http://www.boeing.com/assets/pdf/commercial/airports/faqs/787_composite_arff_data.pdf (last accessed 11/05/16)

[10] "Influence of carbon fibre orientation on reaction-to-fire properties of polymer matrix composites" Eibl, S. *Fire*

Mater. **2012**, *36*, 309-324.
[11] "Dangers relating to fires in carbon-fibre based composite material" Hertzberg, T. *Fire Mater.* **2005**, *29*, 231-248.

[12] "Fire-Protective and Flame-Retardant Coatings – A State-of-the-Art Review" Weil, E. D. *J. Fire Sci.* **2011**, *29*, 259-296.
[13] "Recent developments of intumescent fire protection coatings for structural steel: A review" Mariappan, Thirumal *J. Fire Sci.* **2016**, *34*, 120 – 163
[14] "Reaction-to-fire properties of polymer matrix composites with integrated intumescent barriers" Eibl, S. *Fire Mater.* **2015**, *39*, 658-674.
[15] "The use of fire-retardant intumescent mats for fire and heat protection of glass fibre-reinforced polyester composites: Thermal barrier properties" Kandare, E.; Chukwudole, C.; Kandola, B. K. *Fire and Materials* **2010**, *34*, 21-38.
[16] "Study of fire retardant behavior of carbon nanotube membranes and carbon nanofiber paper in carbon fiber reinforced epoxy composites" Wu, Q.; Zhu, W.; Zhang, C.; Laing, Z.; Wang, B. *Carbon* **2010**, *48*, 1799-1806.
[17] "The effect of fire-retardant additives and a surface insulative fabric on fire performance and mechanical property retention of polyester composites" Kandare, E.; Chukwunonso, A. K.; Kandola, B. K. *Fire Mater.* **2011**, *35*, 143-155.
[18] "A review of current flame retardant systems for epoxy resins" Weil, E. D.; Levchik, S. *J. Fire. Sci.* **2004**, *22*, 25-40.
[19] "Commercial Flame Retardancy of Unsaturated Polyester and Vinyl Resins: Review" Weil, E. D.; Levchik, S. *J. Fire Sci.* **2004**, *22*, 339-350.
[20] "Thermal decomposition, combustion and flame-retardancy of epoxy resins – a review of the recent literature" Levchik, S. V.; Weil, E. D. *Polym. Int.* **2004**, *53*, 1901-1929.
[21] "Synthesis and flammability testing of epoxy functionalized phosphorus-based flame retardants" Benin, V.; Cui, X.; Morgan, A. B.; Seiwert, K. *J. App. Polym. Sci.* **2015**, DOI 10.1002/APP.42296.
[22] "Pyrolysis combustion flow calorimetry" Lyon, R. E.; Walters, R. N. *J. Anal. Appl. Pyrolysis* **2004**, *71*, 27-46.
[23] "Molar Group Contributions to Polymer Flammability" Walters, R. N.; Lyon, R. E. *J. App. Polym. Sci.* **2002**, *87*, 548-563.
[24] "Microcombustion calorimetry as a tool for screening flame retardancy in epoxy" Morgan, A. B.; Galaska, M. *Polym. Adv. Technol.* **2008**, *19*, 530-546.
[25] "Prediction of the burning rates of charring polymers" Stoliarov, S. I.; Crowley, S.; Walters, R. N.; Lyon, R. E. *Combustion and Flame* **2010**, *157*, 2024-2034.
[26] "Flammability and Thermal Analysis Characterization of Imidazolium-Based Ionic Liquids" Fox, D. M.; Gilman, J. W.; Morgan, A. B.; Shields, J. R.; Maupin, P. H.; Lyon, R. E.; De Long, H. C.; Trulove, P. C. *Ind. Eng. Chem. Res.* **2008**, *47*, 6327-6332.
[27] "Correlations between pyrolysis combustion flow calorimetry and conventional flammability tests with halogen-free flame retardant polyolefin compounds" Cogen, J. M.; Lin, T. S.; Lyon, R. E. *Fire Mater.* **2009**, *33*, 33-50.
[28] "A molecular basis for polymer flammability" Lyon, R. E.; Takemori, M. T.; Safronava, N.; Stoliarov, S. I.; Walters, R. N. *Polymer* **2009**, *50*, 2608-2617.
[29] "Evaluation of three flame retardant (FR) grey cotton blend nonwoven fabrics using micro-scale combustion calorimeter" Parikh, D. V.; Nam, S.; He, Q. *J. Fire Sci.* **2012**, *30*, 187-200.
[30] "Material properties and fire test results" Lyon, R. E.; Safronava, N.; Quintiere, J. G.; Stoliarov, S. I.; Walters, R. N.; Crowley, S. *Fire Mater.* **2014**, *38*, 264-278.
[31] "Pyrolysis combustion flow calorimeter: A tool to assess flame retarded PC/ABS materials?" Schartel, B.; Pawlowski, K. H.; Lyon, R. E. *Thermochimica Acta* **2007**, *462*, 1-14.

USING NATURE TO FLAME RETARD POLYOLEFINS

David A. Schiraldi* & Taneisha Deans
Case Western Reserve University, USA

A version of this paper is being submitted to the *Journal of Fire Sciences*, January 2017.

INTRODUCTION

Polyolefins are amongst the most flammable of all polymers, given their minimal degree of oxidation, and high hydrogen-to-carbon ratios. These polymers are, nonetheless, used ubiquitously throughout electronic, packaging, transportation, appliance and home furnishing industry, in many flammability-sensitive applications. In order to make use of polyolefins in such applications addition of one or more endothermic (metal hydrates), flame-quenching (organohalides), and barrier-forming materials (typically phosphorous-based) is required. Many of these flame retarding agents bring with them environmental and toxicological concerns, and the large loadings of these additives generally degrade polymer processing and mechanical properties significantly.

The present research is focused on the use of naturally-occurring materials, especially those from the food processing industry, to replace traditional flame retarding agents in polyolefins. Previous efforts at using bio-based flame retardants have examined, for example, modified starch and sugars, lignin, ramie fiber, tannins, chitosan and chitin. Chitosan and chitin have been investigated as coating for fabrics, foams, and cotton by a layer-by-layer procedure. Sucrose-derived isosorbide was derivatized to produce a polyphosphonate, which was then blended with polylactic acid. In this paper we discuss the use of gelatine and tannins, by themselves and in combination with relatively benign traditional flame retardants, in low density polyethylene. Similar results have been obtained with high density polyethylene and polypropylene, but the one system to be discussed herein is illustrative of what we have learned.

EXPERIMENTAL

Tannic Acid (TA), stearic acid (Str A), poly(tetrafluroethylene) (PTFE; 1 μm particle size), poly(ethylene-graft- meleic anhydride)(PE-g-MA), silicone oil (s.o), and fish gelatine from cold water fish (FG), (Sigma Aldrich); melamine cyanurate (Melapur® MC25;MC;BASF), and magnesium hydroxide-(MH; Vertex® 60SV; J. M. Huber), and alumina (Dispal® 10C; AP; Sassol) and Low density polyethylene (LDPE; Dow; mfi=5.0) were used without further purification

A Haake PolyLab OS RheoDrive 7 batch mixer was used to blend the additives with the polymer. TA, MH and AP were dried at a temperature of 110° C for 24 hours. All the other additives, except silicone oil and PTFE were dried at 60° C for 24 hours under vacuum. These materials were transferred to the mixer in desiccators. Typical composition are given in **Table 1.** The compositions were converted into films using a Carver Model 4122 press, processing at the same temperatures as were used for melt blending, and operating at a pressure of approximatley 1.4 MPa, applied in a stepwise fashion to the polymer melt (the films were melted in the press for 8 minutes without the addition of pressure and 6 minutes with pressure). Each sample was cut into dimensions similar to those required for the UL-94 vertical test: an average thickness of 1.2-1.3mm, length 127 mm, width 12.7 mm. Key samples were also scaled up using a ThermoScientific TSE24MC 24-mm corotating twin-screw extruder with a 40:1 *L/D* ratio. When FG was compounded, all the zones were set to 150° C; all the other systems were processed at 180° C. The extruder screw rotation speed was set to a constant 500 rpm for all experiments. Materials were fed into the extruder at a constant rate of 5.5 kg/h for all experiments using a Brabender volumetric feeder with a single spiral screw configuration. Material was extruded through a three-hole strand die with a hole diameter of 3 mm and collected through a ThermoScientific water bath and pelletizer. Then the pellets are dried overnight in the vacuum oven at 60° C, and were later injection molded.

A Boy model 22s molding machine, operating with a barrel temperature of either 150 °C or 180° C, mold temperature of 100° C, and injection pressure of 17.2 MPa, equipped with a family mold which produced a Type IV tensile bar which conforms to ASTM standard D-683 and a flex bar which can be used for both impact (ASTM D-256) and flexural (ASTM D-790) testing was used in this study.

Table 1

SPECIMENS	TA	FG	MC	MH	STR. A	PE-G-MA	PTFE	SILICA
1	-	50	-	-	4	5	4	-
2	7	42	-	-	4	5	4	-
3	7	-	-	42	4	5	4	-
4	7	-	42	-	4	5	-	2

Flammability testing was carried out in accordance to the UL 94 vertical test, and cone calorimetry. For UL 94, the molded samples were clamped in a metal clamp, then a blue natural gas flame measuring 20 mm was applied at a 90° angle at a distance of 10 mm. Then the fire was applied for 10 seconds and then removed for 3 secs. The flame was reapplied for another 10 seconds. This was carried out for the compression mold samples; injection molded specimens were tested using an SL-S29 UL94 Horizontal-Vertical Flame Chamber (Skyline Instruments Co.). Cone calorimetry was carried out using a BECC model manufactured by INELTEC, Spain, in collaboration with the Polytechnic University of Catalunya. Specimens with a size of 100 mm × 100 mm × 25 mm, weight range from 20 to 25 g, and a thickness of 1/8", were tested under a heat flux of 50 kW/m.

The thermal stabilities of samples were evaluated using a TA instruments TGA Q500 operating under air and nitrogen conditions, without prior drying of the samples. Sample sizes between 5-10 mg were used, and were ramped to 650 ºC at a rate of 10ºC/min.

Mechanical properties were examined in uniaxial tension using an Instron 5565 universal tester. Dog bone samples that were obtained from the injection molding that follow the ASTM-D638V standard was tested at room temperature at a strain rate of 10 mm/min. Five samples were tested for each composition and the average values were obtained.

RESULTS AND DISCUSSION

LDPE was readily compounded, using both batch and continuous twin screw methods, incorporating fish-derived gelatine, tannins, magnesium hydroxide and melamine cyanurate as the primary flame retardants. Processing aids, compatiblizers and antidrip agents, including polytetrafluoroethylene, silicone oil, silica fibers, and poly(ethylene-graft- meleic anhydride) were included in low levels typical of commercial flame retarded engineering plastics (see **Table 1**). The compounds were compression molded and injection molded into test parts used for flammability and mechanical testing (below) and were also evaluated for thermal stability using thermogravimetic analysis (TGA) run under both air and nitrogen atmospheres.

Figures 1 and **2** show typical TGA traces for the LDPE compounds, run under nitrogen and air, respectively. The nitrogen atmosphere TGA tests show the additives to initiate thermal decomposition 20-50 degrees C lower than unmodified polyethylene; the presence of magnesium hydroxide to the composition gives rise to a second decomposition step associated with water loss from this hydrated mineral prior to the main decomposition of the polymer. TGA analysis run under air gives rise to a very different set of results. Flame retardant packages which include fish-derived gelatine and/or tannins show considerably higher thermal stabilities than is seen for the starting polymer itself. The decomposition steps are given in more detail in **Table 2.**

Figure 1

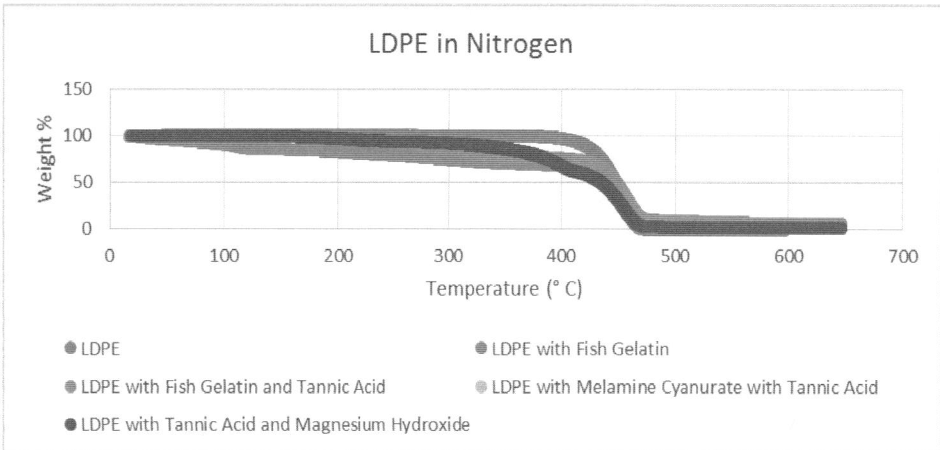

Figure 2

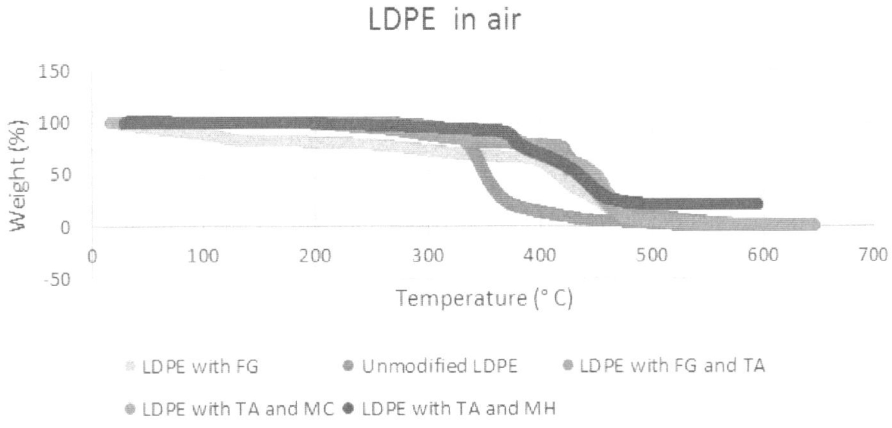

Table 2

Material(s)	Location of Step	$T_{decomp.}$	Char %, Temp (°C)
LDPE	-	298	0, 550
Fish Gelatin	217, 350, 481, 650	217	1.25, 650
LDPE with FG	45, 110, 138, 340, 395,474	340	0.48, 625
Tannic Acid		247	N/A
LDPE with TA, FG	196, 250, 346, 416, 424	416	0.75, 600
$Mg(OH)_2$	242, 310	310	16, 787
LDPE with TA, $Mg(OH)_2$	302, 380, 460	380	20, 594
Melapur MC25	-	275	61, 346
LDPE with TA, MC	254, 352, 411, 479	254	1.3 , 650

It is interesting that fish gelatine and tannic acid, by themselves, begin to degrade at temperature lower than that seen for LDPE polymer, itself, but when combined, these additives delay the main degradation of the polymer, suggesting sacrificial oxidative production. The overall highest level of oxidative stabilization of LDPE appeared to be obtained in the presence of both tannic acid and fish gelatine, this stabilization exceeding that obtained in the presence of traditional magnesium hydroxide or melamine cyanurate FR agents.

UL 94 testing of both compression molded films and injection molded bars were carried out, comparing LDPE to its compounded FR grades which incorporated fish gelatine, tannic acid, magnesium hydroxide and melamine cyanurate. A photo of burned bars from this test is given in **Figure 3**. From this image it is readily apparent that unmodified LDPE burned to the clamp, whereas the FR compounds could be produced such that they self-extinguish, gaining a V(0) rating. 35-50 weight percent FR additives were required in order to obtain V(0) performance, not atypical of traditional brominated and phosphorous-based FR packages in a wide range of engineering plastics. While none of these V(0) compositions were significantly more effective than the others, the combination of fish gelatine plus tannic acid, *Specimen 2* in **Table 1**, appeared to be the superior compound overall, judged by UL 94.

Figure 3

LDPE with:

- **No additive**
- **TA + gelatin**
- **TA + Mg(OH)2**
- **TA+ gelatin**
- **TA + melamine cyanurate**

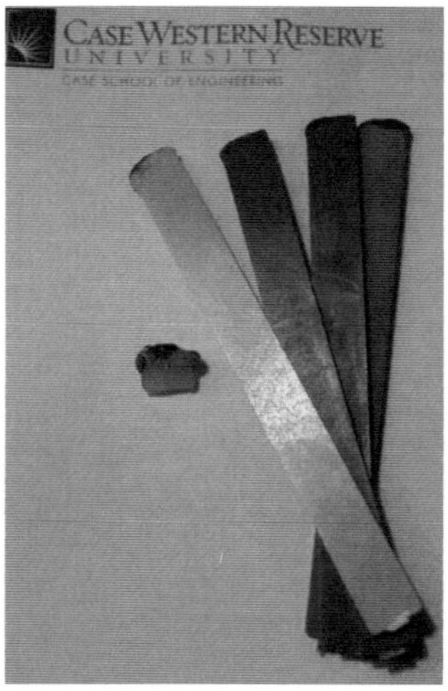

Cone calorimetry was used to investigate the performance of these compounds under a more profound fire condition. A summary of some key parameters from cone calorimetry is given in **Table 3**. see how these materials would perform in a fully developed fire. The time to ignition (TTI), peak heat release rate (PHRR), flame growth rate averages are reported for the best compositions tested (those depicted in **Figure 3**) paint a somewhat different, and more complex picture than was obtained by UL 94. LDPE, itself, has the highest average TTI of the materials tested. So even though the additives provided oxidative stabilization, and gave rise to superior, self-extinguishing behavior, they also led to earlier ignition under cone calorimetry conditions than was observed for the polyolefin without any additives. Essentially no difference in TTI was observed between the three compounded grades tested. The peak heat release value for LDPE incorporating fish gelatine plus tannic acid was slightly higher than that for unmodified polymer, but the combination of tannic acid plus either magnesium hydroxide or melamine cyanurate significantly lowered the peak heat release. FIGRA values for the compounds were somewhat comparable, except for the combination of tannic acid and magnesium hydroxide. So, using cone calorimetry as a screening tool, the combination of tannic acid plus magnesium hydroxide would appear to offer the highest level of fire protection of the materials evaluated.

Table 3.

SAMPLE	AVG. TTI (S)	AVG. PHRR (KW/M^2)	AVG. FIGRA (W/S)
LDPE	52± 6	799 ± 75	7.7 ± 1.0
LDPE WITH FG AND TA	38±5	836 ± 46	8.0 ± 0.7
LDPE WITH TA AND MC	40±6	653 ± 65	4.7 ± 0.5
LDPE WITH TA AND MH	39± 3	487 ± 25	11.0 ± 0.5

The mechanical properties of the above compositions were evaluated, using injection molded test specimens. **Table 4** gives tensile moduli and strength values, evaluated at the minimum necessary additive loadings necessary to obtain UL 94 V(0) performance, for each composition. Striking is that the presence of fish gelatine brings about a significant increase in modulus for the compounded materials, at a very small loss of strength. In the combination of fish gelatine and tannic acid, the modulus increase scales with the gelatine content – this clearly is the dominant factor in determining the increase in mechanical properties. In the other composition, which incorporate tannic acid and either melamine cyanurate or magnesium hydroxide, there is a smaller increase in modulus, and a very modest decrease in tensile strength for the samples. In each case, compounding of additives did decrease the ductility of the LDPE, though the samples given in **Table 4** still exhibited elongations at break of at least 100%, which is quite different from many flame retarded engineering plastics, which often lose virtually all ductility.

Table 4.

System	Modulus (MPa)	Yield Stress (MPa)
Unmodified LDPE	130 ± 10	14.2± 1
1 FG	430 ± 40	9.0 ± 0.4
2 FG + Tannin (TA)	300± 50	13.0 ± 0.6
3 TA+ $Mg(OH)_2$	270 ± 46	12.0 ±0.4
4 TA + Melamine Cyanurate	350 ± 40	12.3 ± 0.2

Combining the thermal and thermo-oxidative stabilities, vertical burn and cone calorimetry results, and mechanical properties, insights as to how these food-derived flame retardant systems act in LDPE is beginning to come together, though the story is certainly not fully developed. Tannins, specifically tannic acid (derived from wine grapes, tree bark, and other naturally-occurring sources) are known antioxidants, and appear to act as sacrificial agents, combusting rapidly to produce char. This oxidation/combustion is evident in the TGA tests run under oxygen, where the tannic acid degrades at a relatively low temperature, and in the cone calorimetry tests, where tannic acid leads to a low time to ignition. In the UL 94 test, however, tannic acid clearly generates char, and leads to self-extinguishing behavior. From the structure of tannic acid, **Figure 4**, it is easy to see how this structure condenses into graphite, a transformation known to occur in the bark of redwood (and other) trees during forest fires.

The role of fish gelatine is less clear, and is the subject of current experimentation. Our hypothesis is that gelatine plays two roles under fire conditions. From its chemical structure (**Figure 4**) it is not difficult to see how this molecule can generate crosslinked structure, much as the denaturing of proteins during heating (the frying of an egg being a good example of a fluid to crosslinked thermoset process), producing a thermoset material which will tend to char, and which will resist flow and dripping under fire conditions. Crosslinking of gelatine also is consistent with the significant increase in modulus seen in polyethylene parts which incorporate this additive – we hypothesize that a low level of crosslinking is obtained during the extrusion and molding steps, a much lower extent of conversion than is seen under fire conditions. From UL 94 testing, it appears that volatile gasses are also evolved from gelatine; nitrogen and carbon dioxide can well be imagined from its chemical structure. The combination of tannic acid and fish gelatine provides fuel-rich polyethylene with a ready char former and a potential thermosetting antidrip agent, a potent combination for resisting the ignition under UL 94 conditions. Under the more profound fire conditions encountered during cone calorimetry, these materials ultimately do serve as fuel and are burned. Tannic acid, in its role as a char former, appears to broadly improve the performance of magnesium hydroxide and melamine cyanurate in LDPE compounds. The levels of mechanical property and thermal property enhancements in these gelatine-free compounds is lower, but for applications where large fires may be encountered (modelled in cone calorimetry experiments), these compositions may be superior to those which include gelatine. *While not presented in this talk, similar results have now been obtained for high density polyethylene and for polypropylene, albeit with modestly different loadings of the same FR additives.*

Figure 4.

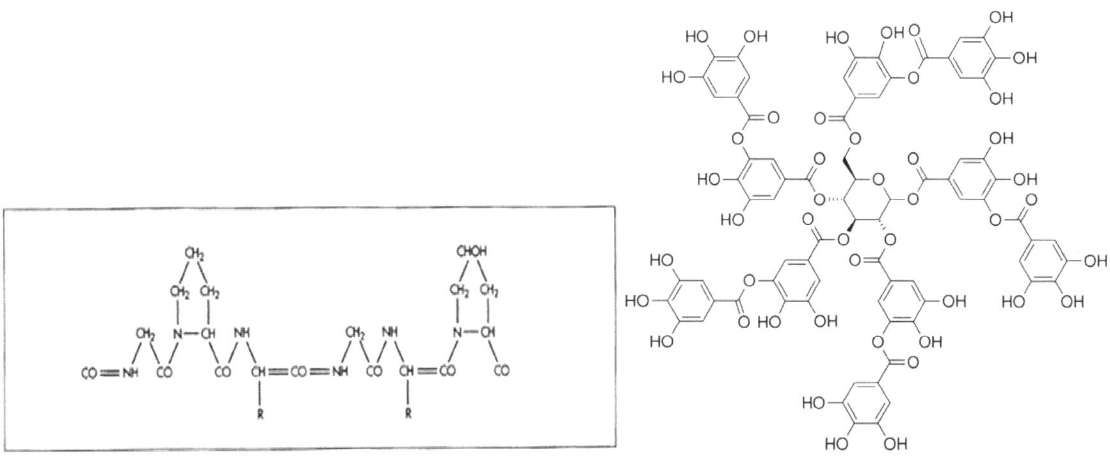

SUMMARY

Bio-based gelatine and tannic acid have been shown to serve as useful flame retardants for low density polyethylene (and in fact also for high density polyethylene and for polypropylene). These bio-based additives can be used by themselves or together, or in combination with other, know, non-halogenated flame retardants, including magnesium hydroxide and melamine cyanurate. The tannic acid appears to serve primarily as a sacrificial antioxidant and char former for the polyolefins under fire conditions. We hypothesis that the gelatine acts primarily as a thermosetting, crosslinking agent which prevents dripping under fire conditions, but may also generate inert gasses which reduce polymer flammability. For ignition source fires, such are modelled by the UL 94 test, the gelatine plus tannic acid FR package, at approximately 35 wt% loading, appears to provide optimum performance in LDPE. For more profound fires, as modelled by cone calorimetry, the combination of tannic acid and magnesium hydroxide, again at about 35 wt% appears to be optimal. Current work continues to explore the mechanism of gelatine in polyolefin flammability, and is expanding the use of these materials to other polymer systems.

PLANT-BASED FLAME RETARDANT COATINGS FOR FLEXIBLE POLYURETHANE FOAM

Douglas M. Fox,* Hussain Khalfan, Noy Kaufman
Department of Chemistry, American University, 4400 Massachusetts Ave, NW, Washington, DC, 20194, USA.

John R. Shields, and Rick D. Davis
Fire Research Division, Engineering Laboratory, National Institute of Standards and Technology, 100 Bureau Dr., Gaithersburg, MD, 20899, USA.

* - Guest Researcher at NIST

INTRODUCTION

Reducing the fire risks associated with upholstered furniture remains to be a challenge. Fires result in more than 20 times the deaths and injuries as all other natural disasters combined, the United States has one of the highest death and property loss rates due to fire among all industrial nations, and building fires represent one-third of all fires and 80% of the deaths related to fires.[1] In the recent past, brominated flame retardants have been added to flexible foam to reduce the fire risks associated with open flame sources.[2] In 2014, California changed their fire regulation for residential upholstered furniture (known as TB 117), eliminating the open flame tests.[3] Although this has reduced the regulatory need that was most often addressed by adding flame retardants, the development of effective flame retardants with low environmental and health safety (EHS) risks continues to be important. Regardless of the current regulations, open flame ignition sources and transitions from smoldering to flaming remains a significant fire risk in residential dwellings.[4] In 1993, fires that started in lounge areas, sleeping rooms, and structural areas had the highest percentage of fires that led to death, and those starting in the cooking, lounge or sleeping areas constituted the highest number of fires and fire related deaths.[5] And, a recent study by NIST determined that many of the 100 lives that perished in the 2003 fire at The Station nightclub in Rhode Island could have been prevented by use of flame retarded sound proofing insulation.[6] Researchers determined that the tested flame retarded materials would likely have withstood the 15 second impingement of pyrotechnics, which was the cause of the blaze. To address these needs, we have been developing flame retardant coatings based on natural materials. In previous studies, we have used sodium polyborate (a compound formed from boric acid and sodium sulfate) as a flame retardant, clay as a char enhancer, and carbohydrates as coating thickeners and coating binders.[7, 8] In this study, we examine the use of lignosulfonate and phytic acid as flame retardants, clay as a char enhancer, and plant proteins as coating binders. Flame spread and cone calorimetry experiments show that this is a promising approach.

EXPERIMENTAL*

Wheat gluten was obtained from either Bob's Red Mill (hpWG, 75 % to 80 % mass fraction protein) or Arrowhead Mills (WG, 60 % mass fraction protein) and zein protein (Z) was obtained from MP

* The policy of NIST is to use metric units of measurement in all its publications, and to provide statements of uncertainty for all original measurements. In this document however, data from organizations outside NIST are shown, which may include measurements in non-metric units or measurements without uncertainty statements.

Biomedicals. Sulfonated lignin (LS, REAX88B) was provided by Mead Westvaco Corporation, sodium montmorillonite (NaMT) was provided by Rockwood Specialties, Inc, and ammonium polyphosphate (APP, Exolit AP-422) was provided by Clariant. Tannic acid (ACS reagent) and phytic acid (PA, 50 % mass fraction in water) was obtained from Sigma Aldrich. 2-Ethylhexyl diphenyl phosphate (E, 90 % to 95 % mass fraction) was acquired from Scientific Polymer Products, Inc. Glacial acetic acid (ACS reagent), glycerol (G, ACS reagent), sodium hydroxide (ACS reagent pellets), and sodium sulfite (anhydrous, 98.5 % mass fraction) were acquired from Fisher Scientific. Ammonium hydroxide (28 % to 30 % mass fraction) was obtained from BDH Chemicals. Ethanol (200 proof) was acquired from Koptec. Type I ultrapure water was produced using a PURELAB Flex system from Veolia.

Open cell, polyester, polyurethane foam (PUF) containing 1% carbon black was obtained from UFP Technologies (Newburyport, MA) and stored in a climate-controlled room with no direct sunlight exposure. Foam was cut with an electric knife into 50 mm x 100 mm x 100 mm pieces for cone calorimetry and with a heated wire cutter into 12.5 mm x 50 mm x 150 mm for UL-94 flame tests. Small pieces, approximately 2 mm x 2 mm x 2 mm, were cut with scissors for thermogravimetric analysis and microcombustion calorimetry.

Foams were coated using a single dip procedure. Coating components were added to a 60 % mass fraction ethanol (aqueous) solution one component at a time, allowing time for dissolution or a stable suspension. The dissolution was aided by adding either sodium hydroxide or glacial acetic acid. A small amount of tannic acid (0.5 % mass fraction of the dry coating) was added as an antifungal compound. NaMT was added as a 4 % mass fraction solution prepared at least 24 h prior and APP was added as a 5 % mass fraction solution in 2 M NaOH at least 48 h prior to the coating preparation. Foam was soaked in solution 3 times and manually squeezed to remove excess solution. After the last soak, the foam was squeezed until there was a 300 % by mass solution pickup. The wet foam was placed on a rack and dried at 50 °C overnight. The codes used to describe the coatings used in this study are shown in Table I. The values represent the percent mass fraction of components added to the solution. The percent mass fraction of each component in the dried coating on the foam may differ slightly due to different selectivities towards foam.

Table I. Composition of dispersions/solutions (% mass fraction) used for foam coating

Code	Protein		Flame retardant			Plasticizer		Additive	
	WG	hpWG	LS	PA	APP	G	E	TA	NaMT
WG-10G	90					10		0.4	
WG-10LS	90		10					0.4	
WG-10LS-3G	87		10			3		0.4	
WG-10LS-10G	80		10			10		0.4	
WG-10PA-10G	80			10		10		0.4	
hpWG-10LS-10G		80	10			10		0.4	
hpWG-20LS-20G		60	20			20		0.4	
hpWG-20LS-20E		60	20				20	0.4	
hpWG-20LS-20E-1C		59	20				20	0.4	1
hpWG-20LS-10PA-20E-1C		49	20	10			20	0.4	1
hpWG-20LS-20E-6C		54	20				20	0.4	6
hpWG-18LS-18PA-18E-6C		40	18	18			18	0.4	6
hpWG-20LS-5APP-20E-6C		49	20		5		20	0.4	6

Coated foams were analyzed using thermogravimetric analysis (TGA), microcombustion calorimetry (MCC), UL-94 flame spread, and cone calorimetry. TGA was conducted on 4 mg to 6 mg samples using platinum pans under a flow of nitrogen. Samples were equilibrated at 90 °C for 30 minutes then heated to 800 °C at 10 °C/min. The microcombustion calorimetry (MCC) samples (5 mg ± 0.1 mg) were tested with a Govmark MCC-2 microcombustion calorimeter at 1 °C/sec heating rate under nitrogen from 200 °C to 600 °C using method A of ASTM D7309 (pyrolysis under nitrogen). Three replicates were tested per sample. The total heat release (THR) and heat release capacity (HRC) had uncertainties of $\sigma = \pm0.7$ kJ/g and $\sigma = \pm11$ J/g·K, respectively. Horizontal burn tests were conducted in a fume hood using UL-94 standard procedures for foamed plastic materials. Coated foam was placed on a wire mesh and burned using a natural gas burner with a curved wing tip. A set of five specimens for each sample was burned and the reported burn rates have an uncertainty of $\sigma = \pm 0.2$ mm/min. Cone calorimetry was conducted according to a standard testing procedure (ASTM E-1354-07) using a prototype instrument at the National Institute of Standards and Technology. The sample (10 cm × 10 cm × 5 cm) was placed in heavy-gauge aluminum foil (Reynolds Heavy Gauge Aluminum Foil) pan, such that only the top surface of the sample was exposed to the cone heater, operating at 35 kW/m^2 heat flux. Ignition was activated by a spark igniter. The tests were conducted on two samples (unless there was a > 10 % discrepancy, in which case a third sample was tested) and standard measurement uncertainty was ±10% of the reported reduction values and ±2 s in time.

RESULTS AND DISCUSSION

The MCC analysis was conducted for a few coatings that were prepared as films and several coated foams. The total heat released, heat release capacity, temperature of peak heat release, residue, and total heat released relative to the amount of material actually pyrolyzed are shown in Table II.

Table II. Flammability properties of protein based coatings using microcombustion calorimetry

Sample	THR (kJ/g)	HRC (J/g·K)	T_{peak} (°C)	char (mass %)	*THR_{rel} (kJ/g)
WG-10G	13.2	146	271	14	15.3
LS	1.6	18	245	64	4.4
WG-10PA-10G	9.6	129	253	20	12.1
WG-10LS	11.7	101	296 / 352	25.1	15.6
WG-10LS-10G	12.1	107	351	22.9	15.7
WG(acid)-10LS-10G	11.2	101	307 / 349	25.4	15.0
hpWG-10LS-10G	12.3	116	354	21.4	15.6
Uncoated foam	25.5	518	286 / 395	0	25.5
PUF+WG-10LS-10G	21.1	397	251 / 424	9.3	23.3
PUF+WG-10LS-3G	21.4	366	292 / 424	10.0	23.8

* - THR relative to amount of material actually consumed = THR / (1 – char fraction)

From the MCC data, it is quite evident that lignosulfonate has very low heat released when pyrolyzed, which is why we chose to use this natural material as the main component in the flame retardant coating. The plasticizer does increase THR of the mixture. Dissolving the gluten in acid resulted in some clumping, which required straining the final solution through cheesecloth, but did not change the flammability of the film much. Using a higher protein gluten source resulted in severe clumping in acidic solutions and the film produced in the basic solution had a flammability similar to the lower protein gluten. The dried coatings on foam added about 30 % mass for all samples. The flammability reduction of the coated foams is between 15 % and 30 % relative to uncoated foam and appears to be a simple law of averages, with the coating representing almost all of the residue after pyrolysis.

In our previous work, we found that the addition of clay was crucial in forming a protective layer for the foam.[8] Enough clay was added to the solutions to comprise 1 % or 6 % mass fraction of the dried coating. Additionally, the protein content was reduced in favour of higher flame retardant contents. This required additional plasticizer to keep the coating flexible. EDP was used in place of G to reduce the water sensitivity of the coating. Coated foams were tested using UL-94 flame spread and cone calorimetry and the results are shown in Table III and Figure 1.

Table III. Heat release and flame spread properties of gluten coated foams containing clay

Sample	t_{ign} (s)	PHRR (kW/m²)	AHRR (kW/m²)	THR (MJ/m²)	char (mass %)	HB rate (mm/s)
Uncoated PUF	3	580	374	30.8	2.0	3.68
PUF+hpWG-10LS-10G	4	395	301	35.1	3.9	n.t.
PUF+hpWG-20LS-20G						2.37
PUF+hpWG-20LS-20E						1.99
PUF+hpWG-20LS-20E-1C	4	453	291	34.6	3.6	1.53
PUF+hpWG-20LS-10PA-20E-1C	6	485	292	32.5	10.2	1.32
PUF+hpWG-20LS-20E-6C	3	441	233	32.9	7.4	n.t.
PUF+hpWG-18LS-18PA-18E-6C	3	401	235	32.2	11.8	n.t.
PUF+hpWG-20LS-5APP-20E-6C	4	277	73	28.9	18.8	1.15

n.t. – not tested

Switching from glycerol as a plasticizer to 2-ethylhexyldiphenylphosphate not only reduced the hydrophilicity, but also reduced the flame spread of the coated foam. This is likely a combination of the phosphorus and the phenyl rings, which increase char to provide a heat barrier. As shown in Table II and Figure 1, it appears that these coatings do increase the heat release peak associated with the depolymerization of the foam, especially as the plasticizer content increases. However, the heat released during the second combustion stage (pyrolysis of the polyol) is reduced by 50 %. As a result, PHRR is reduced by 22 % to 32 %. The use of LS as a char forming flame retardant was not as effective on foams as it was in epoxy composites.[9] Although some of this may be due to the addition of a binding agent (protein in this case) and plasticizer, which increases the fuel content of the material, much is probably due to the large surface area of the foam and the inability to form a thick enough char layer to prevent decomposition of the underlying polymer. Similar to what we observed in carbohydrate based coatings, the addition of phytic acid increases PHRR, but reduces THR. However, the rate of flame spread is reduced with the addition of the phosphate. The addition of clay does not affect the PHRR much, but it does reduce the AHRR and THR. The clay hinders the collapse and liquefaction of the foam, which significantly reduces the heat release rate of the second peak. Surprising, the addition of clay also reduced the flame spread rate. In thermoplastic composites, nanoparticles often result in an increase in flame spread rate. The addition of a small amount of ammonium polyphosphate completely prevented the collapse of the foam and produced a stable char barrier. As a result, the PHRR was reduced over 50 % and the AHRR was reduced by 80 % compared to the uncoated foam. The flame spread was confined to the surface of the foam, leaving nearly two-thirds of the specimens unburned. Solubility tests on the coated foam and examination of other hydrophobic proteins are currently under investigation.

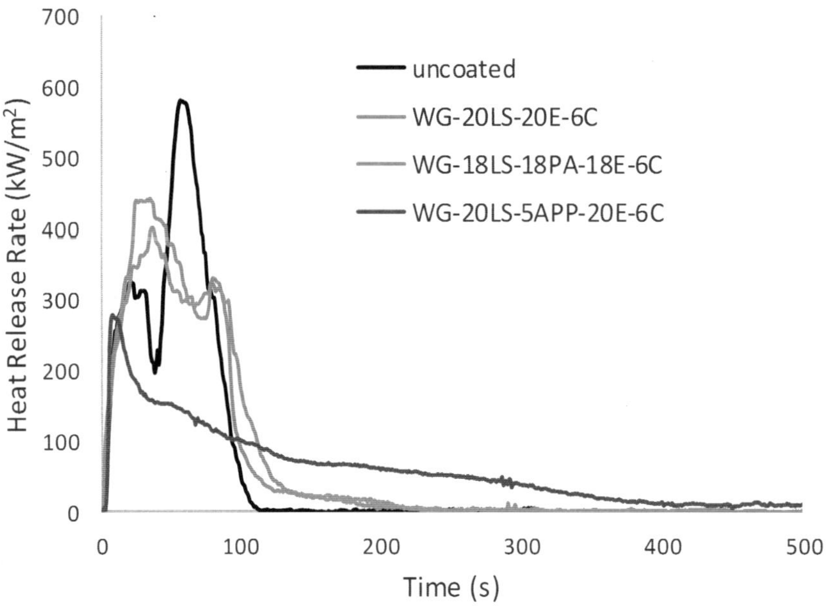

Figure 1. Cone calorimetry of gluten coated foams.

CONCLUSIONS

The use of lignosulfonate as a flame retardant and wheat gluten as a water insoluble binding agent in coatings appear to be a promising halogen-free approach to reduce the heat release rate and horizontal flame spread rate of flexible foam. Peak heat release rates are reduced by over 20 % when using a coating consisting of 20 % mass fraction lignosulfonate. The addition of clay and small amounts of ammonium polyphosphate can produce a stable char that prevents foam collapse, reduces polyol combustion, and slows flame spread. Addition of phytic acid appears to have minimal effects on the heat release of foams in these systems, but does reduce the rate of flame spread. Reducing the protein content in favour of flame retardant components may further reduce the flammability of the foams while maintaining water resistance.

ACKNOWLEDGEMENTS

This work was funded in part by NIST – EL Extramural Fire Research Grant #70NANB14H307, part of the 2014 Measurement, Science, and Engineering Research Grants Program. This material is based upon work supported by the NIST – EL Extramural Fire Research Grants Program. Research was carried out at the National Institute of Standards and Technology (NIST), an agency of the U. S. government and by statute is not subject to copyright in the United States. Certain commercial equipment, instruments, materials or companies are identified in this paper in order to adequately specify the experimental procedure. Such identification is not intended to imply recommendation or endorsement by the National Institute of Standards and Technology, nor is it intended to imply that the materials or equipment identified are necessarily the best available for this purpose.

REFERENCES

1. FEMA. Fire in the United States 2003 - 2007, U.S. Fire Administration & National Fire Data Center, 2009.
2. Stapleton, H. M.; Klosterhaus, S.; Eagle, S.; Fuh, J.; Meeker, J. D.; Blum, A.; Webster, T. F.,

Detection of Organophosphate Flame Retardants in Furniture Foam and US House Dust. Environ. Sci. Technol. 2009, 43, p7490.
3. Stae of California. TB 117 - 2013: Requirements, Test Procedure and Apparatus for Testing the Smolder Resistance of Materials Used in Upholstered Furniture. State of California Department of Consumer Affairs, 2013.
4. Putzeys, O. M.; Fernandez-Pello, A. C.; Rein, G.; Urban, D. L. The piloted transition to flaming in smoldering fire retarded and non-fire retarded polyurethane foam. Fire and Materials 2008, 32, p485.
5. Hasofer, A. M.; Thomas, I. Analysis of fatalities and injuries in building fire statistics. Fire Saf. J. 2006, 41, p2.
6. Grosshandler, W.; Bryner, N.; Madrzykowski, D.; Kuntz, K. Report of the Technical Investigation of The Station Nightclub Fire, National Institute of Standards and Technology, 2005.
7. Davis, R.; Li, Y.-C.; Gervasio, M.; Luu, J.; Kim, Y. S. One-Pot, Bioinspired Coatings To Reduce the Flammability of Flexible Polyurethane Foams. ACS Applied Materials & Interfaces 2015, 7, p6082.
8. Fox, D. M. unpublished results, 2016.
9. Fox, D. M.; Kaufman, N.; Woodcock, J.; Davis, C. S.; Gilman, J. W.; Shields, J. R.; Davis, R. D.; Matko, S.; Zammarano, M. Epoxy Composites Using Wood Pulp Components as Fillers. In Composites from Renewable and Sustainable Materials, InTech, 2016, p199.

PHYSICAL CHARACTERIZATION OF CARBONACEOUS PRODUCTS FROM FIRE AND FIRE RETARDANTS: ASSESSMENT OF THE IMPACT ON FIRE PERFORMANCE

Gizem Okyay, Anil Naik, Pauline Tranchard, Séverine Bellayer, Maude Jimenez, Fabienne Samyn and Serge Bourbigot*
R_2Fire group, UMET-ISP, CNRS UMR 8207, ENSCL, University of Lille, Villeneuve d'Ascq, France

INTRODUCTION

In fire research and safety, it is crucial to understand the kinetics and processes behind the fire phenomenon and the response of materials exposed to it. One should examine the two components of fire: the flame itself and the materials burning in order to assess the fire performance. Those two components can be examined separately or together, depending on the topic and the research interests such as fire spread[1], chemical and physical kinetics[2,3], etc.

Burning of materials, flame retarded or not, leads to the release of heat and combustion products and in particular, carbonaceous products such as soot, ash and char in gaseous and condensed phases. In fire research, it is crucial to study the interactions of this evolved heat with its surroundings, which can become complex due to the kinetics and all modes of heat transfer involved. The response of material to heat is a function of its properties, such as the chemical composition, thermal decomposition, emissivity, morphology, etc. Those properties are eventually time dependent due to the evolution of chemistry and structure throughout burning. Hence, a way to investigate the fire performance of a material is to quantify the above mentioned properties as a function of time when possible.

The objective of this research is to improve the overall understanding of our fire testing and modeling tools, in order to propose better solutions to make fire barrier materials. Here, the morphology and the thermal conductivity are chosen as variables/parameters to examine fire behavior and fire performance. The studied materials and the testing conditions, further detailed in the materials and methods section, have been chosen based on the accumulated experiences and the know-hows of our team. The focus of this study is on the morphology of carbonaceous products issued from burning and it was decided to consider separately the phenomena occurring in the flame and in the material itself.

For the fire tests where radiation dominates (e.g. cone calorimeter test), the emphasis is given on the gaseous phase and flame (i.e. its soot particulate media) because soot is known to be the main heat source and sink of flame emissions[4]. It is also a pollutant and affects the toxicity. The particulate media is extracted and analyzed from the burning of gaseous decompositions of EVA (Ethylene-Vinyl Acetate) and its fire retarded formulation EVA/ATH (EVA loaded with ATH, aluminum trihydroxide mineral filler).

* Correspondence: serge.bourbigot@ensc-lille.fr

For the fire tests where convection and conduction dominate (e.g. furnace tests), the emphasis is given to the condensed phase. Intumescent coatings, and more precisely intumescent chars, which act as heat barrier between the flame and the substrate, are investigated. Knowing that their thermal conductivity plays an important role[5], their expansion pattern and thermal conductivity of the expanded char versus temperature are of interest.

MATERIALS AND METHODS

Study of soot particles in gaseous phase burning

According to state of the art[6,7], thermal degradation of polymeric materials can produce combustible hydrocarbon gases such as methane, propane, ethane, ethylene, etc. Those gases are known to produce nano and micron sized black carbon soot upon burning. Hence, it is possible to examine the decomposition history by observing those carbonaceous particles/residues during fire tests.

EVA copolymer was selected for the study of particulate media because it is a well-known material in fire research and the aim is also to complement the existing studies of our team[6,7,8,9]. Both EVA and EVA/ATH samples were tested, because the effect of flame retardant system on the emitted particulate media should play a significant role. The particulate media was formerly tested by filtering exhaust gas and by low pressure impactor techniques. The chemical composition of the gas phase was investigated by thermogravimetric analysis (TGA)/Fourier Transform Infrared (FTIR) and mass loss cone calorimeter (MLCC) also coupled to FTIR methods.

In this study, an innovative approach is being developed, never used by now in flame retardancy research: the soot particles and agglomerates morphology are studied directly from microscopy imaging with subsequent image processing. The aim is to retrieve information on the particle size (i.e. primary particle size, which is not be confused with the agglomerate size) to observe visually the agglomeration phenomenon and to visualize the time-dependent formation history of particles from our reference polymeric materials.

- Tested specimens:

Specimens used are: pure EVA (containing 28 wt % of vinyl acetate) and EVA/ATH (containing 65 wt% ± 5 wt%). Samples of size 10x10x0.3 cm^3 were prepared by extrusion and by molding for MLCC test, using the materials and methods detailed and verified in former studies[6].

- Fire testing:

The reaction to fire of materials were performed on a mass loss cone calorimeter (MLCC) from Fire Testing Technology (FTT) according to standards ISO 13927 or ASTM E906. The external heat source is only radiative in MLCC; hence the combustion products are intrinsic to the burning of the material, contrary to experiments where the material is subjected to an external flame. For example, in UL94 and UL1709 tests, it is highly probable to collect particles and burnt gases from the external heat source and not only from the material itself. MLCC tests were performed at mild fire scenario of 35 kW/m^2 incident radiative heat flux, with 25 mm separation distance between the heat source and sample surface. At least two MLCC experiments were performed on each material in order to ensure repeatability within the error margins of +/- 10%.

- Particle sampling:

Soot was sampled by using the particle deposition from flame, technique formerly applied and verified in combustion research[10]. The specimens were probed on Transmission Electron Microscopy (TEM) grids (lacey and holey carbon film covered, square mesh copper grids, provided by Agar Scientific). The times of particle sampling were determined according to the characteristic times of the heat release rate (HRR) in MLCC experiments: just after the time to ignition, just after the maximum value of the peak of HRR, and at the end of test before flameout. In this study, only the latter two sampling times will be presented because they lead to highest amount of soot agglomerates.

- Microscopy:

Soot imaging was performed using Scanning Electron Microscopy (SEM) in transmission mode (SEM/STEM). STEM images were taken using a JEOL 7800 FEG LV scanning electron microscope at 30 kV and 7 mA, using a retractable bright and dark field SEM STEM detector completed with Deben Gen5 electronics and a 12 position TEM grid holder.

- Morphology analysis:

Numerical analyses were performed using the electron microscopy recordings, following the methodologies explained and verified in previous researches[11]. Particle counting and the size measurements were performed over more than 20 microscopy recordings for each material type. ImageJ[12] and its plugins were used for image processing and quantitative measurements.

Study of char from intumescent coatings

Former works on intumescent char structure focus on the expansion as a function temperature, on the possible synergies between ingredients of coatings[13] and on kinetics of degradation[14]. Nevertheless, the char morphology should be examined for further consideration, for example in numerical simulations and for better interpretation of thermal conductivity measurements[15].

In this paper, focus is given on the testing and analyses methods of the geometry of the expanded char. The repartition and size of bubbles/pores play an important role for fire properties of intumescent char[15], but it is difficult to integrate such information in modeling and simulation tools, either numerically or experimentally. Therefore, the aim of this study is to determine, if possible, a mathematical formulation of the inner structure with a perspective on its effect on thermal conductivity.

- Intumescent char specimens:

Two types of intumescent chars were examined: one obtained from a commercial epoxy based intumescent coating and the other from a silicon based intumescent coating developed in our lab and described elsewhere[16]. The coated steel plates were subjected to UL1709 fire testing to obtain expanded char specimens. The intact intumescent chars were then analyzed by X-ray tomography, and then cut for visual inspection.

- X-ray Tomography:

X-ray micro-computed tomography (μCT) was applied on chars collected from fires tests (UL1709). This is a non-intrusive technique to observe the inner morphology without structural damage. The tomography was performed using the micro-tomography setup (UltraTom from RX Solutions) at the ISIS4D X-ray μCT platform (Villeneuve d'Ascq, France)[17].

The tomography set-up consists of two X-ray tubes (micro and nano focus), a sliding and rotating stage holder, a flat panel detector (1920*1496 px – 127 μm/px – 0.2 to 60 frame/s), a linear detector (2560 px – 200 μm/px – 0.2 to 60 frame/s), a CCD camera (4000*2624 px – 11.8 μm/px – up to 3.4 frame/s) and an image intensifier.

The char samples were placed on styrofoam holders and then mounted on the rotating stage to minimize the signal noise due to the holder. The epoxy based char had to be placed with its steel plate, while the silicon based char was easily detached from its steel plate and placed on styrofoam directly. Samples were rotated by 360° with an angular step of 0.25°. The subsequent tomography volumes were reconstructed from 1440 projection images and visualized using X-Act (from RX Solutions): reconstructions have a resolution of 81μm/voxel in 3D space.

- Morphology analyses:

Image stacks of the inner structure were generated by slicing the reconstructed 3D volume. Those image stacks were re-processed to adjust gray signal levels to ensure the maximum visibility of carbon elements, noting that low atomic number elements interact only little with X-rays, hence tend to produce low gray levels in stack images[13]. Slice images were binarized for fractal analysis. ImageJ[12] software was used for the image processing. Quantitative analyses of mass density autocorrelations were performed using fractal analysis: Box-counting technique, previously applied on combustion soot[11], was re-written as C++/Matlab algorithms, and applied on μCT image slices of chars.

- Thermal conductivity measurements:

Thermal conductivity was measured from ambient to 500°C using a Transient Plane Source method (TPS2500S) from THERMOCONCEPT (Merignac, France). A thin hot disk sensor is placed between two pieces of sample acting both as heat generator and sample temperature gauge. An adequate sensor is placed between two 10 mm thick disk of 30 mm² diameter of the sample. To ensure a good contact between the sensor and the sample, the sensor was placed during manufacturing process between both sample disks. Then, the sample was degraded to get a thermally stable char. For each temperature, the measurements were performed as soon as the sample temperature was down to +/- 0.5°C of the desired temperature. The furnace was directly connected to a nitrogen flow and all experiments were carried out in an inert atmosphere to prevent oxidation of the sensor and of the char samples. The experiments were run as duplicated and were repeatable within 5-20% margin error.

RESULTS AND DISCUSSION

Particulate media in gaseous phase

Mass loss cone calorimeter experiments were performed on EVA and EVA/ATH. The resulting HRR curves are presented in Fig.1(a) and Fig.1(b) respectively: error margin of the experimental setup is +/- 10%. The probing time is less than or equal to one second. Hence, and as evidenced by the HRR curves, the effect of the soot probing during experiments is negligible. The soot sampling is repeated at the same specific times after ignition, on two different specimens: the results are similar, hence the repeatability of experiments is ensured. This is expected, because the thermal decomposition pathway of the specimens should remain constant throughout the same experimental conditions.

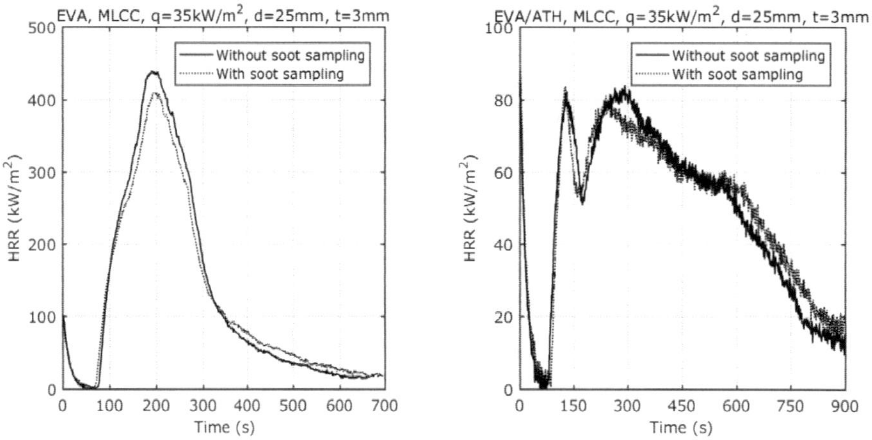

(a) (b)

Figure 1. HRR curves from MLCC experiments on (a) EVA and (b) EVA/ATH

First qualitative observations on the extracted particles from burning EVA formulations are presented in Fig. 2. On Fig.2a, the soot dispersion on TEM grid from EVA is presented: variable sizes of particles and aggregates are observed. On Fig.2b, the soot dispersion on TEM grid from EVA/ATH is presented: tiny aggregates are observed, presumably with some residues of aluminum from ATH. The preliminary observations indicate a higher density of particles inside the gaseous volume for the non-flame retarded (FR) system.

In the FR system the soot emissions are somehow inhibited. This latter was expected because the incorporation of ATH reduces the release of flammable gases thanks to high loading (65 wt% ATH decreasing the polymer content) and to water release (dehydration of ATH), to decrease the system temperature (high endothermic effect due to the dehydration of ATH), hence leading to overall reduction of the formation kinetics of soot[6,18].

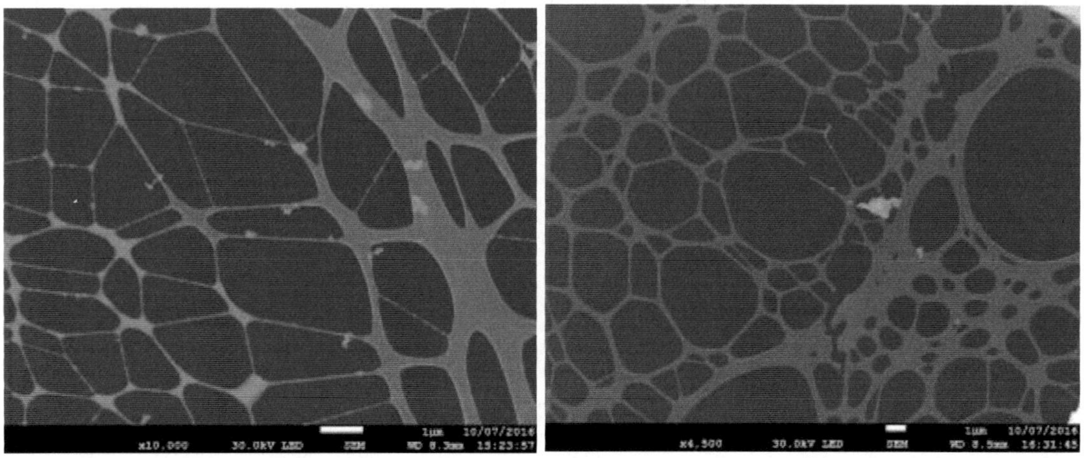

(a) (b)

Figure 2. SEM/STEM images of particles collected from evolved gases of MLCC experiments of (a) EVA and (b) EVA/ATH, to make a qualitative comparison of density of particulate media.

In the next paragraphs, both EVA and EVA/ATH soot agglomerates are investigated separately.

- Investigation of soot from EVA as a neat polymer:

Images of soot extracted from EVA burning are presented in Fig.3a and Fig.3b.: they correspond to different probing times during MLCC test. The agglomeration pattern seems to be independent of probing time. As observed on Fig.3a, soot is in a coated form[19] when extracted just after the peak of HRR. It is believed that some pieces of material are detached and melted on the soot substrate to form coated aggregates, presumably with some other decomposed material condensed on the agglomerates issued from the gaseous decomposition of EVA[6,17]. Aggregates are bare when extracted at the end of experiment as illustrated in Fig.3b.

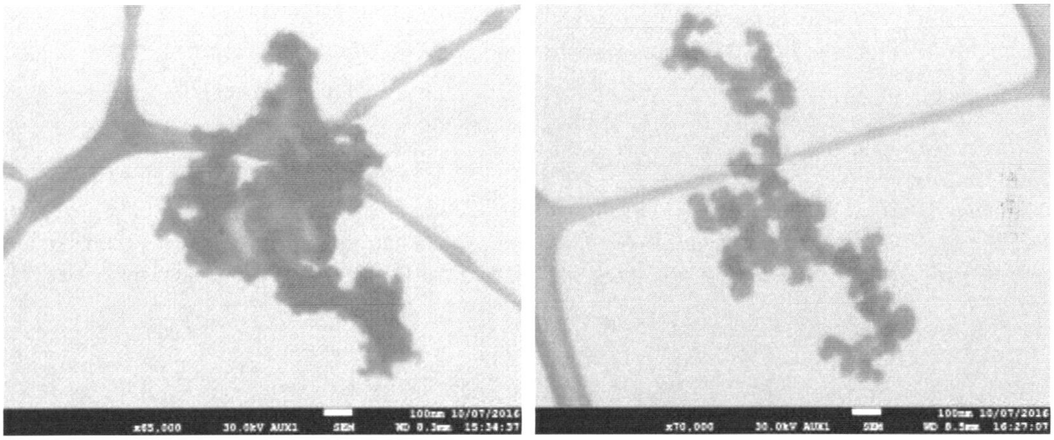

(a) (b)

Figure 3. Example of SEM/STEM images of soot agglomerates collected (a) just after the peak of HRR and (b) before flameout from MLCC experiments of EVA.

The diameters of primary particles are measured on microscopy recordings in order to investigate the size distribution function. According to the probability density function of EVA soot, given in Fig.4, the particle size distribution is multimodal noting that the normal distribution (mean size of 29.5 nm; standard deviation of 11.6 nm) is plotted as a comparison to the real distribution. The modes are dependent on the type of hydrocarbons burning simultaneously.

Based on literature review on combustion soot[11]: small particle size (around 20-25 nm) is characteristic of methane combustion; medium particle size (around 30-35 nm) is characteristic of ethylene and ethane combustion; larger particle sizes are generally issued from heavier hydrocarbon burning like propane, toluene, etc. It is believed that the highest peak in Fig.4 belongs to the burning of methane, whereas it is difficult to distinguish between ethane and ethylene peak. The existence of a third peak is novel compared to the results of smoke particles[18], and indicates traces of soot issued from heavier hydrocarbons and slower reactions. This will be further discussed in a separate paper, together with complementary chemical analyses on the gaseous phase and with potential dependencies on combustion kinetics.

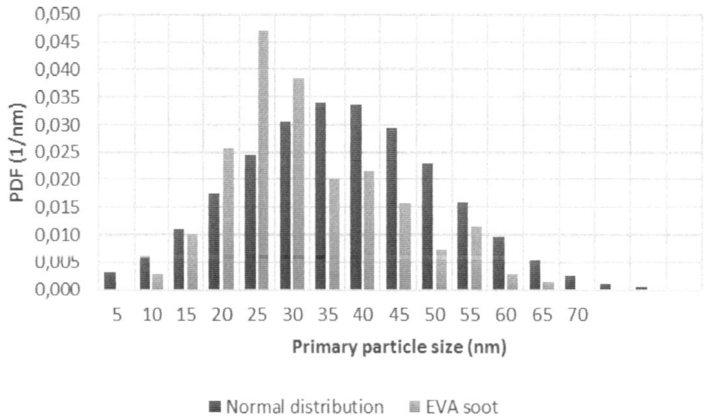

Figure 4. Particle size analysis for agglomerates from EVA burning.

- Investigation of soot from EVA/ATH as a FR system:

Images of soot extracted from EVA/ATH burning are presented in Fig.5. Soot samples probed just after the peak of HRR are coated with a thin layer, as observed in Fig.5a. The agglomeration pattern seems to remain the same according to the bare soot sample presented in Fig.5b. Those results are similar to those obtained for pure EVA.

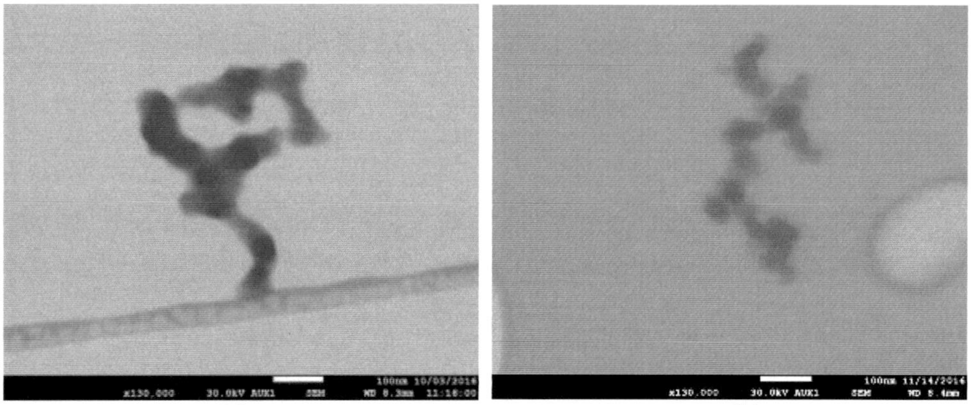

(a) (b)

Figure 5. Example of SEM/STEM images of soot agglomerates collected just after the peak of (a) HRR and (b) before flameout from MLCC experiments of EVA/ATH specimens.

According to the probability density function of EVA/ATH soot given in Fig.6, the particle size distribution is unimodal around 35-40 nm for EVA/ATH soot. The normal distribution (mean size of 34.4 nm; standard deviation of 8.9nm) is plotted as a comparison to the real multimodal distribution. The result is in accordance with the previous results on EVA/ATH decomposition where the propane release is favored with a mass density greater than the overall release of methane and ethane during controlled atmosphere MLCC tests[6]. Existence of lighter hydrocarbons can justify the skewness of the particle size distribution function.

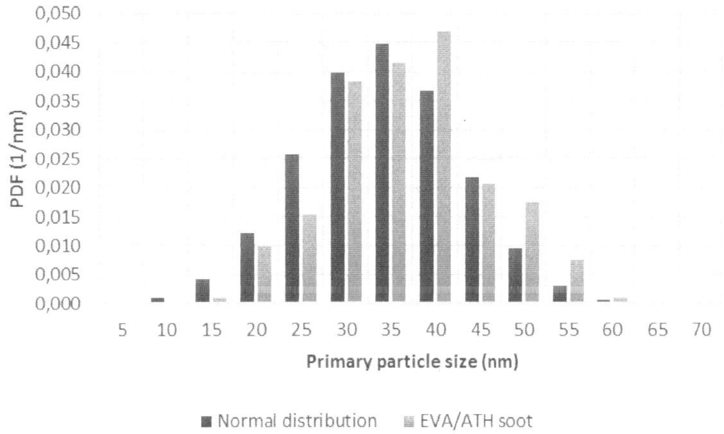

Figure 6. Particle size analysis for agglomerates from EVA/ATH burning.

To sum up, according to the observations on pure EVA and FR EVA/ATH, the volume density and polydispersity of particles is reduced in the flame retarded system. The mean particle size is however found to be larger for this FR EVA/ATH system and this leads to two hypotheses: the mass density of heavier hydrocarbons can be higher in the gaseous decomposition phase and the residence time of soot is longer following the slower reactions due to FR effect.

Some of the particle size measurements, both for EVA and EVA/ATH burning, have an uncertainty due to the coating of aggregates: the coating layer can impact the measurements by up to 5 nm to 10 nm depending on the image resolutions. Another source of error can be the usage of SEM/STEM, due to its lower precision compared to TEM.

The nanoscale morphology of particles cannot be observed in STEM. Nevertheless, the aggregation of primary particles can be observed with preliminary measurements on primary particle size. Also, the specimen is less susceptible to be damaged and polluted, with the usage of lower electron acceleration voltages. Therefore, the measurements are less affected by the deposition of a pollution layer and the restructuring of agglomerates, under the energy of the incident electron beam, justifying our study using SEM/STEM. Error margins (which are dependent on microscopy conditions and on user experience in image processing) will be investigated by future studies on sampling and imaging protocols.

Intumescent char

In the condensed phase, the focus is given to the structure of the expanded char. The final morphology of the expanded intumescent char of epoxy and silicon based coatings submitted to fire test are given in Fig.7(a) and Fig.7(b) respectively. The thickness of the epoxy based char is higher and less uniform compared to the silicon based char. The real life images are visually compared to the reconstructed numerical volumes from tomography illustrated in Fig.8.

(a) (b)

Figure 7. Intumescent char formed from intumescent coatings on steel plates, with (a) epoxy based and (b) silicon based formulations, after UL1709 hydrocarbon fire test.

As it can be seen in Fig.7a and Fig.8a, some part of the information is lost for the epoxy based char, because the steel plate could not be detached without damaging the sample. The steel plate scattered the X-rays resulting in the noisy signals in tomography at the lower part of the char. Hence the quantitative image analyses were performed by suppressing this noisy region from the reconstructions of the X-ray μCT.

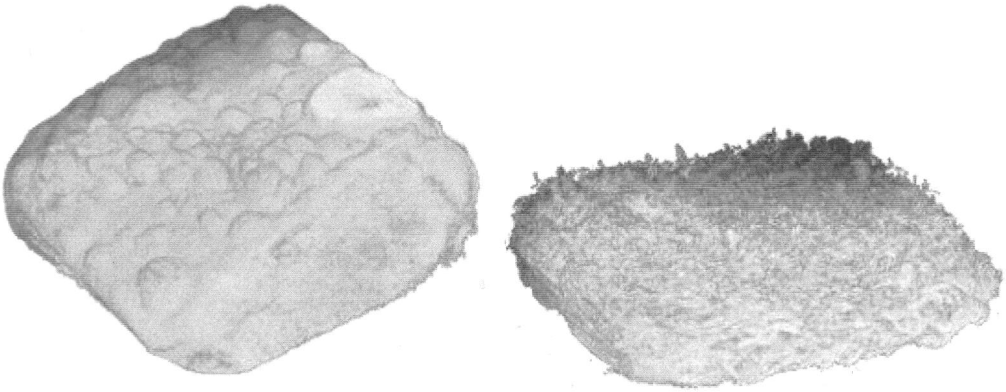

(a) (b)

Figure 8. 3D geometries reconstructed from X-ray μCT of the (a) epoxy based char, (b) silicon based char.

In Fig. 9, the vertical slices were plotted after the X-ray μCT reconstruction. The epoxy based char in Fig.9a contains bubbles of different sizes with a non-uniform distribution. The silicon based char in Fig.9b contains more uniformly distributed voids in the vertical direction.

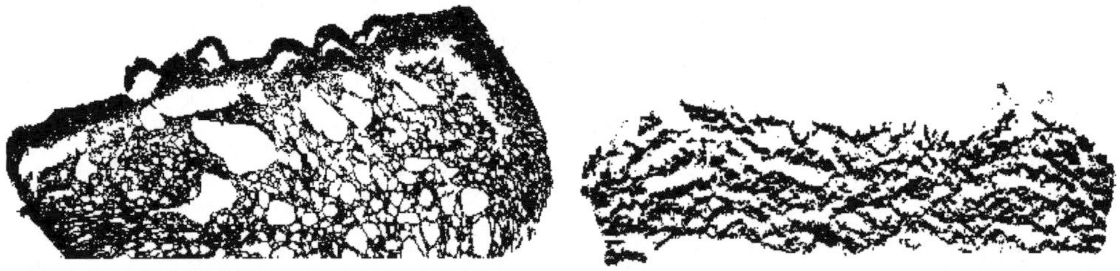

(a) (b)

Figure 9. Binarized slice images of the inner structures of (a) epoxy based char, (b) silicon based char.

Numerical analyses were then performed to investigate the possible existence of a density relationship.

- Numerical analyses on morphology:

Box counting operation was performed in 2D space on tomography slice images of Fig.9 and plotted in Fig.10. The analysis was performed in macro and mesoscale (1 pixel = 81 microns). The computations were repeated for different vertical slices of the char to check any anisotropy inside the material. Starting from the outer vertical face of each char (visible side on the images of Fig.7), different slices were chosen at distances of 3 to 7 centimeters: The captions in Fig.10 are named accordingly and "5cm" corresponds to the middle vertical slice. As the analyses were performed in 2D space, the slope of the space-filling curve corresponds to the maximum slope of 2.

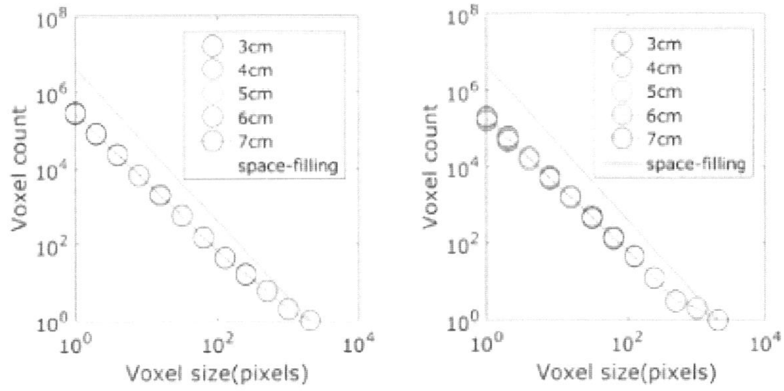

(a) (b)

Figure 10. The fractal analyses: (a) for epoxy based char, (b) for silicon based char.

The preliminary results indicate three main outcomes:

- The fractal dimension is similar for both chars looking at the average slopes of the curves of Fig.10a and Fig.10b.
- The distribution of void spaces, in each vertical slice, is more uniform for silicon based char. This is indicated by the uniformity of the slope of the log-log curves (except at the very macroscopic level). Indeed, the bubbly distribution of the epoxy based char is highly non-uniform from top to bottom, even with visual observations of Fig.10a. Hence, one can say that this mathematical description of morphology is consistent with observations.
- When different vertical slices of the same material are compared, the distribution of void spaces is less uniform for silicon based char of Fig.10b. The analyses of different slices overlap for epoxy based char. For silicon based char, the values start to diverge from mesoscale (medium voxel size) to micro scale (small voxel size). This difference is due to different intumescence mechanisms of both formulations. The epoxy-based formulation is a conventional intumescent system involving charring reactions from the ingredients contained in the material. The silicone-based formulation is based on expandable graphite. The rapid sublimation of sulfuric acid inserted in the graphite platelets creates growing 'worms' forming the intumescent char. It will be commented in future works.

- Experimental results on thermal conductivity:

Thermal conductivity of silicone/modifier and epoxy/modifier were measured from ambient temperature up to 500°C. For epoxy based char, it varies from 0.14 to 0.5 W/m.K depending on the temperature. For the silicon based char, the measurements were more complicated due to an unstable temperature in the sample. Nevertheless, the thermal conductivity varies from 0.33 to 0.5 W/m.K from the ambient to 300°C. Both chars exhibit a low thermal conductivity (lower by 1 W/m.K) and so, heat insulating properties. These preliminary results are in the same order of magnitude as those found in the literature[5,15]. Slight differences between both char thermal conductivities may be explained by differences of expansion and porous media into both chars. To go further in the understanding, morphology analyses were performed to explain it. Based on these preliminary results, measurements will be further affined and then discussed in future work.

CONCLUSIONS AND PERSPECTIVES

In this study, the physical properties of carbonaceous products were examined in gaseous and condensed phases of materials exposed to fire. The focus was given to the morphology of soot (gaseous phase) for EVA and FR EVA and chars (condensed phase) for intumescent coatings. The abundance of soot in the gaseous phase is evidenced for EVA formulations. It is also evidenced that for the FR EVA/ATH system, soot formation is inhibited. The preliminary results on soot morphology evidenced hydrocarbon combustion and the results were consistent with our previous work and with literature on soot from hydrocarbon combustion. The results are to be further affined with future TEM imaging of soot and chemical analyses of specimens.

Concerning condensed phase analyses, the preliminary results on the intumescent char indicated the possible existence of a mass-density autocorrelation for the description of the inner structure. Therefore, the fractal analysis of char morphology is promising for future analyses and correlations to thermal properties, because it can reduce the complex morphology to simpler mathematical descriptions. The investigations of correlations between the morphology and thermal conductivities are in the perspectives.

ACKNOWLEDGMENTS

This work has received funding from the European Research Council (ERC) under the European Union's H2020 – the Framework programme for Research and Innovation (2014-2020) / ERC Grant Advances Agreement n°670747 – ERC 2014 AdG/FireBar-Concept. It has benefited from the research facilities of University of Lille: ISIS4D platform (X-ray µCT), Centre Commun de Microscopie (SEM/STEM, TEM), HPC cluster of CRI (numerical analysis). Dr. Grégory Hauss is gratefully acknowledged for his technical assistance and scientific guidance on X-ray µCT. Johan Sarazin and Benjamin Dewailly are gratefully acknowledged for their technical assistance on MLCC specimen preparations.

REFERENCES

[1] Rothermel, R. C. (1972). A mathematical model for predicting fire spread in wildland fuels.

[2] Lyon, R. E. (1998). Pyrolysis kinetics of char forming polymers. Polymer Degradation and Stability, 61(2), 201-210.

[3] Lyon, R. E. (2000). Heat release kinetics. Fire and Materials, 24(4), 179-186.

[4] Sacadura, J.F. (2005). Radiative heat transfer in fire safety science, Journal of Quantitative Spectroscopy and Radiative Transfer, 93 (1–3), pp.5-24.

[5] Staggs, J.E.J. (2010). Thermal conductivity estimates of intumescent chars by direct numerical simulation, Fire Safety Journal, 45(4), pp.228-237.

[6] Girardin, B., Fontaine, G., Duquesne, S., Försth, M., & Bourbigot, S. (2015). Characterization of Thermo-Physical Properties of EVA/ATH: Application to Gasification Experiments and Pyrolysis Modeling. Materials, 8(11), 7837-7863.

[7] Ngohang, F. E., Fontaine, G., Gay, L., & Bourbigot, S. (2015). Smoke composition using MLC/FTIR/ELPI: Application to flame retarded ethylene vinyl acetate. Polymer Degradation and Stability, 115, 89-109.

[8] Carpentier, F., Bourbigot, S., Le Bras, M., Delobel, R., & Foulon, M. (2000). Charring of fire retarded ethylene vinyl acetate copolymer—magnesium hydroxide/zinc borate formulations. Polymer Degradation and Stability, 69(1), 83-92.

[9] Siat, C., Le Bras, M., & Bourbigot, S. (1998). Combustion behaviour of ethylene vinyl acetate copolymer-based intumescent formulations using oxygen consumption calorimetry. Fire and materials, 22(3), 119-128.

[10] Okyay, G., Héripré, E., Reiss, T., Haghi-Ashtiani, P., Auger, T., & Enguehard, F. (2016). Soot aggregate complex morphology: 3D geometry reconstruction by SEM tomography applied on soot issued from propane combustion. Journal of Aerosol Science, 93, 63-79.

[11] Okyay, G. (2016). Impact of the morphology of soot aggregates on their radiative properties and the subsequent radiative heat transfer through sooty gaseous mixtures (PhD dissertation, Université Paris Saclay, CentraleSupélec, France) – www.theses.fr/2016SACLC031

[12] Schneider, C. A., Rasband, W. S., & Eliceiri, K. W. (2012). NIH Image to ImageJ: 25 years of image analysis. Nature methods, 9(7), 671-675.

[13] Muller, M., Bourbigot, S., Duquesne, S., Klein, R., Giannini, G., Lindsay, C., & Vlassenbroeck, J. (2013). Investigation of the synergy in intumescent polyurethane by 3D computed tomography. Polymer degradation and stability, 98(9), 1638-1647.

[14] Jimenez, M., Duquesne, S., & Bourbigot, S. (2009). Kinetic analysis of the thermal degradation of an epoxy-based intumescent coating. Polymer Degradation and Stability, 94(3), 404-409.

[15] Staggs, J. E. J. (2011). Numerical characterization of the thermal performance of static porous insulation layers on steel substrates in furnace tests. Journal of Fire Sciences, 29 (2), 177-192.

[16] Gardelle, B., Duquesne, S., Vandereecken, P., & Bourbigot, S. (2013). Characterization of the carbonization process of expandable graphite/silicone formulations in a simulated fire. Polymer degradation and stability, 98(5), 1052-1063.

[17] Limodin, N., Rougelot, T., & Hauss, G. ISIS4D – In Situ Innovative Set-Ups under X-ray microtomography – http://isis4d.univ-lille1.fr/

[18] Ngohang, F. E., Fontaine, G., Gay, L., & Bourbigot, S. (2014). Revisited investigation of fire behavior of ethylene vinyl acetate/aluminum trihydroxide using a combination of mass loss cone, Fourier transform infrared spectroscopy and electrical low pressure impactor. Polymer Degradation and Stability, 106, 26-35.

[19] Shaddix, C. R., Palotás, Á. B., Megaridis, C. M., Choi, M. Y., & Yang, N. Y. (2005). Soot graphitic order in laminar diffusion flames and a large-scale JP-8 pool fire. International Journal of Heat and Mass Transfer, 48(17), 3604-3614.

FIRE RESISTANCE

DETERMINATION OF THERMOCHEMICAL PROPERTIES OF STONE WOOL INSULATION MATERIALS

N. Nagy[a] , M. J. DiDomizio[a] , E. J. Weckman[a] , R. Roos[b]
[a]Department of Mechanical and Mechatronics Engineering, University of Waterloo,
200 University Avenue West, Waterloo, ON, N2L 3G1, Canada
[b] Technical Innovations Department, Roxul Inc.
8024 Esquesing Line, Milton, ON, L9T 6W3, Canada

INTRODUCTION

Fire safety is a critical aspect of a building design. Consequently, the National Building Code of Canada (NBCC) stipulates that construction assemblies within Canadian structures must adhere to stringent requirements in terms of their fire performance, with specific requirements set based on the intended application of the structure[1]. In terms of design, fire performance of the assembly is dictated by key factors such as the constituent materials and methods of construction. Once engineered, a near full-scale assembly must be tested to determine its fire resistance rating, which is ascertained using a standard fire performance test such as CAN/ULC S101[2]. Since these tests involve the exposure of full size assemblies to fire, they are destructive. As such, the process for determining these ratings can be both time consuming and expensive. This motivates research towards the development of computer models that can be used to estimate the fire performance of built assemblies early in the design process.

Development of a physically sound model for fire resistance of a construction assembly must begin with detailed investigation into the response of single materials to exposures typical of those seen in fires. Such data provides the input necessary to develop tools to simulate heat and mass transfer through each material in the assembly as it is heated in a fashion representative of a fire resistance test. There are many challenges associated with accurately modelling the response of even single materials as their temperatures increase, however. These include complexities in characterizing thermal decomposition processes or changes to the detailed internal structure of the materials on heating, combined with additional uncertainty in thermo-chemical material properties such as thermal conductivity, density, emissivity and specific heat capacity as functions of temperature and level of decomposition. Research presented in this paper provides some of the background data necessary for the creation of a comprehensive fire model that includes property data measured at expected fire temperatures, rather than estimated on the basis of simplified, steady state assumptions as is currently the norm.

One of the components of many construction assemblies is thermal insulation. These materials are designed to simultaneously optimize thermal insulation, acoustic insulation, permeability, and fire resistance, and consequently tend to have fairly complex properties. The particular materials of interest in this study are Roxul ComfortBatt® and Roxul Safe® insulation products, which are stone wool insulation materials used in commercial, industrial, residential, and marine applications. In general terms these stone wool products are composed of basalt rock and slag, held together by a cured urea extended phenolic formaldehyde (CUEPF) binder[3]. They are reported to have very high melting points, approximately 1177°C[3], and varying densities depending on the specific product. Construction assemblies containing stone wool products have been demonstrated to provide improved fire resistance in many situations, but in comparison to studies into the thermal behavior of wood or gypsum board, surprisingly few detailed evaluations have focused on evaluation of the thermal response of individual insulating materials or insulation in wall assemblies[4-7]. Comparative correlations exist for a limited number of properties[4-9] and specific studies into crystallization[10,11] or specific heat have been conducted[12]. Nonetheless, critical gaps still exist in definition of the high temperature thermophysical properties for these materials, as well as in understanding key phenomena, such as time-varying thermal penetration, formation of void spaces, shrinkage/expansion, transport of water vapour, potential modes of decomposition, or brittleness, friability and failure, that occur in insulating materials under fire exposure. Additional thermophysical property data, particularly high temperature values, are also needed as input to multiphysics models for prediction of the behaviour of the insulation materials in fire situations. The objective of the present study, therefore, is to extend previous results and further explore decomposition and thermal behaviour of stone wool under exposure to temperatures and incident heat flux characteristic of fires.

METHODOLOGY

The stone wool (MW) products tested during this study are Roxul Safe® which has a reported density of 72 kg/m^3 [13] and Roxul ComfortBatt® which has a reported density of 32 kg/m^3 [14] Both products are composed of 94-99% fibre and 1-6% CUEPF binder[15]. Bulk properties of each material were determined based on cubic specimens cut from larger insulation samples. The mass of each cube was measured and the bulk volume was calculated from the measured dimensions. Bulk density was then found by dividing the measured mass of a specimen by its bulk volume. Porosity was measured using the Archimedes' water displacement principle. A sample was placed in a known volume of water, compressed to remove air pockets, and the volume of water displaced was recorded as the skeletal volume of the sample. Porosity (void fraction) was calculated by dividing pore volume for each sample, calculated as the difference between bulk and skeletal volume, by the bulk volume of that sample.

As an initial step towards determining the impact of heating on the stone wool insulation materials, 5cm x 5cm x 5cm cubic specimens of each insulation were placed in an industrial furnace and exposed to constant temperatures of 500°C and 750°C for 10, 20 and 30 minutes. The mass and measured dimensions of each sample were recorded before and after exposure. After heating, the surface of each sample was visually inspected to assess the level of degradation, then the heat exposed samples were sectioned and any changes to the interior of the material noted.

Thermogravimetric analysis (TGA), differential scanning calorimetry (DSC), FTIR reflectance and small-scale inverse heat conduction experiments were used to better understand thermal response, and where possible quantify additional thermochemical properties, of the stone wool insulation materials under exposure to heat. Both TGA and DSC were carried out in inert atmospheres to study properties during thermal pyrolysis of the insulation, as well as in air to better understand the response of each material when heated in an oxidizing atmosphere such as would be encountered during a fire.

Samples of bulk insulation material were powdered for TGA analysis using a mortar and pestle. The powdered samples were allowed to equilibrate to lab conditions for a minimum of 24 hours and were then analyzed using the TA Instruments Q500 TGA apparatus at the University of Waterloo. For each test, a 10 mg specimen was placed in an inert tray and the change in mass of the sample was measured at a predetermined heating rate of 2, 5, or 10°C/min in air, or 5°C/min in nitrogen, across the range of temperatures from 30°C and 800°C. Mass loss rate for each case was computed numerically using the TA Universal Analysis software. The precision of the mass balance in the TA Q500 is rated as ±0.01%, while isothermal temperature accuracy and precision were ±1°C and ±0.1°C respectively. To determine how variations in mass-loss data related to specimen size and preparation, TGA results were also compared to the results from the tests in which the 5cm x 5cm x 5cm cubic specimens of each insulation were heated in an industrial furnace at 500°C and 750°C for 10, 20 and 30 minutes.

Differential scanning calorimetry (DSC) measurements were conducted on powdered specimens of both insulation materials using a NETZSCH DSC 404 C differential scanning calorimeter. Samples, each weighing approximately 10mg, were heated at rate of 10°C/min across the range of temperatures from 30-1200°C in air or an inert (argon) environment to examine the heat flow into or out of each specimen over time as it was heated. The mV/mg output signal was recorded as a function of temperature and used with reference DSC data from a sapphire standard to calculate the specific heat capacity of the specimen according to Equation 1:

$$c_{p,s} = c_{p,st} \frac{D_s W_{st}}{D_{st} W_s} \qquad [1]$$

where c_p is the specific heat capacity, D is the vertical displacement measurement taken from a DSC thermal curve, and W is the mass; the subscripts s and st refer to the test specimen and the sapphire standard respectively. The temperature and heat flow calibration methods followed ASTM E967[16] and ASTM E968[17] respectively, while the ASTM E1269 method was used for determining specific heat capacity[18].

Measurements of spectral reflectance across the range of 2 to 25 μm were made on samples of each insulation using a Surface Optics Corporation (SOC) 400T Infrared (FTIR) Reflectometer. Before testing, the specimens were conditioned for a minimum of 24 hours to approximately 20°C and 30% relative humidity. Values of total near-normal emittance were measured on three different samples across the full spectral range at a temperature of 20°C according to Method C in ASTM E408 – 13[19] in which the reflectivity of the material is determined by comparison to a NIST traceable reference standard and emissivity determined by subtracting the reflectance at each temperature from 100%. Since the insulation specimens were opaque, transmittance was assumed negligible in the calculation. Emissivity was calculated as a function of temperature using a linear estimation, so high temperature values do not account for possible non-linearities, nor do they account for chemical changes that occur with increasing sample temperature. Accuracy of the instrument is rated at ±1%, with additional uncertainty arising via the limitations noted above, as well as in use of Method C for determination of emittance.

Finally, a preliminary series of instrumented heat transfer experiments was carried out on the Roxul Safe® insulation towards development of a protocol for using a modified cone calorimeter experiment to gain insight into thermal response and heat penetration of larger specimens. Disc-shaped specimens, 230 mm in diameter and between 54 and 56 mm thick, were cut from slabs of insulation and conditioned in the lab for 24 hours or more. Each disc was then mounted on a 230 mm diameter disc of 7.7 mm thick concrete board, placed on the load cell of the cone calorimeter and exposed to a constant incident heat flux of 25, 35, or 50 kW/m^2 for a minimum duration of one hour or until measured temperatures approached steady state values. Throughout each test, mass loss of the specimen during heating was recorded via the cone calorimeter load cell. The temperature of the unexposed side of the specimen was also measured via two thermocouples affixed to the top of the concrete board (under the specimen) at r = 0 (the centreline) and at a radial position of r = 70 mm. An additional thermocouple was placed on the edge of the specimen at a position 27 mm below the top surface to gauge the level of heat loss through the side of each sample. Sample surface temperatures were monitored using a FLIR infrared camera at measurement points chosen to correspond to the thermocouple mounting locations on the unexposed side (r = 0 and r = 70 mm). Two tests were conducted on each sample: in the first, the conditioned virgin specimen (never exposed to heat) was exposed to the given level of incident heat flux; in the second, the same specimen was exposed to the same heat flux for a second time using the same methodology to ascertain potential changes in thermal properties of a previously heat exposed sample.

RESULTS

Virgin, as-supplied Roxul Safe® and Roxul ComfortBatt® were cut into cubic specimens and the bulk properties at room temperature were measured as described above. Results are summarized in Table 1. Difficulty in measuring the volume of these insulating materials leads to fairly high values of standard deviation; however, the average values of bulk density of 71.6 kg/m^3 obtained for Roxul Safe® and 30.4 kg/m^3 for Roxul ComfortBatt® are very close to reference values of 72 kg/m^3 [13] and 32 kg/m^3 [14] respectively as expected. Following determination of the density of each sample, the average mean porosity for the virgin materials was determined using the Archimedes method. Porosity of Roxul Safe® was found to be 0.96; as expected, that for Roxul ComfortBatt® was higher at 0.98.

Table 1: Measured values of bulk material properties for the stone wool insulation

Material	Roxul Safe	Roxul Comfort-Batt
Bulk Density [kg/m^3]	71.6 ± 8.4	30.4 ± 6.1
Porosity	0.96 ± 0.006	0.98 ± 0.004

The impact of heating on the samples was examined by heating cubic specimens in a constant temperature furnace for 10, 20 and 30 minutes. Representative results are shown in Figures 1a) through c) and 1d) through f) for samples of ROXUL ComfortBatt® exposed at 500°C and 750°C for 10, 20 and 30 minutes respectively[20]. For exposures to 500°C (Figure 1a) – c)), there is no evidence of damage on the outside of the specimens, and sections indicate that there is only relatively minor damage visible on the interior. As the exposure temperature increases, significant internal degradation begins to occur, as seen in Figure 1d) through f). The large spherically shaped cavitation seen in the bottom specimens suggests that an exothermic reaction involving the binder may occur. This is also supported by temperatures measured in the interior of the insulation during heating which, for many samples were observed to rise higher than the temperatures of exposure.

Figure 1: Cubic furnace specimens of Roxul ComfortBatt® heated for various times

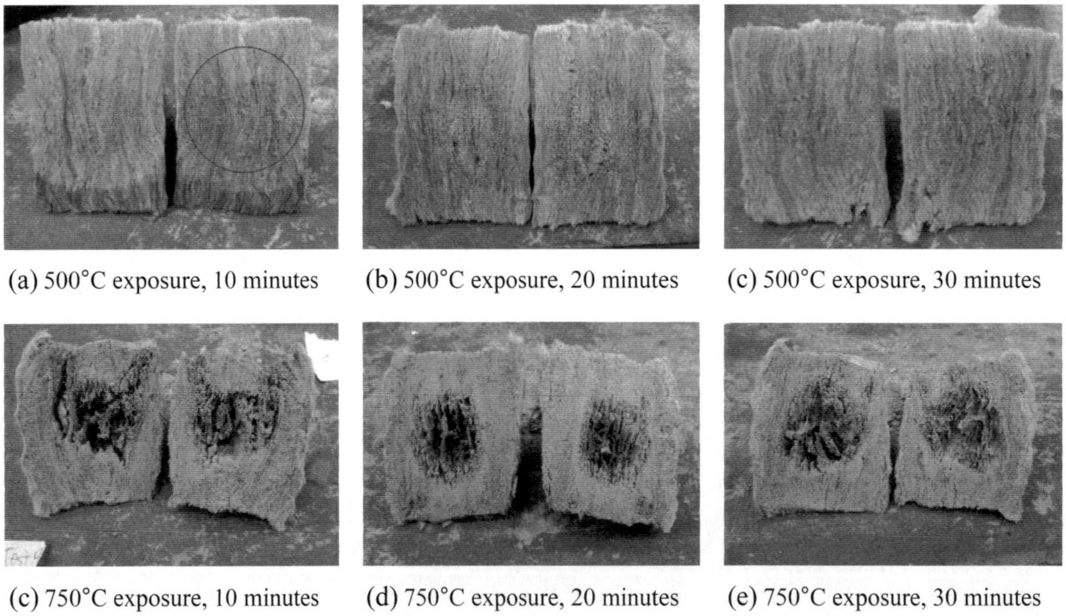

(a) 500°C exposure, 10 minutes (b) 500°C exposure, 20 minutes (c) 500°C exposure, 30 minutes

(c) 750°C exposure, 10 minutes (d) 750°C exposure, 20 minutes (e) 750°C exposure, 30 minutes

One plausible explanation is that the CUEPH binder exothermically decomposes at a critical temperature. Due to the insulating qualities of the surrounding material, the excessive heat is trapped internally and if the reaction produces a volume of pressurized hot gas at a rate high enough that it cannot equilibrate with the surroundings, it will push out isotropically from the centre of the specimen against any incoming gas and the large cavity in the centre may form. Further, the striations and damage through the material to the exterior in Figure 2 suggest that, due to the porous nature of the material, the cavity may fill with hot gases which would be at approximately the same temperature as the surrounding furnace gases. In contrast, for the samples exposed to a lower temperature, Figure 1, the rate of reaction may be slow enough that it equilibrates with the outside as it occurs. To further determine the impact of such internal changes during heating, an attempt was made to measure changes in porosity between room temperature and heat exposed samples. It was found, however, that the resolution of the Archimedes method was not sufficiently high to discern any difference in porosity arising due to the observed cavitation or other heat induced effects. Thus further investigation was required to more systematically study processes that occur in the stone wool insulation during exposure to different levels and times of heating[21].

As one alternative, TGA was conducted on 10 mg powdered specimens heated at several different heating rates for additional insight into the mass loss and possible decomposition modes for each material as a function of temperature. A set of typical results are plotted in Figure 2a) and 2b) for Roxul Safe® and Roxul ComfortBatt® respectively with average event onset temperature across several samples found for each heating rate as summarized in Table 2.

Figure 2: TGA Results: Percent mass remaining as a function of temperature for various heating rates in air and nitrogen purge gas

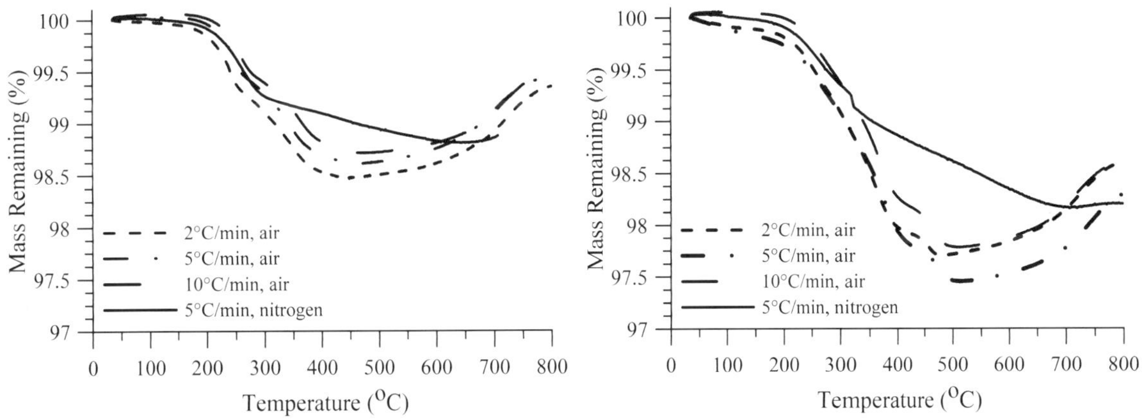

(a) Roxul Safe®

(b) Roxul ComfortBatt®

Table 2. TGA Event onset temperatures for the stone wool samples

Material	Average Initial Onset Temperature (◦C)	Average Secondary Onset Temperature (◦C)	Average Tertiary Onset Temperature (◦C)	Average Final Onset Temperature (◦C)
Roxul Safe®	223	304	329	421
Roxul ComfortBatt®	215	303	352	451

The TGA results in Figure 2 show that there is little impact of heating rate on mass loss rate for either stone wool insulation. For both materials, mass loss transitions begin at temperatures just above 200°C, followed by several periods characterized by different rates of mass loss until maximum values of mass loss are reached at temperatures between 400–500°C when tested in air (specific value of the minimum in the TGA trace depends on heating rate and material under study). Upon further heating, it appears that the materials begin to react with the air since the mass climbs in value until the final test value of 800°C is reached. In contrast, when the stone wool is tested in an inert environment, the initial marked increase in mass loss again begins at around 200°C, but instead of the subsequent changes in slope of mass loss versus temperature seen in the oxidizing environment, the rate of mass loss from both materials in nitrogen remains linear between approximately 300°C and 700°C and then levels off until the end of the heating period.

The transitions in mass loss can be more clearly seen in Figure 3 which shows the change in mass remaining with heating as a function of the temperature for Roxul Safe® heated at a rate of 2°C/min in air in contrast to 5°C/min in a nitrogen environment. The average mass loss and temperatures at which these transitions occur are further summarized in Table 3.

Figure 3: TGA Results: Change in percent mass remaining with temperature as a function of temperature in air and nitrogen purge gas

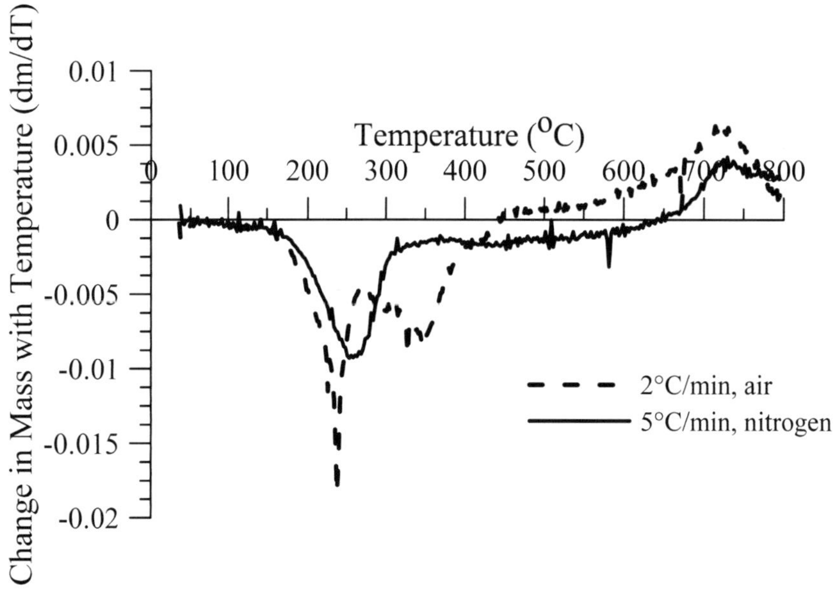

Table 3. TGA Mass loss summary for the stone wool samples

Material	Average Mass Loss (Air, %)	Average Mass Loss (N2 , %)	Temperature of max dm/dT (Air, °C)	Temperature of max dm/dT (N2,°C)
Roxul Safe®	1.412	1.126	246	264
Roxul ComfortBatt®	2.284	1.848	368	264

The table shows that the rate of change of mass of these insulating materials will be at a maximum for temperatures between 250–370°C. Interestingly, the temperature at which the maximum value of dm/dT occurred was the same for both materials when heated in nitrogen, but significantly different when they were heated in air. This would indicate that while the two materials may undergo very similar thermal degradation processes in the absence of oxygen, they may degrade somewhat differently than one another in air. Further, the average mass loss in nitrogen was only 80% of that in air, consistent with the observation that oxygenated pyrolysis did occur. Interestingly, Roxul ComfortBatt® exhibited greater average mass loss in both environments, perhaps indicating that it contained more binder than Roxul Safe®.

Values of mass loss measured during the small-scale (TGA) results (Table 3 above) were compared to values of mass loss obtained during larger scale furnace and cone calorimeter tests to determine how mass loss might change with specimen size and/or preparation. Values of mass loss in Roxul ComfortBatt® samples measured after long duration heating at constant furnace temperatures are plotted in Figure 4 as a function of the exposure temperature. A second order polynomial fit to the data is also shown. The mass loss initially increases as a function of temperature, leveling off at around 4% for temperatures above 550°C. Average values of mass loss measured during cone calorimeter heating of Roxul ComfortBatt® samples for 1 hour surpassed 2% on initial heating and continued to increase even after 1.5 hours of exposure. These values are consistent with those obtained in both TGA and furnace heating tests, particularly when combined with the observation that under exposure to incident flux levels of 50 kW/m^2, the test samples heated to average steady state temperatures of around 345°C. It is likely that similar decomposition reactions are taking place despite the different specimen preparation and heating rates. It was interesting to note that in subsequent cone calorimeter tests on previously heat-exposed samples, the average mass loss was less; it plateaued at approximately 1.5%. This indicates that the decomposition reactions were already at least partially complete during the first exposure of a sample. The measured mass loss in the second round of may have been due to residual moisture being driven off the already heat exposed insulation samples.

Figure 4: Total percent mass loss as a function of furnace temperature for Roxul ComfortBatt®

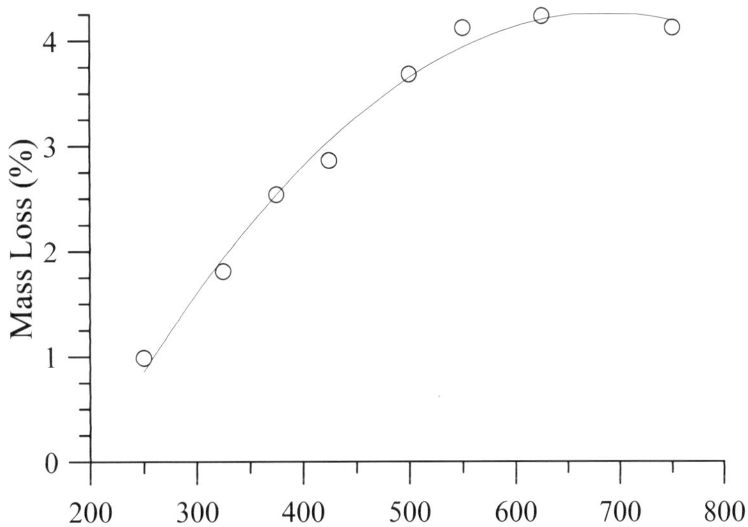

It was reported by Poljanšek[22] that a similar binder to that used in the stone wool samples here loses as much as 80% of its mass by the time it is heated to 600°C. Since the binder content of the present materials is listed as between 1- 6% by mass[13,14], the mass loss of these samples due to decomposition of the binder alone could then be between 0.8 - 4.8%, values very similar to those reported in all of the tests here.

To further investigate details of temperature induced activity, as well as to derive preliminary estimates of the specific heat capacity as a function of temperature, for Roxul Safe® and Roxul ComfortBatt® samples, DSC measurements was carried out in both air and nitrogen environments. Figures 5a) and b) present typical results for Roxul Safe® and Roxul ComfortBatt® respectively.

Figure 5: DSC Results: Specific heat as a function of temperature in air and argon environments

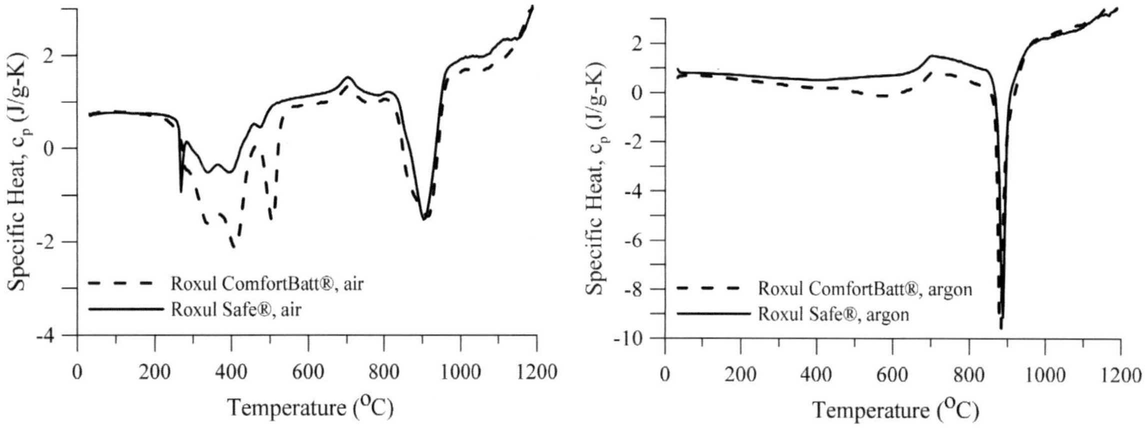

a) Specific Heat in Air

b) Specific Heat in Inert Argon

Based on TGA results, the temperatures at which major thermal effects (thermal transitions or chemical reactions) occurred are documented in Table 2. The data from the DSC results shown in Figure 5 are consistent with those values, also indicating that some form of reaction or transformation takes place in both materials in air beginning at temperatures of around 200°C. Further, exothermic or endothermic reactions are indicated by peaks in a DSC plot, and can be discerned by a positive or negative concavity, respectively. From Figure 5a), it appears that there are multiple exothermic reactions taking place at temperatures between 200 and 550°C. This corresponds to the temperature range in which multiple mass loss events occurred (Table 2) in the TGA tests, and to the temperature range where there was the maximum change in mass loss with temperature (Table 3). In the inert argon environment, on the other hand (Figure 5b)), the DSC data confirm temperature induced activity again beginning at approximately 200◦C but levelling off until temperatures rise to above 640◦C. Finally, there appears to be some form of reaction or material transformation that takes place at 800◦C in air, and a little later perhaps, nearer to 900◦C, in an inert environment. To date then, the combined results have discerned the significant chemical and physical processes occurring during heating of the stone wool insulating materials; however, more testing and data analysis is required to deconvolute, for example, the potential roles of moisture transfer or binder decomposition in heat and mass transfer in the material.

To further delineate the thermal properties of the two stone wool insulation materials and develop input properties for future modelling efforts, values of near total normal emissivity were determined for both

ROXULSafe® and ROXUL ComfortBatt® at temperatures of 273K and 1273K. These are summarized in Table 4. At room temperature, the emissivities of the two materials are quite similar with measured values of 0.968 and 0.978, respectively, but the values change differently as a function of increasing temperature. While the emissivity of ROXULSafe® decreases by approximately 21% to a value of 0.782 as it heats to 1273K, that of ROXUL ComfortBatt® decreases around 28% to a value of 0.724. These marked changes in value with temperature indicate that high temperature emissivity values should be taken into consideration when examining heat transfer analysis involving stone wool insulating materials.

Table 4. Measured values of near normal total emissivity for the stone wool insulation

Temperature (K)	Emissivity	
	Roxul Safe	Roxul ComfortBatt
273	0.968	0.978
1273	0.782	0.724

As a final stage in the present work, several heat conduction experiments were run to develop a modified cone calorimeter experiment with which to gain insight into the thermal response and heat penetration into larger specimens of stone wool insulation. Temperatures were measured at various locations of interest; steady state temperatures values attained after 1 hour of exposure to various levels of incident radiant heat flux are summarized in Table 5. These were then used in a global energy balance on the system, to deduce values of thermal conductivity from measurements.

Table 5. Steady State Temperatures Measured During Heat Conduction Experiments (C)

Incident Flux (kW/m^2)	Round Number	r = 0 mm		r = 70 mm		
		TTOP	TBOT	TTOP	TBOT	TEDGE
50	1	612	155	531	92	70
50	2	602	156	545	87	55
35	1	527	137	453	83	52
35	2	526	132	443	79	45
25	1	473	110	401	69	42
25	2	458	105	397	68	43

As can been seen from the Table, at steady state there was a notable gradient in temperature radially from the centre of the sample with edge temperatures higher than the ambient air temperature (20–22°C) but notably less than temperatures at r = 70 mm and on the centreline. The percent differences in temperature between the centreline and at r = 70 mm were 15% on average at the top surface, and 50% on average at the bottom surface making it difficult to apply a simple 1-dimensional model of heat transfer to the system. Instead, in this preliminary study, values of thermal conductivity were estimated via a global energy balance, configured to account for the radial gradient in temperature by using 13 concentric annular sections across the sample. Heat flux incident on the surface of each annular section was modelled accounting for the specified input heat exposure from the calorimeter modified by appropriate shape factors to connect the cone flux and the annuli on the sample surface. Surface radiative effects were estimated using linearly extrapolated values of average steady state surface temperature for each annular section with temperature dependent estimates of emissivity obtained via the measurements above. Convective heat losses from the perimeter of the sample were estimated using the average temperature difference between the edge and ambient, with an convective heat transfer coefficient of 10 W/m^2-K. Similarly, estimates of heat losses from the top of the sample were based on average temperature differences between the top surface and ambient. Convective heat transfer coefficients on this surface were varied between 10 and 15 W/m^2-K to study the impact on estimates of thermal conductivity related to varying the magnitude of the surface convective effects. Values of thermal conductivity estimated based on the experimental data for exposure of the sample to a heat flux of 50 kW/m^2, with the above global energy balance, ranged between a high value of 0.26 W/m-K for a value of convective heat transfer coefficient of 10 W/m^2-K to a low value of 0.068 W/m-K for a value of 15 W/m^2-K. Given that the average steady state temperature of the sample was around 350°C, these values bracket the values of 0.089 W/m-K to 1.1 W/m-K reported in the literature for similar stone wool materials.

CLOSING COMMENTS

The present results provide additional insight into the temperature dependent behaviour of two examples of stone wool insulating materials. In initial experiments, significant cavitation and high interior temperatures were observed in some specimens during and after exposure to heat while in others there was little evidence of cavitation, suggesting that exothermic reactions, possibly involving the organic binder, were occurring inside the samples under certain heating rates or as the insulation material reached a certain internal temperature. A combination of thermogravimetric analysis (TGA) and differential scanning calorimetry (DSC) both in air and in inert atmospheres, pointed to different levels of exothermic reaction associated with various temperature ranges. The onset of reaction began at around 200°C in both atmospheres, with multiple reactions continuing across the range between 200°C and 550°C in air. There was evidence of another distinct transformation at higher temperatures of around 800°C in both air and the inert environment. Additional work is needed to better understand the nature of these transformations or reactions in order to incorporate key transient phenomena into any future advanced models for the thermochemical behaviour of stone wool at elevated temperatures.

Values of near total emissivity for both stone wool insulating materials were obtained using an FTIR reflectance method, followed by the standard approach of linear extrapolation to high temperature values. Estimated values of emissivity were found to vary by over 25% from room temperature to 1273K, with differing values seen for ROXULSafe® and ROXUL ComfortBatt® at the higher temperature. Thus, the temperature dependence of emissivity should be accounted for in situations where temperatures increase to the high values typically found in fire situations.

Small-scale heat conduction experiments were carried out to determine steady state temperatures across the surface and on the bottom of larger disc-shaped insulation specimens subjected to varying levels of radiant heating via a cone calorimeter. Temperature data, coupled with a model of input heat energy from the cone and a global energy balance on the sample, facilitated confirmation of values for thermal conductivity as a function of temperature for the materials under study. In future work, by combining information deduced from the TGA and DSC analyses with measured values of temperature dependent emissivity and thermal conductivity, additional insight can be gained into the thermal response of the stone wool materials under exposure to heat; information that is needed to refine multiphysics models for prediction of the behaviour of the insulation materials in fire situations.

REFERENCES

1. National Building Code of Canada. Canadian Commission on Building and Fire Codes, Institute for Resaerch in Construction, National Research Council, Ottawa, ON, 2010.
2. CAN/ULC S101, Standard Methods of Fire Endurance Tests of Building Construction and Materials, Underwriters' Laboratories of Canada, Toronto, ON, 2014.
3. Roxul ComfortBoard IS Technical Product Information. Roxul Inc. Accessed March 31, 2015, from <http://www.roxul.com/files/RX-NA_EN/pdf/Technical%20Data%20Sheets-%20updated/ Building %20Envelope/ComfortBoard%20IS.pdf>, 2014.
4. Just, A, Schmid, J and Konig, J 2012 'Post-protection effect of heat-resistant insulations on timber-frame members exposed to fire', *Fire and Materials,* vol. 36, pp.153–163.
5. Frangi, A, Schleifer, V and Hugi, E 2012 'A New Fire Resistant Light Mineral Wool', *Fire Technology,* vol. 48 pp.733–752.
6. Takeda, H and Mehaffy, JR 1998 'Wall 2D: a model for predicting heat transfer through wood-stud wall exposed to fire', *Fire and Materials*, vol. 22, pp.133–140
7. Clancy, P 2001 'Advances in Modelling Heat Transfer Through Wood Framed Walls in Fire', *Fire and Materials*, vol.25, pp.241–254.
8. Keerthan, P and Mahendran, M 2013 'Thermal Performance of Composite Panels Under Fire Conditions Using Numerical Studies: Plasterboards, Rockwool, Glass Fibre and Cellulose Insulations', *Fire Technology*, vol. 49, pp.329-356.
9. Benichou, N, Sultan, M, MacCallum, C and Hum, J 2001 'Thermal Properties of Wood, Gypsum and Insulation at Elevated Temperatures', Institute for Research in Construction - National Research Council of Canada, Ottawa, Canada.
10. Moesgaard, M, Pedersen, HD, Yue, YZ and Nielsen, ER 2007 'Crystallization in stone wool fibres', *Journal of Non-Crystalline Solids,* vol. 353, pp.1101–1108.
11. Smedskjaer, MM, Solvang, M and Yue, Y 2010 'Crystallisation behaviour and high-temperature stability of stone wool fibres', *Journal of the European Ceramic Society* vol. 30 pp.1287–1295.
12. Ohmura T, Tsuboi M, Onodera M, and Tomimura T 2003 'Specific heat measurement of high temperature thermal insulations by drop calorimeter method', *International Journal of Thermophys*ics, vol. 24, pp.559–575.
13. Roxul Safe Technical Product Information. Roxul Inc. Accessed March 16, 2016, from <http://www.roxul.com/files/RX-NA_EN/pdf/Technical%20Data%20Sheets-%20updated/ Building%20Envelope/ROXUL_ SAFE.pdf>, 2015.
14. Blanket (and Batt) Insulation (COMFORTBATT). Roxul Inc. Accessed March 31, 2015, from <http://www.roxul.com/files/RX-NA_EN/pdf/specs/en-ca/072116%20-%20Blanket%20Insulation % 20REV%202013-10-28%20ComfortBatt.pdf>, 2013.
15. Roxul Material Safety Data Sheet. Roxul Inc. Accessed March 16, 2016, from <http://www.roxul.com/files/RX-NA_EN/pdf/MSDS%20and%20Safety%20Bulletin/Roxul %20Material%20Safety%20Data%20Sheet%2002-13-14.pdf>, 2014.
16. ASTM Standard E967-08. Standard Test Method for Temperature Calibration of Differential Scanning Calorimeters and Differential Thermal Analyzers, ASTM International, West Conshohocken, PA, 2008.
17. ASTM Standard E968-02. Standard Practice for Heat Flow Calibration of Differential Scanning Calorimeters, ASTM International, West Conshohocken, PA, 2002.
18. ASTM Standard E1269-11. Standard Test Method for Determining Specific Heat Capacity by Differential Scanning Calorimetry, ASTM International, West Conshohocken, PA, 2011.
19. ASTM Standard E408-13. Standard Test Methods for Total Normal Emittance of Surfaces Using Inspection-Meter Techniques, ASTM International, West Conshohocken, PA, 2013.
20. Nagy, N, DiDomizio, MJ and Weckman. EJ 2015 'Investigation of behaviour of stone wool insulation in a tube furnace at various temperatures', *Proceedings of Spring Technical Meeting of the Canadian Section/Combustion Institute*, Saskatoon, May.

21. Nagy, N, DiDomizio, MJ and Weckman. EJ 2016 'Assessment of the thermal properties of insulation materials'. *Proceedings of Spring Technical Meeting of the Canadian Section/Combustion Institute*, Waterloo, May; modified for Interflam 2016, 14th International Fire Science and Engineering Conference, Interscience Publications, pp. 547-553.
22. Poljanšek, I, Šebenik, U and Krajnc, M 2006 'Characterization of phenol–urea–formaldehyde resin by inline FTIR spectroscopy', *Journal of Applied Polymer Science*, vol. 99, no. 5, pp.2016–2028

AUTHOR INDEX

Grayson Franks Ltd
1 West Yard
Guilford Grove Greenwich
London SE10 8JT
United Kingdom

CURRAN ASSOCIATES INC.
proceedings.com

ISBN 978-1-5108-4674-6
9 781510 846746

15th International Conference and Exhibition on Fire and Materials 2017

San Francisco, California, USA
6 – 8 February 2017

Volume 2 of 2

ISBN: 978-1-5108-4674-6

Printed from e-media with permission by:

Curran Associates, Inc.
57 Morehouse Lane
Red Hook, NY 12571

Some format issues inherent in the e-media version may also appear in this print version.

Printed by Curran Associates, Inc. (2017)

For permission requests, please contact Grayson Franks Ltd
at the address below.

Grayson Franks Ltd
1 West Yard
Guilford Grove Greenwich
London SE10 8JT
United Kingdom

Phone: 44 208 692 5050
Fax: 44 208 692 5155

office@intersciencecomms.co.uk

Additional copies of this publication are available from:

Curran Associates, Inc.
57 Morehouse Lane
Red Hook, NY 12571 USA
Phone: 845-758-0400
Fax: 845-758-2633
Email: curran@proceedings.com
Web: www.proceedings.com

Fire and Materials 2017

15th International Conference

San Francisco, California, USA

6th – 8th February 2017

Published by

interscience communications

London, UK

www.intersciencecomms.co.uk

The organisers wish to thank and acknowledge the companies for their sponsorship

Proceedings of the Fire and Materials 2017 Conference

held at the Hyatt Centric Hotel Fisherman's Wharf Hotel, San Francisco, California, USA
6 – 8 February 2017

A conference organised by
Interscience Communications Ltd
1 Burt Ash Lane
Bromley, Kent BR1 4DJ
England

Fire and Materials 2017
1004 pp

Fire and Materials 2017 Programme Committee

Stephen Grayson
Interscience Communications, UK
Vytenis Babrauskas
Fire Science & Technology Inc, USA
Marcelo Hirschler
GBH International, USA
Anja Hofmann
BAM, Germany
Marc Janssens
SWRI, USA
Patrick Van Hees
Lund University, Sweden

Fire and Materials 2017 Scientific Committee

Serge Bourbigot
PERF, France
Karen Boyce
Ulster University, UK
Steven Craft
CHM Fire Conslt, Canada
Nicolas Dembsey
WPI, USA
Eric Guillaume
Efectis, France
Tuula Hakkarainen
VTT, Finland
Brian Lattimer
Jensen Hughes, USA
Christine Lukas
Dow Chemical Co, UK
Richard Lyon
FAA, USA
Samuel Manzello
NIST, USA
Alexander Morgan
Univ of Dayton Research Inst, USA
Bernhard Schartel
BAM, Germany
Bjorn Sundstrom
SP Tech Research Inst, Sweden
Colleen Wade
BRANZ, New Zealand
Koichi Yoshida
RIME, Japan

To select a paper click on title, or use "Page Thumbnails" on the left hand navigation bar

TABLE OF CONTENTS

FIRE DYNAMICS

UPHOLSTERED FURNITURE

COMPARTMENT FIRES

FIRE HAZARD

IGNITION AND PYROLYSIS MODELING

American Chemistry Council®

North American Flame Retardant Alliance

Focused on Fire Safety, Regulation, and Public Policy

The American Chemistry Council's North American Flame Retardant Alliance (NAFRA) supports a robust and transparent regulatory system that provides both strong fire protection and chemical safety. NAFRA's members represent the cutting edge of fire safety chemistry and technology and are dedicated to protecting consumers through the safe and effective use of flame retardants. Our work includes:

- Conducting original, peer-reviewed research;
- Enhancing awareness of fire safety and the efficacy of flame retardants;
- Supporting risk-based policies and regulations based on sound science; and
- Establishing constructive dialogue with flame retardant users and other stakeholders.

ALBEMARLE®
Fire Safety Solutions

Great Lakes SOLUTIONS
a Chemtura business

ICL Industrial PRODUCTS

EXPERIMENTAL ANALYSIS OF STONE WOOL SANDWICH COMPOSITES EXPOSED TO CONSTANT INCIDENT HEAT FLUXES AND SIMULATED PARAMETRIC FIRES

B.Andres[1,2], J.P. Hidalgo[3,4], L.Bisby[3] & P. van Hees[2]
[1]Danish Institute of Fire and Security Technology, Denmark
[2]Division of Fire Safety Engineering, Lund University, Sweden
[3]School of Engineering, The University of Edinburgh, UK
[4]School of Civil Engineering, The University of Queensland, Australia

ABSTRACT

An experimental analysis of the fire behavior of stone wool composites with stainless steel and gypsum plasterboard linings was performed. The focus was to characterize the micro-scale behavior of stone wool, and identify the main processes occurring in stone wool composites when exposed to heat. Thermogravimetric analysis was applied to two different wools with varying densities and binder content in air and nitrogen atmospheres at 5, 10, and 20 K/min. Results show a low percentage of mass loss that could be linked to the low fraction of organic binder contained in the wool. Intermediate-scale composite assemblies were tested using the Heat-Transfer Inducing Rate System (H-TRIS). Specimens were composed of stainless steel-stone wool, and gypsum plasterboard-stone wool. Tests were performed at (1) constant low incident heat flux (7 kW/m²), (2) constant high incident heat flux (60 kW/m²), and (3) simulated parametric incident heat exposure. The results of these tests help to identify the relevant phenomena that should be taken into account when modelling building assemblies with stone wool cores (i.e. pyrolysis and combustion of the binder, water transport through the cross-section, and burning of paper).

INTRODUCTION

Building fire barriers are compartment elements intended to limit the spread of fire and smoke to neighboring spaces. They play a major role in the overall response of a building in the event of a fire, providing the necessary time for a safe evacuation. Moving towards explicit performance-based design of buildings there is a need to develop improved models that can predict the behavior of building elements under different potential fire scenarios. Modeling techniques to predict the fire behavior of building barriers can provide tools for safer building designs and support product development for the building products manufacturing industry. Many attempts have been made in the history of fire safety engineering to model the behavior of building barriers in standard fire tests [1–3]. Nevertheless these models are typically empirically fitted to reproduce specific results from standard furnace tests. Models are consequently scenario dependent, and attempting to model exposures from other fire scenarios presents various challenges [4]. The study presented in this paper is part of the FIRETOOLS project[5], the main goal of which is to develop prediction methodologies for the fire behavior of building elements using a scaling approach. In particular, the methodology aims to predict the behavior of specific composite building barriers under different heating scenarios, reducing the large amount of prescriptive testing needed and providing a better understanding of the relevant phenomena.

Building fire barriers are often sandwich constructions composed of a core insulation material and covering panels. Common core materials with enhanced thermal and/or acoustic properties are mineral wool, rigid polyisocyanurate foam (PIR), and rigid polyurethane foam (PUR). Covering panels provide strength and stability to the construction and may be steel plates, gypsum plasterboards, chipboards, etc. These individual materials react to fire in very different ways, and when placed together in a construction the interactions can lead to different and complex behaviors. There is therefore a need to better characterize material behaviors and the interaction between them in composite structures when subjected to different heating exposures. For instance, understanding thermal contact, evaporation, and transport of water through adjacent materials, or internal pyrolysis and combustion within components, could lead to improved predictive models.

Within the FIRETOOLS framework, the work performed for the current paper consists of characterizing the fire behavior of gypsum-stone wool and steel-stone wool sandwich composite panels. The main goal is to identify and isolate the phenomena occurring in the materials and their effects when interacting with other materials in a composite construction, thus providing basic knowledge for model development. The reactions occurring in the stone wool were identified using micro-scale thermogravimetric (TGA) testing. Further tests on intermediate-scale composite specimens were performed under constant and variable incident radiant heat fluxes using the novel testing methodology (H-TRIS[6]). These tests can be regarded as more representative of full-scale behavior under different heat exposures than smaller scale tests in a cone calorimeter or fire propagation apparatus.

The results presented herein will also serve as basis for future validation of a multi-scale modeling methodology, where material properties at small or medium scale are used to model full-scale behavior of building barriers. The methodology aims to predict performance in various heating scenarios, providing a rational, physics based tool for performance-based design. Hence, the experiments are performed under constant and simulated parametric incident heat exposures to ensure the models are not only able to predict standard ISO-834[7] fire resistance tests results, but also more realistic fire scenarios including cooling phases.

MATERIALS AND METHOD

The main goal of this study is to identify and isolate phenomena occurring within gypsum plasterboard-stone wool and steel-stone wool sandwich composites when exposed to severe heating conditions. The methodology consists of micro-scale thermogravimetric testing of the stone wool under air and nitrogen atmospheres to characterize and quantify the thermal degradation reactions taking place. Intermediate-scale tests were performed on steel-stone wool composites to investigate the reactions experienced by the wool when exposed to: (1) a low level of irradiation (7 kW/m^2), (2) a high level of irradiation (60 kW/m^2), and (3) a simulated parametric type heating exposure. The aim of the experimental program was to isolate the exothermic or endothermic reactions within the range of temperatures studied. Further tests were performed on gypsum-stone wool composites where the aim was to characterize other phenomena occurring in this case, for instance burning of the paper lining of the gypsum plasterboards, water transport along the building assembly, and cracking of the board which could increase the heat exposure to the insulation material. Figure 1 shows a schematic of the multi-scale experimental approach followed.

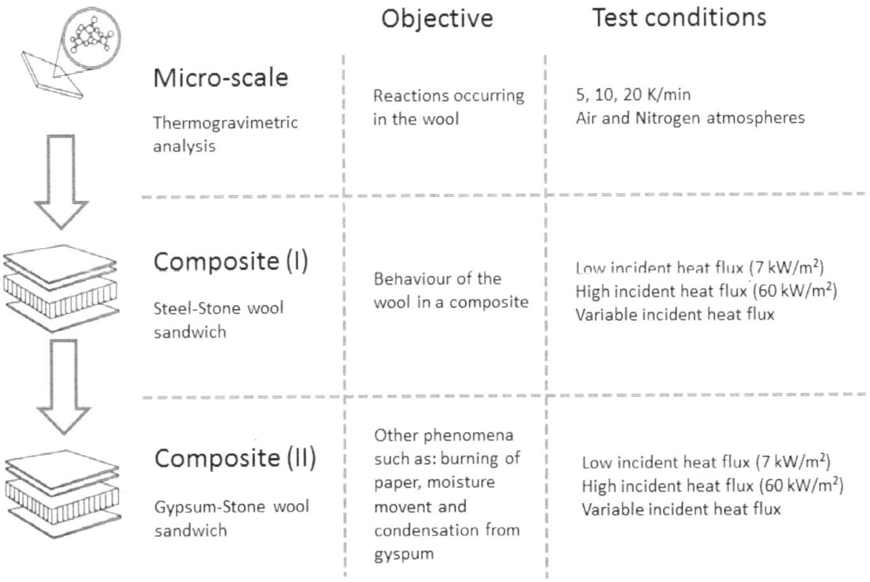

Figure 1: Multi-scale experimental methodology to characterize the fire behavior of composite panels consisting of a stone wool core, and linings such as steel or gypsum.

Materials

Two common sandwich constructions were chosen for this study; these correspond to steel sheet with stone wool insulation (S-SW) and gypsum plasterboards with stone wool insulation (G-SW). The G-SW construction is commonly found in office or residential buildings, whereas the S-SW composite is more common in industrial or commercial buildings. The two-construction choice was made simply so that different common applications would be represented in the work.

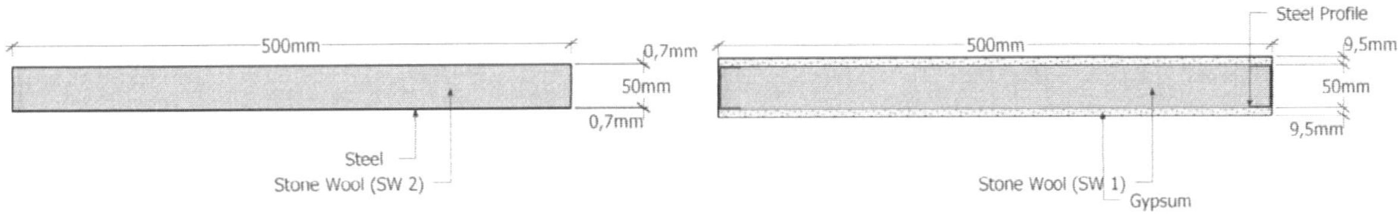

(a) (b)

Figure 2: Cross-section of the specimens tested: (a) S+SW2, (b) G+SW1.

The S-SW (Figure 2 (a)) construction was composed of 0.7 mm thick stainless steel plates and 50 mm stone wool insulation (SW2 in Table 1). Stainless steel is a highly thermally conductive material, with a thermal conductivity value varying from 15-30 W/m·K as a function of temperature[8]. Therefore, for common scenarios with convective heat transfer coefficient varying in the expected natural range (4-60 W/ m^2·K), the Biot number[9] is sufficiently low that a thermally thin behavior can be assumed, thus with no thermal gradient through the cross-section. Stone wool is a permeable material composed of rock fibres and a small percentage of organic binder held together with water repellent oils [10,11]. Although stone wool is non-combustible[12] (Euroclass A1 or A2), the small percentage in weight of binder undergoes an exothermic reaction during heating in the presence of oxygen, thus potentially increasing the temperature within the insulation[13–15]. The G-SW (Figure 2 (b)) in the current study was composed of glass fiber reinforced gypsum plasterboard (Type EH2 according to EN 520[16]) and 50 mm of stone wool insulation (SW1 in Table 1). The thickness of the gypsum boards was 9.5 mm including their paper lining. Gypsum plasterboards are composed of calcium sulfate dihydrated that calcines in a two-step endothermic reaction process[17]. In the first reaction, calcium sulfate dehydrates to hemihydrate form releasing 75 % of its water content (Equation 1), the remaining water is released in the second reaction (Equation 2). The former reaction occurs between 100-150 °C, and the second after 200 °C depending on the heating conditions[18,19]. The endothermic nature of the reactions delays the transfer of heat through the neighboring materials when used in composites, this being a primary reason for its extensive use in buildings. Relevant material properties at ambient temperature are given in Table 1.

$$1^{st}: CaSO_4.H_2O + Q_1 \rightarrow CaSO_4.\tfrac{1}{2}H_2O + \tfrac{3}{2}H_2O \qquad [1]$$

$$2^{nd}: CaSO_4.\tfrac{1}{2}H_2O + Q_2 \rightarrow CaSO_4. + \tfrac{1}{2}H_2O \qquad [2]$$

Table 1: Ambient temperature material properties for the materials used in the current study.

Material	Density (kg/m^3)	Thermal Conductivity (W/m·K)	Water Content (%)	Binder Content (%)
Gypsum	650 ± 5	0.19	18.4	-
SW1	36.8 ± 1.2	0.037	-	2.5 ± 0.06
SW2	105.1 ± 2.4	0.041	-	4.36 ± 0.06
Stainless Steel	7805	15	-	-

Micro-Scale Thermogravimetric Analysis

Thermogravimetric analysis (TGA) was performed on samples from the two different stone wools used in this study, to identify their thermal degradation reactions (e.g. pyrolysis and oxidation reactions). Tests were performed using a METTLER-TOLEDO TGA/DSC 1 apparatus (Figure 3 (a)). Samples of 11.4 ± 3 mg were cut and placed in alumina oxide sample holders with care to avoid contamination. This sample mass is rather large considering the recommendations for TG testing [20], nevertheless this was necessary since only a small percentage of binder in the wool would thermally degrade in the range of temperatures tested (20-800 °C). The samples were gently crushed into the sample holder to have a compact mass (Figure 3 (b)). An empty sample holder was also tested simultaneously to obtain a baseline measurement and to control the sample heating rate. Tests were performed under both air and nitrogen atmospheres using heating rates of 5, 10, and 20 K/min.

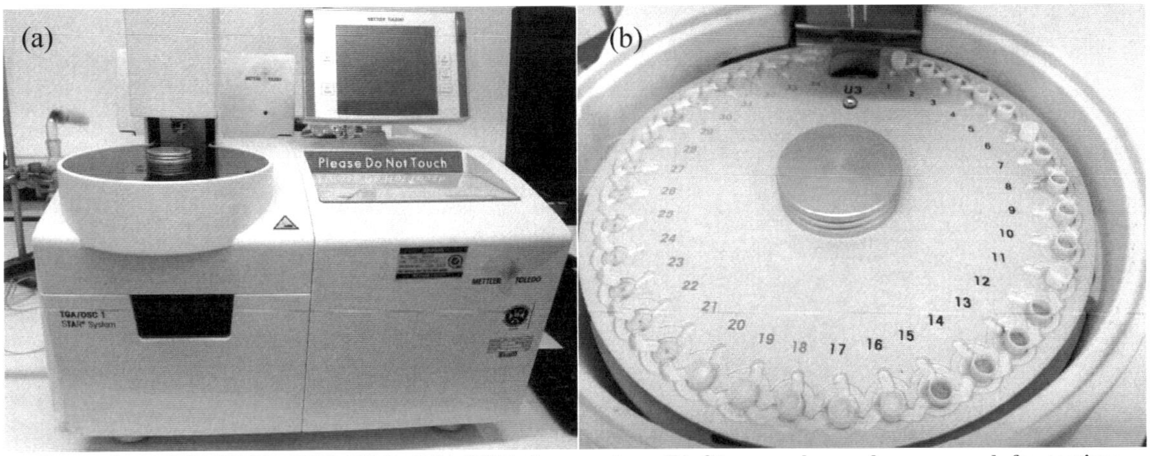

Figure 3: (a) METTLER-TOLEDO TGA/DSC 1 apparatus. (b) Stone wool samples prepared for testing.

Heat-Transfer Rate Inducing System (H-TRIS)

A novel fire testing methodology, the Heat-Transfer Rate Inducing System (H-TRIS)[6], was used to expose intermediate-scale composite samples to both constant and variable incident radiant heating exposures. The size of the specimens tested (500 mm^2 surface area) was sufficient to represent heat transfer *system* behavior. Hence this approach can be used as an extrapolation tool for full-scale fire tests, limited to 1D heat transfer through the composite panels. Discontinuities in the assemblies, such as joints or studs, are not accounted for in these tests. The H-TRIS consists of four propane-fired radiant panels placed on a linear motion system. The incident radiant heat flux to the tested specimen is automatically controlled by the calibrated relative distance between the radiant panels and the surface of the sample. Almost any incident radiant heat exposure-time history can be achieved with H-TRIS, limited only by the size and type of radiant panels, and the minimum distance to the tested surface[6]. The net heat delivered to the specimen is defined as a fraction of the absorbed incident radiant heat flux to the sample's target surface minus the losses at the exposed surface due to convection and radiation (see Equation 3):

$$\dot{q}''_{net} = \alpha \cdot \dot{q}''_{in} - \varepsilon \cdot \sigma \cdot (T^4 - T^4_{amb}) - h_c \cdot (T - T_{amb}) \qquad [3]$$

where $\dot{q}''_{net}$ is the net heat flux, α is the absorptivity of the specimen, $\dot{q}''_{in}$ is the incident heat flux, ε is the emissivity of the material, σ is the Stefan-Boltzmann constant, T is the temperature of the exposed panel, T_{amb} is the ambient temperature, and h_c is the convective heat transfer coefficient. An additional benefit of the H-TRIS methodology is that allows a careful understanding of the thermal boundary conditions. This reduces the uncertainties linked to the modeling and provides a cleaner data set for model validation purposes. Furthermore, the exposed surface can be directly observed during testing, so it is possible to detect phenomena occurring that can influence the heat transfer through the cross-section (e.g. cracking of the boards or burning of the paper linings).

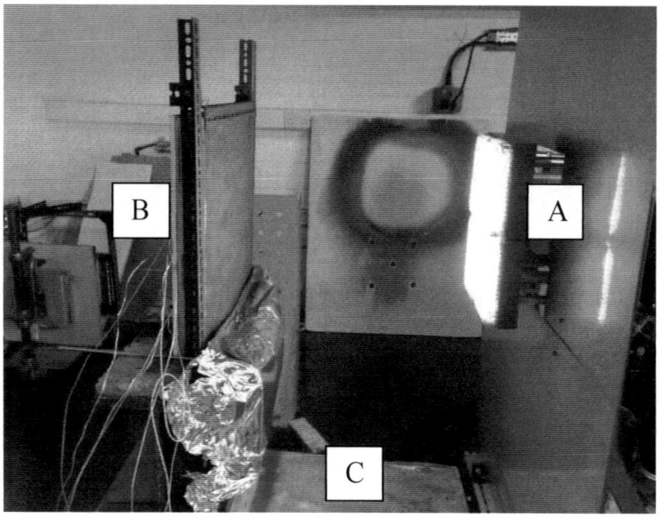

Figure 4: The Heat Transfer Inducing System (H-TRIS): (A) Radiant Panels (B) Specimen Tested (C) Motion System.

Tests were performed under three different heating conditions:
(1) Constant low incident heat flux (7 kW/m^2);
(2) Constant high incident heat flux (60 kW/m^2); and
(3) Variable incident heat flux (simulated parametric fire curve).

The constant incident heat fluxes were used to identify phenomena occurring (or not) at high and low heating rates (e.g. exothermic reactions in the wool, burning of the paper in gypsum, etc.), based on prior thermal calculations. A comparatively low incident heat flux (7 kW/m^2) was selected so as to reach temperatures below those sufficient to cause the exothermic reactions in the wool. By contrast, a relatively high incident heat flux (60 kW/m^2) was chosen as it ensured that the reactions in the wool would occur. Applying a variable external heat flux allowed investigating the response of the composites under a range of more 'realistic' thermal exposures, comparable to a hypothetical real compartment fire scenario. The exposure in this case was calculated using the Swedish Method[21] for a temperature-time history in a standard compartment of 20 m^2 floor area, 32 MJ/m^2 fire load, and standard openings of windows and doors (equivalent to an opening factor of 0.04). The net heat flux (Equation 4) in this case was calculated as the net convective and radiative fluxes due to the fictitious (predicted) room gas temperature. The incident heat flux time-history for H-TRIS was set accounting for the losses due to radiation and convection at the target surface of the specimen (Equation 3).

$$\dot{q}''_{net} = \varepsilon \cdot \sigma \cdot (T^4_{room} - T^4) + h_c \cdot (T_{room} - T) \qquad [4]$$

The surface of the specimens was painted with solar collector coating of known absorptivity (0.92), in order to reduce the uncertainty in the analysis by a better characterization of the boundary conditions. Temperatures were measured through the cross-sections and on the unexposed faces of the specimens (Figure 5) using Type K thermocouples; these have a precision of ± 2.5°C. Sheathed thermocouples were used to measure the temperatures within the stone wool, and thermocouples welded to a copper disk were used for the surface of the covering panels. The temperature on the exposed surface was also measured using a FLIR® A320 thermal imaging camera. At least two repeat tests were performed for each of the assemblies for each of the three heating conditions used. Further tests would be recommended for assessing repeatability.

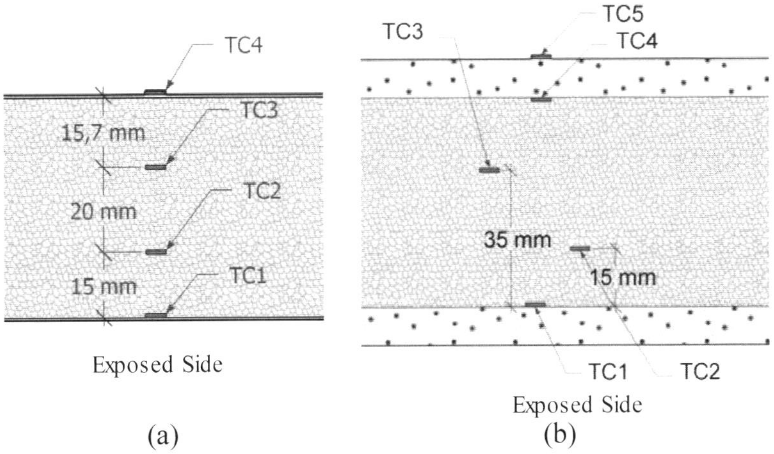

Figure 5: Position of the thermocouples through the cross-section during testing: (a) Steel+SW2, and (b) Gypsum+SW1

TEST RESULTS

Micro-Scale TGA Results

Figure 6 and Figure 7 show the mass loss (TGA) and the differential thermogravimetric curves (DTG) for SW1 and SW2 tested in air and nitrogen atmospheres. Thermogravimetric curves plot the normalized mass loss as a ratio of the initial mass. Differential thermogravimetric curves are obtained as the first derivative of the mass with respect to temperature. A positive DTG value indicates loss of mass.

The mass loss for SW1 in nitrogen atmospheres (Figure 6 (a)) is in the range 1-2.5 % at 800 °C reference temperature for the three different heating rates. For an air atmosphere (Figure 6(c)) at 500 °C reference temperature the mass loss is 3 % for two of the heating rates and 1.5 % at 10 K/min heating rate. This difference could be due to the inherent heterogeneity of stone wool, meaning that some samples might have a higher content in combustible binder than others. From 500-700 °C a plateau is observed, which is followed by an increase of mass. Figures 6 (b) and (d) show the DTG under nitrogen and air respectively. One peak is observed between 250 °C and 350 °C in nitrogen atmosphere. The peak is moving slightly towards higher temperatures with increased heating rates. However, in air atmosphere two sequential peaks are observed at 250-350 °C and 400-500 °C. The first peak is likely caused by pyrolysis of the binder whereas the second could be due to the oxidation of the binder residue. Between 700-800 °C a peak of weight gain is observed when tested in air; this could be due to a crystallization process in the fibers[10]. The slight increase in weight at the beginning of the tests for some of the specimens is likely due to a difference in the buoyancy effect between the baseline and the sample tested.

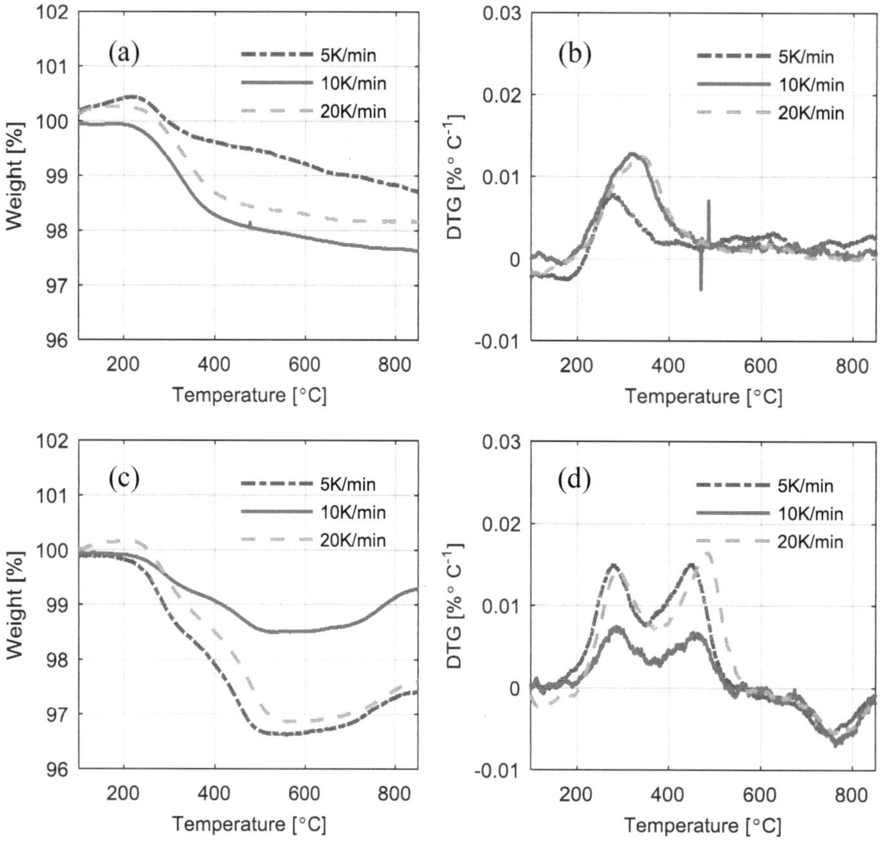

Figure 6: Thermal decomposition of SW1 at different heating rates: (a) TG loss in nitrogen, (b) DTG in nitrogen, (c) TG in air, and (d) DTG in air.

SW2 experiences a higher mass loss compared to SW1, being 2-3.5 % in nitrogen atmosphere at 800 °C and 5-6.5 % in air atmosphere at 500°C reference temperature (Figure 7). This is due to the fact that SW2 has a much larger density and binder content than SW1. The same pattern, with a plateau after 500 °C and a slight increase of mass after 700 °C, is observed. Similar behavior is also observed in the DTG curves for SW2 with respect to DTG SW1. However, while the reactions are all happening at the same temperatures, in the latter case the magnitude of the second peak is greater.

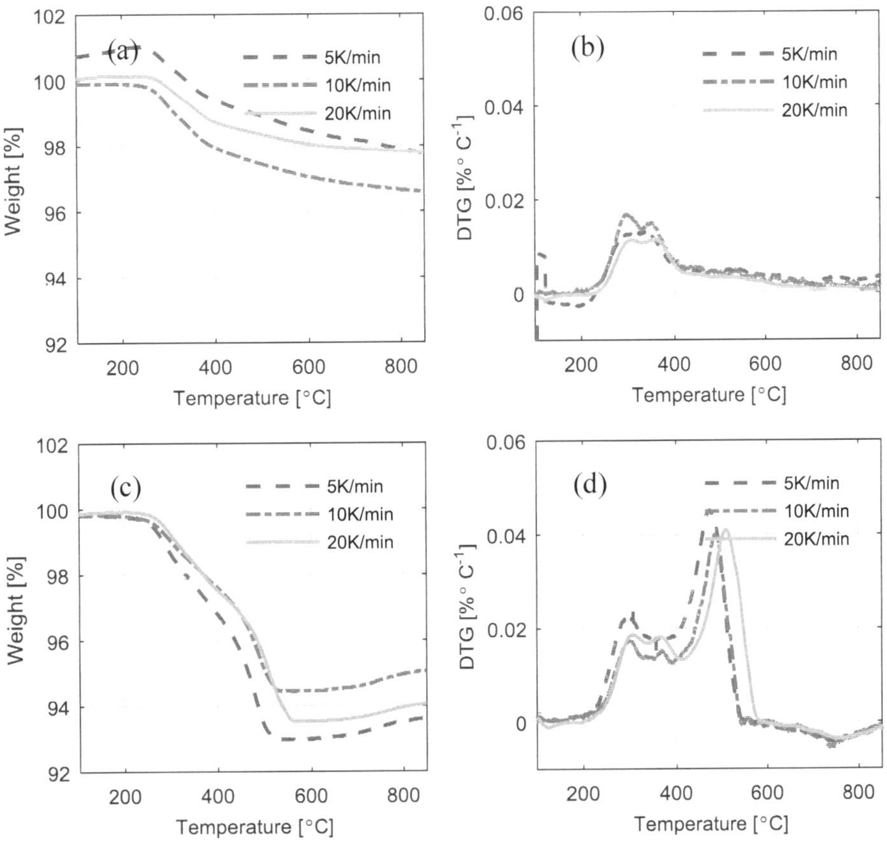

Figure 7: Thermal decomposition of SW2 at different heating rates: (a) TG in nitrogen, (b) DTG in nitrogen, (c) TG in air, and (d) DTG in air

H-TRIS Test Results

Figure 8 to Figure 10 show the temperature measurements for S+SW2 and G+SW1 composites under 7 kW/m², 60 kW/m², and simulated parametric incident heating exposure, respectively. The temperature was measured on the unexposed side of the exposed panels (Exp), at 15 and 35 mm deep in the stone wool and at the unexposed side of the unexposed panels (Unexp) as shown in Figure 5. Additional measurements were also made at the exposed side of the unexposed board for the G+SW1 sandwich panels (i.e. at 45 mm).

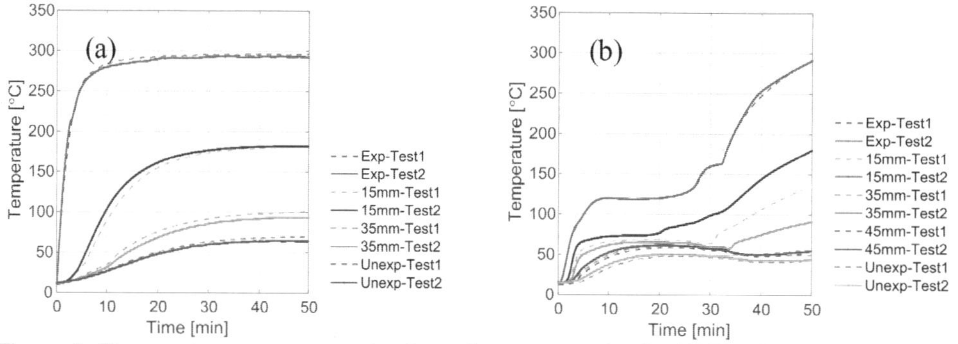

Figure 8: Temperature measurements along the cross-section for 7 kW/m² incident heat flux: (a) S+SW2, and (b) G+SW1.

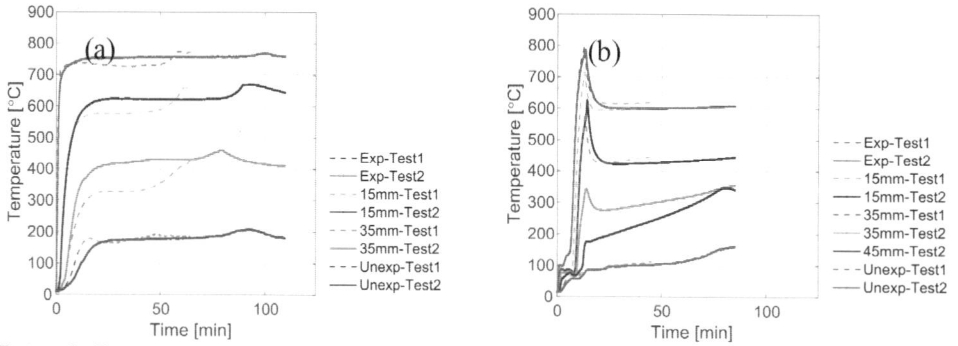

Figure 9: Temperature measurements along the cross-section for 60 kW/m² incident heat flux: (a) S+SW2, and (b) G+SW1.

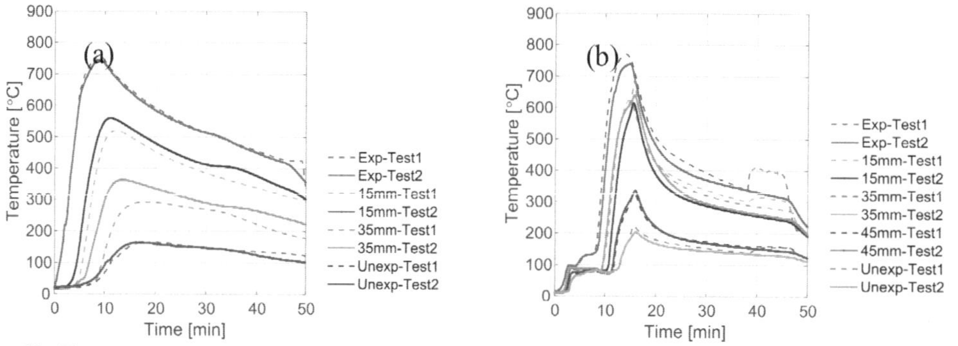

Figure 10: Temperature measurements along the cross-section for a parametric like incident heat flux: (a) S+SW2, and (b) G+ SW1.

Due to the thermally thin behavior of the stainless steel mentioned above, for S+SW2 constructions the Exp temperature can be assumed to be the thermal boundary at which the panels are exposed. The variability between the measured temperatures in the different tests is shown in Table 2 as the percentage difference between the readings for different time points of the tests. In general, tests showed good repeatability, especially for the low incident heat flux. The position giving more disperse values are the ones corresponding to thermocouples embedded in the wool (15 mm and 35 mm), which is linked to the uncertainty on the exact position of the thermocouple, as it was difficult to ensure it during the test. The thermocouples at 15 and 35 mm in Test 1 G+SW1 60 kW/m² (Figure 9 (b)) appear to be closer to the exposed board, giving high differences in the temperature readings between Test 1 and Test 2.

Table 2: Percentage of variation between the measurements in the tests presented herein.

		10 min				20 min				40 min			
		Exp (%)	15mm (%)	35mm (%)	Unexp (%)	Exp (%)	15mm (%)	35mm (%)	Unexp (%)	Exp (%)	15mm (%)	35mm (%)	Unexp (%)
S+SW2	7 kW/m²	1.3	14.2	9.5	5.7	0.9	4.1	9.9	4.9	1.0	1.1	6.3	6.0
	60 kW/m²	0.8	2.0	23.2	42.9	2.6	6.5	19.9	5.8	4.0	7.0	22.1	2.2
	Variable heat flux	1.2	11.5	57.0	26.4	1.5	5.6	12.9	2.9	1.4	10.1	15.9	7.9
G+SW1	7 kW/m²	0.1	9.4	20.3	12.0	0.4	10.5	7.4	5.2	2.3	26.2	32.3	3.4
	60 kW/m²	3.9	46.6	57.2	10.3	2.7	29.0	34.8	5.0	2.8	29.1	33.8	6.0
	Variable heat flux	22.3	35.7	1.8	5.2	5.9	9.0	1.3	7.0	19.3	18.8	8.4	11.6

Figure 8-10 (a) shows the temperature distribution for S+SW2 at the different heating rates. The exposed temperature measurements show an instant initial temperature rise and a later increase smoothly following the heating exposure. No reactions can be observed at 7 kW/m², thus achieving an inert behavior. Nevertheless, for 60 kW/m² incident exposure, an increase on the temperature can be observed after reaching the steady state (50-60 minutes). This bump in the temperature-time curve could be due to the wool binder burning. Given that the wool is porous, the heat generated by the combustion of the binder could diffuse and be transferred through the cross-section, thus explaining the delay in time on the readings of the different thermocouples. This assumption implies also that the heat created within the wool is travelling in both directions. The open edges of the samples could have an influence on the available oxygen to undergo the reaction and different behavior could be expected in sealed specimens. In Figure 11 (a) some residue from the stone wool binder burning can be observed on the exposed side of the unexposed panel, as well as from the discoloration of the wool (Figure 11 (b)) after the test. A weak bump may also be observed for the simulated parametric incident heat flux (Figure 10 (a)), although in this case it could be masked by a cooling down effect.

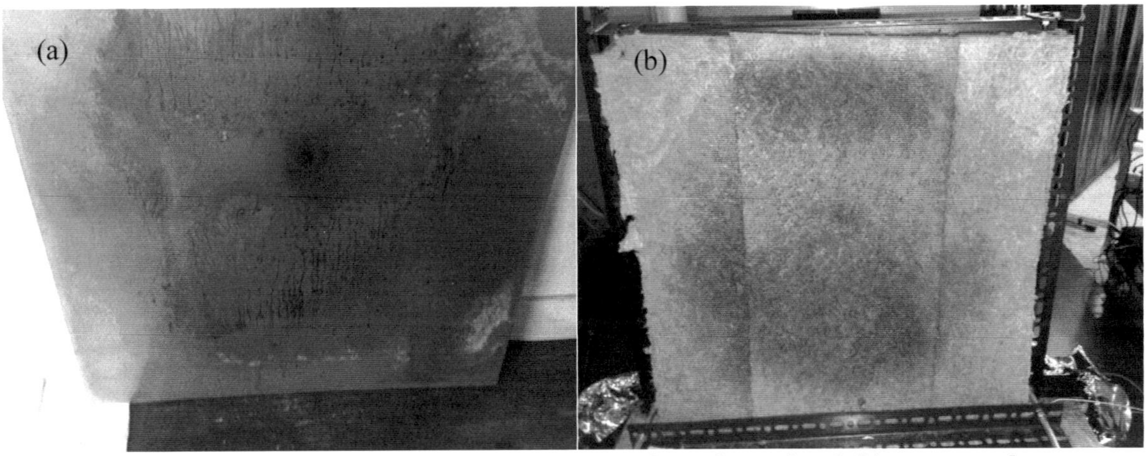

Figure 11: S+SW2 60 kW/m²: (a) unexposed steel panel and (b) stone wool.

Figure 8-10 (b) shows the temperature distribution for G+SW2. All the figures show a water plateau near 100 °C due to the evaporation of the water from the gypsum plasterboard. The length of the plateau depends on the heating conditions, being longer for less severe heating exposures (Figure 8 (b)). For the exposed temperature of the low incident heat exposure (Figure 8 (b)) the two step calcination reactions of the gypsum plasterboard can be observed. After the gypsum has undergone the full calcination process, there is a very rapid increase of the temperature. A peak can be observed between 10-25 minutes in Figure 9 (b), which could be linked to combustion of the paper linings on the unexposed side of the exposed board. Also, few millimetre wide cracks appear on the exposed boards at 10 minutes, potentially increasing the heat exposure through the samples. Thermocouples at 35 mm and 45 mm show a steady increase of temperature after their peak. This could be due either to the burning of the paper on the unexposed board or to combustion of the binder. Figure 12 (a) shows the exposed side of the unexposed board and the burning of the paper around the position of the thermocouple and the unexposed side of the stone wool (Figure 12(b)), with certain discoloration in the same area. Regarding the simulated parametric incident heat flux (Figure 10 (b)), the peak due to the burning of the paper and peak of the heating overlap, with a peak on the unexposed side of the board. Furthermore, in Test 1 of the parametric like incident heat exposure, the board fell off and led to a bump in the temperature profiles around 35 minutes although this heat wave did not have a great impact on the unexposed temperature measurements.

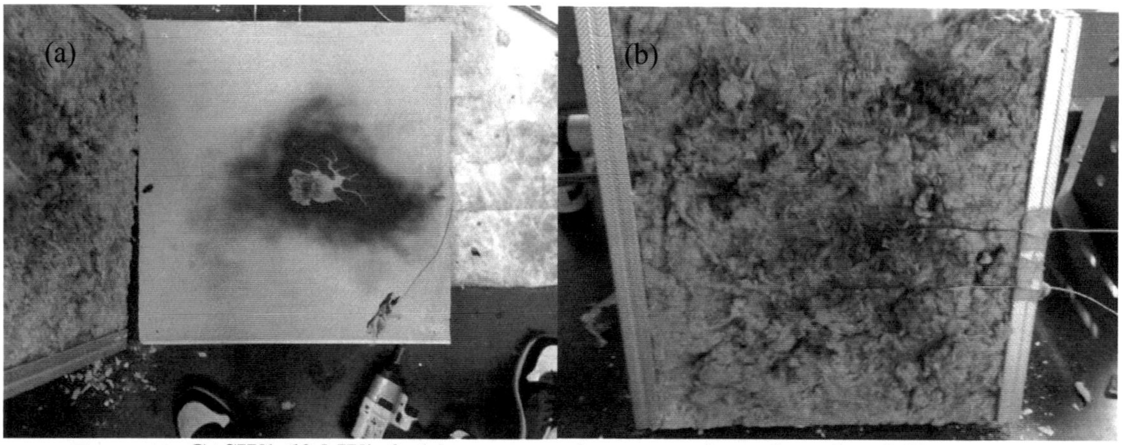

Figure 12: G+SW1 60 kW/m²: (a) unexposed gypsum plasterboard and (b) stone wool.

DISCUSSION AND CONCLUSIONS

Stone wool is a fibrous inorganic material with a low content of organic binder. Two different types of stone wool with varying densities and binder content were subject to a thermogravimetric analysis to characterize the behavior with increasing heat. Results showed a mass loss varying between 1.5-3 % for the low density stone wool (SW1) and 5-6.5 % for the high density stone wool (SW2) in air atmospheres. The total mass lost in the tested samples varied due to the heterogeneous nature of the material, meaning that the samples tested in micro-scale could contain more or less binder. Therefore, further tests are required to develop a better statistical distribution and understanding of the data. The differential thermogravimetric curves show one or two peaks in nitrogen and air atmospheres respectively (Figure 6 and Figure 7). This indicates that the second peak could be linked to an oxidation process from the binder residue after pyrolysis, or from some other subcomponent in the wool.

H-TRIS results have shown the composite behavior of gypsum-stone wool (G+SW) and steel-stone wool (S+SW) composites. Tests performed with S+SW composites provide information on the behavior of the stone wool in larger scales with certain limitation on the air entrainment. The stainless steel panel can be regarded as a thermally thin material, thus the temperature of the exposed panel can be assumed as the boundary temperature for the stone wool. Tests at low incident heat exposures (7 kW/m²) have shown a smooth temperature rise following the exposure, with minor differences between tests. In the high heating exposures (60 kW/m²) test results present a bump in the temperature readings after a steady state has been reached. This occurs at slightly different times between tests. The cause of this temperature increase is not clear, although it could be associated to the combustion of the binder occurring at different locations in the stone wool. The fact that it occurs after reaching the steady state could be due to lack of appropriate conditions for the reaction to occur, for instance oxygen. This would be a main difference between the micro-scale testing (TGA) where all the air necessary for the reaction is provided, and the H-TRIS where conditions are restricted to assembly configuration and lab environment. Due to the permeable nature of stone wool, the heat transfer has been previously identified as a combination of conduction through the fibers and gas, plus radiation and convection through the porous medium[11]. The significant influence on the convective term could explain that the transfer of heat generated by the binder reaction. A lesser bump is observed in the simulated parametric fire curve that could be due to the binder effect. The behavior of the wool in G+SW composites is masked by the behavior of the gypsum. The calcination process of the gypsum generates a plateau in the temperature measurements, delaying the heat of transfer through the construction. In the high heating exposures (60 kW/m²) a great peak in temperatures is observed before the steady state is reached. This peak is probably linked to combination of phenomena such as the combustion of the paper facing, cracking of the exposed board, and combustion of the binder. The simulated parametric fire exposure is a more complex scenario. After the calcination plateau, the different phenomena occurring overlap with the peak incident exposure.

The results presented herein serve as an initial basis for future validation of a multi-scale modelling approach for the behavior of building barriers. Phenomena such as pyrolysis and combustion of the binder, burning of the paper, and water transport have been identified as relevant for successful thermal modeling. The results also show the need of further analysis of convective processes inside porous materials, because air entering through gaps or joints will affect the burning processes in intermediate and large scales. The test results at different heat exposures can be used to assess applicability of the models to different fire scenarios, leading to a scenario independent model. Nevertheless further work is recommended to address the uncertainties linked to the position of the thermocouples embedded in the wool. Likewise additional tests to assess repeatability would be beneficial, especially under severe heating exposures.

ACKNOWLEDGMENTS

This study is part of a collaborative project between Lund University and the Danish Institute of Fire and Security Technology called FIRETOOLS. The authors acknowledge the European Union's Seventh Framework Program for the support under grant no. 316991. Authors gratefully acknowledge the FIRETOOLS group and the BRE Centre for Fire Safety Engineering at The University of Edinburgh, particularly Michal Krajcovic for his assistance in the lab. Special thanks to Rockwool International A/S.

REFERENCES

1. Takeda, H. A model to predict fire resistance of non-load bearing wood-stud walls. *Fire and Materials* **27,** 19–39 (2003).
2. Keerthan, P. & Mahendran, M. Numerical modelling of non-load-bearing light gauge cold-formed steel frame walls under fire conditions. *Journal of Fire Sciences* **30,** 375–403 (2012).
3. Craft, S., Isgor, B., Mehaffey, J. & Hadjisophocleous, G. Modelling Heat And Mass Transfer In Wood-Frame Floor Assemblies Exposed To Fire. *Fire Safety Science* **9,** 1303–1314 (2008).
4. Andres, B. & van Hees, P. Experimental and Thermal Analysis of Wall Assemblies Exposed to Standard and Parametric Fires. *Proceedings of CONFAB* (2015).
5. van Hees, P. *et al.* Simulation of Fire Technical Properties of Products and Construction Barriers to Support Efficient Product Developement in Industry. *Proceedings of 13th international Fire and Engineering Conference Interflam* (2013).
6. Maluk, C., Bisby, L., Krajcovic, M. & Torero, J. L. A Heat-Transfer Rate Inducing System (H-

TRIS) Test Method. *Fire Safety Journal* (in press) (2016). doi:10.1016/j.firesaf.2016.05.001
7. International Organisation for Standardisation. ISO 834–1:1999: Fire Resistance Tests – Elements of Building Construction – Part 1: General Requirements. (1999).
8. CEN. Eurocode 3: Design of Steel Structures. Part 1-2: General Rules - Structural Fire Design. (2005).
9. Drysdale, D. *An Introduction to Fire Dynamics*. (John Wiley & Sons, 2011). doi:0.1002/9781119975465
10. Hidalgo, J. P. Performance based methodology for the fire safe design oof insulation materials in energy efficient buildings. (The University of Edinburgh, 2015).
11. Dyrbøl, S. Heat transfer in Rockwool modelling and method of measurement. (Technical University of Denmark, 1998).
12. EN ISO 1182: Reaction ro fire tests for products. Non-combustability test.
13. Olsen, H., Sjostrom, J., Jansson, R. & Anderson, J. Thermal Properties of Heated Insulation Materials. *Proceedings of 13th international Fire and Engineering Conference Interflam* 1049–1060 (2013).
14. Sjostrom, J. & Jansson, R. Measuring thermal material properties for structural fire engineering. in (15th International Conference on Experimental Mechanics, 2012).
15. Hidalgo, J. P., Torero, J. L. & Welch, S. Experimental Characterisation of the Fire Behaviour of Thermal Insulation Materials for a Performance-Based Design Methodology. *Fire Technology* (in press) (2016). doi:10.1007/s10694-016-0625-z
16. EN 520: Gypsum Plasterboard-Definitions, Requirements and Test Methods. (2004).
17. Thomas, G. Thermal Properties of Gypsum Plasterboard at High Temperatures. *Fire and Materials* **26,** 37–45 (2002).
18. Park, S., Manzello, S., Bentz, D. P. & Mizukami, T. Determining Thermal Properties of Gypsum Board at Elevated Temperatures. *Fire and Materials* (2009). doi:10.1002/fam
19. Melinge, Y. *et al.* One-Dimensional-Time Study of the Dehydration of Plasterboards Under Standard Fire Condition (ISO 834): Thermo-Chemical Analysis. *Journal of Fire Sciences* **29,** 299–316 (2011).
20. Vyazovkin, S. *et al.* ICTAC Kinetics Committee recommendations for performing kinetic computations on thermal analysis data. *Thermochimica Acta* **520,** 1–19 (2011).
21. Pettersson, O., Magnusson, S. E. & Thor, J. *Fire Engineering Design of Steel Structures*. (1976).

THE FIRE BEHAVIOUR OF GYPSUM BOARDS INCORPORATING PHASE CHANGE MATERIALS FOR ENERGY STORAGE IN BUILDING APPLICATIONS

Baljinder K. Kandola*, Awni Alkhazaleh and Graham J. Milnes
Institute for Materials Research and Innovation
University of Bolton, Bolton BL3 5AB, UK

ABSTRACT

In order to study the effect of adding organic phase change materials (PCM) in gypsum board on the latter's fire performance, thermal analysis and cone calorimetric techniques have been used. The cone calorimeter has been used in both horizontal as well as vertical orientation using spark and pilot ignition sources. For vertical orientation a special sample holder was constructed 'in house', which had a frame made of 1.5mm mild steel sheet, without a rear central area but covered with a ceramic board with holes in it to enable easy insertion of thermocouples through them for attaching to the rear surface of the sample. For pilot ignition a simple diffusion butane flame of 50mm length was used. The results were analysed in order to choose one best set of conditions for these sample, which turned out to be vertical orientation with spark igniter on all the time. The rest of samples were tested under these conditions and showed that the ignition and sustainment of the flame in these gypsum boards are dependent on the concentration and distribution of the PCM absorbed in gypsum board. While by an addition method required amount of PCM can be added to the gypsum board, with an immersion method the absorption depends upon the porosity of the material and needs to be engineered to get samples of low flammability.

INTRODUCTION

Thermal energy storage is a key technology for renewable energy utilization and improvement of the energy efficiency of the heating and cooling processes. Solar energy has enormous potential for the heating of buildings in winter, but is time and weather dependent. One way of utilising the solar energy to improve thermal energy storage efficiency of building materials is to use Phase Change Materials (PCMs) in the latter, which are capable of storing a large amount of latent heat during their phase transitions [1]. Heat is absorbed during the daytime and released at night when the material changes from one state to another (e.g. solid to liquid or vice-versa). Organic PCMs such as paraffins, butyl stearate and capric-lauric acids, active as latent heat thermal energy storage materials, have numerous benefits such as no supercooling, chemical stability, low vapour pressure during the phase change and no phase segregation [2]. The selection of the PCM depends on many factors such as: the melting temperature of the PCM must be within the temperature range of operation, the PCM must be chemically stable and should be non toxic, and should not impact overall cost. Paraffin waxes because of these properties and in particular low cost, have been a popular choice so far among various researchers [3,4,5].

PCMs can be incorporated in several forms to building materials like gypsum or concrete [6,7]. One way is the immersion method, by soaking gypsum blocks into bath of a melted PCM. By optimising the porosity of the gypsum board, the desired amount of PCM can be absorbed. The main disadvantage of this method is that PCM leakage is not avoided. The second method is direct incorporation, where PCMs are added into gypsum during its mixing phase. Since the PCM is uniformly distributed within the board, the PCM leakage is lower than that by immersion method.

The usage of PCMs, in particular organic, can be detrimental to flammability of the construction materials [8,9]. The presence of organic PCMs in building materials increases the risks of fire even when incorporated in a non-flammable matrix such as gypsum or cement. Despite a vast amount of literature available on the energy absorbance efficacy of PCMs in the construction materials (gypsum board and concrete), there are only limited studies on the effects of PCMs on flammability of construction materials [2,4,10,11]. As is well known, in recent years, stringent flammability, safety standards and legislations have been imposed on building materials to protect these buildings from fire hazards [9]. In most of the tests the heat release rate of combustibles and ignition time of materials have to be taken into account in order to satisfy the criteria set for making sustainable buildings [ASTM E119, ISO/TR 834 and BS 476]. The deterministic criteria relate mainly to life safety levels, fire exposure, structural performance, fire growth and spread levels [12]. The aim of this work is to devise a suitable methodology of fire testing them at a micro and laboratory scale, which can be used for screening purpose during material development stage with the potential of passing commercial fire tests at the product development stage in compliance with building codes.

EXPERIMENTAL

Materials

Gypsum powder sourced from British Gypsum, British Gypsum Thistle Multi Finish was used as received. The general composition provided by manufacturer is that it mainly contains calcium sulphate hemihydrate. Other constituents such as clay, limestone (calcium carbonate) and minor amounts of quartz and hydrated lime are also present [13].

Phase change materials:

- Eutectic mixture (EM), prepared by mixing 30 wt-% dodecanoic (lauric) acid (Acros chemicals) and 70 wt-% decanoic (capric) acid (Acros chemicals)
- Paraffin (Pa) n-octadecane (Alfa Aesar 99%)

Sample preparation

Plaster board was prepared by mixing gypsum powder with water in the ratio of 2.5:1 in order to get a well proportioned slurry. Then, the prepared slurry was poured in a wooden mould of dimensions 122mm x 122mm x 12mm and the mould was placed at room temperature for 7 days to dry. The sample is referred to as Gy in this paper.

PCMs were introduced by two methods, direct incorporation and immersion methods. In direct incorporation method required amount of PCM was added to gypsum powder before making slurry and then poured in a mould to dry for 7 days. These samples are referred to as Gy_PCM_Ad, where Ad stands for additive.

For immersion method two sets of samples were prepared. In the first set the gypsum board samples were immersed in a bath of melted PCM, keeping the temperature slightly above its melting point, for 30 minutes. After that period, samples were removed and dried. The samples were weighed before and after immersion, from which the % PCM absorbed could be determined. This type of samples are referred to as Gy_PCM_Imm (Table 1), where PCM is eutectic mixture (EM) or paraffin (Pa) and Imm stands for immersion. With this ~6.5% (w/w) PCM material could be absorbed in the board. In the second set Gy samples were first placed in a vacuum oven for 10 minutes to draw all the air out of the board. The samples were then immersed in a bath of melted PCM for 30 minutes. With this method upto 22.9% (w/w) PCM could be absorbed (Table 1). These type of samples are referred to as Gy_PCM_Im_V, where V stands for vacuum. The compositions and densities of all samples are given in Table 1.

Table 1: Composition of samples

Sample ID	Method of incorporation of PCM	PCM absorbed/added (wt-%)	Sample thickness(mm)	Density (g/cm^3)
GY	-	-	10	1.70
Gy_EM_Ad	Addition	10	10	1.22
Gy_EM_Imm	Immersion	6.5	10	1.50
Gy_EM_Im_V	Vacuum suction and Immersion	18.3-22.9	10	1.96
Gy_Pa_Ad	Addition	10	10	1.60
Gy_Pa_Imm_V	Vacuum suction and Immersion	17.2-18.9	10	1.86

Thermal analysis

The thermal behaviour of the PCMs and the gypsum board containing PCMs has been studied by means of an SDT 2960 simultaneous DTA (differential thermal analysis) – TGA instrument from room temperature to 900 °C using 15 ± 1 mg samples heated at a constant rate of 10 °C/min in air flowing at 100 ± 5 mL/min.

Fire performance

A cone calorimeter (Fire Testing technology, UK) was used to evaluate the fire performance of gypsum board without / with PCMs. The sample holder and methodology were slightly modified than those described in ISO 5660. The limitation in the quantity of materials available led to the use of specimens of reduced size of 75 mm x 75 mm rather than using the normal 100 mm × 100 mm sizes as outlined in ISO 5660. Previously in our laboratories it was noted that the use of reduced size gave similar relative trends to those observed using 100 mm × 100 mm specimens [14]. Firstly one set of samples containing EM (samples Gy, Gy_EM_Imm and Gy_EM_Ad) was tested according to standard ISO 5660 procedure at 50 and 70 kW/m^2. No ignition was observed at 50 kW/m^2, whereas at 70 kW/m^2 ignition occurred in EM containing samples. Limitation of the cone heater prevented testing above this heat flux.

Cone calorimeter was used in two orientations, horizontal and vertical. Two types of ignition sources, spark and pilot ignition were used and samples tested under different conditions. For pilot ignition a simple diffusion butane flame of 50mm length was provided from a 3.2mm outside diameter stainless steel tube burner arrangement

Methodolgy used for two cone orientations is as:

1. Horizontal orientation: A sample holder of 75mm x75mm dimensions was used as in ISO 5660 with specimens wrapped in the usual way and the height adjusted to 25mm from front face of sample to the cold cone heater face. For pilot ignition, the spark igniter was pushed to the side and the HT cables

disconnected; the pilot flame was set, using a retort stand on an adjacent table, the stand was used to hold the pilot flame burner horizontally in a position 10mm above the sample surface with the flame extending from the edge of the sample to slightly beyond the mid-point.

2. Vertical orientation: In this case the cone was moved into vertical orientation and a special sample holder was constructed 'in house' from 1.5mm mild steel sheet; 25mm side right angle pieces were folded and cut to the required length and depth and welded up. This yielded a frame without a rear central area with the view that if this was omitted then temperature measurements could be made from the rear face of the sample. The sample holder was provided with an internal backing board of 6mm thick calcium silicate (Duratec 750) followed by a layer of calcium silicate wool immediately behind the sample which was wrapped in the usual way. Further backing boards were cut and drilled for 1mm diameter mineral insulated thermocouples (type 310 stainless steel sheath). Initially a board was marked with a vertical centre line the position of the sample determined and holes for the thermocouple were drilled to match the centre of the sample and points 10mm above and 10mm below the horizontal centre of the sample. This was later replaced with a board marked for the horizontal centre and drilled on the vertical centre of the sample and 10mm each side of that vertical. The thermocouples were pushed through the calcium silicate wool into contact with the sample wrap and then held in place using wire twists. The pilot ignition flame was kept 10mm from the sample face with the orientation as described previously that is a horizontal tube burner with the flame across the sample face approximately 10mm below the horizontal centre line. Because a high irradiance was applied to the sample, a mild steel shutter was constructed with calcium silicate card insulation on the side facing the cone radiator in order to reduce pre- heating of the sample (and the operator). Spark ignition in this case was as in standard orientation. It was found from initial tests that a drip collector was not required for the materials being studied so it was omitted.

A number of tests were performed on one set of samples (gypsum containing EM) with spark igniter on until ignition, spark igniter on all the time, pilot flame on until ignition, and pilot flame on all the time. From this one condition was selected and used for the rest of samples as described in a later section.

RESULTS AND DISCUSSION

Figure 1 shows TGA and DTA curves of the two PCMs used to make composites. Pa starts losing mass at ~105°C, losing 100% mass by 244 °C. EM's mass loss starts at 130 °C, but the mass loss rate is much higher and by 218 °C all mass is lost. In both cases endothermic peaks can be seen from Figure 1(b) with peak maxima at 193 and 206 °C, respectively, followed by exothermic transitions. The sharp exothermic peak in Paraffin at 232 °C is in good agreement with its auto-ignition temperature of 235 °C, whereas in EM only exothermic base line shift is observed.

The effect of heat on gypsum board is well known from literature.[15] At lower temperatures (100–170 °C) the gypsum loses 75 % of its bound water and is converted from calcium sulfate dihydrate to calcium sulfate hemihydrates. At higher temperatures (130–180 °C) the material loses its remaining bound water and is converted to calcium sulfate anhydrate. From Figure 2a) 16% mass loss is observed between 105 - 180 °C. Both endothermic reactions are seen in Figure 2b) as an endothermic peak initiating at 110 °C with maximum at 158 °C. Another reaction occurs around 700 °C. This reaction, seen as 4% mass loss between 650 - 720 °C is believed to be the decarbonation of calcium carbonate [16]. The endothermic peak of the gypsum representing desorption of water and dehydration in both Py_PCM samples is similar to the control (Gy), indicating that PCM presence does not interfere with gypsum's water release. The exothermic peaks observed for PCMs in Figure 1a) are superimposed by the exothermic base line shift of the gypsum board, indicating that these might not be too much affecting the process. In mass loss curves of PCM containing samples in Figure 2a) the additional mass loss is due to evaporation and combustion of the paraffin and EM.

In conclusion while this study indicates the thermal decomposition behaviour of the two Gy_PCM materials, there is nothing which suggests that the flammability of the mixture would be increased.

Figure 1. TGA and DTA curves of PCMs

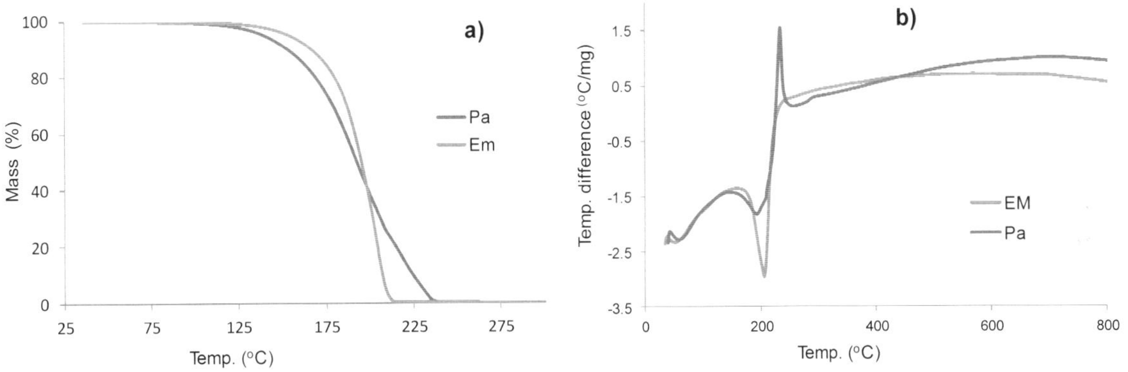

Figure 2. TGA and DTA curves of Gy without / with PCMs incorporated by additive method

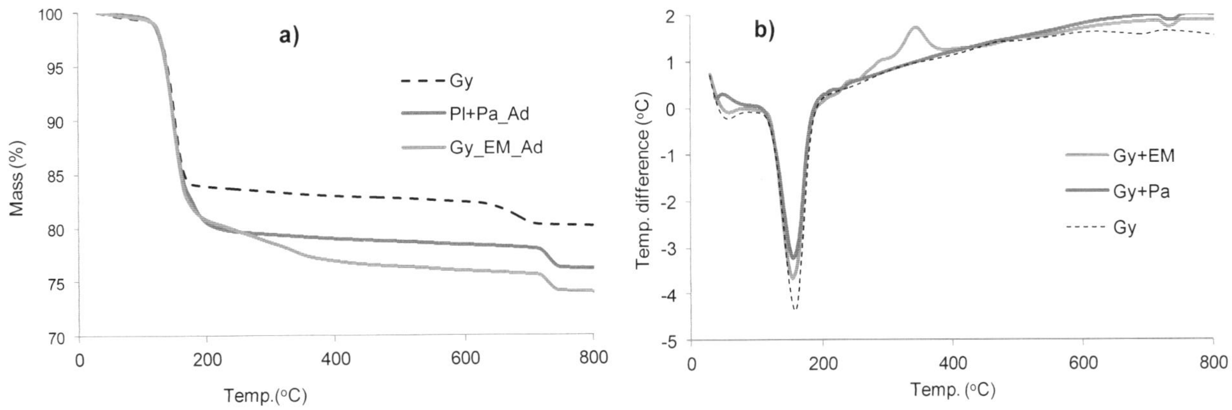

Flammability properties

Establishment of cone exposure conditions

One set of samples, gypsum and gypsum containing eutectic mixture (samples Gy, Gy_EM_Ad and

Gy_EM_Imm) were tested in horizontal and vertical orientations with spark or pilot igniter on until ignition and igniter on all the time to understand their behaviour in different conditions. All results in terms of time-to-ignition (TTI), peak heat release rate (PHRR), total heat release (THR) and total mass loss during the test are presented in Table 2, while selected HRR and % mass versus time curves are presented in Figures 3 – 5.

The control sample, Gy, did not ignite under any condition. Samples with EM with both immersion and additive methods ignited but the burning (appearance of flame) was very low and erratic, i.e. flame will appear when the EM is exposed to surface irradiance and then disappear and come back again. Samples with immersion method also burnt less vigorously despite expected more PCM on the surface due to absorption, which can be explained because of low concentration of the PCM in Gy_EM_Imm (~ 6.5% w/w, Table 1). Moreover, since each specimen was individually immersed in the melted EM bath, there was variation in the concentration of EM in different samples, hence there was variation in cone parameters as well.

Effect of orientation, horizontal versus vertical: In Figure 3a), the effect of orientation, while the spark igniter was on until ignition, on the cone parameter can be seen. It can be seen that in vertical orientation the PHRR was slightly higher than in the horizontal orientation, the difference though is still within the error range of cone tests. There is no significant effect on the mass loss rate, as seen from Figure 3b).

Figure 3. a) HRR and b) mass loss versus time curves for gypsum boards without/with EM on cone exposure at 70 kW/m^2 heat flux, horizontal versus vertical orientation with spark ignition till ignition.

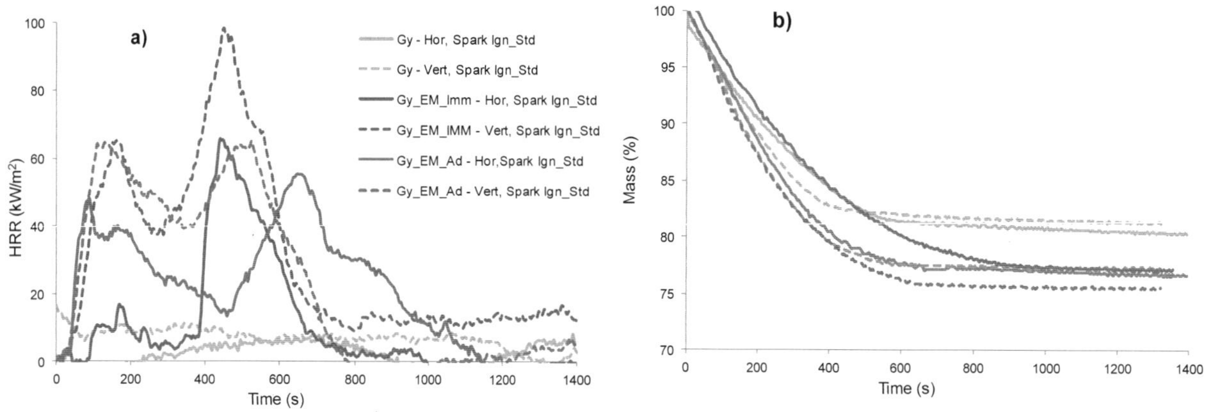

Figure 4. a) HRR and b) mass loss versus time curves for gypsum boards without/with EM on cone exposure at 70 kW/m^2 heat flux, vertical spark ignition versus flame ignition all the time

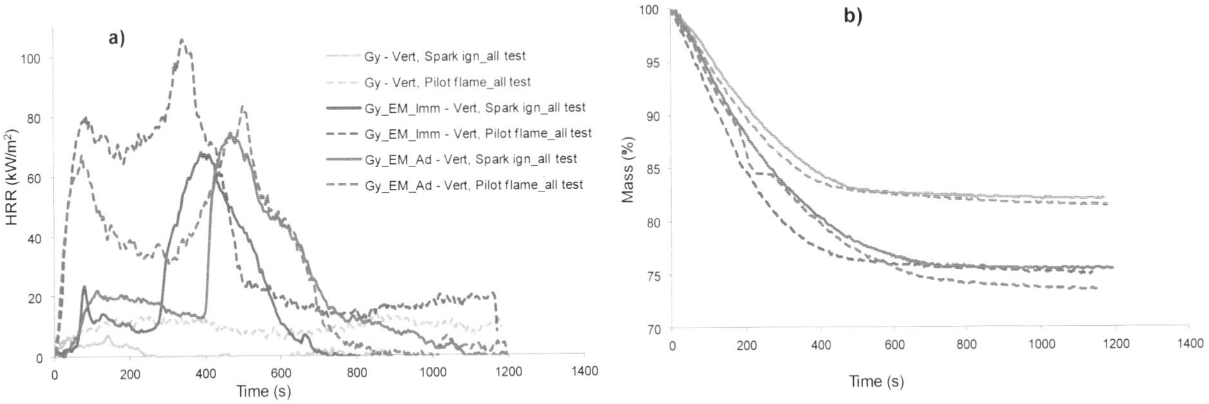

Effect of ignition source, spark versus pilot ignition: The effect of ignition sources on vertically oriented samples is shown in Figure 4. With the pilot ignition it is difficult to judge when the ignition has occurred, hence the pilot flame was left on all the time and to compare with spark ignition, the latter was on all the time as well. It should be noted that the heat release observed due to the pilot flame was compensated in the heat release observed by the cone. As can be seen from results, with pilot ignition there was an early ignition than from the spark ignition. PHRR was also increased, but not significantly. THR was not much different. Mass loss rate is also not much affected.

The reproducibility of the test can be seen from Figure 5 where one of the samples has thermocouples on the reverse side. The results are quite consistent, any variations are due to different concentration of PCM in each specimen plus the normal error range of the cone calorimeter (±10%).

Figure 5. a) HRR versus time curves for gypsum boards without/with EM on cone exposure at 70 kW/m^2 heat flux in vertical orientation with pilot flame ignition all the time for samples without and with thermocouples

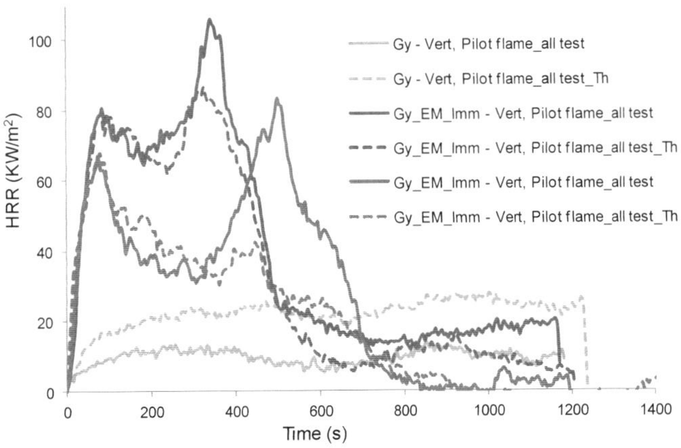

Table 2: Cone calorimetric results of gypsum boards without/with EM at 70 kW/m^2 external heat flux under different orientations and ignition scenarios

Sample	Testing condition	TTI (s)	PHRR (kW/m^2)	THR (MJ/m^2)	Mass loss (%)
Gy	Hor, Spark Ign_Std	-	-	2.7	19.6
	Vert, Spark Ign_Std	-	-	9.8	18.9
	Vert, Spark Ign_all test	-	-	2.1	17.9
	Vert, Pilot flame all test	-	-	11.3	16.2
Gy_EM_Ad	Hor, Spark Ign_Std	47	55.3	28.2	23.4
	Vert, Spark Ign_Std	40	64.6	31.3	24.4
	Vert, Spark Ign_all test	390	73.2	24.2	22.3
	Vert, Pilot flame all test	45	83.3	33.3	24.1
Gy_EM_Imm	Hor, Spark Ign_Std	305	61.7	13.6	23.4
	Vert, Spark Ign_Std	47	95.7	40.2	22.7
	Vert, Spark Ign_all test	245	67.7	16.2	21.1
	Vert, Pilot flame all test	47	105.5	45.8	23.6

Effect of igniter: on until ignition or all the time: There was not much effect of igniter being on all the time or just until ignition as can be seen from results in Table 2.

From these results it can be concluded that in these samples, which are not very flammable, there is not much effect of the orientation, type of igniter and whether it is on all the time or not. In a separate study wood samples were also tested under these conditions to study a flammable material, the results will be published in a forthcoming publication.

Effect of PCMs on flammability of gypsum board

Based on above results all the samples were tested in vertical orientation with spark ignition on all the time. Results are presented in Table 3 and Figure 6. As can be seen from results, both samples of Gy_Pa prepared by immersion or direct incorporation method ignited around 23s. Sample from immersion method have much higher PHRR and THR than by direct incorporation method. This is due to higher PCM concentration (22.9%, Table 1) in the former. Sample of Gy_EM prepared by immersion method ignited at 6s, whereas by direct incorporation did not ignite until 162s. As can be seen both types of Gy_EM samples have lower PHRR and THR than respective Gy_Pa sample. Mass loss results are representative of the PCM concentration in the sample. The samples with additives, Gy_EM_Ad and Gy_Pa_Ad have very similar mass loss rates (Figure 6(b)). Mass loss rate is higher for samples made by immersion method due to higher PCM concentration, these being much higher in Gy_EM_Imm sample than Gy_Pa_Imm.

Figure 6 a) HRR and mass loss curves for gypsum boards without/with PCMs on cone exposure at 70 kW/m^2 heat flux in vertical orientation with flame ignition all the time

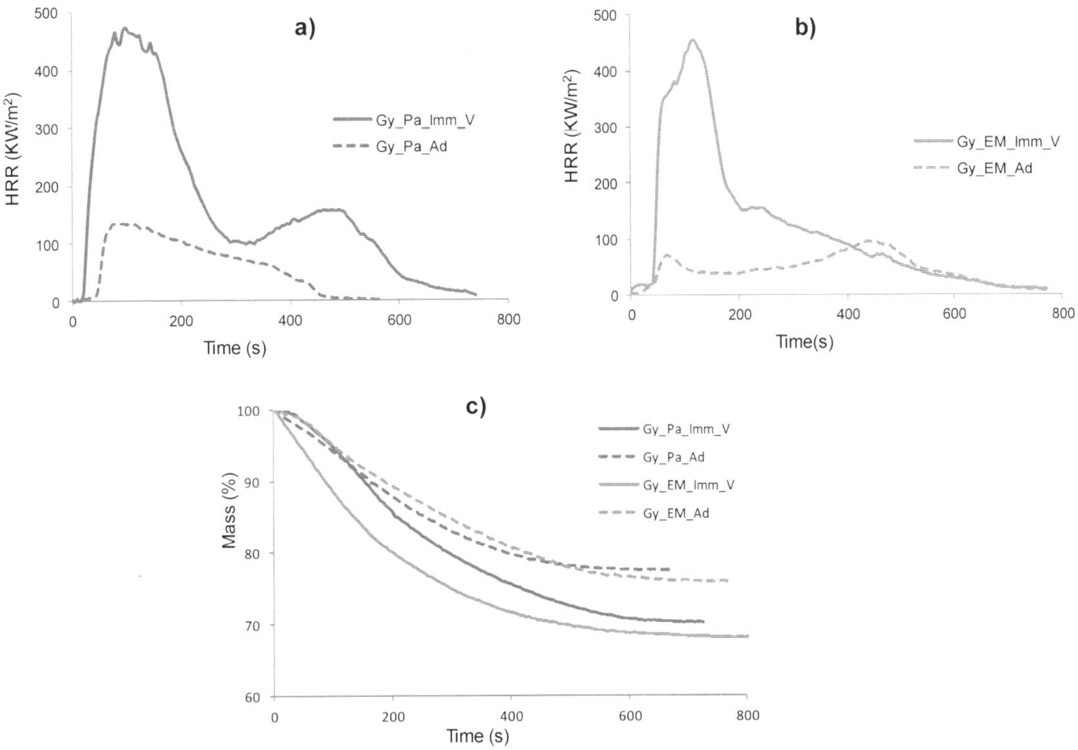

Table 3: Cone calorimetric results of gypsum boards without/with PCMs at 70 kW/m^2 external heat flux in vertical orientation with flame ignition all the time

Sample	TTI (s)	PHRR (kW/m^2)	THR (MJ/m^2)	Residue (%)
Gy	-	-	2.1±1.33	17.3±1.4
Gy_Pa_Ad	26±18	128±15	41.6±6.3	22.7±0.7
Gy_Pa_Imm_V	23±8	537±69	133.1±10	29.7±0.7
Gy_EM_Ad	162±18	82±11	311±4.9	25.7±3.9
Gy_EM_Imm_V	6±2	490±49	101.8±9.6	30.5±3

*Out of three samples, one contained thermocouples attached to the unexposed surface.

The thermocouples embedded on the reverse sides of the sample enabled the measurement of temperature of the unexposed surface as a function of time during the test. As an example results of Gy, Gy_Pa_Ad and Gy_Pa_Imm_V are shown in Figure 7. Trends in rise of temperatures in different types of samples are reflection of heat release rate versus time curves. Gy, which does not burn, has a low temperature rise until 470 s, probably due to loss of water after which voids may be created within the sample, hence coherence is lost and the temperature starts rising, reaching 480 °C after 1000s. Gy_Pa_Ad behaves very similarly to Gy until ~550s despite ignition at 26s, which can be explained due to the reason that ignition occurred on the surface and the burning was very low, but after 550s, the temperature rose more than the Gy, reaching 570 °C after 1000s. However, in Gy_Pa_Imm sample, due to much higher Pa concentration there was an early ignition and burning was persistent, the temperature rise is much higher from 33s onwards, reaching maximum 600 °C after 960s. These results will be used to validate the heat transfer modelling work being carried out in our laboratories.

Figure 7. Temperature versus time curves of the reverse surfaces of gypsum and gypsum containing paraffin samples on cone exposure at 70 kW/m^2 heat flux in vertical orientation with pilot igniter on all the time

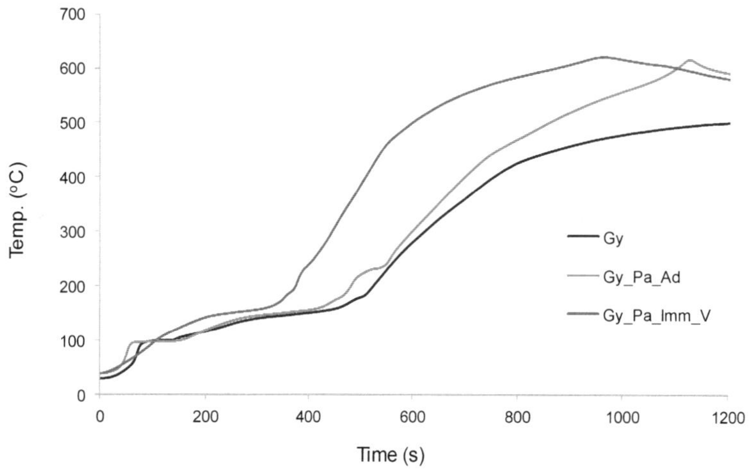

CONCLUSIONS

This work has shown that while required amount of PCM can be added to the gypsum board, with immersion method the absorption depends upon the porosity of the material. By subjecting the sample to vacuum prior to inserting in the molten PCM bath, the air could be extracted and hence porosity increased. The flammability of the materials tested by a vertically oriented sample holder in a vertically oriented cone and using a spark ignition throughout the test showed that the gypsum board does not ignite. Organic PCM such as paraffin and eutectic mixture when present in gypsum board ignite, but the burning is dependent on the concentration as well as dispersion of the PCM. If the PCM is in low concentration, the sample ignites, the flame goes out and then ignites again. However, with immersion method, due to higher concentration of the PCM on the surface of the sample, the ignition was early, flame was persistent and with high peak heat release rate. The measured temperature profiles of samples during the course of the tests give useful insight into the behaviours of the heat exposed samples.

REFERENCES

[1] Sharma, A., Tyagi, V.V., Chen, C.R., Buddhi, D. Review on thermal energy storage with phase change materials and applications, *Renewable and Sustainable Energy Reviews*, 2009; 13, p 318–345.

[2] Cai, Y., Song, L., He, Q., Yang, D., Hu, Y. Preparation, thermal and flammability properties of a novel form-stable phase change materials based on high density polyethylene/poly(ethylene-co-vinylacetate)/ organophilic montmorillonite nanocomposites/ paraffin compounds, *Energy Conversion and Management*, 2008; 49, p2055–2062.

[3] Farid, M.M., Khudhair, A.M., Razack, S.A.K., Al-Hallaj, S. A review on phasechange energy storage: materials and applications, *Energy Conversion and Management*, 2004; 45, p1597–1615.

[4] Haurie, L., Serrano, S., Boscha, M., Fernandezc, A. I., Cabezab, L.F. Single layer mortars with microencapsulated PCM: Study of physical and thermal properties, and fire behavior, *Energy and Buildings*, 2016; 111, p393–400.

[5] Cabeza, L.F., Castell, A., Barreneche, C., de Gracia, A., Fernández, A.I. Materials used as PCM in thermal energy storage in buildings: a review, *Renewable and Sustainable Energy Reviews,* 2011; 15 (3), p1675–1695.

[6] Zhang, Y., Zhou, G., Lin, K., Zhang, Q., Di, H. Application of latent heat thermal energy storage in buildings: State-of-the-art and outlook, *Building and Environment,* 2007; 42, p2197–2209

[7] Bentz, D. P., Turpin, R. Potential applications of phase change materials in concrete technology, *Cement and Concrete Composites*, 2007; 29 (7), p527–53.

[8] Pons, O., Aguado, A., Fernández, A.I., Cabeza, L.F., Chimenos, J.M. Review of the use of phase change materials (PCMs) in buildings with reinforced concrete structures, *Materiales de Construcción*, 2014; 64 (315), e031

[9] Khudhair, A.M., Farid, M.M. A review on energy conservation in building applications with thermal storage by latent heat using phase change materials, *Energy Conversion and Management,* 2004, 45 (2), p263–275.

[10] Banu, D., Feldman, D., Haghighat, F., Paris, J., Hawes, D. Energy-Storing Wallboard: Flammability Tests, *Journal of Materials in Civil Engineering,* 1998; 10 (2), p98–105.

[11] Haurie, L., Mazo, J., Delgadob, M., Zalba, B. Fire behaviour of a mortar with different mass fractions of phase change material for use in radiant floor systems, *Energy and Buildings,*2014; 84, p86–93.

[12] Hadjisophocleous, G.V., Benichou, N. Performance criteria used in fire safety design, *Automation in Construction*, 1999; 8, p489–501.

[13] Thistle MultiFinish Safety Datasheet, http://www.british-gypsum.com/products/thistle-multi-finish.

[14] Biswas, B., Kandola, B.K. The effect of chemically reactive type flame retardant additives on flammability of PES toughened epoxy resin and carbon fiber-reinforced composites, *Polym. Adv. Technol.* 2011; 22 (7), p1192–1204

[15] Paulik, F., Paulik J., Arnold, M. Thermal decomposition of gypsum. *Thermochimica Acta,* 1992; 200, p195-204.

[16] Bénichou, N., Sultan, M.A. Thermal properties of lightweight-framed construction components at elevated temperatures, *Fire and Materials,* 2005; 29 (3), p165-179.

DECOMPOSITION OF LARGE-SCALE MULTICOMPONENT CONSTRUCTION ASSEMBLIES EXPOSED TO REAL FIRES

Matthew J. DiDomizio[a]*, Elizabeth J. Weckman[a]†, Richard Roos[b]‡

[a]Department of Mechanical and Mechatronics Engineering , University of Waterloo, Canada
[b]Technical Innovations Department, Roxul Inc.Canada

ABSTRACT

A new experimental method was developed for the purpose of investigating the decomposition mechanisms of large-scale multicomponent construction assemblies exposed to real fires. The experiment was designed to accommodate a 6 ft × 6 ft (1.8 m × 1.8 m) wall, which was mounted opposite the opening of an insulated compartment. The apparatus allows for direct observation of a wall, from both exposed and unexposed sides, during a compartment fire. A custom wood crib fire was designed in which the peak temperature and total duration were controlled by adjusting the crib configuration and compartment ventilation. This fire developed gas temperatures near the ceiling exceeding 700°C, and burned steadily for approximately 45 min. A calibration test was performed using a steel sheet in place of the wall; temperature measurements were used to calculate the net heat flux to the wall. The fire exposure was found to vary significantly in the vertical direction, as well as in the horizontal direction to a lesser degree. A test was performed on a wall assembly constructed from 2×4 dimensional lumber with standard gypsum board wallcoverings, in which half of the wall was insulated with glass fibre and the other half with stone wool. Temperatures and video on both the exposed and unexposed sides of the wall were recorded. The wall was exposed for a total duration of 3 hours, allowing for investigation of the decomposition of the wall from the onset of the fire through to post-flashover. During this time the gypsum board wallcovering became significantly compromised. Greater temperatures and damage were observed on the glass fibre side of the wall. The method was successful in its objective of providing real-time measurement and observation of the response of a full-scale wall to a realistic building fire. Future work will investigate the response of a variety of multicomponent construction assemblies to various fire exposures.

INTRODUCTION

A typical building contains a number of different construction assemblies which enclose and compartmentalize the structure. Oftentimes these construction assemblies consist of multiple components; for example, a wall might be constructed from a wood stud frame, in-cavity insulation, and a gypsum board wallcovering. The previously mentioned example is classified as "combustible construction" according to the National Building Code of Canada (NBCC)[1]. The use of combustible construction as fire separations in Canada is regulated according to the NBCC's prescribed level of fire protection for the separation of various types of occupant spaces[1]. Fire protection is therefore considered to be a property specific to a given construction assembly, which is colloquially termed "fire resistance".

---

* PhD candidate, corresponding author, matt.didomizio@uwaterloo.ca
† Professor, beth.weckman@uwaterloo.ca
‡ NA Codes, Standards and Fire Safety Manager, richard.roos@roxul.com

Fire resistance can generally be described as the ability of a construction assembly to resist fire, insofar as an assembly with "better fire resistance" is expected to have a greater ability to contain a fire within a structure, thereby limiting fire spread and growth. Generally, this property is quantified by a fire resistance rating, which is developed from a fire resistance test on a representative sample of the construction assembly; in Canada, this test is the CAN/ULC S101[2], which is very similar to the ASTM E119[3] fire test used in the United States. In the fire resistance test a full-scale construction assembly sample is placed in a furnace enclosure. One side of the assembly, the "exposed side", is exposed to heat from a gas burner, and the burner is regulated such that the temperature of furnace gases over time matches the prescribed curve. During the test the temperature of the "unexposed side" is measured at several locations, and the structural integrity of the assembly and any passage of flame to the unexposed side is assessed. The fire resistance rating of a test specimen is calculated to be either the time for a critical temperature rise to be measured on the unexposed side, or the time for the passage of flames to be observed through the wall[2,3].

By its design, the fire resistance test furnace creates a predominantly uniform thermal distribution over the exposed side of a wall specimen at any given time. This contrasts with the reality of a building fire environment, which will give rise to spatially non-uniform temperature distributions on the exposed side of walls. For example, localized fuel sources may lead to exposure levels that vary horizontally over a wall, while a well-developed hot layer may lead to varying exposure levels in the vertical dimension. Given the natural buoyancy of hot fire gases, it can be expected that vertical thermal gradients over the surface of a wall will be particularly significant in building fires.

Furthermore, it is well known that the temporal progression of the rate of heat release in a well-ventilated fire undergoes distinct stages of growth, steady burning, and decay. The same can be expected of the temperature of the gas environment in a building fire. The temperature profile in the fire resistance test, on the other hand, consists only of an initial rapid growth stage followed by a slow growth stage for the entirety of the test, reaching temperatures in excess of 900°C after 60 minutes of exposure[3]. Consequently, the incident heat flux on a wall in the fire resistance is also quite severe, and can be estimated to be approximately 100 kW/m^2 after 60 min assuming a black surface and neglecting convection heat transfer.

While gas temperatures at flashover in a severe building fire can exceed 1000°C with initial rates of growth surpassing even that of the standard fire resistance temperature-time curve, these temperatures can only be sustained while fuel remains and an adequate supply of fresh air is available; oftentimes, this is not the case. In contrast, it is more often the case that a wall exposed to a building fire will experience levels of incident heat flux comparable to those of the fire resistance test only for a short duration of time, followed by prolonged exposure to a post-flashover decay environment. For example, in work by Manzello et al.[4] the temperature of a compartment wall (gypsum board facing material) in a severe fire reached 800°C at the fire's peak rate of heat release of 13.5 MW. Still, even in this case involving three large workstations in a 10.7 m × 7.0 m × 3.5 m tall compartment, the fire reached its peak heat release rate and began to decay after only 23 minutes, at which time the total heat flux incident on the wall was measured to be approximately 160 kW/m^2 using a Schmidt-Boelter heat flux gauge[4].

It is of interest to the fire science community to understand how a more traditional building fire, in particular one with a prolonged post-flashover decay period, might impact a building's constituent construction assemblies. Certainly it has been established that, in general, walls will be affected differently by real fires compared to a furnace test. The two reasons discussed above, spatial and temporal dependence of exposure levels, consider the heat transfer aspect of the problem. In addition, damage such as cracking gypsum board, burning wood studs, and decomposing insulation will significantly influence the overall degradation of the wall. The nature of a fire resistance test, in which the exposed side of a wall is placed in a furnace enclosure, precludes the possibility of observation of these phenomena. These issues must be studied both at the small-scale, in order to understand how these materials thermally decompose, as well as at the large scale, in order to investigate the physical mechanisms for these observed phenomena.

The objective of the present study is to present an experimental method for investigating the decomposition mechanisms of large-scale multicomponent construction assemblies exposed to real fires, as well as the interactions between the construction assemblies and the fire environment. This will be accomplished by designing an experiment suitable for producing a fire environment typical of a building fire, and exposing a reference wall assembly to the conditions of this fire. The experiment must accommodate any number of different wall types and fire configurations. In order to adequately investigate the physical mechanisms of import, the experiment must include instrumentation to support measurements of temperature and heat flux, as well as visual observation of damage to the wall. The fire environment of the test must not only be representative of a typical building fire, but must also be repeatable and well characterized. This paper presents the design for this new experimental method, as well as the results for a single test case.

DESCRIPTION OF THE EXPERIMENT

The essential design component of this experiment was to modify a traditional compartment fire experiment so as to allow one of the compartment walls, identified as the "test wall", to be constructed, instrumented, and replaced each test. A first attempt at this involved the construction of a 3.1 m × 3.1 m × 2.5 m high fire compartment, with the interior insulated in refractory ceramic fibre insulation, within a pre-existing structure[5]. In this previous work one of the compartment walls, measuring 3.1 × 2.3 m, was selected as the test wall and instrumented with more than 100 thermocouples over five stud cavities (spaced 400 mm on centres). The fuel load in this case was a total of 5 softwood cribs, the specifications of which are provided in the Methodology section below, added in stages to the fire compartment. The three-dimensional temperature gradient within the wall for the particular fire exposure was then measured, and assessed.

While this experiment was successful in acquiring a set of temperature data for a large decomposing multicomponent wall, there were some notable limitations. The most significant issue was practicality — it was exceptionally difficult to repeat tests within the confines of this pre-existing structure. Also there was a significant limitation in that real-time observation of the decomposition of the wall was not possible, due to the orientation of the compartment opening with respect to the test wall. Since damage assessment is a critical component of the overall objectives, this would have to be addressed. Furthermore, with the tests completed by late November, the outdoor temperatures in southern Ontario were not ideal for continued fire testing. Still, the methodology showed promise, therefore it was decided to design a new experiment over the winter months that would address these shortcomings.

First, a new fire compartment was designed and constructed. Given the ubiquity of the ISO 9705 Room Corner test[6], which uses a compartment measuring 3.6 m × 2.4 m × 2.4 m tall with a single opening measuring 0.8 m wide by 2.0 m tall, this configuration was selected as a starting point. A standard sea container has interior dimensions measuring 6.1 m × 2.4 m × 2.4 m tall, making it a good candidate for retrofit. Sea containers are modular and extremely flexible in installation, and can easily be installed in a conditioned area if necessary (although in this case the container was located outdoors). The solution adopted here was to retrofit a sea container into three sections, each separated by "frames" fabricated from 10 ga (3.6 mm) steel sheet and 3/16 in (4.8 mm) structural steel square tube welded in-place within the container. A door was cut into the front wall of the container, with dimensions matching that of the ISO 9705 opening. A side section view of the modified container is shown in Figure 1.

A critical component of the apparatus was that there be a system in place for the rapid mounting and instrumentation of a wall assembly. The first frame, which separates the fire section (left) from the transition section (middle), is identified as the “wall frame” as shown in Figure 1. The wall frame was designed to accommodate a 6 ft × 6 ft (1.8 m × 1.8 m) wall, which corresponds to three stud cavities at 600 mm centres, or four stud cavities at 400 mm centres (with an extra 100 mm on either side in the second configuration). The wall frame interior was oversized by 3 in (76 mm) in both dimensions so that it could be lined with 1.5 in (38 mm) of ceramic fibre insulation, limiting heat transfer at the boundaries of the wall. A series of 0.5 in (13 mm) diameter holes, spaced 1 in (25 mm) apart, were drilled in four locations on each of the four sides of the wall frame. These provide a means by which to directly affix a stud wall within the steel frame, and may also be used to route instrumentation cables.

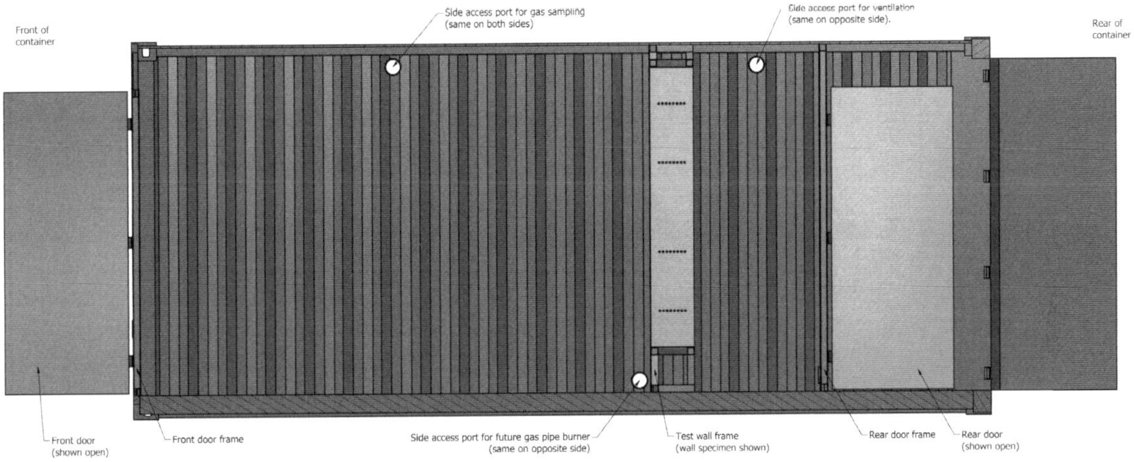

Figure 1: Side section view of the compartment wall fire test apparatus.

The instrumentation section (right) accommodates a data acquisition system, computer, and digital video recorder (DVR). A steel frame separates the instrumentation section, and includes a door for access to the rear of a test wall (this door is closed during a test). As hot gases from a decomposing wall collect in the transition section (middle) during a fire test, the rear frame ensures that instrumentation will be kept safe. The transition section also has ventilation ports at ceiling level so that gases may optionally be exhausted from the unexposed side of the wall.

Finally, the entire interior of the fire compartment was insulated in order to retain heat for prolonged post-flashover exposures, reducing the interior dimensions to 3.5 m × 2.3 m × 2.3 m tall. Insulation panels were constructed from cement board and ceramic fibre insulation and mounted on the walls and ceiling of the fire compartment (the floor was insulated with fire bricks). The panels were supported from the interior using steel struts, also protected with ceramic fiber insulation, which were also used for routing instrumentation. An interior view of the fire compartment, facing the test wall, is shown in Figure 2.

To quantify the severity of the compartment fire, it was necessary to measure the gas temperatures within the compartment. This was accomplished by installing four vertical steel channels at the four corners of the room (covered in insulation for fire protection), each with 8 thermocouples spaced 12 in (305 mm) apart, with the top thermocouple located 1 in (12.7 mm) below the ceiling. Heat flux to particular locations within the compartment can also be measured (e.g. using Gardon gauges), and the apparatus can accommodate lines for cooling water, nitrogen purge, and data within the channels shown in Figure 2, however this instrumentation was omitted in the present work. Instead, as detailed in the following section, a steel calibration wall was used to investigate the exposure level on a wall for a given fire.

Figure 2: Interior of the insulated fire compartment with steel calibration wall in-place.

Even though the wall frame is not directly exposed to fire, being protected by ceramic fibre insulation, it is certainly possible for it to reach temperatures sufficiently great so as to impact the boundary conditions on the wall. Therefore 8 thermocouples were welded directly to the inside of the wall frame, 2 per side, between the steel and insulation lining, in order to measure the temperature of the frame during a test. Furthermore, an additional 8 thermocouples were installed in the transition section, two on the steel surroundings and the remaining six separated into two vertical rakes for measuring gas temperatures. An additional 36 thermocouples were prepared for use in instrumenting a test wall.

Finally, video cameras were used to observe the decomposition of the wall on both the exposed and unexposed sides. On the unexposed side, a camera was positioned on the container floor and directed upward to observe any smoke passage or other events. On the exposed side, a camera was placed at the bottom of the compartment opening, facing the wall. Both cameras were connected to a DVR for continuous recording. IR thermography may also be employed, and in fact was during one test, however it is of questionable relevance to use this on the exposed side of the wall since the camera's line of sight to the wall will be obscured by the fire. With this said, once the fire has entered the post-flashover decay stage, IR thermography is very useful for identifying which, if any, portions of the wall continue to smoulder.

METHODOLOGY

Design Fire

In contrast to a fire resistance test, which utilizes a gas burner, a realistic building fire may involve an assortment of solid fuels such as wood, fabrics, and various polymers. Solid fuels must undergo a heating and volatilization process before contributing significantly to fire growth, and a building fire involving multiple fuel items may have a very long growth period. Therefore, the goal was to design a fire that was relatively slow growing, reaching peak compartment gas temperatures in excess of 600°C, and maintaining a steady rate of heat release for at least 45 min — the reasoning for this is based on the selected wall type, as outlined in the "Wall Test" section below. The fire should then decay steadily over the next two hours or so, allowing for investigation of the wall decomposition during the post-flashover decay stage.

Softwood cribs were selected as the fuel source, the characteristics of which were previously evaluated by Obach[7]. Cribs were constructed using 2 by 2 in nominal spruce furring strips cut to 2 ft lengths and assembled using brads with 6 pieces per layer, and 6 layers thick. The total mass of each crib varied from approximately 14–16 kg, depending on the condition of the constituent lumber and on its moisture content. Accounting for these factors, Obach measured the effective heat of combustion of these cribs to be approximately 14.8 MJ/kg, and found that burning a single crib resulted in a peak rate of heat release of approximately 200 kW (using a large-scale calorimeter). Therefore, an optimal combination of multiple cribs and ventilation control would need to be utilized in order to achieve the objective burn.

Previous work[5] explored the idea of extending the duration of a wood crib fire by intermittently adding additional cribs at regular time increments. While this endeavour was successful in its intent, it required firefighting personnel to actively enter the fire compartment in order to re-fuel the fire, and the resulting temperature-time traces were consequently quite discontinuous. If this experiment is to be used to produce reference data for modelling, the fire progression should ideally be continuous. Therefore, in the present study it was decided to use four cribs in-line, igniting at the end furthest from the test wall using 200 mL of methanol in an aluminum foil pan measuring 8 in × 8 in (203 mm × 203 mm). Cribs were elevated above the container floor using bricks with thickness of 2.5 in (63.5 mm). The first crib was positioned 12 in (305 mm) from the test wall. The remaining cribs were separated by 8 in (203 mm), 4 in (102 mm), and 2 in (51 mm) gaps, respectively. This configuration helped to slow the progression of the fire toward the test wall. The fuel configuration adopted in the present work is shown in Figure 3 at the time of ignition.

In addition to the fuel configuration shown in Figure 3, it was necessary to control ventilation at the compartment opening to prolong the fire duration given the static fuel load. This was accomplished by lowering the soffit from 2.0 m to 1.3 m (2/3 the origin opening size) and placing an angled panel in front of the opening as shown in Figure 4. The intent here was to restrict air entrainment into the compartment near the bottom of the opening, while still allowing adequate exhaust of fire gases from the compartment. A constant angle of 30° was used in the present work, however future studies may dynamically adjust the panel angle for fine control of fire growth. With this fuel configuration and ventilation control the time at which the peak gas temperature was reached was extended to 45 minutes, achieving the objective of approximately 1 hour of significant heat release within the fire compartment.

Calibration Test

Before conducting a test on a real wall assembly, the compartment fire exposure was first characterized using a steel "calibration wall", as shown in Figure 2. This calibration wall was constructed by bolting 18 ga (1.27 mm) steel sheet to a frame of 1-5/8 in Unistrut®. The calibration wall was then bolted directly into the wall frame, with 6.4 mm thick silicone sheet placed between the Unistrut and the wall frame, with thermocouples welded to the edges of the wall (2 per side). These, in combination with the thermocouples welded to the wall frame, effectively measure the temperature differential across the silicone sheet at several locations around the exterior of the wall, thereby affording a means by which to characterize that boundary condition. In addition, 3 sets of 8 thermocouples were welded to the unexposed side of the steel sheet at regular intervals. Figure 5 shows the instrumented steel wall from the unexposed side prior to application of paint, after surface preparation (hence the high reflectivity). At this point both the exposed and unexposed sides of the wall were painted with four light coats of matte black VHT Flameproof® paint, for which the emissive and absorptive properties were previously evaluated[8].

Figure 3: Design fire at initial ignition.

Figure 4: Ventilation control at the opening.

Figure 5: Unexposed side of the steel wall before application of paint.

A number of calibration tests were carried out with the design fire described above. Measured temperatures were used as input to a basic 2D inverse model to determine the net heat flux on the exposed side of the wall. Assuming that the wall is thermally thin and that the steel frame does not add significant thermal mass to the wall, the net heat flux into the wall may be calculated by lumped conservation of energy:

$$q_f'' = \alpha\rho c_p\delta - h_r T_c - h_c T_{\infty,c} + \frac{A_{loss}}{A} q_{loss}'' \ , \qquad \text{Eq. 1}$$

where

$$\alpha = \beta\left(T + \frac{\beta(T - T_i)}{\exp(\beta t) - 1}\right) \ , \qquad \text{Eq. 2}$$

$$\beta = \frac{h_r + h_c}{\rho c_p \delta} \ , \text{and} \qquad \text{Eq. 3}$$

$$h_r = \sigma\left[\frac{1}{\varepsilon} + \frac{A}{A_c}\left(\frac{1}{\varepsilon_c} - 1\right)\right]^{-1} (T + T_c)(T^2 + T_c{}^2) \ . \qquad \text{Eq. 4}$$

In this model ρ is the density of steel, c_p is the specific heat of steel, δ is the thickness of steel, A is the area of the steel, A_c is the area of the surroundings on the cool side, A_{loss} is the area over which heat transfer occurs at the frame, q_{loss}'' is the heat loss per unit area at the frame, T is the measured temperature of the steel wall, T_c is the measured temperature of the surroundings on the cool side, $T_{\infty,c}$ is the measured temperature of gases on the cool side, T_i is the initial steel temperature, ε is the emissivity of the wall, ε_c is the emissivity of the surroundings on the cool side, and h_c is the convection heat transfer coefficient (which was estimated using data from Veloo and Quintiere[9]). The q_{loss}'' term was approximated using the measured temperature differentials and thermal conductivity across the four silicone boundaries. This model was solved numerically using a coarse mesh in which the centre of each area element corresponded to the thermocouple measurement locations on the wall. While this model is certainly a gross simplification, it does allow for a simplistic approximation of the net heat flux over the surface of the wall.

Wall Test

The wall assembly selected in the present work is the reference "W1c" from the NBCC: a wall consisting of a 2x4 wood stud frame spaced 400 mm on centres, in-cavity insulation, and a single sheet of 12.7 mm regular gypsum board on both sides (with joints taped and finished)[1]. In a non-loadbearing application this wall has a prescribed fire resistance rating of 30 min when built using glass fiber insulation, or 45 min when the insulation is mineral fiber processed from stone or slag[1]. This wall assembly would find typical use as an interior partition in residential buildings when sound or thermal insulation is desired. It is an interesting candidate for this test because of the clear distinction in fire resistance rating when the wall is insulated with, for example, glass fiber insulation (30 min) compared to stone wool insulation (45 min). Therefore, in the case presented here, 50% of the wall cavities were insulated with CertainTeed® Sustainable Insulation™ glass fiber insulation, and 50% of the cavities were insulated with Roxul® ComfortBatt® stone wool insulation. The wall was constructed as follows:

- the wood stud frame was constructed in-place inside the wall frame, using 2-1/2 in (63.5 mm) brads, and compressing a layer of ceramic fiber insulation between the outer studs and the steel;
- the exposed-side gypsum board was fastened to the wood frame using 1-3/8 in (34.9 mm) course thread drywall screws spaced 12 in (305 mm);
- 6 thermocouples were attached to the back of the exposed-side gypsum board using masking tape (refer to Figure 6), at the centre of each cavity, spaced 12 in (305 mm) apart;

- insulation batts were installed in each cavity, pressing the in-cavity thermocouples tight between the insulation and the gypsum board, as well as in the small gaps at the sides of the wall;
- the unexposed-side gypsum board was installed flush on the wood stud frame in the same manner, leaving small cut-outs on the bottom for the in-cavity thermocouple leads, which were wrapped in a strip of ceramic fiber insulation and sealed using drywall compound;
- 3 thermocouples were attached to the back of the unexposed-side gypsum board using insulated pads (refer to Figure 7), at the centre of each cavity, spaced 24 in (610 mm) apart; and
- all exposed joints, and screw heads were taped and sealed using drywall compound.

Figure 6: In-cavity thermocouples.

Figure 7: Unexposed side thermocouples.

Once constructed the wall was allowed to acclimate for three days using space heaters to minimize moisture accumulation, and giving sufficient time for the drywall compound to set before being sanded flush. The unexposed side thermocouples were affixed using square pads of ceramic fibre insulation measuring 2 in by 2 in (50.8 mm) by 3/8 in (9.5 mm) thick, and taped down using three 6 inch (152.4 mm) strips of duct tape as shown in Figure 7. Duct tape tends to shrink when heated, maintaining adhesion as the plastic melts around the fibers. This technique proved to be quite effective in the present work.

The test was conducted using the design fire described above. In addition to temperature instrumentation, video cameras recorded the state of the wall from both exposed and unexposed sides. The test was allowed to continue for 3 hours, after which time the observable state of the wall was recorded and the test ended.

RESULTS AND DISCUSSION

Calibration Test

The results of the steel wall calibration test are presented here, and the real wall test results in the following section. Figure 8 shows the average gas temperatures measured in the fire compartment during one of the calibration tests (in actuality three tests were performed) as a function of height above floor level. Temperature of gases in the fire compartment rose steadily for the first 28 min, after which time a rapid increase in temperature was measured as the fire spread toward the wall. At this point all fuel within the compartment was involved, with gas temperatures approaching 700ºC near the ceiling.

Figure 9 shows the average net heat flux to the exposed side of the wall as a function of height above floor level. Note that the net heat flux reported here is considerably less than an incident heat flux, which would be measured by a typical heat flux gauge. Due to the low thermal mass and high emissivity of the steel wall, there was very little time delay in the temperature rise of the steel wall relative to that of the fire gases. Since the net heat flux to the wall was computed from the measured wall temperatures, it is not surprising that the time at which full fuel involvement occurred was resolved with good accuracy. While the spread in compartment gas temperatures was quite significant over the test duration, the difference in net heat flux was relatively minor, again attributed to the low thermal mass of the thin and highly conductive steel. Notably the heat flux near the bottom of the wall became greater than at other locations in the decay phase (after about 45 min), as gas temperatures dropped and the remains of the fire heated the wall directly.

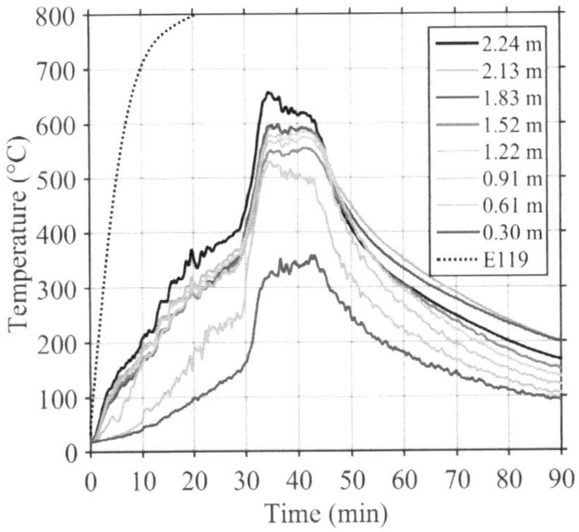

Figure 8: Mean temperature of gases in the fire compartment by height, calibration test.

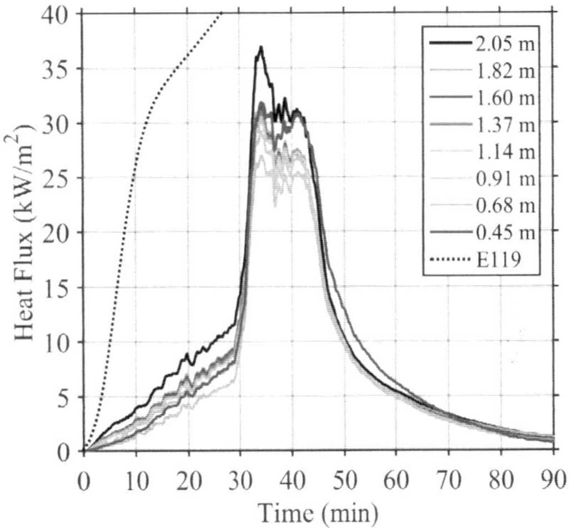

Figure 9: Net heat flux to the exposed side of the wall by height, calibration test.

Also shown is the gas temperature (Figure 8) and net heat flux to the steel wall (Figure 9) if the exposure side gas temperature were equivalent to that of the ASTM E119 fire resistance test furnace. These traces were determined by the same energy conservation model described above, except using the fire temperature (E119) as input for calculating the average temperature of the steel wall, and thus net heat flux into the exposed side. From the above figures, it is clear that the fires used in the present work were significantly less severe than a fire resistance test, as was the resulting net heat flux to the wall.

Figure 10 and Figure 11 show a 2D contours of the net heat flux over the exposed side of the wall after 25 min and 35 min, respectively. The axes in these figures correspond to the edges of the 6 ft × 6 ft (1.9 m × 1.9 m) wall, where the bottom of the wall was located 12 in (0.30 m) above the floor. These figures show that heat flux to the wall before the transition to full fuel involvement was dominated by the developing hot gas layer in the compartment, with a relatively even distribution over the width of the wall and a strong vertical gradient. As the fire spread toward the wall the vertical gradient persisted, and a horizontal gradient also developed. This can undoubtedly be attributed to flame radiation, for which the bottom central region would experience the most significant exposure. Note that these figures show the mean heat flux calculated at the centre node of each element (corresponding to thermocouple locations). With the simplified model adopted here it was not possible to extend calculations to the edges of the wall, though this may be addressed in the future with a more sophisticated inverse model. Despite the coarse resolution, the results presented here provide meaningful insight into the exposure profile.

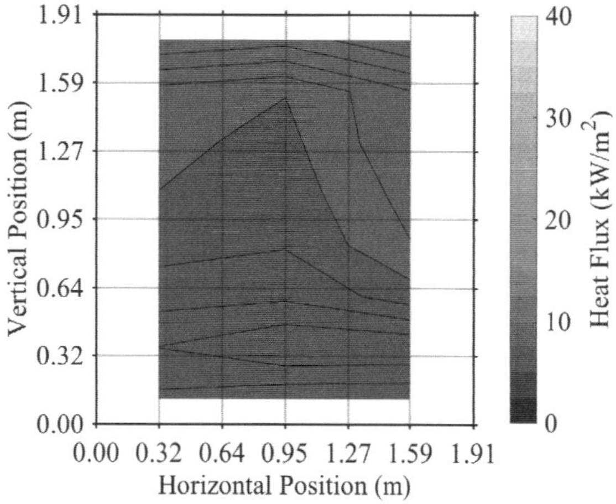

Figure 10: Net heat flux to the exposed side of the wall, 25 minutes after ignition.

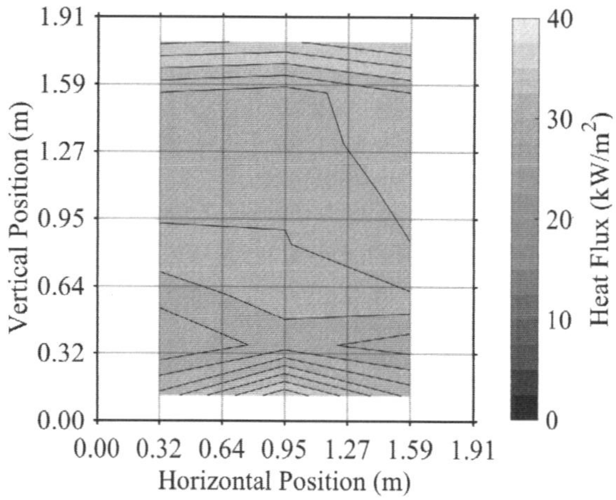

Figure 11: Net heat flux to the exposed side of the wall, 35 minutes after ignition.

Wall Test

The fire environment that evolved during the wall test was similar to that of the calibration test, transitioning to full fuel involvement after approximately 30 min and starting to decay after approximately 45 min. The gas temperatures measured in the fire compartment were slightly greater than those measured in the steel wall tests, reaching a maximum average value of approximately 750°C near the ceiling. Similar levels of net heat flux into the wall, if not slightly greater, can be expected.

Figure 12 shows the temperature measured on the back of the exposed-side wall, averaged between the two leftmost wall cavities (those containing glass fibre insulation), while Figure 13 shows the same for the two rightmost wall cavities (those containing stone wool insulation). The transition to full fuel involvement occurred at 30 min, however the corresponding temperature spikes measured on the unexposed side of the wall lagged behind this by approximately 3 minutes near the top of the wall and 7 minutes near the bottom of the wall, due to the large thermal mass of the gypsum board wallcovering. The peak temperatures measured on the glass fibre side were slightly greater compared to those on the stone wool side, and remained so into the decay stage. Additionally, the glass fibre side of the wall exhibited a secondary rise in temperature after approximately 75 minutes as the gypsum board became compromised and the underlying insulation began to melt, allowing hot gases to penetrate the wall cavities. In this instance, once the gypsum board began to crack, the side containing stone wool insulation was better able to withstand the decay-phase fire environment than the side containing glass fiber insulation.

Figure 14 and Figure 15 show the temperatures measured on the unexposed side of the wall, on the glass fiber and stone wool sides respectively. Temperature at the top thermocouple location was virtually identical on both sides of the wall throughout the test, and at the middle thermocouple location was approximately 10°C cooler on the stone wool side after 90 min. The bottom thermocouple location is particularly interesting as a rapid increase in temperature was measured on the glass fiber side after approximately 75 min. This is likely attributed to the formation of a crack at the bottom-left of the exposed side of the wall, and subsequent direct exposure of the underlying glass fiber insulation to fire gases. This crack can be seen in Figure 16 below, taken at a time corresponding to 60 min after ignition.

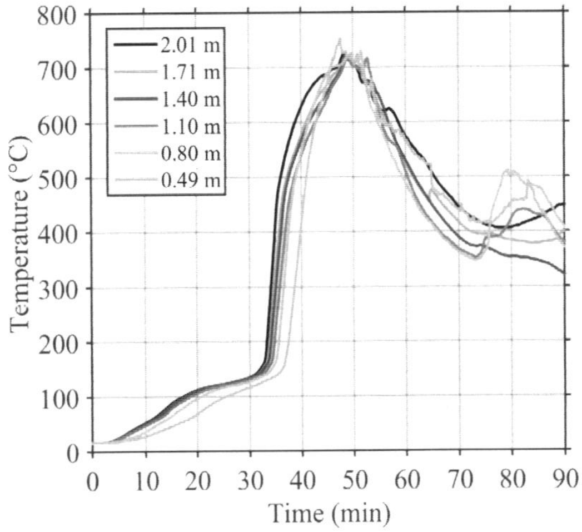

Figure 12: Mean temperature of the exposed side of the wall by height, glass fiber side (left).

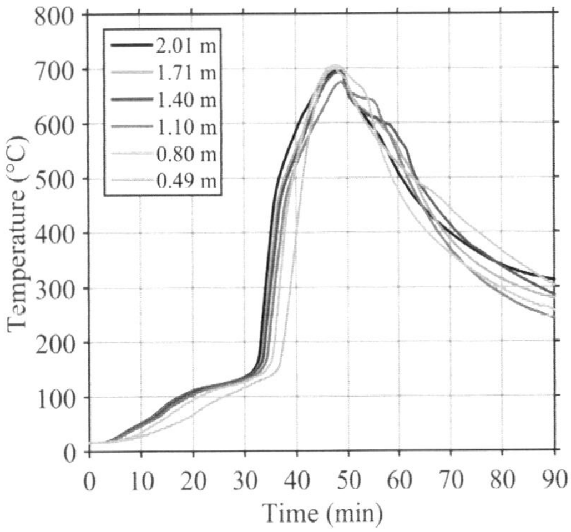

Figure 13: Mean temperature of the exposed side of the wall by height, stone wool side (right).

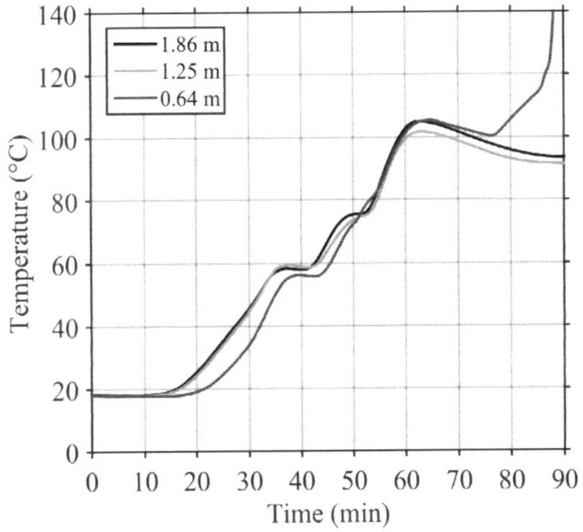

Figure 14: Mean temperature of the unexposed side of the wall by height, glass fiber side (left).

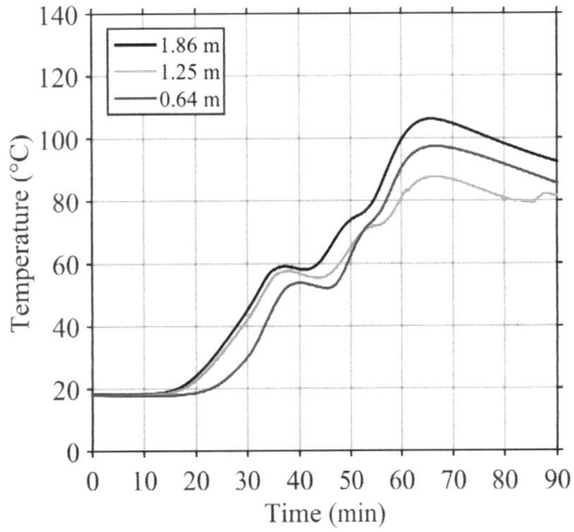

Figure 15: Mean temperature of the unexposed side of the wall by height, stone wool side (right).

The following figures show the condition of the wall as captured by a video camera at 1 hour (Figure 16) and 2 hours (Figure 17) following ignition. By this time the majority of the fuel was burned, but the gas temperatures within the compartment remained very high (Figure 8). After 1 hour the gypsum board was significantly cracked over the entire surface, and the horizontal joint had opened; one of the studs on the glass fiber sides is visibly burning in Figure 16. After 2 hours a significant portion of the exposed side gypsum board had fallen off the wall, more or less evenly on both sides, and localized to the centre of the wall. This is not surprising considering that the fire exposure was most significant at the centreline. It can also be seen in this figure that the glass fiber insulation had burned in the centre of the stud cavities, while the stone wool insulation remained intact.

Figure 16: Condition of the wall after 1 hour (glass fiber on left, stone wool on right).

Figure 17: Condition of the wall after 2 hours (glass fiber on left, stone wool on right).

After 3 hours the test was ended and the panel removed from the compartment opening. Figure 18 shows the final condition of the wall at this time. Despite the relatively cooler environment of the decay-phase compartment, the glass fiber side of the wall continued to decompose significantly more than the stone wool side. Furthermore, the camera installed on the unexposed side of the wall verified that there was no significant smoke leakage through the wall in this time. While the wall would continue to thermally decompose if left unattended, it is clear that this interior partition wall was effective in containing a realistic building fire for a total duration of 3 hours.

Figure 18: Condition of the wall after 3 hours (glass fiber on left, stone wool on right).

One shortcoming of the experimental methodology adopted here was that the exposed side thermocouples did not provide representative measurements of the temperature of the solid material once the wallcovering became compromised. This is to be expected, and it is still beneficial to have temperature measurements as close to the fire side as possible since these will have the fastest response. All the same, it would be interesting to have additional temperature measurements within the insulation layers, which should experience a much more significant rise in temperature than the unexposed side of the wall. The apparatus in its present configuration can accommodate an additional 28 thermocouples without any expansion to the data acquisition system, therefore additional temperature measurements are possible in future work.

Another consideration for future applications of this experiment is to modify the fuel configuration and ventilation so as to produce longer duration fires. It has been shown in the present work that this particular design fire is not capable of burning through the reference W1c wall, which has a stated fire resistance rating of 30 or 45 minutes depending on the insulation, even after 3 hours of total exposure time. The experiment is flexible in that it can accommodate a range of different types of wall assemblies, including those with steel studs or multiple layers of gypsum board for example. Walls with greater fire resistance ratings are intended to be used in spaces that inherently pose a greater risk for a large fire to occur, e.g. those with a greater fire load. Therefore, it is of interest to evaluate the response of such walls to fires in which they are likely to be involved. The apparatus could accommodate this by adjusting the fuel load (e.g. changing the number, size, material, or configuration of wood cribs) or by adopting an entirely different fuel (e.g. a liquid or gas fuel burner). Regardless of the methodology, the fire should burn with some soot formation, as is typical of a solid fuel fire, and develop a distinct hot layer, thereby producing an exposure that is non-uniform over the surface of the wall.

Finally, it should be noted that only a subset of the data measured has been presented in this paper. The numerous temperature measurements provide a good foundation upon which to build a computer model relating the compartment fire environment to the wall, with the ultimate goal being to make predictions of the thermal decomposition of large-scale multicomponent walls.

CONCLUSIONS

A new experimental method was developed for investigating the decomposition mechanisms of large-scale multicomponent construction assemblies exposed to real fires. The experiment was successful in this regard, providing the real time measurements and observations of both sides of a wall as it was exposed to a realistic building fire on one side. Future work will apply this experimental method to a variety of different types of multicomponent construction assembly, perhaps also implementing alternative fire exposures and additional instrumentation to support future modelling efforts.

A methodology was developed for estimation of the heat exposure, in terms of net heat flux to the wall, using a steel sheet in place of a real wall. While the inverse model deployed here was rather simplistic, it was demonstrated to be effective in visualizing the non-uniform heat exposure that a wall experiences in this test. The present work established a foundation from which future work can build upon by adoption of a more sophisticated modelling scheme.

A custom wood crib fire was designed in which the peak temperatures and burning duration were controlled by adjusting the crib configuration and compartment ventilation. It was found that by increasing the spacing between cribs, the rate of spread, and thus the rate of fire growth, could be prolonged. Additionally, by obstructing the door opening it was possible to reduce ventilation to the compartment, ultimately allowing additional control over the fire growth. This combination of controlling fuel configuration and compartment ventilation may be used in future work to produce alternative fires.

REFERENCES

[1] 2010, *National Building Code of Canada*, Canadian Commission on Building and Fire Codes, Institute for Research in Construction, National Research Council of Canada, Ottawa, ON.

[2] 2007, *CAN/ULC-S101-07 Standard Methods of Fire Endurance Tests of Building Construction and Materials*, Underwriters Laboratories of Canada, Ottawa, ON.

[3] ASTM International, 2016, *ASTM E119-16a Standard Test Methods for Fire Tests of Building Construction and Materials*, ASTM International, West Conshohocken, PA.

[4] Manzello, S. L., Gann, R. G., Kukuck, S. R., Prasad, K. & Jones, W. W., 2005, 'Real fire performance of partition assemblies', *Fire and Materials*, vol. 29, no. 6, pp. 351–366.

[5] DiDomizio, M. J., Weckman, E. J. & Roos, R., 2015, 'Experimental Evaluation of Damage and Thermal Penetration in Residential Walls under Realistic Fire Loads', in *Combustion Institute Canadian Section 2015 Spring Technical Meeting*, Saskatoon, SK.

[6] ISO, 2016, *ISO 9705-1:2016 Reaction to fire tests -- Room corner test for wall and ceiling lining products -- Part 1: Test method for a small room configuration*, International Organization for Standardization, Geneva, Switzerland.

[7] Obach, M. R., 2011, 'Effects of Initial Fire Attack Suppression Tactics on the Firefighter and Compartment Environment', University of Waterloo.

[8] DiDomizio, M. J., Wilson, D. P., Weckman, E. J. & Devaud, C., 2016, 'A modified cone calorimeter experiment for study of pyrolysis', *Fire Safety Journal*, Submitted: October 18, 2016, No. FISJ-D-16-00250.

[9] Veloo, P. S. & Quintiere, J. G., 2013, 'Convective heat transfer coefficient in compartment fires', *Journal of Fire Sciences*, vol. 31, no. 5, pp. 410–423.

FIELD TEST METHODS FOR ALUMINUM ALLOY RESIDUAL STRENGTH FOLLOW FIRE

Mark McKinnon[1], Robert Mills[2], Scott Case[2] & Brian Lattimer[1]
[1]Jensen Hughes, USA
[2]Virginia Tech, USA

ABSTRACT

Aluminum alloys are commonly used in structural members for transportation applications. These applications generally lead to difficulty in sampling the material to assess structural strength after exposure to a fire environment. Four commonly used non-destructive field methods to quantify various aspects of the microstructure were used to test each of four aluminium alloys that had been subjected to well-defined fire-like heat exposures up to 500°C. These methods assessed the electrical resistance, ultrasonic attenuation coefficient, and Vickers hardness of each alloy. Destructive tensile tests were conducted on each sample to determine the 0.2% offset yield strength (proof strength) of each alloy and correlations were drawn between each field test measurement and the proof strength. Vickers hardness was shown to have the most reliable correlation with proof strength although the specific relationship was different for each alloy.

INTRODUCTION

Aluminum alloys are commonly used in transportation applications (rail, ships, aircraft) where light-weight materials are needed to meet design requirements. Fires are capable of heating structural aluminum to temperatures (>200°C) that will result in permanent degradation in the mechanical properties. The inability to effectively assess mechanical properties of aluminum alloys that are in use in transportation applications using conventional techniques facilitates the need for non-destructive field techniques to quantify the reduction in mechanical properties of aluminum alloys following a fire to support post-fire inspections, evaluations, and replacement plans. Of particular importance for structural stability considerations is the reduction in proof stress of aluminum as this is generally the designed failure point.

Portable hardness measurement probes have been proposed in previous work[1,2] as a potential field method to indicate the reduction in proof stress due to heating from fire-like conditions. Previous related work indicated that Vickers hardness measurements do not correspond to aluminum proof stress but instead are more closely a measure of ultimate tensile strength. This presents a problem particularly for evaluation of 5xxx series aluminum alloys since their stress-strain curves exhibit more plastic hardening following fire exposure, resulting in a more significant difference between proof stress and ultimate tensile strength. As a result, the decreases in ultimate tensile strength do not correspond well to the proof stress reductions, which increases the uncertainty when attempting to correlate a Vickers hardness at a single force level to proof stress.

Several alternative non-destructive evaluation (NDE) techniques to measure properties that may correlate to proof stress were identified through a literature review, and the techniques that appeared to be most promising were used to characterize several 5xxx series aluminum samples that had been exposed to fire-like temperatures. Two techniques were identified to measure the electrical conductivity in the sample and one technique was found to measure the ultrasonic propagation velocity and the attenuation coefficient of the aluminum samples.

Electrical resistance and its inverse, electrical conductivity, are affected by the grain size, chemical composition, and distribution of defects in an electrically conductive material. The effect of heat and solution treatments on both the strength and electrical resistance of aluminum alloys has been demonstrated by several researchers [3–6]. It has been hypothesized that the electrical conductivity of metals may be affected by changes in the microstructure of aluminum due to exposure to fire-like temperatures.

Sevostianov et al.[7–9] conducted several studies attempting to develop a connection between the strength reduction due to accumulated damage and the electrical resistance variation across a damaged specimen. The authors utilized four-point Kelvin probes to determine the electrical resistance and found that the resistance increased by as much as four times in areas of aluminum samples that had been exposed to cyclic loading in fatigue tests and, as a result, had degraded mechanical properties.

Oppenheim et al. [6] conducted an investigation to correlate mechanical and physical properties of two aluminum alloys that had undergone up to 90 different heat treatments to a maximum solution treatment temperature of 530°C and heat treatment temperature of 165°C. A stated objective of the research was to draw a correlation between the material hardness, electrical conductivity measured with an eddy current tester, and the proof stress of the alloys. The authors were not able to determine a direct correlation between either the hardness and the proof stress or the electrical conductivity and the proof stress for the 6xxx series aluminum, although other research has indicated that it may be possible to develop cross-correlations when analyzing several measured quantities. Research conducted by the Australian Defense Science and Technology Organisation (DSTO) has shown the efficacy of eddy current measurements are a potential NDE method to correlate with changes in microstructural and mechanical properties [10].

Salazar-Guapuriche et al. [11] attempted to correlate the material strength to the hardness and electrical conductivity of several samples of a 7xxx series aluminum alloy that had been treated under a wide range of temper conditions at a maximum temperature of 173°C. The authors concluded that the proof strength, ultimate strength, and temper treatment for 7xxx series aluminum alloys may be estimated when combining information from hardness and electrical conductivity measurements.

Measurements of the time-of-flight and velocity of ultrasound waves have been used widely as a non-destructive evaluation method to detect flaws, study the change in local material properties during plastic deformation, and determine material specimen thickness. The propagating velocity of longitudinal waves travelling through a material may also be affected by the attenuation coefficient, which is directly proportional to the average grain diameter. Since it has been shown that heating a material results in smaller mean grain sizes and a decrease in material strength, it has been hypothesized that this method may be used to assess the decrease in yield strength and proof stress caused by an exposure to fire-like temperatures.

Kwun et al. [12] and Popov et al. [13] used measurements of ultrasonic attenuation coefficients to estimate the grain size of steels and related the grain size to the yield strength. Popov et al. applied this technique to low carbon steels (AISI 1006, 1010, 1015, and 1020) and Kwun et al. applied the technique to several unspecified steel slabs that had been cooled in controlled settings at specified rates to achieve well-defined microstructures. Kwun et al. found that the best correlation for yield strength incorporated information pertaining to the chemical composition of the alloy with the grain size determined through ultrasonic testing. Each investigation resulted in a good correlation between the yield strength predicted using the ultrasound measurements and through destructive tests. Kobayashi [14–17] has developed analytical techniques to correlate ultrasonic measurements to plastic deformation metrics in structural metals, although there has not been a study that investigated the feasibility of ultrasonic testing on aluminum alloys, nor the correlation between sonic transmission properties and material strength.

The non-destructive field methods of detecting changes in the microstructure of metal samples due to exposure to high temperatures that were demonstrated in these previous studies were investigated in the current study. These previous investigations generally either made no attempt to characterize aluminum alloys or lacked direct correlations between each field measurement and the yield stress of the metal. The objective of the current work is to investigate the feasibility of these measurement techniques and to attempt to draw robust conclusions pertaining to potential cross-correlations between electrical conductivity, sonic transmission characteristics, Vickers hardness, and the proof stress of several 5xxx series aluminum alloys.

METHODS AND MATERIALS

Four different NDE techniques were investigated to determine the most appropriate field screening tests for determining the reduction in proof stress of four 5xxx aluminum alloys after various temperature exposures. The instruments included in the study were a four-point Kelvin probe to measure electrical conductivity, an eddy current tester to determine electrical conductivity, an ultrasonic probe, and a portable Vickers hardness tester (5kgf). These methods were selected based on a review of the literature which indicated the potential existence of a relationship between each field measurement and material strength. The measurements of electrical conductivity and ultrasonic attenuation coefficient were conducted exclusively in this work and tensile testing and determination

Materials

Four 5xxx series aluminum alloys were used to evaluate the nondestructive techniques. These alloys were designated by 5456-H116, 5086-H116, 5454-H32, and 5454-H34. The samples were machined to the standard sheet-type specimen size for tensile testing. The gauge section of each sample (25.4 mm long, 12.7 mm wide, and 6 mm thick) was heated using an induction heater. Samples of each alloy were heated at 25°C/min to 100°C, 200°C, 300°C, 400°C, and 500°C and immediately quenched in water after reaching the desired temperature. An as-received sample was also tested for each alloy. Reported material strengths are the 0.2% offset yield stress (proof stress).

Experimental Methods

There are two dominant field methods for the measurement of electrical conductivity for structural materials. One of these methods uses the eddy current principle to determine the electrical conductivity of the material. Eddy current methods utilize a coil through which an alternating current is passed as shown schematically in Figure 1. The coil is brought close to the conductive material being tested, which induces currents within the material. The eddy current is disrupted when a defect is present in the sample, and, as a result, the impedance in the coil is also disrupted. The same method is used to determine the electrical conductivity through a direct correlation between the impedance of the coil and the conductivity of a reference material.

A four-point Kelvin probe measurement utilizes a micro-ohmmeter and four probes directly in contact with the material. Two probes are responsible for applying a current to the material and the electrical potential is measured across the other two probes to determine the resistance between the two measurement probes. The principle behind this method is illustrated in Figure 2.

Figure 1. Measurement concept schematic for Eddy Current Testing [18].

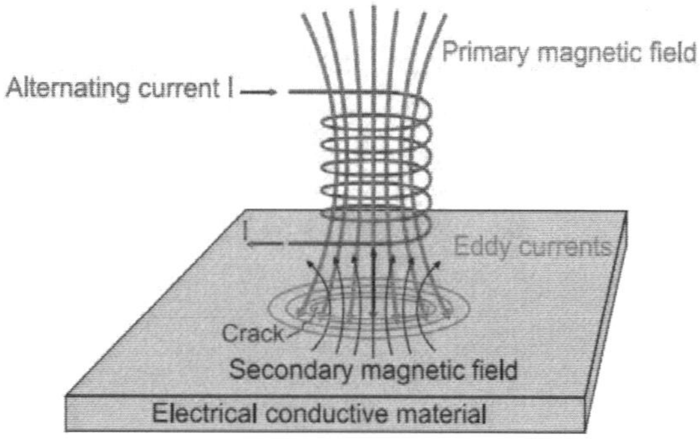

Figure 2. Schematic of principle of operation for four-point resistance measurement [19].

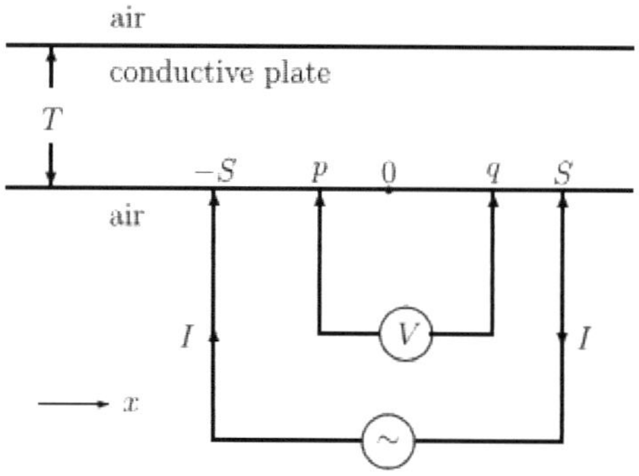

An Olympus Nortec 2000D+ Eddy Current Tester with a 60 kHz conductivity probe that is shown in Figure 3 was used to measure the electrical conductivity in the heated region of all of the samples. All measurements were conducted according to ASTM 1004 [20]. The instrument was calibrated against conductivity standards provided by the manufacturer (30.28% International Annealed Copper Standard (IACS) and 61.28% IACS) and shown in Figure 3b prior to measurement of each alloy and the calibration was verified prior to measurement of each sample. Five measurements were made in the heated region of each sample to accumulate the necessary statistics.

Figure 3. Eddy Current Instrument used to Measure Electrical Conductivity of Sample Specimens: (a) Olympus Nortec 200D+ (b) 60 kHz Probe and Conductivity Calibration Standards.

(a) (b)

An AEMC Model 6250 Micro-ohmmeter and a four-point Kelvin probe, shown in Figure 4, were used to measure the resistance in each sample. The micro-ohmmeter was set to supply 10 A to facilitate the high precision measurement (0.0001 mΩ). The four-point Kelvin probes were affixed to a strip of polymer with a clamp to ensure the relative position between each probe was well-defined and invariant. The tip of each contact point was spring-loaded to ensure all four points made even contact with the material for each measurement. The distance between the measurement probes was 9.85 ± 0.05 mm. The probes were used to measure the resistance in the electrical conductivity calibration standards provided with the eddy current tester to ensure consistently repeatable measurements prior to testing each sample. Five measurements were collected from each sample in which all four probes were within the heated region.

Figure 4. Kelvin Probe used to Measure Electrical Resistance of Samples

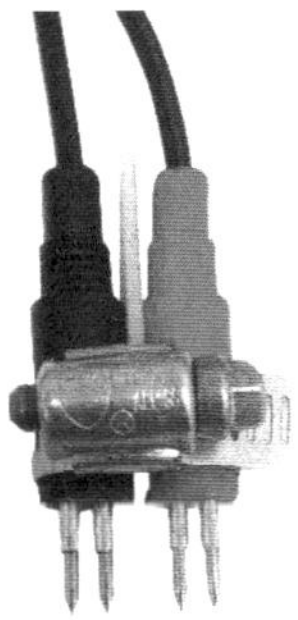

Ultrasonic testing uses electrically produced high-frequency sound waves that pass through the material to achieve a qualitative and quantitative characterization of the material microstructure. Ultrasonic waves that pass through a material are reflected when the wave is incident to a defect or other change in the microstructure. A simple schematic of this phenomenon and the effect of flaw detection on the reflected wave is displayed in Figure 5.

Figure 5. Schematic showing the principle of attenuation of ultrasonic waves and reflection due to defects [21].

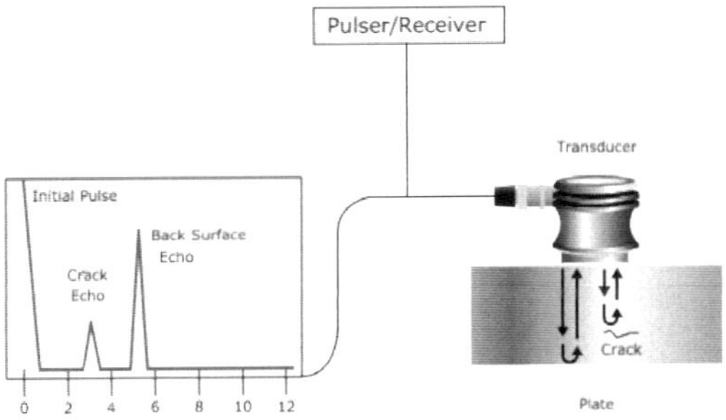

An Olympus XT Ultrasonic flaw detector (shown in Figure 6) was used to measure the longitudinal ultrasonic velocity and the attenuation coefficient of each of the samples. The longitudinal velocity was determined according to ASTM E494. A single probe that was rated for a frequency of 5 MHz was used to generate and receive the ultrasonic pulse using the pulse-echo technique. It was impossible to measure the transverse velocity because of the dimensions of the sample.

Several measurements of the time-of-flight between echoes were made on a 25.4 mm thick 1018 Steel International Institute of Welding (IIW) calibration block to provide a reference measurement. Five measurements of the time-of-flight between reflections of the ultrasonic waves off the back surface of the sample were collected in the heated region of each sample to determine the longitudinal propagation velocity for each sample according to ASTM E494 [22]. Three additional measurements of the difference in magnitude between consecutive echoes were recorded from the heated region of the samples to approximate the ultrasonic attenuation coefficient. The attenuation coefficient was approximated as the decrease in amplitude between consecutive echoes measured in decibels per the sound path length of the ultrasonic wave.

Figure 6. Ultrasonic Flaw Detector used to measure sonic transmission characteristics of sample specimens.

A correlation, presented as Equation 1 [10] has been proposed for the mean grain diameter ($\bar{D}$) as a function of the longitudinal velocity (V_L), transverse velocity (V_T), longitudinal attenuation coefficient (α_L), and the probe frequency (f). Upon determination of the mean grain diameter from the experimental data, a correlation may be made to determine the yield strength (σ_y) of the material using the Hall-Petch relationship [13], provided as Equation 2. In Equation 2, σ_0 and K_y are the friction stress resisting motion of gliding dislocation and the Hall-Petch slope, respectively, and each is a material-dependent constant.

$$\bar{D} = \left(\frac{1125\alpha_L}{4\pi^2 \left(\frac{V_T^4}{V_L^3}\right)\left(\frac{2}{V_L^5} + \frac{3}{V_T^5}\right) f^4} \right)^{1/3} \tag{1}$$

$$\sigma_y = \sigma_0 + K_y \bar{D}^{-1/2} \tag{2}$$

Vickers hardness measurements were performed with a General Electric MIC 10 Micro-Hardness Indenter. This device utilizes a pyramidal tip at 5 kgF to make micro-indentations. The hardness tester was calibrated to 83 Vickers hardness. Five measurements were made on each sample. In the case of the post-fire samples, the measurements were taken within the heated gauge length on both as received and heated samples.

Uniaxial, quasi-static destructive tensile tests were performed at ambient conditions on all of the samples using an Instron 5984 150 kN electro-mechanical testing machine after each NDE technique was evaluated. The tensile tests were performed at a constant displacement rate of 3.048 mm/min, corresponding to an initial strain rate of 10^{-3}/s. Strains were measured using strain gages (small strain analysis) and digital image correlation (large strain analysis).

RESULTS AND ANALYSIS

Destructive Tensile Testing

The data in Figure 7 are the yield strength (proof stress) measured through the tensile testing on a range of different 5xxx series aluminum alloys, which are strain hardened. It is evident in the figure that there are two temperature ranges through which the proof stress decreases. An initial decrease in strength is observed beginning at 150-250°C and is due to recovery, which is the reorganization of the subgrains. At approximately 300°C, the decrease in strength is due to recrystallization, which is the conversion of the primary grain structure into a more equiaxed structure. At 200°C, the 5456-H116 had a drastic drop in yield strength due to recovery. The alloys with lower Mg content exhibited greater drops in yield strength after 300°C due to recrystallization, with full recrystallization occurring by 400°C; thus no drop was seen from 400-500°C.

Figure 7. Residual yield strength (proof stress) of 5xxx series aluminum from tensile testing.

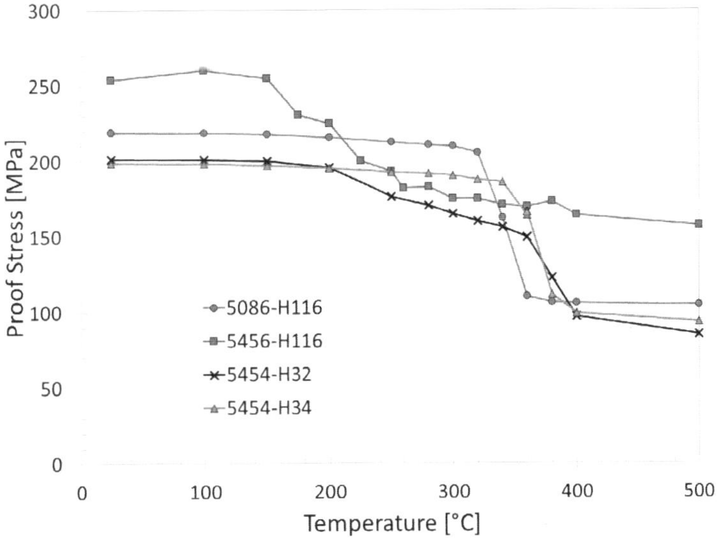

Eddy Current Conductivity Measurement

The electrical conductivity of the four alloys measured with the eddy current method are plotted against the maximum exposure temperature in Figure 8. The error bars in the figure correspond to one standard deviation of five measurement values for each sample. It appears that the conductivity measurement for the as-received 5454-H34 sample is an outlier for that data set because the measurement magnitude is much smaller than all other measured values for that alloy and the conductivity is not expected to change so drastically with a small increase in temperature. There is a relatively weak general trend of decreasing electrical conductivity with increasing exposure temperature.

Figure 8. Mean Electrical Conductivity Measurements Made with Eddy Current Conductivity Probe for: (a) 5456-H116, (b) 5086-H116, (c) 5454-H32, (d) 5454-H34.

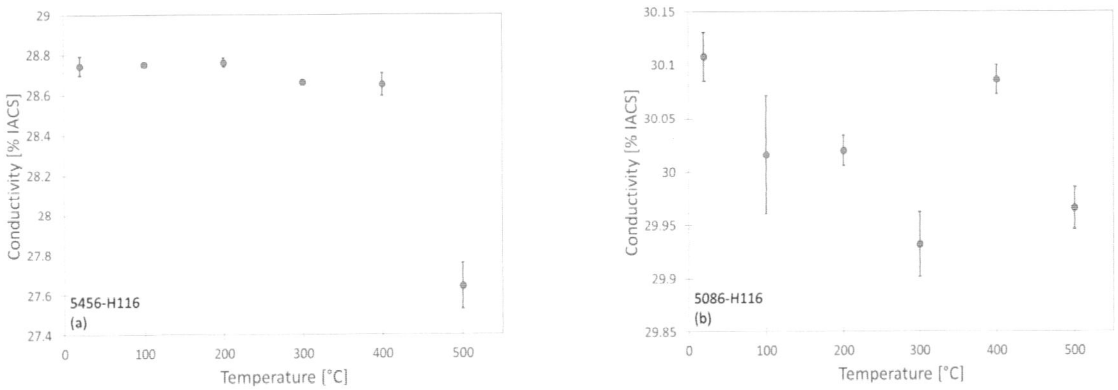

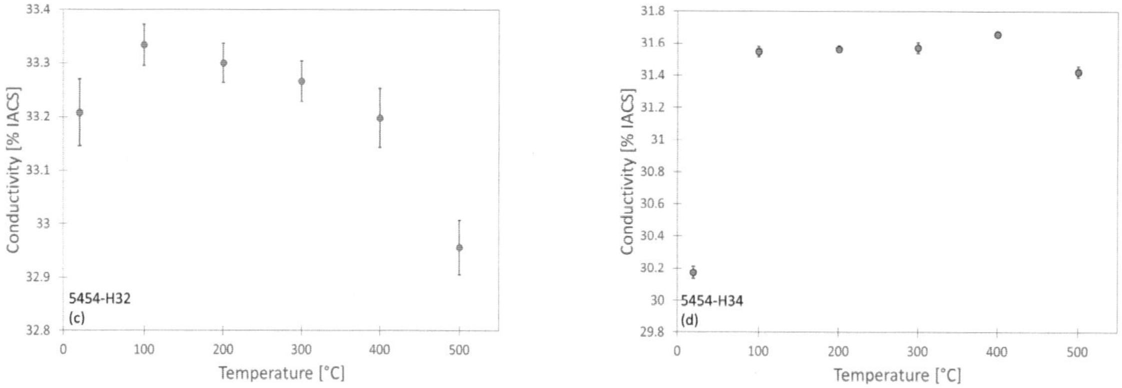

The electrical conductivity measurements are plotted against proof strength determined through destructive testing in Figure 9. It is evident in the figure that there is generally a small decrease in the electrical conductivity as the proof stress decreases for all alloys except the 5086-H116 alloy. It has been noted that eddy current measurements are sensitive to edge effects [20] and the width of these aluminum samples relative to the measurement probe may register edge effects in the conductivity measurement, though the potential effect of interference from the edges on the measurement is uncertain.

Figure 9. Mean electrical conductivity plotted against the proof stress of each sample specimen determined through destructive testing.

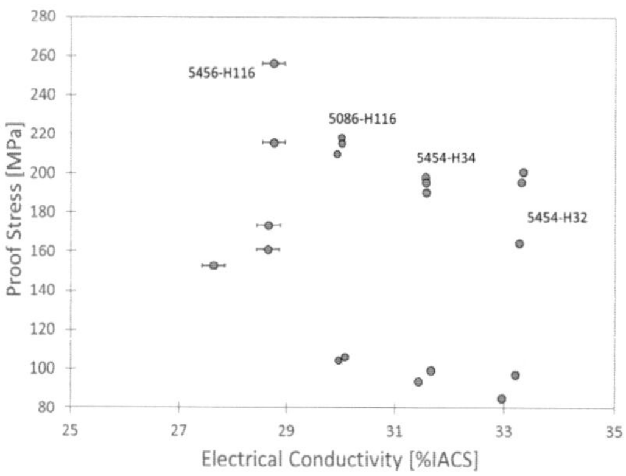

Four-Point Kelvin Probe Resistance Measurement

The mean electrical resistance of each sample as a function of exposed temperature measured using the four-point probe technique is presented in Figure 10. The error bars represent one standard deviation of the data collected on each sample. There is a general trend of increasing resistance with increasing temperature to which the sample is exposed, although the scatter in the measurement makes it difficult to draw any definite conclusions on the existence of a trend because many of the error bars overlap.

Figure 10. Mean Electrical Resistance Measurements Made with Four-point Kelvin Probe for: (a) 5456-H116, (b) 5086-H116, (c) 5454-H32, (d) 5454-H34.

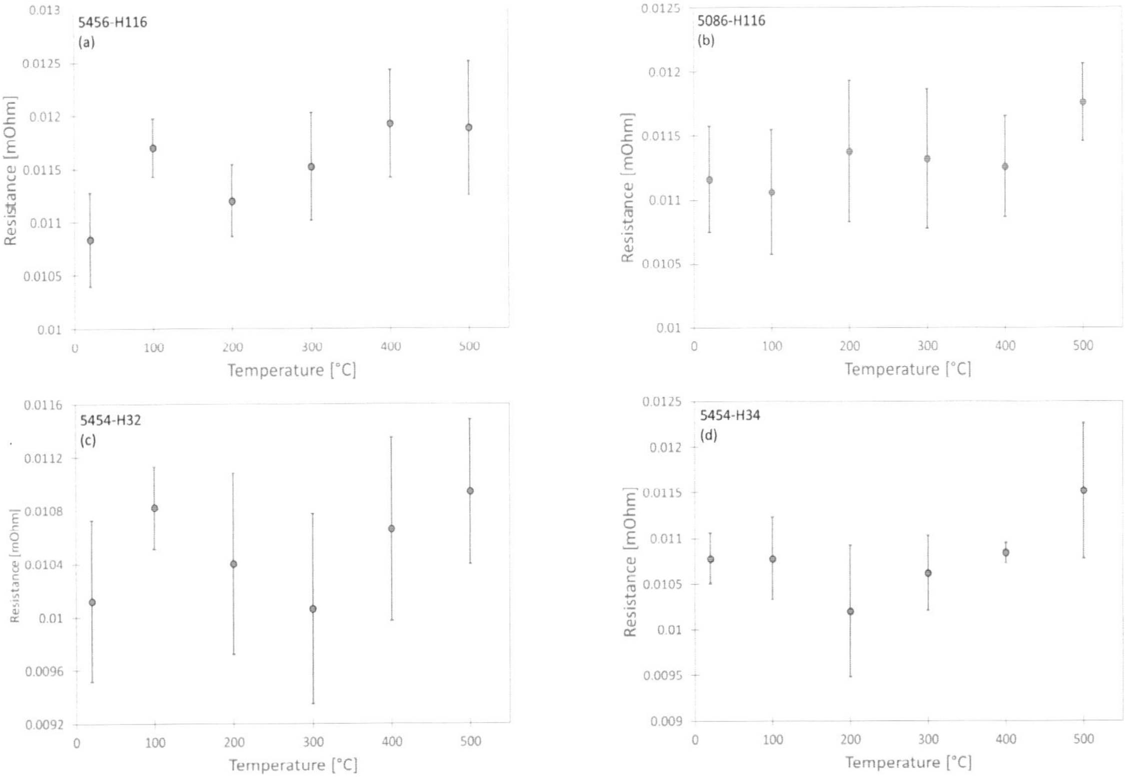

Electrical resistance measurements are plotted against proof strength determined through destructive testing in Figure 11. In general, the mean measured resistance appears to increase slightly with decreasing proof strength. Although, due to the uncertainty in each resistance measurement, there is no clear correlation between the measured resistance and the proof stress for any of the alloys.

Figure 11. Mean electrical resistance measurements plotted against proof stress determined through destructive testing.

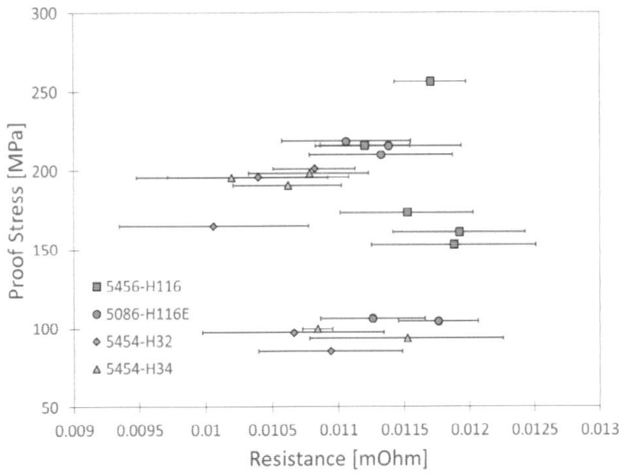

Ultrasonic Testing

The mean longitudinal velocities of sound waves propagating through each sample measured using the ultrasonic flaw detector are plotted against exposure temperature in Figure 12. The error bars in the plots correspond to one standard deviation of the calculated values. It is evident in the figure that the uncertainty in the measurement is relatively large, and there does not appear to be a clear correlation between the exposure temperature and the propagation velocity for any of the tested samples.

Figure 12. Mean longitudinal ultrasonic velocity measurements plotted against specimen exposure temperature for: (a) 5456-H116, (b) 5086-H116, (c) 5454-H32, (d) 5454-H34.

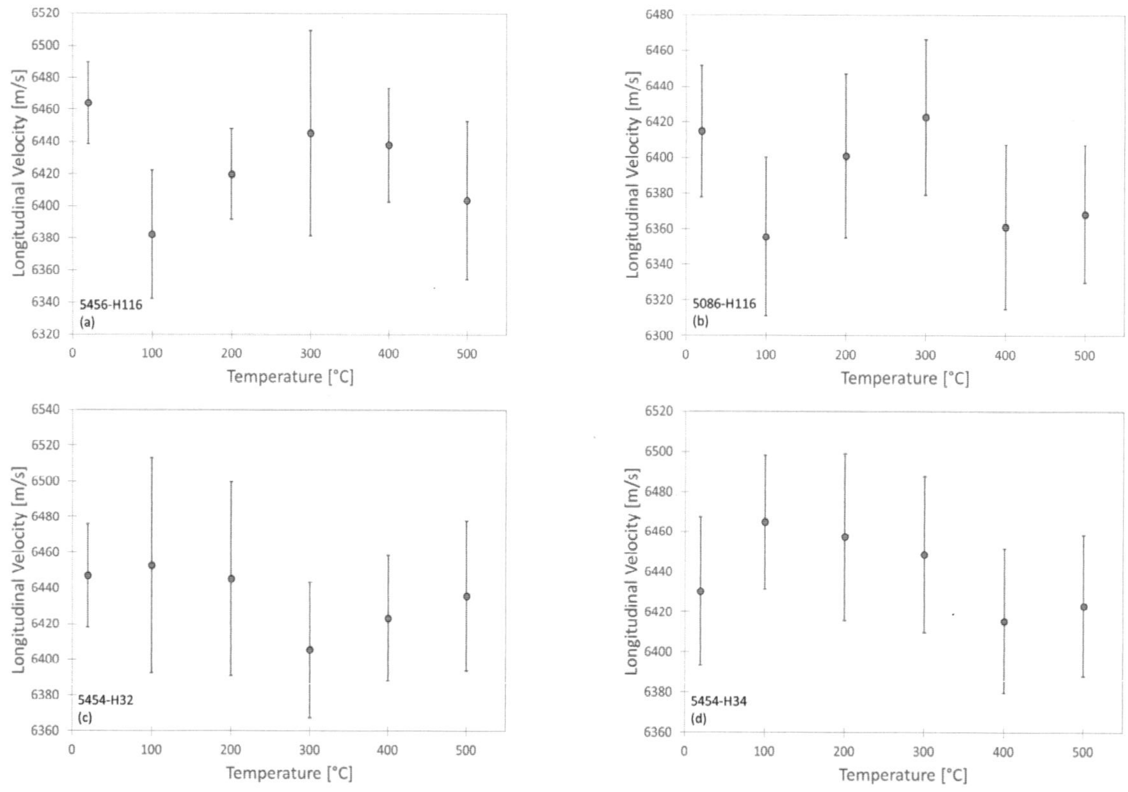

Attenuation coefficient measurements are plotted against the proof stress of each sample measured in destructive testing in Figure 13. There is a weak trend of increasing attenuation coefficient with decreasing proof stress, although this trend is not strong enough to make a conclusive correlation between these two properties. Measurements of velocities and attenuation coefficients are sensitive to edge effects and, due to the width of the samples tested in this work, it is possible that interference from the edges effected the results of the measurements that are presented in this section.

Figure 13. Mean attenuation coefficient calculations plotted against proof stress measured in destructive tests.

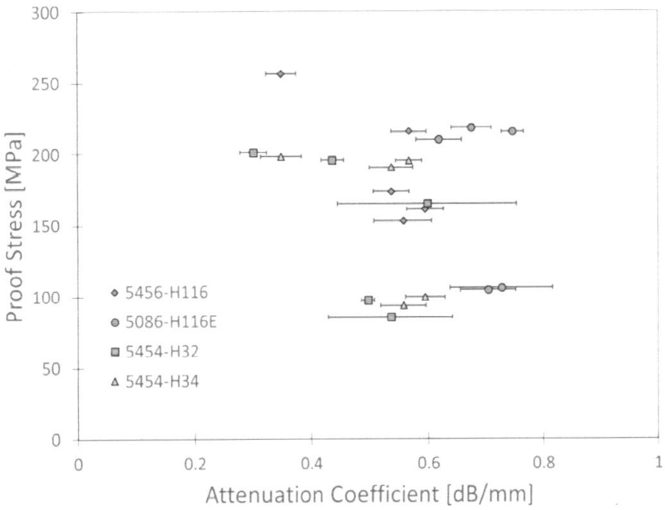

Portable Vickers Hardness Tester

The post-fire hardness results are summarized in Figure 14. Dislocation recovery is suspected to occur in 5XXX series alloys from 150-250°C, while recrystallization occurs from 250-350°C [23]. Figure 15 shows the correlation between the proof strength and Vickers hardness data. In nearly each alloy, there appears to be a correlation between hardness and proof stress but the relationship is different for each alloy.

Figure 14. Vickers Hardness Measurements Plotted against Specimen Exposure Temperature.

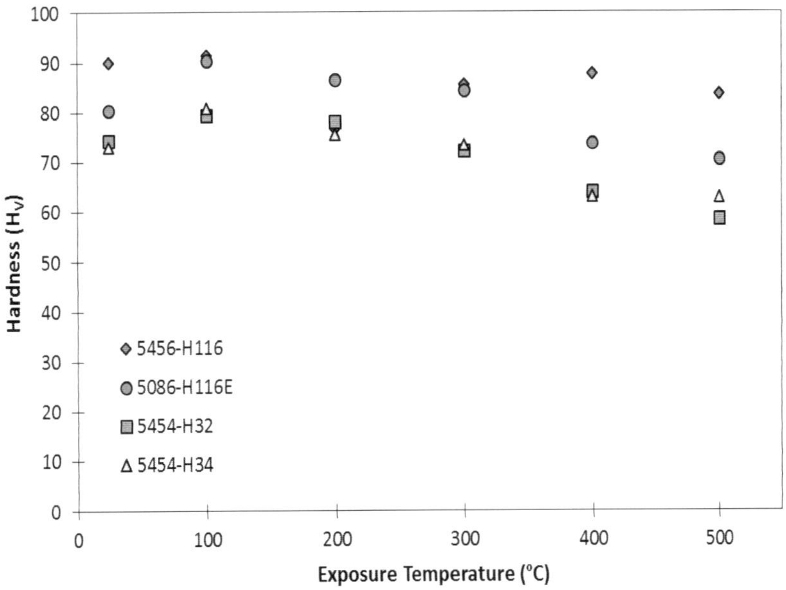

Figure 15. Correlations between post-fire hardness and yield strength.

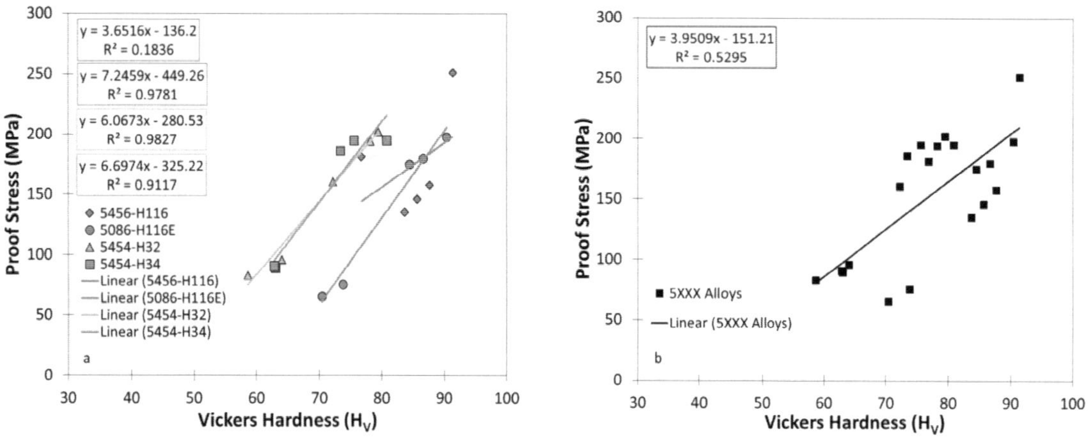

CONCLUSIONS

NDE testing as well as destructive tensile testing were conducted on four 5xxx aluminum alloys subjected to various heating programs to simulate exposure to fire-like conditions. The NDE methods included two methods to measure the electrical resistance of the sample, a method to determine the ultrasonic attenuation coefficient, and a portable probe to evaluate the hardness of each sample. No correlation was evident between the proof stress and the four-point Kelvin probe electrical conductivity measurement or the ultrasonic probe measurement. There was a weak trend between electrical conductivity measured with the eddy current method and proof stress at room temperature, but the method was not able to capture the effects of heating on the reduction in proof stress. The Vickers hardness was shown to have a linear relationship with material proof stress, but the relationship did appear to be alloy dependent. Based on these results, it appears that measurement of Vickers hardness is the best off the shelf NDE method to screen the residual strength of materials after fire exposure. However, this requires calibration with the alloy being evaluated. Based on the limited data available, extracting samples to confirm the proof strength is recommended.

ACKNOWLEDGEMENTS

The authors would like to thank the Office of Naval Research for funding this effort under Grant N00014-14-1-0608, Scientific Monitor Dr. Tom Fu.

REFERENCES

1. Hamada, A. S., Haggag, F. M. & Porter, D. A. Non-destructive determination of the yield strength and flow properties of high-manganese twinning-induced plasticity steel. *Mater. Sci. Eng. A* **558,** 766–770 (2012).
2. Cao, Y. P. & Lu, J. A new method to extract the plastic properties of metal materials from an instrumented spherical indentation loading curve. *Acta Mater.* **52,** 4023–4032 (2004).
3. Starink, M. J. & Wang, S. C. A model for the yield strength of overaged Al–Zn–Mg–Cu alloys.

Acta Mater. **51,** 5131–5150 (2003).
4. Tariq, F., Naz, N., Baloch, R. A. & Faisal. Characterization of material properties of 2xxx series Al-alloys by non destructive testing techniques. *J. Nondestruct. Eval.* **31,** 17–33 (2012).
5. Ahmadi, S., Arabi, H. & Shokuhfar, A. Formation mechanisms of precipitates in an Al-Cu-Li-Zr alloy and their effects on strength and electrical resistance of the alloy. *J. Alloys Compd.* **484,** 90–94 (2009).
6. Oppenheim, T. *et al.* On the correlation of mechanical and physical properties of 6061-T6 and 7249-T76 aluminum alloys. *Eng. Fail. Anal.* **14,** 218–225 (2007).
7. Sevostianov, I., Zagrai, A., Kruse, W. A. & Hardee, H. C. Connection Between Strength Reduction, Electric Resistance and Electro-Mechanical Impedance in Materials with Fatigue Damage. *Int. J. Fract.* **164,** 159–166 (2010).
8. Sevostianov, I. & Kachanov, M. Local minima and gradients of stiffness and conductivity as indicators of strength reduction of brittle-elastic materials. *Int. J. Fract.* **164,** 147–154 (2010).
9. Armstrong, T. W. & Sevostianov, I. Evluation of strength of plane microcracked structural elements via electrical resistance scanning. *Int. J. Fract.* **185,** 171–178 (2014).
10. Mathys, Z., Jeleniewski, V., McCarthy, P., Pedrina, L. & N. Goldsmith. Microstructural analysis and measurement of heat affected aluminium. (2014).
11. Salazar-Guapuriche, M. a., Zhao, Y. Y., Pitman, A. & Greene, A. Correlation of Strength with Hardness and Electrical Conductivity for Aluminium Alloy 7010. *Mater. Sci. Forum* **519–521,** 853–858 (2006).
12. Kwun, S. I., Hong, S. T. & Choo, W. Y. Ultrasonic nondestructive evaluation of microstructure and strength. *J. Mater. Sci. Lett.* **19,** 1453–1456 (2000).
13. Popov, A., Kavardzhikov, V. & Pashkouleva, D. Non-destructive evaluation of the yield stress for low carbon steel by ultrasound measurements. *Russ. J. Nondestruct. Test.* **49,** 328–333 (2013).
14. Kobayashi, M. Ultrasonic nondestructive evaluation of microstructural changes of solid materials under plastic deformation—Part II. Experiment and simulation. *Int. J. Plast.* **14,** 523–535 (1998).
15. Kobayashi, M. Ultrasonic nondestructive evaluation of microstructural changes of solid materials under plastic deformation—Part I. Theory. *Int. J. Plast.* **14,** 511–522 (1998).
16. Kobayashi, M. *et al.* Ultrasonic nondestructive material evaluation method and study on texture and cross slip effects under simple and pure shear states. *Int. J. Plast.* **19,** 771–804 (2003).
17. Kobayashi, M. Analysis of deformation localization based on proposed theory of ultrasonic wave velocity propagating in plastically deformed solids. *Int. J. Plast.* **26,** 107–125 (2010).
18. Garcia-Martin, J., Gomez-Gil, J. & Vazquez-Sanchez, E. Non-destructive techniques based on eddy current testing. *Sensors* **11,** 2525–2565 (2011).
19. Bowler, N. & Huang, Y. Electrical conductivity measurement of metal plates using broadband eddy-current and four-point methods. *Meas. Sci. Technol.* **16,** 2193–2200 (2005).
20. ASTM E1004. Standard test method for determining electrical conductivity using the electromagnetic (Eddy-current) method. 3–7 (2009). doi:10.1520/E1004-09.
21. Center, N. R. Basic Principles of Ultrasonic Testing. *NDT Course Material* (2001). Available at: https://www.nde-ed.org/EducationResources/CommunityCollege/Ultrasonics/cc_ut_index.htm.
22. Qualification, T. P. & Personnel, T. Standard Practice for Measuring Ultrasonic Velocity in Materials 1. *Annu. B. ASTM Stand.* 1–15 (2011). doi:10.1520/E0494-10.
23. Summers, P. T., Case, S. W. & Lattimer, B. Y. Residual mechanical properties of aluminum alloys AA5083-H116 and AA6061-T651 after fire. *Eng. Struct.* **76,** 49–61 (2014).

COMPARING TIMBER ADHESIVE SHEAR STRENGTH PROPERTIES AFTER FIRE DAMAGE

Hailey Quiquero [i] & John Gales [ii]
[i] Graduate student in Structural Engineering, Carleton University, Ottawa, Canada
[ii] Assistant Professor in Fire Safety Engineering, Carleton University, Ottawa, Canada

ABSTRACT

Due to the rapid growth of the engineered timber industry worldwide, a deeper knowledge of the high temperature material properties and behaviour in fire is required. The primary difference between mass timber and engineered timber is the addition of an adhesive through which several laminates of timber are built up into larger sections. Adhesives may degrade at lower temperatures than timber which can lead to delamination during or after a fire. In order to fully understand the performance of engineered timber materials after extreme heating, it is necessary to study the behaviour of these adhesive layers during in-fire and post-fire exposure, the latter being the primary focus of this paper. Current standards do not look at the performance of wood adhesives beneath the char layer in burned engineered timber, but previous tests have shown that this may introduce complex failure mechanisms. To study this issue, samples with two different types of adhesives were heated with various incident heat flux exposures and heating durations followed by mechanical shear testing. A cone calorimeter was used for controlled heating exposure, and the residual shear strength of the glue lines was measured through an innovative mechanical apparatus. Digital image correlation technology was used to assess the deformation (slip) of the adhesive layer as load was applied. It was found that the residual shear strengths were highly variable but did show a slight correlation with the heating duration. Additionally, the slip measurements of the shear line showed that generally, the unheated adhesives were more ductile than the heated samples which exhibited brittle failure.

INTRODUCTION

Undoubtedly, the use of engineered timber has greatly increased over the past decade, with societal drives for sustainable and aesthetic buildings, research advancements and building code amendments. Such buildings are beginning to use massive engineered timber elements such as glued laminated timber (Glulam) and cross-laminated timber (CLT). This is primarily driven through sustainability measures as mass timber takes time to grow, whereas timber used in engineered cases can be harvested earlier and quickly re-grown. Currently, numerous timber building are planned or being constructed, with several competing for the title of tallest wood building in the world. Recently, the University of British Columbia constructed a wood residence building which took the lead with 18 storeys, but this will not remain the tallest as other projects have already been proposed, approved and begun constructed.

Currently, the methods for evaluating the fire performance of exposed engineered timber can be analogous to that of solid massive timber, in which a sacrificial char method is employed. Using previously calibrated charring rates, the cross-sectional area loss of wood over the desired fire resistance period could be estimated for un-encapsulated timber. That char depth thickness is then added to the minimum required cross-sectional area to resist the structural loading during a fire. This method works quite well as a means of fire protection of exposed wood structural elements (negating spread effects), but does not consider a holistic understanding of the composite material behaviour. The introduction of adhesive bonds into the wood section may have a complex influence on the performance of engineered timber, and may differ from solid wood in elevated heat exposures as lower temperatures may have an effect on the polymeric resin materials. The topic of the effects of different adhesives in fire-damaged engineered wood products materials has been identified as needing research by a colloquium of international material experts at the National Institute of Standards and Technology fire safety roadmap workshop [1]. This study begins to serve this discussion with reference to Glulam and its typical adhesives.

BACKGROUND

As a relatively nascent material that is rapidly increasing in use and function, engineered timber performance must be understood in detail. This entails a thorough understanding of the effects of adhesive and how it differs from solid-sawn lumber.

Engineered Timber Adhesives

Timber adhesives have been under development for several decades, with improving properties as science has gained understanding in optimizing polymer properties – they are however, like any polymer, very complex and require detailed consideration when one wishes to analyse their in-fire performance [2]. Currently, there are numerous types of adhesives available to laminated timber manufacturers. This study is focused on two of the most commonly used adhesives at present, polyurethane and resorcinol resins. Polyurethane, commonly abbreviated to PUR, is a newer adhesive which is beginning to see industry wide usage as it becomes more economical [3]. The second adhesive, resorcinol, is a phenol-formaldehyde formulation commonly referred to as PRF. This adhesive has been used in manufacturing wood bonds in several applications for decades. In this study, the PUR samples were bonded by a Glulam manufacturer with a proprietary adhesive, while the PRF samples were prepared by the authors on site.

Standards on Adhesives Performance at Elevated Temperatures

In standards, the consensus is that the adhesive should be of higher strength than the timber itself, so that the material strength properties can be based on those of the wood. This allows easy access to the practitioner in the design of engineered wood, which is largely akin to the design of solid-sawn lumber, with some different modification factors. Like in all material designs, these factors adjust material strength for an acceptable probability of failure, accounting for uncertainties in the product manufacturing.

As there are many different types of engineered wood products available and numerous types of adhesives, several different adhesive standards exist in Canada and internationally. In the authors' opinion, the standards are somewhat disjointed, and many refer to different high-temperature tests and acceptance criteria. Standards exist for adhesives on their own and for the combination of wood and adhesives, and may take the form of a product standard or a manufacturing standard. An outline of some of the standards concerning timber adhesives at elevated temperatures are shown in Figure 1.

The Canadian wood design standard CSA-O86 currently references to the Glulam product and manufacturer standards [4]. These, in turn, cite the CSA-O112-M series of standards for the evaluation of timber adhesives which was originally developed in the late 1970s [5]. The series gives an overview of test procedures in O112.0, and then defines acceptance criteria for several different types of adhesives in O112.1-8 (O112.7 referring to PRF). O112.9 [6] and O112.10 [7] are newer standards which were developed to provide manufacturers with the ability to use newer adhesives not included in the earlier series, after the Canadian federal government funded a wood research initiative [3]. The main test method for evaluating the strength of adhesives in this series is a shear test by compression loading, with specimens that are 50 mm squared by 40 mm thick, with a slight offset on either side of the shear plane. This test is akin to the ASTM D905 test for evaluating the shear strength of adhesives [8].

Figure 1: Overview of Standards in Canada related to Timber Adhesives at Elevated Temperatures

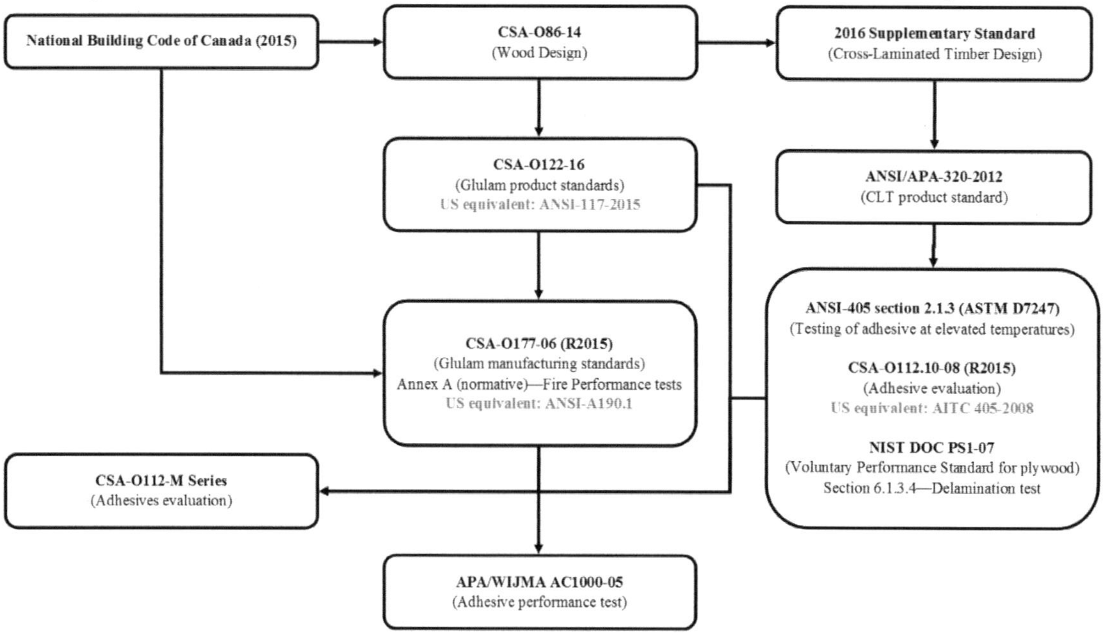

The 2016 Supplementary Standard for CSA-O86 will give added guidance for cross-laminated timber (CLT), and points to ANSI/APA-320 for increased adhesive standards [9]. This document gives provisions for Canadian adhesives to meet ASTM D7247 through ANSI-405 [10]. The ASTM D7247 standard test is meant to evaluate the shear strength of adhesives at elevated temperatures [11], using an almost identical test specimen as in the CSA-O112-M series and ASTM D905. The test procedure involves heating several specimen sets of bonded and solid control samples to a bondline temperature of 220°C for a specified duration of time which depends on the product. Following the heating, the residual shear strength ratio for the bonded specimens must be at least equivalent to the lower 95% confidence interval for the results of the solid wood control specimens. ANSI/APA-320 also specifies that Canadian adhesives must meet section 6.1.3.4 of the NIST Voluntary Performance Standard DOC PS1 [12]. This test uses a Bunsen Burner flame exposure on a sloped surface of the specimen where several glue lines are exposed. After the test, the degree of delamination of the glue lines is visually inspected.

CSA-O177, the "Qualification Code for Manufacturers of Structural Glued-Laminated Timber" gives extra requirements for adhesives meeting CSA-O112.9 [13]. Adhesives following this standard for evaluation of structural wood adhesives for exterior exposures must follow additional tests outlined in normative Annex A of the standard. In this section, a manufacturer has two options of test paths. In the first method, the products must pass a small scale flame test akin to the DOC PS1 Bunsen Burner test, to assess delamination and glue line opening. Additionally, the products must pass APA/WIJMA AC1000, which was the first standard test meant to assess the performance of adhesives at a temperature near the ignition point of wood [14]. It uses the same test procedure as ASTM D7247, which was originally published one year after APA/WIJMA AC1000. As of yet, the authors are not aware of any standard tests that evaluate the performance of adhesive beneath the char layer in burnt engineered timber. Alternatively, in lieu of these two tests, the manufacturer may perform a full-scale fire test in conformance with CAN/ULC S101 to prove adequate fire performance. Adhesives meeting CSA-O112.7 (PRF) inherently meet bondline fire performance requirements.

MOTIVATION

As aforementioned, with the use of engineered timber rapidly increasing globally –even in the un-encapsulated condition, it is important to increase the data available with respect to every aspect of such materials and their behaviour during and after heat exposure. The primary difference between engineered timber and solid massive timber is the introduction of adhesive lines within the structural section. Previous tests by the authors have shown that this introduction may have a complex effect on the failure mechanisms of these materials after an extreme heating event [15]. Moreover, the adhesive standard tests that have been thoroughly conducted by labs and manufacturers do not test the effect of fire on the adhesive below the char layer. As the previous tests suggested that the adhesive within the internal unheated wood zone may have been affected, the goal here is to study that effect in detail. A deeper understanding of the adhesive contribution to the behaviour of engineered timber, specifically Glulam, after fire damage is of importance.

METHODOLOGY

In order to achieve the objective of this research study, to investigate the behaviour of Glulam adhesives beneath the char layer, a test programme was developed in which numerous samples of Glulam adhesive lines were tested under various conditions.

Specimen Preparation

Specimens were prepared in the lab from Glulam products made of spruce-pine-fir species, primarily of Black Spruce, which is the most commonly used species group available in Canada. Glulam blocks were prepared with a thickness of one laminate (40 mm), and with an area of 100 mm by 100 mm. The specimens used differed from the standard shear test specimens used in CSA-O112 to fit in the cone calorimeter apparatus. The blocks were cut from the larger Glulam billets, ensuring that an adhesive line was located along the centreline of the sample for shear testing. The majority of samples were cut around manufacturer made glue lines which used a PUR adhesive, and additionally limited samples with lab-prepared PRF adhesive lines were produced.

Cone Calorimeter Heating

A cone calorimeter was used as a heat source with a constant incident heat flux. Samples were wrapped in aluminium foil to concentrate the heating to the top surface, and were placed in holders with insulation beneath the specimen. The tests done are listed in Table 1 and Table 2.

Table 1: Incident Heat Flux and Heating Duration of PUR adhesive samples

Heat Flux (kW/m^2)	-	50	50	50	50	30	30	30	30
Heating Duration (mins)	-	3	6	10	15	3	6	10	15
Number of Samples	2	2	2	2	2	2	2	2	2

Table 2: Incident Heat Flux and Heating Duration of PRF adhesive samples

Heat Flux (kW/m^2)	-	50	50	50	30	30	30
Heating Duration (mins)	-	3	6	10	3	6	10
Number of Samples	1	2	2	2	1	1	1

An incident heat flux of 50 kW/m^2 was chosen as a typical heat exposure [16] and 30 kW/m^2 was chosen as a comparative heat flux exposure. Exposure times were limited to three, six, ten and fifteen minute durations to observe the varying performance of the remaining adhesive with different quantifiable charring depths. Two samples from the PUR set and one sample from the PRF set were left unheated as control specimens.

This testing procedure also differs from some of the standard tests with regards to the heating, as many standards for evaluating adhesives at elevated temperatures specify a target bondline temperature which must be sustained throughout heating. They are typically done in an oven, where the entire specimen sees uniform heating, and they are tested while still at the elevated temperature. For the purposes of the study, one-sided extreme exposure was used to actually char the wood and leave an unheated wood and adhesive region below the pyrolysis zone. Additionally, tests were done once the specimens had been cooled to mimic the previous tests.

Shear Test Apparatus

A novel shear test apparatus was modified from previous shear analyses of concrete [17], to create a setup compatible with the sample dimension requirements for the cone calorimeter. This non-standard test method was used because the testing objective was to perform an exact comparative basis of the remaining shear capacity of several different samples tested under the same setup.

Figure 2: Shear plane shown on unheated and heated samples (left) and shear apparatus test setup (right)

Digital Image Correlation

Digital Image Correlation (DIC) was used to monitor the deformations of the specimens as they were tested in shear. This is a relatively nascent technique, which was previously proven accurate for measuring displacements and strain in wood specimens [18,15]. The accuracy of the program has been quoted as 0.1 pixels [19], and a past study comparing the accuracy of the DIC measured strains to a bonded 60 mm strain gauge on wood determined an accuracy of about 0.15 pixels [20]. The technique commonly uses a validated pixel-tracking software, GeoPIV8, to locate user-specified locations on a series of high-resolution photographs. In these tests, the slip was measured at 3 points along the shear line: at the top, middle and bottom of the specimens. A sample of the pixel patches used to track the slip are shown in the right-hand image in Figure 2.

RESULTS AND DISCUSSION

Each of the heated samples were exposed to a constant heat flux of 30 kW/m^2 or 50 kW/m^2 in the cone calorimeter for the durations listed in Table 1 and Table 2. No spark was used to ignite the specimens, but rather they were allowed to self-ignite to be more representative of reality in real construction. Sample heat release rate (HRR) curves of four samples are shown in Figure 3.

Figure 3: Sample heat release rate curves for samples with 10 minute heating duration

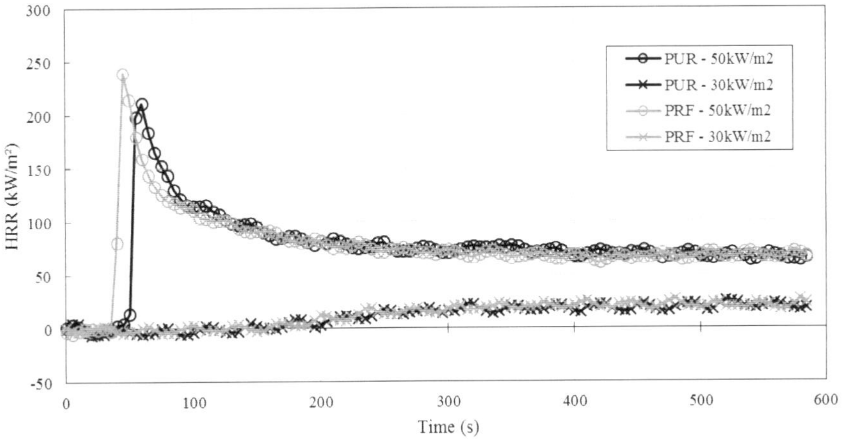

In the 50 kW/m^2 tests, the specimens ignited at an average of 34 seconds after heat exposure began. Divided into adhesive types, PUR ignited around 37 seconds on average, while the average time that PRF samples began flaming was at around 30 seconds. This behaviour can be seen in the trends in Figure 3 above, where the 50 kW/m^2 PRF sample reaches its peak HRR before that of PUR. This can be attributed to the fact that the PRF adhesive used had a much lower flash point than the PUR adhesive - 67°C versus 93°C, respectively as quoted by the manufactures. This indicates that the PRF adhesive may give off sufficient vapours to ignite at a lower temperature than the PUR.

In the 30 kW/m^2 tests, typically the specimens did not ignite at all. Smouldering was seen on the test specimens after about 3 minutes of heat exposure. On the PUR specimens that were tested at 30 kW/m^2 for 15 minutes, flaming was seen after about 7-7.5 minutes of heat exposure. However, the 10 minute samples did not ignite.

Average charring rates over the total heating duration for each sample were computed and are plotted in Figure 4 below versus their respective heating duration.

Figure 4: Averaged charring rates versus heating duration of all test specimens

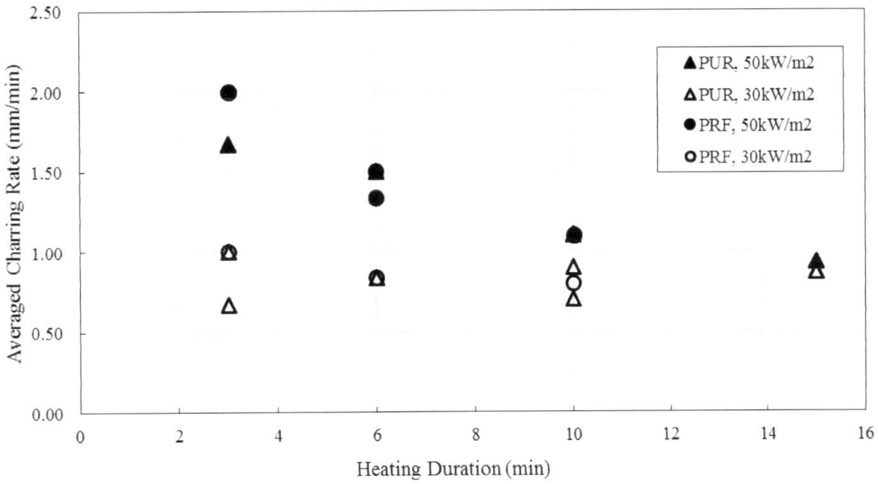

This plot shows a negative correlation between the averaged charring rates and the heating duration, indicating that the charring rates slow down as the heating time increases. This is to be expected as the timber begins to char and build up a protective insulating layer to protect the inner wood and slow the char rate. The values appear to be approaching the recommended value of 0.7 mm/min as the heating duration increases. Additionally, 50 kW/m^2 can be seen as a more severe heat exposure than the standard fire for which the code charring rates are calibrated. The cone test utilises incident heat exposure, whereas a standard fire test would compensate for the heat exposure being emitted thereby reducing the incident heat exposure as timber can give off heat.

It has also been observed that the charring rates on smaller specimens are more severe than on larger specimens, likely due to larger thermal inertia [15]. The 30 kW/m^2 samples do not show a decrease in charring rate with time possibly because they do not build up a sufficient char layer to protect the inner wood, however the charring rates are smaller with the lower incident heat flux and only smouldering fire.

The peak loads of each specimen compared to heating duration were also examined and can be seen in Figure 5 and Figure 6.

Figure 5: Peak loads versus heating duration for PUR test specimens

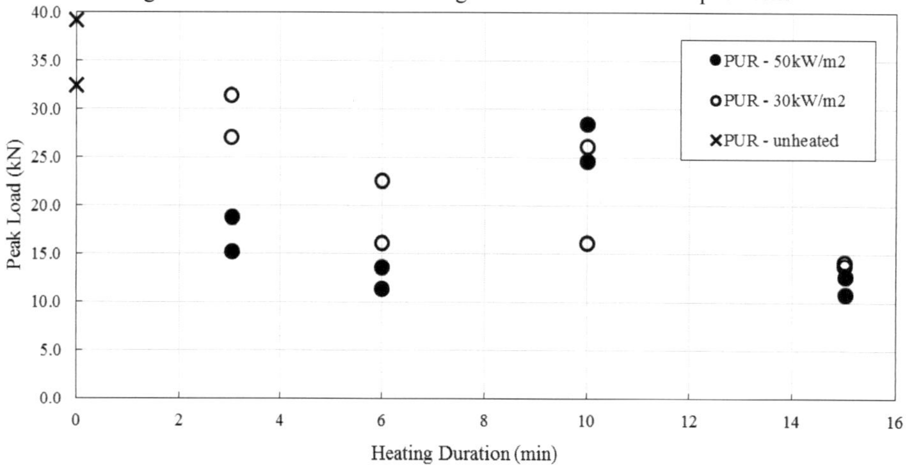

Figure 6: Peak loads versus heating duration for PRF test specimens

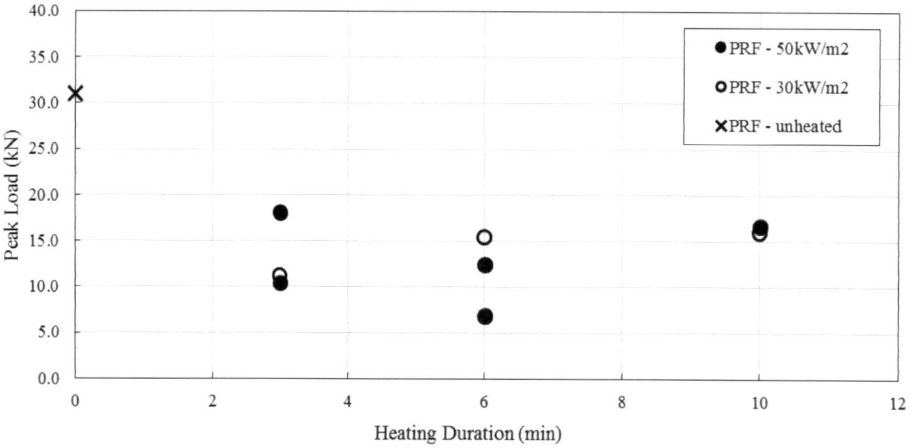

As can be seen in the above figures, the peak loads for all of the specimens were highly variable. In both cases, there appears to be several outliers at the 10 minute heat duration in an otherwise linear trend correlating loss of strength to heating duration and therefore loss of cross-section. The peak loads in Figure 6 for the PRF samples seem to have had a greater loss of strength than the samples in the PUR plot (Figure 5). It is important to note that the PRF samples were the ones with adhesive bonds prepared in the lab, not by the manufacturer. In these samples, glue lines could be considered thicker as craftsmanship was not equivalent to a factory produced bondline, however this craftsmanship used can be considered of typical research practice particularly for proprietary engineered timber system manufacture. This highlights the importance of manufacturing standards and good detailing on engineered wood products, to be used in structural applications and in lab testing which may be used to inform standards.

The residual shear capacity for each specimen was compared to the control specimen shear capacity (the average of the two samples for PUR) to get a percentage of remaining shear capacity in each sample. This was then compared to the remaining shear area based off of the char depth measured at the centreline of each specimen. The results are plotted in Figure 7, where the red dashed line represents a perfect correlation between the residual shear capacity ratio and the remaining shear area, should sacrificial char layers be considered wholly accurate and not account for adhesive degradation. The reader is cautioned however to consider the size of the samples tested when interpreting Figure 8 for application. The plot demonstrates that the typical specimen lost more shear strength than expected compared to the reduction in area.

Figure 7: Remaining shear capacity in each specimen versus remaining shear area

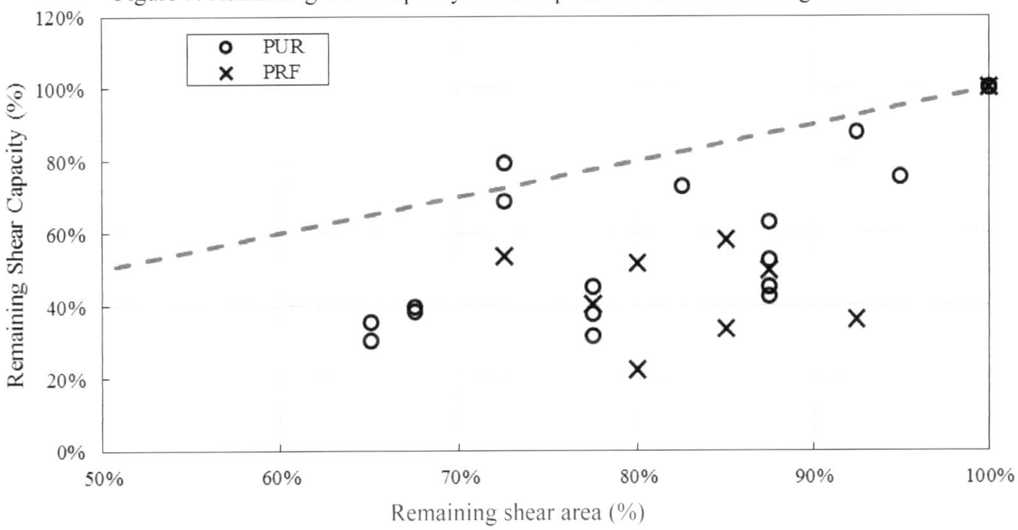

The shear stresses in the specimens were also calculated using the peak load recorded over the reduced shear area beneath the char depth. Figure 7, shows that the shear stresses experienced by the unharmed adhesive layers were again highly variable, although char depths increased almost linearly with heating duration. This indicates that the adhesive may be affected by temperatures within the unheated wood region that are too low to degrade the wood itself. This phenomenon is typically what is studied in standard adhesive evaluation tests in which the specimens are heated to temperatures below the wood pyrolysis temperature range.

Figure 8: Shear stress of all samples versus heating duration

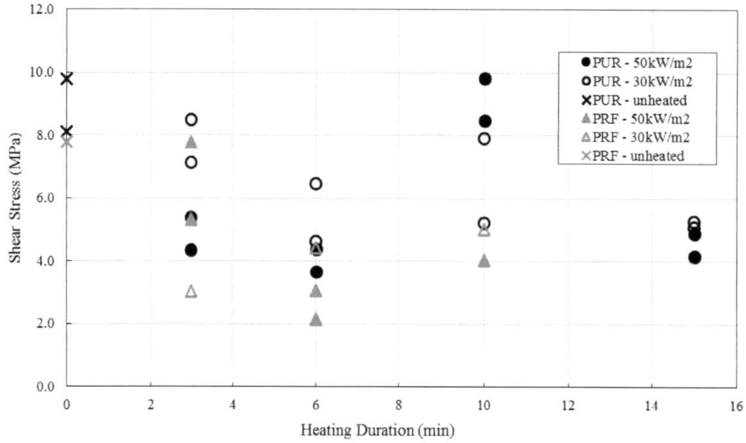

The slip between the two shear surfaces at failure was measured for several samples using the Digital Image Correlation technique. The PRF unheated sample had an average slip of 0.20 mm at failure. One of the unheated PUR samples had an average slip of 0.29 mm, and one had a slip of 0.09 mm. The latter is close to the average failure slips of the heated samples, which had a mean value of 0.10 mm and a standard deviation of 0.05 mm. Sample plots of slip versus load comparing heated and unheated samples of PRF and PUR are shown in Figure 9 and Figure 10.

Figure 9: Sample slip versus load trends for two PUR samples

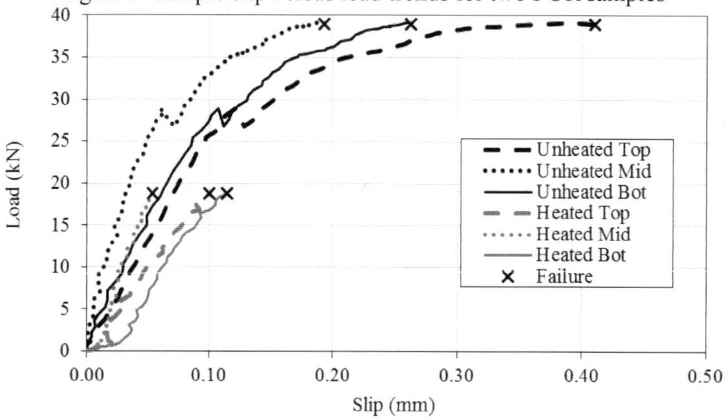

Figure 10: Sample slip versus load trends for two PRF samples

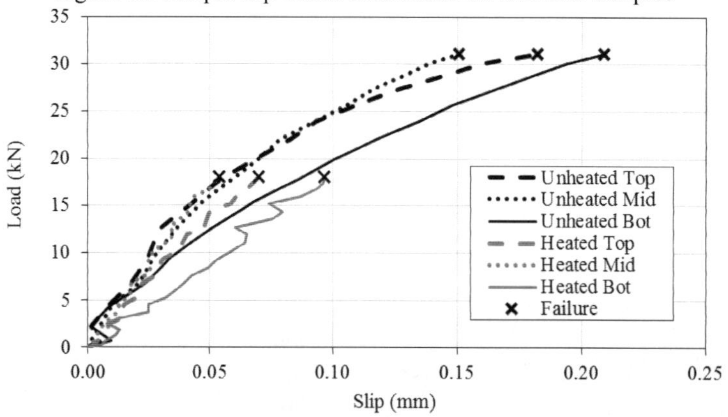

The trends show that the heated samples are follow similar trends to the unheated samples, however they exhibit a more brittle adhesive failure. This could be due to adhesive damage from the elevated temperatures as resins cannot typically be melted and reformed like some other polymers, but rather they degrade. This behavior could explain the reduction in residual shear stress seen in Figure 7.

However, it is important to note that the slip data collected is quite sensitive to error when dealing with such small displacements. As aforementioned, the previously determined error in DIC on wood specimens was 0.15 pixels. For these test images 0.15 pixels corresponds to about 0.0065 mm. Additionally, it is expected that there is some sensitivity to other errors in the measurements [18]. The authors suggest that the values reported be considered over 92.5% accurate when these issues were taken into account (user error for example).

PRELIMINARY CONCLUSIONS AND FUTURE WORK

The primary goal of this study was to investigate into the performance of adhesives in burnt Glulam specimens, specifically the glue lines beneath the pyrolysis zone. In the sacrificial char method, this wood is assumed to be unaffected by the extreme heating. In previous tests involving Glulam, however, it was suggested that the presence of these adhesives after heating introduces a more complex behaviour of the specimen that is not fully captured in this methodology [15]. The test results indicate that the residual strength of the adhesives are quite variable on such small-scale samples when actual pyrolysis of the timber is involved. Future replica tests would be useful in determining the extent of such variability and what uncertainties are involved. The tests do indicate, however, that the adhesive beneath the char layer is somewhat affected by the heating exposure as the shear stresses varied and the specimens exhibited a more brittle behaviour. This is in line with previous tests in which the behaviour of some burned Glulam samples indicated a higher stiffness and slightly lower strength than their unburned counterparts.

These preliminary results demonstrate the importance of investigating the behaviour of the adhesives within the unheated wood regions of burnt samples of engineered timber. Such sporadic results also indicate that more tests are required to be done, and show the importance of quality in professionally prepared engineered timber adhesive bonds. It is important to note once again that the specimens used in this study were very small compared to Glulam structural members which would be used in the modern high-rise timber buildings being proposed. Such members could have cross-sectional dimensions upwards of half a metre, and would not lose nearly as much strength as the unheated region would be much larger with higher reserve strength. If we are to continue innovating with wood and using it as a sustainable alternative building material, we must ensure that the complexities of engineered timber in fire are wholly understood.

ACKNOWLEDGEMENTS

The authors would like to thank NSERC for their funding through the CGS-M and Discovery programs, Ba Lam-Thien and George Hadjisophocleous for their support of the project.

REFERENCES

[1] NIST. (2015). International R&D Roadmap for Fire Resistance of Structures Summary of NIST/CIB Workshop. NIST Special Publication 1188, Maryland, US. p. 131.

[2] Gales, J., Nagy, N., Weckman., B., Gaudreault, C. and Smith, H. (4-6 July 2016). Improving Fire Safety Of Glass Fibre Reinforced Polymers For Bridge Infrastructures. Interflam 2016: 14th International Conference and Exhibition on Fire Science and Engineering. Royal Holloway College, Windsor, UK. pp. 747- 752.

[3] Value to Wood Program. (2010). Research and Development Summary: New adhesive standards help Canadian manufacturers, [online] Quebec: FPInnovations, pp. 1-4. Available at: http://www.valuetowood.ca/html/english/utilities/contact.php [Accessed 4 Dec. 2016].

[4] Canadian Standards Association. (2014). CSA-O86: Engineering design in wood, 10th ed. Toronto: CSA Group.

[5] Canadian Standards Association. (1977). CSA-O112-M: Standards for Wood Adhesives. Toronto: CSA Group.

[6] Canadian Standards Association. (2010). CSA-O112.9: Evaluation of adhesives for structural wood products (exterior exposure). Toronto: CSA Group.

[7] Canadian Standards Association. (2008). CSA-O112.10: Evaluation of Adhesives for Structural Wood Products (Limited Moisture Exposure). Toronto: CSA Group.

[8] American Society for Testing and Materials. (2008). ASTM-D905: Standard Test Method for Strength Properties of Adhesive Bonds in Shear by Compression Loading. Pennsylvania: ASTM International.

[9] American National Standards Institute and The Engineered Wood Association. (2012). ANSI/APA 320: Standard for Performance-Rated Cross-Laminated Timber. Washington: ANSI/APA.

[10] American National Standards Institute and The Engineered Wood Association. (2008). ANSI 405: Standard for Adhesives for Use in Structural Glued Laminated Tımber. Washington: ANSI/APA.

[11] American Society for Testing and Materials. (2016). ASTM-D7247: Standard Test Method for Evaluating the Shear Strength of Adhesive Bonds in Laminated Wood Products at Elevated Temperatures. Pennsylvania: ASTM International.

[12] NIST. (2007). Voluntary Product Standard PS1: Structural Plywood. Maryland: NIST.

[13] Canadian Standards Association. (2006). CSA-O177: Qualification Code for Manufacturers of Structural Glued-Laminated Timber. Toronto: CSA Group.

[14] Yeh, B., Herzog, B. and Williamson, T. G. (2005). Adhesive Performance at Elevated Temperatures for Engineered Wood Products. [online] International Council for Research and Innovation in Building and Construction. Available: https://www.apawood.org/Data/Sites/1/documents/technicalresearch/paper-2005-cib-w18---adhesive-heat-durability.pdf [Accessed 4 Dec. 2016].

[15] Quiquero, H., Gales, J. and Hadjisophocleous, G. (4-6 July 2016). Behaviour of Char Layer in Fire-Damaged Box Section Timber Beams. Interflam 2016: 14th International Conference and Exhibition on Fire Science and Engineering. Royal Holloway College, Windsor, UK. pp. 1063-1074.

[16] American Society for Testing and Materials. (2016). ASTM-E1354: Standard Test Method for Heat and Visible Smoke Release Rates for Materials and Products Using an Oxygen Consumption Calorimeter. Pennsylvania: ASTM International.

[17] Smith, H. K. M., Reid, E. R. E., Beatty, A. A., Stratford, T. & Bisby, L. A. (2011). Shear strength of concrete at elevated temperature in Applications of Structural Fire Engineering: Focusing on utilisation of software tools. Prague, pp. 133-138.

[18] Gales, J. and Green, M. (2-4 February 2015) Optical Characterization of High Temperature Deformation in Novel Structural Materials. Proceedings of the 14th International Conference on Fire and Materials. San Francisco, CA. pp. 626-640.

[19] Gales, J., Bisby, L. and Stratford, T. (2012). New Parameters to Describe High-Temperature Deformation of Prestressing Steel Determined Using Digital Image Correlation. Structural Engineering International, 4: pp. 476-486.

[20] Quiquero, H. and Gales, J. (1-3 June 2016). Behaviour of Fire-Damaged Engineered Timber Beams. In CSCE 2016, 5th International Structural Specialty Conference. London, ON.

USING CFAST AND FDS TO EVALUATE THE RESPONSE OF CLT IN NATURAL AND STANDARD FIRES

Marc Janssens and Jason Huczek
Southwest Research Institute, USA

ABSTRACT

A large-scale room test was conducted to evaluate the performance of CLT construction protected with two sheets of 16 mm type X gypsum board in the pre-flashover, post-flashover and cool-down phases of a severe living room fire. Since minimal or no charring of the CLT was observed in the room test, additional small-scale furnace tests were performed on gypsum-protected CLT specimens using standard and simulated natural fire exposures.

CFAST was used to quantify the thermal environment and exposure conditions of walls and ceiling in the room test. The pyrolysis model in Fire Dynamics Simulator (FDS) was used to simulate the response of the protected CLT in the room fire test and the furnace experiments. The input parameters for the pyrolysis model were based on small-scale measurements and thermal property data reported in the literature for CLT and gypsum board. With these input parameters, FDS generally under-predicts the ability of the gypsum board to protect the CLT. Possible reasons for the discrepancies are discussed, and recommendations are made for future work to address the limitations of the present approach.

KEYWORDS: CFAST, Cross-Laminated Timber, FDS, Natural Fires, Tall Wood Buildings

INTRODUCTION

Wood is a renewable resource when produced by sustainable forestry techniques. It has significant advantages over other construction materials in terms of the impact of its use on the environment. This has been demonstrated, in recent life cycle assessment studies[1,2].

In recent years the renewable and sustainable nature of the materials has resulted in the construction of an increasing number of mid- and high-rise wood buildings in Europe and other parts of the world. These structures use cross-laminated timber (CLT) panels and other types of massive engineered wood members such as glued-laminated (glulam) beams and columns. A substantial amount of fire testing and computer fire modeling has been (and continues to be) conducted to demonstrate that tall wood buildings can provide an acceptable level of safety in case of fire.

U.S. building codes currently do not allow wood structures higher than four stories. However, several tall wood buildings up to 12 stories in height have been, or are now being constructed based on case-by-case acceptance by the local authority having jurisdiction (AHJ). This paper first summarizes the results of a room fire test and four furnace tests that were conducted to support a more general acceptance of engineered wood products such as CLT in tall buildings in the U.S. The paper then describes an initial attempt at modeling the performance of CLT protected with type X gypsum in the tests. CFAST was used to quantify the thermal environment and exposure conditions of walls and ceiling in the room test. The pyrolysis model in Fire Dynamics Simulator (FDS) was used to simulate the response of the protected CLT in the room fire test and the furnace experiments. The paper concludes with suggestions for future work to address these limitations of the present approach.

EXPERIMENTAL

Room Test

In September 2015 Southwest Research Institute® (SwRI®) conducted two full-scale room tests for the American Wood Council (AWC) to evaluate the performance of CLT and nail laminated timber (NLT) construction protected with two layers of 16-mm type X gypsum board when exposed to the thermal environment of a severe living room fire. In the first test the walls of the test room were constructed from CLT panels, which consisted of five layers of spruce-pine-fir (SPF) lumber stacked at right angles and glued together into nominally 0.175 m thick panels. The ceiling consisted of 2 by 6 Southern pine floor joists nailed together one-by-one using 89 mm long 3.33 mm diameter nails. Both walls and ceiling of the test room for the second test were constructed from CLT panels.

The interior dimensions of the test room were approximately 4.46 × 3.25 × 2.38 m. The front narrow wall of the compartment was provided with an open doorway, measuring 1.87 × 2.07 m. Concrete blocks were placed on top of the ceiling to impose a distributed structural load of 1.9 kN/m^2. No additional load was imposed on the walls of the structure.

The room was generously furnished with typical residential non-FR treated living room furniture as shown in Figure 1. The corresponding fire load density was equal to the 94th percentile of the fire load density frequency distribution for living rooms developed from surveys by the National Research Council of Canada[3]. The intent was to maximize the intensity of the fire at the expense of the duration of exposure since an earlier furnace test demonstrated that a loaded CLT wall panel protected with a single layer of 16 mm Type X gypsum board is capable of achieving a 3-hour ASTM E119 fire resistance rating.

The test was started by igniting a sofa seat cushion with a small flame source (BS 5852 Source 2). Flashover occurred in approximately four minutes (see Figure 2), and the peak heat release was measured at approximately 4.9 MW. However, this was thought to be a low value because the hood failed to capture all fire effluents during the peak burning period. To more accurately estimate the peak heat release rate, CFAST 7[4] was used to adjust the heat release rate curve to bring the predicted upper layer temperature in agreement with the measured temperature. The measured and adjusted heat release rates are shown in Figure 3. This figures also illustrates how the fuel load in the CFAST calculations was separated into furniture upholstery and (primarily) cellulosic materials. The total heat released by the upholstery (area under the triangle in Figure 3) is consistent with the total mass and assumed heat of combustion of the upholstery. This figure indicates that the true peak heat release rate was slightly in excess of 8 MW.

Figure 4 compares the room temperatures measured at five locations 0.1 m below the ceiling to the standard ASTM E119 and UL 1709 (hydrocarbon) time-temperature curves. The figure indicates that the UL 1709 curve is more representative of the upper layer temperatures that were recorded in the test than the ASTM E119 curve. However, with the exception of some heavy pieces of furniture (a bookcase and an armoire in the rear corners, and a heavy bookcase in the front left corner), the contents were largely consumed in less than 15 minutes. The test was terminated after 2 hours and 15 minutes to ensure that any continued heating and potential charring of the wood structure during the cool-down phase was captured.

The highest temperature between the two layers of gypsum board was recorded at the ceiling (231°C). The thermocouple temperatures between the base layer gypsum board and the CLT did not exceed 100°C at any time during the test. The gypsum board face layer sustained limited damage, except for the pieces that fell from the ceiling during the test and some pieces that fell down during cool-down (see Figure 4). The base layer remained largely intact although there was some noticeable damage in the three corners where solid furniture items burned for a long time. The wood structure was also intact except for the same three corners, where some minor superficial charring or discoloration was observed after post-test removal of the gypsum board.

Figure 1. Furnished Room for the Second AWC Room Test.

Figure 2. Second AWC Room Test Close to Flashover (4 Minutes).

Figure 3. Second AWC Room Test: Measured and Adjusted Heat Release Rates.

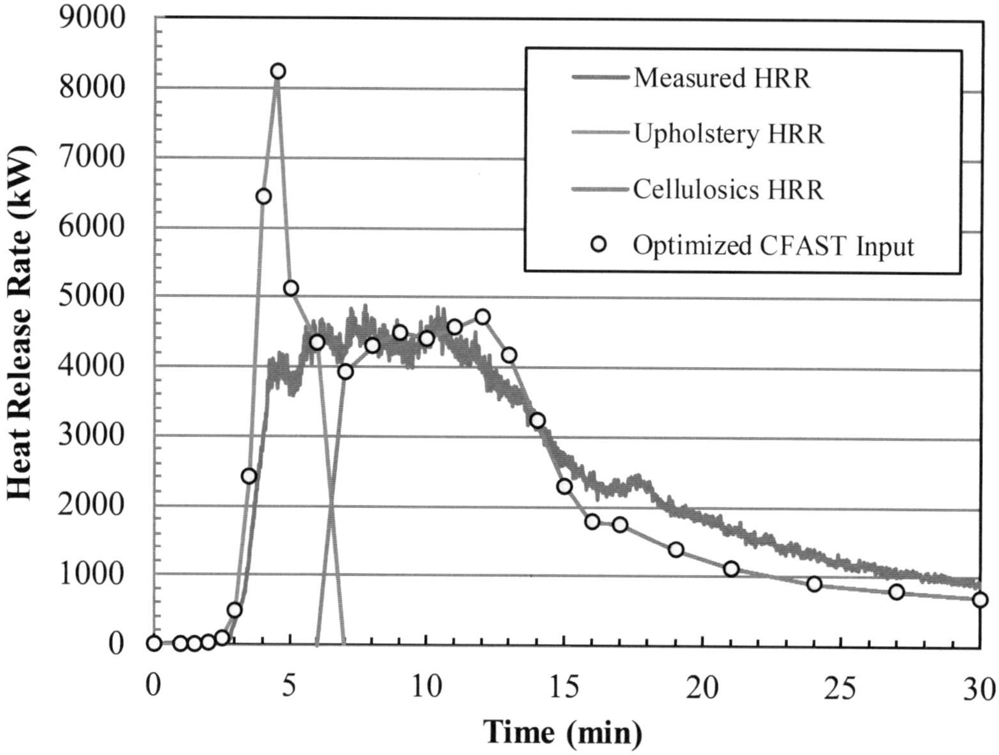

Figure 4. Second AWC Room Test: Measured Hot Gas Layer Temperatures.

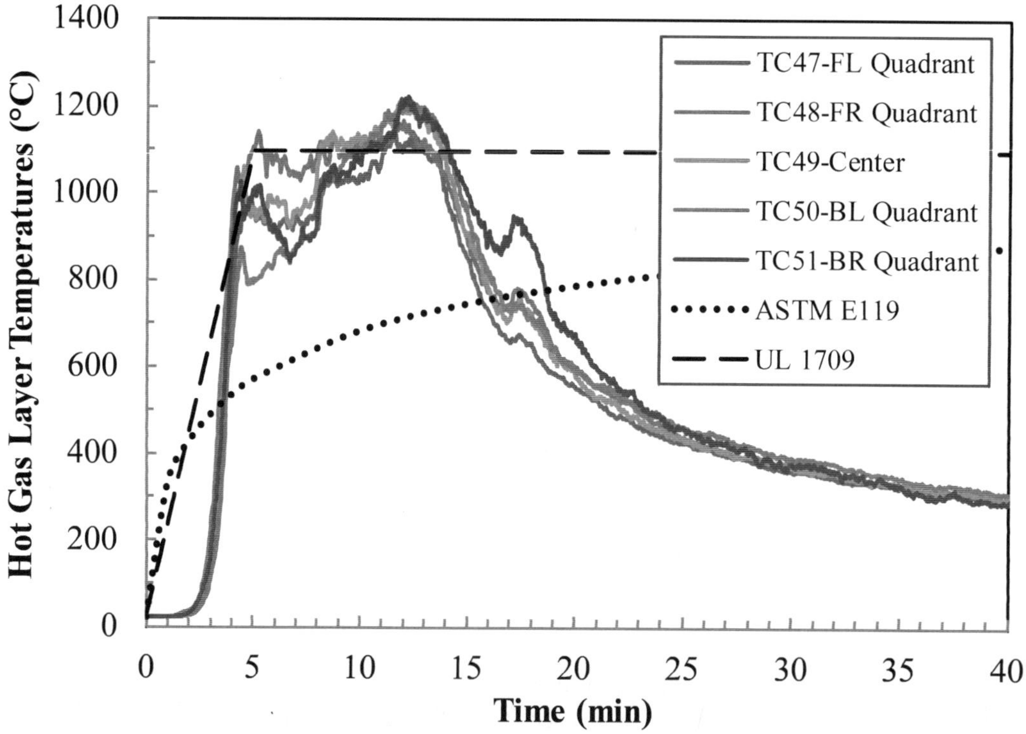

Furnace Tests

Since there was minimal charring very limited information was gained from the room tests about the performance of the wood structure during cool-down. A series of fire resistance furnace tests were therefore conducted in which 0.6 × 0.6 m slabs of CLT and NLT protected with one or two 16 mm layers of type X gypsum board were exposed to the standard ASTM E119 and high intensity UL 1709 time-temperature curves. In these tests the furnace burners were turned off shortly after the char front progressed 13 mm into the wood to simulate cool-down following room burnout.

The ASTM E119 and UL 1709 time-temperature curves are shown in Figure 5. In a UL 1709 test the furnace temperature is ramped to 1073°C above ambient in 5 minutes, and then maintained at that temperature. The UL 1709 curve was chosen because it is reasonably representative of the severe thermal environment that was observed in the room tests, as shown in Figure 4. The modified ASTM E119 and UL 1709 tests were conducted in the small horizontal furnace at SwRI. Four 0.6 × 0.6 m slabs, two CLT slabs and two NLT slabs, were exposed in each test as shown in Figure 5. The test matrix is presented in Table 1. In Test 1 the specimens were lifted from the furnace at the end of the ASTM E119 exposure, which resulted in accelerated cooling. Test 1 is therefore of limited value and was repeated. In the remaining three tests the test frame was left on the furnace during the cool-down, i.e., after the burners were turned off.

Test No.	Thermal Exposure	Gypsum Board Protection
1	ASTM E119	Single 16 mm layer
2	ASTM E119	Single 16 mm layer
3	UL 1709	Single 16 mm layer
4	UL 1709	Double 16 mm layer

1. Small-scale Fire resistance Furnace Test Matrix.

The test specimens were instrumented with thermocouples to measure the temperature along the centerline at the exposed surface of the gypsum, in between the layers of gypsum and wood, and at 13 mm, 25 mm, and 38 mm from the surface of the wood surface facing the furnace. To minimize conduction losses along the thermocouple wires, thermocouples were inserted in holes that were drilled from the edge to the center of each CLT or NLT slab. Since the distance between the edges and the center is 0.3 m and the drill bit is approximately 1 mm in diameter, this technique introduced significant uncertainty in the exact location of the thermocouple junction, which partly explains the scatter in the temperature data presented below.

Figures 6-8 show examples of the measured temperatures for protected CLT specimen exposed to each of the three conditions; ASTM E119 and single layer gypsum protection (Test 2), UL 1709 and single layer gypsum protection (Test 3), and UL 1709 and double layer gypsum protection (Test 4). Figure 6 indicates that CLT protected with a 16 mm layer of type X gypsum board and exposed to the standard ASTM E119 fire continues to char, but that the char front eventually ceases to advance (in this case at 80 min, when the char depth is about 13 mm). This same behavior is expected for the UL 1709 exposure with single double layer of gypsum, although this is not evident from Figures 7 and 8 as the two tests were terminated prematurely.

The temperature and charring rate data obtained from the small-scale furnace test in themselves are not very useful for structural fire design purposes because they are a function of when the cool-down period is initiated. However, the intent was to use the data from these tests for model development and validation. An initial attempt at modeling the performance of protected CLT in the small-scale furnace tests is discussed below.

Figure 5. Test Specimens at End of Test 4.

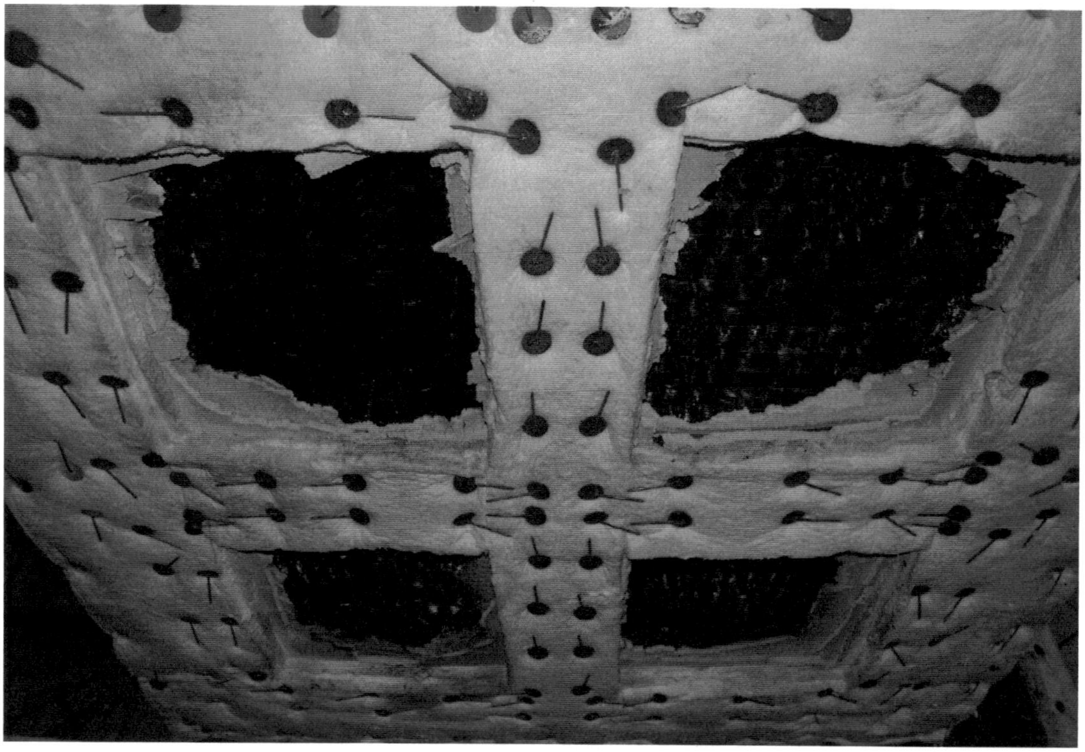

Figure 6. Measured and Calculated Temperatures for Test 2.

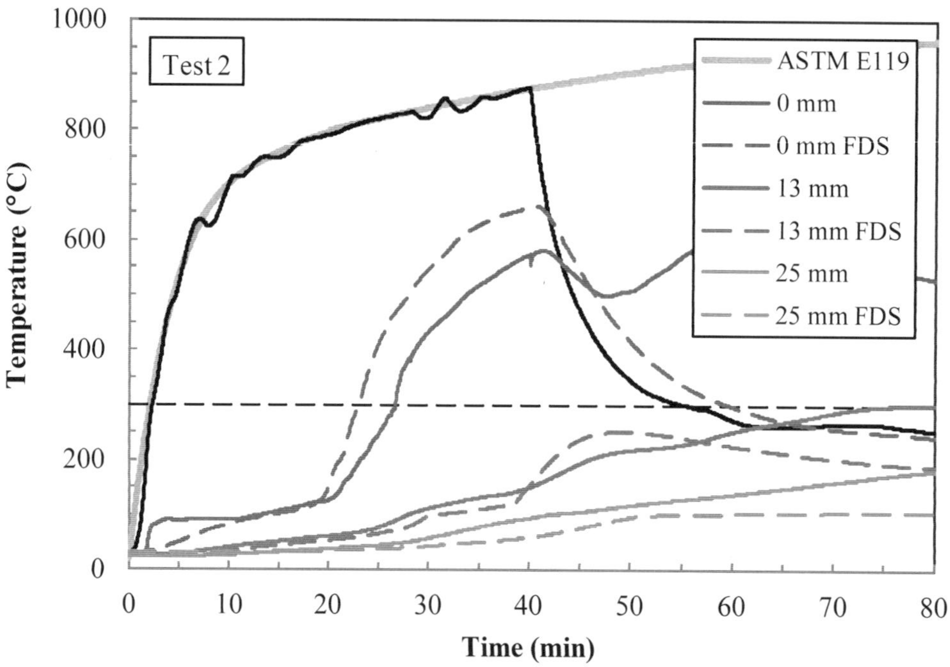

Figure 7. Measured and Calculated Temperatures for Test 3.

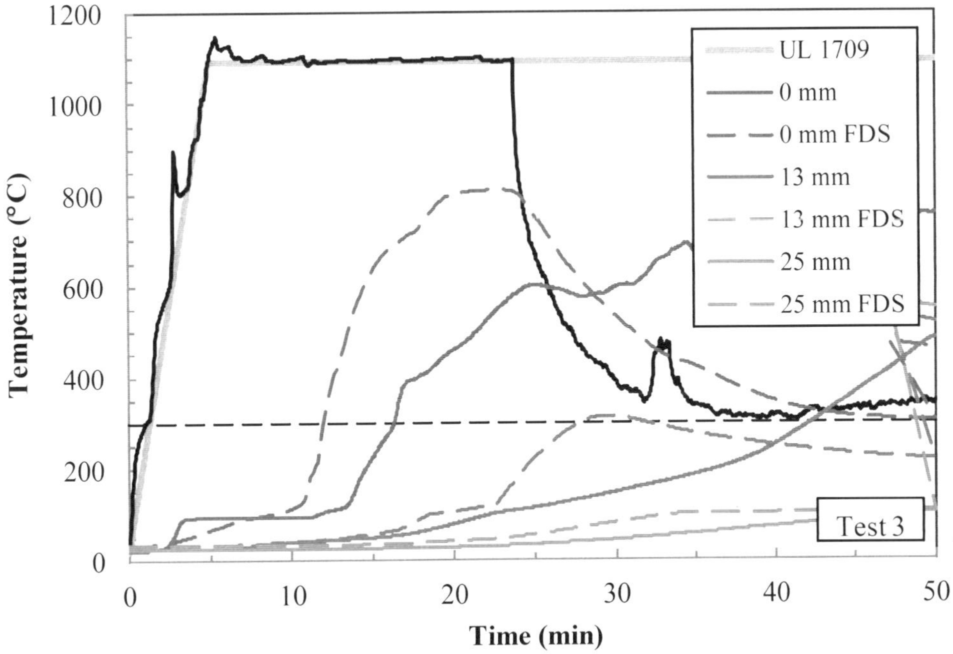

Figure 8. Measured and Calculated Temperatures for Test 4.

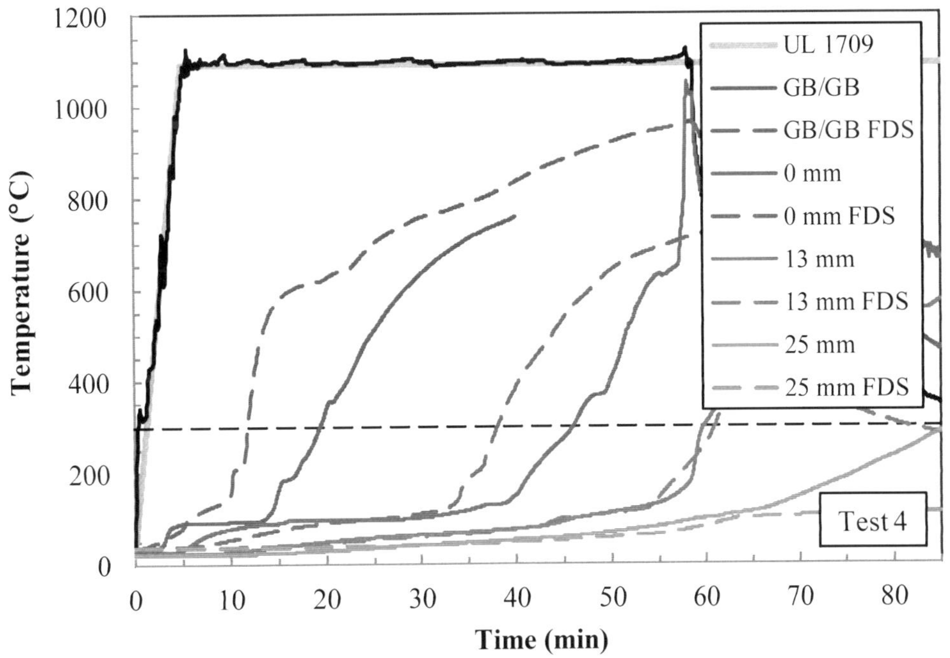

ROOM TEST MODEL CALCULATIONS

Approach

The approach to calculate the response of the protected CLT structure in the room tests involved two steps. In the first step CFAST 7.0 was used to determine the heat flux evolution to the walls and ceiling of the compartment. Two heat flux targets were located in the back wall at 0.6 and 1.8 m above the floor, respectively. A third heat flux target was placed in the center of the ceiling. CFAST was used to calculate the total "gauge" flux, $\dot{q}''_t$, to targets at the three specified locations. The resulting gauge fluxes are shown in Figure 9.

The total gauge flux to a target calculated by CFAST is the net heat flux to the target assuming that its surface remains at the initial or ambient temperature, T_a. It consists of a radiative and a convective component and can be expressed as follows:

$$\dot{q}''_t = \dot{q}''_{t,r} + \dot{q}''_{t,c} = \varepsilon\dot{q}''_r - \varepsilon\sigma T_a^4 + h(T_g - T_a) \qquad [1]$$

where

$\dot{q}''_t$ = total net gauge flux to the target (kW/m^2);
$\dot{q}''_{t,r}$ = radiative component of the gauge flux (kW/m^2);
$\dot{q}''_{t,c}$ = convective component of the gauge flux (kW/m^2);
ε = surface emissivity of the target;
$\dot{q}''_r$ = incident radiant heat flux (kW/m^2);
σ = Stefan-Boltzmann constant ($5.67 \cdot 10^{-11}$ $kW/m^2 \cdot K^4$)
T_a = ambient temperature (K);
h = convection coefficient ($kW/m^2 \cdot K$); and
T_g = gas temperature at the target (K).

Figure 9. Gauge Heat Fluxes from CFAST Simulation of the Second AWC Room Test.

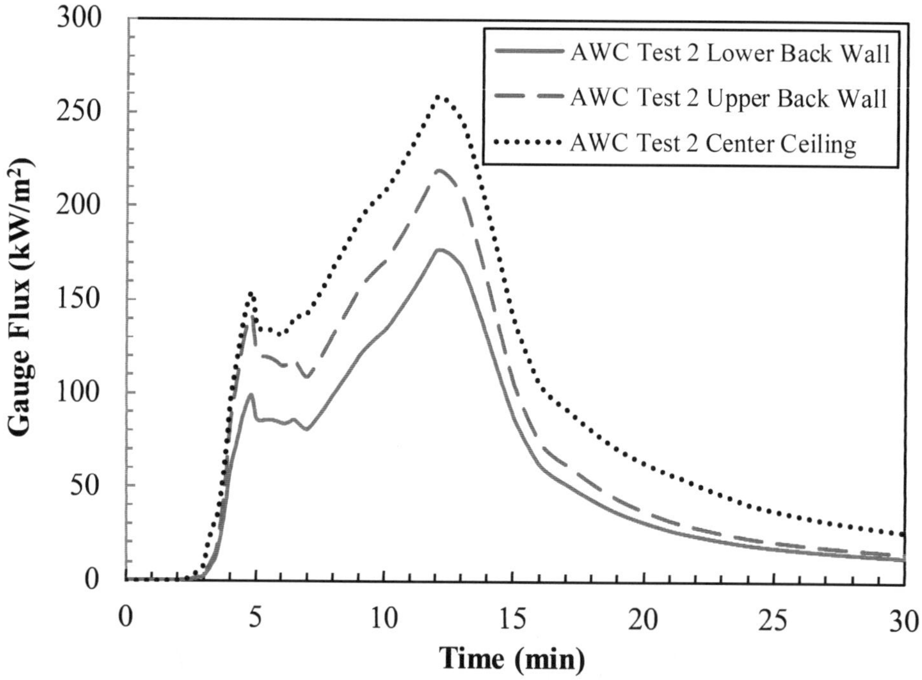

The surface emissivity of the target was set equal to that of gypsum board (0.9). T_a is equal to the air temperature at the start of the test (22°C or 295 K). The convection coefficient was determined to be approximately 0.025 kW/m^2·K.

In the second step the CFAST heat flux predictions were used as input into the pyrolysis model of FDS 6.2[5] to estimate the temperature at various distances from the exposed surface time of the protected CLT. The gas phase calculations were turned off to accelerate the FDS runs. FDS assumes that the solid is exposed to a user-specified incident radiative heat flux, $\dot{q}_r''$, and may be subject to convective heating or cooling. The latter is quantified based on a user-specified gas temperature, T_g, and convection coefficient, h. Hence, the net heat flux at the surface in FDS can be written as

$$\dot{q}_{net}'' = \varepsilon\dot{q}_r'' - \varepsilon\sigma T_s^4 + h(T_g - T_s) \qquad [2]$$

where

$\dot{q}_{net}''$	=	net heat flux at the exposed surface of the solid in FDS (kW/m^2);
T_s	=	exposed surface temperature of the solid (K); and
T_g	=	gas temperature (K).

T_s in Equation 2 is calculated by FDS, while $\dot{q}_r''$, h, and T_g are specified by the user. Unfortunately, only the gauge flux $\dot{q}_t''$ is known (from the CFAST calculations), while $\dot{q}_r''$ and T_g are not. However, after substituting Equation 1 into Equation 2, the latter can be rearranged as follows:

$$\dot{q}_{net}'' = \varepsilon\left[\frac{\dot{q}_t''}{\varepsilon} + \sigma T_a^4\right] - \varepsilon\sigma T_s^4 + h(T_a - T_s) \qquad [3]$$

Hence, FDS can be used to calculate the temperature at different depths from the exposed surface by specifying an incident radiative heat flux at the exposed surface equal to the term in brackets in Equation 3, and a gas temperature equal to T_a.

Material Properties

The material properties for the CLT and type X gypsum board that were used in the FDS calculations are given in Table 2. The densities of the CLT and gypsum board were measured, as well as the moisture content of the CLT. The remaining properties were based on literature values with some adjustments, as discussed below.

Adjusted CLT Properties (ρ≈365 kg/m^3, MC≈12%)				**Adjusted Gypsum Board Properties (ρ≈615 kg/m^3)**			
T (°C)	**k (W/m·K)**	**T (°C)**	**c_p (kJ/kg·K)**	**T (°C)**	**k (W/m·K)**	**T (°C)**	**c_p (kJ/kg·K)**
20	0.100	20	1.53	0	0.215	20	0.95
200	0.125	99	1.77	70	0.170	95	0.95
350	0.070	100	16.28	75	0.170	125	30.60
500	0.090	120	14.56	600	0.245	155	0.75
800	0.350	121	1.90	700	0.280	200	0.75
1200	1.500	200	1.79	1200	0.455	210	23.26
		250	1.35			220	0.70
		300	0.48			1200	0.70
		400	0.34				
		500	0.35				
		600	0.35				
		1200	0.32				

2. Thermal Properties for CLT and Gypsum Board Used in the FDS Calculations.

The thermal conductivity of CLT at 20°C and 200°C was based on measurements conducted at SwRI according to ASTM C518. Values at higher temperatures were taken from Eurocode 5. The specific heat values for CLT were also based on the recommendations for softwoods in Eurocode 5. However, since the density of a solid cannot be varied in FDS unless kinetic parameters for thermal degradation reactions are specified, the density was maintained at 365 kg/m^3 and the specific heat was adjusted as follows:

$$c_{p,adj}(T) = c_p(T)\frac{\rho(T)}{\rho(T_a)} \tag{4}$$

where

$c_{p,adj}(T)$	=	adjusted specific heat at temperature T used in FDS (kJ/kg·K);
T	=	temperature (°C or K);
$c_p(T)$	=	softwood specific heat at temperature T from Eurocode 5 (kJ/kg·K);
$\rho(T)$	=	softwood density at temperature T from Eurocode 5 (kg/m^3); and
T_a	=	ambient temperature (20°C or 293 K).

This approach is justified because ρ and c_p do not appear separately in the heat conduction equation, but only as a product. Equation 4 ensures the product of ρ and c_p, referred to as the volumetric specific heat, is identical as when the temperature dependent density and specific heat from Eurocode 5 were used.

The thermal conductivity of type X gypsum board at 20°C and 70°C was based on measurements conducted at SwRI according to ASTM C518. Values at higher temperatures were based on data published by Manzello et al.[6] The specific heat values for type X were based on the values recommended by Hopkin et al.[7], and adjusted according to Equation 4 with $\rho(T)$ from the same reference.

Selected Results

Figure 10 shows the calculated temperatures between the gypsum board base layer and the CLT at the three locations. It was mentioned earlier that the measured temperature between the gypsum board and the CLT did not exceed 100°C at any time during the 2 hour 15 minute test. Although the temperatures in Figure 10 do not exceed 100°C in 30 minutes, all appear to be on a trajectory that leads to a maximum temperature that exceeds the 100°C threshold, but will not result in temperatures that will initiate charring of the CLT. The latter is consistent with post-test observations. Figure 11 indicates that at least two layers of 13 mm type X gypsum board are needed to avoid charring during the fully developed stage of the fire, but may not avoid charring during a long cool-down.

FURNACE TESTS MODEL CALCULATIONS

The approach to calculate the response of the protected CLT specimens is similar to that employed for the room test, except that in this case the boundary conditions at the exposed surface in the FDS calculations were determined directly from the measured furnace gas temperature. Equation 3 was modified as follows:

$$\dot{q}''_{net} = \varepsilon_{eff}\left[\sigma T_f^4 + \frac{h}{\varepsilon_{eff}}(T_f - T_a)\right] - \varepsilon_{eff}\sigma T_s^4 + h(T_a - T_s) \tag{5}$$

Where T_f is the measured furnace gas temperature, and ε_{eff} is the effective emissivity (0.9). The results of the FDS temperature calculations for the furnace tests are shown as dashed lines in Figures 6-8. In general, FDS appears to predict an accelerated temperature rise compared to the measurements.

Figure 10. FDS Gypsum Board/CLT Interface Temperatures at Target Locations.

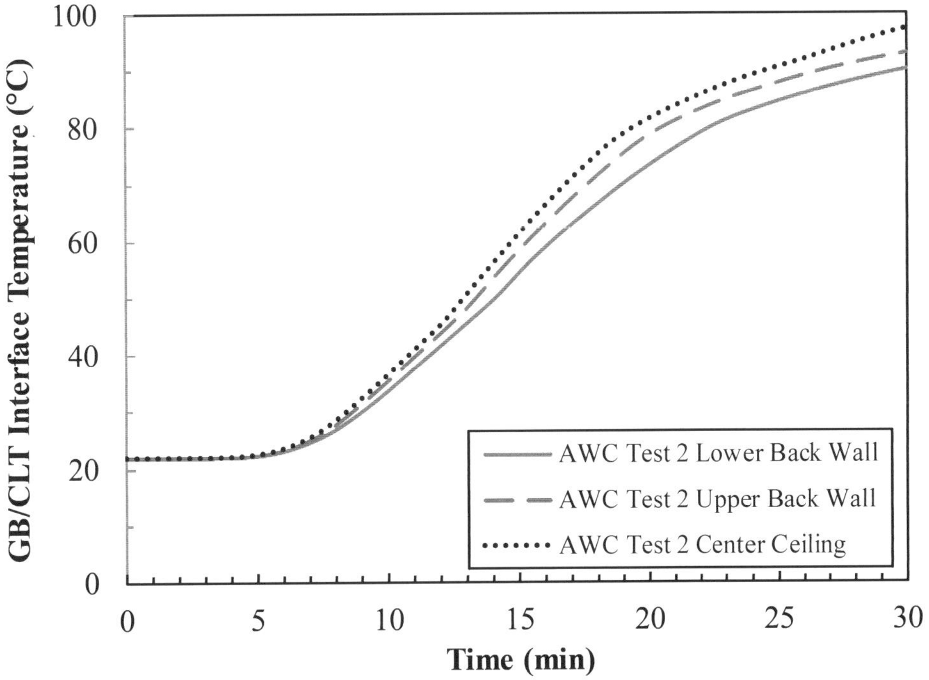

Figure 11. FDS GB/CLT Interface Temperatures at Ceiling for Various Board Thicknesses.

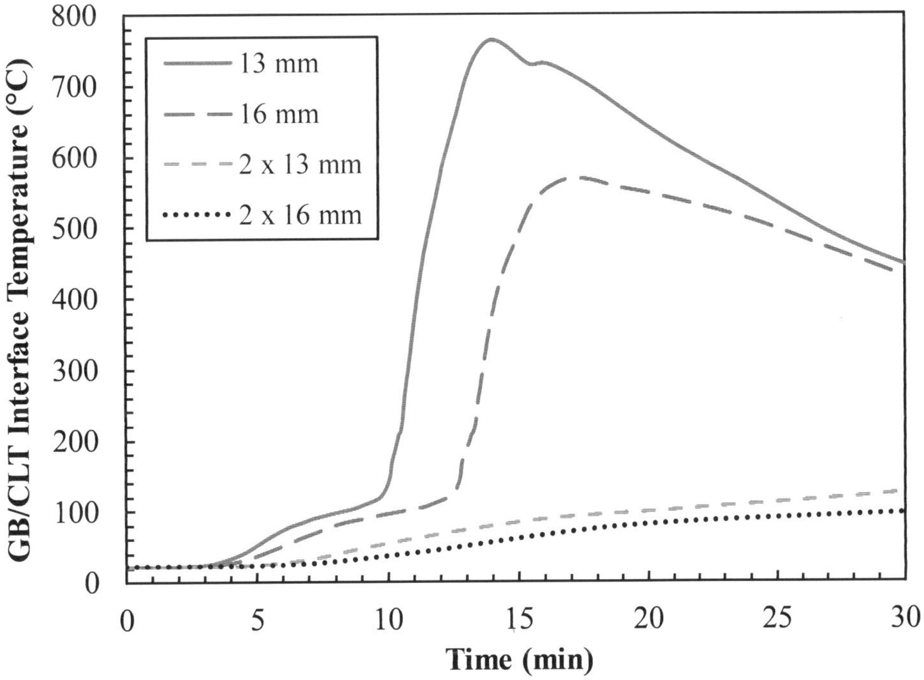

RECOMMENDATIONS FOR FUTURE WORK

The following can be explored to improve agreement between predictions and measurements:

- Both the type X gypsum board and the CLT had a relatively low density (615 and 365 kg/m^3, respectively). Thermal properties for these materials appear to be predominantly based on experimental data for specimens with higher densities. An analysis could be performed to determine which of the properties the model calculations are most sensitive to. Alternatively, thermal conductivity seems to have the largest scatter in the literature data. Parameter estimation techniques could be used to determine apparent thermal conductivities for the two materials that optimize agreement between model predictions and measurements.

- Strictly speaking, thermal properties that are only a function of temperature may not be valid during the cool-down period for materials that decompose and/or lose mass when heated. This problem can be addressed by using a detailed pyrolysis model that captures the irreversible thermal degradation reactions. Since TGA data have been obtained for the gypsum board and CLT that were used in the room and furnace tests, it is worth exploring whether the predictions can be improved by using more detailed pyrolysis models that include chemical kinetics.

- Existing room test data for protected and unprotected CLT could be used to further explore the validity of the two-step CFAST-FDS approach used to model the performance of protected CLT in the second AWC room test. In addition, existing furnace data could be used to evaluate the predictive capability of FDS in conjunction with adjusted literature property data.

ACKNOWLEDGEMENTS

The authors would like to express their gratitude to the American Wood Council for sponsoring the room fire tests. In addition, the authors would like to acknowledge the financial support of the Advisory Committee for Research at Southwest Research Institute for the furnace experiments, small-scale material characterization tests, data analysis, and computer modeling.

REFERENCES

1. Lippke, B., J. Wilson, J. Perez-Garcia, J. Bowyer, and J. Meil, "CORRIM: Life-Cycle Environmental Performance of Renewable Building Materials", *Forest Products Journal*, Vol. 54, 8-19, 2004.
2. John, S., N. Nebel, Perez, N., and A. Buchanan, "Environmental Impacts of Multi-Storey Buildings Using Different Construction Materials", University of Canterbury, Christchurch, New Zealand, Research Report 2008-02, 2009.
3. Bwalya, A., G. Lougheed, A. Kashef, and H. Saber, "Survey Results of Combustible Contents and Floor Areas in Canadian Multi-Family Dwellings", *Fire Technology*, Vol. 47, 1121-1140, 2011.
4. Peacock, R., P. Reneke, and G. Forney, "CFAST–Consolidated Model of Fire Growth and Smoke Transport (Version 7), Volume 2: User's Guide", National Institute of Standards and Technology, Gaithersburg, MD, NIST Technical Note 1889v2, 2015.
5. McGrattan, K., H. Hostikka, R. McDermott, J. Floyd, C. Weinschenk, and K. Overholt, "Fire Dynamics Simulator User's Guide", National Institute of Standards and Technology, Gaithersburg, MD, NIST Special Publication 1019, 2015.
6. Manzello, S., W. Grosshandler, and T. Mizukami, "Furnace Testing of Full-Scale Gypsum Steel Stud Non-Load Bearing Wall Assemblies: Results of Multi-Laboratory Testing in Canada, Japan, and USA", *Fire Technology*, Vol. 46, 183–200, 2010.
7. Hopkin, D., T. Lennon, J. El-Rimawi, and V. Silberschmidt, "A Numerical Study of Gypsum Plasterboard Behaviour Under Standard and Natural Fire Conditions", *Fire and Materials*, Vol. 36, 107–126, 2012.

FIRE-STRUCTURE SIMULATIONS USING FULLY COUPLED THERMO-MECHANICAL SHELL ELEMENTS

Paul A Beata* & Ann E Jeffers
University of Michigan, USA

ABSTRACT

A practical approach to simulating fire-structure interaction problems involves performing a computational fluid dynamics (CFD) simulation to model the fire evolution in the vicinity of a structure and then sequentially coupling the output to a thermal model of the structure for a subsequent heat transfer analysis. A problem arises when the analyst needs to model the deformation in the structure due to the thermal exposure, which would require re-meshing in the finite element analysis (FEA) setting and thus another coupling between the thermal and mechanical FEA models. However, employing thermo-mechanical shell elements allows the user to perform this coupled analysis in a single FEA model without re-meshing and by solving for temperatures and displacements simultaneously. The current study shows the accuracy and usefulness of this approach, combined with the fire-structure coupling methods developed in previous work by the authors, for realistic structural components. Implications of such an approach to the fire-structure problem are reduced computational demand and simplification of the FEA model development as the two field solutions may be compressed into one model.

INTRODUCTION

To better understand how structures respond under a wide range of possible fire hazards, there has been an increasing interest in modeling the thermal boundary conditions at the fire-structure interface using CFD software, such as Fire Dynamics Simulator (FDS) developed by NIST. For example, Zhang et al.[1] used adiabatic surface temperatures (AST) as the transfer data from FDS to an FEA model in ANSYS for thermo-mechanical analysis. Luo, Xie and DesJardin[2] provided a method for performing 2D fluid-structure coupling based on a level-set function approach, specifically for use in the fire-structure problem. Chen, Luo, and Lua[3] developed a two-way coupling approach with a customized version of FDS and linked it with the FEA software package Abaqus, which formed the basis of the Abaqus Fire Interface Simulator Toolkit (AFIST) developed by the company Global Engineering & Materials, Inc. Tondini, Vassart, and Franssen[4] employed a partitioned solution approach to the fire-structure analysis in their development of an interface between CFD fire simulation and FEA heat transfer models for applications to compartment fire models. Finally, Silva, Landesmann, and Ribiero[5] presented the Fire-Thermo-Mechanical Interface tool used for processing fire simulation results and transferring AST measurements to an FEA model.

The fact that researchers, including the authors of this current study, need to perform subsequent heat transfer and structural simulations is at least in part due to the fact that the FDS software for fire simulation is not typically employed for its conduction analysis in the solid domain. This results in users applying a wide range of assumptions regarding the representation of the boundary conditions (e.g., AST vs. incident heat flux vs. net heat flux, etc.), and the method for resolving the discrepancies in the mesh (e.g., sampling heat fluxes vs. averaging heat fluxes) and the time step.

Presently, it is not known what the ramifications of these assumptions are. Yu and Jeffers[6] showed that sampling boundary conditions in time can have a significant impact on accuracy due to heavy oscillations in heat fluxes that come from a CFD model, and Beata and Jeffers[7] similarly showed that sampling data in space can have consequences if the solid and fluid models have significantly different meshes. However, there are few "best practices" that have been identified in the literature for performing a coupled fire-structure simulation using a combined CFD-FEA approach. Moreover, the verification and validation of existing models is inconsistent, making reproducibility of results a major concern.

In general, the transfer of any CFD output representing the thermal boundary conditions as the input of the FEA model for heat transfer is a key aspect of providing an accurate characterization of the non-uniform thermal boundary observed in localized fire scenarios. At its most fundamental level, the fire-structure coupling analysis is a fluid-structure interaction (FSI) problem that is, for many practical applications, assumed to be a one-way coupling such that the structural deformation and the temperature field of the structure do not influence the fire simulation. Moreover, for the fire-structure simulation problem, each system is typically solved separately using what would be considered a partitioned approach in the general FSI context. Thus, the analysis consists of a fire simulation, which is conducted using CFD, and then a thermo-mechanical simulation of the structure, which may be conducted using FEA. In Fig. 1 below, a diagram of this process shows the individual analysis components.

Figure 1: Overview of the three-stage simulation process

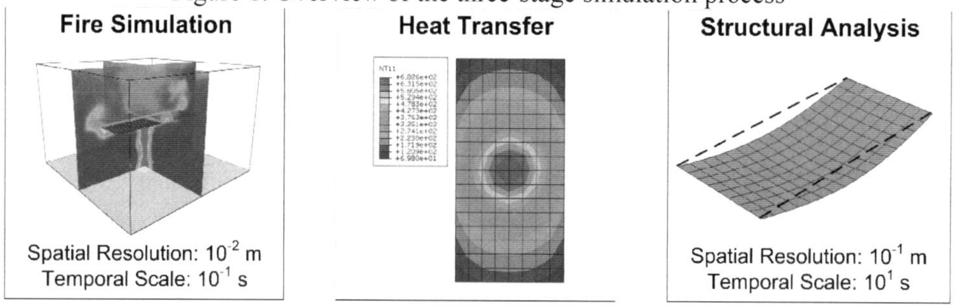

One aim of the fire-structure simulation problem is to characterize the thermal state of the structure as well as the deformation in an accurate and efficient procedure. For thin-walled structures such as floor slabs in buildings or structural panels in naval vessels, finite-element approaches based on shell theory are the preferred means to analyze the structural response. The benefit of shell elements from a computational perspective is the reduction of the degrees of freedom (DOF) in structures with larger planar dimensions than the thickness dimension. Shells overcome the computational restrictions posed by 3D continuum-based elements, which may demand an excessively fine mesh of elements to capture bending behaviours. A similar restriction has been encountered in the 3D conduction heat transfer analysis to evaluate the temperature field in the structure. Some researchers have used 3D continuum finite elements to calculate temperatures and then mapped the temperature field to the structural shell elements. Others (including the authors) have explored thermal shell formulations that capture the 3D conduction more efficiently.

In the present work, fully coupled thermo-mechanical shell elements are used to simultaneously solve the temperature and displacement fields using the same macro-element mesh (i.e., made up of thermo-mechanical shell elements) to model the coupled response of thin-walled structures. The conduction heat transfer analysis is performed using the thermal shell element that was previously developed by the authors[8]. This thermal shell element is capable of capturing the temperature variation through the thickness of the elements by resolving the thickness dimension into a collection of independent layers connected by shared temperature nodes. A piece-wise linear temperature field is defined in the thickness direction while the planar temperature field in a layer is interpolated from nodal temperatures. The theory presented by Kanok-Nukulchai[9] for a degenerated bilinear shell element was adapted here for the coupled temperature-displacement analysis. At present, the model is limited to small-displacement analysis and linear elasticity, although the model can readily be extended for nonlinear analysis, which is a necessary extension for structural fire applications. A selective reduced integration technique presented by Hughes[10] is employed for reducing the shear locking effects in shell elements.

The thermal shell element presented by Jeffers and Beata[8] and the classical mechanical shell element presented by Kanok-Nukulchai[9] were combined using a virtual work approach for finite-element coupled thermoelasticity in order to produce the element formulation discussed in this current study. The process for formulating the coupled-element system of equations was used to develop the non-symmetric element arrays of the coupled thermoelastic solid by starting from a temperature-dependent constitutive law. In this approach, layer-wise temperatures were considered DOF on the left-hand side of the system of equations as opposed to thermal strain forcing terms on the right-hand side. The coupled thermo-mechanical shell element was subsequently implemented in Abaqus as a user-defined element.

The virtual work approach to generate the coupled-element matrices was essential for solving for the temperatures and displacements simultaneously in the shell (i.e., in the same time increment of the analysis) using the non-symmetric matrix solver in Abaqus. The result was a fully coupled thermo-mechanical shell finite element with a bilinear temperature field in the plane of the element, a piece-wise linear temperature field through the thickness, a bilinear displacement field at the mid-surface, and containing four corner nodes while capable of any number of temperature layers. The total number of DOF is dependent on the number of layers selected for the analysis; the default value of five layers is typical. The transfer process to map nodal temperatures from the thermal model to mechanical model, as seen in other attempts, was eliminated with this solution to the coupled thermo-mechanical problem.

In addition, previous work by the authors which focused on coupling the fire simulation (fluid domain) to the structural model (solid domain) was updated for use in the thermo-mechanical shell element modeling approach[7]. Thus, earlier computing frameworks which focused on sequentially performing a three-stage analysis, namely the fire simulation, heat transfer, and structural analysis, as shown in Fig. 1, can now be reduced to a two-stage approach with the heat transfer and structural analyses coupled into the thermo-mechanical shell element model. The thermo-mechanical shell element, coupling approach, and examples are provided here to showcase the performance of this simulation approach.

METHODS

In the current study, two shell finite-element theories were coupled to provide a thermo-mechanical shell element formulation for use in linear thermoelastic analyses. The formulation of a fully coupled thermo-mechanical shell element was designed for 3D transient temperature-displacement analyses and was implemented as a user element in Abaqus. The main formulation for each of the individual shell theories used can be found in the literature. Specifically, the *thermal* shell involves a layered formulation for the 3D conduction analysis[8], while the original degenerated *mechanical* shell element is based on equivalent single layer (ESL) theory in which the element properties and arrays are computed only at the mid-surface of the element[9]. In the coupled formulation presented here, consideration was given for linking the various layer temperatures from the thermal shell to the mechanical shell. The coupling was based on the principle of virtual work and uses descriptions of the finite-element approximation of the shell geometry and DOF from each individual formulation directly.

The final coupled shell element employs a bilinear displacement field approximation, a bilinear temperature field within the plane of each layer, and a piece-wise linear temperature field through the element thickness for representing temperatures within the element domain in the mechanical stiffness equation. As a result, only four nodes need be defined by the user for each element and the calculation of nodal direction vectors was handled automatically in the Fortran implementation. Additionally, previous work related to the fire-structure coupling was referenced here to describe the process of transferring computed FDS results to the FEA model[7].

Thermal Shell Element

The thermal shell element[8] provides a finite-element approach for solving the 3D heat conduction equation within a solid body:

$$\frac{\partial}{\partial x}\left(k_x\frac{\partial T}{\partial x}\right)+\frac{\partial}{\partial y}\left(k_y\frac{\partial T}{\partial y}\right)+\frac{\partial}{\partial z}\left(k_z\frac{\partial T}{\partial z}\right)+Q=\rho c\frac{\partial T}{\partial t} \quad [1]$$

Here, k_x, k_y, and k_z represent the thermal conductivity in each principal direction; Q is the internal volumetric heat source; ρc is the material constant defining storage rate based on the product of density (ρ) and specific heat (c); and T is the unknown temperature field within the solid medium. Derivatives present in Eq. 1 are with respect to the global x, y, and z directions and also with respect to time t. The element was discretized into layers (Fig. 2a) in which the temperature in the thickness direction was lumped at each layer and the temperature in plane was approximated as bilinear or biquadratic. Using a combination of finite-element and control-volume approaches, the authors[8] showed that the governing equations for the thermal shell element could be expressed in the form

$$\mathbf{C}_T\dot{\mathbf{U}}_T+\mathbf{K}_T\mathbf{U}_T=\mathbf{R}_T \quad [2]$$

Here, $\mathbf{C}_T$ is the specific heat matrix, $\mathbf{K}_T$ is the total conductivity matrix, $\mathbf{R}_T$ contains the thermal forcing terms, and the array $\mathbf{U}_T$ is a vector that contains all the temperature DOF in the system. For the sake of brevity, the details of the formulation are not repeated here. In terms of input, the definition of the shell geometry requires only specification of the nodal coordinates at the mid-surface of the domain of the shell (Fig. 2b); the normal direction vectors (i.e., the vectors normal to the mid-surface) were computed internally in the Fortran implementation.

Figure 2: Thermal shell element (a) layered-temperature view and (b) single-node view of the thickness direction vector and mid-surface

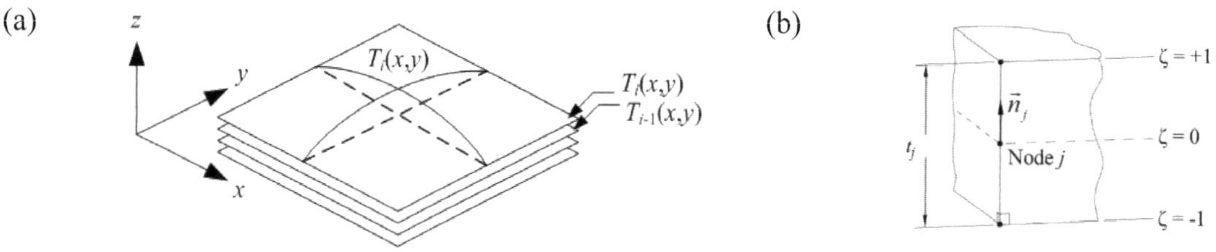

Thermoelastic Coupling Approach

For thermoelastic coupling, the formulation begins with a statement of virtual work for a solid deformable body:

$$\int \delta\boldsymbol{\varepsilon}^T\boldsymbol{\sigma}\,dV=\int \delta\mathbf{u}^T\mathbf{F}\,dV+\int \delta\mathbf{u}^T\boldsymbol{\Phi}\,dS \quad [3]$$

The expression given above includes both virtual displacements $\delta\mathbf{u}$ and virtual strains $\delta\boldsymbol{\varepsilon}$, signified by the δ present in each of these terms of the equality. The remaining components are the stresses $\boldsymbol{\sigma}$, the body forces **F**, and the surface tractions **Φ**. Left-hand-side terms include the single volume integral which is integrated over the whole domain of the elastic body. The right-hand-side also has one volume integral for the body forces terms as well as a surface integral for capturing the applied traction forces on the surface of the body. That surface is a particular subset of the full element domain on which a boundary condition, namely the traction forces **Φ**, has been specified.

The formulation of the coupled shell requires the linear-elastic constitutive law which takes into consideration thermal strains, partitioned into membrane and shear terms (in a local coordinate system at the individual element level):

$$\begin{Bmatrix} \sigma_{11} \\ \sigma_{22} \\ \sigma_{12} \end{Bmatrix} = \begin{bmatrix} D & \nu D & 0 \\ \nu D & D & 0 \\ 0 & 0 & G \end{bmatrix} \begin{Bmatrix} \varepsilon_{11} - \alpha\theta \\ \varepsilon_{22} - \alpha\theta \\ \varepsilon_{12} \end{Bmatrix} \Rightarrow \boldsymbol{\sigma}_m = \mathbf{D}_m(\boldsymbol{\varepsilon}_m - \boldsymbol{\varepsilon}_{th}) \qquad [4]$$

$$\begin{Bmatrix} \sigma_{13} \\ \sigma_{23} \end{Bmatrix} = \begin{bmatrix} fG & 0 \\ 0 & fG \end{bmatrix} \begin{Bmatrix} \varepsilon_{13} \\ \varepsilon_{23} \end{Bmatrix} \Rightarrow \boldsymbol{\sigma}_s = \mathbf{D}_s \boldsymbol{\varepsilon}_s \qquad [5]$$

Consistent with Kanok-Nukulchai[9], the σ_{33} stress term is restricted to zero and that is maintained through this constitutive law which specifically does not include that stress component in the above equations. In the expression of the stress-strain relation, material properties were considered uniform through the thickness for this initial development. Therefore, the constant D can be computed as $E / (1 - \nu)^2$ where E is the modulus of elasticity, ν is Poisson's Ratio, G is the shear modulus, and f is a shear correction factor of 5/6 which is the typical value for such shell elements. Note that the thermal strain term ε_{th} is a product of the coefficient of thermal expansion α and the temperature rise $\theta = T - T_0$, where T_0 is the reference temperature representing a state of zero thermal stress. Note that thermal expansion only affects the σ_{11} and σ_{22} stresses.

Definition of the Coupling Terms

The remainder of the formulation presented here seeks to establish the thermoelastic stiffness components that are produced by the temperature change. As the temperature change only appears in the σ_{11} and σ_{22} stresses, the thermoelastic coupling term can be isolated from the virtual work equation. In a typical finite element, strains **ε** are interpolated from nodal displacements **u** according to the strain-displacement matrix **B**, i.e., **ε** = **Bu**. In the original formulation by Kanok-Nukulchai[9], the strain-displacement matrix was decomposed into membrane $\mathbf{B}_m$ and shear $\mathbf{B}_s$ terms to simplify the notation. Keeping the original notation, the coupling term that arises from the substitution of the constitutive law in Eqs. 4 into the virtual work equation in Eq. 3 is:

$$-\int (\mathbf{B}_m^I)^T \mathbf{D}_m \mathbf{A} \theta \, dV^{(e)} \qquad [6]$$

The superscript I in Eq. 6 denotes the nodal contribution; thus, I ranges from 1, 2, …, n number of nodes. Note that the thermal strain has been rewritten as $\mathbf{A}\theta$ where $\mathbf{A} = [\alpha \ \alpha \ 0]^T$ and θ is the temperature difference (a scalar). The negative sign in the front of Eq. 6 is to preserve the sign of this term which originates on the left-hand-side of Eq. 3 and is a result of the constitutive expression which subtracts the thermal strains, as can be seen in Eq. 4.

Eq. 6 is separated into terms that appear on the left-hand side of the element equations (i.e., a coupling stiffness $\mathbf{K}_C$ multiplied by the nodal temperatures $\mathbf{U}_T$) and a forcing term on the right-hand side of the element equations including the reference temperature, which is expressed as $\mathbf{R}_0$. Writing the equation of motion for the mechanical shell element to include the coupling terms that arise from Eq. 6 gives:

$$\mathbf{K}_M^{IJ}\,\mathbf{U}_M^J + \mathbf{K}_C^{IL}\mathbf{U}_T^L = \mathbf{R}_M^I + \mathbf{R}_0^I \qquad [7]$$

Note that $\mathbf{K}_M$ is the elastic stiffness for the shell element, $\mathbf{U}_M$ is the vector of nodal displacements and rotations, and $\mathbf{R}_M$ is the vector of nodal forces. These terms remain unchanged from the original formulation by Kanok-Nukulchai[9] and are denoted by superscripts I and J, which each range from nodes 1, 2, …, n. The superscript L that is appended to the coupling terms corresponds to 1, 2, …, NL number of layers in the layered heat transfer element. Eq. 7 gives the governing finite-element equation for a deformable elastic body with a consideration for temperature variations through the volume of the body as provided by the thermoelastic constitutive law of Eqs. 4-5.

By bringing the thermal shell element (Eq. 2) together with the mechanical shell that includes thermoelastic coupling (Eq. 7), the full system of coupled equations for the thermo-mechanical shell element can finally be expressed in the following matrix-vector form:

$$\begin{bmatrix} [0] & [0] \\ [0] & \mathbf{C}_T \end{bmatrix}\begin{bmatrix} \dot{\mathbf{U}}_M \\ \dot{\mathbf{U}}_T \end{bmatrix} + \begin{bmatrix} \mathbf{K}_M & \mathbf{K}_C \\ [0] & \mathbf{K}_T \end{bmatrix}\begin{bmatrix} \mathbf{U}_M \\ \mathbf{U}_T \end{bmatrix} = \begin{bmatrix} \mathbf{R}_M + \mathbf{R}_0 \\ \mathbf{R}_T \end{bmatrix} \qquad [8]$$

The exact form of $\mathbf{K}_C$ and $\mathbf{R}_0$ arises from the assumptions that are embedded in the thermal and mechanical shell formulations. As these terms are unique to the coupled formulation, they are described here in more detail. Recall α, D, and v were defined previously; they remain as constants in the present work and thus can be factored out of the volume integral in Eq. 6; these terms are represented by the scalar $\kappa = -\alpha D(1 + v)$ in Eq. 9 below. Moreover, an isoparametric finite element approach is used to map the element from global (x, y, z) coordinates to a tri-unit domain in natural coordinates (ξ, η, ζ) using the Jacobian transformation $|\mathbf{J}|$:

$$\kappa \int_{-1}^{1}\int_{-1}^{1}\left\{\int_{-1}^{1}\begin{bmatrix} \left(\mathbf{B}_{1m}^I\right)^T \\ \zeta\left(\mathbf{B}_{3m}^I\right)^T \end{bmatrix}\begin{bmatrix} 1 \\ 1 \\ 0 \end{bmatrix} T\,|\mathbf{J}|\,d\zeta\right\} d\xi\, d\eta \qquad [9]$$

Three main features should be noted from Eq. 9:

1. The remaining 3×1 array, namely $[1\ 1\ 0]^T$, in the integrand effectively sums columns one and two of its pre-multiplier, the $\mathbf{B}_m$ matrix, which has been split into its two sub-components using the following relation to explicitly factor out the ζ component: $\mathbf{B}_m = [\mathbf{B}_{1m}\ \ \zeta\mathbf{B}_{3m}]$.
2. The fiber integral (i.e., the integral for through-thickness integration) was highlighted using the curled braces above as this integrand is important for handling the layer discretization.
3. The temperature DOF will be introduced through the T variable above in the next section; as it appears here for now, T is simply the continuous field variable $T = T(\xi, \eta, \zeta)$ in the shell element domain: the finite-element interpolation has not been introduced yet in this presentation.

To arrive at the final form for $\mathbf{K}_C$, the temperature field approximation from the thermal shell element[8] is introduced in place of the continuous temperature field variable T in Eq. 9. The nodal temperatures $\mathbf{U}_T$ are then factored out of the integrand, leaving $\mathbf{K}_C$ remaining. For simplicity, $\mathbf{K}_C$ is conveniently partitioned as follows: $\mathbf{K}_C = [\mathbf{X}\ \ \mathbf{Y}]$, where the IK^{th} contribution of the two sub-matrices $\mathbf{X}$ and $\mathbf{Y}$ are given as:

$$\mathbf{X}^{IK} = \kappa\Delta \int_{-1}^{1}\int_{-1}^{1} \boldsymbol{\beta}_1^I N^K \left|\mathbf{J}\right| d\xi\, d\eta \quad [10]$$

$$\mathbf{Y}^{IK} = \frac{\kappa\Delta}{2} \int_{-1}^{1}\int_{-1}^{1} \boldsymbol{\beta}_3^I N^K \left|\mathbf{J}\right| d\xi\, d\eta \quad [11]$$

The notation of $\boldsymbol{\beta}^I$ signifies the product of $\mathbf{B}_m \cdot [1\ 1\ 0]^T$ for each sub-matrix of the $\mathbf{B}_m$ matrix. Δ is the thickness of a single lamina, which is assumed constant. N^K is the Kth interpolation function for interpolating the temperature field. The K superscript ranges from 1, 2, …, n nodes and enters Eqs. 10-11 by inserting the basis function for interpolating temperatures within a thermal layer L. Summation over the K nodes is implied by this expression resulting in the full $\mathbf{K}_C\mathbf{U}_T$ product. Thus, these left-hand-side terms of the governing equation serve to pre-multiply the temperature DOF of each thermal layer, as in Eq. 8; this fact completes the governing equation for coupled thermoelasticity and provides a unified, simultaneous solution of the temperature and displacement fields in the solid body.

Fire-Structure Coupling

To handle the transfer of temporal data from the FDS fire to the FEA model comprised of thermo-mechanical shell elements, the methods by Yu and Jeffers[6] were employed to use a subcycling technique allowing multiple steps of the fluid domain solution to be time-averaged and passed into the solid domain. Specifically, the net heat flux along a structured grid in the FDS model was passed from the fire domain to the FEA model containing the structure.

Handling non-uniform boundary conditions (such as a net flux) over the surfaces of solid structures can be accomplished within the FEA solution framework once the computed surface fluxes are transferred from the fire simulation and mapped to the FEA mesh. Numerical integration is the basis for computing finite-element arrays, both left-hand-side and right-hand-side terms, in any FEA simulation. Traditionally, Gaussian quadrature has been used to compute element arrays by integrating through finite-element volumes and surfaces. Gaussian quadrature is very accurate for the integration of smooth functions and especially polynomials, using a small number of sampling points over an element surface and/or through the volume.

However, for the case of highly non-uniform surface conditions, as in the case of fire simulations, other methods exist which may be more appropriate for handling discretely measured data points as opposed to smooth functions; for example, Beata and Jeffers[7] handled integration of element arrays at the interface with the trapezoid rule for numerical integration by using as many available discrete data points as were available at the element boundary. The trapezoid rule approach uses additional integration points in the element surface integrals to compute equivalent nodal fluxes from the non-uniform fluxes available at the surface. This approach was employed for the fire-structure simulations presented here.

RESULTS

To demonstrate the capabilities of the fire-structure coupling framework using the thermo-mechanical shell element, a series of tests have been included here. First, the thermo-mechanical shell element was used in a thermoelastic verification problem. Second, the shell and fire-structure coupling methods were used to model a flat plate structure exposed to a localized fire simulated by FDS. Finally, the third case demonstrates the ability of the coupling procedure to incorporate multiple interfaces, as in the case of the I-beam model presented here.

Verification of Thermo-mechanical Shell

The first example using the thermo-mechanical shell element is a verification problem for computing the stress in a hollow cylinder subjected to a higher temperature (200 °C) on the inside surface and a lower temperature (100 °C) on the outer surface. These conditions on each of the surfaces were provided by specifying a temperature boundary at the surface of the shell finite element. The test was defined in the Abaqus Verification Guide (Section 1.3.17) [11]. In Fig. 3 below, the full cylindrical structure is shown and the single-element mesh required for this benchmark is also highlighted as only one element was used for this test. Additionally, a steady state condition was assumed according to the documentation.

Figure 3: Verification test for computing the thermal stress in a hollow cylinder

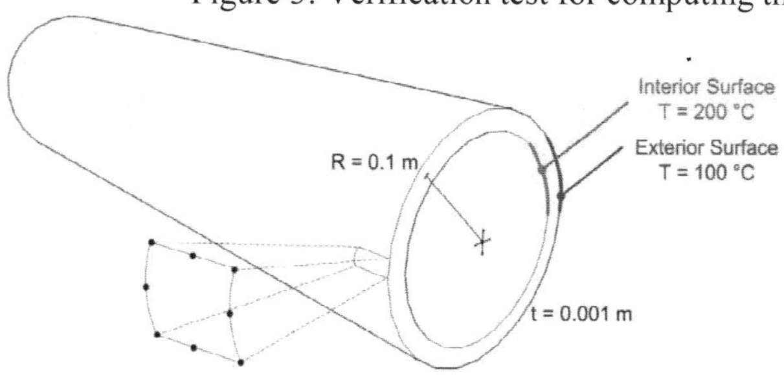

Density:	7800 km/m^3
Conductivity:	52 W/(m·K)
Specific Heat:	586 J/(kg·K)
Expansion Coef.:	1.2×10^{-5} / °C
Young's Modulus:	2.0×10^{5} MPa
Poisson's Ratio:	0.3
Reference Temp.:	0 °C

The size of the element used in this test corresponds to one which subtends an angle of 11.25° at the center of the cylindrical shell opening; thus it is consistent with a scenario in which 32 elements compose the circumferential direction. Proper boundary conditions were specified for the stress analysis such that the rotations about the circumferential direction were constrained but the cylinder was allowed to freely expand along its axial dimension, in accordance with the original test.

A reference solution was given in the Abaqus Verification Guide for this problem[11]. The theoretical stress at the outer and inner surfaces of the cylindrical structure should be ±171.43 MPa, with the sign depending on the surface. Results computed by Abaqus using the Fortran implementation of the proposed thermo-mechanical shell element matched this reference solution exactly. In fact, using the analytical solution for this problem to compute the stress with more significant digits, the thermo-mechanical shell element demonstrated sufficient accuracy, showing an absolute error of $O(10^{-13})$ when comparing computed stresses with the analytical solution.

Application with FDS Data: Flat Plate

With the successful verification of the thermo-mechanical shell element in the previous verification test, the Fortran implementation of this element was upgraded to handle non-uniform thermal boundary conditions consistent with previous work by the authors but now in the setting of coupled temperature-displacement analyses using Abaqus. For this application, a flat plate was exposed to a localized fire computed using FDS; specifically, FDS was used to compute a non-uniform surface flux field which was applied as a net flux boundary condition in the FEA model on the top and bottom surfaces of plate.

All the properties are defined in Fig. 4 below and the definitions of the boundary conditions for the top and bottom surfaces of the plate come from the computed net flux results from FDS. The "edges" of the plate were thermally insulated; this is signified by the shaded markings around the perimeter of the plate in Fig. 4 as well. Note that the displacement boundary conditions represent a simply supported restraint: pinned connections and roller connections at the mid-surface along the full 1.0-m widths at opposite ends of the plate, respectively.

Figure 4: Test setup for the flat plate exposed to a local fire in FDS

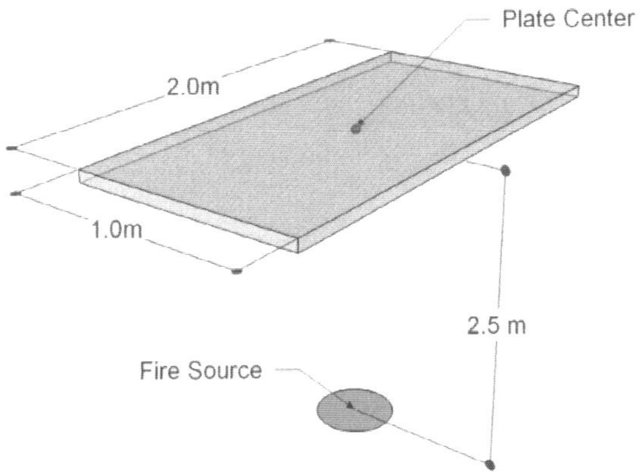

Density:	2000 km/m³
Conductivity:	2 W/(m·K)
Specific Heat:	1000 J/(kg·K)
Expansion Coef.:	1.0×10^{-5} / °C
Young's Modulus:	2.0×10^{4} MPa
Poisson's Ratio:	0.2
Reference Temp.:	20 °C
Plate Thickness:	5 cm
Heat Release Rate:	500 kW
Burning Area:	0.16 m²
Fire Duration:	10 min
FDS Cell Size:	10 cm

Employing the custom user element in Abaqus containing the Fortran implementation of the thermo-mechanical shell element, a coupled temperature-displacement analysis was carried out. The analysis used the fire-structure coupling approaches described previously to map the measured fluxes from the FDS simulation to the FEA model for each time increment (using bilinear interpolation at integration points on the shell element surfaces). Note that a time step size of 10 sec was chosen for the data transfer process from the FDS to the Abaqus model based on the temporal subcycling methods presented earlier.

A comparison of the temporally varying results for the plate center location are shown next (i.e., at the center location in the plane of the plate, as displayed in Fig. 4 above). First, Fig. 5 shows the temperature-time results when using four different meshes in the FEA model. Nodal temperatures at the center of the plate were recorded at the bottom surface, which was facing the fire in the FDS model, and the middle surface, which is the center of mass of the plate structure. Figure 6 shows the corresponding transverse displacements of the plate center with respect to time.

Figure 5: Mid-surface and bottom-surface temperature of the plate center point over time

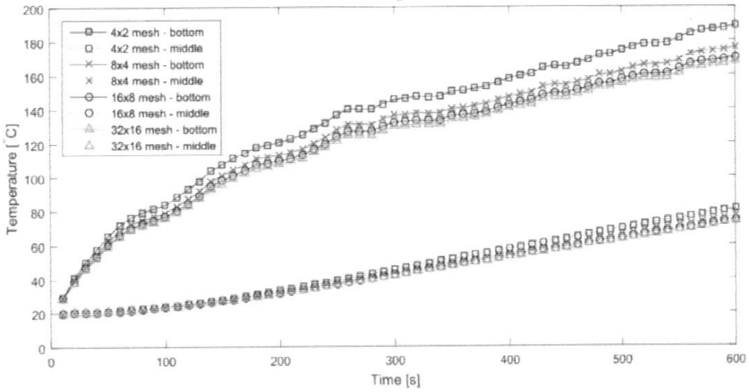

Figure 6: Transverse displacement of the plate center point over time

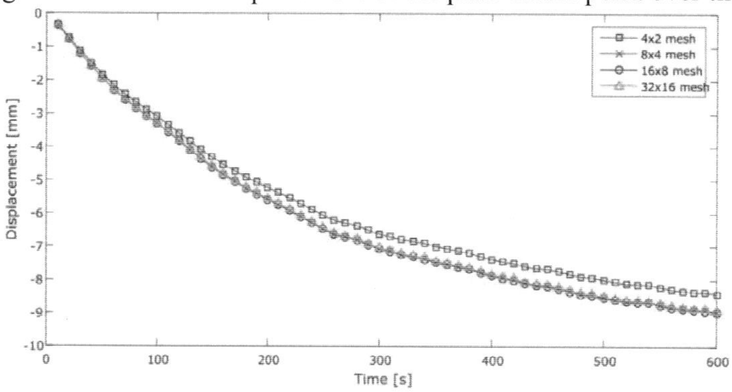

The following contour plots demonstrate the computed results in the plane of the plate at the final time step. In Fig. 7, the mesh for each test is shown followed by the final temperature state and displacement field in the mid-plane of the plate once the simulation had completed.

Figure 7: Results from the plate fire test using FDS data; (a) four mesh configurations used; (b) contour plots of the final temperature state (in °C) at the mid-surface of the plate with maximum and minimum highlighted; (c) corresponding transverse displacement at the final time step (in mm), with maximum.

(a)

(b)

(c)

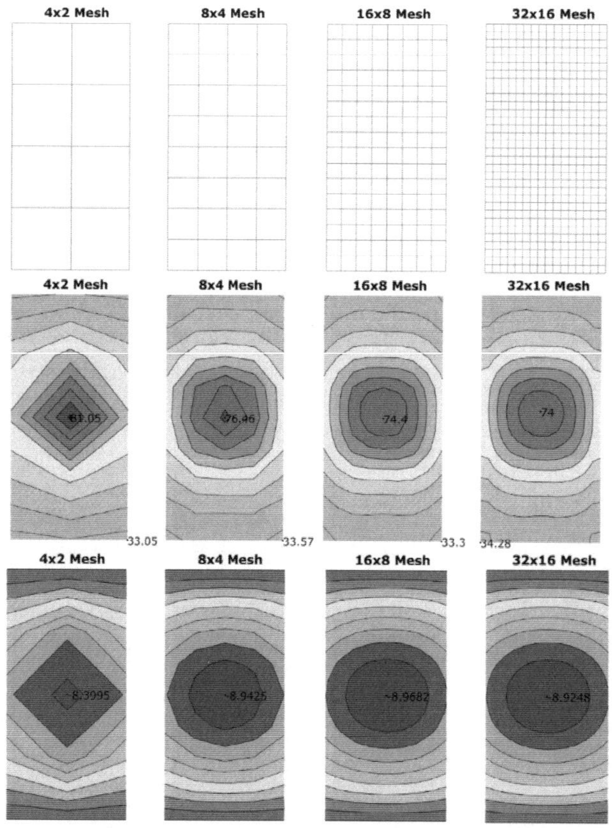

In the contour plots of Fig. 7, the bilinear interpolation of the coarser meshes (such as 4×2 and 8×4) is fairly noticeable as large linear segments can be seen in the temperature and displacement plots for those meshes. As desired, using a higher mesh resolution improves the smoothness of the contour plot.

Simulation times for these models were recorded in two parts: first, preprocessing time of the FDS data and second, the actual coupled fire-structure thermo-mechanical simulation. The preprocessing step, much like the fire simulation, was only needed one time; this is one benefit of the coupling procedure used which allows various FEA meshes to map the FDS output data individually. Preprocessing performed by MATLAB required only 1.5 sec for this test. The time required for each transient temperature-displacement simulation in Abaqus was very reasonable; for increasing mesh resolution, the simulation times were 2.8, 4.8, 14.8, and 56.1 sec, respectively.

Application with FDS Data: I-Beam Model

For the final application, the fire-structure coupling method was extended to handle multiple interfaces, as in the case of the I-beam scenario shown in Fig. 8 below. This problem presents the need to map fluxes measured in the FDS fire simulation to multiple surfaces of the beam flanges and webs. The particular shape selected for this test was a $W18 \times 35$ wide-flange steel section (converts to $W460 \times 52$ for the metric version of this section). Dimensions and properties were defined in Fig. 8 and the fire source was modeled using FDS just as in the plate fire problem previously, with the fire source offset from the beam center, as depicted in the figure. Additionally, at the mid-span of the beam, a concentrated load of 3.0 kN was applied. The beam was simply supported with a pinned connection at one end and a roller at the other, as in the previous application.

Figure 8: Test setup for the I-beam exposed to FDS fire

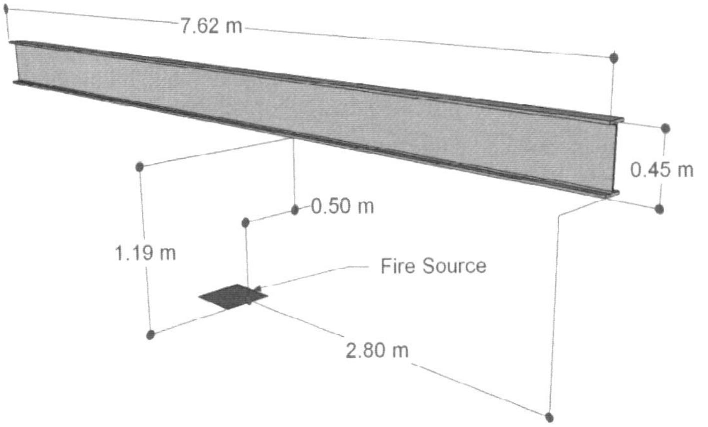

Density:	7833 km/m^3
Conductivity:	54 W/(m·K)
Specific Heat:	465 J/(kg·K)
Expansion Coef.:	1.0×10^{-5} / °C
Young's Modulus:	2.0×10^{5} MPa
Poisson's Ratio:	0.2
Reference Temp.:	20 °C
Flange Thickness:	10.8 mm
Web Thickness:	7.6 mm
Heat Release Rate:	110 kW
Burning Area:	0.16 m^2
Fire Duration:	10 min
FDS Cell Size:	10 cm

The purpose of this test was to demonstrate the ability of the thermo-mechanical shell element in conjunction with the fire-structure coupling techniques to model a realistic structural component (namely, a wide-flange beam) exposed to a localized fire simulated by FDS. For example, Fig. 9 shows the final temperature field of the beam with the localized heating accounted for in the region nearest the fire. Additionally, the displacement field is displayed next to it in the same image. Note that in this exploratory example the 3-kN load applied at the mid-span of the beam had a more significant effect on the final displacement state.

Figure 9: Results for the mid-surface temperature (left) and transverse displacements (right) in the beam model at the end of the fire event

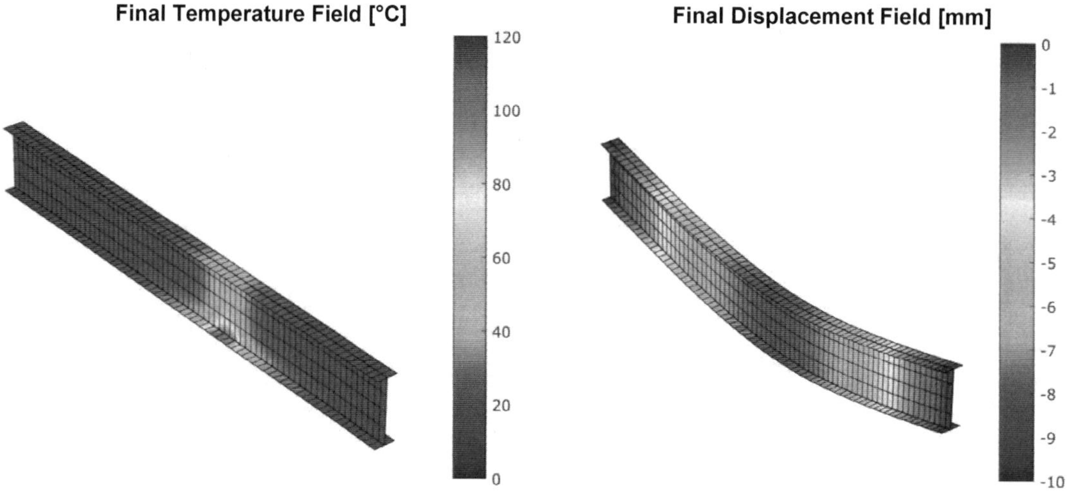

One of the benefits of this thermo-mechanical shell analysis combined with the fire-structure coupling methods is the ability to perform the heat transfer and structural analysis using the same mesh and in a single analysis (i.e., a single coupled temperature-displacement analysis). In the case of this beam application, the analyst may wish to refine (or coarsen) the mesh in order to investigate the convergence of their finite-element model. It is important to note that the FDS fire simulation need only be completed one time for a given CFD mesh and the data can subsequently be mapped to the finite element model. For this particular mesh of 576 elements and nearly 6,000 thermal and mechanical DOF, less than two minutes (105 sec) were needed for analyzing the response to this set of data from the FDS simulation using a single computing workstation.

CONCLUSIONS

In order to improve the state of fire-structure simulation methods, an approach using coupled thermo-mechanical shell elements was presented in the current study. By developing the finite-element equations for a coupled thermo-mechanical shell element and subsequently implementing the formulation as a Fortran subroutine, the presented shell element showed beneficial results from a computational modeling perspective. The thermal stress cylinder problem demonstrated the accuracy of the coupled formulation; the plate exposed to the local fire showed how the shell could be efficiently linked with fire simulation data; and finally the beam problem provided a practical application of the methods for a typical structural component. While this approach has shown promising early results, additional consideration and future research will be needed to apply these methods in scenarios with more structurally significant fire exposure. For such cases, the analyst must consider the effects of temperature-dependent material properties and the potential for large displacements or buckling.

REFERENCES

[1] Zhang, C., Silva, J., Weinschenk, C., Kamikawa, D. and Hasemi, Y. (2015). Simulation Methodology for Coupled Fire-Structure Analysis: Modeling Localized Fire Tests on a Steel Column. Fire Technology, 52(1), pp.239-262.

[2] Luo, C., Xie, W. and DesJardin, P. (2009). Fluid-Structure Simulations of Composite Material Response for Fire Environments. Fire Technology, 47(4), pp.887-912.

3 Chen, L., Luo, C. and Lua, J. (2011). FDS and Abaqus coupling toolkit for fire simulation and thermal and mass flow prediction. In: Proceedings of the IAFSS 10th International Symposium on Fire Safety Science, International Association for Fire Safety Science.

4 Tondini, N., Vassart, O. and Franssen, J. (2016). Development of an interface between CFD and FE software. In: Proceedings of the 7th International Conference on Structures in Fire.

5 Silva, J., Landesmann, A. and Ribeiro, F. (2016). Fire-thermomechanical interface model for performance-based analysis of structures exposed to fire. Fire Safety Journal, 83, pp.66-78.

6 Yu, X. and Jeffers, A. (2013). A comparison of subcycling algorithms for bridging disparities in temporal scale between the fire and solid domains. Fire Safety Journal, 59, pp.55-61.

7 Beata, P. and Jeffers, A. (2015). Spatial homogenization algorithm for bridging disparities in scale between the fire and solid domains. Fire Safety Journal, 76, pp.19-30.

8 Jeffers, A. and Beata, P. (2014). Generalized shell heat transfer element for modeling the thermal response of non-uniformly heated structures. Finite Elements in Analysis and Design, 83, pp.58-67.

9 Kanok-Nukulchai, W. (1979). A simple and efficient finite element for general shell analysis. International Journal for Numerical Methods in Engineering, 14(2), pp.179-200.

10 Hughes, T. (2000). The finite element method. 1st ed. Mineola, NY: Dover Publications.

11 Abaqus (2013). Abaqus Documentation, Dassault Systèmes, Providence, RI, USA.

FDS-FEM SIMULATION METHOD FOR STRUCTURAL FIRE EIGINEERING DESIGN: VALIDATION AND A CASY STUDY

Chao Zhang[1,2] & Hong-Xia Yu[3]
[1]National Institute of Standards and Technology, MD 20899-8666, USA
[2]Tongji University, Shanghai, 200092, China
[3]Arup International, Beijing, China

ABSTRACT

The Fire Dynamics Simulator (FDS) is a large-eddy simulation (LES) code, used in fire engineering for modelling the gas phase environment (temperature, heat flux, velocity, species concentrations, etc.) in fires. This paper discusses the application of the FDS for structural fire engineering analysis and design. In particular, a FDS-finite element model (FEM) coupling method was used to model a localized fire test on a real-scale steel beam recently conducted at the National Fire Research Laboratory (NFRL) of the National Institute of Standards and Technology (NIST). The concept of adiabatic surface temperature was used for transferring data from FDS to FEM. A case study on using FDS for structural fire engineering design in an airport project at Beijing China was also presented.

INTRODUCTION

The engineering design practice in China and other countries usually includes fire safety design of steel structures in large volumes. Prescriptive method in fire code usually assumes confined combustion and is not applicable to large volumes. Fire engineering method, as an alternative, is usually adopted and the Fire Dynamics Simulator (FDS) is a numerical tool used in fire engineering design. FDS is a free and open source software developed by the National Institute of Standards and Technology (NIST) for fire related simulations [1]. It has been used in industrial projects for modeling fire spread and smoke movement in sport stadiums, transportation terminals, and atriums in large buildings. In recent years, there has been increased research in the application of FDS for structural fire analysis [2]. Fire-structure interface tools for transferring data from FDS to particular FEM (finite element method) codes (such as ANSYS, ABAQUS, SAFIR) have been developed. However, there is a lack of validation of FDS for structural fire analysis, mainly due to the limited available test data on structures exposed to realistic fires.

The reasons why lacking test data of structures in realistic fires might be: (1) realistic (compartment) fire tests are difficult to control due to the uncertainties of many parameters affecting the fire behavior, which include fire load, combustion, ventilation, compartment size and geometry, and thermal properties of compartment boundaries [3]; and (2) realistic fire tests require unique testing facilities and instruments, e.g. cone calorimeter is needed to measure the heat release rate which is the most important parameter to characterize the severity of a real fire [4]. Most of the structural fire tests reported in literature are conducted in an enclosed furnace and some are conducted in real compartments (e.g., the Cardington full-scale fire tests [5]) but without measuring the heat release rate. Well controlled, detailedly measured realistic fire test data are essentially required for validating analytical methods and numerical tools for scientific fire safety design.

The National Institute of Standards and Technology added a unique facility named the National Fire Research Laboratory (NFRL), which will serve research needs of studying the response of real-scale structural systems to realistic fires as well as other fire research topics such as advanced firefighting technologies, engineered fire safety, material flammability reduction and wildland-urban interface fire research [6]. The unique facility enables large-scale experiments using fires up to 20 MW and would contribute to the technical basis for performance-based design methodologies for structures exposed to fire. As part of commissioning the structural-fire test capabilities at the NFRL, a series of tests were conducted on loaded structural steel beams exposed to a localized fire [7]. This paper presents the integrated fire-structure simulation of one of those tests. In particular, FDS was used to simulate the localized fire behavior and the thermal boundary conditions at the exposed surfaces of the test specimen, and the FEM program ANSYS (version 16.1) was used to do the heat transfer-thermo-mechanical analysis to get the temperature and structural responses. A case study on using FDS in structural fire engineering design was also provided.

INTEGRATED FIRE-STRUCTURE SIMULATION APPROACH

Fire Dynamics Simulator is a large-eddy simulation (LES) code [1]. LES is a technique used to model the dissipative processes (viscosity, thermal conductivity, material diffusivity) that occur at length scales smaller than those that are explicitly resolved on the numerical grid. In FDS, the combustion is based on the mixing-limited, infinitely fast reaction of lumped species, which are reacting scalars that represent mixtures of species. Thermal radiation is computed by solving the radiation transport equation for gray gas using the Finite Volume Method (FVM) on the same grid as the flow solver. FVM is based on a discretization of the integral forms of the conservation equations. It divides the problem domain into a set of discrete control volumes (CVs) and node points are used within these CVs for interpolating appropriate field variables. The governing equations are approximated on one or more rectilinear grids. Obstructions with complex geometries are approximated with groups of prescribed rectangles in FDS. One-dimensional (1D) heat conduction is assumed for solid-phase calculations. Note that a three-dimensional heat conduction feature is under development in FDS [8].

FDS currently has no solver for solid mechanics and coupling methods between FDS and other codes with solver for solid mechanics (usually FEM) are used to do fire-structure simulations. Principally, there are four FDS-FEM approaches that may be used in a structural fire engineering design of steel structures:

- Approach 1: FDS models with including steel members are used. The steel temperatures predicted by FDS are imported to FE structural models.
- Approach 2: FDS models with omitting steel members are used. The fire (gas) temperatures predicted by FDS are used in simple thermal models (e.g. the Eurocode 3 equations) to calculate the steel temperatures. The calculated steel temperatures are imported to the FE structural models.
- Approach 3: FDS models with including steel members are used. The thermal boundary conditions are imported to FE thermal models to do three dimensional (3D) heat conduction analyses. The steel temperatures predicted by the FE thermal model are mapped to the FE structural models.
- Approach 4: FDS models with including steel members are used. The real time thermal boundary condition data predicted by FDS are imported to the FE thermal model to predict the steel temperatures which are then mapped to the FE structural models to predict deformations, strains, stress, etc. The deformations are transferred and used to update the geometry of the FDS model for the next time increment calculation.

FDS currently includes an output quantity of adiabatic surface temperature (AST) calculated according to the concept proposed by Wickstrom [9]. Consider an ideal adiabatic surface exposed to a heating condition; the net heat flux to the surface is by definition zero, thus

$$\varepsilon_{AS}(\dot{q}''_{in} - \sigma T_{AS}^4) + h_{c,AS}(T_g - T_{AS}) = 0 \tag{1}$$

where ε_{AS} is emissivity of the adiabatic surface; T_{AS} is temperature of the adiabatic surface or adiabatic surface temperature; $h_{c,AS}$ is film coefficient between the adiabatic surface and the surrounding gas; T_g is the surrounding gas temperature; and σ is the Stefan-Boltzmann constant. From Eq. (1), the incident radiative flux to a surface can be calculated from an adiabatic surface temperature,

$$\dot{q}''_{in} = \frac{h_{c,AS}(T_{AS} - T_g)}{\varepsilon_{AS}} + \sigma T_{AS}^4 \tag{2}$$

Consider a real surface exposed to the same heating condition, the net heat flux to the surface can be calculated by

$$\dot{q}'' = \varepsilon_s \sigma (T_{AS}^4 - T_s^4) + \frac{\varepsilon_s}{\varepsilon_{AS}} h_{c,AS}(T_{AS} - T_g) + h_c(T_g - T_s) \tag{3}$$

If the emissivity of the adiabatic surface is taken as the emissivity of the real surface ($\varepsilon_{AS} = \varepsilon_s$), and the film coefficient between the adiabatic surface and the surrounding gas is equal to the film coefficient between the real surface and the surrounding gas ($h_{c,AS} = h_c$), we get

$$\dot{q}'' = \varepsilon_s \sigma (T_{AS}^4 - T_s^4) + h_c(T_{AS} - T_s) \tag{4}$$

Eq. (4) shows that the net heat flux to a surface can be approximately calculated by using a single parameter T_{AS}. Obviously, when exposed to a same heating condition, the surface temperature of a real body cannot exceed the surface temperature of a perfect insulator or AST.

MODELING A LOCALIZED FIRE TEST ON A STEEL BEAM

Experimental description

Figure 1a shows a photograph of the considered localized fire test on a steel beam conducted at the NFRL. The structural testing system composed of the W16x26 beam specimen, reaction frames, and HSS (hollow structural section) loading beams. All the components of reaction frames were designed in accordance with the ANSI/AISC-360 (2010) [10]. The length of the beam specimen was 6.17 m. The beam specimen was simply supported with the bearing-to-bearing length of 5.87 m. The beam specimen was placed at 1.67 m above the strong floor and were loaded by means of two HSS loading beams placed at 1.22 m away from the center of the specimen. The ends of the beam specimen and the regions in contact with the HSS loading beams were laterally and rotationally restrained. The lengths of the HSS loading beams were 6.71 m. The lengths of the cover plates welded to the HSS section were 3.81 m. Two holes were dug on the HSS tube to apply pull force through the high-strength robs. The distances from the holes to the center of the loading beam were 2.74 m. The fuel delivery system consisted of two natural gas burners with a nominal flame zone of one square meter. The distance from the lower flange of the beam specimen to the strong floor was 1.6 m. The assembled burner was placed 1 m below the lower flange of the specimen. Figure 1b shows the heat release rate (HRR) of the assembled burner for the considered test.

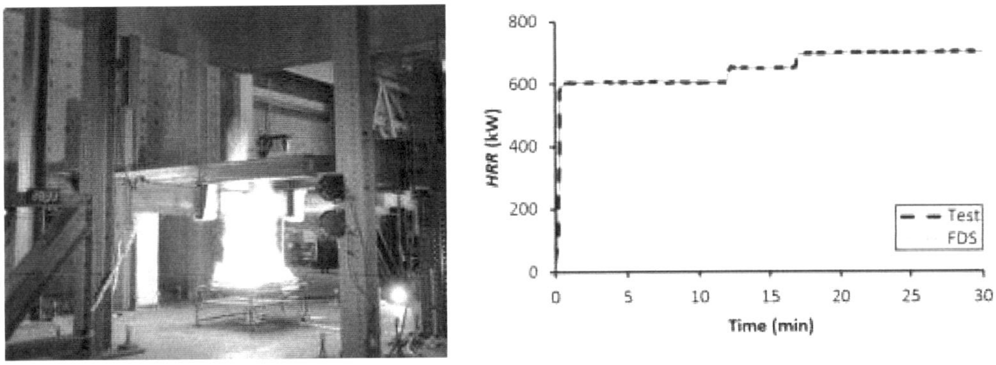

Figure 1 A photograph of the considered localized fire test (left) and the *HRR* curve used (right)

Numerical model

Figure 2a shows the FDS numerical model for the localized fire test. The computational domain had dimensions of 7.2 m (X) × 1.2 m (Y) × 3.6 m (Z). The side walls and top boundary of the domain were assumed to be open during the fire test. The grid size used is an important numerical parameter in CFD (computational fluid dynamics) because of its impact on numerical accuracy. Sensitivity analyses were conducted to determine a proper mesh so that a balance between simulation time and numerical accuracy could be achieved. For a computational time of 1843 seconds, the running for the FDS model uses 12 CPU cores in the FDS simulation cluster maintained by NIST and takes 5 day 1 hours and 8 minutes (physical time). Information about the sensitivity analyses and detailed description of the FDS numerical model are reported in [11] and are not presented here for brevity. Figure 2b shows the FEM structural model for the test setup. The beam specimen and the loading beams were meshed with the structural shell element SHELL181 in ANSYS. The FEM thermal model only included the beam specimen meshed with the thermal shell element SHELL131 in ANSYS and was not presented. The thermal properties specified in the Eurocode 3 [12], the thermal expansion coefficient recommended by NIST TN 1681 [13] and the stress-strain model for structural steel developed by Luecke et al. [14] were used in the FEM numerical models. The Luecke et al.'s model was selected among several models available for its better representation of the tensile coupon test data [15]. The measured yield strength for the steel of the beam specimen is 440 MPa. A detailed description of the FEM models is reported in [11] and are not presented here for brevity.

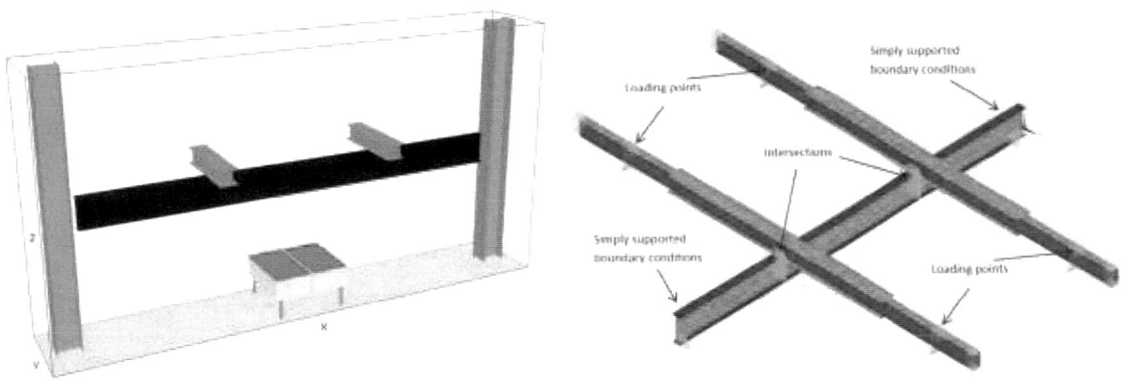

Figure 2 FDS (left) and FEM (right) numerical models

Numerical results

The beam test was conducted in two steps as follows: i) The HRR of the burner was increased to a target magnitude of 700 kW and maintained constant throughout the test (Fig.1). ii) After the maximum temperature at the fire-exposed cross sections reached the steady-state condition, two point loads were programmed to increase at a nominal rate of 2 kip/min (8.90 kN/min) simultaneously until failure occurred.

Figure 3a-c shows the numerical results for steel temperatures at the mid-section of the beam specimen. The measured temperatures were also plotted for comparison. The lower flange (LF) and upper flange (UF) temperatures are taken as the average temperatures measured by thermocouples placed at symmetric positions (Fig.3d). The plotted web temperatures were the temperatures at the mid of the web. The predicted bottom flange and the web temperatures agreed well with the test data, while the predicted top flange temperature was slightly higher. The differences among the predicted and measured maximum temperatures for bottom flange, web, and top flange were 635 °C - 654 °C = -19 °C (-2.9%), 397 °C - 437 °C= -40 °C (-9.4%), and 411 °C - 350 °C = 61 °C (+17.4%), respectively. Fig.4 compares the predicted and measured flame behaviors.

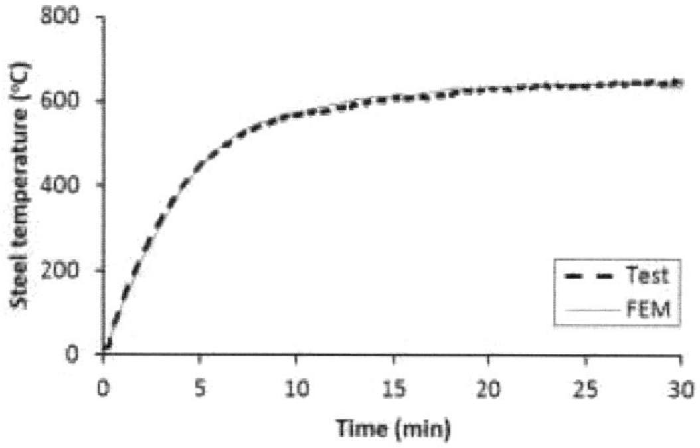

(a) bottom flange temperature

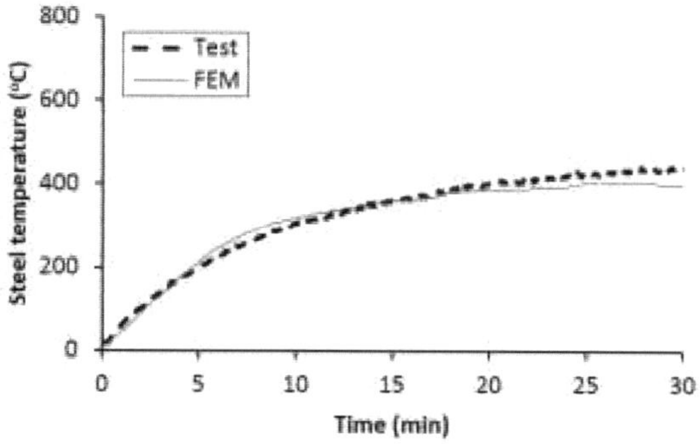

(b) web temperature

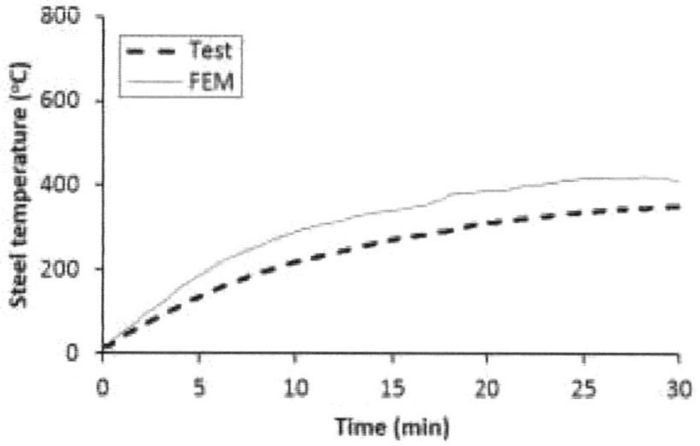

(c) top flange temperature

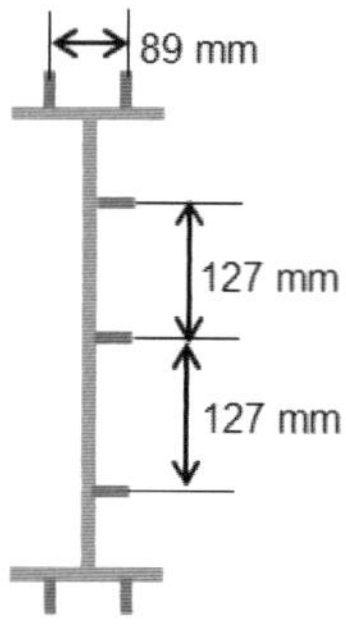

(d) location of thermocouples

Figure 3 Comparison between numerical results and test data

Figure 5a compares the observed and predicted deformations of the beam specimen. Fig.5b compares the measured and predicted deflections at the mid-section of the beam specimen. The beam specimen failed by lateral torsional bucking at 30.6 mins after ignition and the measured peak reaction force was 100 kN. The predicted failure time was 30.8 mins and the predicted peak reaction force was 109 kN.

Figure 4 Predicted and measured flame behavior

(a)

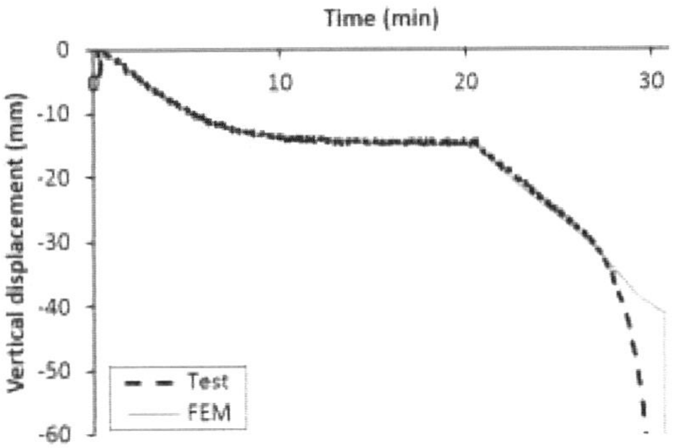

(b)

Figure 5 Numerical results for the localized fire test on a steel beam

CASE STUDY

Project overview

As shown in Fig. 6, the project was an airport terminal in Beijing, China with a site area of 70 ha. The building includes five above-ground floors, where only F1 and F2 occupies nearly the whole site area. F3 to F5 are raised floors with significant recession from the floor below. A giant roof was constructed to cover the entire building and created a huge indoor open space below the roof. In the central region, the roof was supported mainly by eight C-shaped columns with a span of over 100 m between the columns. The roof adopted a special grid system, which gradually mixed with the columns seamlessly.

In this case study, a fire was assumed to be located around a C-shaped column, and some of the column members are engulfed by the flame. The relative location of the fire and the column is shown in Fig. 7, where the cylinder shows the envelope of the flame. The fire size was assumed to be 20 MW and according to the SFPE handbook, the flame diameter is 3.6 m and the flame height is 5 m.

Figure 6 Bird-eye view of the airport terminal

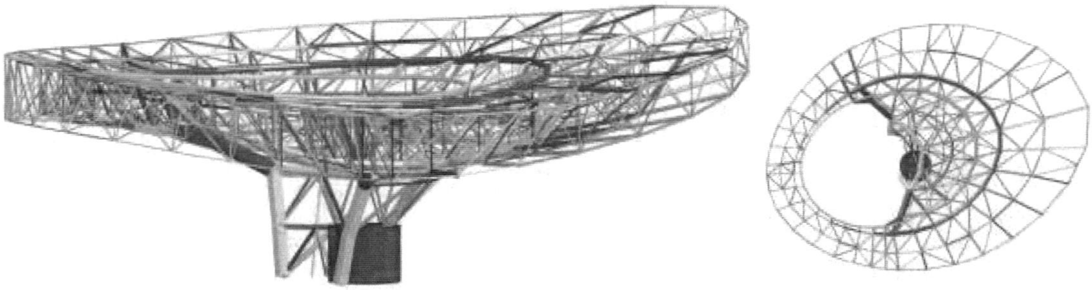

Figure 7 Location of the fire relative to the C-shaped column

Structural fire engineering Design

Design based on empirical correlations

An approach based on empirical equations are developed to calculate the temperature of the steel members when subjected to the localized fire. This approach was based on the theory proposed in the Eurocode 1[16] for external fire, but with the flame shape replaced by that of a localized fire.

According to the Eurocode 3 [12], the increase of the temperature of a steel member when subjected to incoming heat flux can be calculated according to:

$$\Delta T_s = \frac{A_m/V}{c_s \rho_s} \dot{q}_{net} \Delta t \tag{5}$$

where A_m/V is the shape factor of the member; c_s and ρ_s are the specific heat and density of steel; $\dot{q}_{net}$ is the incoming heat flux.

The heat flux includes heat transfer by radiation and convection. Here, radiation considers the radiation from the flame and the convection is the heat transfer between the steel member and its surrounding air.

Assuming at any time t, the flame temperature is T_f, the air temperature is T_g and the steel temperature is T_s, the incoming heat fluxes by radiation ($\dot{q}_{net,r}$) and convection ($\dot{q}_{net,c}$) are

$$\dot{q}_{net,r} = \Phi \cdot \sigma \cdot \left[(T_f + 273)^4 - (T_s + 273)^4 \right] \tag{6}$$

$$\dot{q}_{net,c} = h_c \cdot \left[T_g - T_s \right] \tag{7}$$

For simplicity, the flame temperature is assumed to be constant at 834 °C. The air temperature at the location of interest can be calculated according to the empirical equations by the SFPE handbook [16].

To calculate the configuration factor, the fire source was simplified as a cylinder with its side surface and the top surface as radiation panels. The radius R and height H of the cylinder are the radius and height of the fire plume calculated according to SFPE handbook [17]. The fire radiation model is shown in Fig.8, where the view factor for any location of the steel member can be calculated from its relative location to the fire source (D_h, D_v). The view factor will be the sum of the view factors from all the surfaces that are viewable from the steel member. The top surface is simplified as a square plane of $2R \times 2R$, which is perpendicular to the receiving surface. To consider the radiation from the side surface, a discretization method was employed, with the viewable surface divided into a number of finite plates, each with a width of $R \cdot d\theta$, and its relative angle to the receiving surface θ. Assuming there are n plates on the side surface, the total configuration factor for the steel section will be

$$\Phi = \Phi_{top} + \sum_{i=1}^{n} \Phi_i \tag{8}$$

The view factor from any radiating plate to a receiving surface can be calculated according to the Eurocode 1 [16].

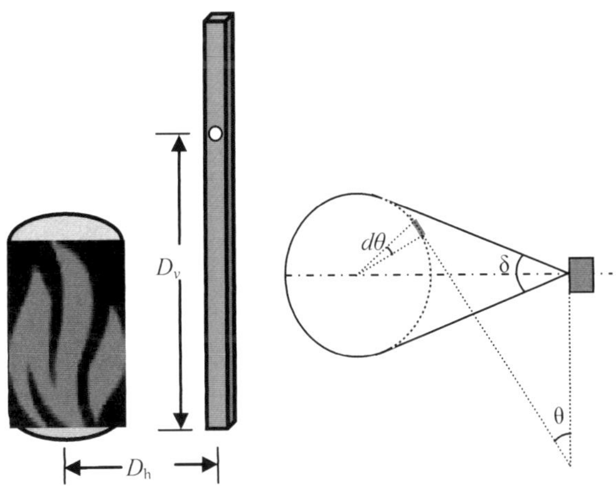

Figure 8 Fire radiation model

Temperature of the steel structures calculated from equation (5) is shown in Fig.9 when the fire duration is taken as 90 min. The structural response was then analyzed using finite element program LS-DYNA. During the analysis, mechanical load was applied and maintained constant. The temperature loading was then gradually applied. At 90 min, the structure shows no significant deformation or no signs of failure. Plastic strains were observed on members that are directly engulfed by flame as shown in Fig.10.

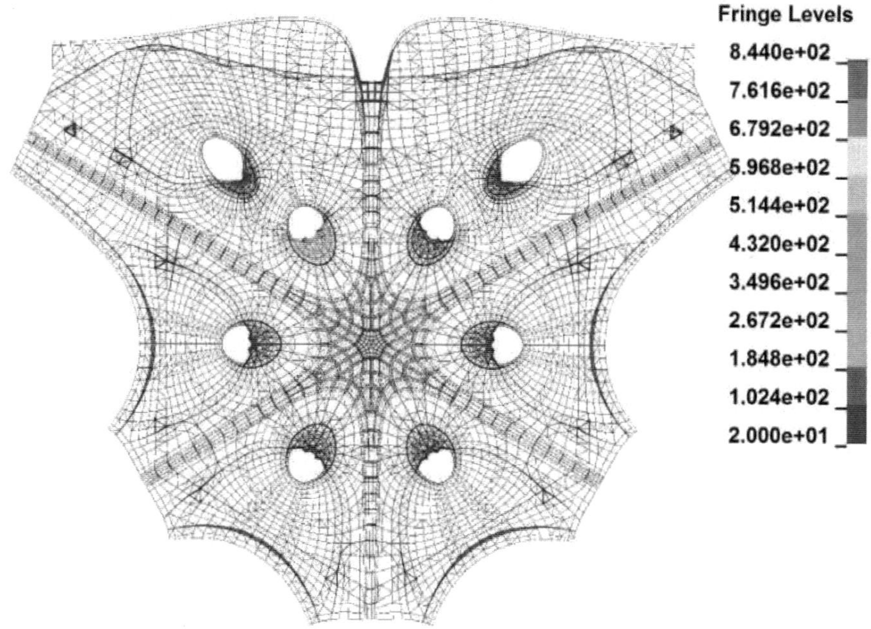

Figure 9 Temperature of the structure from empirical equations

Figure 10 Plastic deformation according to temperature from empirical equation

Verify design by FDS

The sophisticated fire model, FDS, was used to check the design based on empirical correlations. Fig.11 shows the FDS numerical model of the C-shaped column analyzed in the previous section. A square fire source with the same heat release rate as in the empirical formula calculations was considered. Fig.12 shows the distribution of ASTs predicted by FDS. The maximum AST on the exposed steel members is about 540 °C which means, the simplified approach has been over-conservative, by 1) assuming the flame temperature is uniform at 834 °C and by 2) assuming the steel temperature equals to the flame temperature around it. As discussed earlier, when exposed to a same heating condition, the temperature of a steel member cannot exceed the AST. Therefore, the design based on the empirical correlation is conservative.

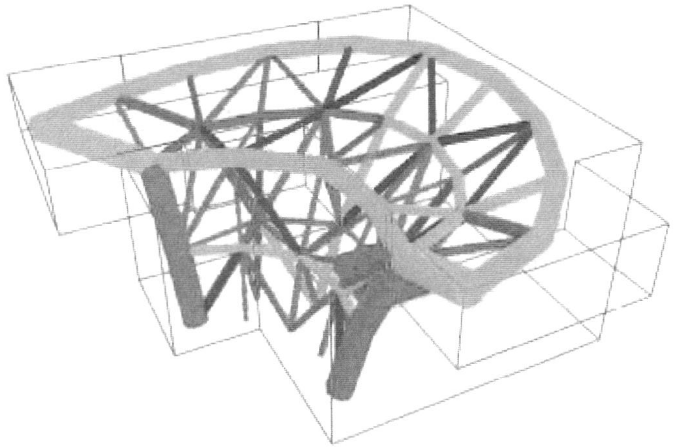

Figure 11 FDS numerical model for a C-shaped column adjacent to a fire source

Figure 12 FDS predicted AST distributions

CONCLUSION

The structural fire safety design practice is moving from prescriptive codes toward performance-based fire safety design, which requires sophisticated numerical models to adequately predict the performance of structures under realistic fire scenarios. Engineering structures in many situations are subjected to low fire risk, typically unlikely to go to flashover. In those cases, with fire risk analysis and subsequent structural fire engineering analysis, a large amount of fire protection may be saved. This paper applied a FDS-FEM coupling approach to successfully predict the fire-thermal-structural behaviors of a steel beam exposed to combined structural load and localized fire. The ratios of difference between predicted and measured values for failure time, failure temperature and failure load can be as small as 0.7% (0.2 min), 2.9% (19 °C) and 9% (9 kN), respectively. The FDS-FEM approach uses general models (e.g. material models) and assumptions (e.g. initial imperfection assumptions), which is a general approach that can be used to study the complex behavior of structures in realistic fires. Besides, using the concept of adiabatic surface temperature, the case study demonstrates the application of FDS to verify the structural fire engineering design based on empirical correlations.

Disclaimer Certain commercial entities, equipment, or materials may be identified in this document in order to describe an experimental procedure or concept adequately. Such identification is not intended to imply recommendation or endorsement by the National Institute of Standards and Technology, nor is it intended to imply that the entities, materials, or equipment are necessarily the best available for the purpose.

Authors' contributions This paper represent a true joint effort between the two listed authors. Yu wrote the case study section and Zhang wrote the remaining sections.

REFERENCES

[1]McGrattan K., Hostikka S., McDermott R., Floyd J., Weinschenk C., Overholt K. Fire Dynamics Simulator, User's Guide. National Institute of Standards and Technology, Gaithersburg, Maryland, USA, and VTT Technical Research Centre of Finland, Espoo, Finland, sixth edition, September 2013.

[2]Zhang C., Silva J.G., Weinschenk C., Kamikawa D., Hasemi Y. Simulation Methodology for Coupled Fire-Structure Analysis: Modeling localized fire tests on a steel column. Fire Technology, 2016;52:239-262.

[3]Drysdale D. An introduction to fire dynamics. 2nd edition. John Wiley and Sons, 1999.

[4]Babrauskas V., Peacock R.D. Heat release rate: The single most important variable in fire hazard. Fire Safety Journal, 1992;18:255–272.

[5]Kirby B.R. Large scale fire tests: the british steel european collaborative research programme on the bre 8-storey frame. Fire Safety Science, 1997;5:1129-1140.

[6]Bundy M., Hamins A., Gross J., Grosshandler W., Choe L. Structural fire experimental capabilities at the NIST national fire research laboratory. Fire Technology, 2016;52:959-966.

[7]Choe L., Ramesh S., Zhang C., Gross J.L.. The performance of structural steel beams subjected to a localized fire. Proceedings of the 9th International Conference on Structures in Fire, July 8-10, Princeton, USA, 2016.

[8]Zhang C., McDermott R., Yu H.X., Choe L., Vanella M., Forney G., Gross J. (2016b). "Fire dynamics simulations for structural fire analysis: engineering demand and recent development.", Fire Safety Science. (under review).

[9]Wickstrom U., Duthinh D., McGrattan K.B. Adiabatic surface temperature for calculating heat transfer to fire exposed structures. In Proceedings of the 11the International Interflam Conference, pages 943-53, London, England, 2007.

[10]AISC. Steel Construction Manual, 14th edition. American Institute of Steel Construction (AISC), 2010.

[11]Zhang C. Integrated fire-structure simulation methodology for predicting the behavior of structuresin realistic fires. Structure Congress, Denver, Colorado, 2017.

[12]BSI. Eurocode 3: Design of steel structures- Part 1-2: General rules-Structural fire design, British Standard Institution, 2005.

[13]Phan L.T., McAllister T.P., Gross J.L., Hurley M.J. Best practice guidelines for structural fire resistance design of concrete and steel buildings. NIST technical note 1681, National Institute of Standards and Technology, Gaithersburg, Maryland, 2010

[14]Luecke W.E., Banovic S.W., McColskey J.D. High-temperature tensile constitutive data and models for structural steels in fire. NIST technical note 1714, National Institute of Standards and Technology , Gaithersburg, Maryland, 2011

[15]Choe L., Zhang C., Luecke W.E., Gross J.L., Varma A.H. Influence of material models on predicting the fire behavior of steel columns.Fire Technology, 2016.

[16]BSI. Eurocode 1: Actions on structures- Part 1-2: General actions-Actions on structures exposed to fire, British Standard Institution, 2002.

[17]SFPE Handbook of Fire Protection Engineering, Third edition, 2002.

ENGINEERED WOOD PRODUCTS

LAMINATED VENEER LUMBER PLATED CONNECTIONS IN FIRE

Arlin Otto [i], John Gales [ii], and George Hadjisophocleous [iii]
[i] Research Student, Architectural Conservation and Sustainability Engineering, Carleton University, Canada
[ii] Assistant Professor in Fire Safety Engineering, Carleton University, Canada
[iii] Professor in Fire Safety Engineering, Carleton University, Canada

ABSTRACT

Society is constantly being pushed in a more sustainable direction, which has resulted in substantial advances in Engineered Wood Product (EWP) technologies in the construction industry. Engineered wood products have many benefits including a high strength to weight ratio, carbon sequestration as well as ease of construction as many EWP systems are prefabricated. The purpose of this research was to produce material data sets for future modelling of Laminated Veneer Lumber (LVL) to aid in a more robust numerical modelling effort of connection systems. The research was performed at small scale using a Cone calorimeter and LIFT apparatus and material focused. Additional to this testing, advanced instrumentation was utilised which considered imaging technology utilising narrow-spectrum illumination to selectively filter out the light from the flame in order to study the underlying charring mechanisms associated to flame spread. The technology was found to be promising. Preliminary conclusions show that external connections may perform better than internal connections with short term heating applications.

INTRODUCTION

Engineered wood products have become a competitive option worldwide due to emerging technology and building code advances. Laminated Veneer Lumber (LVL) is a common engineered wood product due to its strength in comparison to conventional dimensional lumber. As building codes allow for larger scale structures to be constructed out of wood, the use of LVL is expected to increase. In standard building applications with LVL members, connections are commonly fabricated out of steel. This steel plating is present either externally surrounding the member, or internally in a concealed-type connection. The former is the focus of this study. The behavior, both structurally and thermally, of these composite-type connections when exposed to fire is relatively unknown [1]. For this reason, there is a need to research and for implementation of fire-safe and resilient LVL structural components usage in construction industry. This paper begins to investigate the thermal response of the interface between the steel and LVL elements through focus on the development of charring through flame spread and concentrated heating. The research provides a necessary prerequisite for a more complete structural and study of connections in fire.

BACKCRGOUND AND MOTIVATION

With the development of mid-rise (6+ storeys) and high-rise (8+ storey) timber construction in Canada, there has been a surge in supporting research focused on assessing how timber behaves in real fires [2]. At the time of publication, the tallest timber building to be constructed in Canada was Brock Commons at the University of British Columbia. The building is an eighteen story hybrid mass timber structure built primarily out of engineered wood products (EWP) or commonly referred to as engineered timber. Engineered timber being high quality timber glued using adhesives to build up a larger section. The construction methodology is sustainable in the sense that mass timber is not needed as trees can be grown more quickly, harvested responsibly and built up into a larger and more defect free structural member. This building in British Columbia has pushed the boundaries for mass timber construction and code compliance through the alternative solutions measures in the Canadian building code. To enable alternative solutions in future timber infrastructure for fire design, significant modelling and testing may be required.

Figure 1: Brock Commons Building in British Columbia

The bulk of testing in Canada performed for alternative solution development has focused on engineered timber testing with standard fire furnace tests [2] and comparison with more conventional materials like steel and concrete. While these standard tests in Canada, have been informative for purposes of generating valuable information towards the development of novel alternative solution development for floors, walls and beams, they have not necessarily improved our understanding of the underlying mechanisms and theory for which can govern engineered timber in fire – particularly the role of connections. This theme has been identified as a research need and interest by many international research bodies [3, 4].

The underlying 'thermo- physical' mechanisms in particular to fire spread and char formation under non-standard formations are critical when considering how a real fire interacts with timber connections, particularly those with external plated connections. The addition of steel and associated heat transfer only complicates these analyses.

The research herein involves testing LVL sections using a Lateral Ignition and Flame Spread Test (LIFT) apparatus [5] and Cone Calorimeter [6] to study flame propagation and heat penetration through and beneath the steel plates used commonly in connections. This testing style was deemed appropriate in light of more difficult to perform standardized fire testing. These tests were performed in initiating construction of a robust numerical model for evaluating timber connections for future work which can be later used in analyzing full-scale tests, or within possible performance based analyses – the data can provide a basis for future model verification and validation.

The cone calorimeter tests were used to quantify the effect of the steel plate on pyrolysis and charring. Alternatively, the LIFT apparatus was used to quantify the influence of the geometry of the steel plate section on retarding a spreading flame as may be seen in real construction. These apparatuses were located at Carleton University Campus Fire Safety Engineering Lab.

Novel to this study is the testing of a new Digital Image Correlation technique which was adapted within the LIFT apparatus testing herein [5, 7]. As the envelope for timber construction is pushed, in particular with the emerging EWP technologies, our current methodology for testing and understanding these assemblies in fire needs to keep pace with these innovations.

METHODOLOGY

A combination of testing was performed on LVL samples, primarily in two phases. First to consider a controlled heat source and study char formation followed by a controlled flame spread and study of its interaction with the timber material. These two phases were chosen as they can represent timber material assemblies in heat exposures similar to real fires. LVL samples were prepared with and without steel plating screwed into the surface at a sufficient depth. All LVL samples had a moisture content between 5.5 and 5.9% by mass.

Quantification of Char

For both phases, the char depth of these test specimens were determined visually by measuring the unheated material thickness from the samples after the applied heat flux and comparing this to the original sample thickness. This technique ensures that the char depth is measured accurately as the mass loss that occurs may make visual estimation of char on its own challenging.

Controlled Heat Flux Testing

The first phase of this project was to study samples of LVL with and without steel plating to determine the influence of the steel plates on char and pyrolysis depth. See Figure 2. This was completed using a Cone Calorimeter to induce a constant and incident heat flux on the samples at fixed durations of time typical to previous testing by others for comparison. Standard incident heat fluxes of 30kW/m^2 and 50kW/m^2 were used to simulate two severities of heat fluxes.

Figure 2: Plated and Un-plated Sample LVL sections

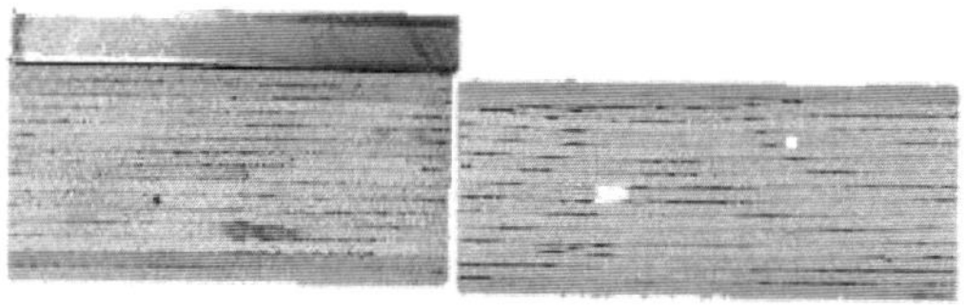

The LVL for these samples was obtained from a manufacturer in Ottawa, Ontario. The adhesive used in this LVL is proprietary to the company. This is quite common with EWP products. A series of 44.5mm by 240.0mm Grade 2.0E single ply LVL beams were acquired and cut down to 100mm x 100mm x 44.5mm test samples. The samples have a coating on the exterior faces for weathering, which is green in colour, and this coating was left on for the testing. See Table 1 for test schedule.

Table 1. Test Schedule for Applied Heat Flux Testing

Sample	Plate Thickness [mm]	Applied Heat Flux [kW/m^2]
A30-C-1, A30-C-2, A30-C-3	n/a	30
A30-4.7-1, A30-4.7-2	4.7	30
A30-9.5-1, A30-9.5-2	9.5	30
A30-12.7-1, A30-12.7-2	12.7	not tested
A50-C-1, A50-C-2, A50-C-3	n/a	50
A50-4.7-1, A50-4.7-2	4.7	50
A50-9.5-1, A50-9.5-2	9.5	50
A50-12.7-1, A50-12.7-2	12.7	50

The samples were tested for 10 minutes at either 30kW/m^2 or 50 kW/m^2 incident heat flux in accordance to Table 1. The time is common in other cone testing by other researchers and can also be used for comparison with other EWP testing done by the authors and collaborators. The specimens were prepared to standardized test procedures where adjacent faces were covered in foil to reduce any subsequent charring which could influence the recorded char depth along the sides among other functions. After testing, the samples were cut so that the char depth in the center could be measured, as the worst case char depth is assumed to would be located in the center of the sample. This area will be exposed to the most uninterrupted heat flux. The samples 12.7 mm samples were not tested at the 30kW/m^2 as the 9.5 mm plated samples would later show no observable char or pyrolysis formation for the given test times.

Flame Spread Heat Flux Testing

The LIFT apparatus is being used for phase two of the project to understand the behavior under a defined length of steel used in a LVL external plated connection (see Figure 3). The test program for Phase two is designed to study the minimum length required in an external style connection to disrupt flame spread across (or beneath) connections in the case of a primary structural system of timber. This experimental plan was to observe if a severe localized fire could spread over the connection and to study how the char and flame spread front develop for this material.

In consideration of typical conventions in practice, external steel plates varied in thickness and length. This variability in plate thickness allowed for the consideration of geometry effects of external connections to either impede flame as well as to study heat conduction through a composite connection.

Figure 3: Typical steel tie plate style external connection

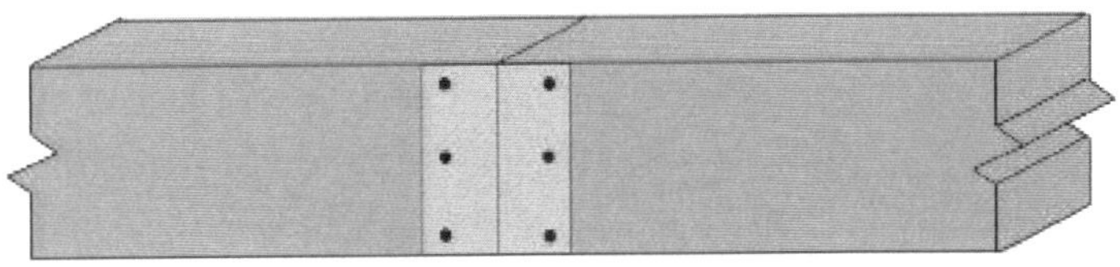

In this sense, the pyrolysis and charring of the LVL under the steel plates is also studied herein. The char depth was measured using the same process as the control heat flux and the samples were cut down the centerline to measure the depth, as per the standard ASTM 2013 E1321 [5]. These tests are underway at the time of publication.

Table 1. Planned LIFT Test Schedule

Sample	Plate Thickness [mm]	Plate Length [mm]
B-C-1	-	-
B-C-2	-	-
B-C-3	-	-
B200-4.7-1	4.7	200
B300-4.7-1	4.7	300
B400-4.7-1	4.7	400
B200-9.5-1	9.5	200
B300-9.5-1	9.5	300
B400-9.5-1	9.5	400
B200-12.7-1	12.7	200
B300-12.7-1	12.7	300
B400-12.7-1	12.7	400

Testing using the LIFT deviated slightly from the ASTM standard as it was performed for research purposes to investigate common external connections in such mass timber applications it was deemed an acceptable deviation.

An imaging technique, using narrow-spectrum illumination and bandpass optical fitering, developed at the National Institute of Standards and Technology (NIST) National Fire Research Laboratory (NFRL) was used for this set of tests using the LIFT apparatus (see [8] for previous application where objects engulfed in clean-burning (natural gas) diffusion flames were observed. This test series is ongoing as the technique is still being refined. The technique aims to allow heated LVL timber to be viewed with reduced obstruction by flames induced by the radiant heater of the LIFT apparatus. A Light Emitting Diode (LED) light that emitted light in a narrow spectrum near the bottom of the visible range (approximately 450 nm) was used to illuminate the sample. The LED light had sufficient power (95 watts) to overcome the radiation in this portion of the electromagnetic spectrum emitted by the flames. A bandpass optical filter that allowed illumination from the LED light to pass, but filtered out electromegatic radiation above and below the target frequence was adapted to fit the lenses on a Canon EOS 5Ds Mark III DSLR camera. This camera has a 50 Megapixel resolution (images of 5792 x 8688 pixels) which allows it to be placed farther away from the sample during testing. This placement eliminates any risk of damage to the camera, lenses and parts due to heating and radiation from the LIFT apparatus. The resolution of the camera still provides detailed resolution of the timber material while it is flaming. To assess the technology an 'adaptive' flame spread analysis was performed by the authors which considered correlation between sequenced images by color intensity change being the marker for char front development (not flame spread). This was opposed to the traditional image correlation measurement through photo sequencing which relies on overall pixel movement [6]. The set up can be viewed in Figure 4.

Figure 4: LIFT Apparatus and Imaging Setup in Carleton Campus Fire Lab

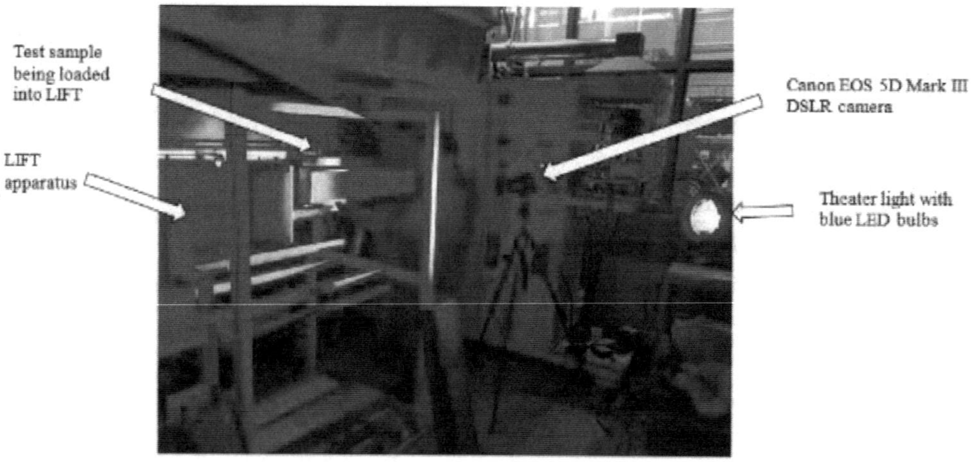

RESULTS AND DISCUSSION

The results presented herein are preliminary as the authors are currently developing additional data sets and expanding and refining their test procedures to holistically consider this topic and to have adequate data to enable model verification and validation.

Controlled Heat Flux Testing

The standardized char rate for LVL members is 0.65mm/min which is similar to other EWPs [2]. Averaged char rates herein were found slightly in excess to these for even low non-standard heat flux exposures but comparable to those seen by the authors for other short term heat exposures [9]. This may be of consequence of the lower moisture contents observed for the LVL samples tested (~5%) and of the use of non-standard heat fluxes outside standard fire testing (ie., not the E119 fire), and smaller developed char layering for low heating times of 10 minutes.

Figure 5: Plate thickness versus Char Depth and Mass Loss for 10 minute exposure

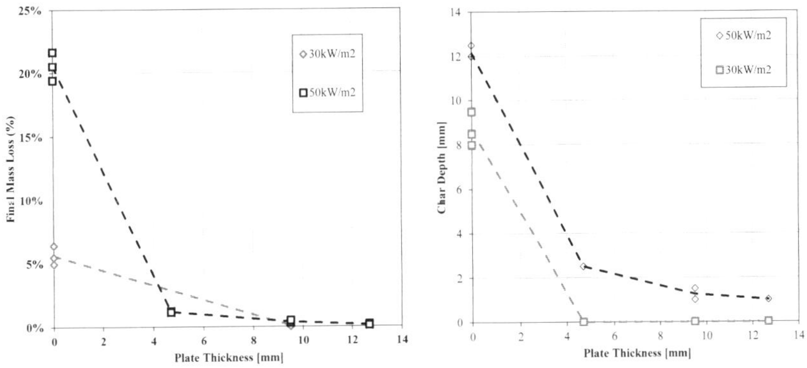

By increasing the thickness of the plate there was a quantifiable difference in depth of charring. The addition of a 4.7mm plate to the LVL caused a 19% mass loss reduction in the 50kW/m^2 heat flux case. The use of external plating in the 30kW/m^2 case has a lessened impact overall as the mass loss was roughly a quarter of that in the 50kW/m^2. The mechanisms are deemed to be related to steel reflectivity, heat transfer and lack of oxygen all contributing to a slowing of pyrolysis and progression to char. This data shows that the relationship between mass loss and heat flux is not linear at least for the testing conducted to date. Utilizing these material trends and developing associated computational tools, as is being considered in industry, could aid in creating better connections. As connections are often the weakest building component in fire, this consideration of the underlying thermal mechanics of the composite materials heightens the overall performance of the building.

The data seems to suggest that structural stability on the basis of the minimal char alone is likely to remain in the connection for the test configuration configured. Its representation of reality though can be open for discussion. Additional testing at different test durations would be essential in order to develop more comprehensive models. The data does suggest that external connections would perform better than an internal connection because the steel may delay the heat transfer to the timber.

Flame Spread Heat Flux Testing

The LIFT apparatus was used to study the influence of steel plating on flame spread therefore an 18-minute basis was used for the testing. Eighteen minutes was chosen arbitrarily as this was the time for the fire to spread during the first unexposed test from one end to the other. As these tests are underway, caution is given to the reader in formulating specific conclusions on the data presented particularly in reference to adhesive performance. Though preliminary insights are considered herein.

The behaviour of plated LVL sample was unique, as the progression of charring along the length of the timber was reduced. As can be seen below in Figure 6, the black-hashed line represents the start point and end point of the steel plating. The top sample has a reduction in char depth directly before the start of the steel plate and the initial char under the plate is reduced by approximately 60% compared to the uncovered samples. The behaviour observed reflects the idea of the steel plate working as a heat sink. This comparative test showed that the sample stopped charring approximately 150mm into the steel plate and that the plate was successful at inhibiting flame spread with this length.

The imaging technique developed at NIST was used on the control tests and showed promising results for the filtration of the flame front in the testing. Figure 7 demonstrates the technique results in the first control tests. The image on the left shows an overall view of the test setup, which was taken with a non-lensed camera, and the flame can be observed. The second image (on the right) shows an image taken from the double-lensed camera where the flame is filtered out successfully.

Figure 6: Char Depth of Sample Compared to Induced Heat Flux

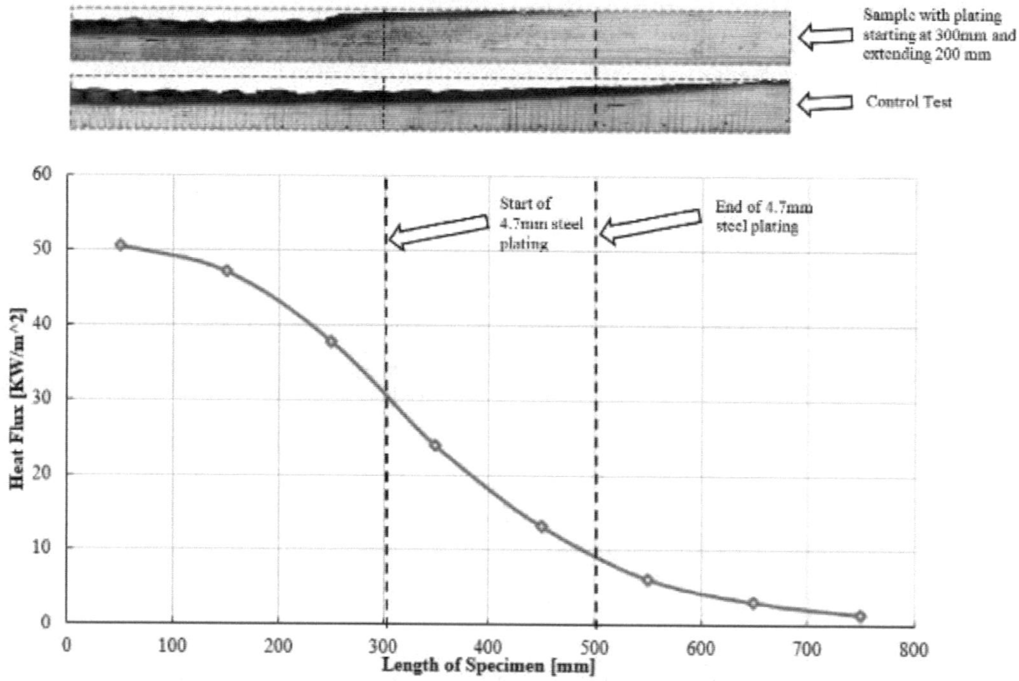

Figure 7: LIFT Testing and Imaging Technique Result

By using narrow-spectrum illumination and optical filtering, the char front progression could be directly observed, rather than the traditional method of tracking the flame front. The below table compares the measured char front to the measured flame spread as is traditionally measured through the naked eye. Since the flame front is fairly turbulent, measured by the naked eye, it could be beneficial and important to consider a spread rate from char front as opposed to the flame front – providing the mechanisms from char to ignition are rationally understood. Figure 8 provides a preliminary analysis of char spread estimated through image correlation based upon color change. Beyond lift testing, the ability to filter the emission of light from fire could allow the measurement of in-fire mechanical properties of polymers and other engineered materials – which the authors are also currently studying.

Figure 8: Char spread measurement

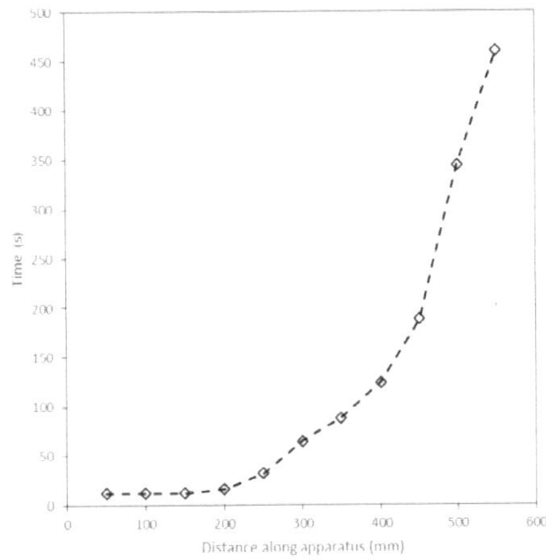

PRELIMINARY CONCLUSIONS AND FUTURE WORK

Engineered timber is gaining competitive economic popularity in the construction industry and this leads to timber performance in fire being more intensely studied. This means that evaluation and modelling techniques for timber in fire are being globally scrutinized for usability and accuracy. This need for more realized fire modelling was the basis for this testing. This testing demonstrates the use of the new imaging technology in fire engineering to understand the development of the char layer rather than the development. The program herein is ongoing and designed to study the optimal configuration required in an external style connections in LVL to disrupt flame spread and reduce char depth from fire exposure. Initial conclusions show that external connections perform well in short term heating applications leading to the structural stability of the structure during egress in the case of fire. Timber structures have proven resiliency despite the ever-changing standards for testing. As these testing methods change we only become more aware of how industry changing engineered wood construction is.

ACKNOWLEDGEMENTS

The authors wish to thank NSERC Canada under grant 2015-05081and Carleton University's I-CUREUS for supporting this research. Considerable thanks to the National Fire Research Laboratory at the National Institute of Standards and Technology, USA for loaning equipment for image correlation studies. In particular to Matthew Hoehler, and Chris Smith.

Many thanks to Carleton university staff and faculty: Jason Arnott, Stanley Conley, Stephen Vickers, Steve Tremblay, Pierre Trudel and Ba Lam-Thien. Carleton University students: Hailey Quiquero, Hailey Todd, Sydney Van Bakel, Josh Woods, Mina Li, Julia Dalphy, Don Jackson and Jeffery Tolton.

REFERENCES

[1] C. Maraveas, K. Miamis and E. Matthaiou (2015). Performance of Timber Connections Exposed to Fire: A Review. Fire Technology. 51, 1401–1432.

[2] FPInnovations, 2014, Technical Guide for the Design and Construction of Tall Wood Buildings in Canada, FPInnovations, Pointe-Claire, QC

[3] Yang, J. C. et al., 2015, International R&D Roadmap for Fire Resistance of Structures: Summary of NIST/CIB Workshop, National Institute of Standards and Technology, Maryland, US.

[4] Hopkin, D., Schmid, J., Friquin, K., 2016. Timber Structures Subject to Non-Standard Fire Exposure – Advances and Challenges. WCTE. 12 pp.
[5] ASTM 2013, E1321- Standard Test Method for Determining Material Ignition and Flame Spread Properties. ASTM International. 16 pp.
[6] ASTM 2015, E84- Test Method for Surface Burning Characteristics of Building Materials ASTM International. 22 pp.
[7] Gales, J., and Green, M. 2015. Optical Characterization of High Temperature Deformation in Novel Structural Materials. Proceedings of the 14th International Conference on Fire and Materials. San Francisco, CA. 626-640.
[8] Smith, C., and Hoehler M. 2017. Imaging through open flames using narrow-spectrum illumination. (pending title and submission date).
[9] Quiquero, H., and Gales J. 2017. Comparing timber adhesive shear strength properties after fire damage. Fire and Materials Conference. 11pp.

FULL SCALE TESTS ON THE PERFORMANCE OF HYBRID TIMBER CONNECTIONS IN REAL FIRES

Samuel B. Amankwah Boadi* & George Hadjisophocleous
Carleton University, Canada

ABSTRACT

Connections are essential elements in any building system as they contribute significantly to its structural stability. Connections in hybrid timber construction would normally fail due to loss of strength in the wood member. To investigate the structural resistance of unprotected hybrid steel-column and glulam-beam shear tab connections when exposed to a real fire, six (6) full scale tests were carried out in this study. This paper presents and discusses the results obtained during the tests that include temperature profiles at sections of the timber beam close to the connections and the mid-span, as well as the failure modes and charring rates for each shear tab connection assembly. The influence of parameters of load ratio and the configuration of the shear tab connection considered are also presented.

Keywords: hybrid connections, fire resistance, charring rate, real fires, failure times, load ratio

1 INTRODUCTION

Connections perform the important role of transferring loads from one structural member to the other (beam to beam, or column) in a structural building system. The ability of any structural system to perform its intended function depends on the strength and stiffness properties of the structural members, and the connections between those members[1]. Connections are therefore designed based on realistic assumptions of the distribution of internal forces in the connecting members. They however constitute weaker links in a structural framing system especially when exposed to fire due to the numerous variables that will influence their structural integrity. This problem becomes critical when considering hybrid construction of timber and steel as the properties of these two materials have to interact to resist the physical and fire loads. Glulam, a commonly used engineered wood product chars at temperatures close to 300°C. The strength and stiffness of structural steel is greatly reduced at temperatures close to 600°C. Obviously these two materials react differently to extreme temperatures thereby complicating the structural stability of systems made using these two materials in fire. As such, an understanding of the behaviour of such systems such as connection assemblies involving these two materials is paramount for fire and life safety considerations[2].

Shear tabs are a common type of simply supported connections in structural steel framing due to their ease of fabrication and erection[3]. They are employed in hybrid construction involving timber and steel. Their configuration contributes significantly to the performance of structural hybrid assemblies due to the ability of the good conductor steel to transfer heat to the connected wood members, and ultimately leading to reduction in strength of the connection system.

Prescriptive-based design has been the approach for fire safety design of structures. In Canada, design requirements for fire safety are found in Division B, Part 3, Fire Protection, Occupant Safety and Accessibility, of the National Building Code of Canada (NBCC) [4]. The NBCC has no detailed engineering design provisions for the design of fire resistance of timber connections[5]. Designs are normally based on experience[6] and prescriptive fire resistance ratings of 45 min or 1 hour for combustible construction[4]. Due to limitations on the use of these prescribed values of fire resistance ratings for highly industrialised and complex structures, the use of performance-based design is on the rise. Using this approach, structural systems are designed using fire safety engineering principles and shown to satisfy requirements and provisions in the building codes. In order to fully comprehend the behaviour of hybrid connections and to complement the performance-based design approach, studies have been conducted by various researchers [2, 7-13] on the behaviour of timber connections in fire. To add to the existing research on connections in fire, this paper sought to investigate the behaviour of unprotected shear tab connection assemblies in a real fire.

2 EXPERIMENTAL PROGRAM

For the purpose of this study, two glulam beams were tested for each shear tab connection type of Exposed (EX), Seated (SE) and Concealed (CN), at the Full Scale Fire Research Facility of Carleton University in Almonte, Ontario. The structural assemblies were tested in a furnace especially designed for studying the fire performance of connection assemblies. The interior dimension of the furnace measures 2700 mm x 2700 mm x 2200 mm high. The walls are heavily insulated with a 25-mm thick Fibrefrax® mineral blanket. The furnace is equipped with two propane line burners and a blower to supply air and to maintain a positive pressure in the room during testing. The temperature in the furnace can be monitored and controlled to replicate any time-temperature curve.

2.1 Testing Facilities

2.1.1 Temperature measuring devices in the furnace

Six (6) shielded rod thermocouples and four (4) plate thermometers are installed to measure the temperature in the furnace. These measured temperatures are used to monitor and control fuel supply to the burners of the furnace to follow the required time-temperature curve. As required by ASTM E119[14] and CAN/ULC S-101[15] Standards, the shielded thermocouples are made of No. 18 gauge type K thermocouple wires enclosed in a porcelain tube.

2.1.2 Load application system

A loading frame is available in the furnace to support the connection assembly and apply the desired load. The load is applied using a hydraulic cylinder, Model RD1610 from Enerpac®. A 250KN capacity load cell (MTS® 661.22D-01) is employed to measure the load. An exterior view of the furnace with the attached testing frame is shown in Figure 1. Figure 2 shows the configuration of a typical connection assembly under testing conditions.

Figure 1: Furnace with testing frame

Figure 2: Configuration of a typical tested assembly

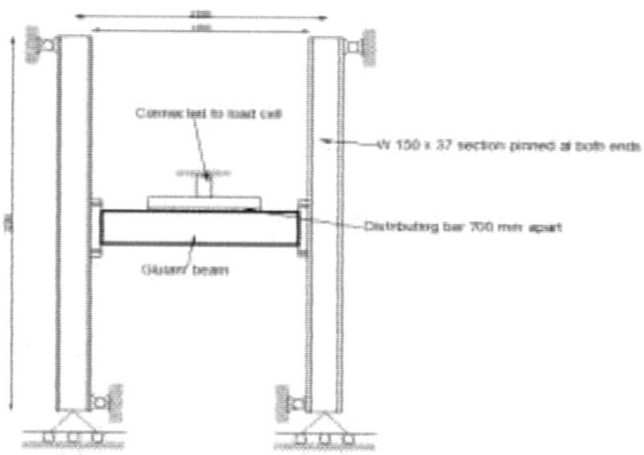

2.2 Specimen Details and Instrumentation

The tested connection assemblies consisted of a 1880 mm long glulam beam connected to two 3200 mm long pin-pin steel columns (W 150 x 37) using a 9.5 mm thick shear tab plate (Grade 300W). The beam was connected to the columns using 4-12.7 mm diameter Grade A325 bolts. Two glulam beam cross-sectional dimensions of 140 mm wide x 191 mm deep, and 137 mm wide x 222 mm deep were studied. As shown in Figure 2, the steel columns are pinned to prevent development of moments at the ends. To mimic the presence and the effect of a floor slab system on the beams, the top surface was protected with a 25 mm thick layer of Fibrefrax® mineral blanket. As such, all the beams were tested with three sides exposed to the furnace temperatures. The steel columns and the loading frame in the furnace were also heavily insulated. A typical assembly in the furnace showing the insulated columns and top surface of the beam can be seen in Figure 3.

Figure 3: Typical assembly set up for testing in the furnace

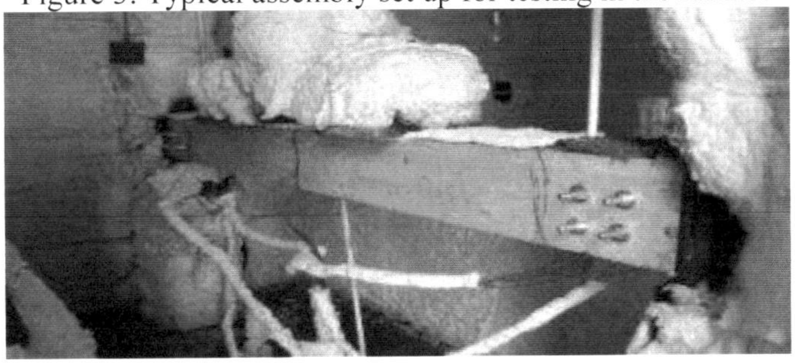

Table 1 shows the details of each assembly tested with information on the shear tab connection used and the load applied.

Two load ratios of 60% and 100% were considered for the series of tests conducted. These load ratios are fractions of the factored resistance found for each connection type assembly under ambient conditions using provisions in the CAN/CSA O86-09[16]. Detailed design drawings of the shear tab plates are shown in Figures 4-6. All dimensions are in millimetres.

Table 1: Details of each assembly tested

Assembly Number	**Size of Beam**				**Connection Type**	**Load**		**Bolt**	
	Width (mm)	Depth (mm)	Length (mm)	Moisture Content (%)		LR (%)	Applied Load (KN)	Diameter (mm)	Length (mm)
1	137	222	1880	10	Exposed	100	98	12.7	175
2	137	222	1880	9	Exposed	60	59	12.7	175
3	140	191	1880	10	Concealed	60	43	12.7	175
4	140	191	1880	9	Concealed	100	71.8	12.7	175
5	137	222	1880	9	Seated	60	70	12.7	175
6	137	222	1880	11	Seated	100	118	12.7	175
All bolts are Grade A325									

Figure 4: Exposed (EX) shear tab connection

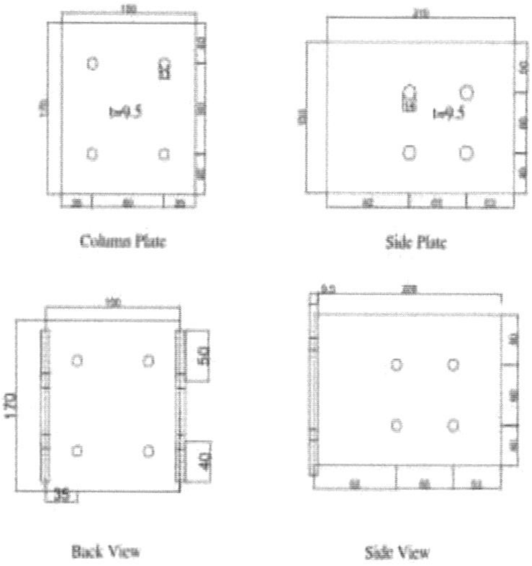

Figure 5: Concealed (CN) shear tab connection

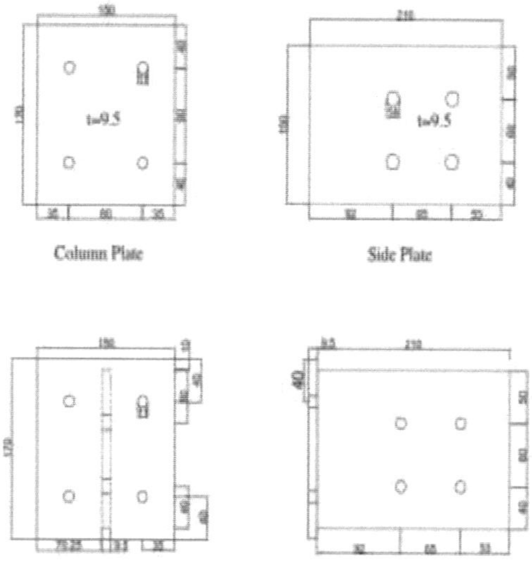

Figure 6: Seated (SE) shear tab connection

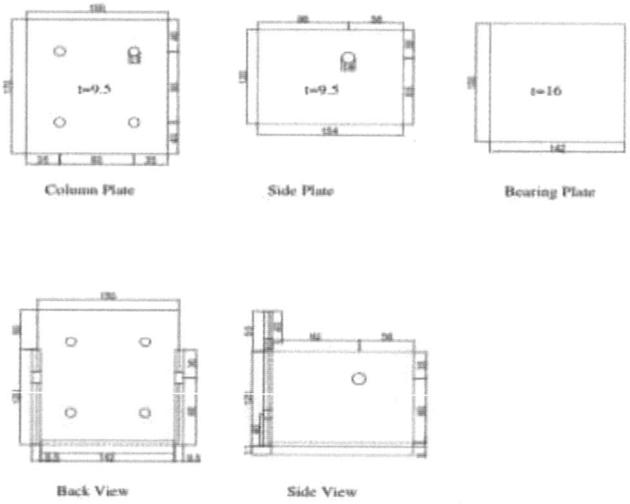

2.2.1 Beam instrumentation

The glulam beam specimens were instrumented with thin metal-sheathed and mineral-insulated (MSMI) type K thermocouples at various depths. The thermocouples were installed from all faces of the beams as detailed in Figures 7-9. The installed thermocouples are represented with dots and numbers in the figures. Each face of the beam is given a different colour representation and marked with the letters F, R, T and B for thermocouples installed from the front, rear, top and bottom faces respectively. In order to measure temperatures of sections of the wood close to the connections, thermocouples were installed at 220 mm from the ends of the beam. Others were also installed in the mid-span to measure and compare the difference in temperature distribution to that at the connection.

As shown in Figures 8 and 9, the thermocouples were installed at 20 mm and 40 mm depths to measure the change in temperature with time in the glulam beams. Others were also located within the inner core of the beam.

Figure 7: Instrumentation of a typical beam specimen with location of installed thermocouples

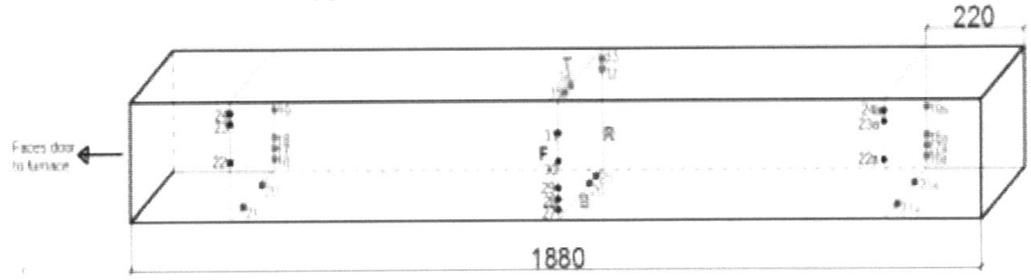

Figure 8: Plane in a typical beam showing the position of thermocouples in the 137 mm x 222 mm specimens

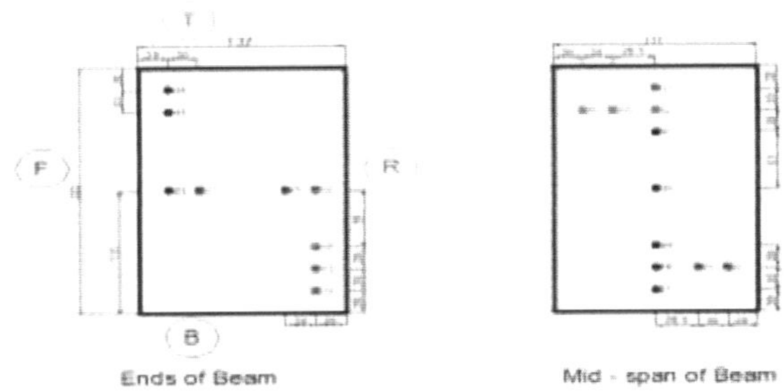

Figure 9: Plane in a typical beam showing the position of thermocouples in the 140 mm x 191 mm specimens

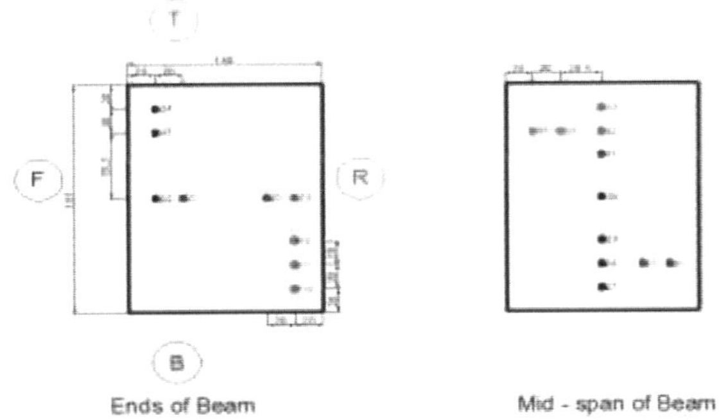

Representation of the thermocouples in the various plots of temperature with time is formed based on the face of the beam from which they were installed, and the number and the depth of installation of the thermocouples. For instance, a thermocouple with the identification TF-24-20mm represents thermocouple number 24 installed at a depth of 20 mm from the front face of the beam. Similarly, TB-21-111mm represents readings recorded by thermocouple 21 installed at a depth of 111 mm from the bottom face of the beam. All the thermocouples are therefore given the simple representation shown below;
T(face of beam)-Number-Depth of installation from the face of the beam
The temperature profiles recorded were used to compute the charring rate and depth of the beams.

2.3 Fire Exposure Model

The real fire curve employed in this study was derived by Aguanno[17]. It is based on a room fire test conducted by McGregor[18] in 2012 on two fully furnished CLT rooms of size 4.5 x 3.5 x 2.5 m high built in the Full Scale Fire Research Facility of Carleton University in Almonte, Ontario. Both rooms were lined with 3-ply CLT panels, with the floor made of hardwood on a cement board. The second room however had an extra material in the form of a 2-layer 12.7 mm Type C gypsum board protecting the CLT panels. The contents in the room amounted to a total fuel load of 560 MJ/m^2.

The temperature reading from these two tests were used to design the real fire/non-standard fire curve. As shown in Figure 10, the temperature readings in the gypsum lined room began to decay after 20 minutes of exposure. That of the unlined room, however, continued to rise incessantly as the CLT walls and ceiling contributed to the fire load until it was manually extinguished[18]. Aguanno[17] therefore modelled a real fire curve that incorporated behaviours involving a blend of steady burning for some time and a decay phase as seen in both tests. A slower decay phase initiated at 1100°C to 300°C in a time interval of 2 hours was used.

Figure 10: Fire temperatures as observed in both protected and unprotected CLT room tests

The real fire exposure model was developed using the formulations in Equation [1].

$$
\begin{aligned}
&\mathbf{T = 126.446t^2 + 18} && \mathbf{for\ 0 \le t < 2.2\mathit{min}} \\
&\mathbf{T = 1100 - \left(\frac{t-24}{1.005559}\right)^2} && \mathbf{for\ 2.2 \le t < 24\mathit{min}} \qquad [1] \\
&\mathbf{T = 1100} && \mathbf{for\ 24 \le t < 36\mathit{min}} \\
&\mathbf{T = 1881.12 - 130.186\sqrt{t}} && \mathbf{for\ 36 \le t \le 200min}
\end{aligned}
$$

where T = furnace temperature and t = exposure time in minutes.
The use of this real fire curve was deemed appropriate as both CLT and glulam beams are combustible, and likely to be found together in a typical residential room.

2.4 Testing Procedure

The glulam beams and the connection types were fabricated in the atrium of the Carleton University Research Facility in Almonte, Ontario. The connections were then bolted to the insulated steel columns by means of a shear tab plate with 4-12.7 mm diameter bolts.

The beam to be tested was placed in the furnace and bolted to the connections. The thermocouples in the beam were hooked to the data acquisition system. Prior to igniting the burners, the beam was subjected to a two-point constant gravity load at 1/3 points along its length. This load was applied at least 30 minutes before the fire resistance test, and maintained until failure of the assembly. A hydraulic jack connected to the load cell was used to apply and maintain the constant load. The load was transmitted through two distributing bars 700 mm apart, and attached to a vertical steel post. The burners were ignited, and the assembly was subjected to the modelled real fire curve until failure. Failure of the assembly is characterised by a sudden drop in the applied load, and the inability of the specimen to sustain the desired load. During testing of an assembly, the furnace temperature readings were monitored and controlled to follow the modelled real fire curve. At failure of an assembly, the propane burners and blowers were shut down. Enough jet of water was then sprayed on the assembly immediately after shut down of the burners to prevent further charring and smouldering combustion of the beam.

After the fire resistance test, the assembly was inspected and observed in the furnace. It was then taken out, and the layer of char formed was brushed off. Measurements of the residual cross-section were then taken at various sections along the length of the beam for charring rate calculations.

3 EXPERIMENTAL RESULTS

Figure 11 shows the adequacy and accuracy of the furnace in replicating the modelled real fire curve for all the assemblies tested. The average furnace temperature readings were monitored and controlled to follow the desired real fire curve.

Figure 11: Average furnace temperatures recorded by the plate thermometers during testing of all the connection assemblies

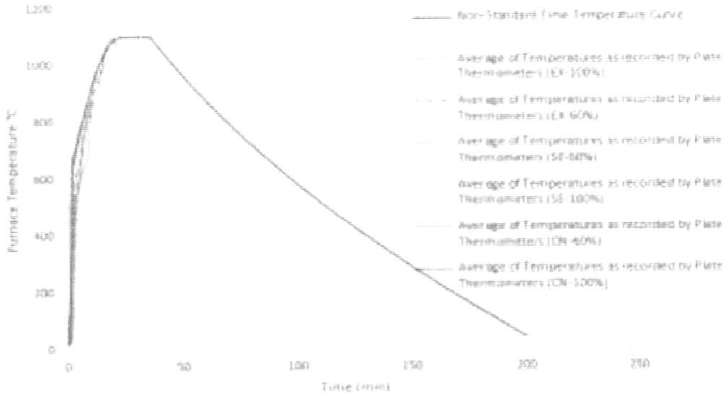

Loading of each assembly was done manually as each assembly was subjected to a constant two point load at least thirty minutes before commencing the fire resistance tests. The loads applied were based on the calculated load ratios from the design capacities of the assemblies. Figure 12 demonstrates how close the applied loads were to the desired loads for each connection assembly tested. The applied loads were measured from the beginning of the fire resistance tests. The initial drop in the applied load curve at the start of the test for the Exposed (EX) and Seated (SE) Connection Assemblies under 100% load ratio is as a result of oil leakage in the hydraulic cylinder of the load application system. As stated in Section 2.4, failure of an assembly was characterised by a sudden drop in the applied load.

Figure 12: Load curve and characterisation of failure

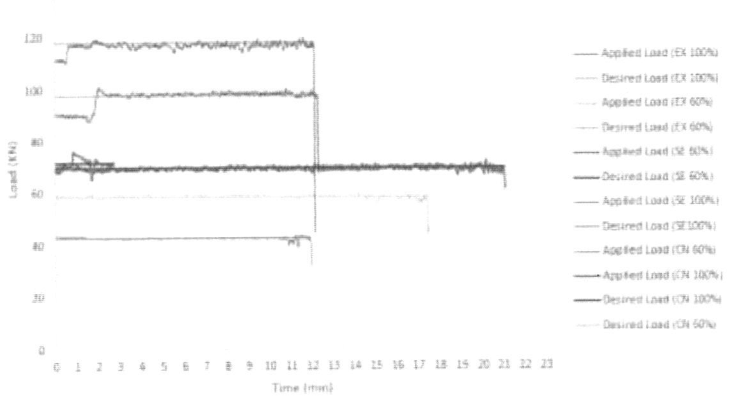

3.1 Mode of Failure

Failure of the connection assemblies was a result of loss of strength in the wood members due to thermal degradation of the wood at elevated temperatures. Failure of all the assemblies except SE-100% and SE-60% were sudden and brittle.

Assembly EX-100% was subjected to a total applied load of 98 KN corresponding to its full factored resistance. Failure of the assembly involved both splitting at the joints and tensile cracks at the mid-span of the glulam beam. Visual observations made during testing showed bending failure at the mid-span (Fig. 13). Post fire test observations however revealed the hidden line split initiated at the left connection end. This shear split was concealed by the side brackets and the layer of char formed. As shown in Figure 13, the splitting failure of Assembly EX-100% along the grain of the glulam beam, occurred just above the upper row of the bolt holes. The split was slightly above the glue line of one of the laminates. After clearing off the layer of char, it was observed that flexural cracks formed prior to failure at the finger joint between laminates (Fig. 14 and 15).

Figure 13: Failure mode of Assembly EX-100%

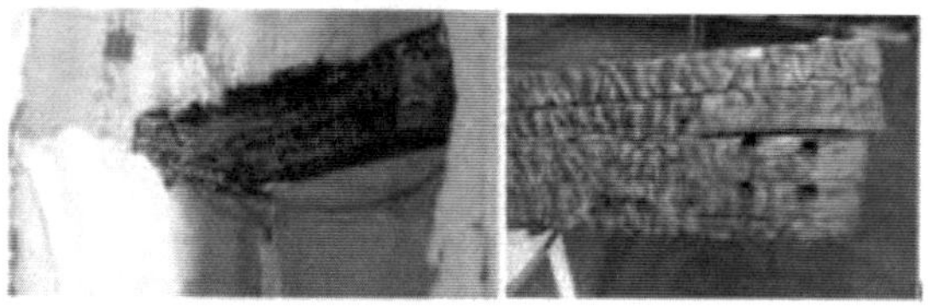

Figure 14: Flexural crack observed in Assembly EX-100%

Figure 15: Flexural crack in Assembly EX-100%

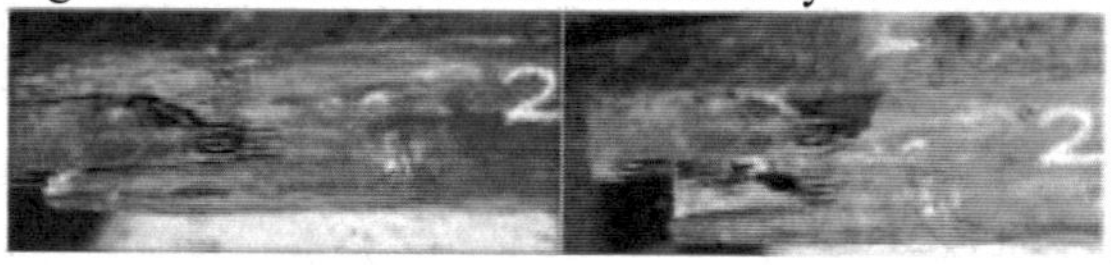

Figure 16: Splitting failure of Assembly EX-60%

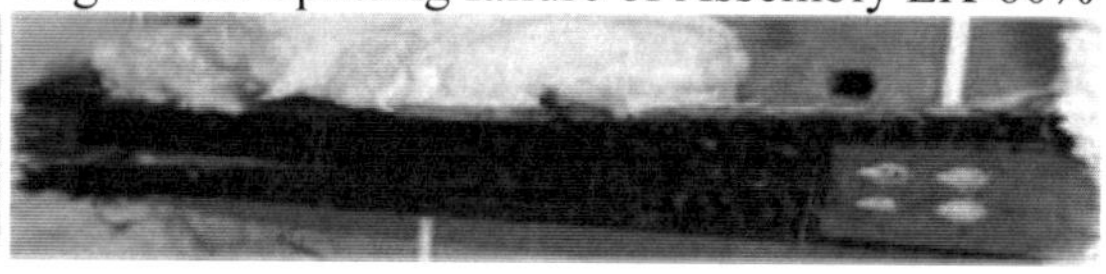

A 59 KN load corresponding to a 60% load ratio was applied to Assembly EX-60%. The mode of failure for this assembly was a longitudinal split initiated at the connection end below the bottom row of bolts. The split was wide enough to expose the core of the glulam beam to the high furnace temperature (Fig. 16).

The Seated connection assembly under 100% load ratio (SE-100%) was subjected to a load of 118 KN. The assembly failed in flexure, with a resulting fracture in a finger joint holding the laminates together in the tensile region of the glulam beam (Fig. 17 and 18).

Figure 17: Assembly SE-100% at failure

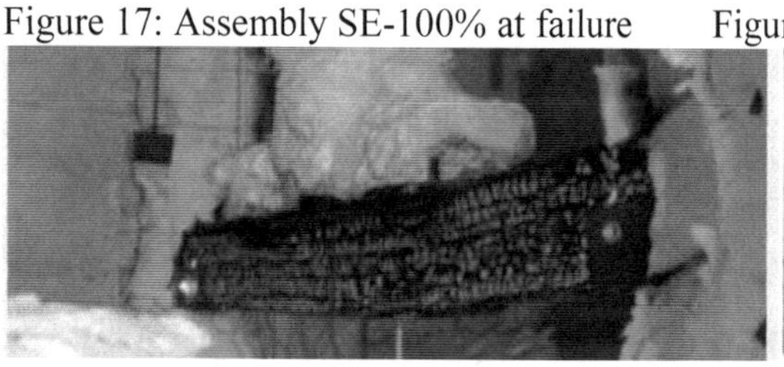

Figure 18: Fracture at finger joint causing failure of SE-100%

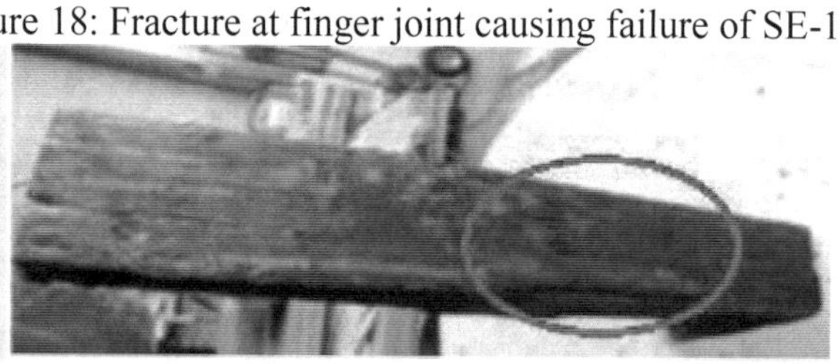

Under 60% load ratio, the seated connection assembly failed in flexure with an accompanying twisting of the glulam beam under loading (Fig. 19). In the absence of any lateral restraint, lateral torsional buckling of the glulam beam commenced as the sides and bottom of the beam lost most of its cross-section in the elevated temperature causing it to twist in a manner similar to that experienced by a steel T-beam with a thinner web and no restraint.

Figure 19: Assembly SE-60% at failure

Figure 20: Splitting failure of Assembly CN-100%

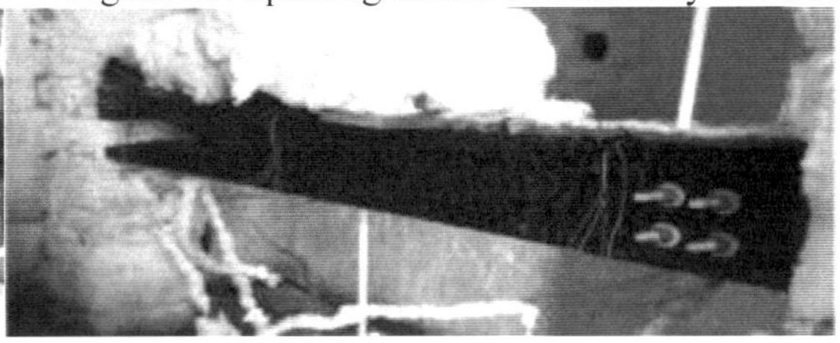

The Concealed connection assembly under 100% load ratio (CN-100%) was subjected to a total applied load of 71.8 KN. As shown in Figure 20, failure of the assembly was characterised by a sudden longitudinal split under the bottom row of bolts. The split profile was around a knot close to the joint of the beam.

An applied load of 43 KN corresponding to a 60% load ratio was imposed on the CN-60% assembly. The assembly also failed in splitting at the connection ends. As shown in Figure 21, the shear split was initiated above the top row of bolts.

Figure 21: Splitting failure of Assembly CN-60%

Table 2 summarises the failure modes of the connection assemblies tested.

Table 2. Main failure modes of each connection type

Connection Type	Failure Modes
Concealed (CN)	Longitudinal split at the joints
Exposed (EX)	Longitudinal split at the joints and rupture of a finger joint in the tensile region of the beam a mid-span
Seated (SE)	Bending failure with subsequent twisting when the full load is applied.

3.2 Relative Failure Times of the Connection Assemblies

In this section, the influence of the load ratios of 60% and 100% on the fire resistance ratings of the connection assemblies is presented. The load ratio is a measure of the fraction of the total factored design load of the assembly in the event of a fire. Buchanan[1] noted that most buildings usually have about 50% of their load capacity present in the event of a fire. This could even be less in structures purposefully designed to resist more catastrophic events such as earthquakes. Lower load ratios result in better fire performance of the assemblies as it can be seen in the results presented in the underlying section.

Figure 22 shows the failure times in minutes of the connection assemblies tested for each shear tab connection under load ratios of 60% and 100%. The significant impact of the load ratio on the failure times is evident. An increase in the load ratio led to a significant decrease in the time to failure of each tested assembly.

Figure 22: Failure times of the connection assemblies under load ratios of 60% and 100%

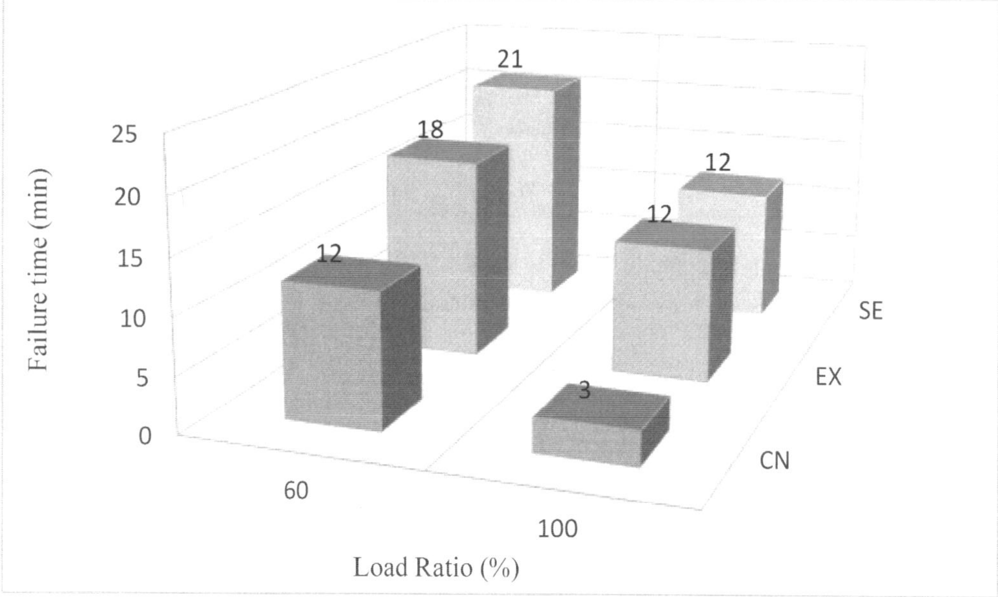

For the lower load ratio of 60%, the seated connection provided the best performance with a failure time of 21 minutes. The times to failure of the assemblies were improved in the order of concealed to exposed and to seated. The concealed connection assembly under 60% load ratio collapsed after 12 minutes of testing. Compared to the concealed connection assembly, the seated connection assembly offered 75% improvement in the failure time under 60% load ratio. At a failure time of 18 minutes, the exposed connection assembly also offered an extra 50% increase in the time to failure compared to the concealed assembly. It was therefore concluded that, the connection assemblies with exposed steel side plates performed better under loading and under exposure to the modelled real fire curve than the one with a slotted-in steel plate. The presence of the steel plate in the concealed connection assemblies, and possible leakage of hot gas temperature to the wood surface in the notch, initiated charring at the inner core as the gas temperature increased prior to failure of the assembly. Heat transfer to the concealed plate through the bolts, in addition to charring initiated from the exposed sides of the wood member at the joint is believed to be responsible for the relatively poor performance of the concealed connection assembly in fire. The increase in time to failure of the connection assemblies with exposed steel side plates can be attributed to the partial protection provided by the side plates in the growing phase of the fire. The thicker the side plate, the longer the time taken for the heat front to travel to the wood surface in contact.

3.3 Thermal Response

The temperature build up in the glulam beams at sections close to the connection ends and at the mid-span for the connection assemblies is plotted as a function of the exposure time in Figures 23-34. Generally, the temperature profiles at the various depths were higher for sections closer to the connection ends due to the ability of the steel plates to conduct and transfer heat to the wood in contact. In all figures, the temperature profiles follow a similar trend with a steady incipient phase within the first minute of testing. In all figures, the temperature growth within the inner core of the beam remained slightly above ambient as the layer of char formed during testing offered protection to the sections underneath. This was evident in thermocouples installed 111 mm and 68.5 mm deep for the beam sections 137 x 222 mm and 140 x 191 mm respectively. The plotted temperature profiles are useful in computing the average charring rate of the glulam beams for each connection assembly.

Figure 23: Temperature profile at connection end of beam as a function of exposure time (EX-100%)

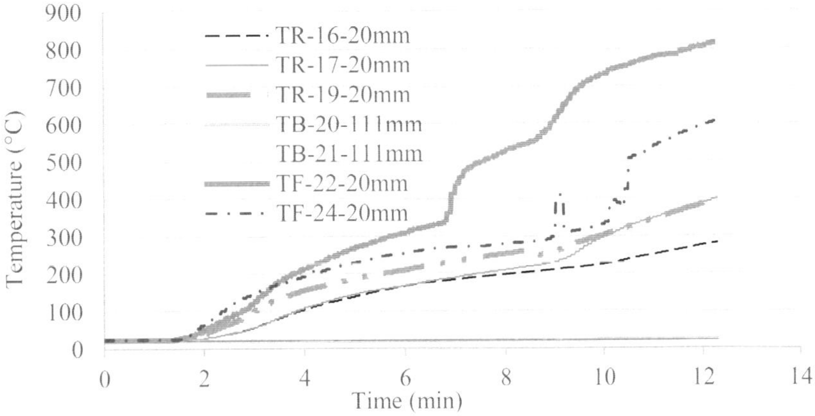

Figure 24: Temperature profile at mid-span of beam as a function of exposure time (EX-100%)

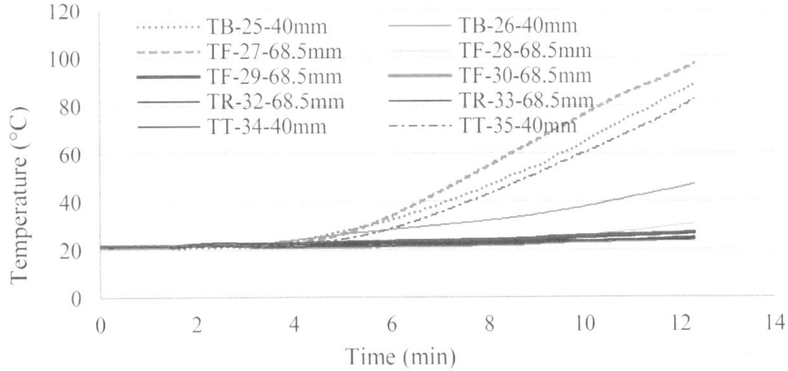

Figure 25: Temperature profile at connection end of beam as a function of exposure time (EX-60%)

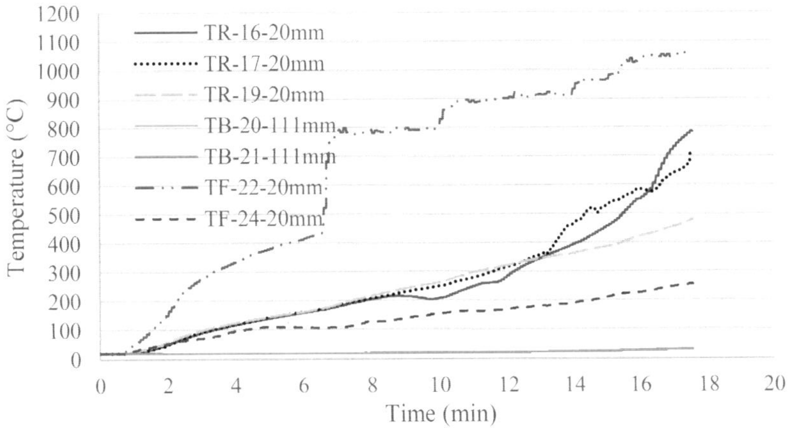

Figure 26: Temperature profile at mid-span of beam as a function of exposure time (EX-60%)

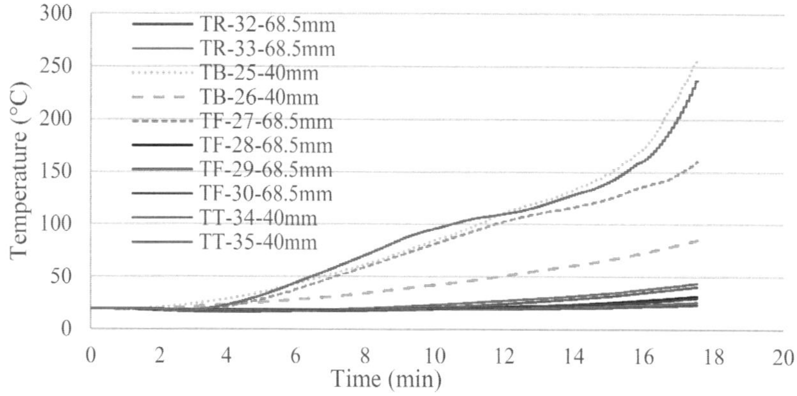

Figure 27: Temperature profile at connection end of beam as a function of exposure time (SE-100%)

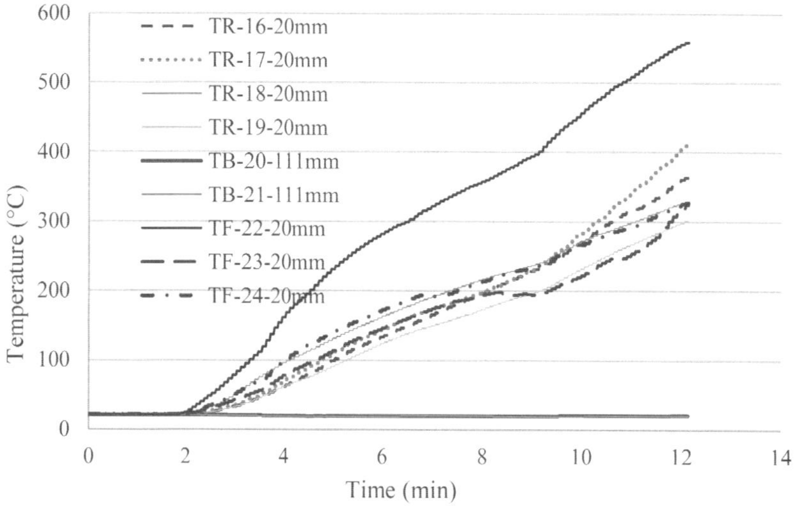

Figure 28: Temperature profile at mid-span of beam as a function of exposure time (SE-100%)

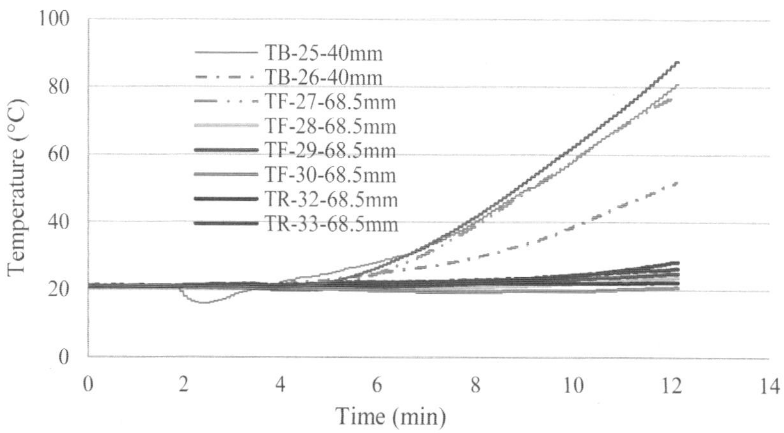

Figure 29: Temperature profile at connection end of beam as a function of exposure time (SE-60%)

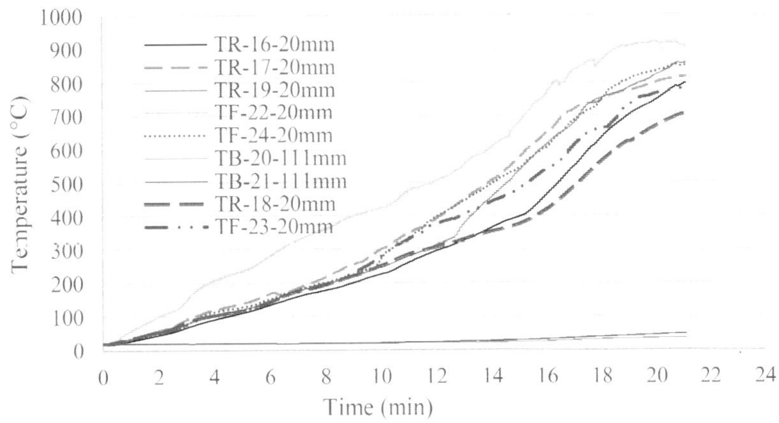

Figure 30: Temperature profile at mid-span of beam as a function of exposure time (SE-60%)

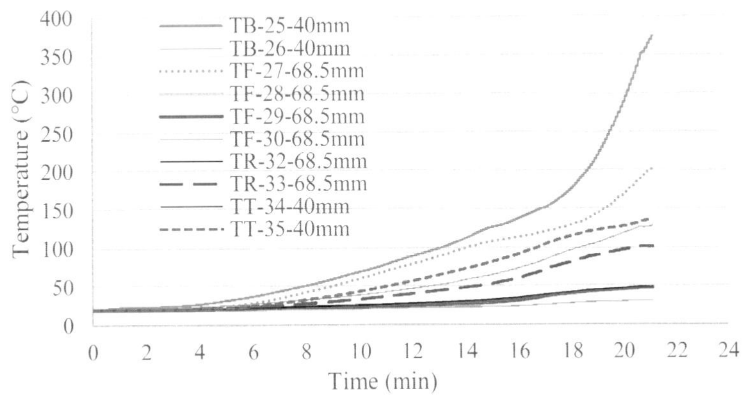

Figure 31: Temperature profile at connection end of beam as a function of exposure time (CN-100%)

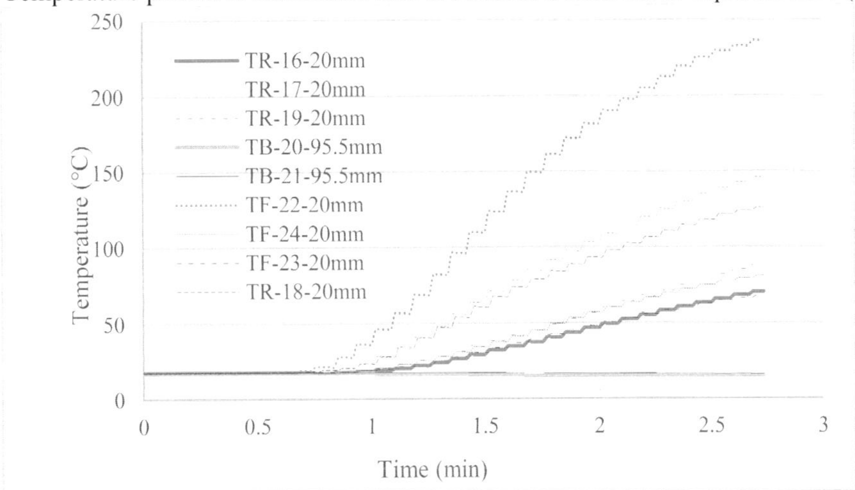

Figure 32: Temperature profile at mid-span of beam as a function of exposure time (CN-100%)

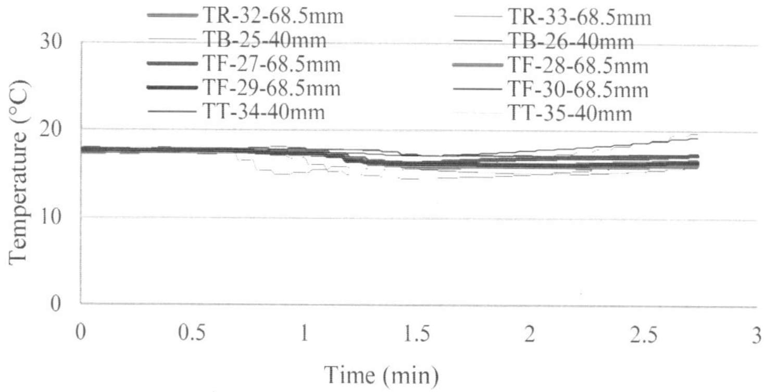

Figure 33: Temperature profile at connection end of beam as a function of exposure time (CN-60%)

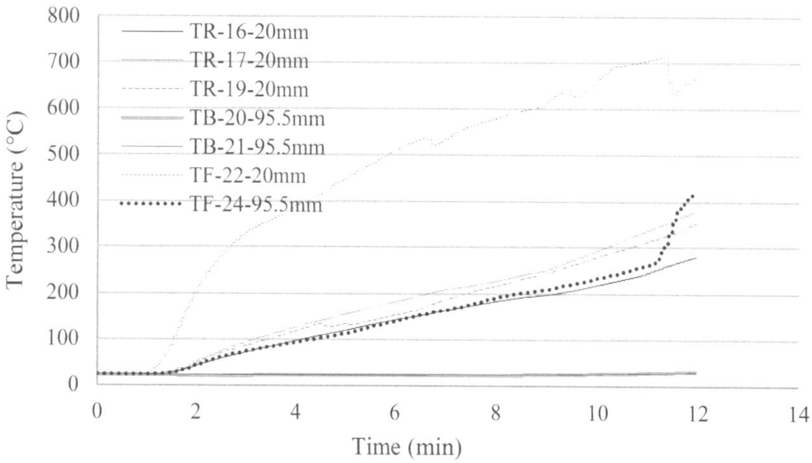

Figure 34: Temperature profile at mid-span of beam as a function of exposure time (CN-60%)

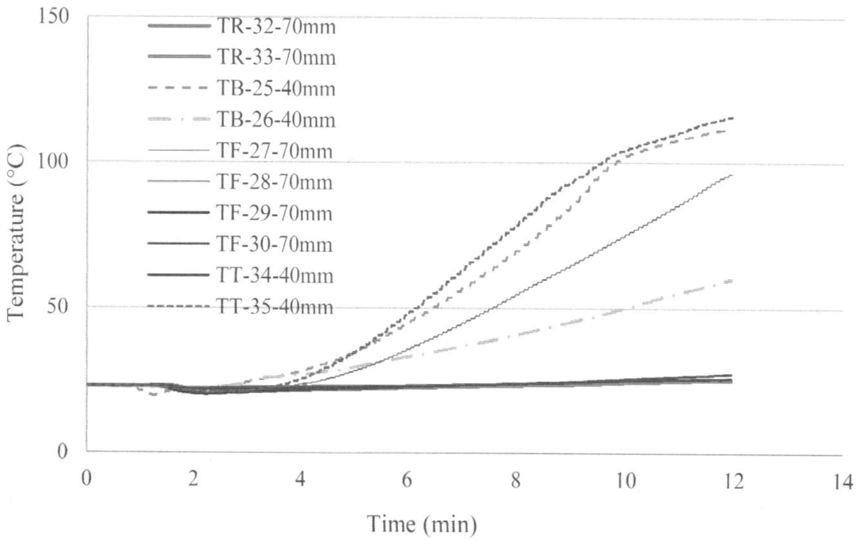

3.4 Charring Rate (mm/min)

Charring of the connected glulam beam member in all the tested assemblies leads to loss of cross-section, loss of strength and ultimately failure of the assembly. To estimate the charring rate in mm/min of the tested assemblies, two main methods were employed.

- Temperature profile estimates of the charring rate (mm/min) based on the time taken for the 300°C isotherm to travel to the depth of installed thermocouples in the glulam member. Estimates from this method was referred to as the Average Charring Rate. It is significantly dependent on the performance of the installed thin-wire thermocouples.
- Charring rate estimates from the measured residual cross-section of the glulam beams after clearing the char layer off tested samples. Estimates from this approach was referred to as the Real Charring Rate (mm/min). It is estimated over the failure time of the assembly and provides the most accurate results[19].
-

Table 3 presents the charring rate values (mm/min) computed at sections of the beams close to the connections as functions of the load ratios.

Table 3: Computed average and real charring rates (mm/min) based on the temperature profiles within the wood member and the residual cross-sections after testing respectively

Connection Type	Charring Rate (mm/min)			
	60% LR		100% LR	
	Average Charring Rate	Real Charring Rate	Average Charring Rate	Real Charring Rate
Exposed Assembly (EX)	2.51	1.04	2.05	1.16
Seated Assembly (SE)	1.94	1.16	1.92	1.01
Concealed Assembly (CN)	2.64	1.39	-*	2.60

*CN-100% assembly recorded temperatures lower than 300°C at the depth of the installed thermocouples. As such average charring rate computations could not be made for this assembly (-)

Direct measurements of the real charring rate at sections of the beam closer to the connections produced values exceeding 1 mm/min. The seated connection assemblies recorded a mean real charring rate of 1.16 mm/min and 1.01 mm/min under 60% and 100% load ratios respectively (Table 3). This assembly offered the best failure time with the least charring rate values, losing 24 mm of its cross-section to the non-strength bearing char layer in the assembly SE-60%. At the connection ends, lower mean real char rate values of 1.09 mm/min and 1.1 mm/min were recorded for the seated and exposed connections respectively. For the concealed connection, a higher real charring rate averaging 2.0 mm/min was measured.

4 CONCLUSIONS

Based on the results and the observations made, the charring rates of the glulam beam connected to the connections in the real fire tests exceeded the value of 0.65 mm/min obtained in standard fire tests, and employed in the design of wood members for fire safety. Charring rate values reported in this study exceeded 1.0 mm/min with a higher charring depth of 24 mm[20].

The load ratio also contributed significantly to the capacity of the assemblies in the fire. Decreasing the load ratio from 100% to 60% resulted in an increased failure time of the tested assemblies. The seated connection had a 75% increase in its failure time, with the exposed connection reporting an increase of 50%. Based on the analysis of the results obtained, the seated connection assembly was found to offer the best fire resistance, with failure times that were 17% and 75% higher than that of the exposed and the concealed connection assemblies respectively, under a load ratio of 60%.

ACKNOWLEDGEMENT

This research was funded by the Natural Sciences and Engineering Research Council (NSERC) of Canada and FPInnovations. We are grateful to them for their support.

REFERENCES

1. Buchanan, A. H. (2002). Structural Design for Fire Safety. John Wiley and Sons Ltd. Chichester, England.
2. Mehaffey, J. R. (2003) Fire performance of heavy timber connections. In: Report No. 3223, Forintek Canada Corp. Ottawa, ON, Canada.
3. Sherman, D. R., Ghorbanpoor, A. (2002). Design of Extended Shear Tabs, Final Report. American Institute of Steel Construction.
4. NBCC (2010). National Building Code of Canada 2010, Canadian Commission on Building and Fire Code, National Research Council of Canada, Ottawa, ON, Canada.
5. Mehaffey, J. R. (2003) Fire performance of heavy timber connections. In: Report No. 3223, Forintek Canada Corp. Ottawa, ON, Canada.
6. Wood Design Manual (2010). The complete reference for wood design in Canada, Canadian Wood Council.
7. Erchinger, C., Frangi, A., and Mischler, A. (2005). Fire behaviour of multiple shear steel-to-timber connections with dowels.Working Commission W18—Timber Structure, Meeting 38, International Council for Research and Innovation in Building and Construction, Karlsruhe, Germany.
8. Erchinger, C., Frangi, A., and Mischler, A. (2006). Thermal investigations on multiple shear steel-to-timber connections. Proceedings of the World Conference on Timber Engineering, Portland, OR, U.S.A.
9. Frangi, A., and Mischler, A. (2004). Fire tests on timber connections with dowel type fasteners. Working Commission W18—Timber Structures, Meeting 37, International Council for Research and Innovation in Building and Construction, Edinburgh, U.K.
10. Laplanche, K., Dhima, D., and Racher, P. (2006). Thermo-mechanical modelling of the timber connection under fire using 3D finite element model. Proceedings of the World Conference on Timber Engineering, Portland, OR, U.S.A.
11. Schabl, S., and Turk, G. (2006). Coupled heat and moisture transfer in timber beams exposed to fire. Proceedings of the World Conference on Timber Engineering, Portland, OR, U.S.A.
12. Ali, S. G., Akotuah, A. O., Erochko, J., and Hadjisophocleous, G. V. (2014). Performance of hybrid timber connections with full-scale tests and finite element modelling. In: 8th International Conference on Structures in Fire, 511-519.
13. Peng, L., Hadjisophocleous, G., Mehaffey, J., and Mohammad, M. (2010). Fire resistance performance of unprotected wood-wood-wood and wood-steel-wood connections: A literature review and new data correlations. Fire Safety Journal, 45, 392-399.
14. ASTM E119. (2000). Standard Test Methods for Fire Tests of Building Construction & Materials. American Society for Testing and Materials.
15. CAN/ULC S-101. (2007). Standard Methods of Fire Endurance Tests of Building Construction & Materials. Underwriter's Laboratories of Canada. Scarborough, Canada.
16. CAN/CSA-O86-09 (2010). Engineering Design in Wood. Canadian Standards Association. ON, Canada.
17. Aguanno, M. (2013). Fire resistance tests on cross-laminated timber floor panels: an experimental and numerical analysis. Department of Civil and Environmental Engineering: Carleton University.
18. McGregor, C. (2012). Contribution of cross-laminated timber panels to room fires. Department of Civil & Environmental Engineering: Carleton University.
19. Tsai, W. K. (2010). Charring rates for different cross-sections of laminated veneer lumber. In: Fire Engineering Research Report 10/5. Department of Civil Engineering, University of Canterbury, NZ.
20. Lange, D., Bostrom, L., Schmid, J., and Albretsson, J. (2014). Charring rate of timber in natural fires. In: 8th International Conference on Structures in Fire. Shanghai, China. 521-528.

PERFORMANCE OF BEAM-TO-COLUMN CONCEALED CONNECTIONS IN STANDARD FIRES

Christian Gonzalez Crespo[1], Osama Salem[2] and George Hadjisophocleous[1]
[1]Carleton University, Canada
[2]Lakehead University, Canada

ABSTRACT

The majority of structural load-transferring zones in timber buildings mainly rely on steel connectors. These zones are susceptible to loss of strength under fire conditions, which negatively affects their ability to support applied loads. To investigate the performance of concealed steel-timber connections under exposure to elevated temperatures that follow the standard time-temperature curve (CAN/ULC S-101), beam-to-column assemblies were subjected to static transverse loads equivalent to 30% and 100% of their design moment capacity. This paper discusses the results obtained from these fire resistance tests including the connection failure modes and charring rates.

Keywords: fire performance, standard fire, charring rates, timber connections, parallel-to-grain resistance

1 INTRODUCTION

1.1 Background

In North America, wood is a widely popular construction material for single family houses and low-rise buildings. Building codes, however, place a lot of restrictions on wood construction mainly because wood is a combustible material. In order to overcome these barriers, fire safety studies have focused on developing a thorough understanding of fire resistance of wood structures. The studies have indicated that even though wood is a combustible material, its fire performance is predictable. As the temperature of the wood increases, it becomes a multi-layer material with deteriorated mechanical properties[1]. Wood charring occurs at about 300°C; however, irreversible decomposition starts at 200°C[2]. The char acts as an insulating layer that helps in delaying the transfer of heat to the interior core of the member. Since the charring rate of wood under standard fire exposure is constant, this rate is used to design wood members to achieve the fire-resistance rating required. In modern timber construction, steel plates and fasteners are normally used as load-transferring components in most glued-laminated (glulam) beam-to-column connections. This combination of materials in relation to their behaviour at elevated temperatures, makes the study of this type of connections complex. Although codes and standards provide guidelines for the design of connections at normal temperatures, there is no well-established method to design timber connections for fire resistance.

Empirical research has identified these load-transferring zones, such as beam-to-column connections, to be more susceptible to failure in a fire event. The main reason is that steel plates and fasteners have high thermal conductivity, which transfers more heat to the core of the wood members[3]. Additionally, there is some discrepancy in the reported charring rates of glulam members. Md Daud et al.[4] performed fire tests in accordance to (BS EN 1363-1:2012)[5]. The tested specimens were glulam members made of Keruing and

Malagangai. The glulam members were graded as hardwood, and had a density of 800 kg/m^3. They reported a charring rate of 0.44 mm/min, and indicated that charring rate depends on the density of the wood. Yang et al.[6] analyzed the charring rates of glulam members using the standard fire test (CNS 12514)[7], and the cone calorimeter method (ISO 5660)[8]. Five different types of softwoods with densities less than 500 kg/m^3, were studied. Their investigation concluded that two factors cause higher charring rates: lower density and longer fire exposure time. The Eurocode 5[9] recommends design charring rates of 0.64 mm/min and 54 mm/min for softwoods and hardwoods, respectively.

Although, there is a general agreement on the minimum accepted temperature for charring of wood, strength loss at high temperatures is still under investigation. Correia et al.[10] conducted a series of experiments on load bearing capacity and embedment strength parallel-to-grain. Tests at 20°C, 50°C, 100°C, 150°C, and 230°C were conducted for (one dowel embedment strength) and (one and three bolts load bearing capacity) in accordance to (NF EN 383)[11]. The latter was assembled as per (EN1995-1-1)[12]. All the specimens were preheated for 150 minutes prior to testing, and loaded parallel-to-grain. Their results showed that embedment strength loss started at 50°C. The reported strength losses at 100°C and 230°C were 30.6% and 35.6%, respectively. Load bearing capacity loss started at 50°C for tests with one and three bolt joints. The maximum loss for one-bolt joint occurred at 50°C with a strength reduction of 47.2%. For the three-bolt joint, the maximum loss bearing capacity occurred at 100°C with a strength reduction of 41.2%.

Literature shows different types of tests related to glulam connections performance in fires. Ali et al.[13] reported on the fire performance of hybrid steel-timber connections. Their tests were carried out under the standard time-temperature curve (CAN/ULC S-101)[14]. The configuration of the test assemblies consisted of glulam beams connected to steel columns. Bending and shear stresses were attained by loading the beam perpendicular to grain on the top side with two-point loads at one-third and two-thirds along the length of the beam. For both sets of tests, connections were composed of two steel plates welded together to form a t-shape connector section. The end of the steel connector was inserted into the beam section, and then secured in place using a number of bolts running across the width of the beam. The other end of the steel connector was fixed to the column using long steel bolts. Test results indicated failure by splitting of the beam section in a brittle fashion at the bottom row of bolts. Palma et al.[15] tested glulam beam to glulam column connections loaded in shear under fire conditions, and reported that failure was a result of beam splitting. They also found that a thicker plate between the beam and the column led to faster failure times.

1.2 Objectives and Scope

This paper presents the results of large-size experiments carried out with the objective of gathering experimental data to assist in developing methodologies for the design of timber connections. Test assemblies were made of glulam beams and columns fixed together using concealed type connections.

2 MATERIALS AND METHODOLOGY

The beams and columns used were made of 38mm thick glued-laminations of Canadian Douglas Fir-Larch 20f-E. The glulam sections were stored for a week at approximately 0°C (32°F) prior to testing. The specimens underwent a moisture content drop from 13% - 15% at manufacturing to 9% - 10% at the moment of testing. Steel plates of 300W grade along with A325 bolts were used in all test assemblies.

2.1 Furnace Description and Test Setup

Fire resistance experiments were conducted at Lakehead University's Fire Testing and Research Facility that accommodates a large-size fire testing furnace. This furnace consists of large steel-reinforced insulated chamber with interior dimensions of 3.4 m x 1.8 m x 1.6 m high. The furnace has several openings at its top, bottom and side walls. The two side openings in the furnace floor were used to insert two 3.0 m-apart steel columns Figure 1a. The state-of-the-art furnace is fully-automated to follow any desired time-temperature fire curve including the standard ones, such as (CAN/ULC S-101)[14].

Each test assembly was composed of two glulam beams – 171 mm (width) x 228 mm (depth) x 1400 mm (length) connected to the opposite sides of a glulam column – 190 mm (width) x 215 mm (depth) x 800 mm (height). The two beam sections were fixed to the centre column using concealed steel connectors. At the beam ends connected to the middle timber column, the beam-to-column connections were free to displace vertically, but restrained against rotation. At the far ends, the beams were completely restrained against both rotation and displacement. Static loads were applied on the top side of the column to exert cantilever effect that produced equal but opposite moment reactions at the beam-to-column connections, Figure 1b. Fire insulation blankets were placed on the supporting steel columns including the far ends of the beams. During testing, the entire glulam beam-to-column test assembly was exposed to elevated temperatures on four sides.

Figure 1a: Furnace Sectional Elevation

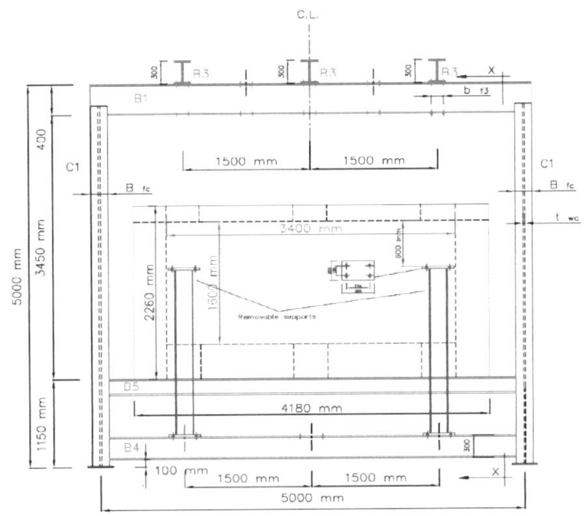

Figure 1b: General Test Setup

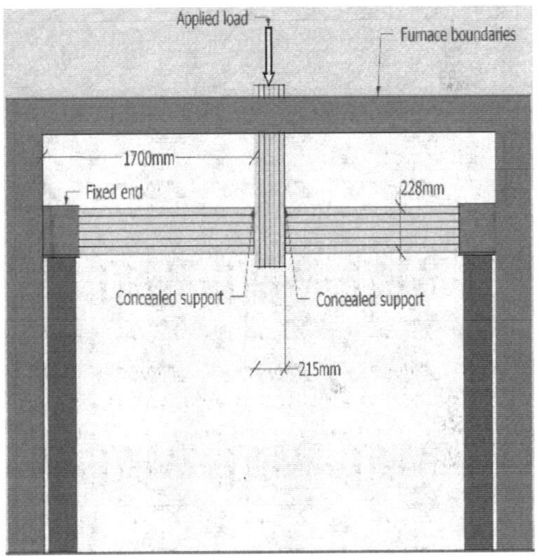

The concealed connection was inserted into the end of the glulam beams and fastened with four – 19.1 mm diameter bolts. The base of the concealed connection was attached to the short glulam column with four – 19.1 mm diameter bolts, as shown in Figure 2. The concealed connector was made of 9.5 mm steel plates welded together in t-shape section. The dimensions and bolt spacing are shown in Figure 3.

Figure 2: Beam-to-column connection details

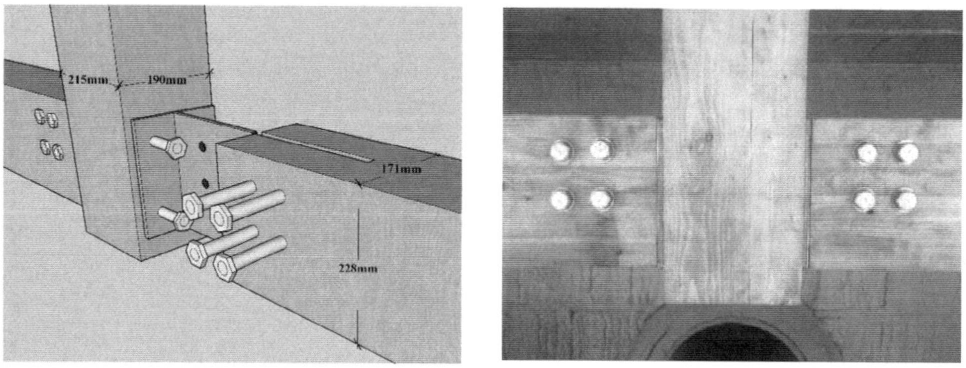

Figure 3: Concealed steel connector dimensions

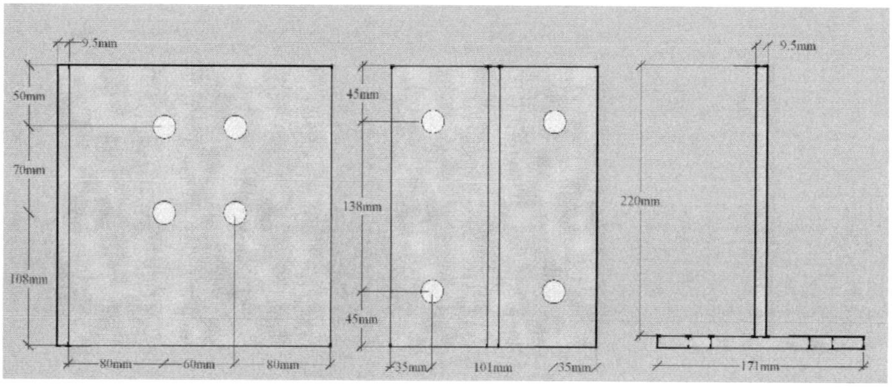

a) Side b) Front c) Top

2.2 Load Application and Instrumentation

The applied load was based on the governing failure mode calculated as per the provisions of (CAN/CSA O86-09 Canadian Handbook of Wood Design 2010 edition)[16]. Equation 1 determines the governing failure mode in parallel-to-grain row-shear resistance.

$$PR_{rT} = \Sigma(PR_{ri}) \quad [1]$$

$$PR_{ri} = \phi_w PR_{rij\,min} n_R (K_D K_{SF} K_T)$$

$$PR_{rij} = 1.2 f_V K_{ls} t n_c a_{cr\,i}$$

Note: The following factors were used: long load duration (K_D =0.65), dry service condition (K_{SF}=1), untreated K_T=1), and member loaded surface (K_{ls} = 1).

Two linear variable differential transducers (LVDTs) were placed outside the furnace, but connected to the top side of the beams, at the right and left of the column to measure vertical displacements. The temperatures of the test assembly were measured with shielded K-type thermocouples at different locations: welded on the bolts, and inserted into the beam and column sections. Fire insulation blankets were used to provide added protection to the thermocouples, only the ends were left exposed.

3 NORMAL TEMPERATURE TESTING

A test was conducted at ambient temperature. The glulam beam-column test assembly had a moisture content of about 10% with some visible cracks in the parallel to grain direction prior testing. During the test, the load was gradually increased until the beams failed by splitting parallel-to-grain along the bottom/tension row of bolts at about 14.4 kN. This value is approximately 29% more than the calculated design value of 11.14 kN. The split propagated passed the back bolt in the bottom row. The beam side of the bolt heads had a wider splitting propagation than the side of the nuts, as shown in Figure 4. The measured rotation at failure was 38 degrees.

Figure 4: Parallel-to-grain splitting failure at 14.4 kN

4 FIRE RESISTANCE TESTING

Two beam-column assemblies were tested at elevated temperatures of a standard fire. Several thermocouples were placed at various depths and heights inside the beam and column sections, as shown in Figure 5. Additionally, two thermocouples were welded on the beam bolts, at the head (M1, M3, M5, M7), and at 39 mm (M2, M4, M6, M8). Three thermocouples were welded on the column bolts, at the head (N1, N4), at 54 mm (N2, N5), and at108 mm (N3, N6).

Figure 5: Thermocouple location on beam and column

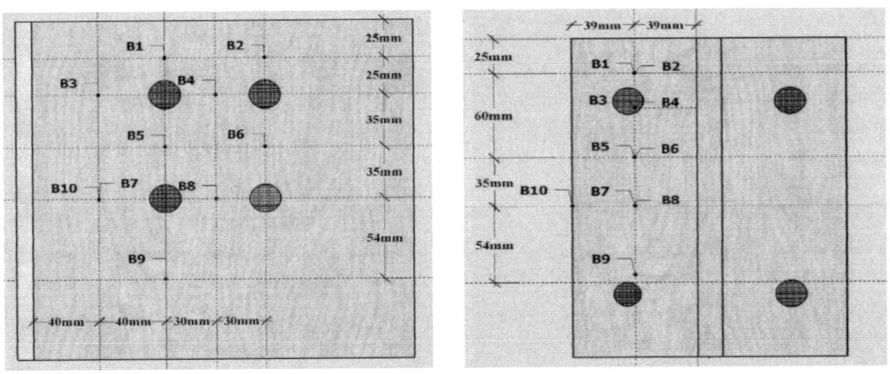

a) Beam side view

b) Beam front view

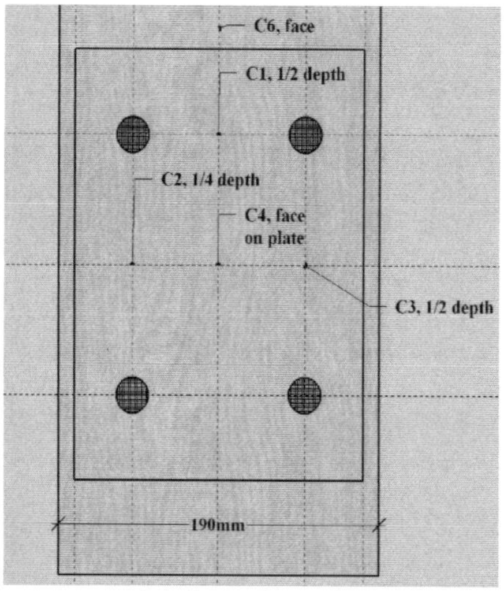

c) Column front view

4.1 Assembly Loaded at 30% of the Connection Design Capacity

For this test, the assembly was subjected to a constant vertical load of 3.36 kN, which is equivalent to a 30% of the design capacity.

4.1.1 Beam Analysis

Figure 6 shows the temperature vs. time profile of the plate thermometer, which was located near the connection, a thermocouple placed on the surface of the beam, and the thermocouples placed inside the beam section. Due to the presence of flames, the temperatures measured by the plate thermometer and the surface thermocouple reached higher values than the standard temperatures. The exposed face of the beam (B10) reached 300°C in approximately one minute. The temperatures along the bottom row bolts at 39 mm (B7) from the exposed surface reached 300°C in approximately 31 minutes. Based on this, the calculated charring rate was 1.28 mm/min. It was observed that at the concealed plate side of the beam, the wood had a lower charring rate. Visual observation agreed with most of the calculated charring rate. Figure 7 shows images taken at 10 minutes (a) and 40 minutes (b) of the fire test. In the former photo, a small portion of the beam is charred and the steel plate is visible at the bottom side. At 40 minutes, the wood between the top row of bolts and the beam's top side (originally at 50 mm) was consumed. At 31 minutes, the beams were not able to resist shear parallel-to-grain along the bottom row of the bolts, which caused splitting failure. At that time the beams had an approximate 40% loss of their cross-sectional area. The rotation of the beam end connected to the column increased gradually with temperature rise, as shown in Figure 8. Despite the splitting failure at the beam-to-column connection, the assembly continued to support the applied load for another 20 minutes due to the fixed conditions at the beam far ends.

Load redistribution occurred at the beam-column interface, transforming the connection system from moment-resistant to shear resistant. As a result, the governing failure mode became row shear perpendicular-to-grain resistance. Figure 9 shows the various temperatures measured of the steel bolts in the beam section. For instance, (M2, M4, M6 and M8) reached temperatures ranging from 600°C to 650°C at the failure time. The entire test assembly finally collapsed at 51 minutes of fire exposure; however, the beam-to-column connections failed much earlier. The observed failure patterns can be seen in Figure 10.

Figure 6: Temperature profile of beam – (30% LR)

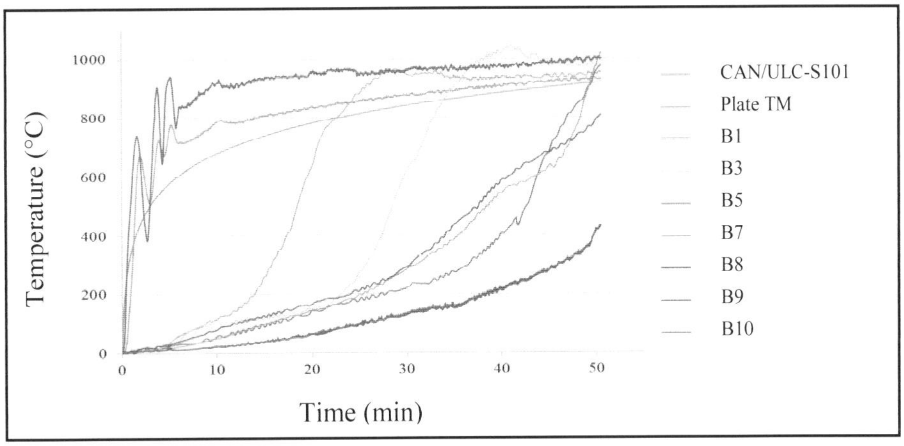

Figure 7: Images of fire testing progress – (30% LR)

a) Taken at 10 minutes

b) Taken at 40 minutes

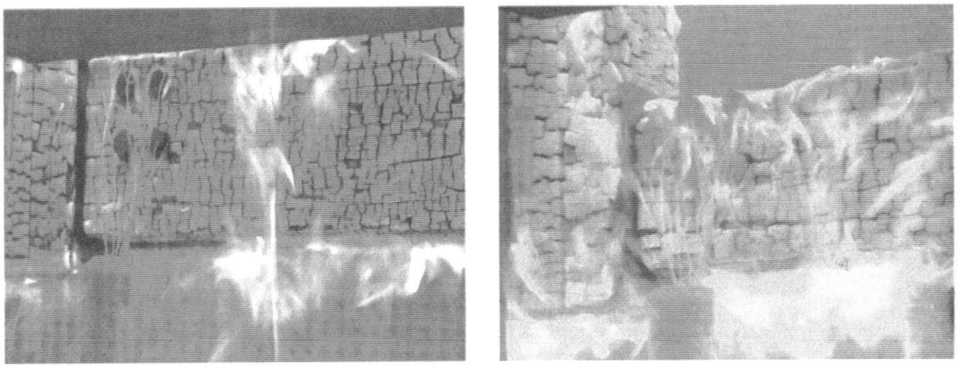

Figure 8: Beam-end rotation over time – (30% LR)

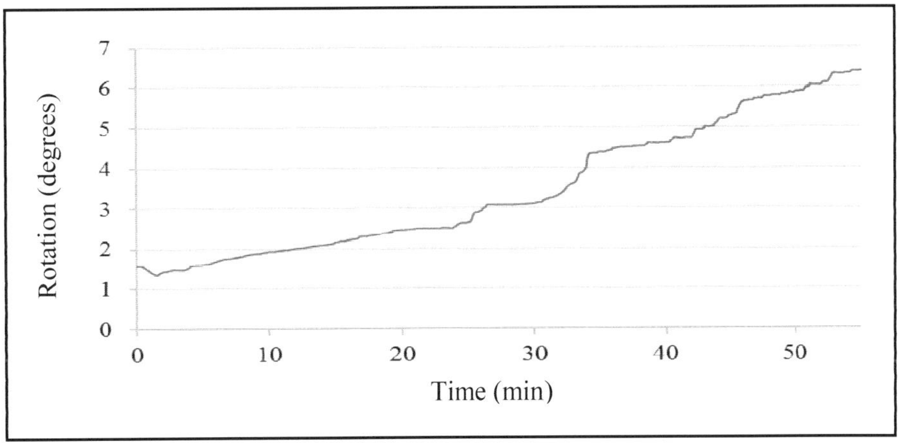

Figure 9: Temperature profile of steel bolts in the beam section – (30% LR)

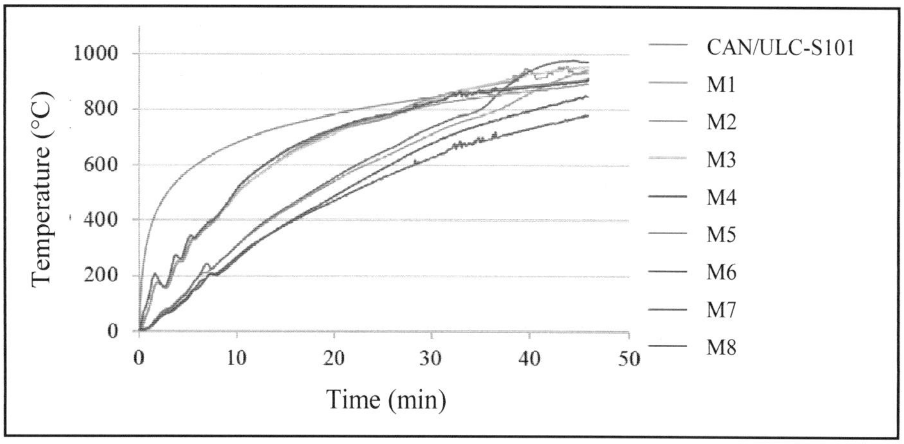

Figure 10: Connection failure modes

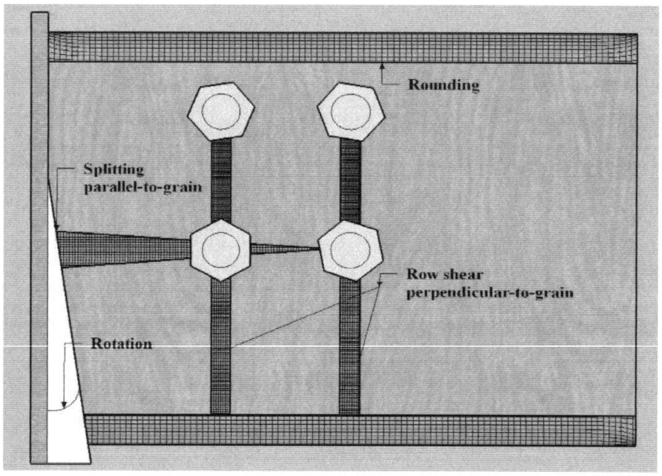

4.1.2 Column Analysis

Column temperatures at different depths are shown in Figure 11. The t-shape steel connectors in contact with the column sides did not transfer significant heat to the core of the column. At the points in direct contact with the steel plates, temperatures increased linearly. At 31 minutes (beam splitting time), the temperature of the column face in contact with the t-shape steel connectors had a temperature of approximately 550°C (C4). At 50 minutes (collapsing time), the temperature of this section matched the standard fire temperature of about 900°C.The maximum temperatures recorded at depths of 54 mm (C2) and 108 mm (C3) inside the column section were only 100°C and 40°C, respectively. The steel bolts inside the column reached higher temperatures faster. Figure 12 shows the temperature profiles of the upper and lower bolts. At 31 minutes, the average temperature of the bolts was about 600°C, which resulted in some charring of the wood in contact with the bolts. This may have caused less vertical displacement and consequently reduced rotation of the beams.

Figure 11: Temperature profile of column section – (30% LR)

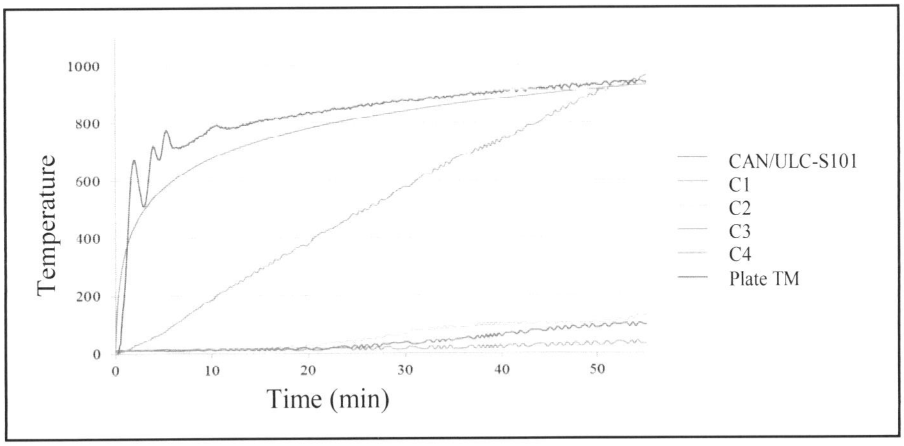

Figure 12: Temperature profile of bolts in the column section – (30% LR)

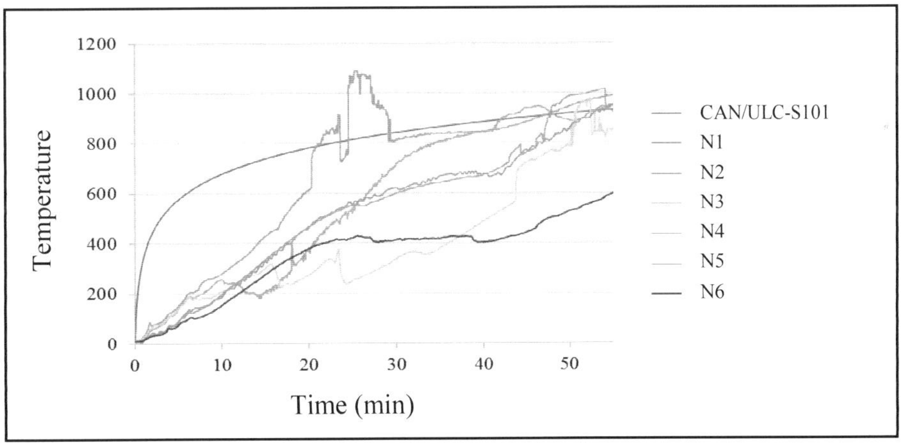

4.2 Assembly Loaded at 100 % of the Design Capacity

For this test the assembly was subjected to a constant vertical load of 11.14 kN, which is equivalent to 100% of the connection design capacity.

4.2.1 Beam Analysis

Figure 13 shows the temperature vs. time profile of the thermocouples placed inside the beam sections at 100% load ratio. The exposed face thermocouple (B10) reached 300°C in approximately one minute. As with the tests with 30% load ratio test, failure occurred at the bottom row of bolts. Points located at 39 mm from the beam surface (B7 and B8) reached 300°C in about 25 minutes. At these points, the calculated charring rate was about 1.53 mm/min. This high charring rate may be due to the cracks observed in the glulam beam prior the fire tests. Splitting along the bottom row of bolts occurred in the parallel-to-grain direction at about 15 minutes. At the time of connection failure, the width of the beam reduced to approximately 130 mm. However, after splitting parallel-to-grain occurred, the entire assembly did not collapse. Again, the load transferring mechanism in the connection changed from moment-resistant to shear-resistant. Also, the governing failure mode became shear row perpendicular-to-grain along the bolts. Pronounced bending of the beam close to mid-span and large parallel-to-grain splits, were observed, as shown in Figure 14. Beam rotation at the support was 8° at 15 minutes (beam splitting time) and 12° at 30 minutes (assembly collapse time).

At 15 minutes, the temperature of the bolts at the centreline of the side wood members reached approximately 400°C, see Figure 16 (M2, M4, M6, and M7). However, at the same depth between the bottom row bolts, the temperature was only 65°C (B7). Perpendicular-to-grain row shear resistance along the bolts contributed to an additional 15 minutes before collapsing.

Figure 13: Temperature profile of beam section – (100% LR)

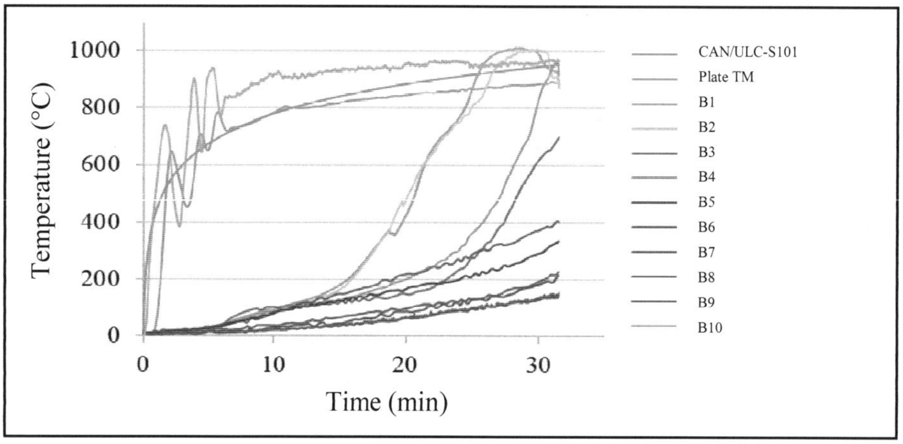

Figure 14: Image taken at 15 minutes – (100% LR)

Figure 15: Beam-end rotation over time – (100% LR)

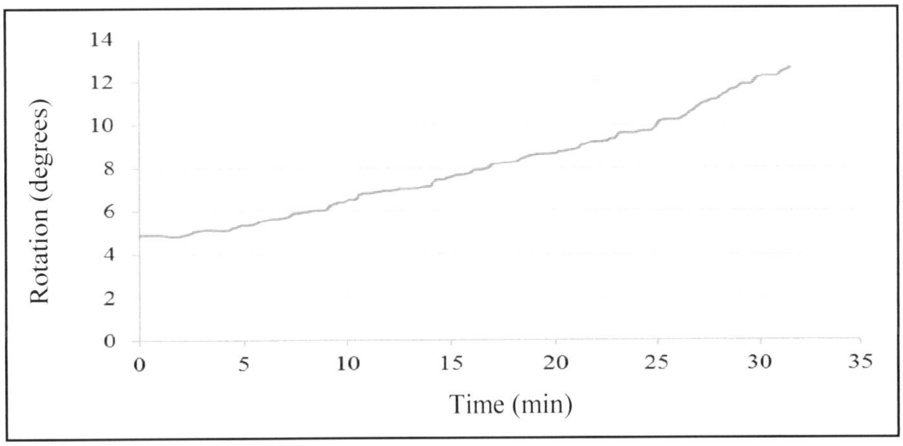

Figure 16: Temperature profile of bolts in beam section – (100% LR)

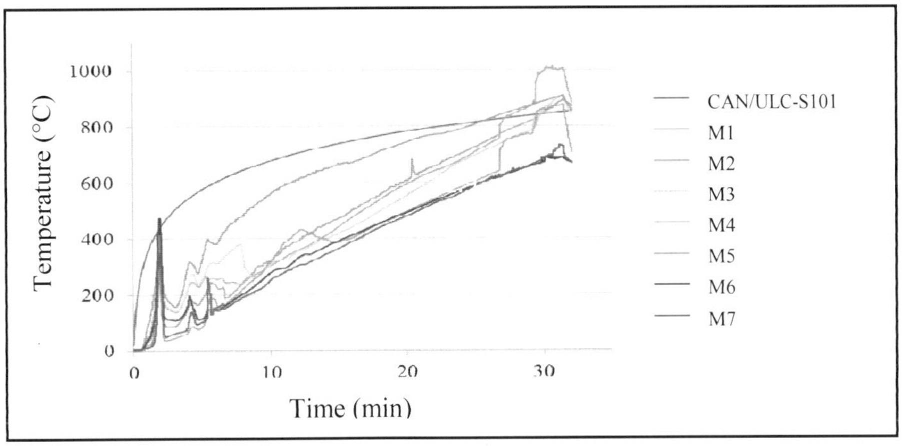

4.2.2 Column Analysis

As with the earlier test, the column performed better than the beams. Column temperatures at different depths are shown in Figure 17. The sides of the column in contact with the t-shape steel connectors had a linear increase in temperature (C4). At 31minutes (collapsing time), temperature reached 560°C. The steel connectors in contact with the column sides did not have a significant contribution to the temperatures across the column section. The steel bolts inside the column had average temperatures of about 300°C at 15 minutes, and 800°C at 31 minutes, see Figure 18. There was minimal vertical relative displacement of the column bolts inside their holes, which was observed afterward.

Figure 17: Temperature profile of bolts in beam section – (100% LR)

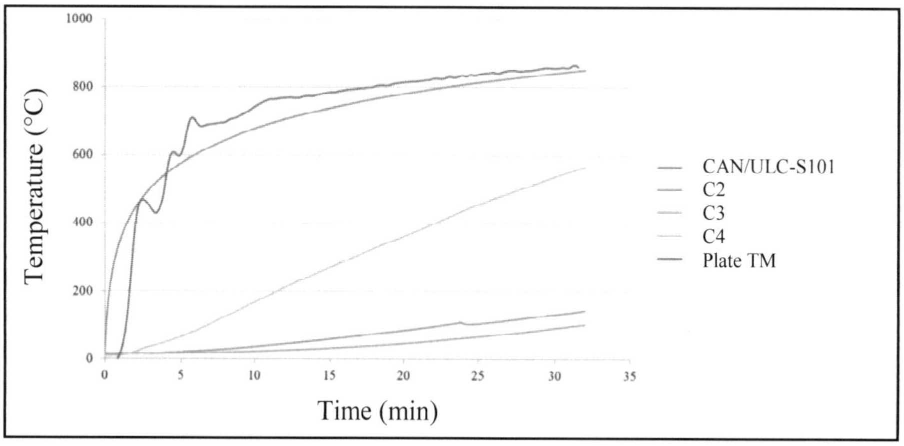

Figure 18: Temperature profile of bolts in column section – (100% LR)

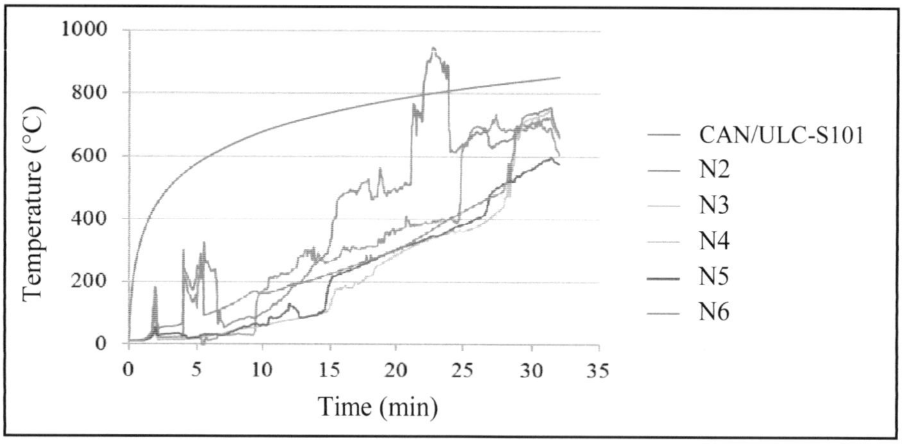

5 DISCUSSION

The location of the governing connection failure, which is aligned with the locations of thermocouples (B7 and B8), was selected for charring rate calculations. Experimentral charring rates based on times required to reach 300°C are shown in Table 1. Beam's charring layer thickness formation times required for failure are presented in Table 2. Beams' charring thickness required for failure was scaled from the ambient temperature ultimate capacity, described in Section 3.

Table 1: Charring rates measured at section B7.

	Time to reach 300°C (min)	Charring rate (mm/min)
30% Design (3.4 kN)	30.5	1.28
100% Design load (11.14 kN)	25.5	1.53

Table 2: Percentage difference between experimental failure times and computed failure times based on charring rates

	Load-based half-beam thickenss for failure (mm)	Charred thickeness for failure (mm)	Time required based on charring rate (min)	Experimental failure time (min)	Difference (%)
Ulitimate experimental capacity (14.2kN)	78	N/A	N/A	N/A	N/A
30% LR (3.4 kN)	18.5	59.5	46.5	31	- 40
100% LR (11.14 kN)	61.2	16.8	10.9	15	31

There is a difference between the experimental failure times and the calculated based on the residual cross section thickness at failure. The 30% LR test failure time was 40% lower than the failure time based on charring rate, while the 100% LR test failure time was 31% higher. The larger difference of the 30% LR test could be a result of the effect of the connection plates in contact with the beam as the test lasted for a longer time. Literature indicates that significant strength loss in the parallel-to-grain direction can occur at temperatures as low as 50°C[10]. Further investigation of the specimen type under study (Douglas Fir-Larch 20f-E) and thermal degradation of mechanical properties in the load direction applied is required.

6 CONCLUSIONS

Hybrid steel-timber connections were tested at normal temperature (20°C) and under exposure to standard fire (CAN/ULC-S101) at 30% LR and 100% LR. Wood splitting along the bottom bolt row was the first observed mode of failure in all three tests. This was in good agreement with the failure mode predicted from the design calculations, discussed in Section 3 of this paper. Splitting failure occurred at about 31 minutes and 15 minutes for the 30% LR and 100% LR tests, respectively. The values were lower than the predicted failure times based on charring rates. This shows that thermal degradation of wood at temperatures between 20°C and 300°C, has an important role in predicting accurate failure times. Having both beam far ends restrained against rotation contributed to the overall stability of the assembly. After splitting failure, both 30% LR and 100% LR tests, allowed for an extra 20 and 15 minutes, respectively, before collapsing. Additionally, steel connectors slightly reduced the charring rates of the adjacent wood section compared to other areas of the glulam members that were directly exposed to elevated temperatures. During both fire tests, the charring rates were higher than the 0.65 mm/min reported in the literature[9].

7 REFERENCES

1 Park, W. C., Atreya, A., and Baum, H. E. (2007). Determination of Pyrolysis Temperature of Charring Materials. In: Report No. 60NANB3D1086, Building and Fire Research Laboratory of the National Institute of Standards and Technology. Gaithersburg, MD, USA.

2 White, R.H.; Dietenberger, M.A. 2001. Wood products: thermal degradation and fire. Encyclopedia of materials: science and technology. [S.l.]: Elsevier Science Ltd, c2001: Pages 9712-9716

3 Buchanan, H., "Fire Performance of Timber Construction," Progress in Structural Engineering and Materials, 2, Issue 3, 2000.

4 Md Daud A.F., Ahmad Z., Hassan R. (2015) Charring Rate of Glued Laminated Timber (Glulam) Made from Selected Malaysian Tropical Timber. In: Hassan R., Yusoff M., Alisibramulisi A., Mohd Amin N., Ismail Z. (eds) InCIEC 2014. Springer, Singapore

5 BS EN 1363-1:2012, Fire resistance tests. General requirements. BSI Standard

6 Yang, T., Wang, S., Tsai, M., and Lin, C. (2009). The charring depth and charring rate of glued-laminated timber after a standard fire exposure test. Building and Environment. Vol. 44, Issue 2, 231-236.

7 CNS Standard, CNS 12514. Fire resistant examination for building structural element. Bureau of Standards, Metrology and Inspection, Ministry of Economic Affairs (MOEA), Taipei, Taiwan, 2002.

8 International Standard Organization, ISO 5660. Fire tests—reaction to fire—rate of heat release from building products, 1993.

9 Eurocode 5: Design of Timber Structures. Part 1-2: General rules structural fire design. ENV 1995-1-2. Brussels: European Committee for Standardization. 1994.

10 Correia ND, Rodrigues JPC, Moraes PD (2010) Timber Bolted Joints Subjected to High Temperatures, Sixth International Conference on Structures in fire, East Lansing, USA, June 2-4

11 CEN, 2004. "Eurocode 5: Design of timber structures—part 1-1: General common rules and rules for buildings," Brussels.

12 Norme Française, 1993. "NF EN383 - Détermination de caractéristiques de fondation et de la portance locale d'éléments d'assemblages de type broche".

13 Ali, S. G., Akotuah, A. O., Erochko, J., and Hadjisophocleous, G. V. (2014). Performance of hybrid timber connections with full-scale tests and finite element modelling. In: 8th International Conference on Structures in Fire, 511-519.

14 CAN/ULC S-101. (2007). Standard Methods of Fire Endurance Tests of Building Construction & Materials. Underwriter's Laboratories of Canada. Scarborough, Canada

15 Palma, P., Frangi, A., Hugi, E. Cachim, P., and Cruz, H. (2014). Fire resistance tension beam-to-column shear connections. In: 8th International Conference on Structures in Fire, 577-584.

16 CAN/CSA-O86-09 (2010). Engineering Design in Wood. Canadian Standards Association. ON, Canada.

RATIONALIZATION OF CROSS-LAMINATED TIMBER DESIGN STANDARDS

Steven T. Craft*
CHM FIRE CONSULTANTS LTD
Ottawa, Canada

Ineke Van Zeeland
Canadian Wood Council
Ottawa, Canada

ABSTRACT

With increased interest in building larger and taller timber buildings, the use of timber panel products such as cross-laminated timber (CLT), laminated veneer lumber (LVL), and nail-laminated timber (NLT) have seen an increase in use. Standard fire-resistance tests conducted on CLT panels have demonstrated the wall and floor assemblies can provide excellent fire resistance when either left exposed or, increasingly so, when protected by gypsum board. In the last few years, the design standards in the US (NDS) and Canada (CSA O86-14) have both added design criteria for CLT. In addition to the regular design provisions for structural design, the standards have included informative design criteria for standard structural fire resistance.

This paper provides some of the background rationale for the design methodology and provisions and compares standard fire-resistance test results with predictions of standard fire-resistance times using the design standards. Excellent agreement is shown between the test results and the fire-resistance times calculated in accordance with the design standards.

INTRODUCTION

Around the world, developers, architects and engineers are becoming increasingly interested in building and designing tall mass-timber buildings. Tall mass-timber buildings in North America are commonly classified as buildings that are 7 storeys and greater, and are built with wood products such as glued-laminated timber (glulam) or other structural composite lumber (SCL) beams and columns and cross-laminated timber (CLT) or laminated veneer lumber (LVL) panels. These structural wood products tend to have a significant amount of inherent fire resistance due to their cross-sectional area and the relatively slow rate of charring when exposed to fire. The newest product of those mentioned above, CLT, has become increasingly common as part of mass-timber buildings over the last 8 years.

Cross-laminated timber (CLT) comes in large, solid panels consisting of layers of kiln-dried lumber glued together. The product standard for CLT for both Canada and the US, ANSI/APA PRG-320 *Standard for Performance-Rated Cross-Laminated Timber*,[1] defines CLT as "a prefabricated solid engineered wood panel made of at least three orthogonally bonded layers of solid-sawn lumber or structural composite lumber (SCL) that are laminated by gluing of longitudinal and transverse layers with structural adhesives to form a solid rectangular-shaped, straight, and plane timber intended for roof, floor or wall applications." The panels are constructed such that each layer of lumber within the panel is oriented perpendicular to the next layer thereby providing strength and dimensional stability in both directions of the panel. Typical panel layups are in odd number of layers such as 3 plies, 5 plies and 7 plies typically providing thickness of approximately 105 mm, 175 mm and 245 mm, respectively, when the typical ply thickness of 35 mm is used. ANSI/APA PRG 320 contains many performance requirements that CLT panels must meet, including minimum elevated temperature performance criteria that the adhesives must meet.

One of the main advantages of using CLT panels is that they can be prefabricated prior to arriving on site, including preparation using CNC equipment that allows all necessary openings and joint details to be cut into the panel before being shipped to site. This can allow for very fast construction times – for example, some high-rise buildings have been built at a rate of one to two floors per week. Panels are often available in dimensions on the order of 3 m by 20 m.

To help designers utilize this relatively new product, the American Wood Council published design criteria for CLT in the 2015 edition of the National Design Specification (NDS) for Wood Construction[2]; and, the Canadian design standard, CSA O86-14 *Engineering design in wood,*[3] was updated in 2016 to include CLT. As part of the design standards, methodologies are now also available for calculating the standard fire resistance of CLT panels used in horizontal or vertical applications.

Nevertheless, there continues to be concern by building code committees and Authorities-having-jurisdiction (AHJs) when discussing mass-timber high-rise building design for fire resistance. This paper intends to provide the background information to the Canadian and US fire resistance design methodologies to demonstrate that the design methods that have been standardized do an excellent job of predicting the standard fire resistance of CLT wall and floor assemblies.

As mentioned, design methods for the calculation of standard fire-resistance ratings for CLT have been standardized in both Canada and the US. While the methods are similar, there are some differences such as the type of char rates (which affects the numerical value of the char rates used), how the effect of lamination thickness on the char rate is addressed, and how the loss in strength of heated wood beyond the char front is accounted for. For instance, the char rates used in the 2015 NDS and CSA O86-14 design standards are different for all wood products groups, as the NDS uses a non-linear char rate, while CSA O86-14 uses a system of linear ones. The two methods of calculating standard fire resistance are described in much greater depth later in this paper.

It is important to understand that the approach used by the two calculation methodologies to determine the fire-resistance rating of an assembly or structural element is not the same as what is used in regular, ambient-temperature, structural design. That is, in Canada the methodology used for regular structural design is limit states design (LSD) and in the US it is load and resistance factor design (LRFD), which use a probabilistic approach, whereas the two methods to calculate standard fire-resistance ratings are deterministic. While a move to probabilistic design for fire resistance would be a great step forward, the nature and cost of testing for fire resistance makes this approach unattainable at this point for any material, whether it is concrete, steel or timber. Furthermore, the deterministic approach is supported by the approach used in prescriptive building codes, which require a single successful fire-resistance test to demonstrate performance. Therefore, an engineering-based calculation methodology that has been validated based on a number of tests is likely better than relying on a single test.

The calculation methods used to determine the fire-resistance rating of timber members and assemblies in the Canadian and US design standards for wood construction are intended to predict the structural fire resistance as determined using the standard fire-resistance test methods CAN/ULC-S101-14[4] and ASTM E119[5], which are the test standards required for the determination of fire-resistance ratings in the National Building Code of Canada[4] (NBC) and the International Building Code[5] (IBC) in the US, respectively. Therefore, the calculations conducted in accordance with the design standards are simply predicting what the result of a full-scale standard fire-resistance test would be when carried out in accordance with the test standard using the standard fire exposure. Similar to all other construction materials, if the fire exposure changes, such as occurs in fire compartment experiments or in a real fire event, we can expect the actual performance of the construction to be different than in the standard test, including apparent charring rates.

FIRE-RESISTANCE TESTS ON CLT ASSEMBLIES

A large number of standard fire-resistance tests conducted in accordance with CAN/ULC-S101-14[6] and ASTM E119[7] have been completed over the last 6 years in Canada and the United States. Tests have been completed by research organizations such as the National Research Council of Canada (NRC) and FPInnovations, industry associations including the Canadian Wood Council and the American Wood Council, as well as manufacturers. Of these tests, some of the results are proprietary, and therefore, the test reports are not available publicly. Other tests have tested specific combinations of CLT, lumber framing or furring, insulation and gypsum board, which the design standard calculation methodologies are not able to directly address. The design standard CSA O86-14 is currently able to determine the fire resistance of CLT assemblies when left exposed or when protected by fire-rated gypsum board either directly applied or applied to relatively thin furring material. The NDS currently is limited to CLT assemblies that are left exposed (i.e. not protected by gypsum board). Therefore, only those tests with exposed CLT or CLT protected by gypsum board directly applied or fastened to furring without insulation can be predicted directly using the design standards. A total of 12 full-scale tests have been identified as meeting these criteria and are listed below in Table 1.

FPInnovations, in cooperation with the NRC, completed 8 full-scale fire-resistance tests in accordance with CAN/ULC-S101-14[6] and ASTM E119[7] on CLT wall and floor assemblies[8]. The CLT assemblies were sourced from two Canadian manufacturers and tested in varying thicknesses, with and without fire-rated gypsum board protection. A summary of the 8 tests including the assembly details, the applied load and the fire-resistance time achieved are included in Table 1. Also included in Table 1 are two tests completed by the Canadian Wood Council[9,10], one test completed by the American Wood Council[11] and one test completed by Nordic Strucutres[12].

Of the twelve tests available for comparisons, five tests were conducted with the CLT assembly exposed, five tests with a single layer of gypsum board protection and two tests with two layers of gypsum board protection on the exposed surface.

CANADIAN DESIGN STANDARD, CSA O86-14 (Update No. 1)

In May of 2016, an update (Update No. 1) was published to the design standard, CSA O86-14, *Engineering design in wood*[3], which added structural design criteria for CLT to the design standard. In addition to regular structural design criteria such as resistance in bending and compression, a calculation methodology was added to Annex B of the standard, which addressed design for structural fire resistance.

The design standard's fire-resistance calculation methodology uses what is called the "effective cross-section method". The "effective cross-section" of a member is used along with the full strength properties of the wood to calculate the member resistance after some duration of standard fire exposure. The effective cross-section is determined by subtracting the depth of char from the original cross-section as well as an additional "zero-strength layer" depth. The zero strength layer accounts for the various degrees of loss of strength of the portion of wood heated beyond the char front by subtracting an additional 7 mm of cross-section beyond the char depth, effectively assigning the layer zero strength. After subtracting the char layer and the zero strength layer, the effective cross-section remains and the capacity is calculated using the strength of the wood under non-fire conditions. The residual resistance of the timber element can then be calculated using the effective cross-section and compared to the imposed loads. If the resistance after a set period of standard fire exposure is greater than the imposed loads, it can be concluded that the structural fire resistance is equal to or greater than the duration of fire exposure used in calculating the effective cross-section. The effective cross-section method has been validated previously for use to predict the fire resistance of large solid-sawn timbers, glulam beams and columns, and timber decks (tongue-and-groove)[13].

Table 1. Summary of fire-resistance tests completed on CLT wall and floor assemblies.

Test	Assembly Details	Total Load (applied + self-weight)	Load Ratio‡	Fire Resistance
FPI-1	3-ply (3x38mm) floor assembly - E2 Stress Grade (pre product standard) - 2 layers of 12.7 mm Type X gypsum board	3.4 kPa M = 9.5 kN.m/m	31 %	*Failure did not occur, test stopped at 77 min
FPI-2	3-ply (3x38mm) wall assembly - E2 Stress Grade (pre product standard) - 2 layers of 12.7 mm Type X gypsum board	335 kN/m	72 %	106 minutes (structural failure)
FPI-3	5-ply (5x35mm) floor assembly - El Stress Grade - No gypsum board	12.5 kPa M = 35.0 kN.m/m	57 %	96 min (integrity failure)
FPI-4	5-ply (5x35mm) wall assembly - El Stress Grade - No gypsum board	333 kN/m	33 %	113 min (structural failure)
FPI-5	3-ply (3x35mm) floor assembly - V2 Stress Grade - 1 layer of 15.9 mm Type X gypsum board	2.94 kPa	67 %	86 min (integrity failure)
FPI-6	5-ply (5x35mm) floor assembly - V2 Stress Grade - 1 layer of 15.9 mm Type X gypsum board	8.27 kPa M = 23.1 kN.m/m	93 %	124 min (integrity failure)
FPI-7	7-ply (7x35mm) floor assembly - V2 Stress Grade - No gypsum board	15.6 kPa M = 43.7 kN.m/m	92 %	179 min (Structural failure)
FPI-8	5-ply (5x21mm) wall assembly - V2 Stress Grade - No gypsum board	73.3 kN/m	31 %	57 min (structural failure)
CWC-1	3-ply (3x35mm) wall assembly - V2 Stress Grade - 1 layer of 15.9 mm Type X gypsum board	120 kN/m	60%	76 min (structural failure)
CWC-2	3-ply (3x35mm) wall assembly - V2 Stress Grade - 1 layer of 15.9 mm Type X gypsum board	200 kN/m	100%	66 min (structural failure)
AWC-1	5-ply (5x35mm) wall assembly - El Stress Grade - 1 layer of 15.9 mm Type X gypsum board	129 kN/m	13 %	185 min (structural failure)
INT-1	3-ply (3x35mm) wall assembly - El Stress Grade - No gypsum board	296 kN/m	73 %	32 min

‡ The load ratios are based on CAN/ULC-S101-14 and CSA O86-14 using limit states design and slightly different ratio values are expected if the assembly capacity was based on the 2015 NDS.

The design standard's fire-resistance calculation methodology uses what is called the "effective cross-section method". The "effective cross-section" of a member is used along with the full strength properties of the wood to calculate the member resistance after some duration of standard fire exposure. The effective cross-section is determined by subtracting the depth of char from the original cross-section as well as an additional "zero-strength layer" depth. The zero strength layer accounts for the various degrees of loss of strength of the portion of wood heated beyond the char front by subtracting an additional 7 mm of cross-section beyond the char depth, effectively assigning the layer zero strength. After subtracting the char layer and the zero strength layer, the effective cross-section remains and the capacity is calculated using the strength of the wood under non-fire conditions. The residual resistance of the timber element can then be calculated using the effective cross-section and compared to the imposed loads. If the resistance after a set period of standard fire exposure is greater than the imposed loads, it can be concluded that the structural fire resistance is equal to or greater than the duration of fire exposure used in calculating the effective cross-section. The effective cross-section method has been validated previously for use to predict the fire resistance of large solid-sawn timbers, glulam beams and columns, and timber decks (tongue-and-groove)[13].

With respect to the loads to be used in calculating the fire resistance, the same approach is used as for full-scale fire-resistance testing. In Canada, that means the specified loads are used to calculate the fire resistance (i.e. Dead + Live due to gravity loads) as specified in CAN/ULC-S101-14[6].

As previously mentioned, the intent of the standard is to provide the same fire-resistance rating that would be obtained by running a full-scale standard fire-resistance test. Therefore, as opposed to regular design where 5th percentile strength and stiffness values are used to calculate the member or assembly's factored resistance, the fire resistance is determined using average strength values. The strength values from the design standard are multiplied by a modification factor to convert from the 5th percentile strength value to the average strength value. Similarly, the resistance factor, which is typically 0.8 for compression members and 0.9 for bending members, which is used to account for variability in resistance, is assigned a value of 1.0 for calculating fire resistance. Lastly, the strength value is increased slightly based on duration of load factor (for short-term loading), since a fire event occurs over a relatively short period of time and, for shorter durations of loading, timber members exhibit greater strength.

The char rate used in calculating the fire resistance value is a critical factor since it determines the remaining cross-section that is available to resist the applied load. Most wood products experience similar char rates and, for design purposes, use the same char rate. However, the char rates used for CLT panels in this methodology are different since it has been widely published that some adhesives allow for falling-off of the CLT plies as each one becomes fully charred[8,14,15]. This is particularly important in calculating the char rate since the loss of the protective char layer has the effect of initially increasing the char rate on the remaining cross-section. The CSA standard addresses this influence by prescribing two char rates. If the char depth does not reach the first adhesive bond interface between plies during the fire duration, the standard one-dimensional char rate of 0.65 mm/min is to be used. However, if the char depth passes the first adhesive bond interface during the fire exposure, the notional char rate of 0.8 mm/min is to be used. This notional char rate is used to account for the initial increase in char rate that occurs when each lamination falls off, exposing non-charred wood that now has no insulating char layer to protect it. The use of the notional (i.e. effective) char rate of 0.8 mm/min was a simple solution to address the increased char rates associated with CLT while being conservative enough to be applicable to all lamination thicknesses currently permitted under the ANSI/APA PRG 320 CLT product standard[1]. This is because CLT panels with thinner layers tends to char at a higher effective rate than panels with thicker laminations when the adhesive allows the charred layers to fall off. This simple approach of using a higher notional charring rate, which implicitly accounts for the falling-off of charred layers, was chosen since it reflects the current practice for other wood products in the standard to use a notional char rate to account for various effects that effectively reduce the cross-section of the wood member. This includes glued-laminated timber, where a slightly higher notional char rate implicitly accounts for corner rounding of beams and columns.

Once the char depth has been determined in accordance with the Annex B methodology, an additional 7 mm zero strength zone is subtracted to account for the loss of strength of the heated wood beyond the char front. This zero strength zone depth is the same value used for all other wood products such as solid sawn timber and glued-laminated timber.

One aspect that differentiates calculating the fire resistance of CLT apart from other engineered wood products is the need to calculate section properties such as the location of the neutral axis, the moment of inertia in order to calculate the section modules (for beams), and slenderness (for columns) at each iteration since the cross-sections are not homogeneous. The strength and stiffness of the cross-plies of the CLT panel are much lower than the longitudinal plies and, therefore, for fire-resistance design, are often ignored as a conservative approach. The need to recalculate the location of the neutral axis, moment of inertia and other properties at each iteration is demonstrated in Figure 1. The cross-section on the left of Figure 1 is a 5-ply CLT panel before fire exposure, which has the neutral axis located at mid-depth, making the calculation of the section properties relatively straight forward. The cross-section on the right of Figure 1 is a 5-ply CLT panel with some duration of fire exposure (from the "bottom" side, in the Figure), causing half of the first ply to be charred. In this case, the neutral axis starts to migrate towards the top of the panel (the opposite side from the fire exposure) as soon as the strength properties of the bottom ply are affected. This moving of the neutral axis requires the moment of inertia and section modulus or slenderness to be calculated at each fire exposure time for which the capacity of the assembly is sought.

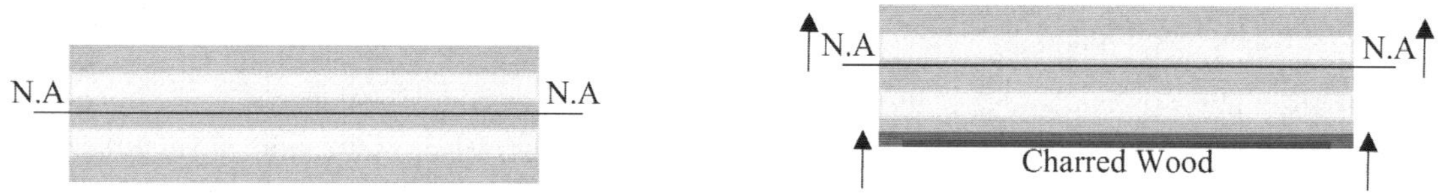

Figure 1. CLT cross-section (5-ply) with movement of neutral axis (N.A) indicated as CLT panel starts to char.

Gypsum board protection is common on CLT assemblies and the CSA O86-14 design standard's Annex B includes information in order to account for fire-rated gypsum board protection. The times in the Annex are based on research conducted by Dr. White at the US Forest Products Laboratory, where tests were conducted on mass-timber beams with gypsum board protection[16] (with the exception of the times for two layers of 12.7 mm Type X gypsum board, which are based on tests FPI-1 and 2[8] listed in Table 1). The Annex includes times which can be added to the fire-resistance time calculated for the CLT panel alone when one or two layers of 12.7 mm or 15.9 mm Type X gypsum board is applied. The following table summarises the times to be added to the fire resistance of CLT associated with gypsum protection.

Table 2. Additional fire resistance provided by gypsum board as specified in CSA O86-14

Gypsum Board Protection	Number of Layers	Additional Fire Resistance
12.7 mm Type X	1	15
	2	60
15.9 mm Type X	1	30
	2	60

US DESIGN STANDARD, 2015 NDS

The National Design Standard for Wood Construction (NDS) published by the American Wood Council included CLT for the first time in the 2015 edition. With respect to fire resistance, the NDS approach is very similar to that of the Canadian Design Standard previously described, as the effective cross-section method is used.

However, one significant difference between the methods is in the char depth determination. While the method in Annex B of the Canadian standard, CSA O86-14, uses a linear char rate for CLT irrespective of lamination thickness (within the thicknesses currently permitted by the ANSI/APA PRG 320 CLT product standard[1], the NDS method uses a non-linear equation that is a function of the lamination thickness and proximity to the lamination bond line.

The equation in the 2015 NDS for calculating the char depth for CLT is as follows:

$$a_{char} = 1.2\left[n_{lam}\, h_{lam} + \beta_n \left(t - \left(n_{lam} t_{gi}\right)\right)^{0.813}\right] \quad [1]$$

$$t_{gi} = \left(\frac{h_{lam}}{\beta_n}\right) \quad [2]$$

$$n_{lam} = \frac{t}{t_{gi}} \quad [3]$$

where:
a_{char} = effective char layer thickness (in.)
β_n = nominal char rate (in./hr.)
t = exposure time (hr.)
t_{gi} = time for char front to reach glued interface (hr.)
h_{lam} = lamination thickness (in.)
n_{lam} = number of laminations charred (rounded to lowest integer)

Equation [1] above uses a non-linear char rate, which increases significantly after passing a bond line to reflect the increased char rate when a lamination of CLT falls off. Once the char rate passes a bond line in the CLT, Equation [1] provides for a char rate of 3.8 mm/min in the first minute, reducing to 1.4 mm/minute in the next minute and 1.1 mm/min in the minute after. The char rate then slowly decreases to approximately 0.6 mm/minute as the char front progresses through the CLT ply before reaching the next bondline and increasing again. Also, note that the 1.2 value in equation [1] effectively increases the total effective char depth by 20%, which historically accounts for the loss of strength of the heated wood beyond the char front in a similar manner to the use of a zero-strength zone depth.

A comparison between the char depth used in CSA O86-14 and the 2015 NDS for CLT is shown in Figure 2. As described earlier, CSA O86-14 uses a char rate of 0.65 mm/min if the char depth does not reach a glue line (indicated as CSA O86(3-ply) in Figure 2) and 0.8 mm/min if the char depth passes at least one glue line during the fire exposure duration (CSA(CLT)). Two char depths are provided using the 2015 NDS, one with 19 mm thick laminations (2015 NDS(19 mm)) and one with 35 mm thick laminations (2015 NDS(35 mm)). The 35 mm lamination thickness represents the most common thickness used in North American CLT while the 19 mm thickness provides a thinner thickness to demonstrate the impact of lamination thickness on the effective char depth following the NDS method. It is evident from Figure 2 that the simplified char rate used in CSA O86-14 provides a relatively conservative estimate of char depth compared to the more involved calculation in the 2015 NDS. Because the calculation is more onerous in the 2015 NDS, meaning a spreadsheet is necessary to calculate the effective cross-section, a table is provided in the standard for common lamination thicknesses and exposure durations which is reproduced below in Table 3.

Besides the differences in the approach to char rate, the general approach is consistent between the two design standards. For example, similar to the CSA O86-14 design standard method, the 2015 NDS method uses an adjustment factor to convert from design strength to mean strength for the design of fire resistance.

One item that is not currently included in the 2015 NDS is guidance on the contribution of gypsum board to the fire resistance of mass-timber members, including CLT panels. Without this, the methodology is currently limited due to not being able to calculate the fire resistance of CLT assemblies which have gypsum board protection. This is reflected below in the comparison to test data where only 5 of the 12 assemblies can be compared with the 2015 NDS predictions since the remaining 7 had gypsum board protection of one or two layers. It is expected that guidance on the contribution of gypsum board to the fire resistance of mass-timber members will be added in the near future and some work is already underway.

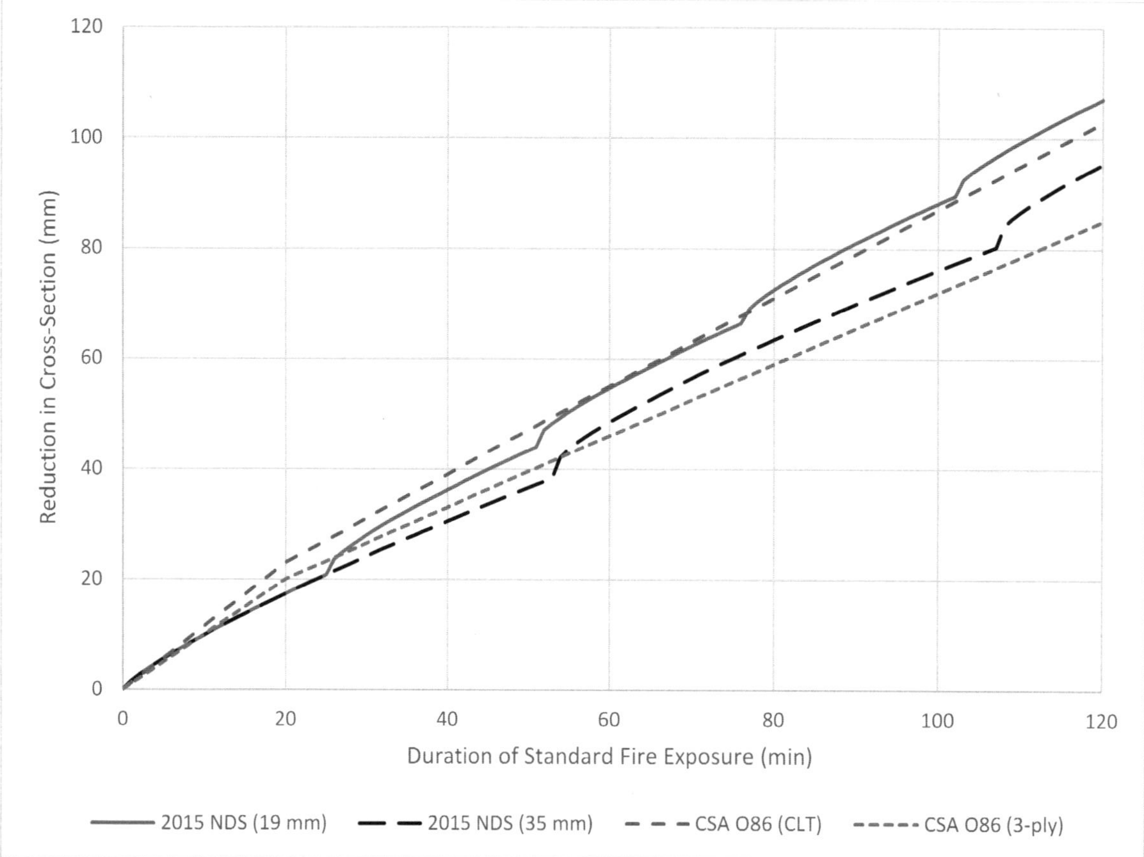

Figure 2. Comparison between char rates specified in CSA O86-14 for CLT and 2015 NDS for CLT with 19 mm thick and 35 mm thick laminations.

Table 3. Effective char depth for CLT as specified in 2015 NDS[2].

Required Fire Endurance (hr.)	Lamination Thickness, h_{lam} inches (mm)								
	5/8 (16)	3/4 (19)	7/8 (22)	1 (25)	1-1/4 (32)	1-3/8 (35)	1-1/2 (38)	1-3/4 (44)	2 (51)
1-hr.	2.2 (56)	2.2 (56)	2.1 (53)	2.0 (51)	2.0 (51)	1.9 (48)	1.8 (46)	1.8 (46)	1.8 (46)
1½ -hr.	3.4 (86)	3.2 (81)	3.1 (79)	3.0 (76)	2.9 (74)	2.8 (71)	2.8 (71)	2.8 (71)	2.8 (71)
2-hr.	4.4 (112)	4.3 (109)	4.1 (104)	4.0 (101)	3.9 (99)	3.8 (97)	3.6 (91)	3.6 (91)	3.6 (91)

Note: Dimensions in millimetres have been added to Table 3 in parentheses.

COMPARISON OF DESIGN STANDARDS AND TEST RESULTS

As previously discussed, a total of 12 tests are available for comparison to the predictions of the design standard methods. Table 4 summarizes a comparison between the results of the two design standard methodologies with the 12 test results. Fire-resistance times can be calculated for all 12 tests using the CSA O86-14 Annex B method since it includes information regarding the contribution of gypsum board to fire resistance. Conversely, comparisons to the NDS method are limited to 5 of the 12 tests due to no provisions for the protection provided by the gypsum board being currently provided with that method. For the most part, the design standards do an excellent job of predicting the fire resistance of CLT assemblies. For the instances where the fire resistance is over-predicted, it is limited to just a couple of minutes, which is within the repeatability of the test method.

Table 4. Summary of fire resistance tests completed on CLT wall and floor assemblies.

Test	Assembly Details	Fire Resistance	CSA O86-14, Annex B	NDS-2015‡
FPI-1	3-ply (3x38mm) floor assembly - E2 Stress Grade - 2 layers of 12.7 mm Type X gypsum board	*Failure did not occur, test stopped at 77 min (However, Test 5 got 86 min with one layer of GB)	101 min	N/A
FPI-2	3-ply (3x38mm) wall assembly - E2 Stress Grade - 2 layers of 12.7 mm Type X gypsum board	106 min (structural failure)	96 min	N/A
FPI-3	5-ply (5x35mm) floor assembly - E1 Stress Grade - No gypsum board	96 min (integrity failure)	98 min	114 min§
FPI-4	5-ply (5x35mm) wall assembly - E1 Stress Grade - No gypsum board	113 min (structural failure)	106 min	111 min
FPI-5	3-ply (3x35mm) floor assembly - V2 Stress Grade - 1 layer of 15.9 mm Type X gypsum board	86 min (integrity failure)	66 min	N/A
FPI-6	5-ply (5x35mm) floor assembly - V2 Stress Grade - 1 layer of 15.9 mm Type X gypsum board	124 min (integrity failure)	120 min	N/A
FPI-7	7-ply (7x35mm) floor assembly - V2 Stress Grade - No gypsum board	179 min (Structural failure)	102 min	178 min
FPI-8	5-ply (5x21mm) wall assembly - V2 Stress Grade - No gypsum board	57 min (structural failure)	58 min	61 min
CWC-1	3-ply (3x35mm) wall assembly - V2 Stress Grade - 1 layer of 15.9 mm Type X gypsum board	76 min (structural failure)	67 min	N/A
CWC-2	3-ply (3x35mm) wall assembly - V2 Stress Grade - 1 layer of 15.9 mm Type X gypsum board	66 min (structural failure)	63 min	N/A
AWC-1	5-ply (5x35mm) wall assembly - E1 Stress Grade - 1 layer of 15.9 mm Type X gypsum board	185 min (structural failure)	174 min	N/A
INT-1	3-ply (3x35mm) wall assembly - E1 Stress Grade - No gypsum board	32 min	32 min	31 min

‡ NDS calculated fire-resistance values taken from American Wood Council's Technical Report 10[13].

§ American Wood Council's Technical Report 10 does not compare this value to the test result since the test result has been reported as integrity failure while the calculation is for structural failure.

A graphical comparison between the test results and the calculated fire resistance based on CSA O86-14 is provided in Figure 3. The calculated fire resistance is in excellent agreement with the test results with the exception of test FPI-7 where the CSA O86-14 method is significantly conservative. This large difference between 179 minutes in the test method and 102 minutes calculated is due to the nature of how CLT assemblies experience a reduction in capacity as exposed to fire. While the fire is causing a lamination of the CLT in the longitudinal direction to char, the strength of the panel is decreasing. However, while a lamination of the CLT in the perpendicular direction is charring, there is no decrease in resistance of the panel. In the case of FPI-7, the failure is calculated to occur when the char front has progressed approximately half way through the third ply while the CLT assembly during the standard test did not fail until the char front reached the fifth ply. The cross-ply between the third and fifth longitudinal plies takes 44 minute to char through based on the char rate or 0.8 mm/min, which accounts for significant portion of the discrepancy.

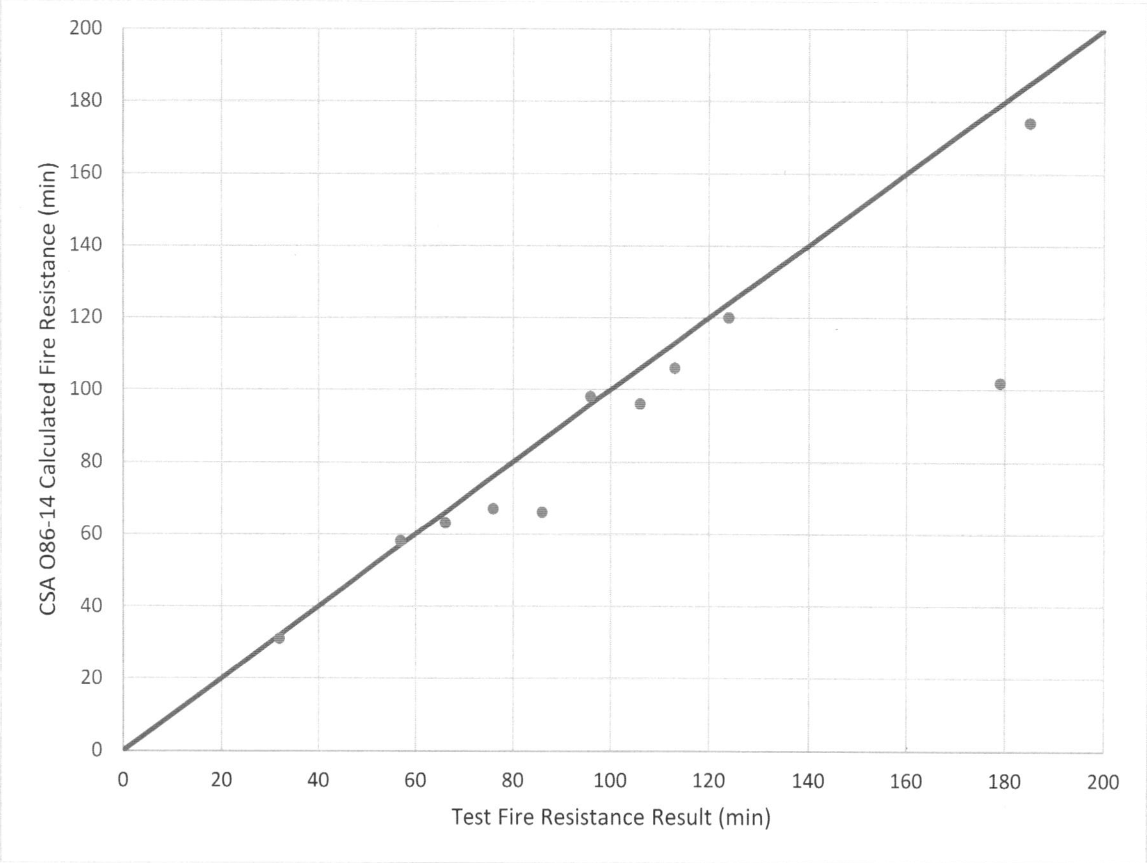

Figure 3. Comparison between CSA O86-14 calculated fire resistance and fire resistance test results.

Based on the excellent agreement between the test data and the calculated fire resistance in accordance with CSA O86-14, it is believed that the reliability of the fire-resistance rating is greater when based on the design standard methodology compared to the building code requirement of completing a single successful fire-resistance test. The calculation methodology has the benefit of being validated based on a larger number of tests as well as the ability to consider the specific construction details in the determination of fire resistance. For example, the fire resistance of a beam with a 20 m long span can be calculated complete with consideration for lateral restraint while the alternative is to base the design on test method which is capable of testing a 5-m-long span where the lateral restrain conditions likely will not be accurately reproduced. Similarly for columns, some designs include eccentricity in the column connections which can be accounted for when calculating the fire resistance using the design standard but typically cannot be addressed when relying on standard fire-resistance tests. For these reasons, it is believed that designing CLT assemblies for fire resistance in accordance with the respective design standard methods provides a more reliable fire resistance design when the methodology has been validated as presented in this paper.

Lastly, connections are a critical component of any structure and the fire resistance of connections is no less important than the fire resistance of the structure and assemblies. The CSA O86-14 standard provides guidance in the design of connections stating that provided the steel components of a connection are embedded within what ends up being the effective cross-section (within the non-charred cross-section which remains), the steel is considered protected. Similarly, the 2015 NDS specifies connectors and fasteners shall be protected from fire exposure by wood, fire-rated gypsum board or any approved coating.

In the case of cross-laminated timber assemblies, it is generally relatively easy to conceal fasteners. Fasteners connecting panels are typically self-taping screws embedded in the CLT, which behave very well when exposed to fire. Note that all the fire resistance tests listed in Tables 2 and 3 include self-taping screws connecting the CLT panels together. With attention to detail, the fasteners used in connecting CLT panels to each other, as well as to post-and-beam structures, can easily meet the fire-resistance rating required.

CONCLUSIONS

It is evident from the comparison between the predictions of the design standards' methodologies and the twelve standard fire-resistance tests that the agreement between the two is excellent. In fact, it is expected that predictions using the design standard methodologies provide a more reliable method of determining fire resistance since the methods have been validated based on the results of many fire resistance tests while the building codes in Canada and the US will accept the fire resistance of an assembly or structural element based on a single test. Additional advantages to using the calculation methods specified in the design standards includes the fact that the actual dimensions, loads and loading conditions can be directly accounted for when the fire resistance is calculated, as opposed to the implicit restrictions and limitations when using the standard fire resistance test furnace. Considerations such as point loads on a floor assembly or eccentricity in a wall assembly are not part of standard fire-resistance testing, however they do exist in buildings. By using the design standards' methods to calculate fire resistance, the specific loads, dimensions, and other structural considerations can be included in the determination, resulting in designs which have a higher reliability in terms of fire resistance.

REFERENCES

1. ANSI/APA PRG 320-2012, "Standard for Performance Rated Cross-Laminated Timber" APA-The Engineered Wood Association, Tacoma, WA.
2. National Design Specification for Wood Construction (NDS). 2015. American Wood Council. Leesburg, VA.
3. CSA O86-14 (Update No. 1 – May 2016), "Engineering design in wood", CSA Group, Mississauga, Canada.
4. 2015 National Building Code of Canada (NBC), National Research Council of Canada, Ottawa, ON.
5. 2015 International Building Code (IBC), International Code Council, Washington, DC.
6. CAN/ULC-S101-14, Standard Methods of Fire Endurance Tests of Building Construction and Materials. Scarborough, ON: ULC.
7. ASTM E119-16a. Standard Test Methods for Fire Tests of Building Construction and Materials, ASTM, Philadelphia.
8. Osborne, L. Dagenais, C and Benichou N. Preliminary CLT Fire Resistance Testing Report. Joint publication between FPInnovations and the National Research Council. FPInnovations Project No. 301006155, July 2012.
9. Intertek Test Report prepared for Canadian Wood Council, CAN/ULC-S101. Test Report No. No. 100585447SAT-002B. December 30, 2013.
10. Intertek Test Report prepared for Canadian Wood Council, CAN/ULC-S101. Test Report No. No. 101700231SAT-003_Rev.1. November 24, 2014.
11. NGC Testing Services Test Report for American Wood Council, ASTM E119-11a. Test Report No. WP-1950. October 15, 2012
12. Intertek Test Report prepared for Nordic Engineered Wood, CAN/ULC-S101. Test Report No. No. 100668113SAT-002A_Rev.1. July 26, 2012.
13. Technical Report No. 10 – Calculating the Fire Resistance of Exposed Wood Members (2016).

American Wood Council. Leesburg, VA.
14. Frangi, A., Fontana, M., Hugi, E. and Jobstl, R. (2009) "Experimental analysis of cross-laminated timber panels in fire", *Fire Safety Journal*, vol. 44, pp. 1078-1087.
15. Craft, S.T., Desjardins, R. and Mehaffey, J.R. (2011) "Investigation of the Behaviour of CLT Panels Exposed to Fire", Proceedings of the Twelfth International Conference Fire and Materials held 31st January – 2nd February, 2011.
16. White, R. (2009) Fire Resistance of Wood Members With Directly Applied Protection. Proceedings of the Eleventh International Conference Fire and Materials held 26-28th January, 2009.

WILDLAND

STATISTICAL ANALYSIS ON FIREBRAND GENERATION FROM STRUCTURAL FUELS

Faraz Hedayati & Aixi Zhou

University of North Carolina at Charlotte, Charlotte, NC, USA

ABSTRACT

In large scale fires, spot ignitions away from the fire front caused by firebrands are of a major mechanism in spreading of fire. The firebrand phenomenon can be described in three phases namely, generation, transport and ignition. In the last fifty years, although a lot of studies have been conducted, the generation phase is probably the phase that we have the least knowledge about. In this study, available experimental data from 1965 to 2015 about the generation of firebrands from structural fuels were reviewed and categorized into three sections (Outdoor Whole Buildings, Indoor Whole Buildings, and Indoor Structural Components) and analyzed. The novelty of this study is presenting a Probability Density Function (PDF) for each experiment which considers the correlation between different parameters such as traveling distance, mass, surface area etc. Based on the correlation factor and the previous observations, we employed the multivariate truncated Gaussian distribution to find the PDF for the brands and presented a statistical frame work to approach the firebrand phenomenon. The obtained PDFs can provide us with valuable information about the total heat flux of a firebrand shower as well as the probability of having specific sizes or masses among the whole firebrand population. The results show that the correlation factor between mass and projected area of the brands are much stronger than that of between distance from the firebrand and projected area. Moreover, it was observed that the distribution of brands at closer distances to the fire was affected by the brand lofted from the corners and at the further distances, the whole building generates the brand like a flat wall, regardless of the shape. Finally, the effects of roof size and the pressure underneath the roof on the generation were discussed.

INTRODUCTION

Spreading of fires occurs with three main mechanisms; direct flame contact, radiant/convective heat flux and firebrands. The first two mechanisms get weakened with distance noticeably. For example, it has been reported that the direct flame mechanism is weak enough that it is incapable of initiating another structure when the distance is more than 30.5 meters (100 feet) [1]. On the other hand, firebrands can initiate a spot fire miles away from the fire line. These flying hot particles can be either burning combustible materials, like vegetation, or high temperature non-combustible materials, like metals, depending on their origin at which they have been lofted.

Firebrands play an important role in fire spread in large scale fires, such as fires in the wildland and Wildland Urban Interface (WUI). In order to mitigate the fire hazards caused by firebrands, it is necessary to understand the firebrand (or spotting) phenomenon. The firebrand phenomenon can be described in three steps: firebrand production, firebrand transport, and firebrand ignition of recipient fuel. Several recent review articles (Koo et.al.[2]; Mell et.al.[3]; Manzello et.al.[4]; Caton et.al.[5]) may serve as a good source of information for the spotting phenomenon. From the literature, considerable work on firebrand transport has been conducted, but much less has been done to understand firebrand production and firebrand ignition of fuel. Firebrand production is the first step and the basis for understanding the subsequent transport and ignition processes. For example, by estimating the heat flux that a firebrand shower can deliver to the recipient fuel, we can predict the growth and spread of large scale fires more accurately.

The motivation of the current paper is to fill the gap for the generation phase by reviewing all the available experimental data in the last 50 years and statistically analyzing them. To achieve this goal, a comprehensive statistical analysis was conducted for each available data set by presenting a probability density function (PDF) and calculating the mean, variance and median for mass, projected area and flying distance. Moreover, a confidence interval has been presented for each study to have an estimation of the population mean. For studies after 2000, we contacted the authors and, upon responses, requested for raw data of their studies. When raw data could not be obtained from the authors, we extracted experimental data from tables or figures in the articles. Table 1 represents all the studies and the details for the structural fires.

Table 1. Summary of firebrand generation experiments using structural fuels

Experiment in Literature	**Structure ID**	**Testing conditions**[1] wind speed, Temperature, relative humidity	**Height** (stories)	**Collection Method**
Outdoor Whole Building (OWB) Experiments				
Vodvarka (1969)[6], Star Center	OWB1	5.3-8.9 m/s, 1.6 °C, 89%	2	N.A.
Vodvarka (1969)[6], Batavia	OWB2	4.4-8 m/s, 21.1 °C	1	
Vodvarka (1969)[6], Heartview	OWB3	8.9-11.1 m/s, 4.4 °C	1	
Vodvarka (1969)[6], Building 68-1	OWB4	0.9-1.7 m/s, 28.8 °C, 65%	2	Polyethylene sheets
Vodvarka (1969)[6], Building 68-2	OWB5	0.4-1.3 m/s, 30 °C, 70%	1.5	
Vodvarka (1969)[6], Building 68-3	OWB6	2.6-3.5 m/s, 28.3 °C, 70%	1.5	
Vodvarka (1970)[7], Building 69-1	OWB7	< 2.2 m/s, 21.1 °C, 75%	2.5	Plastic sheets
Vodvarka (1970)[7], Building 69-7	OWB8	2.2-15.6 m/s, 13.8 °C, 45%	2	
Suzuki et al (2012)[8], at 4m	OWB9a	5.7 m/s, 25 °C	2	Water pans
Suzuki et al (2012)[8], at 18m	OWB9b	5.7 m/s, 25 °C	2	
Indoor Whole Building (IWB) Experiments				
Yoshioka et al. (2004)[9], wet	IWB1a	2-4 m/s	1	Trays
Yoshioka et al. (2004)[9], dry	IWB1b	2-4 m/s	1	
Suzuki et al., 2014[10]	IWB2	6 m/s	1	
Indoor Structural Component (ISC) Experiments				
Waterman (1969)[11], all	ISC1	6.9-9.1 m/s	N.A.	Mesh screen
Suzuki et al. (2013)[12], walls	ISC2	6 m/s		Water pans
Suzuki et al. (2013)[12], corners, 6 m/s	ISC3a	6 m/s		
Suzuki et al. (2013)[12], corners, 8 m/s	ISC3b	8 m/s		

This paper is organized as the following. First, the statistical approach and the assumptions used for this study is described and the benefits of the introduced PDFs and their applications are discussed. Then, a summary of research related to firebrand production from structural fuels is presented in a table. After this, a detailed rendering of analyzed data for physical and combustion characteristics of firebrand generation, as well as factors affecting firebrand generation are provided. Finally the conclusions are drawn.

STATISTICAL APPROACH

As mentioned earlier, the generation/distribution phase of the firebrands have not been studied in deep yet. Usually the sample size of the available data is not large enough that helps us to make a firm conclusion. Only recently, Zhou et.al. 2015[13], conducted a comprehensive experiment with a large sample size that studied wind speed effects on the distribution of the embers. They concluded that the mass distributed under a normal distribution curve. Their conclusion makes sense since based on the Central Limit Theorem, when the sample size is large enough, the distribution gets closer to the Gaussian distribution. In fact, the Gaussian distribution is considered the most important distribution of all probability distributions in many practical engineering problems (Bird, 2014[14]).

From physical point of view, the distribution of the firebrands depends on both fuel characteristics (fuel moisture content, mass, surface area, aerodynamic shape) and environmental characteristics (wind speed, humidity, and temperature). The novelty of this study is to consider the effects of each parameter as well as their mutual correlation with each other on the final distribution.

Considering the previous conclusions about the distribution of the embers and bearing in mind the mutual effects of each parameter on the other one, we suggest the multivariate normal distribution can be a good fit to the experimental data. In addition, since all the physical quantities of interest like mass and surface area cannot have a negative value, whenever it was necessary, we modified our model by truncating the Gaussian curve. This can be easily done by setting the values in the negative range to zero and normally scaling the PDF in the positive range of physical quantities in a way that the total double integral equals to unity.

In Figures 1-4, we have evaluated the accuracy of our proposed PDF for the distribution of the firebrands with the experimental data. The bars represents the experimental data and the curve represents the model. For the sake of brevity, the validating curves have been presented for one study in each category of OWB, IWB and firebrand generators. In worth noting that since the volume beneath a standard PDF is unity, we have multiplied the curve by a constant to scale it up over the bars. Since 2D graphs are usually easier to visualize, we only present the 3D graphs for validation and the rest are depicted by 2D contour graphs.

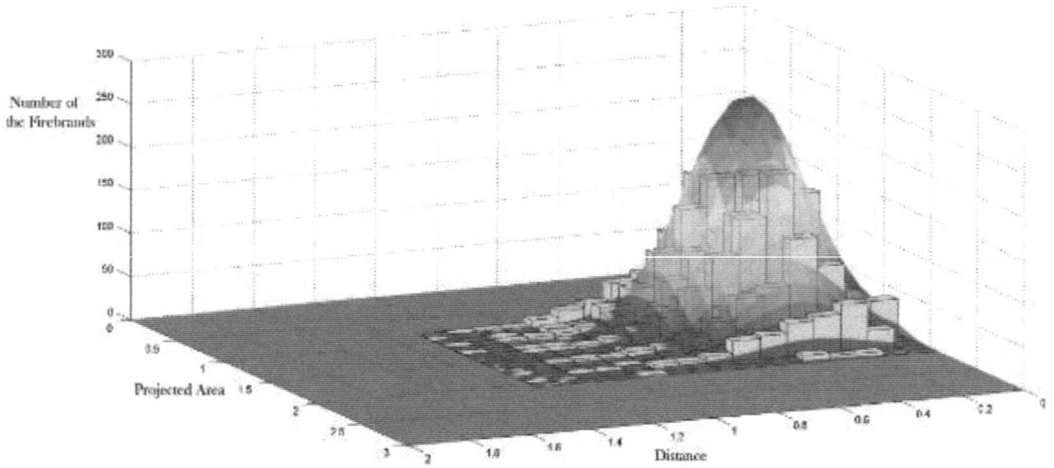

Figure 1. Validation Curves for distance and projected area at 0 wind speed (Zhou et al., 2015)

The mathematical formulation for such a distribution with p variables is as the following:

$$P(x_1, x_2, x_3, ..., x_p) = (\frac{1}{2\pi})^{p/2} |\Sigma|^{-\frac{1}{2}} e^{\left[-\frac{1}{2}(x-\mu)' \Sigma^{-1}(x-\mu)\right]} \quad [1]$$

where $\mu = (\mu_1, \mu_2, \mu_3, ..., \mu_p)$ is the vector of means and Σ is the variance-covariance matrix. For a two dimensional distribution, the above equation can be simplified to the following one:

$$P(x,y) = \frac{1}{2\pi\sigma_1\sigma_2\sqrt{1-\rho^2}} e^{\left[\frac{-1}{2(1-\rho^2)}\left[\frac{(x-\mu_1)^2}{\sigma_1^2} - \frac{2\rho(x-\mu_1)(y-\mu_2)}{\sigma_1\sigma_2} + \frac{(y-\mu_2)^2}{\sigma_2^2}\right]\right]}$$

$$\rho = cor(x,y) = \frac{\mathrm{cov}(x,y)}{\sigma_1\sigma_2} \qquad [2]$$

where μ is the mean, σ is the standard deviation and ρ is the correlation factor. The latter term presents how strong the two quantities are correlated regardless of their relation. Table 2 presents the values of the parameters in Equation 2, which have been obtained after the analysis for each structure.

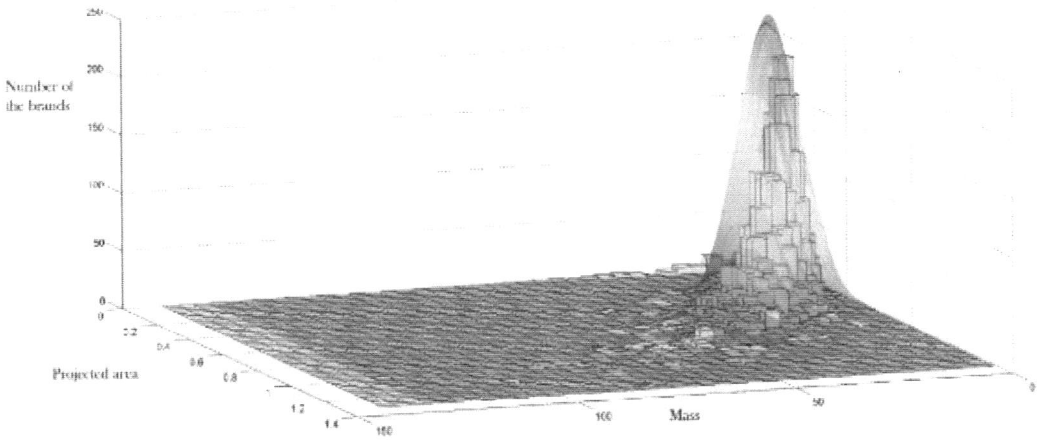

Figure 2. Validation Curves for mass and projected area at 0 wind speed (Zhou et al., 2015)

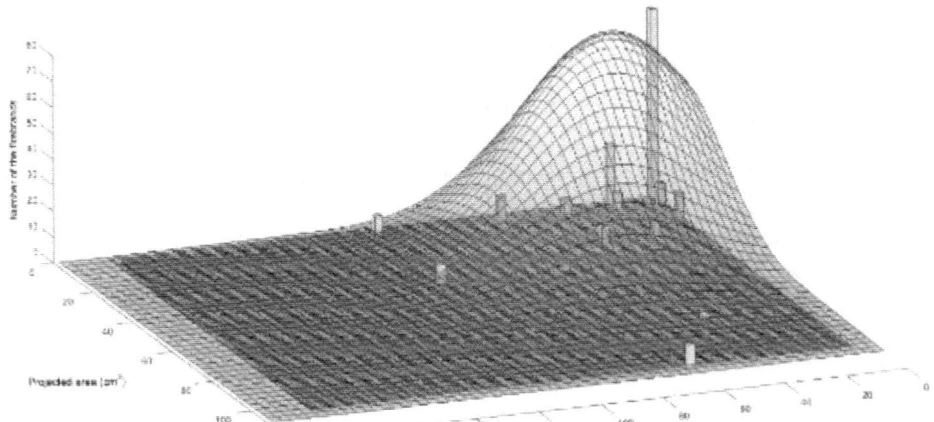

Figure 3. Validation for Batavia Building, distant and projected area (Vodvarka 1969)

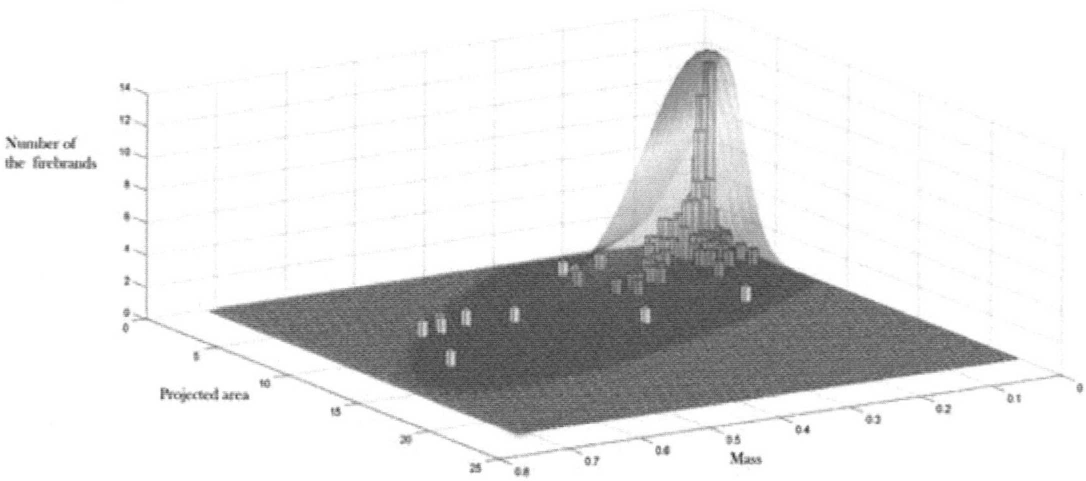

Figure 4. Validation for mass and projected area (Suzuki et al., 2014)

Statistical approach is the best way to study the problems whose nature is stochastic. Wherever a system includes numerous particles, like the firebrand phenomenon, it is practically impossible to study each particle individually. As discussed earlier, in the current survey, we employed a bivariate Gaussian distribution as PDF to statistically model firebrands regardless of their formation mechanisms and transport. In practice, the PDFs can be used to estimate the probability and distribution of firebrands landing on locations of concern. There are several criteria that can evaluate the ignition of a material such as critical surface temperature, critical heat flux, and critical mass loss rate. The PDFs can help us to predict the critical values that can initiate a fire.

Table 2. Parameters for firebrands in each experiment obtained from statistical analysis

ID	Flying Distance x (m)			Projected Area a (cm^2)			Mass m (g)			$\rho_{x,a}$	$\rho_{m,a}$
	μ_x	σ_x	$\tilde{x}$	μ_a	σ_a	$\tilde{a}$	μ_m	σ_m	$\tilde{m}$		
OWB1	210.91	32.88	182.88	71.96	60.44	58.06				-0.49	
OWB2	29.51	31.41	15.24	12.94	23.25	0.25				0.41	
OWB3	103.3	32.73	76.20	24.66	25.71	25.80				-0.11	
OWB4	30.14	12.43	21.33	2.22	4.41	1.35	N.A.			0.13	N.A.
OWB5	49.81	11.11	50.29	1.88	3.49	1.35				0.04	
OWB6	46.07	12.62	45.72	1.10	1.22	0.45				0.03	
OWB7	56.10	32.40	38.1	6.34	8.32	2.71				-0.15	
OWB8	45.04	22.06	47.24	2.85	1.02	2.71				-0.04	
OWB9a				2.03	2.70	1.23	0.06	0.10	0.03		0.89
OWB9b				4.50	11.25	1.92	0.12	0.16	0.04		0.51
IWB1a				0.83	0.57	0.62	0.01	0.02	0.01		0.96
IWB1b				1.50	1.26	0.62	0.06	0.07	0.02		0.99
IWB2	N.A.			3.26	4.93	1.74	0.67	0.67	0.07	N.A.	0.98
ISC1				N.A.			0.12	0.05	0.11		N.A.
ISC2				2.27	2.87	1.31	0.16	0.33	0.05		0.59
ISC3a				3.13	3.19	2.18	0.30	0.44	0.17		0.94
ISC3b				4.52	4.08	3.11	0.56	0.72	0.30		0.86

With the knowledge of firebrand surface temperature and heat flux as well as total potential heat energy, one can check whether the recipient fuel reaches these critical values and estimates the probability of ignition from firebrand attacks. The expected value and variance of the whole population for any quantity can be obtained with the help of the PDF. For example, if one intents to calculate the average heat energy (E) from the whole population of firebrands, obtaining a function for a single brand, the expected E value of the entire brand population can be calculated as the following:

$$\langle E \rangle = \int_{-\infty}^{\infty}\int_{-\infty}^{\infty}...\int_{-\infty}^{\infty} E_{one-brand}(x,y...)\ PDF(x,y...)\ dxdy... \qquad [3]$$

where x, y and z can be different quantities of interest such as mass, projected area, or any other parameter. Based on available data, we presented the PDFs that are functions of mass, area and distance. Thus, once a function for any desired physical quantity based on the area and mass or flight distance of a firebrand is known, the average value of the entire population, or in a specific range, can be estimated. This value can predict the initiation phase of a fire, either starting from smoldering and then transitioning to flaming or starting with flaming directly.

Integrating the PDF between two desired boundaries (instead of minus infinity to infinity) enables us to calculate the probability of having specific size and mass of the brands in the whole population. This will help us to evaluate how fatal the brands can be depending on the situation. For example, small brand can penetrate through the ventilation system in the house and the can easily start a fire if they land on the carpet. On the other hand, they are more likely to burnout during the flight. Thus, in one case they can be extremely dangerous and the other case they are not a big threat. So, based on the characteristics of the area, urban or wildland, one can have a reasonable estimation of the lethality of the brands.

RESULTS AND DISSCUSIONS

Out of 13 different experiments listed in Table 1, some recorded mass and projected area and some measured the flying distance and projected area. In order to facilitate the visualization of the graphs, we present the contour plots along with the 3D graphs. Due t page limit, only some of the graphs corresponding to the selective experiments listed in Table 1 are presented in each category of OWB, IWB and ICS.

To draw a general view of the distribution of the firebrands, Figure 5 is presented at the first in this section. As can be seen, the histograms of the mean and the median for all the experiments have been depicted. The figure suggests that unlike smaller brands, the frequency of finding brands larger than 6 square centimeters does not vary with the size. Moreover, the frequency decreases approximately under an exponential function ($f(a) = 8e^{(-a+3)} + 1$).

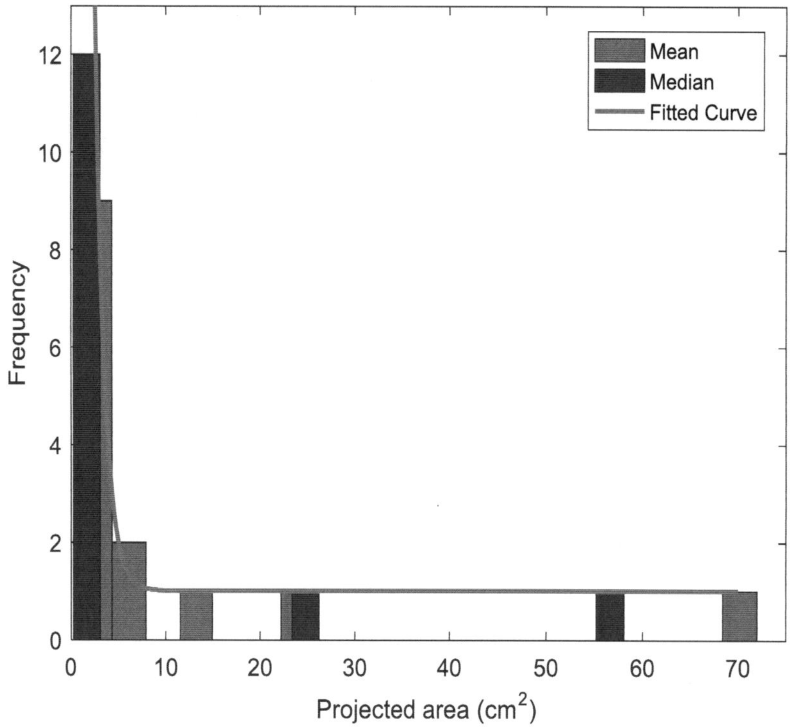

Figure 5. Histograms of mean and median projected area values for all OWB experiments.

Based on Figures 6a and b, for OWB1 the brands could travel further away from the fire front comparing to the other two. This may show the effects of height and the used materials in the building on the generation of the embers; OWB1 wind speed was not the highest one but still the brands could travel further away. Regarding Figures 7a and 7b, one can realize that the correlation between mass and area is stronger at closer distances to the fire front. Needless to mention that the smaller minor diameter of the oval shows stronger relation between the parameters. Finally, considering Figures 8a and 8b, it is understandable that the brands generated in ISC3b, had more variety in the size and mass comparing to the other two which may be the result of the wind speed. Moreover we can conclude that for the corner assemblies, the correlation between mass and projected area is higher than the wall.

More details can be found in Table 2 where the mean, variance, median, correlation factors for mass, flying distance and projected area for all the studies have been summarized.
The effects of roof size and the pressure underneath the roof is presented in Table 3. The general trend shows that the number of the firebrand per unit area decreased when the area of the firebrand collection sheet decreased, except for the last one. In order to see the effect of internal pressure on firebrand generation, we categorized the data into two groups according to pressure levels: a low pressure group from 0.13 to 0.15 inch (0.33-0.38 cm) of water, and a high pressure group 0.49 to 0.59 inch (1.24-1.49 cm) of water. It can be concluded that higher pressure led to more firebrands, in average 19,967 firebrands for the high pressure group versus 6,344 for the low pressure group. Moreover, the average mass of firebrand in the high pressure group (2389.10 g) is significantly higher than the low pressure group (743.30 g).

Finally, In order to assess the accuracy of the experiments, in figure 9, the area and mass of the brands generated in the experiments OWB9, IWB2 and ISC2 are compared. Considering the figure, one can realize that the results for IWB2 and ISC2 are very similar. However, when it comes to real fires, the results for mass are very far. For the projected area, an overlap between the results can be seen.

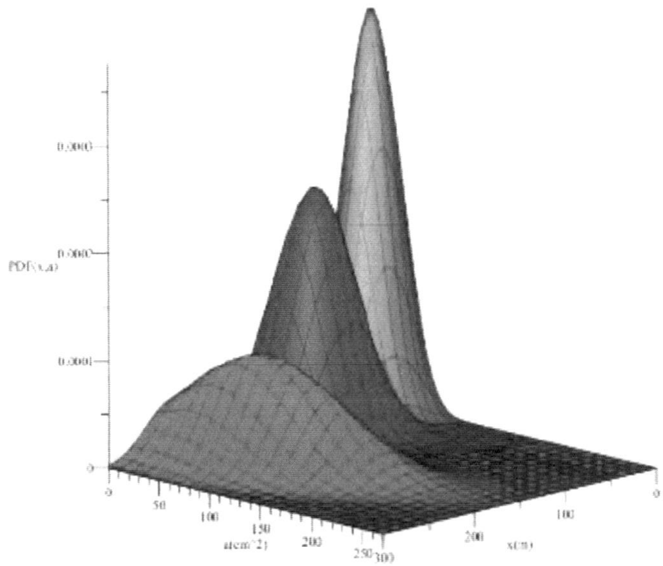

Figure 6a. 3D PDFs for OWB1-3

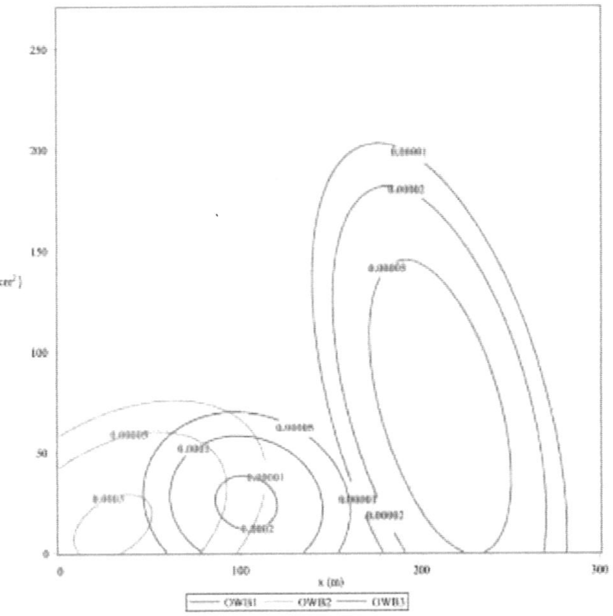

Figure 6b. 2D contour plots for OWB1-3

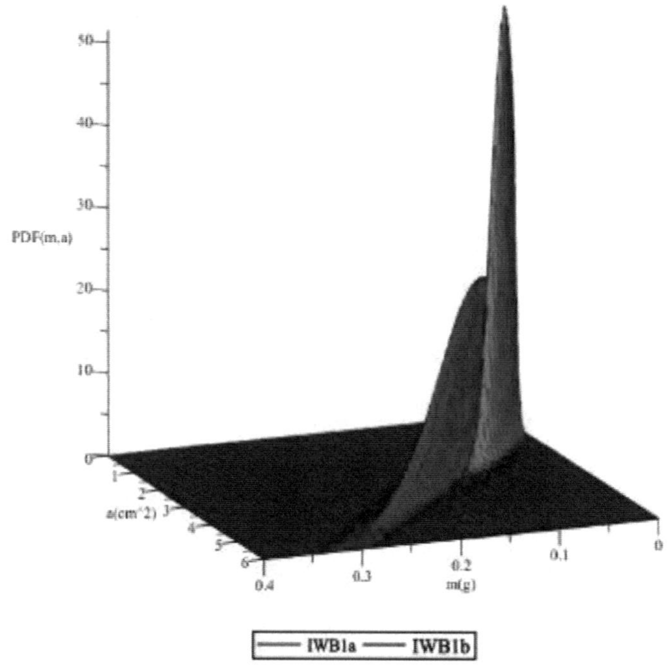

Figure 7a. 3D PDFs for IWB1a, IWB1b

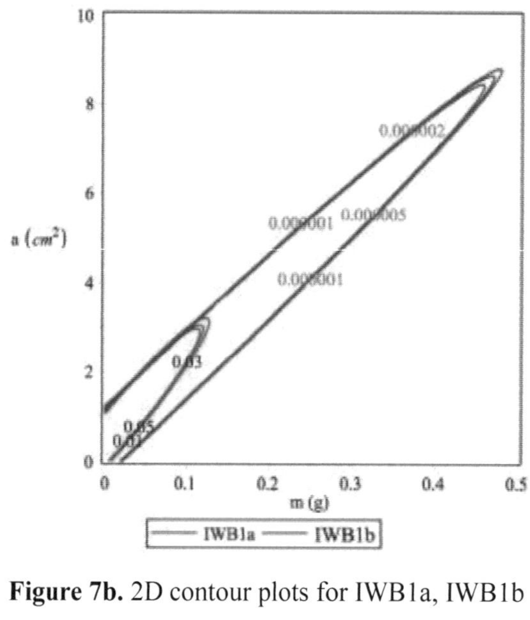

Figure 7b. 2D contour plots for IWB1a, IWB1b

Table 3. Summary of **ISC1** firebrand information

Sheet area (m^2)	Frequency of associated sheet	Total number of brands	Brand per unit area
8.45	8	202740	2999
6.41	2	3177	248
5.94	24	220527	97
5.66	3	9952	586

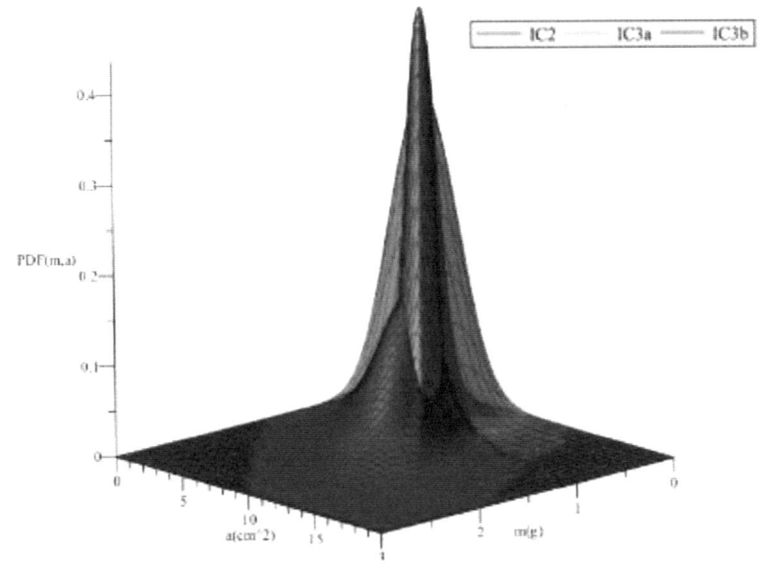

Figure 8a. 3D PDFs for ISC2, ISC3a and ISC3b

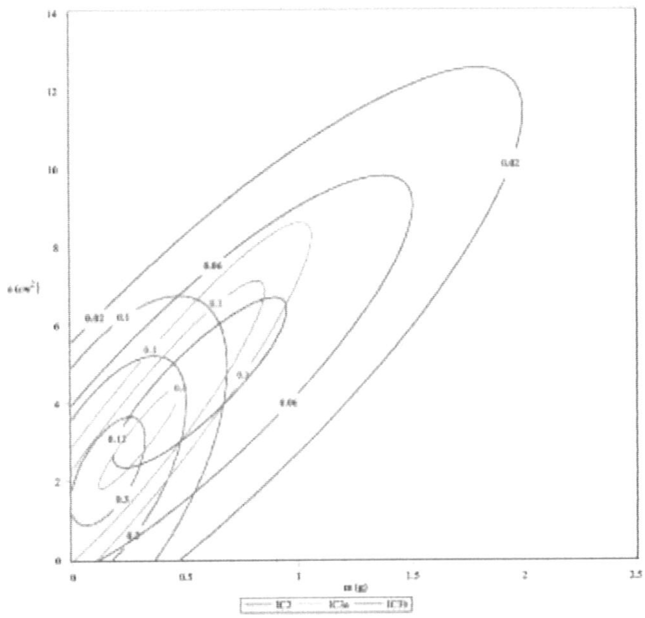

Figure 8b. 2D contour plots for ISC2, ISC3a and ISC3b

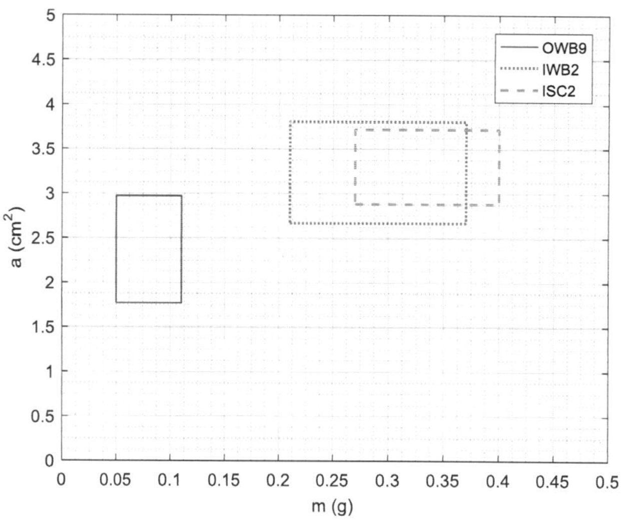

Figure 9. Comparison of the ranges of the mass and size populations of firebrands in the OWB9, IWB2 and ISC2 experiments based the 99% confidence interval

CONCLUSION AND SUMMARY

In order to build a statistical framework for the generation and distribution of the firebrands, the truncated bivariate Gaussian distribution was employed. We figured out that most of the parameters are correlated and their mutual effects on each other need to be considered in the analysis. The obtained probability density functions can provide us with valuable information about the probability of having specific size of the brands as well as the expected value and variance of a quantity of interest for the whole population. Some of the key conclusions are listed below:

- Using the proposed PDFs one can calculate the expected value and variance for any specific quantities of interests.
- Based on available firebrand data from structural fuels, the correlation factor between mass and projected area is stronger than that of distance and projected area of the brands.
- Stronger wind carries the brands to further distances, however, its effects may be overshadowed by construction height and aerodynamic shape of the brand. Stronger wind leads to more scattered brand distribution as well.
- The pressure underneath the roof plays a very important role; the higher it is, the larger the generated brands are and the number of generated brands increases markedly with it. This pressure depends on undivided heights inside the structures.
- Avoiding sharp edges in the constructions (corner assemblies) can reduce the size/mass of the generated brands.
- Based on the 95% confidence interval for the total population of the firebrands in outdoor whole buildings flying distance falls in the range of 48.3 and 50.2 meters from the origin and the projected area lies between 14.7 and 17.6 square centimeters. For indoor whole buildings, the projected area is between 2.5-3.2 square centimeters and mass is between 0.2-0.3 grams.

ACKNOWLEDGEMENT

We would like to thank Dr. Samuel Manzello from Fire Research Division, National Institute of Standards and Technology (NIST) in the United States and Dr. Sayaka Suzuki from Large Fire Laboratory, National Research Institute of Fire and Disaster in Japan who generously shared their valuable experimental results with us. The authors would like to thank the Joint Fire Science Program for sponsoring this project (Project ID # 15-1-04-4). Any opinions, findings, and conclusions or recommendations expressed in this material are those of the authors and do not necessarily reflect the views of the Joint Fire Science Program.

References

[1]Cohen J., (2008) The Wildland-Urban Interface Fire Problem, forest history today, 20

[2]Koo E., Pangi P., Weise D., Woycheese J., (2010) Firebrand and spotting ignition in large scale fires, International journal of wildland fires, 19, 818

[3]Mell W., Manzello S., Maranghides A., Butry D., Rehm R., (2010)the wildland urban interface fire problem- current approaches and research needs, International journal of wildland fires, 19, 238

[4]Manzello S. L., Shields J. R., et al, (2008) On the development and characterization of a firebrand generator, Fire Safety Journal 43, 258

[5]Caton S. E., Hakes R. S. P., Gorham D.J., Zhou A., Gollner M.J., (2016) A Review of Pathways for Building Fire Spread in the Wildland Urban Interface Part I: Exposure Conditions, Fire Technology, 1-45

[6] Vodvarka F., (1969), Firebrand field studies, IITRI final technical report J6148

[7] Vodvarka F.J., (1970). Urban burns – Full-scale field studies, Final Technical Report J6171, IIT Research Institute, Chicago, IL, January 1970.

[8]Suzuki S., Manzello S. L., et al, (2012) Firebrand generation data obtained from a full-scale structure burn, International Journal of Wildland Fire, 21, 961–968

[9] Yoshioka H., Hayashi Y., et al, (2004) Real-Scale Fire Wind Tunnel Experiment on Generation of Firebrands from a House on Fire, Fire Science and Technology Vol.23 No.2 142

[10]Suzuki S., Brown A., et al, (2014) Firebrands generated from a full-scale structure burning under well-controlled laboratory conditions, Fire Safety Journal 63, 43–51

Suzuki S., Manzello S. L., (2016) Firebrand production from building components fitted with siding treatments, Fire Safety Journal 80, 64

[11]Waterman T. E., (1969) experimental study of firebrand generation, Final Technical Report, Project J 6130

[12]Suzuki S., Manzello S. L., et al, (2013) The size and mass distribution of firebrands collected from ignited building components exposed to wind, Proceedings of the Combustion Institute 34 2479

[13]Zhou K., Suzuki S., Manzello S. L., (2015) Experimental study of firebrand transport, Fire Technology 51, 785

[14] Bird J., (2014) Engineering mathematics, seventh ed., Taylor and Francis, New York

LABORATORY STUDIES ON THE GENERATION OF FIREBRANDS FROM CYLINDRICAL WOODEN RODS

Sara Caton[a,b], Raquel Hakes[a], Matthew Weston-Dawkes[a], Ali Tohidi[a], Michael Gollner[a], and Nelson Bryner[b]

[a]Department of Fire Protection Engineering, University of Maryland, College Park, MD, 20742, USA

[b]Fire Research Division, National Institute of Standards and Technology, Gaithersburg, MD, 20899, USA

ABSTRACT

Firebrands, pieces of burning material that break off of structures or vegetation and are transported far from the fire front, are the primary cause of home ignitions during wildland urban interface (WUI) fires. The purpose of this work is to study the breakage process to better understand how firebrands are produced. Cylindrical wooden dowels are used as a simplistic representation of vegetative fuels in WUI fires. The dowels were exposed to various heating conditions and then three-point bending tests were performed on each dowel to study the effect of combustion on mechanical properties. It has been found that there are two distinct regimes that describe the breakage. The size of the dowel and the final density of the dowel both control the transition between regimes. A scaling analysis was performed to show that the two regimes and the transition point are the same for all species tested. Predictions of wind velocities needed to produce the measured critical stresses were calculated, because a connection needs to be drawn between the combustion and the wind to fully understand firebrand generation. Dowels of diameter less than 6 mm were predicted to break off at reasonable velocities, consistent with previous measurements of firebrands from WUI fires, which were generally less than 5 mm in diameter.

INTRODUCTION

The term wildland urban interface (WUI) describes a geographic location where humans and their communities meet or comingle with undeveloped wildland areas. Although the concept of WUI fire risk is not new, it is a problem that is often overshadowed by the risk of structural fires, resulting in a large gap of knowledge and multiple areas of research needs in WUI fire research. Over the past few decades, fires in WUI communities have been increasing in size and severity, resulting in large-scale WUI disasters internationally every year[1], and this trend is projected to continue to increase due to climate change, wildfire management policies, and an increase in rural development[2].

There are three mechanisms for fire to spread into WUI communities and ignite homes: direct flame contact, radiation, and firebrands. Firebrands are transport downstream of the fire and then cause smaller spot fires ahead of the main fire front. They can also directly ignite structures or structural components. Case studies of past fires have found that firebrands are the leading cause of home ignitions in the WUI, and they are often cited for at least 50% of ignitions, if not more[3-5]. Although it is evident that firebrands play a large role in fire spread and the destructiveness of a WUI disaster, there is limited research on firebrands and their processes.

Exposure from firebrands can be broken down into three processes: generation, transport, and ignition. Of these three processes, firebrand transport has received the most attention. It is known that firebrands can be lofted large distances to ignite spot fires, and there is a large body of work in the literature on modeling the lofting and transport process. Although firebrand ignition of fuel is a very complex and stochastic process, there has been an effort to understand the ignition of fuel beds and some structural components, i.e. roofs and decks, from firebrands. There are also studies on the generation of firebrands, but almost all of these studies are focused on collecting firebrands downstream to determine size (area and mass) and distribution of typical firebrands. A detailed summary of previous studies conducted on the firebrand processes can be found in the literature[3], which illustrates that information on how and when firebrands are generated or break off of the burning material appears not to have been addressed.

The focus of the current research is to study the generation of firebrands from wildland fuels, like conifers and shrubs, in their simplest configuration. Experiments were conducted with uniform wooden dowels at the small scale to better understand the effects of thermal degradation on characteristics responsible for breakage. Forces were measured using a three-point bending test in order to calculate the critical stresses required for failure. In real WUI fires, these forces come from the influence of wind and weight of the vegetation, which is estimated as well. Various wood species and sizes were used to understand the effect of the initial properties of the vegetation on breakage mechanics. The experiments here do not apply to those firebrands produced from tree bark or structural fuels, which likely degrade and break by different mechanisms than the configuration studied here.

LITERATURE REVIEW

The majority of firebrand generation studies in the literature are primarily focused on collecting firebrands produced from vegetation, structural components, full-scale structures, and WUI fires to observe size and distribution and have found that the majority of firebrands are very small. For vegetative fuels, Manzello et al.[6] studied firebrand generation from Douglas fir trees that were 2.6 m and 5.2 m tall and ranged from 10% - 50% moisture content. All firebrands that were collected in these experiments were cylindrical with an average diameter of 3 mm and length of 40 mm for the 2.6 m tall tree and 4 mm diameter and 53 mm long for the 5.2 m tall tree. Manzello et al.[7] also studied firebrand generation from Korean pine trees. These experiments also produced solely cylindrical firebrands of similar size and mass, with the average firebrand diameter measuring 5 mm and average length measuring 40 mm.

Size distributions of firebrands have also been measured in real WUI fires. The first study of this kind was during the Angora fire in California in 2007[8,9]. Data was collected after the fire by analyzing holes that were created in a trampoline from brands melting the canvas of the trampoline. This study found that 95% of the burn holes had an area of less than 1 cm^2. Observations were made about firebrand size distributions after the Bastrop Complex fire, which occurred in 2011 in Texas, using the same technique as in the Angora fire study of measuring the sizes of burn holes in trampolines[10]. The size distribution of holes from firebrands in this fire is similar to what was measured after the Angora fire with 90% of the holes less than 0.5 cm^2 in size.

Firebrand generation studies were also performed in the New Jersey Pine Barrens to characterize mass, size, and origin of firebrands in the field[11]. This is one of the first studies that incorporated the importance of characterizing and closely studying the sizes of the source fuel. Pine bark and shrubs were sampled during the tests to determine what size classes were breaking off to become firebrands. After examining all the data, it was found that all branches less than 2 mm in diameter were consumed, 2.01 mm to 4 mm diameter fuels were partially consumed, and fuels greater than 4 mm were not consumed.

Barr and Ezekoye[12] investigated breakage and subsequent lofting of firebrands through a thermal plume. They used a fractal model to describe wildland fuels and paired this with combustion models to create inputs for numerical models. In these experiments, three-point bending tests were performed on Yellow poplar rods with diameters of 4.8 mm, 6.4 mm, and 7.9 mm. There were three conditions for the tested dowels: ambient conditions (14°C, 90% relative humidity), oven heated at 101°C for 1 hour, and oven heated at 250°C for 1 hour. The results from these experiments show a somewhat linear trend between the flexural strength and density; however, the rate of change of the strength with respect to the density varies for the three different heating conditions.

Tohidi et al.[13] proposed a mechanical model for firebrand breakoff due to shear stress associated with an internal bending moment that exceeds the maximum bending moment the specimen can withstand. In this model, the bending moment is created due to the material weight and drag forces, which can be produced in the vertical direction generated by buoyancy from the fire plume or the horizontal direction from ambient winds. This is a simplistic model that does not account for charring or burning of the material, which causes degradation and weakening. From this model, it was determined that fuel with large aspect ratios (long and thing) are more susceptible to firebrand break off. The studies by Barr and Ezekoye[12] and Tohidi et al.[13] both propose a beginning approach to include generation of firebrands into an initialization model, which has been missing up to this point. This information is necessary for better prediction of firebrand transport, and data needs to be collected to inform these models.

EXPERIMENTAL SETUP

In this study, experiments were conducted with cylindrical wooden dowels of three different wood species: birch, oak, and poplar. These species were chosen because they provide a range of material properties and were able to be machined into small wooden dowels. The average initial densities measured before testing were 610 kg/m^3, 700 kg/m^3, and 540 kg/m^3 for the birch, oak, and poplar, respectively. Because density is an important factor in the strength properties of materials and Barr and Ezekoye[12] found a relationship between strength and density, a range of initial densities was studied to better understand the role of density in firebrand generation. The density of Douglas fir and pine species, which are more typical in WUI fires, overlap with the lower range of densities tested in this study.

The dowels tested were cylindrical in shape and had diameters ranging from about 3 mm to 13 mm (1/8 in to 1/2 in). Previous work has found that the average firebrand diameter ranges from 3 mm to 5 mm[6,7,11], which motivated the dowel sizes used in these experiments. Cylindrical shaped dowels were chosen because this is the shape primarily found during firebrand production studies involving vegetative fuels[6,7]. This research is focused on the production of firebrands from natural fuels, like conifers or shrubs, which contain some branching members without leaves or needles. The reason for this is to start with modeling the simplest form of firebrand generation and to physically understand what is happening on the small scale. All of the dowels were cut to 10.16 cm (4 in) in length for the three-point bending tests. Moisture contents were measured with a moisture content analyzer which dried the fuel over a load cell and were found to range from 5% - 8% on a dry basis for all three species. Drying the dowels to lower moisture contents was tested but did not have a significant effect on the results of the bending tests.

Two modes of heating were used to produce thermal degradation before performing the three-point bending tests: a hot plate and a propane flame. The hot plate tests were divided into two groups, so that the effects of increasing the exposure time and increasing the temperature could be made. One set of tests was performed at a constant heating temperature of 350°C with increasing times of 10 min, 15 min, and 20 min. A second set of tests was performed on the hot plate with a constant exposure time of 15 min and increasing temperatures of 350°C, 375°C, and 400°C. This temperature range was used because the char layer begins to form in this range. A third group of bending tests were performed on dowels that were exposed to a propane flame created by a large Bunsen-type burner to induce flaming combustion, which was missing in the previous heating method. While the temperatures produced by this burner were likely higher than those produced in a real wildland fire, they were much more representative of the temperatures and heating rates expected than hot plate testing. The effect of the flaming time was observed by increasing the flaming time in 5 s intervals from 10 s to 20 s. Details describing these three groups of tests with the species and sizes used is given in Table 1.

Table 1. Summary of details of the heating methods and dowels used with each test condition.

	Group 1	Group 2	Group 3
Heating Mode	Hot Plate Constant T = 350°C Times: 10, 15, 20 min	Hot Plate Constant t = 15 min Temps: 350, 375, 400°C	Propane Flame Flaming times: 10, 15, 20 s
Species	Birch (B), Oak (O), Poplar (P)	Birch (B)	Birch (B), Oak (O), Poplar (P)
Diameter (mm)	12.7 (B) 9.525 (B, O, P) 6.35 (B, O, P) 4.76 (B) 3.175 (B)	12.7 (B) 9.525 (B) 6.35 (B) 4.76 (B) 3.75 (B)	12.7 (B, O, P), 9.525 (B, O, P), 6.35 (B, O, P),

In order to ensure that all pyrolysis and combustion stopped and the heating times were as accurate as possible, the dowels were placed in a 22.9 cm x 33 cm x 6.35 cm (9 in x 13 in x 2.5 in) aluminum box after each exposure time. The box was lined with kaowool ceramic insulation fiber, so that the dowels would not fracture when placed inside and nitrogen gas was injected through an inlet hole in the top of the box. The nitrogen gas acted as an inert to extinguish any combustion processes, flaming or smoldering, that were still occurring.

A PASCO Materials Testing System was utilized with the bending accessory to perform three-point bending tests on all of the heated dowels. A set of three-point bending tests was also conducted on virgin dowels of all the species to provide data on initial mechanical properties. The length span between the supports, L_S, is 8.65 cm (3.4 in), which means that supports are each located 0.76 cm (0.3 in) from the edge of the 10.16 cm (4 in) long dowel. The Materials Testing System's software measures the force applied to the specimen through the plunger by a 7,100 N load cell in the base of the machine, and the position of the tip of the plunger is measured when the software is recording data by an optical encoder.

RESULTS

Comparisons between failure mechanisms as a result of the two heating methods, the hot plate and the propane flame, were made through visual observations of the breakage of the dowels during the three-point bending tests in order to understand what breakage mechanisms were occurring during the different tests. The majority of the dowels that were heated on the hot plate experienced fibrous breakage during the three-point bending tests. The splintering of the fibers during this type of breakage often occurred at a 45° angle, signifying yielding, and the location of the applied force was obvious because the dowels did not snap into pieces. This breakage mechanism was dominant for all three species. However, birch dowels were tested with a larger range of diameters, and it was found that 4.76 mm (3/16 in) diameter dowels fractured in a fibrous manner as well as abruptly in a brash failure. The dowels tested with diameters below 4.76 mm (3/16 in) had all brash failures ending with the original dowel broken into at least two pieces. This indicates that the diameter of the dowel contributes to the mechanics of the breakage; yet, it is not clear if the transition occurs for all species at a similar size, because oak and poplar were not tested on the hot plate with diameters in the lower range.

In the tests with a propane flame as the heat source, there is a wider range of breakage mechanisms over shorter exposure times. For example, Figure 1 shows the results after the bending tests for 6.35 mm (1/4 in) diameter birch dowels after 10 s and 20 s exposures. In this case, the 10 s exposure resulted in a

Figure 1. Birch dowels with a diameter of 6.35 mm (1/4 in) heated in the flame for 10 s (left) and 20 s (right), illustrating that the breakage mechanism is dependent on heating time.

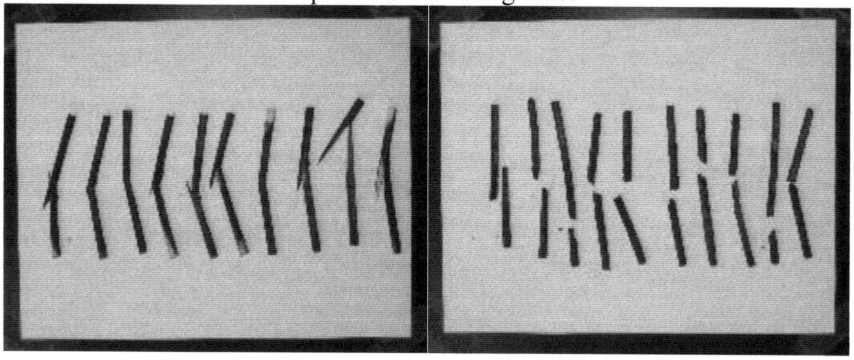

fibrous fracture for the dowels, but after 20 s exposure the dowels all failed in a brittle manner. The 15 s burning time of this size dowel produced a mixture of the two breakage conditions. The three-point bending tests on the oak dowels of this size resulted in the same trend as the birch. However, brash, complete failure was observed for the 6.35 mm (1/4 in) diameter poplar dowels even after only 10 s flaming time. The two larger diameters 9.525 mm and 12.7 mm (3/8 in and 1/2 in, respectively) were observed to break in the fibrous method, with fractures occurring at 45° angles and fibers splintering apart, for all three species, even when the exposure time was 20 s. These tests show that the size of the dowel and the exposure time are controlling factors of the breakage mechanism, a result which was not as clear in the hot plate tests. Not only did the flaming tests allow for more realistic fire conditions to be achieved in a repeatable manner, but they also allowed for a wider distribution of breakage mechanisms to occur for the given species over the same amount of changing variables.

The maximum forces that each dowel could withstand prior to breakage were studied as a function of the initial diameter of the dowels for all the tests. Ten dowels were tested for each testing scenario for both the hot plate and flaming tests, and the mean of the maximum force for each testing condition was calculated to find the average value associated with the test parameters. The results from the hot plate tests showed that increasing the temperature had a larger effect on the value of the maximum force than increasing the exposure time. The results for the flaming tests are shown in Figure 2 for all three species to demonstrate the differences for the species. In the birch tests, the maximum forces decreased with increasing time for each diameter. The results of the oak tests demonstrate that increasing the exposure time from 15 s to 20 s did not significantly change the maximum force measured for any of the sizes. The poplar dowels lost a significant amount of strength after burning for 10 s, but increasing the exposure time beyond 10 s only created small decreases in the ultimate force.

As discussed previously, Barr and Ezekoye[12] performed strength tests on oven-heated dowels and found a linear relationship between the flexural strength and density of the dowel. A comparison between their findings and the results of this research is made to determine if a linear relationship is an accurate description of the breakage behavior. The flexural strength can also be described as the maximum strength before rupture, which is given the term modulus of rupture (*MOR*). The *MOR* can be calculated by converting the maximum force measured to the maximum stress induced on the dowel by following expression:

$$MOR = \frac{8F_{max}L_S}{\pi D^3}, \tag{1}$$

Figure 2. Maximum force as a function of initial diameter for the birch (top left), oak (top right) and poplar (bottom) for the flaming tests.

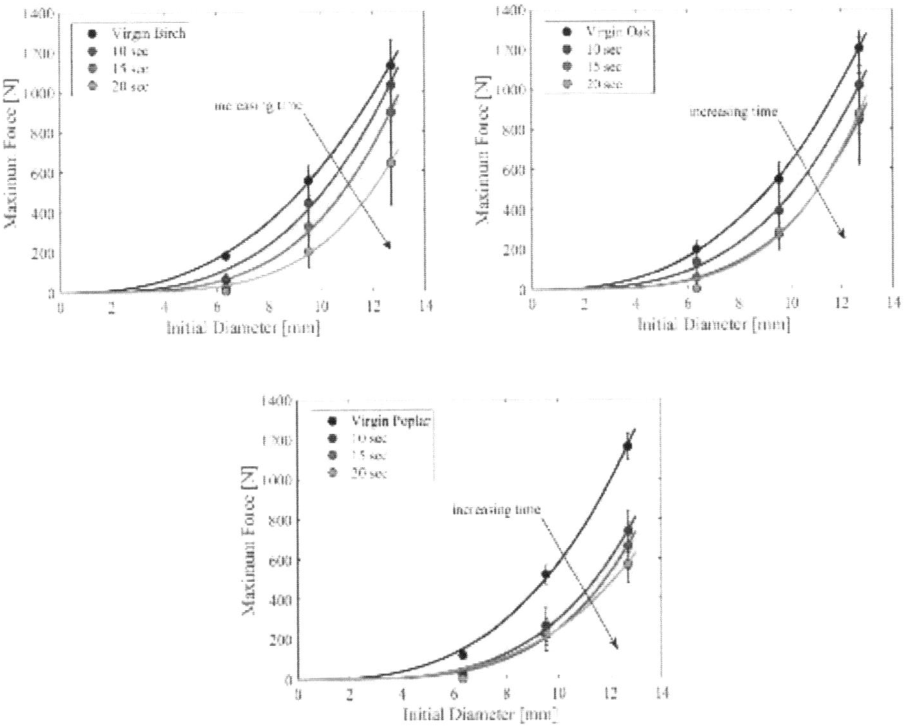

where F_{max} is the maximum force, L_S the length span between supports, and D the dowel diameter. Due to the short exposure times in these experiments, the volume of the dowels is mostly unchanged after heating, so the density at breakage is calculated using mass after heating and the initial volume.

The hot plate tests resulted in a similar trend to the findings of Barr and Ezekoye[12]. The *MOR* decreases with decreasing density in a somewhat linear relationship; however, the results show that the linear relationship changes for different species, which indicates that how sharply the strength decreases due to density will vary dependent on species. The failure data from the flaming tests produces valuable insight into the reduction of strength that is missing in the other heating method. When flaming combustion occurs, two distinct regimes controlled by the density can be observed, as shown in Figure 3. The regime with higher densities was fitted with a linear fit to make a comparison to the results of Barr and Ezekoye[12], which is shown by the dashed line in Figure 3. The flaming tests illustrate that any specimen with a final density below 300 kg/m^3 had a significant reduction of strength and was easily breakable. Above this density, a relatively linear trend is observed to be dependent on species, which was also seen in the hot plate tests. This provides an initial connection between the effects of combustion on structural properties, and shows the importance of including flaming combustion.

Since the data for the flaming tests resulted in a more comprehensive analysis of the relationship between structural properties and combustion effects, these results were studied in more detail. To gain a deeper understanding of the structural breakage process, stress, σ, and strain, ε, relationships were calculated for each bending test with the following equations:

$$\sigma = \frac{8FL_S}{\pi D^3}, \quad (2)$$

$$\varepsilon = \frac{6D\delta}{L_S^2}, \quad (3)$$

where F is the applied force, L_S the length span between supports, D the diameter, and δ the deflection. The modulus of elasticity, E, for each test can be found from the slope of the linear portion of the stress-strain curves from Hooke's Law, and the *MOR* is represented by the maximum point on these curves. Generally, for all three species tested, the *MOR* decreased with increasing flaming times and increased with increasing diameters, as shown in Table 2. Using the *MOR* from these tests as a comparative measure of strength indicates the relatively oak is the strongest and poplar is the weakest. Comparing literature values of *MOR* for these species provides this same trend[14].

Figure 3. Modulus of rupture as a function of density for the flaming tests compared to the results from Barr and Ezeykoye[12].

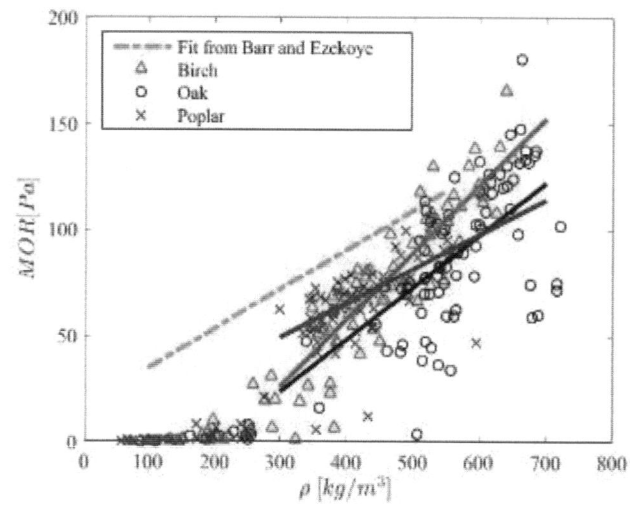

Table 2. Values for the *MOR* for the three species, diameters, and flaming times. The *MOR* decreases with increasing flaming times and increases with increasing diameters.

	Birch *MOR* **(kPa)**			**Oak** *MOR* **(kPa)**			**Poplar** *MOR* **(kPa)**		
D **(mm)**	10 s	15 s	20 s	10 s	15 s	20 s	10 s	15 s	20 s
6.35	46,356	10,351	2,797	110,830	44,654	2,271	22,119	2,480	-
9.525	104,650	76,071	45,239	92,205	63,459	60,847	60,901	53,338	58,695
12.7	101,480	86,697	64,530	99,699	79,390	80,393	72,684	67,709	55,340

Figure 4 gives the stress-strain curves for the flaming tests of the poplar dowels. Each curve is an average for the respective testing conditions. The effect of increasing the flaming time is illustrated for the three different diameters. It was found that flaming times of 20 s produced at least a 33% loss of strength for the

12.7 mm (1/2 in) diameter dowels and at least a 53% loss for 9.525 mm (3/8 in) dowels. The 6.35 mm (1/4 in) diameter dowels resulted in about a 90% strength reduction for the 20 s exposure. The birch and oak dowels resulted in similar trends. It can also be seen that the *MOR* occurs at decreasing values of ε for decreasing diameters. This indicates that less deflection is needed for the smaller dowels to fracture than the larger dowels. Figure 4 also demonstrates that the modulus of elasticity decreases as the flaming time increases.

Figure 4. Stress-strain curves for the 12.7 mm (1/2 in) diameter (top left), 9.525 mm (3/8 in) diameter (top right), and 6.35 mm (1/4 in) diameter (bottom) poplar dowels.

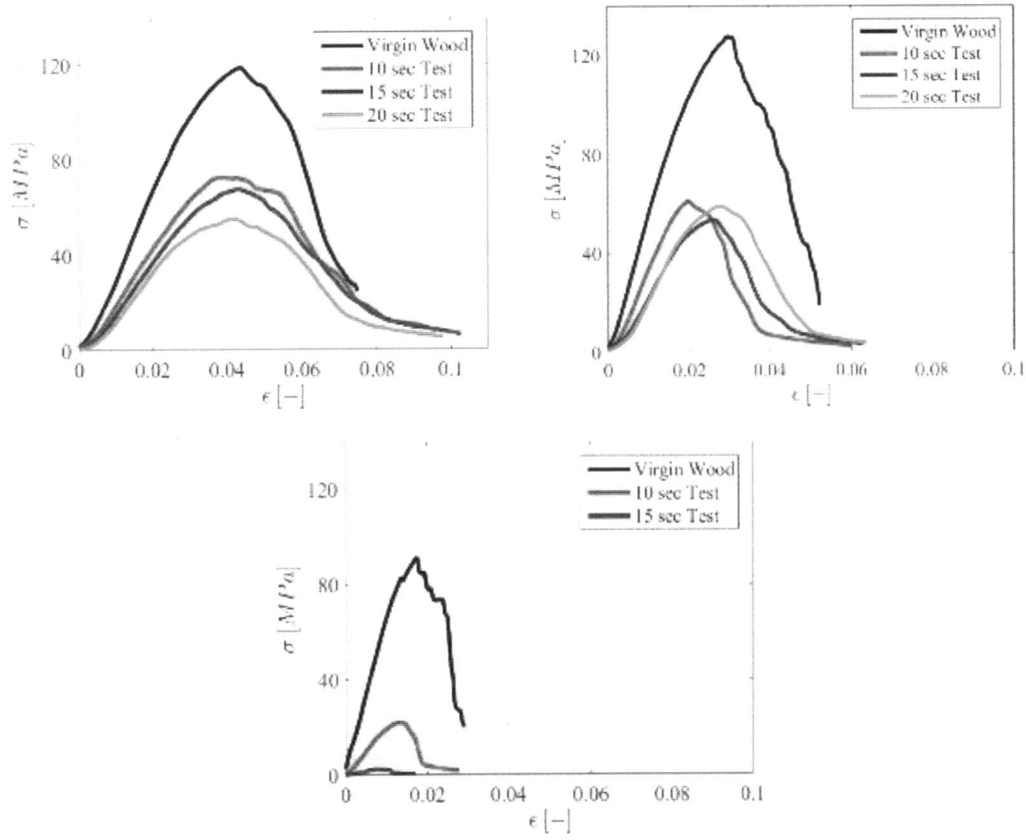

ANALYSIS

The results of the three-point bending tests produced valuable knowledge on how combustion processes affect the strength and breakage mechanics of cylindrical wooden dowels. The results, however, are specific to the species and sizes being tested, so it is beneficial to use non-dimensional scaling to find a more universal relationship to describe the breakage phenomena of firebrands. The critical, or maximum, breakage force should be scaled by using the controlled variables and the measured parameters from the experiments and by incorporating structural and combustion properties to form a model of the important parameters. This can be shown by combining the controlling parameters in the following relationship:

$$F_{max}=f\left(\rho_0, L_0, D_0, v, \alpha, \dot{m}, E, \rho_\infty\right), \tag{4}$$

where F_{max} is the critical force for breakage, ρ_0 the initial species density, L_0 the initial length of the dowel, D_0 the initial dowel diameter, ν Poisson's ratio, α species thermal diffusivity, $\dot{m}$ the mass loss rate, E the modulus of elasticity, and ρ_∞ the density of the air. These controlling parameters can be grouped into four separate categories. L_0 and D_0 strictly describe the geometry of the dowel; these two parameters can be paired with ρ_0 to describe the initial conditions and characteristics of the dowel. Including ν and E is important because these parameters define the fracture mechanics. $\dot{m}$ and ρ_∞ can be grouped together to describe the burning, and α is a description of thermal properties. This simple model is developed to first understand how these values interact before creating a more generalized and detailed model that incorporates the added effects of wind.

The Buckingham Pi theorem allows for manipulation of formulated Π groups to form two main nondimensional parameters that are meaningful and descriptive of the problem that needs to be modeled. The two final nondimensional parameters used in this firebrand breakage model are

$$\Pi_1^* = \frac{F_{max} L_0 \nu}{E D_0^3}, \quad \Pi_2^* = \frac{\dot{m} \alpha \rho_0}{E D_0^3 \rho_\infty}. \tag{5}$$

Π_1^* is a nondimensional formulation of the transverse strain due to bending. If ν is removed from Π_1^*, this parameter is converted to the axial strain in bending. For this analysis, Π_2^* is defined as ratio of the capacity for the material to burn over the capacity of the material to fail.

As indicated in Figure 3, Figure 5 illustrates that there are two distinct regimes that define the effects of combustion on the breakage mechanics in these experiments. There is one regime, shown by the horizontal green line in Figure 5, that is controlled by the burning rate parameter, Π_2^*, and a second regime, shown by the vertical yellow line, that is the controlled by Π_1^*, the nondimensional strain. The nondimensional strain created by the maximum force is constant until a critical value of about 2.2 x 10^{-12} is achieved by the burning

Figure 5. Nondimensional scaling of the influences of combustion on the fracture mechanics shows two distinct regimes controlled by dowel size and heating time

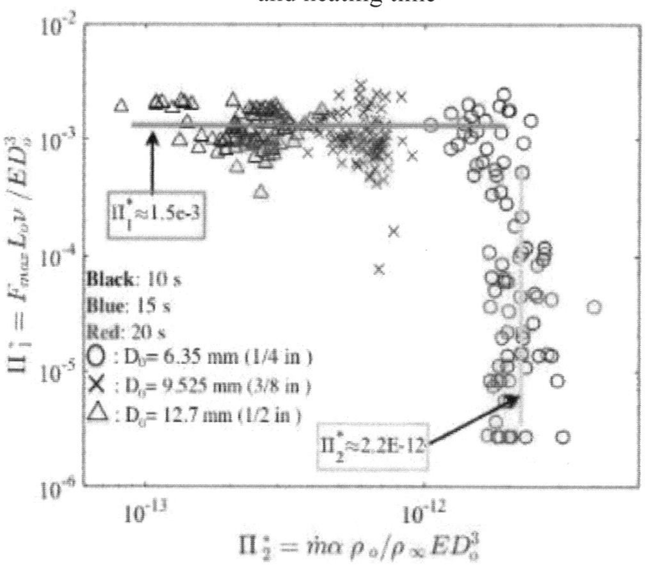

rate parameter. At this value, the critical breakage forces drop off significantly, indicating that this regime is more easily susceptible to breakage.

It is evident from Figure 5 that the diameter size is a controlling factor that determines the transition. The 6.35 mm (1/4 in) diameter dowels, shown by the open circles, are all in the regime that is dominated by Π_1^* meaning that they fractured the easiest and are not dependent on $\dot{m}$ because of their size. The flaming time, shown by the changing colors, also shows an intuitive trend. For the 12.7 mm (1/2 in) diameter dowels the Π_2^* values increase with increasing flaming times, which makes sense as $\dot{m}$ is a controlling parameter in this regime. A trend dependent on exposure time can also be seen in the other regime demonstrating that the dowels were more susceptible to fracture after longer amounts of time. The species are not distinguished in Figure 5 because this relationship is independent of the species used.

All of the results and analysis thus far have been focused on strength and breakage properties of wood based solely on the effects of degradation due to combustion. The effects of degradation and wind, however, must be studied in conjunction to create a comprehensive firebrand breakage model. Wind has a critical role in the process of firebrand generation for two main reasons. The first is that it imposes drag forces on the burning vegetation, which have the ability to cause fracture. It also has the potential to alter the burning rate, which will influence the degradation process. Consequently, it is necessary to study the coupling effect of the wind and degradation to fully understand the details of firebrand production.

An estimation of the wind speeds needed to cause the values of *MOR* measured in the three-point bending experiments can be made by simplifying the breakage model presented by Tohidi et al.[13], which was described earlier, and the correlation between drag coefficient and Reynolds number for a cylinder in cross-flow. For this discussion, the model for *MOR* is simplified to

$$MOR = \sqrt{\left(\frac{8\rho_\infty C_D U_h^2 \eta^2}{\pi}\right)^2 + \left(4\rho_0 g L\eta\right)^2}, \tag{6}$$

where C_D is the drag coefficient, U_h the horizontal wind speed, η the aspect ratio (L/D), and g gravity. The term that included the effect of the vertical drag forces is neglected for this analysis, so that an estimation of the horizontal wind speeds needed to cause the *MOR* can be found. This equation can be re-written to solve for the horizontal wind velocity, U_h, as

$$U_h = \left[\frac{\sqrt{\pi^2\left(MOR^2 - \left(4\rho_0 g L\eta\right)^2\right)}}{8\rho_\infty C_D \eta^2}\right]^{1/2}. \tag{7}$$

This equation relates the horizontal wind velocity to the *MOR* and stresses caused by the weight of the dowel. The weight is included because in real WUI scenarios branches are not simply supported, instead they resemble a cantilever beam, and the force from the weight creates a moment. This equation was solved using an iterative process, where a Reynolds number, *Re,* is calculated from estimating the wind speed and the associated drag coefficient is found from the relationship of C_D and *Re* over a smooth cylinder.

This calculation was performed for the three species, but only used the data from tests with a flaming time of 20 s. In reality, vegetation will have much longer exposure times than this, so the two lowest flaming times are not considered. Figure 6 shows the results of this analysis. The length of the dowel was increased in increments of 5 cm starting from 10 cm, which was the length of the dowels tested. This illustrates that longer dowels require lower wind speeds to cause fracture, which occurs because increasing the length increases drag force. As expected, decreasing the diameter decreases the wind speed needed for breakage because the smaller diameters have lower *MOR*s. It can also be seen that the species plays a small role in determining the critical winds speeds. The dashed black line in Figure 6 represents wind velocities common during WUI fires, about 15 m/s, and indicates that the wind speeds predicted for the 6.35 mm (1/4 in) diameter are reasonable velocity predictions. A previous study[11] in the New Jersey Pine Barrens with a similar wind speed (gusts up to ~7 m/s) found that fuels with diameters greater than 4 mm were not consumed (i.e. broken into firebrands), consistent with our results. The larger diameters, therefore, predicted unrealistic values needed for breakage. This is an indication that these sizes do not fracture until they degrade down to a critical size or do not break off at all.

Figure 6. Estimated wind velocities from the *MOR* values from the 20 s flaming tests. The dashed black line represents typical wind velocities during WUI fires

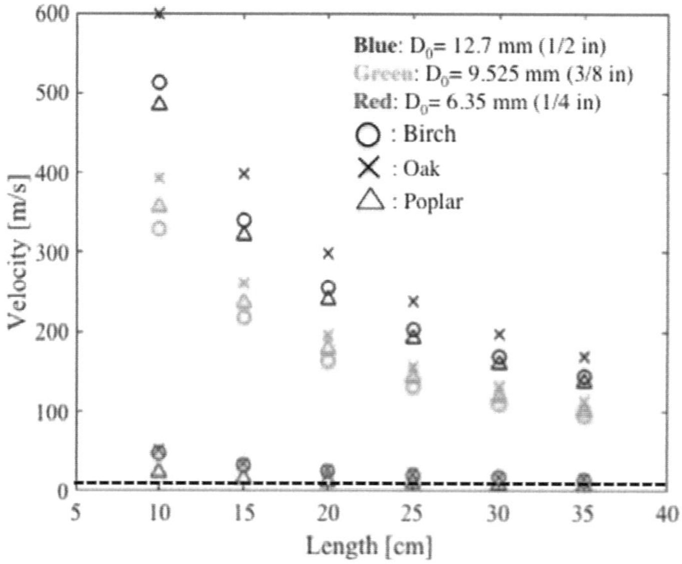

There is relatively little degradation of firebrands while they are in transport because smoldering combustion is slow, so the sizes of the collected firebrands are a reasonable estimation of the size at breakage. The results from the studies discussed in the literature review demonstrate that the majority of firebrands found are very small[6-11]. The 6.35 mm (1/4 in) diameter dowels tested in the current study are slightly larger than the average size collected. However, the wind speeds reported in the previous work are lower than what was predicted to be necessary to cause breakage for the 6.35 mm (1/4 in) diameter dowel, which suggests why these sizes were not observed. This simplified model predicts reasonable wind velocities to fracture 6.35 mm (1/4 in) diameter dowels after a flaming time of only 20 s. Increasing the heating time to more realistic times experienced by burning vegetation during a WUI fire will decrease the *MOR* and the diameter, which will result in lower wind speeds needed to cause fracture.

CONCLUSION AND FUTURE WORK

This work developed a repeatable method for testing the effects of combustion on the strength of small cylindrical wooden dowels. After different heating modes were tested, it was found that using a flame created the most realistic conditions and produced a larger range of results. This was beneficial because similar work in the literature was missing this aspect. Three-point bending tests provided an efficient method for gaining knowledge on multiple aspects of the failure mechanics.

It was found that there were two main mechanisms for failure. These were abrupt, complete failure or fibrous, splintering failure. The 6.35 mm (1/4 in) diameter dowels were the only size to experience abrupt failure when the flaming conditions were present; however, larger sizes experienced this type of failure on the hot plate. If the flaming times were increased in these tests, the larger sizes could have experienced abrupt failure. The dowel size, exposure time, and species were all controlling parameters for the fracture mechanism observed.

For all species tested, it was similarly found that the breakage is controlled by two distinct regimes. The density is one of the parameters that controls these regimes. If the density falls below 300 kg/m^3, the *MOR* is very small indicating that the stress needed for breakage is small. Above this value, there is a somewhat linear relationship where *MOR* increases with increasing density, but the species controls the rate of increase. All species saw a similar trend where the effect of the heating time was dependent on the size of the dowel. The smallest diameter was always most affected by the heating time, and the *MOR* decreased with increasing heating time.

Predictions of the wind speeds associated with the critical stresses measured were calculated, because the connection between how heating and wind effects breakage is vital. These predictions suggest that the 9.525 mm and 12.7 mm (3/8 in and 1/2 in, respectively) diameters would not break from the heating exposure in these experiments, because the wind speeds predicted were not realistic. The wind velocities predicted for the 6.35 mm (1/4 in) diameter dowels are more reasonable, in comparison, even with the short exposure times. The dowels tested were shorter than most branches, so the length was increased to show the effect of increasing drag. The theory provides that increasing the length of the dowel results in wind speeds that could be experienced in WUI fire events for the 6.35 mm (1/4 in) diameter dowels.

There still needs to be an effort to further understand the connection between wind and combustion. Small-scale experiments that study the breakage of dowels in a wind tunnel are needed to determine the critical wind velocity and exposure time that results in breakage. This can be combined with the knowledge developed in this research on the effect that combustion has on the fracture mechanics to develop a comprehensive model of firebrand generation. Experiments involving real WUI fuels should also be included in future work to ensure that the relationships developed in this work scale in the same manner, especially on branched vegetation. Combining these effects with the breakage mechanics and developing an accurate, comprehensive firebrand generation model will allow for better computational modeling of the spread of WUI fires by firebrands. It could also help to determine if certain areas are at a higher risk of firebrand attack depending on the surrounding fuel and fire conditions. Although more work on the effect of the wind is still needed, this study is the first step in understanding the controlling parameters governing firebrand production.

REFERENCES

[1] Grishin, A.M., Filkov, A.I., Loboda, E.L., Reyno, V.V., Kozlov, A.V., Kuznetsov, V.T., Kasymov, D.P., Andreyuk, S.M., Ivanov, A.I. and Stolyarchuk, N.D., 2014. A field experiment on grass fire effects on wooden constructions and peat layer ignition. *International journal of wildland fire*, *23*(3), pp.445-449.

[2] Cohen, J.D., 2000. Preventing disaster: home ignitability in the wildland-urban interface. *Journal of forestry*, *98*(3), pp.15-21.

[3] Caton, S.E., Hakes, R.S., Gorham, D.J., Zhou, A. and Gollner, M.J., 2016. Review of pathways for building fire spread in the wildland urban interface part I: exposure conditions. *Fire Technology*, pp.1-45.

[4] Cohen, J.D. and Stratton, R.D., 2008. Home destruction examination: Grass Valley Fire, Lake Arrowhead, California.

[5] Maranghides, A. and Mell, W.E., 2009. *A case study of a community affected by the Witch and Guejito Fires*. National Institute of Standards and Technology. Building and Fire Research Laboratory.

[6] Manzello, S.L., Maranghides, A. and Mell, W.E., 2007. Firebrand generation from burning vegetation. *International Journal of Wildland Fire*, *16*(4), pp.458-462.

[7] Manzello, S.L., Maranghides, A., Shields, J.R., Mell, W.E., Hayashi, Y. and Nii, D., 2009. Mass and size distribution of firebrands generated from burning Korean pine (Pinus koraiensis) trees. *Fire and Materials*, *33*(1), pp.21-31.

[8] Manzello, S.L. and Foote, E.I., 2014. Characterizing firebrand exposure from wildland–urban interface (WUI) fires: results from the 2007 Angora fire. *Fire Technology*, *50*(1), pp.105-124.

[9] Foote, E.I., Liu, J. and Manzello, S.L., 2011, February. Characterizing firebrand exposure during wildland urban interface fires. In *Proceedings of Fire and Materials 2011 Conference, Interscience Communications, London,* pp. 479-492.

[10] Rissel, S. and Ridenour, K., 2013. Ember production during the Bastrop complex fire. *Fire management today*, *72*(4), pp.7-13.

[11] El Houssami, M., Mueller, E., Filkov, A., Thomas, J.C., Skowronski, N., Gallagher, M.R., Clark, K., Kremens, R. and Simeoni, A., 2016. Experimental procedures characterising firebrand generation in wildland fires. *Fire Technology*, *52*(3), pp.731-751.

[12] Barr, B.W. and Ezekoye, O.A., 2013. Thermo-mechanical modeling of firebrand breakage on a fractal tree. *Proceedings of the Combustion Institute*, *34*(2), pp.2649-2656.

[13] Tohidi, A., Kaye, N. and Bridges, W., 2015. Statistical description of firebrand size and shape distribution from coniferous trees for use in Metropolis Monte Carlo simulations of firebrand flight distance. *Fire Safety Journal*, *77*, pp.21-35.

[14] Ross, R.J., 2010. Wood handbook: Wood as an engineering material.

WIND EFFECTS ON FLAME SPREAD AND EMBER SPOTTING NEAR A STRUCTURE

Kathryn M. Butler, Erik Johnsson, Marco Fernandez, Mariusz Zarzecki, Eric Auth, and Glenn Forney
Fire Research Division, Engineering Laboratory
National Institute of Standards and Technology
Gaithersburg, MD 20899-8662 USA

ABSTRACT

In wildland-urban interface (WUI) fires, combustible materials pose a potential threat to contiguous or nearby structures. Flame spread and firebrand spotting have been studied in a series of field experiments involving mulch beds under a variety of wind conditions and separation distances from a small structure. The rate of flame spread in a mulch bed was observed to slow significantly as the fire approached the structure, with flame contact with the structure occurring only after the gap was jumped by firebrand spotting. Computational fluid dynamics (CFD) modeling was used to gain insight into the flow field. A vortex forming near the base of the structure in the wind field was found to be responsible for changing the flame spread from a concurrent flow situation to opposed flow as it approaches the structure.

INTRODUCTION

In wildland-urban interface (WUI) fires, combustible materials pose a potential threat to contiguous or nearby structures. The wind field around a structure affects the behavior of a fire in its vicinity, including the flame spread over fuels near the structure and the trajectory and accumulation of firebrands. Fire propagation over mulch and other ground fuels is strongly affected by the flow field near the ground. In concurrent flame spread, with the wind flow in the same direction as the flame spread, the rate of flame spread is strongly affected by radiative and convective heat transfer between the flame and the unburned fuel bed.[1] As the wind (or slope) increases, the angle between the flame and the unburned fuel decreases, and the dominant mechanism of preheating changes from radiation to convection as buoyancy forces become weaker than inertial forces.[2,3] For opposed flow, or creeping, flame spread, gas phase chemical kinetics play a leading role, and the heat transfer between the flame and unburned fuel is limited to a small region in which the material is heated to pyrolysis temperatures.[1,4] Opposed flow flame spread is therefore much slower than concurrent flame spread. For concurrent flow, the flame spread rate over packed natural fuel beds has been found to be roughly linear with wind speed.[5]

The flow fields around surface-mounted rectilinear objects have been studied by a number of researchers. The shed structure used in this set of experiments closely resembles a cube, which has been well studied. Hunt et al.[6] developed a sketch of the flow pattern around a cube oriented perpendicular to the flow, based on wind tunnel experiments and topological considerations. At high Reynolds number ($Re \approx 10^4$), a vortex wrapped around the structure near the ground in a horseshoe pattern is observed upstream of the object. Note that a ground fire approaching a structure under these conditions would go from a region of concurrent flow to a region of opposed flow as the flames encounter the vortex. Depending on the flow parameters, additional smaller counter-rotating vortices may add to the complexity of the upstream flow. These vortex patterns were confirmed by Martinuzzi and Tropea[7] in flow visualization experiments at $Re \approx 10^5$. The upstream separation distance was found to be about 0.8 times the height of the object H for a cube. A numerical study of a wall-mounted cube in channel flow by Sedighi and Farhadi[8] showed that the number of upstream vortices decreases as Re increases from 1000 to 40 000. The upstream separation distance reached a maximum of about 1.8 H at Re =10 000 and a minimum of 0.9 H at Re = 40 000.

This paper reports on a series of field experiments on flame spread and firebrand spotting over shredded hardwood mulch under a variety of wind conditions and separation distances from a nearby structure. A simulation of the flow fields in this set of experiments is used to provide insight into the characteristics of the flame spread rates. The experiments described in this paper are part of a larger study that includes multiple mulch types and a variety of fence types placed within the mulch bed.

EXPERIMENTS

A series of experiments was performed to observe the behavior of fire over a bed of shredded hardwood mulch. Both the movement of the flame front and spotting from firebrands generated by the mulch were of interest. Researchers from the National Institute of Standards and Technology (NIST) carried out these experiments at the Frederick County Fire and Rescue Training Facility in Maryland, U.S.A.

Experimental setup

Figure 1 shows the configuration for this set of experiments. The mulch bed consisted of a 0.05 m thick layer of shredded hardwood mulch packed within a steel pan 0.91 m wide and 3.4 m long, with 0.04 m high sides. Since a single pan of the required length was not available, the mulch pan consisted of two overlapping pans nested together, with the end pieces folded down to make a single unit. The mulch bed was arranged perpendicular to the side wall of a small structure 2.44 m on a side. A sacrificial piece of T1-11 plywood siding covered the bottom 1.2 m of the wall facing the mulch bed. The separation distance d_{sep} between the mulch bed and the structure was varied from 0 m to 6 m. In some cases, a separate pan 0.46 m wide and containing a 0.05 m thick layer of shredded hardwood mulch was placed along the base of the structure, perpendicular to the mulch pan, in order to observe ignitions from firebrands generated by the burning mulch. In these cases, the plywood layer over the wall of the structure was replaced with a non-combustible surface (cement board) for better protection.

Figure 1. Experimental setup

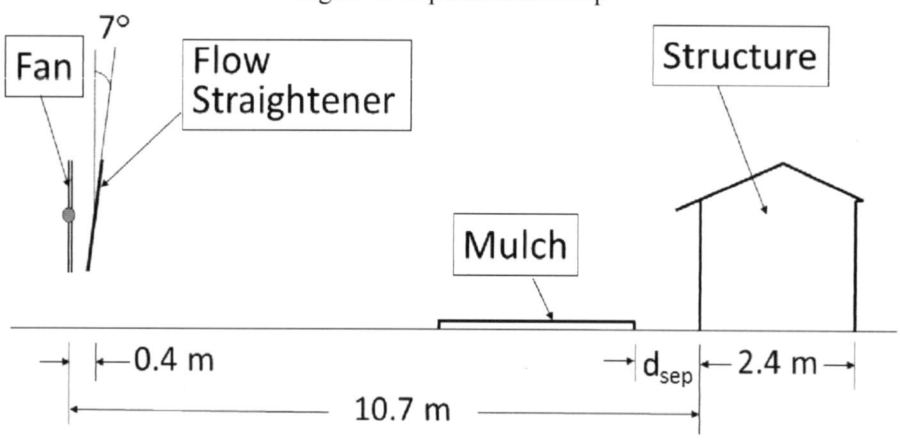

The wind field was generated by a large fan (a gas-engine powered propeller) 2.1 m in diameter located 10.7 m from the structure. A flow straightener tilted at 7° directed the flow toward the mulch bed. Figure 2 shows the experimental setup from a viewpoint to the right and behind the fan. Wind speed was measured and recorded using a set of 13 bidirectional pressure probes in an array mounted 1.22 m in front of the mulch bed. Five probes were arranged vertically along the centerline of the experiment, and 5 probes (including the one along the centerline) were located in a horizontal line 1.22 m above the ground. The other 4 probes were located along top and bottom rods of the probe array.

Figure 2. Experimental setup showing fan and flow straightener aiming wind flow toward mulch pan and structure

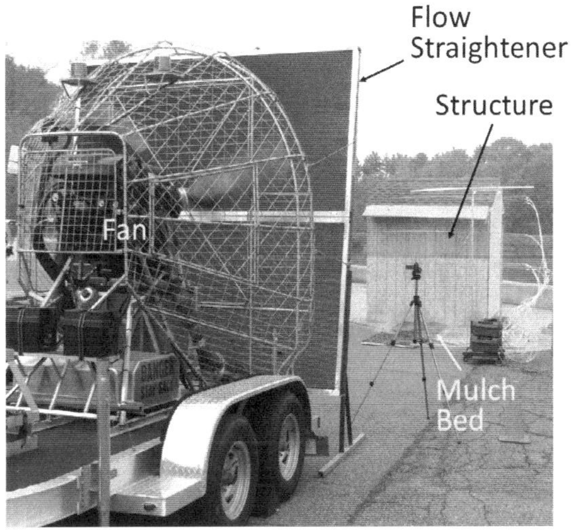

Figure 3 shows average wind velocity profiles from bidirectional probes arranged along vertical and horizontal lines. The plots show values at each probe time-averaged over the entire experiment, from sample experiments conducted on different test days. The fan was nominally set to deliver steady wind speeds of 6 m/s (blue), 10 m/s (green), and 14 m/s (red). The plots in Fig. 3a show that the value of average velocity at the lowest location was the most repeatable, that the repeatability of the average velocity worsened with height, and that the value of average velocity at the highest location dropped significantly. Except for one case with a wind speed setting of 14 m/s, the vertical profile was reasonably uniform from 0.3 m through 1.7 m in height. Assuming that the horizontal profile in Fig. 3b was symmetric, the flow speed was reasonably uniform over the 0.91 m width of the mulch bed.

In order to maintain the specified wind over the entire length of the mulch bed and to the wall of the structure, experiments were carried out only when the speed of the ambient wind was lower than 1/3 times the required wind flow from the fan.

Figure 3. Average wind velocity profiles from bidirectional probe array a) along vertical line at centerline of mulch bed and b) at 1.22 m height above ground

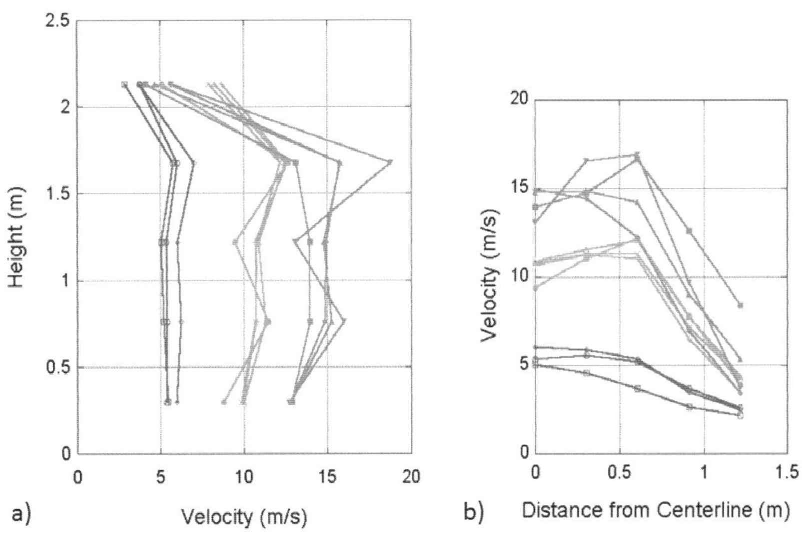

The configurations in Fig. 4 show the four separation distances that were tested: $d_{sep} = (0, 0.3, 0.9, 1.8)$ m. Without a separate mulch bed sitting at the base of the structure, the zero separation distance case was the only configuration that allowed direct contact between the burning mulch and the wall of the structure. The flames over the mulch bed were not high or intense enough to cause ignition by radiation. Therefore, without contact between the burning mulch and the structure, firebrands were the only way to ignite the plywood layer. This required the firebrands to either be caught in a crevice underneath the structure or accumulate at its base.

Figure 4. Top view of mulch bed layouts

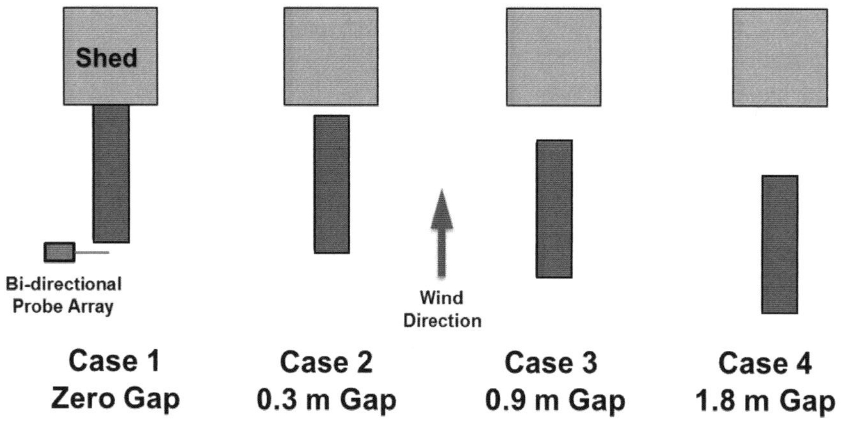

The behavior of the mulch bed after ignition was captured in four cameras; two were located with a view of the mulch bed from the right or left side, and two were positioned on either side of the wind machine with a near end-on view of the mulch bed. At least one camera had a clear view of the digital clock on the right side of the structure that was the official timekeeper of the experiment.

Experimental procedure

Before each experiment, the mulch was dried on pallets in the sun or in aerated barrels until samples taken from multiple sites measured less than 7 % moisture content using a moisture analyzer. The mulch bed was prepared by positioning the steel pan in front of the structure and filling the pan with an even layer of shredded hardwood mulch, which was lightly compressed by foot. Then the mulch was ignited with a propane burner placed on the mulch bed near the end closest to the fan. The burner was left in place for 1½ min to allow the fire to become established. After the burner was removed, the engine driving the fan was started and brought up to speed. The progress of the fire was observed until ignition or charring was observed somewhere along the outer sacrificial layer of the structure, at which point the mulch and all firebrands were extinguished with water.

EXPERIMENTAL ANALYSIS

The modes of fire progression over the mulch bed can be separated into flame spread, which moves the fire continuously from the point of ignition into unburned material, and firebrand spotting, which ignites new fires at some distance away from the fire front.

In every mulch experiment, the flame spread was observed to slow significantly as the flame front approached the structure wall, with the first ignition of the plywood layer resulting in each case from a firebrand landing in and igniting the mulch at a point much closer to the wall than the fire front. Figure 5 shows a particularly clear example of this phenomenon. The photo at 9:00 shows the position of the flame front that has stalled near the wall of the structure. At this time a firebrand has just landed very close to the wall and ignited the mulch. From this second point of ignition, flames spread toward the wall and eventually ignite the plywood. In the other direction, within 12 s the flames from this second ignition point have burned in a line back toward the original flame front, as shown in the top right photograph. The flame front then spreads laterally from the burned line, while the original flame front continues to move forward very slowly.

Other phenomena were observed to affect the rate of flame spread in the mulch experiments. In many experiments, a firebrand ignited the mulch downstream of the main flame front, creating a separate burning region that expanded with time until the flame front merged with the upstream edge of the main flame front. This resulted in jumps in the flame front location.

Figure 5. Mulch experiment in 10 m/s imposed flow, showing the evolution of the flame front due to spotting from a firebrand igniting mulch next to the wall

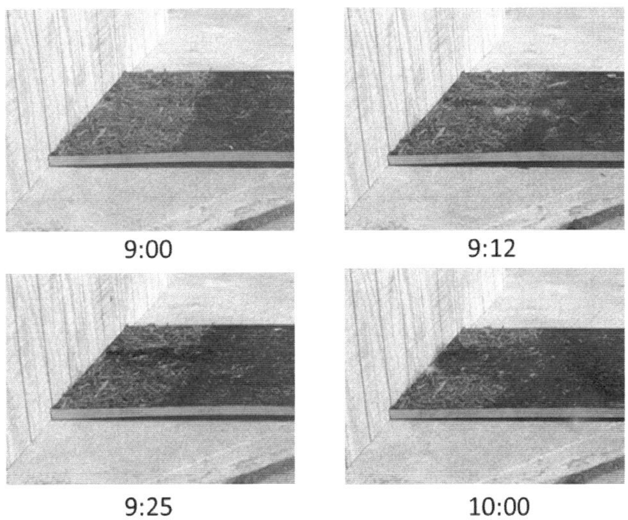

An assessment of the number of minutes between the time of removal of the propane burner and the time when the flame front reached the end of the mulch bed resulted in the plot of test duration as a function of fan wind speed and separation distance shown in Fig. 6. This indicates that experiments take less time (the flame spread is faster) as the wind speed increases and as the separation distance increases from 0 m to 1.8 m. Some anomalous results are seen for the zero separation distance cases.

Figure 6. Test durations for all shredded hardwood mulch experiments

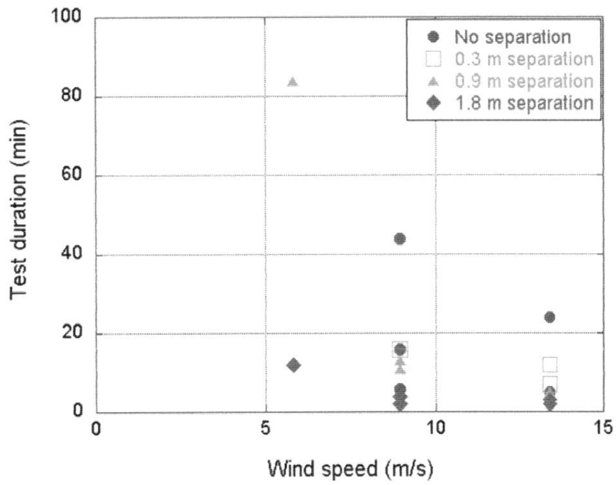

Further investigation of the behavior of the flame front required more detailed analysis of the video recordings of the experiments and a qualitative look at the flow field using computational fluid dynamics (CFD).

Video analysis

The position of the flame front as a function of time for a given experiment was determined from a set of frames extracted from the video recorded by the right or left camera or both. Each camera view was fixed in place during the experiment after having been adjusted beforehand to capture the field of interest. To measure the position of the flame front in a given frame, a physical scale was developed using a set of perspective lines created for each camera view. Figure 7 gives an example, with lines marked off in units of feet. The set of perspective lines in this frame make it possible to estimate distances from the structure wall, including the position of the flame front closest to the wall. The lines were generated using MATLAB* software by selecting two points on top of the mulch at each end of the mulch pan and setting a physical scale from a known distance. In this case, the corner of the white square marking on the asphalt in the foreground was known to be 3 ½ ft (1.07 m) from the wall. In later experiments, markings added to the pavement made it easier to determine both the physical scale and the angle of perspective lines along the length of the mulch bed.

For camera views in which the upstream end of the mulch pan was not in the frame, as was the case for some of the experiments with 0.9 m and 1.8 m separation distances, the perspective line away from the structure was established using other information. On a sunny day, such as that shown in Fig. 7, the shadow cast by the bidirectional probe array could be used, assuming that the array had been positioned parallel to the wall. In cases in which the mulch pan was separated from the wall, the location of the perspective line along the wall at mulch height was estimated.

Figure 7. Perspective lines to track the flame front superimposed on a video frame

* Certain commercial products are identified in this paper in order to specify the experimental procedure adequately. Such identification is not intended to imply recommendation or endorsement by the National Institute of Standards and Technology, nor is it intended to imply that the products identified are necessarily the best available for the purpose.

The boundary between burned and unburned mulch was considered to be where the color of the mulch changed from brown to black. Flames attached to the mulch were observed in some video frames but were too sporadic to serve as an indicator of flame front location. The flame front was typically not a straight line. The leading edge of the flame front was therefore defined as the point of the continuous burn pattern closest to the structure. On occasion, burn spots appeared in the mulch bed downstream of the flame front due to firebrands. When a spot became connected to the main burned region, the location of the flame front jumped to the leading edge of the spot, which was then incorporated into the main burning region. For the plots in this paper, the initial time t = 0 s was taken as the time when the propane burner was removed.

The measurement of the flame front location was subject to uncertainty from both setting the physical scale and determining the leading edge of the flame front. Sources of uncertainty were all Type B, with standard uncertainty based on scientific judgment. The perspective line at mulch height along the wall could be incorrect by ± 0.5 cm for zero separation cases, due to uneven piling of mulch against the wall and distortions in the mulch pan, and as much as ± 2 cm for experiments with nonzero separation distances, for which the position was estimated. The accuracy of the points defining the perspective line far from the structure was estimated as ± 1 cm when the end of the mulch pan was visible in the frame and ± 4 cm when it was not. Other minor sources of uncertainty included lens distortions in the camera, which may throw off the assumption of linearity used to create the set of perspective lines, and wind vibrations that caused the image to move under the superimposed lines.

Determining the location of the flame front was challenging in many cases. The contrast between burned and unburned areas depended on the lighting on sunny or cloudy days and on the presence of smoke and flames in the image. Frames in which the smoke or flames completely obscured the flame front were skipped. The inhomogeneity of the shredded hardwood mulch also added to the difficulty of determining the location of the flame front. This source of uncertainty was estimated to be on the order of ± 3 cm.

The combined standard uncertainty in the flame front location, calculated as the square root of the sum of squares from all of these sources, was ± 5.5 cm. An analysis of the videos from both right and left cameras from a single experiment provides a test of this estimate of uncertainty. Examples will be pointed out in the plots in the next section.

Flame front locations

Figure 8 shows the flame front location as a function of time for an experiment in which the mulch bed was separated from the structure by a distance of 0.9 m and subjected to an imposed wind set to 6 m/s. The flame spread rate is given by the slope of this plot. This plot demonstrates the main features observed in many of these mulch experiments. Initially there is a start-up period, during which the burning region expands from the area under the propane burner toward the edges of the pan. The spread rate of the flame front toward the wall of the structure is slower during this time period, which in this case is from 0 s to about 500 s. The flame spread rate then increases, and is constant or gradually decreasing until it reaches a location, here about 1.6 m from the wall, where there is an abrupt decrease. The flame spread rate is then constant at this lower value except for locations where firebrand spotting causes a jump in the location. In this case, a change in the flame spread rate due to spotting does not occur until the flame front is nearly at the end of the mulch bed at 0.9 m from the wall.

Figure 8. Flame front location with time for separation distance of 0.9 m and wind speed of 6 m/s

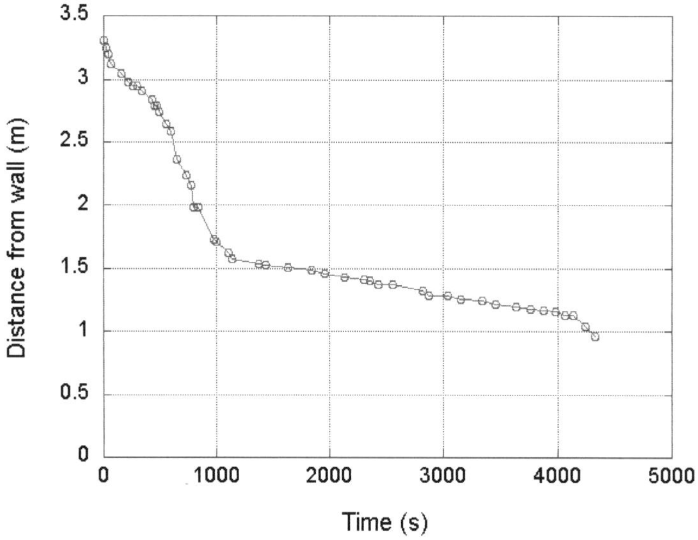

Figure 9 shows flame front locations relative to the structure wall for a set of experiments at various separation distances, with the fan set to deliver a wind speed of 10 m/s. Note that the behavior for 0.3 m and 0.9 m separation distances is similar. In both cases, the flame spread rate is relatively constant until the flame front reaches about 1.4 m from the wall in the 0.3 m separation case and about 1.7 m from the wall in the 0.9 m separation case. At this point, the flame spread rate decreases abruptly. For the 1.8 m separation case, the flame spread rate is faster than for either 0.3 m or 0.9 m separation cases. It is decreasing as it reaches the end of the mulch pan at 1.8 m from the wall. The two experiments at 0 m separation from the wall show widely disparate flame spread rates, with one (#2) showing the same rate at the 1.8 m case until the flame front stalls at 0.4 m from the wall and the other (#8) showing the same rate as the final rate of 0.3 m and 0.9 m separation cases. The reasons for this behavior are not yet understood and will be studied further.

Right (R) and left (L) videos have been analyzed for 0.3 m, 0.9 m, and 1.8 m cases. The maximum difference between measured flame front locations at a given time is roughly 10 cm, which is in line with uncertainty estimates.

Figure 9. Flame front location with time as a function of separation distance for 10 m/s wind speed

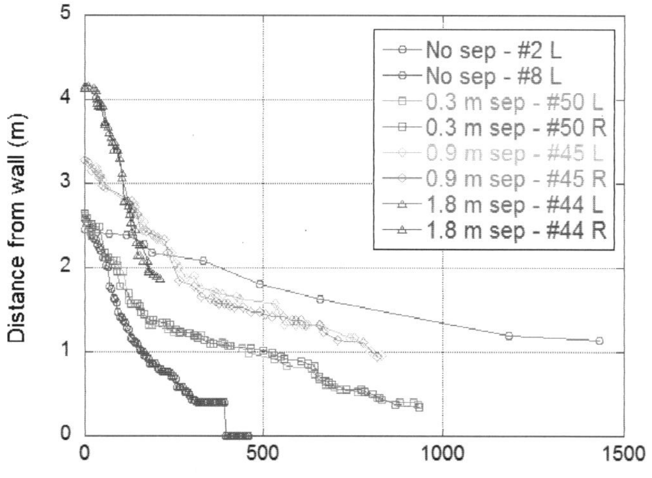

Flame front locations for experiments at various separation distances at a wind speed of 14 m/s are plotted in Figure 10. Again, the behavior for 0.3 m and 0.9 m separation distances is similar. Here, the flame spread rate decreases more gradually from its maximum value after the start-up period, until the flame front slows significantly at a distance between 1.3 m and 1.7 m from the wall. The flame spread rate for the experiment at 1.8 m separation is comparable to the 0.3 m and 0.9 m separation cases at the same distance from the wall. Just as in the 10 m/s wind cases in Figure 9, the 0 m separation case behaves differently, showing a much slower flame spread rate throughout.

Figure 10. Flame front location with time as a function of separation distance for 14 m/s wind speed

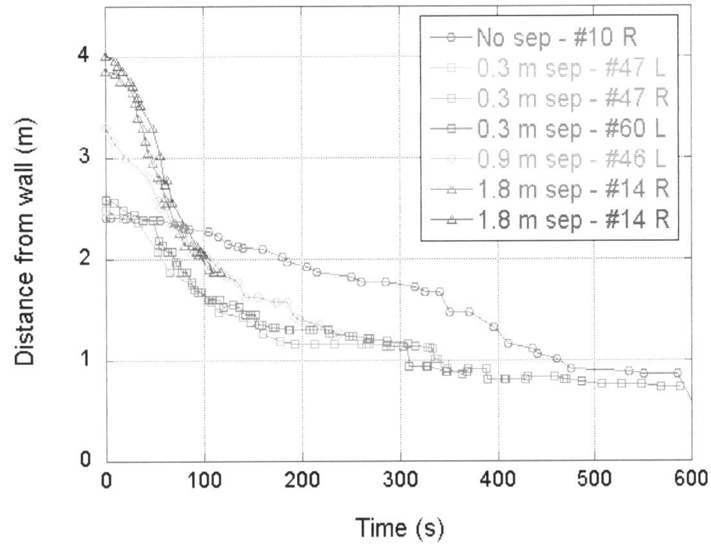

Figure 11 plots the flame front location with time for experiments at the same separation distance of 0.9 m and at three different wind speed settings. The flame spread rate after the start-up period (when the location of the flame front is about 2.8 m from the wall) is a function of wind speed, with slower flame spread rate corresponding to slower wind speed. In all cases, the flame spread rate changes abruptly at a distance of about 1.6 m from the wall.

Figure 11. Flame front location with time as a function of wind speed for 0.9 m separation distance

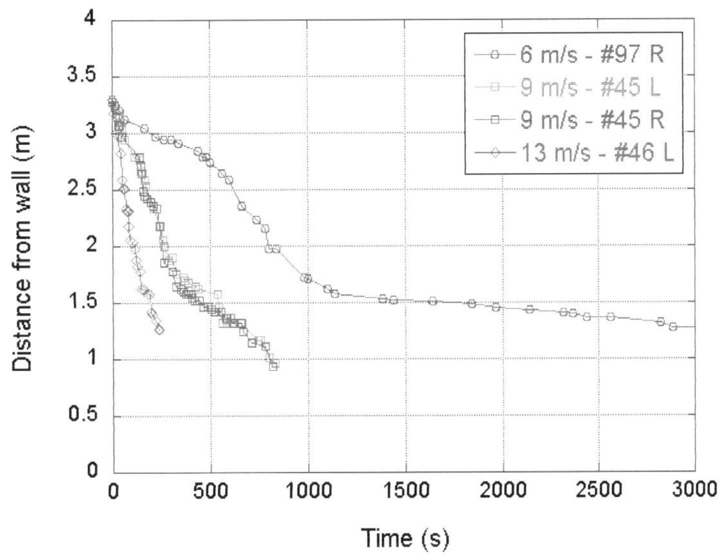

Figure 12 shows the flame front location with time for experiments at three different wind speed settings, all at a separation distance of 1.8 m. The flame spread rate after the start-up period is a function of wind speed, with slower flame spread rate corresponding to slower wind speed. Although the flame spread rate is slowing as the flame front reaches the end of the mulch, no abrupt decrease is observed.

Figure 12. Flame front location with time as a function of wind speed for 1.8 m separation distance

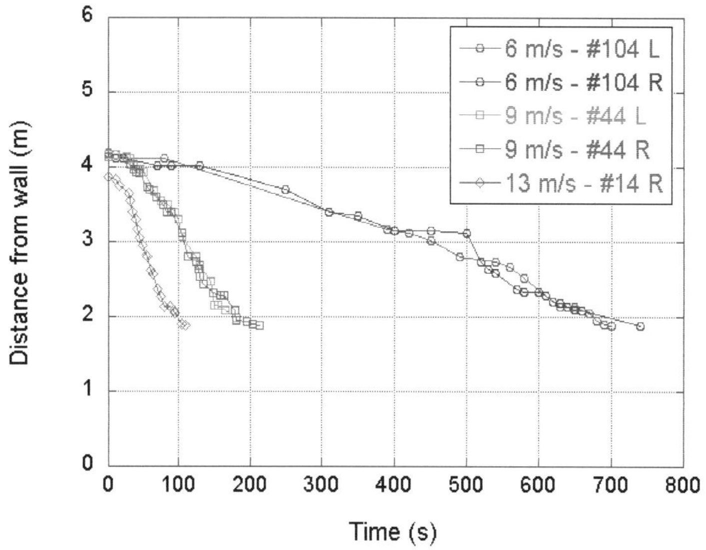

COMPUTATIONAL ANALYSIS

The observation during this set of field experiments that the flame front consistently slowed as it approached the structure suggested that the flow field could be playing an important role. To obtain insights into the fire behavior, the experimental setup was modelled using the NIST Fire Dynamics Simulator (FDS),[9] a computational fluid dynamics (CFD) software program typically used for calculating fire-driven fluid flow in enclosures. As an initial exploration into the mulch bed experiments, FDS was used for calculating fluid flow only (no fire) in an exterior computational space with open boundaries. Subgrid mixing due to turbulence was represented by a Smagorinsky large eddy simulation (LES) model. Recent validation work[10] has showed that FDS is capable of good results when compared with experiments for similar problems, including building ventilation, wind impinging on the exterior of a structure, and transport of pollutants in an urban environment.

Flow field results

In the model of the mulch test, illustrated in Figure 13, the fan and shed structure were represented as obstacles, shown in brown. Since the intent of the model was to provide insight rather than a quantitative comparison, the propeller-driven flow field and the geometries of the structure and the fan/flow straightener system were not represented in detail. The fan was simulated by a square block 1.8 m on a side and 0.5 m thick with a steady velocity boundary condition directed toward the structure at an angle of 7° to the horizontal. The distance of the fan from the structure was 10.7 m in agreement with the experimental setup. Boundary conditions at the exterior of the domain were open. Resolution in x, y, and z directions was 0.1 m.

Instantaneous velocity vectors in Fig. 13 illustrate some of the features of the wind flow, which is highly turbulent. Figures 13a and b show instantaneous velocity vectors along the center plane of the experimental geometry (side view) and in a plane 0.1 m above the ground (top view), respectively. The vectors are colored by wind speed, with red corresponding to the maximum value of 9 m/s. The velocity flow field shows that a vortex forms near the ground in front of the structure and wraps around the sides, consistent with the horseshoe vortex observed for a surface-mounted cube in a flow field found in the literature.[6] The behavior of the wind near the ground is of particular interest for the mulch experiments. The side view in Figure 13a shows the vortex near the ground in front of the structure. In the top view in Figure 13b, the wind velocity just above the ground is high and directed toward the structure at distances between one and two structure heights (about 2 m to 7 m) from the structure. Within one structure height distance (about 2 m) from the structure, the wind velocity is lower and directed toward the fan.

The flow field strongly influences the behavior of firebrands generated in the mulch bed. Firebrands near the ground experience the opposing flow from the vortex, which acts to keep them away from the structure and directs them into the high velocity flow around the left or right side of the structure shown in Fig. 13b. Firebrands that are lofted to an intermediate height of 0.1 m to 1 m may be caught in the vortex flow and deposited close to the wall of the structure, where they can ignite combustible wall material or mulch adjacent to the wall. Firebrands that approach the structure at a height greater than a meter may enter an updraft that transports them over the top of the structure; open vents could allow these firebrands to enter the structure.

Figure 13. FDS model results showing instantaneous velocity vectors. Side view (a) shows velocities along the center plane, and top view (b) shows velocities in a plane 0.1 m above the ground.

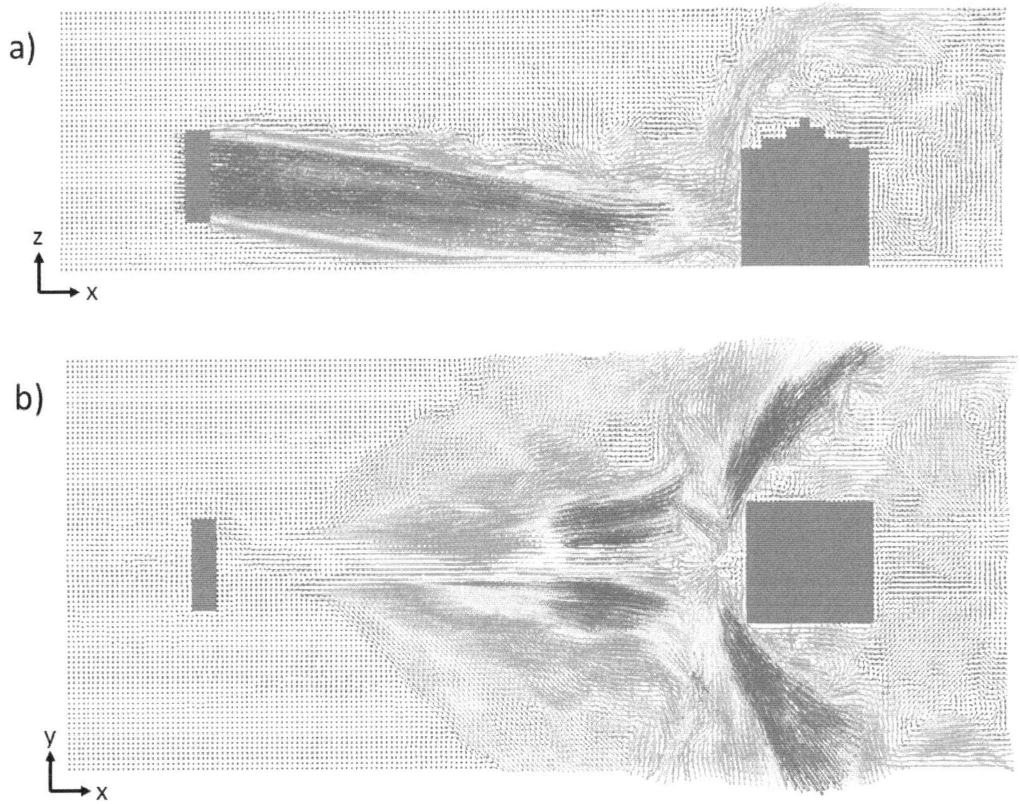

Figure 14 shows the x component of velocity v_x, in the direction toward or away from the structure, averaged over time. The size of the vortex is approximately 0.7 times the dimensions of the structure, roughly consistent with the literature on flow around surface-mounted cubes at high Re.[7,8] Between the fan and the structure, the wind is directed toward the structure (positive values of v_x) except for a region close to the ground just in front of the structure, where v_x is strongly negative (blue). Depending on how far away from the structure the mulch bed is ignited, the flames spreading toward the structure may initially experience concurrent flow, with wind velocity in the same direction as the flame spread. As the flame front enters the vortex region, however, the flame spread changes to opposed flow mode and slows considerably. If a firebrand ignites the mulch within the vortex region, as in Figure 5, the flame spread from the point of ignition back to the main flame front is concurrent with the vortex flow field, and is thus rapid.

Simulations at imposed wind speeds from 6 m/s to 14 m/s indicated that the vortex size is relatively independent of wind speed in this range. Vortex intensity and concurrent and opposed flow velocities are strongly affected by wind speed.

Figure 14. FDS model results showing time-averaged values of velocity v_x in the x direction. Side view (a) shows v_x along the center plane, and top view (b) shows v_x in a plane 0.1 m above the ground

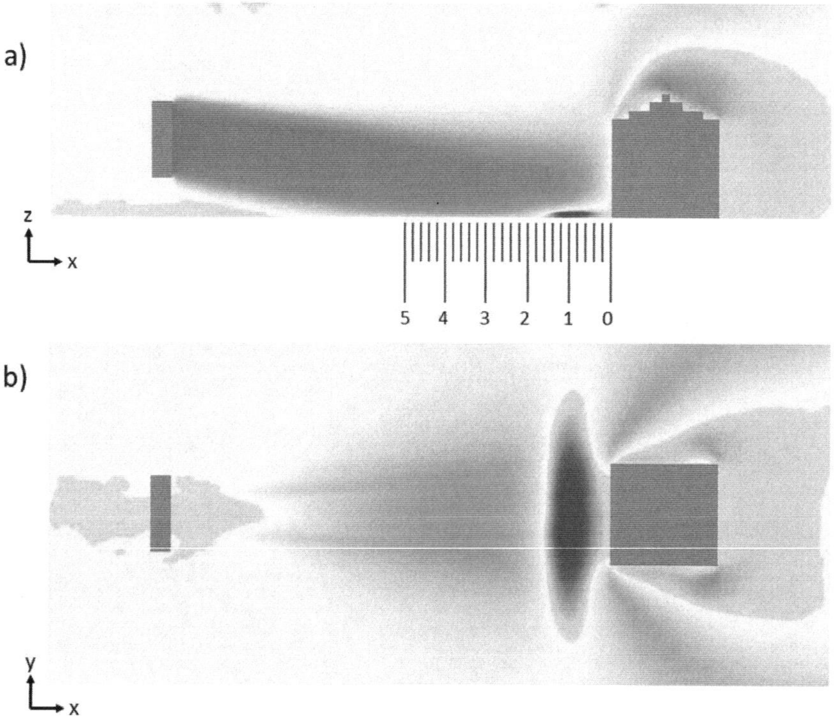

Figure 14a includes a scale marked from 0 m to 5 m away from the structure. This figure can be compared with Figures 8 through 12 showing flame front location with time to understand the changes in flame spread rate. For the mulch pan separated from the structure by 1.8 m, the flame spread takes place completely outside of the vortex. However, the concurrent velocity slows as the flame approaches the end of the mulch pan, which could explain the gradual decrease in flame spread rate. The flame fronts in mulch pans separated from the structure by 0.3 m and 0.9 m encounter the vortex as they approach the building, causing the flame spread to go from concurrent to opposed flow mode. This accounts for the abrupt change in flame spread rate at around 1.7 m from the structure. The flame front in a mulch pan that is adjacent to the structure with no spacing is primarily within the vortex. The lack of a transition in flame spread rate for a burning region ignited just outside of the vortex requires further study, as does a zero separation case that shows rapid flame spread.

CONCLUSIONS

Field experiments have been carried out on flame spread in mulch pans at a variety of separation distances from a small building and at a variety of imposed wind speeds. The flame spread rates for mulch pans separated at 0.3 m and 0.9 m from the structure undergo an abrupt decrease between 1.3 m and 1.7 m from the structure. Simulations demonstrate that this change is due to the presence of a vortex near the ground in front of the structure, which causes the flame front to transition from concurrent to opposed flow modes. Flame spread rates for mulch pans at 1.8 m separation from the structure begin to slow near the end of the mulch pan but do not enter the vortex and do not change abruptly. Flame spread rates at a distance greater than the height of the structure are dependent on wind speed.

This work demonstrates how field experiments and modelling can lend insight into the physics underlying a recognized WUI fire hazard.

REFERENCES

[1] Fernandez-Pello, A.C. and Hirano, T. (1983). Controlling mechanisms of flame spread. Combustion Science and Technology 32, pp. 1-31

[2] Pagni, P.J. and Peterson, T.G. (1973). Flame spread through porous fuels. In Proceedings of the Fourteenth Symposium (International) on Combustion, 20-25 August 1972, University Park, PA, pp. 1099-1107. (The Combustion Institute : Pittsburgh, PA).

[3] Nelson Jr., R.M. (2015). Re-analysis of wind and slope effects on flame characteristics of Mediterranean shrub fires. International Journal of Wildland Fire 24, pp. 1001-1007. doi :10.1071/WF14155.

[4] Delichatsios, M.A. (1996). Creeping flame spread : Energy balance and application to practical materials. In Proceedings of the Twenty-Sixth Symposium (International) on Combustion, 28 July-2 August 1996, Naples, Italy, pp. 1495-1503. (The Combustion Institute : Pittsburgh, PA).

[5] Catchpole, W.R., Catchpole, E.A., Butler, B.W., Rothermel, R.C., Morris, G.A. and Latham, D.J. (1998). Rate of spread of free-burning fires in woody fuels in a wind tunnel, Combustion Science and Technology 131, pp. 1-37.

[6] Hunt, J.C.R., Abell, C.J., Peterka, J.A. and Woo, H. (1978). Kinematical studies of the flows around free or surface-mounted obstacles; applying topology to flow visualization. Journal of Fluid Mechanics 86(1), pp. 179-200.

[7] Martinuzzi, R. and Tropea, C. (1993). The flow around surface-mounted, prismatic obstacles placed in a fully developed channel flow. Journal of Fluids Engineering 115, pp. 85-92.

[8] Sedighi, K. and Farhadi, M. (2006). Three-dimensional study of vortical structure around a cubic bluff body in a channel. Facta Universitatis : Mechanical Engineering 4(1), pp. 1-16.

[9] McGrattan, K., McDermott, R., Weinschenk, C., Overholt, K., Hostikka, S. and Floyd, J. (2013). Fire Dynamics Simulator User's Guide. NIST Special Publication 1019, 6th ed. National Institute of Standards and Technology, Gaithersburg, MD.

[10] McGrattan, K., Hostikka, S., McDermott, R., Floyd, J., Weinschenk, C., and Overholt, K. (2014). Fire Dynamics Technical Reference Guide, Volume 3: Validation. NIST Special Publication 1018, 6th ed. National Institute of Standards and Technology, Gaithersburg, MD.

EXPOSING FENCING ASSEMBLIES TO FIREBRAND SHOWERS CHARACTERISTIC OF BURNING STRUCTURES

Samuel L. Manzello
Fire Research Division
National Institute of Standards and Technology (NIST)
Gaithersburg, MD USA 20899
samuelm@nist.gov

Sayaka Suzuki
Large Fire Laboratory
National Research Institute of Fire and Disaster (NRIFD)
Chofu, Tokyo, JAPAN

Ichiro Hagiwara
Department of Fire Engineering
Building Research Institute (BRI)
Tsukuba, Ibaraki, JAPAN

ABSTRACT

Wildland-Urban Interface (WUI) fires have devastated communities on multiple continents. Large outdoor fires in Japan mainly occur due to post-earthquake fires, which may result in severe urban fire spread. Once structures are ignited in these fires, firebrand production from structures becomes a key factor in fire spread. The authors are constructing a database of firebrand production from burning structures, as there is a little quantifiable data available in the literature. Based on post-fire disaster investigations, fencing assemblies are known to be prone to ignition in WUI fires, and may provide pathways to structure ignition. In this work, a comparison of ignition results from full-scale fencing assembly experiments conducted using the Building Research Institute's (BRI) full-scale wind tunnel facility, to mock-ups of full-scale fencing assemblies using the recently developed experimental capability at the National Research Institute of Fire and Disaster (NRIFD) are discussed. In both experimental facilities, the fencing assemblies were exposed to firebrand showers using custom built continuous-feed firebrand generators with size and mass distributions similar to the structure firebrand database mentioned above. The purpose was to determine if similar fencing ignition vulnerabilities were observed for reduced sized experiments. These results are required to guide the necessary configurations of fencing assembly mock-ups that can potentially be used in standard laboratory test methods. The results of this study will be presented.

INTRODUCTION

Recently, there have been significant number of wildfires that spread into communities, in Argentina, Australia, Brazil, Canada, Chile, France, Greece, Portugal, Spain, and the USA. Such fires are referred to as Wildland-Urban Interface (WUI) fires. Japan is another country that has been subject to large scale outdoor fires, but these fires have been initiated by earthquakes. A prominent mechanism of structure ignition in both WUI and urban fires is the production of firebrands, also referred to as embers[1].

Post-fire studies suggest that attached building components may provide a pathway for structures to ignite in WUI fires[2]. Examples of attached building components include decking and fencing assemblies. A recent study by the authors demonstrated that fencing assemblies may be ignited by firebrand showers, but the capability of the ignited fencing to transfer the fire to an adjacent structure, such as wall assembly, was not considered[3]. This was done intentionally, since the *first step* was to determine if firebrand showers could even ignite various wood fencing types.

The focus of this paper is to present a comparison of ignition results from full-scale fencing assembly experiments conducted using the Building Research Institute's (BRI) full scale wind tunnel facility, to mock-ups of full-scale fencing assemblies using the recently developed experimental capability at the National Research Institute of Fire and Disaster (NRIFD). In both experimental facilities, the fencing assemblies were exposed to firebrand showers using custom built continuous-feed firebrand generators with size and mass distributions similar to the structure firebrand database developed by the authors.

It is of interest to determine if similar fencing assembly ignition vulnerabilities were observed for reduced sized experiments. Such data are required to guide the necessary configurations of fencing assembly mock-ups that can potentially be used in standard laboratory test methods. Furthermore, obtaining insights into ignition behaviour from reduced sized experiments is a desirable tool. It is hoped that these studies will also provide guidance for further full-scale experimental campaigns focused on structure ignition by ignited fencing assemblies, such as those conducted at the Building Research Institute (Japan), and recent studies conducted at NIST[4] to investigate fire spread along fencing assemblies using a burner (without firebrand attack).

EXPERIMENTAL DESCRIPTION

Full-Scale Experiments at the Building Research Institute (BRI)

The interested reader is referred elsewhere for a comprehensive description of the redesigned full-scale continuous feed firebrand generator system used here[5]. Namely, the feeding system was redesigned to be able to generate firebrand size and mass distributions using larger wood chips in an effort to produce a wider range of exposure conditions, as compared to prior studies[3]. Recent experiments have evaluated the performance of full-scale roofing components exposed to structure firebrand showers using this redesigned firebrand generator[6]. For completeness, a terse overview is provided presently.

The device consisted of the main body and continuous feeding component. The main body of the generator was the same as prior investigations (Dragon component); the feeding system was rebuilt. The feeding system was connected to the main body and was equipped with two gates to prevent fire spread. A blower was connected to the main body and the purpose of the blower is to loft and control the combustion state of the firebrands. A difficulty when constructing this device was designing a completely contained feeding system shielded from the wind tunnel flow.

The feeding system consisted of a pneumatic cylinder coupled to a cylindrical container where wood chips were stored (see **Figure 1**). Below the wood chip storage area, a plate was installed that allowed variation in the volume of wood chips to fall from the storage area to the first gate. This plate was set precisely to allow a specific mass of wood chips to fall into this volume. When the air pressure was applied, the sliding rod of the pneumatic cylinder moved forward and separated the wood chips from the storage area to the first gate, where they were then dropped into the second gate that leads to the Dragon where they are ignited (see **Figure 1**). The gate system was required to contain the fire from spreading from the Dragon to the feed system and each gate was driven by pneumatic cylinders as well.

For all experiments, Japanese Cypress wood chips were used to produce firebrands. These were provided from a supplier and upon arrival, these chips were filtered using a 10 mm mesh to remove very fine wood pieces. The chips were also oven dried, as they were shipped under wet conditions. The sizes of wood pieces were selected to produce firebrands with larger projected area at a specific mass than our prior studies using continuous firebrand generation[3]. As structure firebrand production is an important factor in WUI and urban fire spread, the authors have begun to develop a database of firebrand production from burning structures/structure components, and have demonstrated the ability to generate firebrands similar to structure firebrands from this database[5]. The fencing assemblies, described below, were exposed to these structure firebrand showers.

As in prior experiments using the NIST Dragon, the new experimental device was installed inside the test section of BRI's Fire Research Wind Tunnel Facility (FRWTF). The facility was equipped with a 4 m diameter fan to produce the wind field. The cross section of the FRWTF is 5 m wide by 4 m high. The maximum wind speed available using the FRWTF is 10 m/s. The leading edge fencing assemblies were placed at a distance of 3.25 m downstream of the full-scale Continuous Feed Firebrand Generator.

Redwood lattice fencing assemblies were constructed for the experiments (see **Figure 2**). These were custom fabricated by using two 1.22 m in height, 2.44 m in width, redwood lattice pieces, and held together using a 2 x 4 wood array (2 x 4 boards are 38 mm by 89 mm). While two-sided Redwood lattice fencing assemblies were used in this study, this type of fencing assembly may also be used in a one-sided configuration. Experiments are also underway to investigate these differences between one-sided and two-sided configurations, both at NIST and BRI/NRIFD. To simulate fine fuels that may be located near fencing assemblies, mini-pine bark mulch was used. Mini-pine bark nugget mulch was placed around the fencing assemblies at a depth of 51 mm (both sides). The mini-pine bark nuggets were oven dried and the density was 0.15 g/cm^3. Pine bark nugget mulch is known to be harder to ignite than other types of mulch, so this mulch type was selected to be able to clearly see the dynamic ignition process. These redwood lattice fencing assembly/mulch beds were exposed to firebrand showers. Further details are shown in **Figure 2.**

Figure 1 Picture of full-scale continuous feed firebrand generator installed at BRI's FRWTF. Japanese Cypress wood chips are feed into the device to produce firebrand showers similar to burning structures.

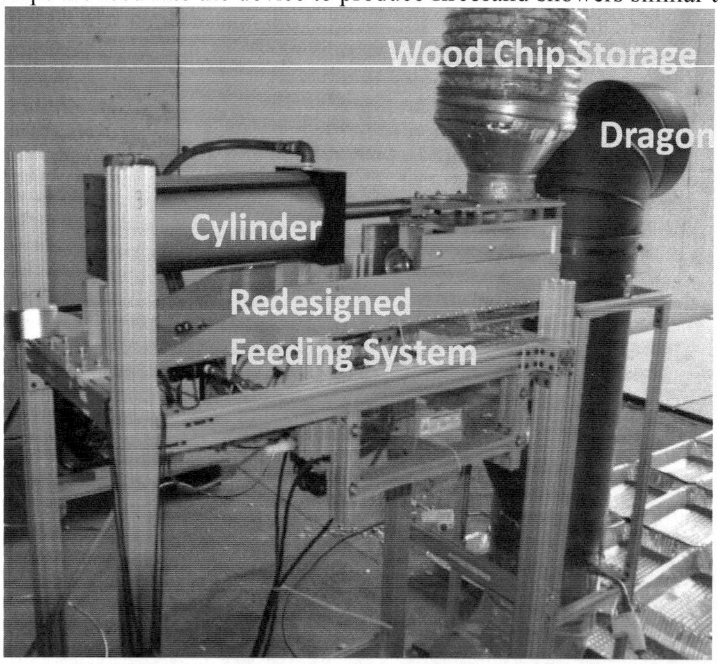

Figure 2 Picture of full-scale Redwood lattice fencing assembly with mulch bed.

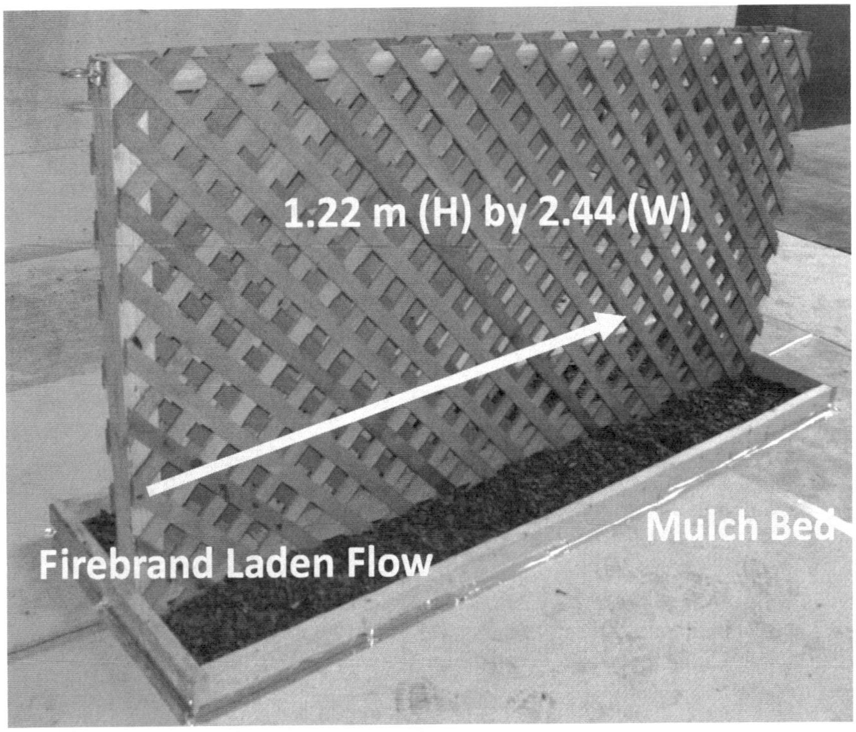

Reduced-Scale Experimental Facility at National Research Institute of Fire and Disaster (NRIFD)

A reduced-scale continuous-feed firebrand generator was used to generate firebrand showers (see **Figure 3**). A conveyor was used to feed wood pieces continuously into the device. For all tests, Japanese Cypress wood chips were used to produce firebrands. These same size wood pieces were used in the full-scale fencing assembly studies (above) and have been shown to be within projected area/mass of burning structures. Specifics of the wood feeding rate are provided below. Recently, this facility has been used to expose mock-ups of full-scale roofing assemblies to firebrand showers and the interested reader is referred elsewhere for a more comprehensive description of the National Research Institute of Fire and Disaster's facility (NRIFD)[7].

Since the base of the fan used to generate the wind field is located 1.6 m from the floor, the conveyor was placed under a custom stage designed for experiments when using the NRIFD wind facility (see **Figure 3**). The flow field was measured to be within ± 10 % over a cross-section of 2.0 m by 2.0 m.

Mock-ups of full-scale Redwood lattice fencing assemblies were constructed for the experiments. The fencing assemblies were custom fabricated by using two, 0.61 m in height, 1.22 m in width, redwood lattice pieces, and held together using a 2 x 4 wood array (see **Figure 4**). To simulate fine fuels that may be located near fencing assemblies, mini-pine bark mulch was used. Mini-pine bark nugget mulch was placed around the fencing assemblies at a depth of 51 mm (both sides). The mini-pine bark nuggets were oven dried and the density was 0.15 g/cm^3. The leading edge fencing assemblies were placed at a distance of 1.5 m downstream of the reduced-scale Continuous Feed Firebrand Generator in an effort to have the arrival number flux of firebrands similar to the full-scale experiments.

Figure 3 Schematic of reduced-scale continuous feed firebrand generator installed at NRIFD (side view). The diameter where the firebrands exit the firebrand generator (Dragon) was 10 cm. The stage dimensions were 5.5 m wide, 6.4 m long, with a height of 1.6 m.

Figure 4 Picture of mockup of full-scale Redwood lattice fencing assembly with mulch bed. The dimensions were 0.61 m (H) and 1.22 m (W).

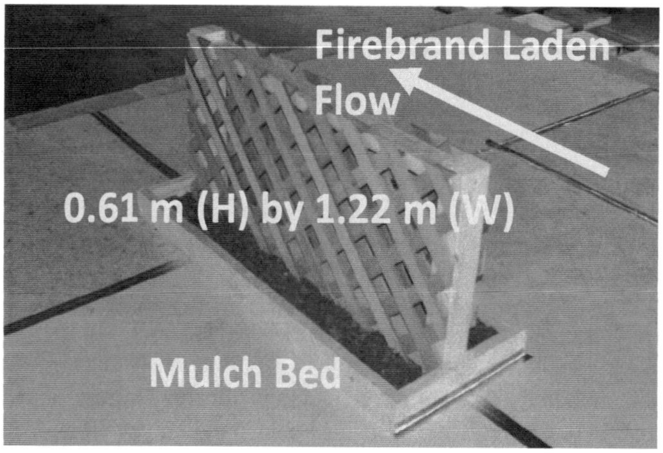

RESULTS AND DISCUSSION

Full-Scale Fencing Assemblies

It was found that 170 g ± 30 g, fed into the Dragon every 15 s, provided an adequate firebrand generation rate to ignite fencing assemblies. A supply of 170 g corresponded to approximately 560 wood pieces deposited every 15 s. The number flux (number of firebrands generated/m^2s), at the exit of the device, was measured at a feeding rate of 170 g every 15 s (680 g/min). To determine the number flux, the number of firebrands was counted at every frame of the video recording, and summed every second. Based on the analysis, the number flux reached a steady value of 530/m^2s 150 s after feeding began. Mass flux data (mass of firebrands generated/m^2s) were calculated by multiplying the number flux and the average mass of each firebrand at a feed rate of 170g every 15 s.

The projected area of the firebrands directed at the fencing assembly/mulch bed was also quantified using a series of separate experiments where a water pan array was placed downstream of the firebrand generator. The firebrands were extracted, dried, and the mass (using balance) and projected area were determined using image analysis methods. **Figure 5** displays the projected area of the generated firebrands for various wind speeds. In **Figure 5**, data are presented for *both* the full-scale and reduced-scale firebrand generators. The reduced-scale firebrand generator data, however, is described below. Image analysis software was used to determine the projected area of a firebrand by converting the pixel area using an appropriate scale factor. More details on the methodology used are available elsewhere[6]. The standard uncertainty in determining the projected area was ± 10 %. The mass of each firebrand was measured by a precision balance. The standard uncertainty in the firebrand mass was approximately ± 1 %. The mean mass and standard deviation of each firebrand were obtained and observed to be 0.03 g ± 0.02 g. Therefore, the mass flux of generated firebrands was calculated to be 16 g/m^2s for the full-scale firebrand generator.

Figure 5 Size and mass distribution of generated firebrands for full-scale and reduced-scale firebrand generators (comparison at 6 m/s, and 9 m/s).

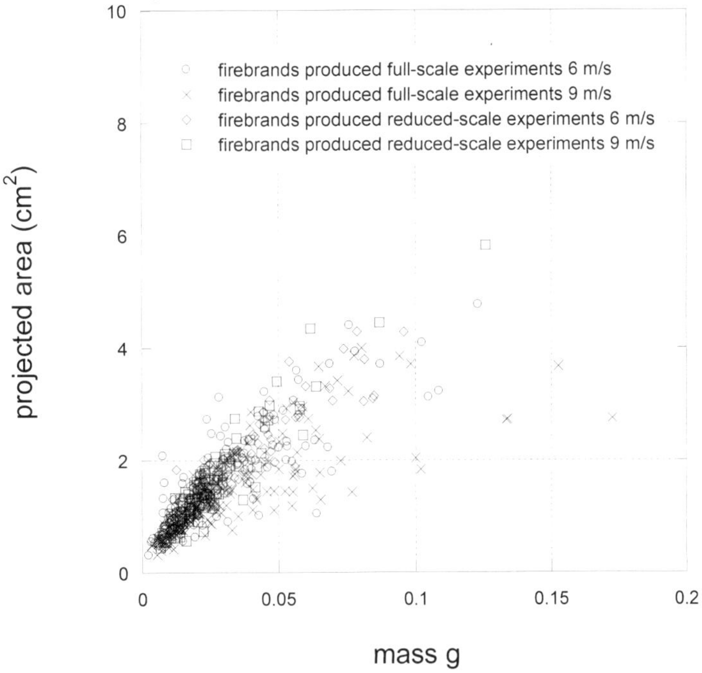

Experiments were conducted for wind speeds of 4 m/s, 6 m/s, and 9 m/s. In all cases, the evolution of the ignition process was similar. Specifically, glowing firebrands ignited the mulch beds via smoldering combustion, and then the mulch transitioned to flaming combustion. These flaming mulch beds then ignited the redwood lattice fencing assemblies. **Figure 6** displays the temporal evolution of the ignition process for a wind speed of 4 m/s. At the highest wind speed considered (9 m/s), once the fencing assembly was fully involved in flaming combustion, significant firebrand production from fence combustion was observed (see **Figure 7**; firebrands were observed to be transported outside the wind tunnel).

Figure 6 Temporal evolution of Redwood lattice fencing assembly/mulch bed ignition for wind speed of 4 m/s. The top panel displays smoldering ignition in the mulch bed from firebrand attack. The smoldering mulch bed has transitioned to flaming ignition, and the Redwood lattice fencing assembly is ignited (middle panel). In the bottom panel, the ignition has spread to the entire Redwood lattice fencing assembly.

Figure 7 Redwood lattice fencing assembly/mulch bed combustion with a wind speed of 9 m/s. Significant firebrand production may be observed.

Not all the firebrands that are generated by the Dragon arrive at the fencing assembly/mulch bed. The average number of firebrands arriving on the mulch beds was counted as a function of wind speed. To simplify the analysis, firebrands were counted on the left-hand side of the mulch bed, and these values were doubled for the entire assembly. Based on these analysis, the average total arrival number flux of firebrands was $4/m^2s$ (average total arrival number flux based on results for 4 m/s and 9 m/s wind speeds). As described below, efforts were made to try to have a similar firebrand arrival number flux for the mock-up experiments.

Mock-Ups of Full-Scale Fencing Assemblies

To expose the mockups of full-scale fencing assemblies to firebrand showers, a reduced-scale continuous-feed firebrand generator was used. A conveyor was used to feed wood chips continuously into the reduced-scale firebrand generator. The conveyor belt was operated at 1.0 cm/s, and wood chips were put on the conveyor belt at 12.5 cm intervals. The wood feed rate was fixed at 80 g/min, near the upper limit of reduced-scale firebrand generator. These same size wood chips were used in the full-scale roofing assembly study. The sizes of Japanese Cypress wood chips used for firebrand generation (before combustion) were 28 ±8 mm (length), 18 ±6 mm (width), and 3 ±1.0 mm (thickness) (average ±standard deviation), respectively. **Figure 5** displays the projected area of the generated firebrands at 6 m/s, and 9/s using the same analysis methods described for the full-scale continuous feed firebrand generator (see above).

Similar to the full-scale experiments, the wind speed was varied from 4 m/s to 9 m/s. In all cases, the evolution of the ignition process was similar to the full-scale fencing assembly experiments. Specifically, glowing firebrands ignited the mulch beds via smoldering combustion, and then the mulch transitioned to flaming combustion. These flaming mulch beds then ignited the redwood lattice fencing assemblies.

Figure 8 displays the temporal evolution of the ignition process for a wind speed of 4 m/s. At the highest wind speed considered (9 m/s), once the fencing assembly was fully involved in flaming combustion, significant firebrand production from fence combustion was also observed (see **Figure 9**).

Figure 8 Redwood lattice fencing assembly/mulch bed combustion with a wind speed of 4 m/s. The dynamics of the ignition process were similar to the full-scale experiments. The top panel displays smoldering ignition in the mulch bed from firebrand attack. The smoldering mulch bed has transitioned to flaming ignition, and the Redwood lattice fencing assembly is ignited (middle panel). In the bottom panel, the ignition has spread to the entire Redwood lattice fencing assembly.

Figure 9 Mockup of full-scale Redwood lattice fencing assembly/mulch bed combustion with a wind speed of 9 m/s. Significant firebrand production may be observed.

The average number of firebrands arriving on the mulch beds was counted as a function of wind speed. To simplify the analysis, firebrands were counted on the left-hand side of the mulch bed, and these values were doubled for the entire assembly. Based on these analysis, the average total arrival number flux of firebrands was 4.5/m^2s (average total arrival number flux based on results for 4 m/s and 9 m/s wind speeds).

It is important to note that similar ignition behaviors were observed between the full-scale fencing assemblies and the mock-up of full-scale fencing assemblies. It is believed that this was observed since the following parameters were matched: size and mass of firebrands that are generated, the arrival number flux of firebrands, wind speeds, and same ignitable materials. Additional experiments are required for other fencing assembly types to further verify these important findings.

SUMMARY

The authors are constructing a database of firebrand production from burning structures, as there is a little quantifiable data available in the literature. Fencing assemblies are known to be prone to ignition in WUI fires, and may provide pathways to structure ignition. In this work, a comparison of ignition results from full-scale fencing assembly experiments conducted using the Building Research Institute's (BRI) full scale wind tunnel facility, to sections of full-scale fencing assemblies using the recently developed experimental capability at the National Research Institute of Fire and Disaster (NRIFD) was presented. In both experimental facilities, the fencing assemblies were exposed to firebrand showers using custom built continuous-feed firebrand generators with size and mass distributions similar to the structure firebrand database mentioned above.

Firebrand showers resulted in ignition of the full-scale redwood lattice fencing assemblies with mulch beds present. In all cases, the evolution of the ignition process was similar to the full-scale fencing assembly experiments. Specifically, glowing firebrands ignited the mulch beds via smoldering combustion, and then the mulch transitioned to flaming combustion. These flaming mulch beds then ignited the redwood lattice fencing assemblies.

It is believed that ignition similarities were observed since the following parameters were matched: size and mass of firebrands that are generated, the arrival number flux of firebrands, wind speeds, and same ignitable materials.

Naturally, further analyses of these interesting findings are in progress. In particular, a well-defined approach to quantify the flame spread along the fencing is required. In the case of firebrand showers, ignition may be observed at multiple points along the fencing assembly/mulch bed due to the dynamic firebrand attack, in contrast to recent NIST experiments[4], where the fencing assembly/mulch bed was always ignited at the leading edge using a burner. Experiments are also required for other fencing assembly types to further verify these important findings.

These results will provide guidance for further full-scale experimental campaigns focused on structure ignition by ignited fencing assemblies, such as those ongoing at the Building Research Institute (Japan), and those also currently in progress at NIST using burners for ignition, as compared to firebrand showers.

The ability of a given burning fencing assembly to ignite adjacent wall assemblies will be considered in future experiments. Furthermore, obtaining insights into ignition behaviour from reduced sized experiments is a desirable tool, as understanding firebrand attack on buildings is a complex and difficult task.

ACKNOWLEDGEMENTS

Mr. Marco Fernandez of NIST's Engineering Laboratory (EL) is acknowledged for shipping materials for the experiments.

REFERENCES

1. Manzello, S. L., Yamada, T., Jeffers, A., Ohmiya, Y., Himoto, K., and Fernandez-Pello, A.C., Workshop for Fire Structure Interaction and Urban and Wildland-Urban Interface (WUI) Fires – Operation Tomodachi Fire Research, *Fire Safety Journal*, 59 (2013) 122-131.

2. Maranghides, A., McNamara, D., Vihnanek, R, Restaino, J., and Leland, C., A Case Study of a Community Affected by the Waldo Fire – Event Timeline and Defensive Actions NIST TN1910, 2015.

3. Suzuki, S., Johnsson, E., Maranghides, A., and Manzello, S.L., Ignition of Wood Fencing Assemblies Exposed to Continuous-Wind Driven Firebrand Showers, *Fire Technology*, 52 (2016) 1051-1067.

4. Johnsson, E., and Maranghides, A., Effects of Wind Speed and Angle on Fire Spread along Privacy Fences NIST TN 1894, 2016.

5. Manzello, S.L., and Suzuki, S., Generating Firebrand Showers Characteristic of Burning Structures, *Proceedings of the Combustion Institute*, 36 (2016), published online.

6. Suzuki, S., Nii, D., and Manzello, S.L., The Performance of Wood and Tile Roofing Assemblies Exposed to Continuous Firebrand Assault, *Fire and Materials* (2016), published on-line.

7. Suzuki, S., and Manzello, S.L., *Fire Safety Science* 12 (2017), in review.

ACCUMULATION AND CHARACTERIZATION OF EMBERS DEPOSITED ADJACENT TO BUILDINGS

Christine D. Standohar-Alfano*, Stephen L. Quarles, & Murray J. Morrison
Insurance Institute for Business & Home Safety, Richburg, SC

ABSTRACT

As communities continue to develop into the wildland-urban interface (WUI), there is increased interest in understanding building vulnerabilities to wildfires. Due to the importance of embers (firebrands) in causing structural ignitions, this facet of wildfire exposure receives significant attention. Recent ember exposure experiments at the Insurance Institute for Business & Home Safety (IBHS) Research Center aimed to identify wind speeds and directions which attributed to high values of ember accumulation. This was achieved with the use of water-filled pans placed along a one-story residential building where embers were quenched and collected. In addition to quantifying ember accumulation, a subset of the captured embers were characterized by mass and surface area. The final aspect of this work involved a qualitative analysis of ember accumulation on a smooth deck surface. While not included in the original experimental design, this portion was included due to the difficulty of having embers settle on a smooth surface immediately adjacent to the building. This was primarily due to the recirculation of flow on the windward side of the structure. While embers were unable to settle immediately adjacent to the building, they did tend to accumulate approximately 1.5 m (5 ft) from the building edge. These results highlight the importance of maintaining a non-combustible zone at least 1.5 m (5 ft) from a home or business. Embers that are able to settle adjacent to the building can result in direct ignition if combustible material or detailing is used in construction. Likewise, embers that settle and accumulate outboard from the building can ignite combustible features, such as vegetation. This can result in a direct flame and/or radiant heat exposure to the building.

INTRODUCTION AND MOTIVATION

On average, the United States experiences over 73,000 wildfires annually[1]. Fortunately, the majority of wildfires do not threaten the built environment; however, as development in the wildland-urban interface (WUI) continues, more homes and businesses are becoming susceptible to losses due to wildfire. Likewise, wildfire occurrence and frequency is increasing worldwide with major wildfires occurring outside of normal fire season, exhausting firefighting resources[2].

One of the main challenges in designing buildings for wildfire exposures is that they must be able to resist three exposures including flame impingement, radiant heat, and ember (firebrand) attacks. The latter has been identified as the most important cause of wildfire spread in the WUI and accounts for between 65%–90% of structural igntions[3,4]. The risk of ignition from ember attacks depends on several factors including the number and size of the embers, duration of the exposure, and the amount of combustible materials or debris in the vicinity of the building. In cases where buildings are constructed with combustible materials, embers can cause direct ignition to the building by igniting exterior structural or aesthetic components or through entry into building openings such as vents. However, it is more common for embers to result in indirect ignition of buildings. In this case, embers can accumulate near the building and ignite combustible materials such as mulch, vegetation, or accumulated debris. Once that material ignites, it can result in a direct flame or radiant heat exposure to the building. Thus, guidance suggests that home and business owners maintain a non-combustible zone immediately adjacent to the home of at least 1.5 m (5 ft).

There is a direct link between building survival in a wildfire and the building materials used to construct the home and attachments. Burning buildings can also be a source of embers, which poses a threat to downwind structures and vegetation. Thus, by preventing the ignition of buildings, a reduction of fire spread in WUI communities can be anticipated, especially in locations with a high density of closely spaced homes. Developing an understanding of the potential for near-building ember accumulation, as a function of wind speed and direction and building design, would be an important step in fully understanding the vulnerability of buildings to wind-blown embers. Results from recent experiments which evaluated the accumulation of wind-blown embers near a building and ember accumulation on attached decks will be discussed.

EXPERIMENTAL PROCEDURE

Ember exposure experiments were performed at the Insurance Institute for Business & Home Safety (IBHS) Research Center located in Richburg, South Carolina. This is a unique research facility that contains a large-scale wind tunnel that is capable of testing full-scale one- and two-story residential and small commercial buildings, allowing for realistic investigations into structural performance under high winds. The facility is able to generate wind speeds in excess of 71.5 m/s (100 mph) using a 105-fan array. Likewise, the wind tunnel is able to generate a constant wind speed or replicate observed, fluctuating wind records from hurricanes, thunderstorms, or open country conditions. For the recent ember exposure experiments, the open country wind record was used and scaled to generate wind time histories defined as medium and high for the purposes of the wildfire experiments. The medium and high wind speed traces were fluctuating records with an average wind speed of 10.3 and 17.4 m/s (23 and 39 mph), respectively. Figure 1 shows the medium and high wind speed traces used in the experiments. In addition, an idle fan speed was used which ranged between 4.5–5.4 m/s (10-12 mph). The idle wind speed is the minimum capacity of the fans and is not a fluctuating record.

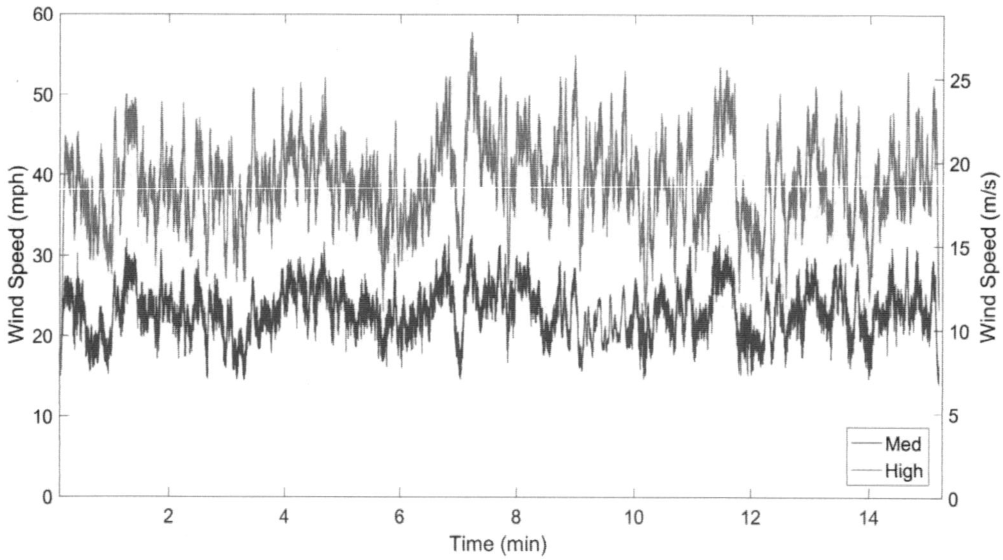

Figure 1. Medium and high wind speed records used in the ember exposure experiments.

The wind tunnel also has a large turntable with a radius of 16.8 m (55 ft) which can rotate 360° allowing for the full rotation of the test building. Since buildings in the wind tunnel are positioned on the turntable, the effect of wind speed *and* direction on the deposition of embers can be evaluated by rotating the test building.

The ember exposure experiments used a custom-made system to deliver fuel and generate embers. The fuel was a mixture of southern yellow pine wood chips and commercially available wooden dowels commonly used in furniture manufacturing, with a ratio of 80% and 20%, respectively. The raw materials were dried to a moisture content of less than 10%. The fuel was delivered into the generators by an auger feed line shown in Figure 2A. Upon entry into the body of the generator, the fuel collected on a metal grate on top of a burner (not shown). A vertically oriented fan was placed at the bottom of the generator underneath the burner indicated in Figure 2 at B. The fan introduced a vertical component of wind which pushed embers up and out of the exhaust duct shown in Figure 2C. A total of ten generators were manufactured and placed at equally spaced increments in front of the fan array so that embers could be introduced into the wind stream. An overview of the auger system, generators, and building location is shown in Figure 3.

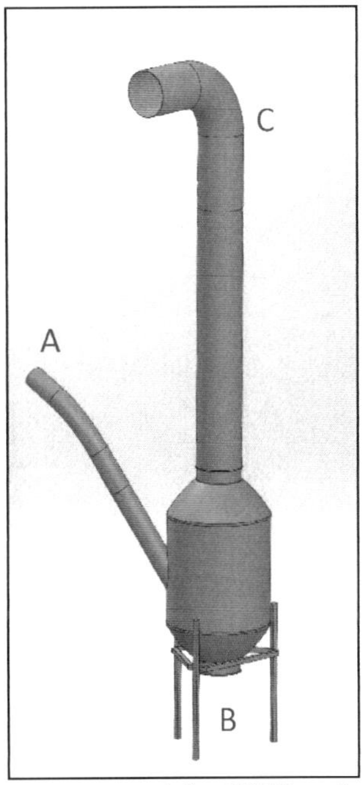

Figure 2. Schematic of the ember generators used for IBHS ember experiments. A is the auger feed line which introduced the fuel into the generator; B is the location of the vertically oriented fan; and C is the exhaust duct.

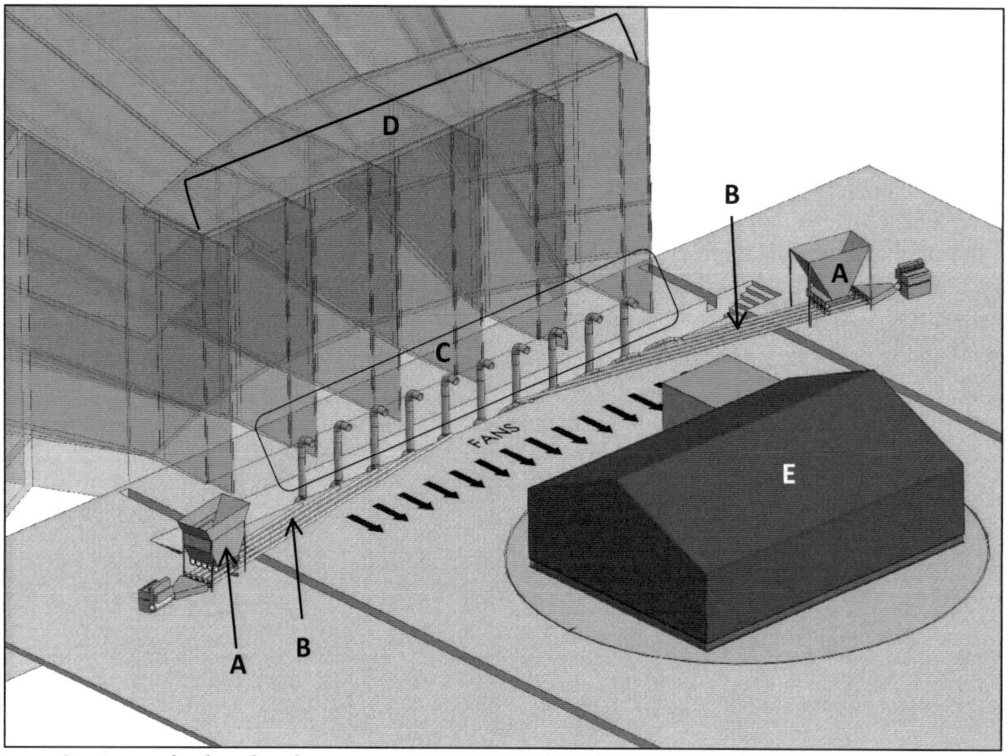

Figure 3. Auger feed and ember generation system. The left and right hoppers are shown in A; B indicates the auger feed lines which deliver fuel to the ten generators indicated in C; D is the fan array behind the generators; and E is the test building.

Prior to testing, the fuel was placed in the hoppers shown in Figure 3A. On the underside of the hoppers were five augers (not shown) which were used to feed the raw material into the auger feed lines. When active, the auger delivered fuel into the auger feed lines shown as B in Figure 3. These feed lines brought the fuel to the generators shown in Figure 2A and Figure 3C. As described earlier, the custom-made generators burned the fuel and a vertically oriented fan pushed the burning material up and out of the exhaust duct. Using the wind traces shown in Figure 1, the fan array (Figure 3D) created the wind flow in the wind tunnel. The test building shown by Figure 3E was then subjected to a wind-blown ember exposure. The duration of the exposure depended on which aspect of the experiment was being investigated (i.e., accumulation or ignition potential of decking).

In order to evaluate ember accumulation in the vicinity of a building, a full-scale residential steep-slope roof building was used in the wind tunnel as shown Figure 3E. The structure was one-story with a building footprint of 9.1 x 12.2 m (30 x 40 ft). Furthermore, since re-entrant corners have been known to be problematic in terms of vertical flame growth, a 3 x 3 m (10 x 10 ft) flat-roof cube was placed on one corner of the long axis, as shown in Figure 4. The building was covered with non-combustible material to minimize the opportunity for ignition of the test building. The orientation of the building was varied throughout the experiments with the use of the turntable. This allowed for an assessment of the impact of wind direction on ember accumulation.

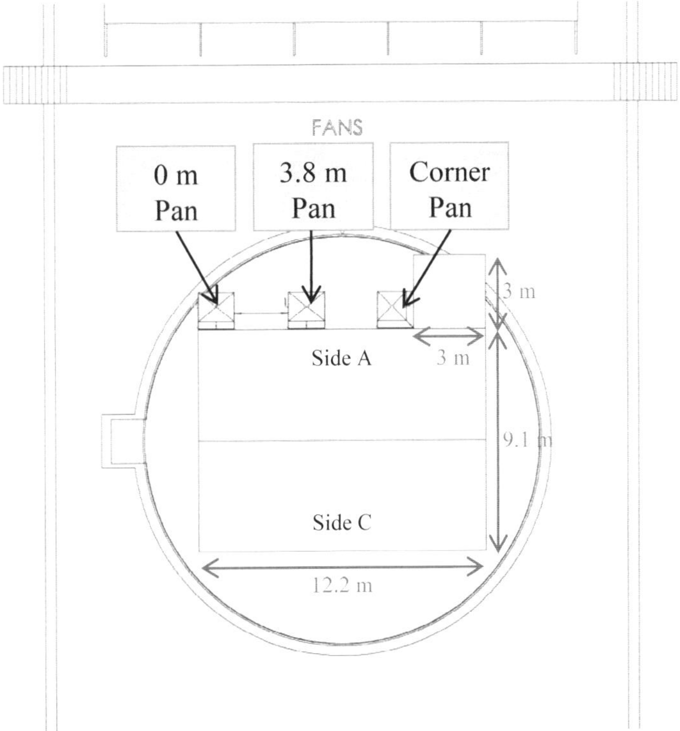

Figure 4. Plan view of the test building used in ember exposure experiments.

NEAR-BUILDING EMBER ACCUMULATION STUDY

The first aspect of the ember exposure experiments involved a quantitative analysis of ember accumulation. The goal of this work was to understand the impact of wind speed and direction on accumulation in the vicinity of the test building. This was done using water-filled pans placed at various locations adjacent to the structure. The building was then subjected to an ember exposure for 15 minutes. The building was originally oriented in the configuration shown in Figure 4. In order to assess the impact of wind direction on ember accumulation, the building was rotated both clockwise and counter-clockwise in 15° increments up to 30°. Additional orientations were investigated as part of the experiment, but will not be discussed here.

Water-filled pans, shown in Figure 4 along Side A, were placed adjacent to the building and were used to quench embers as they landed. A total of three pans were used. The first pan was placed at the end of the building to assess ember accumulation as wind flowed around the sharp edge of the building corner. A second pan was placed along the windward wall, approximately 3.8 m (12.5 ft) from the edge to assess accumulation away from the edge of the wall. The last pan was placed in the re-entrant corner since previous experiments had resulted in rapid flame growth in this area. Thus, this region was identified as a critical location to assess ember accumulation.

The water-filled pans measured 1.5 x 1.5 m (5 x 5 ft) and were divided into two sections. A 0.3 m (1 ft) portion of the pan was placed immediately adjacent to the building, defined as the *wall* section. This area captured embers that were able to land next to the building and could pose a threat to a building via direct ignition. Similarly, the *field* section was the remaining portion of the pan that spanned 0.3–1.5 m (1–5 ft) from the building. At this location, embers would not pose a threat to the building from direct ignition. Rather, any combustible material in this region could ignite from ember accumulation and result in a direct flame or radiant heat exposure.

Results from the accumulation study are shown in Figure 5. As expected, there was variability in the accumulation depending on orientation, wind speed, pan section (wall or field), and pan placement along the wall of the test building. The orientation of the test building, as shown in Figure 4, was designated as 90°. A clockwise rotation of the turntable allowed for an assessment of angles less than 90°. Similarly, for the larger angles investigated, the turntable was rotated counter-clockwise.

For the orientations shown in Figure 5, the 60° orientation had the highest accumulation for both the medium and high wind speeds. At this orientation, the two water-filled pans placed at the edge and near the midsection (0 m and 3.8 m, respectively) were closer to the ember generators and the re-entrant corner was positioned so that embers were naturally caught in the corner pan. For this reason, accumulation values were relatively high. Likewise, at 105° and 120°, ember accumulation decreased. For these orientations, the cube used to simulate the re-entrant corner did provide some shelter for locations on the leeward side of the cube where the water-filled pans were placed. Results also indicated that regions immediately adjacent to the building resulted in the largest ember accumulation. This could be problematic, especially if the building was constructed with combustible material.

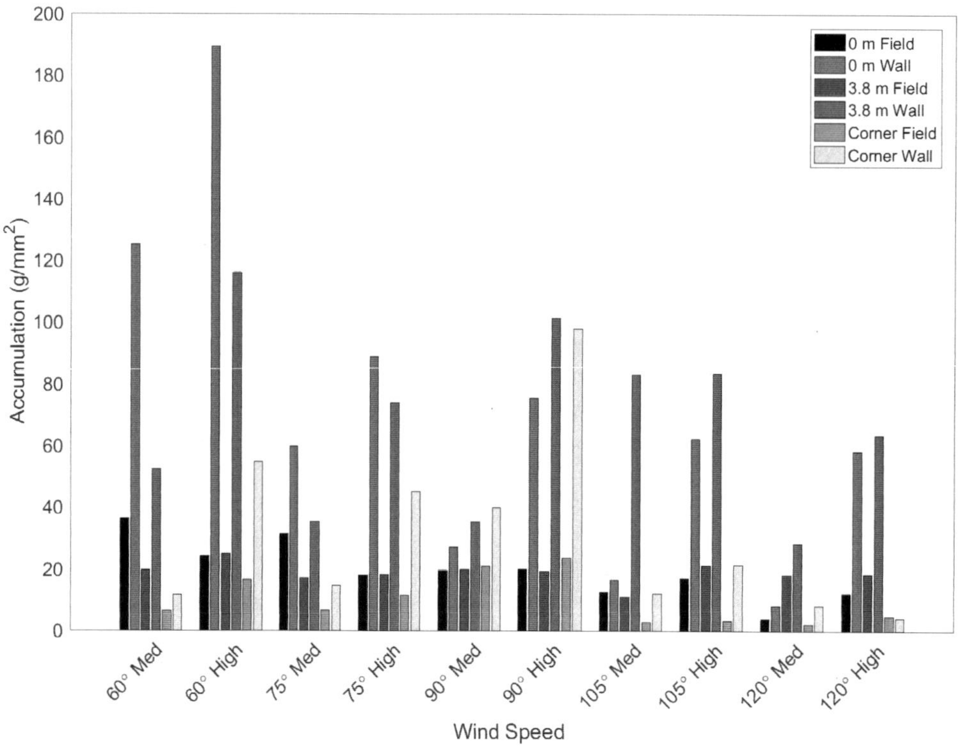

Figure 5. Ember accumulation in water-filled pans.

QUALITATIVE ASSESSMENT OF EMBER ACCUMULATION ON DECKS

The ember exposure experiments at IBHS also included an assessment of ember accumulation on deck surfaces. This aspect was motivated in part due to the destructive nature of the Waldo Canyon Fire of 2012 in

Colorado Springs, Colorado[5]. Wood decks were observed to be vulnerable to ignition from ember exposures. Similarly, ember exposure is not currently considered in Chapter 7A of the California Building Code[6] for decking material. In this code, product acceptance for most combustible deck boards is based exclusively on an underdeck flame impingement exposure[7]. Whereas a flame impingement exposure for an attached deck is possible during a wildfire, an ember exposure is almost certain when a home is threatened by a wildfire. One of the goals of this research was to determine if an underdeck flame impingement exposure provided sufficient information to describe performance from an ember exposure. In order to determine ignition potential of decks, the location of maximum ember accumulation was evaluated.

While the primary goal of the deck research was to evaluate ignition potential, in order to achieve this, an understanding of ember accumulation patterns on deck surfaces was necessary. The original experimental design called for 1.5 x 1.5 m (5 x 5 ft) decks to be placed at selected locations immediately adjacent to the test building. However, due to the complexities of near-building flow, and the surface roughness characteristics of the test decks, the original deck setup had to be modified. It was observed that the test matrix, as designed, was not appropriate since the embers were unable to accumulate on the deck surface without being blown off in the recirculation along the windward wall. Thus, a modification to the text matrix was warranted and a qualitative aspect of ember accumulation on decks was conducted.

In order to test ignition potential of decks, the location where ember accumulation would occur had to first be identified. This was done qualitatively with the use of two large decks that were constructed and attached to the A and C sides of the test building (refer to Figure 4). The two decks are shown below in Figure 6 with Side A on the left and Side C on the right. The deck shown in Figure 6a went around a portion of the cube so that accumulation in the re-entrant corner could be assessed. The large portion of the deck measured 6.1 x 4.6 m (20 x 15 ft) whereas the small portion along the cube was 3 x 1.5 m (10 x 5 ft). The deck placed on Side C of the building (Figure 6b) measured 6.1 x 3.7 m (20 x 12 ft) and was centered along the midpoint of the wall length.

Figure 6. The large deck used for qualitatively determining ember location when the house was oriented with a.) Side A towards the fans; and b.) Side C towards the fans.

The building was then rotated at various orientations and subjected to a 15-minute ember exposure at both medium and high wind speeds. The location of maximum ember accumulation was visually observed and documented. It was determined that at medium wind speed, with the house oriented with Side C towards the fan, the most significant ember accumulation occurred. The area of maximum accumulation was located approximately 1.2 m (4 ft) from the edge of the building and continued to the end of the deck as shown in Figure 7. Using these results, the deck test matrix was modified as described below.

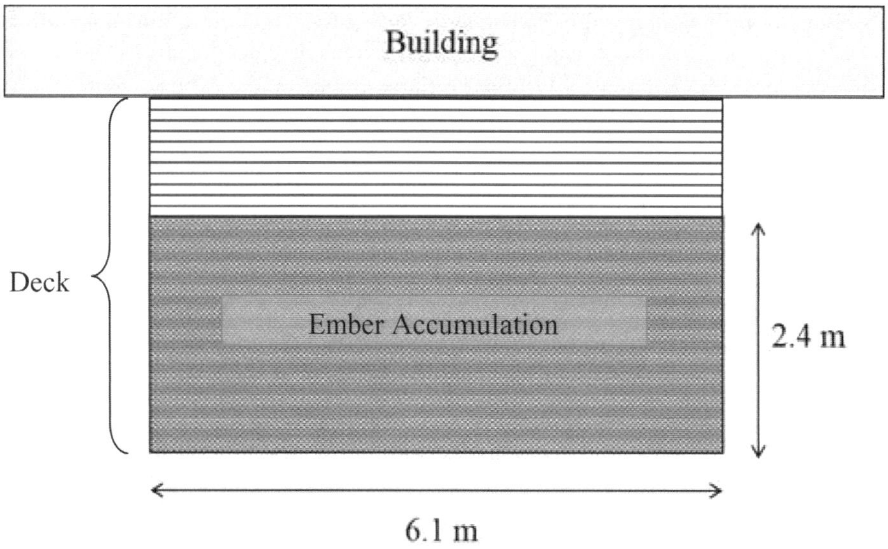

Figure 7. The location of maximum ember accumulation, indicated with shading, on the large deck attached to Side C.

Within the large deck shown in Figure 6b, a 1.5 x 1.5 m (5 x 5 ft) opening was created on one end to allow the smaller deck test specimens to be positioned. This opening was placed approximately 1.5 m (5 ft) from the test building since embers caught in the recirculating airflow were able to settle and accumulate at this point. A framed, movable wall was used on the same side as the opening to simulate a re-entrant corner. The experimental setup for the deck experiments is shown in Figure 8. Since the balance of the large deck was constructed using non-fire-retardant-treated deck boards, it was covered with a panelized gypsum product to minimize ignition when exposed to embers. Results of the ignition potential of various deck boards will be discussed in future reports.

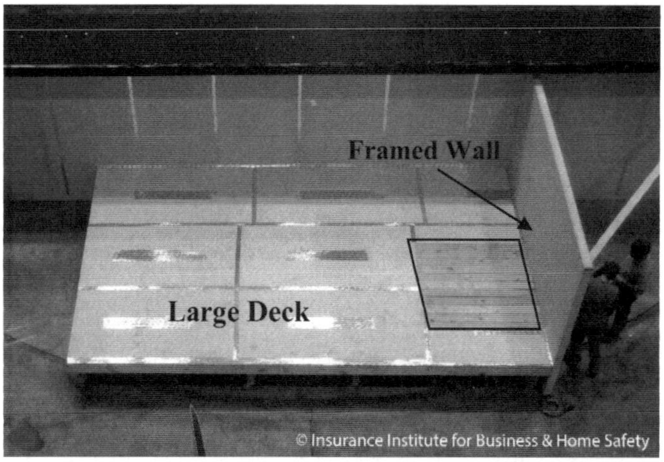

Figure 8. Test setup for ember exposure on decks. The framed wall could be moved in and out of test position to insert and remove the test deck.

EMBER CHARACTERIZATION

In addition to quantifying ember accumulation, characterization of individual embers was also performed by measuring mass and surface area. The goal of this portion of the work was to add to the knowledge base regarding the quantification of ember exposure since this is an area of needed research[8]. Due to the large number of orientations investigated, only a sample of embers from the 90° orientation were measured. The information on ember characteristics can ultimately be used in the future to estimate an ember mass flux (g/m²s) which will aid in evaluating the ignition potential of common exterior-use construction materials and contribute to the development of the proposed WUI hazard index [9]. This is an active area in wildfire research so that emergency managers and incident command can quantify expected fire severity based on ember and fire exposure.

The characterization of embers collected at 90° are shown in Figure 9. The relationship between mass and surface area was relatively linear and as the surface area increased, so did mass. There did not appear to be any strong dependency between wind speed and ember surface area or mass. Likewise, although not shown, the orientation of the building did not appear to have an important impact on ember surface area or mass. This was not surprising since fuel type plays a larger role in ember characteristics than building orientation. Since the fuel used to create the embers was not varied throughout this series of experiments, ember characterization between orientations should be relatively similar. Likewise, as data regarding ember characteristics from active wildfires becomes available, a comparison can be made to the embers created with the IBHS generators. Using that information, the generators and/or fuel can be modified to better replicate embers collected in the field.

Embers collected and characterized during this experiments were in general agreement with data published by the National Institute of Standards and Technology (NIST)[8] in that most embers were small (i.e., low mass and small surface area). It should be noted that the embers analyzed during these experiments were overall smaller than those reported by NIST.

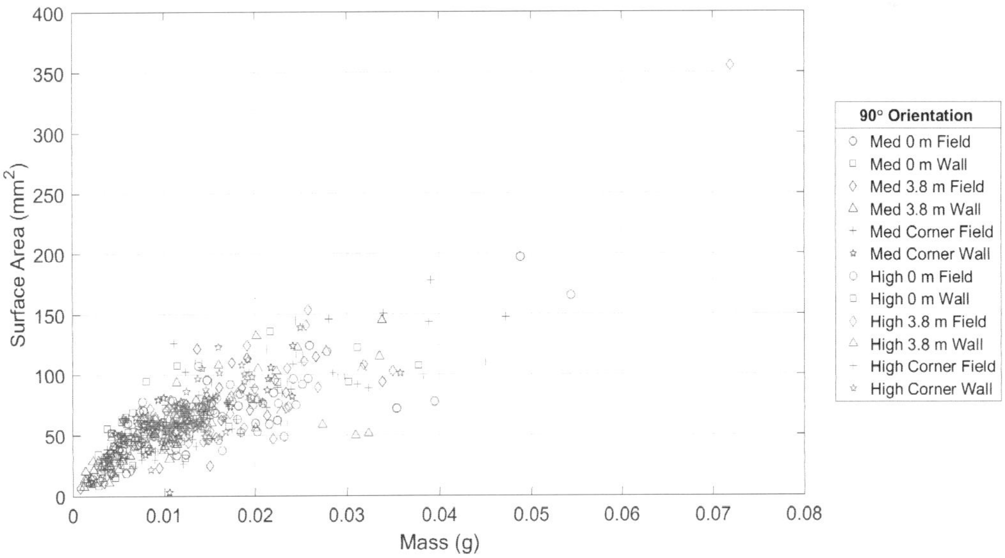

Figure 9. Mass versus surface area for embers collected from the ember generators when the test building was rotated to a 90° orientation.

SUMMARY

The recent ember exposure experiments at IBHS focused on ember accumulation and characterization. Embers have been shown to be responsible for 60%–90%[3,4] of building ignitions from wildfires, so there is significant interest in understanding the location of ember accumulation near a home or business. This included an investigation on the sensitivity of ember accumulation to wind speed and orientation. A subset of the embers collected for the study was also characterized by mass and surface area.

The accumulation of embers was estimated using water-filled pans. The embers were able to either settle in the pans on their downward trajectory from the generators or were caught in the recirculation and forced down into the pans where they were quenched. This resulted in the area of maximum accumulation immediately adjacent to the wall. Likewise, as the wind speed increased, ember accumulation also increased, especially for the wall section of the water-filled pans.

These results highlight the importance of maintaining a non-combustible region immediately adjacent to a home or business. In cases where buildings are constructed with combustible materials, embers can cause direct ignition to the building. However, it is more common for embers to result in indirect ignition of buildings. In this case, embers can accumulate near the building and ignite combustible materials such as mulch, vegetation, or debris that is able to accumulate. Once that material ignites, it can result in a direct flame or radiant heat exposure to the building. By maintaining defensible space within 1.5 m (5 ft) of the building, embers may be able to accumulate, but will likely not result in structural ignition.

The ember characterization portion of these experiments indicated a linear relationship between surface area and mass. There was not significant variation in either characteristic between the different orientations, wind speeds, or pan locations investigated. This is due to the fact that ember characteristics are driven by fuel type. For these set of experiments, the fuel was kept consistent, thus significant variations in ember characteristics were not anticipated.

In addition, this was the first time the embers generated at the IBHS Research Center were characterized. Thus, this highlights the capability of IBHS to provide a methodology to develop critical ember data for improved understanding of ember characteristics. Likewise, using the ember characterization data, a comparison of IBHS laboratory-generated embers with those collected in situ from an active wildfire can be performed. Using that information, the IBHS generators and/or fuel can be modified to better replicate observations from the field.

An additional aspect of the ember exposure experiments was on deck performance, specifically with respect to ignition potential. This was partially due to the destructive nature of the 2012 Waldo Canyon Fire in Colorado. This fire resulted in the complete loss of over 300 homes and it is hypothesized that fences and decks contributed to fire spread. While decks have been included in the California Building Code[6], there is limited information on their vulnerability to ember exposures. However, prior to investigating ignition potential of decks from embers, a better understanding of ember deposition on decks had to be determined. This was achieved qualitatively with the use of a two large decks placed along the test building.

One interesting observation from this set of experiments was the importance of surface roughness. For instance, with the deck experiments, the recirculation along the windward wall of the test building captured embers and forced them to settle outboard of the wall. Thus, with smooth surfaces, it would be difficult to have significant ember accumulation immediately adjacent to the building. However, with the water-filled pans, the highest accumulation was in the portion of the pan against the wall rather than in the field section. In these cases, the embers caught in the recirculation were forced down and landed in the water-filled pans where they were quenched and unable to blow away. This is similar to ground conditions with a rough surface, such as mulch, where embers can land and have less likelihood of settling outboard of the building. Thus, if a home or business includes rough ground conditions adjacent to the building, it is important that the materials used are non-combustible. Bark mulch or pine straw can be easily ignited from ember deposition which can result in flame contact and/or elevated amounts of radiant heat to the building.

REFERENCES

[1] United States Forest Service, (2016). Managing Land - Fire. [online] Available at http://www.fs.fed.us/managing-land/fire. [Accessed 29 June 2016].

[2] Short, John Rennie, (2015). The West is on fire – and the US taxpayer is subsidizing it. The Conversation. [online] Available at https://theconversation.com/the-west-is-on-fire-and-the-us-taxpayer-is-subsidizing-it-47900. [Accessed on 20 June 2016].

[3] Potter, Mark and Leonard, Justin, (2010). Spray System Design for Firebrand Attack – Research Findings and Discussion Paper. CSIRO – Sustainable Ecosystems, Report No. EP103159, 1-27.

[4] Quarles, Stephen L., Valachovic, Yana, Nakamura, Gary M., Nader, Glenn A., and De Lasaux, Michael J., (2010). Home Survival in Wildfire-Prone Areas: Building Materials and Design Considerations. University of California Agriculture and Natural Resources. ANR Publication 8393.

[5] Maranghides, Alexander, McNamara, Derek, Vihnanek, Robert, Restaino, Joseph, and Leland, Carrie, 2015. A Case Study of a Community Affected by the Waldo Fire – Event Timeline and Defensive Actions. NIST Technical Note 1910. National Institute of Standards and Technology.

[6] California Building Code, 2009. Chapter 7A [SFM] Materials and Construction Methods for Exterior Wildfire Exposure. [online] Available at: http://www.fire.ca.gov/fire_prevention/downloads/ICC_2009_Ch7A_2007_rev_1Jan09_Supplement.pdf. [Accessed 14 July 2016.]

[7] SFM 2009, Materials and Construction Methods for Exterior Wildfire Exposure. Decking SFM Standard 12-7A-4.

[8] Manzello, Samuel L., Suzuki, Sayaka, and Hayashi, Yoshihiko, 2011. Summary of Full-scale Experiments to Determine Vulnerabilities of Building Components to Ignition by Firebrand Showers. NIST Special Publication 1126. National Institute of Standards and Technology.

[9] Maranghides, Alexander and Mell, William, 2012. Framework for Addressing the National Wildland Urban Interface Fire Problem – Determining Fire and Ember Exposure Zones using a WUI Hazard Scale. NIST Technical Note 1748. Gaithersburg, MD: National Institute of Standards and Technology.

UNDERSTANDING IGNITION SUSCEPTIBILITY OF WILDLAND-URBAN INTERFACE FUELS TO FIREBRAND ATTACK

Raquel S P Hakes, Matthew J Weston-Dawkes, Sara E Caton, Michael J Gollner & Jiann Yang
University of Maryland, College Park, USA
National Institute of Standards and Technology, USA

ABSTRACT

Structures in the wildland-urban interface often ignite as a result of firebrands, burning embers which are launched from the fire plume and can travel over several kilometers. While some previous research has been conducted to understand the generation, lofting, and transport of firebrands, knowledge of the ignition process of materials by firebrands on a small scale is lacking. Most small scale testing of firebrand ignition has been restricted to vegetative fuels, which are more porous than solid WUI fuels, such as decks, porches, and rooftops. The overarching goal of this research is to understand ignition by firebrands of wood, plastic and composite assemblies attached to structures. In order to de-couple the firebrand heating from the ignition process, experiments were conducted to determine the heat flux imparted from a pile of firebrands to an inert board. Trends were found which related heat flux produced to the diameter of the firebrands and the sizes of firebrand piles, by comparing temporal and spatial heat flux variations, as well as the net heat produced over a full test.

INTRODUCTION

Over the past several years, the number of devastating wildland-urban interface (WUI) fires has increased drastically. The WUI is defined as the location where urban areas connect or intermix with wild vegetation, posing potential fire risks. Although the number of wildland fires has stayed approximately constant over the past several decades, the area that these fires burn has increased significantly[1] and more of these fires are impinging on WUI areas. The result of these interactions is that an average of 2,000 structures in the United States are lost annually to WUI disasters[2]. Similar statistics are reflected in the loss of structures in other countries internationally[3].

Structures are lost in WUI fires when the fire is able to ignite the home. Ignition can be a result of radiation from the main fire front (uncommon due to short exposure times), direct flame contact, if ornamental vegetation or other flammable material can form an unbroken path to the structure, or firebrand exposure[1]. Firebrands, or burning embers, are small pieces of burning material generated from vegetation or structures in the fire, then lofted ahead of the fire front. These firebrands can travel up to 9 km ahead of the fire front[4] and may ignite structures several hours after the fire has passed. Firebrands are thought to be responsible for the majority of structure ignitions in WUI fires[5].

The process by which firebrands ignite WUI fuels, such as decks or rooftops, is not well-understood. Firebrands often land in a smoldering state and tend to form piles when they deposit on solid WUI fuels; however, the processes governing the transition from smoldering firebrands to fuel ignition are currently unknown, as are the interactions resulting from the piled formation.

Previous research on ignition has primarily focused on either large scale ignition (e.g. whether a structure or structural component will ignite in response to a firebrand shower)[6,7,8,9] or small-scale tests with small numbers of firebrands or hot particles on loose vegetative fuels[10,11,12]. In a series of studies, Manzello et al. and Suzuki et al. have used a firebrand generator (the National Institute of Standards and Technology, NIST, Fire Dragon) in a large-scale wind tunnel to produce firebrand showers to expose building components, including fences, rooftops, and full structures[7-9]. This work has highlighted the tendency of firebrands to form piles, particularly in corners or crevices[8], though they have also been observed on top of decks[7].

A number of small-scale studies on vegetative fuels have also been conducted. In these studies, smoldering or flaming firebrands or hot particles are placed into a bed of fuel, and ignition or non-ignition is typically recorded. Most of these studies have used single firebrands, sometimes up to four firebrands, rather than firebrand piles[10-12]. This configuration is different than what occurs on structures, with vegetative fuels being much less dense than solid WUI fuels and the small number of larger brands not replicating the interaction and exposure from piles of firebrands.

Relatively few studies using structural fuels have been conducted. Early experiments by Dowling[13] looked at ignition of timber by firebrands produced in Australia. Ignition occurred more commonly in gaps and crevices where firebrands could accumulate[14]. Despite this previous study, there are few studies that focus on small-scale ignition of solid WUI fuels. One glaring gap is the lack of quantitative information necessary to define ignition limits. In radiative ignition of fuels, there are clearly defined heat fluxes used for ignition; however, for potentially conduction-dominated ignition by piles of firebrands there is no such scale. It is also believed that these fuels may ignite differently as vegetative fuels are more porous than solid fuels, such as deck boards. The porosity allows firebrands to firmly lodge themselves in the fuel, thus increasing the contact between fuel and ignition source. On a solid fuel, there is poor contact between the smoldering firebrands and the recipient fuel. This study will seek to address these gaps by quantitatively measuring heat fluxes from piles of firebrands to an inert surface. In the future, these will be connected with ignition studies.

EXPERIMENTAL SET-UP

An experimental set-up was developed to isolate the heating from firebrands and quantify this heating so that it can be connected to ignition properties on solid fuels. Firebrands were fabricated in this study from birch wooden dowels of initial diameters 6.35 mm, 9.52 mm, and 12.7 mm and a fixed length of 25.4 mm. Birch was chosen for this study as it was readily available in the sizes of firebrands typically observed from WUI fires, however its density is slightly higher than that of softwoods (~600 kg/m^3 vs. 350-530 kg/m^3)[14] more commonly found in WUI fires.

For ignition, manufactured brands were placed on a metal mesh and exposed to a propane Bunsen-type burner flame. Once the wood pieces achieved flaming (~ 30 s), the propane flame was turned off and the firebrands were allowed to burn freely in air until flaming combustion had ceased and all brands were glowing (~ 150-200 s depending on pile size), as shown in Figure 2. This technique was found to be the most reliable and repeatable method for firebrand production. Smoldering firebrands were then deposited on a sheet of ceramic fiber insulation board outfitted with heat flux, temperature, and mass loss sensors.

Heat flux was measured using two separate techniques. Single point heat flux measurements were made directly below the center of the firebrand pile using a calibrated 1.27 cm (0.5 in) Schmidt-Boelter water-cooled heat flux gauge. In separate tests, qualitative spatially-resolved heat flux measurements were made using thin-skin calorimeters. The thin-skin calorimeters were fabricated by spot-welding a thermocouple to the back of a thin (0.05 mm) sheet of Inconel metal. The front side of the metal was painted black and placed beneath the firebrand pile. Heat fluxes were calculated using the measured temperature difference over time conducted through the Inconel[15]. Both heat flux measurement set-ups are shown in Figure 1 below.

Figure 1. Schematics of experimental set-up for both water-cooled heat flux gauge measurements (Set-up 1) and thin-skin calorimeter heat flux measurements (Set-up 2). The propane burner and data acquisition (DAQ) system were used for both tests.

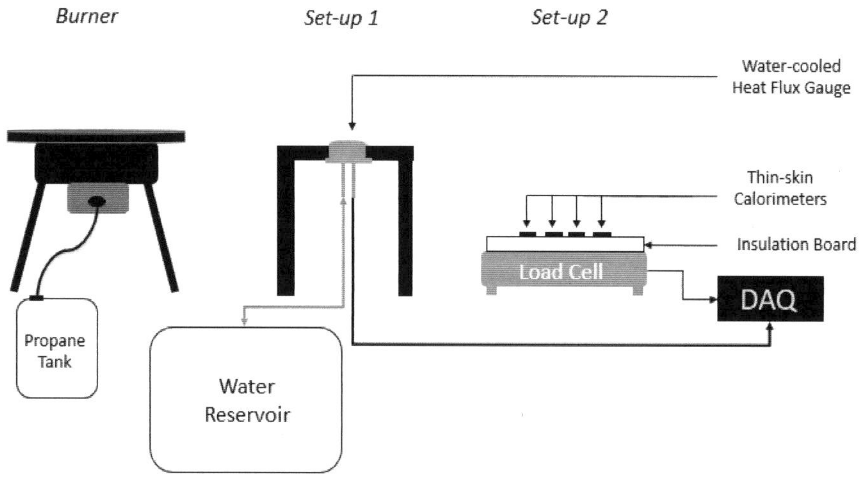

Both heat flux set-ups were also instrumented with temperature measurements. Temperatures beneath the firebrand pile were measured using a single 0.25 mm wire diameter K-type thermocouple placed on the surface of the ceramic insulation board, so that it touched firebrands when placed on top of the board. The mass loss was measured as a function of time using a Mettler Toledo load cell with a resolution of $\pm$ 0.01 g. Mass data was only recorded for the tests using thin-skin calorimeters, as the water lines for the cooled gauge interfered with mass measurements. For all tests, the ambient air temperature was also recorded. Data was recorded by a National Instruments data acquisition system at 10 Hz.

For these experiments, the diameter of the brands used and the ultimate size of the pile (number of brands) were varied. Firebrand length, wood species, and moisture content (MC) were kept constant throughout all tests. All firebrands were fabricated from 25.4 mm (1 inch) long cylindrical birch dowel pieces, dried for approximately 6 hours at 103°C (per ASTM D4442 [16]) resulting in a MC of 0 $\pm$ 0.5%. Pile sizes varied from 0.5 – 25 g before burning, which translated to ~ 0.1 – 2.1 g initially deposited on the ceramic insulation board. The number of brands was varied from 1 – 40 for the 6.35 mm firebrands, 1 – 21 for the 9.52 mm brands, and 1 – 10 for the 12.7 mm brands. These numbers corresponded to analogous mass piles.

RESULTS AND DISCUSSION

Previous literature on firebrand ignition[10-12] has focused almost solely on testing ignition from a single firebrand. While a few tests have used a small number of brands in a pile (up to four), no small-scale studies have investigated larger pile sizes[10]. In this study, the number of brands in a pile was increased up to 10 – 40 brands (equivalent masses for different diameters).

Overview of Experiments

Once firebrands ceased flaming and were deposited on top of the heat flux gauges, they began a glowing phase of smoldering combustion, which continued for a long duration, up to 600 s for the largest piles. Photographs of smoldering firebrands of varying pile sizes placed on the water-cooled heat flux gauge are shown in Figure 2 below. The photographs of the single firebrands show the water-cooled heat flux gauge. For a single brand, placement directly on top of the gauge allowed for good contact between the firebrand and the heat flux gauge. Despite the gauge being water cooled, it can be seen that sections of the brand still glow significantly on top of the cooled section of the sensor. On the other hand, piles of firebrands (seen in the lower rows of Figure 2) are subject to different configurations upon being deposited. For larger piles, these configurations repeatedly result in full coverage of the heat flux gauge; however, for smaller piles, such as the three brand piles on the second row, pile configuration may allow for exposed portions of the heat flux gauge, depending on how brands land next to the senor. All of these configurations can result in variations of contact and view factor between the brands and the gauge, which can change heat flux results.

Figure 2. Photographs of firebrand experiments 30 seconds after being placed on the sensor. From left to right, diameter of the brands placed increases and from the top down, the mass of firebrands initially deposited on the sensor increases.

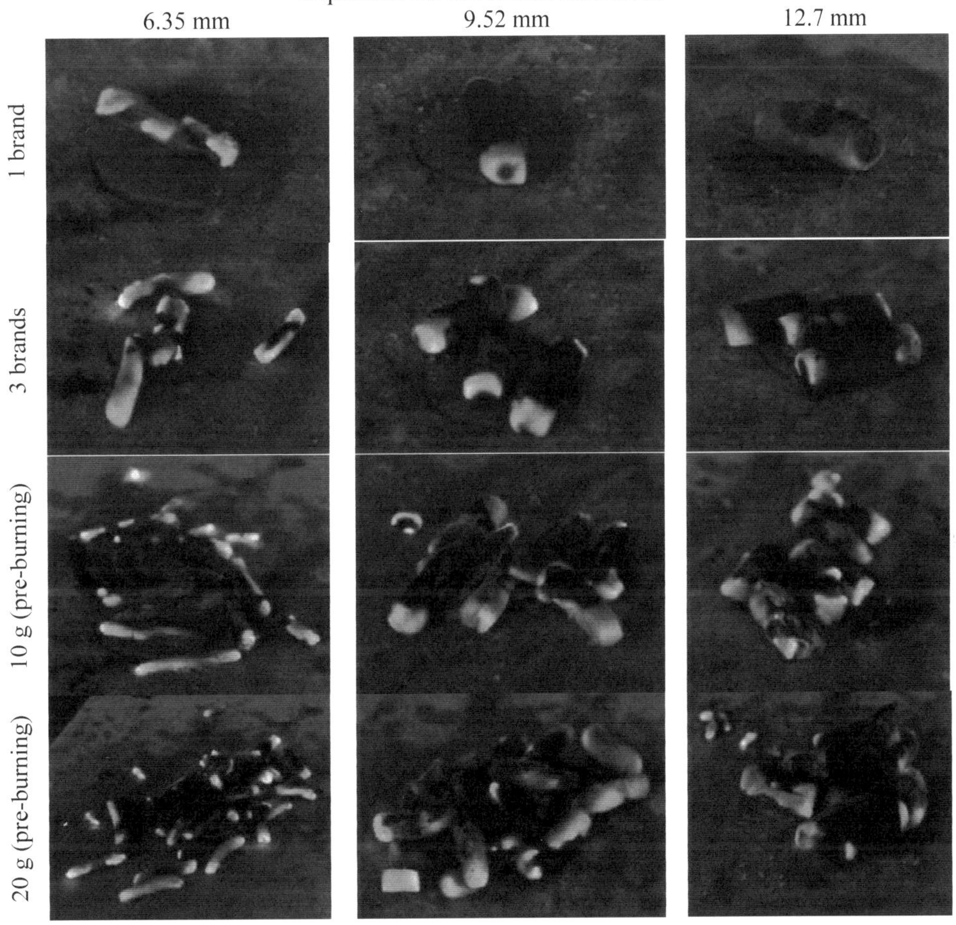

The images in Figure 2 show the firebrands in the first thirty seconds of the test, where an existing char layer is undergoing smoldering combustion over most of the surface of the brands. As the test proceeds, glowing continues over the char layer, but more of the brands progressively turn to ash. This ash layer increases contact between the brands and surface; however, it also acts as an insulator between the hot brands and the gauge surface. Later in the experimental time, piles of ash were in fact observed to produce very low heat flux levels, despite the presence of a still-glowing core.

Multiple heating mechanisms therefore act to heat both recipient fuels and the firebrand pile. The glowing in the photos above indicates the presence of radiation, as do spatial distributions of heat flux shown later in this paper. Re-radiation within piles also serves to keep the piles at higher temperatures for longer. The direct contact between firebrands and the recipient material (ceramic fiber insulation board in these tests, or solid fuel in a WUI fire) allows for potentially conduction-dominated heat transfer.

Heat Flux Variations between Experiments

For all experiments on the water-cooled heat flux gauge, the heat flux was recorded as a function of time and the tests were repeated several times until an average trend was observed. Figure 3 shows results from the smallest and largest mass tests, the 6.35 mm single brand and 12.7 mm ten brands (2.1 g), respectively. Individual tests are shown in Figure 3 as dashed lines, with the average of all tests shown as a dark solid line, and the standard deviation from the average indicated with the gray shading. Despite nearly twice as many tests run for the single 6.35 mm brand as for the 12.7 mm pile, the single firebrand tests produce a significantly larger variation in heat flux from test to test. For the single brand tests, the peak heat flux varied between tests by as much as 20 kW/m^2, while the standard deviation averaged around 7-8 kW/m^2. On the other hand, for the large pile of 12.7 mm brands, peak heat fluxes were within 6 kW/m^2 of one another, and standard deviation averaged closer to 3-4 kW/m^2. These trends indicate that larger piles of multiple brands can produce more repeatable heat flux results, while single brands experience larger amounts of variation. This result suggests that using piles for ignition experiments, a more realistic configuration, may have more information to offer on heat fluxes from firebrands, as well as ignition as a result of piles configurations.

A slight change in contact from how the firebrand is placed on or near the heat flux sensor can have a drastic effect on the measured heat flux. Because there can also be considerable variability between samples of wood used during tests, stochastic effects of ignition and burning behavior mean each test will produce slightly different burning, contact, and ultimately heat flux results. These differences become smoothed over when multiple firebrands are used in a test, not only producing more realistic results, but also more repeatable experiments when searching for trends in the data.

Figure 3. Heat flux over time for 6.35 mm (left) and 12.7 mm (right) diameter firebrands. The 6.35 mm diameter test shows 1 brand, the lowest mass tested, while the 12.7 mm diameter test shows 2.1 g, the highest mass tested. Individual, averaged, and the standard deviation of heat flux are shown. Note the time scale between the tests is different so that variability can be observed.

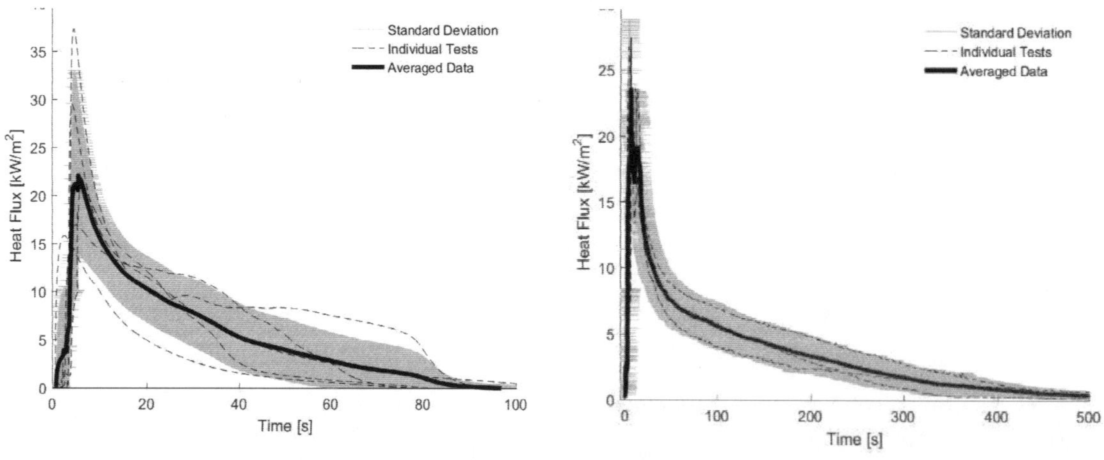

Spatial Distribution of Heat Flux

Thin-skin calorimeters were used to map the spatial variation of heat flux over the 13.5 cm^2 area covered by the sensor array, with each sensor's surface area measuring 1 mm^2. The spatial variation of heat flux is important as it may play a role in whether a fuel sample will ignite due to firebrand exposure. For instance, a high heat flux over a very small area may not induce ignition, while a moderate heat flux over a larger area might produce sufficient pyrolyzate to cause flaming ignition. The actual relationship for this combined conduction and radiation process over area is unknown at this time, but it is expected to play a role and will be important to model in the future.

Figure 4 shows heat fluxes averaged over the entire test time, and so heat fluxes represent qualitative spatial trends rather than specific heat fluxes for a given time in the test. Similar to the observation from photographs that the way in which firebrands fall has a large impact on the heating, Figure 4 shows that heating was not constant spatially throughout the pile. It is likely that the configuration of firebrands within the pile influences the areas of higher and lower heating. Additionally, it is worth noting that the firebrand piles only covered the middle portion of the thin-skin array and the lower left side; however, heat fluxes are measured even at the edge of the array where no firebrands were located. Thus the brands produced sufficient radiation to heat the outer thin-skins. The effect of radiation away from the pile may influence ignition, particularly in L-shaped or corner configurations, such as a deck near the siding of a house.

Figure 4. Spatial distribution of averaged heat flux for piles of similar mass, with diameters of 6.35 mm (30 brands, left) and 12.7 mm (8 brands, right). Higher heat fluxes are represented by yellows and lower heat fluxes are represented by blues.

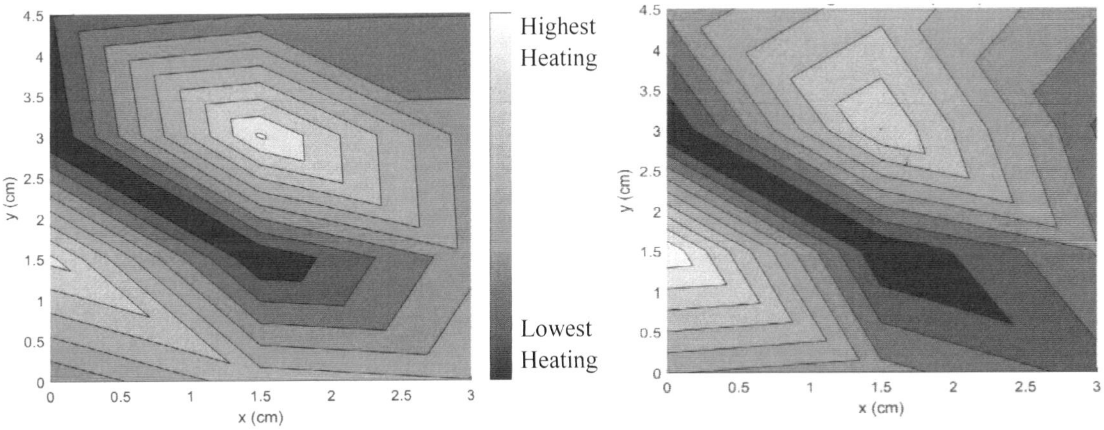

Temporal Distribution of Heat Flux

All test configurations were repeated in order to produce averaged heat flux curves, as shown in Figure 3. These averages were used to investigate trends between mass and diameter. Figure 5 shows comparisons of temporal variation in heat flux produced by different initially deposited masses for each diameter tested. As can be expected, larger piles of firebrands deposited on the heat flux gauge result in a longer decay time for the heat flux, varying from about 100 s to go to zero heat flux for a single firebrand to about 600 s for the largest piles studied (1.7 – 2.1 g). For the 6.35 mm diameter tests, there is variation in the curves between different masses; however, as the diameter increases, the variation between curves of different masses appears to collapse, so that different masses produce similar heat flux curves.

Figure 5. Averaged heat fluxes over time for three diameters, increasing diameter from top to bottom, and different pile masses: (top) 6.35 mm, 0.1 g, 0.2 g, 0.5 g, and 1.0 g; (middle) 9.52 mm, 0.4 g, 1.0 g, 1.2 g, 1.7 g; (bottom) 12.7 mm 0.8 g, 1.9 g, and 2.1g.

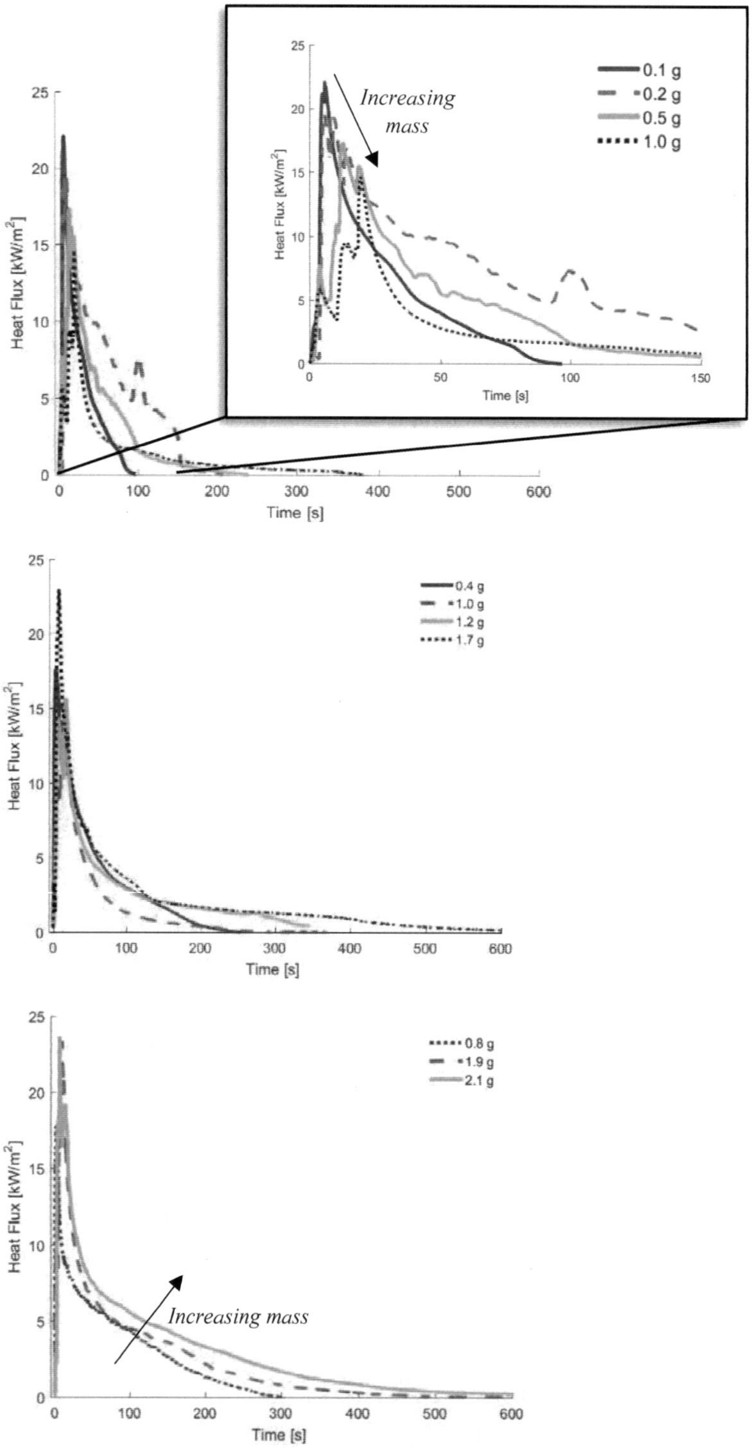

The trends for different diameters also vary. For the smallest diameter, the peak heat flux is highest for the single firebrand (0.1 g) tests and decreases for the piled brands with larger masses. This high peak heat flux for the single brand may be due to the ease of attaining good contact between the brand and gauge for a single piece of smoldering wood. Additionally, the heat flux peak occurs progressively later for an increasing mass of firebrands; however, this could be due to the slightly longer time required to deposit a pile rather than a single brand. For firebrand piles, brands are poured onto the ceramic fiber insulation board, then pushed to the center, while single brands can be directly placed atop the gauge.

For the 9.52 mm diameter brands, unlike for the 6.35 mm brands, the largest mass pile produced the highest heat flux peak, as well as the highest heat fluxes throughout the tests. The heat fluxes for the other pile sizes did not follow a clear trend. For the 12.7 mm diameter brand piles, the largest pile again produced the highest peak heat flux, similar to the 9.52 mm diameter brands; however, the smaller pile heat fluxes also produced a trend for this larger diameter. The largest mass pile produced the highest heat fluxes throughout the entire test, and the heat flux decreased with decreasing pile size. Because of the large standard deviation between tests, as was shown in Figure 3, the averaged trends in Figure 5 are mostly within the standard deviation of the other averages. More tests will be conducted to improve this variability and further investigate these trends.

Net Heat Imparted

In order to compare the full range of tests conducted, a net heat imparted, or simply net heat, was calculated for each test. The net heat was defined as the total heat imparted from the firebrand pile to the heat flux gauge throughout the entire duration of the test, e.g. the area under the heat flux curves shown in Figure 3. The net heat was calculated only for the single-point tests using the water-cooled heat flux gauge, and thus does not take into account the total area heated.

Figure 6 shows this net heat imparted as a function of initial brand diameter for different size piles. There is a wide range of variability, as shown by error bars in the figure; however, there may be a slight upward trend with increasing diameter. This trend may potentially be due to the increasing diameters corresponding to generally increased masses of firebrand piles, but is well within the error bars for tests and will require more tests to understand.

Figure 6. Net heating throughout test as a function of diameter. Net heat received increases with increasing diameter.

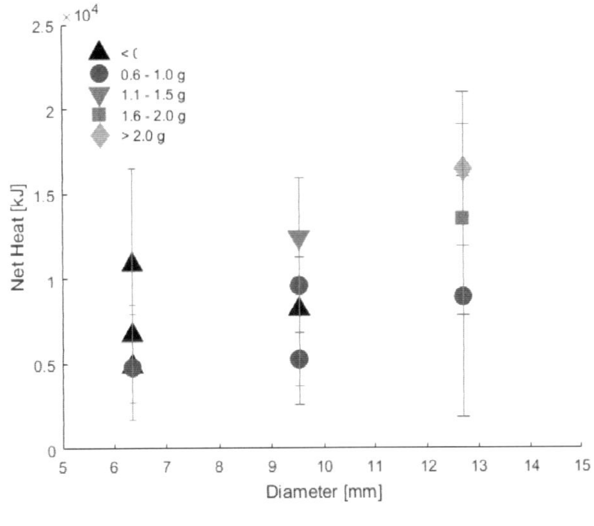

Figure 7 shows the same averaged net heating values, but plotted as a function of initial mass of the firebrand pile deposited on the sensor. Here, we see a much clearer trend of increasing net heat received by the sensor with increasing pile size. This trend is clearest in the largest diameter brands (12.7 mm) which show an increasing trend beyond the error bars. This is expected as larger piles should smolder for a longer period, imparting more heat over the test duration, potentially increasing the probability of ignition.

Figure 7. Net heating as a function of mass initially deposited on the inert board. While the two smaller diameters produce similar amounts of heat, regardless of mass, the heating produced by the 12.7 mm firebrands increases with increasing initial mass.

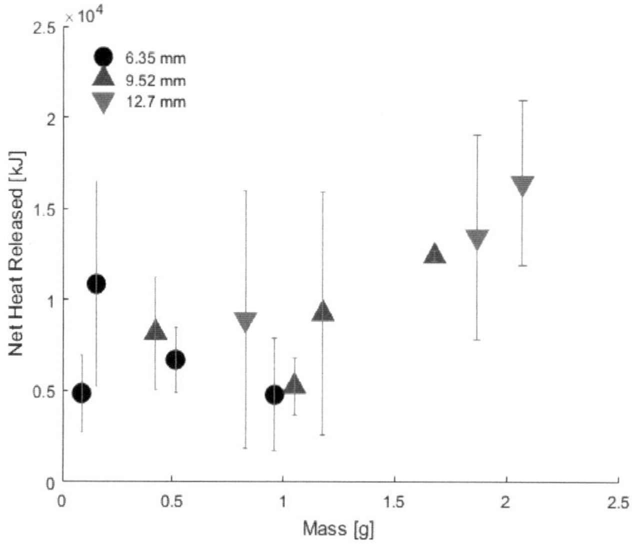

Figure 8. Net heating as a function of number of brands deposited on the inert board.

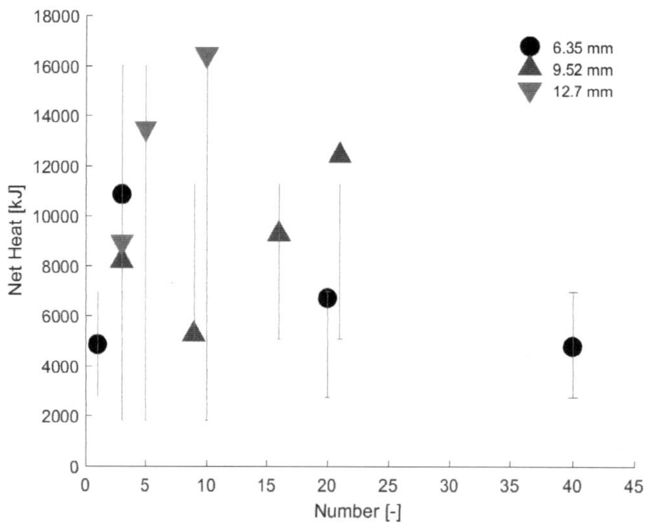

Figure 8 shows the same averaged net heating values but now plotted as a function of number of brands deposited on top of the sensor. Here, we see a very poor trend between the net heating value and number of brands. Depending on the diameter, the trend may increase or decrease with the number of brands, which indicates that there is not necessarily a universal trend for these initial tests. With clearer results seen when results are plotted as a function of mass in Figure 7, the number of firebrands does not appear to be a good metric for comparison.

Discussion

There is not currently a metric for comparing critical ignition parameters for recipient fuels as a result of firebrand exposure; however, it may be possible to use metrics from radiative ignition to formulate a method of comparing firebrand heat fluxes. For radiative ignition to occur, it is necessary for a critical heat flux to be sustained for a certain time. As part of the International Crown Fire Modeling Experiments, Cohen[17] performed ignition experiments adjacent to a large fire front (20 x 50 m^2) exposing a small wooden structure to incident radiation heat flux as a function of time. He found that a critical heat flux must be sustained for a given amount of time to produce ignition. For example, wood in his tests exposed to 31 kW/m^2 ignited after 60 s, while a higher exposure of 40 kW/m^2 would have take fewer than 30 s to ignite the same material. It will likely be necessary to define a similar criterion for firebrand ignition. The net heating parameter is used as one method of comparing heat fluxes, but does not take into account the time at a given heat flux. As we better understand the process of ignition – whether the firebrand pile transitions to flaming and ignites the recipient fuel below or whether the recipient fuel begins smoldering and both fuel and firebrand pile transition to flaming – it will be possible to define a more appropriate parameter, such as time at a critical heat flux.

CONCLUSIONS

Previous research has focused on experiments with one or a few firebrands on a porous fuel, rather than a pile of brands on a solid fuel, which is more like conditions expected during ignition of structural components in WUI fires. Experiments here have shown that it is possible to quantify heating from firebrands to a solid surface and that trends can be obtained from these measurements. It was found that the heat flux from a single firebrand is incredibly variable, suggesting a potential reason that previous experiments have resulted in wide variability for ignition thresholds. A pile of firebrands, on the other hand, started to smooth out variations between individual brands, and clearer trends began to be observed.

Peak heat fluxes measured from both individual and larger piles of firebrands do not seem to vary that widely, peaking on average on the order of 23 kW/m^2, though individual tests could see peaks up to 37 kW/m^2 for short times. The change with piles of firebrands is that the heat flux decay takes much longer for larger mass piles, resulting in much greater net heating to the target fuel.

Results using the net heat imparted from firebrands reveal that mass is an important parameter to characterize deposited firebrands. Determining such a parameter is important as firebrand "flux" has been identified in the literature as a key parameter to measure exposure from WUI fires[5] and the quantity of "flux" has yet to be defined. "Flux" could include mass, surface area, or number of firebrands. While only measuring the number of firebrands would be desirable for using video analysis, it appears that this is not a good metric. In order to connect the flux of firebrands to the ignition potential on a structure, knowledge of the mass and/or size of firebrands deposited in a specific location as a function of time appears to be necessary.

There is still much to be explored in the area of firebrand ignition. A wider range of diameters, fuels and masses needs to be tested. Ignition of actual WUI fuels needs to be connected to the values of heat flux measured here and a theory for ignition, using these values, developed. The influence of ambient wind and configuration of the pile and fuel are also important and will be explored in further studies.

REFERENCES

[1]Caton, S.E., Hakes, R.S.P., Gorham, D.J., Zhou, A., and Gollner, M.J. Review of Pathways for Building Fire Spread in the Wildland Urban Interface Part I: Exposure Conditions. *Fire Technology*, [online]. Available at: http://link.springer.com/article/10.1007/s10694-016-0589-z [Accessed 13 April 2016].

[2] Bailey, D. (2013). *WUI Fact Sheet.* 1st ed. [pdf] International Association of Wildland Fire. Available at: http://www.iawfonline.org/pdf/WUI_Fact_Sheet_08012013.pdf

[3] Grishin A.M., Filkov A.I., Loboda E.L., Reyno, V.V., Kozlov, A.V., Kuznetsov, V.T., Kasymov, D.P., Andreyuk, S.M., Ivanov, A.I., and Stolyarchuk, N.D. (2014). A field experiment on grass fire effects on wooden constructions and peat layer ignition. *International Journal of Wildland Fire* 23, pp. 445-449.

[4] Maranghides A., McNamara D., Mell W., Trook, J., and Toman, B. (2013). *A case study of a community affected by the Witch and Guejito fires: Report #2—evaluating the effects of hazard mitigation actions on structure ignitions*. [online] Gaithersburg: National Institute of Standards and Technology, pp. 1-47. Available at: http://nvlpubs.nist.gov/nistpubs/TechnicalNotes/NIST.TN.1796.pdf [Accessed 12 Jan. 2015].

[5] Mell W.E., Manzello S.L., Maranghides A., Butry, D., and Rehm, R.G. (2010) The wildland–urban interface fire problem – current approaches and research needs. *International Journal of Wildland Fire* 19, pp. 238-251.

[6] Manzello, S.L., Park, S. Suzuki, S., Shields, J.R., and Hayashi, Y. (2011). Experimental investigation of structure vulnerabilities to firebrand showers. *Fire Safety Journal* 46, pp. 568-578.

[7] Manzello, S.L. and Suzuki, S. (2014). Exposing Decking Assemblies to Continuous Wind-Driven Firebrand Showers. In: *Fire Safety Science – Proceedings of the Eleventh International Symposium*. [online] Lund: International Association for Fire Safety Science, pp. 1339-1352, Available at: http://www.iafss.org/publications/fss/11/1339# [Accessed 9 Dec. 2016].

[8] Manzello, S.L., Suzuki, S., and Hayashi, Y. (2012) Enabling the study of structure vulnerabilities to ignition from wind drive firebrand showers: A summary of experimental results. *Fire Safety Journal* 54, pp. 181-196.

[9] Manzello, S.L. and Suzuki, S. (2012). The new and improved NIST Dragon's LIAR (Lofting and Ignition Research facility). *Fire and Materials* 36(8), pp. 623-635.

[10] Manzello, S.L., Cleary, T.G., Shields, J.R., and Yang, J.C. (2006). On the ignition of fuel beds by firebrands. *Fire and Materials* 30, pp. 77-87.

[11] Viegas, D.X., Almeida, M., Raposo, J., Oliveira, R. and Viegas, C.X. (2012). Ignition of Mediterranean Fuel Beds by Several Types of Firebrands. *Fire Technology* 50(1), pp. 61-77.

[12] Hadden, R.M., Scott, S., Lautenberger, C., and Fernandez-Pello, C. (2010). Ignition of Combustible Fuel Beds by Hot Particles: An Experimental and Theoretical Study. *Fire Technology* 47(2), pp. 341-355.

[13] Dowling, V.P. (1994). Ignition of timber bridges by bushfires. *Fire Safety Journal* 22(2), pp. 145-168.

[14] Forest Products Laboratory, (2010). Wood Handbook: Wood as an Engineering Material. [online] Madison, United States Department of Agriculture, pp. 2-4 – 2-18, Available at: http://www.woodweb.com/Resources/wood_eng_handbook/wood_handbook_fpl_2010.pdf [Accessed 8 Dec. 2016].

[15] American Society for Testing and Materials, (2015). ASTM E459: Standard Test Methods for Direct Moisture Content Measurement of Wood and Wood-Based Materials. West Conshohocken: ASTM International, pp. 1-5.

[16] American Society for Testing and Materials, (2016). ASTM D4442: Standard Test Method for Measuring Heat Transfer Rate Using a Thin-Skin Calorimeter. West Conshohocken: ASTM International, pp. 1-6.

[17] Cohen, J. (2004). Relating flame radiation to home ignition using modeling and experimental crown fires. *Canadian Journal of Forest Research* 34, pp. 1616-1626.

IGNITION OF ATTIC INSULATION FOAMS SUBJECTED TO FIREBRANDS

Eric Hamann, Babak Bahrani, Jacob Kadel, Mark Stickles, and Aixi Zhou
University of North Carolina at Charlotte, Charlotte (UNC Charlotte), NC, USA

ABSTRACT

Residential attics are usually not protected against fire incidents. Mostly, attics are insulated with different types of foams. The purpose of this study was to investigate the combustibility characteristics of polyurethane attic insulation foam against various testing parameters with respect to firebrand attacks. For this study, three different types of foams were used: class A, class B, and untreated foam specimens. Experiments consisted of generating different firebrands (in size and shape) using an ember generator and placing them against the foam sample at different positions (top and side) and combustion modes (flaming and smoldering), in presence and absence of wind. In addition to single firebrand tests, ignition of foam samples subjected to the accumulation multiple firebrands was also investigated. Our results showed that flaming firebrands had a higher ignition potential in comparison to smoldering firebrands. The ignition potential of foams exposed to single embers was significantly lower in comparison to accumulated embers on the foam. Firebrands placed on the side of the foam sample had a greater effect of foam ignition than firebrands placed on top of the foam samples. The larger the firebrand was, the greater the ignition potential of the foam. The presence of wind at first aided the ignition of the foam samples at lower wind speeds (0-3.5 m/s), whereas speeds greater than 3.5 m/s reduced ignition chances. Overall, the fire retardant treated foams sufficiently prevented ignition or inhibited flame spread.

INTRODUCTION

An estimated 30 deaths, 125 injuries, and $477 million dollars in property loss was recorded from 10,000 residential building attic fires each year in the US from 2006 to 2008[1]. Attics usually do not have smoke alarms or heat detectors, and attic fires are considered as non-confined fires which makes them more dangerous for both residents and firefighters. The three major heat sources contributing to ignition in residential building attic fires are[1]: heat from powered equipment, heat from smoldering objects, and heat from chemical/natural heat sources. Among all the above ignition sources, more than 20% attributed to heat from embers or smoldering objects. Insulation materials (including thermal/acoustical and wire/cable) accounted for 36% of the first items ignited in attic fires[1]. Thus, it is desirable to investigate the ignition and fire properties of these insulation materials. Typical insulation forms and their materials in attics are listed in Table 1. Due to this arising problem, researchers at the University of North Carolina at Charlotte conducted a study which aimed to examine the ignition potential of attic insulation materials from burning embers. The material for this study was the polyurethane (PU) foam type of insulation that have been used for spray application or in rigid board.

Table 1. Typical Insulation Forms and Materials in Residential Building Attics

Form	***Material***
Loose Fill (Blown-In)	Cellulose (e.g., recycled papers), fiberglass, minerals (rock/slag)
Spray Foam (Closed-Cell)	Polyurethane (PU), Polyisocyanurate (PIR), phenolic, cementitious
Batt (or Roll, Blanket)	Fiberglass, mineral (rock, slag) wool, plastic or natural fibers
Foam Board (Rigid Foam)	Polystyrene (PS), Polyurethane (PU), Polyisocyanurate (PIR)

BACKGROUND

Firebrands are burning embers that are considered the primary source of ignition in the wildland-urban interface (WUI) [2-6]. These firebrands can directly ignite combustibles using radiant and conductive heat[2]. There are two different firebrand shapes used for this experiment: disk and cylinder. Disks represent firebrands generated from structural fuels and cylinders represent firebrands generated vegetation fuels[3-6]. The firebrand wood that was used for this experiment was Southern Yellow Pine (SYP) because it was most commonly used to construct houses in wildfire prone areas in the south east and eastern states of US, they can be both structural fuels and wildland fuels in case of a wildland or WUI fire, and they are most readily available to the research team. For wildland (or vegetation) fuels, laboratory tests performed by Manzello et al[2] that found for a 2.6 m Douglas-fir the average firebrand size was 4 mm in diameter and of 53 mm in length. This was used as a rough estimate for the cylinder shapes to be tested from vegetation. The masses that was found of those firebrands were approximately 3.5 to 3.7 grams which is around the size of the largest ember tested in this experiment. The smallest firebrand that their experiment produced was 0.3 grams. A recent study[6] reviewed all available experimental data on firebrands generated from structural fuels since the 1960s. A summary of the size (represented as the projected area a in cm^2) and frequency of firebrands from structural fuels is shown in Figure 1. This figure shows that the "frequency of finding brands larger than 6 cm^2 does not vary with the size"[6]. The firebrands for this experiment were based on the literature studies mentioned above in order to create coverage over a majority of sizes. According to Figure 1, the following four sizes were used in this study according to their projected areas represented in different regions on the fitted curve: small (1.0-2.0 cm^2), medium (2.0-6.0 cm^2), large (6.0-15.0 cm^2), and extra-large (larger than 15.0 cm^2).

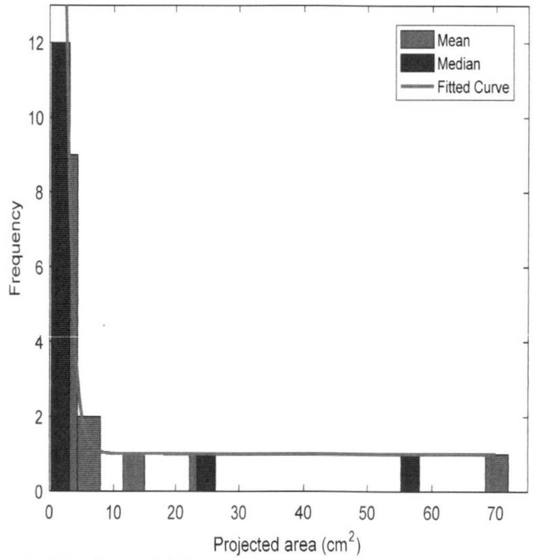

Figure 1. Firebrand Size and Frequency of Structural Fuels[6]

It is important to also realize that the surface area of a firebrand will decrease from the initial surface area and mass to something significantly smaller over time due to its ongoing combustion process. The firebrands that were used to potentially ignite the insulation foam were all within the parameters set by references 2-6 firebrand mass vs. percentage data with a majority of the firebrand's mass's being between 0.1g and 0.3 g and are realistic of a traveling ember that could potentially land on insulation foam in an attic[2-6]. Final projected area is an interesting parameter when dealing with fire embers because it is difficult to determine the surface area of a burnt ember that may not resemble an actual shape. Therefore, through examining the burning process and the mass loss characteristics of firebrands, it was possible to control the times of burning and smoldering to achieve virtually any of the sizes and masses listed within this manuscript. The majority of firebrands are relatively small and do not pose a threat to ignite the insulation foam (from an individual small firebrand), therefore the research team tested larger embers (with their initial sizes shown in Table 2) which in turn created a higher safety factor for the foam.

Table 2. Pre-Burnt Ember Sizes (Southern Yellow Pine)

Fuel Type	*Length [cm]*	*Diameter [cm]*	*Exposed Surface Area [cm²]*	
Disk *(Structural Fuels)*	1	1.5	1.76	(small)
		2.5	4.9	(medium)
		3.6	10.2	(large)
		4.6	16.6	(extra-large)
Cylinder *(Vegetation Fuels)*	1.5	1.5	1.9	(small)
	3.8	1.5	5.7	(medium)
	2.5	4.6	11.5	(large)
	3.8	4.6	17.5	(extra-large)

Ember sizes directly relate to the mass loss rate, temperature, heat flux, and heat release rate of embers. It is vital to properly decide what sizes should be tested and determined accurately. This research mostly required the use of large disks (4.6cm in diameter) in regard with the higher emitted heat from such samples. It is important proving the validity of the experiments, and in this case, however unlikely it is that an ember of that size would travel through the air and land inside an attic on insulation foam. The firebrands are heated due to the temperature and heat capacity of the brand after landing, the potential of heat due to exothermic reactions inside the firebrand, and the possibility of a flaming firebrand which can cause direct flame contact[2,7].

Heat transfer from glowing firebrands is due to convection, radiation, and conduction (when direct contact exists). When wind is introduced, the heat release rate and rate of smoldering will increase and as will the rate of heat loss due to convection. The better the contact that the firebrand shares with the bed of fuel, the better the conductive heating. However, with less contact, there is more radiation within the contact distance[7]. Also, when there is space in between, there is more oxygen supply which in turn will increase the possibility to change the firebrand from smoldering to flaming[7]. Thus it is important to investigate the effects of the modes of combustion (flaming and smoldering), firebrand positions (on top or side), and wind speeds on firebrand ignition of insulation foams. It was within the study's aspirations to determine whether foams would ignite via various testing parameters being: single vs. multiple firebrands, top vs. side orientation of firebrands, flaming vs. smoldering modes against the foam samples, and wind vs. no wind. Ignition for this study was defined as the foam sample sustaining flames for 4 or more seconds.

EXPERIMENT DESIGN

The objective of this study was to examine the ignition of polyurethane (PU) foam insulation materials subjected to heat from burning firebrands. This was achieved through studying the ignition potential and ignition characteristics of PU insulation foams under firebrand attacks in a controlled laboratory conditions. The research was accomplished by constructing a small firebrand generator and wind generator, fire ember size determination, firebrand combustion characterization, and performing foam specimen tests. Multiple tests were performed to examine the ignition capability on class A, class B, and untreated polyurethane spray foams via various parameters including number of firebrands, contact orientation, combustion modes, and wind speeds. These parameters against each of the foam sample types are shown in Tables 3, 4, and 5. Polyurethane (PU) spray foam was chosen because they are one of the most common materials used in attics for insulation, as well as the foam was supplied to the research team for desired analysis. The foams used in this study include: ASTM E84 Class-A fire retardant, Class-B fire retardant foam, and untreated foam. The contents of the fire retardant are unknown, but it was assumed that there are brominated, and phosphoric based retardants due to the charring and inhabitation of chemical reactions. The reason for these different experimental parameters was to test for random conditions regarding firebrand attacks on attic insulation foam.

Fire embers from the generator were characterized by their size (mass and geometry), and were divided into four groups according to their sizes: small, medium, large, and extra-large, per Table 2. The pre-burnt fire ember sizes were determined through previous studies producing a wide range of ember size distribution for the tests. Fuel for embers were cut from Southern Yellow Pine pieces of lumber. A vertical, fixed drill was used to cut the lumber into desired shapes[2-6]: (a) disks, representing structural fuels, and (b) cylinders, representing vegetation fuels, where a disk would have the circular part of the ember facing down and the cylinder would have the thickness or sides of the circle facing down. From that contact area with the surface, an appropriate heat flux could be measured to determine the total heat flux to the ignition surface. The sizes were classified based on surface area of the disk, and cylinder, and are shown in Table 2. These embers were applied to the PU foam insulation materials. A wind generator was also constructed to provide wind during testing. Fire ember size selection was based on the frequency of previous experimental data, as shown in Figure 1. The research team performed multiple experiments to test the ignition capability of smoldering and flaming embers on polyurethane spray foams. From these experiments, a better understanding of the foams combustion properties was achieved.

Wooden Dowel Preparation

The wooden dowel samples for this test were made from both 2.5 cm x 9 cm x 305 cm and 4 cm x 19 cm x 244 cm. Southern Yellow Pine boards. Using differing drill bit sizes of 51 mm, 25 mm, 19 mm, the wood samples were created in dimensions that either represent disk shape (structural fuels) or cylinder shape (vegetation fuels). The structural fuel and vegetation fuel size ranges can be found in Table 2.

Moisture Content Monitoring

All specimens were dried to 0% moisture content. The method for drying the ember samples are in comply with the ASTM D4442-15 standard[8]. This procedure requires that the samples are stored in vapor-tight containers. Samples were weighed using a balance with a minimum balance readability of 1 mg with a precision level of 0.01%. The oven was preheated to a temperature of 103 ± 2°C. Based on observations, it took at least approximately 6 hours for a specimen to reach a final 0% moisture content; depending on the size and initial moisture content. In the beginning stages of observing the moisture content loss, the weight of the group of samples were measured every 3 hours as recommended by the ASTM D4442-15 standard. The moisture content (MC) was considered 0% by using the following equation[8]:

$$MC\% = [(A - B)/B] \times 100 \qquad [1]$$

Where A = Initial Mass [g], and B = Oven-Dried (Final) Mass [g].

Fire Ember Generation

Once the wood samples were dried to 0% of moisture content, they were placed in the same spot on the testing tray. With an open flame at a point directly above the burners' gas release point, the gas knob is then fully opened to create a steady heat flux. The ignition source is flaming from natural gas flame with steady maximum flame heat flux of 90-95 kW/m^2 and maximum flame temperature of 800-810 °C. Using a stopwatch, multiple times were taken regarding each tested dowel sample. Recording was started once the wooden dowel was exposed to the flame from the burner. The recorded time was the initiation of flaming (time-to-ignition). The second recorded time was when the wooden sample ceased flaming and began to smolder (flameout time). This was the point at which the samples were considered as smoldering embers, and were placed quickly on the scale to monitor the mass loss in 1-second time intervals. The last recorded time was the when the ember was no longer smoldering, recorded as the *burnout time*. Note that an ember is established to be at the burnout stage once the mass loss rate becomes steady state with respect to its weight[7].

Foam Specimen Testing Method

The ASTM E84 Class-A fire retardant, Class-B fire retardant foam, and untreated foam were tested to observe their performance against various burning/ smoldering fire embers. Both burning and smoldering fire embers were placed on different positions: on top or beside the foam. Note that 48 tests were conducted for each experiment. Tests were performed under a laboratory hood with a ventilation fan producing approximately 0.81m/s wind.

PROCEDURE AND METHODS

Fire Ember Generator

The purpose of development of the fire ember generator was to produce fire embers that could be used to perform fire tests on various foam samples. The generator was constructed using 304 16 gage stainless steel. This steel was chosen with regards to the corrosion and thermal resistance since high temperatures were to be achieved inside the generator. Figure 2 and Figure 3 show the actual photo and a schematic of the fire ember generator.

Figure 2. Fire Ember Generator in Hood Apparatus

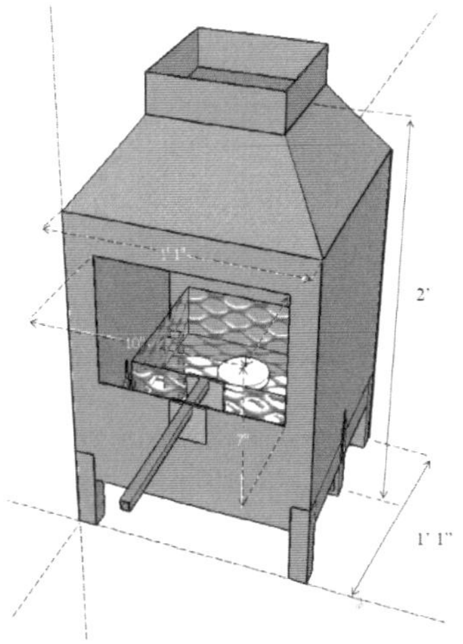

Figure 3. Fire Ember Generator Dimensions

As seen from the image above, the ember generator stands 61cm (24 in) high, 33cm (13 in) wide, and 13cm (5 in) long. The window is 17.8cm (7 in) high, and 25.5cm (10 in) wide. The dimensions of the generator were chosen based on the hood apparatus dimensions at the testing facility as shown in Figure 4a. The generator could not be any taller than 117cm (46 in) since an ample space was needed for combustion gases to escape the top of the generator.

The generator was also equipped with a circular bayou classic BG12 cast-iron burner. The burner was located directly in the middle of the generator and was the only heat source used to heat the wood specimens as shown in Figure 4b. It measures 14cm (5.5 in) in diameter having an air shutter and spring to control air flow to the burner. The burner produced an average heat flux of 92.6 kW/m^2 (90-95 kW/m^2) to the tray, where the tray was located 7.5 cm (3 in) above the burner inside of the ember generator. The average heat flux was measured using a MEDTHERM 64 Series Gardon Type Sensor, of various points around the hot red ring on the ember generator tray by the burner. The burner was supplied by natural gas at 13.8 kPa (2 PSI). The maximum flame temperature was in the 800-810 °C range as measured by a thermocouple at various locations.

The generator was also equipped with a T310 stainless steel woven wire screen tray. T310 stainless steel was chosen with regard to the good resistance against the temperature range of 800°C-900°C. The screen was then fabricated to 21.5cm (8.5 in) wide, 21cm (8.3 in) long, and 7.6cm (3 in) tall to accommodate the wood specimens when burned.

Wind Generator

The wind generator was created to simulate wildfire conditions including high or low wind speeds. The simulator was built in order to provide more oxygen to the fuel therefore increasing the heat release rate of the ember. Based of experimental videos, there is a clear difference with and without wind. When wind is present the ember glows a brighter red and ignition of the fuel is more likely to occur and decreases the time to ignition as well. The generator was built using a 15cm (6 in) inline with a 15 cm (6 in) to 10 cm (4 in) nozzle and a 10 cm (4 in) to 7.6 cm (3 in) nozzle attached to increase and direct air flow. The maximum speed for the wind generator was approximately 13 m/s at 30 cm (1 ft.) from the edge of the duct reducer. Presence of wind at first aids the ignition of the foam sample, and as the foam is further affected by the wind, decreases its ignition. It was noted that this transition wind speed was approximately 3.5 m/s.

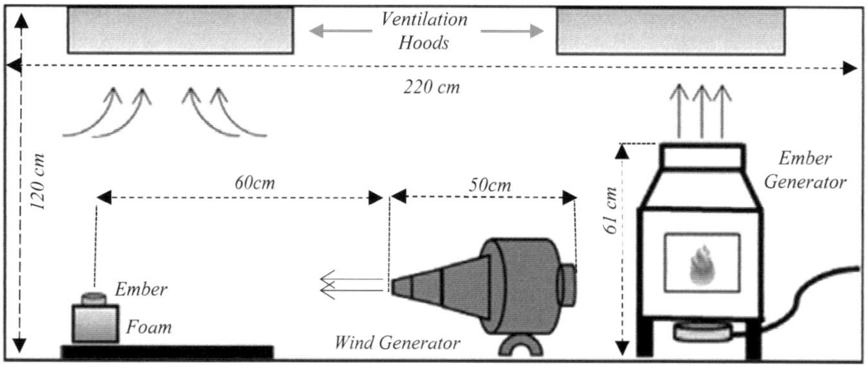

Figure 4a. Testing Zone Schematic

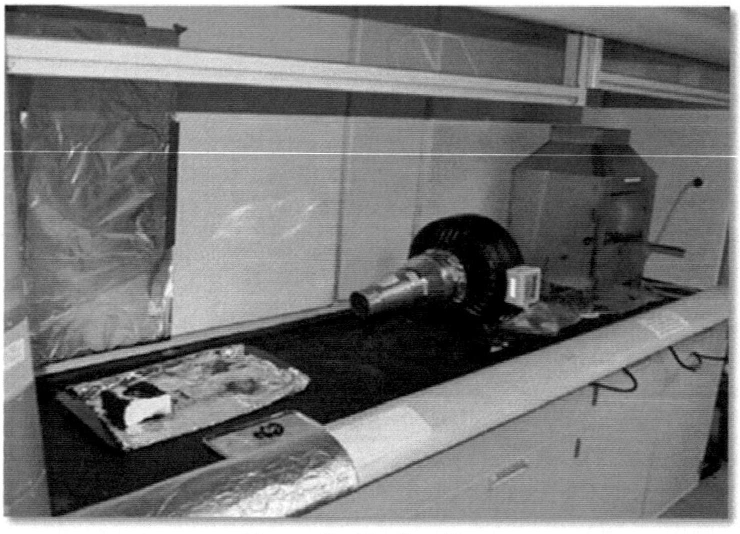

Figure 4b. Testing Zone

RESULTS

Observations from the tests are summarized in Table 3, Table 4, and Table 5. It is important to understand that ignition is defined by sustained flaming of the foam specimen for 4 or more seconds.

Table 3. Various Experimental Parameters vs. Class A Insulation Foam Ignition

Firebrand	*Orientation*	*Flame/Smoldering*	*Wind (2 m/s)*	*Ignition*
Single Large Disk	Top	Flaming	*	0 out of 3
		Smoldering	*	
	Side	Flaming	*	1 out of 3
				2 out of 3
		Smoldering	*	0 out of 3
Multiple Large Cylinder	Top	Flaming	*	
				2 out of 3
		Smoldering	*	0 out of 3
				2 out of 3
	Side	Flaming	*	0 out of 3
				1 out of 3
		Smoldering	*	2 out of 3
				0 out of 3

Note: *Asteroid (*) indicates either flaming or smoldering was observed in presence of 2m/s wind.*

Table 4. Various Experimental Parameters vs. Class B Insulation Foam Ignition

Firebrand	*Orientation*	*Flame/Smoldering*	*Wind (2 m/s)*	*Ignition*
Single Large Disk	Top	Flaming	*	1 out of 3
		Smoldering	*	0 out of 3
	Side	Flaming	*	1 out of 3
		Smoldering	*	0 out of 3
Multiple Large Cylinders	Top	Flaming	*	1 out of 3
		Smoldering	*	1 out of 3
				2 out of 3
	Side	Flaming	*	1 out of 3
		Smoldering	*	0 out of 2
				0 out of 3

Note: *Asteroid (*) indicates either flaming or smoldering was observed in presence of 2m/s wind.*

Table 5. Various Experimental Parameters vs. Untreated Insulation Foam Ignition

Firebrand	*Orientation*	*Flame/Smoldering*	*Wind (2 m/s)*	*Ignition*
Single Large Disk	Top	Flaming	*	3 out of 3
		Smoldering	*	0 out of 3
	Side	Flaming	*	3 out of 3
		Smoldering	*	1 out of 3
				0 out of 2
Multiple Large Cylinders	Top	Flaming	*	3 out of 3
		Smoldering	*	
	Side	Flaming	*	
		Smoldering	*	1 out of 1
				0 out of 2

***Note:** Asteroid (*) indicates either flaming or smoldering was observed in presence of 2m/s wind.*

Tables 3, 4, and 5 show the results of all the test conducted with various experimental parameters against the foam samples. (It is important to note from tables 3 and 4 that some of the experimental parameter test were not able to be run 3 times. This was due to the fact in that there were no more untreated samples nor class B samples for testing at the time of submission of this manuscript.):

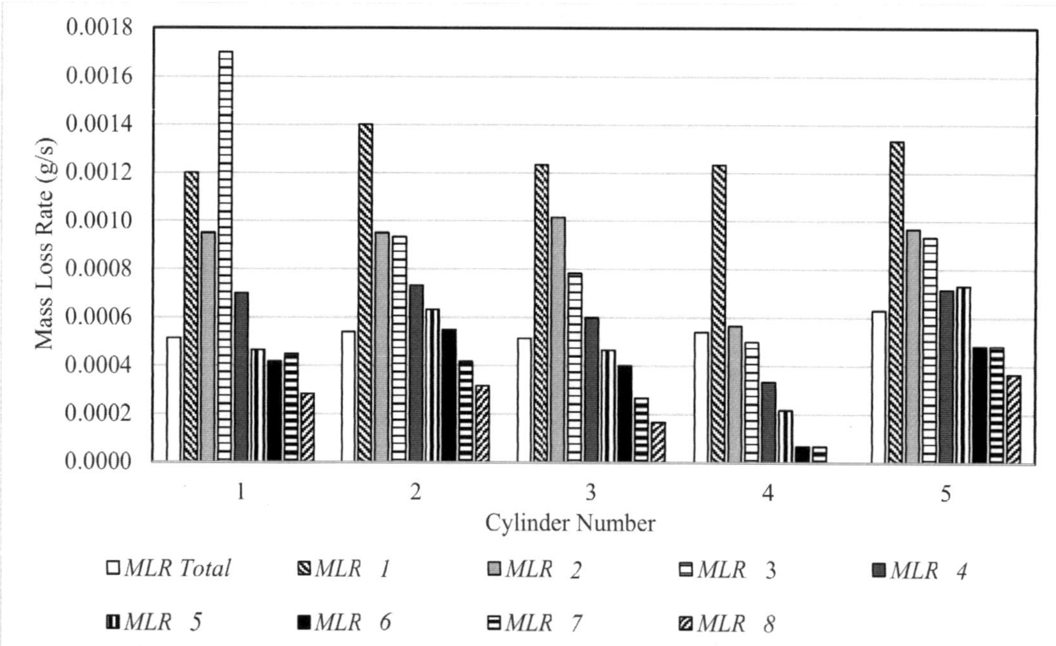

Figure 5. Mass Loss Rate

Figure 5 shows the mass loss rate for 5 separate medium cylinder samples. The highest mass loss rate was observed in all the tests within 60s after ignition was initiated. As the burning continued, the corresponding heat flux emitting from each of the embers was also decreased. As can be seen for sample 4, this sample reached burnout in the least amount of time which is represented by the missing MLR 8 bar for sample 4.

Figure 6 shows that the average mass of the medium cylinder wooden dowels started from 0.6 grams and dropped 0.3 grams right before the burnout was observed. The burnout time that occurred for these 5 samples varied from an approximate 400 seconds to 735 seconds. Compared to Figure 7, the large cylinder wooden dowels originally had an average initial mass of 4.5 grams. After an averaged approximate 450 seconds, the mass then decreased to of 3.9 g.

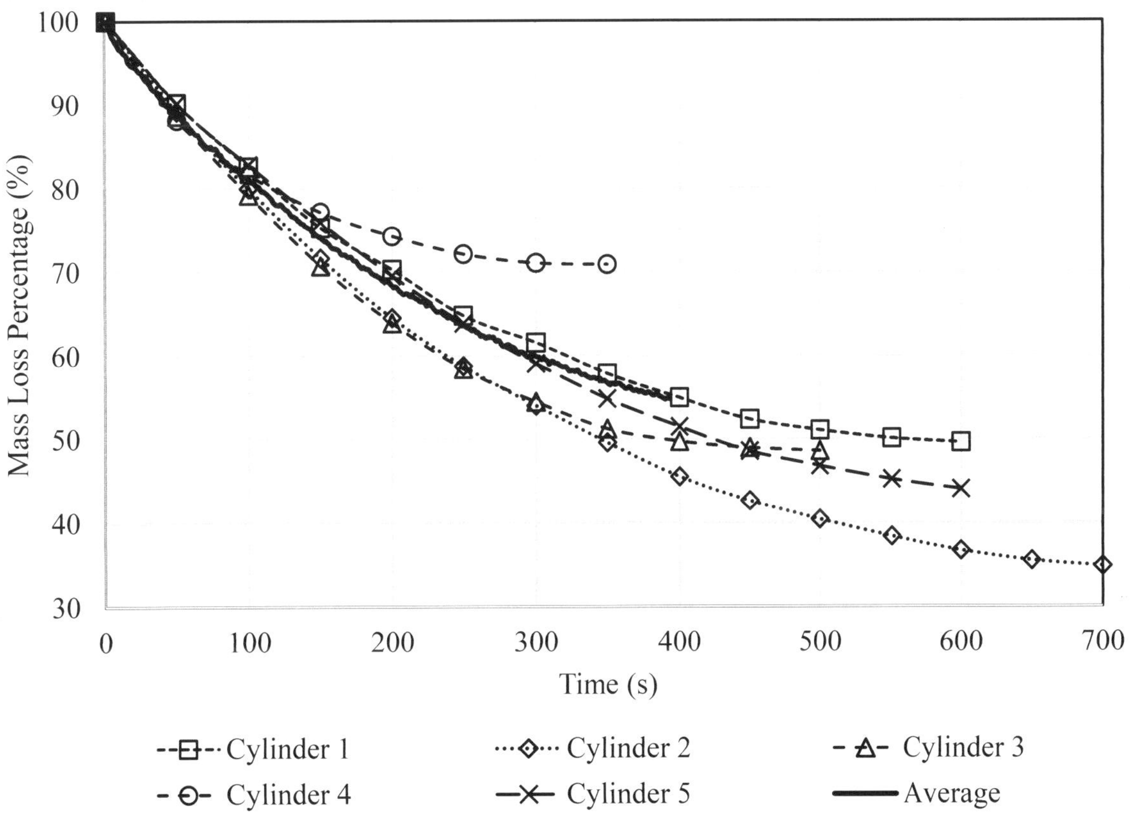

Figure 6. Medium Cylinder Remaining Mass vs. Time

Figure 8 shows the mass loss for the large smoldering cylinder embers. The average initial mass of these embers were 4.5-5 grams, and the average final mass after burnout was 3.5 to 4 grams. The mass loss was taken until the embers reached a state of burnout which took a range of time from 350 s to 550 s.

It is evident that the large cylinders had a longer ignition time, flame burnout time, and smoldering time compared to the medium cylinders. This could be simply the result of size, where smaller objects tend to ignite faster compared to larger objects because there is less available surface area and fuel.

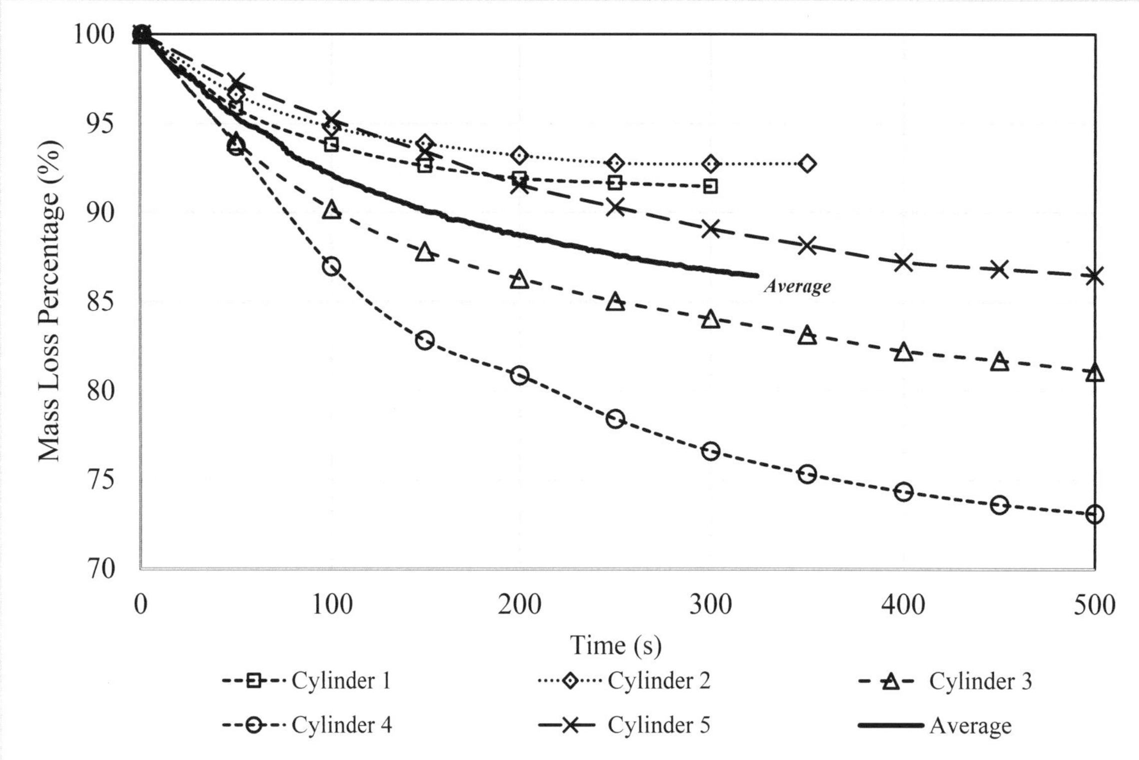

Figure 7. Large Cylinder Mass vs Time.

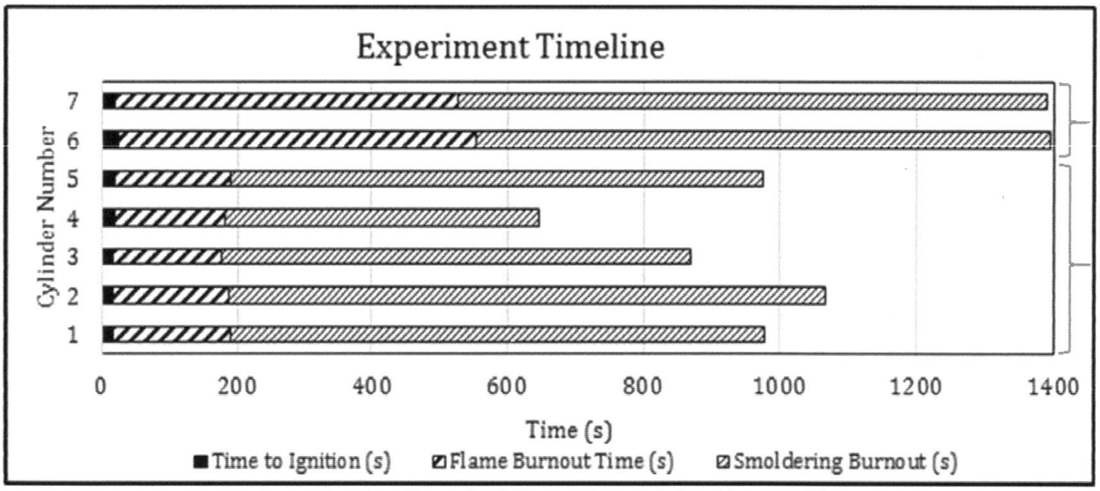

Figure 8. Timeline of Different Stages of Ember Ignition

**Note that cylinders 6 and 7 are the large sized embers to show comparison*

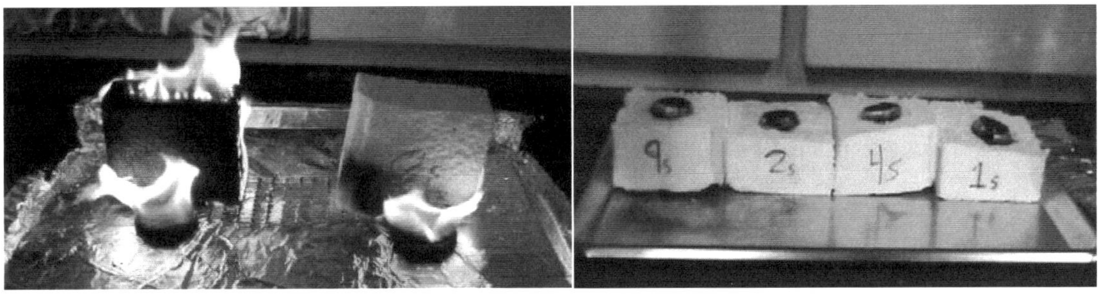

(a) (b)

Figure 9a. Single Flaming Ember Foam Tests Untreated vs. Treated Foam
Figure 9b. Single Ember Smoldering on All Treated Foams after 60 s, No Wind

Figure 9a displays two large disk flaming embers placed at a vertical position to the foam samples to better show the flame spread. The foam samples used in this experiment were untreated (left), and Class A fire retardant foam (right). The flaming embers were placed at the foot of each foam at approximately the same time. It is evident that the treated foam on the right inhibited the flame spread compared to the untreated foam on the left. The flame on the left was able to travel all the way up and to the sides of the foam, which would not be ideal for any attic fire incident. The treated foam on the right deterred the flame spread so that it only traveled barely half way up the foam. Therefore, the treated foam worked as intended in inhibiting flame spread when exposed to direct flame impingement from the flaming embers.

From Figure 9b, it's evident that no ignition occurred on the foam 60s after starting the test. Ignition would never occur in this particular test which ran for about 5 minute's time. The foam used in this experiment was all treated foam; 1s, 2s, and 4s were all Class-A ASTM E84 foam. The 9s foam was a Class-B fire retardant foam. The fire ember used in the experiment was the large disk.

(a) (b)

Figure 10a. Smoldering Ember on Untreated Foam after 180s
Figure 10b. Smoldering Ember on Untreated Foam after 60s

Figure 10a and Figure 10b shows that no ignition occurred throughout the test. The experiment was conducted with untreated foam and one single large disk. Throughout the test, it is noticeable that the disk started to sink into the foam displaying a reaction between the ember and the foam, but no ignition occurred. On further investigation, once the foam specimen was cut open, it was shown that there were craters the size of the embers underneath that area in which the smoldering ember was placed. Once the smoldering ember had reached the burnout stage, crackling sounds could be heard for minutes after from the foam specimens as the foam specimen materials were still reacting to the effects of heat.

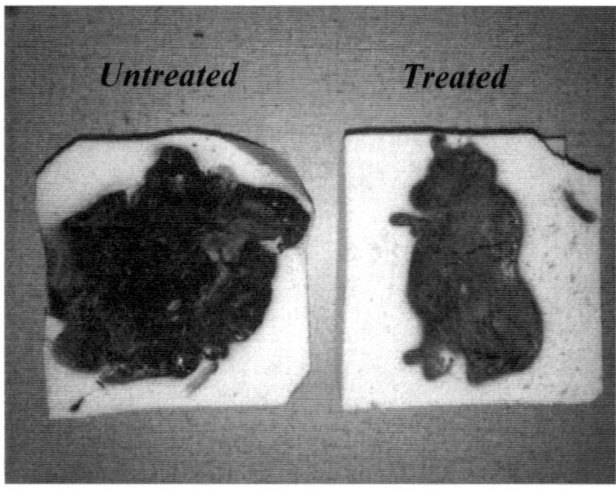

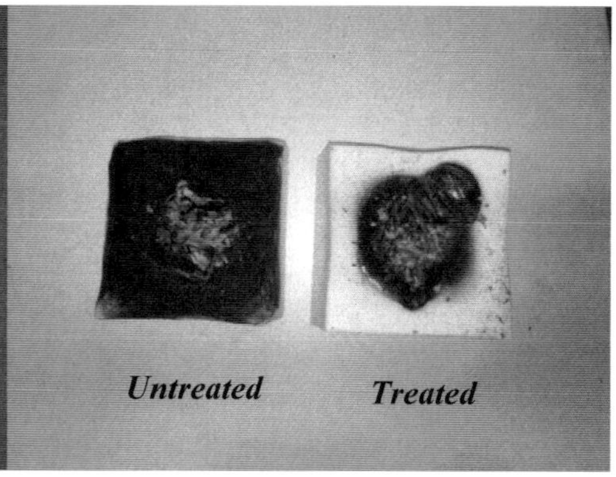

(11) (12)

Figure 11. Untreated Foam (Left) vs. Treated Foam (Right) - Multiple Smoldering Embers
Figure 12. Woodchips Accumulation- Untreated Foam (Left) vs. Treated (Right)

Figure 12 displays a test where two smoldering ember cylinders were positioned side by side and placed simultaneously on top of the treated and untreated foams, where ignition occurred in both tests. As seen in Figure 12, treated foam inhibited flame spread better than the untreated foam. In both, foam ignition occurred, but the flame spread was a lot more severe in the untreated versus the treated foam. The smoldering embers used in the tests were large cylinders. Figure 11 emphasize that more severe damage to the foam occurred on the untreated foam from the smoldering wood chips. Ignition did not occur in either of the tests but the treated foam performed better when exposed to the smoldering wood chips.

(a) (b) (c)

Figure 13a. Class A Treated Foam Large Disk 4 Meter per Second Wind 80 Seconds into Test
Figure 13b. Class B Foam Large Disk 4 Meter per Second Wind 80 Seconds into Test
Figure 13c. Untreated Foam Large Disk 4 Meter per Second Wind 80 Seconds into Test

Figure 13a, 13b, and 13c show large disk placed on a Class-A, Class B fire retardant foam, and untreated foam in presence of 4 m/s wind. No ignition occurred for the full duration of each test which ended once the ember burned out. There was a slight increase in char distance (Figure 13c) which was expected as it was the untreated foam. Note that the presence of wind was causing the ember to glow more brightly and rise in temperature due to the increased supply of oxygen to the reaction zone. This increase in oxygen develops a more intense heat flux and surface temperature exposure from the ember. Through multiple test at various speeds, it was seen that the wind first increases ember temperature and heat flux resulting in more foam ignition, but at the 3.5 m/s wind, then begins to decrease ignition chances.

CONCLUSIONS

The ignition of PU insulation foam subjected to firebrand attacks was investigated through experiments under controlled laboratory conditions. Three different types of foams (class A, class B, and untreated) were used in the study. Experiments consisted of generating different firebrands in size and shape and placing them against the foam sample at different positions and different combustion modes, in presence and absence of wind. In addition to single firebrand tests, ignition of foam samples subjected to the accumulation of multiple firebrands was also investigated. Some key results include the followings:

- Flaming embers in ambient conditions had a greater effect on foam ignition in comparison to smoldering firebrands in a similar condition
- Ignition potential of foams exposed to single embers was significantly lower in comparison to accumulated embers on the foam.
- Firebrands placed on the side of the foam sample had a greater effect of foam ignition than firebrands placed on top of the foam samples
- The larger the firebrand was, the greater the ignition potential of the foam.
- Wind speeds from 0 m/s to approximately 3.5 m/s induced a greater chance for foam ignition whereas speeds greater than 3.5 m/s then reduced ignition chances.
- FR treated foam samples showed better resistance to firebrand ignition than untreated foams.

Overall, the fire retardant foam achieved its goal of preventing/delaying ignition or inhibiting flame spread. Further testing will include using accelerated aging methods to investigate the ignition of aged foam specimens.

ACKNOWLEDGEMENTS

The authors would like to thank the Joint Fire Science Program (JFSP Project #15-1-04-4, for firebrand research) and ICL Industrial Products (ICL-IP, for PU foam research) for sponsoring this project. Any opinions, findings, and conclusions or recommendations expressed in this material are those of the authors and do not necessarily reflect the views of JFSP, ICL-IP, or UNC Charlotte.

REFERENCES

1. USFA (2011). Attic Fires in Residential Buildings, Topical Fire Report Series, 11(6), 2011, US Fire Administration (USFA) National Fire Data Center, Emmitsburg, MD, p 12. Available at: https://www.usfa.fema.gov/downloads/pdf/statistics/v11i6.pdf [Last Access: Nov. 3, 2016].
2. CATON, SARA ET. AL. "Review of Pathways for Building Fire Spread in the Wildland Urban Interface Part 1: Exposure Conditions". Fire Technology (2016): 15.
3. MANZELLO, S. L., MARANGHIDES, A., SHIELDS, J. R., MELL, W. E., HAYASHI, Y., & NII, D. (2009). Mass and size distribution of firebrands generated from burning Korean pine (<i>Pinus koraiensis</i>) trees. Fire and Materials. 33, 21-31.
4. SUZUKI, S., BROWN, A., MANZELLO, S. L., SUZUKI, J. & HAYASHI, Y. 2014. Firebrands generated from a full-scale structure burning under well-controlled laboratory conditions. Fire Safety Journal, 63, 43-51.
5. SUZUKI, S., & MANZELLO, S. L. (2016). Firebrand production from building components fitted with siding treatments. Fire Safety Journal. 80, 64-70.
6. HEDAYATI, F, & ZHOU, A. 2017 'Review and Analysis of Experimental Data on Firebrand Production from Structural Fuels': *Proceedings of the 15th (2017) Fire and Materials International Conference*, San Francisco, CA.
7. BAUM, H., & ATREYA, A. (2014). A Model for Combustion of Firebrands of Various Shapes. Fire Safety Science. 11, 1353-1367.
8. ASTM D4442-15 Standard Test Methods for Direct Moisture Content Measurements of Wood and Wood-Based Materials, ASTM International, West Conshohocken, PA, 2015, https://doi.org/10.1520/D4442-15.

IGNITION OF WILDLAND FUELS BY HOT METAL PARTICLES AND DROPLETS

Erik Christiansen & James Karnesky
Exponent Failure Analysis Associates

ABSTRACT

A series of experiments was carried out in order to investigate the incendivity of small masses of metal applied to wildland fuels. Specifically, this study focused on a comparison of molten versus solid materials. Tests were conducted using typical wildland fuels found in Southern California which were conditioned to summertime temperature and humidity conditions. The materials used to attempt ignition were particles or droplets of aluminum, copper and glass which were heated to a range of temperatures and dropped into the fuel bed. The experimental results demonstrate that when molten metal is dropped onto a bed of heterogeneous wildland fuels, the liquid metal conforms to surface irregularities, promoting efficient conduction of heat from the metal to the fuel. Conversely, a hot particle in solid form is more likely to land in a position where it does not make suitable contact with the fuel and consequently may lose a larger percentage of its heat to air in void spaces, reducing the energy available to heat the fuel. This study finds that molten droplets have a greater likelihood to ignite typical wildland fuels as compared to solid particles with similar total energy and material properties such as thermal conductivity and heat capacity.

INTRODUCTION

Hot surface ignition is a common initiator of wildland fires.[1,2] In particular, heated metal particles or droplets may be generated from a variety of sources including the failure of overhead powerlines, hot work, and friction or impact (e.g rock strikes) involving metal equipment. Previous studies[3-8] have investigated the ignition of granular cellulosic fuel beds from heated metallic particles, as well as the generation of these particles. These studies have reported that molten aluminum spheres have greater incendivities than steel spheres of the same diameter. It was hypothesized that "this is most likely the result of their increased bulk energy due to the heat of fusion and potentially also the different impact dynamics of molten objects."[8]

The present study aims to investigate the potential for hot particle ignition of realistic wildland fuels. To this end, dry vegetation of varying composition was collected from a park in coastal southern California (Figure 1). Metals selected for testing as ignition sources included copper and aluminum; these materials are typical of electrical conductors in overhead power lines. In order to isolate the effect of latent heat of fusion as well as the ability of molten material to conform to an irregular surface, borosilicate glass was also tested. This glass has a similar energy content to the aluminum at the tested temperatures, but is dramatically different in rheology.

Figure 1. Area of fuel collection.

EXPERIMENTS

Various wildland fuels (e.g. twigs, leaves, brush, pine needles, and decomposing wood) were gathered from a coastal area in Southern California known to be susceptible to wildfires. These materials were conditioned using an environmental chamber to summertime temperature and humidity conditions typical for the location from which they were collected. They were then arranged into a fuel bed similar to the typical ground cover observed during the summer months at the collection location as shown in Figure 2.

Figure 2. Fuel bed prior to material application.

The collected fuels were conditioned in an environmental chamber to conditions typical for the area of collection during the summer months. They were then removed from the environmental chamber and placed in a fuel bed on top of a container of soil. Ignition materials (copper, aluminum, or borosilicate glass) were placed in crucibles and heated in a furnace to the desired temperature, ranging from 670 to 1000°C. In each test, a crucible was removed from the furnace and its contents poured from a height of approximately 50 cm onto the fuel bed, which was observed for ignition.

Immediately after the heated material landed in the fuel bed, one of the following three events would occur:

1. No observed ignition, either in the form of glow or smoke.
2. Smoldering combustion was initiated. Smoke and glowing embers were observed, but not flames.
3. Direct ignition of flaming combustion.

For several of the trials which resulted in smoldering combustion, a fan was used to apply a known wind speed to the fuel bed, and the reaction was allowed to either self-extinguish or transition into flaming combustion (Figure 3).

Figure 3. Fuel bed immediately following transition from smoldering to flaming combustion.

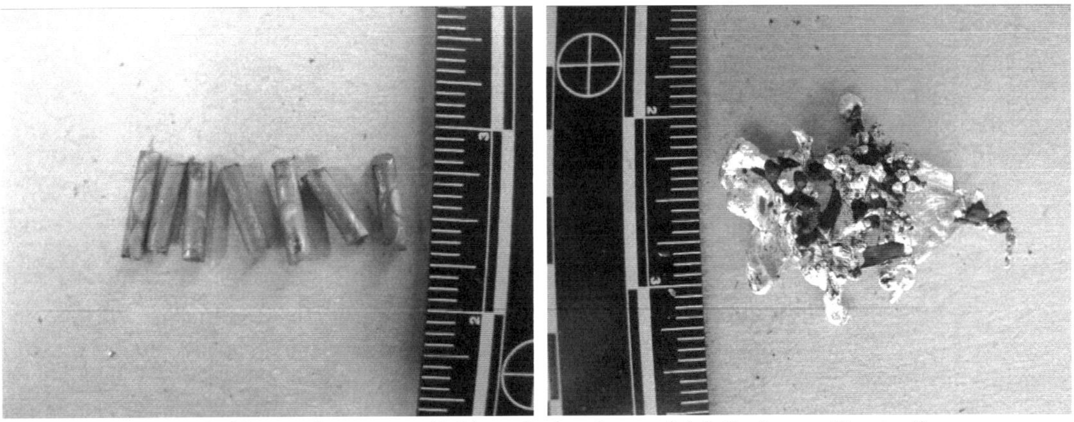

Figure 4. Example copper (left) and aluminum (right) pieces after testing.

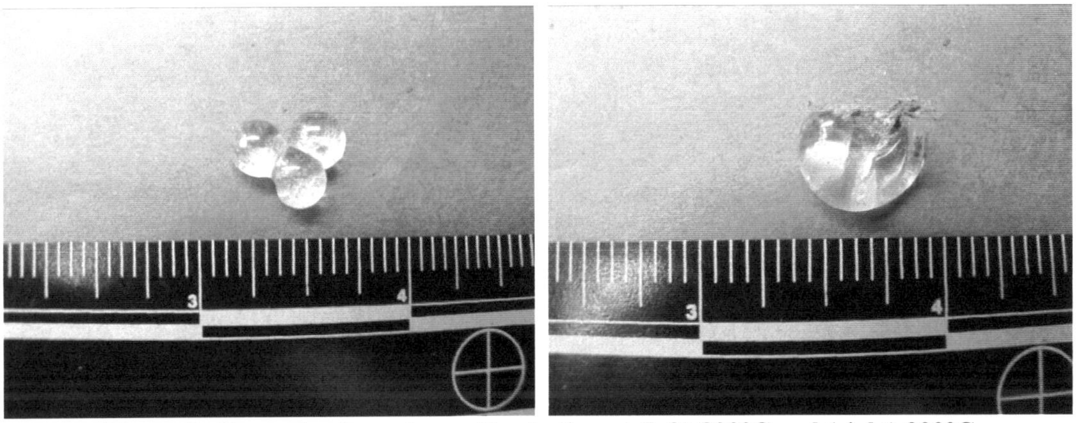

Figure 5. Example glass pieces after testing at (left) 800°C and (right) 900°C

Copper Tests

Table 1 contains a summary of the tests performed with copper particles. All copper testing was conducted at 1000°C. At this temperature, copper is a solid. The copper pieces consisted of single strands from copper conductors cut to approximately 1 cm in length. The crucible used in the first test contained seven of these pieces. During this test it was observed that the pieces spread out to different areas of the fuel bed, essentially acting as independent ignition sources, and the subsequent 28 test crucibles each contained a single piece.

In most of the tests (19 out of 29), the copper caused the fuel to begin smoldering. The ten tests during which no smoldering combustion was observed are highlighted in grey in the table. In fourteen of the tests, a fan was positioned to provide an airflow of 1.3 m/s over the fuel bed. Such wind velocities can encourage the transition from smoldering combustion to flaming combustion, however this transition was not observed during any of the copper tests.

Table 1. Copper tests.

Test	Temp (°C)	Total Mass(g)	Fan	Notes
1	1000	8	no	Smoldering
2	1000	1.2	no	no ignition
3	1000	1.1	no	Smoldering
4	1000	1.2	no	Smoldering
5	1000	1.2	no	Smoldering
6	1000	1.2	no	no ignition
7	1000	1.2	no	Smoldering
8	1000	1.2	no	no ignition
9	1000	1.6	no	Smoldering
10	1000	1.6	no	Smoldering
11	1000	1.7	no	Smoldering
12	1000	1.7	no	Smoldering
13	1000	1.6	no	Smoldering
14	1000	3.3	no	Smoldering
15	1000	1.4	1.3 m/s	no ignition
16	1000	0.9	1.3 m/s	no ignition
17	1000	0.8	No	no ignition
18	1000	0.9	1.3 m/s	smoldering
19	1000	1	1.3 m/s	smoldering
20	1000	0.9	1.3 m/s	no ignition
21	1000	1.1	1.3 m/s	smoldering
22	1000	0.5	1.3 m/s	smoldering
23	1000	0.8	1.3 m/s	smoldering
24	1000	1.2	1.3 m/s	no ignition
25	1000	0.7	1.3 m/s	no ignition
26	1000	1.6	1.3 m/s	smoldering
27	1000	1.8	1.3 m/s	smoldering
28	1000	0.5	1.3 m/s	no ignition
29	1000	1.5	1.3 m/s	smoldering

Aluminum Tests

Table 2 contains a list of the tests performed using aluminum. The aluminum particles were cut from conductors and connectors used in overhead power lines. All testing was conducted above the melting point of aluminum (660°C), and ranged from 670-1000°C. After the solid oxide layer is broken, the molten aluminum is a freely flowing liquid which conforms to the shape of whatever object it is poured over (see Figure 6). All of the tests using aluminum resulted in smoldering combustion. For temperatures of 800°C and higher, direct ignition of flaming combustion resulted from first contact with the ignition source. For temperatures at which this did not occur, a fan was used to provide an airflow of between 0.4 and 1.3 m/s over the fuel bed. This airflow resulted in the transition to flaming combustion in several cases, including the lowest temperature tested, which was only 10°C higher than the melting point of aluminum. This indicates that under the right conditions, molten aluminum at *any* temperature may result in a wildfire. Tests which resulted in flaming combustion (either promptly or as a result of transition from smoldering) are highlighted in yellow in the table.

Table 2. Aluminum Tests

Test	Temp (°C)	Total Mass(g)	Fan	Notes
1	1000	3.2	no	Smoldering
2	1000	2.7	no	Flaming ignition on contact
3	1000	4.5	no	Flaming ignition on contact
4	1000	4.1	no	ignition on contact -> smolder
5	1000	4.3	no	Flaming ignition on contact
6	800	3.1	no	Flaming ignition on contact
7	800	4.4	no	Flaming ignition on contact
8	800	3.6	no	smoldering
9	800	4	no	smoldering
10	800	3.9	no	Flaming ignition on contact
11	800	3.6	no	smoldering
12	700	3.8	no	Smoldering
13	700	2.3	no	Smoldering
14	700	2.3	no	Smoldering
15	700	1.3	0.9-1.3 m/s	Smoldering
16	700	3.7	0.9-1.3 m/s	Smoldering, transition to flames
17	700	3.6	0.9-1.3 m/s	Smoldering
18	700	1.8	0.4-0.9 m/s	Smoldering
19	700	5	0.4-0.9 m/s	Smoldering
20	700	3.7	0.4-0.9 m/s	Smoldering
21	700	2.9	0.4-0.9 m/s	Smoldering
22	700	3.1	0.4-0.9 m/s	Smoldering, transition to flames
23	670	2.5	0.4-0.9 m/s	Smoldering -> brief flames -> smoldering
24	670	1.5	0.4-0.9 m/s	Smoldering
25	670	2.4	0.4-0.9 m/s	Smoldering
26	670	2.3	0.4-0.9 m/s	Smoldering
27	670	3.3	0.9-1.3 m/s	Smoldering

28	670	4.1	0.9-1.3 m/s	Smoldering then flaming ignition
29	670	4.8	no	Smoldering
30	670	5.3	0.9-1.3 m/s	Smoldering
31	670	4.2	0.9-1.3 m/s	Smoldering
32	670	4	0.9-1.3 m/s	Smoldering
33	670	3.7	0.9-1.3 m/s	Smoldering
34	800	1	no	Smoldering
35	800	1	no	Smoldering
36	800	1	no	Smoldering
37	800	1	no	Smoldering
38	800	1	no	Smoldering
39	800	1	no	Smoldering
40	800	1	no	Smoldering
41	800	1	no	Smoldering
42	800	1	no	Smoldering
43	800	1	no	Smoldering
44	800	1	no	Smoldering
45	800	1	no	Smoldering
46	900	1	no	Smoldering
47	900	1	no	Smoldering
48	900	1	no	Flaming ignition on contact
49	900	1	no	Smoldering
50	900	1	no	Flaming ignition on contact
51	900	1	no	Flaming ignition on contact

Figure 6. Aluminum particle immediately after a pour. Note that it has flowed over and around the leaf.

Glass Tests

Table 3 contains a list of tests performed using borosilicate glass. The glass particles began as beads approximately 0.5 cm in diameter. Three beads were placed in each crucible to achieve a total mass of approximately 1 gram of glass. Glass tests were conducted at 800 and 900°C. At 800°C, the glass beads largely retained their forms, but in some of the tests, the beads fused together into a single three-bead piece. At 900°C, the beads softened to a taffy-like consistency. Rather than flowing like the liquid aluminum when poured over the fuel bed, the globs of molten glass retained the structure they had in the crucible (Figure 7). All of the glass tests resulted in smoldering combustion.

Table 3. Glass tests.

Test	Temp (°C)	Total Mass(g)	Fan	Notes
1	800	1	no	Smoldering
2	800	1	no	Smoldering
3	800	1	no	Smoldering
4	800	1	no	Smoldering
5	800	1	no	Smoldering
6	800	1	no	Smoldering
7	800	1	no	Smoldering
8	800	0.7	no	Smoldering
9	800	1	no	Smoldering
10	800	1	no	Smoldering
11	800	1	no	Smoldering
12	900	1	no	Smoldering
13	900	1	no	Smoldering
14	900	1	no	Smoldering
15	900	1	no	Smoldering
16	900	1	no	Smoldering
17	900	1	no	Smoldering

Figure 7. Glass particle immediately after 900°C test. Its structure does not conform to the fuel.

DISCUSSION

Smoldering combustion was achieved for all materials tested, indicating that under the right circumstances, fires may result from any one of them. The results demonstrate that of the three materials tested, aluminum has the greatest propensity to ignite the fuel bed; this is consistent with the findings of previous researchers. Direct comparison between the glass tests and aluminum tests numbered 34-51 is of interest. These tests were conducted with similar masses of material, under the same temperatures. Moreover, the materials have similar energy contents.

A key result of these experiments was that at 1000°C and equal metal masses, aluminum was found to ignite the fuel bed into immediate flaming combustion, while the copper trials were found to ignite either smoldering combustion or no observable reaction. In all cases for which copper ignited smoldering combustion, the reaction failed to transition into open flames. This would suggest that aluminum's higher heat capacity, coupled with the fact that aluminum is liquid at this temperature (copper is solid), is critical to the difference in incendivity between the two materials. Figure 4 shows copper and aluminium material after testing. It is easily seen in the photograph that the liquid aluminum flowed around the fuel sample before it resolidified

Table 4 contains relevant thermal properties of the materials used in the ignition experiments. The available energy content per unit mass of a metal at elevated temperature is $c_p\Delta T$, where c_p is the average specific heat over the temperature range, and ΔT is the temperature difference from a reference condition. If a phase transition is present within this temperature range, the energy becomes $c_p\Delta T+\Delta h_{fus}$, where Δh_{fus} is the heat of fusion of the material. Thus a molten material will have more energy available for ignition than a solid material with a similar mass and specific heat capacity at the same temperature. Furthermore, the liquid material will deform to make greater thermal contact with the fuels than the solid material. Both of these factors result in a greater incendivity for the molten material. Of the three ignition sources tested, only aluminum underwent a distinct phase transition, and carried the latent heat of fusion. This coupled with the higher specific heat results in an energy content nearly three times greater than that of copper at the test temperatures. The borosilicate was raised above its glass transition temperature, which does not result in a latent heat. At high temperatures, the specific heat of the borosilicate is higher than that of aluminum. These are competing effects when it comes to the energy content of the material, with the result that the energy contents of the glass and aluminum are within a few percent of one another at the tested temperatures. With similar energy contents and similar masses, aluminum was found to have a greater incendivity than the glass. This is most likely due to dramatically different contact resistances. The aluminum is freely flowing and conforms to the material it comes into contact with, maximizing the area available for heat transfer. The taffy-like glass retains its structure when dropped on the fuel bed, leaving greater void space rather than conforming to the surface irregularities of the fuel.

Table 4. Thermal properties of ignition materials.

Material	STP C_p (J/g-K)	High T C_p (J/g-K)	Avg C_p (J/g-K)	Δh_{fus} (J/g)	h(800 K) (J/g)	h(900 K) (j/g)	h(1000 K) (J/g)
Aluminum	0.897[12]	1.18[9]	1.0	397[9]	900	1000	1100
Copper	0.385[12]	0.44[12]	0.4	N/A	320	360	400
Borosilicate Glass	0.84[11]	1.4[10]	1.1	N/A	880	990	1100

CONCLUSIONS

Molten aluminum droplets were found to be a competent ignition source of typical Southern California wildland fuels at any temperature above the melting point of the material. This finding was contrasted with solid copper particles within a similar temperature range, which exhibited lesser incedivity. However, it was noted that copper particles of the have approximately one third the total available energy as aluminum droplets with the same mass.

In order to isolate the effects of the liquid state of the material from energy content, additional experiments were conducted with borosilicate glass. This material has similar thermal conductivity and almost identical energy capacity as aluminum at the temperatures tested. The primary difference between the two materials as ignition sources is that the glass is solid or taffy-like while the aluminum is a free flowing liquid. A comparison of the results show that the aluminum also has a greater incendivity than the glass, indicating the importance of heat transfer from the material to the fuel. The free flowing nature of the liquid allows the aluminum to conform to surface irregularities, resulting in a lower contact resistance between the materials.

REFERENCES

1. Schwarzenegger, A., Chrisman, M., Grijalva, R., and Dargan, K. Power Line Fire Prevention Field Guide. Cal Fire. November 2008.
2. Guide to Wildland Fire Origin & Cause Determination. PMS 412-1. NFES 1874. Natioanl Wildfire Coordinating Group. April 2016.
3. Stokes, A.D. Fire Ignition by Copper Particles of Controlled Size. Journal of Electrical and Electronics Engineering Australia. Vol. 10. 1990.
4. Rowntree, G.W.G. and Stokes, A.D. Fire Ignition by Aluminum Particles of Controlled Size. Journal of Electrical and Electronics Engineering Australia. Vol. 14. 1994.
5. Rallis, C. J. & Muissa Mangaya, B. Ignition of Veld Grass by Hot Aluminium Particles Ejected from Clashing Overhead Transmission Lines. Fir Technology 38 :81-92. 2002.
6. Russell, B. D., Benner, C. L., and Wischkaemper, J. A. Distribution Feeder Caused Wildfires : Mechanisms and Prevention. 65th Annual Conference for Protective Relay Engineers. IEEE. 2012.
7. Tse, S. D. & Fernandez-Pello, A. C. On the Flight Paths of Metal Particles and Embers Generated by Power Lines in High Winds—a Potential Source of Wildland Fires. Fire Safety Journal 30 :333-356. 1998.
8. Zak, C., Urban, J., Tran, V., & Fernandez-Pello, C. Flaming Ignition Behavior of Hot Steel and Aluminum Spheres Landing in Cellulose Fuel Beds. Fire Safety Science 11 :1368-1378. 2014.
9. Chase, M.W., Jr., NIST-JANAF Themochemical Tables, Fourth Edition, J. Phys. Chem. Ref. Data, Monograph 9, 1998.
10. Shibata, H., Suzuki, A, & Ohta, H. Measurement of Thermal Transport Properties for Molten Silicate Glasses at High Temperatures by Means of a Novel Laser Flash Technique. Materials Transactions 46 :8 :1877-81. 2005.
11. Pegasus Glass. Properties of Borosilicate (PYREX 7740) Glass. PYREX Technical Information.
12. CRC Handbook of Chemistry and Physics. 85th ed. CRC Press: Boca Raton, FL, 2004-2005.

STATISTICAL ANALYSIS ON THE EFFECTS OF NATURAL WEATHERING ON THE PERFORMANCE OF INTUMESCENT COATINGS USING ANOVA

Vahid Hemmati[1], Babak Bahrani[1], Aixi Zhou[1], and Stephen L. Quarles[2]
[1] University of North Carolina at Charlotte, Charlotte, NC, USA
[2] Insurance Institute for Business and Home Safety (IBHS), Richburg, SC, USA

ABSTRACT

Intumescent fire retardant coatings are one of the most frequent passive fire protection methods used for structures in case of wildland or Wildland-Urban Interface (WUI) fires. When exposed to a heating source, they form a thick layer of carbon-rich char on the exterior surface of a structure and can delay or event prevent the structure's ignition. These coatings are economical and easy to apply. However, the performance of such systems is affected by weathering, as aging affects the intumescence mechanism and fire protection effectiveness. Moreover, coatings may act as an additional source of fuel to the structures after weathering. In this study, three intumescent coatings were selected from the market to compare their fire protection performance after weathering. A bench-scale cone calorimeter was used to collect data for fifteen different classes of samples. The classes include the samples which were exposed to natural weathering for 0, 3, 6 and 12 months in two different weathering orientations (north and south). Then, the specimens were randomly picked up from each class and were tested using the cone calorimeter. The Analysis of Variance (ANOVA) method as a statistical tool was then employed to analyze the differences among the classes. First, the uncoated and coated specimens were compared. In all cases, coatings showed a significant protection. Then, coated specimens with different weathering intervals were compared. The results showed that not all coatings preserved their protective effectiveness after 3 months of weathering exposure. In some cases, there were no statistical evidence of performance difference between coated and uncoated specimens just after 3 months.

INTRODUCTION

Wildland-Urban Interface (WUI) fires have imposed much property damage and many human casualties around the world, particularly in the recent several decades[1]. Various fire protection systems have been used to mitigate fire damage from WUI fires. In general, these fire protection systems can be divided into two categories: passive and active. The Active Fire Protection (AFP) systems are mostly used just in the time of incidents to manage and finally extinguish the fire. The main characteristic of AFP systems is their requirement of physical movement or action. Various systems from smoke/fire alarming system to sprinklers and extinguishers are categorized as AFP systems. Passive Fire Protection (PFP) systems are those in which there are no physical movement required at the time of incident. These systems work simply by their existence in the incident scene, such as fire walls, fire protection coating layers, fire protection wraps, or wetting agents. Proper design of structures and using fire-resistance materials in structures can be considered as passive fire protection. Intumescent coating is one of the most frequent PFP systems used to protect structures. The intumescent coating treatment is not only relatively simple to apply, but also it is considered as an economical PFP system. Intumescent coatings swell 20-50 times their original thickness, and form a carbonaceous char layer which delays or prevents ignition, reduces the flame spread and decrease the fire growth at the early stages of fire[2-4]. Smoldering combustion can also be contained in presence of carbon char[5].

An intumescent coating system usually consists of three main components: 1) a carbon source, which acts as the main component for the char formation, 2) an acid source, which acts as a catalyst in simultaneous expansion, and 3) a blowing agent, which produces gas[6]. Some systems consider binders as another major component, where they keep all the components together during the intumescence process[7]. Intumescence mechanism can be divided into chemical and physical aspects. A series of physical and chemical reactions occur as such coatings are exposed to a heat source. Chemical mechanism takes place in 6 main steps: binder softening, acid release, carbonization, gas formation, foaming, and solidification[8-9.] Physical mechanism includes the cooling effect, dilution, and protective layer formation[10].

Despite numerous studies on intumescent coatings on steel substrates and impregnated coatings on woods, the weathering effects on the performance of the intumescent coating applied on wooden substrates has not been specifically studied [8, 17-20]. The effects of weathering on intumescent coatings include visual characteristics (such as fading, loss of gloss, chalking, and crack formation) and mechanical properties (such as weight loss, surface roughening, embrittlement, and impact strength)[8, 14]. In addition, combustion properties such as time-to-ignition and heat release rate along with intumescence properties like time-to-intumescence and intumescence height will also be affected after weathering exposure[8].

In terms of fire protection effectiveness, there are two major weathering concerns which may challenge the advantage of using intumescent coating as a passive fire protection method: 1) weathering reduces the fire protection efficiency in a way that the coatings show no protection after a certain period of time, and 2) the intumescent coating may act as an extra fuel to structures after exposed to weathering conditions.

Primary weathering parameters considered in weathering tests are solar energy (radiation), moisture, and temperature, alongside with atmospheric contaminants as secondary parameter[8]. There are three main ways to perform weathering durability tests: natural weathering, accelerated natural weathering, and accelerated laboratory (artificial) weathering. According to available standards such as ASTM D2989, ASTM D7032, and ASTM D6662, constructions should maintain their performance against weathering conditions[11-12]. For this study, outdoor natural weathering method was used to examine the non-concentrated sunlight effects along the other weathering parameters in pre-defined weathering orientations[14] (north and south) to examine combustion properties of specimens.

The goal of this research was to determine the efficiency of intumescent coatings in natural weathering conditions. To achieve the goal, three commercially available intumescent coatings were selected to compare their fire protection performance after weathering. A cone calorimeter was used to collect data for fifteen different classes of samples. The classes include the samples which were exposed to natural weathering for 0, 3, 6 and 12 months in two different weathering orientations (north and south). Then, the specimens were randomly picked up from each class and were tested using the cone calorimeter. Measured parameters included Time-to-Ignition (TTI), Heat Release Rate (HRR), Mass Loss Rate (MLR), and Effective Heat of Combustion (EHC). Next, the Analysis of Variance (ANOVA) method as a statistical tool was employed to analyze the differences among the classes. First, the uncoated and coated specimens were compared. Then, fire properties of coated specimens with different weathering intervals were compared. The focus of this article is to present results on TTI data.

EXPERIMENT DESIGN

The implementation of ANOVA as statistical method develops a significant tool in comparing measured data. Despite of the popularity of this method in quality control, the implementation in weathering effects on intumescent coatings is almost a new application for the method.

For this study, three different commercially available intumescent systems (designated as A, B, and C) were selected. Specifications of coating can be seen in Table 1[9].

Table 1. Intumescent Coatings Specifications

Coating System	Description (Per Manufacturer's Instructions)	Primer Number of Layers	Coating Number of Layers	Coating Dry Film Thickness [m]
A	Water-based intumescent coating (w/formaldehyde, non-flammable)	2	1	1.016×10^{-4}
B	Latex-based intumescent coating	0	1	2.286×10^{-4}
C	Water-based intumescent coating, non-combustible	2	2	3.048×10^{-5} to 4.572×10^{-5}

The substrate was AC grade southern yellow pine 4-ply plywood with a nominal thickness of 12.5 mm (1/2-in). Samples for each coating were processed from three separate plywood panels, each measuring 2490 mm×1170 mm (98-in×46-in). Each panel was cut into 121 pieces in 200 mm×100 mm (8-in×4-in). A total of 363 specimens out of three panels were prepared. The specimens were selected using a random allocation process, where each 100 mm×200 mm (4-in×8-in) sample was given a number that corresponded to a weathering orientation (North or South), radiant heat exposure level (30, 50 and 70 kW/m^2), and weathering time. The results used in this paper are from 0, 3, 6, and 12 months of exposure. Three replications were prepared for each treatment combination. Based on the number of available samples, an average of 32 specimens were considered for each weathering period. All specimens were finished with one of the three selected intumescent coatings. The C-grade side and the four edges of the plywood substrate were painted with one layer of regular, semi-gloss white paint and primer. The purpose of this primer coat was to provide some protection to water absorption, thereby minimizing the associated dimensional change that could occur. The front face (A-grade side) of the specimens was finished with one of the three intumescent coatings. A regular brush was used to apply the coatings per the suppliers' instructions.

Specimens were conditioned (temperature of 23±3°C and relative humidity of 50±5%) and tested using a cone calorimeter in comply with ASTM E1354[15]. Specimens were tested in three different radiative heat flux levels, representing different scenarios in case of a WUI fire: low (30 kW/m^2), medium (50 kW/m^2), and high (70 kW/m^2). Procedures for preparation, conditioning, test procedures and data collection were according to ASTM E1354 and discussed in detail in[8-9]. Uncoated specimens were tested alongside the coated ones, and collected data have been analyzed using the one-way ANOVA method. Analyzing data could statistically prove whether coatings are proper treatments after weathering exposure or not. Additionally, how long the protection and efficiency of such coatings would last.

DATA ANALYSIS

One-way ANOVA is a popular method in many scientific and engineering fields. In most cases, ANOVA method is applied to study the effectiveness of a specific treatment. The effectiveness of a suggested treatment is expected to shift the mean value of response variable in a random- sample set from an entire population[16].

In this research, the intumescent coating is suggested as a treatment to improve fire protection of wood structures. The fire protection properties, called the *response variables* in ANOVA terminology, includes TTI, HRR, PHHR, MLR, and EHC. Weathering can be considered as a treatment as well. This treatment may affect the response variables in either positive, negative, or even neutral way. One-way ANOVA can be used as one of the best statistical tools to detect the effects of weathering on the performance of intumescent coatings. The null hypothesis is that that the mean (average value of the dependent variable) is the same for all groups. A P-value of 0.05 (i.e., a 95% confidence interval) was used as a threshold value to reject the null hypothesis.

MATLAB program has subroutine for one-way ANOVA. The MATLAB ANOVA subroutine that used in this research not only calculates the ANOVA table, but also plots the Box-Plot.

RESULTS

The measured fire property data included TTI, HRR, PHHR, MLR, EHC, and several others. Due to length limitation, only TTI data are presented here. The measured average TTI values are shown in Table 2. Baseline data (uncoated, non-weathered specimens) had TTI values of 75s for 30 kW/m^2, 20s for 50 kW/m^2, and 12s for 70 kW/m^2. Values in Figures 1 to 31 are normalized average TTI values generated using MATLAB ANOVA subroutine.

Table 2. Average Time-to-Ignition (TTI) Values

Coat.	HF	W	O	TTI	Coat.	HF	W	O	TTI	Coat.	HF	W	O	TTI
A	30	0	-	**381**	B	30	0	-	**468**	C	30	0	-	**268**
		3	N	**109**			3	N	**66**			3	N	**250**
			S	**105**				S	**73**				S	**199**
		6	N	**119**			6	N	**82**			6	N	**255**
			S	**111**				S	**61**				S	**211**
		12	N	**69**			12	N	**66**			12	N	**157**
			S	**74**				S	**52**				S	**153**
	50	0	-	**69**		50	0	-	**26**		50	0	-	**292**
		3	N	**29**			3	N	**42**			3	N	**48**
			S	**28**				S	**18**				S	**39**
		6	N	**42**			6	N	**18**			6	N	**45**
			S	**40**				S	**31**				S	**34**
		12	N	**29**			12	N	**17**			12	N	**74**
			S	**31**				S	**17**				S	**61**
	70	0	-	**39**		70	0	-	**23**		70	0	-	**19**
		3	N	**17**			3	N	**14**			3	N	**13**
			S	**16**				S	**16**				S	**14**
		6	N	**27**			6	N	**16**			6	N	**28**
			S	**19**				S	**16**				S	**19**
		12	N	**20**			12	N	**12**			12	N	**11**
			S	**18**				S	**7**				S	**12**

**Note)* HF: Heat Flux Level [kW/m^2], W: Weathering Period [Month(s)], O: Weathering Orientation,

Low Heat Flux Level (30 kW/m^2)

Figure 1 shows the effect of coating on TTI under low heat flux exposure. The calculated P-Value (shown as "Prob>F" in the figure) of much less than 0.05 rejects the null hypothesis. The mean values of TTI for coated specimens increased more than 60%. These results indicate that the coatings can be considered as an effective treatment. Among the three intumescent coatings, Coating C showed the best performance.

Another interesting result is the variation of error bars. The uncoated specimen data showed more repeatability due to relatively smaller error bar. The increasing of standard deviation in all coating treated specimens can be interpreted as the result of coating process.

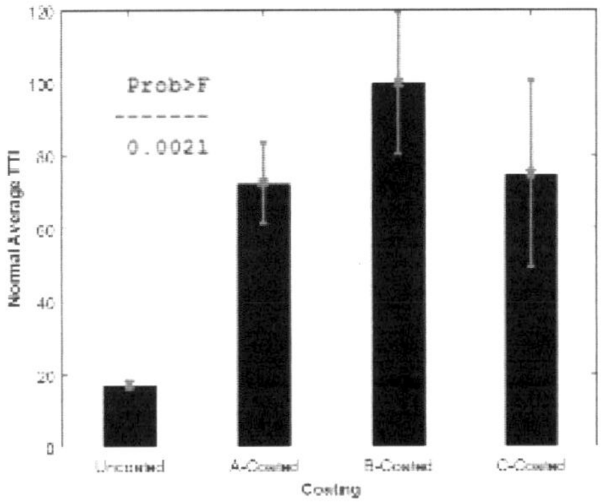

Figure 1. Effect of Coating Application on TTI- Low Heat Flux

Coating A at low heat flux level

Figure 2 and Figure 3 show the normalized average TTI at different time intervals for Coating A at low heat flux level. Data of each installation orientations, northward and southward, are plotted in separate graphs. The calculated P-Values confirm the effects of weathering with remarkable certainty of more than 99%. In both cases, the measured TTI dropped to less than 40% of its initial values just after 3 months of weathering exposure. Another decline in TTI is seen just after 12 months. In addition, there were no clear signs of different responses in specimens in north and south orientations.

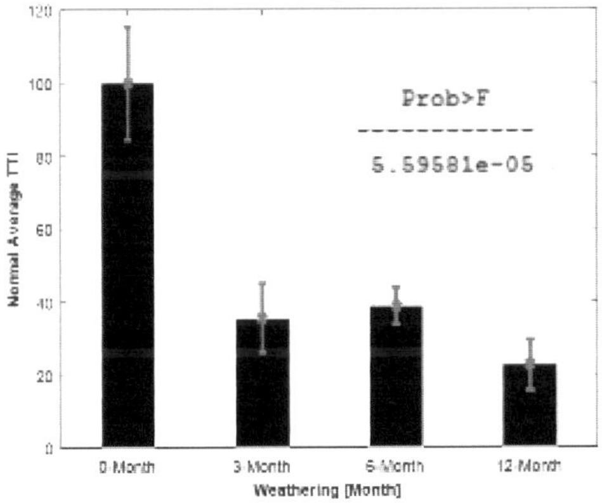

Figure 2. Weathering Effect on Type A (North, Low Heat Flux)

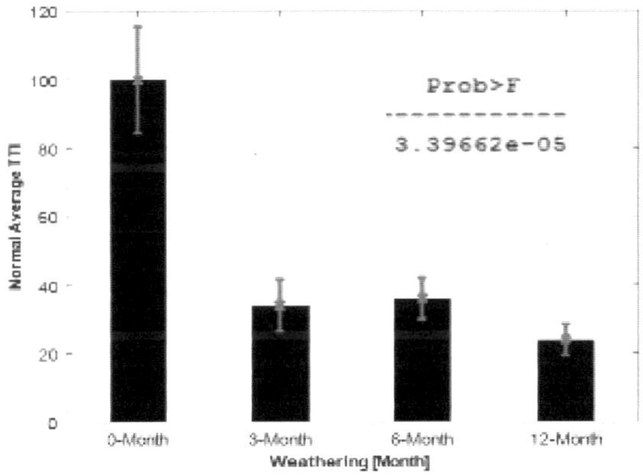

Figure 3. Weathering Effect on Type A (South, Low Heat Flux)

Comparisons of TTI values at low heat flux level for the weathered A-coated specimens and uncoated samples are shown in Figure 4 and Figure 5. The P-Values reject the null hypothesis. In addition, the plotted error bars clearly show the distinct responses after 3 and 6 months weathering exposure. Also, the similarity in the responses of uncoated and 12-months weathered coated is shown. These results indicate that the A type specimens provided a better protection than the uncoated for at least 6 months. This advantage is not seen after 12 months weathering.

The similarity in the behavior of the southward and northward installed specimens shows that the weathering orientation was not a major factor in the weathering process.

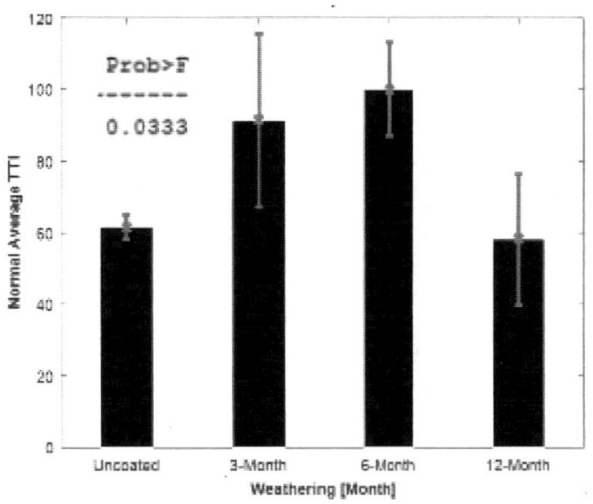

Figure 4. Comparison- Uncoated and Weathered Specimens (Type A, North, Low Heat Flux)

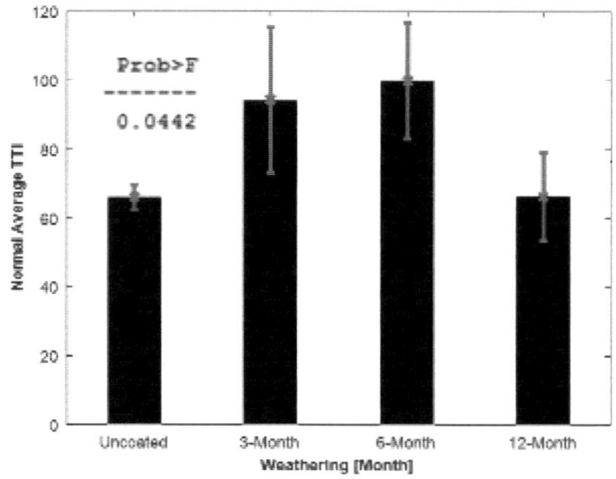

Figure 5. Comparison- Uncoated and Weathered Specimens (Type A, South, Low Heat Flux)

Coating B at low heat flux level

Figure 6 and Figure 7 show the normalized TTI values at low heat flux level at different time intervals for Coating B. The calculated P-Values confirm the effects of weathering with remarkable certainty of more than 99%. These P-value for both types are in the same range, indicating a significant change due to weathering. However, in both orientations, the measured TTI for Coating B dropped to less than 20% of its initial values just after 3 months of weathering. A major difference between Coating A and Coating B is the TTI values after 6 months. In Coating A, a new trend of decline was seen after 6 months. This decline was not observed in Coating B. It could be a sign that the damage after 3 and 6 months weathering was enough to destroy all protection of the coatings, and no protection remained afterward. In addition, there is no clear sign of different responses in southward and northward specimens.

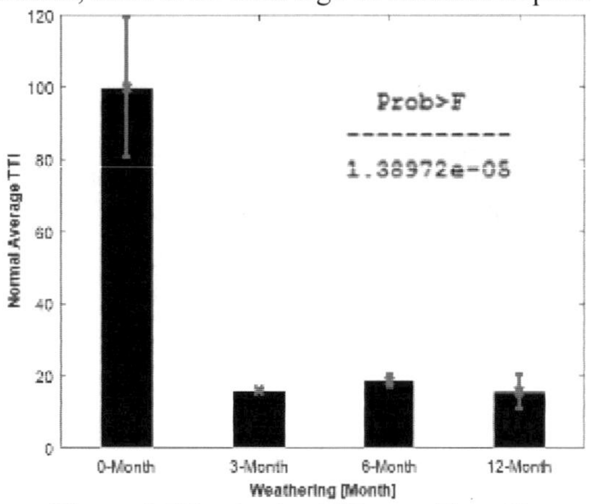

Figure 6. Weathering Effect on Type B (North, Low Heat Flux)

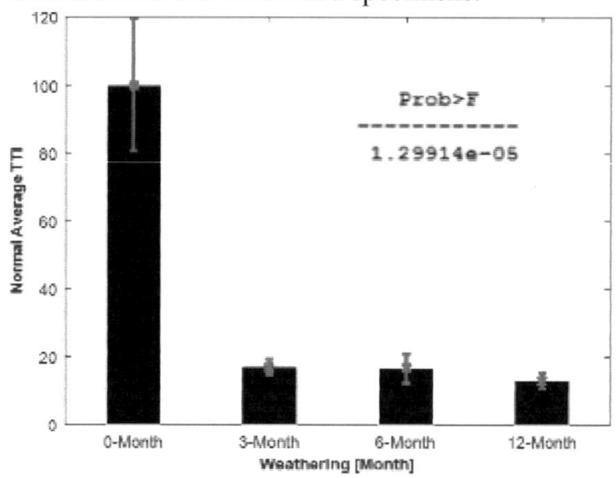

Figure 7. Weathering Effect on Type B (South, Low Heat Flux)

To investigate the advantage of deploying the coating treatment, the weathered B-coated specimens were compared to uncoated samples. The comparisons are shown in Figure 8 and Figure 9. The calculated P-Values confirm the null hypothesis. In addition, the plotted error bars show that there is no distinct response in coated and uncoated specimens. These results show that the B type coating failed to provide a better performance in comparison to uncoated specimens just after 3 months of weathering. It was shown in Figure 1 that coatings A and B performed almost the same before weathering. The results of weathered specimens show the advantage of using Coating A over Coating B for low heat flux level when weathering is involved.

The similarity in behavior of the southward and northward specimens shows that weathering orientation was not a key factor in the weathering process.

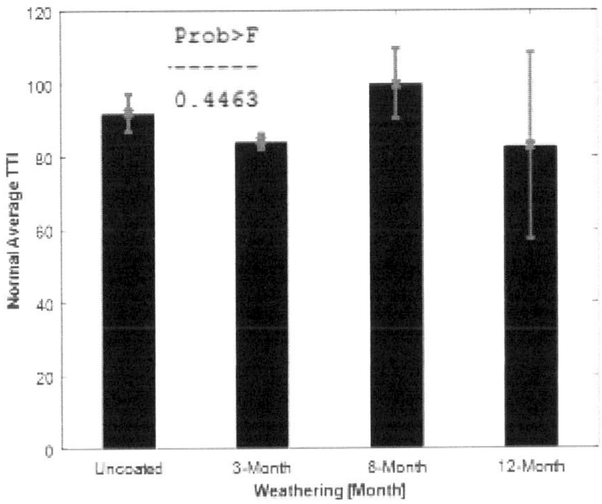

Figure 8. Comparison- Uncoated and Weathered Specimens (Type B, North, Low Heat Flux)

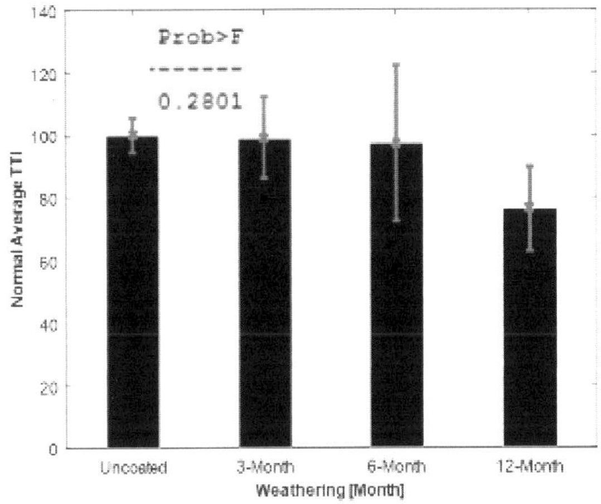

Figure 9. Comparison- Uncoated and Weathered Specimens (Type B, South, Low Heat Flux)

Coating C at low heat flux level

Figure 10 and Figure 11 show the normalized TTI values at low heat flux level at different time intervals for Coating C. The calculated P-Values for both weathering ordinations are greater than 0.05. These P-values confirm the null hypotheses. Therefore, there is no statistical evidence that weathering changed the response variable, TTI, even after 12 months of weathering. The ANOVA concludes that Coating C can resist the weathering much better than the other two types. In addition, Coating C showed a better protection before weathering exposure. In addition, bar charts in Figure 10 and Figure 11 show that the average TTI values of weathered specimens drop around 25%, which is not greater than standard deviation of data. Therefore, the ANOVA results are in consistence with the bar charts.

Similarity in the behavior of the southward and northward specimens shows that the weathering orientation was not a major factor in the weathering process.

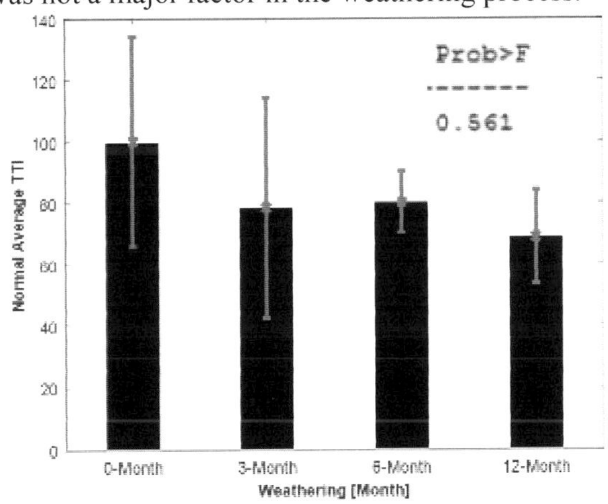

Figure 10. Weathering Effect on Type C (North, Low Heat Flux)

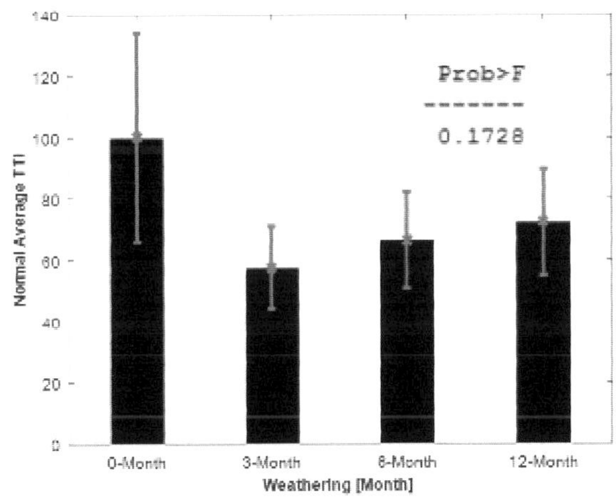

Figure 11. Weathering Effect on Type C (South, Low Heat Flux)

Medium Heat Flux Level (50 kW/m^2)

Figure 12 shows the effect of coatings on TTI values under medium heat flux exposure. The P-value strongly *rejects* the null hypothesis. The bar chart shows that Coating C performed significantly better in comparison to the other two. Performance of Coating A and Coating B were compared to the uncoated specimens. The results are shown in Figure 13. The P-value rejects the null hypothesis, and the performance of Coating A was much better than the uncoated and Coating B specimens. Coating B did not show any advantage over the uncoated specimens based on the P-value shown in Figure 14. The P-value in this case is greater than 0.05, and per ANOVA analyses, there is *no statistical evidence* that Coating B performance is different from the uncoated specimens. This behavior was not seen at low heat flux exposure. In low heat flux level, all the three types of coating provided protection before weathering exposure.

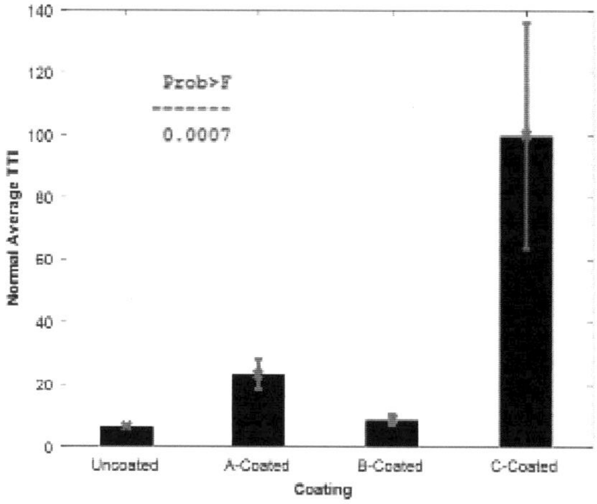

Figure 12. Effect of Coating Application on TTI- Medium Heat Flux

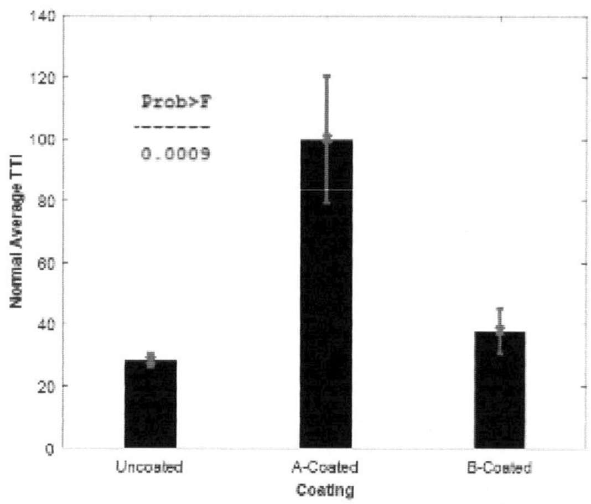

Figure 13. Effect of Coating Application (Types A and B) on TTI- Medium Heat Flux

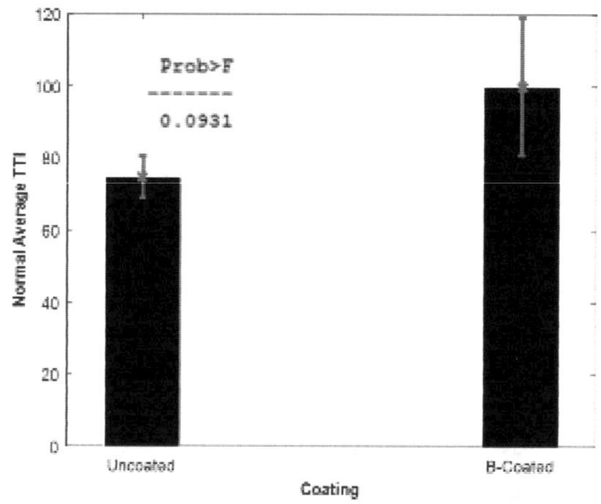

Figure 1. Effect of Coating Application (Type B) on TTI- Medium Heat Flux

Coating A at medium heat flux level

Figure 15 and Figure 16 show the normalized TTI values at medium heat flux level at different time intervals for Coating A. The calculated P-Values confirm the effects of weathering on the performance of Coating A. In both cases, the measured TTI values dropped to nearly 40% of its initial values just after 3 months of weathering. It seems the TTI values stayed stable after 3 months in a way that the measured

TTI in 12 month-weathered specimens were almost equals to the 3 month-weathered specimens. In addition, there is no clear sign of different responses regarding weathering orientations.

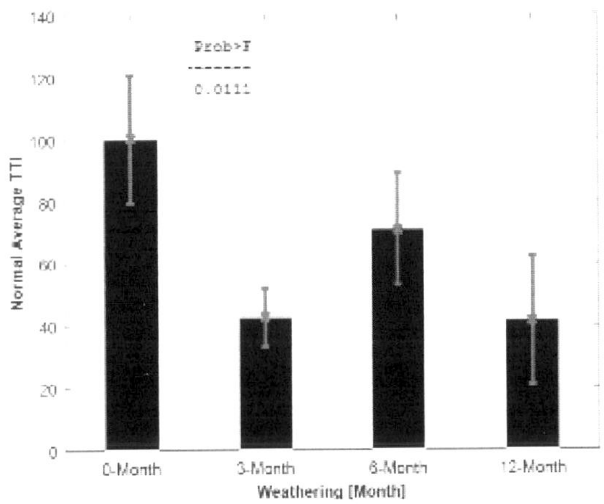

Figure 15. Weathering Effect on Type A (North, Medium Heat Flux)

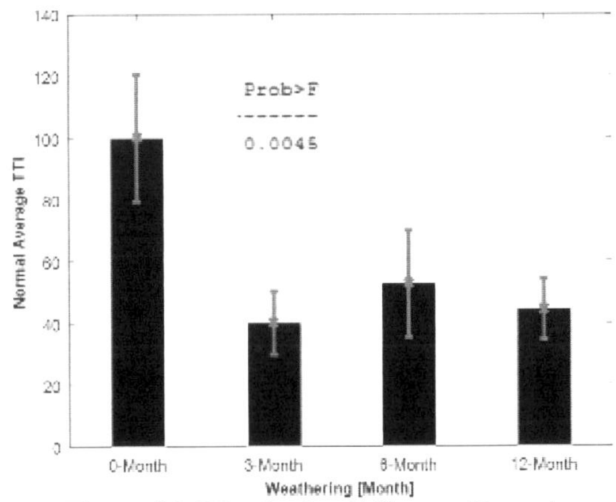

Figure 16. Weathering Effect on Type A (South, Medium Heat Flux)

The weathered A-coated specimens were compared to uncoated samples in Figure 17 and Figure 18. Data of 6 months weathered specimens were excluded in Figure 17. The P-Values confirm the null hypothesis. Based on statistical analysis, the use of Coating A is pointless just after 3 months of weathering exposure at medium heat flux level, although A showed resistance against weathering for at least 6 months at low heat flux level. The similarity in the behavior of southward and northward specimens shows that the weathering orientation was not a major controlling factor in the weathering process.

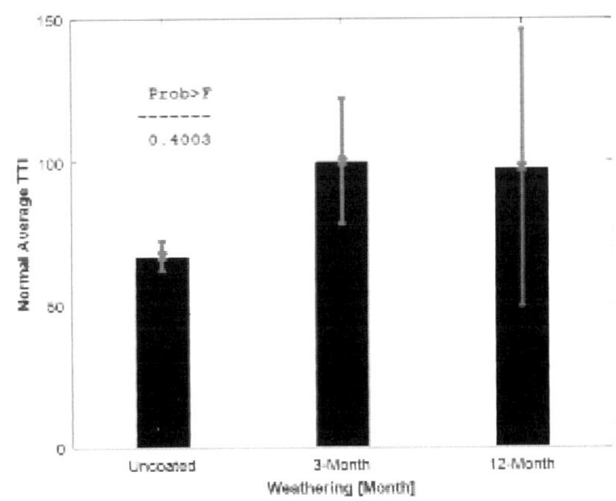

Figure 17. Comparison- Uncoated and Weathered Specimens (Type A, North, Medium Heat Flux)

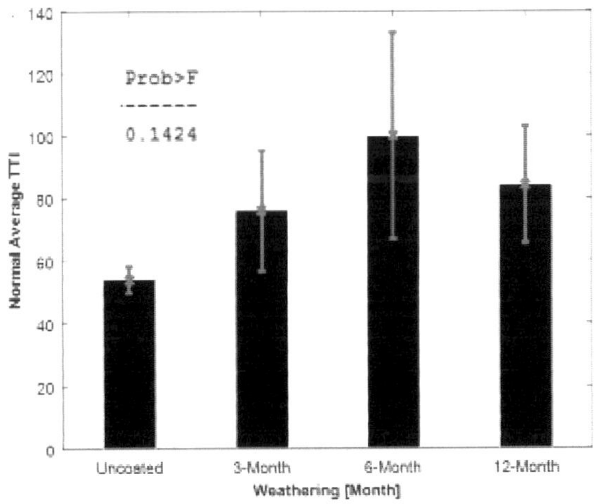

Figure 18. Comparison- Uncoated and Weathered Specimens (Type A, South, Medium Heat Flux)

Coating B at medium heat flux level

The results of TTI values for Coating B are presented in Figure 19. The P-value confirms the null hypothesis. Statistically, there is no evidence between Coating B and the uncoated specimens regardless of weathering exposure. This conclusion was made from results plotted in Figure 14.

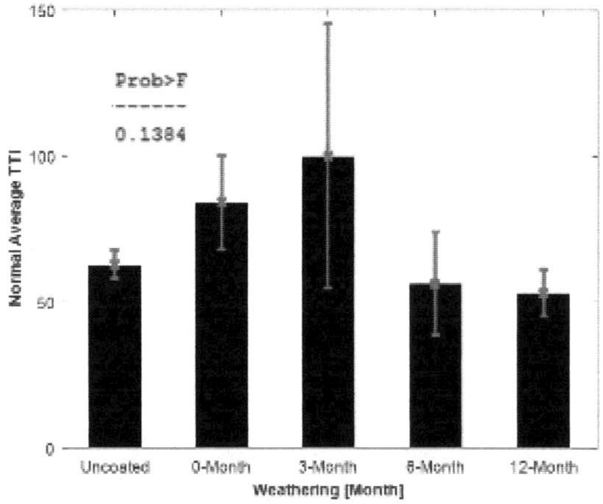

Figure 19. Performance of Type B in Medium Heat Flux

Coating C at medium heat flux level

Figure 20 and Figure 21 show the normalized TTI values at different time intervals for Coating C. The P-Values confirm the effect of weathering on the performance of type C coating. In both cases, the measured TTI values dropped to less than 20% of its initial values just after 3 months of weathering. The decline in TTI was not seen after 3 months. It seems that the TTI was not further reduced after 3 months in a way that the TTI values measured in the 12 month-weathered specimens were almost equal to those measured in the 3 month-weathered specimens. In addition, there is no clear sign of difference due to weathering orientation.

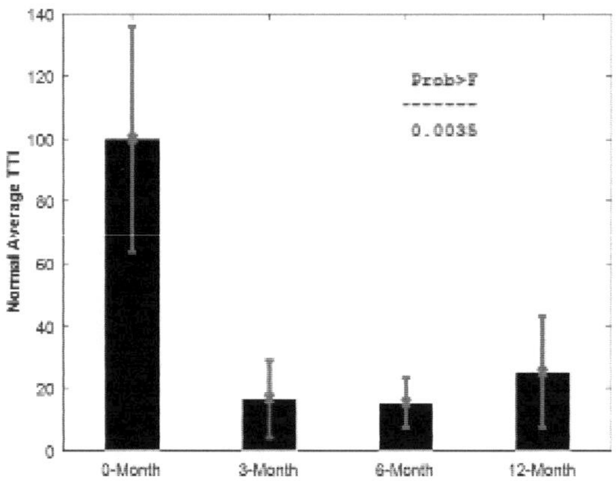

Figure 20. Weathering Effect on Type C (North, Medium Heat Flux)

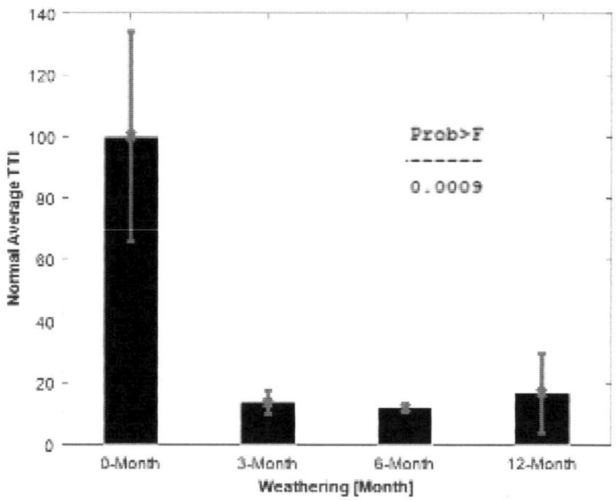

Figure 21. Weathering Effect on Type C (South, Medium Heat Flux)

The weathered C-coated specimens were compared to uncoated samples in Figure 22 and Figure 23. The P-Values confirm the null hypothesis. Based on statistical analysis, the use of Coating C is pointless after 3 months of weathering at medium heat flux level, however Coating C showed resistance against weathering for at least 12 months at low heat flux level. Similarity in the behavior of the southward and northward specimens shows that the weathering orientation was not a controlling factor during the weathering process.

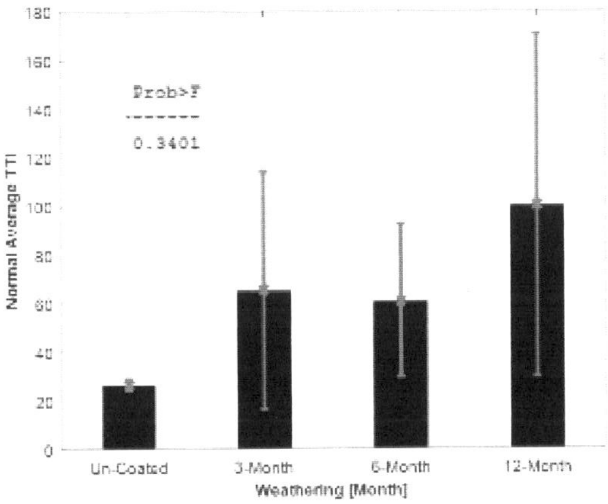

Figure 22. Comparison- Uncoated and Weathered Specimens (Type C, North, Medium Heat Flux)

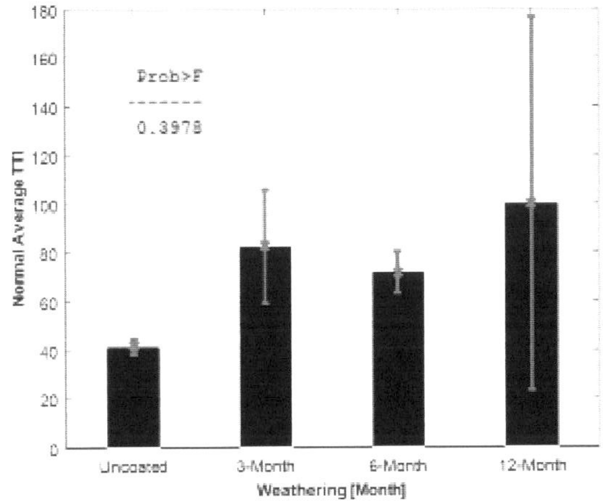

Figure 23. Comparison- Uncoated and Weathered Specimens (Type C, South, Medium Heat Flux)

High Heat Flux Level (70 kW/m^2)

Figure 24 shows the effect of applying intumescent coatings on TTI values under high heat flux level. The P-value rejects the null hypothesis. It shows that, at high heat flux level, Coating A performed significantly better than the other two types. To evaluate the effectiveness of Coating B and Coating C, the performance of these two types are compared to the uncoated specimens as shown in Figure 25. The P-value confirms the null hypothesis. It is interesting that Coating C, which showed the best performance at low and medium heat flux levels, performed poorly at high heat flux level. Coating B, which performed well at low heat flux level, showed poor performance at both medium and high heat flux levels.

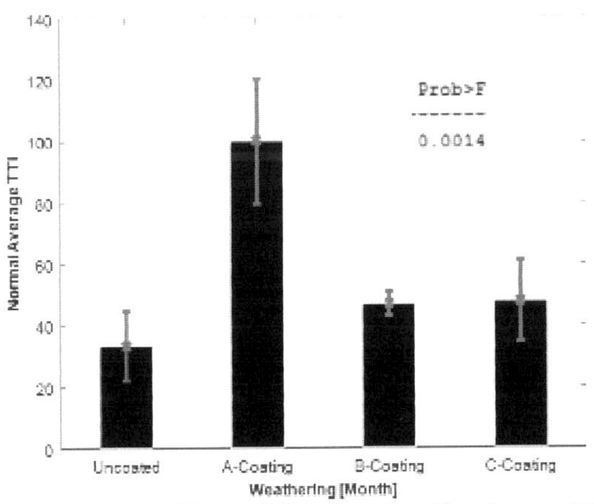

Figure 24. Effect of Coating Application on TTI- High Heat Flux

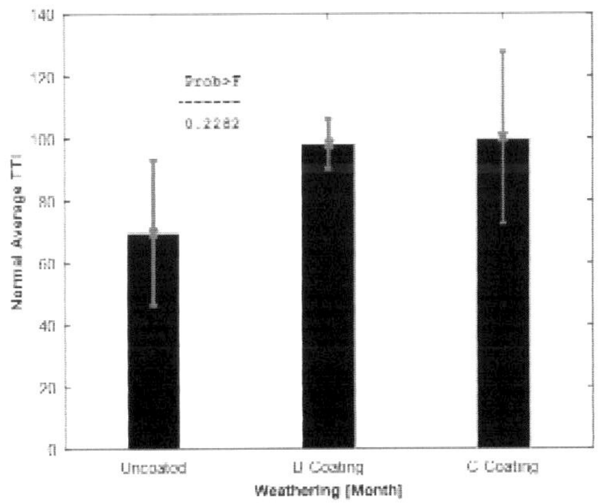

Figure 25. Effect of Coating Application (Types B and C) on TTI- High Heat Flux

Coating A at high heat flux level

Figure 26 and Figure 27 show the normalized TTI values at different time intervals for Coating A at high heat flux level. The P-Values confirm the effect of weathering on performance of Coating A specimens. In both cases, the measured TTI values dropped to around 40% of its initial values after 3 months of weathering. Further decline in TTI was not seen after 3 months. It seems that the TTI values remained stable after 3 months in a way that the measured TTI from the 12 month-weathered specimens were almost equal to those from the 3 month-weathered specimens. In addition, there is no clear sign of difference regarding weathering orientation.

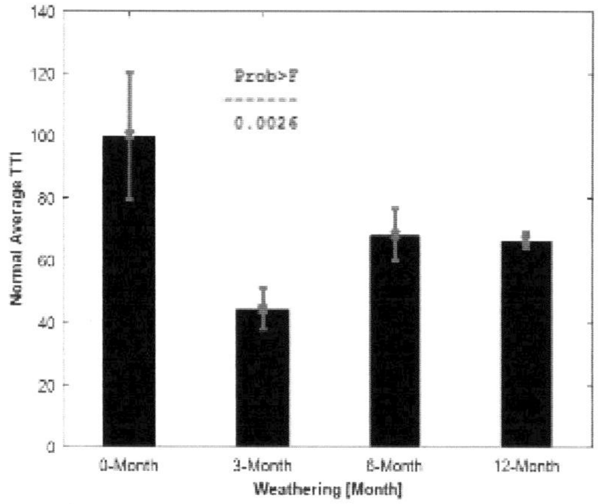

Figure 26. Weathering Effect on Type A (North, High Heat Flux)

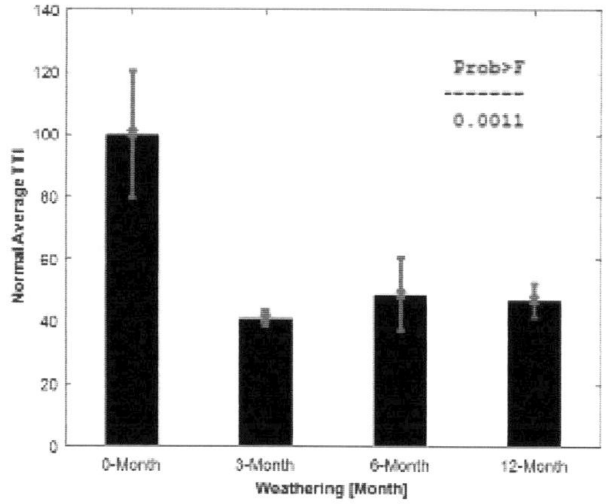

Figure 27. Weathering Effect on Type A (North, High Heat Flux)

The comparisons of weathered A-coated specimens and the uncoated samples at high heat flux level are shown in Figure 28 and Figure 29. The calculated P-Value in Figure 28 (which belongs to northward specimens) rejects the null hypothesis. However, the P-Value in Figure 29 (which belongs to southward specimens) confirms the null hypothesis. For the first time, the results are different due to different weathering orientations. More degradation and decrease in protection were observed in the southward specimens.

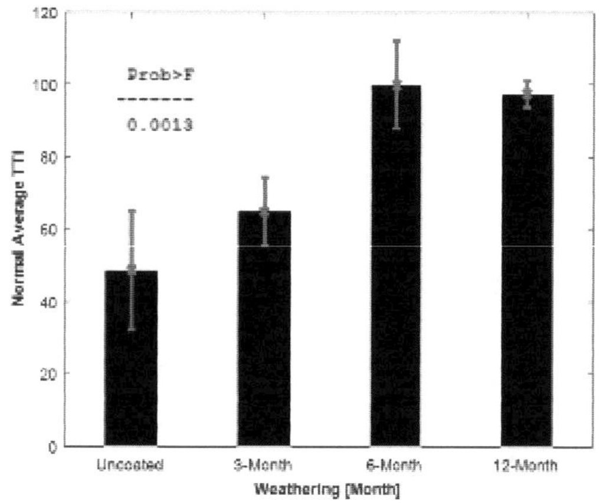

Figure 28. Comparison- Uncoated and Weathered Specimens (Type A, North, High Heat Flux)

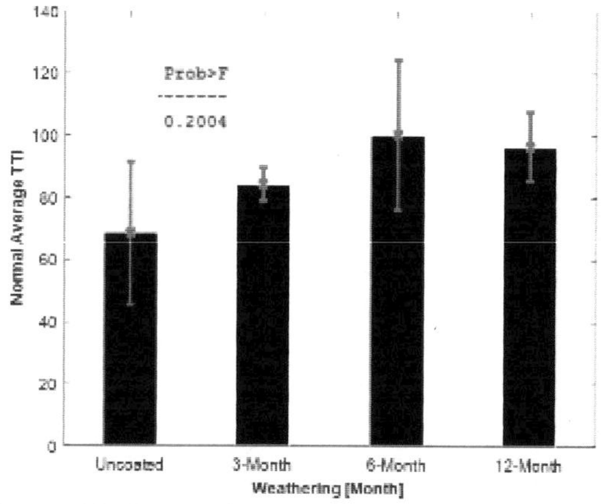

Figure 29. Comparison- Uncoated and Weathered Specimens (Type A, South, High Heat Flux)

Coatings B and C at high heat flux level

The comparisons of TTI values at high heat flux level in coatings B and C are presented in Figure 30 and Figure 31. The P-values confirm the null hypothesis. Statistically, there is no evidence among B, C, and the uncoated specimens. This same conclusion was made from results shown in Figure 25.

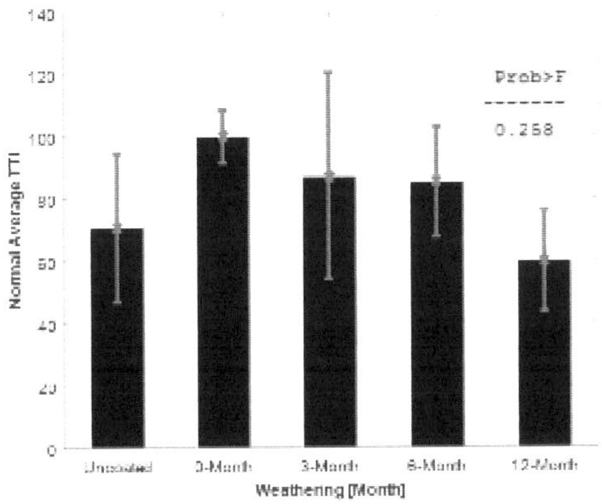

Figure 30. Performance of Type B in High Heat Flux

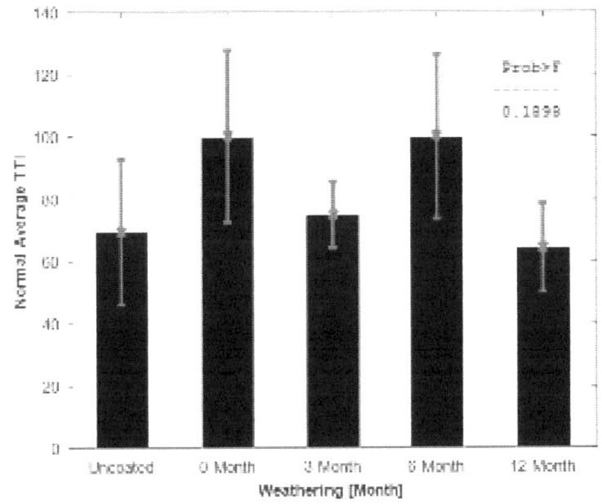

Figure 31. Performance of type C in High Heat Flux

CONCLUSIONS

This study evaluated the effectiveness of intumescent coatings exposed to natural weathering conditions for 12 months. The ANOVA method was employed to conduct data analyses. The following conclusions can be made based on the findings from this study:

- Weathering exposure reduced the effectiveness of fire protection of intumescent coatings. Degradations in coatings were observed in all three types. However, some types can preserve their protection longer.
- Heat flux levels played an important role in the performance of intumescent coatings. Information on heat flux range for activation and performance of intumescent coatings would be helpful for consumers. The intumescence mechanism sequence was likely affected by increasing exposure heat flux level.
- Weathering orientation showed much less effect on the performance of intumescent coatings in comparison to other parameters. Weathering orientation was not a controlling factor during the weathering process.

ACKNOWLEDGEMENTS

The authors would like to thank the IBHS Research Center for sponsoring this project. Any findings, opinions, conclusions, and recommendations expressed in this paper are those of the authors and do not necessarily represent the views of UNC Charlotte and IBHS.

REFERENCES

1 INTERNATIONAL ASSOCIATION OF WILDLAND FIRE (IAWF), 2013. WUI Fact Sheet [Online]. Available: http://www.iawfonline.org/pdf/WUI_Fact_Sheet_08012013.pdf [Accessed October 25, 2016].

2 LAWRENCE, C., 2014. High Performance Textiles and Their Applications: Elsevier, pp. 161.

3 SAXENA, N., & GUPTA, D., 1990. Development and Evaluation of Fire Retardant Coatings. Fire Technology, 26(4), 329-341.

4 MORGAN, A. B., & WILKIE, C. A. (2014). The Non-halogenated flame retardant handbook. http://public.eblib.com/choice/publicfullrecord.aspx?p=1666531. Pp. 293-346.

5 BEYLER, C. L., & HIRSCHLER, M. M., 2002. Thermal Decomposition of Polymers. SFPE Handbook of Fire Protection Engineering, 7, 113-123.

6 BISSCHOFF, J., 2000. *Oxygenated Hydrocarbon Compounds as Flame Retardants for Polyester Fabric.* MS Thesis, University of Pretoria, Pretoria, South Africa.

7 LABUSCHAGNÉ, F. J. W. J., FOCKE, W.W., and Mentz, J.C., 2001. South Africa Patent No. 2002/4876 University of Pretoria, Pretoria, South Africa.

8 BAHRANI, B. 2015. *Effects of Weathering on Performance of Intumescent Coatings for Structure Fire Protection in the Wildland-Urban Interface*. MS Thesis, University of North Carolina at Charlotte.

9 BAHRANI, B., ZHOU, A. & QUARLES, L. S. 'Effects of Weathering on Performance of Intumescent Coating', Proceedings of *Society for the Advancement of Material and Process Engineering* (*SAMPE*) Conference and Exhibition, Available from: SAMPE Digital Library. [23-26 May 2016]

10 BOURBIGOT, S., & DUQUESNE, S., 2007. Fire Retardant Polymers: Recent Developments and Opportunities. Journal of Materials Chemistry, 17(22), 2283-2300.

11 INTERNATIONAL WILDLAND-URBAN INTERFACE CODE (IWUIC), 2015. International Code Council (ICC). pp. 19-20, [Last access date: November 12th, 2016] From: http://codes.iccsafe.org/app/book/toc/2015/ICodes/2015%20IWUIC%20HTML/index.html

12 WHITE, J. R. & TURNBULL, A. 1994. Weathering of polymers: mechanisms of degradation and stabilization, testing strategies and modelling. Journal of Materials Science, 29, 584-613.

13 ASTM STANDARD G113, 2014, "*Standard Terminology Relating to Natural and Artificial Weathering Tests of Nonmetallic Materials*," ASTM International, West Conshohocken, PA, 2014, DOI: 10.1520/G0113-14.

14 SEARLE, N. D., 2009. Natural and artificial weathering of sealants. Handbook of Sealant Technology (CRC Press, New York, 2009), pp. 93-142.

15 ASTM STANDARD E1354, 2015, "*Standard Test Method for Heat and Visible Smoke Release Rates for Materials and Products Using an Oxygen Consumption Calorimeter*," ASTM International, West Conshohocken, PA, 2015, DOI: 10.1520/E1354-15A.

16 PAGANO, R. R. (2013). Understanding statistics in the behavioral sciences. Australia, Wadsworth/Cengage Learning.

17 HARADA, T., MATSUNAGA, H., KATAOKA, Y., KIGUCHI, M., & MATSUMURA, J. (2009). Weatherability and combustibility of fire-retardant-impregnated wood after accelerated weathering tests. Journal of Wood Science: Official Journal of the Japan Wood Research Society. 55, 359-366.

18 DANILIUR, A. (2012). New trends in wood coatings and fire retardant. EUROPEAN COATINGS JOURNAL. 20-25.

19 ROBERTS, T., SHIRVILL, L., WATERTON, K., & BUCKLAND, I. (2010). Fire resistance of passive fire protection coatings after long-term weathering. Process Safety and Environmental Protection. 88, 1-19.

20 VAHABI, H., SONNIER, R., & FERRY, L. (2015). Effects of ageing on the fire behaviour of flame-retarded polymers: a review. Polymer International. 64, 313-328.

MODELLING IN WFDS OF A WILDFIRE SCENARIO

Johan Anderson, David Lange, Anders Lönnermark & Pierrick Mindykowski
SP Safety: Fire Research
SP Technical Research Institute of Sweden
Borås, Sweden
Johan.Anderson.br@sp.se
David.Lange@sp.se
Anders.Lonnermark@sp.se
Pierrick.Mindykowski@sp.se

ABSTRACT

In this paper a coarse graining process is used to subsequently model large wildland fires, starting from a model of a single tree. The models are created using Wildland Urban Interface Fire dynamics Simulator (WFDS), and it is here found that reasonable fire spread in small forests can be obtained although the results are quite dependent on grid resolution as well as moisture content. In most realistic scenarios the computational volume is rather large yielding massive amounts of data. In using WFDS a rather small grid size is needed to appropriately model the fire spread this will be a severely limiting factor in creating large models.

INTRODUCTION

The issue of Wildfire is a problem in many countries all over the world and gets increasing attention in relation to climate effects. Wildfires can affect the society in many different ways and can result in cascading effects. For that reason the EU project CascEff decided to select wildfire as one of the scenarios to include when looking at the risks for cascading effects between different systems and different societal functions. When studying the wildfire scenario it was also decided to look at the possibilities for modelling the wildfire as such and how to draw conclusions on the risks for other systems within or in the vicinity of a wildfire.

Empirical, semi-empirical and physics based models of forest fires have been under development since 1940 [1-10]. Empirical models or conceptual models from experience and intuition from past fires can partly be used to anticipate fires in the future and can thus provide with valuable information regarding fighting woodland fires. Most semi-empirical fire spread equations have been developed for Australasian fuel complexes that can be used for quick estimation of fundamental parameters of interest such as fire spread rate, flame length, and fire intensity of surface fires. A number of key assumptions are built in to these semi-empirical models such as, assuming a representative location, wind and terrain slope. The most complete models combine CFD models with a wildfire component while allowing the fire to have a feedback loop with the atmosphere. These models include NCAR's Coupled Atmosphere-Wildland Fire-Environment (CAWFE) model, WRF-Fire at NCAR and University of Colorado Denver which combines the Weather Research and Forecasting model with a spread model using the level set method and the WUI (Wildland Urban Interface) Fire Dynamics Simulator (WFDS) [11-13]. All these complex modelling tools have different emphases and have been applied to better understand the fundamental aspects of fire behaviour.

In this particular work the wildfire modelling has been performed using the extended WFDS (Wildland Urban Interface FDS) built on the regular FDS model described in [11]. WFDS was originally created for simulating stationary outdoor fires. However, starting with the 5th official release of FDS, WFDS was expanded to contain routines which are aimed at predicting wildland fire scenarios. The effect of the solid fuel such as wood, grass, pine needles, i.e. vegetation, is represented by a series of correlations [12]. This includes representing phenomena such as heat sources and sinks from the solid as well as the mass flow of pyrolysis gases from the solid. The physics for the gas phase is solved by the regular FDS. In this paper, the adopted scenario is the Skatås wildfire which occurred over an extended forest area, meaning that some simplifications are needed in order to have a tractable model. The Skatås forest is located around lakes Stora Delsön and Lilla Delsjön east of the city of Gothenburg on the west coast of Sweden, see Figure 1. The initial event was a forest fire, which was started by children who were playing with fire. The incident started in the forest area just north of the transition between the two lakes, Stora Delsjön and Lilla Delsjön. The red "x" in Figure 2 denotes the starting location of the incident. The initial impact was a forest fire which spread throughout the forested area. The present approach is a coarse graining process, starting with a single tree burning including pyrolysis and detailed thermal modelling on a rather small grid, and then making simplifications to be able to consider a large area with randomly distributed trees on a coarse grid. In this process we will consider a meso-scale model with approximately 100 identical trees randomly distributed over approximately 1200 m^2 while keeping a simplified combustion model and using a 1 m × 1 m × 1 m grid. Furthermore, in order to model a large area of the size of the scenario containing thousands of trees, bulk obstructions with tree properties are created that are larger than regular trees however they contain the same amount of fuel and also representing a similar impedance to the flow of hot gases created by the wind and the fire.

In this work a part of a real wildland fire has been modelled in order to investigate if WFDS can provide qualitative agreement with incident reports and in the future evaluate the likelihood of possible other severe outcomes with small variations in the data. The recorded weather data only provides with wind direction and speed at the start of the fire however the wildfire continued for several hours. Furthermore, the model is simplified to only accommodate for one type (pine) and one particular shape of tree.

DESCRIPTION OF THE EVENT

The two bodies of water are connected by a narrow channel. The two lakes and the surrounding forest are a popular destination for locals and tourists for water sports such as fishing, canoeing and kayaking, as well as hiking, running, mountain biking, picnicking and sunbathing. The lakes are a part of the potable water supply for the city of Gothenburg and act as a reservoir for the city. Water is pumped to the lakes from the Göta älv, the river which flows into the estuary south of which Gothenburg is built. South of the two lakes is the Riksväg 27 highway, which runs from the city out to the airport and then east towards the city of Borås. The Riksväg 27 is an important commuter and transportation highway for the region. South of the highway from Delsjön is a water treatment plant which treats the water in Stora Delsjön prior to distribution around the city. There is one other water treatment plant in the north of the city closer to the Göta älv. There are dirt roads throughout the forest. These are mainly used for running, hiking and for mountain biking.

Figure 1. The Skatås forest to the east of the city of Gothenburg and around the lakes Stora and Lilla Delsjön.

There is a telecommunications mast near to the location of the fires ignition. This includes a small office amongst other buildings as well as a mast which is 300 m high. The mast is anchored to the ground with steel cables, which also have hydraulic dampers. Although the area around the telecommunications mast is secure the vegetation does continue throughout the compound (see Figure 3). There are also buried power lines within the forest. Although these were not affected by the fire there is a risk that they could have been damaged during either the fire or as a result of fire fighting operations.

Figure 2. The Skatås forest to the east of Gothenburg around the lakes Stora and Lilla Delsjön.

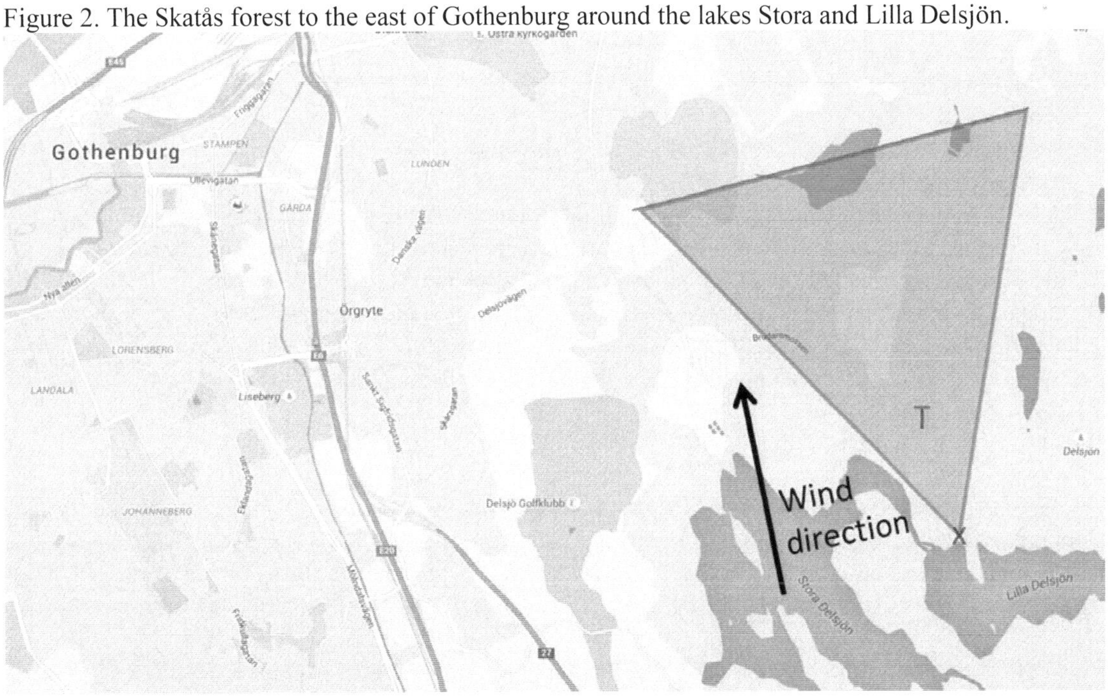

The wind speed on the day of the incident was 15 m/s and the wind direction was SSE. The weather on the day of the incident was sunny and 15 °C. Before the event there had been a generally dry climate for a few weeks contributing to the low moisture content of the fuel and rapid fire spread. The total area directly affected by the fire was 2.88 km^2. The wind direction and the approximate area directly affected by the fire are shown in Figure 2. The first call to the emergency services was made at 11:24 local time on the 29th of April in 2008. The point of ignition was at the transition between the two lakes of Stora Delsjön and Lilla Delsjön. The point of ignition is about 2.8 km from the city of Gothenburg.

Figure 3. The telecommunications mast is located in a secure compound. However, vegetation does continue through the compound and the cables for the mast extend towards the perimeter.

REAL AND POSSIBLE OUTCOMES OF THE WILDFIRE

The real consequences of the incident were limited. 2.88 km^2 of the forested area was affected by the fire, and one fire fighter was injured. There was no interruption to service in the hospital or the telecommunications mast, however workers at the telecommunications mast were evacuated and the hospital did close external ventilation and make plans for evacuation.

Possible consequences include:

- injuries to members of the public
- contamination of the drinking water supply
- ecological damage
- interruption to usability of the Riksväg 27 highway
- damage to the Riksväg 27 highway
- damage to the telecommunications mast and interruption to or reduced telecommunications service
- evacuation of the hospital as a result of smoke spread over the urban area, interruption to provision of medical care as a result of reduced number of hospital beds and surgical theatres
- damage to power lines and interruption to power supply to some neighborhoods
- social consequences including reduced access, delays in collecting children from day care or returning home from school or work
- evacuation of residential areas
- spread of the fire to the wildland urban interface and subsequent damage to property

All of these possible consequences were prevented by either the actions of the fire and rescue service or the conditions (wind conditions) on the day.

MODELLING OF WILDFIRES IN WFDS

In the present work the wildfire modelling has been performed using the extended WFDS (Wildland Urban Interface FDS) built on the regular FDS as discussed previously [8, 12]. In WFDS the effect of solid fuel such as wood, grass, pine needles is included by a series of experimentally determined correlations

[12,13]. This includes representing phenomena such as heat sources and sinks from the solid as well as the mass flow of pyrolysis gases from the solid. The physics for the gas phase is solved by the regular physics modelling framework in FDS. The model will hereafter be referred to as WFDS. In order to have a tractable model in WFDS while still being representative of the involved physics some key assumptions was needed. The Skatås wildfire occurred over an extended area which means that some simplifications on the size of the models are needed however also reducing the complexity of burning bio-materials is needed. The present approach is a coarse graining process, starting with a single tree burning including pyrolysis and detailed thermal modelling on a rather small grid, and then making simplifications to be able to consider a large area with randomly distributed trees on a coarse grid. In this process we will consider a meso-scale model with approximately 100 identical trees randomly distributed over approximately 1200 m^2 while keeping a simplified combustion model and using a 1m x 1m x 1m grid. Furthermore, in order to model a large area of the size of the scenario containing thousands of trees, bulk obstructions with tree properties are created that are larger than regular trees however they contain the same amount of fuel and also representing a similar impedance to the flow of hot gases created by the wind and the fire. In WFDS, two approaches for modelling vegetation have been developed which are called the fuel element and the boundary fuel models. The fuel element method can be used to represent surface or raised vegetation. The boundary fuel method is used to represent surface fuels only and was designed to operate at coarser grid resolutions than the fuel element model. The current version of WFDS uses fuel elements to represent vegetation. The boundary fuel element was used in earlier versions of WFDS and is being added to the current version.

One particular difficulty in wildfire CFD modelling is treating the interaction between vegetation and the turbulent flow of the gas phase components. In WFDS, the effect of the vegetation on the solution of the momentum equation is represented in the form of a body force (Fd), called the drag force [14, 15], which approximates the drag created by vegetation. The second particularity with WFDS is the use of the fuel model. In the fuel model the following parameters have to be set:

VEG BULK DENSITY: Bulk density of vegetation in kg/m^3. Default is 0.3 kg/m^3. This can be determined from field measurements of the fuel loading (kg/m^2) divided by the height of the vegetation (m). It is the mass of dry vegetation divided by the bulk volume containing that vegetation.
VEG BURNING RATE MAX: Maximum value allowed for the rate that fuel gases can be created per unit volume in a grid cell, kg/(m^{3}s). Default is 0.4 kg/(m^{3}s), based on Douglas fir tree burning experiments. For some problems this bound is needed to avoid too rapid of a burning rate. Whenever possible it should be based on experimental measurements.
VEG CHAR FRACTION: Fraction of virgin dry virgin vegetation that becomes char. Default value is 0.25.
VEG DEHYDRATION RATE MAX: Maximum value allowed for the loss of moisture during the thermal degradation of the vegetation in kg/(m^{3}s). Default is 0.5 kg/(m^{3}s).
VEG DENSITY: Density of vegetative fuel in kg/m^3. Default is 540 kg/m^3.
VEG DRAG COEFFICIENT: Non-dimensional multiplicative factor used in the drag model. Default value is 1.
VEG INITIAL TEMPERATURE: Initial temperature of vegetation in °C. Default value is 20 °C or TMPA which is set in the MISC namelist (see FDS user guide about TMPA).
VEG MOISTURE: Fraction of moisture on a dry mass basis (mass of moisture in vegetation / dry mass of vegetation). Default value is 10.
VEG REMOVE CHARRED: Either .TRUE. or .FALSE., default is .TRUE.. Specifies if, once the thermal degradation has reduced the vegetation to pure char, the fuel element should be removed (.TRUE.) or kept. If the vegetation is kept then it participates in the computation as a source of drag and head sink (or source) through radiative and convective heat transfer. Currently the thermal degradation model does not include char oxidation (smoldering combustion).
VEG SV: Surface-to-volume ration of the vegetation element in 1/m. Default value is 4000 1/m.

In general detailed measurements have be performed for the particular trees in the region of interest however there are a few studies already present such as the NIST Douglas fir tree experiments [14] that will be used in the current model.

SINGLE BURNING TREE MODEL

The single burning tree model uses the fuel element model in WFDS where an ignitor fire starts a fire on the ground a few meters from the tree. The HRRPUA of the fire is 260 kW/m^2 on 0.2 m × 2.0 m area. This case is built on the example files available for the WFDS [14, 16] however it is customised to northern European condition in terms of moisture content etc. The vegetation data of the tree crown is based on results in Ref. [14]. The WFDS input variables for our model are summarized here for ground element such as grass:

- VEG_SV=4000,
- VEG_MOISTURE=0.17,
- VEG_CHAR_FRACTION=0.20,
- VEG_DRAG_COEFFICIENT=0.375,
- VEG_DENSITY=514,
- VEG_BULK_DENSITY=0.626,
- VEG_BURNING_RATE_MAX=0.45,
- VEG_DEHYDRATION_RATE_MAX=0.45.

However, the tree will be considered pine trees with properties according to Ref. [14]. In Figure 4 a snapshot of the model are shown. The tree with an ignitor fire to the left is displayed. Utilizing this model while varying the grid size from 10 cm × 10 cm × 10 cm to 5 cm × 5 cm × 5 cm yields a slight difference. However, as is displayed in Figure 5 the moisture content of the ground material (grass) is rather important for the fire spread rate. This effect will be important in modelling a larger scenario where the fire will consume the ground material as it progresses yielding a fire spread time constant that differs with moisture content. Note that ground vegetation with large surface to volume fraction as used for Australian grass (VEG_SV) may burn too fast to allow for ignition of trees, c.f. the green line in Figure 5.

Figure 4 The geometry of the tree with an ignitor fire on the left with grass material on the ground.

Figure 5 The heat release rate as a function of time for varying degree of moisture content. Where not stated, VEG_SV = 4000.

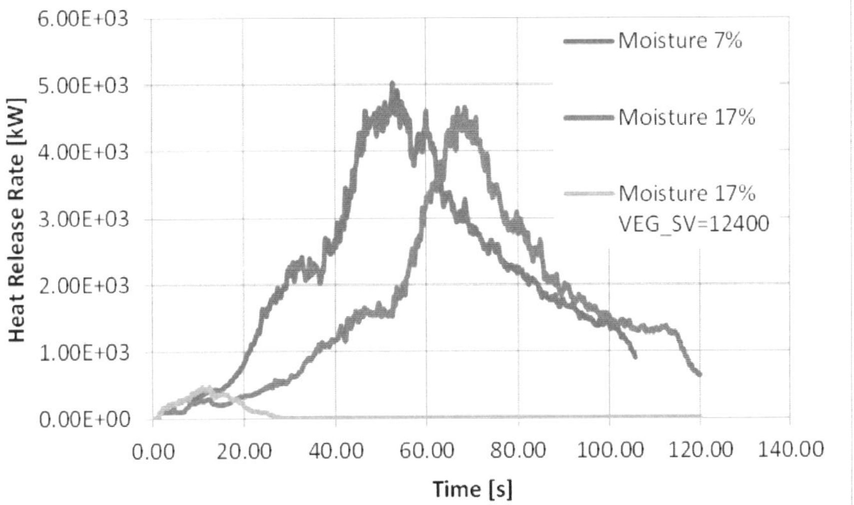

A FIELD OF SEVERAL BURNING TREES

The field is populated with 93 randomly placed identical trees in a domain that is 40 m × 32 m × 20 m. All trees are 6 m high and the base of the crown is 3 m wide with grass covered ground. The physical properties are similar to those used in the single tree model presented in Figure 7. A wind with wind velocity 2 m/s in the positive x direction is applied on the surface at minimum x.

Every single tree is modelled using the same properties as follows:

- VEG_INITIAL_TEMPERATURE=20.,
- VEG_SV=4000.,
- VEG_MOISTURE=1.0,
- VEG_CHAR_FRACTION=0.25,
- VEG_DRAG_COEFFICIENT=0.375,
- VEG_DENSITY=520.,
- VEG_BULK_DENSITY=2.2,
- VEG_BURNING_RATE_MAX=0.4,
- VEG_DEHYDRATION_RATE_MAX=0.4,
- VEG_REMOVE_CHARRED=.TRUE..

Where the stem is modelled separately as

- VEG_SV=3.,
- VEG_MOISTURE=1.0,
- VEG_DRAG_COEFFICIENT=0.375,
- VEG_DENSITY=520.,
- VEG_BULK_DENSITY=520.

In Figure 7 a snapshot of the temperature distribution in the small part of the forest after 36 s is displayed. Here it is observed that an increased moisture level prolongs the fire spread using a 0.5 m × 0.5 m × 0.5 m grid. However, larger grids may have a much smaller heat release rate compared to the current simulation since the average temperature of the larger cells will be smaller meaning that the trees will not catch fire and only the ground vegetation will be completely burnt out. Decreasing the grid only has a limited effect on the results. This may be very important for the large wildfire scenario since only a coarse grid will be practical to use. The heat release rate as a function of time for two different values of the moisture content is presented in Figure 8.

Figure 7 The geometry of the field with randomly distributed trees.

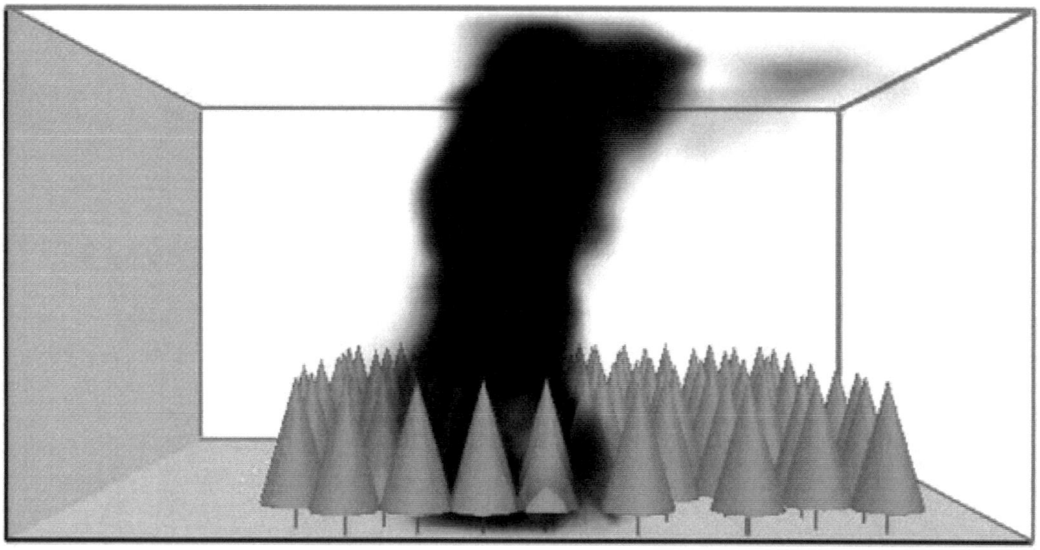

Figure 8 The heat release rate as a function of time for two different values of the moisture content.

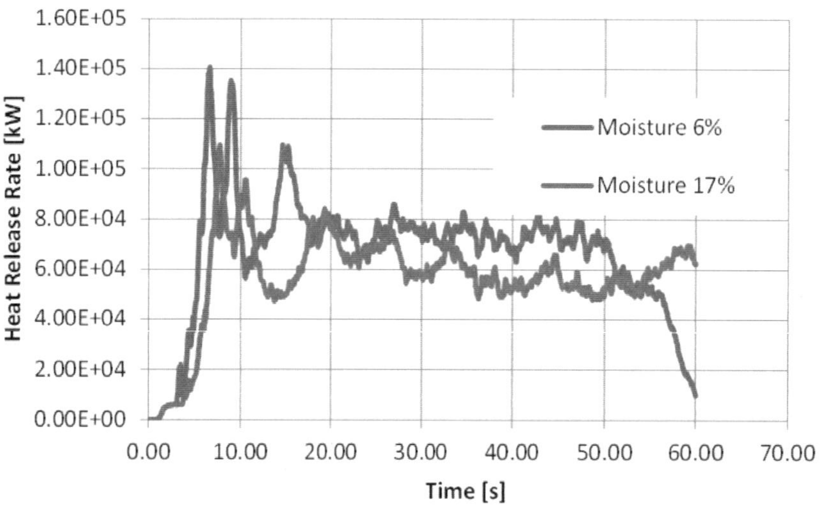

THE WILDFIRE SCENARIO

The wildfire scenario is modelled using a similar approach as in the previous model however the natural amount of trees is impractical to implement as well as requiring very small grid sizes in comparison to the size of the model. In the scenario a slope upwards to the telecom tower is presently modelled through different levels in the models where ground vegetation is present on all surfaces even the vertical surfaces. Using a 2 m × 2 m × 2 m grid results in no fire spread at all since the grid is too coarse whereas halving the grid seems the to support fire spread. One other important factor is the tree density; the tree density will affect the flows and wind close to the ground and may obstruct the fire spread. To illustrate this we have set-up a model including 5000 trees on a 100 m x 200 m piece of land covered by grass c.f. Figure 9, and 10 and 11. Although a fire starts using an ignitor fire for 27 s and spread with the wind for a while it later stops due to flow interference caused by the density of the trees, see the time evolution of the heat release rate in Figure 10. A comparison is done for lowered moisture content as well increased VEG_SV as used for Australian grass. Note that there is plenty of combustible material left to burn in the high moisture content case whereas in the low moisture content case most of the trees take part in the fire.

Figure 9 A medium sized model 100 m × 200 m × 30 m with 5000 randomly distributed trees.

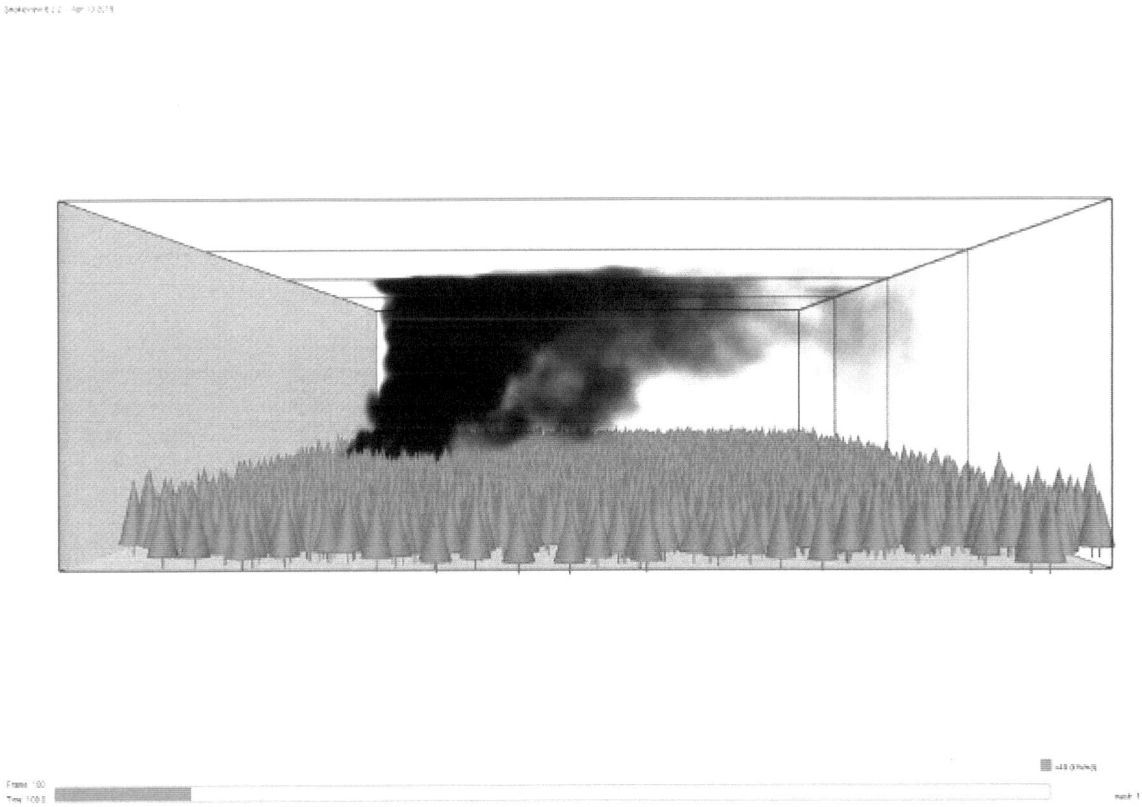

Figure 10 A comparison of the heat release rate as a function of time.

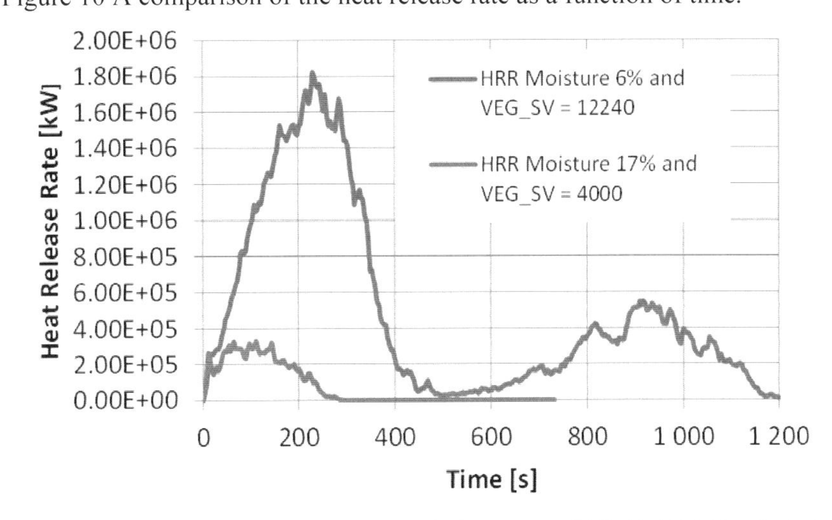

The next step is to model the proximity to the telecom mast in order to investigate the possible adverse effects of the forest fire on the mast itself but also on the supports cables with hydraulic dampers to the tower that may be affected earlier in the development. The telecom mast is visible in Figure 11 in the model as a thin tower that measures 2 m × 2 m × 30 m.

Figure 11. The model of the neighbourhood of the telecom mast (shown in red).

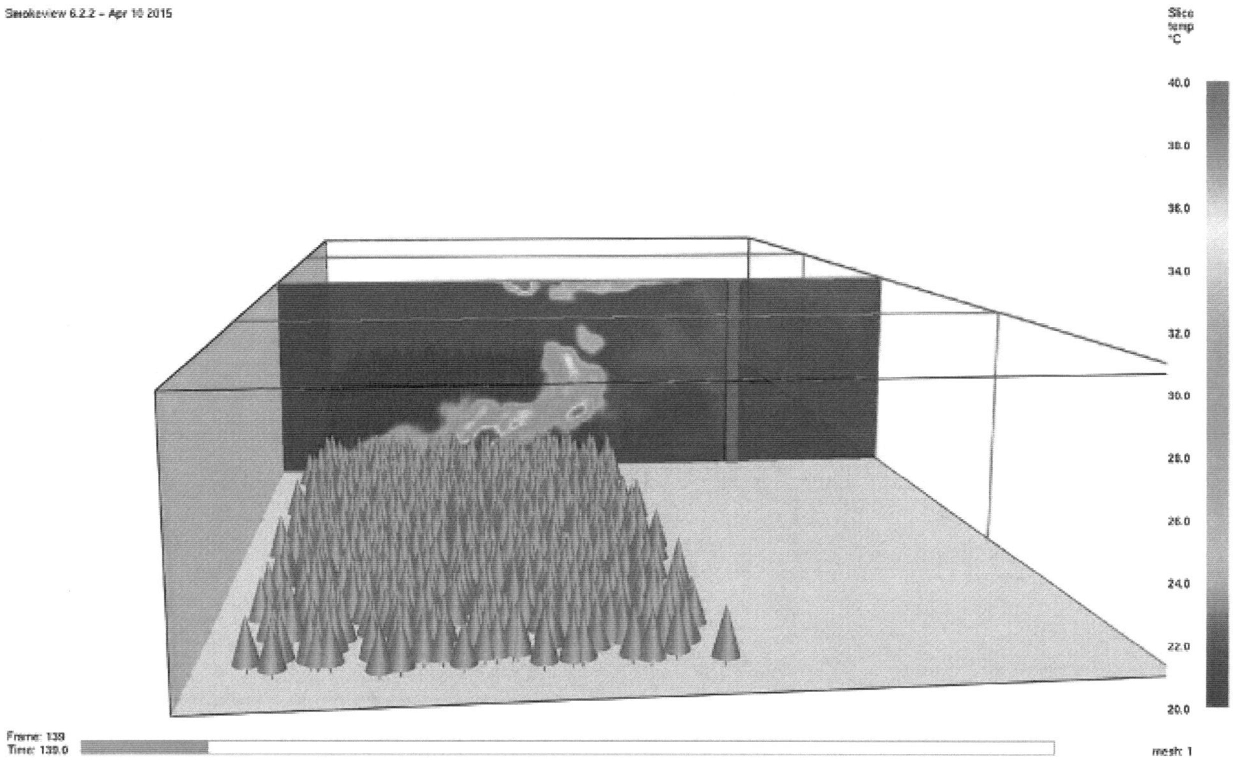

The model is similar in size as the previous medium scale forest fire however the area around the tower is cleared from trees. It should be noted that the tree density is around the half of the density of the previous model. Due to stability problems with the model the wind speed is only 2 m/s however in the actual scenario it was around 15 m/s; this may cause hot gases to pass close by to the tower. In this model the cleared safety distance from the forest was large enough to keep down the temperatures in the vicinity of the tower.

In addition, one large model was created that spanned approximately 1000 m x 200 m x 120 m where the total height of the model is decreased along the slope up to the telecom mast. The telecom mast is also visible in Figure 11 of the previous model as a thin tower that measures 2 m x 2 m x 30 m. One further simplification is to utilize larger but fewer trees however this may have an impact on the fire spread and in future studies this simplification has to be more closely evaluated. Modelling forest fires on these rather large areas is very demanding computationally and generates an enormous amount of data. For the fire spread model to at least function a comparably small grid size is needed otherwise the spread stops or is not well represented which means that these models will become huge. Unfortunately, the current case stopped due to limited disc memory and thus no results are shown from this model.

DISCUSSION

In this work the WFDS model is evaluated as a tool for fire spread estimation in rural/woodland areas. It is found that comparing to Ref. [14] that the fire spread from the ground vegetation to a crown fire the overall trends are reproduced. It has been noted that the performance of WFDS is better for dry fuels whereas the performance elsewhere is mixed. It is found that reasonable fire spread in small forests can be obtained although the results are quite dependent on grid resolution as well as moisture content. In most realistic scenarios the computational volume is rather large yielding massive amounts of data. In this work a part of a real wildland fire has been modelled in other to investigate if WFDS can provide qualitative agreement with incident reports and in the future evaluate the likelihood of possible other severe outcomes with small variations in the indata. The recorded weather data only provides with wind direction and speed at the start of the fire however the wildfire continued for several hours. There are several likely sources of errors such as the particular drag and heat transfer models used in WFDS, the simulated wind field, thermal degradation of the materials, spatial and size distributions of the vegetation. Furthermore, the model is simplified to only accommodate for one type (pine) shaped and shape of trees.

In using WFDS a rather small grid size is needed to appropriately model the fire spread this will be a severely limiting factor in creating large models. In comparison with the actual incident, and addressing the question of whether or not WFDS is a suitable tool for modelling such a large incident, the following comments are made:

- Given the wind speed measured on the day, the WFDS software would require a significantly smaller cell size to have modelled the incident as it occurred. The practicalities of this are rather difficult because of the size of the model. The cell size of the models presented was of the order of 0.5 m to 1.0 m cubed. For the larger models this results in a total number of cells of between 24 and 100 million. These were run on 9 cores. The authors did try to run smaller celled models on as many as 43 cores but ran into memory issues as a result of the cluster which was being used. As a result of this the wind speed in the models is generally much lower than it was on the day.
- The impact of cell size on flame spread rate as discussed above is also significant, leading to reduced burning rates.

Given these computational limitations it is not practical to model the wildfire scenario in its current state. The computational effort required for the wind speed and which would result in a satisfactory fire spread rate is simply too high.

ACKNOWLEDGEMENT

The research leading to these results has received funding from the European Union Seventh Framework Programme (FP7/2007-2013) under grant agreement n° 607665.

REFERENCES

1. Pastor, E., Zarate, L., Planas, E. and Arnaldos, J., Mathematical models and calculation systems for the study of wildland fire behaviour. Progress in Energy and Combustion Science, 29:139–153 (2003).
2. Finney, M. A., FARSITE: Fire area simulator-model development and evaluation. Res. Pap. RMRS-RP-4, Ogden, UT: U.S. Department of Agriculture, Forest Service, Rocky Mountain Research Station. 47 p., http://www.farsite.org (1998).
3. Tymstra, C., Bryce, R. W., Wotton, B. M. and Armitage, O. B., Development and structure of Prometheus: the Canadian wildland fire growth simulation Model. Inf. Rep. NOR-X-417. Nat. Resour. Can., Can. For. Serv., North. For. Cent., Edmonton, AB.. PROMETHEUS (2009)
4. Coen, J. L., Simulation of the Big Elk Fire using coupled atmosphere-fire modeling. International Journal of Wildland Fire, 14(1):49--59 (2005).
5. Mandel, J., Beezley, J.D., Coen, J. L. and Kim, M., Data Assimilation for Wildland Fires: Ensemble Kalman filters in coupled atmosphere-surface models, IEEE Control Systems Magazine 29, Issue 3, June 2009, 47-65.
6. Sun, R., Krueger, S. K., Jenkins, M. A., Zulauf, M. A. and Charney, J. J.. The importance of fire-atmosphere coupling and boundary-layer turbulence to wildfire spread. Int. J. of Wildland Fire,18(1) 50–60 (2009).
7. Linn, R., Reisner, R., Colman, J. J., and Winterkamp, J.. Studying wildfire behavior using FIRETEC.

Int. J. of Wildland Fire, 11:233—246 (2002).
8. Mell, W., Jenkins, M. A., Gould, J. and Cheney, P. A physics-based approach to modelling grassland fires. Intl. J. Wildland Fire, 16:1--22 (2007).
9. Coen, J. L., Clark, T. L., and Latham, D.. Coupled atmosphere-fire model simulations in various fuel types in complex terrain. In 4th. Symp. Fire and Forest Meteor. Amer. Meteor. Soc., Reno, Nov. 13-15, pages 39—42 (2001).
10. Clark, T. L., Coen, J. L., and Latham, D.. Description of a coupled atmosphere-fire model. International Journal of Wildland Fire, 13:49--64 (2004).
11. K. McGrattan et al, FDS User Guide, NIST Special Publication 1019, Sixth Edition (2016).
12. Mell, W., User Guide to WFDS
13. Mueller, E. W., LES Modelling of Flow through Vegetation with Applications to Wildland Fires, MSc thesis, Worcester Polytechnic Institute
14. Mell, W., Maranghides, A., McDermott and R., Manzello, S. L., Numerical simulation and experiments of burning douglas fir trees. Combustion and Flame. 156, pp. 2023-2041 (2009).
15. Grishin, A. M., Mathematical modeling of forest fires and new methods of fighting them, Publishing House of the Tomsk State University, Tomsk, Russia, 1997.
16. http://www.fs.fed.us/pnw/fera/wfds/index.

FIRE HAZARD

RRECOMMENDATIONS FOR DOCUMENTATION OF REACTION-TO-FIRE PROPERTIES OF MATERIALS IN THE OIL AND GAS INDUSTRY

Karolina Storesund* & Anne Steen-Hansen
SP Fire Research AS, Trondheim, Norway

ABSTRACT

This paper presents a set of recommended reaction-to-fire test methods and related criteria for different construction products and furnishing products for use on offshore petroleum industry facilities. The system was first developed and published in 2008, as a response to performance based fire safety regulations. It has since then been extensively used by petroleum companies operating on the Norwegian shelf. The recommendations have been revised and updated recently to take revisions and developments of regulations, standards and test methods into account.

These recommendations can be used as a part of the risk management system where the fire safety in high risk application areas shall be controlled, and are not limited to application of materials on facilities in the offshore petroleum industry only.

INTRODUCTION

Fire is a major threat to on- and offshore oil- and gas facilities and every year, a number of fire related incidents occur. A facility in this context is an installation, plant and other equipment for petroleum activities (excluding supply and support vessels or ships that transport petroleum in bulk). On the Norwegian shelf the requirements for materials used follow from the Facilities regulations[1]. These are performance based regulations and come with guidelines including recommended pre-accepted solutions. The Facilities Regulations describes selection of materials in or on facilities, and the guidelines specifies in more detail which test methods that should be used to document reaction to fire properties[2]. This list of test methods has not been updated for many years, it contains several outdated methods and several references are not stating failure criteria where this is necessary. Therefore SP Fire Research (previously SINTEF NBL) developed recommendations on how to interpret the regulations, which are now in frequent use within the industry[3,4]. This paper presents our updated recommendations for testing and documentation of materials used on oil- and gas facilities on the Norwegian shelf. The recommendations may also be useful for other industrial applications worldwide, both on- and offshore.

The offshore petroleum industry is an area where the consequences of a fire may be very large. Fires on offshore facilities may start in several types of equipment and areas. An analysis of 985 fire incidents registered in the Norwegian Petroleum Safety Authority's database between 1997 and 2014 showed that approximately 67 % of the incidents are registered as fires or explosions that did not involve hydrocarbons[5,6]. About one third of the fires took place in so called *ancillary system*, which are areas in support of the production areas, and cover a wide range of characteristics. Most of the platform systems that are not categorised as belonging to the main process are ancillary systems, which could explain the large share of incidents. Six of the 985 incidents were categorized at the highest severity level, having a large potential for serious accident or fatalities – three of these occurred on offshore facilities. 66 of the incidents occurred in living quarters. Thus, severe incidents related to fire do not occur very often. However, some incidents reported could have developed into disastrous events, which indicate the need for continued work to increase safety and a focus on barriers preventing possible consequences of an escalating incident.

Therefore the safety regime on offshore installations has to be very strict, and barrier management is one of the strategies to obtain a high safety level. According to the Norwegian Petroleum Safety Authority, *Barrier management includes the processes, systems, solutions and measures which must be in place to ensure the necessary risk reduction through the implementation and follow-up of barriers*[7]. A proper selection of materials in components, construction and interior is an important barrier preventing fires from starting, and preventing small fires from developing into larger fires that will be difficult to control. The regulations for health, safety and environment (HSE) in the Norwegian petroleum sector contain mainly risk- and performance based requirements[8]. Risk management is defined in the international standard ISO 31000 as *coordinated activities to direct and control an organization with regard to risk* [9]. Risk based safety management should include all safety related activities required to manage an organisation, and includes both human resources and material goods. The principles of risk based safety management are usually applied at a general level in an organization, but may very well include detailed information like reaction-to-fire behaviour of construction products in areas with a high probability of fire. According to ISO 31000, risk management is based on the best available information. The work presented here is an effort to give updated and relevant guidelines on how fire risk can be managed with regard to selection of materials.

REGULATIONS, STANDARDS AND GUIDELINES FOR THE NORWEGIAN SHELF OIL- AND GAS FACILITIES

The regulations for the petroleum activity under the jurisdiction of the Norwegian Petroleum Directorate and the Petroleum Safety Authority consist of a resource management part and a Health Safety and Environment (HSE) part.

The Norwegian Environment Agency, the Norwegian Directorate of Health and the Norwegian Petroleum Directorate and the Petroleum Safety Authority cooperate on regulating HSE for the petroleum activities on the Norwegian Shelf. So the different HSE regulations are established with basis in a number of legislative acts concerning e.g. petroleum activities, health and working environment, pollution control, product control, etc. There are a number of regulations with relevance for HSE, of which the facilities regulations with guidelines are central for reaction-to-fire properties of materials in offshore applications. Since the regulations are performance based there are also several additional guiding documents with the aim of stating specific requirements, e.g. the NORSOK standards and other specifications.

The facilities regulations with guidance

The facilities regulations describe principle requirements including reaction-to-fire properties for materials used on offshore installations on the Norwegian shelf[1]. Material requirements are only mentioned in the Facilities regulations §12, where it is stated that:

"Materials to be used in or on facilities shall be selected considering

...

d) fire-resistance properties".

The guidelines to the regulations elaborates on this by stating that when using combustible materials, the spread of flames, smoke production, heat release and smoke toxicity shall be limited. They also list a selection of fire test methods to be used to determine the technical fire qualities of materials, including ISO, IMO and NT Fire methods[2]. These particular methods are recommended to use but not required.

However, the guidelines do not state the failure criteria when using the test methods. Only a part of the methods in the guidance list have defined criteria for use for assessment of the materials. Many of the methods are connected with parallel classification systems. Some have different criteria for different areas of application. In addition, the list is due for an update with regards to test method edition. Some of the methods, such as ISO 5660-1[10], are normally used for evaluating the fire performance and not for classification or rating. Where there is a system with criteria for rating, these are commonly not related to offshore applications, but e.g. ships and buildings. This means that for each project an evaluation will be required to assess whether a material should be considered suitable for offshore use or not.

Regulations of 31 January 1984 No. 227 on precautionary measures against fire and explosion on mobile offshore units[11]

These regulations apply to mobile platforms and drilling ships registered in a Norwegian ship registry. They are placed under the jurisdiction of the Ministry of Trade, Industry and Fisheries and are maintained by the Norwegian Maritime Authority. The regulations contain a number of requirements related to fire safety but for material application, chapter 4 *"Fire safety measures"* is the most relevant. The requirements have many references to the IMO Fire Test Procedure Code (FTPC)[12] but they also include a few specific requirements:

- Furniture with drawers or cabinets or cupboards shall be made of certified, non-combustible material, except that the surface of such furnishings may be a combustible veneer not more than 2 mm thick.
- Free-standing furniture such as chairs, sofas, beds and tables shall be made with frames of non-combustible material.

The NORSOK standards

The NORSOK standards were the result of an initiative in 1993 established by the petroleum industry and Norwegian authorities. The standards were developed in order to ensure satisfactory safety, value creation and cost efficiency for the operations in the oil and gas industry on the Norwegian shelf, closing the gap between international standards and the Norwegian requirements[13]. The NORSOK standards are owned by the Norwegian Oil and Gas Association, the Federation of Norwegian Industries, and the Norwegian Shipowners' association, and the standards are developed by experts from the Norwegian petroleum industry. There are both general and specific requirements for materials stated in different NORSOK standards, mainly concerning non-combustibility, smoke production and toxicity and spread-of-flames. The requirements are considering both living quarters and processing areas.

The most relevant NORSOK standards with regards to material fire safety are:

- NORSOK S-001 Technical safety[14]
- NORSOK C-002 Architectural components and equipment[15]
- NORSOK R-004 Piping and equipment insulation[16]

These have various references to e.g. the former IMO Fire Test Procedure Code IMO Res. MSC 61(67)[17], which was replaced in 2010, the guidance to the facilities regulation, and different references to the European testing and classification system for construction products.

Other guiding documents

The operators, e.g. the Norwegian petroleum company Statoil, have their on technical specifications. These are generally based on their risk analyses and their interpretation of the Norwegian regulative. Survey organisations, e.g. DNV GL, have also defined their set of requirements for material certification. NORSOK S-

001 is also referring to a report with recommendations that has been published by SINTEF NBL (now SP Fire Research AS)[4].

An overview of regulations, standards and guidelines is shown in Figure 1 below.

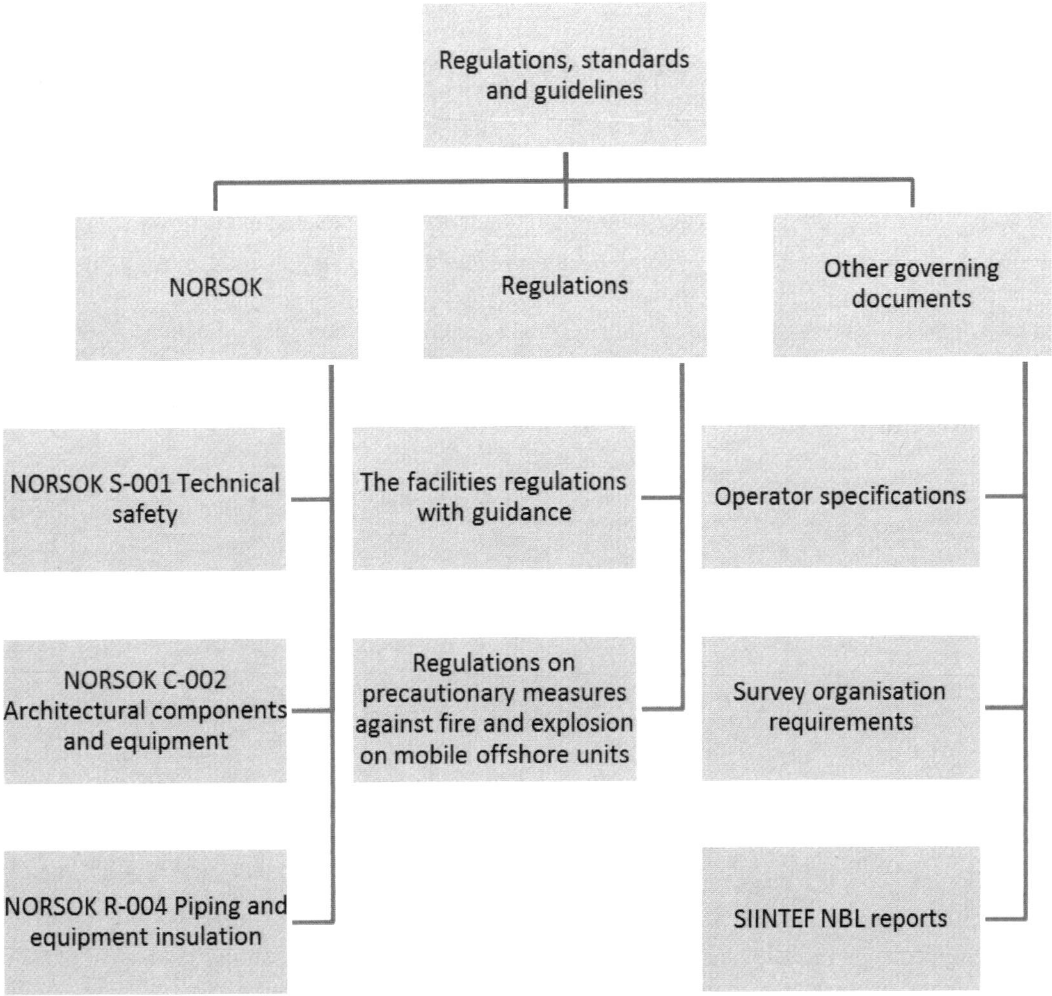

Figure 1 Regulations, standards and guidelines for materials in petroleum industry offshore applications.

BASIC REQUIREMENTS TO COMBUSTIBLE MATERIAL IN OFFSHORE APPLICATIONS

The use of combustible materials offshore can have several positive aspects for different applications. These advantages may include lower procurement costs, lower maintenance requirements, lower weight and easier installation. The use of combustible materials can also have safety related advantages[18]. When combustible materials are applied offshore, it is important to have control over which products are being used, what applications, and the quantities of products that can be permitted.

It is natural to set different requirements for combustible materials depending on the application on the offshore installation. Fire risk will be considerably different in living quarters as compared to areas near or in the process areas. Possible fire scenarios will be different both in terms of ignition sources, fire development, the spread of fire and smoke and possible consequences. We have therefore added the following philosophy as the basis for assessment of the requirements that should apply to combustible materials offshore:

Requirements for living quarters

- Combustible materials used in living quarters should be of such a fire technical quality that a fire does not spread from the room of origin.
- A fire in the living quarters shall not normally provide flashover, and material properties should be of such a quality so that the emergency team will have a high probability of controlling the fire.

Requirements for process areas

- A fire shall not be able to occur in combustible materials on offshore installations without the contribution of fire in petroleum products.
- Combustible materials should not substantially worsen the conditions during a fire in petroleum products.

Critical reaction-to-fire properties

During a fire, people will be affected by the heat from the fire, toxic gases from the smoke and reduced visibility due to dense smoke.

The following are some material properties that are critical for fire development and the spreading of fire and smoke:

- Heat release
- Smoke production
- Production of toxic gases
- Spread of flame
- Production of burning droplets and pieces

Individuals will be affected by heat from the fire, toxic gases and reduced visibility. There are a number of test methods that measure one or several of these factors.

Heat exposure may be caused either by direct contact to hot fire gases or radiation from flames or hot gases. Breathing air of high temperatures can cause painful damages to the upper respiratory tract. Heat flux of approximately 20 kW/m^2 would result in pain and burns within few seconds.

Some toxic gases lead to reduced consciousness and eventually to asphyxiation (especially CO, CO_2 and HCN), while irritating gases (mainly HCl, NO_X, NH_3, SO_2, HF, HBr and acrolein) can irritate eyes and upper respiratory tracts making it difficult for victims to escape.

The smoke will also reduce the visibility which will hinder efficient escape.

Loose fittings, especially upholstered furniture and mattresses, will have large impact on the fire development during fire in the living quarters. The selection of fire safe materials for furniture is an example of passive fire safety measures that will increase the fire safety in rooms. Fire alarms and automatic extinguishing systems would be the most relevant active fire safety measures in these areas.

DOCUMENTATION OF FIRE SAFETY

Documentation of fire safe materials for offshore applications can be prepared in several ways. Results from fire testing according to a standardized method are a commonly used starting point. The documentation can be in the form of a report in which methods and results are described, or it can be designed as an assessment on the basis of test results. Such assessments can be made in relation to a recognized classification system or it may be assessed with respect to the product's usefulness in a given application.

Non-combustible materials and materials with limited combustibility

Both the guidelines to the Facilities Regulations and NORSOK S-001N indicate that materials primarily should be non-combustible. The test method for documentation of non-combustibility, ISO 1182[19], is used in many countries, and the same method is used for buildings, ships and offshore.

Documentation of limited combustibility primarily involves the determination of a material's calorific value which can be performed with ISO 1716[20]. Neither ISO 1182 nor ISO 1716 contain any criteria for the assessment of the test results. IMO 2010 FTPC Part 1 however describes testing according to ISO 1182 and states criteria for the approval as a non-combustible material according to IMO[12]. The European classification standard EN 13501-1[21] uses ISO 1182 and ISO 1716 as basis for the classification of Euroclass A1 (non-combustible) and A2 (limited combustibility).

ISO 5660-1[10], the cone calorimeter test, can also be used to document the combustibility of different products.

Applications

Materials for insulation and sound absorption for offshore applications shall be non-combustible.

Test methods for combustible materials

If, for technical and practical reasons, combustible materials are selected, they should show properties of limited flame spread, low heat release, low levels of smoke production, and the smoke shall be of low toxicity. Non-combustible materials will fulfil these requirements, but other materials must be tested with the relevant test methods.

IMO 2010 FTPC Part 5 Test for surface flammability[12], IMO 2010 FTPC Part 2 Smoke and toxicity test[12] and test and classification according to EN 13501-1 are suitable for documentation for most surface applications.

Table 1 below presents our recommendations for selection of test methods and minimum criteria for documentation of combustible materials for the stated areas or applications, and is based on the different requirements in the regulations mentioned above. Where several test methods are listed for a specific application, these should be understood as alternatives to one another. Non-combustible materials will be acceptable for all applications and should be documented by IMO 2010 FTP Code Part 1, ISO 1182, Euroclass A1 or A2-s1,d0 (EN 13501-1) or ISO 5660-1 as described in Table 2.

Table 1 Recommendations for selection of test method and criteria

APPLICATION	RECOMMENDED TEST METHODS AND CRITERIA WHEN USING COMBUSTIBLE MATERIALS
Thermal insulation and sound absorption materials	– Non-combustible only – ISO 5660-1. Test conditions and criteria are listed in Table 2.
Passive fire protection of structures and equipment	– IMO 2010 FTPC Part 5 and Part 2. Moderate fire risk: criteria for floor coverings and primary deck coverings. Significant fire risk: consider criteria for bulkhead, wall and ceiling linings. – Euroclass B-s1,d0 (EN 13501-1) – IMO 2010 FTPC Part 10[12] (ISO 9705[22]) – ISO 5660-1. Test conditions and criteria are listed in Table 2. In case the materials are used outdoors, in areas where the risk for exposure to smoke is small, lower requirements on smoke production may be considered.
Surface materials on walls and ceilings	– IMO 2010 FTPC Part 5 and Part 2. Moderate fire risk: criteria for floor coverings and primary deck coverings. Significant fire risk: consider criteria for bulkhead, wall and ceiling linings. – Euroclass B-s1,d0 (EN 13501-1) – IMO 2010 FTPC Part 10 (ISO 9705) – ISO 5660-1. Test conditions and criteria are listed in Table 2. In case the materials are used outdoors, in areas where the risk for exposure to smoke is small, lower requirements on smoke production may be considered.
Tarpaulins	– IMO 2010 FTPC Part 5 and Part 2, criteria must be determined in accordance with the application area and amount of material used – Euroclass B-s1,d0 (EN 13501-1) – IMO 2010 FTPC Part 10 (ISO 9705)
Noise reduction curtains	– IMO 2010 FTPC Part 5 and Part 2, criteria for bulkhead, wall and ceiling linings – Euroclass B-s1,d0 (EN 13501-1) – IMO 2010 FTPC Part 10 (ISO 9705) – ISO 5660-1. Test conditions and criteria are listed in Table 2.
Flooring and primary deck coverings	– IMO 2010 FTPC Part 5 and Part 2, criteria for flooring or primary deck coverings – ISO 5660-1. Test conditions and criteria are listed in Table 2. Euroclass D_{fl}-s1 (EN 13501-1) can be acceptable in living quarters, subject to evaluation
Gratings	– IMO 2010 FTPC Part 5 and Part 2, criteria for primary deck coverings – ISO 5660-1. Test conditions and criteria are listed in Table 2.

Table 1 (continued) Recommendations for selection of test method and criteria

Insulation for pipes and ducts	– NT FIRE 036[23], class PI – IMO 2010 FTPC Part 5 and Part 2, criteria for primary deck coverings – Euroclass B_L-s1,d0 (EN 13501-1) – ISO 5660-1. Test conditions and criteria are listed in Table 2.
Electric cables	Selection of class must be made based on area of application as well as other available fire safety measures. In high risk applications Euroclass A_{ca} (EN 13501-6[24]) is recommended. Dependent on the amount of exposed area of cables, Euroclasses C_{ca}-s1,d1 or D_{ca}-s2,d2 would normally be acceptable in living quarters[25].
Materials for the construction of lifeboats	– IMO MSC/Circ.1006[26]
Draperies, curtains and other supported textiles in living quarters	– NT FIRE 043[27], class I. Acceptable for all areas. – IMO 2010 FTPC Part 7[12]
Upholstered furniture in living quarters	– BS 5852, crib 7[28]. Acceptable for all areas. Testing with smouldering and small flaming ignition sources (e.g. IMO 2010 FTPC Part 8[12], or EN 1021-1[29] and EN 1021-2[30]) can be acceptable, subject to evaluation with respect to hazard area and available additional fire protective measures)
Mattresses in living quarters	– BS 6807, crib 7[31] (or equivalent method). Acceptable for all areas. – SS 876 00 10[32], or equivalent with a 30 kW burner ignition source. Acceptable for all areas. Testing with smouldering and small flaming ignition sources (e.g. IMO 2010 FTPC Part 9[12], or EN 597-1[33] and EN 597-2[34]) can be acceptable, subject to evaluation with respect to hazard area and available additional fire protective measures)
Materials for furniture and fixtures in living quarters	– IMO 2010 FTPC Part 5 and Part 2, criteria for bulkhead, wall and ceiling linings – Euroclass B-s1,d0 (EN 13501-1) – IMO 2010 FTPC Part 10 (ISO 9705 or ISO 5660-1 as relevant) – ISO 5660-1. Test conditions and criteria are listed in Table 2. These materials can be used in both smaller components and larger surfaces. Therefore the choice of test method should be determined with respect to the application area and amount of material used.

The cone calorimeter (ISO 5660-1) is also a useful tool to be used for documentation of the combustibility of different products. Table 2 below is based upon results from a literature review comprising literature from several countries on assessment of materials for different applications[35,36,37,38,39,40,41,42,43,44].

Table 2 Recommended criteria related to testing according to ISO 5660-1

<table>
<tr><th>Product</th><th>Heat flux density level [kW/m^2]</th><th>Test duration [s]</th><th>t_{ign} [s]</th><th>$HRR_{avg,300s}$ [kW/m^2]</th><th>HRR_{max} [kW/m^2]</th><th>THR [MJ/m^2]</th><th>SPR_{avg} [s^{-1}]</th></tr>
<tr><td>Non-combustible materials and materials with limited combustibility

Thermal insulation and sound absorption materials</td><td>50</td><td>900</td><td>≥ 150</td><td>-</td><td>≤ 10</td><td>≤ 10</td><td>≤ 0.17</td></tr>
<tr><td rowspan="6">Surface materials on walls and ceilings

Passive fire protection of structures and equipment

Noise reduction curtains

Insulation for pipes and ducts

Pipes of plastic and composite materials

Materials in ventilation ducts</td><td colspan="7">Either</td></tr>
<tr><td>50</td><td>900</td><td>≥ 150</td><td>-</td><td>≤ 50</td><td>≤ 10</td><td>≤ 0.17</td></tr>
<tr><td colspan="7">Or</td></tr>
<tr><td>50</td><td>900</td><td>≥ 150</td><td>≤ 50</td><td>≤ 65</td><td>-</td><td>≤ 0.17</td></tr>
<tr><td colspan="7">Or</td></tr>
<tr><td>75</td><td>900</td><td>≥ 90</td><td>≤ 100</td><td>≤ 100</td><td>-</td><td>≤ 0.17</td></tr>
<tr><td>Flooring and primary deck coverings in process areas

Gratings</td><td>50</td><td>900</td><td>≥ 150</td><td>-</td><td>≤ 100</td><td>≤ 25</td><td>≤ 0.3</td></tr>
<tr><td>Flooring in living quarters</td><td>50</td><td>900</td><td>≥ 150</td><td>-</td><td>≤ 100</td><td>≤ 25</td><td>≤ 0.17</td></tr>
<tr><td>Materials for furniture and fixtures in living quarters</td><td>50</td><td>1200</td><td>≥ 20</td><td>-</td><td>≤ 60
NB!
Average of 30 s</td><td>≤ 25</td><td>≤ 0.57</td></tr>
<tr><td>Upholstered furniture, mattresses and bedding components</td><td colspan="7">The cone calorimeter is not best suited but may be an alternative*)</td></tr>
<tr><td>Electric cables

Tarpaulins

Draperies, curtains and other supported textiles in living quarters</td><td colspan="7">The cone calorimeter not suited</td></tr>
</table>

*) The research program *Combustion Behaviour of Upholstered Furniture (CBUF)* in the early 1990's specified a method of testing materials in upholstered furniture and mattresses in the cone calorimeter[45]. The heat flux density level was 35 kW/m^2 and the test duration was set to flameout of mass loss below 150 g/m^2 during 1 minute.

SUMMARY

SP Fire Research is in a continuous dialogue with industry and suppliers about documentation, test methods and criteria to use for installations in the North Sea, assisting the actors to fulfil the regulations and uphold a satisfactory level of fire safety. It is up to the different operators to determine what documentation they require for different products, based on the requirements in the Facilities Regulations and the NORSOK standards.

We have developed guidance and recommendations for suitable documentation requirements for materials on oil and gas facilities with regards to fire performance. The recommendations are based on an evaluation of fire risk and hazards in different areas on offshore installations combined with our knowledge on reaction-to-fire properties and a sound application of the selected test methods. These are currently in frequent use by the oil and gas operators and its suppliers. They cover different products and applications on board the facilities and consider different levels of risk. Many products can also be documented by use of the cone calorimeter, and a set of test parameters and test criteria are suggested, based on existing and relevant regulations, previous research and testing experience.

REFERENCES

[1] Petroleum Safety Authority Norway, Norwegian Environment Agency, Norwegian Directorate of Health and the Norwegian Food Safety Authority: Regulations relating to design and outfitting of facilities etc. in the petroleum activities (The Facilities Regulations). Last amended 2015.

[2] Petroleum Safety Authority Norway, Norwegian Environment Agency, Norwegian Directorate of Health and the Norwegian Food Safety Authority: Guidelines regarding the Facilities Regulations. Last updated 2015.

[3] Kristoffersen, B., Steen-Hansen, A., 2005. Branntekniske krav til materialer i boligkvarter offshore (No). Report NBL A05103. SINTEF NBL, Trondheim, Norway.

[4] Steen-Hansen, A., Kristoffersen, B., Stensaas, J.P., Hofstad, G.E., 2008. Branntekniske krav til materialer offshore (No). Report NBL A08122. SINTEF NBL, Trondheim, Norway.

[5] Storesund, K., Steen-Hansen, A., Hertzberg, T., Sesseng, C., 2016. Shelf analysis. Fire Risk Management, March 2016.

[6] Steen-Hansen, A., Storesund, K., Hertzberg, T.,. Analysis of fire related incidents in petroleum activities on the Norwegian shelf., in: Nordic Fire & Safety Days, Book of Abstracts, p. 51. Presented at the Nordic Fire & Safety Days, 16-17 June 2016, Copenhagen, Denmark.

[7] Petroleum Safety Authority Norway, 2013.Principles for barrier management in the petroleum industry. Stavanger, Norway. http://www.ptil.no/getfile.php/PDF/Barrierenotatet%202013%20engelsk%20april.pdf

[8] Petroleum Safety Authority Norway About the HSE regulations. Stavanger, Norway. http://www.psa.no/about-the-hse-regulation/category929.html

[9] ISO 31000:2009. Risk management - Principles and guidelines. International Organization for Standardization, Geneva, Switzerland.

[10] ISO 5660-1:2015 Reaction-to-fire tests - Heat release, smoke production and mass loss rate - Part 1: Heat release rate (cone calorimeter method) and smoke production rate (dynamic measurement). International Organization for Standardization, Geneva, Switzerland.

[11] Norwegian Ministry of Trade, Industry and Fisheries: Regulations of 31 January 1984 No. 227 on precautionary measures against fire and explosion on mobile offshore units. Oslo, Norway.

[12] IMO Resolution MSC.307(88), IMO 2010 Fire Test Procedures (FTP) Code, 2012. International Maritime Organization, London, UK.

[13] NORSOK Direktiv A-001N:2016 - Retningslinjer for utvikling og utforming av NORSOK standarder. Standards Norway, Oslo, Norway.

[14] NORSOK STANDARD S-001 Technical safety. Edition 4, February 2008. Standards Norway, Oslo,

Norway.
[15] NORSOK STANDARD C-002 Architectural components and equipment Edition 4, September 2015, Standards Norway, Oslo, Norway.
[16] NORSOK STANDARD R-004 Piping and equipment insulation, Rev. 3, August 2006, Standards Norway, Oslo, Norway.
[17] IMO Resolution MSC.61(67). Adoption of the international code for application of fire test procedures. Adopted on 5 December 1996. International Maritime Organization, London, UK.
[18] Steen-Hansen, A., Landrø, H., Opstad, K., Schmidt Pedersen, K., 1993. Anvendelse av brennbare materialer offshore. Brennbare materialers bidrag til brann. SINTEF report no. STF25 A93048. SINTEF NBL as, Trondheim, Norway.
[19] ISO 1182:2010-Reaction to fire tests for building products - Non-combustibility test. International Organization for Standardization, Geneva, Switzerland.
[20] ISO 1716:2010 - Reaction to fire tests for building products - Determination of the gross heat of combustion (calorific value). International Organization for Standardization, Geneva, Switzerland.
[21] EN 13501-1:2007+A1:2009 Fire classification of construction products and building elements - Part 1: Classification using data from reaction to fire tests, CEN, European Committee for Standardization, Brussels, Belgium.
[22] ISO 9705-1:2016. Reaction to fire tests - Room corner test for wall and ceiling lining products - Part 1: Test method for a small room configuration. International Organization for Standardization, Geneva, Switzerland.
[23] NT Fire 032 Edition 2 Upholstered furniture: Burning behaviour - Full scale test. 1988. NORDTEST, Espoo, Finland.
[24] EN 13501-6:2014 Fire classification of construction products and building elements - Part 6: Classification using data from reaction to fire tests on electric cables, CEN, European Committee for Standardization, Brussels, Belgium.
[25] Steen-Hansen, A.E., Storesund, K., Försth, M., Strömgren, M., Jensen, B.V., Bluhme, D., 2013. Conditions for Nordic harmonisation of fire classification of cables. Proposal of implementation of the new European classification system in the building regulations. Report no. NBL A13111. SINTEF NBL, Trondheim, Norway.
[26] IMO MSC/Circ.1006 - Guidelines on fire test procedures for acceptance of fire-retardant materials for the construction of lifeboats. 08-Jun-1998. International Maritime Organization, London, UK.
[27] NT FIRE 043. Large, free-hanging curtain and drapery textiles: Heat release, fire spread and smoke production - Full scale test. 1989. NORDTEST, Espoo, Finland.
[28] BS 5852:2006 Methods of test for assessment of the ignitability of upholstered seating by smouldering and flaming ignition sources, BSI, British Standards Institution, London, UK.
[29] EN 1021-1:2014 Furniture - Assessment of the ignitability of upholstered furniture - Part 1: Ignition source smouldering cigarette. CEN, European Committee for Standardization, Brussels, Belgium.
[30] EN 1021-2:2014 Furniture - Assessment of the ignitability of upholstered furniture - Part 2: Ignition source match flame equivalent. CEN, European Committee for Standardization, Brussels, Belgium.
[31] BS 6807:2006 Methods of test for assessment of ignitability of mattresses, upholstered divans and upholstered bed bases with flaming types of primary and secondary sources of ignition. BSI, British Standards Institution, London, UK.
[32] SS 876 00 010:2016 Healthcare textiles - Nursing beds - Burning requirements for mattresses to be used in high risk environment. Swedish Standards Institute, Stockholm, Sweden.
[33] EN 597-1:2015 Furniture - Assessment of the ignitability of mattresses and upholstered bed bases - Part 1: Ignition source smouldering cigarette. CEN, European Committee for Standardization, Brussels, Belgium.
[34] EN 597-2:2015 Furniture - Assessment of the ignitability of mattresses and upholstered bed bases - Part 2: Ignition source: match flame equivalent. CEN, European Committee for Standardization, Brussels,

Belgium.
[35] Richardson, L.R., 1994. Determining Degrees of Combustibility of Building Materials – National Building Code of Canada. Fire and Materials 18, 99–106.
[36]C/VM2 Verification Method: Framework for Fire Safety Design For New Zealand Building Code Clauses C1-C6 Protection from Fire, 2014. https://www.building.govt.nz/assets/Uploads/building-code-compliance/c-protection-from-fire/asvm/cvm2-protection-from-fire-amendment-4.pdf
[37] Hakkarainen, T., Hayashi, Y., 2001. Comparison of Japanese and European Fire Classification Systems for Surface Linings. Fire Science and Technology 21, 19–42.
[38] Sumathipala, K., 2006. Flammability tests for regulation of building and construction materials, in: Flammability Testing of Materials Used in Construction, Transport and Mining. Woodhead Publishing, Cambridge, UK, pp. 217–230.
[39] Sorathia, U., Long, G., Gracik, T., Blum, M., Ness, J., 2001. Screening Tests for Fire Safety of Composites for Marine Applications. Fire and Materials 25, 215–222.
[40] MIL-STD-2031 (SH). Test method standard. Fire and toxicity test methods and qualification procedure for composite material systems used in hull, machinery, and structural applications inside naval submarines, 1991.
[41] Grenier, A.T., Janssens, M.L., Nash, L., 2000. Developing Cone Calorimeter Acceptance Criteria for Materials Used in High Speed Craft. Fire an Materials 24, 29–35.
[42] Peacock, R.D., Bukowski, R.W., Markos, S.H., 1999. Evaluation of Passenger Train Car Materials in the Cone Calorimeter. Fire Mater. 23, 53–62.
[43] Van Hees, P., Hertzberg, T., Steen-Hansen, Anne, 2002. Development of a Screening Method for the SBI and Room Corner using the Cone Calorimeter. Nordtest Project 1479-00 (No. SP Report 2002:11). SP Technical Research Institute of Sweden, Borås, Sweden.
[44] Kristoffersen, B., Steen-Hansen, Anne, A.E., Hakkarainen, T., Östman, B., Johansson, P., Pauner, M.A., Grexa, O., Hovde, P.J., 2003. Using the Cone Calorimeter for screening and control testing of fire retarded woodproducts. Nordtest project 1526-01. Report NBL A03119. Trondheim, Norway.
[45] Sundström, B. (Ed.), 1995. Fire Safety of Upholstered Furniture: the final report on the CBUF research programme. Interscience Communications Ltd, London, UK.

UNREGULATED COMBUSTIBLES: DECORATIVE MATERIALS

Douglas H. Evans[1] & Marcelo M. Hirschler[2]
DHE FPE LLC[1] & GBH International[2], US
([1]: 7260 W. Azure Dr., Suite 140-90; Las Vegas, NV, 89130 - [2]: 2 Friars Lane, Mill Valley, CA, 94941)

ABSTRACT

Building code regulations cover the fire safety of construction materials and products. These regulations apply during design and construction, until the building starts being occupied. On the other hand, it is the fire codes (and, in the US, the life safety code) that regulate the majority of combustibles brought into buildings (or at least into public buildings) that are not building products. However, a major category of combustible products that is responsible for several high-profile fires and remains mostly unregulated, is decorative materials. In 2015 a very deadly nightclub fire occurred in Romania, shortly after an outdoor fire on a rooftop pool deck of a Las Vegas hotel; both involved decorative materials.

Throughout the developed world, combustible interior finishes are regulated for fire performance, particularly because experience has demonstrated that fires involving interior finish can have devastating consequences. However, interior finish, which is "the exposed surfaces of walls, ceilings, and floors within buildings", is sometimes challenging to interpret and/or difficult to fabricate into representative specimens for fire testing. In many environments, various types of decorations are affixed to walls or ceilings. Typically, interior finish materials are tested by assessing flame spread, propensity for flashover (room-corner test), or a heat release property. Thus, when the surfaces are covered with decorations, it often becomes unclear how exactly the interior finish needs to be tested.

In the US, regulation of combustible products in public buildings has long, properly, focused on upholstered furniture and mattresses, as they can release very high heat on burning. However, they are not the sole products with high heat release rates, and decorative items can also be a serious problem.

A key concern is whether the fire hazard of a decorative feature is sufficiently limited such that it need not be regulated. In order to assess the fire hazard associated with a particular item, two aspects need to be considered: the material properties of the item and its overall size. Size also needs to be considered when the decorative feature may become a falling hazard capable of causing injury to occupants or blocking egress paths.

Typical fire safety regulation of decorative materials (when it exists) is based on a small-scale fire test to assess the materials of construction of the item, but ignores the issue of size. Clearly, a branch, or a leaf, of a tree does not have the same potential to cause a large fire as the full tree. Therefore, full scale testing (e.g. using a furniture calorimeter) is much more realistic, although the ignition source typically used is 20 kW.

Active fire protection (such as sprinklers) is often (but not always) required in the type of environments where large quantities of combustible contents (including decorative materials) are present. Depending on the material properties and configuration of the fuel load (e.g. obstructions to sprinklers), this protection may not always be sufficient to ensure appropriate fire safety. This is not addressed fully in this work.

This work discusses the present code requirements (in the US) with regard to decorative materials, including decorative vegetation, based on the latest editions published at time of writing: 2015. It also discusses some notable fire losses, and the appropriate type of fire testing to assist in determining what fire protection measures are reasonable and what level of passive fire protection is prudent.

INTRODUCTION

In the US, fire safety in new buildings is regulated by a building code, typically either the International Building Code[1] (IBC) or the NFPA Building Construction and Safety Code[2] (NFPA 5000). The International Code Council (ICC) has also issued the International Existing Building Code[3] (IEBC) which deals with "the repair, alteration, change of occupancy, addition and relocation of existing buildings, regardless of occupancy". However, combustible products (including decorations and other furnishings) brought into a building for "permanent" use once the building has been completed (unless alterations trigger IEBC involvement) are regulated by the International Fire Code[4] (IFC) or, in the NFPA system, by the NFPA Life Safety Code[5] (NFPA 101) or the NFPA Fire Code[6] (NFPA 1). Most of the information regarding combustible products in existing buildings contained in NFPA 1 is "extracted" from NFPA 101, meaning that the text in both codes is identical and the responsibility of the NFPA 101 committees.

The key products regulated by the IFC and NFPA 101 are interior finish (wall, ceiling and floor), upholstered furniture, mattresses, curtains, drapes and decorations (including decorative vegetation such as Christmas trees). There is no need to explain the meaning of upholstered furniture, mattresses or curtains and drapes. There is a need, however, to distinguish between interior finish and decorations. Both types of products are typically combustible and may be associated with severe fires if left unregulated.

The IFC defines decorative materials as: "All materials applied over the building interior finish for decorative, acoustical or other effect including, but not limited to, curtains, draperies, fabrics, streamers and all other materials utilized for decorative effect including, but not limited to, bulletin boards, artwork, posters, photographs, paintings, batting, cloth, cotton, hay, stalks, straw, vines, leaves, trees, moss and similar items, foam plastics and materials containing foam plastics. Decorative materials do not include wall coverings, ceiling coverings, floor coverings, ordinary window shades, interior finish and materials 0.025 inch (0.64 mm) or less in thickness applied directly to and adhering tightly to a substrate."

The IFC defines interior wall and ceiling finish as: "The exposed interior surfaces of buildings, including but not limited to: fixed or movable walls and partitions; toilet room privacy partitions; columns; ceilings; and interior wainscoting, paneling or other finish applied structurally or for decoration, acoustical correction, surface insulation, structural fire resistance or similar purposes, but not including trim." The IFC also states that: "Interior finish includes interior wall and ceiling finish and interior floor finish."

NFPA 101 defines interior finish as: "The exposed surfaces of walls, ceilings, and floors within buildings.", with interior wall finish being: "The interior finish of columns, fixed or movable walls, and fixed or movable partitions." NFPA 101 does not define decorative materials, unfortunately.

The definitions above indicate that decorative materials can "be applied over", or cover, interior finish but do not, themselves, constitute interior finish. Typically, the fire safety of interior finish is regulated by assessing flame spread and/or heat release of relatively flat materials, using either the Steiner tunnel test (ASTM E84[7], with the criteria shown in Table 1) or the room-corner test (NFPA 286[8]), with the following ***key*** pass/fail criteria: no flashover, the heat release rate cannot exceed 800 kW, the total smoke released cannot exceed 1,000 m^2 and the flame cannot spread to the edges of the test specimen in the room.

Table 1 - Classification for Code Use of Interior Finish Materials or Products Using ASTM E84

	Flame Spread Index	Smoke Developed Index
Class A or Class I	$FSI \leq 25$	$SDI \leq 450$
Class B or Class II	$26 \leq FSI \leq 75$	$SDI \leq 450$
Class C or Class III	$76 \leq FSI \leq 200$	$SDI \leq 450$
Plenum Materials	$FSI \leq 25$	$SDI \leq 50$

However, Figures 1 and 2 show one of the key difficulties when regulating decorative materials. Fig. 1 is a complex interior finish/decorative environment in a restaurant and Fig. 2 is the interior of a Las Vegas casino. It is not immediately obvious how to assess fire safety or even what is to be tested.

Fig. 1 - Environment in a Decorated Restaurant

Fig. 2 – Interior Passageway in a Las Vegas Casino

DECORATIVE MATERIALS AND INTERIOR FINISH

There are similarities and key differences between decorations and interior finish and between fire incidents involving both types of products. Some similarities are: they can both be highly combustible, they are often present in public places with high occupant load and there is usually insufficient care for fire safety in public occupancies. Some key differences are that interior finish is well regulated (at least in the US) while many decorations are mostly not well regulated, particularly when they are placed outside but in the proximity of the building and that interior finish fires are virtually always the result of non-compliance with code requirements but decorations fires are more likely to occur where not regulated.

Decorative materials can be attached to walls or ceilings or can be free-standing, which compounds the problem, as already shown above, of how to assess their fire performance. Note that interior finish materials/products can be "decorative" (See Fig. 3) and that is typically the intention of the architect and interior decorator. However interior finish performs a function: it covers (and at times protects) the wall material. Decorative materials are not functional.

Fig. 3 – Ceiling Above a Lamp

In some instances, entire structures and even entire buildings (perhaps functional and perhaps not) are incorporated inside a space or room. If they are not in contact with walls or ceilings (typical) they are not normally classified (or regulated) as interior wall or ceiling finish and can interfere with sprinkler action.

Fig. 4 - Huts with Thatched Roofs and Artificial Vegetation under a Las Vegas Roof

Fig. 5 - Buildings and Statues Inside Las Vegas Casino

This work was inspired by two large fires in 2015 involving decorations. The installations in both cases were deemed to be compliant with code requirements.

Fig. 6 – July 25, 2015 Fire at Cosmopolitan Hotel, Las Vegas

On July 25, 2015, a fire started at an artificial tree present on an outdoor deck in the Cosmopolitan Hotel in Las Vegas (Fig. 6). Videos posted on social media showed the large fire consuming the poolside tent-like cabanas, imitation palm trees, furniture, and just about everything else on the open-air deck of the hotel, sending thick black smoke billowing high above the area. There were no fire fatalities or serious fire injuries (but two people were treated for smoke inhalation). There were some broken windows, but little other damage to the building but smoke penetrated into some of the hotel rooms. The trunks of the artificial trees present were constructed of a steel pipe covered with foam plastic (maybe 15-20 cm thick) and a rigid plastic molded to simulate a palm tree. The combustible materials would not have met the code fire test requirements for foam plastic when used as interior finish indoors[9, 10]. In particular, foam plastic used indoors in the US must always be protected from the habitable area by a thermal barrier and this was not.

It is interesting to note that building and fire codes adopted by the local jurisdiction (Clark County, NV, including Las Vegas, as explained in work Evans in 2004[11]) did not apply to outdoor decorative vegetation. Local authorities commented after the fire that they felt no policy changes should be recommended because it would be difficult to regulate, identify or standardize such adornments, and could cause an undue burden to the businesses. Further statements included that current requirements were sufficient to keep the fire from spreading beyond the pool's deck where it started. However, code officials are urging hotels to voluntarily test and/or remove highly combustible decorations, but did not give exact details.

On October 30, 2015, a heavy metal band's pyrotechnical show sparked a deadly fire at the Colectiv Bucharest, Romania, nightclub, killing 62 people and injuring 200 of the club's mostly youthful patrons. Witnesses said that there were between 300 and 400 people at the club, housed below grade in a former factory, with only one exit door; people panicked and rushed for the exit. A spark on stage ignited a column covered with or constructed of a polystyrene decorative material (Fig. 7). Fire regulations are said to be "lax" in Bucharest clubs and restaurants and there was no prohibition on using such decorative materials. In comparison, if this had been interior finish in the US, the foam plastic would have been required to be covered with a thermal barrier and also flame retarded (there is no evidence as to the fire performance of the actual polystyrene material involved in this fire).

Fig. 7 – October 30, 2015, Foam Plastic-Clad Column Burning in Bucharest Nightclub

These two examples, which follow a large list of decorative materials fires with multiple fatalities going back decades, show that a significant part of the problem of the fire hazard posed by decorative materials is the fact that regulation is either non-existent or (as will be shown below) inadequate when it exists or even inadequately enforced.

US NFPA 101 CODE REQUIREMENTS FOR DECORATIVE MATERIALS

The Life Safety Code (NFPA 101), adopted by most US states[12], includes requirements for interior finish, interior trim and furnishings and contents (including decorative materials) within Chapter 10. The requirements apply to new and existing buildings but only come into force if they are being called for by the corresponding occupancy chapters. All occupancy chapters are subdivided into those applying to new occupancies and those applying to existing occupancies (which tend to have less stringent requirements). The key occupancies that do (or do not) apply the requirements from Chapter 10 are: health care, detention, assembly, educational, hotels and dormitories, mercantile, business and residential board and care.

The complete requirements for decorative materials in Chapter 10, 2015, are as follows:
"10.3.1* Where required by the applicable provisions of this Code, draperies, curtains, and other similar loosely hanging furnishings and decorations shall meet the flame propagation performance criteria contained in Test Method 1 or Test Method 2, as appropriate, of NFPA701[13], Standard Methods of Fire Tests for Flame Propagation of Textiles and Films."
"10.3.5* Furnishings or decorations of an explosive or highly flammable character shall not be used.
10.3.6 Fire-retardant coatings shall be maintained to retain the effectiveness of the treatment under service conditions encountered in actual use.
10.3.7* Where required by the applicable provisions of this Code, furnishings and contents made with foamed plastic materials that are unprotected from ignition shall have a heat release rate not exceeding 100 kW when tested in accordance with UL 1975[14], Standard for Fire Tests for Foamed Plastics Used for Decorative Purposes, or when tested in accordance with NFPA 289[15], Standard Method of Fire Test for Individual Fuel Packages, using the 20 kW ignition source." *(Note that this is relatively small)*

The asterisks (*) refer to non-mandatory annex notes that read as follows:
"A.10.3.1 Testing per NFPA 701, Standard Methods of Fire Tests for Flame Propagation of Textiles and Films, applies to textiles and films used in a hanging configuration. If the textiles are to be applied to surfaces of buildings or backing materials as interior finishes for use in buildings, they should be treated as interior wall and ceiling finishes .... The test results from NFPA 701 are suitable for classification purposes but should not be used as input into fire models, because they are not generated in units suitable for engineering calculations."
"A.10.3.5 Christmas trees that are not effectively flame-retardant treated, ordinary crepe paper decorations, and pyroxylin plastic decorations might be classified as highly flammable."
"A.10.3.7 Neither UL 1975, Standard for Fire Tests for Foamed Plastics Used for Decorative Purposes, nor NFPA 289, Standard Method of Fire Test for Individual Fuel Packages, is intended for evaluating interior wall and ceiling finish materials. Actual test results for heat, smoke, and combustion product release from UL 1975 or from NFPA 289 might be suitable for use as input into fire models intended for performance-based design."

NFPA 701, a small burner test, includes two methods. Method 1 (the less severe of the two test methods) applies to materials with an areal density ≤ 700 g/m^2 including the following: (1) fabrics or other materials used in curtains, draperies, or other window treatments, (2) single-layer fabrics, (3) multilayer curtain and drapery assemblies in which the layers are fastened together by sewing or other means and (4) where required, fabrics with an areal density ≤ 700 g/m^2 and used in other construction applications. Method 1 does not apply to any material to which Method 2 applies nor to the following: (1) vinyl-coated fabric blackout linings or lined draperies using a vinyl-coated fabric blackout lining, (2) plastic films or (3) decorative materials other than fabrics. Method 2 applies to: (1) decorative materials other than fabrics with areal densities ≤ 700 g/m^2, (2) fabrics, including multilayered fabrics, films, and plastic blinds, with or without reinforcement or backing, with areal densities > 700 g/m^2, (3) vinyl-coated fabric blackout linings and lined draperies using a vinyl-coated fabric blackout lining, (4) plastic films, (5) fabrics, with or without reinforcement or backing, used for decorative or other purposes inside a building or as temporary or permanent enclosures for buildings under construction and (6) fabrics used in the assembly of awnings, tents, tarps, membrane structures or banners. This test assesses flame propagation performance based on mass loss (Test 1) and char length (Test 2), as well as after-flame (for both tests).

The test specimen for Method 1 is 400 mm x 150 mm and methane at a gas flow rate of 1205 mL/min (flame height of 100 mm) is applied for 45 s perpendicularly at a distance of 25 mm from the bottom of the test specimen. The test is "passed" with weight loss of ≤ 40%. The test specimen for Method 2 is 1200 mm x 280 mm and methane at a gas flow rate of 113 L/hr. (flame height of 280 mm) is applied for 2 minutes at an angle of 25° from the vertical at a distance of 100 mm from the bottom of the test specimen. The test (for flat specimens) is "passed" if the char length is ≤ 435 mm. In both tests flaming must cease within 2 seconds of removal of the burner.

UL 1975 and NFPA 289 are both furniture calorimeter tests in which an ignition source is applied to a single decorative product. The ignition source for UL 1975 is a 340 g wood crib and the one for NFPA 289 is a propane gas burner, namely the same one that is used in the room-corner test (NFPA 286). However, codes use the smallest flame intensity referenced in NFPA 289 as the ignition source for decorative materials: the ignition source used is a 20 kW source. Both tests measure heat release.

In NFPA 101, foam plastic fuel packages of various kinds must meet requirements based on UL 1975 or NFPA 289 (either 100 or 150 kW criterion) in assembly occupancies. They also require that "fabrics and films used for decorative purposes, and all draperies and curtains, and similar furnishings" must meet the requirements of 10.3.1, meaning the criteria based on NFPA 701.

However, in all other occupancies these requirements are much milder and occasionally contradictory.

For example, requirements in new and existing health care occupancies allow much more leeway:
"Combustible decorations shall be prohibited in any health care occupancy, unless one of the following criteria is met:
(1) They are flame-retardant or are treated with approved fire-retardant coating that is listed and labeled for application to the material to which it is applied. *(Note that the term flame-retardant is not defined)*
(2) The decorations meet the flame propagation performance criteria contained in Test Method 1 or Test Method 2, as appropriate, of NFPA 701, ...
(3) The decorations exhibit a heat release rate not exceeding 100 kW when tested in accordance with NFPA 289, ...using the 20 kW ignition source.
(4) The decorations, such as photographs, paintings, and other art, are attached directly to the walls, ceiling, and non-fire-rated doors in accordance with the following:
(a) Decorations on non-fire-rated doors do not interfere with the operation or any required latching of the door and do not exceed the area limitations of ... (b), (c), or (d).
(b) Decorations do not exceed 20% of the wall, ceiling, and door areas inside any room or space of a smoke compartment that is not protected throughout by an approved automatic sprinkler system .....
(c) Decorations do not exceed 30% of the wall, ceiling, and door areas inside any room or space of a smoke compartment that is protected throughout by an approved supervised automatic sprinkler system ....
(d) Decorations do not exceed 50% of the wall, ceiling, and door areas inside patient sleeping rooms having a capacity not exceeding four persons, in a smoke compartment that is protected throughout by an approved, supervised automatic sprinkler system ... ."
"Draperies, curtains, and other loosely hanging fabrics and films serving as furnishings or decorations in health care occupancies" *must meet the requirements of NFPA 701.*

In the case of ambulatory health care occupancies, the first of the 4 criteria for decorations is useless, as it states simply the combustible decorations are permitted if "they are flame retardant". In earlier editions of codes stating that a material was "flame retardant" was taken to mean that it met the criteria in NFPA 701.

However, when a reference to NFPA 701 was added explicitly to NFPA 101, the "flame retardant" criterion remained in the code even though there is no defined interpretation of what it means.

In the case of educational, day care and detention occupancies, the only requirements are that draperies, curtains, and other similar furnishings and decorations must meet NFPA 701. Interestingly, detention occupancy requirements state: "Combustible decorations shall be prohibited in any detention or correctional occupancy unless flame-retardant." The sections on hotels and lodging or rooming houses, include the following (helpful) comment: "Furnishings or decorations of an explosive or highly flammable character shall not be used." The section on apartments states even more intriguingly: "Furnishings or decorations of an explosive or highly flammable character shall not be used outside of dwelling units." Hotels also require that "new draperies, curtains, and other similar loosely hanging furnishings and decorations" meet the criteria of NFPA 701. Those requirements are not included for lodging or rooming houses and apply only to those residential board and care occupancies that are not sprinklered. By default, all other occupancies (including mercantile and business) are silent on requirements for any decorations.

Mercantile occupancies regulate materials in plastic signs and kiosks. If the signs are not foamed plastics they must meet the requirements for light transmitting plastics, which are based on ignitability per ASTM D1929[16] and both flame spread and smoke release per ASTM E84. If the signs are foamed plastics they must release less than 150 kW when tested to UL 1975 or NFPA 289. In kiosks, foam plastics are regulated via UL 1975 or NFPA 289 ($\leq$ 100 kW) and textiles and films are regulated via NFPA 701.

It should be noted that NFPA 1 (Fire Code) contains requirements for decorative vegetation, natural and artificial, which are missing from NFPA 101. It states that the presence of natural combustible vegetation, including natural cut Christmas trees, is severely limited. In NFPA 1, natural Christmas trees are prohibited inside assembly, board and care, detention, dormitories, educational and hotel occupancies. In other occupancies the amounts permitted are to be regulated by the authority having jurisdiction (AHJ) based on the provision of adequate safeguards in terms of the quantity and nature of the combustible vegetation. With regard to artificial vegetation and artificial Christmas trees, they must be labeled or otherwise identified or certified by the manufacturer as being fire retardant and the "fire retardance" *(note that this term is a deprecated term in the fire safety community but it is widely used, since it cannot be quantified and simply means that a fire retardant was added)* must be demonstrated for each individual decorative vegetation item in an approved manner. Only listed electrical lights and wiring are permitted on natural or artificial combustible vegetation and no electrical lights are permitted on metal artificial trees.

Both NFPA 101 (and NFPA 1 with regard to combustible vegetation) contain requirements that decorations cannot obstruct exits or means of egress, irrespective of their fire properties.

US IFC CODE REQUIREMENTS FOR DECORATIVE MATERIALS

The International Fire Code (IFC) is the most widely adopted fire code in the US because it works consistently in conjunction with the IBC, which is adopted in virtually every state. This means that there is some inconsistency when jurisdictions adopt both NFPA 101 and the IFC (which some do). The IFC contains all requirements for interior finish, decorative materials and furnishings in chapter 8. The same chapter also includes the applicability of each requirement.

Combustible decorative materials, other than decorative vegetation, must meet, for all occupancies: (a) furnishings or decorative materials of an explosive or highly flammable character shall not be used, (b) fire-retardant coatings in existing buildings shall be maintained so as to retain the effectiveness of the treatment under service conditions encountered in actual use and (c) there is no limit to the permissible amount of noncombustible decorative materials.

In other than detention occupancies, curtains, draperies, fabric hangings and other similar combustible decorative materials suspended from walls or ceilings must comply with either NFPA 701 or NFPA 289

(20 kW ignition source and 100 kW criterion) and cannot exceed 10% of the specific wall or ceiling area to which they are attached.

The code has some exceptions, addressing some specific occupancies, as follows (not verbatim language):

1. In auditoriums of assembly occupancies the permissible amount of curtains, draperies, fabric hangings and other similar combustible decorative material suspended from walls or ceilings can be up to 75% of the aggregate wall area, if the building is sprinklered.

2. In residential dormitories, within sleeping units and dwelling units, the permissible amount of curtains, draperies, fabric hangings and other similar decorative materials suspended from walls or ceilings can be up to 50% of the aggregate wall areas, if the building is sprinklered.

3. In business and mercantile occupancies, the amount of combustible fabric partitions suspended from the ceiling and not supported by the floor must comply with NFPA 701 or NFPA 289 (as above) but the amounts are not limited.

In assembly occupancies, foam plastics (either exposed foam plastic materials or unprotected materials containing foam plastic used for decorative purposes or stage scenery or exhibit booths) must comply with either NFPA 701 or NFPA 289 (20 kW ignition source and 100 kW criterion), except for individual items with less than 0.45 kg of foam plastics or trim that meets the general foam plastic trim requirements. In these same occupancies, motion picture screens must comply with NFPA 701 or with a Class B in the ASTM E84 test and "imitation leather or other material consisting of or coated with a pyroxylin *(which is based on nitrocellulose)* or similarly hazardous base" is not permitted.

In places of religious worship the IFC states: "wood used for ornamental purposes, trusses, paneling or chancel furnishing shall be allowed" without limit. In educational occupancies, IFC exceptions permit significant display of artwork and teaching materials in corridors and classrooms.

Natural cut trees are not allowed in ambulatory care facilities, assembly, educational, institutional, mercantile and residential facilities (other than dwellings), except that they are allowed in most places if the entire building is sprinklered. Artificial decorative vegetation must meet the requirements of NFPA 701 or of NFPA 289 (20 kW ignition source, 100 kW criterion).

No unlisted electrical wiring or lighting is permitted on natural cut trees or on artificial decorative vegetation and the use of electrical wiring and lighting on artificial trees constructed entirely of metal is prohibited.

Just like in NFPA 101, there are statements prohibiting decorations from obstructing means of egress.

TESTING OF COMBUSTIBLE DECORATIVE VEGETATION

California codes[17] have included, for many years, a requirement to test (and certify) flame retardant treatments that are commercialized to improve the fire performance of natural vegetation (typically Christmas trees). This test is a modification of an old edition of NFPA 701, but is based on the so-called "small scale" test (neither Test 1 nor Test 2), a test which has been eliminated from NFPA 701 since the 1989 edition because it was shown not to predict full scale fire behavior[18]. Reality is, however, that multiple fabrics are commercially available accompanied by certificates saying that they comply with the NFPA 701 test (meaning the above-mentioned *small scale test*). The California test for natural decorative vegetation consists of applying a 38 mm Bunsen burner flame for 12 seconds to the bottom of a vertical specimen (branch of Douglas fir treated with the flame retardant and left to dry). The material is deemed to have satisfactory performance if there is no flame spread from the area in contact with the test flame and after-flame does not exceed 10 seconds. Experience indicates that the test is not very reliable.

In both ASTM and NFPA, activity is underway to develop, or update, fire test methods for combustible decorative vegetation. At NFPA, modifications are being considered to NFPA 289 to specifically address testing of Christmas trees in full scale and at ASTM a brand new test is being developed.

In the work underway for NFPA 289, trees are exposed to a propane gas burner and measurements are made of heat and smoke release, flame spread and mass loss. Artificial Christmas trees are to be tested by exposing the tree (placed on a load cell) to a 20 kW gas burner placed within 25 mm of the base of the tree, for 15 min. For natural Christmas trees the recommendation is to investigate the effect of the flame retardant treatment by placing the treated tree (on a load cell) to the same gas burner but at a linearly increasing flow rate ranging from 0 to 20 kW until 2.5 minutes and then decreasing it again linearly until there is no flow rate at 5.0 minutes. The vertical centerline of the burner is located 150 mm inside the outermost portion of the tree and within 100 mm vertically from the bottom branches. At ASTM, the concept is similar to the large scale NFPA test for natural Christmas trees but it is accompanied by a small scale test similar to the California one.

DISCUSSION

In 2016 an NFPA article[19] and a presentation at the NFPA June conference[20] discussed this problem and showed that it is a fire safety issue. It was pointed out that the deadliest nightclub fire in US history is related to combustible decorative features, although an exact cause of ignition is under debate. On November 28, 1942, fire destroyed the Cocoanut Grove club in Boston, killing 492 people. Witnesses said the fire started in an artificial palm tree, while others said it began in a cloth covering the ceiling.

It is the opinion of the authors that the requirements for decorative materials in the US are inadequate to ensure proper fire safety. The following is a list of key issues with the present requirements:

- Requirements typically involve small scale tests or tests with a low heat/flame input: the key test used for decorative materials (NFPA 701, even in its new versions, Test 1 and Test 2) applies a relatively low heat input and is likely to be inappropriate for large items because they involve high fuel loads that the test cannot assess; furthermore the test may not even "fail" them.
- It is a good idea to have "bench-scale"(or small-scale) testing initially conducted to determine if adverse behavior of the specific material can be predicted under actual fire conditions. However, failure to achieve ignition in small-scale tests is not a substantial proof of adequate fire performance, since many materials incapable of reaching a self-supporting fire in bench scale test configurations prove to be very hazardous when subjected to larger scale testing. Thus, successful outcomes in small scale tests need to be followed up be larger scale or higher heat input fire tests.
- There is no regulation of a number of products, such as the statues or small (non-occupiable) indoor buildings described above. There is little, if any, regulation of large items that are not based on textiles or foam plastics, as evidenced by the facts (shown above) that many decorative products are not covered by regulations.
- There is little, if any, regulation that provides a size limitation for decorative items: NFPA 701 tests relatively small test specimens and cannot determine whether a larger version of the same material or product which may be suitable as a small decoration is equally suitable as a large hanging item, for example.
- There is virtually no regulation based on fire hazard: the results based on NFPA 701 cannot be used for conducting fire hazard assessments. The results based on heat release (from UL 1975 or NFPA 289) can be used for such purposes but the code requirements simply assess a single item (and not the largest possible version of such an item) and assume that if one item is satisfactory, multiple items of the same type will be equally safe. That is manifestly untrue.

- Decorative materials need to be assessed for fire safety as a function of the fire hazard that can result when they burn, which is a function of the type of material, as well as its size and location.
- There is virtually no regulation of exterior combustible decorative items placed in close proximity to buildings. The presence of decorative materials or products in the vicinity of buildings with combustible components is a growing practice. In fact, it is not uncommon to have combustible components (perhaps including decorative products) arranged near or attached to non-combustible walls, such as overhangs or sidewalls. Of course the fire hazard is even larger in the presence of combustible walls, which are very common.
- "Bench-scale"(or small-scale) testing should initially be conducted to determine if adverse behavior of the specific material can be predicted under actual fire conditions. Failure to achieve ignition in small-scale tests is not a substantial proof of adequate fire performance, since many materials incapable of reaching a self-supporting fire in bench scale test configurations prove to be very hazardous when subjected to larger scale testing.
- The use of active fire protection (such as sprinklers) is not properly considered. In particular, issues to consider include the proximity of decorations to fire sprinklers and their potential obstruction of sprinkler discharge, as well as exterior fires large enough to adversely affect the building. Also, thin combustible ceilings can be a concern regarding sprinkler protection because there is the potential for adverse interaction with the sprinklers.

CONCLUSIONS

The modern world uses an increasing amount of combustible decorative materials, both inside buildings and in their proximity. Building codes regulate building materials and not contents and furnishings, which are regulated by a fire code (or a life safety code). However, since contents (including decorative materials) are easy to bring into buildings that have already received a certificate of occupancy, there is more enforcement associated with a building code than a fire code or a life safety code. Perhaps it might be worth considering the following rule of thumb.

When should a furnishing or content (such as a decorative material) be regulated as a building material? If a single individual can pick it up and carry it to its final location, it is likely just a decorative material. On the other hand, if a combustible item requires a crew of people to haul it to its final location for assembly, the decorative feature should be regulated in the same way as other building materials.

Decorative materials are widely found in public buildings (and in residential construction) and fires associated with them can have very severe consequences but present regulations are still insufficient.

REFERENCES

[1] IBC (International Building Code), International Code Council, Washington, DC, US.

[2] NFPA 5000, NFPA Building Construction and Safety Code, National Fire Protection Association, Quincy, MA, US.

[3] IEBC (International Existing Building Code), International Code Council, Washington, DC, US.

[4] IFC (International Fire Code), International Code Council, Washington, DC, US.

[5] NFPA 101, NFPA Life Safety Code, National Fire Protection Association, Quincy, MA, US.

[6] NFPA 1, NFPA Fire Code, National Fire Protection Association, Quincy, MA, US.

[7] ASTM E84, "Standard Test Method for Surface Burning Characteristics of Building Materials", ASTM International, West Conshohocken, PA, US.

[8] NFPA 286, "Standard Methods of Fire Tests for Evaluating Contribution of Wall and Ceiling Interior Finish to Room Fire Growth", National Fire Protection Association, Quincy, MA, US.

[9] Earl, T.T. and Hirschler, M.M., "Regulation of foam plastic insulation by US construction codes", in Fire and Materials Conf., San Francisco, CA, Feb. 2-4, 2015, pp. 529-542, Interscience Communications, London, UK.
[10] Hirschler, M.M., "Fire Testing of Interior Finish", Fire Protection Engineering Fall 2004, # 24, pp. 12-17, 2004.
[11] Evans, D.H., "Unique Interiors on the Las Vegas Strip", Fire Protection Engineering Fall 2004, # 24, pp. 4-11, 2004.
[12] Hirschler, M.M., "Codes & standards development in the USA: How does it work?", Proc. Interflam 2016, pp. 1175-1186, Egham, Surrey, UK, July 4-6, 2016, Interscience Communications, London, UK.
[13] NFPA 701, "Standard Methods of Fire Tests for Flame Propagation of Textiles and Films", National Fire Protection Association, Quincy, MA, US.
[14] UL 1975, "Standard for Fire Tests for Foamed Plastics Used for Decorative Purposes", Underwriters Laboratories, Northbrook, IL, US.
[15] NFPA 289, "Standard Method of Fire Test for Individual Fuel Packages", National Fire Protection Association, Quincy, MA, US.
[16] ASTM D1929, "Standard Test Method for Determining Ignition Temperatures of Plastics", ASTM International, West Conshohocken, PA, US.
[17] California Code of Regulations, Title 19 (Public Safety), Chapter 8 (Regulations Relating to Flame-Retardant Chemicals, Fabrics and Application Concerns), "Flame Retardant Regulations", Section 1264 (Test Requirements for Interior Flame-Retardant Chemicals – When applied to Materials other than Fabrics) and Section 1237 (Fire Resistance Test), http://www.oal.ca.gov/ccr.htm
[18] Belles, D. W. and Beitel, J.J., "Do multiple-layer draperies pass the single layer fire test?", Fire Journal, pp. 25-30, 90-91, Sept./Oct. 1988.
[19] Smith, A., "Looks Nice. Burns Hot. How much of a fire hazard is posed by decorative features inside and outside buildings?" NFPA Journal, May-June 2016.
[20] Evans, D.H. and Hirschler, M.M., "NFPA 101, Life Safety Code, Combustible Decorative Features and the Las Vegas Strip", Session T44 at NFPA Conference and Expo, June 13-16, 2016, Las Vegas, NV.

FIRE HAZARD ANALYSIS OF INDOOR LARGE-SCREEN DIGITAL MEDIA SYSTEM

Xiaolei Chen, PhD
California State University, Los Angeles, USA
Frank Wang, PE
JENSEN HUGHES, USA

INTRODUCTION

Large-screen digital media systems (LSDMSs) consist of hundreds and thousands of pixel modules that are typically made of polymer materials, and have been widely adopted outside the buildings to decorate metropolises nowadays. With emerging technical innovations, more and more buildings seek to elevate their occupants' indoor experience with these innovative digital media features. Installation of LSDMSs inside buildings is becoming one of the most fantastic means to create a one-of-a-kind experience, generate a revenue platform, develop a unique dynamic brand expression, as well as provide an intelligent interior space. For instance, the new Tom Bradley International Terminal (TBIT) opened at Los Angeles International Airport (LAX) in September 2013 features the most advanced multi-media environment "Integrated Environmental Media System (IEMS)" that includes seven architecturally-scaled media features designed to enhance a particular aspect of the passenger departure or arrival itineraries, as shown in Figure 1 [1].

Figure 1 – IEMS Installed in the new TBIT at LAX [1]

PTO/Xiaolei Chen

However, with the extensive application of such indoor LSDMS, attention has been drawn to potential fire hazards posed by these systems, i.e., the thermal and toxic threat to which building environment and occupants may be exposed.

As it is known, electrical equipment installed in buildings shall comply with relevant codes and standards, such as International Building Code (IBC), International Fire Code (IFC), UL 94 and NFPA 70. Requirements in these codes and standards are largely limited to electrical system wiring, power and clarification. Currently there are few guidelines with respect to special electronic equipment such as LSDMSs. In practice, some manufacturers state that their LSDMSs have UL 94 flammability rating certification such as HB or V-0 rating and are considered as electrical devices. However, UL 94, the Standard of Tests for Flammability of Plastic Materials for Parts in Devices and Appliances, which includes small scale tests that evaluate the flammability of polymeric (plastic) materials in response to a small, open flame or radiant heat source under controlled laboratory conditions, is typically used for parts in devices and appliances. The scope of UL 94 clearly indicates that it does not cover polymeric materials used for building construction, finishing or contents such as wall and floor coverings, furnishings or decorative objects. The results of most UL 94 flammability tests are not applicable to materials whose thickness exceeds 13.0 mm, or whose surface area exceeds 1 square meter (approximately 10 square feet) [2].

Since current codes and standards do not regulate and are not applicable for the use of LSDMS inside buildings, the point at which low voltage electronic equipment transitions from standard electrical equipment to a unique interior feature is totally subject to the interpretation of authority having jurisdiction (AHJ) and application of alternate design methodology using performance-based engineering analysis methods.

METHODOLOGY OF FIRE HAZARD ANALYSIS

In consideration that current code and standard requirements might not be appropriate for or cannot be easily applied to LSDMS, it is imperative to conduct an evaluation on this type of systems to determine its potential fire hazards. This study proposes using the following testing standards as a basis of fire hazard analysis for indoor LSDMS.

- ASTM E84-12, *Standard Test Method for Surface Burning Characteristics of Building Materials* to determine flame spread and smoke development behavior.

- ASTM E1354-11b, *Standard Test Method for Heat and Visible Release Rates for Materials and Products Using an Oxygen Consumption Calorimeter* to determine combustibility, heat release rate per unit area (kW/m^2 or Btu/s.ft^2), fire growth and smoke toxicity of materials.

With data collected from the two ASTM tests, we are able to obtain the properties and characteristics of LSDMS in a fire. The potential fire hazards of indoor LSDMS can be evaluated with a performance-based design methodology through combination of an engineered fire hazard analysis and a Fire Dynamics Simulator (FDS) [3] simulation with associated baseline evaluation criteria.

In order to study the fire hazards of indoor LSDMS, two design fire scenarios are established: one is a regular atrium fire normally used for an atrium smoke management system design and the other is an indoor LSDMS fire. These two design fire scenarios are simulated and compared to facilitate the identification of fire hazards brought by the indoor LSDMS.

FIRE HAZARD ANALYSIS

Fire hazard analysis of indoor LSDMS in this study is performed on material *combustibility*, smoke *toxicity* and fire *spreadability*.

Material Combustibility

Typically, a LSDMS is installed within large spaces such as atria, covered mall buildings, airport terminals and concourses, arenas and stadiums, etc. to provide an intriguing environment as well as present relevant information to occupants in the buildings. According to IBC and NFPA 101, if an indoor LSDMS is able to meet Class B requirements, i.e. its flame spread index is not greater than 75 and its smoke-developed index is less than 450, such media system may be classified as an *interior finish* material. Otherwise they shall be treated as *combustible fuel loads*. Based on the two ASTM testing outcomes, we should be able to conclude whether a LSDMS should be treated as interior finishes or combustible fuel loads.

Currently LSDMS products on the market are typically made of one or multiple polymer materials [4, 5, 6]. Based on available testing data of LSDMS, both of their flame spread index and smoke-developed index have greatly exceeded the code requirements for being defined as interior finishes. Furthermore, the heat release rate (HRR) per unit area for LSDMS can be as high as 600 kW/m^2. Therefore, LSDMS would normally be classified as combustible materials and treated as fuel loads.

Smoke Toxicity

When a LSDMS is considered as fuel loads according to two ASTM testing results, it is imperative to analyze the effects of smoke toxicity caused by such LSDMS fire on the proposed fire protection systems, i.e., smoke management, fire alarm and sprinkler systems, as well as the safe evacuation of occupants in buildings. In this study, a LSDMS product example is used in FDS modeling simulations to analyze smoke toxicity.

As indicated earlier in this paper, two design fire scenarios, a regular atrium fire and an indoor LSDMS fire, are simulated to perform the smoke toxicity analysis. In accordance with most of current codes and standards, such as IBC, UBC and NFPA 101, a smoke management system is required for an atrium. Such smoke management system should be able to maintain a height of the lowest accumulating smoke layer at least 1.83 meters (6 feet) above any walking surface that forms a means of egress within the atrium such that occupants would not be exposed to an unsafe (smoke toxic) environment.

An atrium with geometry of 182.88 meters (600 feet) (length) by 60.96 meters (200 feet) (width) by 24.384 meters (80 feet) (height) is modeled to analyze smoke toxicity of the LSDMS example. The plan and section views in Figures 2 and 3 exhibit the modeled atrium with the highest occupied floor on Level 4. In addition, a LSDMS with dimensions of 9.144 meters (30 feet) wide by 18.288 meters (60 feet) tall is assumed to be installed at the west side of the atrium, as shown in Figures 2 and 3.

Figure 2 – Plan View of the Modeled Atrium

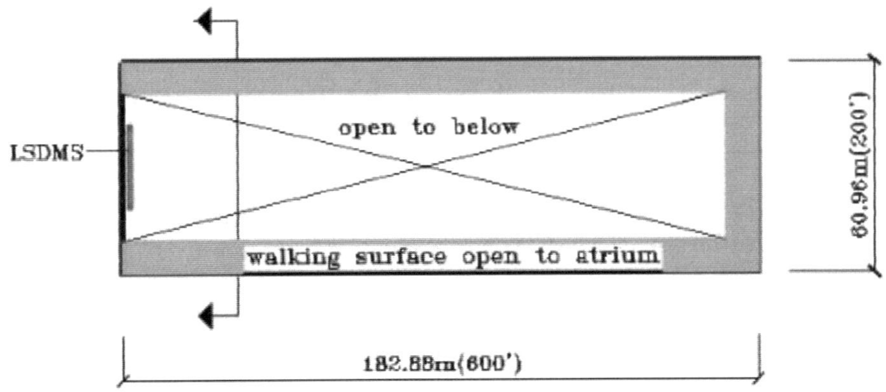

Figure 3 – Section View of the Modeled Atrium

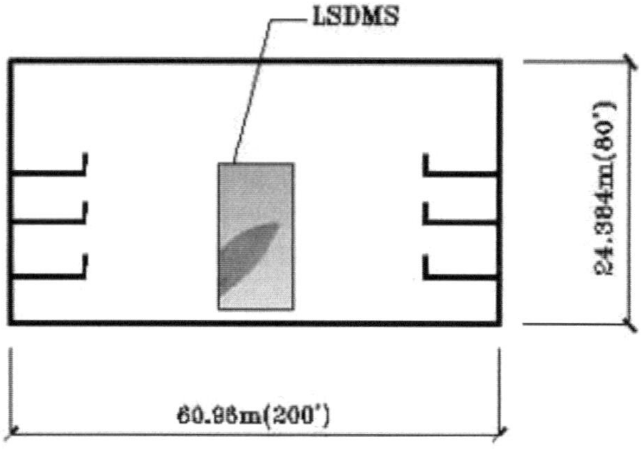

In order to compare effects of smoke toxicity in the two design fire scenarios, a couple of baseline evaluation criteria are established. The evaluation criteria are values with which the performance of the proposed design can be measured, compared and deemed successful (i.e., demonstrate equivalency to the code requirements). Numerous tenability criteria have been suggested and used in published fire hazard analyses and test studies over the past two decades. Table 1 presents the three evaluation criteria adopted in this study. They are used to analyze the modeling simulation results for the determination of the smoke layer interface (i.e., maintaining the height of the smoke layer at 1.83 meters above the Level 4 walking surface [7, 8]). This study only considers the effect of acute (short-term) exposures to toxic products, other than the chronic (long-term) effects. Fire Dynamics Simulator (FDS) version 6.4.0 [3] developed by National Institute of Standards and Technology (NIST) is used to determine the smoke layer height and evaluate the thermal and toxic threat to which building occupants are exposed.

Table 1 – Evaluation Criteria

Parameter	Description of Criterion	Value of Criterion
Temperature	Limit temperature to a maximum of 60°C (140°F) at a height of 1.83 meters (6 feet) above Level 4 walking surface [9, 10]	60°C
Visibility	Maintain visibility of at least 10 meters (33 feet) at a height of 1.83 meters (6 feet) above Level 4 walking surface [11]	10 meters
Carbon Monoxide (CO)	Limit CO concentration to 220 ppm at a height of 1.83 meters (6 feet) above Level 4 walking surface [12, 13]	220 ppm
Duration	Evaluate trial for 20 minutes [7, 8]	20 minutes

In addition, a design fire of 5,275 kW (5,000 Btu/s) with a fast T-square fire growth rate is used to design the regular atrium smoke management system and to evaluate smoke toxicity effects of the LSDMS fire on the proposed fire protection systems provided in the atrium.

For the LSDMS, its fuel characteristics, such as heat release rate per unit area, fire growth rate and fuel properties, including soot yield, carbon monoxide (CO) yield, heat of combustion, etc. can be obtained from ASTM E1354-11b testing report, and the fire spread speed can be roughly derived from ASTM E84-12 testing results. Figure 4 presents the fire development (i.e., HRR vs. time) of these two types of design fires used in this study. Table 2 summarizes other parameter values associated with these two design fires.

Table 2 – Summarized Input Parameters of Design Fires

Parameter	Regular Atrium Design Fire	LSDMS Design Fire
Soot Yield (kg/kg)	0.05	0.08
CO Yield (kg/kg)	0.04	0.032
Heat of Combustion (kJ/kg)	18,000	20,400

Figure 4 – Fire Development of Regular Atrium and LSDMS Fires

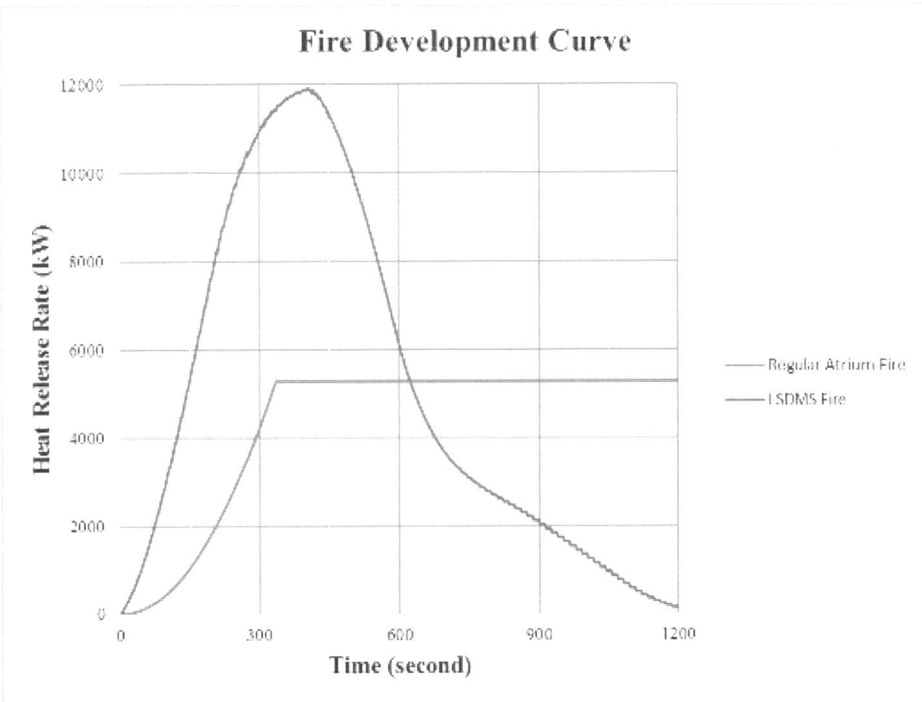

In addition to the evaluation criteria and design fire parameters, the following assumptions are included in the study. A default visibility factor of 3 for non-illuminated objects (i.e., light-reflecting signs) is used in the study per Mulholland, SPFE Handbook [9]. A total smoke exhaust airflow capacity of 396.44 m^3/s (840,000 ft^3/min) is provided at the top of the atrium and natural makeup air supply with a total free opening area of 356.75 square meters (3,840 square feet) is provided at north and south sides of the atrium on Level 1. Note that the smoke exhaust airflow capacity and makeup air opening area come from the regular atrium smoke management system design.

In this study, two base design fire scenarios are analyzed to evaluate smoke toxicity of the LSDMS fire on the fire protection systems that are designed to handle a regular atrium fire, as shown in Table 3.

Table 3 – Proposed Design Fire Scenarios

Fire Scenario	Sprinkler Controlled	Fire Growth Rate	Max. HRR (kW)	Smoke Exhaust Rate (m^3/s)
Regular Atrium Fire	No	Fast, T-Square Unsteady	5,275	396.44
LSDMS Fire	No	Fast - Ultrafast, T-Square Unsteady	12,000	396.44

The evaluation criteria listed in Table 1 to determine the smoke layer interface at 1.83 meters (6 feet) height above Level 4 are compared in two design fire scenarios, as shown in Figures 5 and 6. These two figures present the outcomes of visibility, a dominant criterion among all the three evaluation criteria to justify the smoke layer height and toxicity.

Figure 5 – Visibility isosurfaces indicating where visibility is 10m (black line shows 1.83m above Level 4)

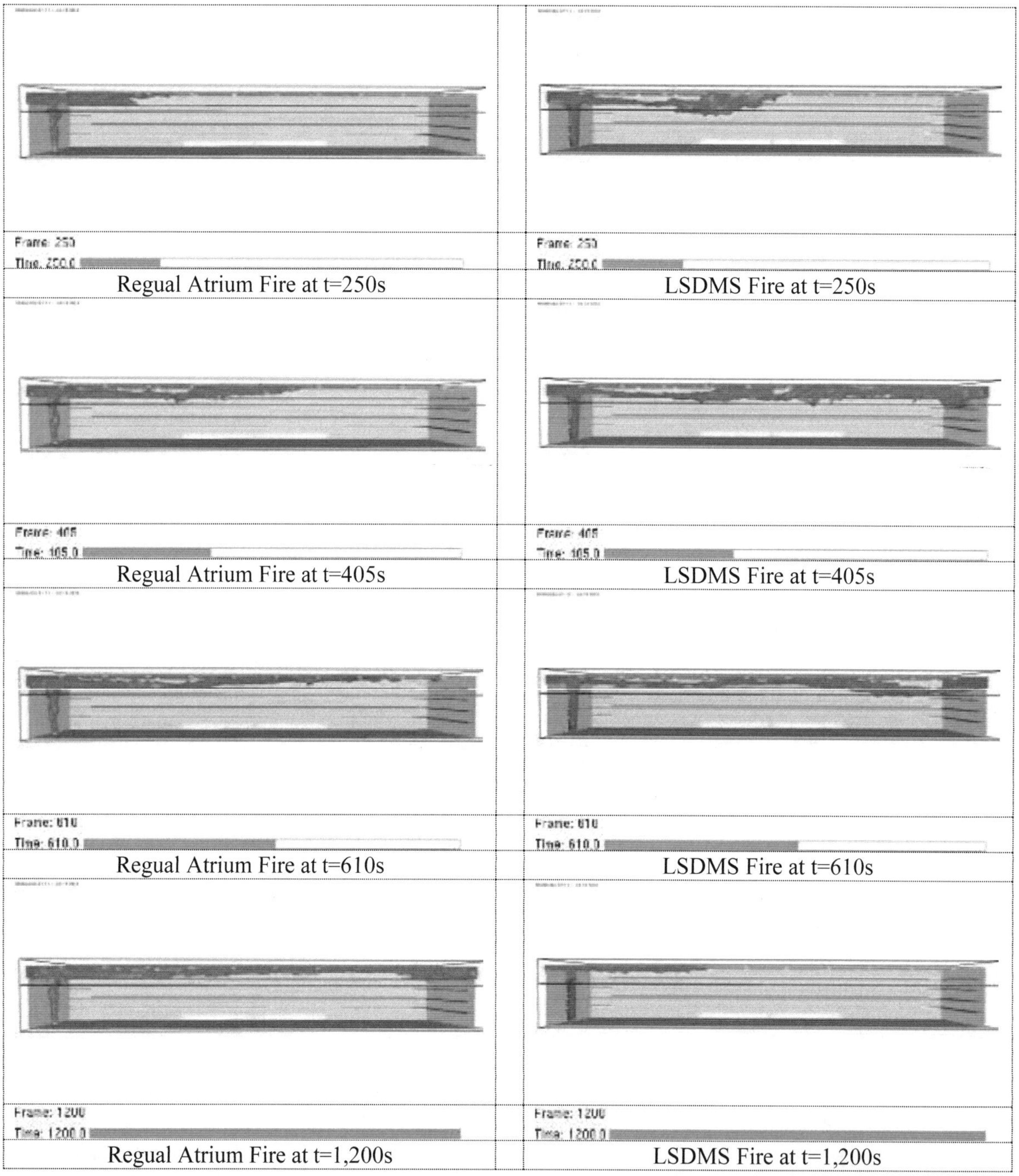

Regual Atrium Fire at t=250s | LSDMS Fire at t=250s

Regual Atrium Fire at t=405s | LSDMS Fire at t=405s

Regual Atrium Fire at t=610s | LSDMS Fire at t=610s

Regual Atrium Fire at t=1,200s | LSDMS Fire at t=1,200s

Figure 6 – Visibility horizontal slices indicating where visibility is 10m at 1.83m above Level 4

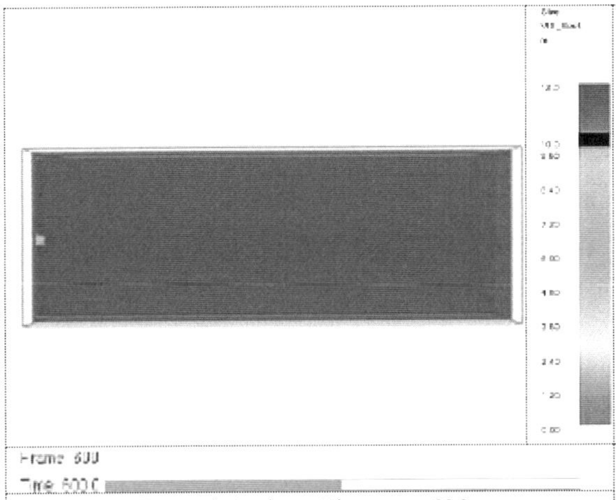

Regual Atrium Fire at t=600s

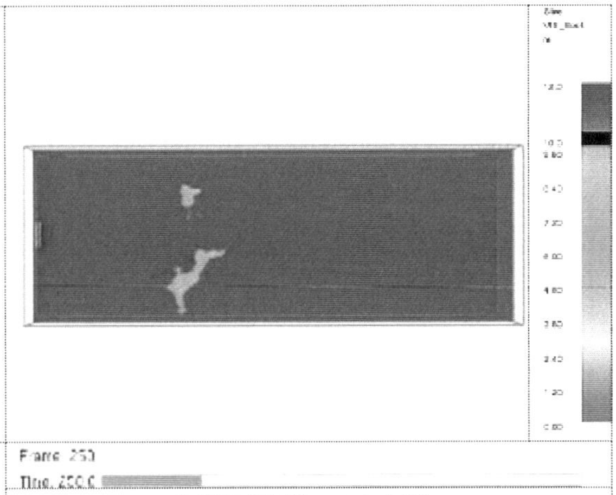

LSDMS Fire at t=250s

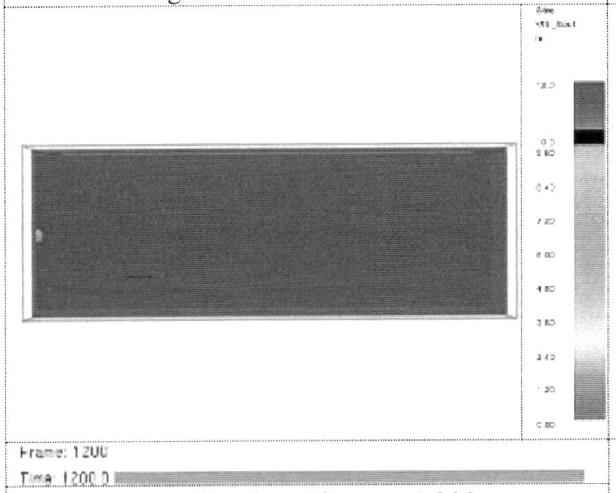

Regual Atrium Fire at t=1,200s

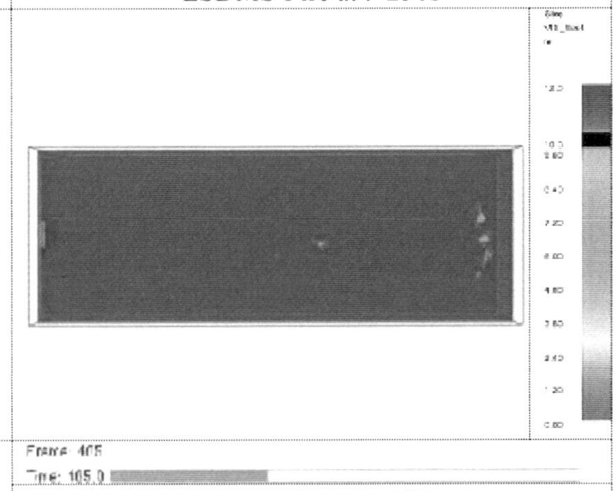

LSDMS Fire at t=405s

In the LSDMS fire scenario, the simulation results show that the smoke layer height starts descending below 1.83 meter on Level 4 at around 250 seconds and an untenable condition lasts up to approximately 610 seconds after ignition of the LSDMS fire. After that, the smoke layer height maintains above 1.83 meter on Level 4. In contrast, in the regular atrium fire scenario, the smoke layer height is able to maintain above 1.83 meter on Level 4 for the whole simulation duration of 1,200 seconds (20 minutes). Furthermore, it is observed that in the LSDMS fire scenario, fire grows faster and smoke spreads more rapidly throughout the atrium when compared with the regular atrium fire scenario, as shown in Figure 5.

Based on the fire modeling analysis results, smoke toxicity caused by the LSDMS fire adversely affects not only the fire protection systems (i.e., smoke management, alarm detection, etc.) designed and installed in accordance with current codes and standards, but also the safe evacuation of occupants inside the building.

According to the outcomes of smoke toxicity analysis, additional fire life safety mitigation features, such as increase of the smoke exhaust capacities, control of the LSDMS fire development, early detection of the fire alarm system, etc., need to be provided to compensate the deficiencies and discrepancies of the currently proposed fire protection systems. This study proposes the following approach as shown in Figure 7: using a water curtain system immediately installed at the top of the LSDMS to control the fire development, and providing an additional beam detector right above the LSDMS to achieve early detection of the fire.

Figure 7 – Additional Fire Protection Mitigation Features at LSDMS

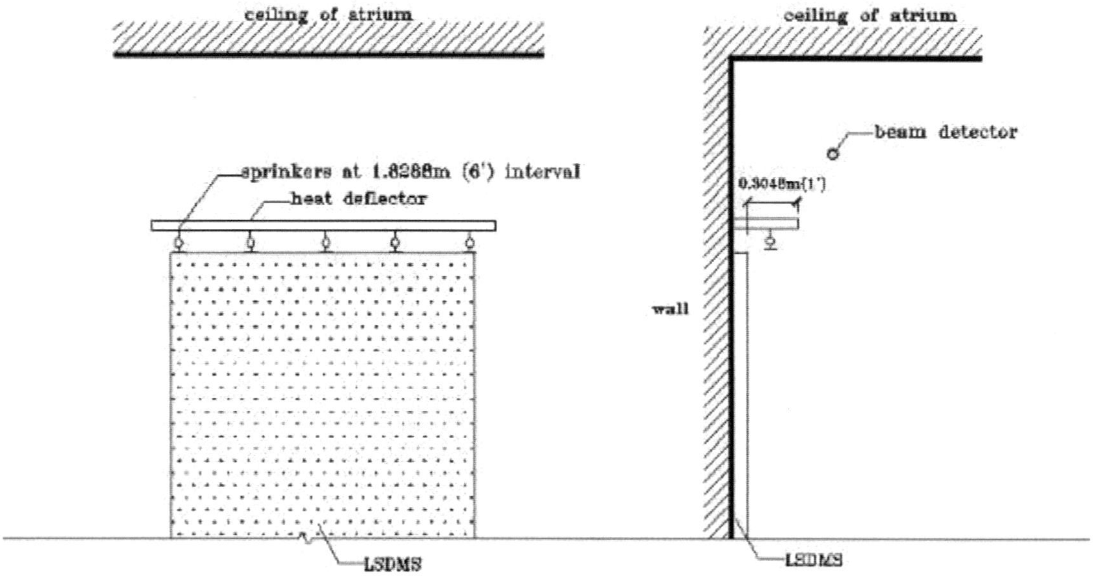

Based on the proposed additional fire life safety mitigation features, the LSDMS fire development (i.e., HRR) is able to be controlled as presented in Figure 8. The associated fire modeling analysis results are shown in Figure 9.

Figure 8 – Fire Development of Regular Atrium and Sprinkler-Controlled LSDMS Fires

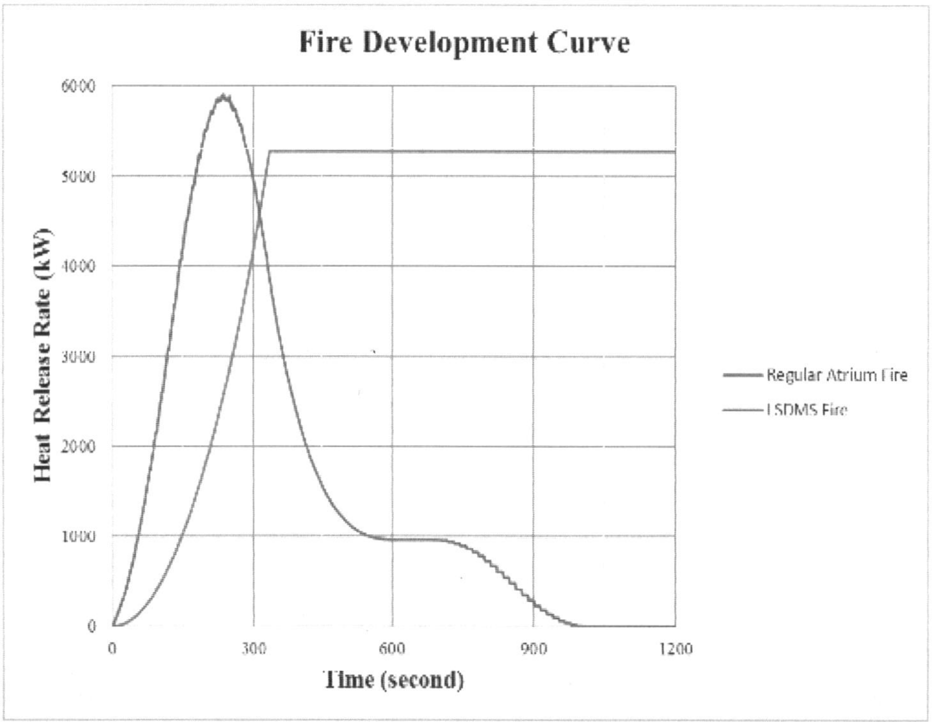

Figure 9 – Visibility isosurfaces indicating where visibility is 10m (black line shows 1.83m above Level 4)

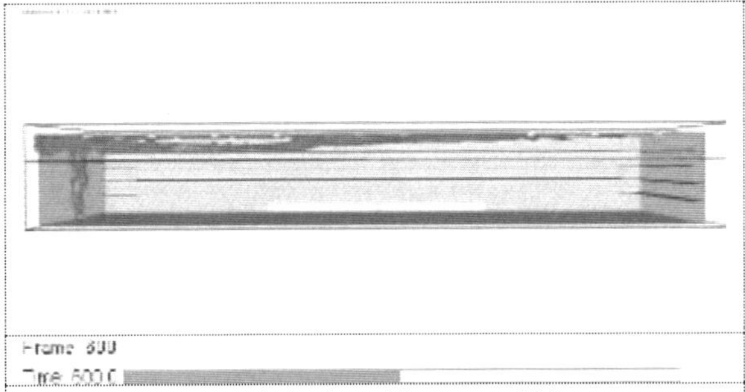

Regual Atrium Fire at t=600s

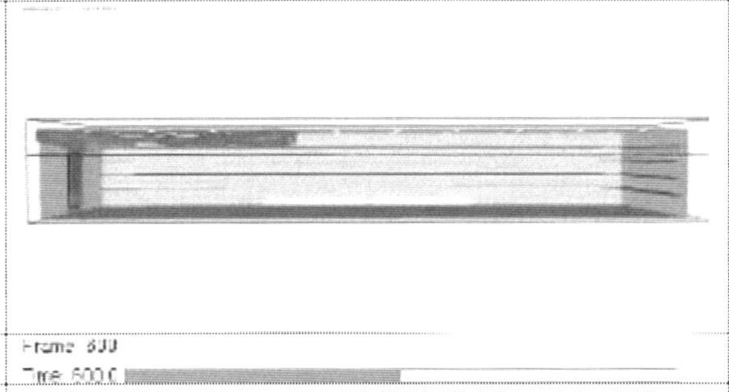

Sprinkler-Controlled LSDMS Fire at t=600s

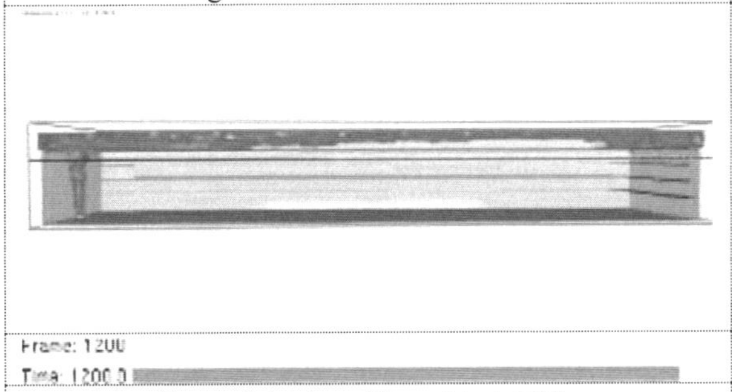

Regual Atrium Fire at t=1,200s

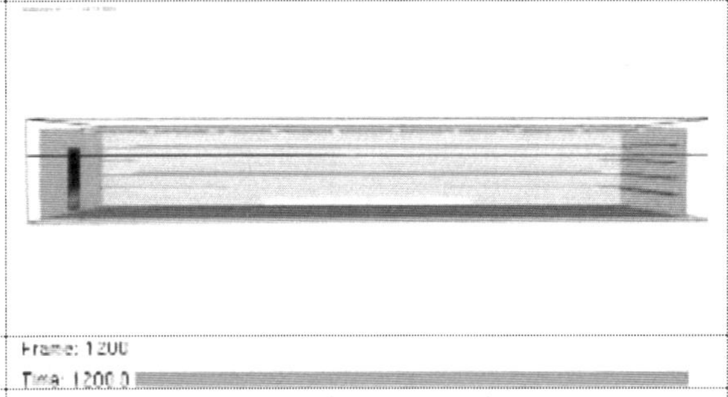

Sprinkler-Controlled LSDMS Fire at t=1,200s

The simulation outcomes demonstrate that upon provisions of the additional fire protection mitigation features and systems, a tenable environment is able to be maintained and the adverse effects brought by the indoor LSDMS can be eliminated.

Fire Spreadability and Egress Disability

In comparison with the regular atrium fire burnt on a horizontal surface, the indoor LSDMS is installed on a vertical surface and its surface flame typically spread vertically at a faster velocity and with a higher heat release rate. Therefore, it is of importance to evaluate the effects of radiant heat flux to the built environment (fire spreadability) as well as to the building occupants in proximity to such a fire (egress disability).

Fire spread can occur when radiant heat flux exceeds critical heat flux of adjacent items. According to Appendix 3 of 5th Edition SFPE Handbook [14], the critical heat flux for wood and most polymer products is approximately 10 kW/m^2. Furthermore, radiant heat flux can cause skin pain of occupants when the difference between the rate of supply heat to the skin surface exceeds the rate at which heat is conducted away from the skin [9, 15]. According to Chapter 63 of 5th Edition SFPE Handbook [9], a fairly obvious tolerance limit for skin exposure to radiant heat is around 1.7 – 2.5 kW/m^2.

In order to analyze the effects of radiant heat, thermocouples are placed in the FDS model at different distance from the burning surface to which LSDMS is attached. These thermocouples are to specify the output of gas phase scalar quantities (e.g., radiative heat flux) at a particular time and at a particular spot within the modeling environment.

When comparing the FDS thermocouple outcomes as shown in Figure 10 with the critical heat flux of 10 kW/m^2, It is found that items (i.e., sofa, table and chair) located within about 4 meters (13 feet) from the LSDMS may be ignited by the LSDMS fire. However, with provisions of additional fire protection mitigation features, the distance is reduced to about 2 meters (7 feet), as presented in Figure 10.

Figure 10 – Ignition Radiant Heat Flux by LSDMS Fires

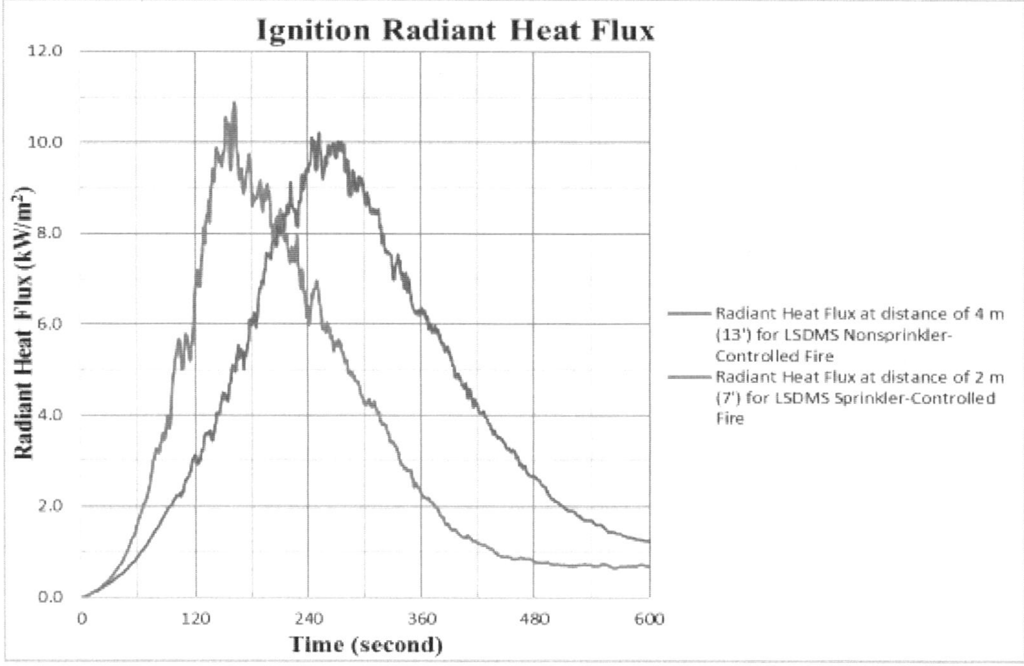

Similarly, when comparing the FDS thermocouple outcomes as shown in Figure 11 with the tolerant radiant heat flux of 1.7 – 2.5 kW/m^2, it is found that occupants located within about 12 – 9 meters (40 – 30 feet) from the LSDMS fire may feel skin burn. However, with provisions of additional fire protection mitigation features, the distance is reduced to approximately 8.5 – 6 meters (28 – 20 feet) as presented in Figure 11.

Figure 11 – Skin-Burn Radiant Heat Flux by LSDMS Fires

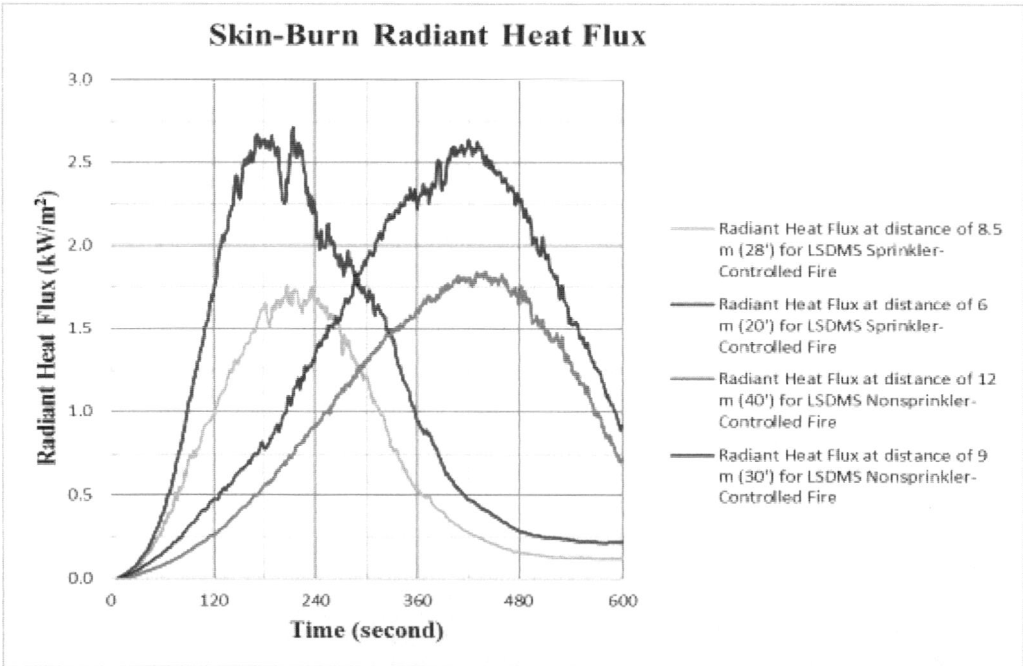

DISCUSSIONS

Compared with regular fires, fires caused by LSDMS typically grow faster and generate more heat (a higher peak heat release rate). Therefore, the fire protection systems, such as a smoke management system, designed on a basis of regular fires, are not able to handle fire hazards caused by LSDMS fires. Consequently, the deficiencies of the malfunctioned fire protection systems may cause occupants to be exposed to an unsafe built environment and fail to safely evacuate or relocate.

Furthermore, due to the vertical burning pattern, LSDMS fires usually have stronger fire spreadability when compared with regular fires. Potentially, fire spread throughout the built environment may increase the risk of occupants being exposed to a severely hazardous environment. In addition, the vertical burning pattern of LSDMS fires can cause skin burn of occupants located within a broad range around the fires. This implies that if exits are allocated in proximity to LSDMS, occupants may avoid using these exits due to such heavy radiant heat flux generated from burning LSDMS (egress disability).

Another finding from this study is that LSDMS fires are typically fuel-controlled fires and associated fire development experiences growth, peak and decay phases as shown in Figures 4 and 8. Such fuel characteristic of LSDMS can be utilized to control the fire size of LSDMS and further minimize effects of associated fire hazards if appropriate fire protection mitigation features are provided.

CONCLUSIONS

This study proposed the fire hazard analysis methodology using ASTM E84-12 and ASTM E1354-11b testing standards to evaluate fire hazards of indoor LSDMS. Through the preliminary engineering analysis, the study outcomes support UL's clarification on the applications of UL 94, i.e. UL 94 certification does not cover polymeric materials used for building construction, finishing or contents. The application of UL 94 flammability testing results to indoor LSDMS is inappropriate. Therefore, any indoor LSDMS should be evaluated through well-established fire hazard analysis methodologies.

Second, fire hazards caused by indoor LSDMS fires have severe impacts on the fire protection systems designed on the basis of current codes and standards. Due to absence of clearly defined regulations and guidelines, extra attentions should be paid on the design of fire life safety systems when a LSDMS is installed inside a building. As presented in this study, offering additional fire protection mitigation features will help control the fire development and further resolve the adverse effects brought by a LSDMS fire.

Third, as a result of the parametric study on effects of radiant heat flux from a LSDMS fire, when a LSDMS is proposed within a building, the interior design (i.e., furniture layout and decoration locations) and the emergency evacuation plan for the building should consider the effects of fire spreadability and egress disability caused by LSDMS fires.

This study performed a preliminary analysis on fire hazards of indoor LSDMS. The preliminary analysis outcomes apply to the fire hazards in a relatively large volume space with provisions of a smoke management system. A number of other factors, such as varieties on space volume and geometry, variations on LSDMS dimension, presence/absence of smoke management system, need to be further quantified on how they affect the fire hazards of indoor LSDMS.

In addition, the fire propagation rate obtained from ASTM E84-12 represents a *horizontal* flame spread rate. Since indoor LSDMSs are normally installed on vertical surfaces, the application of ASTM E84-12 fire propagation rate may not be accurate. Development of an appropriate engineering methodology to determine the vertical fire propagation rate for burning LSDMS will be necessary in future studies.

It is anticipated that based on this preliminary and further studies, prescriptive code compliance will be quantified and relevant design guidelines can be established to eventually resolve the fire life safety deficiencies brought by indoor LSDMS.

REFERENCES

[1] New Tom Bradley International Terminal Integrated Environmental Media System, Los Angeles World Airports, Los Angeles, USA.
Available: https://www.lawa.org/uploadedFiles/LAXDev/News_for_LAXDev/06 IEMS Fact Sheet FINAL.pdf

[2] Underwriters Laboratories, (2007) "Understanding UL 94 Certifications and Limitations", *The Code Authority newsletter, Issue 3*. Available: www.ul.com

[3] McGrattan, Kevin (ed.), (2016) *Fire Dynamics Simulator version 6.4.0*, National Institute of Standards and Technology.

[4] Daktronics, Inc. Available: http://www.daktronics.com/en-us/products/video-displays

[5] EKTA Company. Available: http://ekta-led.com/en/products/led-panels/equipment-for-rent

[6] Standard Vision, LLC. Available: http://www.standardvision.com/#content-anchor

[7] International Code Council, (2014) *2015 International Building Code Section 909.8*, ISBN-13 9781609834685, International Code Council, Inc.

[8] National Fire Protection Association, (2014) *NFPA 101: Life Safety Code Section 8.6.7 (5) 2015 Edition*, NFPA: Quincy, Massachusetts, USA.

[9] Purser, D.A., McAllister, J.L., (2016) "Chapter 63 Assessment of Hazards to Occupants from Smoke, Toxic Gases, and Heat", *SFPE Handbook of Fire Protection Engineering Fifth Edition*, pp. 2308 – 2428. New York: Springer Science +Business Media.

[10] Klote J.H., Evans D.H., (2007) *A Guide to Smoke Control in IBC 2006*, pp. 6-1 – 6-14, International Code Council, Inc.

[11] Yamada, T., Akizuki, Y., (2016) "Chapter 61 Visibility and Human Behavior in Fire Smoke", *SFPE Handbook of Fire Protection Engineering Fifth Edition,* pp. 2181 – 2206, New York: Springer Science +Business Media.

[12] Committee on Toxicology, National Research Council, (2010) "Chapter 2 – Carbon Monoxide", *Acute Exposure Guideline Levels for Selected Airborne Chemicals: Volume 8*, pp. 49 – 143, National Research Council of the National Academies.

[13] National Fire Protection Association, (2013) *NFPA 130: Standard for Fixed Guideway Transit and Passenger Rail Systems Annex B Section B.2.1.2 2014 Edition*, NFPA: Quincy, Massachusetts, USA.

[14] Hurley, M.J., (ed.) (2016) "Appendix 3: Fuel Properties and Combustion data", *SFPE Handbook of Fire Protection Engineering Fifth Edition,* pp. 3437 – 3475, New York: Springer Science +Business Media.

[15] Christopher, J.W. and Nicholas, A.D., (2016) "Chapter 68. Effects of Thermal Radiation on People: Predicting 1st and 2nd Degree Skin Burns", *SFPE Handbook of Fire Protection Engineering Fifth Edition,* pp. 2705 – 2737, New York: Springer Science +Business Media.

TRAVELLING FIRE SPREAD BETWEEN VEHICLES IN CAR PARKING BUILDINGS

Mohd Zahirasri Mohd Tohir*[1] & Michael Spearpoint[2]
[1]Universiti Putra Malaysia, Malaysia
[2]University of Canterbury, New Zealand

ABSTRACT

Previous work to develop a risk-based vehicle fire spread model for car parking buildings has assumed that vehicles burn simultaneously such that the fire severity is represented by the summation of the peak heat release rate of each vehicle. However, this assumption appears to be highly unlikely particularly where a large number of vehicles are involved. This paper seeks to improve on the previous assumption by examining the growth characteristics of vehicle fires rather than only considering the combination of the peak heat release rate values. It is expected that in a multiple vehicle scenario that after the fire grows in the first vehicle, it will spread to a second vehicle and by the time the fire spreads to subsequent vehicles it is possible that the fire in the initial vehicles will be declining. This is a phenomenon that can be regarded as a form of travelling fire in which a fire in a large space burns over limited area at any one time. To analyse these fire growth characteristics a multiple vehicle fire tool has been developed. The main feature of the tool is its ability to conduct a probabilistic analysis of multiple vehicle fire spread on a single row arrangement of parked vehicles. In this work, simulations are set up considering a scenario in car parking building in which 29 vehicles are lined up in a single row. It is found from the simulations that a travelling fire in a multiple vehicle fire scenario can occur such that the peak heat release rate approaches a steady-state value rather than continuing to increase as more vehicles are considered.

INTRODUCTION

Vehicle fires in car parking buildings can impact on the life safety of the vehicle occupants as well as the building occupants who are in the vicinity of the fire. Vehicle fires in car parking buildings can also result in material losses in terms of the vehicles itself, to the building structure as well as to neighbouring property. Previously there have been several significant fires in car parking buildings involving multiple vehicles and in some cases these have led to fatalities. For example, in 2006 there was a car park fire in Bristol, United Kingdom where 22 vehicles were destroyed in the incident and one person died in the occupancy above the car park[1]. Therefore, it is useful to understand the risks of vehicle fires and the need to potentially reduce their likelihood of occurrence and/or mitigate the severity if a fire does occur.

The expectation in a multiple vehicle scenario is that after a fire grows in the first vehicle ignited, it will spread to subsequent vehicles by which time it is possible that the fire in earlier ignited vehicles is in decline. This is a phenomenon is known as a travelling fire[2] in which a fire in a large space burns over limited area at any one time. The time to ignition of subsequent vehicles depends on a number of factors including the heat release rate profile of the already burning vehicle, the distance between vehicles, the materials that can likely ignite on the target vehicles etc. These characteristics are not fixed but will vary depending on a specific scenario.

Several previous studies have assumed the time to ignition of a subsequent vehicle after the first vehicle has ignited and the design fire of the vehicles can be represented deterministically. For example, in the multiple vehicle simulation study by van der Heijden[3] it was assumed that the second vehicle will burn 10 min after the ignition of the first vehicle and the third vehicle will burn 5 min after the second car starts burning. Similarly in a multiple vehicle fire study by Collier[4] the time of ignition and the heat release rate curve suggested by Schleich[5] was adopted so that the time to ignition of each subsequent vehicle was 12 min. Other work which uses similar assumptions suggested by Schleich includes that by Jansen[6], Nordijk and Lemaire[7] and de Feijter and Breunese[8].

Consequently, there are two major components in performing a simulation of multiple vehicle fires, namely (1) the selection of the design fire for the vehicles involved and, (2) the time to ignition of subsequent vehicles after earlier vehicles have ignited. These two components have been addressed by Tohir et al.[9] leading to the work described here in which a probabilistic design fire and ignition prediction method have been combined to create a simulation tool which is able to forecast the overall heat release rate history for a multiple vehicle fire scenario.

The findings in this work are important in determining an overall heat release rate history for a multiple vehicle fire scenario. In the earlier fire risk analysis by Tohir and Spearpoint[10] one element required is the fire severity. This element is represented by the peak heat release rate such that the higher peak heat release rate the greater the fire risk level. As an initial approximation Tohir and Spearpoint[10] assumed that all vehicles in a cluster burn simultaneously, where a cluster is taken to be a group of closely neighbouring vehicles. However, this assumption appears to be highly unlikely when compared to the statistics discussed in Tohir and Spearpoint[10] where most of the reported cases resulted in only one vehicle being involved. Even for cases of more than single vehicle involved, it would seem very unlikely that they would catch fire and burn at their peak heat release rate simultaneously.

MULTIPLE VEHICLE FIRE SCENARIO TOOL

Vehicle fleet distribution

In this work vehicles are categorised by their curb weight according to the ANSI classification system[11] as shown in Table 1.

Table 1. Single passenger vehicle ANSI classification by curb weight.

Classification	Curb weight
Passenger car: Mini	1500 – 1999 lbs (680 – 906 kg)
Passenger car: Light	2000 – 2499 lbs (907 – 1134 kg)
Passenger car: Compact	2500 – 2999 lbs (1135 – 1360 kg)
Passenger car: Medium	3000 – 3499 lbs (1361 – 1587 kg)
Passenger car: Heavy	≥ 3500 lbs (≥ 1588 kg)
Minivan/MPV	Unspecified
SUV	Unspecified

Vehicle fleet distribution is an available input for the simulation tool as a process of selecting the classification of each vehicle. Figure 1 shows a vehicle fleet distribution obtained from Tohir and Spearpoint[10] based on the data from the European Union and USA. In addition, a survey of almost 5,000 vehicles in New Zealand by Anderson and Bell[12] is also shown for comparison.

Figure 1. Distribution of vehicle population curb weights, standard deviation indicated for Tohir & Spearpoint[10] values.

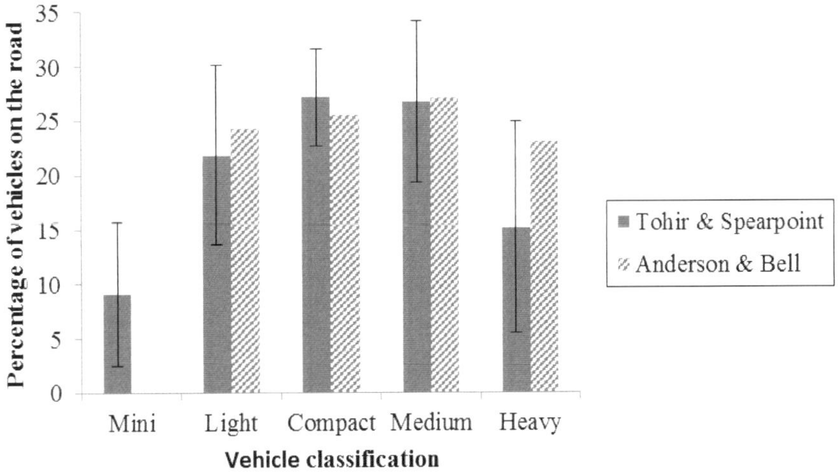

Characterisation of design fire curves

Tohir and Spearpoint[13] discuss the characteristics of single passenger vehicle design fires based on experimental results from the literature. As mentioned previously, design fires for single passenger vehicles that are widely adopted in the literature are often either represented by a specified heat release rate history or as mathematical functions with fixed coefficients irrespective of the size (curb weight) of the vehicle.

The design fire in this work, presented in terms of heat release rate as a function of time, is represented by mathematical functions for the growth and decay phases. Therefore, in constructing the design fire, information regarding the peak heat release rate, fire growth coefficient and fire decay coefficient must be determined. For the simulation tool used in this study the growth is be defined as a t-squared function such that

$$\dot{Q}(t) = \alpha_{peak} t^2 \qquad (t \leq t_{peak}) \qquad [1]$$

where t_{peak} is the time at which the maximum heat release rate is reached and the decay is defined by an exponential function is such that

$$\dot{Q}(t) = \dot{Q}_{max} e^{\beta_{exp}(t - t_{peak})} \qquad (t \geq t_{peak}) \qquad [2]$$

Figure 2. Characterisation of a vehicle design fire using the peak growth and exponential decay methods.

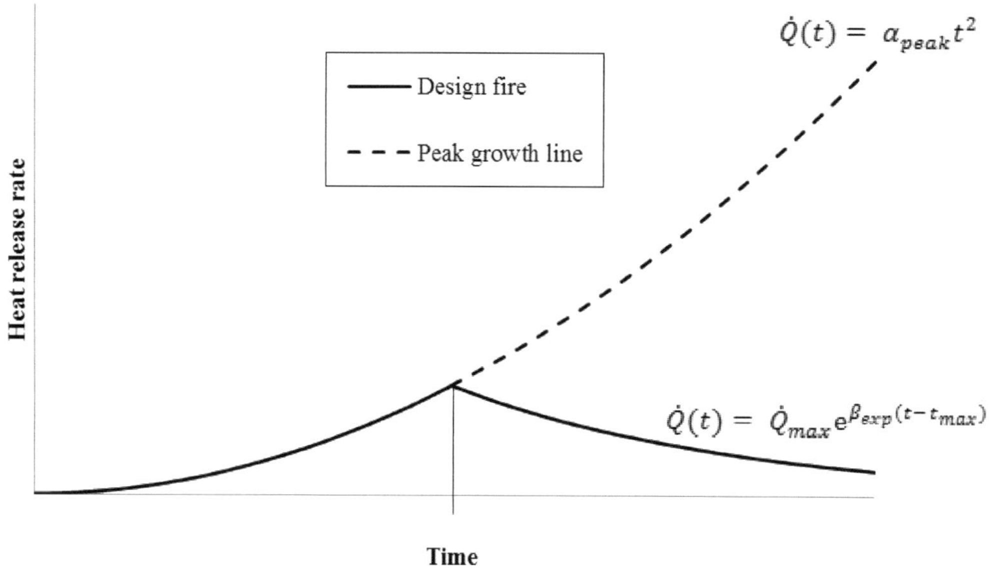

Figure 2 illustrates the combination of Equation 1 and Equation 2 to form a single vehicle design fire curve. Fixed values for the coefficients can be used to match the design fire curve to individual experiments published in the literature. To apply the design fire curve in the simulation tool, values for the peak heat release rate, fire growth coefficient and fire decay coefficient parameters for different curb weight classifications come from sets of probability distributions introduced in Tohir and Spearpoint[13]. Table 2 outlines the distribution statistics for peak heat release rate, fire growth coefficient and decay coefficient for every classification. The parameter κ is the shape parameter, which determines the shape of a distribution function and parameter θ is the scale parameter, which determines the position of the data distribution along the x-axis.

Table 2. Summary of the single vehicle distribution analyses for peak heat release rate, fire growth coefficient and decay coefficient parameter values.

		Peak heat release rate, $\dot{Q}_{peak}$ (kW)		**Fire growth coefficient, α_{peak} (kW/min²)**		**Fire decay coefficient, β_{exp} (min⁻¹)**	
Distribution shape		Weibull		Gamma		Weibull	
Distribution parameters		**κ**	**θ**	**K**	**θ**	**κ**	**θ**
Class	**Mini**	5.19	3809	1.39	11.86	0.93	0.17
	Light	1.66	5078	1.23	14.78	1.21	0.11
	Compact	2.40	4691	1.18	5.14	3.93	0.08
	Medium	3.18	7688	2.24	2.75	1.38	0.11
	Heavy	3.11*	8723*	1.51*	1.82*	1.86*	0.11*

*extrapolated values

Tohir and Spearpoint[13] found that due to a lack of data, the Minivans/MPVs and SUVs classifications had to be included in the appropriate vehicle classification by using their known curb weights rather than by identifying them as separate categories. Also due to lack of data, the distribution parameter values for the Heavy classification were extrapolated from the distribution parameters of the lighter weight classes. Since there was no clear trend from the lighter weight class values an average value for the shape parameter (κ) is used.

However, for the scale parameter (θ), different approaches were taken for each distribution. For the peak heat release rate distribution an increasing linear trend was used for the extrapolation as the curb weight class increases. For the fire growth coefficient, a decreasing exponential trend was used for the extrapolation, since it was the best fit as the curb weight class increases. For the fire decay, it was evident from the analysis that there was no obvious trend, thus a value of 0.11 was assumed for the Heavy class fire decay coefficient using the average values of the lighter weight classes.

Ignition prediction

A method to predict the time to ignition of a target vehicle, given another is already burning is discussed in Tohir et al.[9]. This approach uses the flame radiation point source model (PSM) for predicting heat flux from a burning vehicle, and uses the flux-time product (FTP)[14] as the ignition criterion for the prediction for the time to ignition of the target vehicle. The use of the PSM and FTP are consistent with the work of Baker et al.[15] and is incorporated into the B-RISK fire model[16].

The work by Tohir et al.[9] suggests that 0.3 be used for the radiative fraction in the PSM for the prediction of heat flux. In addition it has been shown that assuming a material representative of vehicle bumper trim gives a reasonable match with experimental measurements using an FTP value of 21862 $(kW/m^2)^n \cdot s$, a power law index of 2 (thermally thick) and a critical heat flux of 3.1 kW/m^2. These attributes are used as the specified values for the simulation tool applied to this work.

Effective distance

The effective distance is the radial distance from the heat source to the nearest point on the target vehicle. Using the ignition prediction study by Tohir et al.[9] an approximation for the location of the heat source position is suggested as the at the far end of the burning vehicle, away from the target vehicle. The effective distance is best explained in Figure 3 where a is the width of the burning vehicle and as a simplification it is assumed that each vehicle is parked in the middle of its parking space such that the distance to the edge of the parking space is b. Thus the distance from the burning vehicle to the target vehicle is $2b$ and the effective distance is simply $a + 2b$.

Figure 3: Illustration of the effective distance.

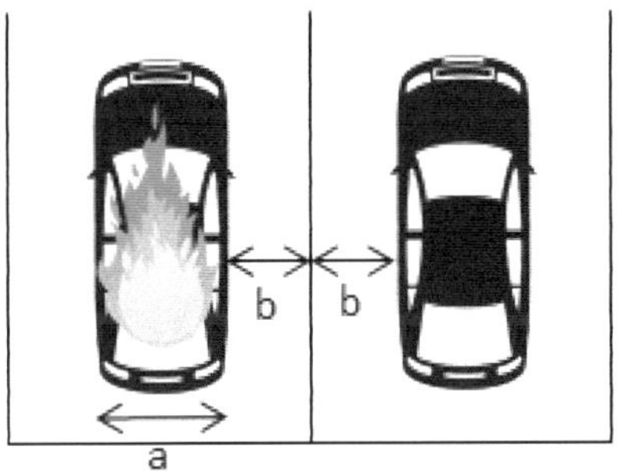

Simulation tool

The simulation tool used in this work combines the selection of single passenger vehicle design fires and time to ignition predictions to simulate a multiple vehicle fire scenario. The tool has been written in Microsoft Excel and Visual Basic Application (VBA) to achieve the combination of the two components. This is then coupled with @RISK software for generating random numbers from different sets of probability distributions. The overall simulation process is shown in Figure 4.

Figure 4. Simulation process.

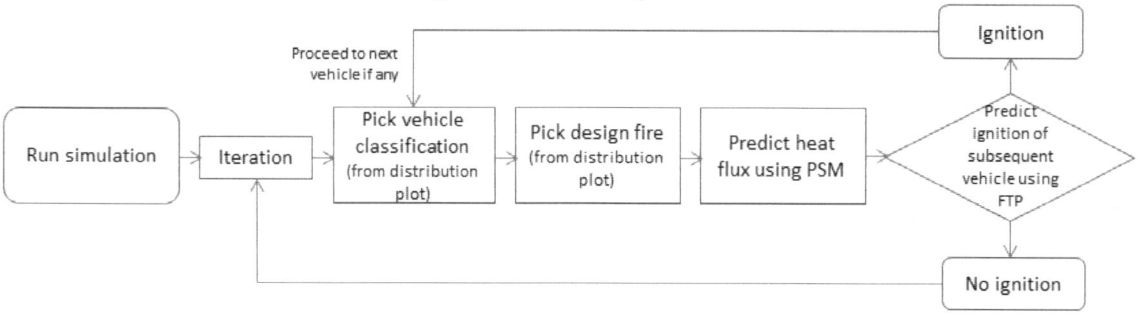

Prior to running the simulation tool, the user enters the attributes desired, i.e., number of iterations, classification of vehicles and the effective distance. The user has the option to set the classification of vehicles manually or to randomly select the probability distribution of from one of the two pre-defined vehicle fleets described above.

On starting a simulation, the tool selects vehicle classification for the first vehicle either from the vehicle fleet distribution or using the specified classification if it has been assigned manually. The tool then constructs a design fire for the first vehicle using a random sampling from the fire growth coefficient; peak heat release rate and fire decay coefficient probability distributions using statistics shown in Table 2. Here the tool uses the Monte Carlo feature in the @RISK software to select the random samples from the probability distributions. The design fire for the first vehicle is used to predict the heat flux received by the second vehicle using the PSM. If the second vehicle receives sufficient radiant energy from the first vehicle (i.e., when the accumulated FTP exceeds the FTP threshold value) then the second vehicle is ignited. The tool proceeds to select the classification for the second vehicle and consequently chooses its design fire using the same procedure as before. The ignition process is repeated until no further vehicles ignite or all of the vehicles in the cluster are involved in the fire. Once the design fire curves for each ignited vehicle have decayed to zero the tool proceeds to the next iteration until the defined number of iterations have been completed.

It is assumed that all the vehicles potentially receive heat flux from any burning vehicle, which disregards the possibility of a blocking body in between. This assumption follows Baker et al.'s[16] assumption in the creation of the B-RISK software. For example, as shown in Figure 5, Vehicles 2 and 3 receive heat flux from Vehicle 1, although Vehicle 2 is supposedly blocking the radiation from Vehicle 1. After Vehicle 2 ignites, Vehicle 3 receives heat flux from Vehicles 1 and 2 until either one of the heat sources burn out. This also occurs for the next vehicle, and so on.

Figure 5. Example of three-vehicle fire spread.

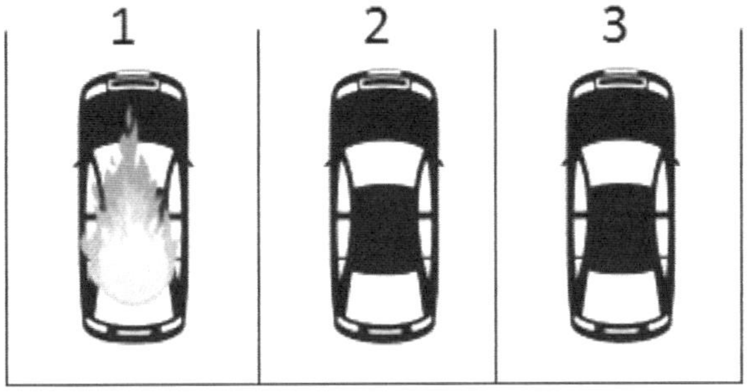

The tool includes a number of other simplifications. It does not consider any other effects such as radiation from the underside of the hot upper layer during the fire and other compartment effects. It is assumed that there is no external wind effect on the spread of fire which may be the case for car parking buildings with open sides and also there is no manual or automatic fire suppression.

METHODOLOGY

For the purpose of getting the multiple vehicle fire scenario overall heat release rate history some assumptions for the input parameters are required as summarised in Table 3. For the vehicle width, an assumed value of 1.6 m is used based on the average width of several vehicles in the Passenger car: Mini classification. This assumed width is chosen as to produce the lowest reasonable effective distance. The distance between vehicles is taken as 0.7 m as discussed by de Feijter and Breunese[8] as a generic separation distance. The addition of the vehicle width and the distance between vehicles equates to 2.3 m which is taken as the effective distance. The effective distance is essentially the parking width, and from Tohir[10], the range of parking space widths from the literature ranges from 2.2 – 3.2 m. Thus the 2.3 m effective distance is at the lower end of the range and represents a worst case scenario in terms of potential fire spread.

Table 3. Parameters for the multiple vehicle scenario simulation.

Parameter	Value
Vehicle width	1.6 m
Distance between vehicles	0.7 m
Effective distance	2.3 m
Parking width	2.3 m
No. of iterations	10,000 times
Distribution of vehicle fleet	European & USA

Another important assumption for the analysis is the arrangement of the vehicles during the fire. This is illustrated in Figure 6 which is an example of five-vehicle cluster scenario. To maximise the consequence, it is assumed that in each simulation iteration the first vehicle ignited is the one in the middle of the row so that the fire can spread in the two directions. Since the simulation tool is limited to assuming the vehicles are arranged in a single row starting the fire in the middle also somewhat represents a two-row fire. To allow fire spread to an even number of vehicles to the left and right of the initial vehicle only odd number cluster sizes are used. For this work, the number of vehicles involved in a scenario is varied using odd numbers between 1 and 29 where 29 is the maximum odd number of vehicles permitted in the tool.

Figure 6: Example of five-vehicle cluster scenario.

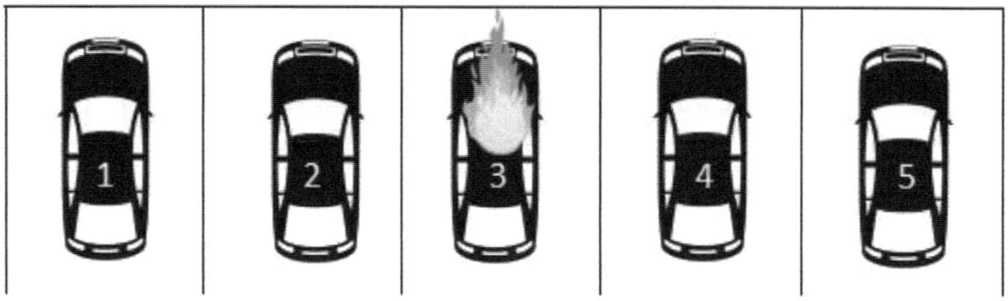

RESULTS

Simulations for each cluster size consist of 10,000 iterations and the results are presented in the form of a plot of peak heat release rate against vehicle cluster size as shown in Figure 7. The plot shows the average peak heat release rates ($\dot{Q}_{peak,ave}$) with the maximum and minimum values shown by the range bars. As the number of vehicles in a cluster increases the average and maximum peak heat release rate also increase whereas the minimum shows very little change. A logarithmic trend line fitted through the average peak heat release rate values, with an R^2 value of 0.99, is given as

$$\dot{Q}_{peak,ave} = 6197 \ln(n) + 6158 \qquad [3]$$

where n is the vehicle cluster size. This equation can be used to extrapolate the average peak heat release rate for larger vehicle cluster sizes for further analysis. The trend supports the hypothesis of a travelling fire in that by the time the fire spreads to subsequent vehicles the heat release rate for the previous vehicles is already declining.

Figure 7: Peak heat release rate for vehicle cluster sizes using the fire spread simulation tool.

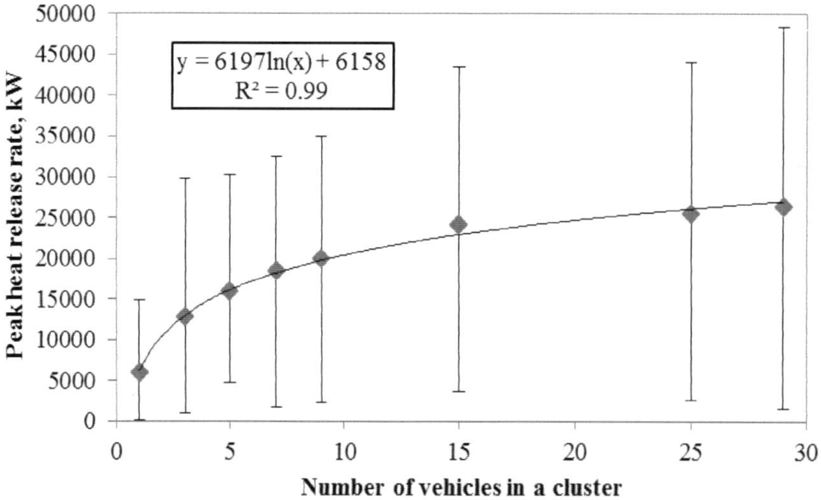

Using Equation 3, the expected average peak heat release rate for 1 to 29 vehicles in a cluster can be compared with a plot where it is assumed that all vehicles in the cluster catch fire simultaneously taken from the work by Tohir and Spearpoint[2]. For example, when there are three vehicles in a cluster, it was assumed by Tohir and

Spearpoint[2] that all three vehicles catch fire simultaneously, thus tripling the peak heat release rate due to the combination of each individual vehicle heat release rate curves.

The comparison of the plots is shown in **Error! Reference source not found.**. The red squares are the average peak heat release rate for all vehicles in the cluster where they are assumed to catch fire simultaneously while the blue diamonds are the peak heat release rate using Equation 3. It is clear that when the cluster size is small the average peak heat release rate are similar but the difference between the two predictions increase as the number of vehicles in a cluster increases. This means that the use of the fire growth characteristics as a consequence component in the probabilistic quantitative risk analysis model will significantly change the fire risk level as demonstrated in **Figure 9**. It is evident from the figure that the fire risk level difference obtained using data from Tohir and Spearpoint and by using the data using Equation 3 could reach to up to 3.5[th] order of significance.

Figure 8: Comparison Equation 3 with Tohir and Spearpoint[2] for average peak heat release rate versus cluster size.

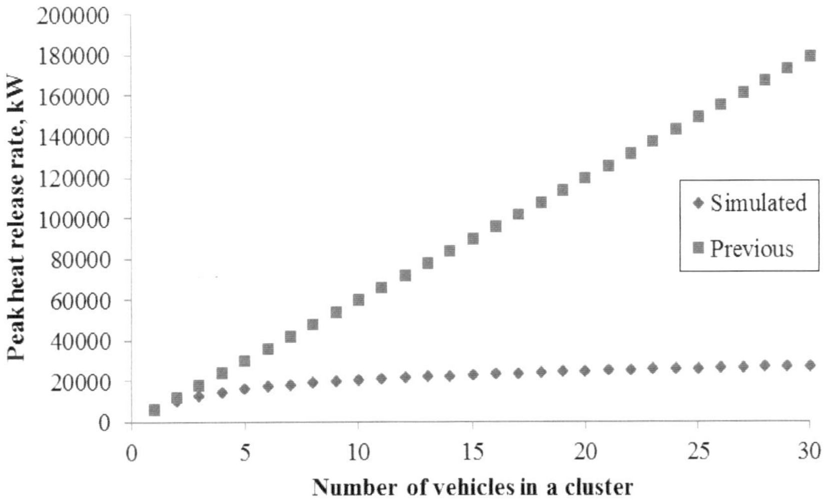

Figure 9: Comparison of the fire risk level from previous work and current analysis

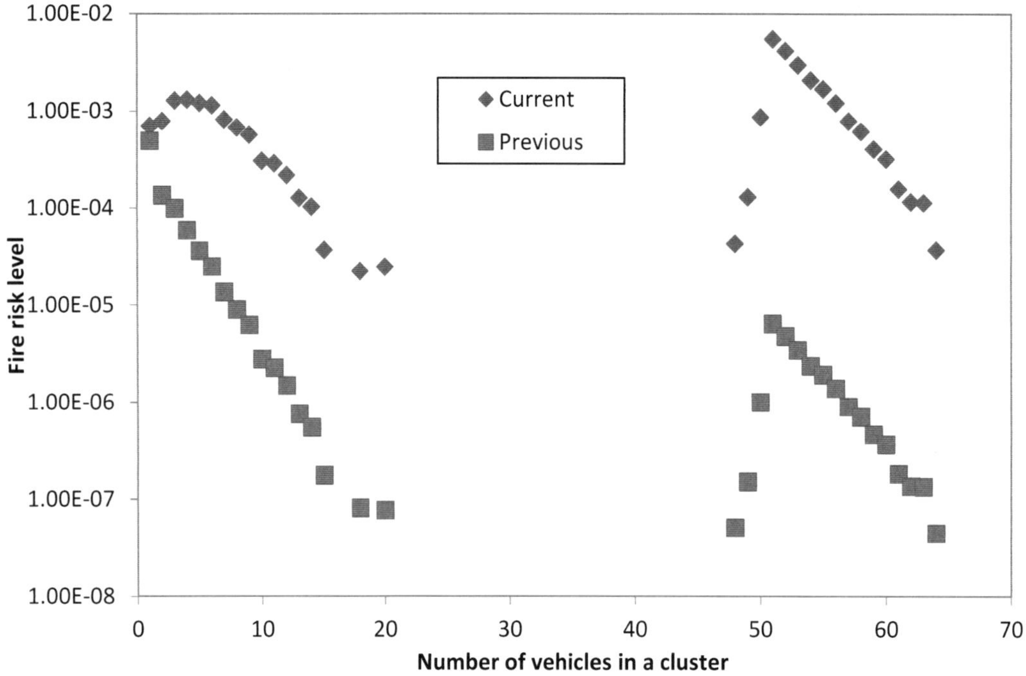

CONCLUSION

A probabilistic multiple vehicle fire simulation tool for car parking buildings has been created in the work by Tohir et al. This tool has the ability to create design fire curves for different vehicle classifications and predict the ignition of neighbouring vehicles using the PSM and FTP calculations. A series of simulations using the tool for different vehicle cluster sizes has been carried out to predict the peak heat release rate during a fire. The simulations indicate that the fire moves from one vehicle to another while the earlier ignited vehicles run out of fuel. This results in the average peak heat release rate increasing as a logarithmic trend as the cluster size increases and thus suggests that a travelling fire in a multiple vehicle fire scenario can occur. Therefore, the application of the fire growth characteristics to determine the fire severity component of a fire risk analysis will be reduced when compared to the previous assumption that all vehicles simultaneously ignite and burn with their peak heat release rate.

REFERENCES

[1] Shipp, M. (2010). Fire spread in car parks, Technical Report BD2552, Building Research Establishment (BRE), London,United Kingdom.

[2] Stern-Gottfried, J. and Rein, G. (2012). Travelling fires for structural design–Part I: Literature review. Fire Safety Journal, 54, 74-85.

[3] van der Heijden, M. G. M. (2010). Heat and smoke removal in semi-open car parks. Master's thesis, Eindhoven University of Technology, Eindhoven, Netherlands.

[4] Collier, P. C. R. (2011). Car parks - Fires involving modern cars and stacking systems. Building Research Association New Zealand (BRANZ), New Zealand, BRANZ Study Report 255.

[5] Schleich, J. B., Cajot, L. G., Pierre, M., and Brasseur, M. (1999). Development of design rules for steel structures subjected to natural fires in closed car parks. European Commission, Luxembourg.

[6] Jansen, D. (2010). Autobranden in parkeergarages - brandscenario's, brandsimulaties en de gevolgen voor constructies. Postgraduate thesis, Universiteit Gent.
[7] Noordijk, L. and Lemaire, T. (2005). Modelling of fire spread in car parks. HERON, 50(4).
[8] de Feijter, M. P. and Breunese, A. J. (2007). Investigation of fire in the Lloydstraat car park, Rotterdam. Efectis Nederland BV, Postbus, Netherlands, 2007-Efectis-R0894(E).
[9] Tohir, M.Z.M., Spearpoint, M., and Fleischmann, C.M., Prediction of time of ignition in a multiple vehicle fire spread experiment. Submitted to Fire & Materials.
[10] Mohd Tohir, M. Z. and Spearpoint, M. (2014) Development of fire scenarios for car parking buildings using risk analysis, Fire Safety Science, 11, 944-957.
[11] American National Standards Institute, Manual on Classification of Motor Vehicle Traffic Accidents, ANSI D16.1-2007, 2007.
[12] Anderson, C. M. and Bell, N. M. (2014). Analysis of vehicle distribution in parking buildings. presented at the Civil and Natural Resources Engineering Research Conference, University of Canterbury, New Zealand, vol. 2 pp. 65 – 72.
[13] Mohd Tohir, M.Z, and Spearpoint, M. (2014). Simplified approach to predict heat release rate curves from multiple vehicle fires in car parking buildings. Proc. 3rd International Fires in Vehicles Conference, Berlin, Germany.
[14] Smith, E. E. and Satija, S. (1983). Release rate model for developing fires. Journal of Heat Transfer, 105(2); 281-287.
[15] Baker, G., Collier, P. C. R., Wade, C., Spearpoint, M., Fleischmann, C. M., Frank, K., and Sazegara, S. (2013). A comparison of a priori modelling predictions with experimental results to validate a design fire generator submodel. 13th International Fire and Materials Conference, San Francisco, CA, pp. 449 - 460.
[16] Wade, C. (2004). BRANZFIRE Technical Reference Guide. Building Research Association of New Zealand (BRANZ), Judgeford, New Zealand, BRANZ Study Report 92.
[16] Baker, G. B., Spearpoint, M. J., Fleischmann, C. M., and Wade, C. A. (2011). Selecting an ignition criterion methodology for use in a radiative fire spread submodel. Fire and Materials, 35(6).

IGNITION AND PYROLYSIS MODELING

Determining Ignition Temperature Using Dynamic Thermal Analysis

Richard E. Lyon and *Natallia Safronava
Aviation Research Division
Federal Aviation Administration
Atlantic City International Airport, NJ

*Technology and Management International, LLC, Toms River, NJ

ABSTRACT

The surface temperature at piloted ignition in a fire test is derived from gas phase criteria for the initiation of flaming combustion of solids and related to the temperature of a milligram sample in thermal analysis by the requirement for controlled heating. It is found that initiation of sustained piloted ignition in a fire test and the onset of thermal decomposition in thermal analyses occur at similar temperatures during transient heating of combustible solids, and that these temperatures are coincident when the sample is in thermal equilibrium with the heat source. Under these conditions, the surface temperature at the initiation of sustained flaming combustion depends only on the ignition criteria in the fire test and the thermokinetic properties of the solid. Thermal analysis temperatures from derivative thermogravimetry and microscale combustion calorimetry are in reasonable agreement with measured ignition temperatures in standard fire tests for 16 polymers.

INTRODUCTION

Ignition of liquids occurs when a fluid is heated to a uniform temperature at which the equilibrium vapor pressure in air reaches the lower flammability limit of the fuel gases. In a closed-cup experiment, these conditions are well defined and the flash point temperature is calculable from the system properties. An approximation of this closed environment test for solids is the vertical tube furnace test for ignition temperature of plastics[1]. In open-cup measurements of liquid ignition temperature, the fuel diffuses away from the surface producing a concentration gradient that monotonically decreases with height, so the ignition temperature depends on the position of the igniter and the characteristics of the airflow. The analog of this open-cup test for liquids is a fire calorimeter [2] or ignitability test [3] for solids. In these fire tests, a sample is exposed to radiant heat on one side and ignition occurs at the surface in the gas phase when the solid thermally decomposes to gaseous fuels that mix with air at an igniter. Transient ignition occurs when the fuel/air mixture first reaches the lower flammability limit or a critical energy density at the igniter [4,5]. Sustained ignition commences when the fuel/air mixture increases beyond the lower flammability limit at the igniter and the combustion energy density increases to a level that is sufficient to sustain a diffusion flame. Once ignition is sustained, the additional heat flux provided by the flame increases the burning rate to a quasi-steady value with a near-stoichiometric fuel/air ratio. Since ignition is a gas phase phenomenon, a critical mass or energy flux at solid/air interface is the generally accepted criterion for the onset of flaming combustion of solids [4,5]. Unfortunately, these gas phase phenomena are difficult or impossible to quantify as criteria for piloted ignition.

A condensed phase criterion for piloted ignition has been proposed [6-11] in which the heated solid reaches a surface temperature at which thermal decomposition to gaseous fuel proceeds at a rate that is sufficient to maintain a flame at the surface. The time to reach this surface temperature at ignition (ignition temperature) is important because it is the rate-limiting step for ignition and flame spread. However, the surface temperature at ignition is known to depend on the ventilation rate, the location and strength of the igniter and the radiant heat flux incident on the surface so, unlike the gas phase criteria, the condensed phase criterion of a critical surface temperature at the onset of flaming combustion is a test parameter [6-13]. Although straightforward in principle, surface temperatures at piloted ignition are difficult to measure, so they are often inferred from unsteady heat transfer analyses of material response in standardized fire calorimeter tests [4,14] or measured in separate bench scale ignitability tests [1,3,13].

In this paper we show that the initiation of flaming combustion in a fire calorimeter and the onset of thermal decomposition in laboratory thermal analyses occur at similar temperatures for transient heating of combustible solids, and that these temperatures are coincident when the sample is at thermal equilibrium with the heat source. Because ignition is a gas phase phenomenon, the surface temperature of the solid at ignition depends on the flows and geometry of the fire test as well as the thermal stability and combustion properties of the material.

APPROACH

The rate of conversion of solid to gaseous fuel in a single step, thermal decomposition process can be expressed,

$$\frac{d\alpha}{dt} = k(T)f(\alpha) \tag{1}$$

In Equation 1, α is the fraction of the solid that has been converted to gaseous fuel at temperature T, and $k(\mathrm{T}) = A\exp[-E_a/\mathrm{RT}]$ is a global rate constant for pyrolysis having the Arrhenius form, R is the gas constant, $f(\alpha)$ is the reaction model and $A(\mathrm{s}^{-1})$ and E_a(J/mole) are kinetic parameters called the global frequency or pre-exponential factor and activation energy for pyrolysis, respectively. Typically, $f(\alpha) = (1-\alpha)^n$ where n is an additional kinetic parameter called the reaction order [15]. In derivative thermogravimetry (DTG), experiments are conducted at a constant rate of temperature rise, dT/dt = constant = β (K/s) and the instantaneous mass of a milligram sample m is measured as a function of time/temperature. If m_0 is the initial mass of sample, m_c is the residual mass after pyrolysis and the char yield/pyrolysis residue is constant, $m_c/m_0 = \mu$, the available fraction of solid fuel converted to gaseous products at temperature T is,

$$\alpha = \frac{m_0 - m}{m_0 - m_c} = \frac{1 - m/m_0}{1 - \mu} \qquad \text{(DTG)} \tag{2}$$

In microscale combustion calorimetry (MCC) the combustion energy (heat) release rate per unit initial mass of sample, $\dot{Q} = dQ/dt$ (W/g) is also measured at a constant heating rate β. In MCC, the extent of conversion at temperature T is,

$$\alpha = \frac{\frac{1}{\beta}\int_{T_0}^{T}\dot{Q}(\theta)d\theta}{\frac{1}{\beta}\int_{T_0}^{\infty}\dot{Q}(T)dT} = \frac{Q}{Q_\infty} \qquad \text{(MCC)} \tag{3}$$

The solution to Equation 1 for the extent of conversion of solid to gaseous fuel in a first order process where $f(\alpha) = (1-\alpha)$, at constant heating rate is [15],

$$\alpha = 1 - \exp\left[-\frac{1}{\beta}\int_{T_0}^{T} k(\theta)d\theta\right] \approx 1 - \exp\left[-\frac{RT^2 k(T)}{\beta(E_a + 2RT)}\right] \tag{4}$$

Equation 4 shows that the extent of conversion of solid to gaseous fuel at a given heating rate depends only on the instantaneous temperature of the sample, so the rate of conversion of solid to gaseous fuel at β is,

$$\frac{d\alpha}{dt} = \beta\frac{d\alpha}{dT} \equiv k(T)(1-\alpha) = k(T)\exp\left[-\frac{RT^2 k(T)}{\beta(E_a + 2RT)}\right] \tag{5}$$

Of particular interest is the fractional conversion rate of solid to gaseous fuel at temperature T as $\beta \to \infty$ or as $\alpha \to 0$. Under these conditions, the conversion rate depends only on the sample temperature and the kinetic parameters A and E_a, which are assumed to be material properties,

$$\frac{d\alpha}{dt} = k(T)\exp[-0] \approx k(T) = A\exp\left[-\frac{E_a}{RT}\right] \tag{6}$$

The scalar conversion rate in nonisothermal analysis (Equation 6) is related to the surface flux of mass or energy in a fire test by the condition for controlled heating of a thermal analysis sample,

$$Bi = \frac{(dT/dx)_{internal}}{(dT/dx)_{external}} = \frac{hV}{\kappa S} = \frac{hm}{\rho\kappa S} \le 1 \tag{7}$$

In Equation 7, dT/dx are the internal and external temperature gradients at the sample surface, Bi is the Biot modulus, h is the overall heat transfer coefficient from the furnace to the sample of mass m, density ρ, thermal conductivity, κ, surface area S and volume V. When $\alpha \ll 1$, the mass, surface area and volume of the sample are essentially unchanged from the initial values, so $m(0) \equiv m_0 = \rho V_0$, $S = S_0$, and the relationship between the specific mass loss (fuel generation) rate at the onset of thermal decomposition and the surface mass flux $\dot{m}''$ (kg/m^2-s) of a milligram-sized sample in constant heating rate thermal analyses under anaerobic conditions is,

$$\dot{M}(T) \equiv (1-\mu)\frac{d\alpha}{dt} = -\frac{\dot{m}(T)}{m_0} = -\frac{\dot{m}(T)/S_0}{\rho\kappa/h} \geq \frac{\dot{m}''}{\rho\kappa/h} \tag{8}$$

The specific heat release rate analog of Equation 8 for a MCC sample in terms of the energy flux, $\dot{q}''$ (W/m^2) is,

$$\dot{Q}(T) = Q_\infty \frac{d\alpha}{dt} \geq \frac{\dot{q}''(T)}{\rho\kappa/h} \tag{9}$$

The gas phase criterion for piloted ignition is either a concentration of fuel in air that is above the lower flammability limit (LFL) at the igniter, or a combustion energy density that exceeds 1.9 MJ/m^3 in the fuel/air mixture [5]. If the minimum mass flux at ignition in a particular fire test is MLR*, the corresponding fractional mass loss rate in DTG is,

$$\dot{M}(T_{ign}) = \dot{M}^* = \frac{MLR^*}{\rho\kappa/h} \tag{10}$$

Likewise, if the minimum heat release rate per unit sample area (energy flux) at ignition in a particular fire test is HRR*, the corresponding specific heat release rate in the MCC is,

$$\dot{Q}(T_{ign}) = \dot{Q}^* = \frac{HRR^*}{\rho\kappa/h} \tag{11}$$

The DTG and MCC temperatures at $\dot{M}^*$ and $\dot{Q}^*$, respectively, depend on the conditions of the fire test in which MLR* and HRR* were measured, so they are not strictly material properties. Consequently, these thermal analysis temperatures depend on the fire test as well as the thermal stability parameters (A, E_a), and thermochemical properties (μ, Q_∞) of the material by Equations 6, 8 and 9,

$$T_m = \frac{E_a}{R\ln\left[A/\dot{M}^*\right]} \quad \text{or,} \quad T_q = \frac{E_a}{R\ln\left[AQ_\infty/\dot{Q}^*\right]} \tag{12}$$

Equation 12 shows that the DTG temperature T_m for a particular fire test is a function of the thermal stability parameters, A and E_a, while the MCC temperature T_q also includes the combustion energy of the material, Q_∞.

In thermal analyses, heat transfer is primarily by radiant energy exchange between the furnace at temperature T_f and the sample at temperature T, and the temperature difference, $\Delta T = (T_f - T)$ is small compared to T or T_f to ensure accurate heating rate control. For a typical polymer thermal decomposition temperature, $T = 773$K (450°C), the effective heat transfer coefficient in the TA test is approximately, $h \approx 4\sigma T^3 = 100$ W/m^2-K, which satisfies the inequality of Equation 6 for initial polymer sample dimensions typical of thermal analysis, $V_0/S_0 < 10^{-3}$ m.

Equations 10 and 11 provide a physical basis for using DTG or MCC to estimate the ignition temperature of a solid sample in a fire test for which MLR* or HRR* are known. Table 1 is a summary of MLR* and HRR* for piloted ignition in a standard bench scale fire calorimeter test in which the ventilation rate is prescribed and the position and energy of the igniter are fixed [2]. The thermal analysis criteria in Table 1, $\dot{M}*$ and $\dot{Q}*$, were computed for MLR* and HRR* at transient and sustained ignition using Equations 9 and 10 with, h = 100 W/m^2-K and typical polymer properties, ρ = 1300 kg/m^3 and κ = 0.24 W/m-K [14]. A distinction is made between transient ignition, which occurs at the LFL or when E = 1.9 MJ/m^3, and sustained ignition, which occurs at fuel concentrations between the LFL and UFL or when $E \approx 6$ MJ/m^3 [5].

Table 1. Specific Mass Loss Rate ($\dot{M}*$) and Specific Heat Release Rate ($\dot{Q}*$) in Thermal Analysis Corresponding to the Critical Mass Flux (MLR*) and Energy Flux (HRR*) at Transient and Sustained Ignition[5] in a Cone Calorimeter[2].

	Cone Calorimeter			DTG		MCC	
Mode of Ignition	MLR* (g/m^2-s)	HRR* (kW/m^2-s)	*Surface Temp.* (°C)	$\dot{M}*$ (s^{-1})	*Sample Temp.* (°C)	$\dot{Q}*$ (W/g)	*Sample Temp.* (°C)
Transient	1.0 ±0.4	21±6	T_{flash}	3x10^{-4}	T_{m1}	7	T_{q1}
Sustained	3.2 ±1.2	66 ±17	T_{ign}	1x10^{-3}	T_{m2}	20	T_{q2}

EXPERIMENTAL

Materials

Polymers tested in our laboratory were unmodified, natural materials containing little or no additives obtained in extruded sheet form from commercial sources. All gases were ultra high purity grades (>99.999%) from local suppliers.

Methods

Thermogravimetric analysis (TGA) of ≤5 mg samples was performed in a commercial instrument (TGA/DSC 1, Mettler-Toledo) under a nitrogen purge gas flow rate of 50 cm^3/min and at heating rates, β = 3, 10, 30 and 60 K/min according to a standard procedure [16].

Microscale combustion calorimetry was performed on samples weighing ≤5 mg according to a standard method [17].

Fire calorimeter experiments were conducted in a cone calorimeter (Cone 1, Fire Testing Technologies) according to a standard method using an edge frame holder [2] on 3-mm thick samples over a range of external heat flux.

Numerical simulations of DTG experiments were performed using the one-dimensional pyrolysis model ThermaKin [18] for thermally thin (< 5 mg) samples having typical, constant polymer properties listed in Table 2 [14]. The sample temperature at the specific mass loss rates $\dot{M}$ in Table 1 was computed for specified constant heating rates of 1,3,10,30,60,100 and 300 K/min.

ThermaKin simulations of the burning rate of a typical polymer (Table 2) in a fire calorimeter under standard conditions [2] were also conducted for samples of thickness, L = 3, 6, 12 and 24 mm at external radiant heat fluxes, $\dot{q}''_{ext}$ = 20, 40, 60 and 80 kW/m^2 for a perfectly insulated rear face boundary condition. The surface temperatures at transient ignition (T_{flash}) and sustained ignition (T_{ign}), corresponding to the mass flux at transient ignition MLR* = 1 g/m^2-s and the mass flux as sustained ignition, MLR* = 3.3 g/m^2-s [5] were computed for each sample thickness L and external heat flux, $\dot{q}''_{ext}$. In order to obtain a representative surface temperature for comparison to experimental measurements using thermocouples, the surface temperature T_s at each MLR* was obtained from the ThermaKin temperature profile $T(x)$ as a kinetically weighted, integral average over the sample thickness,

$$T_s\big|_{MLR^*} = \frac{\int_0^L A\exp\left[-E_a/RT(x)\right]T(x)\,dx}{\int_0^L A\exp\left[-E_a/RT(x)\right]dx} \tag{13}$$

Table 2. Constant Properties Used for Numerical Simulations of Typical Polymer [14]

Property	Symbol	Value
Inert/Char Mass Fraction	μ	0
Thermal Conductivity	κ	0.24 W/m-K
Density	ρ	1300 kg/m^3
Heat Capacity	c_p	2300 J/kg-K
Arrhenius Frequency Factor	A	9.5x10^{13} s^{-1}
Arrhenius Activation Energy	E_a	198 kJ/mole
Surface Emissivity	ε	0.95
Surface Reflectivity	R	0.05

Ignition temperatures: Surface temperatures of polymers at sustained, piloted ignition T_{ign} were measured in our laboratory using a thermocouple or were inferred from critical heat flux measurements and transient heat transfer analysis [5,14,19,20]. Sustained ignition temperatures for polymers were also obtained from the literature [6-13]. Between 2 and 7 values of T_{ign} for each polymer obtained by direct [1-3] or indirect [4,14] methods were averaged to obtain mean values ± one standard deviation for comparison to DTG (T_{m2}) and MCC (T_{q2}) estimates.

RESULTS AND DISCUSSION

Figure 1 contains representative data from the MCC showing the location of T_{q1} and T_{q2} for polystyrene/PS at $\dot{Q}* = 7$ W/g and $\dot{Q}* = 20$ W/g, corresponding to the HRR* for transient and sustained ignition, respectively, in the cone calorimeter.

Figure 1.

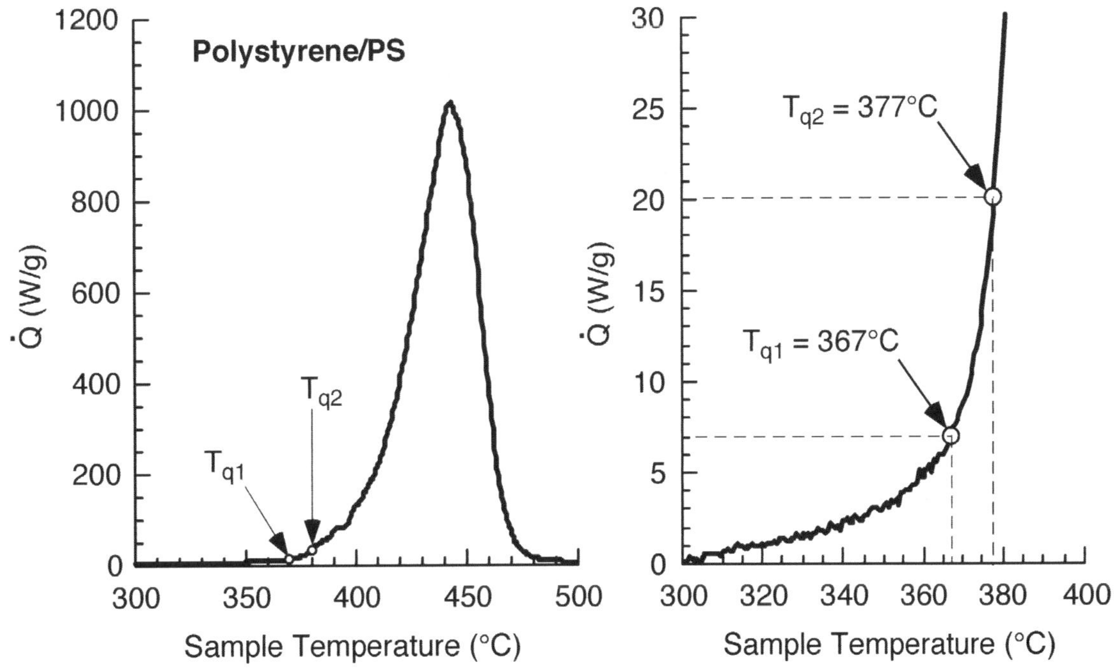

The mass of PS that is lost up to temperature T_{q2} in Figure 1 is less than 1% of the original sample mass based on the integrated heat release rate histories, so $\alpha < 0.01$ and $m(T_b)/m_0 \approx 1$ as per Equation 6. Note that the MCC temperatures corresponding to transient (T_{q1}) and sustained (T_{q2}) ignition occur within a very narrow temperature interval, 10°C in this case.

Figure 2 is DTG data for T_{m1} and T_{m2} for several of the polymers in Table 3 versus heating rate. It is seen that T_{m1} (transient ignition temperature) and T_{m2} (sustained ignition temperature) are essentially independent of heating rate at large β and for $\alpha << 1$, as per Equation 5. Data from MCC are similar in magnitude and trend to the DTG data. Temperatures T_{m1} and T_{m2} at $\beta \leq 3$ K/min are less reliable because of balance sensitivity at these low fractional mass loss rates.

The DTG data in Figure 2 indicate that T_{m1} and T_{m2} are independent of heating rate at $\beta > 30$ K/min. These experimental results are consistent with ThermaKin simulations of a DTG experiment on a 5 mg sample of a polymer having the kinetic parameters in Table 2. Shown as points in Figure 3 are T_{m1} and T_{m2} at specific mass loss rates, $\dot{m}/m_0 = 3\text{x}10^{-4}$ s^{-1} (transient ignition) and $1\text{x}10^{-3}$ s^{-1} (sustained ignition). Lines in Figure 3 were calculated using Equation 4 with the same kinetic parameters of Table 2.

Figure 2. DTG Sample Temperatures Corresponding to Transient Ignition (T_{m1}) and Sustained Ignition (T_{m2}) Versus Heating Rate for Several of the Polymers in Table 3.

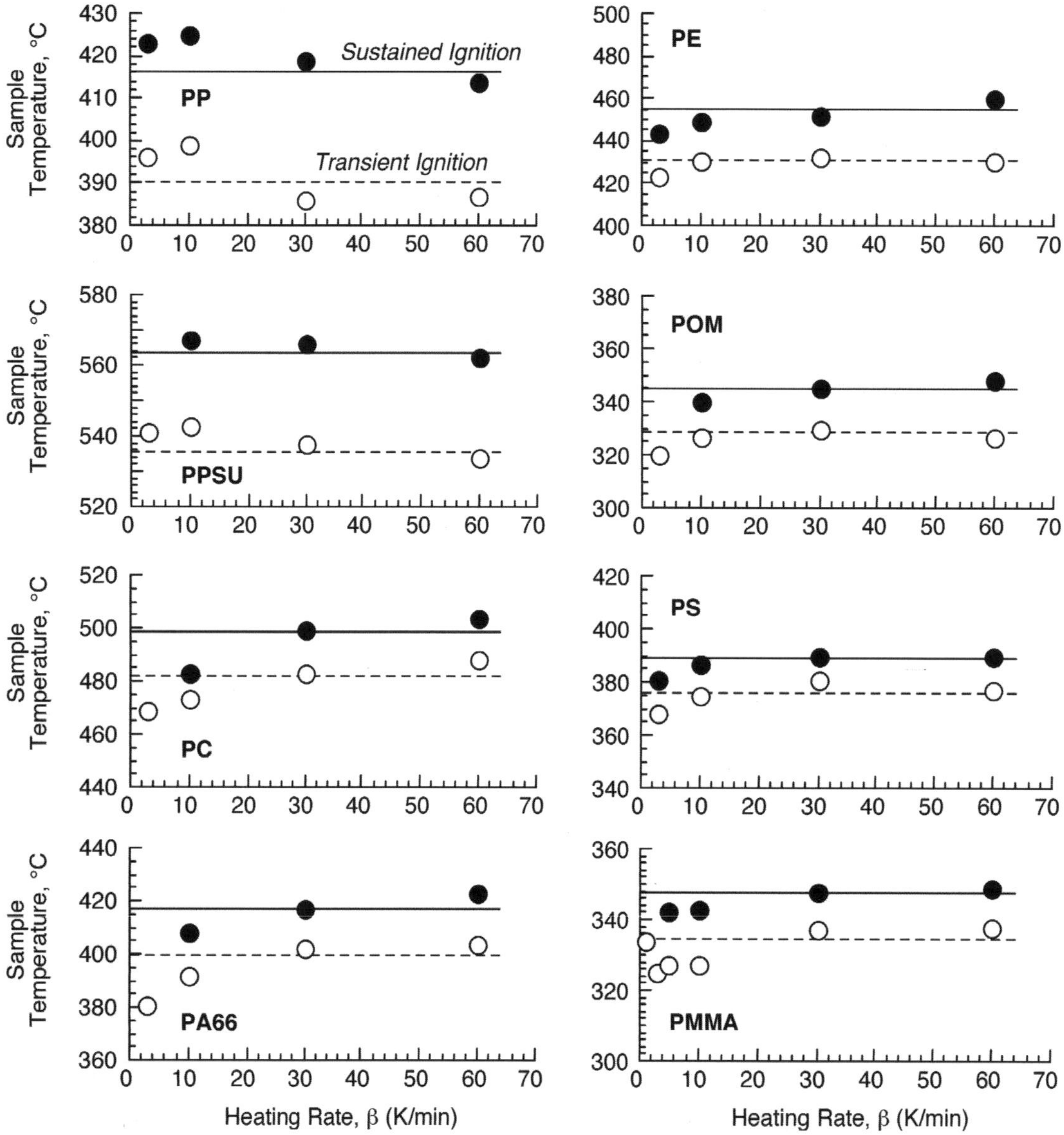

Table 3 contains TA sample temperatures, T_{m1} and T_{q1}, at the mass and energy fluxes for transient ignition in Table 1, as well as nonisothermal analysis sample temperatures, T_{m2} and T_{q2}, for sustained ignition using the criteria in Table 1. Table 3 also contains the extent of conversion α_{ign} at T_{q2}, T_{m2}. The average value of the extent of conversion in Table 3 at T_{q2} (neglecting PVC) is $\langle\alpha_{ign}\rangle = 1.7\% = 0.017 << 1$, as required by Equations 6-8.

Figure 3. Simulated Sample Temperatures T_{m1} and T_{m2} Versus Heating Rate for the Typical Polymer of Table 2. Points are ThermaKin Simulations. Lines are From Equation 5.

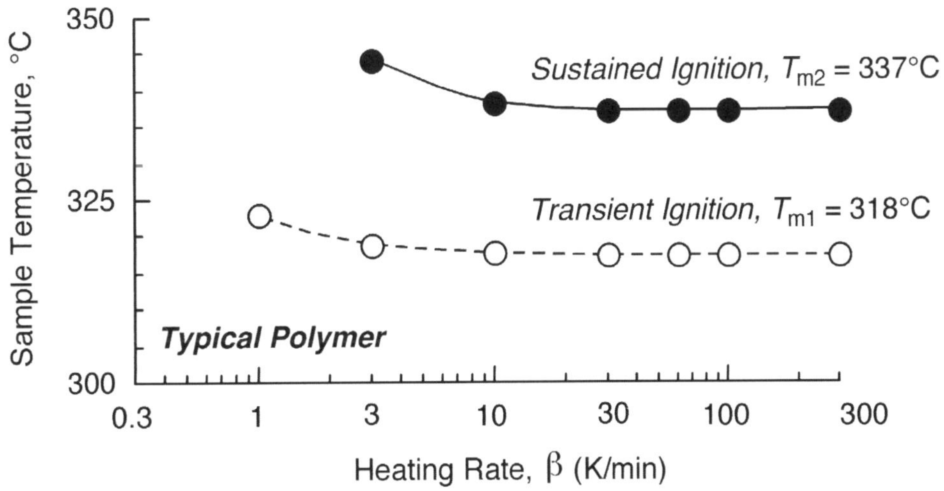

Table 3. TA Sample Temperatures Corresponding to Transient (T_{q1}, T_{m1}) and Sustained (T_{q2}, T_{m2}) Ignition In Bench Scale Fire Tests Compared to Measured Surface Temperatures at Sustained Ignition.

	Thermal Analysis Data					Measured Values
	Transient Ignition		*Sustained Ignition*		α_{ign}	*Sustained Ignition*
	MCC	DTG	MCC	DTG		
	T_{q1} (°C)	T_{m1} (°C)	T_{q2} (°C)	T_{m2} (°C)	(%)	T_{ign} (°C)
POM	331	326	352	344	0.3	306 ±25
PMMA	328	331	336	346	0.8	336 ±42
PP	392	392	406	420	0.6	358 ±39
PS	367	375	377	387	0.6	363 ±3
PET	406	386	415	407	1.2	391 ±23
ABS	372	387	384	408	0.7	402 ±11
PE	433	429	449	458	0.6	403 ±56
HIPS	383	390	396	409	0.8	405 ±7
PVC	276	243	305	251	5.5	419 ±55
PA66	391	395	422	416	1.8	451 ±33
PC	488	478	499	495	1.4	474 ±31
PEI	525	508	536	523	2.4	528 ±29
PVDF	470	425	480	439	6.3	531 ±158
PPSU	538	576	565	609	3.5	543 ±37
PEEK	563	560	570	571	1.0	553 ±18
PPS	471	475	499	491	3.2	564 ±39

Figure 4 is a plot of the measured ignition temperature T_{ign} versus the MCC and DTG estimates T_{q2} and T_{m2}, respectively, for the polymers in Table 3. Linear regression of T_{ign} on T_{q2} forced through the origin gives a slope of 1.00±0.03 and a correlation coefficient R = 0.84 for the MCC data in Table 3. Results for DTG are a slope of 1.00±0.04 and a correlation coefficient of R = 0.67. Despite the large uncertainty of the measured T_{ign} and the PVC outlier, the MCC temperature T_{q2} is a somewhat better predictor of T_{ign} than the DTG estimate, T_{m2}. The average difference between the thermal analysis estimates T_{m2}, T_{q2} and the measured surface temperatures at ignition for each polymer is 31°C, which is also the uncertainty of T_{ign}, .

Figure 4. Bench-Scale T_{ign} Versus Thermal Analysis Temperatures T_{m2}, T_{q2} Corresponding to Sustained Piloted Ignition.

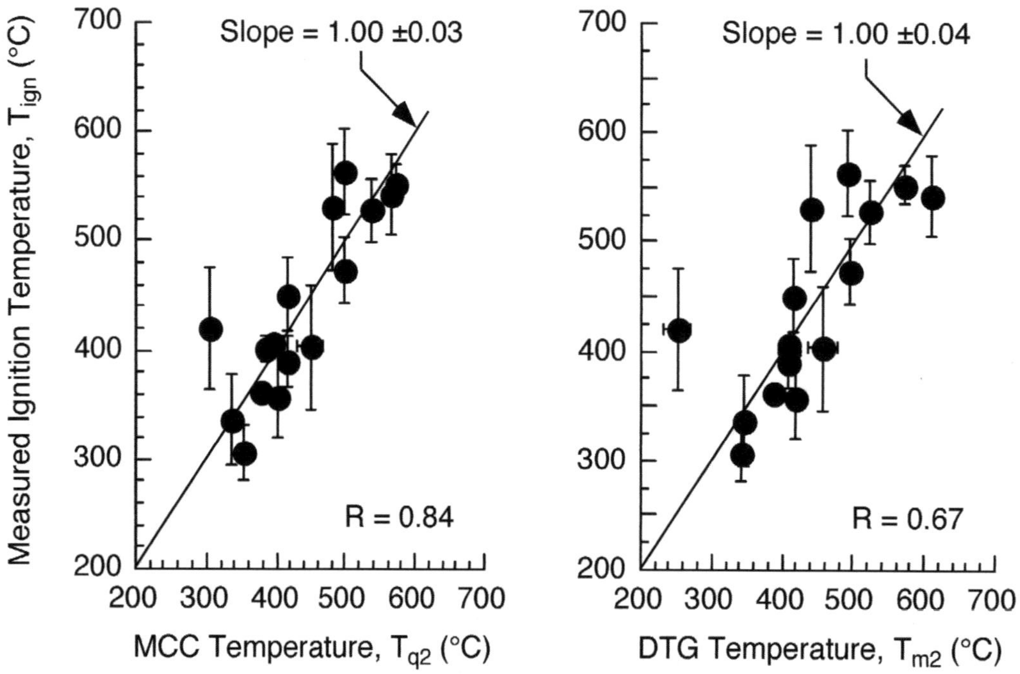

Figure 5 shows ThermaKin simulations of the surface temperatures at the mass fluxes MLR* for transient and sustained ignition from Table 1 for the typical polymer of Table 2. The critical external heat flux for ignition was obtained by an iterative procedure such that $\dot{q}''_{ext}$ = CHF = 10 kW/m^2 at MLR = MLR*. Plotted at CHF in Figure 5 are the DTG temperatures of Figure 3. Figure 5 shows that the simulated ThermaKin surface temperatures at transient and sustained ignition are substantially independent of sample thickness, as indicated by the superposed open circles at each $\dot{q}''_{ext}$ for the different thicknesses. However, the surface temperatures at MLR* in the ThermaKin simulations (Equation 13), are seen to increase somewhat with external heat flux, as is observed experimentally [6-12]. This increase in the surface temperature at MLR* is kinetic compensation for the decrease in the pyrolysis volume at high $\dot{q}''_{ext}$. Another important result in Figure 5 is that the surface temperature at ignition approaches the thermal analysis value (Equation 12) as $\dot{q}''_{ext} \rightarrow$ CHF and the sample approaches thermal equilibrium with the radiant heat source. Consequently, T_{ign} obtained from a surface energy balance at CHF should be in reasonable agreement with measured and TA estimated values, as has been observed [14].

Figure 5. ThermaKin Calculation of the Surface Temperatures at Ignition for a Typical Polymer Over a Range of External Heat Flux and Thickness, L = 3, 6, 12 and 24 mm.

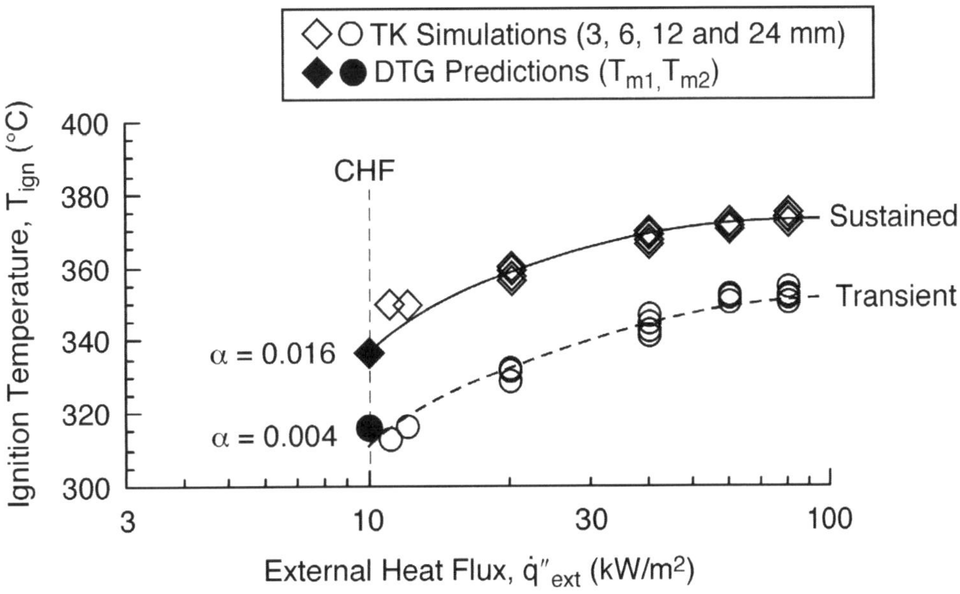

CONCLUSIONS

It is found that the onset of flaming combustion near the critical heat flux in a fire test and the onset of thermal decomposition in a thermal analysis experiment occur at the same temperature for combustible solids, and that this temperature is determined by the thermal stability of the material and the gas phase criteria for ignition in the fire test. Thermal analysis estimates of the surface temperature at piloted ignition in standard fire tests are in reasonable agreement with measured values for 16 polymers.

REFERENCES

[1] ASTM Standard D1929. (2016). Test Method for Determining Ignition Temperature of Plastics, American Society for Testing and Materials, West Conshohocken, PA.

[2] ASTM Standard E1354. (2013). Test Method for Heat and Visible Smoke Release Rates for Materials and Products Using and Oxygen Consumption Calorimeter, American Society for Testing and Materials, West Conshohocken, PA.

[3] ISO 5657. (1997). Reaction to Fire Tests: Ignitability of Building Products Using Radiant Heat Source, International Standards Organization, Geneva, Switzerland.

[4] Drysdale, D.D., (1998). *An Introduction to Fire Dynamics*, 2nd Edition. Chichester, England: John Wiley & Sons.

[5] Lyon, R.E. and Quintiere, J.G. (2007). Piloted Ignition of Combustible Solids. *Combustion and Flame*, 151, 551-559

[6] Rasbash, D.J., Drysdale, D.D., and Deepak, D. (1986). Critical Heat and Mass Transfer at Pilot Ignition and Extinction of a Material. *Fire Safety Journal*, 10, 1-10.

[7] Tian, N. and Zhou, A. (2016). Sensitivity and Uncertainty Analysis for Ignition of Fiber-Reinforced Polymer Panels, *Fire and Materials*, 40, 396-415.

[8] Thomson, H.E., and Drysdale, D.D. (1987). Flammability of Plastics. Part 1. Ignition Temperatures, *Fire & Materials*, 11(4), 163-172.

[9] Drysdale, D.D. and Thomson, H.E. (1989). Flammability of Plastics. Part 2. Critical Mass Flux at the Fire Point, *Fire Safety Journal,* 14, 179-188.

[10] Rasbash, D.J., and Drysdale, D.D. (1983). Theory of Fire Processes. *Fire & Materials*, 7, 79-88.

[11] Thomson, H.E., and Drysdale, D.D. (1988). An Experimental Evaluation of Critical Surface Temperature as a Criterion for Piloted Ignition of Solid Fuels. *Fire Safety Journal*, 13, 185-196.

[12] Drysdale, D.D., and Thompson, H.E. (1994). The Ignitability of Flame Retarded Plastics. Proceedings of the Fourth International Symposium on Fire Safety Science, Ottawa, Canada, 196-204.

[13] Babrauskas, V. (2003). *Ignition Handbook*, Issaquah, WA, Fire Science Publishers.

[14] Safronava, N., Stoliarov, S.I., and Lyon, R.E. (2015). Effect of Moisture on Ignition of Polymers, *Fire Technology*, 51(5), 1093-1112.

[15] Lyon, R.E., and Safronava, N. (2013). Comparison of Direct Methods to Determine *n*-th Order Kinetic Parameters of Solid Thermal Decomposition for Use in Fire Models, *Journal of Thermal Analysis and Calorimetry,* 114, 213–227.

[16] ASTM Standard E2550. (2011). Test Method for Thermal Stability by Thermogravimetry, American Society for Testing and Materials, West Conshohocken, PA.

[17] ASTM Standard D7309. (2013. Test Method for Determining Flammability Characteristics of Plastics and Other Solid Materials Using Microscale Combustion Calorimetry, American Society for Testing and Materials, West Conshohocken, PA.

[18] Stoliarov, S.I. and Lyon, R.E. (2008). Thermo-Kinetic Model of Burning, Final Report DOT/FAA/AR-TN08/17, Federal Aviation Administration.

[19] Oztekin, E.S., Crowley, S.B., Lyon, R.E., Stoliarov, S.I., Patel, P., and Hull, T.R. (2012). Sources of Variability in Fire Test Data: A Case Study on Poly(aryl ether ether ketone)(PEEK), *Combustion and Flame*, 159, 1720-1731.

[20] Lyon, R.E. (2004). *Handbook of Building Materials for Fire Protection*, C.A. Harper, ed., Chapter 3, Plastics and Rubber, New York, NY, McGraw-Hill.

IDENTIFICATION OF THE SOLID FUEL AUTO-IGNITION CHARACTERISTICS: DIRECT NUMERICAL SIMULATION OF THE PMMA AUTO-IGNITION UNDER CONE CALORIMETER CONFIGURATIONS

Simon Roblin[*,1,2], Thomas Rogaume[2], Franck Richard[2] & Arnaud Trouvé[3]
1: Calyxis, Pôle d'Expertise du Risque, France
2: Institut P', UPR 3346 CNRS, France
3: University of Maryland, USA
*Corresponding author: simon.roblin@ensma.fr

ABSTRACT

Fire propagation on the scale of buildings and cities is a major stake. It is conditioned by the ignition of solid fuels in rooms adjacent to the one where the disaster originally takes place. The ignition is so piloted by the initiation of the combustion reaction of the gaseous mixture stemming from the thermal decomposition of the condensed phase induced by heat transfer.
Two types of ignition are defined in the literature: piloted-ignition related to the presence of a hot spot and auto-ignition resulting from the thermal runaway within the gas phase. The auto-ignition plays a major role in the context of fire spread between rooms. However, this process has been very little experimentally studied, because of its complexity, and only theoretical analyses were lead concerning the phenomena which take place during solid fuels auto ignition.
The aim of the present study is to characterize auto-ignition regimes according to various solid typologies (regarding to thermal and kinetic behaviour) in order to understand better their processes and their occurrence conditions. Thereby, this fine understanding allows to develop global models of fire spread for a deterministic consideration of the fire hazards at urban scale.
The brief and local character of the auto-ignition requires the choice of a complete resolution for flows, transfers and chemistry. Thus, the Direct Numerical Simulation (DNS) was selected to capture the phenomena, with the introduction of a fine and non-infinitely fast chemistry of thermal decomposition and combustion.
The results show that the ignition time decreases linearly with the density as the mass flux. On the contrary, the ignition surface temperature increase and it leads to two different regimes : the first one correspond to a gas phase ignition while the second is piloted by the solid surface. The formers are describing and confirm those exposed by the literature.

KEYWORDS: Auto ignition; Solid Fuels; Direct numerical simulation; Combustion; Flammability; Pyrolysis; Chemical kinetics; Stoichiometry; Density.

INTRODUCTION

The development of large scale simulation tools on wild fires or floods is actually possible by the increasing of the computing resources. In front of this observation, the scientific fire community and the fire services are trying to develop fire spread models in urban scales to optimise the location of fire brigades and to manage fire with real time simulation during operations. This kind of tools are currently based on probabilistic methods and only some integrate global deterministic variables such as critical times, critical temperatures or critical heat flux are used.
The fire propagation between premises is conditioned by material ignition in the rooms adjacent to the disaster. Two types of ignitions are defined in the literature, auto-ignition and piloted ignition, and this

study will focus on the first one. Indeed, the second one was very studied by the introduction of apparatus like the cone calorimeter. The auto-ignition dependency to the experimental conditions makes difficult the phenomenon characterization and only few studies coupled to homogeneous combustion theories lead to a good interpretation of it. In our literature review, many criterion was found to define auto-ignition. Even though some of them are difficult to measure experimentally, many involved condensed phase parameters and they have a major impact on the auto-ignition which takes place in the gaseous phase. So, analysis of solid fuels auto-ignition needs a complete resolution of the flow, kinetics and thermal conditions both in the solid and in the gaseous phase. The present study proposes to handle it with a Direct Numerical Simulation approach in the well-known condition of the cone calorimeter. The NIST's Fire Dynamics Simulator (FDS) is used to lead the calculus required.

The first part of this paper tackles the different auto-ignitions theories in the gaseous phase and with the introduction of the solid fuels variables. After this, the numerical setup and models are described. FDS abilities to simulate the phenomenon are tested. Then, a sensitivity analysis is lead to find better mesh size (in both phases) in order to lock onto the ignition. Finally, results are presented with a discussion about the identified ignition regimes.

AUTO-IGNITION AND SOLID TYPES

Auto ignition theories

Auto-ignition theories were identified in the homogeneous combustion literature (Glassman & Yeter 2008) and authors like Quintiere (Quintiere 2006) or Fernandez Pello (Fernandez Pello 2011) give us mechanisms to understand solids auto-ignition. The Semenov theory is the simple one to understand the process of gases auto ignition. Two approaches were found to characterize it: the thermal one and the chemical one. In the first, the auto ignition occurrs when the thermal balance becomes positive regarding the heat loss around the reactor. So, the ignition takes place by the thermal runaway arising from the oxidation reaction. In a chemical approach, the ignition occurs when the number of radicals exceed the number of termination reaction in the reactor. Therefore, a chemical runaway takes place in the gas phase and is at the origin of the fire tetrahedron concept.

Concerning solid fuels auto-ignition, Quintiere (Quintiere 2006) describes it through a chronological approach which considers the sum of three characteristics times:

$$t_{ig} = t_{py} + t_{mix} + t_{chim} \quad (1)$$

The ignition time t_{ig} corresponds to the sum of a pyrolysis time t_{py} (the time required to start a sufficient gaseous emission level), a mixing time t_{mix} (characterising the mixing of species to reach a flammable mixture concentration) and a chemical time t_{chim} (necessary to start the combustion reaction in the gas phase and directly related to the Semenov's theory).

Niioka, Fernandez-Pello and Torero (Niioka et al. 1981; Chao & Fernandez Pello 1997; Torero 2016) propose an additional approach to explain the phenomenon. In fact, the ignition takes place over the solid surface and they introduce the notion of critical Damkölher number to reach the ignition point in the gas phase (this number compares the turbulence charasteristic time to the chemical one). Thus, two kinds of auto-ignition are identified for a solid exposed to an external heat flux. A first one takes place in the gas phase and is initiated by the thermal (or chemical) runaway between the species. The gas absorptivity plays a key role in this type of auto ignition and it leads to a faster heating of the gas phase than the solid one. The second type of auto ignition corresponds to a reaction initiation by solid surface which heats faster than the gas phase. So, the condensed phase seems to play a major part in the auto-ignition and it is necessary to characterize it. Moreover, in his works related to the ignition, Boonmee (Boonmee 2004) gives a state of art of studies which introduce auto-ignition criterion. More of them are related to the solid phase and proves its impact on the process. To identify the solid phase impact, a solid classification appears to be indispensable.

Solid types

Pyle and Zaror (Pyle & Zaror 1984) give us this kind of classification with two parameters: the thermal thickness and the chemical kinetics of the condensed phase. Thermal thickness is given by the Biot number and characterizes the presence (if $Bi \gg 1$) or not ($Bi \ll 1$) of a temperature gradient inside the solid phase. It is given by:

$$Bi = \frac{hL_C}{\lambda_s} \quad (2)$$

With h, the heat transfer coefficient, L_C the characteristic length and λ the thermal conductivity of the material.

As kinetics is not stationary, it is traduced by a quotient of two characteristic times, the conduction one et and pyrolysis one. These times are given by Pyle and Zaror as:

$$t_{cond} = \frac{\rho C_p}{\lambda_s \sigma_s^2} \ (3) \text{ and } t_{py} = \frac{1}{k} = \frac{1}{Ae^{-\frac{E_a}{RT}}} \ (4)$$

With C_p, the specific heat, $\sigma_s$the specific surface of the material, A the pre exponential factor, E_a the activation energy, R the ideal gas constant and T, the temperature.

Then, the Biot number and the quotient of t_{cond}/t_{py} leads to the determination of four solid typologies which are presented in the figures 1. For the figures 1(a) and 1(b), there is no temperature T gradient inside the condensed phase. Solids are considered as thermally thin but we can notice a change in their kinetics traduced by $\dot{m}$. For the case (a), $Bi \ll 1$ and $t_{cond}/t_{py} \gg 1$, the chemical reaction is fast regarding to the thermal transfer and if the pyrolysis temperature is reached, the reaction takes place in the entire solid volume. In (b), $Bi \ll 1$ and $t_{cond}/t_{py} \ll 1$, the characteristic reaction time is great behind the heat transfer one and the process is dominated by the kinetic which takes place at the solid temperature, uniform and equal to the external temperature. Then, the chemical process is isothermal. For the figures 1(c) and 1(d), temperature gradients are important in the solid phase. In 1(c), the reaction time is low compared to the heat transfer one and the reaction zone could be assimilated to a "thin pyrolysis front". The degradation process is dominated by the heat transfer. For the 1(d) case, the reaction velocity is low regarding the thermal wave propagation inside the condensed phase. The pyrolysis front and the reaction are not heterogeneous and it could not be considered as homogeneous.

Figure 1: Solid typologies according to Pyle and Zaror; (a) thermally thin, low kinetic; (b) thermally thin, fast kinetic; (c) thermally thick, low kinetic; (d) thermally thick, fast kinetic

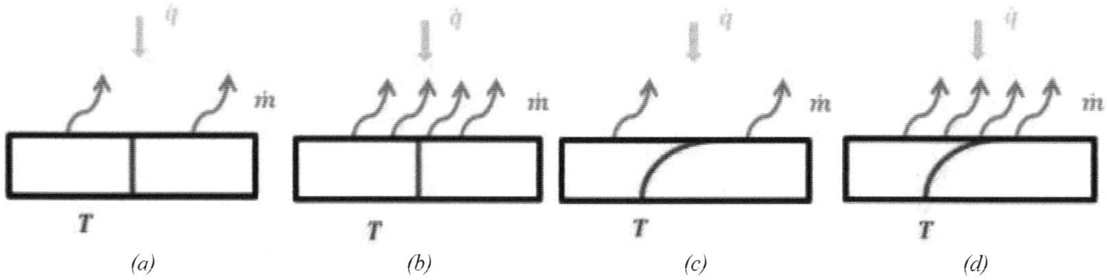

Considering equations (2) and (3), it appears that the density could modify the characteristic time of the conduction. This study will focus on thermally thick solids and a wide range of density will be tested to identify the impact of the condensed phase on the ignition process.

NUMERICAL SETUP

Description of the model

A full resolution of the flow is need to describe auto ignition. In fact, the phenomenon is brief and very local as we find in the existent literature about auto-ignition. The chemical processes in both phases, transport and heat transfer have to be modelled in a precise way. To ensure this objective, the Direct Numerical Simulation with a full resolution of the Navier Stokes equation is required and was chosen.

A well-known material is selected to model auto-ignition of a solid. Polymetyl methacrylate was very studied and a wide range of analysis were lead about its decomposition (Luche et al. 2011; Li & Stoliarov 2013) and about the combustion of its gas phase (composed by the methyl methacrylate). This material was then selected in this study and various recent studies about it are reviewed to implement physical and chemical parameters in both phase.

In the fire engineering community, the NIST's Fire Dynamics Simulator (McGrattan et al. 2013) is a very important tool to model fires in large scales. This software is also a reference in the research community. It offers a Direct Numerical Simulation mode which leads to a complete resolution of the flow and, depending to the user, fine kinetics models in the condensed and the gaseous phases.

Two single step models were selected to represent both pyrolysis and combustion. The first one represents the PMMA decomposition as its monomer follows this kind of reaction:

$$PMMA + heat \rightarrow MMA \ (5)$$

The reaction rate is directly dependant of the temperature as it is shown in the following equation:

$$r_{\alpha\beta} = \left(\frac{\rho_s}{\rho_s(0)}\right) A \exp\left(-\frac{E_\alpha}{RT_s}\right) \ (6)$$

The pyrolysis is resolved in one dimension.

After the decomposition into the monomer, the MMA reacts with the oxygen from the air in this proportions:

$$C_5H_8O_2 + 6O_2 \rightarrow 5CO_2 + 4H_2O \ (7)$$

This reaction rate is calculated by FDS like:

$$r_{F,i} = -k_i \prod C_\alpha^{a_{\alpha,i}} \ (8) \text{ with } k_i = A_i T^{n_i} \exp\left(-\frac{E_{a,i}}{RT}\right) \ (9)$$

Cone Calorimeter Apparatus

The cone calorimeter corresponds to a standard and is used all around the word to characterize solid mass loss rate, combustibility parameters, heat release rate, and piloted ignition time. Moreover, this apparatus is commonly used in a research way to sense the auto-ignition phenomenon. Then, the solid is exposed to an external heat flux and starts its decomposition until his auto-ignition. A horizontal configuration was chosen for this study.

To be sure that FDS could handle the auto-ignition, a pre-study was led in a 1D configuration but this paper is only focussed on the 2D cases.

Solid sizes (10 cm × 10 cm) in cone calorimeter lead to large simulation time and it is impossible to model it with the DNS mode. The width (in 2D) was reduced to 2 cm and to ensure a fast ignition, the prescribed heat flux is 100 kW.m^{-2}.

Sub-models of pyrolysis and combustion

Data has to be selected in the both phases to deal with the polymer degradation and to lead to the auto-ignition of the gaseous mixture. The selected model in the solid phase was extracted from Stolariov's works of the University of Maryland (Li & Stoliarov 2013; Korver 2015). The solid phase parameters are presented in table 1. This model includes the PMMA's glass transition.

Properties	**Values**	**Units**
Emissivity	0,95	-
Specific heat	$0,6 + 3,6.10^{-3}T$	kJ.kg^{-1}.K^{-1}
Thermal conductivity	$(T < 378\ K)\ 0,45 - (3,8.10^{-4})T$	W.m^{-1}.K^{-1}
	$(T > 378\ K)\ 0,27 - (2,4.10^{-4})T$	
Initial density	1155	Kg.m^{-3}
Pre-exponential factor	$8,6.10^{12}$	s^{-1}
Activation energy	$1,88.10^{5}$	J.mol^{-1}

In the gas phase, the selected model is from Korver (Korver 2015) and it was refined by Stoliarov (Li et al. 2014; Leventon et al. 201). It is given in table 2. The kinetics parameters (A and Ea) were found in Tsai (Tsai et al. 2001) and (Di Blasi et al. 1991).

Properties	**Values**	**Temperature range (K)**	**Units**
Specific heat	$(3,66.10^{-9})T^3 + (1,47.10^{-4})T^2 + (3,6754.10^{-1})T - 18,68$	-	kJ.kg^{-1}.K^{-1}
	0,015	$T < 373$	W.m^{-1}.K^{-1}

Thermal conductivity	$(1{,}85.10^{-8})T^2 + (5{,}64.10^{-5})T - 0.0086$	$373 < T < 1000$	
	$0{,}066$	$T > 1000$	
Dynamic viscosity	$9{,}44.10^{-6}$	$T < 373$	kg.m^{-1}.s^{-1}
	$(5{,}09.10^{12})T^2 + (3{,}11.10^{-8})T - 1{,}45.10^{-6}$	$373 < T < 1000$	
	$2{,}45.10^{-5}$	$T > 1000$	
Diffusivity	$(3{,}796.10^{-6})T^{1{,}75}$	-	m^2.s
Pre-exponential factor	$175{,}5$	-	s^{-1}
Activation energy	$4{,}9.10^{17}$	-	J.mol^{-1}

Density modification

In the equation (3), t_{py} appears to be conditioned to the material density. This parameter plays a key role to modify the solid typology regarding to the Figure 1. The diminution of the density decrease the conduction characteristic time and could lead to a decomposition piloted by the pyrolysis processes.In this study, few densities have been tested and they are presented below in table 3. For guidance, a material has been associated to densities. This study will focus on the densities which are representatives of solids materials, so volume masses will be considered until $\rho/50$.

Denomination	**Density (kg.m^{-3})**	**Materials (as an indication)**
ρ	1155	PMMA
$\rho/1{,}25$	924	Polyethylene, rubber
$\rho/1{,}50$	770	Various woods
$\rho/1{,}75$	660	
$\rho/2$	577,50	
$\rho/3$	385	
$\rho/4$	288,75	Cork
$\rho/6$	192,50	Balsa wood
$\rho/10$	115,50	Rock wool
$\rho/20$	57,75	Cotton, expanded polystyrene
$\rho/40$	28,875	
$\rho/50$	23,10	
$\rho/100$	11,55	Liquids/gas
$\rho/200$	5,775	
$\rho/250$	4,62	
$\rho/300$	3,85	
$\rho/350$	3,30	

SENSITIVITY ANALYSIS TO THE MESH

Solid phase

Five grid meshes were tested for the 1D resolution of the condensed phase: 0.2 mm, 0.1 mm, 0.04 mm, 0.02 mm and 0.01 mm. Over the simulations, a numerical artefact was identified on the mass flux. The shrinking of the first fuel cell leads to a decrease of the mass flux and it is not representative of the physic (figure 2). To solve this problem, a refinement of the grid size inside the solid phase has been done until the complete disappearance of the artefact. The study demonstrates that a 0.01 mm grid size is required, so it was selected.

Figure 2: Mass flux function of time for different grid sizes

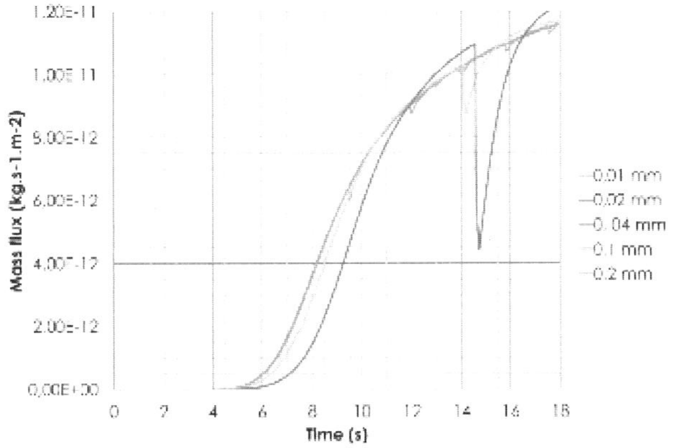

Gas phase

Local and global values have to be checked in the gas phase. In this sensitivity analysis, a convergence through the results is researched. Four mesh sizes were tested for global values. The ignition corresponds to the first heat release of the combustion reaction so the total Heat Release Rate (HRR) (heat release integral over the gas phase domain) appears to be a coherent indicator as a global value. Three convergences have been shown in the figure 3 between the following grid sizes: 0.1×0.1 mm^2, 0.2×0.2 mm^2 and 0.5×0.5 mm^2. These three mesh sizes were selected and tested for local values in the gas phase: the temperature 1 cm height over the solid surface (figure 4) and the incident heat flux at the solid surface (figure 5).

Figure 3: Heat Release Rate function of time for different grid sizes

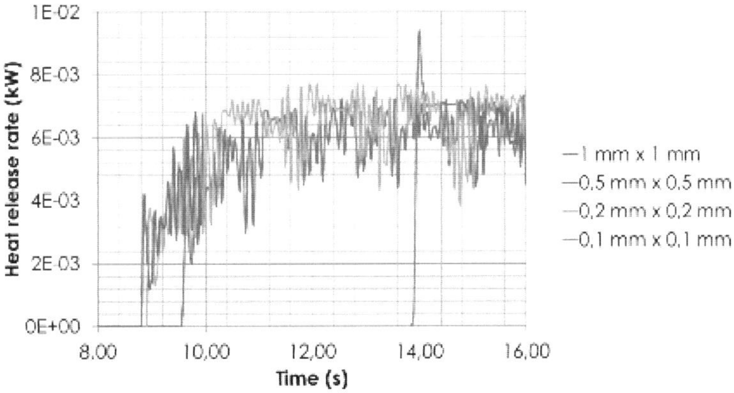

Through the figures 4 and 5, two values have been checked to identify the convergence between mesh sizes. The ignition time is the first one. It is possible to see that the ignition time is approximately the same for 0.1×0.1 mm^2 and 0.2×0.2 mm^2 (and less than one second later for 0.5×0.5 mm^2). The values taking by the heat flux before and after the ignition are also important to be checked and it can be seen that they are convergent for a 0.1×0.1 mm^2 and a 0.2×0.2 mm^2 meshes. To minimize the calculation duration, our choice was oriented on the 0.2×0.2 mm^2 grid size for our simulation in the gas phase.

Figure 4: Temperature 1 cm over the solid surface function of time for different grid sizes

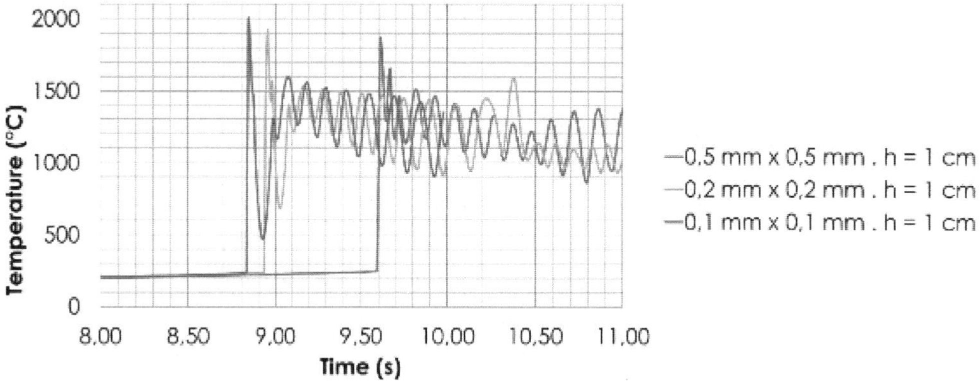

Figure 5: Incident heat flux at the solid surface function of time for different grid sizes

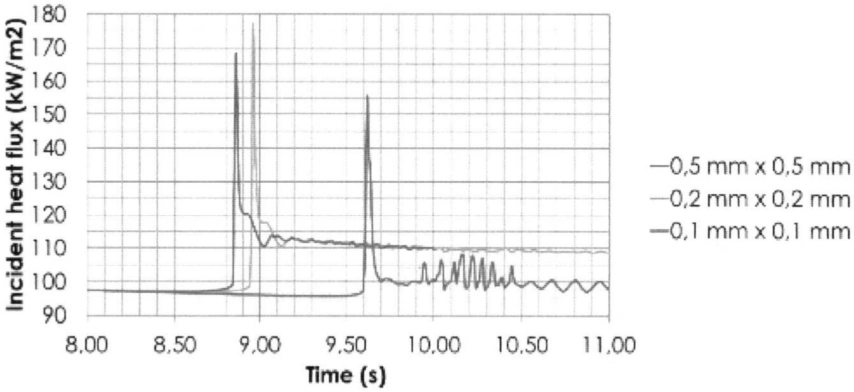

RESULTS AND DISCUSSION

As we have already seen in the previous part of this paper, local and global results were found and we have to deal with them in our analysis.

Global results

The ignition time is the first parameter easily available through the simulation. It was chosen to determine it with the total HRR, integral of the heat release on all over the gas phase domain. The ignition time function of the density is presented in the figure 6.
It is possible to note the impact of the density reduction, which generates a decrease of the ignition time. Quintiere (Quintiere 1998) proposes an equation (10) to calculate the ignition times of thermally thick solids.

$$t_{ig} = c\left(k\rho C_p\right)\left[\frac{(T_{ig}-T_{\infty})}{\dot{q}''}\right]^2 \quad (10)$$

This equation was plotted in the Figure 6 to analyse the coherence of this formula (based on a conduction model which assume that the solid processes are dominant) compare to the very fast phenomena in the gas phase. A good agreement between the empirical method of Quintiere is shown with our DNS model. The agreement between this relation and our simulation is good because we assume that fire is very well ventilated, it could not lead to the same results with a variation of the oxygen concentration in the domain. To conclude on the ignition time, we have shown with our results that the Quintiere's formula could be used considering thermally thick solids, with kinetics piloted by the heat transfer or by the pyrolysis and in well ventilated conditions.

Figure 6: Ignition time function of densities

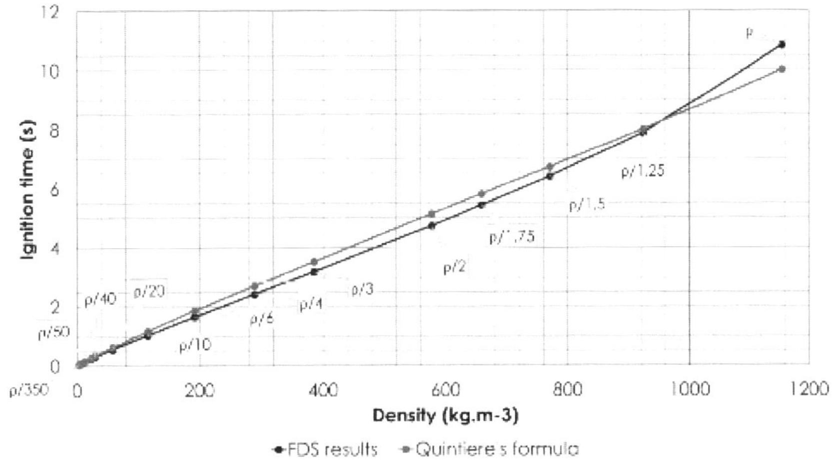

Auto-ignition depends both on temperature condition at the solid surface and on the quantity of pyrolysate emitted by the condensed phase. For the ignition time steps, those values are plotted in the figure 7. The density diminution generates an increase of the ignition surface temperature while the mass flux decreases. When the surface temperature is high, the ignition takes place very quickly and it can explain the decrease of the mass flux.

Figure 7: Ignition surface temperature and mass flux function of densities

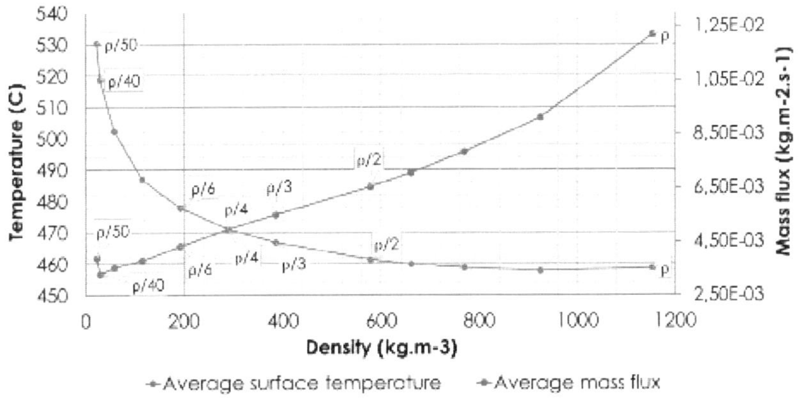

All the global and averaged results are presented taking into account the local variation of heat flux, temperature and mass flux on the solid width. These variations are very small and are negligible in the study of global analysis of solid results. We are going to see them in the next part of this paper.
Two representative densities were selected to lead this study and they correspond to the extreme cases presented in the figure 7: ρ and $\rho/50$.

Local results

Solid phase

The solid surface temperature is given by the figure 8 for the two considered densities. It is possible to see that the pre-ignition temperature is constant on the width of the surface of the solid. So, the consideration of an average temperature is representative of the surface conditions. The reaction rate is temperature dependant and the mass flux is calculated through it. The mass flux could also be considered as constant on the surface. We could note that the temperature levels are higher for the lower density. For this same density, the ignition generates a global increase of the temperature surface while it is not the case for the high density. The higher temperature is generated by the reduction of the density and consequently the thermal inertia. The global increase shows a global phenomenon (in opposition to the local one observe at high densities)

Figure 8: Temperature at the solid surface

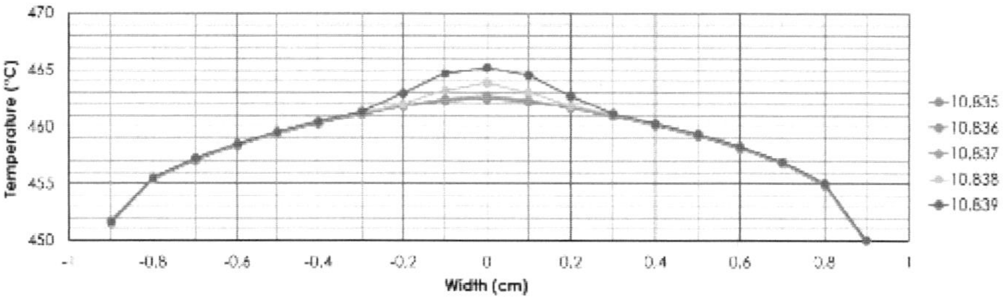

$\rho = 1155\ kg.m^{-3}$

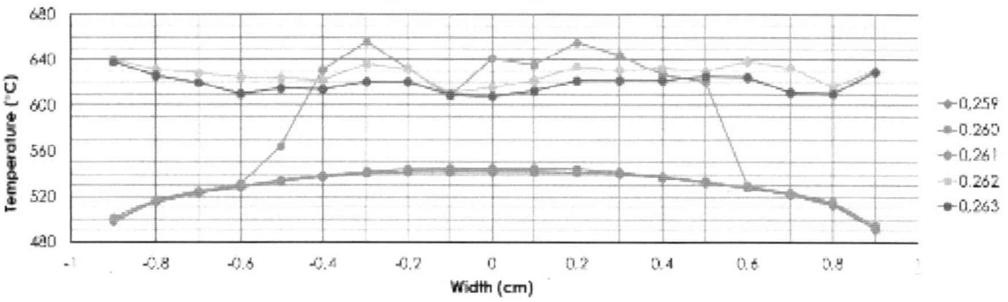

$\rho/50 = 23.1\ kg.m^{-3}$

Gas phase

The first parameter to check in the gas phase is the mixture composition. It is possible to describe it in terms of mixture fraction, Z.

For the following reaction:

$$C_5H_8O_2 + 6O_2 + 24N_2 \rightarrow 5CO_2 + 4H_2O + 24N_2$$

The mixture fraction could be noted like:

$$Z = \frac{sY_{C_5H_8O_2} - Y_{O_2} + Y_{O_2}^0}{sY_{C_5H_8O_2}^0 + Y_{O_2}^0}$$

With s, a stoichiometric quotient:

$$s = \frac{\nu'_{O_2} W_{O_2}}{\nu'_{C_5H_8O_2} W_{C_5H_8O_2}} = \frac{6 \times 32}{1 \times 100} = 1{,}92$$

By convention, $Z = 1$ for a mixture only constituted by fuel and $Z = 0$ if it is only constituted by oxidizer. In the reaction zone, all the fuel and oxidizer are consumed and it is possible to define a stoichiometric mixture fraction, Z_{st}, as:

$$Z_{st} = \frac{Y_{O_2}^0}{1{,}92\ Y_{C_5H_8O_2}^0 + Y_{O_2}^0} = \frac{0{,}2}{1{,}92 + 0{,}2} = 0{,}094$$

For the present mixture of MMA and air, the stoichiometric mixture fraction is about 0.1.

Figure 8 shows the evolution of the mixture fraction for the two cases mentioned before regarding to the temperature evolution. The five graphs represent the two time steps just before ignition, the ignition time step and the two time steps just after. Through the five time steps corresponding to the initial density, we can see that the reaction starts for a mixture fraction of 0.65. This value is six times superior to the stoichiometric mixture fraction. Blue points represent the mixture fraction and the temperature in the centre of the gas phase domain . As it can be shown in t_4 and t_5, the reaction is localised in the centre of the domain. For a density divided by 50, the reaction takes place around 500°C too but the major part of the domain is below the stoichiometric mixture fraction. Few points are over this limit but the maximum mixture fraction is around 0,2. In t_5, it is possible to note that the maximum temperature is reached for the stoichiometric mixture fraction which is led to a greater level of temperatures (more than 2000°C). For both densities, the mixture is not a limiting parameter.

Figure 9: Conditioned temperatures

$\rho = 1155\ kg.m^{-3}$ $\rho/50 = 23.1\ kg.m^{-3}$

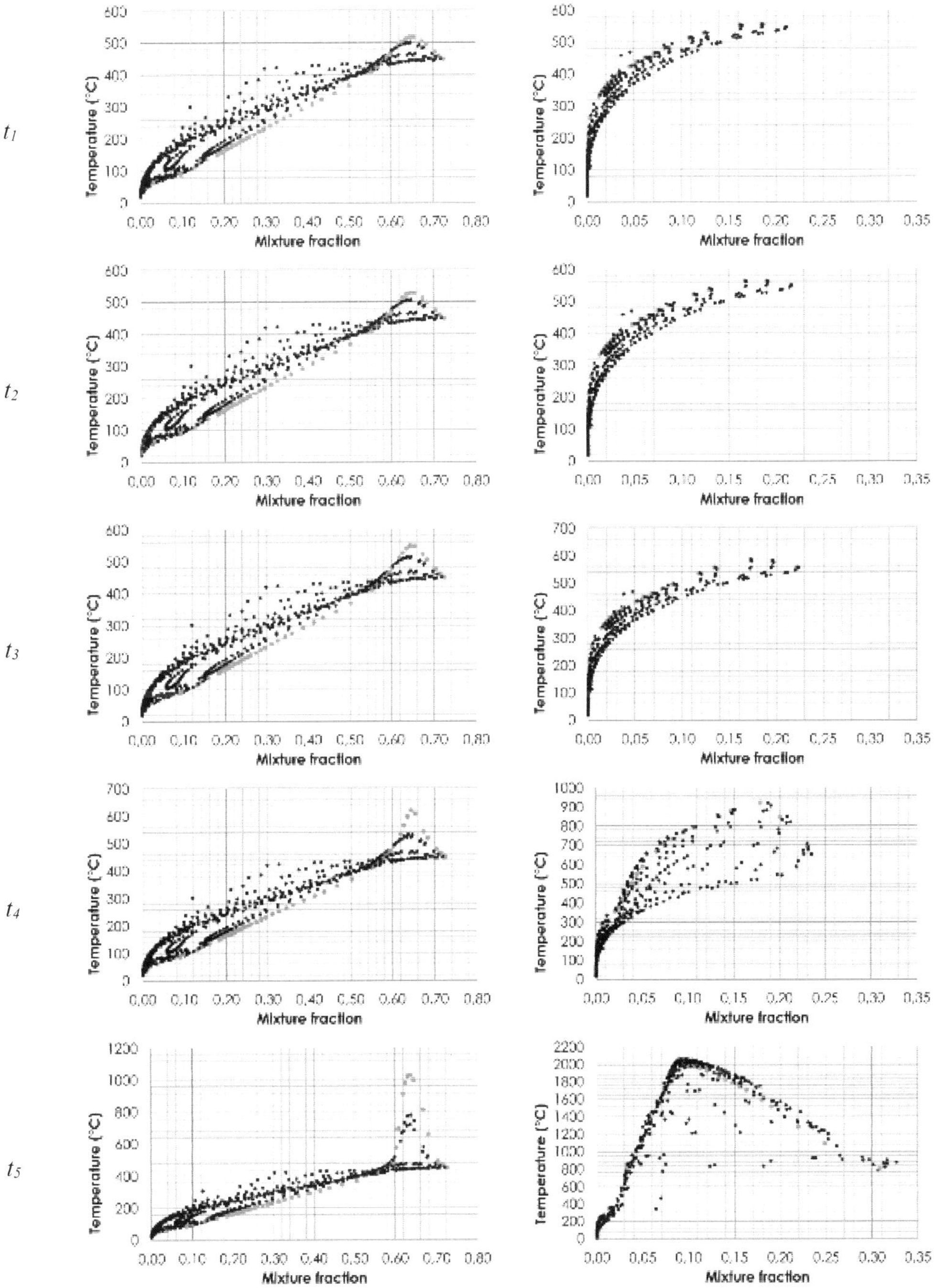

In fire safety engineering and safety in general, it is common to use the flammability limits notions to verify the flammability of a mixture. The MMA's flammability limits are calculated for a temperature of 550°C (mixture temperature just before the ignition) and the Lower Flammability Limit (LFL) is found to be 1.2% vol. and the Upper one (UFL) to be 17,3 % vol. Prior to ignition (t_2), the flammability limits are plotted to get a spatial representation of the mixture. They are given in figure 9 with the reaction rate, to determine where the ignition takes place. LFL is given by the white line and UFL by the red one. The reaction takes place in a fuel rich zone for the initial density, superior to the UFL. This shows representative problem induced by flammability limits when the occurrence conditions are far from the experimental ones. Furthermore, the temperature dependency equations (known as Burgess Wheeler equations) are empirical and this case is far from experimental conditions of homogeneous combustion.

Figure 10: Flammability limits location and reaction rate (arbitrary units)

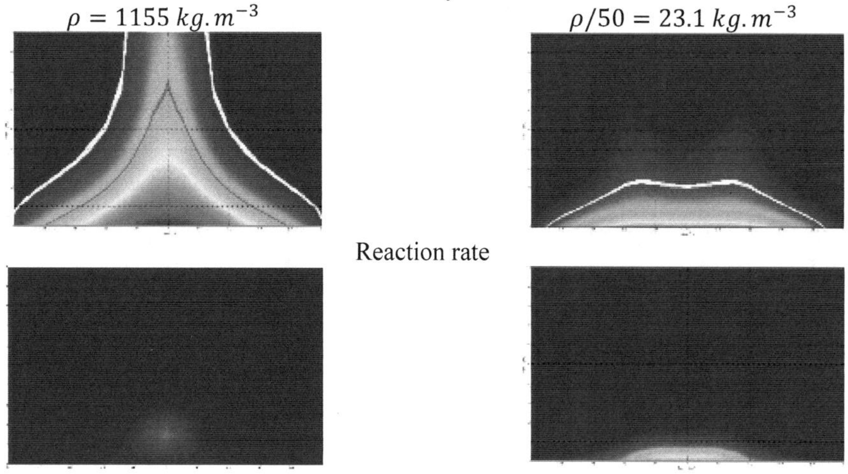

Torero (Torero 2016) indicates that auto-ignition takes place when a critical Damköhler is reached in the gas phase and it could be interesting to evaluate it before ignition. For the three times steps corresponding to the ignition, Damköhler number is high, traducing an infinitely fast chemistry compare to the residence time.

Figure 11: Damköhler number over the solid centre

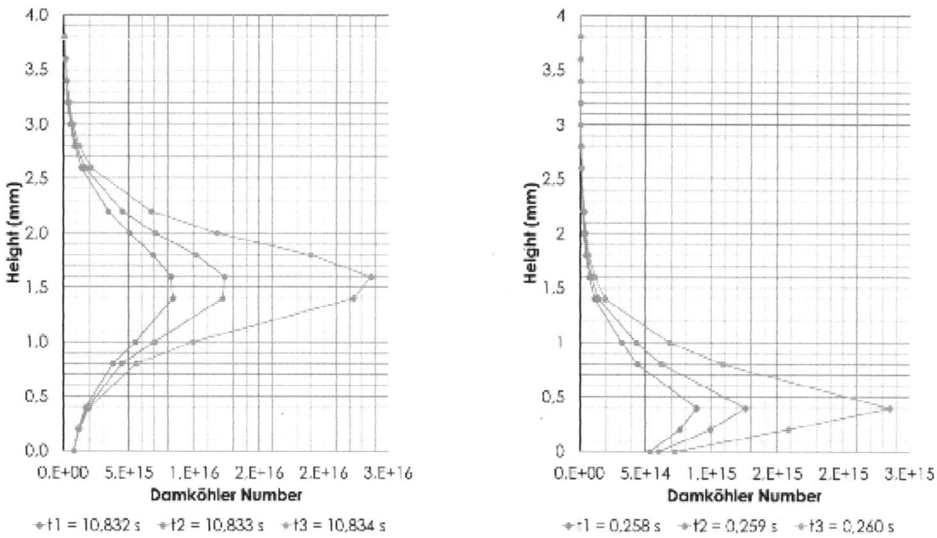

Those mixing and reacting conditions let the reaction takes place in the gas phase and two kinds of ignition are identified. For a high density, the reaction takes place over the solid surface, 1.5 millimetres in the gas phase. The ignition is piloted by the convective transfers in the gas phase.
For low density, the ignition is piloted by the solid surface (hotter than the other case). The solid pilots the phenomenon and the ignition time is equal to the sum of the conduction time and of the pyrolysis one.

Figure 12: Gas phase reaction rate

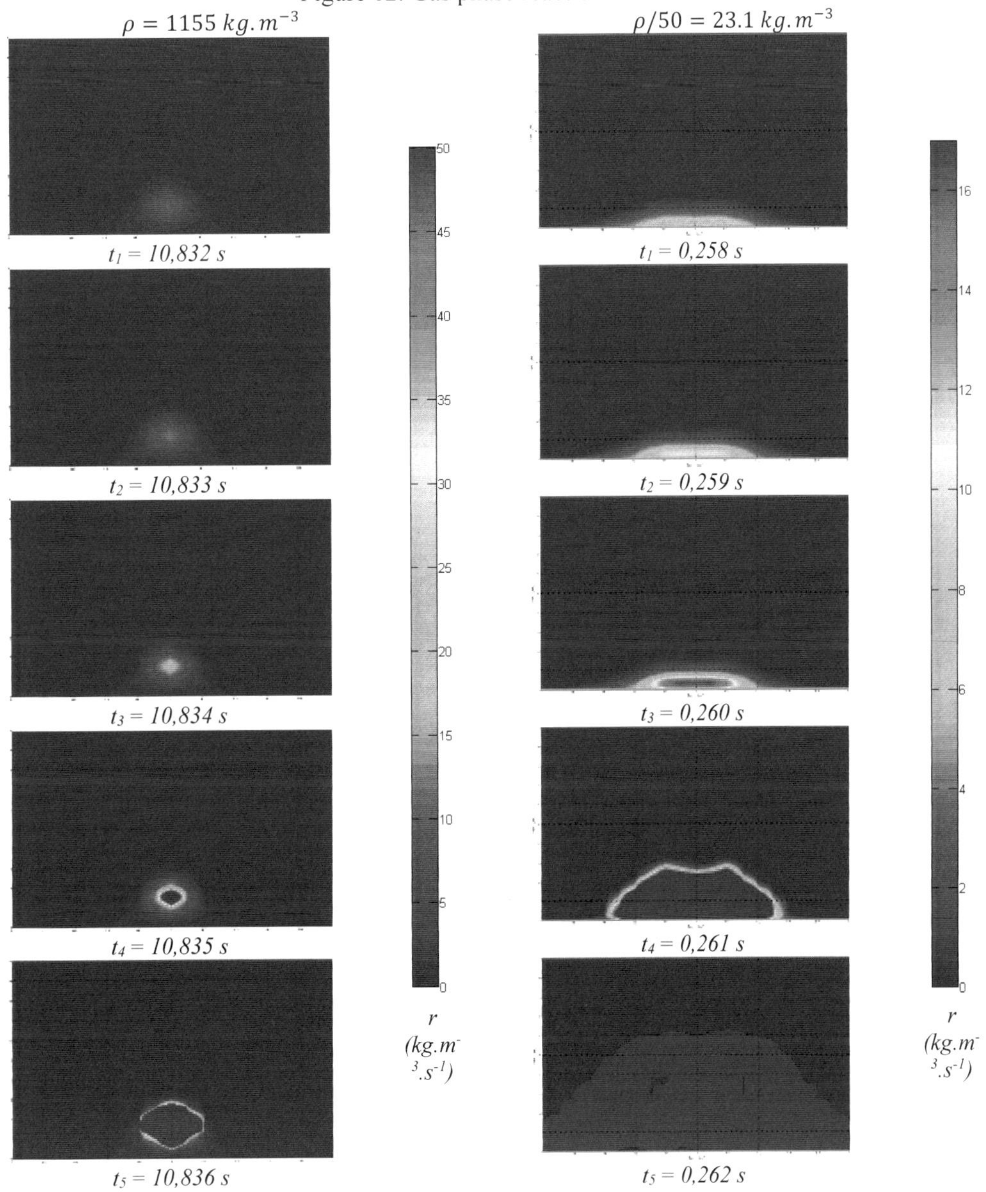

CONCLUSION

The present study is focussed on the auto-ignition of solid fuel in order to clearly establish the characteristics of this process. The DNS mode of Fire Dynamics Simulator has then be chosen and applied to the study of the PMMA. A pyrolysis and a combustion models have been implemented while the characterization of the influence of the solid properties on the auto-ignition has been determined by the variation of the density.

This work leads to the identification of four solid types and tries to explore the comportment differences in case of their auto-ignition. Two auto-ignition regimes are identified and correspond to the description done by Torero (Torero 2016). The first one corresponds to a gas phase ignition where the limiting parameter seems to be the convective heat transfer in the gas phase. The second one is piloted by the solid phase which heat faster than the gas phase and the phenomena is limited by the solid phase.

ACKNOWLEDGMENTS

The authors thank:

- Calyxis for its financial support
- The Institut des Risques Industriels Assurantiels et Financiers and the Hestia Platform for its technical and human support.
- Computations have been performed on the supercomputer facilities of the Mésocentre de calcul de Poitou Charentes.
- This work pertains to the French Government program "Investissements d'Avenir" (LABEX INTERACTIFS, reference ANR-11-LABX-0017-01

REFERENCES

Di Blasi, C. et al., 1991. Numerical model of ignition processes of polymeric materials including gas-phase absorption of radiation. *Combustion and Flame*, 83, pp.333–344.

Boonmee, N., 2004. Theorical and experimental study of autoignition of wood. *NIST Special Publication 1019*.

Chao, Y.H.C. & Fernandez Pello, A.C., 1997. Forced ignition of a solid fuel in a turbulent boundary layer oxidizing flow. In *Physical and Chemical Aspect of Combustion - A tribute to Irvin Glassman*. pp. 409–431.

Fernandez Pello, A.C., 2011. On Fire Ignition. In *Fire Safety Science - Proceeding of the Tenth International Symposium*. pp. 25–42.

Glassman, I. & Yeter, R.A., 2008. *Combustion - Fourth Edition*,

Korver, K.T., 2015. *A Generalized Model For Wall Flame Heat Flux During Upward Flame Spread On Polymers*,

Leventon, I.T., Li, J. & Stoliarov, S.I., 2015. A flame spread simulation based on a comprehensive solid pyrolysis model coupled with a detailed empirical flame structure representation. *Combustion and Flame*, 162(10), pp.3884–3895.

Li, J., Gong, J. & Stoliarov, S.I., 2014. Gasification experiments for pyrolysis model parameterization and validation. *International Journal of Heat and Mass Transfer*, 77, pp.738–744.

Li, J. & Stoliarov, S.I., 2013. Measurement of kinetics and thermodynamics of the thermal degradation for non-charring polymers. *Combustion and Flame*, 160(7), pp.1287–1297.

Luche, J. et al., 2011. Characterization of thermal properties and analysis of combustion behavior of PMMA in a cone calorimeter. *Fire Safety Journal*, 46(7), pp.451–461.

McGrattan, K. et al., 2013. Fire Dynamics Simulator, Technical Reference Guide, Volume 1: Mathematical Model. *NIST Special Publication 1018*, 1, p.175. Available at: http://www.thunderheadeng.com/wp-content/uploads/2013/08/FDS_Technical_Reference_Guide.pdf.

Niioka, T., Takahashi, M. & Izumikawa, M., 1981. Gas -Phase Ignition of a Solid in a Hot Stagnation-Point Flow. *Journal of Chemical Information and Modeling*.

Pyle, D.L. & Zaror, C.A., 1984. Heat transfer and kinetics in the low temperature pyrolysis of solids. *Chemical Engineering Science*, 39(1), pp.147–158.

Quintiere, J.G., 2006. *Fundamentals of Fire Phenomena* John Wiley.,

Quintiere, J.G., 1998. *Principles of Fire Behavior* Delmar, ed., Alar Elken.

Torero, J.L., 2016. Flaming Ignition of Solid Fuels. In *SFPE Handbook of Fire Protection Engineering*. pp. 633–661.

Tsai, T.H. et al., 2001. Experimental and numerical study of autoignition and pilot ignition of PMMA plates in a cone calorimeter. *Combustion and Flame*, 124, pp.466–480.

SIMULATION OF MICRO-SCALE FIBER REINFORCED POLYMER RESIN AND ADDITIVE SPECIMENS FOR KINETIC MODELING

Nicholas A. Dembsey[1], William Kreysler[2], Krista Anderson[1]
[1]WPI, 100 Institute Road, Worcester, MA 01609 USA
[2]Kreysler & Associates, 501 Green Island Road, American Canyon, CA 94503 USA

ABSTRACT

There is strong interest currently in the use of new materials such as fiber reinforced polymers (FRPs) in modern building construction. In the future computer simulation of FRP pyrolysis under fire conditions will allow for economical optimization. Computer simulations require input data such as material thermal properties and decomposition kinetic properties. Decomposition properties are typically determined from micro-scale specimens that are decomposed in the TGA and DSC. An important assumption about the specimen behavior is that the specimen is in thermal equilibrium with the apparatus furnace. In previous work the results for mixtures of resin and additive in the TGA showed potentially non-equilibrium behavior. To investigate the potential for non-equilibrium behavior of these micro-scale specimen mixtures their pyrolysis behavior in the TGA was simulated using the general pyrolysis program Gpyro.

INTRODUCTION

There is strong interest currently in the use of new materials such as fiber reinforced polymers (FRPs) in modern building construction. The interest in FRP is due to its aesthetics, shape variability shape, strength-to-weight, durability, insulation and low carbon footprint characteristics. The use of FRP in modern buildings is regulated by codes such as the International Code Council's International Building Code (IBC) [1]. The IBC requires that materials and their assemblies for a specific application meet specified reaction to fire and fire resistance requirements.

Currently development of fire safe FRP materials and systems is conducted via standard fire tests as specified in the IBC. This physical modeling approach does not allow for economical optimization of the FRP materials and assemblies. In the future computer simulation of FRP pyrolysis under fire conditions will allow for economical optimization. Computer simulations require input data such as material thermal properties and decomposition kinetic properties.

Decomposition properties are typically determined from micro-scale specimens that are decomposed in Thermogravimetric Analysis (TGA) and Differential Scanning Calorimetry (DSC) [2,3,4,5] under a constant heating rate. An important assumption about the specimen behavior is that the specimen is in thermal equilibrium with the apparatus furnace, e.g. the specimen temperature is uniform and the same as that of the furnace [6,7,8]. Estimation of decomposition kinetic properties are typically conducted under this assumption. If this assumptions does not hold then it is unclear how the property estimations will be affected.

Previous Work

Commercially available unsaturated polyester resin (Hetron™ 650 T20) [9] liquid and additive alumina trihydrate (ATH) power were mixed in various proportions to create a series of micro-scale specimen "drops" that were tested in the TGA and DSC [10]. The following mixtures of resin to additive mass ratios were studied: 100:33, 100:66, 100:100, and 100:130. Pure resin and pure additive were tested as well.

The mass loss and mass loss rate (MLR) results for these mixtures from the TGA showed potentially non-equilibrium behavior. The degree of this behavior was a function of the level of ATH used in the specimen mixtures.

Current Project

The current project investigated the potential for non-equilibrium behavior of resin and ATH mixture micro-scale specimens in the TGA [11]. Since measurement of micro-scale specimen temperatures is challenging at best, we chose to simulate the pyrolysis behavior of a micro-scale specimen in the TGA using the general pyrolysis program Gpyro [12].

GPYRO MODELING

Gpyro is a generalized pyrolysis model that can be applied to a wide variety of condensed phase fuels. The model simultaneously calculates the condensed phase mass conservation, gas phase mass conservation, condensed phase species conservation, and condensed phase energy conservation equations. Extensive details are given in Reference [12].

The resin and ATH mixtures are considered to be homogeneous. Gas phase decomposition products are assumed to be released instantaneously. Local thermal equilibrium is assumed between the condensed phase and the decomposition products. As the specimen decomposes its volume is assumed constant which means that the final residue forms a porous char.

In the previous work involving micro-scale specimens of resin and ATH, the specimens tested in the TGA were "drops" up to 10 mg in mass and up to 1 mm in radius [10]. The actual heating of these specimens in the TGA furnace is a complex 3D heat transfer problem involving the geometry of the furnace, the furnace heating elements, the flow rate of purge gas (nitrogen), the specimen pan and the specimen itself. To solve the 3D problem is beyond the scope of this work and not warranted at this time given the simplified kinetic models currently used to simulate the decomposition of specimens.

Simplified TGA Micro-scale Specimen Model

The "drop" specimen was assumed to be symmetrical and exposed to symmetrical heating. The "drop" was then modeled as a 1D (slab) with a half-width of 1 mm (20 nodes). The exposed face of the specimen was assumed to have a constant net heat flux boundary condition. The back face of the specimen was assumed to be perfectly insulated consistent with the assumed symmetry.

To simulate the heating rates (10 to 45 $°Cmin^{-1}$) of the micro-scale specimens in the TGA, the appropriate net heat flux boundary condition must be established for the 1D model. To do this the measured TGA MLR curves from the previous work were used [10]. This 3D data was then converted into an equivalent 1D form. From the corresponding TGA MLR curves an average normalized MLR was determined. The mass of the drop was known and the surface area of the drop was estimated based on the specimen radius. For each resin and ATH mixture tested a heat of decomposition was determined. The net heat flux absorbed by the 1D specimen model then is the product of the mixture heat of decomposition, the drop mass and the average normalized MLR. This product is then divided by the specimen surface area to determine the net heat flux. For the range of specimen mixtures and heating rates tested in the TGA the net heat flux ranges from 100 to 800 Wm^{-2}.

The specimen is assumed to consist of four species: resin, resin residue, ATH, and ATH residue. Each species has its own thermal properties. The resin and ATH each undergo their own one-step decomposition reaction starting with the virgin material, and producing gas and residue as decomposition products. The decomposition is modeled as an nth order reaction, see Equation 1 below. Each decomposition reaction is endothermic and its energetics is characterized by a heat of decomposition. All properties were determined in the previous work [10].

$$\frac{d(1-\propto)(t)}{dt} = \left[A\exp\left(-\frac{E_a}{RT(t)}\right)\right](1-\propto(t))^n \quad [1]$$

$$\frac{dT(t)}{dt} = \beta \quad [2]$$

where:

$\alpha = 1 - (m/m_o)$	E_a = activation energy (kJ/mol)
t = time (minutes)	R = ideal gass constant (8.314 J/molK)
m = mass of sample (kg)	T = temperature (K)
m_o = initial mass of sample (kg)	n = exponent
A = pre – exponential constant (1/s)	β = heating rate (K/min)

Input Parameters

Other than the input parameters described above or noted below, Gpyro default parameters were used. Tables 1 and 2 below show the input parameters for each species. These values were determined based on previous work [10,13,14].

Table 1: Species Thermal Properties [13,14]

	k [$Wm^{-1}K^{-1}$]	ρ [kgm^{-3}]	c [$kJkg^{-1}K^{-1}$]	εε [-]
Hetron	0.23	1200	1.40	0.84
Hetron Residue	0.19	253	1.90	0.90
ATH	1.22	2300	1.20	0.81
ATH Residue	0.24	1558	1.20	0.89

Table 2: Species Decomposition Properties [10]

	$Log_{10}Z$ [s^{-1}]	E [$kJmol^{-1}$]	ΔH [$kJkg^{-1}$]	n [-]
Hetron to Hetron Residue	13.7	195	172	1.125
ATH to ATH Residue	11.4	140.1	1000	1.24

RESULTS

Gpyro simulations were conducted for all the resin and ATH mixtures using the simplified TGA micro-scale specimen model. The mixtures considered are shown in Table 3. Net heat fluxes in the range 100-800 Wm^{-2} were used to produce heating rates of 10 to 45 °$Cmin^{-1}$. Consistent with the general nature of the pyrolysis model Gpyro, a wide range of simulated results can be output from the model.

Table 3: Material Simulations for Gpyro

Mixture Ratio Hetron : ATH	Material Label
100 : 33	Hetron+25%ATH
100 : 66	Hetron+40%ATH
100 : 100	Hetron+50%ATH
100 : 130	Hetron+57%ATH

General Specimen Behavior

Typical Gpyro output is shown in Figures 1 to 3 for the mixture Hetron+25%ATH. The simulated MLR shows two peaks as would be expected for the decomposition of the ATH and of the resin. As can be seen for various times throughout the simulated TGA test, specimen temperature, bulk density, porosity, reaction rates, and species mass fractions are uniform (flat profiles). Similar results are seen for all mixtures and all heating rates. The uniform behavior of the specimens appears consistent with the equilibrium assumption in TGA testing.

Figure 1: Hetron+25%ATH micro-scale specimen, heating 300 Wm^{-2} or 10 $°Cmin^{-1}$; MLR history; temperature, bulk density and porosity profiles

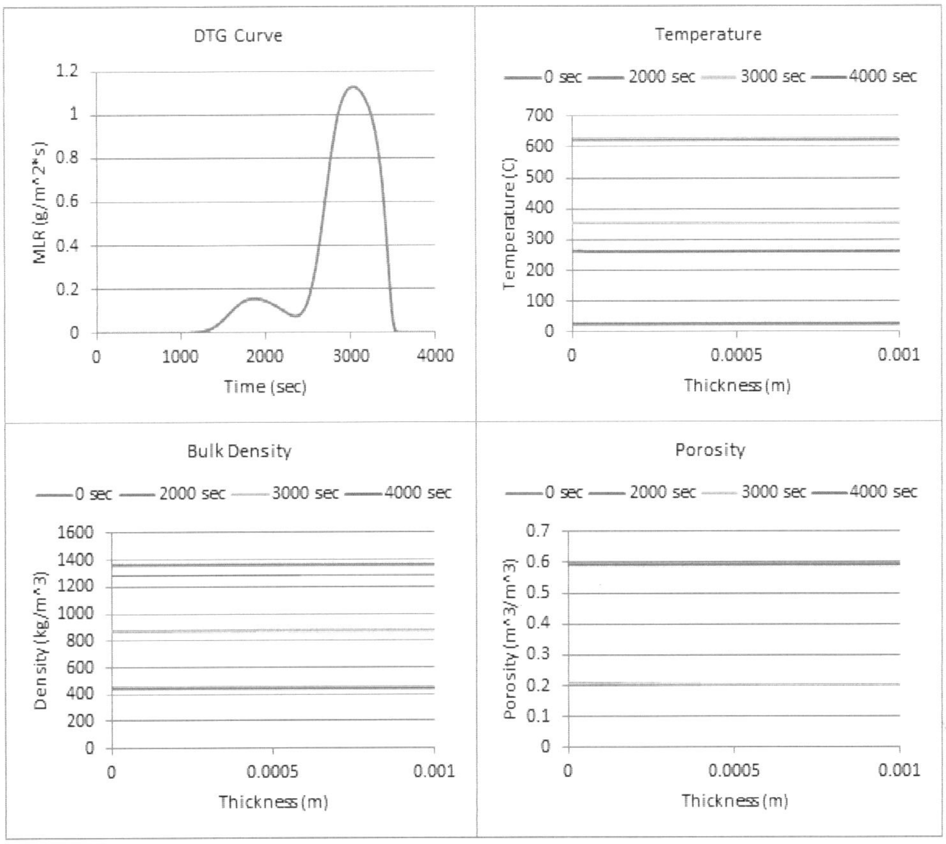

Figure 2: Hetron+25%ATH micro-scale specimen, heating 300 Wm^{-2} or 10 $°Cmin^{-1}$; reaction rate profiles

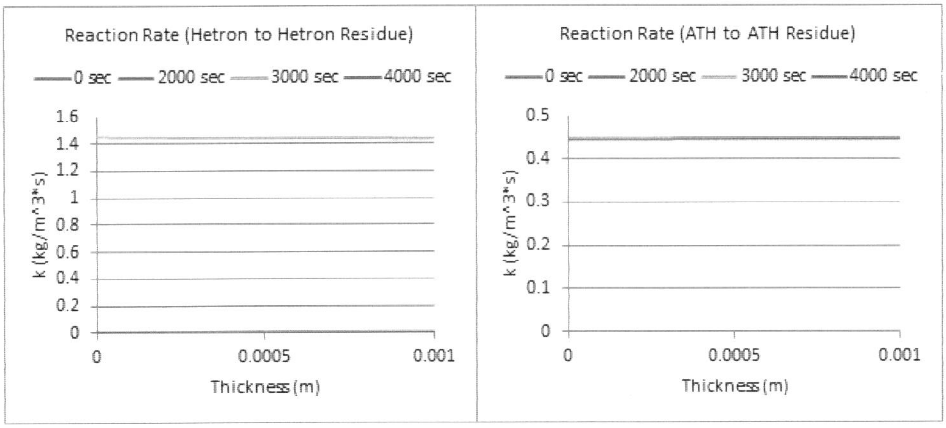

Figure 3: Hetron+25%ATH micro-scale specimen, heating 300 Wm^{-2} or 10 °$Cmin^{-1}$; species profiles

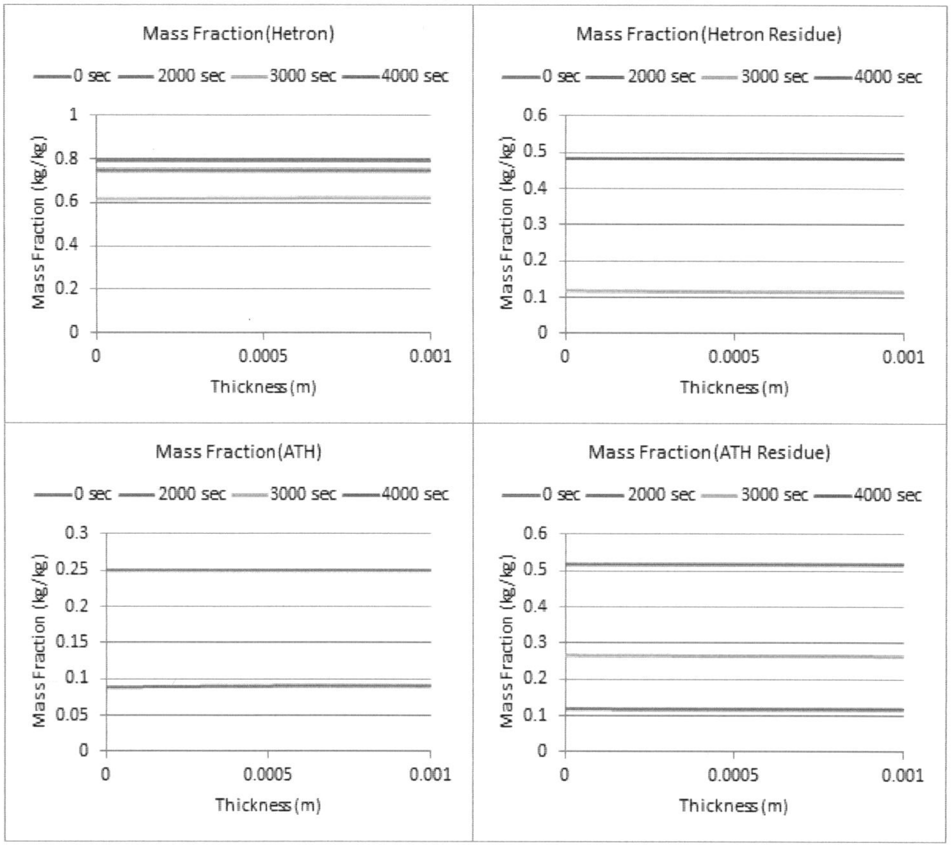

Comparison of Simulations and TGA Measurements

The Hetron+25%ATH mixture MLR simulations were compared to the corresponding TGA measurements for 4 heating rates as shown in Figure 4. Similar results are found for the other mixtures. The TGA MLR measurement had to be converted to a 1D equivalent so it could be compared to the corresponding simulation. To convert the TGA measurement to 1D, the normalized MLR measurement was multiplied by the specimen mass and then divided by the specimen surface area based on the "drop" radius. This then converted the TGA normalized MLR measurement in 3D with units s^{-1} to 1D with units $gm^{-2}s^{-1}$. The simulated heating rates are within 5 °C of the rates used in the TGA.

The magnitude of the simulated peaks for both the ATH (smaller peak) and the resin (larger peak) are similar to that measured in the TGA when converted to a 1D equivalent for all four heating rates. The width and the timeframe of the simulated peaks vary significantly from the TGA measurements. This is likely due to the fact that the simulations are pure 1D and that the TGA specimens are actually 3D in the way they are heated. The fact that the simplified 1D micro-scale specimen model can produce qualitatively consistent results in comparison with the actual TGA measurements suggestions that the modeling results can provide useful insight.

Figure 4: Hetron+25%ATH micro-scale specimen, MLR comparison between TGA measurements (°Cmin^{-1}) and simulation (Wm^{-2})

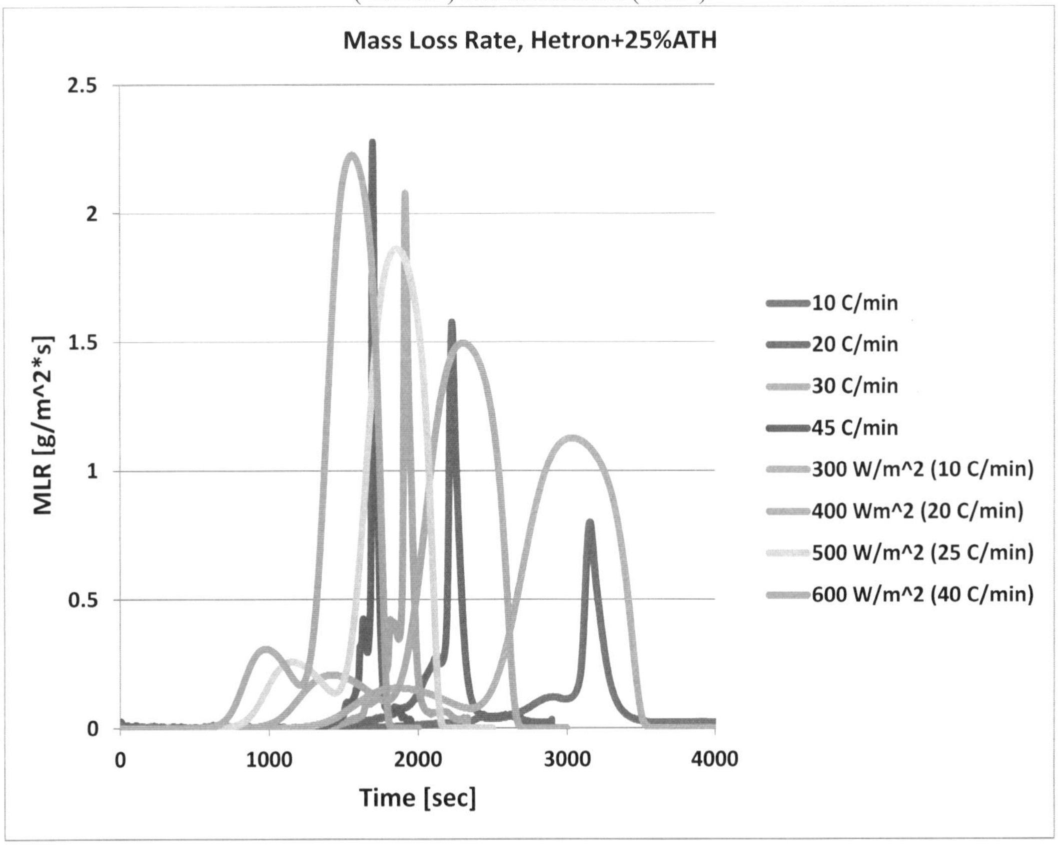

To more clearly understand if the resin and ATH mixtures might undergo non-equilibrium behavior the simulated specimen temperature can be compared to the TGA furnace temperature for the Hetron+25%ATH mixture, see Figure 5. Similar results are seen for all the mixtures and all the heating rates. As can be seen in Figure 5, over time the simulated specimen temperature deviates from the furnace temperature. For the 10 °Cmin^{-1} heating rate, the deviation starts to become significant at 20 minutes (1200 s) corresponding to the beginning of decomposition as seen in Figure 4. For the 30 °Cmin^{-1} heating rate, the deviation starts to become significant at 10 minutes (600 s) corresponding to the beginning of decomposition as seen in Figure 4. The magnitude of the maximum temperature deviation is approximately 200 °C for both heating rates. As the simulated decomposition MLR decreases at later times (higher temperatures) towards the end of the test the specimen temperature again tracks the furnace temperature.

The likely reason for the non-equilibrium behavior of the specimen is due to the large heat of decomposition of the ATH. This behavior has been noted by others [15]. Although the Gpyro simulations are only qualitatively consistent, they indicate that for resin and ATH mixtures that there is potential for non-equilibrium behavior.

Figure 5: Hetron+25%ATH micro-scale specimen simulated temperature deviation from TGA furnace temperature

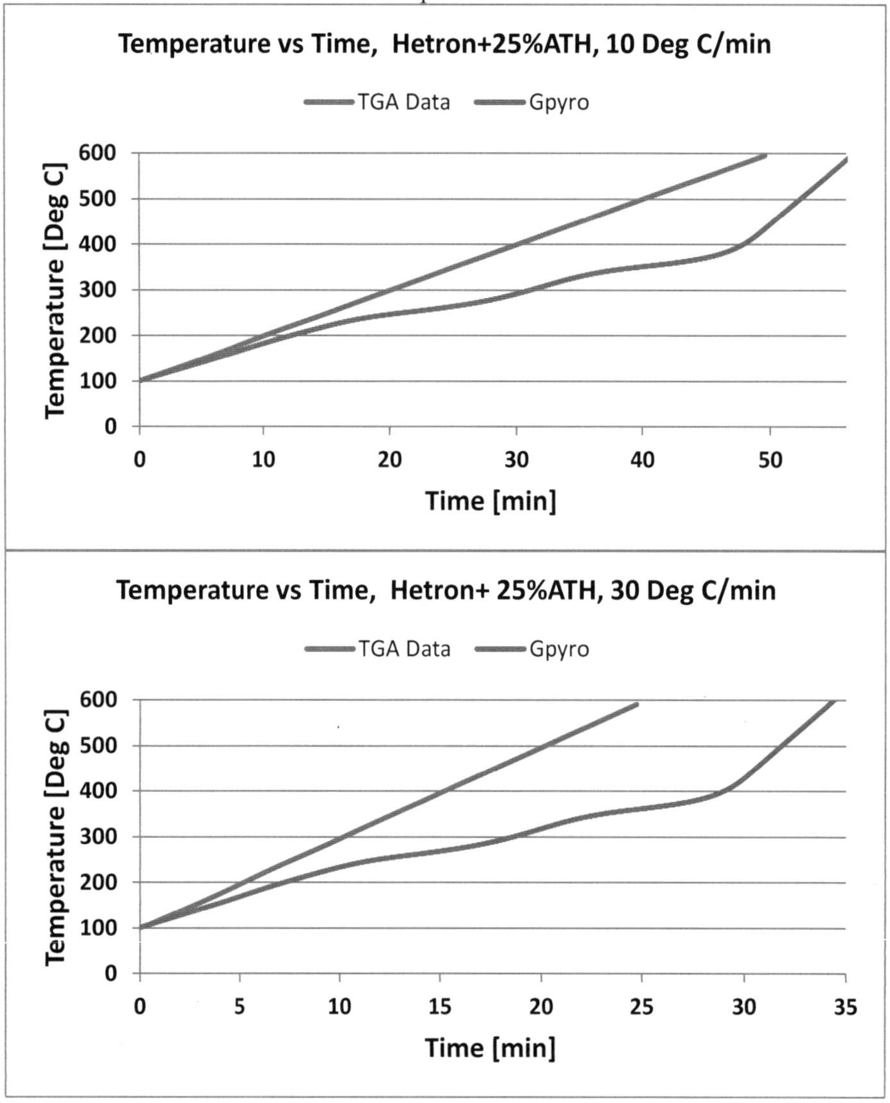

ACKNOWLEDGEMENTS

The authors would like to thank Victoria Grimes, Zachary Stewart and Shuyang Wang for their assistance in support of the Gpyro modeling work. The authors would like to thank Randall Harris, WPI Fire Lab Manager, for his assistance with related lab work.

CONCLUSIONS

There is strong interest currently in the use of new materials such as fiber reinforced polymers (FRPs) in modern building construction. The use of FRP in modern buildings is regulated by codes such as the International Code Council's International Building Code (IBC) [1]. Currently development of fire safe FRP materials and systems is conducted via standard fire tests as specified in the IBC. This physical modeling approach does not allow for economical optimization of the FRP materials and assemblies. In the future computer simulation of FRP pyrolysis under fire conditions will allow for economical optimization. Computer simulations require input data such as material thermal properties and decomposition kinetic properties.

Decomposition properties are typically determined from micro-scale specimens that are decomposed in TGA and DSC [2,3,4,5] under a constant heating rate. An important assumption about the specimen behavior is that the specimen is in thermal equilibrium with the apparatus furnace, e.g. the specimen temperature is uniform and the same as that of the furnace [6,7,8]. Estimation of decomposition kinetic properties are typically conducted under this assumption. If this assumption does not hold then it is unclear how the property estimations will be affected.

Results

Previous work with resin and ATH micro-scale specimens in the TGA showed potentially non-equilibrium behavior [10]. To better understand micro-scale specimen behavior in the TGA, a general pyrolysis program Gpyro [12] was used to simulate specimen behavior in the TGA. The results of the simulations reported here show that the micro-scale specimens have uniform characteristics such as temperature, bulk density, reaction rate and species mass fractions throughout the timeframe of the TGA test. The simulated MLRs are showed to be qualitatively consistent with those measured in the TGA. The likely reason for only qualitative agreement in MLR is due to the 1D simulation in comparison to the 3D measurement. The simulated specimen temperature deviates significantly (up to 200 °C) from that of the TGA furnace temperature. The likely reason is due to the large heat of decomposition of the ATH.

Future Work

The effect of the non-equilibrium micro-scale specimen behavior should be investigated in terms of its effect on the kinetic properties estimation process. The simulated micro-scale specimen temperature histories can be used in place of the TGA furnace temperature histories to develop MLR vs. "actual" specimen charts. These charts then can be analyzed using standard procedures to estimate updated properties. These updated properties then can be compared to current properties which are based on furnace temperatures. Both sets of properties can be used in the program Gpyro to investigate the effects of these sets of kinetic properties on pyrolysis simulation of FRP involving resin and ATH.

REFERENCES

1. "2012 International Building Code," International Code Council, Birmingham, AL, Country Club Hills, IL and Whittier, CA, 2012.
2. ASTM D3850-12, "Standard Test Method for Rapid Thermal Degradation of Solid Electrical Insulating Materials By Thermogravimetric Method (TGA)," ASTM International, West Conshohocken, PA, 2012.
3. "Thermogravimetric Analysis (TGA) A Beginner's Guide," Perkin Elmer, Inc. 2010. Product website. http://www.perkinelmer.com/CMSResources/Images/44-74556GDE_TGABeginnersGuide.pdf
4. ASTM D7426-08, "Standard Test Method for Assignment of the DSC Procedure for Determining Tg of a Polymer or an Elastomeric Compound," ASTM International, West Conshohocken, PA, 2013.
5. ASTM D3418-12e1, "Standard Test Method for Transition Temperatures and Enthalpies of Fusion and Crystallization of Polymers by Differential Scanning Calorimetry," ASTM International, West Conshohocken, PA, 2012
6. Lyon, R.E., Safronava, N., Senese, J., & Stoliarov, S., "Thermokinetic Model of Sample Response in Nonisothermal Analysis," *Thermochimica Acta* 545 (2012): 82-89.
7. Comesaña, R., Gómez, M. A., Álvarez, M. A., & Eguía, P., "Thermal Lag Analysis on a Simulated TGA-DSC Device.," *Thermochimica Acta* 547 (2012): 13-21.
8. Narayan, R. & Antal, M.J., "Thermal Lag, Fusion, and the Compensation Effect during Biomass Pyrolysis," *Industrial & Engineering Chemistry Research* 35(5) (1996): 1711-1721.
9. "Ashland 4107 V3 F2, Hetron FR 650 Series Flame Retardant Resins," Ashland, Inc., 2014
10. Dembsey, N.A., Gillespie, B., Long, M., McMillan, N., Walde, C., & Kreysler, W., "Pyrolysis Simulation of Fiber Reinforced Polymer (FRP) Composites: Challenges of Zero-Dimensional Testing of Resin and Additive Mixtures to Measure Kinetic Parameters," *Proceedings of Fire and Materials 2015, 14th International Conference*, Interscience Communictions, Hyatt Hotel Fisherman's Wharf, San Francisco, CA, USA, 2-4 February 2015.
11. Anderson, K., Grimes, V., Stewart, Z., and Wang, S., *Micro-scale Testing of Fiber Reinforced Polymer Resin and Additive to Evaluate Kinetic Parameters for Pyrolysis Modeling*, Worcester Polytechnic Institute, 2015.
12. Lautenberger, C.W., *A Generalized Pyrolysis Model for Combustible Solids*, University of California, Berkeley, 2007.
13. Kim, E., & Dembsey, N.A., "Evaluation of Pyrolysis Parameters for Fiberglass Reinforced Polymer Composites Based on Multi-objective Optimization," *Fire and Materials* 39(8) (2015). [DOI: 10.1002/fam.2262]
14. Kim, E. & Dembsey, N.A., "Parameter Estimation for Comprehensive Pyrolysis Modeling: Guidance and Critical Observations," *Fire Technology* 5(2) (2014): 443-477.
15. Stoliarov, S.I., & Li, J., "Parameterization and Validation of Pyrolsis Models for Polymeric Materials", *Fire Technology* 52(1) (2016). [DOI: 10.1007/s10694-015-0490-1]

NUMERICAL STUDY ON THE INFLUENCE OF IN-DEPTH RADIATION IN THE PYROLYSIS OF MEDIUM DENSITY FIBREBOARD

Guoxiang Zhao*, Tarek Beji, Davood Zeinali, Bart Merci
Dept. of Flow, Heat and Combustion Mechanics, Ghent University, Ghent, Belgium

ABSTRACT

In order to accurately predict the flame spread along a combustible wall, a better understanding of the burning behaviour of a material exposed to external radiation is required. This work focuses on simulating the pyrolysis behaviour of the Medium Density Fiberboard (MDF) material. The numerical simulation tool used in this study is the Fire Dynamics Simulator (FDS 6.2.0). A one-dimensional heat transfer solver that includes in-depth radiation is employed. The simulation results were compared with the experimental data, including the mass loss rate, and the surface temperature. The base case does not show satisfactory results for the time to reach the first peak and the value of that peak in the mass loss rate curve when compared to experimental data. The influence of the material properties and model parameters on the pyrolysis behaviour of MDF has been investigated in detail through a sensitivity analysis. For the considered range of parameters, the most significant influence on the time to the first peak comes from the emissivity, followed by the thermal conductivity, specific heat, and moisture content. For the peak mass loss rate, the most significant influence comes from the absorption coefficient, followed by the through-thickness density, the moisture content, and the specific heat. Only when proper values of these parameters were employed in the simulation, the onset of the pyrolysis process can be reasonably predicted. Through a simple trial and error procedure, a set of 'optimized' values of parameters including the in-depth absorption coefficient of char was obtained, which results in better agreement with experimental data.

1. INTRODUCTION

In the configuration of flame spread along a vertical combustible wall, thermal radiation of gas-phase combustion products drives the pyrolysis process by heating the unpyrolyzed material surface to pyrolysis temperature. Therefore, a better understanding of the burning behaviour of a material exposed to external radiation is required. A lot of effort has been devoted to study the burning behaviour of both charring and non-charring materials in the literature [1-5]. It has been demonstrated that numerical methods are suitable to predict the mass loss rate and burning rate for non-charring materials. However, the predicted results highly depend on the value of the input parameters, which include the thermal properties of the material and other parameters used in the pyrolysis model (e.g., kinetic parameters).

In-depth absorption of radiation has been reported to have a large influence on the burning behaviour of non-charring materials such as PMMA (Polymethylmethacrylate) [6-7]. However, the influence on the burning behaviour of charring materials (e.g., wood-based products) has not been extensively studied, presumably because opaque conditions are assumed. Nevertheless, in the case of cracks or delamination, in-depth-like radiation absorption may occur.

In this paper we present numerical simulations of MDF pyrolysis using a one-dimensional heat transfer solver that includes in-depth radiation transport. A one-step finite rate reaction is assumed; the virgin material is converted to char and the rest is released as pyrolysate. The values of the kinetic parameters and char yield are estimated from Thermogravimetric Analysis (TGA) test results reported in [3]. The thermal properties of virgin MDF and char are taken from the literature. More details on the pyrolysis model are provided in the numerical model section.

The goal of this study is to predict the burning behaviour of MDF, with an emphasis on the importance of in-depth radiation. The numerical simulation tool used in this study is the Fire Dynamics Simulator (FDS 6.2.0), which is developed and maintained by the National Institute of Standards and Technology (NIST) [8]. The physical processes occurring in the solid phase are solved as a one-dimensional problem with a one-step pyrolysis model. The input parameters include the material properties, kinetic parameters (e.g., pre-exponential factor and activation energy) properties and boundary conditions. A sensitivity analysis is conducted to study the influence of these input parameters on the burning behaviour of MDF.

2. EXPERIMENTAL CONFIGURATIONS

Small scale tests have been conducted by FM Global in the Fire Propagation Apparatus (FPA) [2]. The MDF sample used in these tests are manufactured by SPANOLUX [9]. Figure 1 shows the schematic of this configuration. The back side of the sample is actually insulated with Cotronics ceramic paper, which has a relatively low thermal conductivity (0.028 W/(m·K)). That is why an adiabatic boundary condition is used in the numerical simulations. The dimensions of the sample are 80 mm×80 mm ×18.4 mm (height).

The tests have been conducted in nitrogen atmosphere in order to eliminate uncertainties related to gas-phase combustion. During these tests, the mass loss rate, surface and back side temperatures have been measured under three constant external heat fluxes, namely 25, 50, and 100 kW/m^2. More details can be found in reference [2].

Figure 1 Schematic of the MDF sample in FPA test.

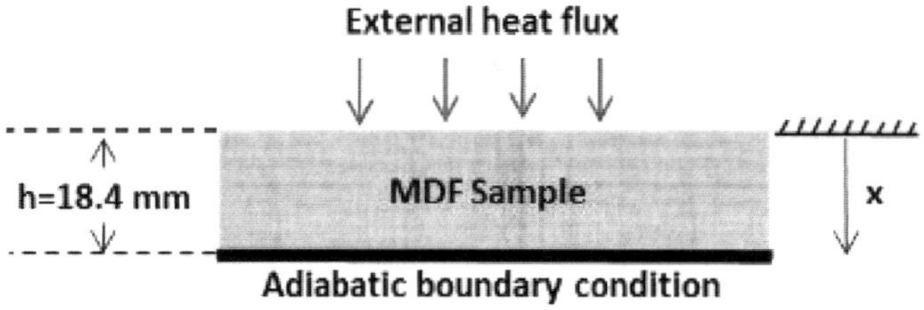

3. NUMERICAL MODEL

As mentioned in the introduction, numerical techniques have been employed in this study. In this section, the numerical model and input parameters will be discussed in detail.

3.1 Pyrolysis model

The pyrolysis process is simplified as a one-step finite rate reaction. Only three species are treaded here, namely the virgin solid, char, and pyrolysate. For the solid reaction, the virgin is converted to char and releases pyrolysate, as shown in the following reaction.

$$\text{Virgin} \rightarrow (1-\nu_c)\text{Pyrolysate} + \nu_c \text{Char} \qquad [1]$$

where ν_c is the yield of char. Since neither shrinkage nor swelling is modelled here (i.e., constant volume), the yield of char can be calculated as $\nu_c = \rho_c / \rho_v$, where ρ_v and ρ_c are the densities of respectively the virgin and the char.

With this one-step pyrolysis model, the pyrolysis rate of the sample is calculated using a temperature-dependent Arrhenius expression. The mass conservation equations for virgin and char are expressed as follows:

$$\frac{dY_v}{dt} = -AY_v \exp(-\frac{E}{RT}) \qquad [2]$$

$$\frac{dY_c}{dt} = -v_c \frac{dY_v}{dt} \quad [3]$$

where Y_v and Y_c are the mass fractions of virgin and char respectively. The initial value of the virgin mass fraction $Y_{v,0}$ is equal to 1. T is the temperature (in Kelvin) within the solid. R is the universal gas constant, 8.314J mol^{-1} K^{-1}. A is the Arrhenius pre-exponential factor, with units of s^{-1}. E is the activation energy, with units of kJ/kmol.

It should be noted that the kinetic parameters are not available for most real materials. In practice, the values of A, E and v_c can be obtained from bench-scale measurements like TGA under an inert atmosphere. These measurements consist of heating the sample at a fixed rate, $\dot{T}$, and measuring the residual mass and the reaction rate (in s^{-1} or min^{-1}) as a function of the temperature (see example in Fig. 2).

Figure 2 Typical TGA results for MDF pyrolysis at a heating rate of $\dot{T}$ = 5 K/min [8]. (a) black line represents the normalized residual mass (b) blue line represents the reaction rate

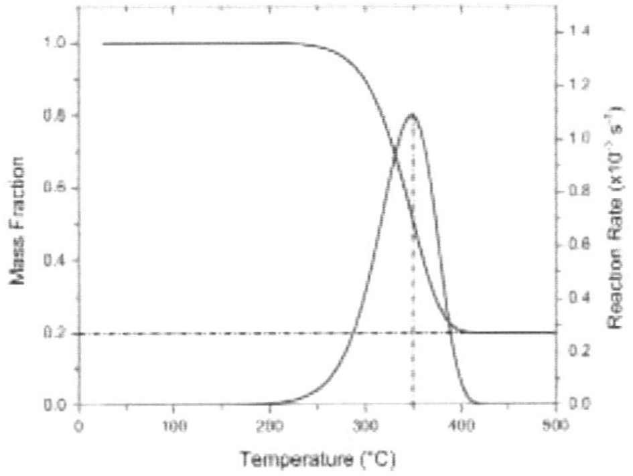

The results in Fig. 2 show a residual mass of 21% at high temperatures. This indicates that we can take $v_c = 0.21$. The peak reaction rate $\gamma_p = 0.0011\,\text{s}^{-1}$ occurs at a temperature $T_p = 350\,°\text{C}$. The blue curve in Fig. 2 can also be characterised by a pyrolysis range (ΔT), which is the approximate width of this curve (in degree Celsius or Kelvin). Assuming a triangular shape for the curve, this width can be estimated as [8]:

$$\Delta T = (1 - v_c) Y_{v,0} \frac{2\dot{T}}{\gamma_p} \quad [4]$$

Based on the obtained value of T_p, ΔT and v_c from TGA tests, the kinetics parameters E and A can be estimated as [8]:

$$E = \frac{2e(1 - v_c) R T_p}{\Delta T} \quad [5]$$

$$A = \frac{e \gamma_p}{Y_{v,o}} \exp\left(\frac{E}{R T_p}\right) \quad [6]$$

In this work, the estimated values for A and E are 358000 s^{-1} and 89100 kJ/kmol respectively.

3.2 Heat transfer within the solid

The 1-D heat transfer equation, which accounts for the in-depth radiation absorption with an energy source term, reads:

$$\rho c_p \frac{\partial T}{\partial t} = \frac{\partial}{\partial x}\left(k \frac{\partial T}{\partial x}\right) + \dot{q}_{s,c}''' + \dot{q}_{s,r}''' \quad [7]$$

where ρ, c_p and k are respectively the density, specific heat and conductivity of the solid. The variables $\dot{q}_{s,c}'''$ and $\dot{q}_{s,r}'''$ denote respectively a chemical and a radiative source term. Eq. [7] is solved using a uniform mesh with a cell size $\Delta x =$ 0.376 mm.

The chemical source term $\dot{q}_{s,c}'''$ is modelled using an Arrhenius expression:

$$\dot{q}_{s,c}''' = -\rho_v A Y_v \exp(-\frac{E}{RT}) H_p \qquad [8]$$

where H_p is the heat of reaction in kJ/kg. A positive value of H_p means that the pyrolysis reaction is endothermic, while a negative value of H_p means that the reaction is exothermic.

The radiative absorption term $\dot{q}_{s,r}'''$ is a function of absorption coefficient (κ) and emissivity (ε).For an opaque material, the thermal radiation is absorbed within an infinitely thin layer at the solid surface, and the corresponding absorption coefficient is infinitely large. For some other materials the radiation can penetrate to some finite depth, and the absorption coefficient value is smaller than infinity.

In order to solve the radiative absorption term in the heat conduction equation, a two-flux model is employed. This model is based on the Schuster-Schwarzschild approximation [5], which assumes that the intensity is constant inside the 'forward' and 'backward' hemispheres. Accordingly, the radiative source term $\dot{q}_{s,r}'''$ is taken as the sum of the 'forward' $\dot{q}_r^+(x)$ and 'backward' $\dot{q}_r^-(x)$ flux gradients.

$$\dot{q}_{s,r(x)}''' = \frac{d\dot{q}_r^+(x)}{dx} + \frac{d\dot{q}_r^-(x)}{dx} \qquad [9]$$

The forward radiative heat flux into the solid is computed as follow:

$$\frac{d\dot{q}_r^+(x)}{dx} = 2\kappa(\sigma T_{(x)}^4 - \dot{q}_r^+(x)) \qquad [10]$$

A similar formula can be given for the 'backward' direction. The boundary condition for $\dot{q}_r^+(x)$ at the solid surface is given by:

$$\dot{q}_r^+(0) = \varepsilon\dot{q}_e'' + (1-\varepsilon)\dot{q}_r^-(0)) \qquad [11]$$

The boundary condition at the front surface of the solid reads:

$$-k\left.\frac{\partial T}{\partial x}\right|_{x=0} = \varepsilon\dot{q}_e'' - \dot{q}_c'' - \varepsilon\sigma\left(T_s^4 - T_\infty^4\right) \qquad [12]$$

On the right hand side, $\dot{q}_e''$ is the incident heat flux, $\dot{q}_c''$ is the convective heat flux. It is calculated with the following expression:

$$\dot{q}_c'' = h(T_s - T_\infty) \qquad [13]$$

where T_s and T_∞ are the surface temperature of the solid and the temperature of the surrounding air; h is the convective heat transfer coefficient, which is calculated based on a combination of natural and forced convection correlations.

For the back-side, a perfectly insulated condition is applied:

$$-k\left.\frac{\partial T}{\partial x}\right|_{x=h} = 0 \qquad [14]$$

3.3 Thermal properties

The properties of MDF and the model parameters will be discussed in this section, including the density, thermal conductivity, specific heat, heat of pyrolysis, emissivity, and the absorption coefficient. In general, determining these parameters is challenging due to their dependency on temperature, test conditions, and uncertainties related to the measurement method.

3.3.1 Densities of virgin MDF and char

It has been reported in the literature [2, 9-11] that the density of the MDF is not uniform along the thickness of the sample, namely due to the hot-processing operation during its manufacturing. It is highest at the surface of the sample and lowest at mid-depth. The MDF density profile suggested in [2] (see parabolic profile in Fig.3) is expressed as:

$$\rho_v(x) = \frac{3\rho_{v,bulk}}{2+\xi}\left[4(1-\xi)\left(\frac{x}{h}\right)\left(1-\frac{x}{h}\right)+\xi\right] \qquad [15]$$

where the bulk density is $\rho_{v,bulk} = 605$ kg/m^3 (as measured), and the ratio, ξ, of the maximum density to the minimum density in the sample panels is estimated in reference [2] to be 1.61 that also agrees with values observed in the literature, ranging between 1.50 and 1.75 [9] for different MDF panels.

Figure 3. Through-thickness density profile for both uniform density and non-uniform density cases.

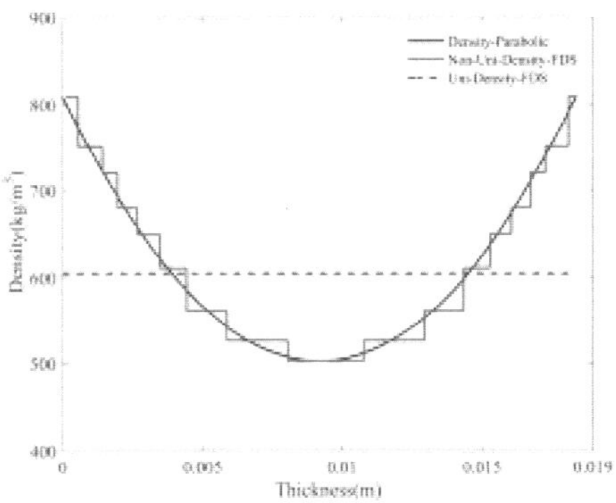

Both the uniform and non-uniform density cases have been considered in this work. In FDS, the solid can be represented with different layers, and each of the layers having its thermal properties can undergo reactions. For the non-uniform density case, 17 layers have been used to represent the sample (see Fig. 3). Each layer has a constant density. The minimum value is 502.92 kg/m^3, i.e. at the middle of the sample, while the maximum value is 809.46 kg/m^3 ,i.e. at the front and back surface.

The density of char was calculated based on the assumption that the volume of the solid sample is fixed during the burning process. Considering a value of 0.21 for the char yield [3], a char density of 127 kg/m^3 is approximated.

3.3.2 Thermal conductivity

Although the thermal conductivity depends on time-varying temperature as well as the local density and constant values are still widely used in pyrolysis models (for example [14-16]). Here, taking the range of 0.08 to 0.41 W/(m·K) for wood material as reference [13], a constant value of 0.18 W/(m·K) is considered for MDF.

The char thermal conductivity of MDF should be similar as that of softwood, since their densities are similar. In [17, 18], the value for wood char is reported within a range of 0.05 to 0.175 W/(m·K) for temperatures between 300 K and 873 K. Since for the most time during the pyrolysis process, the sample temperatures range from 700 K to 1000 K, a constant value of 0.17 W/(m·K) is considered appropriate..

3.3.3 Specific heat capacity

The specific heat capacity of wood based materials depends on the temperature and moisture content of the material, but is independent of density and species [19]. According to a literature survey by Gronli [13], specific heat capacities used for unreacted wood material in pyrolysis models vary from 1.5 to 2.51 kJ/(kg·K), whereas Gupta et al.[18] report that values from 0.67 to 2.5 kJ/(kg·K) have been used. In this study, a constant value of 1.58 kJ/(kg·K) is considered.

The reported values in the literature for specific heat capacity of wood char range from 0.67 to 1.35 kJ/(kg·K) at room temperature [13]. If the temperature increases up to a range of 523 K – 673 K, the specific heat capacity can increase up to 1.9 kJ/(kg·K). Here, the constant value of 1.4 kJ/(kg·K) is used.

3.4 Heat of pyrolysis

The heat of wood pyrolysis reported in the literature [21] ranges from endothermic (370 kJ/kg) to large exothermic values (-1700 kJ/kg), depending on the experimental conditions. The heat of pyrolysis is not available for MDF used in this study. Here a value of 0 kJ/kg is first considered and other possible values will be discussed in the sensitivity study.

3.5 In-depth radiation parameters

The values for the emissivity and absorptivity are assumed to be the same in FDS by default. It has been reported in [1] that the effective emissivity of virgin material is a function of temperature. For a temperature range up to 1500 K, the corresponding emissivity ranges from 0.6 to 0.9. The effective emissivity of char, as reported in [1], is approximately0.86 and relatively independent of temperature over a wide temperature range (300-3000K). In the base case, values of 0.86 for both the virgin and char emissivity have been used.

The absorption coefficient of MDF, κ, has been hardly mentioned or discussed in the literature. However, for polymeric materials such as PMMA, it has been investigated in the literature. In [5] a wide range of possible values for κ (from 333 m^{-1} to 2000 m^{-1}), is examined for black PMMA. In our study, first the default FDS value of 50000 m^{-1} is considered for both the virgin material and char, which is deemed suitable for opaque materials. As a consequence, the in-depth radiation is effectively not considered in the equations. Table 1 shows the summary of input parameters used in the base case.

Table 1 summary of input parameters for base case as well as the optimized case.

Parameter	Unit	Base case	Optimized case	Range
Density of virgin MDF	[$kg.m^{-3}$]	605	non-uniform	-
Density of char	[$kg.m^{-3}$]	127	127	-
Residual fraction	-	0.21	0.21	-
Moisture content	[%]	0	6	-
Heat capacity of virgin MDF	[$kJ.kg^{-1}.K^{-1}$]	1.58	1.58	[0.67, 2.5]
Heat capacity of char	[$kJ.kg^{-1}.K^{-1}$]	1.4	1.4	[0.67, 1.9]
Thermal conductivity of virgin MDF	[$W.m^{-1}.K^{-1}$]	0.18	0.18	[0.08, 0.41]
Thermal conductivity of char	[$W.m^{-1}.K^{-1}$]	0.17	0.17	[0.05, 0.18]
Emissivity of virgin MDF	[-]	0.86	0.6	[0.6, 0.9]
Emissivity of char	[-]	0.86	0.86	-
Heat of pyrolysis	[$kJ.kg^{-1}$]	0	0	[-1700, 370]
Reaction order	[-]	1	1	-
Pre-exponential factor	[s^{-1}]	3.58×10^5	3.58×10^5	-
Activation Energy	[$J.mol^{-1}$]	8.91×10^4	8.91×10^4	-
Absorption coefficient for virgin	[m^{-1}]	50000	50000	-
Absorption coefficient for char	[m^{-1}]	50000	2000	-

4. SMALL-SCALE SIMULATION RESULTS

First, the values for input parameters in Table 1 are applied in the simulations, which is the base case simulation. These results of this simulation case, including mass loss rate and front surface temperature, are presented in the following section.

Moreover, a sensitivity study has been conducted for the different model parameters, namely, for the thermal conductivity, specific heat capacity, heat of pyrolysis, emissivity, through-thickness density profile, and absorption coefficient. Consequently, a set of 'optimized' parameters have been determined through a simple trial and error procedure, shown in Table 1 as the optimized parameters.

4.1. Base case simulation results

4.1.1. Mass loss rate

The mass loss rate results are presented in Fig.4 (a-c). Three external heat flux values have been considered, namely 100, 50, and 25 kW/m^2.

There are two peaks in the mass loss rate curves for different external heat flux values. The first peak arises due to a high temperature reached at the surface. After the pyrolysis process, a char layer is formed at the surface of the sample. This insulating char layer decreases the gasification rate. As a result, the mass loss rate drops to a lower level. The second peak is mainly due to the heat accumulation at the back side.

The simulated values for the first peak are significantly lower than the measured values with relative deviations of about 21%, 50% and 63 % for respective heat flux exposures of 25, 50, and 100 kW/m^2. Furthermore, the time to reach the first peak is shorter in the predicted profiles by about 34 s, 107 s and 381 s for respective heat flux exposures of 25, 50, and 100 kW/m^2. Deviations are particularly pronounced for the lower external heat flux value (25 kW/m^2). One possible reason could be that the one-step reaction is too strong a simplification for the complex phenomena taking place during the pyrolysis process. The consequence of possible shortcomings in the chemistry is expected to be more pronounced as the pyrolysing zone occupies a larger volume in the material, which is the case for lower heat fluxes. Fig. 3(d) illustrates this, showing the through-thickness temperature evolution for heat flux values of 100 and 25 kW/m^2. The thickness in the solid occupied by temperature range [200°C - 400°C], corresponding to the pyrolysing zone (Fig. 2), is indeed much wider for 25 kW/m^2 than for 100 kW/m^2 (0.0081m and 0.0022 m respectively).

Figure 4. Comparisons between experimental data and predicted results for (a) MLR at 100kW/ m^2, (b) MLR at 50kW/ m^2 (c) MLR at 25kW/ m^2 and (d) through-thickness temperature at exposures of 100 and 25 kW/ m^2.

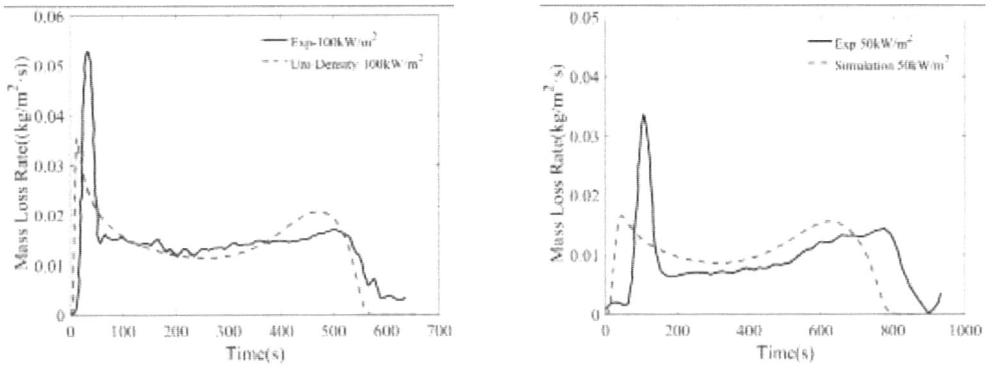

(a) (b)

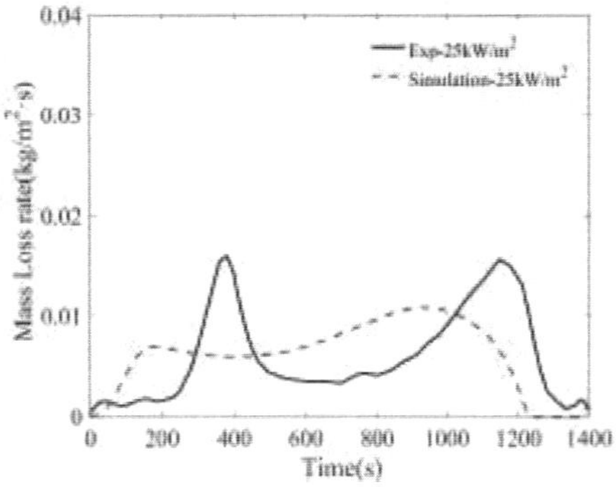

(c)

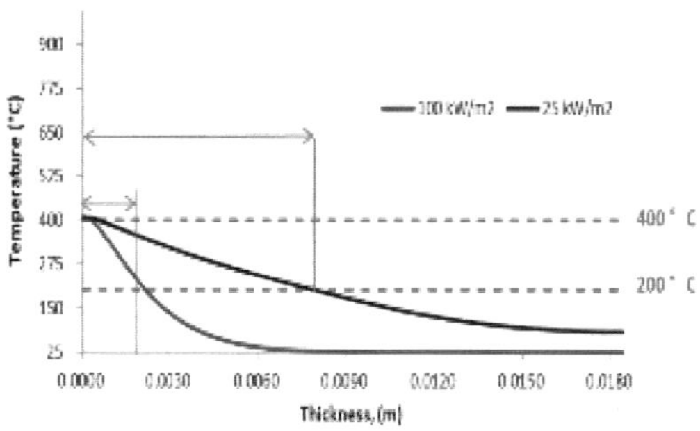

(d)

In order to quantitatively indicate the difference between the experimental data and the predicted results, the time, t_p, to reach the peak, as well as the value of the peak, $\dot{m}''_p$, are listed in Table 2. For the time to reach the peak MLR, the predicted values are much smaller than the experimental results, namely by about 50 to 65%). For the value of the peak, larger deviations are observable for lower heat flux exposures.

Table 2 Characteristic values of t_p and $\dot{m}''_p$ from the experiments and the predictions with FDS.

Cases	t_p-exp(s)	t_p-sim (s)	ε_{tp}	$\dot{m}''_p$-exp(kg/(sm^2))	$\dot{m}''_p$-sim(kg/(s/m^2))	ε_{mp}
100 kW/m^2	34.3	12.0	-65.0%	0.052	0.041	-21.1%
50 kW/m^2	106.8	40.8	-61.8%	0.034	0.017	-50.0%
25 kW/m^2	380.8	186.0	-51.1%	0.016	0.006	-62.5%

4.1.2. Front temperature

The experimental and simulated results for the surface temperature are shown in Fig.5. The temperatures are overpredicted during the initial heating-up period, roughly in the first 70s, 130s, and 400s at heat flux rxposures at 100, 50, and 25 kW/ m^2. This is in accordance with the MLR results displayed in Fig.4. After the initial heating-up period, the temperature results for the heat flux exposures at 100 and 50 kW/m^2 cases are well predicted. However, the temperature at the end of the 25 kW/m^2 case is slightly underpredicted.

Figure 5. Surface temperature comparison between the experiment and simulation for three heat flux exposures.

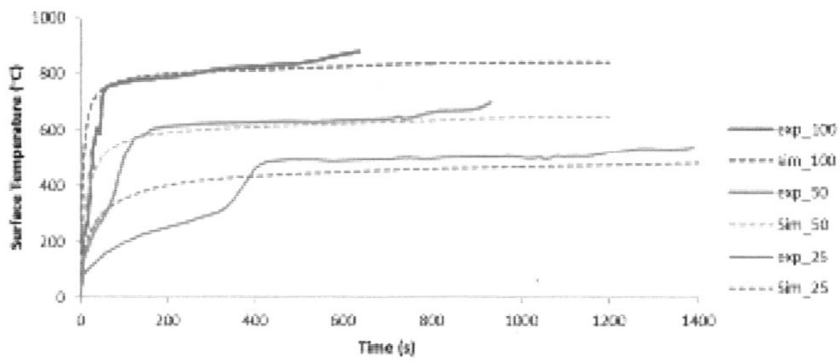

4.2. Influence of input parameters

A sensitivity analysis has been conducted to check the influence of the input parameters on the burning behavior. These parameters include the thermal properties of MDF, namely the conductivity, the specific heat capacity, the heat of reaction, moisture content, emissivity, and the absorption coefficient. The range of these parameters has been determined based on the values reported in the literature, while the applied external heat flux is 50 kW/m^2.

4.2.1. Influence of specific heat capacity

A sensitivity analysis has been conducted for evaluating the influence of the specific heat capacity. Three values have been investigated, namely 1, 1.5, and 2 kJ/(kg·K). The results are illustrated in Fig. 6. With a higher specific heat capacity, more time is required to reach the first peak (e.g., 14 % difference between 1.5 and 2 kJ/(kg.K). Furthermore, the peak value is significantly lower (e.g., 28 % difference between 1.5 and 2 kJ/(kg.K). Moreover, the front surface temperature for the higher specific heat capacity case is lower than that of the case with lower heat capacity value.

Figure 6. Influence of specific heat on the (a) mass loss rate and (b) surface temperature.

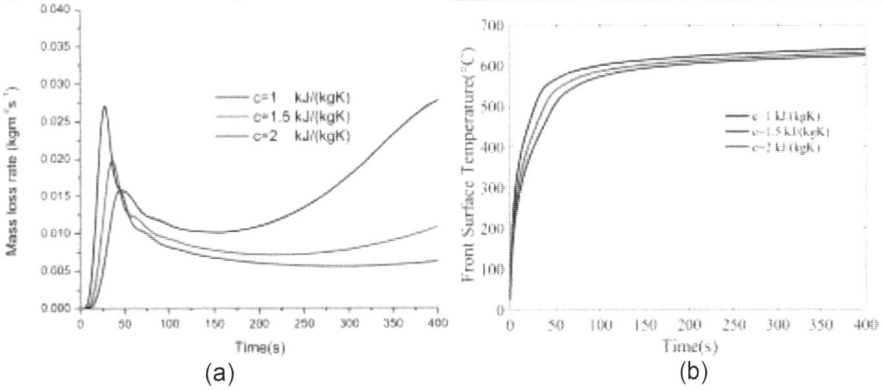

4.2.2. Influence of thermal conductivity

A separate sensitivity analysis has been conducted for the thermal conductivity. Three values have been investigated, namely 0.3, 0.2, and 0.1 W/(m·K). The results are demonstrated in Fig.7. When increasing the thermal conductivity, the thermal diffusivity ($k / \rho c$) increases. Thus, heat can penetrate deeper into the material in the same time period, thus the pyrolysis process at the surface is slow down during the initial heating-up period.

Figure 7. Influence of thermal conductivity on the (a) mass loss rate and (b) surface temperature.

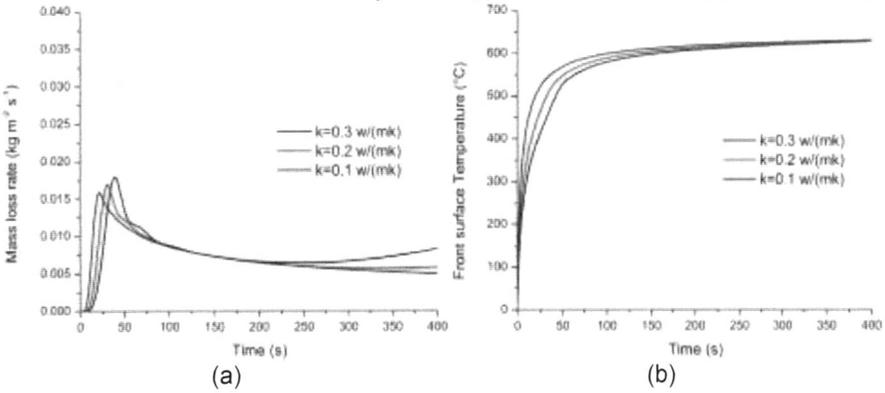

When increasing the thermal conductivity of the virgin material from 0.1 W/(m·K) to 0.3 W/(m·K), the time to reach the peak increases by 31.8% and the value of the peak increases by 6.4%. Moreover, the surface temperatures take more time to reach the experimental values when the conductivity is higher, as heat is conducted into the material more easily.

4.2.3. Influence of the heat of pyrolysis

The heat of pyrolysis does not have a significant influence on the pyrolysis process. Increasing H_p from 0 to 100 kJ/kg results in differences less than 5% in the mass loss rates and the front surface temperatures.

4.2.4. Influence of the moisture content

Moisture content (MC) is a measure of the amount of water within a material. MC is expressed as the percentage of the mass of the material that contributes water content. In this study, the original MC of the sample was approximately 6%.

In order to check the influence of the MC, a second Arrhenius equation for the water evaporation was added. Again Eqs. 2, 3, and 4 were employed to estimate the kinetic parameters *A* and *E*. The peak reaction rate $\gamma_p = 0.0016\,s^{-1}$ occurs at a temperature $T_p = 100\,°C$. A value of 5 K/s is used for the heating rate. Three MC values have been considered, namely 0, 5, and 10%. The comparison of simulation results is displayed in Fig. 8.

Figure 8. Influence of MC on the simulated (a) mass loss rate and (b) surface temperature.

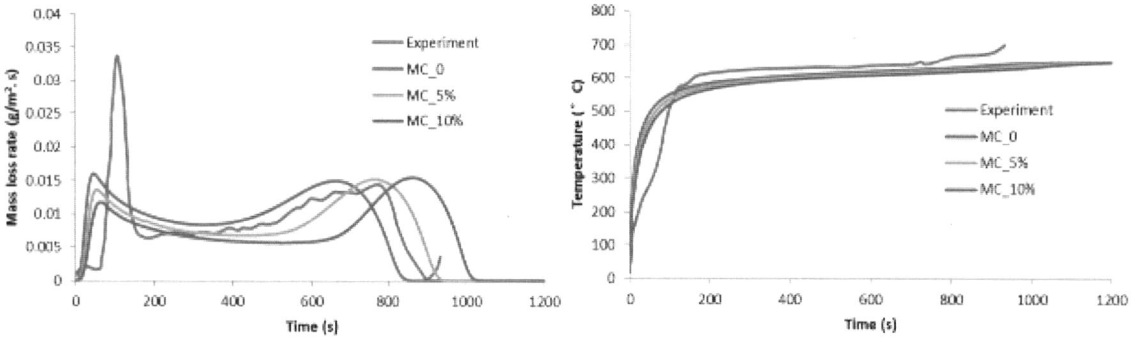

(a) (b)

Increasing the MC from 0 to 5%, the peak value $\dot{m}''_p$ decreases by 16.73%, while the time to reach the first peak t_p increases by 17.53%. Moreover, the overall reaction time increases by 20%. This influence can be explained in terms of the amount of energy required for the evaporation process, i.e. more water or moisture content means that the sample needs more energy for water evaporation. Obviously, the MC has a significant influence on the burning behaviour. The optimized value of 6% MC will be used hereafter.

Up to this point, the results of the aforementioned sensitivity analysis show that the most significant influence on the peak value for the mass loss rate is caused by the specific heat capacity. Nevertheless, none of these results showed satisfactory results for the time to reach the peak value of the burning rate when compared to experimental data, namely, the predicted time to peak is significantly lower than the measured value.

4.2.4. Influence of the emissivity

In light of the above, another sensitivity analysis has been conducted for the effective emissivity of virgin. The considered range of the emissivity rangs from 0.5 to 0.8. The comparisons are shown in Fig. 9. For the boundary condition, we are assuming the absorptivity is equal to the emissivity. When decreasing the emissivity (thus also absorptivity), less energy is received at the sample surface. Therefore, more time is required to reach the first peak in MLR and thus the maximum front surface temperature.

Figure 9. Influence of emissivity on the (a) mass loss rate and (b) surface temperature.

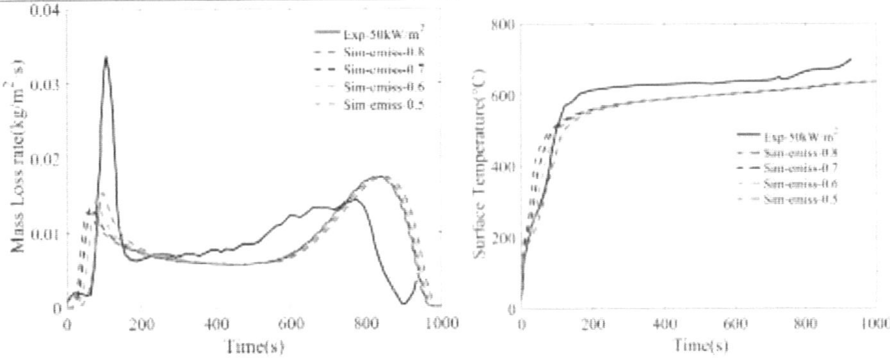

(a) (b)

When decreasing the emissivity of virgin from 0.86 to 0.6, the time to reach the first peak increases by 116%, and the peak value increases by 11.24%. Based on these results, a value of 0.6 for the emissivity of the virgin material is suggested. Consequently, emissivity can be identified as the first parameter that plays a role in the timing of the peak mass loss rate.

4.2.5. Influence of the absorption coefficient

Figures 10 and 11 display the results for the influence of the absorption coefficient on the mass loss rate (a) and front surface temperature (b), fixing the emissivity of the virgin material at 0.6, and the emissivity of char at 0.86. In general, both the absorption coefficient of virgin and char have significant influence on the mass loss rate curve. In Fig. 10, when decreasing the values of the absorption coefficient of virgin from 50,000 m^{-1} to 30 m^{-1} leads to a significant increase in the peak MLR and a delay in the time to reach the peak. Specifically, the values of the absorption coefficient of virgin between 30 m^{-1} and 700 m^{-1} show the most significant influence. The peak value is improved by 48.25%, and the time to reach the first peak is improved by 18.42%.

Figure 10. Influence of absorption coefficient of virgin on the (a) mass loss rate and (b) surface temperature.

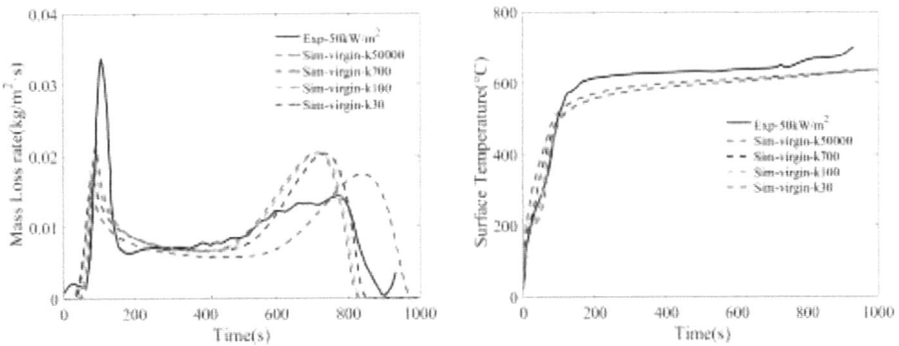

(a) (b)

Figure 11. Influence of absorption coefficient of char on the (a) mass loss rate and (b) surface temperature.

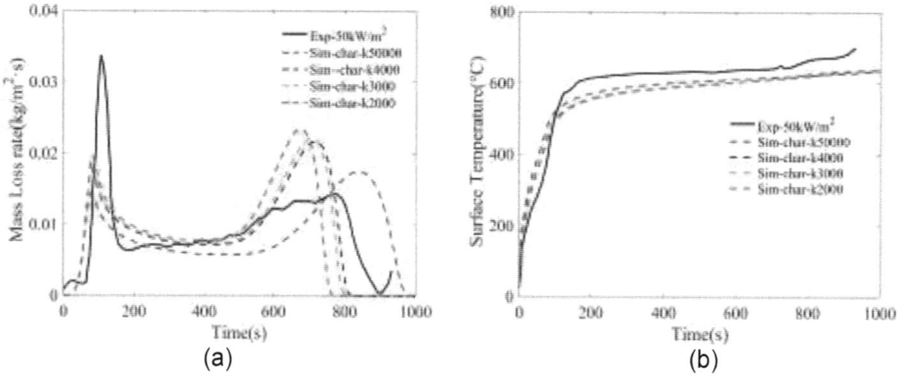

As indicated in Fig. 11 (a), the considered range of the absorption coefficient of virgin between 2000 m^{-1} and 4000 m^{-1} shows significant influence. The peak value is increased by 39%, while the time to reach the first peak is not improved very noticeably. Furthermore, the absorption coefficient of char has significant influence on the second peak. This could be explained by fact that the top layers of the sample are covered with char. And with the increase of the absorption coefficient of char, the radiation can penetrate deeper into the sample, which accelerates the reaction. Thus the overall reaction time decreases. However, this does not significantly influence the surface temperature, as shown in Fig 11 (b).

As mentioned in the introduction, the in-depth-like radiation absorption may occur in case of cracks or delamination. Hence, it is reasonable to assume that the material is opaque until it starts charring. Based on this consideration, values of 50000 m^{-1} and 2000 m^{-1} are suggested for the absorption coefficient of virgin and char, respectively.

This finding should be taken into account with caution and must be further investigated. It highlights, nevertheless, the possible importance of in-depth radiation in the pyrolysis modelling of MDF and more generally wood-based materials.

4.2.6. Non-uniform through-thickness density profile

As mentioned in the previous section, the density is not uniform through the sample thickness. In fact a high density value is observed on the surface. With a higher density, a higher peak value is expected. The non-uniform through-thickness density profile used here is the one mentioned in section 3.3.1. The results are shown in Fig. 12. Two heat flux values have been considered here.

Figure 12. Influence of the through -thickness density profile on the mass loss rate for heat flux exposures of (a) 50 kW/m^2, (b) 25kW/m^2.

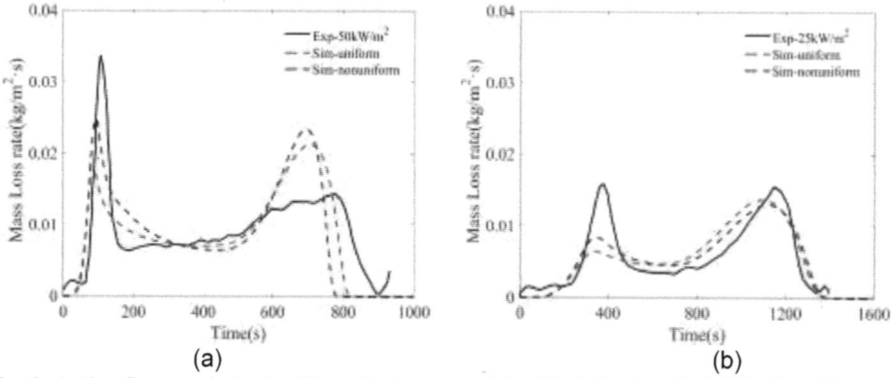

It is notable that, the first peak is significantly improved by 29.84% for 50 kW/m^2 and by 31.74% for 25 kW/m^2 when a non-uniform density profile is applied. Therefore, a non-uniform density profile is suggested for the optimized case.

The impact of parameter variations on t_P (time to peak mass loss rate) and $\dot{m}''_p$ (peak mass loss rate) is quantitatively shown in Table 3 for cases with 50 kW/m^2 heat flux exposure. For the considered range of the parameters, the most significant influence on the time to the first peak comes from the emissivity, followed by the thermal conductivity, specific heat, and moisture content. For the peak mass loss rate, the most significant influence comes from the absorption coefficient, followed by the through-thickness density, the moisture content, and finally the specific heat capacity.

Table 3 The impact of parameter variations on t_P and $\dot{m}''_p$

Parameter	Original Value	Changed value	dt_P(%)	$d\dot{m}''_p$ (%)
Specific Heat capacity of virgin	1.5	2	27.62	-14.47
Thermal conductivity of virgin	0.2	0.1	31.80	6.40
Absorption coefficient of char	50000	2000	8.00	39.00
Heat of pyrolysis	100	50	0	-4.91
Emissivity	0.86	0.6	116.00	11.24
Moisture content	0%	5%	17.53	16.73
Density	uniform	Non-uniform	0	29.84

4.3. Optimized values for parameters and corresponding results

In order to achieve a better estimation of the burning behavior for MDF material, a simple trial and error optimization procedure has been followed to predict the MLR from the experiment. Consequently, a set of parameters has been obtained (see Table 1), and these simulation results are demonstrated in this section as well.

Figures13 shows the simulation results using these optimized values of parameters. These results include the mass loss rate and the front surface temperature for three different external heat flux exposures. In general, as expected a better agreement is achieved when using optimized values of model parameters. For the predictions of the mass loss rate, the onset of the pyrolysis was well captured, and significant improvements were also achieved in terms of the value of the first peak in the mass loss rate curve. For the front surface temperature, the predictions of the initial heating-up period (roughly in the first 70s, 130s, and 400s at the heat flux exposures of 100, 50, and 25 kW/m^2 respectively) were notably improved and fit the experimental data better than those in the preliminary simulations.

Figure 13. Comparisons between experiment and simulation using optimized parameter values (a) mass loss rate, (b) surface temperature.

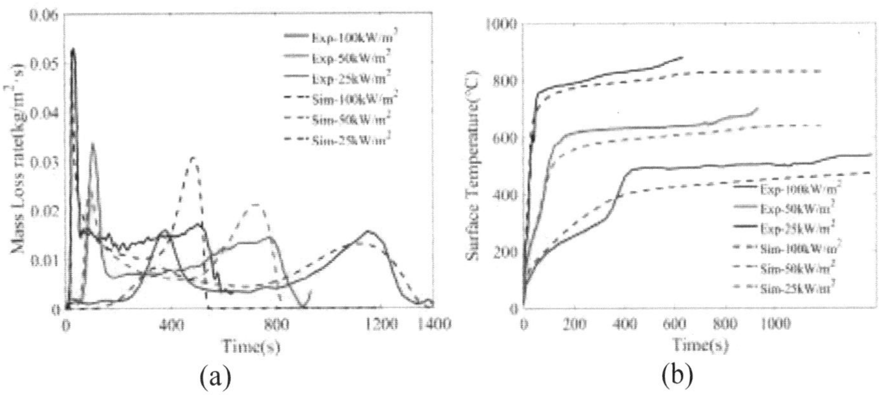

(a) (b)

CONCLUSIONS

In this work, pyrolysis behaviour of the MDF material has been studied numerically using a one-dimensional heat transfer solver, and a one-step Arrhenius type pyrolysis model. The numerical simulation tool used in this study is the Fire Dynamics Simulator (FDS 6.2.0). Within this model, the in-depth radiation has been taken into consideration. The values of the kinetic parameters and char yield are estimated from TGA test results reported in [3]. The thermal properties of virgin MDF and char are taken from the literature for the base case. It is assumed that no shrinkage and swelling occur during the whole process.

The base case does not show satisfactory results for the time to reach the first peak and the value of that peak in the mass loss rate curve when compared to experimental data. The predicted time to peak and the value of that peak are significantly lower than the measured value, namely by about 50 to 62%). The influence of the material properties and model parameters on the pyrolysis behaviour of MDF has been investigated in detail through a sensitivity analysis. The parameters include thermal conductivity, specific heat, heat of reaction, the emissivity, absorption coefficient, moisture content, and through-thickness density profile.

Based on the results of the sensitivity analysis study, it can be concluded that the main parameters influencing the t_p are the emissivity, thermal conductivity, specific heat, and moisture content. For the peak mass loss rate, the absorption coefficient of char, through-thickness density, moisture content, specific heat of virgin, and emissivity are the parameters showing most significant influence. Hence, only when proper values of the parameters were employed in the simulation, the onset of the pyrolysis process can be reasonably predicted.

Through a simple trial and error procedure, a set of 'optimized' parameter values has been obtained (see Table 1). During this optimization procedure, the parameters showing significant influence have been considered, including the in-depth radiation, which shows better agreement with experimental data.

ACKNOWLEDGEMENT

The authors would like to thank the China Scholar-ship Council (CSC) and BOF co-funding from Ghent University for their financial support to the first author, Grant NO.201306420001. Dr. Tarek Beji is a Postdoctoral Fellow of the Fund for Scientific Research-Flanders (Belgium).

REFERENCES

1. Marcos Chaos. Spectral aspects of bench-scale flammability testing: application to hardwood pyrolysis. pp. 165-178. Fire safety science proceedings of the Eleventh International Symposium, 2014.

2. Agarwal, G., Chaos, M, Wang, Y. Zeinali, D., and Merci, B., Pyrolysis Model Properties of Engineered Wood Products and Validation Using Transient Heating Scenarios, Interflam 2016, 14th International Conference and Exhibition on Fire Science and Engineering, Royal Holloway College, UK, 2016.

3. Kai-Yuan Li, Xinyan Huang, etc. Pyrolysis of Medium-Density Fiberboard: Optimized Search for Kinetics Scheme and Parameters via a Genetic Algorithm Driven by Kissinger's Method. Energy & Fuels. Combustion 2014, 28 6130-6139.

4. Florian Kemple, B. Schartel, G.T. Linteris, A. Hofmann-Böllinghaus. Prediction of the mass loss rate of polymer materials: Impact of residue formation. Combustion and Flame 159(9):2974-2984.September 2012.

5. Nicolas Bala, Guillermo Rein. On the effect of inverse modelling and compresation effects in computational pyrolysis for fire scenarios. Fire safety journal Volume 72, February 2015, Pages 68–76.

6. Fenghui Jiang, J.L. de Ris, M.M. Khan b. Absorption of thermal energy in PMMA by in-depth radiation. Fire Safety Journal 44 (2009) 106– 112.

7. Gregory Linteris, etc. Absorption and reflection of infrared radiation by polymers in fire-like environments. Fire and Materials, 2012.

8. McGrattan, K., Randall McDermott, etc. Fire dynamics simulator user's guide, National Institute of Standards and Technology Report NIST special publication 1019. (2016).

9. Wang, X., Mohammad, M., Hu, L.J., Salenikovich, A., Evaluation of density distribution in wood-based panels using X-ray scanning, *Proceedings of the 14th International Symposium on Nondestructive Testing of Wood*, University of Applied Sciences Eberswalde, Germany, 2005.

10. MDF material manufacturer : http://www.spanolux.com/en.

11. Kaiyuan.Li etc. Determining thermal physical properties of pyrolyzing new Zealand medium density fibreboard. Chemical engineering science 95 (2013) 211-220.

12. Cai, Z., Muehl, H.J., Winandy, J.E., Effects of pressing schedule on formation of vertical density profile for MDF panels, 40th International Wood Composites Symposium, Seattle, Washington, 2006.

13. Gronli, M.G. Theoretical and experimental study of the thermal degradation of biomass. Doctoral dissertation. Trondheim. Norwegian University on Science and Technology, Faculty of Mechanical Engineering, Dept. of Thermal Energy and Hydropower. Norway.p339.

14. Wang,X., Kersten, S.R.A., Prins, W.P.m. Biomass pyrolysis in a fluidized bed reactor. Part 2: Experimental validation of model results. Industrial &Engineering Chemistry Research. 2005. Vol.44 pp. 8786-8795.

15. Younsi, R., Kocaefe, D., Poncsak, S. and Kocaefe, Y. Thermal modelling of the high temperature treatment of wood based on Luikov's approach. International Journal of Energy Research. 2006. Vol.30, pp.699-711.Franks, C., Fire and Materials 2017, Interscience Communications Ltd, London, UK, p1- p2.

16. Saastamoinen, J., Simplified model for calculation of devolatilization in fluidized beds. 2006. Fuel. Vol. 85, pp. 2388-2395.

17. Brown, L. E.. An experimental and analytical study of wood pyrolysis. Doctoral dissertation. The University of Oklahoma.1972.

18. Gupta, M., Yang, J., Roy. C. Specific heat and thermal conductivity of softwood bark and softwood char particles. Fuel. Vol. 82, pp. 919-927.2003.

19. Simpon, W. and TenWolde, A. Physical properties and moisture relations of wood. In Forest products laboratory. 1999. Wood handbook.-Wood as an engineering material.

20. Siau, J.F. Transport process in wood. Springer-verlag, Berlin. 1984.

21. A.F. Roberts. The heat of reaction during the pyrolysis of wood. Combustion and Flame. Vol 17 pp. 79-86. 1071.

22. D. Zeinali, etc. Computational analysis of pyrolysis and flame spread for MDF panels placed in a corner configuration. 8th ISFEH Proceedings, (2016).

MODIFIED HEAT OF COMBUSTION AS A TOOL FOR EVALUATION OF BUILDING PRODUCTS COMBUSTIBILITY

Jadwiga Fangrat[1] & Marc L. Janssens[2]
[1]Instytut Techniki Budowlanej, Filtrowa 1, 00-611, Warsaw, Poland, e-mail: j.fangrat@itb.pl
[2] Southwest Research Institute, 6220 Culebra Road, San Antonio, TX 78238, USA

ABSTRACT

Are building products evaluated properly in terms of fire safety? The authors aim to answer the question. The combustion process of building products is discussed on the basis of experimental data in the context of building fire safety. Data are obtained by means of three standard methods: EN ISO 1716 oxygen bomb calorimeter, EN ISO 1182 small non-combustibility furnace and ISO 5660 cone calorimeter. Various categories of building products were tested: cellulose-based products (particle boards, plywood), concrete, ceramics, insulation materials (thermal and/or acoustic), boards (wall/ceiling), mortars, adhesives and thin coatings. The studied products exhibited very different fire properties from non-combustible to highly combustible. In order to more effectively differentiate non-combustible from combustible building products, a modified value of the heat of combustion was calculated using an extensive collection of test results according to the above-mentioned standard methods. A proposal is presented for modifying the criteria for Euroclasses A1 and A2 to obtain a more realistic reaction-to-fire evaluation. The proposed approach relies on the modified heat of combustion as a convenient tool for the fast and cost-effective initial non-combustibility evaluation. The approach appears to be an improvement over the current methods for distinguishing between non-combustible and combustible building products because it is a better representation of the actual physical combustion process.

BACKGROUND

Fires constitute a serious economic and social problem. Fire losses make up approximately 1% of the gross domestic product of each European country. For instance in Poland, where this level is not very different from the European average, there are nearly 35,000 building fires per year (mostly residential buildings) and this number has not fallen in recent years (Fig.1) Over the same period, the number of fatalities in building fires has decreased from 605 in 2005 to 517 in 2013 (Fig.2), and the number of injured in fires has increased[1,2].

The explanation of how fires are initiated and spread inside and between structures is crucial for the fire safety of buildings. Some parts of a building product tend to sustain combustion and generate heat in a fire, while others are not involved. Consequently, the fire behaviour of a building product is strongly affected by its chemical composition, particularly the organic content [3–8, 9]. Also, the internal structure and the condition of the surface exposed to the ignition source have an influence on the fire characteristic of the building product due to the effect on the heat transfer in the reaction zone. Products tending to be involved in combustion are included to combustibles, while others, are defined as non-combustible. The lack of involvement in combustion does not automatically mean that mechanical and chemical properties are preserved as fire exposure may result in strength reduction, appearance of defects and cracks, dripping and falling residue, due to, firstly, water evaporation and then gasification, etc. Moreover, toxic or harmful substances may be present in the non-combustible products of thermal decomposition. Even non-combustible (steel, concrete) load-bearing construction elements may lose fire resistance, which could affect the safety of building occupant dueing evacuation and that of emergency personnel during intervention.

Figure 1.
Statistical data concerning the number of fires of buildings in Poland in the period of 2010-2014.
Own study based on The National Headquarters of The State Fire Service database of 2015 [1,2].

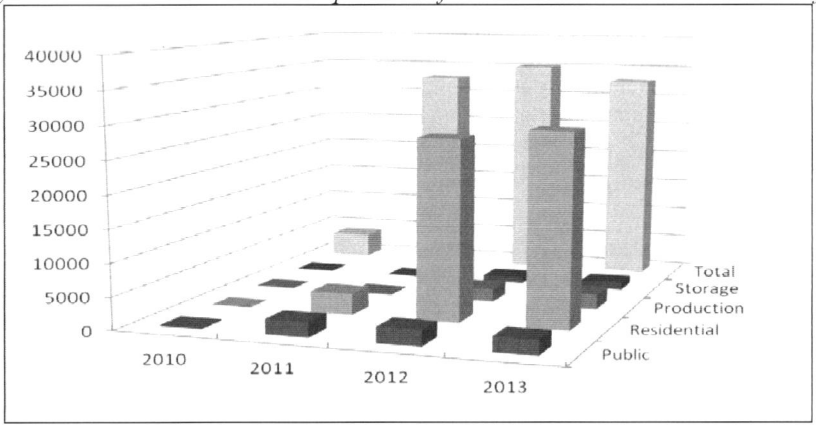

Figure 2.
Statistical data concerning the number of fatalities in fire of buildings in Poland in the period of 2005-2013. *Own study based on The National Headquarters of The State Fire Service database of 2015* [1,2].

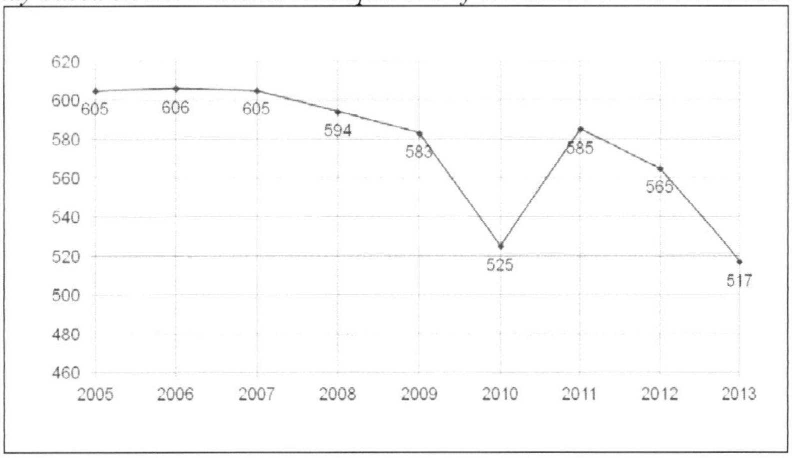

The process of fire is so complex that it is impossible to express quantitatively all the variables necessary to design or estimate functional properties with an accuracy that will make their application credible. Products of combustion in a fire are chemical compounds and thermal energy released in an uncontrollable way. The formation of these combustion products has a negative impact on structures, people and the environment. Therefore, it is necessary to analyse the issue regarding fire safety of buildings in connection with the combustion processes that occur during a fire, involving construction products containing combustible organics.

Environmental aspects give rise to new issues related to the fire safety of buildings and their significance will grow in the future. In light of the following, it is particularly important to formulate the appropriate requirements for construction products in a precise and relatively simple way, in order to differentiate the non-combustible and combustible products in various fire initiation conditions. It becomes necessary to develop or modify the requirements of fire safety to implement modern environmentally-friendly products in the construction industry.

BUILDING PERFORMANCE OF PRODUCTS VERSUS FIRE SAFETY

Fire may burn with or without flames (smouldering or glowing). Flameless combustion results from the deficiency of oxidiser in the reaction zone [10-13]. This situation is frequently observed in the conditions of real fire, particularly in the decay phase. Sometimes flameless combustion may occur in buildings without being noticed for a long time (several days), creating a fire hazard– this applies particularly to insulation made of sandwich panels[13]. Combustible products become fuel during the fire. Fire causes their thermal and chemical degradation. Even non-combustible materials may pose a threat to the evacuees and rescue teams due to the loss of their properties, which affects the fire safety of a building (*e.g.*, high-strength concrete, steel). Addition of retardants may have short-term effects in the fire conditions and cause environmental and health problems for people evacuated from the building. It is an issue which is often omitted or downplayed in a general fire safety strategy.

Transition from smouldering to flaming combustion is a critical point in fire development. The long-term flameless combustion pertains mainly to combustible insulation products of low density and high porosity. The additional amount of oxidiser is located inside pores and supports the sustained combustion and its propagation. Fire is a source of convective flows inside buildings due to the temperature gradients created. Transition from smouldering to flaming combustion strongly depends on flow over the combustible surface. A minimum velocity of 2 m/s is necessary to undergo transition to flaming in case of forward smouldering under increased forced flow. Opposed-flow smouldering propagation does not occurs at flow velocities below 5 m/s[12]. Even non-combustible products may cause a danger in fire situation. Even if a given material does not burn in a fire it does not necessarily preserves its chemical and physical properties. Usually, the following phenomena occur: the loss of strength, appearance of flaws and cracks, dripping and falling of combustion residue, initially water evaporation and then gasification. In the case of load-bearing construction elements, even for those made of non-combustible materials, there is a possibility of loss of structural capacity, which is dangerous for the whole construction, evacuation of occupants and fire-fighter intervention[14]. Problems related to building products are not sufficiently addressed in fire safety research and strategy. The current approach of using "global parameters" does not increase knowledge in this field and is (and will continue to be) a barrier in future development of new products and their application in construction. It becomes necessary to better connect the results of tests with the real processes which products undergo in a fire. The concept of modified heat of combustion presented below is an attempt to improve this relationship.

Technical aspects of the fire safety of buildings linked to combustion vary depending on the phase of fire. One may distinguish the following[14]:

1) in the phase before flashover:
    - providing means of egress for people and protection against injuries
    - limiting initial flame effects
    - <u>reducing the rate of fire development in the room of fire origin</u>
    - <u>limiting the possibility of fire propagation to other rooms of a building</u>

2) in the phase after flashover:
    - Considering safety of emergency responders
    - Safe evacuation of people from the building

- Limiting fire propagation to other separated areas in the building, including the so-called safe zones
- Fire spread to adjacent objects and structures
- Protection of property and continuation of the building function.

The problems addressed in this study are connected to the issues underlined above. One of the most important fire safety problems is to differentiate between combustible and non-combustible products. For example in the European reaction-to-fire classification system[15], the methods and criteria which are currently used for Euroclasses A1 and A2 (see Table 1) allow for a comparison of raw materials on the basis of arbitrary criteria for the defined classification groups. These criteria are not based on the fire development process. For instance, flashover occurs at room temperatures between 500 °C and 600 °C [16, 17], while the EN ISO 1182 non-combustibility test[18] is performed in the temperature of 750°C. Compared to the range of post-flashover temperatures, the allowed temperature increase, ΔT, is indeed relatively insignificant. However, there is no technical basis for the limiting values.

Table 1.
Exemplary requirements for building products–Euroclasses A1 and A2

Test Method	Criterion	A1	A2
EN ISO 1182	Furnace temperature rise ΔT, °C	≤ 30	≤ 50
	Sample mass loss Δm, %	≤ 50	≤ 50
	Duration of flaming combustion t_f, s	0	≤ 20
EN ISO 1716	Gross heat of combustion HC, MJ/kg	≤ 2,0 ≤ 1,4*	≤ 3,0 ≤ 4,0 *

* For the secondary component of an internally homogenous product.

EXPERIMENTS

Experimental data were obtained for commercial building products obtained from three standard test methods: (a) EN ISO 1716 oxygen bomb calorimeter[19], (b) EN ISO 1182 cylindrical furnace [18] and (c) ISO 5660 cone calorimeter[20] and are used for further theoretical analysis. To better distinguish between non-combustible and combustible materials or different classes of combustibles, a modified heat of combustion was defined and calculated for a few dozen materials tested according to these methods. The products studied exhibited very different mass loss values and a low tendency to auto-ignite in the cylindrical furnace.

(a) The instrument for testing the gross heat of combustion[18] is also known in the power engineering industry as an oxygen bomb calorimeter and has been used for many years to determine the calorific value of fuels. Its basic elements include:
  - high-pressure combustion chamber immersed in a water jacket fitted with an agitator driven by a constant speed motor;
  - instrument for measuring and recording temperature with an accuracy up to ±0.002°C;
  - crucible made of material resistant to heat, corrosion and oxidation, where a test sample is placed
  - ignition source, which is a steel resistance wire connected to an electrical circuit.

  A sample of the material with known mass is burnt in pure oxygen under high pressure. The calorimeter is calibrated with certified benzoic acid, which is also used as a combustion promotor for material with very low organic content. At least three samples of each product are tested. During the test the calorific value of the sample is determined on the basis of the temperature increase in a calorimetric vessel with consideration of heat losses. Modern instruments, including the one used for this study, are fully automated, which considerably increases the accuracy of the test method.

(b) The instrument for testing non-combustibility[8] (Fig. 3) –often referred to as the cylindrical furnace method– consists of a cylindrical section of a ceramic tube with the dimensions of: 150 mm (height) x 75 mm (inner diameter) x 10 mm (thickness) mounted on a conical base for stabilising the air flow and the cover. The furnace is heated electrically to the temperature of 750±10°C. A cylinder-shaped sample with the height of 50 mm and base diameter of 45 mm is placed centrally inside the furnace. The test is repeated for 5 samples of the product. During the test the increase of temperature inside the furnace with a type K thermocouples and the time of stabilised flaming combustion are recorded. Total sample mass loss is also determined.

Figure 3.
Furnace for testing of non-combustibility of building products acc. to [18].
Sample thermocouple is optional.

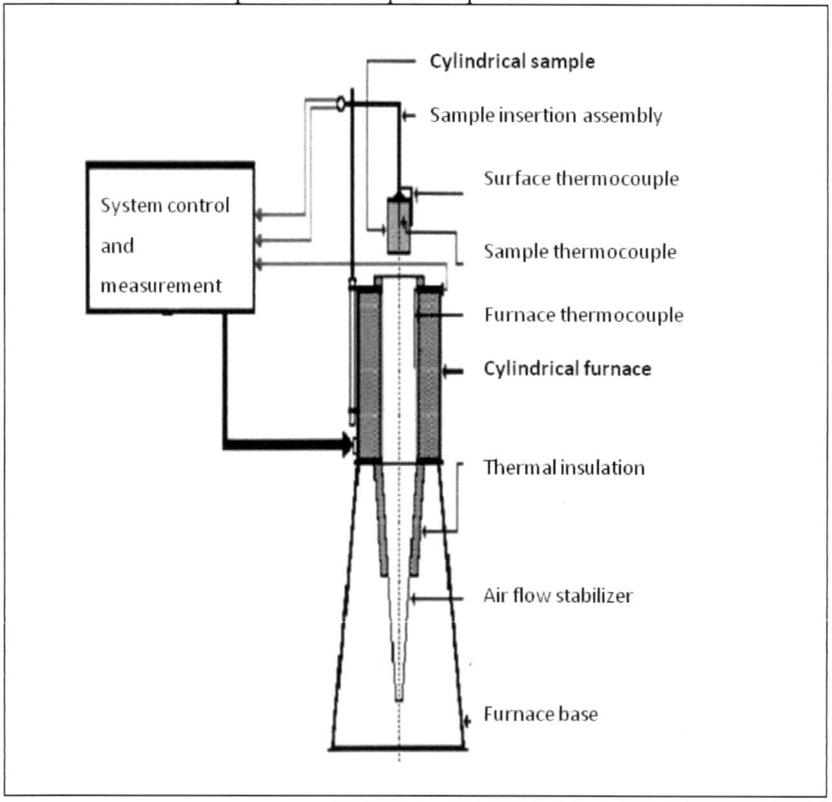

(c) The Cone Calorimeter described in ISO 5660[20] has been used for testing horizontally oriented samples of 10 cm x 10 cm in size, placed in the steel holder under the hood (Fig.4) connected to the exhaust system. The sample is exposed to the external hear flux generated by the electrical cone heater. During the test, the holder with the sample is placed on an electronic load cell, which allows the sample mass loss to be recorded as a function of time. The ignition source is an electrical spark generated at 10 mm above the surface of a sample in its geometrical centre. During 15 minutes (or less if the sample burns out sooner), the following measurements are made: time to ignition, heat release rate (HRR), total heat released (THR), smoke production rate, concentration of O_2, CO, CO_2, flow rate and temperature in the measuring section of the exhaust duct.

The results reported in the next section were obtained from measurements performed on the horizontal samples exposed to the external heat flux of 50 kW/m^2.

Figure 4.
Schematic of the Cone Calorimeter apparatus according to ISO 5660 [20].

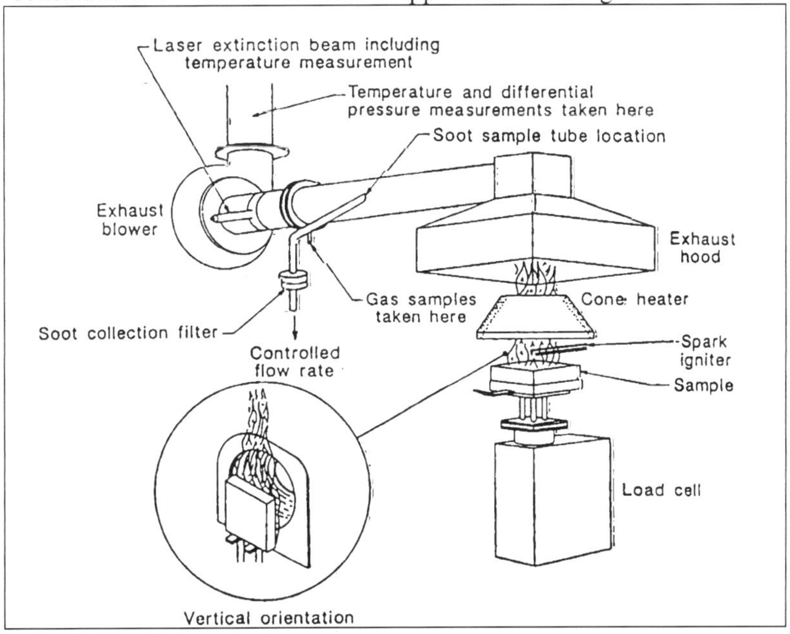

CONCEPT OF MODIFIED HEAT OF COMBUSTION AND ITS VERIFICATION

Cylindrical Furnace Application

Problems with the direct use of heat of combustion data for building materials in real fire analysis are reported in the literature[19, 20]. One of the most common problems occurs for charring materials, for which the heat of combustion of volatiles is not equal to that of the whole material. To address these problems, it was proposed[21] to use the modified heat of combustion defined as follows:

$$MHC = HC \times \Delta m \quad [MJ/kg] \qquad [1]$$

where:

MHC = modified heat of combustion (MJ/kg)
HC = gross heat of combustion (MJ/kg) according to EN ISO 1716[19]
Δm = sample mass loss relative to initial sample mass according to EN ISO 1182 [18]

The MHC concept uses measurements from both the oxygen bomb calorimeter and the cylindrical furnace, and increases the possibility to better distinguish non-combustible (class A1) from combustible products (the remaining classes). This concept was verified on the basis of an analysis of test results for 66 commercial building products, which were grouped into five categories, depending on their intended use in the building[22]: concrete and ceramic, thermal/acoustic insulation materials, covering boards (wall/ceiling), mortars and adhesives, and thin coatings.

The heat of combustion (HC) data (Fig. 5) covered a very wide range; for products under endothermic chemical reactions HC values below zero were found for 18 of 66 samples, mostly in the group of mortars and adhesives.

Figure 5.
Results for heat of combustion for 66 commercial building products; based on previous work[22].

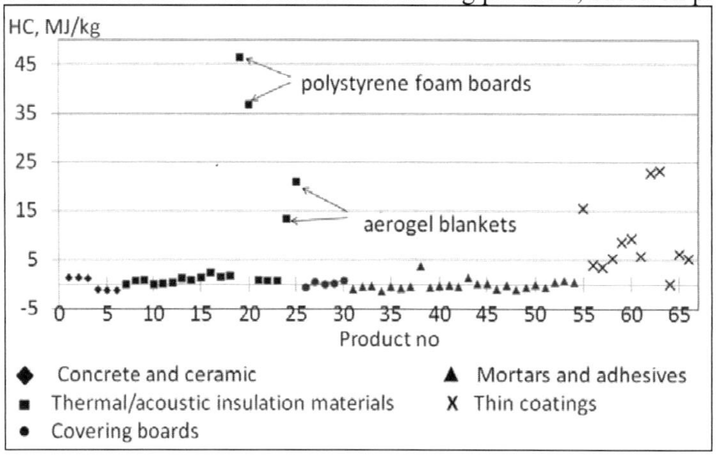

The highest HC value that was measured was approximately 46 MJ/kg for regular polystyrene foam, while for fire retarded polystyrene foam this value was lower and approximately equal to 37 MJ/kg. These HC levels are even higher than reported for real industrial coals for which HC values were found between 7.5 MJ/kg (low rank brown coal[23]) and 28 MJ/kg (anthracite coal[24]). For wood the HC can be as high as 18.5 MJ/kg[25].

A statistically significant correlation was not found between heat of combustion according to [21], the furnace temperature rise ΔT, and the sample mass loss Δm [18]. (Fig. 6).

Figure 6.
Comparison of non-combustibity test results by method [18]. Class A1criteria: ΔT≤30ºC (dashed line), and Δm≤50% [15], based on [22].

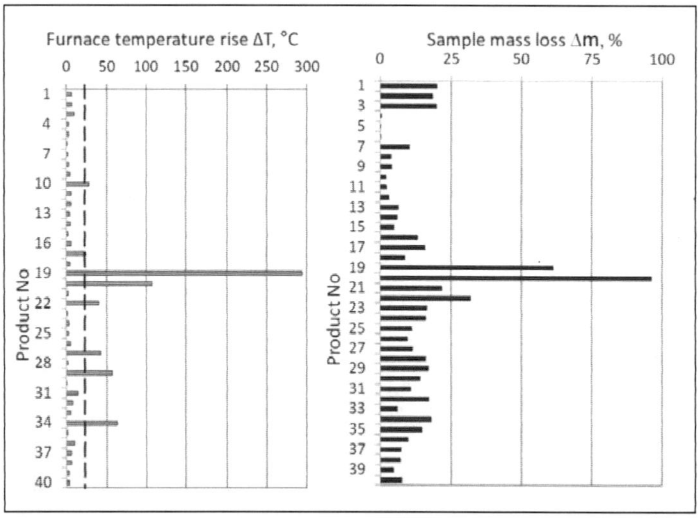

More details on this study are available in another publication[22]. On the basis of experimentally obtained HC and Δm values, the MHC was calculated using formula (1) (Fig. 7).

Figure 7.
Comparison of results with the criteria for Euroclass A1[15]: $\Delta T \leq 30°C$, and HC$\leq$ 2 MJ/kg, based on [22].

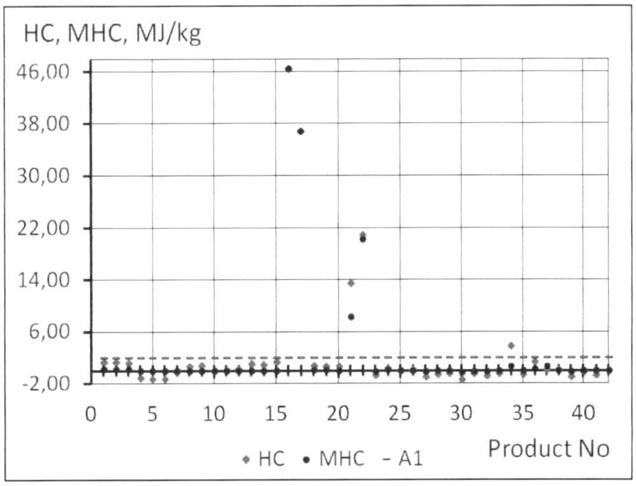

Futhermore, results are presented graphically using criteria of class A1 (Table 1) for ΔT_s and Δm versus HC in Figures 8 and 9 respectively. The horizontal and vertical axes cross at the limit points for the criteria values of ΔT_s and HC (Fig. 8) or Δm and HC (Fig. 9) according to Table 1. The area of each figure is divided into 4 parts. Data points on the left side of the vertical axis correspond to samples which fulfil the heat of combustion criterion (HC $\leq$ 2 MJ/kg) in both figures. Data points below the horizontal axes correspond to samples which fulfil the criterion for furnace temperature of the sample (Fig.8) or mass loss (Fig. 9). The data points in the areas located on the bottom left side and the top right side correspond to samples that would be consistently classified according to both criteria as non combustible (A1) and combustible, respectively.

Figure 8.
Comparison of results with the criteria for Euroclass A1 [15]:$\Delta T \leq 30°C$, and HC$\leq$ 2 MJ/kg, based on [22].

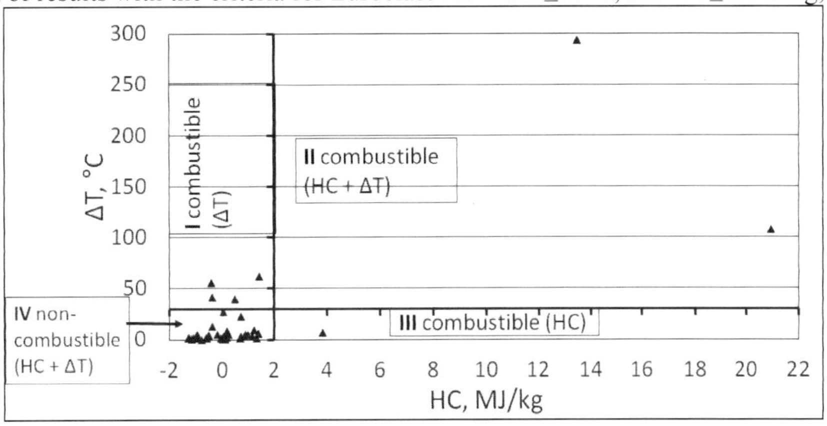

With only four outliers in case of ΔT (see Fig. 8) and no outliers in case of Δm (see Fig. 9), there is excellent consistence of the A1 classification. In other words, method [21] is a very efficient tool for fast and inexpensive screening of non-combustible building products. Without performing more expensive tests it is possible to predict the classification and distinguish between non-combustible and combustible materials. This would be a useful tool for the future design and implementation of new products in the construction industry.

A limitation of the approach is that it would not necessarily classify materials that have an excessive flaming duration in EN ISO 1182 as combustible (t_f in Table 1) [18]. On the basis of results presented in Figures 7, 8 and 9, it is proposed to reanalyzed and modify the classification criteria of non-combustible and combustible products. Moreover, the results show that there was no clear need for two different limits of A1 class for heat of combustion (Table 1) in the CEN system. So the simplification of the classification system seems justified and reasonable.

Figure 9.
Comparison of results with the criteria for Euroclass A1 [15]: $\Delta m \leq 50\%$, and $HC \leq 2$ MJ/kg, based on [22].

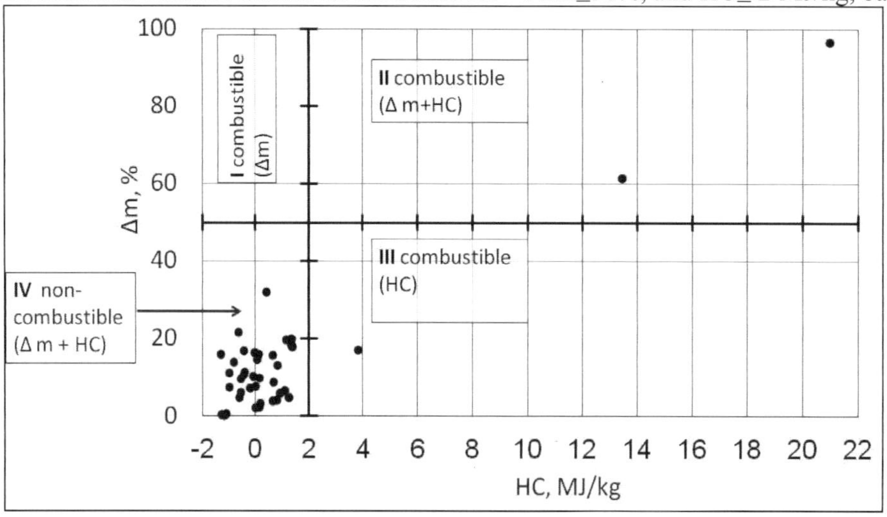

Cone Calorimeter Application

The Cone Calorimeter has the advantage over the cylindrical furnace and oxygen bomb calorimeter tests that materials can be tested under realistic fire conditions in their end-use form. Consequently it may be a more suitable method to evaluate the combustibility of layered materials. In fact the Cone Calorimeter is already used in Japan for that purpose. The Japanese Building Standards Law classifies materials as non-combustible if the peak heat release rate and total heat released in a 20-minute test at 50 kW/m^2 is 200 kW/m^2 or less and 8 MJ/m^2 or less, respectively.

Assuming the (first) peak heat release rate for a homogeneous material is proportional to the maximum temperature rise in the EN ISO 1182 cylindrical furnace, the idea of combining several criteria of the non-combustibility tests could be accomplished by defining a modified peak heat release rate as follows:

$$MPHRR = PHRR \times \Delta m \ (kW/m^2) \qquad [2]$$

Where

MPHRR = modified peak heat release rate (kW/m^2)
PHRR = first peak heat release rate measured in the Cone Calorimeter (kW/m^2)
Δm = total sample mass loss measured in the Cone Calorimeter relative to initial mass (-)

Table 2 provides the modified peak heat release rate for materials of varying degrees of combustibility

Table 2.
Modified peak heat release rate for materials with varying degrees of combustibility*

Material	PHRR (kW/m^2)	Δm (-)	MPHRR (kW/m^2)
Douglas fir plywood	216	0.78	168
FR Douglas fir plywood	106	0.67	71
Type X gypsum board	66	0.05	3.3

* Cone Calorimeter data obtained at a heat flux of 50 kW/m^2

The advantage of the modified peak heat release rate is that the two components, PHRR and Δm, can be measured with relatively high accuracy. The approach, therefore, shows promise in addressing the main problem with the Japanese classification system, i.e., the fact that the accumulation of systematic errors, in particular over a 20-minute period, can easily result in incorrect classification of borderline materials. The latter was the primary reason for the development of the modified Cone Calorimeter described in ISO 5660-4. More work is needed to demonstrate the general applicability of the modified peak heat release rate concept.

CONCLUDING REMARKS

In the course of this study, a new research area was identified within building fire safety, which is essential due to the increasing significance of future environmental requirements. An original approach to the research and analysis aimed at explaining the physical phenomena occurring in construction products in the initial phase of fire development, consequently affecting the level of fire safety in buildings. The primary findings are as follows:

- Development of modified heat of combustion concept as a single quantity, which combines the results of both current non-combustibility test methods. The modified heat of combustion is more suitable for characterizing the fire properties of building materials, than the gross heat of combustion.
- The testing of gross heat of combustion is a very effective way for the preliminary determination related to the fire properties of products; especially those with combustible organic content. EN ISO 1716 is a very convenient and reliable method for time and cost-effective evaluation of building products for their combustibility. On the basis of the single HC value it is possible to predict the fire evaluation particularly to differentiate non-combustible and combustible building products. It would be useful for the future design and implementation of new products in the construction industry. However, one should bear in mind, that the third criterion of auto ignition in the second non-combustibility test EN ISO 1182 could influence the final A1 classification of the building product. The relatively simple and low-cost measurement of gross heat of combustion is proposed for all Euroclasses of building products with different limit values.
- Having such a suitable method, for the purposes of a preliminary assessment of the combustible products with high accuracy, will be important at the stage of design and implementation of new products in the construction industry.
- The similar concept, i.e., the combination of two characteristics that pertain to the combustibility of a material, can also be applied to Cone Calorimeter data. However, more work is needed to.

Further studies should focus on the connection between the results of the experimental tests concerning reaction to fire and the calculations. This will gradually be possible as additional data are analysed and new products emerge.

REFERENCES

1. www.kgsp.pl (2015), [in Polish]
2. Mazur, R.(2014) Statistic analysis of fires in Poland, In: *Red book of fires. Part II. Fire statistic.* 1st ed. Józefów: CNBOP [in Polish].
3. Torero,J.L. (2013) Scaling-up fire, *Proc. Comb. Inst.*, 34, pp. 99-124.
4. Drysdale, D. (1985). *Introduction to fire dynamics*, John Wiley and Sons Ltd.
5. Fernandez-Pello, C.(1995) The solid phase. In: G. Cox, ed., *Combustion fundamentals of fire*, 1st ed. Cambridge, Massachusetts: Academic Press, pp.55-85.
6. Hasemi, Y., (2008). Surface flame spread. In: DiNenno, P.J. ed. *The SFPE Handbook of Fire Protection Engineering*, 4th ed., Quincy: NFPA, , Publication No: HFPE08, pp. 2-278- 2-290.
7. Mouritz, A.P. and Gibson A.G. (2006) *Fire properties of polymer composite materials*, 1st ed. Dordrecht: Springer.
8. Fangrat, J. and Wolański, P. (1991). Major factors influencing flame spreading over solid fuel layer. In: Kuhl, A. ed. *Dynamics of deflagrations and reactive systems: Heterogeneous Combustion.* 1st ed. Washington: American Institute of Aeronautics and Astronautics, pp. 261-274,
9. Fangrat, J. (2012) Effect of increased organic content on fire properties of building products, *Building Materials*, 12, pp. 35-40 [in Polish].
10. Fangrat, J. (2004) Is flameless combustion of importance to fire safety?, *Arch. Combustionis*, 24 (1), pp. 1-4.
11. Ohlemiller, T.J.(1990). Smouldering combustion propagation through permeable horizontal fuel layer. *Comb. Flame*, 81, pp. 341-354,
12. Buckmaster, J. and Lozinski D. (1996). An elementary discussion on forward smoldering, *Comb. Flame*, 104, pp. 300-310
13. Ohlemiller, T.J., (2008). Smouldering combustion. In: DiNenno, P.J. ed. *The SFPE Handbook of Fire Protection Engineering*, 4th ed., Quincy: NFPA, Publication No: HFPE08, pp 2-229- 2-259.
14. Fangrat, J. (2014). How much fire safety in building products? *Building Materials,* 11, pp. 36-39 [in Polish].
15. EN 13501-1 Fire classification of construction products and building elements – Part 1: Classificatin using test data from reaction to fire tests. *CEN,* Brussels.
16. Walton, W.D. and Thomas, P.H. (2008). Estimating Temperatures in compartment fires. In: DiNenno, P.J. ed. *The SFPE Handbook of Fire Protection Engineering*, 4th ed., Quincy: NFPA, Publication No: HFPE08, p. 3-204.
17. Thomas, P.H. (1981). Testing products and materials for their contribution to flashover in rooms. *Fire, Materials*, 5 (3), pp. 103-111.
18. EN ISO 1182 Reaction to fire of building materials – Non combustibility test, *CEN,* Brussels
19. Dietenberger, M. (2002). Update for combustion properties of wood components. *Fire, Materials*, 26, pp.255-260.
20. Madrigal, J., Guijarro, M., Hernando, C., Diez, ., Marino E. (2011). Effective heat of combustion for flaming combustion of Mediterranean forest fuels.*Fire Technology*, 47, pp. 461–474.
21. Fangrat J., (2016) On non-combustibility of commercial building materials, *Fire,Materials*, DOI:10.1002/fam.2369,
22. EN ISO 1716 Reaction to fire of building materials – Determination of gross calorific value, *CEN*, Brussels.
23. Pawlak-Kruczek, H. (2013). Co-firing of biomass with pulverised coal in oxygen enriched atmosphere. *Chemical and Process Engineering*, 34 (2), pp. 215-226.
24. Pawlak-Kruczek, H. (2003). *Problems of the combustion of young low-metamorphism rank fossil fuels*, 1st ed., Wrocław: Wroc. Tech. Univ. [in Polish].

FIRE INVESTIGATION

ARC MAPPING: NEW SCIENCE, OR NEW MYTH?

Vytenis Babrauskas
Fire Science and Technology Inc., San Diego, CA
Dept. of Mechanical and Aerospace Engineering, University of California San Diego, La Jolla, CA

ABSTRACT

Arc mapping was first introduced in the 2001 edition of NFPA 921 and was subsequently expanded so that in the recent editions it constitutes one of the four main methods for determining the origin of a fire. Careful consideration of engineering principles and large-scale experimental studies on the subject indicates that the relevance and prominence of arc mapping as a leading indicator of fire origin is greatly overstated. The technique is valid and applicable only in some very limited scenarios. Yet it has seen very extensive use in recent years by investigators preparing fire reports. In many cases, such attempted use of arc mapping is based on incorrect and invalid hypotheses, which are often implicitly assumed to be true instead of being explicitly stated. The following are myths: (i) An abundance of arc beads at a given locale means that fire originated in that area, while a paucity of arc beads indicates that it did not. (ii) When multiple arcs are present on a circuit, the direction of arcing will necessarily proceed upstream towards the power source. (iii) If an appliance is the victim of a fire, internal arcing will be primarily near the exterior of the unit, while arcing deep inside indicates a fire origin at that place. NFPA is urged to revise NFPA 921 to eliminate arc mapping as one of the four main methods for establishing fire origin, and to subsume it under the more general category of "fire patterns." In addition, it is important that NFPA 921 reduce the implied general utility of the method and provide more explicit information on its interpretation and its limitations and on the circumstances under which it may be a valid method for assisting in the determination of the fire origin.

Keywords: arc mapping; arc beads; electrical short circuits; fire investigation; metallurgical evidence; NFPA 921.

INTRODUCTION

Up until the early 1990s, fire investigation was a notably empirical profession, with very limited scientific foundations. As a consequence, various "rules" were used by practitioners, which were later learned to be myths, not valid precepts of science[1,2,3]. It was already recognized by the mid-1980s that the situation needed to improve, thus in 1985 the National Fire Protection Association (NFPA) established a Technical Committee on Fire Investigations which produced the first edition of NFPA 921—Guide for Fire and Explosion Investigations in 1992, and updated regularly since then[4]. One of the main tasks of the Committee and the Guide has been to replace myths with science-based procedures. While the early origins of arc mapping were in the 1980s, it became a technique in wide usage among fire investigators only since the 2001 edition of NFPA 921. In its recent editions, NFPA 921 progressively increased emphasis on arc mapping as a fire investigation tool. However, this was not accompanied by scientific rigor in establishing a clear understanding of its valid uses, versus its limitations and improper uses. It is the purpose of this study to examine carefully the history of arc mapping and the available experimental validation data, and especially to uncover the underlying hypotheses. It will be shown that some of these, while often tacit and not explicit, are scientifically erroneous or speculative and unsupported, and consequently lead to investigatory conclusions that are not supportable. Recommendations emerge out of this study on improvements for the NFPA 921 treatment of this topic, along with guidance for fire investigators as to what constitutes valid and correct application of arc mapping principles.

When fire engulfs energized wiring, shorting may occur (but does not necessarily have to). If shorting does occur, a detectable artifact (an "arc bead") will not necessarily be created. Laboratory experiments[5] have found that direct metallic shorting is less likely to create identifiable arc beads than is shorting where part of the path is a carbonaceous material. But the latter is often likely to happen when wire insulation gets burned due to a fire. Furthermore, if an arc bead is created, it will not necessarily be located at the point of first fire damage to the cable.

In fire investigation, arc mapping[*] is defined in NFPA 921[6] as "*The systematic evaluation of the electrical circuit configuration, spatial relationship of the circuit components, and identification of electrical arc sites to assist in the*

[*] Sometimes referred to as *arc surveys*, *arc fault surveys*, *arcing fault pattern analysis*, or similar terms.

identification of the area of origin and analysis of the fire's spread." NFPA 921 first referred to arc mapping in the 2001 edition, which did not provide any procedure details, but simply added arc mapping as the fourth source of information in the Origin Determination chapter. Additional material was developed in the 2004, 2008, and 2011 editions, while the 2014 edition greatly expanded the text and for the first time gave a definition of arc mapping, as indicated above. Changes for the 2017 edition have been minimal.

Note that arc mapping is not related to examining the possibility of arcing as a *cause* of fire. An electrical arc may be the initial event starting a fire, but arc mapping is a method suggested for helping identify the area of fire origin, not its cause. The fire cause is defined by NFPA 921 as: "*The circumstances, conditions, or agencies that brought about or resulted in the fire or explosion incident, damage to property resulting from the fire or explosion incident, or bodily injury or loss of life resulting from the fire or explosion incident*." In other words, determining fire cause is establishing what took place to enable the fire to occur. Laboratory examination of molten artifacts may be necessary for this purpose, but this is an undertaking different from that of conducting arc mapping and has already been covered elsewhere[7].

The present paper examines all of the experimental research published on the topic of arc mapping to date. The most important objective has been to examine the hypotheses put forth by various authors on the principles underlying arc mapping and the ways that arc mapping may be used to assist in determining the origin of a fire. The validity of these hypotheses is then examined in terms of the experimental results and in terms of basic principles of electrical and fire science.

HISTORY

The earliest mention of a concept related to arc mapping is found in a 1955 book by Straeter and Crawford[8]. The authors suggested that: "*Condition of wiring after the fire. The amount of destruction of insulation on electric wires can reveal a clear pattern for the area of intense heat. Electric shorts have their physical effects. Investigate to be sure whether the short caused the fire or the fire caused the short. The exact locations of the centers of wire damage, of char, and of other indications may well be compared to answer this question*." And: "*Sequence of shorted electric circuits. The appliances that were stopped by electrical failure may indicate the path of the fire to some degree, especially in a larger building in which there are many circuits and they were put out one at a time. The times at which the clocks stopped could tell which circuits were included first in the fire's path*." The next development was in Japan, where in 1968 Miyake discussed arc evidence at a closed meeting on forensic engineering held by the National Research Institute of Police Science. He then published it in a journal in 1975[9], stating: "*If two points of severing due to shorts are found on an electric cable burned in a fire, it means that the farther point from the power source was damaged by the fire earlier than the other one. This can be used to estimate the place of origin*."

The obscure 1955 and 1975 references were evidently not seen by later workers and, in most literature, the first reference is to a 1983 paper by Delplace and Vos[10], who stated that they had been developing for 12 years a method that they called "systematic mapping of the locations of short circuits." The paper was brief and provided only a single diagram, but it can form the starting point of the present study, since they proposed testable hypotheses how arc mapping might be interpreted. The basic Delplace and Vos hypotheses were:

- "A close parallel exists between the patterns left by combustion and the location of short circuits."
- "A short circuit will normally take place wherever the fire first damages the cable. This gives us the point along the entire length of the circuit earliest affected by the fire."

They also noted that, when several sever-arcs* are found on a particular circuit, the earliest one had to be the farthest downstream from the power source, since the initial sever-arc removes power from the downstream portion, but not the upstream portion. They then offered a hypothetical example of a vehicle fire, which they also considered to be amenable to this analysis: "*If damage by electric arcs is found at the dome light and at the steering wheel, then the fire did not start in the engine compartment, nor in the dashboard*." The paper gave no details of any experiments or analyses that might have been conducted, and the method saw little use during the next two decades, apart from one other forensic paper[11] and the teachings of Robert Svare. Svare is an engineer who taught the technique to numerous individuals in classes on fire investigation, but published only brief summaries[12,13] (Figure 1). Interest within the profession resumed in the late 1990s, after which a number of papers appeared.

* A sever-arc is an arcing location where one or more conductors involved have been severed by action of electrical arcing, that is, the conductor in question is no longer continuous at that location.

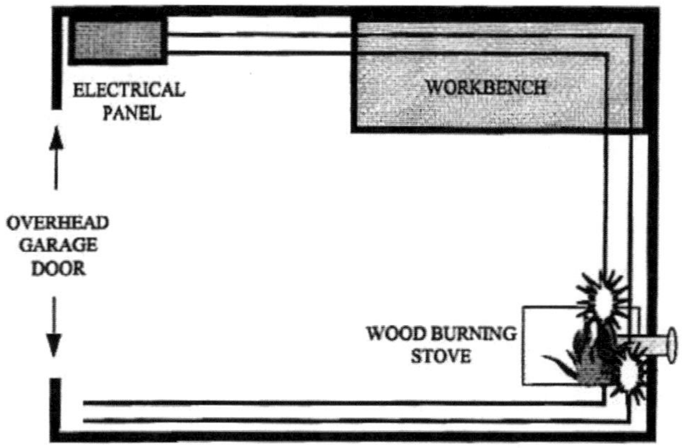

Figure 1 Teaching illustration used by Svare in a 1995 paper. According to the author, this arcing evidence signifies that fire originated at the wood-burning stove.
(Diagram: Robert Svare)

TECHNIQUE

Procedurally, arc mapping is simple. A floor plan has to be made that, if not of the whole structure, encompasses at least all areas potentially likely to be the area of origin. All circuits running through the area need to be traced[14]. The wiring then is inspected for presence of arcs, and those are recorded on the floor plan (map).

Arc mapping is likely to be a tedious process. Unless most of the structure can be excluded on the basis of other considerations, a great deal of wiring may need to be examined. But the examination is likely to be slow, since arc beads can be small and may require use of bare fingers or a sliding a cotton ball in order to detect some of them[15,16]. However, it was also learned that the cotton ball technique does not have value unless charred insulation material is first thoroughly cleaned off conductors; otherwise, cotton balls tend to get stuck on charred debris[17]. For stranded wires, this can be difficult, and ultrasonic cleaning is likely to be needed[17]. In addition, even prior to making this search, it is necessary to trace the relevant circuits[14] (back to a specific circuit breaker in the load center) and in itself this is likely to be a laborious process (and may be impossible if the conductors are broken).

Even though it is sometimes done by electrical or forensic engineers, arc mapping does not require electrical engineering expertise. The only electrical knowledge needed by the individual doing the work is to know the configuration of electrical power cables and the identification of the closed/open/tripped status of circuit breakers. Thus, arc mapping can be done by a competent fire investigator or assistant[18]. Note however that a tripped circuit breaker does not necessarily indicate tripping due to fire-caused short-circuiting. Circuit breakers may have tripped from unrelated causes, prior to the fire, or tripped due to ambient heating of the circuit breaker itself by the fire[19].

GOVERNING PRINCIPLES

The Delplace/Vos hypotheses can be usefully examined as a starting point. The first hypothesis (*A close parallel exists between the patterns left by combustion and the location of short circuits)* can be conceptualized but not readily proven by deduction from any principles of science. However, its validity may be examined by experimental study. The second hypothesis (*A short circuit will normally take place wherever the fire first damages the cable. This gives us the point along the entire length of the circuit earliest affected by the fire.*) needs some detailed scrutiny. The first sentence is either untrue, or a truism. If "first shorts out" is taken as synonymous with "first damages," then it is a truism, making the statement superfluous. But if it is understood that "damage" can entail degrees of severity and is not identical to "shorting," then the sentence is not a general truth. If the fire first causes mild thermal damage that does not result in shorting, then it is incorrect to state "a short circuit will normally take place." Now, consider the second sentence, which, reworded more explicitly, would say that "On any particular circuit, a found arc bead will denote the first place where flames impinged on this circuit." There is no principle of fire science or electrical science that would make this true (or false) for all possible fire scenarios. Potentially, it might be true in a probabilistic sense; it would then be amenable to experimental demonstration, which is considered below.

Carey et al.[25] proposed a simpler hypothesis for arc mapping theory: "*Relating the position of the beads that are formed due to arcing with the layout of the electrical system can help establish the area where the fire first attacked the energized electrical system.*" First, it may be noted that the authors take a weak position: "can help" indicates that, at best, they are anticipating some correlation, rather than describing facts provable from known science. The implied view here is that arcing will line up around some boundary of flaming at some unspecified early time of fire growth. Carey, in fact, put forth such an explicit hypothesis (beading will be "*found on the edge of the initial plumes*") in his 1999 paper[45], but his later work did not support it (nor did any other researcher's). Similarly, Churchward et al.[20] hypothesized that: "*arcing sites will pattern themselves about the area of origin*," while Johnson and Rich[21] hypothesized that "*arc mapping* [will]...*create a boundary around the area of origin*." Taken together, the above hypotheses may succinctly and clearly be worded as:

- Arc sites will be primarily found only in the area of fire origin;

and, possibly:

- Arc sites will be primarily found around the periphery of the area of early fire involvement.

In recent years, fire investigators have been perhaps most likely to adopt hypotheses such as enunciated by Karasinski et al.[22]:

- Initial stages of fire progression
  - Will cause arcing on energized circuits
  - Arcing will typically cause circuits to be de-energized by opening protective devices (breakers, fuses) or severing conductors.
- Later stages of fire progression
  - Further away from the area of origin, there will remain fewer energized circuits, and therefore less arcing.

From fundamental principles, it is useful to consider what can be accepted to be true, solely on the basis of accepted fire science or electrical science:

(1) An arc bead can only be created if the circuit is powered at the time that it gets fire engulfed and shorted*;
(2) Arc beads will not be found on circuits that were de-energized at the time of fire engulfment, irrespective of whether the loss of power was due to tripping of circuit breakers from earlier arcing, severing of conductors due to arcing elsewhere, manual disconnection, or utility outage;
(3) Arc beads are likely (but not guaranteed) to be plentiful in areas where sustained flaming occurred while power was still available; and
(4) Arc beads will not be found on circuits where fire exposure was mild.

The latter two statements stem from the well-accepted facts that applying flames onto energized conductors can burn the insulation off them (in full or in part). Then, conductors with damaged or destroyed insulation and without special means of separation and securement <u>**may**</u> make contact and thereby short out. Energized conductors with damaged insulation may also short out to non-electrical grounded objects. Note that this says nothing about the origin of the fire.

How can then the scheme be applied to fire origin determination? By making an additional hypothesis:

(5) Electric power is available in the area near the fire origin, but is no longer available once flame spread has progressed away from the origin.

There is clearly no principle of science that requires the latter to be true. So next, it must be considered how power becomes lost in structure fires. Several ways are possible:

- *Tripping of circuit breakers*. Shorting may trip a circuit breaker (or open a fuse), but the actual outcome is probabilistic and depends on voltage, available short-circuit current, characteristics of the circuit breaker, and other details of the circuit. Note: if arcing causes conductors to get *welded* together, tripping can be almost certainly assured, provided the conductor size is consistent with the circuit breaker rating and the circuit breaker is not defective. But in some cases, prolonged arcing can occur, and wires of sufficiently small diameter may undergo welding without tripping of circuit breakers.
- *Severing of conductors*. An initial arc may or may not trip a circuit breaker, and it may or may not sever an energized conductor*. But if an energized conductor does get severed, no arcing will be possible

* In some cases, the investigator also needs to be aware that power may be supplied by a flow path that is not the normal one. For example, this can entail back-feeding of a dead circuit through shorting with a still-live one, or operation of an alternate power supply (generator, UPS, etc.).

downstream, although it may remain possible upstream of the sever point. However, no methods exist for calculating if a given event will, or will not sever a conductor, nor which one will get severed if severing occurs. These are probabilistic events.

- *Manual intervention.* Building occupants generally do not respond to fires by turning off electrical circuits, although this may occasionally happen when a fire is seen to be electrical in nature. Standard operating procedures at fire departments vary widely. But fire departments may not shut down electric power unless and until a fire has become large or difficult to control. By that time, fire will have spread greatly beyond the area of origin. Thus, manual intervention is likely to occur only later in the progression of a fire, when fire has progressed greatly beyond any areas potentially to be evaluated as area of origin.
- *Utility power outage.* While this will rarely be an issue, power may simply cease to be available due to a utility outage, unrelated to any of the above actions.

Thus, apart from manual intervention (where timing may sometimes be estimated), or unrelated power outage (which would be exceptionally rare to coincide with a fire event), de-energizing of circuits in a structure fire has to be considered as probabilistic. Given that, it becomes evident that the potential of arc mapping to reveal anything about the area of fire origin is fundamentally probabilistic, and only a sufficient ensemble of experiments can establish utility of the technique. (However, certain deductions involving severed or tripped circuits may sometimes be made reliably, and these deterministic possibilities are considered later.)

The discussion to this point focused solely on establishing a fire origin area smaller than a whole room, assuming the origin is known to be in a particular room. But Carey[45] hypothesized that arc mapping can also "*be used effectively at large fire scenes to identify which compartment the fire started in.*" No research exists on this point, but it is useful to consider the layout of electrical circuits in buildings. Assuming that manual intervention or power outage does not occur, the only way that circuits will be de-energized in regions remote from the area of active burning is by severing of conductors or tripping of circuit breakers. But this will be possible *only if the same circuit's wiring extends from early-burning areas to a remote area in question.* Assuming this is not ancient wiring (where circuits were typically very few), this is unlikely. Modern buildings in North America are commonly wired so that one circuit will serve only a single room, or a few closely-situated rooms or spaces[†]. If that is the case, then there is no physical reason de-energizing of circuits remote from the area of active burning should be expected.

EXPERIMENTAL VALIDATION

As seen above, no principle of science establishes that areas far away from the point of origin will be de-energized and, therefore, bereft of arc beads if fire extends into those areas. But within the general area of origin, will arc beads preferentially occur near the origin on any particular circuit, as opposed to further away? Assuming the initial fire is localized (as contrasted, say, to a vapor cloud ignition), heat fluxes might be expected to decrease monotonically with distance from the fire. Thus, with all else being the same, it might be surmised that arcing would necessarily occur first closest to the fire, and only later additional arcing might occur at farther locations. As will be shown later, the assumption of heat fluxes monotonically diminishing with distance will rarely be valid for real building fires. Apart from the dominant role of local fuel load and local ventilation effects, there are some cable-related reasons why wire shorting potential is *not* likely to bear any simple relation to distance from fire origin:

- The elevation of the cable may change along the run
- There may be more or less thermal protection at various locations along the run
- The stresses in the cable may favor shorting at some locations.

The last factor needs some explanation. Burning away plastic insulation does not, by itself, cause shorting. For shorting to occur, either conductors that are initially a few millimeters apart must move into contact, or else shorting has to involve a carbonized insulation bridge. The propensity for conductors to come into contact will be influenced, among other things, by stresses (bends, kinks, twists, etc.) of the conductors. Once plastic is burned away, conductors may stay in place, or move apart, or move closer together. None of these factors above can be calculated to predict the shorting history of a cable. Shorting may also proceed without metallic contact, due to carbonized insulation bridging the conductors; issues related to this have been discussed by Babrauskas[23]. In view of the above issues—which are not amenable to current-day theoretical or computational analysis—it is important to examine the results of experimental programs.

[*] Note that severing a ground or neutral conductor will not necessarily de-energize the circuit, since arcing can still potentially occur between a live conductor and another conductor that is not at the same potential as the severed conductor.

[†] In the UK, it is common to use 'ring mains.' This entails larger-size branch circuit conductors, with the cable run so as to make a loop, fed from the same circuit breaker at both ends. Such ring mains are likely to traverse sizable areas.

The earliest published studies on this topic gave no useful test data. In 1986, Rothschild[11] published a paper describing his investigation experience, but also provided no experimental results. The first report giving experimental data was published by Carey[24] in 1999. In his first room burn, he found two arc beads, one near the area of origin, the second near a window 3 m away. For a repeat burn, two arc beads were found, both in the general vicinity of fire origin. Carey also conducted another burn, but this was not for the purpose of validating arc mapping. In 2003, Carey supplemented this by another test burn[25] where he found 6 arc beads in the test compartment, 3 close to the origin and 3 at the opposite end of the room. He concluded "*the research has found that the arc mapping methodology has proven to be a reliable indicator of the fire's areas of origin*." This was not justified by the actual data, and a more accurate conclusion would be "*arcing will generally be found in the area of fire origin*." In fact, Carey provided a conclusion very similar to this in his later 2009 study, discussed below.

In 2005 West and Reiter[26], reported on three full-scale room fire tests done for arc mapping validation purposes (Figure 2 shows the detailed results for their third test). In each test, 12 to 15 branch circuits were installed using 14/2 NM cables. These were attached to the ceiling and allowed direct fire impingement. Despite the fact that the tests were run with wiring strung in the open, along the ceiling, in none of the three tests was the arcing clearly localized to the area of origin. While in all cases there was arcing near the area of origin, arcing was also found remote from the area of origin (in some cases, diametrically opposite). For test No. 3 shown in Figure 2, the remote arcing was probably influenced by doorway ventilation, but in other cases no clear reason was seen. Furthermore, in all three tests, the point of ignition was on the sofa, which was, by far, the predominant fuel load item in the test rooms. Thus, the reason for somewhat greater concentration of arc beads in the vicinity of the sofa was most likely due to fuel load, not origin, effects. The authors themselves additionally observed that electrical arcing activity was "greatly influenced by the entrained air velocity," that is, by the ventilation flow patterns.

The largest study of arc mapping was the 2009 Ph.D. dissertation by Carey[14], who conducted 39 compartment burns. This is an order of magnitude greater than presented by any other researcher. However, most of the research was focused on metallurgical studies and only limited conclusions can be drawn from the arc mapping work. The main limitations were: (a) only a single room was studied; the ability of arc mapping to identify the room of fire origin was thus not possible to address; (b) only 4 circuits were used, yet the rooms were notably large; (c) all wiring was run exposed on the ceiling; no protected or low-level circuits were studied; (d) most of the fires were exceptionally fast, flammable-liquid fires, with circuit breakers tripping in 2 – 3 min, and sometimes even around 1 min. Furthermore, 17 of the 39 tests involved scenarios with multiple points of ignition, thus there were only 22 tests where a fire with a single point of origin was configured.

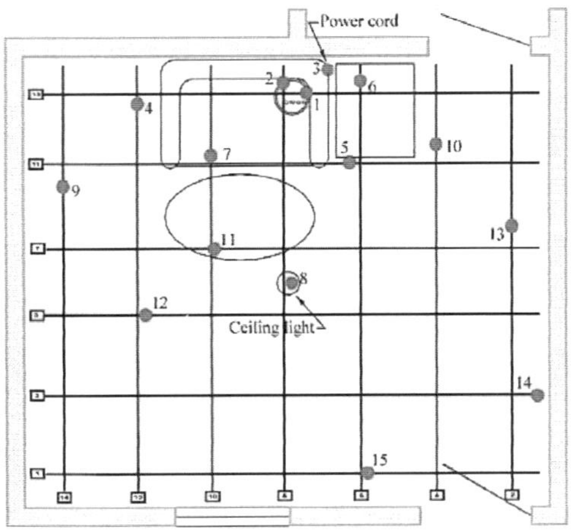

Figure 2 Arcing locations (red dots) found by West and Reiter for their test No. 3; ignition was on the seat of the two-seater couch, indicated by a blue circle
(Diagram: Larry West and David Reiter)

Carey concluded "*Analysis of the three-dimensional data recorded during the experiments and the arcing events has indicated that there is a high probability of arcing damage occurring to electrical conductors located close to the fire's area of origin*." This conclusion is supported by the results presented. However, it does not address the issue of discrimination: given that arc beads are likely to occur near the fire origin, are they rare or unlikely to occur at locales remote from the origin? In fact, an examination of his data shows that these were also plentiful.

Wood and Kimball[27] conducted four full-scale fire tests in a 3.7 × 4.9 m fully furnished bedroom. An array of 14 AWG NM cables was attached to ceiling on a 0.61 × 0.61 m grid; each cable fed by a separate 15 A circuit breaker. While such close spacing will generally not simulate real-life conditions, it does enable arcing locations to be better assessed than the coarse spacings used by Carey. No loads were fed from the cables. In each test, all circuit breakers opened before the fire reached flashover stage. Arcs found were typically between hot and ground, but there were arcs between all three conductors. In most cases, they found that the arcing artifacts were extremely small (1 – 4 mm) and required great effort to locate. The main conclusion that they drew from their work was that arcing is likely to occur in the room of fire origin, but "*no arc mapping patterns emerged from the room fire tests to clearly indicate a point of origin*." But they also added "*all conductors did arc indicating that the conductors may reliably indicate a room or area of origin even in the absence of being able to indicate a point of origin*." The latter conclusion is incorrect, since if arcing (or even copious arcing) suffices to establish the room or area of origin, then buildings with widespread, extensive arcing would necessarily have been victims of multiple areas of origin, which is an unjustifiable conclusion. It bears emphasis that the authors did not run any tests where arcing on the wiring of multiple rooms in a building was studied.

Most experimental studies involved NM cables simply fastened to the bottom of a gypsum wallboard ceiling. This may simulate wiring in unfinished basements or attics, but is not representative of normal wiring conditions in occupied spaces. In 2015, Wheeler[28] was the first researcher to explore experimentally the effect of protective barriers. He conducted three burn room tests using 14 AWG NM cables that were run up one wall, across the ceiling, and down the opposite wall. In test No. 1, the NM cables had no protection. In test No. 2, the cables on the ceiling and on one wall were protected by ½-in (12.7 mm) thick gypsum wallboard, while the cables on the other wall were protected by thin, 1/8-in (3.2 mm) thick wood paneling. In test No. 3, all cables were protected by ½-in gypsum wallboard, ignition was on the floor near a chair, and the fire went through flashover before being extinguished. The results (Figure 3) do not validate the ability of arc mapping to localize an area of origin that is smaller than room-size. Wheeler provided some moderately optimistic conclusions from his tests that were not justified by the data, and for the protected test (No. 3), the results indicate that *arcing was localized to the area of heaviest fuel concentration (the sofa), rather than the area of origin*. It should be noted that even Wheeler's protected experiment did not involve wiring compliant with the electrical code. Sec. 300.4 of the NEC[29] requires that wiring behind a wallboard inside walls have at least a 32 mm (1.25-in) air space between the cable and the back of the wallboard. By omitting this spacing, the cables in Wheeler's tests were more readily susceptible to heating from the fire, thus more likely to arc closer to the location of the dominant fuel load burning in the fire.

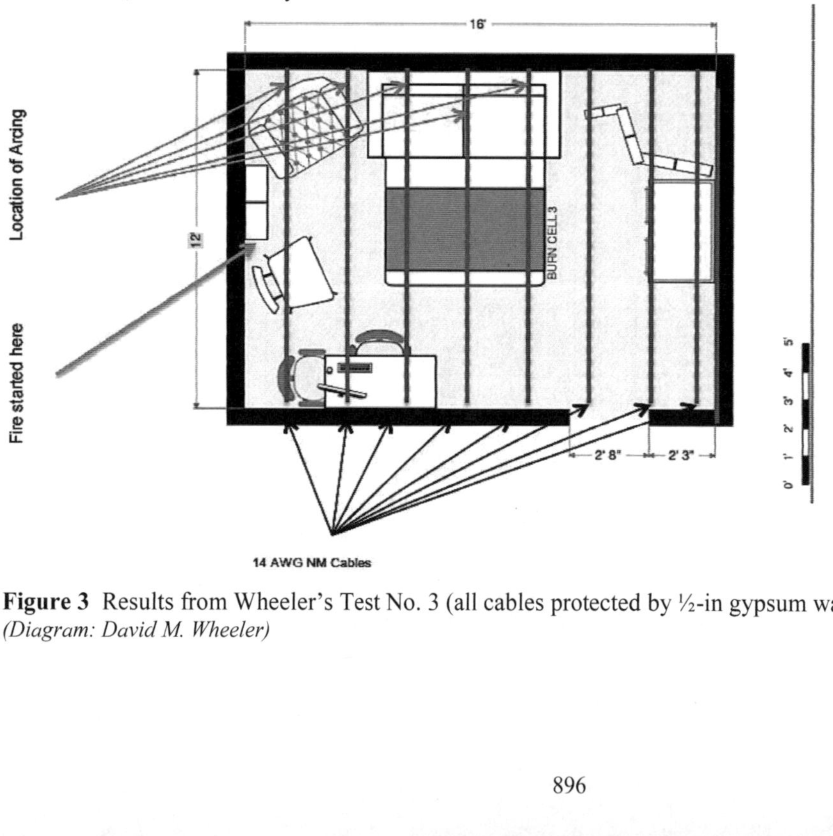

Figure 3 Results from Wheeler's Test No. 3 (all cables protected by ½-in gypsum wallboard)
(Diagram: David M. Wheeler)

MULTIPLICITY OF ARCS

Some investigators used to believe that multiple arcing will not occur on a single circuit, but well-known forensic electrical engineer Bernstein[30] already noted in 1991 that multiple arcs are possible. Rothschild[11] reported finding as many as 4 arcs on one 14 AWG branch circuit, but provided no details. West and Reiter[26] provided the first detailed experimental results on multiple arcs on a single circuit, and Table 1 shows their results; 4.8% of the circuits experienced two arcs and no instances of more than two arcs were found. Carey[14] found that 5.8% of his circuits experienced two arcs (and none three or more). These two studies gave very similar results, and it can thus be concluded that about 5% of solid-core conductor circuits can be expected to show two arcs, and an insignificant fraction may contain a larger number.

Table 1 Results from tests by West and Reiter on fire-induced arcing of 14/2 NM cables

Test No.	Test 1	Test 2	Test 3
circuits with one arc	11	14	15
circuits with two arcs	1	1	0

ARC BEAD TYPE AND SIZE

In his tests, Carey[14] found that 29% of arc sites constituted sever-arcs, while 7% were welded together. For arc sites where severing was found, however, he did not document the fraction where the hot conductor was severed.

West and Reiter[26] analyzed their data for arc bead size, and found beads ranging from very large to barely perceptible. They noted that this was an unexpected finding, since the circuit breakers were identical for all circuits, as was the available short-circuit current, I_{sc}, of 310 – 320 A. These findings indicate that arc bead size is a highly probabilistic quantity. However, there are some known effects on bead size, and Ablenas and Bodzay[5] demonstrated that lower short-circuit currents are more likely to result in large arc beads, than if the current is high and rapid circuit breaker tripping occurs. In their study, they found that, in fires, the short-circuit current is likely to be much lower than I_{sc}, due to shorting occurring with carbonaceous material in the path, not a direct metal-to-metal contact.

ARCING WITH STRANDED WIRES

Branch circuits in 15 or 20 A ratings in North America use solid-core copper wires. But power cords and wiring inside appliances and equipment are likely to use stranded wires. Stranded wires can behave differently from power cables having solid-core conductors for several reasons:

(a) Stranded-wire cords are typically used exposed in the room, not protected by any thermal barrier. Stranded appliance wiring may be protected by an appliance housing, especially if it is steel, but is unlikely to be located behind a gypsum wallboard barrier.
(b) Power cords are likely to be low, or at floor level, as opposed to being high or at ceiling level.
(c) Most stranded cords use only a single layer of insulation, rather than having both wire insulation and a jacket.
(d) Stranded-wire cords are likely to use much smaller size conductors than cords used for building wiring purposes; consequently, overcurrent protection may be much less effective, and multiple arc sites on a single circuit might be readily expected.

Thus, experimental data obtained on solid-conductor power cables may not necessarily directly apply to stranded power cords, and results specific to stranded wires should be considered. In an early paper, Rothschild described a fire where 7 arcs were found on one power cord. Hoffmann et al.[31] conducted a laboratory study on appliance power cords that had 18 AWG copper conductors. A large number of tests (713) were run under several thermal conditions—radiant heating, small gas flame, and a more realistic fire from a small wood crib. The results indicated that 100% of the power cords attacked by a wood crib fire showed arcing damage, while lesser fractions of specimens receiving radiant heat or small burner flames shorted. Johnson and Rich[32,33] ran a single compartment test where they exposed to a room fire an extension cord and three appliance cords plugged into it, arranged at heights ranging from floor to table height. The extension cord was extra-heavy duty (12 AWG) and was powered from a 20 A circuit breaker. The compartment went through flashover, and then was extinguished shortly thereafter. The extension cord showed 4 arcs, one appliance cord had 4 arcs, a second had "multiple" arc marks, while the last one had none. Presumably the extension cord shorted out and severed or tripped the circuit breaker (the authors did not state), thereby de-energizing the last appliance that showed no shorting. The authors also reported a fire case where three appliances were plugged into a circuit-breaker protected power strip. The power strip's cord showed arcing, as did one appliance, but not the other two. In this case, it was found that the power strip's circuit breaker did trip. It can be concluded from this that the likelihood of arcing on power cords is decreased if they are fed by a power strip or an extension cord, since upstream shorting or tripping may act to de-energize plugged in devices. Otherwise, the likelihood is high that energized, stranded cords will get shorted out in a serious room fire, but it should not be inferred that this is a necessary outcome.

Hoffmann et al.[34] also ran room fire tests with energized appliances and showed that fire attack on power cords will often, but not always trip circuit breakers. In some cases, arcing damage to cords occurs, but the affected circuit breaker is not tripped. Hazelwood[35] reported a case where a long extension cord was found to have not only 4 arc locations, but each location was a sever-arc. Such evidence does allow the direction of arcing damage to be established, which necessarily had to progress from the downstream to the upstream end of the cord.

Stahl and Parrott[36] discussed the application of arc mapping to motor vehicles. Vehicles utilize primarily stranded conductors, but also differ from building wiring in the power supply: DC supplied typically at 12 V for passenger vehicles, with higher voltages for heavy vehicles or electric cars. They did no tests, but did give a number of warnings against misuse of the technique for fire investigation in motor vehicles. Their main conclusion was that, given an absence of any laboratory studies applied to motor vehicles, arc mapping cannot be assumed to be a "*dependable, robust technique when translated from structures to vehicles and equipment*."

Lancaster and Meadors[17] studied arc mapping of stranded conductors using three test rooms, fueled by burning wood pallets. Fire was extinguished in the first one after all circuits tripped, while the next two rooms were allowed to burn to the ground. Each test used a number of stranded cords laid out along the floor. Trip times were typically 10 min, as contrasted to 2 – 3 min for solid-core cables stretched across the ceiling in Carey's tests. The work was unfinished, however, and the authors only presented metallurgical results, but not data obtained from the arc mapping itself.

Shanley[37] described using arc mapping as an aid in analyzing a clothes dryer fire, specifically to determine if the fire originated inside the appliance, or outside. No tests were run, and Shanley simply concluded that the fire originated inside, since the heater was energized, and since arcing locations did not correspond to places that would be most vulnerable to fire attack from the outside. Kovarsky[38] also did not perform tests, but reported investigation conclusions from several types of cases, which are generalized and expanded here:

- Arcing was found on the wiring inside an appliance at a location well away from external fire attack, and this is taken to indicate that fire originated internally, near this arcing locale. This scenario is innately questionable, since a fire originating inside an appliance is unlikely to then jump to far-off areas while leaving the near environs of the appliance unburned.
- The preceding scenario is narrowed to specify that, in addition, no arcing is found elsewhere inside the appliance, closer to the external fire. Again, this is a questionable scenario. More likely, a small, self-terminated fire occurred inside the appliance unrelated to the structure fire being investigated.
- Arcing was found inside the appliance at a location close to where an external fire would likely impinge on the device, but far from likely sources of ignition inside the appliance. Kovarsky considered that this allows one to conclude the fire originated externally. However, as discussed below, Hoffmann et al. showed that arcing inside an appliance from an external fire may more frequently be located near the externally heated area, but there is no reliable correlation. Thus, conclusions should not be drawn solely on the basis of distance between the external thermal attack and the location of internal arcing.
- Arcing is found on the cord or plug of the appliance, but no arcing is found inside. Kovarsky considers that this rules out the hypothesis of fire originating inside the appliance. However, this is not a valid deduction from general principles of science, and by itself such a condition does not preclude the origin of the fire being inside the appliance. A fire originating inside the appliance may create arcing close to the origin, inside the appliance, but this is not required. Fire may originate internally at a bad connection or an overheated component, but at a place where wiring cannot readily short out and arc. In such a case, it may well be possible for fire to emerge from the appliance and attack a nearby cord or plug outside. However, any arcing found inside an appliance warrants a detailed laboratory examination of the item.

Hoffmann et al.[39] conducted a test to examine the question whether arcing may occur inside appliances due to an external fire. They set up a single room test where three appliances—a refrigerator, a dishwasher, and a clothes dryer—were exposed to the room fire that was started in a trashcan. The dishwasher showed arc damage at several locations inside the appliance. The refrigerator showed arcing to three different internal circuits, but all close to the same general location. The dryer showed arcing on its power cord and at five different locations in its internal wiring. Two of the 5 locations were near the back of the unit, far away from the room fire that was in front of the appliance. These results indicate that an external fire can readily cause arcing inside energized appliances, but does not support the notion that arcing necessarily will be confined to portions of the interior most directly proximate to the external fire attack. In some cases, external fire attack may be close to the front of an appliance, but internal shorting may occur due to radiant heating from overhead.

The correct conclusion with regards to appliance wiring is that the same principles are obeyed as for room fires. The propensity to arc is related to the propensity for thermal damage, but the latter is controlled by fuel arrangements and ventilation patterns, and not just by time elapsed from start of flaming at a given location.

Hoffmann[40] also presented a scenario where two appliance cords are plugged into an outlet, the circuit breaker has tripped, but only one cord shows arcing (but *not* welded together). Hoffmann suggests that the appliance whose cord does not show arcing can be ruled out as being the origin of the fire. In some cases, an engineering analysis may demonstrate that this is likely to be true, but no fundamental science principles require it to be so. Burning of fuel packages commonly occurs so that, if flames are going to cause cord shorting, this will occur first to the cord of the equipment burning, rather than to a cord of neighboring equipment. But arrangements can also be such that the converse will be true. Thus, a general behavior cannot be assumed in the absence of a specific engineering analysis.

THE ROLE OF APPLIANCE FUSES

Some appliances are required by pertinent product standards to have internal or plug-mounted fuse protection (or, occasionally, circuit breaker protection). In some other cases, protection is not mandatory, but the manufacturer may provide it nonetheless. Does the blowing of the fuse, or the fact that the fuse is intact, indicate anything about the origin of the fire? Kovarsky[38] correctly explained that, in general, no conclusions can be drawn beyond the trivial one that that if a fuse blew due to overcurrent then the appliance was energized at some time prior or during a fire. Note that even the time period for this cannot be established. A fuse may have blown some time prior, and unrelated to a fire. Methods exist to distinguish fuses melted from overcurrent versus external fire heating however[41,42].

ERRONEOUS CONCEPTS

Any concepts associated with arc mapping practice can be rejected without need for considering experimental results or validation, if they are based on an incorrect understanding of electrical science. One such incorrect principle has been succinctly expressed as[43]: "*A basis for arc mapping is the idea that on any given circuit, electrical activity that is furthest from the electrical source (e.g. panel box) occurred prior to activity that is closer to the electrical source.*" The concept was first enunciated by Rothschild[11], then later presented by others[12,38,44,45]. But the early researchers Miyake[9] and Delplace and Vos[10] correctly understood that this is not a general conclusion but only holds if there is a sever-arc—then any electrical activity after that must be upstream of the severing location. And Bernstein[30] explained that: "The arcing farthest from the power source probably occurred first *if arcing closer to the power source caused conductors to separate*." This follows from simple continuity, since current cannot flow beyond a severed place and therefore arcing cannot take place. There is some possibility that arcing will occur at two places simultaneously, thus a more correct version of Bernstein's statement would be: "Arcing downstream of a server point cannot have occurred later than the sever-arc location.*"

Recent authors have generally not explained the basis for their erroneous conception, but Rothschild made his basis clear: he did not differentiate between a bolted short and an arcing short, and thought that, once a short occurs, this means that no current is available to flow downstream of that point. In fact, arcing shorts on 120 VAC, and even 240 VAC, circuits are short-lived even when a circuit breaker is not tripped[47]. The arc may self-extinguish during the next zero-crossing of the current. But magnetic forces tend to push the conductors apart[48], aided by flow of molten material. Thus, an arcing short clears itself and, if the circuit breaker has not tripped and conductor severing has not occurred, the circuit is able to carry current further downstream. Rothschild did correctly point out that, when shorting occurs, current in the conductors upstream will necessarily increase, and this increased current may lead to

* An arc traveling on bare, parallel conductors (e.g., busbars) will move away from the current source (due to Lorentz force), not towards it, but this situation is not applicable to insulated conductors. It also requires that no circuit breaker tripping occur during the travel away from the power source. See Ref. 46.

arcing failures upstream. In another early paper, Sanderson[44] thought that arcing further downstream from the original arc cannot take place since all arcs will be sever-arcs, but this is also not correct.

ARC MAPPING APPLICATIONS WHICH DO NOT REQUIRE VALIDATION

Several arc mapping scenarios follow directly from accepted science principles and, therefore, do not need validation.

(1) Arc marks are found on some conductor(s). This indicates that the circuit was energized at the time of shorting. The conclusion can be reliably made, but will only rarely be of assistance in establishing the origin of a fire.

(2) Multiple arcing, including a sever-arc or a weld-arc. A conductor has multiple arc marks, but one arc location is a sever-arc. Circuit continuity demands that any arc marks downstream of the sever point had to have occurred before (or simultaneous with) the sever-arc[10,30]. Note this specifically refers to situations where the hot conductor is severed. If the neutral or ground conductors are severed, but not the hot conductor, then arcing downstream of the sever point might potentially be possible if the unsevered hot conductor comes into contact with some other grounded objected, e.g., a water pipe. Further analysis should be done to rule out such contact. In the same manner, if an arc location is found where the conductors are welded together, then no significant current can flow downstream of that location. Thus, if a downstream arc location exists, it could not have been formed later than the one at the welded-short location.

(3) Welded-short conductors. Two cords are plugged into one outlet and both cords show arcing, but the conductors on one cord are welded together, while the other cord shows arc marks, but not welded-together conductors. The circuit breaker is tripped. The cord that was welded then necessarily had to be the second event, since if it were first, the circuit breaker would have tripped due to welded-short conditions. This scenario was described by Utt[49].

RELIABILITY OF ARC MAPPING

Churchward and Ryan[15] (Churchward was the Chair of the NFPA Technical Committee on Fire Investigations when arc mapping was first introduced into NFPA 921) pointed out that "*Reliance solely on an arc survey to determine the...proposed area of fire origin places far too much importance on a technique that cannot always be reliable or validated.*" They also noted that if the building is not de-energized relatively quickly, there is likely to be an abundance of arc sites and the imputed 'area of origin' is likely to be huge. They further cautioned that it is not logical to conclude that a circuit where the wire insulation burned up, yet which shows no arcing must have burned only after the power was removed, since exposure of an energized circuit to flaming *may* lead to arcing, but *not inevitably*.

Similarly, Wheeler[28] observed very recently (2015) that "*a great deal has been written about arc mapping,*" however "*there are many factors that influence the reliability of arc mapping* [yet] *very little has been published that specifically addresses such factors or concerns.*" This is actually an excessively optimistic assessment of the reliability of arc mapping. While more experimental studies might reveal additional information, the existing studies already suffice to indicate that, arc mapping is unreliable and unscientific if a hypothesis is made that abundance of arc beads at a given locale means that fire originated in that area, while a paucity of arc beads indicates that it did not. Only in some narrow circumstances, as discussed above, can sound, technically justified conclusions be drawn from the results of arc mapping.

Wheeler's own tests, even though not numerous, have been important for understanding the reliability of arc mapping, since he has been the only researcher to conduct experiments on the effect of wiring behind protective barriers (Novak[50] presented some theoretical modeling of this situation, but not room fire experiments). Analysis of his actual results (but not necessarily his published conclusions) indicates that:

- Significant arcing is likely to be found in areas subjected to ventilation air flows
- Significant arcing is likely to be found in areas of heavy fuel load concentration
- The above two factors tend to dominate the arcing patterns found, and arcing is not likely to be concentrated in the area of fire origin, unless this happens to be an area of heavy fuel load concentration or within the inflow region of incoming ventilation air.

While early researchers hoped that arc mapping would somehow prove to be more reliable than char-depth interpretation, experimental work has not borne this out. Instead, the experimental studies indicate that arc mapping has the same limitations as char-depth studies—local fuel load effects[51] and local ventilation effects[52] are likely to be dominant, and not the location of the fire origin. This is true of *all* the research studies that have been published to date and have been examined here. There is no basis to believe that arc beading will be found to be plentiful in the area of origin, and not present (or sparsely present) in areas away from the area of origin.

Additional considerations with regards to lack of validation are the following:

(1) In most structures, the majority of building wiring is behind protective membranes, not out in the open. Wheeler's study indicates that arcing patterns behind protective membranes do not bear a significant relationship to the area of fire origin. No other researchers have used such realistic test arrangements, but even their results, if interpreted in an unbiased way, do not suggest that arc mapping is a reliable pointer to the location of the origin of a fire. Carey[24] succinctly noted that arc mapping "*is of limited value when the cables are protected from the fire in the initial stages, for example, buried in plaster or installed behind plaster board in cavity walls.*"

(2) Arcing cannot occur where there are no conductors to arc. Most of the experimental studies have been based on a tacit assumption that there is a densely-spaced grid of wiring available for potentially shorting and arcing. This will rarely be true and, in most cases, cable runs are sparsely and irregularly distributed. In other cases, a significant grouping of adjacent, parallel circuit cables may be found, often at locales near the load center[53]. Thus, presence of arc beads may indicate mainly where wiring runs were located, rather than revealing anything useful about the progress of the fire.

(3) Furthermore, runs of branch-circuit wiring are likely to be concentrated below the floor, or in the lowest portions of walls. Generally, much fewer circuits are run along ceilings or in the upper portions of walls. Thus, wiring in practice is most likely to be concentrated along the areas where there is least likely to be early fire damage to it. Consequently, experimental studies based on closely-spaced ceiling grids, when applied to such situations, will necessarily present a positive bias.

(4) Arcing and shorting may preferentially occur at places where there is a kink or bend in the cable, or a similar stress situation that facilitates shorting of heated wires. No study has explored this.

(5) Arcing may occur, but not create an artifact that can be unambiguously associated with arcing[23]. This can also include situations where an identifiable artifact was created, but was subsequently obliterated by fire melting, corrosion, or other fire effects.

(6) Few studies exist for arc mapping pertinent to stranded-wire conductors and none for wires inside conduits. However, there is no basis to believe that arc mapping in those situations would be valid under more scenarios than for branch-circuit wiring in NM cables.

(7) There have been no validation studies of arc mapping in motor vehicles, which also use stranded wires, but typically have low-voltage DC power supplies and sometimes use circuits with no overcurrent protection.

(8) In many cases, the structure will have multiple rooms, and the first task facing the investigator may be to identify the room of origin. No study exists where arc mapping was experimentally studied for a multi-room building. However, there is no basis for assuming that such application would allow valid use of arc mapping under broader circumstances than is justified for single-room scenarios.

PROBABILISTIC EVALUATION

The main scenario for which valid electrical science allows positive conclusions to be drawn from arc mapping is the severed-wire scenario. For this scenario, probabilities of occurrence can be evaluated, at least for solid-conductor branch circuits. Based on the results of West and Reiter[26] and Carey[14] given above, there is approximately a 5% probability of a second arc on any particular circuit that has arced. The probability is 29% that an arc will be a sever-arc. But conclusions can only be drawn if the non-sever arc is downstream of the sever-arc. Since, in general, there is no bias where the second arc occurs, there is a 50% probability for the downstream location. Thus, the probability is $0.05\times0.29\times0.5 = 0.0073$, which is 0.7%, that a circuit with arcing evidence will allow the direction of fire movement to be scientifically established. Limited findings suggest that the likelihood may be much higher for stranded wires, but quantitative estimates are not available.

CONCLUSIONS

Procedurally, arc mapping is a technique for plotting locations of arc marks on a floor plan (map) of a building. As such, it is non-controversial, assuming that the arcs (or absence thereof) are correctly identified and plotted. Conclusions drawn from an arc map, however, may be unreliable and inconsistent with known science, depending on the inferences being drawn.

Arc mapping is a tedious process, since it requires detailed examination of all the conductors in the relevant building area, which is likely to be a painstaking process. In addition, it requires that the circuits be traced and identified, which can be difficult to do. This requirement may also make arc mapping impossible, since there are many fires where, due to the type of damages sustained, circuits cannot be successfully traced. In case of serious fires, structural collapses may produce a jumble of wires with no means of reassembling into their original locations.

Once an arc map is prepared, in a small fraction of cases, reliable conclusions can be drawn from applying basic principles of electricity, especially the fact that arcing cannot occur without current flow, and current cannot flow in circuits beyond a sever point, or if a circuit breaker has tripped. But in the majority of cases, principles of science do not allow conclusions to be drawn from an arc map with regards to the origin of the fire. *In the cases where valid conclusions can be drawn, they only establish the local direction of fire movement at a specific locale, and do not suffice to identify the area of origin of the fire.* Reliable conclusions on the local direction of fire movement can be drawn in those cases where the branch circuit has a sever-arc or a weld-arc, and an additional arc downstream of this location. But experimental data indicate that the probability of a building's branch circuit exhibiting these conditions for allowing a valid arc-mapping conclusion is less than 1%, and is estimated to be 0.7%. This pertains to solid-core conductors; not enough research has been conducted on stranded wire arcing to be able to draw statistical conclusions.

Delplace and Vos proposed the arc mapping method on the basis of two hypotheses: (1) that patterns of arcing will be spatially similar to patterns of combustion damage; and (2) that arcing locations will tend to reflect the early boundaries of flame spread. Hypothesis #1 has been shown to be generally true, but also generally unhelpful. Hypothesis #2 is not supported by principles of science, nor by experimental studies of arc mapping.

Experimental studies have borne out the Delplace/Vos hypothesis that arcing patterns tend to correspond to patterns of combustion damage, specifically intensity of damage, e.g., as found from char-depth studies, at least in a stochastic sense. Consequently, arcing patterns can reflect:

- fuel loading configurations;
- ventilation, specifically, air inflow patterns; and
- duration of burning.

Only the last variable has any relation to the potential of an area being the area of origin for the fire. If fuel loads were uniformly distributed, and combustion air inflows also uniformly and diffusely supplied, then arcing patterns could be expected to reflect the duration of burning and, thereby, point to the area of origin. But such conditions are essentially never true in actual buildings. Thus, there is no valid basis to claim that arc patterns will identify the area of origin. But additionally, fire damage to cabling has a strong stochastic aspect. This includes both the local heat fluxes generated by the fire, and the local susceptibility of the cable to shorting and arcing. The general trends and magnitudes of heat fluxes may be modeled, but due to the turbulent nature of fire and the complexities of local geometry irregularities, precise deterministic estimates cannot be expected. The response of cables to heat is even less amenable to calculation. A limited amount of experimental data exists, but data scatter is large due to complexities such as wire stresses and cable twists.

Inside habitable buildings, wiring is occasionally placed in the open, unprotected. But in most cases, wiring is behind protective layers such as plaster, wallboard, or masonry. Such protective layers will inhibit early thermal attack on the wiring. They will also smooth out the temperature gradients, so that failure of wiring should be expected only when the fire has become widespread, instead of still being localized near the area of origin. Consequently, if the actual wiring is behind protective layers, experiments that are based on unprotected installations innately have a bias towards positive findings.

The totality of published experimental results does not support the ability of arc patterns to identify the area of fire origin. This is especially notable, since most studies have been effectively biased in favor of positive findings, due to lack of thermal protection for the circuits being fire-exposed. And even in the one study that did use thermal protection, the installation was not compliant with the electrical code, and the non-compliance was such as to bias the results towards a positive finding.

In some locales, all wiring is required to be placed in metallic conduits. There have been no experimental data on the arcing patterns in buildings using such wiring methods.

Some authors have published conclusions that are at variance with their data. The fact that some experiments show a presence, or even a copious presence, of arcing in the vicinity of the area of ignition does not logically indicate that arc mapping can distinguish between areas of fire origin and areas where fire did not originate. To be able to make such a claim would require demonstration that arcing is sparse or nil in areas where fire did not originate. However, this is impossible, since there are already sufficient data showing not only that arcing may readily be found in areas not close to the area of origin but that, in some cases, such arcing is more copious than arcing near the origin. The following are myths:

- An abundance of arc beads at a given locale means that fire originated in that area, while a paucity of arc beads indicates that it did not.
- When multiple arcs are present on a circuit, the direction of arcing will necessarily proceed upstream towards the power source.
- If an appliance is the victim of a fire, internal arcing will be primarily near the exterior of the unit, while arcing deep inside indicates a fire origin at that place.

No evidence exists that arc mapping of stranded-conductor wiring can be used to establish the origin of a fire any more so than with solid-conductor wires. However, because arcing on small-diameter stranded wires may be much less likely to trip circuit breakers, it can occasionally be useful towards establishing the *direction* of fire spread in a limited, local area, if multiple sever-arcs are encountered along the length of the same cord.

In fire investigation reports, it is not acceptable for an investigator to report that a conclusion was based simply on "arc mapping." There are a few circumstances where arc mapping may be utilized in a scientifically acceptable manner. But there are many more ways in which arc mapping may be used based on unsupported or disproven hypotheses. Thus, it is essential to identify explicitly the exact hypothesis being invoked, and how the conclusions follow from that hypothesis. A parallel to "photographic evidence" may be helpful. Photographic evidence may reliably enable some conclusion to be drawn, but if that is to happen, the exact items being illustrated in the photographs have to be clearly and explicitly identified. Simply stating that photography was done and that the investigator's conclusions are based on photography is without merit.

NFPA 921 should be revised to eliminate arc mapping as one of the four main methods for establishing fire origin, and to subsume it under the more general category of "fire patterns." In addition, it is important that NFPA 921 reduce the implied general utility of the method and provide more explicit discussion of its limitations and of those circumstances where arc mapping is a valid method for assisting in the determination of a fire origin.

REFERENCES

1. Shanley, J. H. jr., Fire Investigation Change and Evolution, Part 1: An Overview, *Fire Engineering* **147**:1, 42-46 (Jan. 1994).
2. Coleman, R. J., and LaBarbera, B. T., **Arson Investigation...a Team Approach**, California District Attorneys Association, Sacramento (1994).
3. Lentini, J. J., The Mythology of Arson Investigation, pp. 301-312 in *ISFI 2006—Proc. 2nd Intl. Symp. on Fire Investigation Science & Technology*, NAFI, Sarasota FL (2006).
4. Kennedy, P. M., Insights into Fire Investigation from Fifty Years of Professional Practice: What's Wrong? What's Right? What's the Difference, pp. 3-13 in *ISFI 2010—Proc. 4th Intl. Symp. on Fire Investigation Science & Technology*, NAFI, Sarasota FL (2010).
5. Ablenas, F. J., and Bodzay, S. J., Surface Analysis of Beaded Electrical Conductors towards the Determination of Fire Causation (Report 9018 U 810), Canadian Electrical Assn., Montréal (1994).
6. Guide for Fire and Explosion Investigations (NFPA 921), National Fire Protection Assn., Quincy MA (2017).
7. Babrauskas, V., Arc Beads from Fires: Can 'Cause' Beads Be Distinguished from 'Victim' Beads by Physical or Chemical Testing? *J. Fire Protection Engineering* **14**, 125-147 (2004).
8. Straeter, R. L., and Crawford, C. C., **Techniques of Arson Investigation**, R. L. Straeter, Los Angeles (1955).
9. Miyake, K., Fire Investigation—Article No. 5 in a Series of Articles on Scientific Investigation (in Japanese), *Sosa Kenkyu* **24**:1, 45-57 (Jan. 1975).
10. Delplace, M., and Vos, E., Electric Short Circuits Help the Investigator Determine Where the Fire Started, *Fire Technology* **19**, 185-191 (1983).
11. Rothschild, L. I., Some Fundamental Electrical Concepts in Locating the Cause and Origin of a Fire, *National Academy of Forensic Engineers J.* **3**:2, 37-44 (1986).

12. Svare, R. J., Determining Fire Point-of-Origin and Progression by Examination of Damage in the Single Phase, Alternating Current Electrical System, *Journal of People to People, International Arson Investigation Delegation to the People's Republic of China and Hong Kong* (1988).
13. Svare, R. J., Using the Electrical System to Help Reconstruct the Fire Scene, pp. 103-116 in *Proc. Intl. Symp. on the Forensic Aspects of Arson Investigations*, Federal Bureau of Investigations, Washington (1995).
14. Carey, N. J., Developing a Reliable Systematic Analysis for Arc Fault Mapping (Ph.D. dissertation), Univ. Strathclyde, Strathclyde, UK (2009).
15. Sanderson, J. L., Cotton Balls not Just for the Medicine Cabinet Anymore, *Fire Findings* **9**:2, 14 (Spring 2001).
16. Churchward, D. L., and Ryan, R. M., The 'Benefits' of Arcing, pp. 27-38 in *ISFI 2010 – Intl. Symp. on Fire Investigation Science & Technology*, Natl. Assn. of Fire Investigators, Sarasota FL (2010).
17. Lancaster, T., and Meadors, A., Preliminary Analysis of Energized and Non-energized Multi-Stranded Copper Wire within Three Full Scale Compartment Fires, pp. 383-393 in *ISFI 2012—Proc. 5th Intl. Symp. on Fire Investigation Science & Technology*, NAFI, Sarasota FL (2012).
18. Sanderson, J. L., and Dwyer, N. P., Arc Mapping Basics: Inspect, Document and Understand the Pitfalls, *Fire Findings* **13**:2, 9-11 (Spring 2005).
19. Fogle, J. L., In a Fire Scene, What Does a Tripped Circuit Breaker Indicate? *Fire & Arson Investigator* **35**:4, 9-10 (June 1985).
20. Churchward, D. L., Cox, R. M., and Reiter, D., Means to Determine Fire Origin in Residential Structures, *ISFI—Proc. Intl. Symp. on Fire Investigation Science & Technology*, Fire Service College, Moreton-in-Marsh, United Kingdom (2004).
21. Johnson, W., and Rich, L. V., Arc Fault Analysis: Post Flashover Studies of the Power Cord, *2007 Fire and Material Conf.,* Interscience Communications Ltd, London (2007).
22. Karasinski, J., Robin, D., DeMatthies, H., and Hauf, G., Arc Mapping. An Explanation and Example, *Fire Scene* (Newsletter, New York State Chapter of IAAI) (Spring 2009).
23. Babrauskas, V., Short-Circuit Arc Marks, to be published (2017).
24. Carey, N. J., Comparison of Behaviour during Fire: U.K. & U.S. Electrical Wiring Systems, London Fire Brigade (1999).
25. Carey, N. J., Svare, M. J., and Daeid, N. N., Arc Fault Mapping—A Technique to Assist in Identifying a Fire's Area of Origin, presented at The Forensic Science Society, Leicester, England (2003); also *IAAI Annual General Meeting* (2006).
26. West, L., and Reiter, D. A., Full-Scale Arc Mapping Tests, pp. 325-339 in *Fire & Materials 2005*, Interscience Communications Ltd., London (2005).
27. Wood, C. B., and Kimball, A., A Preliminary Report of Full-Scale Arc Mapping Test Results, pp. 615-627 602 in *ISFI 2010 – Intl. Symp. on Fire Investigation Science & Technology*, Natl. Assn. of Fire Investigators, Sarasota FL (2010).
28. Wheeler, D. M., Arc Mapping: Things to Consider, *Fire & Arson Investigator* **66**:1, 18-24 (July 2015).
29. National Electrical Code (NFPA 70), National Fire Protection Assn., Quincy MA (2014); Sec. 300.4.
30. Bernstein, T., Electrical Fires: Causes, Prevention, and Investigation, pp. 116-134 in E. K. Greenwald, ed., **Electrical Hazards and Accidents: Their Cause and Prevention,** Van Nostrand Reinhold, New York (1991).
31. Hoffmann, J. M., Hoffmann, D. J., Kroll, E. C., Wallace, J. W., and Kroll, M. J., Electrical Power Cord Damage from Radiant Heat and Fire Exposure, *Fire Technology* **37**, 129-141 (2001).
32. Johnson, W., and Rich, L. V., Arc Fault Analysis: Post Flashover Studies of the Power Cord, *2007 Fire and Material Conf.,* Interscience Communications Ltd, London (2007).
33. Johnson, W. B., and Rich, L. V., Arc Mapping Follow Up, *Fire Findings* **15**:3, 7-11 (Summer 2007).
34. Hoffmann, J. M., Hoffmann, D. J., Kroll, E. C., and Kroll, M. J., Full Scale Burn Tests of Television Sets and Electronic Appliances, *Fire Technology* **39**, 207-224 (2003).
35. Hazelwood, T. A., The Effectiveness of Arc Mapping to Determine Fire Origin Area, *Claims Magazine* (July 2012).
36. Stahl, D. R., and Parrott, K. D., Applicability and Limitations of Arc Mapping in Vehicle and Equipment Fire Investigation, pp. 617-623 in *ISFI 2012—Proc. 5th Intl. Symp. on Fire Investigation Science & Technology*, NAFI, Sarasota FL (2012).
37. Shanley, J. H., Intra-Appliance (Electric Clothes Dryer) Arc Mapping to Assist Fire Origin and Cause Determinations, pp. 83-89 in *ISFI 2008—Proc. 3rd Intl. Symp. on Fire Investigation Science & Technology*, NAFI, Sarasota FL (2008).
38. Kovarsky, R., Arc Mapping and Appliance Investigations, pp. 293-302 in *Proc. Intl. Symp. on Fire Investigation Science & Technology – ISFI 2016*, Natl. Assn. of Fire Investigators, Sarasota FL (2016).
39. Hoffmann, D. J., Swonder, E. M., and Burr, M. T., Arc Faulting in Household Appliances Subjected to a Fire Test, *Fire Technology* (2016).
40. Hoffmann, D. J., Beyond Tea Leaves—Using Arc Mapping to Pinpoint Origin and Cause of a Fire, *Fire Science and Litigation Seminar*, Defense Research Institute (2013).
41. Buc, E. C., Influence of Overcurrent and Heat on Tin Coated Copper Fuse Elements, pp. 311-324 in *Fire & Materials 2005*, Interscience Communications Ltd., London (2005).
42. Twibell, J. D., and Christie, C. C., The Forensic Examination of Fuses, *Science and Justice* **35**, 141-149 (1995).
43. Kovarsky, R. P., The Use and Misuse of Engineers in Fire Investigation, pp. 371-375 in *ISFI 2012—Proc. 5th Intl. Symp. on Fire Investigation Science & Technology*, Natl. Assn. of Fire Investigators, Sarasota FL (2012).
44. Sanderson, J. L., Arc Mapping May Provide Valuable Clues of Fire's Origin, *Fire Findings* **7**:2, 12-13 (Spring 1999).

45. Carey, N., Powerful Techniques—Arc Fault Mapping, *Fire Engineers J./Fire Prevention* 47-50 (Mar. 2002).
46. Land, H. B. III, The Behavior of Arcing Faults in Low-Voltage Switchboards, *IEEE Trans. on Industry Applications* **44**, 437-444 (2008).
47. Babrauskas, V., **Ignition Handbook**, Fire Science Publishers/Society of Fire Protection Engineers, Issaquah WA (2003).
48. Beavis, E. A., Cable Short Circuit Currents, *Electrical Review* **150**:3874, 385-388 (22 Feb. 1952).
49. Utt, D. P., Bringing Science into the Courtroom, *2009 IEEE Symp. on Product Compliance Engineering*, IEEE (2009).
50. Novak, C. J., An Analysis of Heat Flux-Induced Arc Formation in Residential Electrical Cables (M.S. thesis), Univ. Maryland, College Park (2011).
51. DeHaan, J. D., and Icove, D. J., **Kirk's Fire Investigation**, 7th ed., Brady/Prentice-Hall, Upper Saddle River NJ (2012).
52. Babrauskas, V., Charring Rate of Wood as a Tool for Fire Investigations, *Fire Safety J.* **40**, 528-554 (2005).
53. Goodson, M., Perryman, T., and Colwell, K., Effects of Polyurethane Foam Systems on Wiring Ampacity, *Fire & Arson Investigator* **52**:4, 47-50 (July 2002).

BEWARE THE HIDDEN WOOD PARTICLE BOARD

H. Breulet*, S. Brohez**
(*) ISSeP- Accidental Risks, Liège, Belgium, (**) University of Mons, Belgium

INTRODUCTION

When available, statistics indicate that fires originating in restaurants usually account for the most reported fires for assembly buildings. Not surprisingly, cooking often ranks first amongst the listed fire causes[1 2]. Nonetheless, for such buildings, insurance companies might be prone to search for possible "red flag" (e.g. unsuccessful business) and consider possible arson.

For the restaurant fire described in this paper, we obviously did not rule out any possible cause at first. The refurbishment and transformation of the building in a grill restaurant had just been completed and its owners had planned to open it for year-end holidays, potentially the most prosperous period for restaurant and catering business. Hence, since the fire occurred not long before Christmas, post fire cleaning and restoration work were initiated soon. As a consequence, the fire scene was not totally preserved until we had the opportunity to examine it.

Although the restaurant was not yet open, some kind of "rehearsal" had been organized, with the grill working for hours. The fire was detected during the following day: occupants of the flat on the 1st floor complained about some slight "burn odours" in the morning; later in the afternoon some smoke release was detected outside. The extent of damages was medium, as one might expect considering the time required for the fire brigade to reach the place (around 15 minutes).

Since the examination of the fire scene highlighted a number of non-compliances with standard practice regarding wires and cables mounting (although the electrical system was brand new and had just been inspected by an approved body) several fire investigators favored the hypothesis of an electrical fire cause. However, the thorough search amongst the fire debris enabled to collect other significant pieces of physical evidence, a.o. some fire patterns / partial charring of inside layer of a complex inner wall, which could have been part of the grill enclosure.

Hence at least 2 different hypotheses could be developed and required further investigation: electrical fault inside a partition and heat transfer from the grill fire to inner combustible layer(s) of a wall.

BACKGROUND / ENVIRONNEMENT

The building is a typical rubblestone with 3 levels; the ground level had been recently totally transformed to be used as a grill restaurant. This detached house is located in a rural area.
The restaurant includes the guests area (about 100 m^2), a bar (9 m^2) and a kitchen (32 m^2) with an embedded grill (see Figure 1).

The fire broke out on the eve of the official opening of the restaurant. The previous evening, the grill had been working for several hours (we had to rely on the single testimony of the owners). According to the fire brigade, the fire was mainly located in the eastern part of the bar. In order to prevent re-kindling in the fire, partitions, ceiling and part of the grill wall were dismantled and removed items thrown outside (in the snow).

The possible area of fire origin (see Figure 1) stood between the bar and the grill and included a technical enclosure and a technical conduit. On the northern wall of the restaurant, initially built with stones, a partition (wood particle and /or gypsum boards on metal studs) had been added in order to permit to install utilities (e.g. electric cables, water pipes). The ceiling (one layer of gypsum board on metal studs) acted somewhat as a fire barrier and the fire penetration through the ceiling and floor to the 1st level was pretty much avoided: the fire spread upstairs was restricted to @ ½ m^2 on the northern wall.

The fire developed around 12 hours (actually the estimate ranges from 10 to 16 h) after the grill activity stopped.

Photo 1 : technical enclosure

Photo 2 : grill area

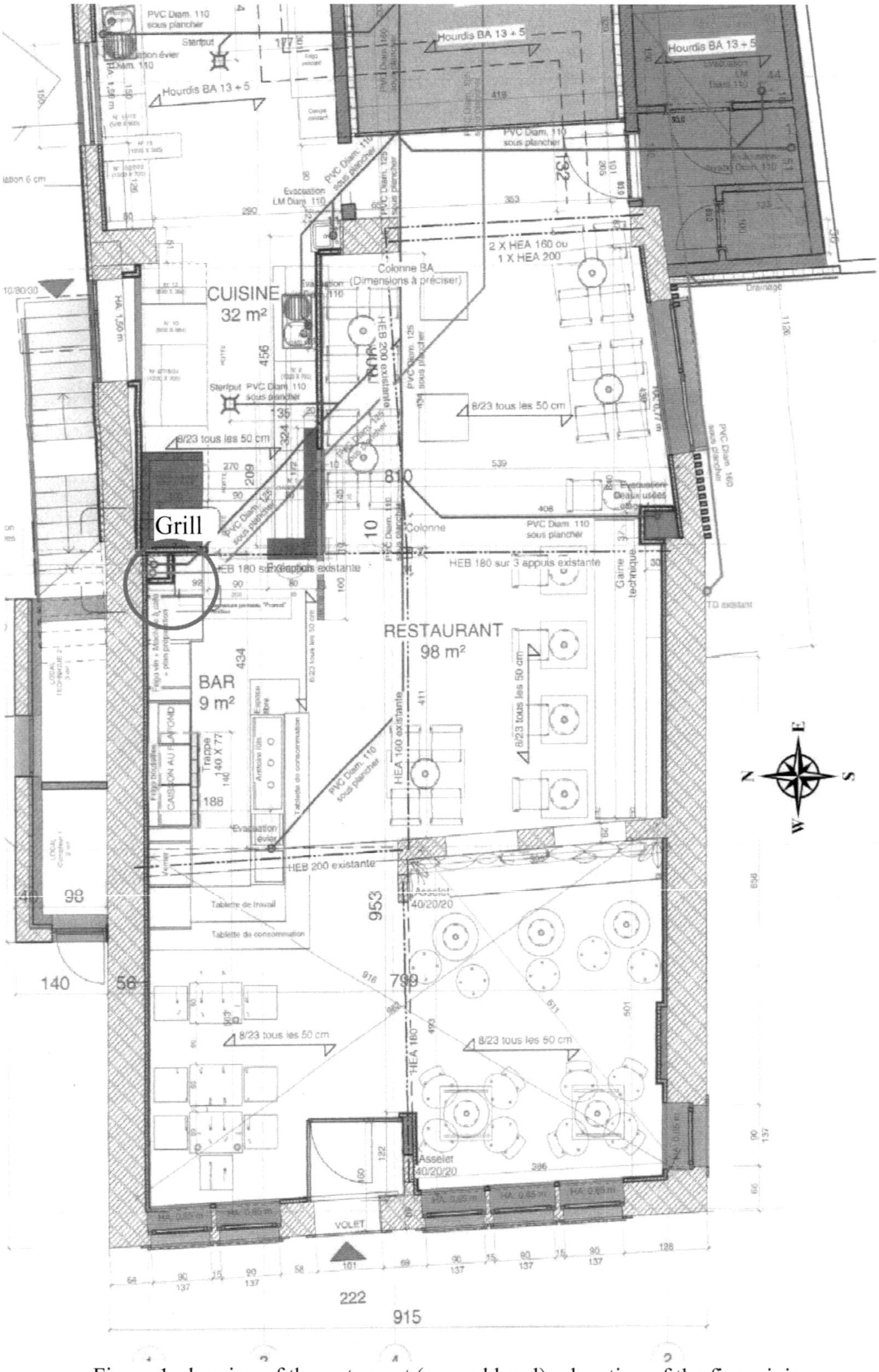

Figure 1: drawing of the restaurant (ground level) – location of the fire origin

BUILDING ELECTRICAL SYSTEM – ELCTRIC FAULT HYPOTHESIS

The building electrical system was a common 3-phase 240/400 V connected to the network via (from upstream): overhead service drop, wheatherhead, service raceway. The restaurant electrical system included a general circuit breaker (63 A), a general ground-fault circuit interrupter (GFCI) with a medium sensitivity (300 mA) and 8 branch circuits (4 3-phases and 4 single-phase).

The electrical system had been inspected and found compliant to national regulations[3] by an approved body[a] about 2 months before the fire. Yet passing cable and wires through metal studs orifices is against reasonable practice since such openings usually have sharp edges likely to damage cable jacket and/or insulation.

Photo 3: example of cables passing through metal studs

Most installed cables were common power cables: FR PVC jacket / filler / XLPE insulation.

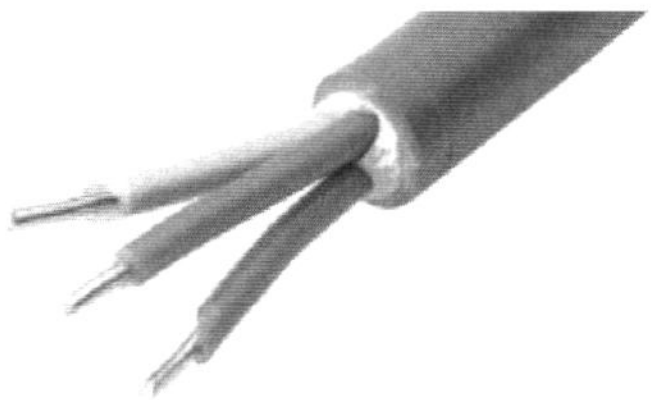

Figure 2: typical power cable installed in the restaurant

All cables present in the area of fire origin or in its vicinity were collected and preserved. Each cable zone passing through an opening in the metal studs was examined using a microscope. Although such examination put into evidence a few gouges (mechanically caused), no single damage caused by electrical energy (e.g. arc damage, short-circuit notch, bead / bead welding,…) was highlighted [4 5].

One must remain careful that having not found damage due to an electrical fault does not suffice to definitely rule out electricity as a possible fire cause: not even a "secondary" electrical damage was found, which could be explained either by the sensitivity of GFCI or by possible loss of electrical parts due to firefighting and/or overhaul.

[a] As it is mandatory for any building electrical system before being used.

GRILL – GRILL WALLS

Next to the kitchen a grill enabled the restaurant to broil pieces of meat. It consisted in 2 walls in "L" , the cooking area and an additional fire area ("basket") used to continuously produce embers to feed to cooking area (see Photo 2 and Figure 3).

From the western wall of the grill only the first layer (grill side) remained. Hence we assumed that its construction was identical to the northern wall. For the latter, we determined its construction by means of a core drilling. This brought to light a complex multi-layer (8 layers!) partition (see Figure 3).

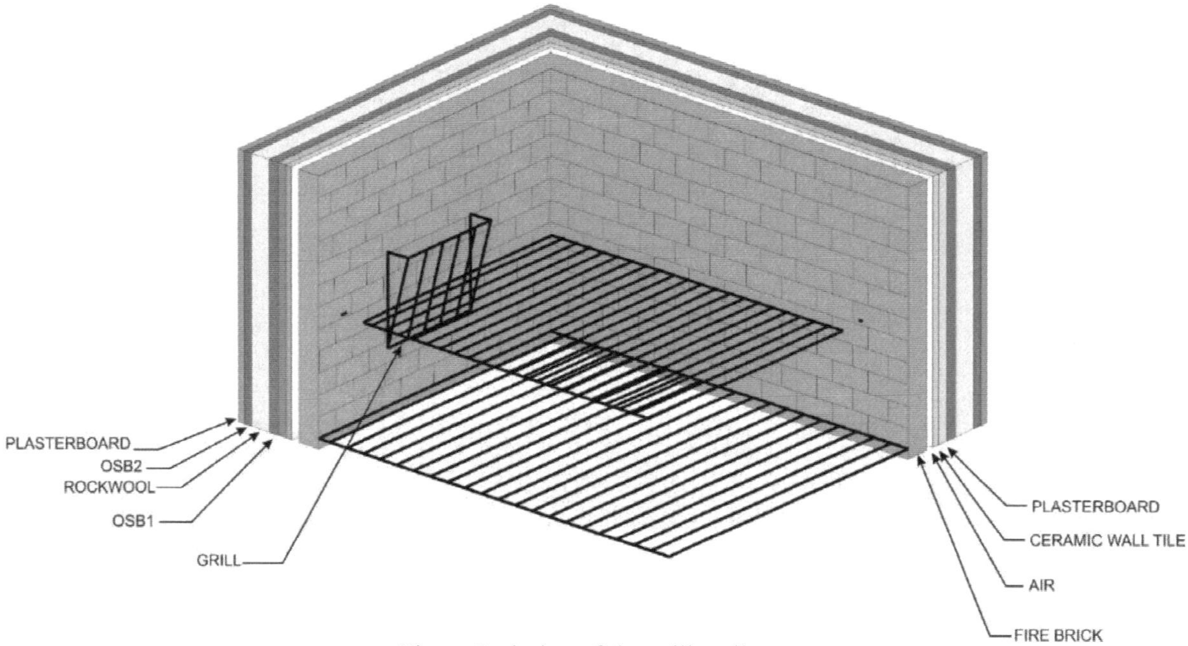

Figure 3: design of the grill walls

In order to be able to apply the equations of the Fourier's law of heat conduction, main thermal characteristics ($\lambda\rho c$ – thermal inertia) of each material were obtained either by measurement or from existing database.

Table 1: layers and their thermal characteristics

Layer nr	Material	Thickness (mm)	λ (W/m,K)	ρ (kg/m³)	c (kJ/kg, K)
1	Fire brick	100	0.96	2200	0.96
2	Air gap	15	0.1	1.2	1.005
3	Ceramic wall tile	8	0.25	1600	0.90
4	Gypsum board	12.5	0.48	1440	0.84
5	OSB 1	18	0.13	640	0.36
6	Mineral wool	60 (1)	0.034	70	2.15
7	OSB 2	18	0.13	650	1.3
8	Gypsum board	12.5	0.17	800	1.09

(1) For our experiment, rock wool 50 mm thick was used.

As aforementioned, the fire in the grill had been burning the day before the fire, for a duration between 5 and 10 h. According the owner's testimony, it worked only with wood logs, although the use of charcoal could not be ruled out (a couple of charcoal bags had been found in the premises).

The first combustible material possibly reached by the heat transfer is the OSB board, aka flakeboard (OSB1 - see Figure 3), i.e. strands of wood with a few % of resin (mix of PF, MUF and PMDI). In Europe, commercial grades of OSB board benefit of the so-called CWFT (Classification Without Further Testing) for CPR (European Construction Products Regulation) with a reaction-to-fire Euroclass ranking of D-s2,d0 [b] [6]. This corresponds to a medium – low level of fire performance [c]. Hence common OSB does not need FR treatment.

HEAT TRANSFER

In order to assess whether critical temperature (in terms of risk of ignition) could had been reached in the 1st board of OSB (OSB1- Figure 3), a two-folded study was undertaken:

1. "Physical simulation: a simplified mock-up of the grill wall was built and temperatures were monitored at different depths within the wall with a fire burning in the grill for a couple of hours;
2. "Mathematical" simulation using the Fourier's law of heat conduction to calculate the heat transfer through the consecutive layers.

1. Physical modeling

A multi layers single wall was built using the same materials as the ones in the actual grill.

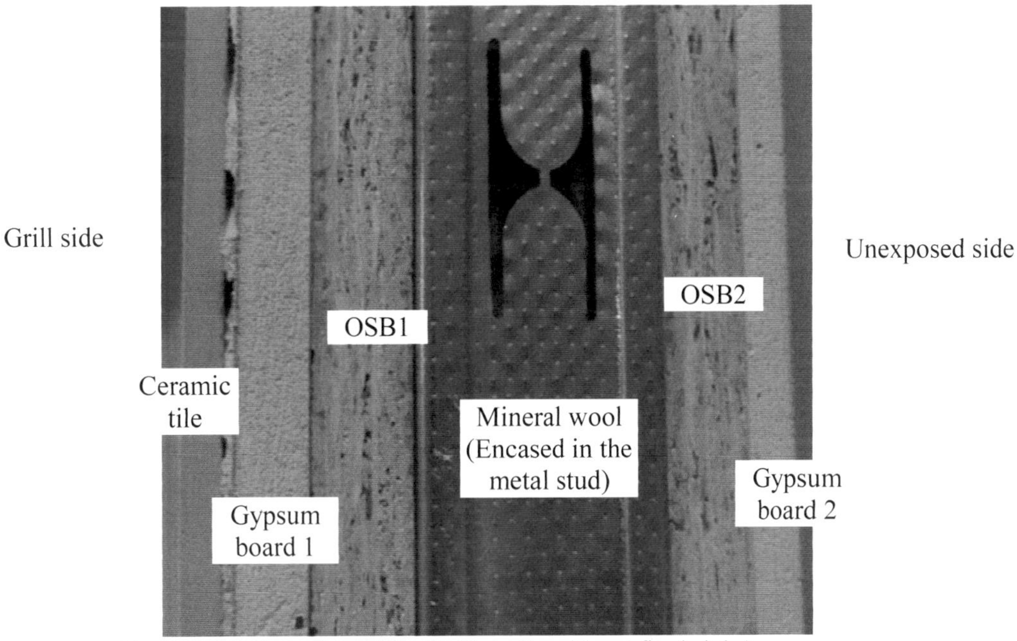

Photo 4: cross section of the built wall (without fire bricks)

[b] Reaction-to fire valid for OSB according to EN 300, with $\rho \geq 600$ kg/m³ and thickness ≥ 9 mm
[c] Euroclass D is supposed to correspond to products with a time to flashover between 2 and 10 min when tested in the Room Corner test (reference test for Euroclasses).

A total of 9 surface temperature probes were embedded in the mock-up, as follows

- T1, T2 and T3 on the unexposed side of the fire bricks;
- T4, T5 and T6 on the close side of OSB1 (i.e. between gypsum board 1 and OSB1);
- T7 & T8 on the close side of OSB2 (i.e. between mineral wool and OSB2). (See Figure 4).

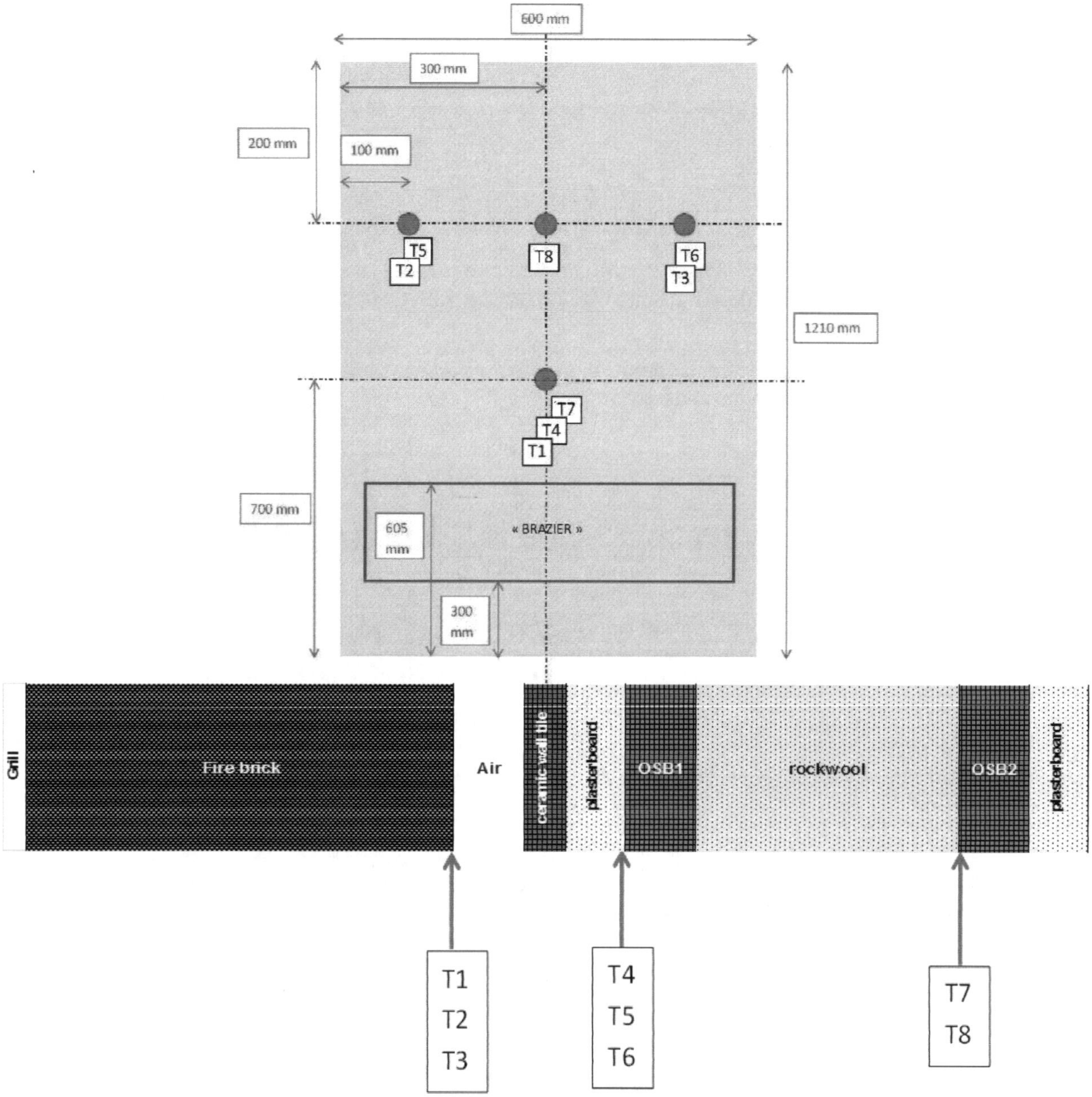

Figure 4 : localization of surface temperature probes

The grill fire was fed with wood logs (mix of birch and oak). The total load of wood was @ 35 kg for 6 hours (hence a fire load ≈ 530 MJ and an average power ≈ 24 kW). The test was stopped once a "thermal runaway" was measured within OSB1.

No surface temperature probe could be properly maintained on the exposed face of the fire bricks. In order to have some estimate of the temperature reached on that face, pyrometers measurements and /or IR camera were targeted the close vicinity of the fire zone.

The temperature measurements are summarized in Table 2.

Photo 5: grill mock-up with fire

Table 2: temperature measurements

	T1	T2	T3	T4	T6	T7	T8
T° max (°C)	332	128	141	331	104	106	72
Time (h)	6h 24 min	7h 54 min	6h 24 min	7h 54 min	7h 54 min	7h 54 min	8h 12 min

Temperature on the exposed face of the fire bricks – measured nearby but outside the fire (flames) zone: @ 500 °C.

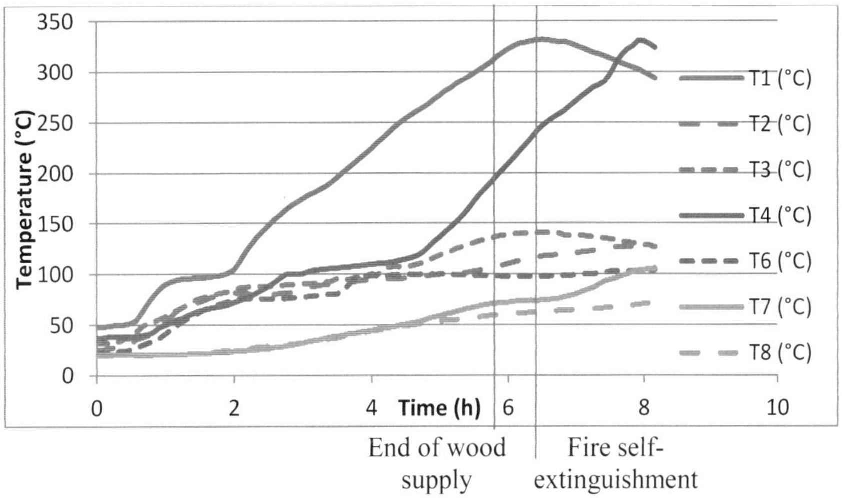

Figure 5 : recorded temperatures

These results are worth a few comments:

- The fire stopped while the temperature of the unexposed side of the fire bricks was still increasing, suggesting that stationary regime had not been fully reached yet;
- The slope of the temperature rise in OSB1 (T4) shows a sudden increase after about 4.5 h. At the same time, the temperature gap at different location in OSB1 (T4 Vs T6) starts to grow significantly;
- The temperature rise in OSB1 (at T4 location) remains sustained long after the fire stops;
- The 2 last points suggest that the temperature increase in OSB1 (at T4 location) cannot be explained solely by the heat transfer (conduction) through the successive layers;
- During our test, no flaming fire was observed within OSB1 (but experiment was conducted with rock wool instead of glass wool).

2. Mathematical modeling

According to the Fourier's law of heat conduction and to the conservation of energy to a differential control volume, an estimate of the temperatures reached in the deeper layers of the wall (including OSB) was obtained (Fourier-Kirshhoff second-order differential equation).

Without any internal heat source, and considering an one-dimensional unsteady-state heat conduction in solid, the conduction equation can be written as follows:

$$\rho c_p \frac{\delta T(t,x)}{\delta t} = \lambda \frac{\delta^2 T(t,x)}{\delta x^2}$$

Where:

ρ = Density ($kg.m^{-3}$)
c_p = Specific heat capacity ($J.kg^{-1}.K^{-1}$)
λ = Thermal conductivity ($W.m^{-1}.K^{-1}$)

The Fourier-Kirshhoff equation was writen for a multi-layer wall, taking into account temperature and heat flow continuity for interface conditions between layers. A finite difference method was used for discretizing the partial differential equation in each layer. A central difference scheme for the computation of temperature at the nodes at time $t+\Delta t$ (Crank-Nicholson method). For the air gap, the radiation heat transfer between two surfaces was taken into account and linearized.

A summary of the obtained results is given in Figure 6, Figure 7 and Figure 8.

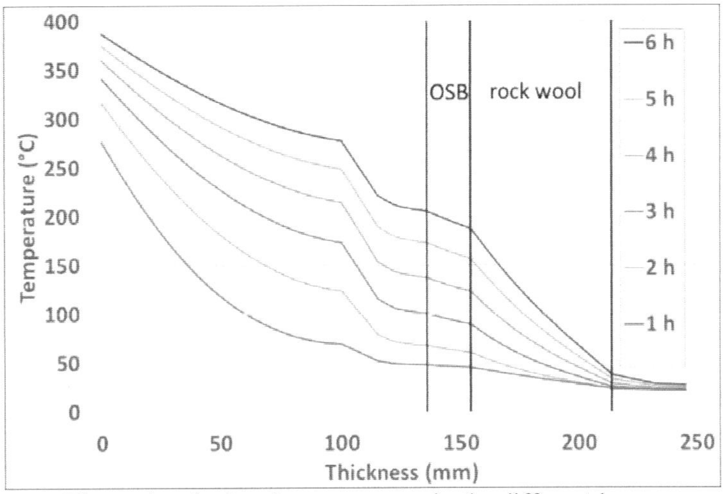

Figure 6: calculated temperatures in the different layers

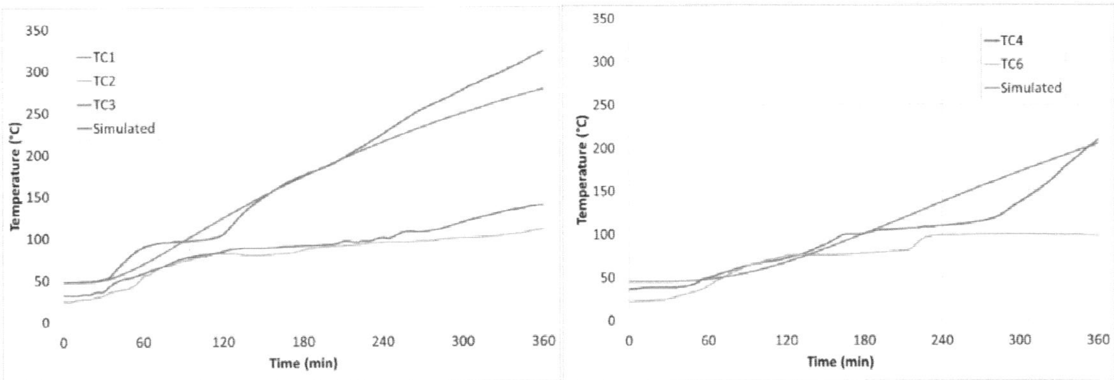

Figure 7: comparison calculated Vs measured (fire bricks – OSB1)

The numerical simulation shows good agreement with the experiment: results are very close for the fire brick layer (unexposed side) and for the 1st particleboard (OSB1). For the latter, beyond 3 h, numerical simulation tends to somewhat overestimate the temperature. This is certainly due to the moisture content in the OSB which leads to a nearly constant temperature of about 100-110°C during approximately 2 hours. This endothermic evaporation was not taken into account in the mathematical simulation. This phenomenon can also be observed for thermocouple TC6 giving a constant 100°C value during a very long time period (Figure 5). At a later stage, measured temperature catches up the calculated one (and even exceeds it) but exothermic decomposition in particleboard is then involved.

Additional simulations were performed in switching the position of OSB1 and the mineral wool (hence with the mineral wool protecting the OSB against the heat transfer, in addition to the gypsum board). As expected, the temperature reached in OSB (OSB1) dramatically drops: from more than 200 °C after 6 h to less than 40 °C (Figure 8). Although this is outside the scope of the fire cause determination, it shows how easily the risk could have been prevented.

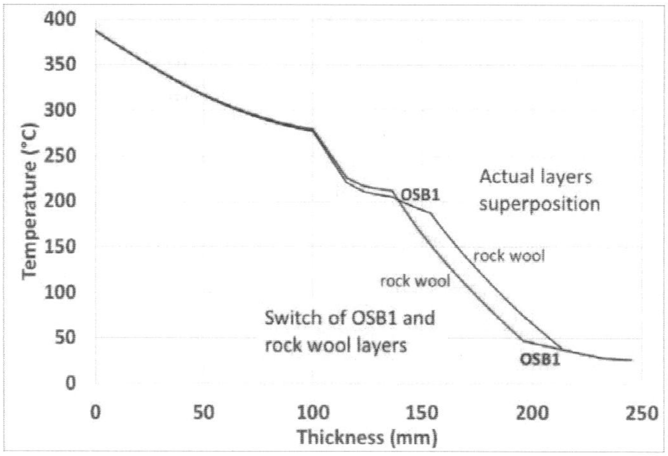

Figure 8: comparison of the temperature reached after 6 h (actual layers superposition Vs superposition with switch of OSB1 and rock wool)

COMMENTS & CONCLUSIONS

Although several unappropriated pathways of cables (through openings in metal studs) were pointed out, a close analysis of all cables and wires located in these critical areas did not highlight any trace of electrical fault.

During the examination of the fire scene, we discovered that the wall of the grill consisted of a complex multi-layer construction, including two wood particleboards. Although the first board was "protected" by several layers (fire brick, air gap, gypsum plasterboard), the question was raised whether a "long-term" heating could eventually lead to critical temperatures in the OSB.

During our experiment carried out using a mock-up of the restaurant grill, temperature higher that 300 °C was reached after a couple of hours. The numerical model has demonstrated reasonable agreement with the experiment but is obviously restricted to the stage without any combustion process.

The propensity of wood particleboards to initiate combustion and possibly ignition had been found similar to solid wood, although much less studies have been dedicated to them [7 8 9].

For the investigated fire, according to testimonies, we should exclude the process of low-temperature long-term ignition of cellulose-based materials[10] (or a similar process due to repeated heating) since the grill had been used once or twice[d]. However the temperatures reached in the first OSB (physical & numerical simulations) are beyond the minimum threshold to initiate a combustion process in a particleboard.

Due to the superposition of layers and the nature of materials constituting them, a 2-step ignition process seems plausible[11]:

1. Under restricted oxygen conditions (due to OSB trapped between the plasterboard and the mineral wool), char is formed locally and a smoldering fire grows. Temperature combined to the thermally insulated environment enables the particleboard to undergo sustained smoldering;

[d] Again this mainly relies on the owner's testimony. Since the building had been recently totally refurbished, it is not possible that the grill had been used a large number of times but a few uses cannot be ruled out.

2. At a later stage (after several hours), transition from smoldering to flaming fire takes place, thanks to change(s) in oxygen availability: increase of the specific area in the charred wood and/or modification of ventilation inside the complex.

During our experiment, we did not reach this transition either because all required conditions (environment, ventilation) were not met or because the test was stopped too early[e]. Nonetheless, the time lap between the moment grill fire stopped and the flaming fire spread, i.e. in the range 10 to 16 h, is consistent with a possible transition from smoldering to flaming fire.

Hence, even if this is not sufficient to build an indisputable evidence of the cause of the fire, it was demonstrated that the hypothesis of the fire initiated in the grill wall would deserve close consideration.

REFERENCES

[1] Evarts, B. Structure Fires In Eating And Drinking Establishments. NFPA, November 2012.
[2] Restaurant Building Fires. Topical Fires Report Series. Volume 12, Isseu1, April 2011.
[3] RGIE – Règlement général des Installations Electriques – AREI - Algemeen Reglement op de Elektrische Installaties. Belgian Royal Decree of March 10, 1981.
[4] NFPA 921 (2014 edition). Guide for Fire and Explosions Investigations. Chapter 9: Electricity and.
[5] Ettling, B. V., "Arc Marks and Gouges in Wires and Heating at Gouges,"*Fire Technology*, Vol. 17, No. 2 (May 1981), pp. 61–68
[6] Commission decision 2003/43/EC establishing the classes of reaction-to-fire performance for certain construction products. OJEC L13/35, January 18, 2003.
[7] Babrauskas, V. Ignition of wood. A review of the State of the Art. Interflam 2001 Proceedings.
[8] Foguenne O. Etude de la réaction au feu de panneaux dérivés du bois. Essais au cône calorimétrique et au SBI. ECAM Thesis, 2002.
[9] Byoung-Ho I. & Al. Evaluating the Flammability of Wood-based panels and gypsum particleboard using a cone calorimeter. Construction and Building Materials 25(7):3044-3050 · July 2011
[10] Babrauskas, V. 'Pyrophoric Carbon' and Long-term, Low-temperature Ignition of Wood. Ignition Handbook, 2003.
[11] Ohlemiller T.J. Smoldering Combustion. Section 2, Chapter 9, SFPE Handbook of Fire Protection Engineering. 3rd Edition, 2002.

[e] For safety reason, since we could not ensure a 24 h supervision, the test had to be stopped after about 12 h.

SIGNATURE RECOGNITION OF ENERGETIC MATERIALS

June Bott, Kevin Lewis, Sebastian Scheiff
CASE Forensics, a Jensen Hughes Company

Jason McCary
Novinium, Inc.

ABSTRACT

The goal of this paper is to provide information that fire investigators can consider at the outset of every investigation both in the interest of safety and to better understand some of the potential fire causes. Fire investigators are trained to recognize and are regularly exposed to fire scenes full of hazardous flammable materials, but unless they have been warned ahead of time, a clandestine drug or explosive laboratory is just that – clandestine. One of our goals is to be able to distinguish some of the traits of explosives production from a drug production lab or even a well-stocked backyard shed. The characteristics of energetic materials are discussed and some of the chemicals and equipment used in their production are described.

BACKGROUND

An explosion is a sudden, violent, uncontrolled release of chemical energy. Energetic materials require two things: an oxidizer and a fuel. The oxidizer and fuel react chemically, and that chemical reaction releases a large amount of energy in the forms of heat, light, sound, and force. Chemically speaking, the fuel is a molecule with a large amount of energy stored in the form of bonds between atoms. The most common fuels are hydrocarbons, which are based on bonds between carbon and hydrogen atoms. The oxidizing agent facilitates the breaking of these bonds. Common oxidizers include ammonium nitrate, hydrogen peroxide, and chlorine-containing chemicals. NFPA 921 is a good resource for a more in-depth understanding the fire science behind explosions.

There are two types of energetic materials. In the first type, the oxidizer and the fuel are two separate chemical compounds that are mixed to form the energetic material. The best-known example of this type of energetic material is a mixture of ammonium nitrate (Figure **1**) and fuel oil, often called ANFO. ANFO will be discussed further in later sections.

The second type contains both the oxidizing agent and fuel within the same molecule. TNT (trinitrotoluene), as seen in Figure **2**, and dynamite are common examples of self-contained energetic chemicals.

What should a fire investigator be looking out for?

In the following section, we will be naming and describing several chemicals that may be used in the production of energetic materials. It is important to note that most of these chemicals are commonplace in the household, in agriculture, and in various commercial industries. The mere presence of these chemicals does not necessarily mean that explosives are being produced in a facility. However, a fire investigator should combine an awareness of these chemicals with other clues in order to judge whether to take extra precautions with a fire scene, or if the situation seems immediately dangerous, to vacate the area and contact the proper authorities.

OXIDIZING AGENTS

Ammonium Nitrate

Ammonium nitrate (AN) is a common and easily sourced oxidizing agent. It is an important agricultural fertilizer. As fertilizer, is typically produced in small, white or off white spherical pellets called prill (Figure **3**). Ammonium nitrate is stable at room temperature and will not readily burn on its own, but can detonate at elevated temperatures and in the presence of certain catalysts and/or combustible materials.[1] Large quantities of AN are used in the agricultural industry as fertilizer and in the mining industry as an energetic material precursor. Explosives-grade AN is made in less dense and more porous prill than agricultural grade. This allows it to combine better with combustible materials, such as fuel oil, to form explosives.

AN has been involved in several deadly explosions, including an explosion at a fertilizer plant in the city of West, Texas in 2013. The explosion at the West Fertilizer Company started as a fire, which the Bureau of Alcohol, Tobacco, Firearms and Explosives ruled was intentionally set.[2] An unknown and uncontrollable series of events caused the AN, which was stored in bins made of combustible materials, to explode.[3]

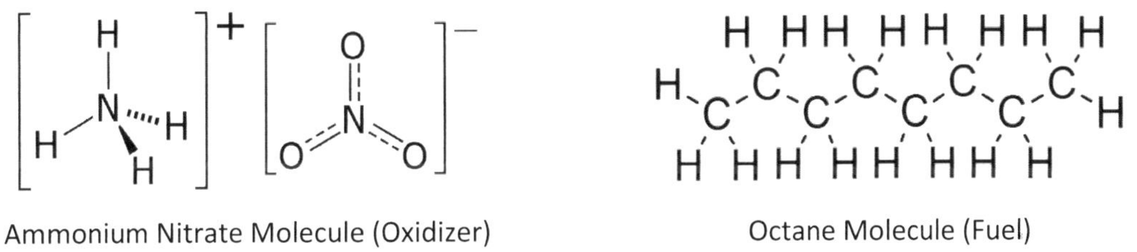

Figure 1: Examples of oxidizer and fuel molecules, such as those that might be found in ANFO explosives (retrieved from https://en.wikipedia.org/wiki/Ammonium_nitrate#/media/File:Ammonium-nitrate-2D.svg and https://en.wikipedia.org/wiki/Octane#/media/File:OctaneFull.png).

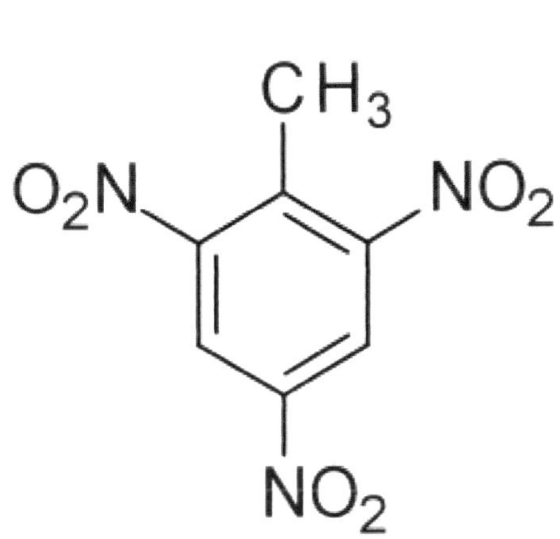

Figure 2: Trinitrotoluene (TNT) molecule (retrieved from https://en.wikipedia.org/wiki/Trinitrotoluene#/media/File:Trinitrotoluene.svg).

Figure 3: Examples of ammonium nitrate prill.

Large stores of AN should be treated with great caution, even if pure AN is non-combustible and not sensitive to shock. The most readily available source of AN is large bags of fertilizer-grade AN. Instant cold packs, such as those that can be purchased at your local drug store, can also be sources of AN. The AN may also be stored loose or in unmarked containers in the form of prill or powder. Ammonium nitrate that has been ground into powder or mixed with combustible fuels is much more dangerous and shock-sensitive than standard high-density fertilizer-grade prill.

Figure 4: Concentrated nitric acid that has been contaminated with nitrogen dioxide, giving it a yellow tint. (Photograph by W. Oelen used under the Creative Commons SA 3.0 license, retrieved from https://en.wikipedia.org/wiki/Nitric_acid#/media/File:Fuming_nitric_acid_40ml.jpg).

Chlorates

Chlorates are chlorine-based oxidizers. They can easily be made from bleach and other pool cleaning materials. Large stockpiles of bleach should be treated with caution. Potassium perchlorate and potassium chlorate are both easily sourced or produced chlorate oxidizers. Chlorine can also be obtained from ordinary salt and from sodium-free salt substitutes such as Lite Salt.

Liquid Oxidizers

Concentrated hydrogen peroxide is another potent oxidizer. It will decompose when exposed to light, so hydrogen peroxide is typically stored in dark-colored opaque containers. Hydrogen peroxide that can be purchased for first aid use at drug stores is generally only a 3% solution. The hydrogen peroxide that is available industrially (such as in the hair care industry) is concentrated at about 30%, which is concentrated enough to use as a precursor chemical. Somewhat simple chemical processes can further concentrate hydrogen peroxide.

Nitric acid is also a strong oxidizer. Nitric acid is a clear to yellow-brown liquid with a viscosity similar to water (Figure 4).

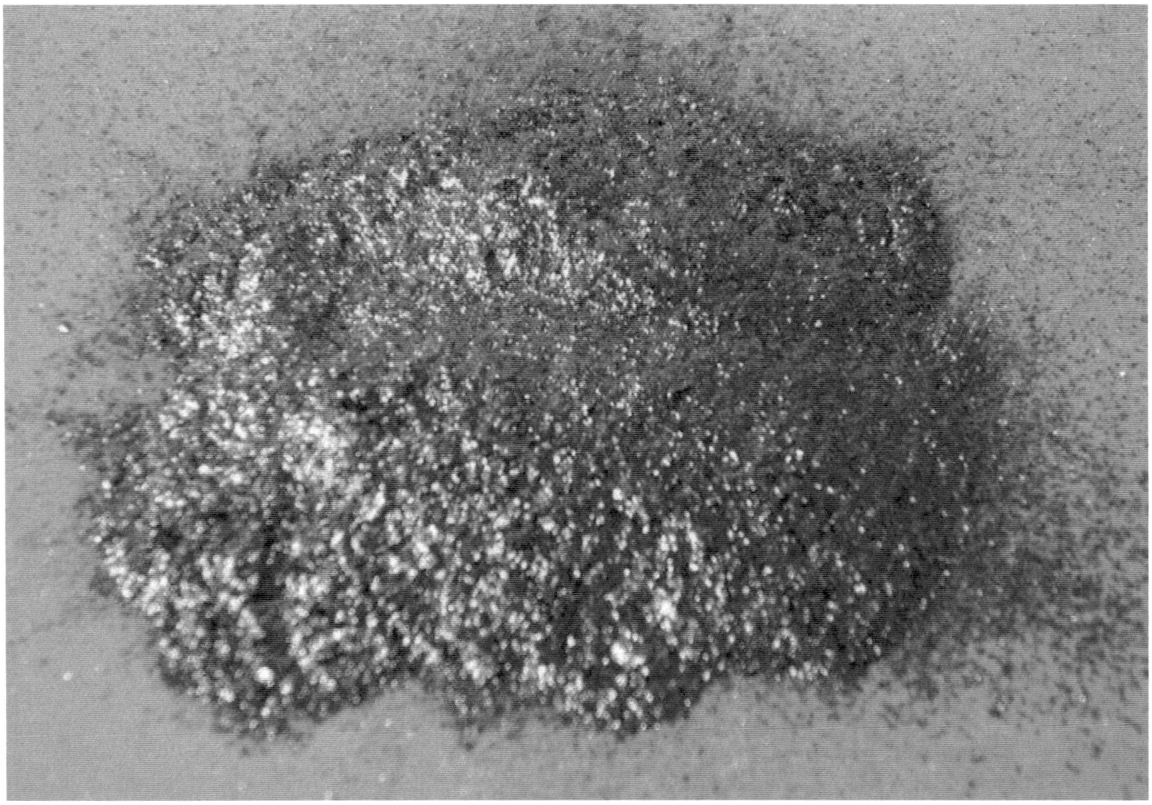

Figure 5: Finely ground aluminum powder (retrieved from https://commons.wikimedia.org/wiki/File:Aluminium_pigment_powder.JPG)

FUELS

Liquid Fuels

Liquid fuels are commonly used in explosives production. For example, ANFO is produced by mixing ammonium nitrate in the form of a powder or low-density prill with fuel oil. Hydrocarbons are the most potent liquid fuels. Examples include diesel fuel, mineral oil, gasoline, heating fuel oil, kerosene, and petroleum jelly.

Solid Fuels

Solid fuels can be just as effective as liquid fuels. Examples of solid fuels include powdered metals (such as aluminum, Figure 5), sugars, and powdered cellulose (finely ground wood). Powdered metals can easily be purchased from fireworks manufacturing suppliers and outdoor sporting retailers. Tannerite®, an energetic material designed for firearms target practice, is a mixture of ammonium nitrate and/or ammonium perchlorate and finely powdered metals, such as aluminum. ATF regulations do not apply to the two components of Tannerite® until they are mixed together to form the binary explosive. Sugars may be found in the form of powdered drink mixes, such as Tang or Kool-Aid.

Hexamine is a solid fuel with a long shelf life that is sold as camping fuel pellets. It is both very flammable on its own and is used as a precursor chemical in the production of other, more volatile/less stable explosive materials.

OTHER TOOLS AND EQUIPMENT

In addition to the chemicals needed to develop explosive materials, a fire scene may also contain a variety of chemical laboratory equipment needed to process the chemicals. Chemicals can be heated using hot plates (both the laboratory variety and, in more improvised labs, the variety used for cooking), heating mantles for round-bottom glassware, and oil or water baths. Chemicals are separated from one another using seperatory funnels, filters, and distillation apparatuses. Vacuum pumps may be used to pull mixtures through filter papers into specially designed flasks. Mortars and pestles may be used to process ammonium nitrate prill into powder on a small scale, while rock tumblers and ball mills can be used to powder materials on a larger scale.

Once the chemicals are produced and processed, they must be packaged and primed. Mixers, such as the large mixers found in commercial kitchens, may be used to combine solid components of explosive materials. Blasting caps, fuses, and electronic ignition systems may all be used to detonate the prepared energetic material. Many of these devices can be themselves be improvised from commercially available materials.

CONCLUSIONS

As with any other fire investigation, numerous clues must be put together and analyzed to make sense of a fire scene containing energetic materials and the equipment and chemicals used to prepare them. As stated previously, the presence of any of the individual chemicals or pieces of equipment at a fire scene is not necessarily cause for alarm. The most important combination to watch out for is the presence of large quantities of both an oxidizer and a fuel. The most important tool, by far, is the investigator's instinct. If any one clue sets off mental alarm bells, it is best to vacate the scene and call for help from the authorities and from people with professional training in handling energetic materials.

REFERENCES

1. Moore, J., Stanitski, C., and Jurs, P. (2008). *Chemistry: The Molecular Science, 3rd Ed.*, Thomson Brooks/Cole, p. 642.
2. Bureau of Alcohol, Tobacco, Firearms and Explosives (2016). "ATF Announces $50,000 Reward in West, Texas Fatality Fire," Retrieved from https://www.atf.gov/news/pr/atf-announces-50000-reward-west-texas-fatality-fire
3. U.S. Chemical Safety and Hazard Investigation Board (2016). "Investigation Report: West Fertilizer Company Fire and Explosion," Retrieved from http://www.csb.gov/west-fertilizer-explosion-and-fire-/

Fire Testing of Aerosols and Propane Bottles

Dan Madsen*, John Barton, Stefan Svensson, Patrick van Hees
Lund University, Division of Fire Safety Engineering, Lund, Sweden.
**Corresponding author email: dan.madsen@brand.lth.se*

ABSTRACT

Aerosol- and propane bottles for consumers, including hairspray, lubricants, etc. were tested for exposure from fire. The tested bottles had a volume between 15- 400 ml of product mixed with propane and bottles with propane only containing 0.35, 0.45, 5 and 6 kg of propane. The aerosol- and propane bottles were exposed to a heptane pool fire, corresponding to properties of common available flammable liquids for sale in stores. The experiments showed that the aerosol bottles exposed to fire exploded within a few minutes only and that this resulted in clouds of burning vapor as well as shrapnel affecting the vicinity. An experiment with several such aerosol bottles exposed to fire on a shelf in an enclosure, showed that unexploded aerosol bottles acted as projectiles when adjacent bottles exploded. Also, the explosion resulted in a high pressure, affecting building components. Experiments with propane bottles exposed to fire showed that such bottles release pressure in different ways, depending on their type and construction. During the experiments, the bottles released their pressure through a pressure valve or melting fuse within a few minutes only. During comparative experiments with propane bottles without safety devices, the bottles exploded within a few minutes only, resulting in large burning vapor clouds and shrapnel being thrown as far as approximately 80 m. Recommendations based on the experiments included suggestions on the maximum amount of combustible liquids which should be stored together with aerosol bottles. Recommendations were also given regarding the relevance of safety devices and securing propane bottles so that they can't move if safety devices activates when bottles are exposed to fire.

INTRODUCTION

Flammable gases such as propane or butane are commonly used as propellant in pressurized aerosol bottles. Such products include hairspray, lubricants, paint and deodorants. When exposed to fire, the pressure may exceed the design pressure of the bottle. Smaller bottles, often found in retail stores, do not have any requirements for safety devices for pressure relief, which is common for larger bottles of propane (so called P6, PC5 and PA6, depending on their construction). When exposed to heat, the pressure increases in the bottle provided that the bottle is not relieving the pressure by leaks or by the bottle's safety devices such as a pressure valve or melting fuse. If the bottle have such a safety device, it can release pressure in a controlled manner. For smaller bottles that are not equipped with safety devices, heating of the bottle may cause the bottle to burst resulting in burning vapor cloud or an explosion. Also, the explosion can generate projectiles, parts from the bottle, flying at great velocity as well as a pressure rise in a storage room.

Knowledge on how much heat from burning flammable liquids is required until an aerosol or propane bottle bursts/explodes or releases pressure by design is limited and is therefore studied in this work. The project was performed in several parts:

1. Aerosol and propane bottles for consumer use (hairspray, lubricants, camping fuel, etc.) exposed to fire.
2. Multiple aerosol bottles for consumer use on a shelf in an enclosure, exposed to fire (single test only).
3. Propane bottles for consumer use (5 – 6 kg) of different type (composite, aluminum, steel), with and without safety devices (pressure valve or melting fuse), exposed to fire.

The overall purpose of the tests was to investigate if there are aerosol bottles that can be excluded from regulations on storage, the importance of safety devices on propane bottles and to get data on the behaviour of aerosols and propane bottles exposed to fire.

EXPERIMENTAL SETUP

In the first experimental setup, aerosol bottles for consumer use, the bottles was placed on a steel structure approximately 50 mm above a poolfire with a diameter of 0.3 meter, containing 1 liter of heptane on a 1 cm water bed. Thermocouples type K where mounted on four sides of the steel structure and an additional thermocouple on the bottle, figure 1.

Figure 1: Single aerosol- and propane bottles on top of a poolfire, here with a bottle of 450 gram propane for camping use.

In the second experimental setup, 6 bottles of hairspray were put on a shelf in an enclosure, similar to that in a retail store, figure 2. The shelf was located in the centre of an enclosure measuring approximately 5×2.5×2.5 m^3 with a hatch measuring 2.5×0.7 m^3 in one wall. Also, the enclosure had a closed hatch covering a 2.4×2.4 m^2 area in the opposite wall, figure 3.

Figure 2: 6 bottles of hairspray on a shelf in an enclosure.

Figure 3: Experimental setup with a vessel of heptane for the pool fire, 6 bottles of hairspray and some smaller containers of nail-polish on the upper shelf.

The tested aerosol and propane bottles in the first and second experimental setup are shown in table 1. The experiments were performed outdoors. The weather conditions during the tests were between 9-11 °C in temperature, wind speed between 5 – 6 m/s, no rain. Consequently, the fire was influenced by wind so that the aerosol- and propane bottles were not fully engulfed in flames.

Table 1. Aerosol- and propane bottles exposed to fire.

Product	**Volume/weight**
Antifreeze lock spray	15 ml
Antifreeze lock spray PTFE	60 ml
Lighter fluid	50 ml
Bicycle tyre fix	50 ml
Silicon spray lubricant	60 ml
Deodorant	150 ml
Shoe deodorant	165 ml
Starter fluid	400 ml
Spray paint	400 ml
Hair spray	500 ml
Propane	300 gr
Propane	450 gr

In the third experimental setup, propane bottles (5 – 6 kg) were placed on a steel structure on top of a poolfire with a diameter of 0.56 m filled with approximately 18 liters of heptane, figure 4. The ignition source was operated remotely. The rate of heat release was approximately 420 kW, determined by measuring mass loss rate from the poolfire separately.

Figure 4: Experimental setup with 5.3 kg propane bottle (composite) on top of a poolfire (test 1, table 2).

A total of 8 tests were made with four types of propane bottles, 4 with and 4 without safety devices. The safety devices was either a pressure valve with an activation pressure of approximately 26 bar or a melting fuse with an activation temperature of about 115 – 130°C. Some of the bottles had both of these devices. See table 2 for details on the tested bottles.

The weather conditions during the experiments were temperatures between 9-11 °C, wind speed between 8 – 10 m/s, no rain.

Table 2. 8 propane bottles exposed to fire to evaluate the necessity of safety devices.

Test	**Material**	**Type/model**	**Volume of bottle [l]**	**Amount of propane [kg]**	**Pressure valve**	**Melting fuse**
1	Composite, one piece	PK5/Ragasco	12.2	5.3	Yes	No
2	Composite, two pieces joined	PK5/Composite Scandinavia	14.3	5.1	Yes	Yes
3	Aluminum	PA6	14,3	5.9	Yes	Yes
4	Steel	P6	14.3	5.9	Yes	No
5	Composite, one piece	PK5/Ragasco	12.2	5.8	No	No
6	Composite, two pieces joined	PK5/Composite Scandinavia	14.3	5	No	No
7	Aluminum	PA6	14.3	5.6	No	No
8	Steel	P6	14.3	5.8	No	No

RESULTS AND ANALYSIS

Aerosol and propane bottles

During the experiments with smaller aerosol and propane bottles, the measurements indicated that the bottles were exposed to a flame temperature of approximately 750 – 800°C.

A total of 33 bottles were tested, of which 30 exploded. One of these unexploding bottles contained lubricant. When the bottles was heated, the fluid started to leak at the top of the bottle through its connection to the nozzle and no flammable gas were exposed to the fire. The two other unexploded bottles contained lighter fluid. The construction of the bottle is such that the bottom is concave so that in case of fire (or otherwise heated), it should be pushed out. During the experiments this happened and the bottle was thrown away from the poolfire without bursting. Table 3 shows the results from the first experimental setup with smaller aerosol and propane bottles (as in table 1).

Table 3: Product, time to explosion and surface temperature for tests with smaller aerosol and propane bottles.

Product and volume/weight	Time to explosion [s]	Surface temperature on bottle at explosion [°C]
Antifreeze spray, 15 ml	70 73	227 184
Antifreeze spray PTFE, 60ml	51 64	165 124
Lighter fluid, 50ml	99	337
Bicycle tyre fix, 50 ml	41 56	139 113
Silicon spray lubricant, 60 ml	80 101	145 145
Deodorant, 150 ml	72 80 83	121 291 263
Shoe deodorant, 165 ml	71 85 109	150 457 98
Starter fluid, 400 ml	81 93 124	141 142 359
Spray paint 400 ml	91 98 111	183 306 321
Hairspray, 500 ml	140 219 156	260 - 190
Propane, 300 gr	69 69 74	125 128 -
Propane, 450 gr	81 85 100	167 141 136

The experiments showed no correlation neither between surface temperature and volume of the bottle, nor between surface temperature and time to explosion. However, time to explosion appeared to be longer for larger bottles which should be an expected results, figure 5. The mean value of time to explosion is 91 seconds with a standard deviation of 34 seconds.

Figure 5: Time to explosion versus volume of bottle

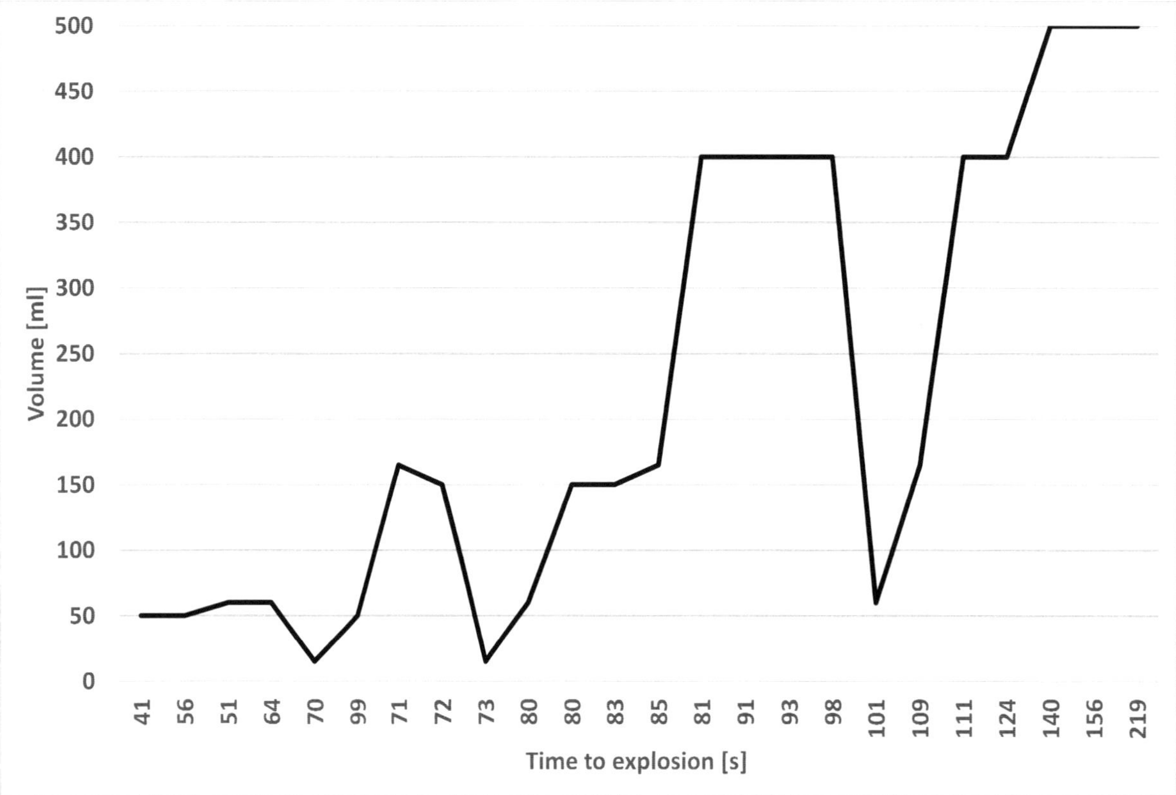

The experiments involving smaller aerosol and propane bottles, 15 – 400 ml, showed that when they burst they result in a subsequent combustion that can be referred to as an explosion. After each experiment, remaining parts and pieces of the bottles could be easily identified, although they in some cases where very small. Apart from the explosion itself, including high temperature, radiation and pressure, such parts and pieces can be expected to cause injuries within at least within a couple of meters from the bottle. The amount of flammable liquid that was burned off at the point of explosion was in the range of 0.07 – 0.36 liters of heptane, which are fairly small amounts of flammable liquids.

Multiple aerosol bottles in an enclosure

The second part of the tests, involving multiple aerosol bottles for consumer use on a shelf exposed to fire in an enclosure (6 400 ml bottles of hair spray), resulted in a in a powerful explosion. 1 – 3 bottles were involved in the explosion, causing the rear hatch of the enclosure to open, although anchored with a large-diameter wire around a steel bar.

The shelves were mounted on a gypsum board that was penetrated by an intact hair spray container which acted as a projectile when adjacent bottles exploded. When several aerosol bottles are stored together and exposed to a burning flammable liquid, it can be assumed that several bottles will participate in the explosion and that adjacent bottles will act as projectiles with sufficient force to penetrate a gypsum board.

Propane bottles with and without safety devices

The third experimental setup included propane bottles for consumer use (5 – 6 kg) of different type (composite, aluminum, steel), with and without safety devices (pressure valve or melting fuse), exposed to fire. Visual observations during and after each test suggests that the bottles were exposed to flames at about half the circumferential surface, as a result of the wind's impact. It has thus been primarily bottom and the leeward side of each bottle that were exposed to flames. In cases where the pressure valve/melting fuse triggered, the assumption can be made that the bottle was then subjected to further heat exposure as a result of the jetflame from the escaping gas. Measurements indicated that the bottles were exposed to a temperature in the order of 800 1000°C.

The measurement on the side of the bottle, on the windward side, suggesting that the bottle has been exposed to a relatively large cooling due to the strong wind.

Test 1

In test 1 a one-piece composite bottle with pressure valve was tested. At 3 min 30 sec after ignition the bottle appeared to be leaking from the valve/valve attachment. At 5 min 48 sec a flow of gas from the lower part of the bottle was observed, probably from a hole due to the construction, figure 6. This was noted primarily as an increase in rate of heat release. Also, this could be confirmed by the temperature measurements. The measurements suggested that gas flows out through the shell of the bottle as it weakens and burns up.

Figure 6: The bottle in test 1 after the test. A hole at the bottom of the bottle can be seen.

Test 2

In test 2 a composite bottle in two parts joined together, with a pressure safety valve and a melting fuse was tested. The melting fuse was activated 57 seconds after ignition, resulting in a jetflame. The flame persisted for about 3 minutes.

Test 3

In test 3 an aluminum bottle with pressure valve as well as a melting fuse was tested. The pressure valve opened intermittently from 1 min 27 sec after ignition approximately every two seconds. After 1 min 42 sec the pressure valve opened intermittently every second up until 2 minutes 2 seconds when the pressure valve open continuously for about 15 seconds after which the melting fuse activated at 2 min 17 sec. It took approximately 7 minutes for the bottle to be emptied of its contents. The release of gas resulted in a reacting force to the bottle, which was secured with chain and three ground rods. When the melting fuse triggered the forces from the escaping gas made the bottle to tip over and push away the poolfire approximately 1 m. The bottle ended up on the ground with the opening for the melting fuse downward, creating a hole in the ground.

Test 4

During test 4 a steel bottle with pressure valve was tested. The pressure valve opened intermittently at 1 min 35 sec after ignition. After this, the pressure valve opened according to table 4.

Table 4: opening times for the pressure valve in test 4.

Time	**Open**
2 min 11 sec	6 sec
2 min 23 sec	1 sec
2 min 35 sec	2 sec
2 min 50 sec	3 sec
3 min 06 sec	2 sec
3 min 21 sec	3 sec
3 min 36 sec	3 sec
3 min 49 sec	3 sec
4 min 03 sec	3 sec
4 min 19 sec	2 sec
4 min 34 sec	2 sec
4 min 51 sec	90 sec

There were strong reaction forces when the pressure valve opened and the bottle would probably have tipped over unless it wouldn't had been attached to the ground. After the test, when the fire was out, a distinct odor of propane could be felt coming from the bottle. It seems as if the pressure valve was damaged during the fire so that not all of the propane escaped. It should be noted that the odorant is very strong and only a small amount of gas it is enough for a strong scent to be felt.

Test 5

In test 5 a composite bottle made in one piece was tested. The bottle had neither pressure valve nor melting fuse.

After ignition the fire was slowly intensified, probably because of the plastic and the leaking gas through the shell. After about 3 minutes a hissing sound could be heard, probably coming from propane leaking through the shell. The bottle did not explode.

Figure 7: Image taken at approximately 4 minutes after ignition, when the fire seemed to be the most intense.

Test 6

In test 6 a composite bottle made out of two parts joined together was tested. The bottle was not provided with neither the pressure valve nor melting fuse. After 2 minutes 5 seconds from ignition a hissing sound was heard and it exploded after 2 min 52 sec. Top of the bottle flew about 80 meters into the air and landed about 70 m from the pool fire, surprisingly enough into the wind. The sound of the explosion was perceived as relatively dull. The fireball was estimated to be about 25 m in diameter.

Test 7

During test 7 an aluminum bottle equipped with neither pressure valve nor melting fuse was tested. The bottle exploded after about 2 min 57 sec. Several parts of the bottle flew in a relatively steep angle, and parts were found up to about 70 m from the primary fire. The bottom part of the bottle flew off and hit a support for a camera in the form of an empty jerry can which was considerably deformed. The camera survived the experience. The explosion also deformed the steel structure supporting the bottle on top of the pool fire and it flattened the pool itself, see figure 8. The sound of the explosion was perceived as relatively dull but with a certain flick. The fireball was estimated to be about 40 m in diameter.

Figure 8: Supporting steel structure and pool after the explosion.

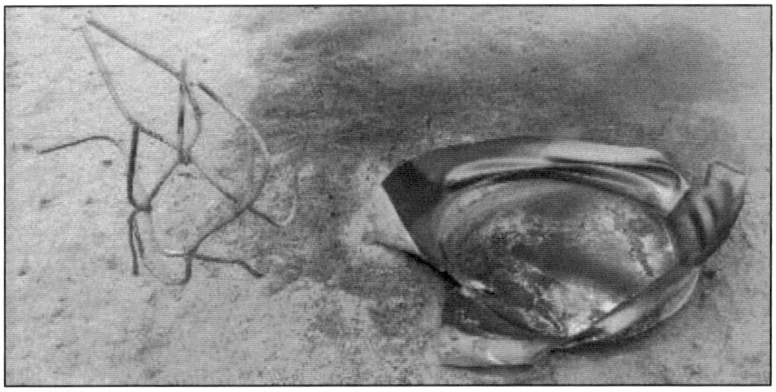

Test 8

In test 8 a steel bottle not equipped with neither pressure valve nor melting fuse was tested. During the first minutes of the test the bottle changed shape and adopted a rounder profile as the pressure increased in the bottle. The bottle exploded after about 4 min 6 sec. Several parts of the bottle flew in a relatively steep angle, where parts were found up to about 70 m from the primary fire, see figure 9. The pool was, again, flattened and the stand was not for further use after the explosion. The sound of the explosion was perceived as sharp and with a loud bang. The fireball was estimated to be about 50 m in diameter.

Figure 9: A collection of pieces after the tested steel bottle in test 8.

Summary of the tests 1 – 8

Table 5, below, summarizes when gas was escaping from the bottles, when pressure valves activated, when melting fuse activated or when the bottle exploded.

Table 5: Summarizing the results from the tests with propane bottles.

Test	Escaping gas [s]	Pressure valve activated [s]	Melting fuse activated [s]	Explosion [s]	Comments
1	210	-	-	-	
2	54	-	57	-	
3	-	87	137	-	
4	-	131	-	-	Opens intermittently for nearly three minutes
5	180	-	-	-	
6	125	-	-	172	
7	-	-	-	173	
8	-	-	-	246	

DISCUSSION

All the tests, except testing multiple aerosol bottles in an enclosure, were performed outdoors thus affected by weather. The test conditions were a bit windy so that the flames did not totally engulf the bottles during the tests. It was estimated that half of the surrounding surface of the bottles was exposed to flames during the tests. Consequently, the other half of the container was cooled by the wind. If the fire tests had been performed without the influence of wind, the sequence of events probably would have been faster

The smaller aerosol and propane bottles were chosen based on their appearance in retail stores. Other types or sizes of aerosol and propane bottles may be present on the market and such bottles may very well have a shorter time to failure and cause a sequence of events that are more severe.

Consequently, this study was performed as single tests of bottles exposed to fire. The sequence of events might differ between individual bottles. Repeated tests of each type of container would be required to determine the repeatability of the sequence of events.

During the tests with smaller aerosol and propane bottles, a fairly small pool fire with a diameter of 0.3 m was sufficient to cause the aerosol bottles to burst or explode. In some cases, the heat exposure led to the pressure being relieved from the container tank as a continuous leakage during the fire test**.** For aerosol bottles with varying internal volumes of 15 – 500 ml, between 0.07 – 0.36 liters of burning heptane was sufficient to initiate a burst or an explosion. The time to explosion for the aerosol bottles varied from 41 to 219 seconds were it was seen that larger bottles resulted in a longer exposure time before explosion. The average time to explosion of the tested aerosol bottles was 91 seconds.

During the tests with the larger (5 – 6 kg) propane bottles, these were affected by fire until they released pressure through leakage from the shell of the bottle, through safety devices or by bursting. The pressure release through leakage that was observed with the composite bottles that was constructed in one piece led to almost no reaction forces during leakage through safety devices. The bottles of aluminum and steel that were equipped with safety devices showed considerable reaction forces as they released pressure through these devices. During pressure release through the melting fuse, one of the bottles pushed away pool fire and tipped over. In a real setting, this would most likely have led to secondary effects, especially if bottles starts to move around as they release pressure. Such secondary effects have not been evaluated in these tests.

There was a significant difference in pressure release for the bottles without safety devices. Bottles made out of aluminum and steel exploded and parts flew up to 80 meters from its original position. The composite bottle that was designed in two parts and assembled by a glue joint also exploded with parts found 70 meters from its original position. However, the composite bottle constructed in one piece released pressure through a more general leakage through its shell, from the valve attachment and from a from a construction hole at the bottom. This resulted in no significant reaction forces and no explosion occurred.

The resulting burning vapor cloud and debris after an explosion may cause harm to people as well as property even without being adjacent to the fire exposed bottle. Consequently, propane bottles for consumer use should be equipped with safety devices (pressure valve or melting fuse or a combination).

CONCLUSION

From these tests of aerosol and propane bottles, the following conclusions can be drawn:

- A small amount of burning flammable liquid only is sufficient to cause aerosol bottles to burst or explode.
- The time it takes for a bottle to burst or explode in a real setting cannot be determined in advance, but the time can be said to be fairly short depending on heat exposure.
- Several bottles stored together may, due to explosion caused by exposure to fire, cause high pressure, high temperature, high levels of radiation as well as projectiles.
- Aerosols containing pressurized flammable gas should be included in regulations on how and how much is allowed for display in retail stores.
- Propane bottles (5 – 6 kg) can release pressure through safety devices within a few minutes only if exposed to fire.
- Propane bottles with safety devices should be fastened to avoid them to tip over or move around in case of pressure relief due to fire exposure.
- Bottles without safety devices perform in different ways when pressure is released, depending on type of construction.
- The time it takes for a bottle to burst or explode in a real setting cannot be determined in advance, but the time can be said to be fairly short depending on heat exposure.
- Safety devices (pressure valve or melting fuse) should be installed on all propane bottles for use by consumers.

REFERENCES

Svensson, S., Madsen, D. *Brandprovning av gasolflaskor, med respektive utan tryckavlastning* (rapport MSB1054). Myndigheten för Samhällsskydd och beredskap. 2016.

Madsen, D., Barton, J. Svensson, S., van Hees, P. *Brandprovning av aerosol- och gasolbehållare* (rapport 3195). Lunds Tekniska Högskola. 2015.

ELECTRONIC BALLAST FIRE ANALYSIS

Sebastian Scheiff, Kevin Lewis, Craig Rice, Karen Rayment, Matt McClelland & June Bott
CASE Forensics, a Jensen Hughes Company

ABSTRACT

In a recent investigation, a fire in a financial institution was caused by a fluorescent UV light used by tellers to check for counterfeit currency. This paper will discuss the post-fire examination of the evidence, and the supporting testing and analyses used to confirm the light as the cause of the fire. This paper will be helpful to scientists exploring fluorescent lighting fires in their own practice.

As part of the investigation, several dozen exemplar UV lights were examined for possible failure modes. During the investigation, unusual heat patterns on the bulb ends, and damage to electronic components of the ballast were observed, warranting further investigation and testing. Upon further testing it was discovered that as the fluorescent UV bulb reached end of life, its startup time would increase, until either the ballast could no longer light the UV bulb or the filament of the lamp broke and the bulb would no longer function at all.

During the startup cycle, while the ballast attempts to strike an arc to ignite the gasses inside the UV bulb tube, higher current and voltage are drawn. As the units age, longer start times damage individual ballast components due to the high heat produced by constant high current during startup cycles, as well as damaging the ends of the UV bulb, until ignition of the nearby plastic housing is achieved.

Damage to the ballast components is most likely due to the higher-than-designed duty cycle. The increase in duty cycle is due to extended startup times in which higher current is drawn.

The ballast type that is the subject of this paper did not have any safety mechanisms built in that would shut down power if lighting of the UV bulb was not achieved within a given time. The ballast design of the subject UV lights is commonly found in compact fluorescent light fixtures. This phenomenon of either "failing cold" at the bulb's end of life, or attempting to strike an arc for an extended period of time is not only found in UV fluorescents, but possibly in all types and spectra of fluorescent lights.

BACKGROUND

The building where the subject fire occurred was a multi-use high-rise in a major metropolitan U.S. city. The direct fire damage was contained to the unit space of the financial institution on the main level. The financial institution tenant construction was completed approximately eight years prior to the fire. There were no sprinklers installed in the financial institution tenant spaces, and the fire detection system was not monitored.

The financial institution purchased many UV-spectrum fluorescent tube lamps and fixtures for use by each teller to detect counterfeit money when bills were scanned underneath a UV desktop fixture.

The last occupant of the unit left at approximately 7 pm on a Friday evening. The unit was vacant from that time until the fire occurred on Sunday at approximately 1:30 pm. The fire was discovered when a passer-by observed smoke when looking in through the front windows.

Surveillance video recorded the teller station leading up to the fire, and a copy of the video was obtained during the fire origin and cause investigation. The subject camera was not pointed directly at the UV light, but just above it, at the service counter where customers walk up to interact with the teller at the subject station. The UV light did not appear to be emitting light until just prior to the fire, when it was observed in the surveillance video to have flickered on and off several times. After visible light was seen flickering from the teller station, smoke was then observed rising from the area.

FIRE ORIGIN DETERMINATION

Fire Damage and Burn Patterns

The area of origin was determined by viewing footage from the extensive video surveillance system, and fire pattern analysis. Since the fire did not achieve flashover, burn patterns were still present and identifiable.

Burn patterns indicated the fire originated at teller station 2. The teller stations were constructed out of particle board and other combustible products. The greatest loss of mass was at station 2 and was consistent with the video surveillance footage. The lack of burning underneath the desktop indicated the fire originated above the work surface.

The items located on top of the work surface were the UV light, a check validator, and a computer monitor. The computer was located below the work surface.

Figure 1: Teller station 2 burn patterns.

Figure 2: Workstation top with UV light (red rectangle) and check validator, a monitor had been moved prior to taking this picture.

FIRE CAUSE DETERMINATION

Evidence Examination

The collected evidence was destructively examined in a laboratory setting. This included all appliances located on the work surface of teller station 2. All ignition sources except for the fluorescent UV light were eliminated. The ballast circuit boards were heavily fire damaged but mostly intact. Some of the electronic components of the ballast circuits were missing.

No obvious signs of failure were found on the Printed Circuit Board (PCB). Arcing was found in several locations on the associated wiring. This included arcing on 120V conductors supplying power to the ballasts as well as high voltage conductors leaving the ballasts and leading to the fluorescent tubes. The arcing indicated that the light was powered on and the ballasts operating at the time of the fire.

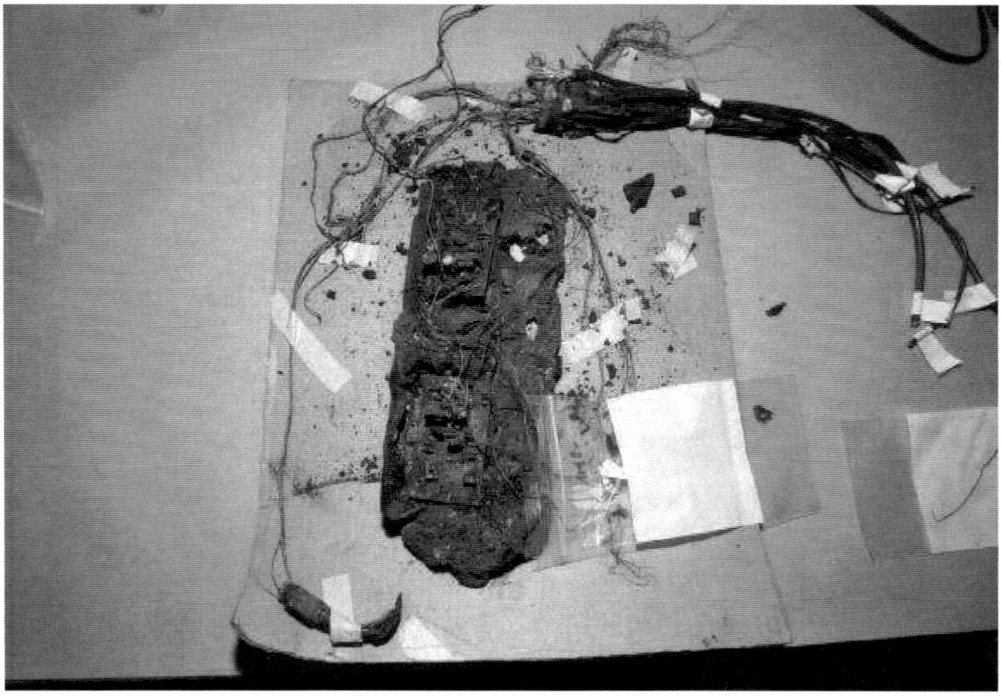

Figure 3: Subject UV light. Both ballast circuits are still present.

Figure 4: Arcing was found in several locations inside the UV light fixture.

The remaining wipe contacts and pin connections for the fluorescent tubes were examined and documented. No apparent failures associated with high resistance connections were found. While the video appeared to show the UV light flickering, it appears power was applied to the ballast and in turn to the UV bulbs. UV light is hard to see unless something is in front of the bulb that glows. The smoke from the initial stages of the fire may have made the light visible giving the effect of the light flickering on and off.

Alternatively, the UV bulbs could have temporarily established an arc and emitted light before losing the arc and not emitting light.

Testing and Analysis

CASE Forensics obtained several dozen exemplar fluorescent UV lights that had been collected by the financial institution from other locations. Each light fixture consisted of two (2) UV bulbs in a plastic housing. Each bulb had its own, independent ballast. A single on/off switch controlled both ballasts. The ends of the fluorescent UV bulbs were covered by a piece of silver plastic.

Initially, all light fixtures were opened and the ballasts inspected for damage. Each ballast was categorized and differences in ballast design and construction were noted. In addition, the lights were tested to determine if both bulbs were working. Thermal imaging was used to determine the operating temperature of each bulb. There were several light fixtures that exhibited abnormal heating at the UV bulb ends. The heat damage ranged from minor discoloration of the plastic bulb caps, to severe deformation and melting of the plastic caps.

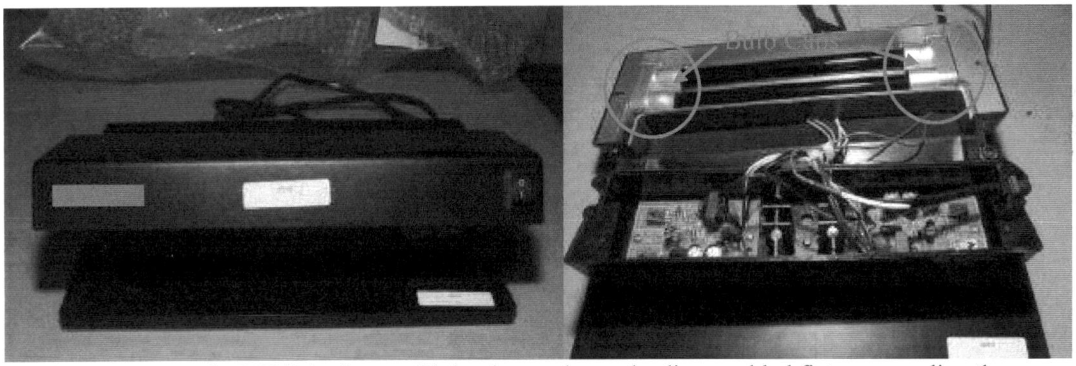

Figure 5: Exemplar UV light fixture. Right picture shows the disassembled fixture, revealing the ballast circuitry.

During the exemplar categorization, it was noted that the ballast had a resistor that could be soldered into the circuit in one of two positions to select whether the light was to be used with 110V or with 220V wall power. The voltage selection resistor provided an 'international feature,' allowing for the fixtures to be easily converted for use with either 110 or 220 Volt mains depending on the region where the fixture was intended to be used.

A schematic was created by tracing out the components and circuit for each of the two international voltage selection variants. It appeared that the 110 Volt version acted as a voltage doubler with a half wave rectifier, while the 220 Volt version was a simple full-wave rectifier.

Figure 6: Selection between 220V and 110V circuit

This appeared to affect the output waveform of the ballast during startup and run time. The frequency of the output, as well as the amplitude, increased.

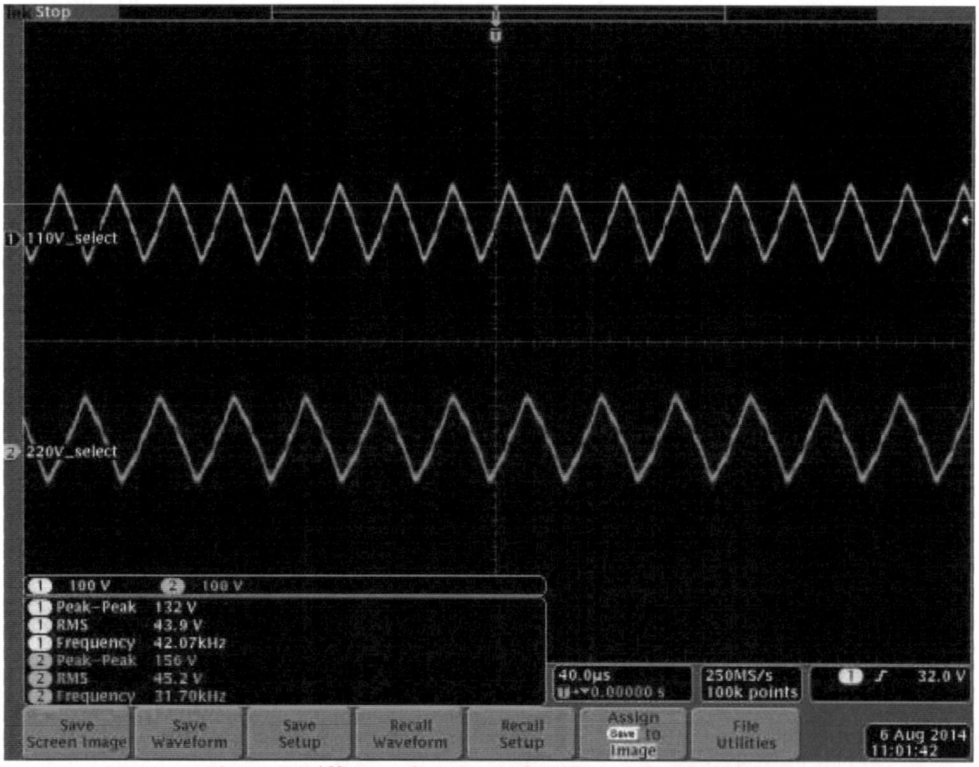

Figure 7: Difference in output of 120V and 220V circuit.

After all the exemplar UV lights were categorized, several were selected for further analysis. During this analysis, it was observed that one of the UV lights (Item 1501) was not emitting light, even though the power to the fixture was switched on. When surveyed by a thermal camera, the lightbulb that was not emitting any visible light was found to be extremely hot at the bulb ends.

The temperatures reached by the powered ends were measured to be more than 400° F using thermal imaging. When looking at the ballast for the hot-ended fluorescent tube, individual components were also running at much higher temperatures than those found in the functioning bulb and ballast in the same fixture. The two ballasts in this fixture were set for 110V operation.

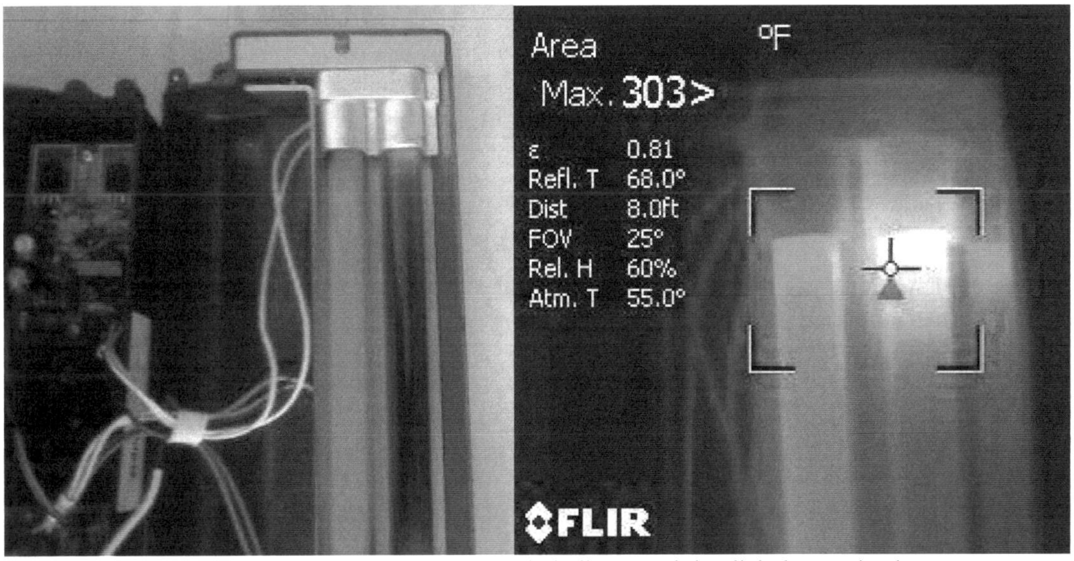

Figure 8: Image and IR caption of UV light bulb not emitting light but getting hot.

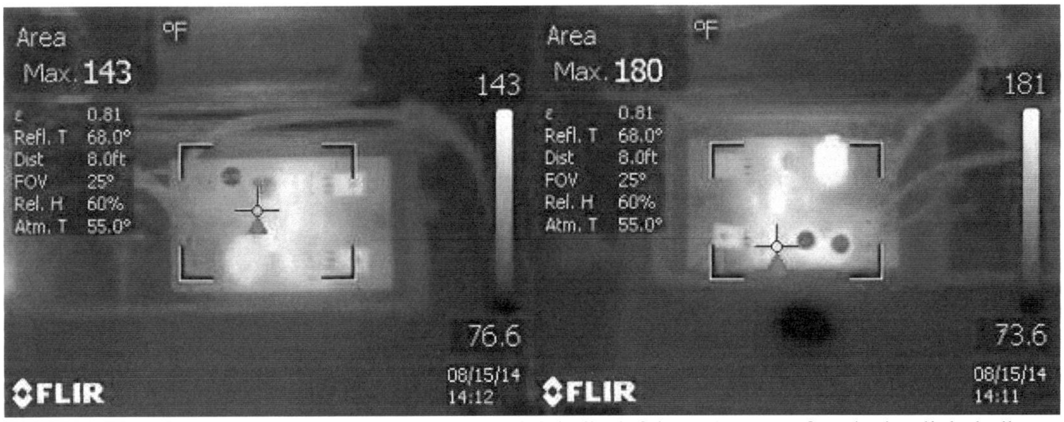

Figure 9: Ballast circuit of normally functioning lightbulb (left image) vs. nonfunctioning light bulb ballast (right image).

When swapping the two bulbs in the fixture, the heating of the bulb ends as well as the ballast circuit appeared to follow the bulb swap. This indicated that the cause of this heating was internal to the bulb and not an issue with the ballast or pin connectors at the end of the bulb. When inspecting the connector pins at the connection points, nothing abnormal was observed that would indicate a poor connection, which can cause higher-than-normal temperatures due to resistive heating.

Voltage and current measurements taken from the ballast circuit with the abnormally hot bulb showed that the ballast continued attempting to start the lamp by continually striking the startup arc, although the gas in the tube would not ignite.

A normal startup sequence is comprised of a short high-voltage, high-current burst that establishes an arc through the base of a fluorescent tube. The current heats up the two elements at each end of the bulb and preheats the emissions mix coated onto the elements, which allows an arc to be established through the gas between the elements. The energized gas emits radiation in two narrow bands of the UV spectrum. For typical lamps intended to produce light in the visible spectrum, coatings on the inside of the glass tube absorb the UV radiation and fluoresce in visible-light wavelengths.[1]

Under normal startup conditions, once the arc in the tube is established, the resistance changes and the voltage and current to the bulb will decrease. In the case of the bulb that did not light, the bulb never established an arc and the ballast never reduced its current output all the while attempting to continually start light.

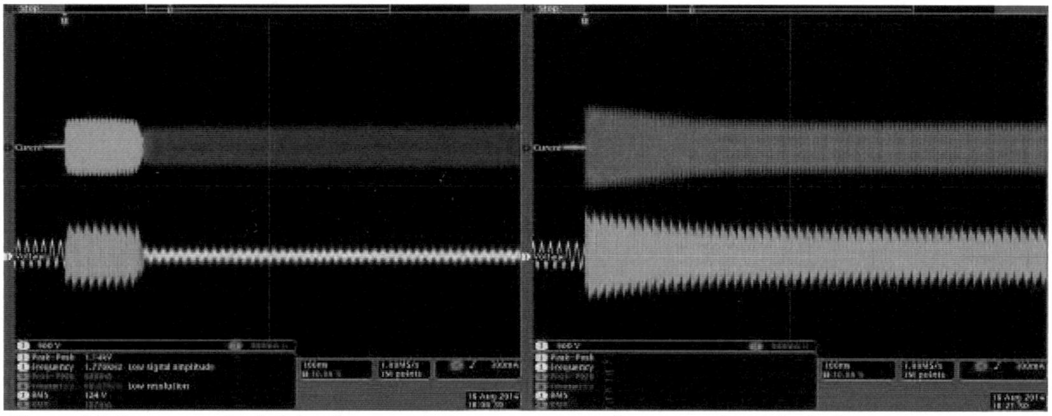

Figure 10: Image on the left is of a correct startup sequence. Image on the right is the startup sequence of the non-functioning bulb.

This continual "startup mode" increased the duty cycle of the individual components of the ballast circuit, causing excessive heating.

During the survey of the exemplar lights, several of the ballasts were found to have components burned out. Incidentally, these components were the same ones found to have excessive heating in the non-functioning UV bulb.

Figure 11: Damaged ballast components.

After the prolonged startup mode discovery, a second survey of the UV lights was conducted. This time, the initial thermal survey of lights was used to identify UV lights that had either deformations of the plastic endcaps, or were exhibiting higher heat signatures at the bulb ends.

The startup current and voltage signatures were analyzed. No other lightbulb was found to be generating excessive heat without showing visible light. During the survey of startup signatures, the startup time was measured and recorded. In general, as the startup time increased, so did the temperatures of the bulb ends. In the case of all these selected lights, the ballast was selected for 220V operation which is different than UV light 1501.

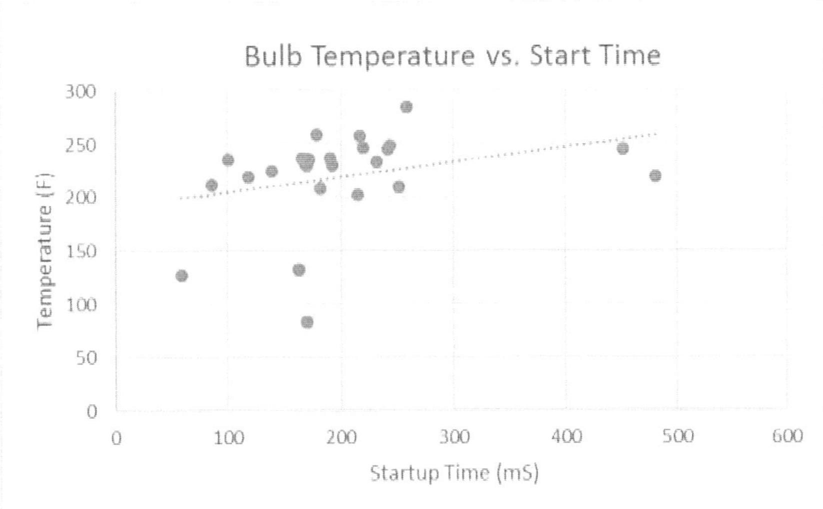

Figure 12: Graphical representation of startup time vs. bulb end temperature.

The plastic housing was analyzed and tested. Using Fourier Transform Infrared Spectroscopy (FTIR), various components of the plastic housing were analyzed and found to be Styrene-Acrylonitrile (SAN)[2] plastic. When subjected to a flame, the plastic did not exhibit any self-extinguishing characteristics and sustained an open flame.

Three of the UV lightbulbs were selected for further analysis: one bulb that was burned out and stopped emitting visible light, a second that had a long startup time, and a third that had a relatively short startup time. Of interest was the condition of the elements at each end of the bulb. The glass tube was carefully sectioned to reveal the elements on each end.

When inspected under a microscope, a clear and progressive depletion of the emissions mix that the element is coated with could be observed, when comparing the three elements. The emissions mix is intended to be consumed over time.

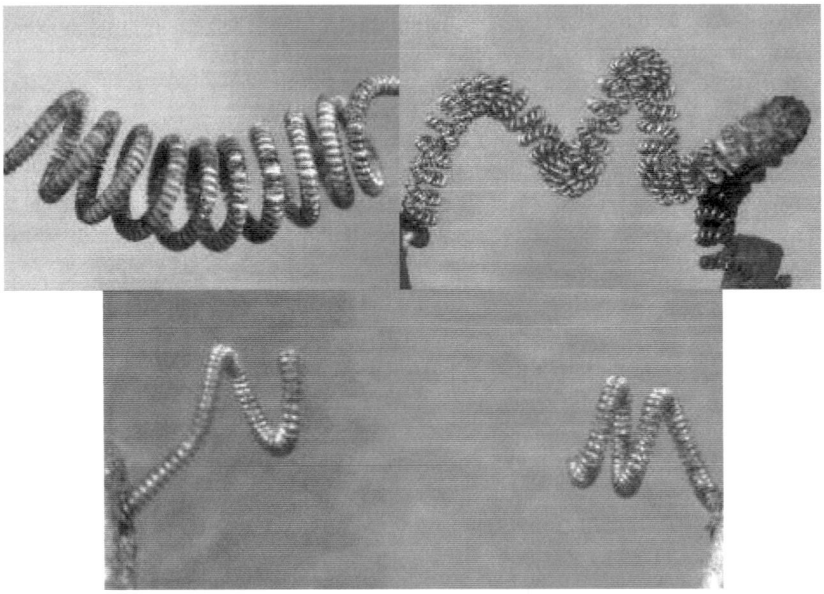

Figure 13:Progression of emissions mix depletion. Fast startup time (top left), slow startup time (top right), and burned out element (bottom).

DISCUSSION

The fire scene examination established that the fire started at a teller station at the financial institution. Video surveillance footage confirmed the fire originating near a UV light located on the work surface. During the examination, arcing was found on the 120V supply wiring as well as the high-voltage wiring for the light bulbs. No failure was identified on the lightbulb connections. While the video surveillance appeared to show the light flickering, physical evidence showed the light to be on at the time of the fire. The appearance of the light flickering could simply be due to the UV light being made visible by the smoke coming out of the light fixture. Another possibility is the fluorescent UV bulb firing intermittently causing it to flicker.

After extensive testing, we established that as fluorescent tube UV bulbs age they lose the emission mix that is necessary to light the fluorescent tube. As the emissions mix is depleted, it takes longer for the element to heat the gasses and establish an arc in the fluorescent tube, if an arc is established at all.

In addition, the elements start running at elevated temperatures and appear to get hotter as the startup time increases. In some cases, the bulb may never establish an arc and the ballast continues to attempt a startup. This increase in duration of the startup sequence overheats the ballast components that were only designed to handle the higher voltage and current of the startup sequence for a short duration.

The subject ballast design does not have any timeout mechanism that stops the ballast from continually trying to start the bulb. In addition, the fluorescent tube ends overheat since the internal coated wire acts as a heating element, which was found to deform, and sometimes melt the plastic components of the fixture.

In this case, the design of the light significantly contributed to the resulting fire. The plastic cover over the ends of the fluorescent tube made it impossible to replace the bulbs, once they burned out, without completely disassembling the entire UV lamp.

Most people would assume that if the bulb does not light that there is no danger. As we have seen in the case of one of the lights (Item 1501) from our study, this may not be true. In addition, the heat damage to the plastic covers over the tube ends is not apparent as the user never has a good vantage point to see it under normal use and operation. The plastic covers are obscured by the lighting hood that shields the users' eyes from direct light from the working UV bulbs, and directs the light downward onto the face of the currency under examination, in normal field use.

During this study, several failure modes were identified. Either the ends of the fluorescent tubes could get hot enough to ignite nearby combustibles, or individual ballast components may fail. The ballast design of this UV light is almost identical to that of a compact fluorescent light bulb. Fires that have been suspected, or even proven to be caused by a CFL, have mostly been blamed on a faulty capacitor in the ballast circuit. However, as we have shown by this work, the failure may originate in the elements of the fluorescent tube in similar fashion as that of the UV light in this case.

The danger of this failure is that most consumers assume that if the bulb does not emit visible light that it has "burned out" and failed in a safe manner - more of a nuisance than a safety concern.

This common assumption can be traced back to the days of incandescent lights - when they burned out (filament broke), current flow would stop. With fluorescent bulbs, when the bulb no longer emits visible light spectra, current can still be flowing at elevated voltages while the ballast continually tries, and fails, to strike a starting arc inside the tube. This phenomenon may be the beginning of a failure that can cause a fire.

CONCLUSIONS

Based on the investigation, the total-loss fire was found to be caused by a Fluorescent UV light at a teller station. Several key pieces of evidence led to the origin and cause determination:

1. The origin of the fire was determined to be the teller station.
2. Surveillance camera footage clearly showed the fire starting at the light.
3. Fire patterns clearly showed the origin on the desk top and not underneath the teller station furniture.

Thorough testing of the hypothesis revealed that a fluorescent tube light can be dangerous, even when it appears to have failed cold, due to the electronic ballast continually operating at elevated current levels without being able to ignite the gas inside the light tube.

REFERENCES

1. Ribarich, T., Thompson, E., & Mathur, A. (2000). Application Note. In International Rectifier (Author), *AN-1013*. Retrieved from http://www.irf.com/technical-info/appnotes/an-1013.pdf
2. Vitanza, A., Scollo, R., & Hayes, A. (1999). Application Note, Electronic Fluorescent Lamp Ballast, STMicroelectronics. Retrieved from http://www.st.com/content/ccc/resource/technical/document/application_note/43/99/f4/24/48/d7/41/7f/CD00003901.pdf/files/CD00003901.pdf/jcr:content/translations/en.CD00003901.pdf

---

[1] Paragon House. (2013). Fluorescent lamp. In *New World Encyclopedia*. Retrieved from http://www.newworldencyclopedia.org/entry/Fluorescent_lamp

[2] *Physical and Chemical Properties of SAN* [Fact sheet]. (2012, December 10). Retrieved December 1, 2014, from AZO Materials website: http://www.azom.com/article.aspx?ArticleID=7959

CASE STUDY: PYROPHORIC IRON REACTIONS IN REFINERY AND PETROLEUM INDUSTRIES

Isabelle Murray, P.Eng., M.A.Sc., CEP Forensic
& Raynald Gauvin, Ph.D. McGill University, Montreal, Canada

ABSTRACT

It is well documented that formation of pyrophoric iron sulphide from rust represents a significant fire and explosion hazard in oil and petrochemical industries. However, many conditions are required to allow growth and development of such pyrophoric iron reactions. More precisely, corrosion of steel of various refinery equipment, piping and storage tanks generates oxides and hydroxides. If the environment also contains hydrogen sulphide as a result of the addition of sulphur acting both as a binder extender and a mixture modifier, it can lead to the formation of pyrophoric iron sulphides and subsequent explosions. However, this represents only part of the problem since pyrophoric material can also create heat which has the potential to ignite residual hydrocarbons associated with the equipment containing the pyrophoric material. Fire and Explosion investigators need to understand the basic formation and oxidation processes of the iron sulphides which could be pyrophoric.

This paper investigates high precision metallurgical techniques that can be applied to detect the presence of iron sulphide from rust and confirm that pyrophoric reaction was related to the cause of the disaster. Analytical methods such as high magnification electron microscopy and XRD analysis, XPS techniques and metallography, together with an evaluation of the oxide layer microstructure, are powerful tools to help Fire and Explosion investigators in their origin and cause analyses. These metallurgical techniques are presented through a case study involving the explosion of a petroleum storage tank that caused severe damage to the tank and a loss of production.

A general literature review of the phenomenon of pyrophoric iron fires together with study of the general risks associated with petroleum products will be addressed in order to determine the chemistry of pyrophorics and the morphology, microstructure and composition of the oxide layers that are formed at the surface of steel equipment. Then, metallurgical analysis techniques will be applied on steel coupons collected from the fire and explosion site to demonstrate the pyrophoric species that had infiltrated the steel surface with respect to this case study. More specifically, this paper will demonstrate that high magnification and precision techniques as well as testing for the chemical liaisons allow identifying the presence of liaisons between sulphur and metal that correspond to the interaction between the iron and the sulphur forming Fe-S, thus pyrophoric oxidation.

INTRODUCTION

It is a fact that steel or iron oxides on the inside walls of molten sulfur, crude oil or asphalt tanks react with H_2S, resulting in the formation of iron sulfides. Under particular circumstances, some types of these compounds cause violent exothermic oxidizing reactions when in contact with air, and consequently lead to an explosion and subsequent ignition of the surrounding materials. During fire and explosion investigations, an examination of petroleum storage components is usually conducted in order to localize the fire origin area, as well as a general overview of all the electrical control systems and devices. Fire and Explosion investigators must be aware that the formation of pyrophoric iron sulfide on the inside walls of tanks constitutes a real problem. More precisely, there are four main conditions where explosions can occur in bitumen petroleum tanks, namely contact with water, cold storage tanks, excessive heating and oxidation reaction.

Metallurgy has evolved into a complete science with methodology, tools and supporting software that are more powerful than ever. Microscopic examination and material detection are insured by scanning and transmission microscopes together with X-Ray photoelectronic and diffraction spectroscopic devices. This paper supports on these high technology metallurgical techniques combined with more basic and conventional ones to determine reproducible and precise elements that are associated with pyrophoric iron reactions.

The first section of this document summarizes the phenomenon associated with pyrophoric reaction while Section 2 presents the general description of the explosion event and subsequent fire, and the selection of the samples collected during our site visit. Section 3 reveals the metallurgical results, including our metallographic and microscopic analyses of the samples using Scanning Electron Microscope (SEM), X-Ray Diffraction (XRD) and X-Ray Photoelectron Spectroscopy (XPS). Our technical discussion in included under Section 4 of this paper. Finally, the last section concludes with our general results regarding the metallurgical characteristics associated with pyrophoric reaction and its relation with the cause of the explosion.

SECTION 1: PHYSICAL PRINCIPLES OF PYROPHORIC OXIDATION OF IRON SULPHIDE

Even though literature indicates that the pyrophoric nature of iron sulfide has been known since 1826, we come to the understanding that the mechanisms involved and the factors affecting the reactivity have not been fully explored[1]. It appears that the pyrophoricity of ferrous sulfide susceptible to cause fire or explosion accidents in the process of petroleum product storage, transportation and refining has arisen wide concerns since the late 1970s, leading to research and various studies[2].

The basic reaction involves carbon steel, the material used to manufacture the storage tanks and equipment. It is expected that such carbon steel will normally corrode and an oxide layer of Fe_2O_3 be formed at the internal surface of the storage device. It also implies hydrogen sulfide products (H_2S) produced by hot petroleum/bitumen products. The latter is heavier than air and known to represent health and safety as well as a fire hazards. Literature indicates that under low oxygen concentration (less than 3%), the hydrogen sulfide (H_2S) from bitumen/petroleum products will react with rust or corrosion deposits that are likely located at welds, around manholes, ventilation conduits and the upper portions of tanks[3]. It should be noted that "virgin" petroleum/bitumen products are frequently doped with polymers in order to change their rheological properties for the desired use. Hence, sulphur is a common element added to bitumen to act as a binder extender and a mixture modifier for road products[4].

This reaction will generate iron sulfides (FeS, Fe_3S_4 and FeS_2) respectively at low, intermediate and high temperatures ranging from 10°C to 100°C. Iron sulfide is one such pyrophoric material that oxidizes exothermically when exposed to air. It is frequently found in iron sulfide scales inside tanks. More precisely, pyrophoric sulfide is formed by the conversion of iron oxide (rust) in iron sulfide, in an oxygen-free atmosphere where hydrogen sulfide gas is present (or where the concentration of hydrogen sulfide (H_2S) exceeds that of oxygen). This rapid exothermic incandescent oxidation is known as a pyrophoric oxidation and it can ignite nearby flammable hydrocarbon-air mixtures. Initially, iron sulfide is formed by the reaction of rust or corrosion deposits with hydrogen sulfide as follows[5]:

$$Fe_2O_3\ (\text{rust}) + 3H_2S = 2\ FeS + 3H_2O + S \qquad [1]$$

Considering this reaction, we understand that the probability of iron sulfide formation increases with high sulfur content of bitumen/petroleum products. Iron sulfide can also be formed under deposits adhered on tank interior walls. Vapor produced by hot bitumen will condensate on internal walls and will generate deposits similar to coke[6].

When the iron sulfide is exposed to air, the process of rapid oxidation of the sulfides to oxides is susceptible to occur, according to the following equations:

$$4FeS + 3O_2 = 2Fe_2O_3 + 4S + Q\ (Q>0) \qquad [2]$$

$$4FeS + 7O_2 = 2Fe_2O_3 + 4SO_4 + Q\ (Q>0) \qquad [3]$$

The heat usually dissipates quickly unless there is an additional source of combustible material to sustain combustion.

Literature also indicates that pyrophoric iron oxidation usually forms when tanks or distillation columns are out of service or opened to air for maintenance[7]. It is basically based on the fact that thin layers of sulfides that are created remained attached or adhered on the internal surfaces of the equipment. Moreover, it is known that this reaction could occur during product transfer operations (fill out) when oxygen is allowed to enter the tank to balance the amount of product transferred, and when process units are stopped for maintenance.

SECTION 2: SAMPLES SELECTION

An explosion followed by a fire occurred inside one of the storage tanks of a company specialized in production and sales of aggregates and hot mix asphalt in Canada. More precisely, the fire was circumscribed inside a petroleum production reservoir characterized by a 12-foot diameter and a capacity to stock up to 100 tons of bitumen products. The tank was made of CSA G40.21 - 44W steel sheets of 0,635 centimeter (1/4 in.) in thickness in the inferior section and 0,475 centimeter (3/16 in.) in the upper sections. The heating of the petroleum product stored in the tank was done by a circulating thermic oil heating mechanism having its energy source located far from the tank.

During our site visit, it was apparent that the top portion of the tank was totally detached and was projected as a consequence of the explosion. No metallic components in the top section were affected by the heat of the fire and showed any signs of melting on both sides. Rather, we located significant heat-induced damage inside the body of the tank. More precisely, our examination of the internal side of the tank showed many areas characterized by oxidation related to corrosion (Figure 1).

Figure 1

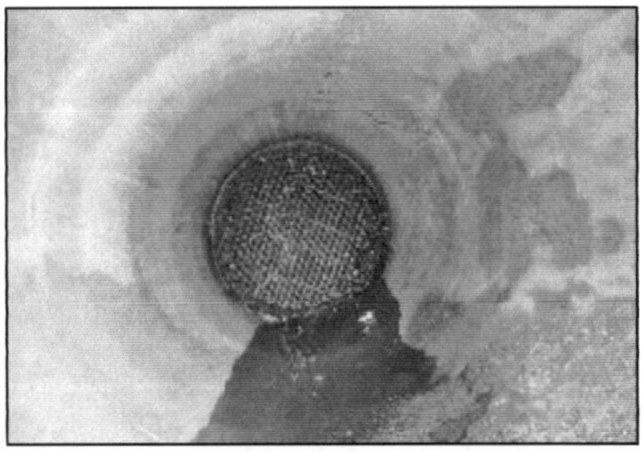

This type of scale was present all around the tank, even though it was installed only 2 years prior to the event. We noted a stain near the base of the tank that was most probably caused by the level of petroleum product in the tank at the time of the explosion and related fire (Figure 2). Moreover, we also noticed the presence of remnants of petroleum products adhered to the surface (Figure 3).

Figure 2

Figure 3

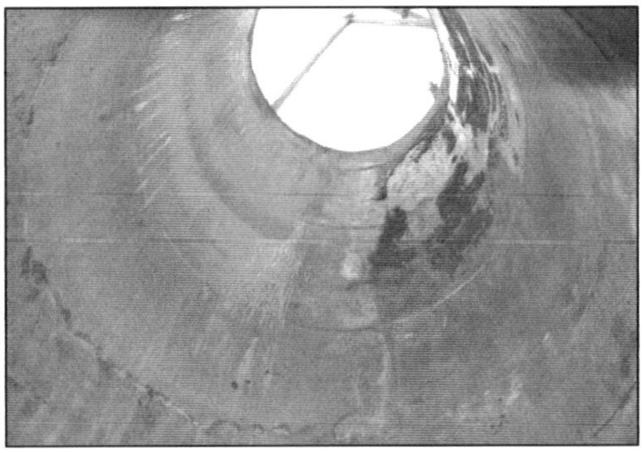

The information gathered indicated that on the morning of the incident, sulphur pellets were poured into the tank. The fill out operation of the storage tank began in the afternoon, and the explosion occurred when 2/3 of the transfer of bitumen product was completed. It should be noted that this operation had to be interrupted for a short period of time due to maintenance issues.

The bottom of the tank was equipped with section of tubes that were parts of the heating system. There were no particular deformations or distortions of these tubes (figure 4). This system was also equipped with many control devices hooked up to control board and a data logger. Our site inspection and testing of all the control devices allowed determining that they were still working properly after the event.

Considering the results of our testing and the parameters and particulars of the event, we collected some residues that were still adhered to the inner walls of the bitumen tank and selected wall portions for sampling and further metallurgical testing (figure 5).

Figure 4

Figure 5

We then identified two areas of interest (A and B) which represented zones without and with deposits, for comparison purposes (figures 6 and 7). Considering the material used to manufacture the tank, the sampling operation was performed using an arc cutting technique In order to avoid heating (and altering/modifying) of the surface layer, the cut was made far from the critical zone used as representative samples from the steel sheets. Subsequently, the arc cut parts were reduced in size with a manual cutting saw to prevent excessive heating.

Figure 6

Figure 7

SECTION 3: METALLURGICAL RESULTS

We collected coupons of both areas for comparison purposes and characterization.

3.1 Metallography

We initially performed metallographies on coupons from samples A and B respectively without and with remaining burnt deposits to allow the examination of the material microstructure as well as the interaction between the steel and the surface, thus the interface.

Our examination of the cross sections from Sample A revealed signs of interaction between the steel and the deposits on the surface. These deposits associated with petroleum products on the steel surface had metallurgically infiltrated the metal (Figures 8 and 9).

Figure 8

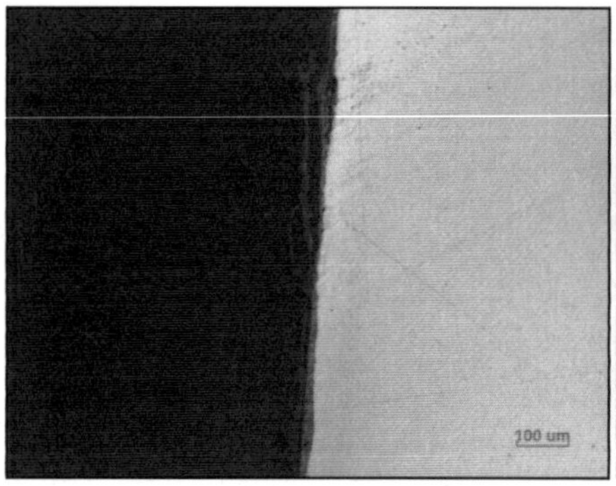

Figure 9

On the other hand, the metallography associated with coupon B showed less signs of reaction between the thick deposit and the metallic surface. We mostly observed signs of generalized corrosion on the surface of the sample (Figures 10 and 11).

Figure 10

Figure 11

3.2 Scanning Electron Microscopy

Both samples A1(from A) and B1(from B) were microscopically examined under a Hitachi SU3500 VP-SEM microscope equipped with an EDS Aztek system from Oxford Instrument. Our analysis confirmed the presence of solidified deposits. More precisely, the oxide layer on sample A1 was mainly composed of iron (Fe) oxygen (O), sulfur (S) and carbon (C) (Figures 12 to 14).

Figure 12

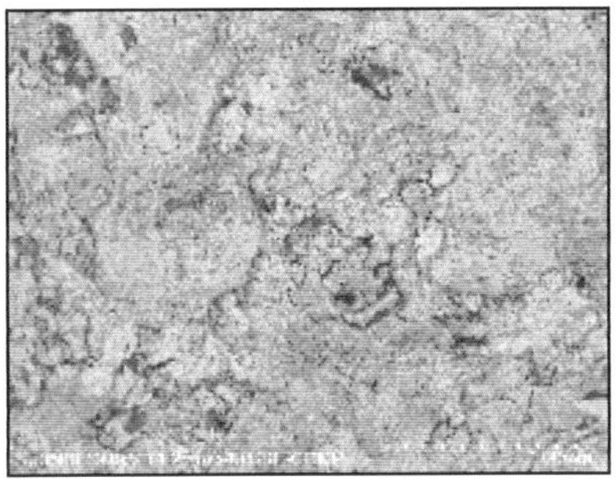

Figure 13

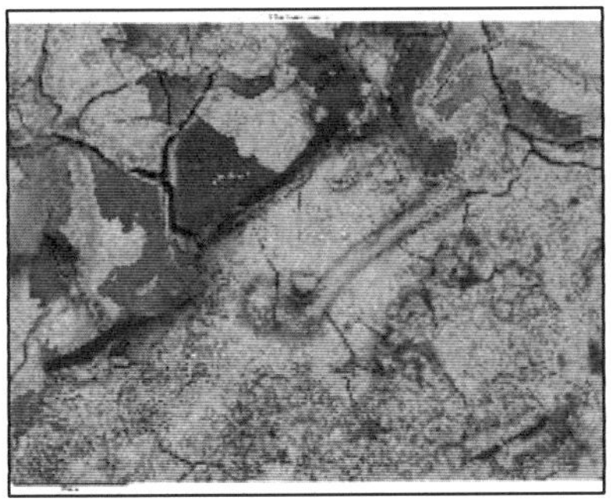

Figure 14

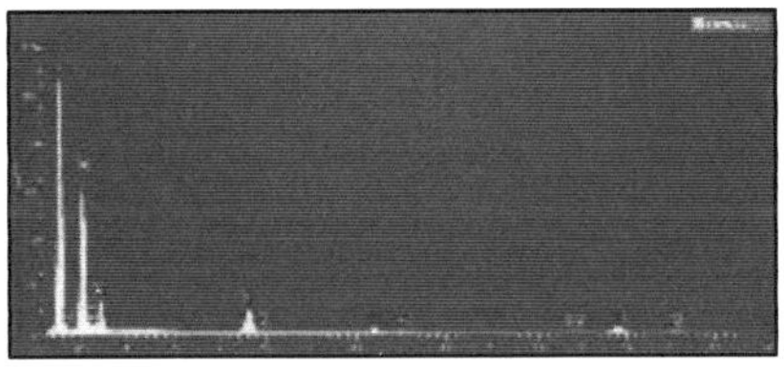

As for the scaling adhered on sample B1, it included essentially carbon (C), sulfur (S) and oxygen (O) (Figures 15 and 16) while the surface under this layer also exhibited iron (Fe) (Figure 17). These results confirmed that the deposit reacted with the surface of the sample.

Figure 15

Figures 16 et 17

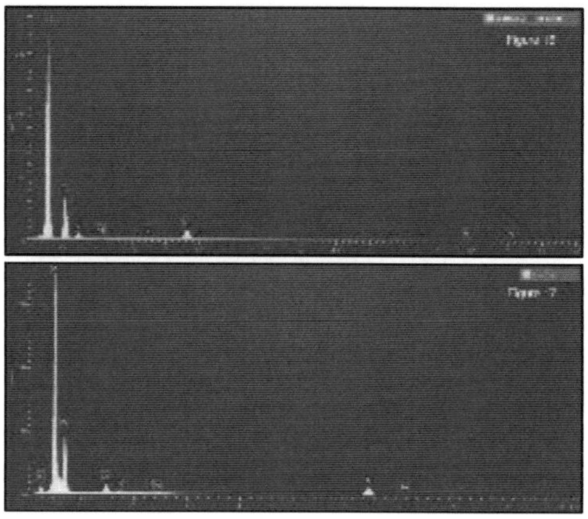

Mapping on the surface of Sample A1 also confirmed the presence of all these elements (Figure 18).

Figure 18

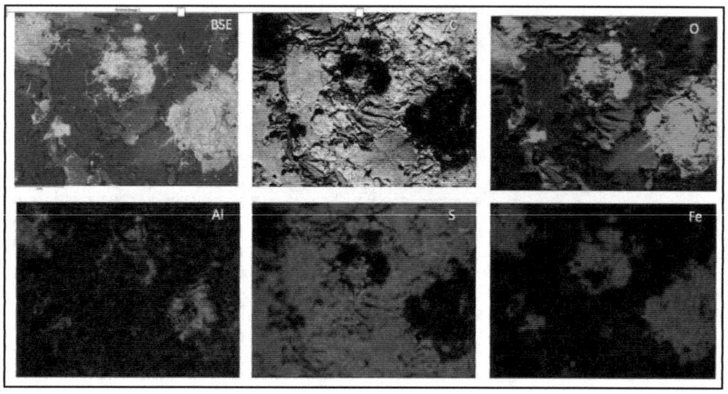

3.3 X-Ray Diffraction (XRD) Analysis

XRD is a non-destructive technique to characterize the crystalline materials. It allows to identify the molecular composition and the 3-dimensionnal structure of materials.

Samples A1 and B1 were examined using this technique. Fe_3O_4, Fe, C as well as FeS were discovered on sample A1(Figure 19) while Sample B1 revealed the presence of Fe_3O_4, Fe and C (Figure 20).

Figure 19 – Sample A1

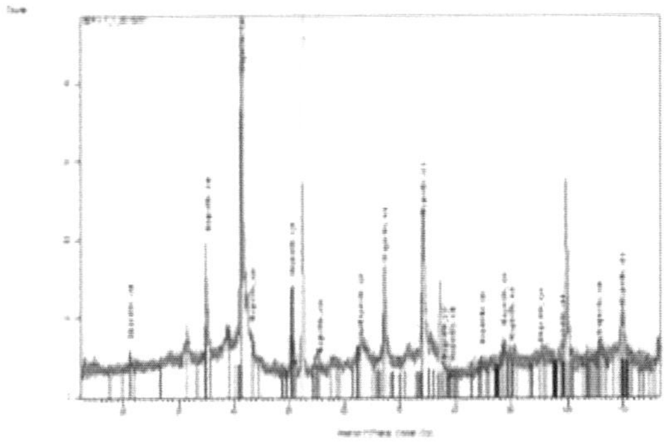

Identified components	Chemical Formula	
Iron Oxide, Magnetite	Fe_3O_4	Major
Iron	Fe	Minor
Carbon	C	Minor
Iron Sulfide	FeS	Presence - traces

Figure 20 – Sample B1

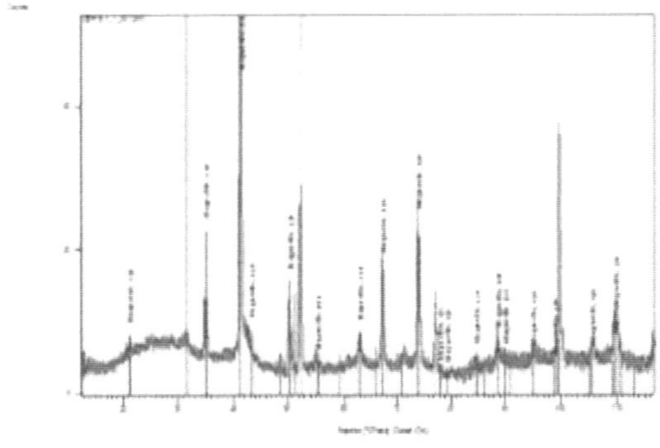

Identified components	Chemical Formula	
Iron Oxide, Magnetite	Fe_3O_4	Major
Iron	Fe	Minor
Carbon	C	Presence - Traces

3.4 X-Ray Photoelectron Spectroscopy (XPS) Analysis

X-Ray Photoelectron Spectroscopy is a very potent technique to quantify all of the elements of the periodic table, with the exception of hydrogen and helium from the surface of the specimen. We performed an XPS analysis using an ESCALAB 3MKII equipment on deposits collected on the surface of Sample A1.

This technique correlates binding energies with associated elements and compounds. More precisely, we noted a binding energy at 164,2 eV for a sulfur with carbon (S-C) liaison. Also, a binding energy at 162.7eV was identified for sulfur, which corresponded to the binding energy of a Sulfur-Metal (S-metal) compound. The only metal found in the layer was iron at 712.9eV for Fe^{3+}. Since iron was the only metal detected, the liaison between Sulfur and a metallic element was most likely attributed to FeS, corroborated with binding energies at 707-708eV for FeS (Figure 21).

Figure 21

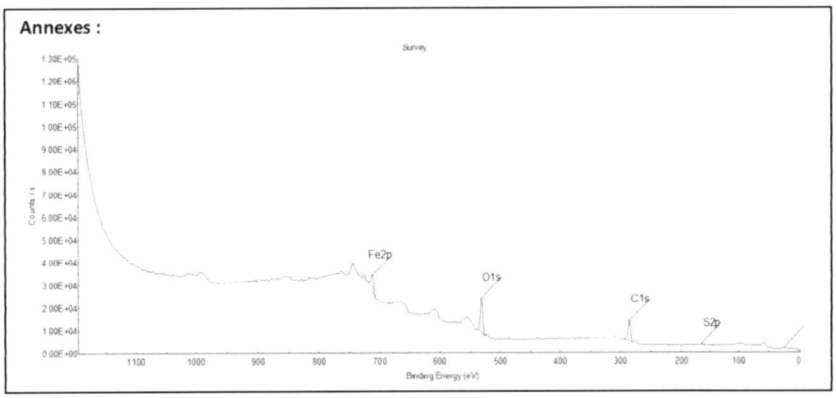

As a result, XPS analysis confirmed the presence of sulphur (S), iron (Fe), carbon (C) and oxygen (O) in the deposits testing. Furthermore, the testing for the chemical liaisons allowed identifying the presence of liaisons between sulphur and iron (Fe-S), the only metal identified in the deposits (Figure 22). It should be noted that the aluminum seen on the SEM spectra was not present in the residues collected from the surface of Sample A1.

Figure 22

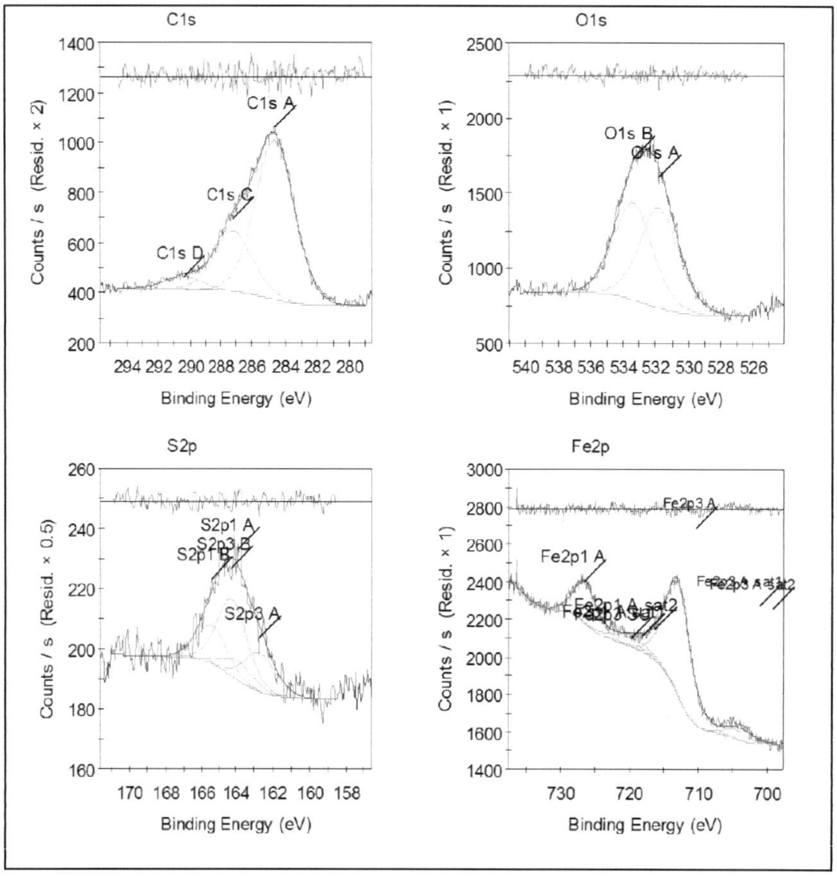

SECTION 4: TECHNICAL ANALYSIS

Our analysis of all the facts gathered allowed us to identify that the tank under study suffered from an explosion followed by a fire inside the tank. In determining the fire/explosion cause, we have considered the following elements known to have the potential to create such phenomenon, namely temperature, water infiltration, mixture, heating system and spontaneous ignition of iron sulfide. Our review of data of monitoring devices did not indicate any unusual elevated temperatures recorded inside the tank prior to the explosion. We also discarded the water infiltration given the explosion happened during the fill out operation. As for the mixture, it was the same as planned. Finally, the heating system located far from the tank had no reported or apparent problems.

The facts reported were to the effect that the production mixing rate could have been influenced by the temporary shutdown of the process during the fill out operation. The mixture was known to content sulphur, and said sulphur was added the morning of the incident. The temporary shutdown could have concentrated the amount of sulphur and increased the hydrogen sulphide concentration.

The metallurgical analysis revealed that bitumen products had infiltrated the steel surface, inside the tank. Microscopic examination confirmed the presence of a solidified deposit that reacted with the surface. More precisely, the metallographies indicated areas where the surface suffered from localized depressions.

The elements identified on the surface, in the areas where the bitumen scaling was burnt away, were composed of iron (Fe), oxygen (O) and sulphur (S). Fe_3O_4, Fe, C as well as FeS were discovered on sample A1 with XRD. In addition, XPS analysis and testing for the chemical liaisons allowed identifying the presence of liaisons between sulphur and another metal. Since the only metal identified in the deposits was iron and considering the energy of the liaison Fe-S, our analysis showed that these liaisons most likely corresponded to the interaction between the iron and the sulphur forming Fe-S. Therefore, all these results are compatible with pyrophoric oxidation.

CONCLUSION

The explosion of a petroleum storage tank caused severe damage to the tank and a loss of production to the company. The explosion was followed by a fire inside the tank.

Our site examinations and analysis of the facts reported revealed that the explosion occurred during the fill out operation of the tank.

Based on our technical investigation and considering the results of our metallurgical analysis, we conclude that the cause of the explosion is directly related to pyrophoric oxidation.

REFERENCES

[1] R. Walker, Pyrophoric oxidation of iron sulphide, Surface and Coatings Technology, 34 (1998)163-175.
[2] P.Li, S.Wang, Z.Zhang, and S.Zhao, Study on the Effect of Water on the Formation and Pyrophoricity of Ferrous Sulfide, Petroleum Science and Technology, 29 :1922-1931, 2011.
[3] Hedlund, F. H., Fire in hot asphalt cargo hold, spontaneous ignition during unloading, Fire in hot asphalt cargo hold, spontaneous ignition during unloading,Loss Prevention Bulletin, 224, 23-24. (2012).
[4] D. Strickland, J. Colange, P. Shaw, N. Pugh, A study of the Low-Temperature Properties of Sulphur Extended Asphalt Mixtures, Shell Sulphur Solutions.
[5] Mukesh Sahdev, Pyrophoric Iron Fires, Safety and Pressure Relief, November 08, 2010.
[6] P.Li, J. Li, S. Zhao, L. Kong and Y. Zhai, Research on the danger of fires in oil tanks with sulfur, Fire Safety Journal 40 (2005) 331-338.
[7] Constantinos Plellis-Tsaltakis, Investigation of a pyrophoric iron fire in a Visbreaker fractionation column provides better cleaning work procedure, Journal of Loss Prevention in the Process Industries, 38 (2015) 268-275.

MEASUREMENTS ON SMOKE GENERATED FROM AN OFFICE FIRE: SOOT MASS CONCENTRATION, WALL DEPOSITION AND PARTICLE CHARACTERIZATION

Louis Decoster[1, 2], Alexis Coppalle[3], François-Xavier Ouf[4], Aurelien Thiry[1]
1Laboratoire Central de la Préfecture de Police (LCPP), France.
2Institut Universitaire des Systèmes Thermiques Industriels (IUSTI), France.
3Complexe de Recherche Interprofessionnel en Aérothermochimie (CORIA), France.
4Institut de Radioprotection et de Sûreté Nucléaire (IRSN), PSN-RES, SCA, LPMA, Gif-Sur-Yvette, 91192, France.

ABSTRACT

A fire test has been carried out in a 30m^2 room containing office furniture. This room was connected to a corridor in which an experimental setup was mounted in order to measure the properties of particles present in the smokes. This study, with a real office fire and the set-up of the measurements, had three main objectives: – Characterise soot particles emitted in smoke – Evaluate soot deposition on a wall – Test and comparison of three soot mass concentration measuring devices. It is found that the particle size increases during the fire. Smallest diameters and highest number concentrations were observed at the beginning of the fire, and as the fire developed, the number concentration decreases while the mass concentration and particles diameter increase. The comparison between the three soot mass concentration measuring devices will be shown and discussed. The method of measuring the deposit, which uses plates at different temperatures, will be described. It will be shown that the comparison between deposits on cold and hot plates provides reliable data useful for the understanding of the deposition process.

INTRODUCTION

Smoke emissions from fires are important to know, in particular for compartmented fires[1]. Flows of smokes contain hot gases and particles that can increase the risk of propagation of the fire outside the source compartment. Smokes reduce the visibility, which makes the evacuation of persons and the intervention of the firemen more difficult. Exposition to toxic gas, CO for example, may be fatal for peoples involved in the accident. At least, smokes confined inside the building cause significant damage that may be difficult to repair.

Smoke is a mixture of gases, aerosols (as water droplets or condensed vapors) and soot particles[1]. In the case of wood or biomass, ash can be also emitted. The quantities and the properties of the particles depend on the ventilation of the flames, which varies during the fire, from well ventilated to under-ventilated if the combustible source is strong enough. The on-line and real-time measurement of aerosols and soot particles properties is an important issue. A lot of works have been performed on car engine emissions, leading to the development and the validation of technologies, allowing the measurement of mass and number concentrations of particles in the engine exhausts[2]. For fire studies, some of them have been applied but not as much as for car emissions. Authors's previous works have shown results obtained at laboratory scale[3-5]. However, the relevance of these technologies needs to be considered analyzed and such tools must be validated for realistic large scale experiments.

Particle measurements have been carried out in a corridor adjacent of an office fire. This work has three objectives: – To characterize soot particles emitted in the smoke – To evaluate soot deposition on a wall – To test and compare three soot mass concentration measuring devices. The article is organized as follows: description of the office fire test, then description of the particle experimental setup, and results are next presented and discussed.

DESCRIPTION OF THE OFFICE FIRE

The 30 m^2 room has a ceiling height of 2 meters. It contains furniture representative of an actual workspace. They are placed in the room, as illustrated in figure 1, so as to form three working spaces delimited by wooden partitions. The fire load therefore consists of the following elements:

- Three wooden desks, with computer, screens, keyboard, mouse.
- Two half-height, wooden, removable partitions.
  - Three storage cabinets against wooden walls.
  - A printer on a table in the center of the room.

Computer hardware consists mainly of plastic and each office has a full paper basket. The desks and cupboards are also filled with paper. The office is instrumented by numerous sensors including thermocouples and cameras. The thermocouples are arranged to form three vertical trees of six sensors in the room which positions are indicated on figure 1.

Figure 1: Dimensions and furnishing of
the office before the fire and localisation of the thermocouples trees

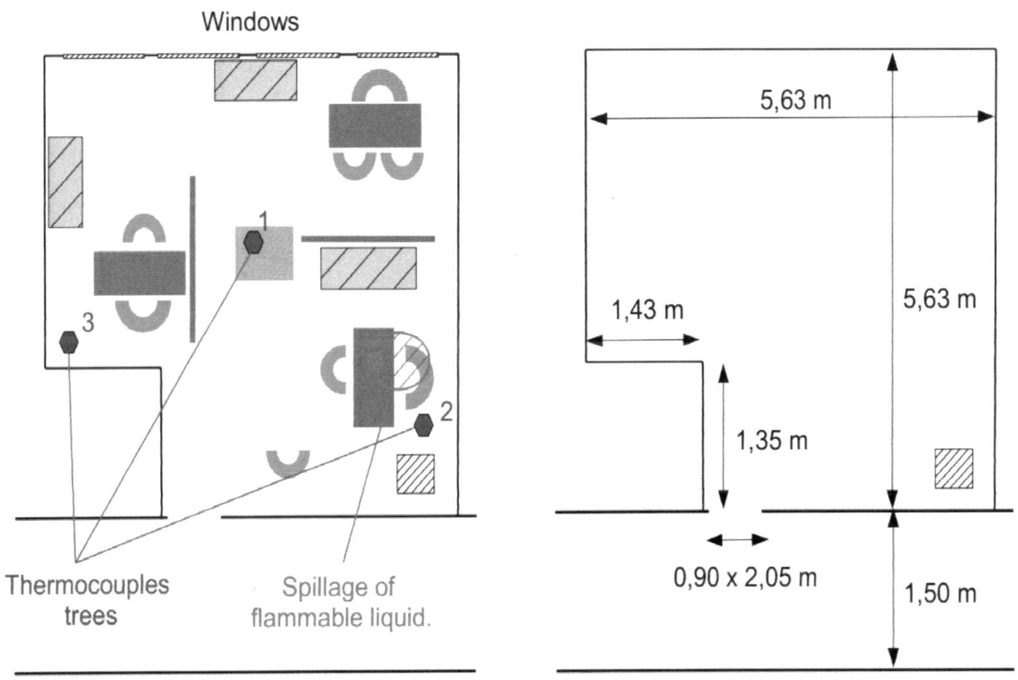

The office door is left open for the duration of the test. The four windows (numbered 1, 2, 3 and 4 from left to right) facing the outside are closed at the beginning of the test. Windows 3 and 4 will break due to the fire respectively at 1015 and 1070 seconds after the fire started. A bowl containing burning gasoline was spilled near a chair to set the office on fire, as shown in figure 1 and figure 2. The initial moment (t = 0 second) corresponds to the appearance of the first flame on the bowl after it was ignited with a flaming torch. The fire lasted approximately 19 minutes (1160 seconds) before the intervention of the firemen and the switch-off of the sensors for safety reasons.

Figure 2: Picture of the furniture close the ignition location (left) and view after the fire (right)

Figure 3 shows the temperatures measured by each of the three thermocouples trees, which positions are given in figure 1. The thermocouple tree number 2 is next to the ignition area whereas the trees number 1 and 3 are further from this zone. The temperatures measured next to the deposition targets, and times corresponding to windows breaking and aerosol sampling are also indicated. Several phases of fire development can be identified. Before 30 seconds, gasoline is burning on the small bowl, involving no significant evolution of the thermocouples' responses. Between 30 and 120 seconds only spilled gasoline is burning. Then the fire intensity slightly decreases until 240 seconds, when the desk and the chair begin to burn. Around 600 seconds, the fire spreads quickly to the melted chair and to the desk, causing the temperatures around the starting area and in the smoke layer to increase. The fire then starts losing power again until the first window (number 2) breaks at 1015 seconds and a second one (number 3) breaks 55 seconds later. With these new sources of air, the fire intensifies and nearly reaches the flash over: the lowest thermocouples show an increasing temperature, even far from the ignition zone (see trees 1 and 3 in figure 3). Firefighters begin to extinguish the fire 1160 seconds (19 minutes) after the fire started, which results in an immediate drop in temperatures in the room.

Figure 3: Temperatures measured in the office

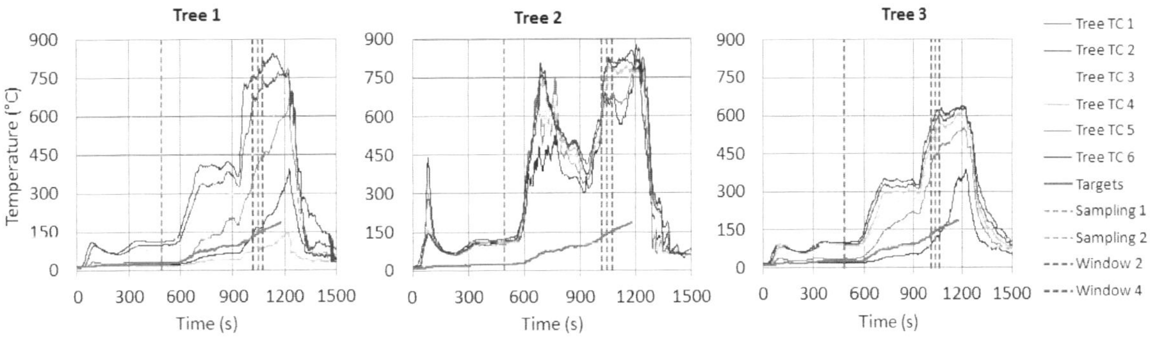

THE PARTICLE MEASUREMENT SET UP

Sampling and dilution

The experiment was set up in the upper section of the corridor (Figure 4). An opacimeter was placed in this corridor for in-situ and real-time analysis of smoke emission rate. In addition to this in-situ device, a stainless steel probe was used for sampling and transporting smoke to the lower stage for performing their analysis using conventional laboratory tools.

Figure 4: Corridor set-up for smoke concentration and deposition measurement

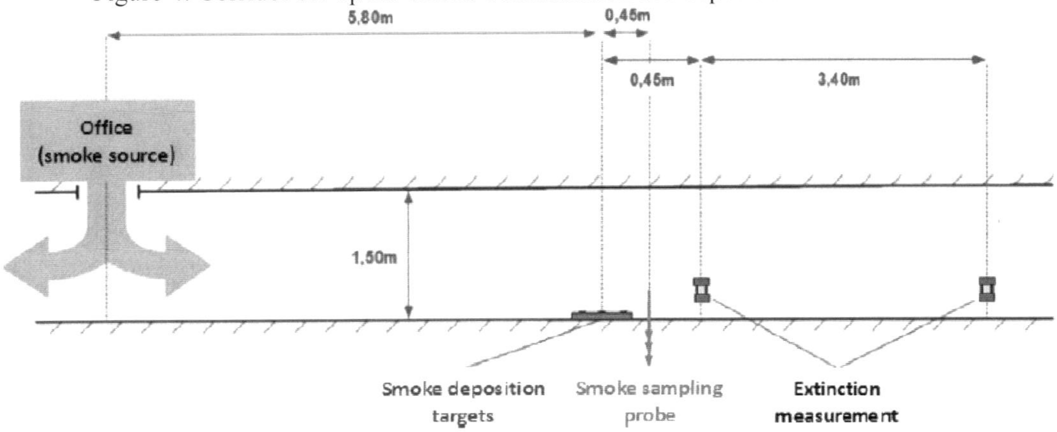

The sampling probe (internal diameter of 8 mm) was inserted through the wall at 2.05 m from the floor and at 45 cm from the aerosol deposition targets. This probe was connected to an ejector type dilution device (DEKATI DI-1000) ensuring a dilution ratio of 1:10 using 2.5 bars of dried and filtered air. For avoiding any clogging of the venture hole of the dilution device by flying ashes, two 90° bended tubes were placed between the sampling probe and the DI-1000. Three on-line measurements devices were mounted downstream the dilution device. Two were devoted to mass concentration analysis (Tapered element oscillating microbalance or TEOM and the Pegasor Particle Sensor or PPS) while the third one (Electrical Low Pressure Impactor or ELPI) was used for measuring the size distribution of aerosol emitted during the fire. In parallel to these on-line and real-time measurement tools, a flow splitter was implemented for sampling on two high efficiency particulate air quartz fiber filters (HEPA 1 and 2) and two transmission electronic microscopy grids (TEM1 and 2) for a posteriori mass concentration, chemical composition and morphological analyses. Figure 5 shows the experimental setup of sampling and characterization of aerosol.

Particle sizes

The Electrical Low Pressure Impactor device (ELPI) is a cascade impactor composed of different successive stages of impaction which collect particles of a specific aerodynamic diameter Da which depends on the geometry of the acceleration stage and the flow rates used. It covers a wide range of aerodynamic diameter from 0.03 μm to 10 μm. Detection of particles is carried out by electrical charging of particles prior to their injection of the cascade impactor and measurement of the corresponding current on each impaction stage by the mean of high resolution electrometers. The particle concentration is then determined using a relationship between the current and the particle number density[6-7]. ELPI provides the total particle number concentration n (number/cm^3) and the size distributions during the test at a frequency of 1 Hz.

Figure 5: diagram of the sampling probe and the dilution system

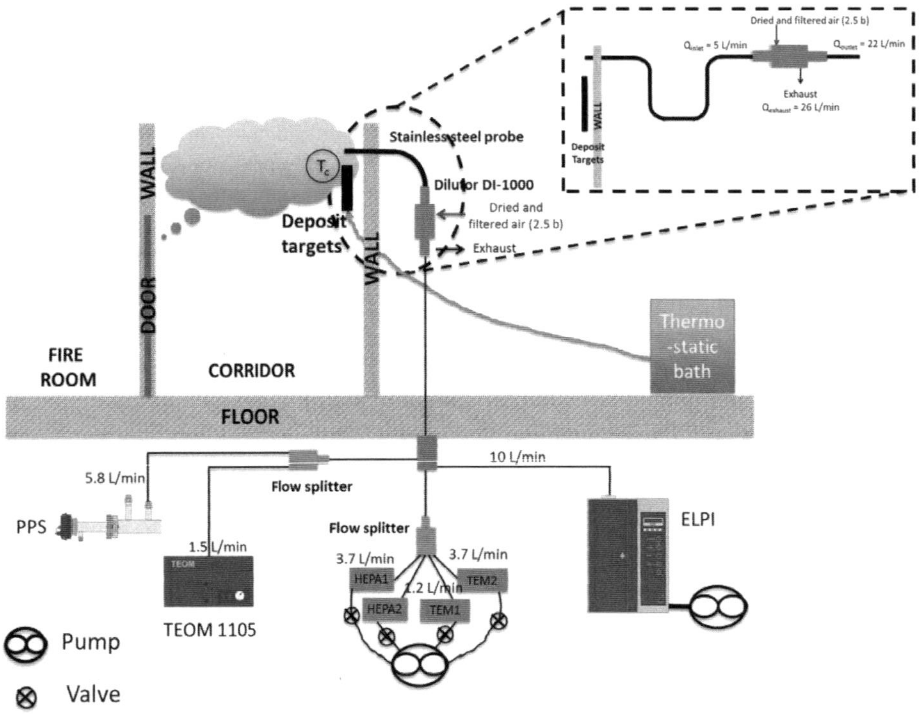

Particle properties

In parallel with the on-line analyses, samples on filters and transmission electronic microscopy grids were made to characterize the composition, of the particles emitted at two characteristic moments of the test (see fig. 3). Quartz fiber filters (Pallflex QAT-UP 2500) of 47 mm of diameter were used for analysing the organic content of the soot particles and for measuring the mass concentration associated to carbonaceous particles. The elementary and organic (EC-OC) content of the soot particles has been carried out using a Sunset Laboratory analyzer[8]. This device consists of an oven where the samples taken on the quartz fiber filters are heated to separate the different types of carbon (elementary and organic). Particles were also collected by filtration on carbon-coated copper TEM grids (Agar Scientific) and post-analysed with one JEOL JEM 3010 electron microscopes.

Particle mass concentration

Three devices exist in the purpose of measuring online the particle concentration. These devices have been already compared in the smoke generated in a small fire compartment[5].

- The opacimeter, simple to build up, provide average values on the optical path, instantaneous. They can be used until 200°C, as the new white light opacimeter[9]. The soot concentration measured by the opacimeter is obtained from measuring the output voltage following the Beer-Lambert law:

$$C_s = \alpha \frac{-\ln(1 - \frac{U_{out}}{U_{max}})}{\sigma_s L} \qquad [1]$$

where Cs represent the soot concentration [g/m3]; α the calibration coefficient [-]; Umax output voltage [V] without sample; σs specific extinction coefficient [m2/g]; L optical path of the device (= 0.05 m). The calibration coefficient is obtained by using reference Kodak wratten neutral filters. The specific extinction coefficient is taken for flaming conditions[10] ($\sigma s = 8.7\ m^2/g$).

- The Pegasor Particle Sensor (PPS), a new system developed recently for automotive application[11]. It provides instantaneous measurements of samples and it can be used until 600°C. Its operating principle is based on the corona effect by electrical charging of the exhaust aerosols. The aerosols are driven into an ion trap by pumping, establishing an electrical voltage proportional to the concentration of aerosol. The calibration proposed by the manufacturers has been done on soot samples in a diesel engine tailpipe, with a 50 nm count median diameter. For such particles, the mass concentration is proportional to the current detected and provided by the ionised particles. The recommended proportionality coefficient is given by L(mg/m3fA) = 6.3e-5 / Qs (stdlpm), Qs being the sample flow rate determined in standard condition of temperature and pressure.

- The Tapered Element Oscillating Microbalance (TEOM)[12], which provides a direct measurement of the mass concentration in a sample flow. It is accurate but more difficult to use compared to the two previous systems, especially because it requires a prior dilution of the smoke. Microbalance (TEOM) measures particle mass PM based on the change in the resonance frequency of a tapered quartz wand as particles accumulate onto a sampling filter attached at its tip[12]. That measurement method is the most direct measure of real mass among the three considered.

Figure 6: pictures of the three different devices used for mass concentration

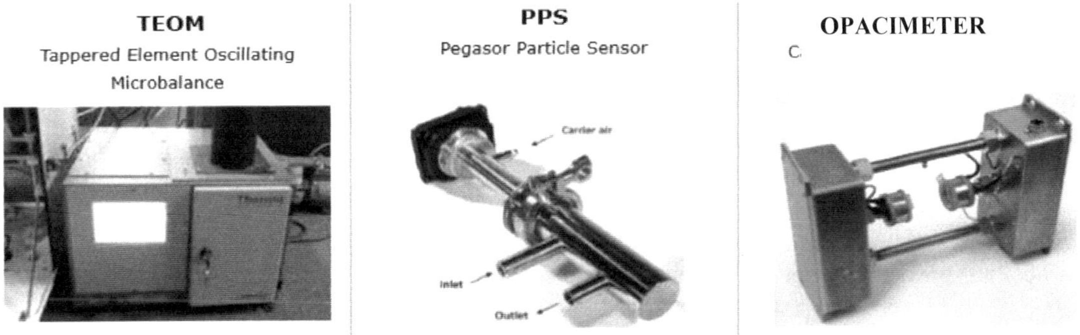

Particle deposition on the walls

A simple device has been built to characterize the deposition of soot on the walls of the corridor. Three 0.18 mm thick aluminium deposition targets are plated to three distinct aluminum supports themselves inserted in a wooden support. This support is attached to the wall at 2.05 m from the floor and close the particle probe (see fig. 4). Two of the aluminium supports (to which we shall refer by supports 1 and 2) are plated to a 10 mm thick wood piece while the third one (support 3) was attached to a heating electrical resistor. Each one of the three fine aluminium plates are weighed before and after the fire, allowing to measure the quantity of soot that deposits during the whole test. The targets are collected one hour after the fire was extinguished, due to operator safety reasons.

Temperature is monitored whit thermocouples at the surface of the three aluminum supports and in the smoke, at 5 cm from the central deposition target, and at the end of the sampling probe. We can verify that the gas sampling temperature is very close to the temperature of the gas near the deposition targets. Supports 1 and 2, which are exposed to similar thermal conditions (gas and backing), have a nearly identical surface temperature. Support 3 however is indeed heated by the resistor at a temperature superior than the gas temperature.

Figure 7 : Temperatures measured on the deposition surfaces and in the gas near the deposition targets.

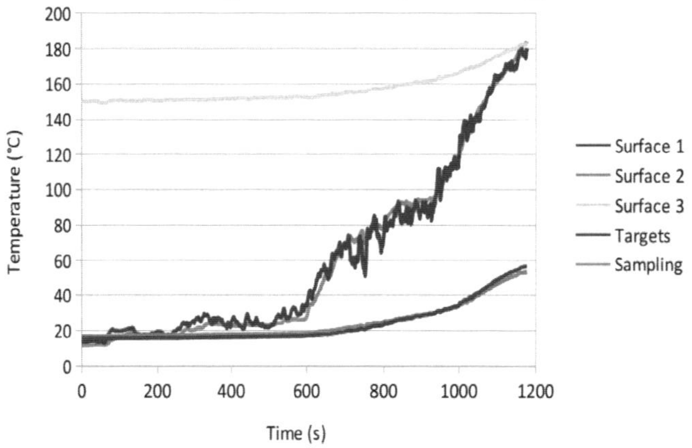

RESULTS AND DISCUSSIONS

Particle sizes and properties

On figures 8 are given the particle size distributions at times taken in each of the three periods of the fire, as shown in figure 3. The bottom graphs present the two parameters that characterise the log normal distribution which fits, at each time, the experimental points, i.e. the count median diameter (CMD) and the geometric standard deviation (GSD)[1]. Figure 9 shows the particle number concentrations n (part/cm3) for three classes for diameters, PM10 for diameters less than 10μm, PM1 for 1μm and PM0.1 for 0.1μm. The first case corresponds to all sizes detected by the ELPI device and PM0.1 the ultrafine particle class. These numbers are calculated from the size distributions reported on fig. 8. PM10 and PM1.0 curves are superposed, meaning that there are a very few particle number with diameters greater than 1μm. The PM1 number is more or less constant during the first period of the fire (t<600s) and then there is a steadily decreasing, on the other hand the PM0.1 number has a well pronounced decrease during the first period. The important feature observed on fig. 9 is that there is in proportion less and less ultrafine particles during the first period of the fire.

Figures 8: left, particle size distributions at several times, measurements (marks) and curve fits (line); right, two parameters of the size distribution, count median diameter and geometric standard deviation.

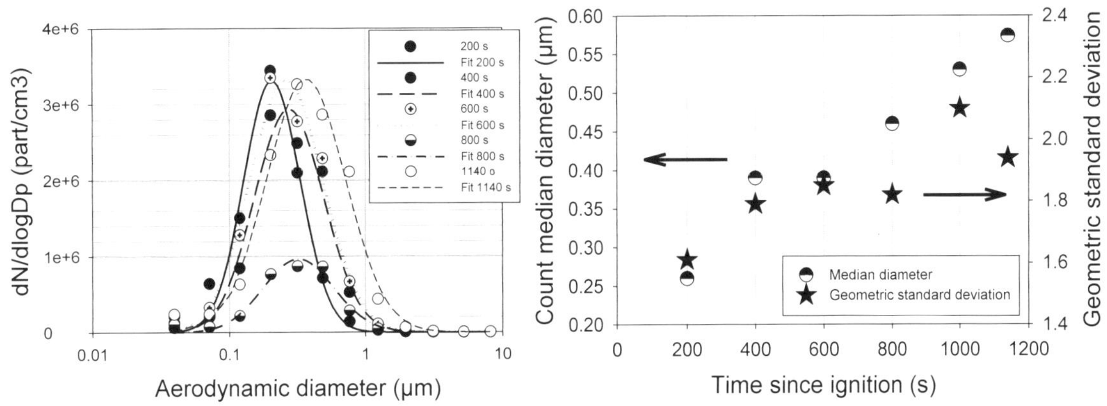

At the beginning of the fire and up to 600s, the count median diameters of smoke particle were lower than 0.4 µm denoting a slight increase as a function of time. This size increase is in agreement with the decrease of number concentration of particles with a diameter lower than 0.1 µm (PM0.1) reported on figure 9. After 600s and due to the increase of the fire intensity, the CMD and GSD significantly increase. However, during the former period, the number is greater than during the second period (600s-1000s). Such overall decrease of number concentration, associated to an increase of CMD and GSD of particle is characteristic of an agglomeration process which can occurred within the fire room and the corridor. Then after 1000s, the fire is strengthened due to window breakages and the number concentration of particles increases again. During the fire test, the particle diameters are ranging from 200 nm to 500 nm, in good agreement with those reported for small and large scale industrial fires[13-14].

One must also notice that the high content of small aerosols (PM0.1) reported at the beginning of the fire (between 100 s and 600 s, see figure 9) may be due to the presence of small water droplets, which result of the condensation of water vapours emitted by the heating of the furniture and material inside the fire room. Goo[15] has determined the particle sizes of wood and PP pellets emitted by a small tube furnace and collected in a mixing chamber. It is interesting to compare the previous results to the present ones, since wood and PP are material that are used for furniture manufacturing. In the Goo's work, both non-flaming and flaming conditions have been used. It has been found a high content of small particles, less than 100nm, for both flaming and non-flaming conditions. One can see on figure 8, that this is in agreement with the present results if time is less than 600s, but this is different for times greater 600s, when the fire intensity becomes important. This difference is more pronounced for wood than for PP. This can be explained by the low temperature in the mixing chamber during the Goo's tests, enhancing the condensation of water vapour and unburnt gases. In our experiment, such droplets may condense in the smoke, which is rather cold at the first moments (before 600 s) and may be progressively evaporated as time and smoke temperature increase.

Figure 9: Particle number concentration during the fire,
PM10-PM1.0-PM0.1 correspond to particles with diameters less than 10-1-0.1 μm respectively.

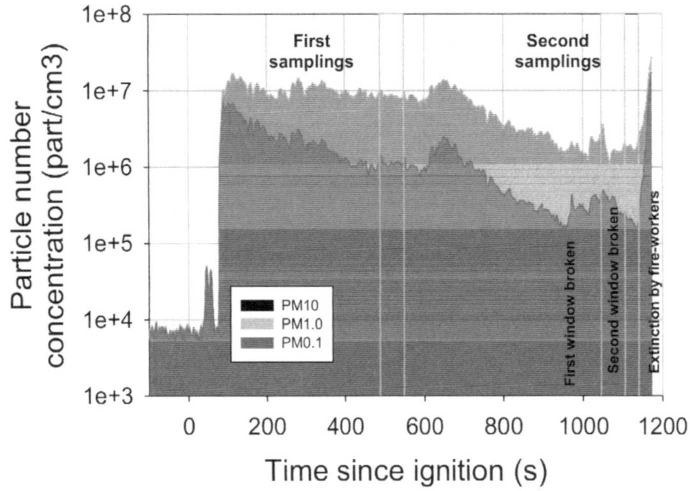

The soot particle contents in elementary EC and organic carbon OC have been analysed on two filters HEPA1 and HEPA2, exposed at 488 s for the former, during the first period of the fire, and at 1045 s for the later during the full fire development. The ratio OC/(EC+OC) has been found equal to 9.8 % +/- 1.4 % and 42.5 % +/- 6.4 % for respectively HEPA1 and HEPA2. The organic content is low during the first period, which corresponds to a well-ventilated fire, and much higher during the full fire period, which correspond to an under-ventilated fire. Figure 10 shows an example of TEM micrograph of a particle sampled on TEM grids during the first sampling period. As shown on this picture, particle denotes a well-known fractal morphology and are mainly composed by carbonaceous nanoparticles with a diameter ranging from 10 to 50 nm. One must also notice that these particles denote a low content of organic content at their surface, confirming the low OC/(EC+OC) reported for the first sampling.

Figure 10: TEM micrographs of particles emitted during the first sampling period (t < 600 s)

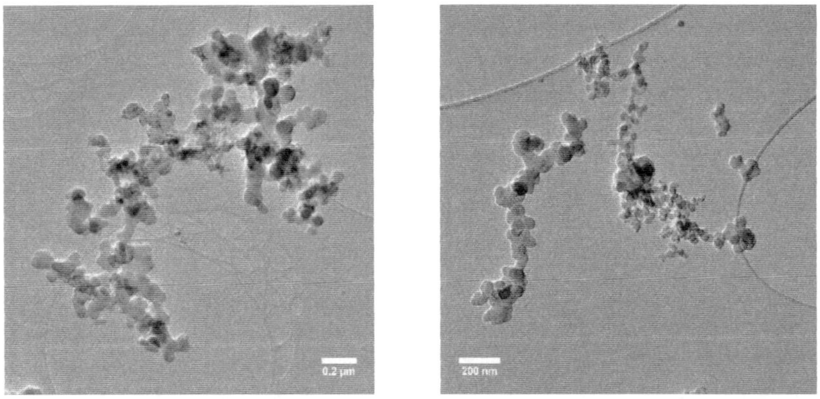

Particle mass concentrations

On figure 11, one can see that the mass concentration Cs curves given by the three devices have the same variations up to 600s. During this first period, Cs reaches a first plateau, which does not last long, and then increases to a second one, which occurs up to 600s. After that, the Cs values measured by the TEOM and the PPS decrease while they increase with the opacimeter. This may be due to a slight contamination and a particle deposition on the windows during the test or a drift of the electronic due to the temperature increase inside the device. The average values on these two plateaux are 90 and 260 mg/m^3 for the TEOM, 80 and 250 mg/m^3 for the opacimeter, 6 and 8 mg/m^3 for the PPS. The conclusion of the comparison is that the opacimeter results, using a specific extinction coefficient value equal σs = 8.7 m^2/g, are in a good agreement with TEOM results. Regarding PPS, this device seems to strongly underestimate the particle mass concentration. This has been already found in a previous study with smokes generated by the combustion of dodecane in a small fire chamber. Further studies have to be done to know how the PPS measurements depend on the particle size and composition but such result is not surprising since the PPS calibration is performed on soot particles emitted by diesel engines and denoting a count median diameter of nearly 50 nm, 10 times lower than particles emitted during our experiment.

Careful examination of the number concentration (figure 9) and of the TEOM mass concentration (figure 11) shows that, during the first period of the fire (<600s), the PM0.1 number and the mass have opposite trends. The first is decreasing while the second is increasing. This is explain by the small size of the PM0.1 particles, less than 0.1 µm, which gives them a small mass compared to ones of larger particles (PM1 or PM10). As the fire increases, the PM0.1 number decreases to the benefits of the larger particles, so the mass concentration increases.

Figure 11: particle mass concentrations measured by the TEOM, PPS and opacimeter devices

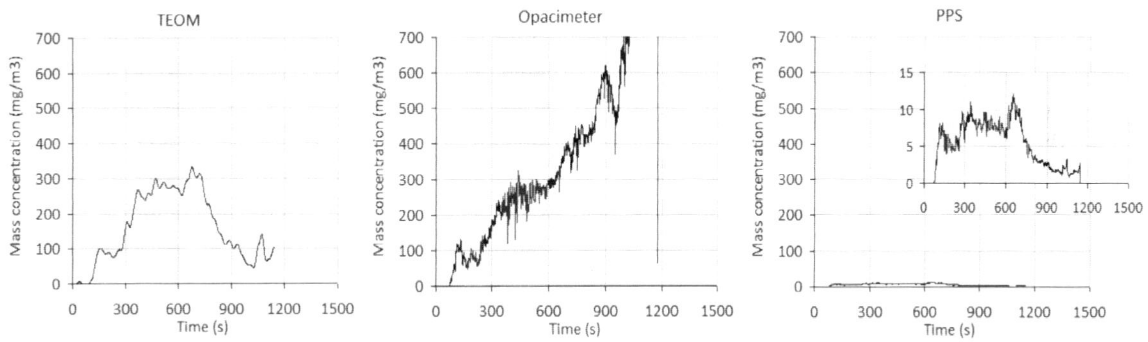

Wall soot deposition

Figure 12 shows pictures of each deposition targets in the order in which they were presented on the plate: Target 1, Target 2 and Target 3. We can see on figure 12 that Target 3, heated to about 200 ° C, has very low traces of aerosol deposition. Only a few large isolated particles can be seen on its surface, which results in a negligible mass weighing. Targets 1 and 2 are covered with a dark and significant deposit.

Figure 12: photo of the targets after the fire

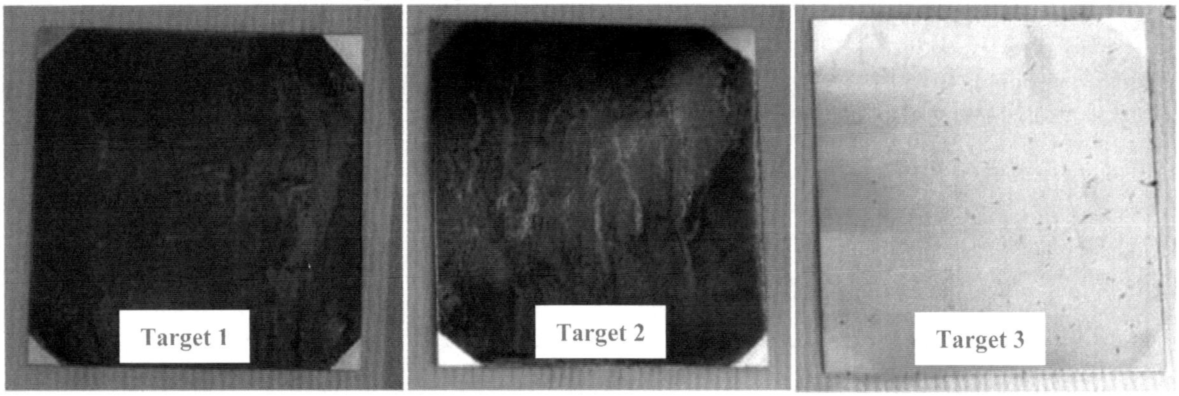

Table 1 shows the results of targets weighing carried out one hour and four days after the end of the fire (the targets were kept in a cold and dark container). It should be noted that the mass lost by the samples during the four days of storage are 11% and 8% which is rather low and may be due to water and highly volatile organic vapors losses. The close results between the targets 1 and 2 are explained by the fact that the temperatures of these two targets during the test are identical. Due to the heating the surface of target 3, the mass deposited is negligible, confirming the major contribution of thermophoresis on the overall deposition of particles during a fire. It is then obvious that the physical description of such phenomena must be improved in the actual soot deposition model, especially by taking into account the specific morphology of soot particles emitted in case of fire[16].

Table 1: Weighing of the deposition targets

Target	One hour after the fire	Four days after the fire
1	11.83 mg	10.53 mg
2	13.88 mg	12.78 mg
3	-0.11 mg	-0.16 mg

CONCLUSION

A particle measurement setup has been installed in the smoke spreading in an adjacent corridor of an office fire. Three on-line mass concentration measurement devices have been compared: a microbalance TEOM, an opacimeter and a PPS. The comparison shows that the opacimeter results, considering a specific extinction coefficient value equal to $\sigma s = 8.7\ m^2/g$, are in close agreement with TEOM results, while PPS strongly underestimates the particle mass concentration. Further studies are in progress to for explaining this discrepancy and for proposing new calibration factors, relevant for fire application. The other results provided by the particle measurement setup show that:

-At the beginning of the fire and compared to the other periods, a high content of small aerosols have been observed, corresponding to low values of the mass concentrations.

-Latter, when the fire source inside the office has a much higher intensity, the size of the particles increases, count mean diameter value at the end of the fire has been found equal to 0.5 μm and the mass concentration close to 250 mg/m^3. Particles present a fractal morphology and are mainly composed by carbonaceous nanoparticles with a diameter ranging from 10 to 50 nm.

-The organic content of the soot particles is low during the first period corresponding to a well-ventilated fire. On the other hand, particles emitted during the full-fire period, associated to under-ventilated conditions, have a significantly higher OC content.

Additional measurements of particles deposition have been also carried out using mass weighing of aluminium targets. Significant masses of particles have been measured and have confirmed the significant contribution of thermophoresis within the overall deposition phenomenon.

REFERENCES

[1] K.M. Butler, G.W. Mulholland, 2004, Generation and transportation of smoke components, Fire Technol. 40 149–176.

[2] B. Giechaskiel et al, 2014, Review of motor vehicle particulate emissions sampling and measurement: From smoke and filter mass to particle number, Journal of Aerosol Science 67 48–86

[3] F.-X. Ouf, V.-M. Mocho, S. Pontreau, Z. Wang, D. Ferry, J. Yon, 2015,Physicochemical properties of aerosol released in the case of a fire involving materials used in the nuclear industry, Journal of Hazardous Materials, 283, 340-349.

[4] A.S.-X. Loo, A. Coppalle, J. Yon, P. Aîné, 2016, Time-dependent smoke yield and mass loss of pool fires in a reduced-scale mechanically ventilated compartment, Fire Safety Journal, 81, 32-43.
Fire Science and Technology 2015

[5] Bellivier A., A. Coppalle, A. S. Loo, J. Yon, L. Decoster, S. Dupont and H. Bazin, 2015, Comparison and Assessment of Particle Mass Concentrations Measurements in Fire Smokes with a Microbalance, Opacimeter and PPS Devices; Fire Science and Technology 2015: The Proceedings of 10th Asia-Oceania Symposium on Fire Science and Technology.

[6] Marjamäki, M., Keskinen, J., Chen, D. R., & Pui, D. Y. H., 2000, Performance evaluation of the electrical low-pressure impactor (ELPI). Journal of Aerosol Science, 31(2), 249–261. http://doi.org/10.1016/S0021-8502(99)00052-X

[7] Ouf, F. X., & Sillon, P. , 2009. Charging Efficiency of the Electrical Low Pressure Impactor's Corona Charger: Influence of the Fractal Morphology of Nanoparticle Aggregates and Uncertainty Analysis of Experimental Results. Aerosol Science and Technology, 43(7), 685–698. http://doi.org/10.1080/02786820902878245

[8] Chow, J. C., Watson, J. G., Chen, L. W. A., Chang, M. C. O., Robinson, N. F., Trimble, D., & Kohl, S. , 1995. The IMPROVE_A temperature protocol for thermal/optical carbon analysis: maintaining consistency with a long-term database. Journal of the Air & Waste Management Association. http://doi.org/10.3155/1047-3289.57.9.1014

[9] Bellivier A., Delorme R., Dupont S., Benhaiem P., Chen Y., LeBars P. and Bazin H., 2011, Study of local aerosol combus-tion concentration with a differential white light opacimeter, EAC 2011, European Aerosol Conference, Manchester, 4-9 sept.

[10] SFPE Handbook of Fire Protection Engineering, 2002, Third Edition, published by the National Fire Protection Association National Fire Protection Association, Inc., One Batterymarch Park Quincy, Massachusetts 02269. chapter 13.

[11] Ntziachristos L., Amanatidis S., Samaras Z., Janka K. & Tik-kanen J., 2013,.Application of the Pegasor particle sensor for the measurement of mass and particle number emissions. SAE 2013-01-1561.

[12] Patashnick H. & Rupprecht E. 1991, Continuous PM-10 meas-urement using the tapered element oscillating microbalance. Journal of the Air&Waste Management Association, 41, 1079–1083

[13] Ouf, F.-X. et al, 2014, Clogging of Industrial High Efficiency Particulate Air (HEPA) Filters in Case of Fire: From Analytical to Large-Scale Experiments. Aerosol Sci. Technol. 48, 939–947

[14] Ouf, F. X., Vendel, J., Coppalle, A., Weill, M. & Yon, J., 2008. Characterization of Soot Particles in the Plumes of Over-Ventilated Diffusion Flames. Combust. Sci. Technol. 180, 674–698.

[15] J. Goo (2015), Study on the real-time size distribution of smoke particles for each fire stage by using a steady-state tube furnace method, Fire Safety Journal, 78, 96-101

[16] E. Brugière, F. Gensdarmes, F.X. Ouf, J. Yon, A. Coppalle, 2014, Increase in thermophoretic velocity of carbon aggregates as a function of particle size, Journal of Aerosol Science, 76, 87-97

ON-SITE INVESTIGATION EXERCISE BASED ON SIX FULL-SCALE LIVE BURNINGS

Aurélien Thiry[1], Eddie Faure[1], Mathieu Suzanne[1], Nicolas Dreuille[1], Bruno Poutrain[2], Olivier Delémont[3], Christophe Remillon[4], Véronique Vidotto[5], Korbinian Pasedag[6]
[1] Laboratoire Central de la Préfecture de Police, Paris, France
[2] Paris Fire Brigade, Paris, France
[3] Ecole des Sciences Criminelles, Université de Lausanne, Switzerland
[4] Institut de recherche criminelle de la Gendarmerie Nationale, Pontoise, France
[5] Institut National de Police Scientifique, Toulouse, France
[6] Büro für Brandschutz, Bergish, Germany

INTRODUCTION

Fire investigation in France involves many actors. They can be distinguished by both time scales and aims relative to their missions. By instance, specialized fire fighters can be first-responders with objectives focusing on fire safety feedback and evidence preservation. Then, depending on the case, investigators from police or gendarmerie may investigate fire scene with evidence collection assignment and help initiating the judiciary process. Between few days and, sometimes, up to one year after the fire, experts from insurances and legal experts may inspect remains of the fire scene for research on responsibilities and last evidences. Such organization is schematically described in the figure 1 below.

Figure 1. Actors involved in fire investigation

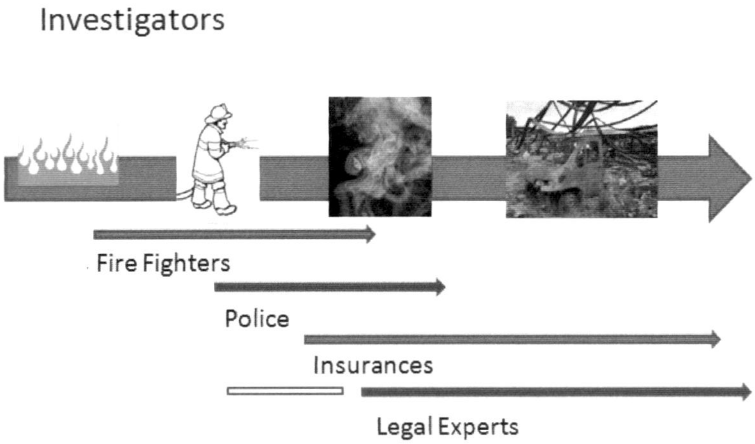

Concerning training of investigators, there is no common framework to ensure reliability and effectiveness. There is also no standards recognized among the different profession exercising post-investigation. By instance:

- Fire fighters missions and training are ruled by an official document from Homeland Security

Ministry[1]: 44h of theory and practice

- Central laboratory of Paris Police Prefecture: internal training including 50 hours of theory and practice, 30 on-duty days as observer and then 1 year of tutoring
- Forensic national institute: 80 hours of external formation including theory and practice, approximately 10 fire scenes as observer
- Gendarmerie (IRCGN): 60 hours of external formation, real training fire scenes and tutoring of less experienced investigators
- Legal experts: recognition of experience and per review for affiliation to a Court
- Institute of Forensic Sciences: specialized bachelor or master degree including around 150 hours of theory and practice and 2 years of tutoring

More generally, there is numerous and distinct reference documents supporting these formations. Many of them are directly inspired by or close to NFPA standards[2,3], reference books related to fire investigation[4,5] or to fire dynamics [6,7].

In this context, Central Laboratory of Paris Police Prefecture regularly organizes live training on real fire scenes equipped with metrology in order to gain experience and knowledge in investigation and to promote exchanges between experts of different backgrounds. This test campaign thus follows Arcueil's one in 2009[8] (5 fire scenes) and Herblay's one in 2012[9] (5 fire scenes).

CONTEXT

The first objective of this test campaign was the formation and the competences evaluation of new investigators of the Central Laboratory and all the investigation first-responders of Paris Fire Brigade. The second objective of the organized exercises was the definition and diffusion, with all the different professional corporations that may be involved in legal investigation, of good practices and the possibility of focusing on training needs.

To allow good conditions of training, 7 real fire scenes were reconstructed. Each was equipped with measurement devices. For all tests, coarse room temperature mapping was performed (between 12 and 24 thermocouples per room) and some specific devices were installed, depending on the fire scenario, to focus on particular aspects of fire development.

Electrical installations were also reconstituted and powered during the tests.

PARTICIPANTS

Participants were as follow and distributed in three 2-days session. The first session was focused on public services experts; the second on legal expert designated by courts; and the third one on private experts. To summarize, it represents a total of:

- 25 military fire-fighters from Paris Fire Brigade
- 20 civilian/professional fire-fighters from territories surrounding Paris and its neighborhood

- 12 Police investigators
- 4 Gendarmerie investigators (military)
- 12 legal experts
- 8 private investigators
- 3 investigators/researchers from Criminalistic Institute of Lausanne, Switzerland
- 1 investigator from Switz police

ORGANIZATION

For each session, 2 days were devoted to the investigation exercises. First session consisted of 6 groups of 4 people with various profiles in order to exchange and compare the different methods of investigation and damage observation. Before each exercise, a team of 2 firefighters from Paris Fire Brigade carried out their technical investigations without the opportunity of moving any material or furniture from the fire scene. Then at the arrival of the groups, they played their own role by informing them of their first observations as it takes place especially on Paris and the suburbs. An external person played the role of police officer by informing the investigators on context, fire testimonies, etc.
Only 6 fire scenes were investigated due to huge damages in one fire scene explained by communication problems with fire-fighters and high thermal stress slowing their approach.
Each group studied for about 1 hour the 6 scenes that was assigned to them. The first day, they were authorized only to evolve through the scenes without removing anything. The second day, they were authorized to move furniture and to remove debris from the floor. They filled out a report containing their main findings, the samples taken and their conclusions. Each day debriefing of the studied scenes common to 3 groups was carried out, in order to promote exchanges between the different groups and participants. During the debriefing, panoramic 360 ° views of each post-fire and pre-fire scene were shown. Various videos showing the development of the fire within each scene and from different angles were viewed as well as the evolutions of the temperature measurements during the development of the fire in the room. These views and technical data gave rise to new exchanges between the participants.

LIVING-ROOM FIRE SCENE-BEFORE

Only one out of the seven fire scenes is described in the present paper. It concerns a living-room and is representative of domestic fires regularly encountered by investigators. The premises chosen to perform the test were located in an old office building made available to the LCPP before destruction. The enclosure of about 18m^2 (2.45 m wide for 7.3 m long and 2.45 m high) had six windows overlooking a roof terrace. It was arranged into a lounge on one side and a dining part on the opposite side. A large sofa was installed transversely to separate these two areas.
The lounge part was equipped with a coffee table, two armchairs (one of which placed near a window), audio-visual appliances currently encountered in a living room and a bookcase (see fig. 2.a). The dining room was equipped with a table, four wooden chairs, a dresser and an office area including a desk, a computer and a chair made up of plastic materials (see fig. 2.b).

Figure 2. View of the scene prior to fire a) lounge part b) dining part

Long curtains were installed on the windows, and wall sockets located along these windows were powered by a generator. The various lamps (halogen or other) as well as the domestic electrical appliances installed in the room were plugged in the aforementioned sockets. The test room door was kept closed during the experiment until fire-fighters intervention.

Experimental set-up

Two thermocouple trees were installed in the living room (see Tr1 and Tr2 in fig.3): 1) centered in the lounge area and 2) next to the dining-room table. Six thermocouples were placed on each tree starting at 0.4m from the floor and fixed every 0.4m afterward.

Figure 3. Experimental set-up

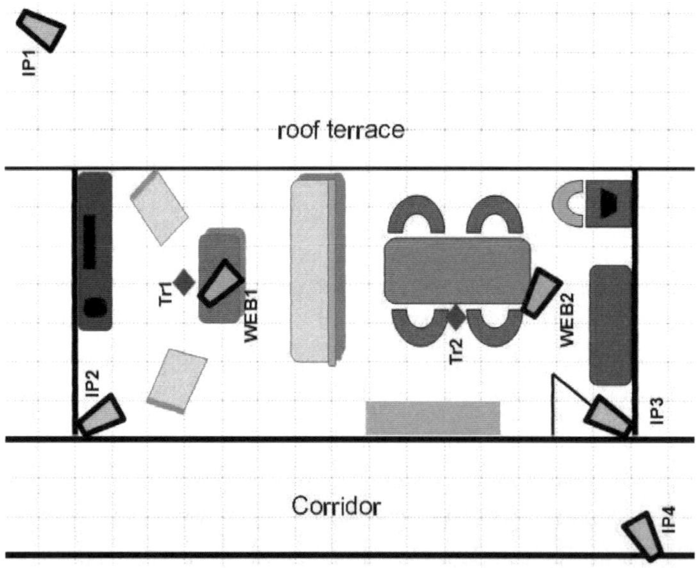

Six cameras were installed as shown in figure 3 ("Web" stands for webcam and "IP" for IP camera), filming the fire development from inside and outside.

LIVING-ROOM FIRE SCENE – AFTER

In order to make fire pattern observations and origin and fire cause determinations easier, the fire scene was entirely preserved by the fire brigade during the extinction operations. In many cases, post-fire investigations are carried out after a partial or complete clearing of the fire scene which makes it more difficult to perform technical investigations. After cooling and ventilation of the room, pictures were taken to freeze the fire scene. It was noticed a striking contrast in the fire degradations between both parts of the enclosure, the lounge and the dining room, with a dividing line located at the back of sofa bed. All the windows of the enclosure were broken but the three windows on lounge side were the most impacted by the fire (see fig. 4).

Figure 4: Exterior view of the living-room

In the lounge part, a clean burn pattern was noticed on the wall at the back of the enclosure on lounge side, which was composed of gypsum blocks. All the other walls, made of steel, were oxidized and deformed. It was particularly verified at the location of the bookcase which was literally burnt to ashes (see fig. 5.a).

Figure 5. Inside damages of the lounge part a) general view, b) upholstered chairs, sofa and table

The ceiling, constituted of metallic plates, was oxidized and partially collapsed. In depth charring was observed on all the furniture containing wood in the lounge part. Furniture was so severely damaged that some pieces had entirely disappeared (e.g. armchair close to the window and coffee, see fig.5.b). A pile of melted and solidified thermoplastic materials only testifies to the presence of a furniture item at the back of the enclosure near the armchair close to the window. The foam of the armchairs and the bed sofa as well as the insulation of the bar refrigerator were entirely consumed (see fig.6). The same observation are made concerning the combustible parts of the appliances and furniture present in this part of the room (see fig.6).

Figure 6. Focus on specific damages in lounge part: chair and refrigerator and TV furniture

Around the dividing line mentioned earlier, a charring gradient was observed on the wooden wardrobe surface located against the longest metallic wall. The deepest and lowest charring levels were found on lounge side. A direct flame contact was visible on the upper part of the wardrobe close to the metallic wall (see fig.7.a).

Figure 7. View of a) wardrobe damages b) dining part of the room

Concerning dining part of the room, metallic walls and suspended ceiling were blackened by soot deposition but not further damaged or oxidized. The paint was however peeling over the right hand side of the desk. The different wooden furniture in the dining room was blackened. Shallow carbonisation traces were visible on the upper part of the furniture near the entrance door. The carpeted floor in this part of the room showed no trace of carbonisation (see fig.7.b).
The fabric covering the chair near the table, and more precisely its back and seat, was carbonized on the side exposed to the sofa. The foam was still present, unconsumed, as the other parts of the chair. The wooden chair next to it also had charring traces on its entire height on the side facing the sofa (see

fig.8.a).

Figure 8. a) Table and chairs b) furniture with laptop

The top of the computer desk and the laptop placed on it had carbonisation traces. Thermoplastic elements in the lower part of the desk were melted (see fig.8.b). The thermoplastic chair in front of the desk was particularly melted and traces of carbonisation under and around the seat were observed (see fig.9). A piece of thermoplastic furniture placed on the ground on the right hand side of the computer desk was melted on its surface (see fig.8.b).

Figure 9. View of the chair in front of the computer desk

INVESTIGATIONS AND FIRE DEBRIS SAMPLE COLLECTION

For this test, the scenario that was established is that of a fire deliberately set and carried out after the spillage of flammable liquid (i.e. gasoline). **One window in two was broken from the outside before the flammable liquid was spilled near the armchair, on the bed sofa and on the thermoplastic chair in front of the computer desk.** The various electrical devices in the room were plugged but switched off or on standby mode. The extinguishing phase started approximately 25 minutes after the ignition of the first area where gasoline was spilled (i.e. near the armchair in the lounge).

The context provided to the investigators for the training was that the fire was detected around 3 am on a hot summer night by a neighbor who heard a distant sound when sleeping. He then smelled smoke and

saw flames coming out through at least one window of the apartment across the street. He also knew that the occupants of the damaged apartment had been on holiday for two weeks. The damage to the room due to the fire is representative of what investigators may encounter (see figs. 5 and 7).
The different groups of investigators worked on this fire scene, taking care of limiting its modifications. Technical investigations of most groups were focused in the lounge area of the living room. It was noticed that a large majority of investigators concentrated their fire pattern analysis where the damage were the greatest. Observations of elements not of little damaged were however performed but briefly, most often due to a lack of time. The last groups that examined this scene paid more attention to all the elements disposed in the room because they had more time to work but they also were allowed to modify the scene and could proceed to a meticulous excavation.

Almost all the investigators identified two fire origins (see fig.10): 1) near the TV cabinet and the armchair near the window and 2) at the bookcase location. However, important discussions were held between participants of each group to determine whether these patterns were related or distinct in terms of origin location.

Figure 10. Fire origins identified by investigators

Depending on the initial training of investigators or their statutes (legal expert, institution, insurance, fire brigade), the search for volatile organic compounds with a PID was not always performed or, when they were negative, did not always lead to fire debris sample collection for analysis in a laboratory specialized in the search for traces of flammable liquids. Some investigators proceeded to a search with PID on all the elements of the room, damaged or not. It is worth noting that detections were negative all over the room except on the floor under the chair in front of the desk, in the dining part. For institutional investigators, this kind of measurement and sampling were generally systematic.
It was observed that investigators with a strong electrical background realized a more careful examination of wall sockets or electrical cables and appliance and wondered more on this topic than more generalist investigators.
Many trainees examined the premises for possible break-in traces. All concluded that the door of the room was closed at the time of the fire but that there was suspicious presence on the ground of many glass fragments characteristics of mechanical shocks. They could not determine whether they were resulting from a break-in or from the intervention of the fire brigade. Only groups with more time and that were able to perform a more thorough excavation of the fire scene discovered many pieces of broken glass with no soot deposition on their surface; they were located directly on the floor, under the rubble generated by the fire.

Concerning investigators conclusions:

- All groups concluded that break-in was certain, based on testimonies, context and glass fragments
- All groups concluded about a point of origin located near the armchair, close to the first window (see fig. 10)
- About 75% of the groups concluded that there were at least two different fire origins
- Among these 75%, only 2 groups concluded correctly about a second point of origin near the sofa
- Only one investigator found the third point of origin, i.e the gasoline spill in front of the computer desk

Measurements

Measurements obtained for three out of the six thermocouples installed on tree 1 and 2 are plotted in figs 11 and 12. Ignition is identified in these graphs by the vertical line at 200s. Maximum temperatures quickly reached a plateau around 350°C which lasted till 1300s. It turned out that the single window panes were reinforced with security films and that temperatures were too low to break them. As a consequence, the door being closed and the window only partially broken to allow ignition, the fire was under-ventilated during this phase and did not get enough oxygen to develop in flash-over. The flame were only located in the lounge, first on the armchairs and then on the sofa but did not spread toward the dining table. The fire located on the chair next to the computer desk did not spread either to the furniture.

As the fire did not spread, it was decided to break the window between the sofa and the computer desk, because it was already very damaged. Then, this opening progressively acted as the main smoke exhaust, oxygen entering through the other openings. Flash-over conditions were quickly reached explaining temperatures measured in tree 1 higher than 1000°C in less than 300s. This was accentuated by the progressive fall of the adjacent window. As soon as it opened, flames went out through this window and shortly after through two others.

Due to ventilation conditions, flash-over did not occur in the dining part of the room despite favorable thermal environment in the top layer.

Figure 11. Temperatures in the lounge part

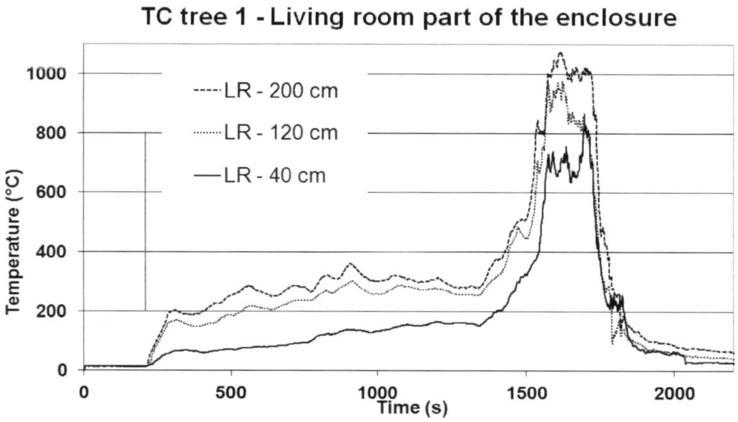

Figure 12. Temperatures in the dining part

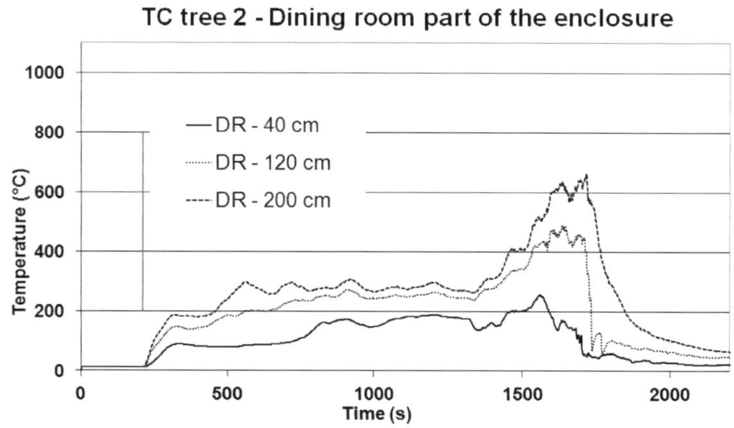

Fig. 13.a below shows the initial stage of the fire, with the 3 ignition points with spill of gasoline. Fig. 13.b shows fully developed stage of the fire, with occurrence of flash-over in the right part f the room.

Figure 13. Exterior view of the fire growth a) initial stage b) fully-developed stage

Fig. 14 focus on the computer desk, which is the ignition point that was missed by a large majority of trainees due to less important damages in this part of the room.

Finally, fig.15 below shows that the area that interested the most the different investigators was not even burnt to ashes during the fire. The total destruction of the wardrobe occurred after extinction, as the result of smoldering process which was not detected during immediate post-fire inspection.

Figure 14. Ignition and spread near the computer desk

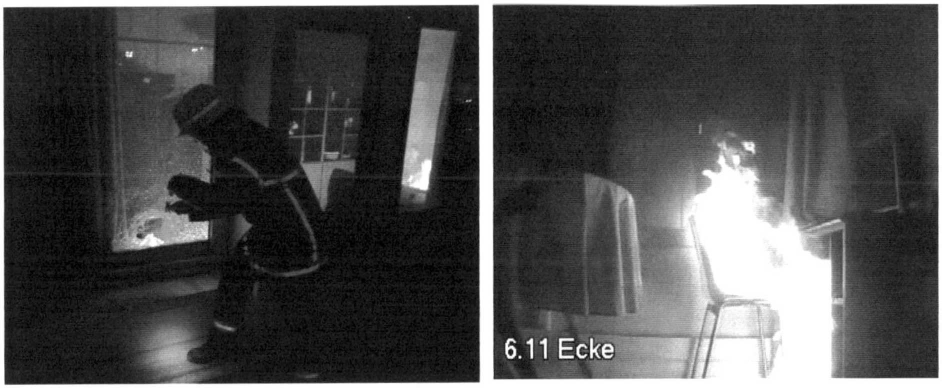

Figure 15. View of the wardrobe after the fire

CONCLUSION

This training highlighted very interesting features concerning investigators methods. First, initial formation has a strong influence on methods employed on a fire scene, and in particular material elements that are meticulously examined. Investigators with strong electrical background focused first on appliances, plugs and electrical installation. Investigators with "materials" background focused on fire patterns and degradation gradient among furniture and walls.

Second, numerous investigators focused on the most damaged areas before trying to consider ventilation of the fire. Such approach led a majority of investigators to determine wardrobe (which was burnt to ashes) as a fire origin while it was located more than 2 meters from the actual fire origins. Nevertheless, such reasoning led a part of them to good deductions in the determination of fire origins, near the chair

and near the sofa. Not considering specific ventilation of this fire led also all the investigators but one to miss unusual degradation and fire patterns near the computer desk, as a third fire origin.
Third, it was important in this case to focus on general features of the fire scene before exploiting fire damages on furniture. By instance, shape and lack of soot deposition on pieces of broken windows were of primary interest as well as reconstitution of ventilation patterns.
To conclude, such exercise involving professional from all the country and of all kind of profession is very useful in order to identify possible improvement in the investigators formation (background and training during career). NFPA 921[2] methodology concerning fire scene investigation mainly focus on the following aspects to conclude on fire origin and cause:

- fire patterns
- fire dynamics
- arc mapping
- context and testimonies

Feedback from all the test campaign (including five other fire scenes) is that, among these 4 points, "Fire dynamics" one should be developed. This field includes, in order of importance, ventilation, fire kinetics, propagation and fire-related phenomena (like smoldering by instance). In addition to that, knowledge in fire dynamics is sometimes difficult to gain by just exploring extinguished fire scenes because it is intrinsically an approach on "frozen" elements. Thus more exercises including video recording and temperature measurements should be organized for a better understanding of time-dependent phenomena.

ACKNOWLEDGMENT

We also wish to thank all the partners involved in the organization of this test campaign; Aéroport de Paris for the availability of the site, the Fire engineering division for the realization of the tests, in particular Delphine Cendra, Sylvie Dupont and Adissa Traoré, and all the investigators involved in the exercises.

REFERENCES

[1] Circulaire Ministérielle n°IOCE1108242C du 23 mars 2011, *Réalisation des missions de recherche des causes et circonstances d'incendie RCCI par les services d'incendie et de secours*.
[2] National Fire Protection Association, 2014. NFPA 921: Guide for Fire and Explosion Investigations, NFPA, Quincy, USA.
[3] National Fire Protection Association, 2014. NFPA 1033: Standard for Professional Qualifications for Fire Investigator, NFPA, Quincy, USA.
[4] Martin JC. 2013. Incendies et explosions d'atmosphère. PPUR, Lausanne
[5] DeHaan J.D., Icove D.J. 2012. Kirk's Fire Investigation, 7th Edition, Brady Fire, Pearson Education Inc., New Jersey
[6] Drysdale D. 1998. An Introduction to Fire Dynamics, 2nd edition, John Wiley & Sons Ltd, New York

[7] Babrauskas V., 2003. Ignition Handbook: Principles and Applications to Fire Safety Engineering, Fire Investigation, Risk Management and Forensic Science, Fire Science Publishers

[8] Bazin, H. and al. Modélisation de scénarios d'incendie et expérimentation des feux réels correspondants dans un immeuble voué à la démolition, Projet National sur l'Ingénierie de la Sécurité Incendie, 2009

[9] Cendra, D. and Dreuille, N., Manoeuvre feux réels Herblay – Formation nationale des sapeurs-pompiers investigateurs, Internal report n°12/6783/RG1, 2012

LATE PAPERS

ABSTRACT

HEAT RELEASE MEASUREMENT AT OXYGEN RICK ATMOSPHERE

Koichi Yoshida, Mr. Yokohama national University
Aki Hosogai, Mr. Japan Aerospace Exploration Agency
Tetsuya Hatayakawa, Mr. Tokyo Systems Vac. Inc.

Heat release measurement technology in reduced oxygen atmosphere using cone calorimeter has been developed, and now it is the stage of ISO standardization (NP 5660-5), aiming at investigating the burning behavior of materials in closed compartment at fires. On the other hand, there is a oxygen rick atmosphere, slightly richer than 20.95% in air, in space crafts. And it is now a matter in aerospace technology and industries to develop methodologies and technologies for investigating and understanding the burning behavior, in particular heat release, of materials in such slightly rich oxygen condition. The experimental study on heat release measurement using cone calorimeter in oxygen rich atmosphere has been initiated, as the first step. This presentation will show the first round results of the experimental study.

ABSTRACT

POLYSTYRENE FOAM INSULATION WITH A SUSTAINABLE FLAME RETARDANT: TRANSFORMING THE MARKET

Christine Lukas [a], Lorraine Ross[b]

Mark W. Beach[d], Inken Beulich[c], John W. Davis[d] Heli Hollnagel[c], John W. Hull[d], Bruce A. King[d], Shari L. Kram[d], Ted A. Morgan[d], Michal E. Porter[d], William G. Stobby[d]

[a] Dow Chemical Company UK Limited, United Kingdom

[b]Intech. Consulting Inc., Gulfport, Florida USA

[c] Dow Europe GmbH, Horgen, Switzerland

[d] The Dow Chemical Company, Midland, 48674 MI, USA

ABSTRACT

In light of increasing emphasis on green buildings, today's construction product marketplace must deliver products that not only perform in their intended use, but must also disclose sustainability attributes. For example, thermal insulation foams are used to satisfy increasing energy efficiency standards. Voluntary eco-label certification programs for environmental and health impacts are now often requested, along with traditional building and fire code compliance reports. However hurdles to adoption of replacement technologies can exist due to the use of simple hazard based approaches which don't adequately acknowledge sustainability improvements.

At the previous conference we introduced the polymeric flame retardant (FR) for polystyrene foam insulation. This polymer (plastic) has a superior, non-hazardous environmental profile and was specifically designed and developed by The Dow Chemical Company to replace the incumbent FR hexabromocyclododecane (HBCD). The polymeric FR solution with the more advantageous profile met the required key performance indicators including low inherent toxicity, being processable in EPS and XPS, maintaining the desired PS foam product properties and economically viable. Licensing the technological development ensures that the innovative solution is available to the whole Polystyrene foam (PS) foam industry.

This paper reports on further experience in the implementation of the polymeric FR which has taken place in the last 2 years. Since the conversion in Japan in 2014, HBCD has been more or less phased out in Europe after the so-called 'sunset date' in August 2015 under the REACH regulation. The majority of the EPS and XPS industry already stopped using HBCD before August 2015 moving to the Polymeric FR as the primary FR of choice.

Gaining this international experience in the production of foams containing the Polymeric FR is very valuable in North America where Canada has confirmed a deadline of 31 December 2016 for the phase-out of HBCD in Polystyrene foams. There is still the need for development of the local mandatory building/fire code certifications and other valued voluntary certifications for specific markets in this geography before the implementation is complete. Indications are nevertheless that North America is transitioning ahead of regulatory deadlines. An update on the transition will be given on how the members of the industry responded to the proposed milestone and in making a step change to a more sustainable flame retardant for PS foams.

Additional discussion will be focused on the various voluntary sustainability certification programs that are used by green rating systems. Some of these programs rely on the so-called "Red List" hazard approach. In this case, scrutiny is a simple check on material content of the product for the presence or absence of "unacceptable red list" chemical categories or technologies. This is a challenge when seeking to gain acceptance and acknowledgement of a significant step change in sustainability and counterproductive to the development of chemicals which are more environmentally suitable materials. It also potentially sets ever increasing sustainability thresholds against providing suitable fire safety solutions and desired energy efficiency goals. In contrast, a certification program based on toxicology principles of risk assessment, rather than blanket "red list" bans, would present a more accurate picture of the improved environmental or health profile of the new product or technology.

In summary we describe the complex journey to a more sustainable non-hazardous flame retardant for Polystyrene foams which is being made available to the whole of the Polystyrene foam industry. This replaces the incumbent technology which has been in use for several decades. We explain the challenges resulting from the use of 'red lists' in voluntary certification and propose a more risk based approach which as a result can better support and accelerate market transformation.

ABSTRACT

Flammability Properties and Radiant Fraction of FRT Wood Plastic Composites using Mass Loss Calorimeter under HRR Hood

By Mark A. Dietenberger, Ph.D., Charles R. Boardman, Nicole Stark, Ph.D.

USFS Forest Products Laboratory*, One Gifford Pinchot Dr., Madison WI

ABSTRACT

A special test arrangement was used to assess the flammability of 4 different wood plastic composites (WPC), most with fire retardants, all of which has a tendency to high smoke production leading to high radiant energy losses to the apparatus walls. The mass loss calorimeter (MLC) was modified to include a thermopile on the exhaust pipe stack to compensate for radiant energy losses to the wall (Figure 1) (Dietenberger, et. al.) A truncated conical electric heater provides a constant heat flux of up to100 kW/m2 onto the test specimen of up to 50 mm thick and a spark pilot igniter. Pure ethylene glycol and methanol for low flame radiant losses and the PMMA and solid polystyrene for high flame radiant losses supplanted the use of methane for calibrating the HRR using both thermopiles. Mass data was processed using a SG digital filter followed by an exponential filter so that the mass loss profile was synchronized the HRR in order to obtain an accurate effective heat of combustion (HOC) with time. Standard materials of methanol, ethylene glycol, PMMA and polystyrene have data listed by Tewarson for the radiant fraction of the chemical heat of combustion, whose values were used in calibration of the modified MLC for estimating radiant heat of combustion of unknown materials.

Figure 1 Modified MLC under FPL's HRR hood surrounded by radiant absorbing walls

The FPL HRR hood apparatus has been calibrated to provide the stoichiometric, chemical, radiant, and sensible heat of combustion as a function of time in which to provide indirect measures for radiant fraction associated with open flame as located under the HRR hood. The well-known properties of propane or methane pool fires and their simulation with FDS provided the verification of HRR hood apparatus calibration, particularly at lower HRR of that associated with MLC test specimen. A completely walled in hood, except for North side doorway, with high emissivity paint coating help achieved a higher accuracy to the radiant fraction measurements. This is in addition with the HRR hood's exhaust pipe instrumented with bidirectional velocity probe, pair of thermocouples, red laser smoke extinction, and extraction for gas analysis of O2, CO2, CO, and H2O with time. Radiant fractions of HRR for methane, methanol, and propane open flames were comparable to the values listed by Tewarson.

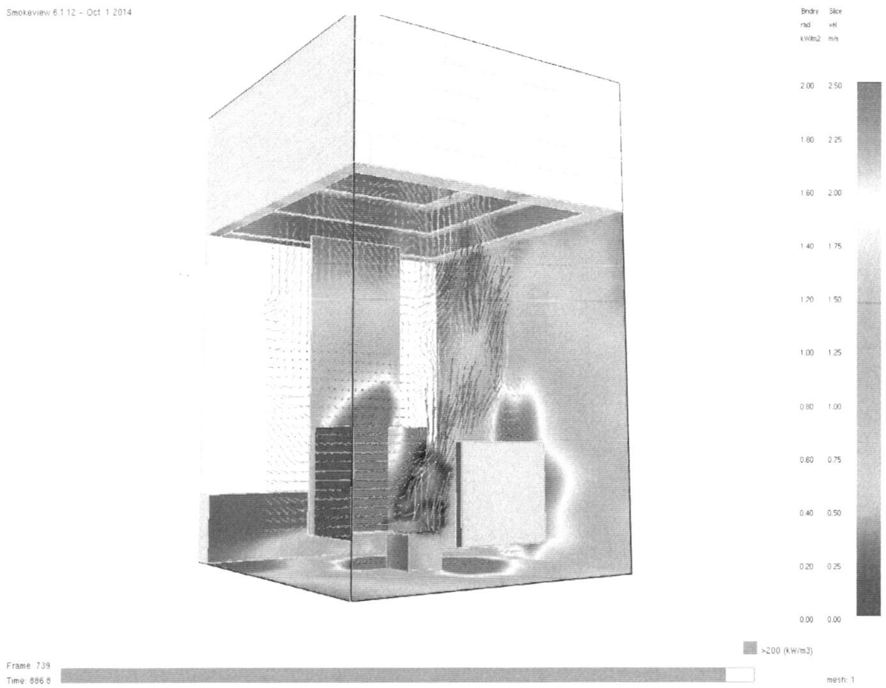

Figure 2 Propane pool fire under HRR hood with opened North wall with FDS rendition of flows and radiant heat fluxes from 30% radiant fraction of HRR measured with advanced O2 consumption method corrected for CO2, CO, H2O and soot measurements

Wood-plastic composite (WPC) boards were manufactured at the Forest Products Laboratory. The base WPCs investigated consisted primarily of high-density polyethylene (HDPE) and wood flour (WF). The HDPE had a 5 melt flow index and was purchased from ExxonMobil (HD 6605.70, Houston, TX). American Wood Fibers supplied 40 mesh, mixed pine WF (AWF 4020, Schofield, WI). To maintain good composite surface characteristics, a lubricant was added to each composite. Struktol Company of America supplied the lubricant (TPW 113, Stow, OH). In addition, two fire retardant systems typical for use in WPCs were investigated: 1) 3:1 ratio of decabromodiphenyl oxide (Saytex 102E, Albemarle Corporation, Baton Rouge, LA) and antimony trioxide (BrightSun HB, China Antimony Chemicals Co., Ltd., Guangxi, China); and 2) ammonium polyphosphate (Exolit AP 422, Clariant Corporation, Charlotte, NC). WPCs without fire retardants had either 50% or 60% by weight WF, 5% lubricant, and the remainder HDPE. Composites with fire retardants incorporated 50% WF, 10% of the fire retardant, 5% lubricant and the remainder HDPE.

Test results with the modified MLC under the HRR hood were comparable with our previous cone calorimeter test results of the same WPC specimen at irradiance of 50 kW/m^2 (Stark, et. al.). We confirm that the addition of WF into PE dramatically improves the fire performance of PE. For example, incorporating 60% wood into PE lowered the peak HRR of PE by 76%. The average HRR decreased, relative to this WF/PE composite, between 19 and 39% when fire retardants were added. Reductions to the estimated radiant HOC were also noted, especially when the wood component was increased.

References. M.A. Dietenberger, & C.R. Boardman, HRR Upgrade to mass loss calorimeter and modified Schlyter test for FR Wood, Fire and Materials Conference, 2013, pp. 25, 1-263

N.M. Stark, R.H. White, S.A. Mueller, T.A. Osswald, Evaluation of Various Fire Retardants for Use in Wood Flour/Polyethylene Composites, Polym Degrad Stab 2010; 95(9) 1903-1910.

AUTHOR INDEX

Grayson Franks Ltd
1 West Yard
Guilford Grove Greenwich
London SE10 8JT
United Kingdom

ISBN 978-1-5108-4674-6

9 781510 846746

CURRAN ASSOCIATES INC.
proceedings.com